SI Base Units

Quantity	Name	Symbol
Length	meter	m
Mass	kilogram	kg
Time	second	s
Temperature	Kelvin	K
Amount of substance	mole	mol
Electric current	ampere	A
Luminous intensity	candela	cd

SI Derived Units

Measurement	Unit	Symbol	Expressed in Base Units	Expressed in Other SI Units
Acceleration		m/s^2	m/s^2	
Area		m^2	m^2	
Capacitance	farad	F	$A^2{\cdot}s^4/kg{\cdot}m^2$	
Density		kg/m^3	kg/m^3	
Electric charge	coulomb	C	$A{\cdot}s$	
Electric field		N/C	$kg{\cdot}m/C{\cdot}s^2$	
Electric resistance	ohm	Ω	$kg{\cdot}m^2/A^2{\cdot}s^3$	V/A
Energy, work	joule	J	$kg{\cdot}m^2/s^2$	$N{\cdot}m$
EMF	volt	V	$kg{\cdot}m^2/A{\cdot}s^3$	
Force	newton	N	$kg{\cdot}m/s^2$	
Frequency	hertz	Hz	s^{-1}	
Illuminance	lux	lx	cd/m^2	
Magnetic field	tesla	T	$kg/A{\cdot}s^2$	$N{\cdot}s/C{\cdot}m$
Potential difference	volt	V	$kg{\cdot}m^2/A{\cdot}s^3$	W/A or J/C
Power	watt	W	$kg{\cdot}m^2/s^3$	J/s
Pressure	pascal	Pa	$kg/m{\cdot}s^2$	N/m^2
Velocity		m/s	m/s	
Volume		m^3	m^3	

GLENCOE

PHYSICS

Principles and Problems

Author

Paul W. Zitzewitz, Ph.D.
Professor of Physics
University of Michigan-Dearborn

Glencoe
McGraw-Hill

New York, New York   Columbus, Ohio   Woodland Hills, California   Peoria, Illinois

A Glencoe/McGraw-Hill Program

Glencoe Physics: Principles & Problems

Program Components

- **Student Edition**
- **Teacher Wraparound Edition**
- **Laboratory Manual, SE and TE**
- **Physics Lab and Pocket Lab Worksheets**
- **Tech Prep Applications**
- **Study Guide, SE and TE**
- **Reteaching**
- **Physics Skills**
- **MindJogger Videoquizzes**
- **Supplemental Problems**
- **Chapter Assessment**
- **Computer Test Bank, Win/Mac**
- **Problems and Solutions Manual**
- **Enrichment**
- **Critical Thinking**
- **Advanced Concepts in Physics**
- **Transparency Package**
- **Transparency Masters and Worksheets**
- **Lesson Plans**
- **Spanish Resources**

Glencoe Science Professional Series:

Performance Assessment in the Science Classroom
Alternate Assessment in the Science Classroom
Cooperative Learning in the Science Classroom
Graphing Calculators in the Science Classroom

Glencoe/McGraw-Hill

A Division of The ***McGraw·Hill*** *Companies*

Send all inquiries to:
Glencoe/McGraw-Hill
936 Eastwind Drive
Westerville, Ohio 43081

ISBN 0-02-825473-2

Printed in the United States of America.

4 5 6 7 8 9 10 027 05 04 03 02 01 00 99

Author

Paul W. Zitzewitz is professor of Physics at the University of Michigan-Dearborn. He received his B.A. from Carleton College and M.A. and Ph.D. from Harvard University, all in physics. His research director at Harvard was Nobel laureate Norman Ramsey. Dr. Zitzewitz has taught physics to undergraduates for 26 years, and is an active experimenter in the field of atomic physics with more than 50 research papers. He has been the president of the Michigan section of the American Association of Physics Teachers, chair of the American Physical Society's Forum on Education, and a member of the American Physical Society's Committee on Education.

Contributing Author

Mark Davids
Physics Teacher
Grosse Pointe South High School
Grosse Pointe Farms, Michigan

Contributing Writers

Linda Barr
Freelance Writer
Westerville, Ohio

Mary Dylewski
Freelance Science Writer
Houston, Texas

Devi Mathieu
Freelance Science Writer
Sebastopol, California

Consultants

Donald J. Bord, Ph.D.
Professor of Physics
The Department of Natural Sciences
University of Michigan-Dearborn
Dearborn, Michigan

David Haase, Ph.D.
Professor of Physics
Director of The Science House
North Carolina State University
Raleigh, North Carolina

Teresa Anne McCowen, M.S.
Chemistry Instructor
Moraine Valley Community College
Palos Hills, Illinois

Safety Consultant

Anne Barefoot
Physics and Chemistry Teacher, Emeritus
Whiteville High School
Whiteville, North Carolina

Reviewers

Kathleen M. Bartley
Westville High School
Westville, Indiana

Stephen C. Benedict, M.A.
Science Department Head
New Albany High School
New Albany, Ohio

John Burger, M.A.
Mansfield High School
Mansfield, Massachusetts

David Capron, M.A.
Suttons Bay High School
Suttons Bay, Michigan

Richard D. Creed, M.A.
Science Department Chair
L.D. Bell High School
Hurst, Texas

Carol G. Damian, Ph.D.
Science Department Chair
Dublin Scioto High School
Dublin, Ohio

Cathy Ezrailson, M.S.
Science Department Head
Oak Ridge High School
Conroe, Texas

Craig Kramer
Bexley High School
Bexley, Ohio

David A. Loeffler
York County Vocational
Technical High School
York, Pennsylvania

John A. Mannette
Bucksport High School
Bucksport, Maine

Tracy W. Phillips
Burbank High School
San Antonio, Texas

Elizabeth Ramseyer, M.S. Ed.
Niles West High School
Skokie, Illinois

Karen A. Russell, M.A.
Marion-Franklin High School
Columbus, Ohio

Joy B. Trauth
Jonesboro High School
Jonesboro, Arkansas

Michael Turner, M.Ed.
Page High School and
Weaver Education Center
Greensboro, North Carolina

Mindy Weaver, M.A.
Jay County High School
Portland, Indiana

Michael D. Wolter, M.Ed.
Science Department Chair
Muncie Central High School
Muncie, Indiana

Elizabeth Woolard
Science Department Chair
W.G. Enloe Magnet High School
Raleigh, North Carolina

Contents *in Brief*

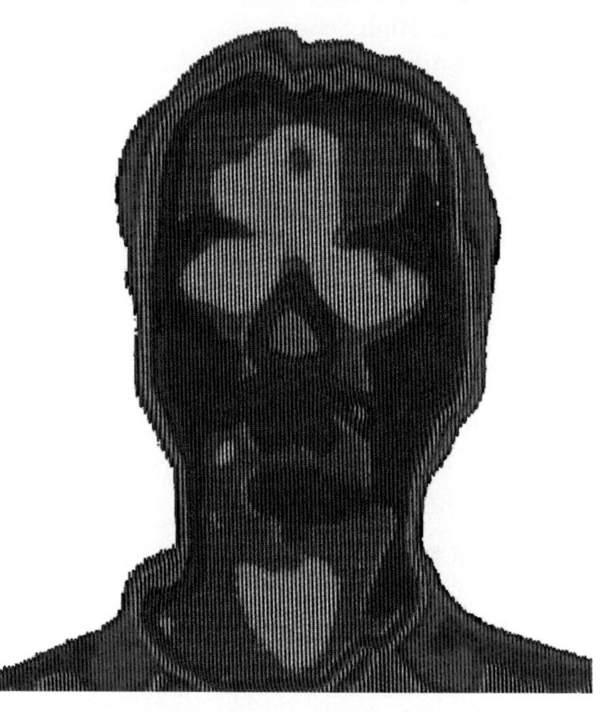

Electricity

Modern Physics

Contents

Contents

Contents

Contents

Electricity

Contents

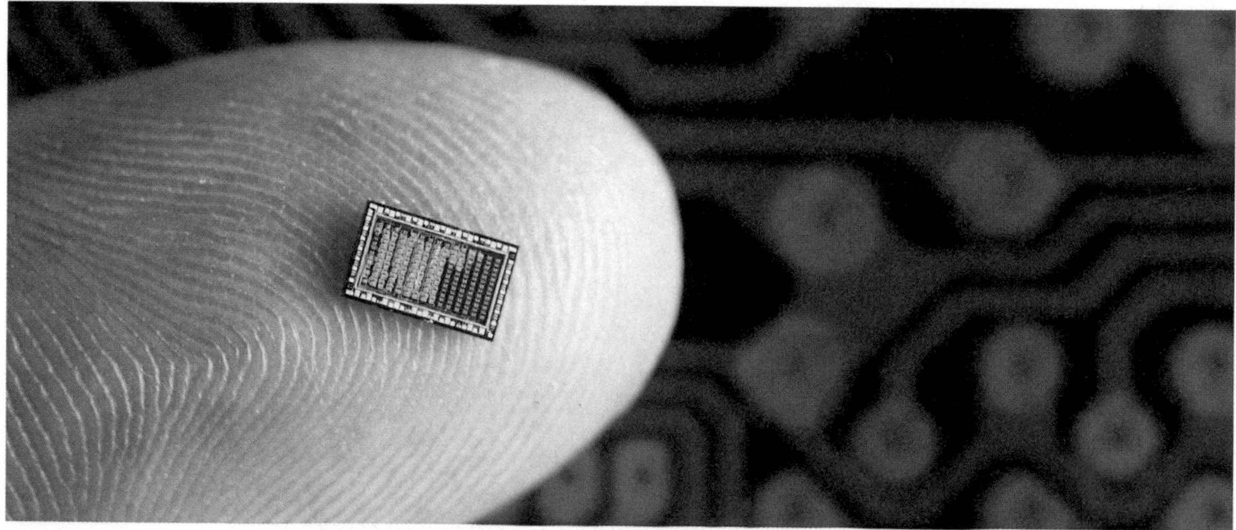

Physics Labs

Pocket Labs

Problem Solving Strategies

interNET CONNECTION

Follow the link on the Glencoe Homepage at **www.glencoe.com/sec/science** to find out more about physics.

Physics & Society

How It Works

Physics & Technology

CONNECTIONS

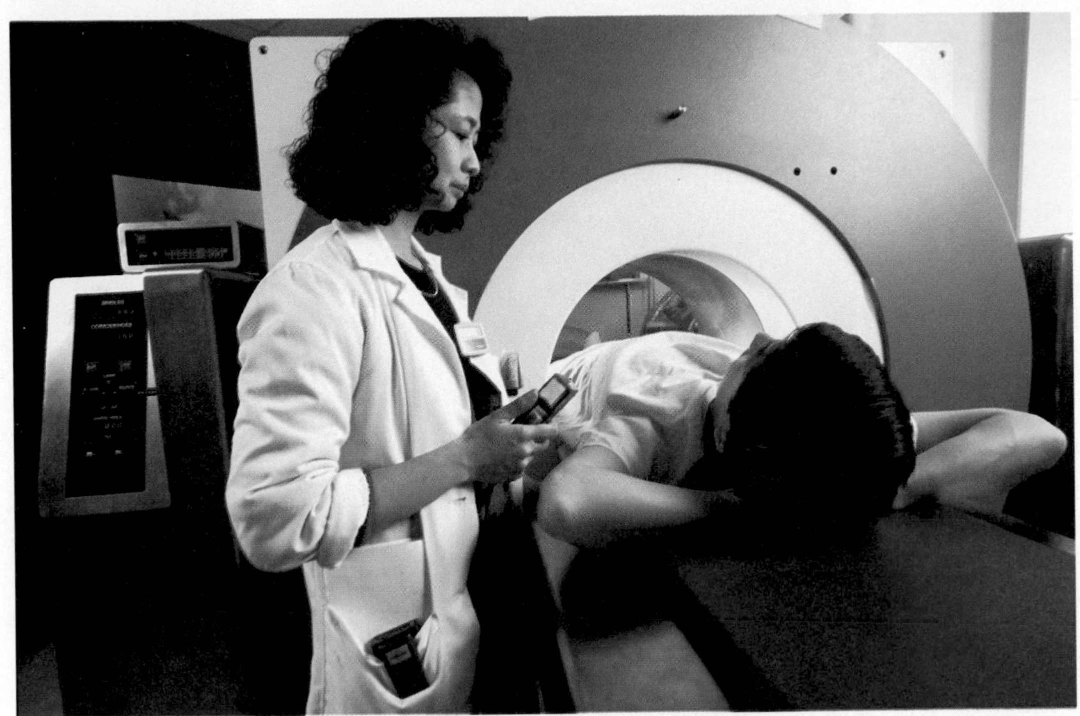

HELP WANTED

Blast Off!

The *Mars Pathfinder* hurtled into space to demonstrate a lower-cost way to deliver scientific equipment to the distant surface of Mars and send back data to Earth. What role does physics play in missions to Mars?

CHAPTER

1 What is physics?

What do you think when you see the word *physics*? Do you recall friends saying how hard it is? Do you think of blackboards filled with equations? The mushroom cloud of the atomic bomb? People in white lab coats? Albert Einstein? Stephen Hawking?

Physics does have a reputation of being difficult. Physics does use mathematics as a powerful language. But physics also involves concepts, ideas, and principles that are expressed in ordinary words.

Yes, the atomic bomb was developed with the aid of many physicists. But so were the computer chips used in PCs and video game systems; the graphite-epoxy materials used in guitars and golf clubs; the CDs on which your favorite music, computer games, and movies are recorded; and the lasers used to play them.

Albert Einstein and Stephen Hawking are examples of remarkable physicists. But so are many men and women who work in universities, two-year colleges, and high schools; at industrial and government labs; at hospitals; and on Wall Street. A physicist could easily be your next-door neighbor—or you!

WHAT YOU'LL LEARN

- You will learn to ask the questions "How do we know?" "Why do we believe it?" and "What's the evidence?" in order to examine and solve problems.
- You will have the satisfaction of understanding and even predicting the outcomes of physical occurrences all around you.

WHY IT'S IMPORTANT

- An understanding of physics will help you make informed decisions as a citizen in an increasingly complex world.

interNET CONNECTION

Follow the link for this chapter on the Glencoe Homepage at **www.glencoe.com/sec/science** to find out more about physics and missions to Mars.

Physics: The Search for Understanding

OBJECTIVES

- **Define** *physics.*
- **Relate** theory, experiment, and applications to the role they play in physics research.
- **Demonstrate** that, while there is no single scientific method, there are common methods used by all scientists.

Physics is a branch of knowledge that involves the study of the physical world. Physicists investigate objects as small as subatomic particles and as large as the universe. They study the natures of matter and energy and how they are related. Physicists and other scientists look at the world around them with inquisitive eyes. Their observations lead them to ask questions about what they see. What makes the sun shine? How were the planets formed? Of what is matter made? Physicists make observations, do experiments, and create models or theories to try to answer these questions. Finding explanations for the original questions often leads to more questions and thus more observations, experiments, and theories. The goal of all scientists is to obtain a compelling explanation that describes many different phenomena, makes predictions, and leads to a better understanding of the universe.

Sometimes the results of the work of physicists are of interest only to other physicists. Other times, their work leads to the development of devices such as lasers, communication systems, computers, and new materials that change everyone's life. As an example of how physics works, let's look at the role of the planet Mars, shown in **Figure 1–1,** in the development of the scientific method and the exploration of Mars.

The Wanderers

Have you ever seen the planet Mars? Mars is among the brightest planets in the night sky. Ancient people were keen observers of celestial objects in order to define the time of year and find the direction of travel on Earth. These observers noticed that five bright "wanderers," or planets, generally followed an eastern course through the constellations, yet, unlike the stars, they also moved westward for periods of time. The deep-red color of one of those planets caused the Babylonians to associate it with disaster and the Romans to name it after their god of war, Mars. Early records of the motion of Mars helped develop the early concepts of the solar system centuries before the invention of the telescope.

FIGURE 1–1 NASA scientists used 102 images taken by the *Viking Orbiter* to form this mosaic of Mars.

Are the stars and planets like Earth?

About 2500 years ago, Greek philosophers tried to determine what the world was made of by making observations of everyday occurrences. Some of these scholars believed that all matter on or near Earth was made up of four elements: earth, water, air, and fire. Each element was thought to have a natural place based on its heaviness. The highest place belonged to fire, the next to air, then water, and, at the bottom, earth. Motion was thought to occur because an element traveled in a straight line toward its own natural place.

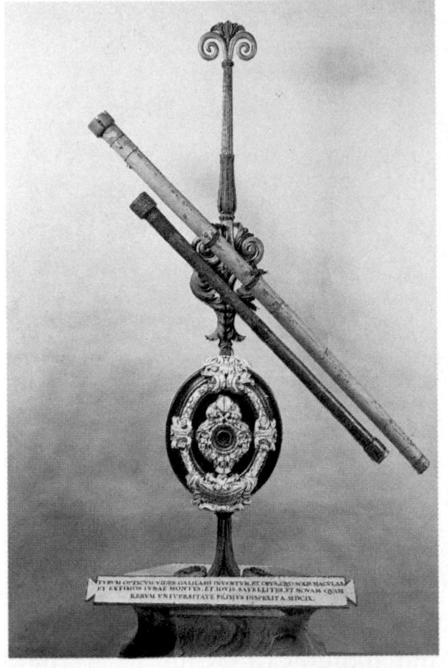

a

b

FIGURE 1–2 The telescope **(a)** and the lens **(b)** with which Galileo first observed the moons of Jupiter are on display at the Museo di Storia della Scienza in Florence, Italy.

Ancient people observed that the sun, moon, stars, and planets such as Mars didn't behave this way. As far as anyone could see, these celestial bodies were perfect spheres and moved in circles about Earth forever. They certainly didn't obey the same laws of motion as objects on Earth, and so it seemed that they couldn't be made of the four elements; rather, they were formed of a fifth element, quintessence.

The writings of these early Greeks, lost to Europe for hundreds of years, were studied and translated by Arabic scholars. In the twelfth century, the writings made their way to Europe and were accepted as truth that did not have to be questioned or tested. One of the first European scientists who claimed publicly that the ancient books were no substitute for observations and experiments was Galileo Galilei (1564-1642).

Galileo and Scientific Methods

In 1609, Galileo built a telescope shown in **Figure 1–2,** powerful enough to explore the skies. He found that the moon wasn't a perfect sphere, but had mountains, whose heights he could estimate from the shadows they cast. He discovered four moons circling the planet Jupiter, that the Milky Way was made up of many more stars than anyone had thought, and that Venus had phases. As a result, Galileo argued that Earth and the other planets actually circled the sun.

As Galileo studied astronomy and the motion of objects on Earth, he developed a systematic method of observing, experimenting, and analyzing that is now referred to as a **scientific method.** Rather than writing his results in Latin, the language of scholars, he wrote them in his native Italian so that any educated person could read and understand them. For these reasons, Galileo is considered to be the father of modern experimental science.

Falling

The Greek philosophers argued that heavy objects fall faster than light objects. Galileo stated that light and heavy objects fall at the same rate. What do you think? Drop four pennies taped together and a single penny from the same height at the same time. Tear a sheet of paper in half. Crumple one piece into a ball. Repeat your experiment with the paper ball and the half sheet of paper. What did you observe each time?

Analyze and Conclude Who was correct, the Greeks or Galileo?

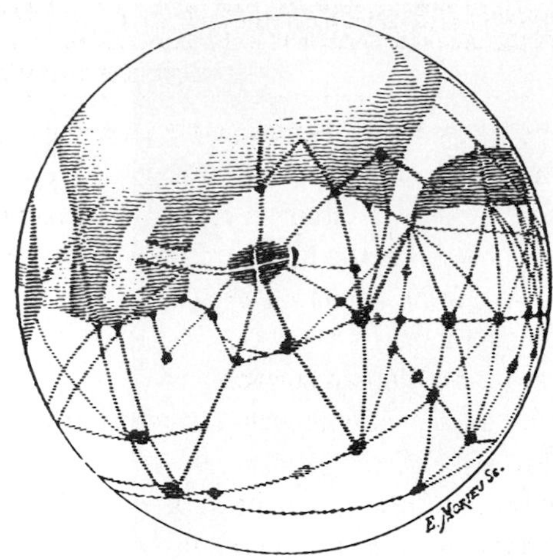

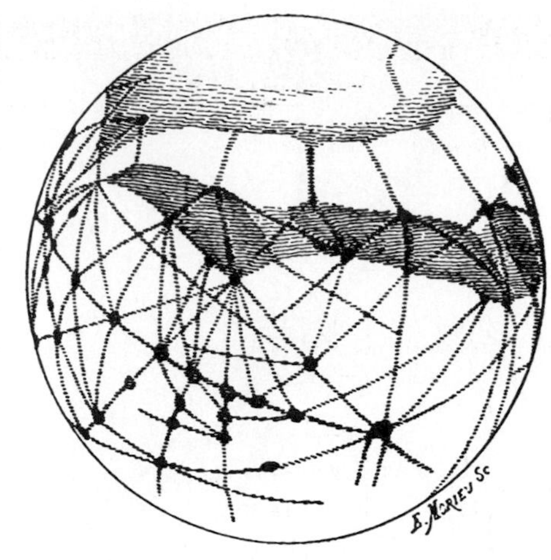

FIGURE 1–3 During the 17th, 18th, and 19th centuries, it was widely believed that Mars was inhabited. Dark linear features were interpreted as canals.

Galileo's methods are not the only scientific method. All scientists must study problems in an organized way. They combine systematic experimentation with careful measurements and analyses of results. From these analyses, conclusions are drawn. These conclusions are then subjected to additional tests to find out whether they are valid. Since Galileo's time, scientists all over the world have used these techniques and methods to gain a better understanding of the universe. Knowledge, skill, luck, imagination, trial and error, educated guesses, and great patience all play a part.

MUSIC CONNECTION

Composer Gustav Holst (1874-1934) is best known for his symphony *The Planets.* Its seven movements represent each planet except Earth and Pluto. (Pluto was not discovered until 1930, 12 years after Holst finished this work.) Holst based *The Planets* on astrology and mythology, not on astronomy. Movement titles include *Mars, the Bringer of War,* and *Mercury, the Winged Messenger.* What do these titles indicate about the studies of mythology, astrology, and astronomy?

Mars in Recent Times

As telescopes improved, Mars became much more interesting to astronomers and to people in general because they thought it looked much like Earth. Astronomers found what appeared to be ice caps that advance and recede, color changes they attributed to vegetation cycles similar to Earth's seasons, and dark areas believed to be seas. To some early observers, strange markings on the surface of Mars, shown in **Figure 1–3,** were mistakenly interpreted as being channels or canals possibly made by intelligent beings. This interpretation became so prevalent that a 1938 radio drama, depicting a Martian invasion of Earth, caused widespread panic in the United States.

When rockets capable of reaching our neighboring planet were developed in the 1960s, both the United States and the former Soviet Union launched a series of probes designed to orbit Mars, take photographs, and land on the planet to return data. A timetable of these probes is shown in **Table 1–1.**

The first of the recent probes, *Mars Pathfinder,* surrounded by protective airbags, bounded down on Mars on July 4, 1997. The entire mission cost less than the production of one Hollywood movie. A 10-kg robot rover, named *Sojourner,* was released to explore nearby rocks. Millions of people used the Internet to retrieve photos directly from the NASA websites into their home computers. The new era of Martian exploration had begun.

TABLE 1–1

	Missions to Mars	Accomplishments
1964	U.S. *Mariner 4*	First photos from 16 898 to 9846 km above surface.
	USSR *Zond 2*	Failed to send back data.
1969	U.S. *Mariner 6*	Examination of Martian equatorial region from an altitude of 3430 km.
	U.S. *Mariner 7*	Examination of the Martian southern hemisphere and south polar ice cap from an altitude of 3430 km.
1971	USSR *Mars 2*	Martian orbit.
	USSR *Mars 3*	Lander on Martian surface.
	U.S. *Mariner 9*	Photographs of entire Martian globe from orbit.
1973	USSR *Mars 5*	Martian orbit.
1975	U.S. *Viking 1*	Panoramic views and close-up photos from the Martian surface.
	U.S. *Viking 2*	Automated experiments on the Martian surface.
1988	USSR *Phobos 2*	Martian orbit.
1993	U.S. *Mars Observer*	Lost during mission.
1996	U.S. *Global Surveyor*	Record of surface features, atmospheric data, and magnetic properties from Martian orbit.
	U.S. *Mars Pathfinder*	Surface landing and release of a mobile vehicle to explore *Ares Vallis.*
		Goals
1998	U.S. *Mars Surveyor Orbiter*	To map Martian surface and weather patterns and act as a communications satellite between landers, rovers, and Earth.
1999	U.S. *Mars Surveyor Lander*	To land at the Martian south pole to sample and analyze the surface.
2001 and beyond	U.S. orbiters and landers	To replace communications relay satellites with new orbiters; to continue analyses of Martian atmosphere and surface with orbiters and landers.
	U.S. sample-and-return spacecraft	To return Martian rock and soil samples to Earth.

Why study Mars?

Mars had seemed in many ways to be similar to Earth. But the probes have confirmed that its climate is very different. Mars is an ideal laboratory for scientists interested in geology and atmospheric physics.

From the study of Mars, scientists may learn more about the types of conditions that could lead to dramatic climatic or atmospheric changes on our own planet, and about the formation and evolution of the entire solar system. These studies may help us understand why Mars grew cold with almost no atmosphere early in its history while Venus and Earth did not.

The search for water is central to future explorations. Mars's northern polar cap contains water in the form of ice, a vital ingredient for future human exploration. Furthermore, there is evidence that gigantic floods helped shape the surface of Mars billions of years ago. What happened to that water? Is it combined with rocks, is it frozen underground, or has it escaped into space?

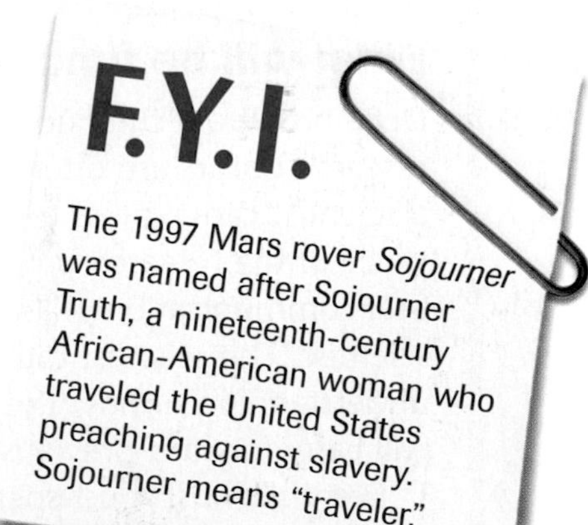

F.Y.I.
The 1997 Mars rover *Sojourner* was named after Sojourner Truth, a nineteenth-century African-American woman who traveled the United States preaching against slavery. Sojourner means "traveler."

Physics & Society

Research Dollars

Some scientific discoveries are made by chance, but most are the result of years of carefully planned research. Most scientists are paid to conduct research—exploring ideas, creating hypotheses, performing experiments, and publishing findings. Professors at universities and their students spend a significant amount of time in the laboratory. Other scientists work for government-funded laboratories or for private companies and spend virtually all of their time doing research.

Who will pay?

Where does the money to pay for this research come from? In the case of university professors, much of the support comes from grants supplied by the government, private foundations, or private companies. Government-funded laboratories receive money from the federal budget. Private companies fund their own projects, often using profits earned by inventions developed in previous, successful research. Sometimes, private companies are hired by government agencies to participate in large projects.

What will be funded?

Decisions about funding with limited research dollars are often based on how well a scientist can express to others the importance of the research project. Written and oral communication skills are vital to every scientist. The scientist must have a good understanding of how the proposed project will carry forward previous research.

Every scientist must spend time studying and evaluating the work of other scientists. He or she must also be able to clearly communicate why the work is needed and who could benefit from its results. Becoming a scientist requires not only an education in a chosen field of study, but also the ability to think critically and communicate effectively.

Who will benefit?

Not everyone agrees about the kinds of research that should be funded, or even how much money should be spent on scientific research in general. This is especially true when research funds come from the government. Taxpayers often disagree with government spending decisions. Some people believe that their money would be better spent on solving more immediate human problems, such as feeding the hungry, sheltering the homeless, and curing disease. Others argue that the benefits humans derive from exploratory research are well worth their cost.

Investigating the Issue

1. **Communicating Ideas** Read several articles from publications such as *Science News* or *Scientific American* about advances in scientific research. Write a brief essay about the research areas that you feel are most interesting or important.
2. **Debating the Issue** Should the U.S. government support research in outer space, or should the money go toward research in areas with more humanitarian applications?

Follow the link for this chapter on the Glencoe Homepage at **www.glencoe.com/sec/science** to find out more about scientific research.

If Mars were a warmer, wetter world early in its history, what happened to cause such climatic devastation and render it lifeless, barren, and frigid? Have there ever been life forms on Mars? Living systems on Mars today would have to be able to survive without oxygen, store water for long periods of time, and live underground or have protection from solar radiation and large temperature fluctuations. In 1976, the *Viking* landers found no evidence of life. But in 1996, scientists claimed to have evidence of primitive life forms in meteorites found in Antarctica that they strongly believed had Martian origins.

Who will study Mars?

The Mars exploration team is made up of many women and men. Some represent the sciences, including physics, chemistry, geology, and astronomy. Others are electrical, mechanical, aeronautic, or computer engineers. Still others are technicians, graphic designers, managers, and administrators. All share some common characteristics. They are curious, creative, and interested in mysteries and in solving problems. They love their work, but they also have many outside interests such as music, drama, sports, and mountain climbing. When they were younger they took science and mathematics courses, but they were also involved in many activities in and out of school.

The members of the Mars exploration team had to join their individual experiences and learn to work together, as shown in **Figure 1–4.** They report that it can be harder to work with a team than on their own, but that the team can do more, and so the rewards can be greater. They also have found that it's more fun when they can share ideas and experiences with others.

HELP WANTED

NASA RESEARCHERS

Seeking researchers for NASA's Jet Propulsion Laboratory (JPL) and other space centers in the United States. Positions are available in engineering, aeronautics, robotics, computer systems, fluid and flight mechanics, chemistry, materials and structures, and telecommunications. Also hiring researchers in lunar and planetary studies, meteorology, radiation, and related areas. Some projects require specialists in physiology, psychology, botany, and biology.
Contact:
JPL Human Resources
4800 Oak Grove Drive
Pasadena, CA 91109

F.Y.I.

Mae Jemison relied on experience as engineer, physician, educator, and first African American woman astronaut to found The Jemison Institute for Advancing Technology in Developing Countries.

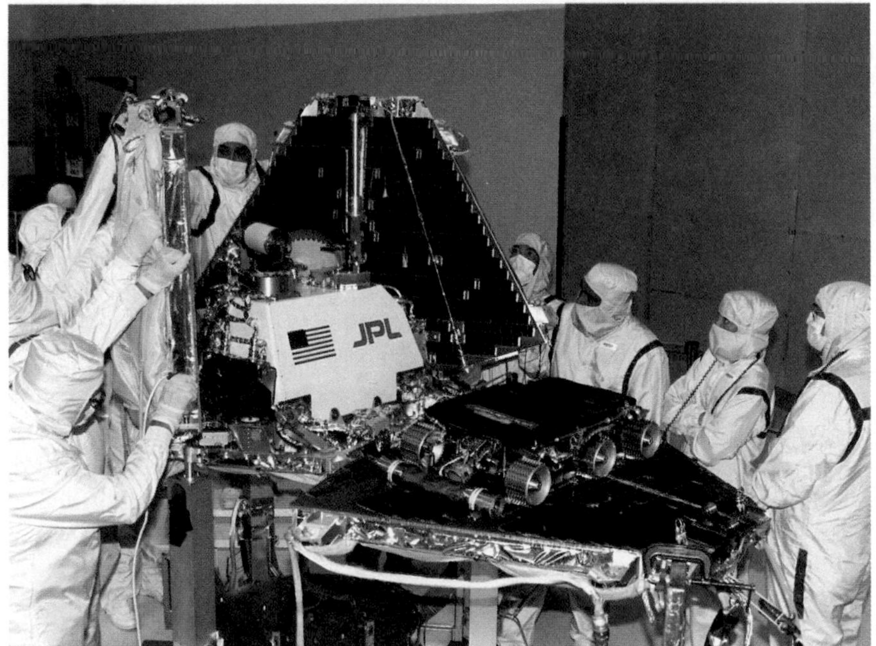

FIGURE 1–4 Jet Propulsion Laboratory scientists prepare the *Mars Pathfinder* for placement atop a Delta II launch vehicle.

There is room on the team for you. Thanks to the Internet, you can send E-mail to the Mars exploration team and ask questions. You also can follow the probes' progress: you can see the photos before they appear in the newspapers and obtain more complete coverage of the results than that which is provided in the ten seconds that fits into local television news. You can choose your course of study so that you can become part of future space exploration.

Is physics important?

Most physicists are not involved directly in the Mars explorations. Most of the people directly involved in the Mars missions are not physicists, nor did they major in physics in college.

Blast Off!

But the Mars missions are based on physics, starting with the design of the rocket engines, the gyroscopic directional controls, and the precision clocks that are needed to indicate where the spacecraft is and how fast it is moving. The solar panels and nuclear electrical sources that keep the probes in contact with Earth during flight are based on physics. Physics is also involved in the design of the cameras, computers, radio transmitters, and receivers that send the photos back to us.

Science and technology constantly interact. Sometimes, scientific results produce new equipment for use outside the scientific community. The efficient design of the mechanical arms that allow the rovers to sample the surface of Mars may be used to make artificial limbs for people with disabilities. Similarly, new equipment produces new scientific results. Advances in computer technology allow faster, lighter computers to be placed on board the spacecraft. As shown in **Figure 1–5,** the applications of such discoveries affect all our lives.

All participants in the Mars missions use the problem-solving skills that they learned in physics and other science courses every day. They can't find answers in the back of a book or by asking a friend! They have learned the skills that enable them to go forward from a predicament to a decision by choosing relevant information, making logical decisions, and applying old applications to new situations and new applications to old situations. Above all, they have learned how to work as a team: dividing the work but making sure that everyone understands, exploring all possibilities but agreeing on one method, and checking to make sure that the problem really was solved. Finally, they have learned how to make presentations, orally or in writing, that communicate what they have learned to their coworkers, their friends, and the general public.

The goal of this course is not to make you a physicist. It is to show you the way that physicists view the world and to give you an understanding of the physical world around you. It may be that you will become interested in a rewarding career in science or technology. Whatever your chosen career, you will be able to make better-informed decisions in an increasingly complex age. You will learn to ask the questions "How do we know?", "Why do we believe it?", and "What's the evidence for that?" when you are presented with new information or new problems.

FIGURE 1–5

We are surrounded in our daily lives by physics success stories. Examine a few highlights below.

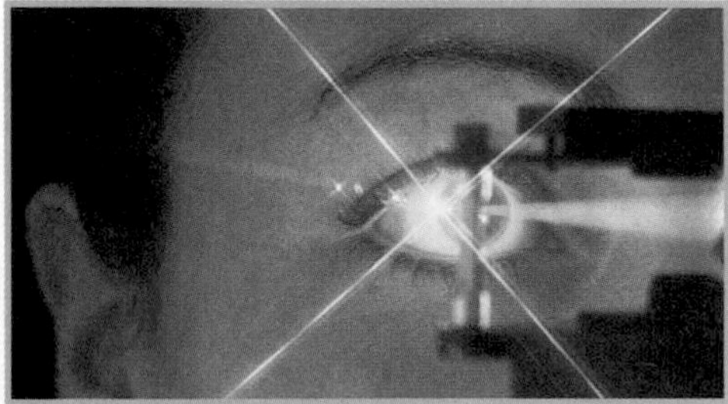

1 Originally thought to serve no useful purpose, lasers are now used in industry and construction, data storage and retrieval, medicine, telecommunications, navigation, and defense.

2 Wartime research into radar and miniature electronics led to the development of microwave ovens.

3 Razor blades are coated with thin film materials using plasma physics techniques. The blade handles are attached by laser welding. Computerized vision systems quality check each batch.

4 Billions of dollars are saved by consumers as automobiles are built with lighter composite materials and polymers, with microcomputers to control fuel injection systems, and with more efficient fuel cells and batteries.

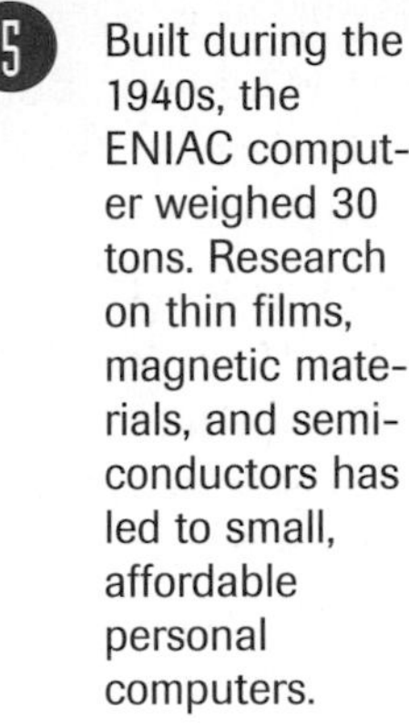

5 Built during the 1940s, the ENIAC computer weighed 30 tons. Research on thin films, magnetic materials, and semiconductors has led to small, affordable personal computers.

6 Energy efficient houses are a result of physics research on heat transfer, thin films, plasma sources, vacuum technology, optics, and new materials.

7 The nanotechnology that built this guitar will allow scientists to study processes and perform functions on a submicroscopic level.

Physics Lab

The Paper Tower

Problem

What is the tallest free standing tower you can construct with a single sheet of paper and 30 cm of cellophane tape?

Materials  

1 sheet of white paper per student
1 sheet of colored paper per group
cellophane tape
scissors

Procedure

1. Each student will receive one sheet of white paper. Use the white sheet to try out various design possibilities. Think creatively.
2. Each lab group will receive one sheet of colored paper to make a competition tower.
3. Before working with the colored paper, examine the designs of each group member.
4. Decide which aspects of each design should be incorporated into your final design. The most important aspects of a winning team are communication and cooperation.
5. Plan ahead. Set a timetable for experimentation and for actual construction. Plan on finishing at least five minutes before the end of the period.
6. Watch your time. Do not fall too far behind schedule.
7. Your tower must be free standing for at least five seconds.
8. Measure the height of your tower before it tips over.

Analyze and Conclude

1. **Forming a Hypothesis** What are some limiting factors to how high a tower can be built?
2. **Analyzing the Results** What were the limiting factors in your tower's construction?
3. **Evaluating the Process** Did your group work well as a team? What could you do differently to be more effective?

Apply

1. What architectural elements have been incorporated into your design?

Data and Observations	
Group	Design

CHAPTER 1 REVIEW

Key Terms
- physics
- scientific method

Summary

- Physics is the study of matter and energy and their relationships.
- Physics is basic to all other sciences.
- A knowledge of physics makes us, as citizens, better able to make decisions about questions related to science and technology.
- Much scientific work is done in groups in which people collaborate with one another.

Reviewing Concepts

1. Define *physics* in your own words.
2. Why is mathematics important to science?
3. Assume for a moment that the theory of matter held by some of the ancient Greeks is correct. How does this theory explain the motion of the four elements?

Applying Concepts

4. Give some examples of applications that resulted from work done by physicists.
5. Give some examples of applications that have resulted from work done by physicists on the exploration of space.
6. Research the aspects of nature investigated by each of the following kinds of scientists: astrophysicists, astronomers, biophysicists, exobiologists, and geophysicists.
7. Some of the branches of physics that you will study in this course investigate motion, the properties of materials, sound, light, electricity and magnetism, properties of atoms, and nuclear reactions. Give at least one example of an application of each branch.
8. What reason might the Greeks have had not to question the evidence that heavier objects fall faster than lighter objects? **Hint:** Did you ever question which falls faster?
9. Is the scientific method a clearly defined set of steps and procedures? Support your answer.
10. Why will the work of a physicist never be finished?

Critical Thinking Problems

11. It has been said that a fool can ask more questions than a wise man can answer. In science, it is frequently the case that a wise man is needed to ask the right question rather than to answer it. Explain.

Going Further

Class Discussion In 1996, scientists reported that meteorites found in Antarctica were actually from Mars, probably ejected from that planet by the impact of a meteor or comet millions of years ago. These meteorites were especially interesting because they contain structures that were interpreted as evidence of simple life-forms.

As a group, brainstorm ways to develop answers to the three questions "How do we know?", "Why do we believe it?", and "What's the evidence for that?" regarding the composition of these meteorites.

interNET CONNECTION

Follow the link on the Glencoe Homepage at **www.glencoe.com/sec/science** to find out more about this chapter.

A Graphic Display

Computers monitor the speed, the location, the fuel consumption, and the environment in and around the spacecraft; the physiological condition of any crew members; and data from experiments. How do scientists use equations and graphs to analyze and display this information?

CHAPTER 2 A Mathematical Toolkit

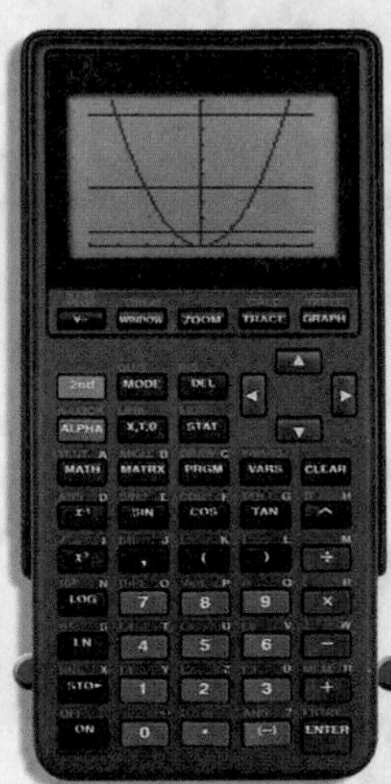

Within your lifetime, humans may set foot on Mars. What will they find when they arrive? Scientists believe that the physical principles learned on Earth will be equally valid on Mars. After all, those principles will have gotten them there. How could they test this hypothesis? They would have to make observations, do experiments, and make measurements. Both qualitative and quantitative experimental results then would be transmitted to Earth using words, diagrams, and mathematics.

Physics is a way of thinking based on experiments with numerical results that can be reproduced by others. Mathematics often has been called the language of physics. In this chapter, you'll find mathematical techniques that will be useful throughout this course. You might think of this chapter as a collection of tools. Appendix A contains more mathematical tools. Look through it now to find out what is there.

WHAT YOU'LL LEARN

- You will perform calculations using SI units and scientific notation.
- You will understand the need for accuracy and precision when making measurements and reporting data.
- You will display and evaluate data using graphs.

WHY IT'S IMPORTANT

- A basic understanding of mathematics is useful not only in the laboratory but also at the shopping mall, on the highway, in the kitchen, and on the playing field.

*inter*NET CONNECTION

Follow the link for this chapter on the Glencoe Homepage at **www.glencoe.com/sec/science** to find out more about use of mathematics in science.

2.1 The Measures of Science

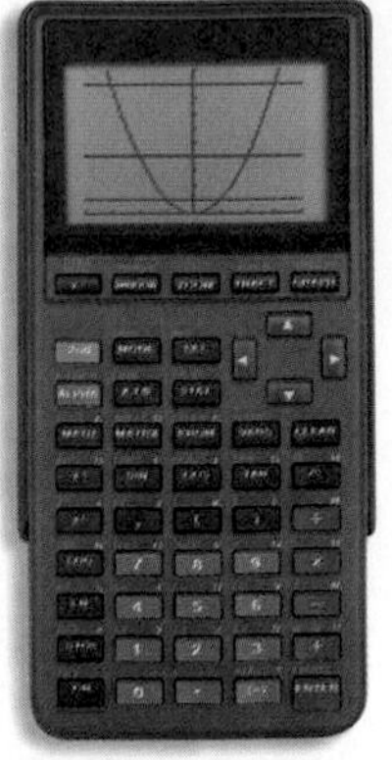

OBJECTIVES

- **Define** the SI standards of measurement.
- **Use** common metric prefixes.
- **Estimate** measurements and solutions to problems.
- **Perform** arithmetic operations using scientific notation.

The first humans on Mars must make the measurements needed to do experiments on Mars. But, what is a measurement? Every measurement is a comparison between an unknown quantity and a standard. If this comparison is to be valid, the measuring device must be compared against a widely accepted standard. For a standard to be useful, it must be practical for the type of measurement being made, readily accessible, reproducible, and constant over time. There must be agreement among users as to what the standard defines.

The Metric System and SI

French scientists adopted the metric system of measurement in 1795. Until that time, communications among scientists had been difficult because the units of measurement were not standardized. Units of measurement had been based on local customs. The **metric system** provides a set of standards of measurement that is convenient to use because units of different sizes are related by powers of 10.

The worldwide scientific community and most countries currently use an adaptation of the metric system to make measurements. The Système Internationale d'Unités, or **SI,** is regulated by the International Bureau of Weights and Measures in Sèvres, France. This bureau and the National Institute of Science and Technology (NIST) in Gaithersburg, Maryland keep the standards of length, time, and mass against which our metersticks, clocks, and balances are calibrated. Because other quantities can be described using combinations of these three units, length, time, and mass are base quantities. The units in which these quantities are measured are thus **base units. Table 2–1** lists the seven base quantities and their units, which are the foundation of SI.

FIGURE 2–1 The Metric Conversion Act became law in the United States in 1975.

TABLE 2–1

SI Base Units

Base quantity	Base unit	Symbol
Length	meter	m
Mass	kilogram	kg
Time	second	s
Temperature	kelvin	K
Amount of a substance	mole	mol
Electric current	ampere	A
Luminous intensity	candela	cd

The SI base unit of length is the **meter,** m, as shown in **Figure 2–1.** The meter was first defined as 1/10 000 000 of the distance from the north pole to the equator, measured along a line passing through Lyons, France. Later, it was more practical to define the standard meter as the distance between two lines engraved on a platinum-iridium bar kept in Paris. Methods of comparing times have become much more precise than those of comparing lengths. Therefore, in 1983, the meter was defined as the distance traveled by light in a vacuum during a time interval of 1/299 792 458 s.

The standard SI unit of time is the **second,** s. The second was first defined as 1/86 400 of the mean solar day. A mean solar day is the average length of the day over a period of one year, approximately 24 hours. It is now known that Earth's rotation is slowing; and days are getting longer; thus that standard is not constant. In the 1960s, atomic clocks were developed that gain or lose only 1 s in approximately 3 000 000 years. The second is currently defined in terms of the frequency of one type of radiation emitted by a cesium-133 atom. NIST adds a leap second every few years as Earth's rotation continues to slow.

The third standard unit, the **kilogram,** kg, measures the mass of an object. The kilogram is the mass of a small platinum-iridium metal cylinder kept at very controlled temperature and humidity. A copy is kept at NIST, as shown in **Figure 2–2.**

A wide variety of other units, called **derived units,** are combinations of the base units. Common derived units include the meter per second, m/s, used to measure speed, and the joule, $kg{\cdot}m^2/s^2$, used to measure energy, as shown in **Figure 2–3.** As you learn the base and derived units, you will find it is useful when solving physics problems to perform dimensional analysis, that is, to treat the units in each term of the equation as algebraic quantities, to help assure the accuracy of an answer.

FIGURE 2–2 The International Prototype Kilogram is composed of a platinum-iridium alloy.

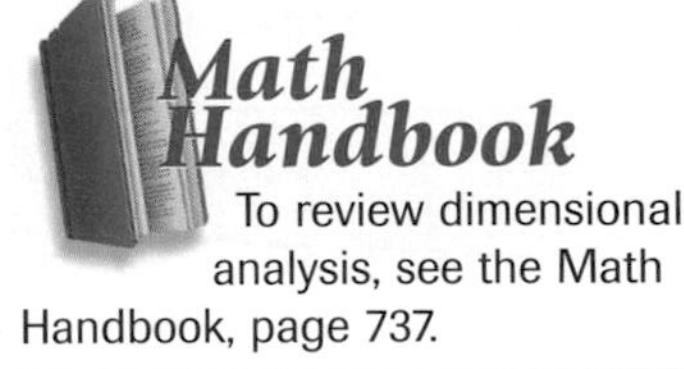

To review dimensional analysis, see the Math Handbook, page 737.

Problem Solving Strategy

Estimates

Does a measurement you made during an experiment or the answer to a problem you solved make sense? Suppose you calculated that a runner was moving at 275 m/s. Is that number reasonable, or could you have made a mistake? To check your results, you often can make a rough estimate.

The ability to make rough estimates without actually solving a problem is a skill you will find useful all your life. Rough estimates give you a hint about how to start working on a solution and whether or not the method has a chance of working.

For example, is 275 m/s a reasonable running speed? Your stride is about 1 m. How many strides can you make in 1 s? A sprinter runs about 100 m in 10 s. A rough estimate of 10 m/s tells you that your answer of 275 m/s is not reasonable.

FIGURE 2–3 The SI derived unit for energy, the joule, is common on food labels in many countries.

HISTORY CONNECTION

How did people measure in ancient times? In Mesopotamia (3500–1800 B.C.), workers built the first cities using cubits (43–56 cm or 17–22 in). They measured weight in shekels (0.5 oz). In the 1400s and 1500s, the Incas used body parts as units of measurement. A span was the length of a man's hand: 20 cm (8 in). A fathom was the width of a man's outstretched arms: 160 cm (64 in). A pace was 1.2 m (4 ft). What problems would result from using these measurement units?

Rough estimates have to start with reasonable lengths and times. How long is a car? How large is a page in this book? How wide is your classroom? You do not have to look for a ruler or meterstick—you already have three built-in measuring sticks. The distance from your nose to the fingertips on your outstretched hand is about one meter. The width of your fist is about 1/10 m, and the width of your finger is about 1/100 m or 1 cm.

How long does it take something to fall? You can count seconds by saying, to yourself, "one chimpanzee, two chimpanzees, and so on" In fact, by placing the emphasis on "**one** chim**pan**zee," you can actually estimate half-seconds quite well. You'll find that it takes about one-half second for an object held high above your head to hit the ground.

With these methods, you can estimate times, distances, and velocities to within a factor of 2 or 3. That is usually good enough to find the worst of your errors.

SI Prefixes

The metric system is a decimal system. Prefixes are used to change SI units by powers of 10. To use SI units effectively, you should know the meanings of the prefixes in **Table 2–2.** For example, 1/10 of a meter is a decimeter, 1/100 of a meter is a centimeter, 1/1000 of a meter is a millimeter and 1000 meters is a kilometer. All of these divisions except the kilometer can be found on a meterstick. **Figure 2–4** shows the vast range of lengths of objects in our universe; commonly used length units also are indicated.

The same prefixes are used for all quantities. For example, 1/1000 of a gram is a milligram, 1/1000 of a second is a millisecond, and 1/1000 of a liter is a milliliter.

TABLE 2–2

Prefixes Used with SI Units

Prefix	Symbol	Multiplier	Scientific notation	Example
femto	f	1/1 000 000 000 000 000	10^{-15}	femtosecond (fs)
pico	p	1/1 000 000 000 000	10^{-12}	picometer (pm)
nano	n	1/1 000 000 000	10^{-9}	nanometer (nm)
micro	μ	1/1 000 000	10^{-6}	microgram (μg)
milli	m	1/1000	10^{-3}	milligram (mg)
centi	c	1/100	10^{-2}	centimeter (cm)
deci	d	1/10	10^{-1}	deciliter (dL)
kilo	k	1000	10^{3}	kilometer (km)
mega	M	1 000 000	10^{6}	megagram (Mg)
giga	G	1 000 000 000	10^{9}	gigameter (Gm)
tera	T	1 000 000 000 000	10^{12}	terameter (Tm)

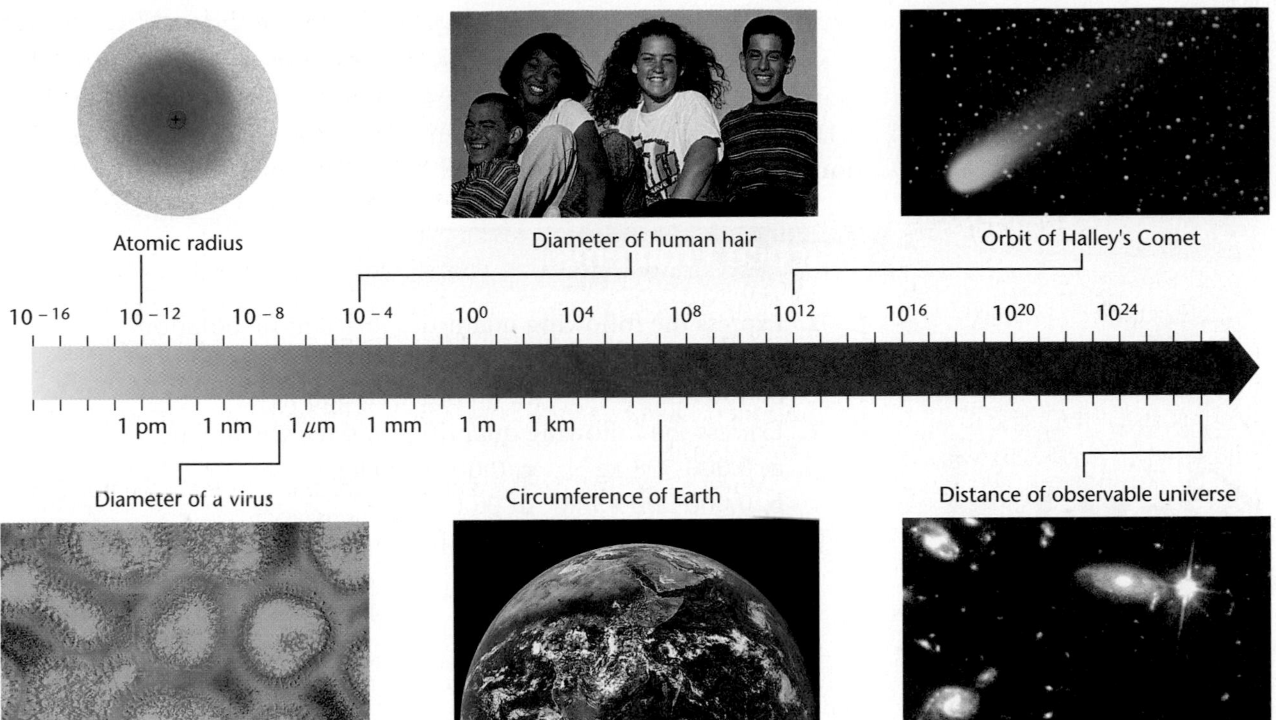

FIGURE 2–4 Objects in the universe range from the very small to the unimaginably large.

Scientific Notation

Many of the numerical values of the multipliers in **Table 2–2** are very large or very small numbers. Written in this form, the values of the quantities take up much space. Such large or small measurements are difficult to read, their relative sizes are difficult to determine, and they are awkward to use in calculations. To work with such numbers, write them in **scientific notation** by expressing decimal places as powers of 10. The numerical part of a quantity is written as a number between 1 and 10 multiplied by a whole-number power of 10.

$$M \times 10^n$$

$1 \leq M < 10$ and n is an integer. To write numbers using scientific notation, move the decimal point until only one non zero digit remains on the left. Then count the number of places you moved the decimal point and use that number as the exponent of 10.

The average distance from the sun to Mars is 227 800 000 000 m. In scientific notation, this distance would be 2.278×10^{11} m. The number of places you move the decimal to the left is expressed as a positive exponent of 10.

The mass of an electron is about

0.000 000 000 000 000 000 000 000 000 000 911 kg.

To write this number in scientific notation, the decimal point is moved 31 places to the right. As a result the mass of an electron is written as 9.11×10^{-31} kg. The number of places you move the decimal to the right is expressed as a negative exponent of 10.

How good is your eye?

The distance from your nose to your outstretched fingertips is about 1 m. Estimate the distance between you and three objects in the room. Have the members in your lab group each make a data table and record their estimates. Verify each distance.

Compare Results Were the estimates reasonably close? Did one person consistently make accurate estimates? What could be done to improve your accuracy?

Scientific notation with calculators

Many calculators display numbers in scientific notation as *M En*. For example, a calculator might show 2.278×10^{11} as 2.278 E11 or show 9.11×10^{-31} as 9.11 E − 31. When you report the results of a calculation, you should write it in normal scientific notation.

Practice Problems

1. Express the following quantities in scientific notation.
   a. 5800 m **c.** 302 000 000 m
   b. 450 000 m **d.** 86 000 000 000 m
2. Express the following quantities in scientific notation.
   a. 0.000 508 kg **c.** 0.0003600 kg
   b. 0.000 000 45 kg **d.** 0.004 kg
3. Express the following quantities in scientific notation.
   a. 300 000 s
   b. 186 000 s
   c. 93 000 000 s

F.Y.I.

Parts used in high-performance car engines must be measured to within 7μm (7×10^{-6} meters).

Converting Units

What is the equivalent in kg of 465 g? You know from **Table 2–2** that 1 kg = 1000 g. Thus, (1000 g)/(1kg) = 1. How can this information be used to convert units?

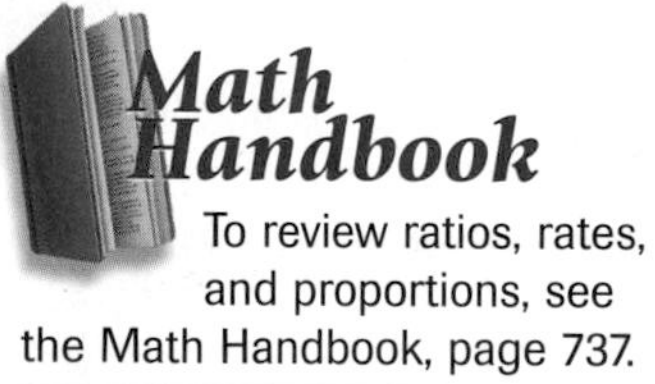

To review ratios, rates, and proportions, see the Math Handbook, page 737.

Problem Solving Strategy

The Factor-Label Method of Unit Conversion

An easy way to convert a quantity expressed in one unit to that quantity in another unit is to use a conversion factor, a relationship between the two units. A conversion factor is a multiplier equal to 1. Because 1 kg = 1000 g, you can construct the following conversion factors.

$$1 = \frac{1\ \text{kg}}{1000\ \text{g}} \text{ or } 1 = \frac{1000\ \text{g}}{1\ \text{kg}}$$

Recall that the value of a quantity does not change when it is multiplied or divided by 1. Therefore, to find the equivalent in kg of 465 g, multiply it by an appropriate conversion factor.

$$465\ \text{g} = (465\ \text{g})\left(\frac{1\ \text{kg}}{1000\ \text{g}}\right) = \frac{465\ \text{g} \times 1\ \text{kg}}{1000\ \text{g}} = 0.465\ \text{kg}$$

Unit labels cancel just like algebraic quantities. If the final units do not make sense, check your conversion factors. A factor may have been inverted or written incorrectly. This method of converting one unit to another is called the **factor-label method** of unit conversion.

Practice Problems

4. Convert each of the following length measurements as directed.
   a. 1.1 cm to meters
   b. 76.2 pm to millimeters
   c. 2.1 km to meters
   d. 2.278×10^{11} m to kilometers
5. Convert each of the following mass measurements to its equivalent in kilograms.
   a. 147 g **b.** 11 Mg **c.** 7.23 μg **d.** 478 mg

Arithmetic Operations in Scientific Notation

Suppose you need to add or subtract measurements expressed in scientific notation, $M \times 10^n$. The measurements must be expressed in the same powers of 10 and the same units.

Example Problem

Addition and Subtraction Using Scientific Notation

Solve the following problems. Express the answers in scientific notation.

a. 4×10^8 m + 3×10^8 m

b. 4.1×10^{-6} kg − 3.0×10^{-7} kg

c. 4.02×10^6 m + 1.89×10^2 m

Calculate Your Answer

Strategy:

a. If the numbers have the same exponent, n, add or subtract the values of M and keep the same n.

Calculations:

4×10^8 m + 3×10^8 m
= $(4 + 3) \times 10^8$ m
= 7×10^8 m

b. If the exponents are not the same, move the decimal to the left or right until they are the same. Then add or subtract M.

4.1×10^{-6} kg − 3.0×10^{-7} kg
= 4.1×10^{-6} kg − 0.30×10^{-6} kg
= $(4.1 - 0.30) \times 10^{-6}$ kg
= 3.8×10^{-6} kg

c. If the magnitude of one number is quite small when compared to the other number, its effect on the larger number is insignificant. The smaller number can be treated as zero.

4.02×10^6 m + 1.89×10^2 m
= $40\,200 \times 10^2$ m + 1.89×10^2 m
= $(40\,200 + 1.89) \times 10^2$ m
= $40\,201.89 \times 10^2$ m
= $4.020\,189 \times 10^6$ m
= 4.02×10^6 m

Practice Problems

Solve the following problems. Write your answers in scientific notation.

6. **a.** 5×10^{-7} kg + 3×10^{-7} kg
b. 4×10^{-3} kg + 3×10^{-3} kg
c. 1.66×10^{-19} kg + 2.30×10^{-19} kg
d. 7.2×10^{-12} kg − 2.6×10^{-12} kg

7. **a.** $6 \times 10^{-8}\ m^2 - 4 \times 10^{-8}\ m^2$
b. $3.8 \times 10^{-12}\ m^2 - 1.90 \times 10^{-11}\ m^2$
c. $5.8 \times 10^{-9}\ m^2 - 2.8 \times 10^{-9}\ m^2$
d. $2.26 \times 10^{-18}\ m^2 - 1.8 \times 10^{-18}\ m^2$

8. **a.** 5.0×10^{-7} mg + 4×10^{-8} mg
b. 6.0×10^{-3} mg + 2×10^{-4} mg
c. 3.0×10^{-2} pg − 2×10^{-6} ng
d. 8.2 km − 3×10^{2} m

Math Handbook

To review the properties of exponents, see the Math Handbook, page 737.

To multiply quantities written in scientific notation, simply multiply the values and units of *M*. Then add the exponents. To divide quantities expressed in scientific notation, divide the values and units of *M*, then subtract the exponent of the divisor from the exponent of the dividend. If one unit is a multiple of the other, convert to the same unit.

Example Problem

Multiplication and Division Using Scientific Notation

Find the value of each of the following quantities.

a. $(4 \times 10^{3}\ \text{kg})(5 \times 10^{11}\ \text{m})$

b. $\dfrac{8 \times 10^{6}\ \text{m}^3}{2 \times 10^{-3}\ \text{m}^2}$

Calculate Your Answer

Strategy:

a. Multiply the values of *M* and add the exponents, *n*. Multiply the units.

Calculations:

$(4 \times 10^{3}\ \text{kg})(5 \times 10^{11}\ \text{m}) = (4 \times 5) \times 10^{3+11}\ \text{kg·m}$

$= 20 \times 10^{14}\ \text{kg·m}$

$= 2 \times 10^{15}\ \text{kg·m}$

b. Divide the values of *M* and subtract the exponent of the divisor from the exponent of the dividend.

$\dfrac{8 \times 10^{6}\ \text{m}^3}{2 \times 10^{-3}\ \text{m}^2} = \dfrac{8}{2} \times 10^{6-(-3)}\ \text{m}^{3-2}$

$= 4 \times 10^{9}\ \text{m}$

Practice Problems

Find the value of each of the following quantities.

9. a. $(2 \times 10^4\ \text{m})(4 \times 10^8\ \text{m})$
b. $(3 \times 10^4\ \text{m})(2 \times 10^6\ \text{m})$
c. $(6 \times 10^{-4}\ \text{m})(5 \times 10^{-8}\ \text{m})$
d. $(2.5 \times 10^{-7}\ \text{m})(2.5 \times 10^{16}\ \text{m})$

10. a. $\dfrac{6 \times 10^8\ \text{kg}}{2 \times 10^4\ \text{m}^3}$ **c.** $\dfrac{6 \times 10^{-8}\ \text{m}}{2 \times 10^4\ \text{s}}$

b. $\dfrac{6 \times 10^8\ \text{kg}}{2 \times 10^{-4}\ \text{m}^3}$ **d.** $\dfrac{6 \times 10^{-8}\ \text{m}}{2 \times 10^{-4}\ \text{s}}$

11. a. $\dfrac{(3 \times 10^4\ \text{kg})(4 \times 10^4\ \text{m})}{6 \times 10^4\ \text{s}}$

b. $\dfrac{(2.5 \times 10^6\ \text{kg})(6 \times 10^4\ \text{m})}{5 \times 10^{-2}\ \text{s}^2}$

12. a. $(4 \times 10^3\ \text{mg})(5 \times 10^4\ \text{kg})$
b. $(6.5 \times 10^{-2}\ \text{m})(4.0 \times 10^3\ \text{km})$
c. $(2 \times 10^3\ \text{ms})(5 \times 10^{-2}\ \text{ns})$

13. a. $\dfrac{2.8 \times 10^{-2}\ \text{mg}}{2.0 \times 10^4\ \text{g}}$

b. $\dfrac{(6 \times 10^2\ \text{kg})(9 \times 10^3\ \text{m})}{(2 \times 10^4\ \text{s})(3 \times 10^6\ \text{ms})}$

14. $\dfrac{(7 \times 10^{-3}\ \text{m}) + (5 \times 10^{-3}\ \text{m})}{(9 \times 10^7\ \text{km}) + (3 \times 10^7\ \text{km})}$

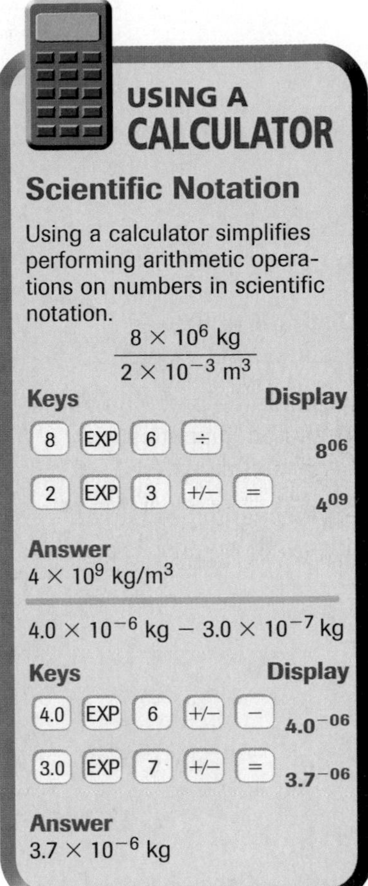

USING A
CALCULATOR
Scientific Notation
Using a calculator simplifies performing arithmetic operations on numbers in scientific notation.
$\dfrac{8 \times 10^6\ \text{kg}}{2 \times 10^{-3}\ \text{m}^3}$
Keys
Display
8 EXP 6 ÷
8 06
2 EXP 3 +/– =
4 09
Answer
$4 \times 10^9\ \text{kg/m}^3$
$4.0 \times 10^{-6}\ \text{kg} - 3.0 \times 10^{-7}\ \text{kg}$
Keys
Display
4.0 EXP 6 +/– –
4.0 –06
3.0 EXP 7 +/– =
3.7 –06
Answer
$3.7 \times 10^{-6}\ \text{kg}$

2.1 Section Review

1.1 A calculator displayed a number as 1.574 E8. Express this number in normal scientific notation.

1.2 Your height could be given either in terms of a small unit, such as a millimeter, or a larger unit, such as a meter. In which case would your height be a larger number?

1.3 Describe in detail how you would measure the time in seconds it takes you to go from home to school.

1.4 **Critical Thinking** What additional steps would you need to time your trip, using one clock at home and one at school?

2.2 Measurement Uncertainties

OBJECTIVES

- **Distinguish** between accuracy and precision.
- **Indicate** the precision of measured quantities with significant digits.
- **Perform** arithmetic operations with significant digits.

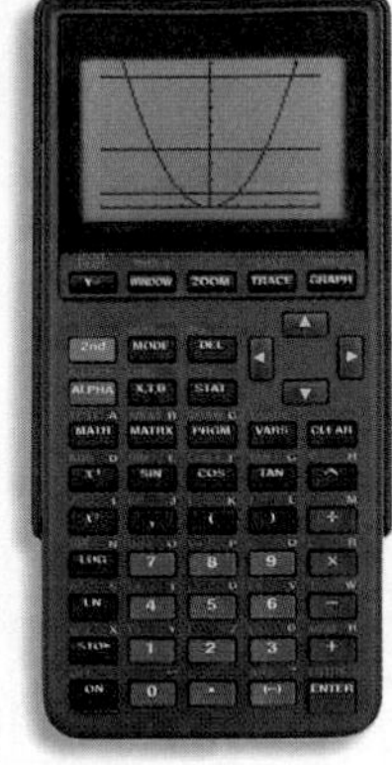

Scientists don't believe the result of an experiment or the prediction of a theory because of the fame of the scientist. Rather, they believe it only when other people have repeatedly obtained the same result. A scientific result has to be reproducible. But no scientific result is perfectly exact. Every measurement, whether it is made by a student or a professional scientist, is subject to uncertainty.

Comparing Results

Before exploring the causes of this uncertainty, let's see how results of experiments along with their uncertainties can be compared. Suppose, for example, three students measured the length of a block of wood. Student 1 made repeated measurements, which ranged from 18.5 cm to 19.1 cm. The average of Student 1's measurements was 18.8 cm, as shown in **Figure 2–5.** This result was reported as (18.8 ± 0.3) cm. Student 2 reported finding the block's length to be (19.0 ± 0.2) cm. Student 3 reported a length of (18.3 ± 0.1) cm.

Could you conclude that the three measurements were in agreement? Was Student 1's result reproducible? The results of Students 1 and 2 overlap, that is, they have the lengths 18.8 cm to 19.1 cm in common. However, there is no overlap and, therefore, no agreement, between their results and the result of Student 3.

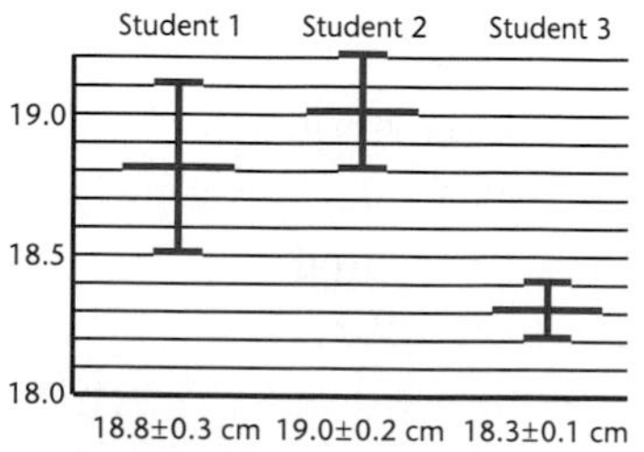

FIGURE 2–5 Three students took multiple measurements of the block of wood. Was Student 1's result reproducible?

Accuracy and Precision

Experimental results can be characterized by their precision and their accuracy. How precise and accurate are the measurements of the three students? **Precision** describes the degree of exactness of a measurement. Student 3's measurements were between 18.2 cm and 18.4 cm. That is, they were within ± 0.1 cm. The measurements of the other two students were less precise because they had a larger variation.

The precision depends on the instrument used to make the measurement. Generally the device that has the finest division on its scale produces the most precise measurement. The precision of a measurement is one-half the smallest division of the instrument. For example, the micrometer in **Figure 2–6** has divisions of 0.01 mm. You can measure an object to within 0.005 mm with this device.

The smallest division on a meterstick is a millimeter, as shown in **Figure 2–7.** Thus, with a meterstick you can measure the length of an object to within 0.5 mm. Would you choose the meterstick or the micrometer to make a more precise measurement?

Accuracy describes how well the results of an experiment agree with

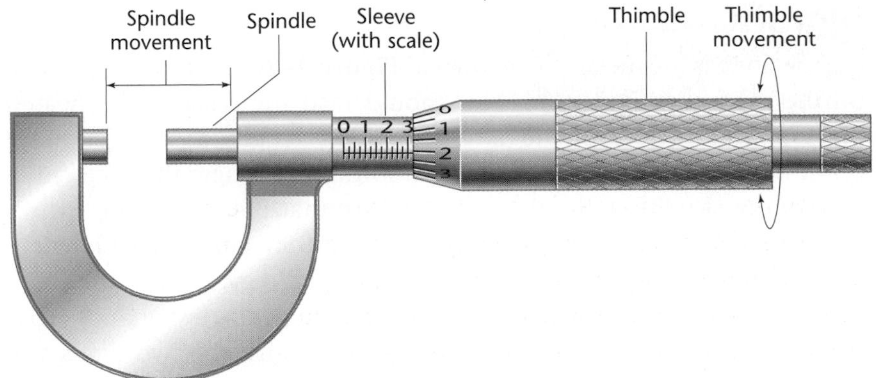

FIGURE 2–6 Micrometers are used to make extremely precise linear measurements.

the standard value. If the block had been 19.0 cm long, then Student 2 would have been most accurate and Student 3 least accurate.

Although it is possible to make precise measurements with an instrument, those measurements still may not be accurate. The accuracy of the instrument has to be checked. A common method is called the two-point calibration. First, does the instrument read zero when it should? Second, does it give the correct reading when it is measuring an accepted standard? The accuracy of all measuring instruments should be checked regularly.

Math Handbook

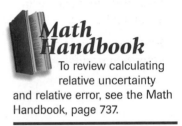

To review calculating relative uncertainty and relative error, see the Math Handbook, page 737.

Techniques of good measurement

To assure accuracy and precision, instruments also have to be used correctly. Measurements have to be made carefully if they are to be as precise as the instrument allows. One common source of error comes from the angle at which an instrument is read. Metersticks should either be tipped on their edge or read with the person's eye directly above the stick as shown in **Figure 2–7.**

If the meterstick is read from an angle, the object will appear to be a different length. The difference in the readings is caused by **parallax,** the apparent shift in the position of an object when it is viewed from different angles.

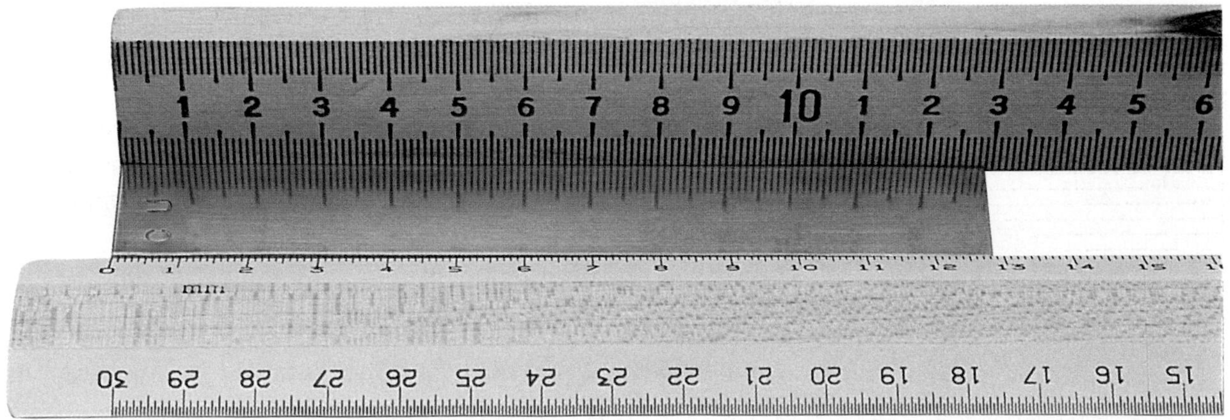

FIGURE 2–7 Good technique while taking a measurement will assure accuracy and precision.

HELP WANTED

ACTUARY

Do you love the challenge of figures and probability? A logical, analytical mind and exceptional knowledge of math and statistics are required for the position of actuary in the corporate headquarters of this major insurance company. A bachelor's degree in math or actuarial science and certification on appropriate examinations are absolutely necessary. For more information contact:
The American Academy of Actuaries
1720 I Street, N.W.
7th Floor
Washington, DC 20006

Significant Digits

How long is the metal strip shown in **Figure 2–8?** The smallest division on the meterstick is 0.1 cm. You should read the scale to the nearest 0.1 cm and then estimate any remaining length as a fraction of 0.1 cm. The metal strip in the figure is somewhat longer than 8.6 cm. If by looking closely at the scale, you can see that the end of the strip is about four tenths of the way between 8.6 and 8.7 cm, then, the length of the strip is best stated as 8.64 cm. The last digit is an estimate. It might not be 4, but it is likely not larger than 5 nor less than 3. Your measurement, 8.64 cm, contains three valid digits: the two digits you are sure of, 8 and 6, and one digit, 4, that you estimated. The valid digits in a measurement are called the **significant digits.** If you were to measure the strip with a micrometer, you might find it to be 8.6365 cm long. This measurement would have five significant digits.

Suppose that the end of the strip were exactly on the 8.6 cm mark. In this case, you should record the measurement as 8.60 cm. The zero indicates that the strip is not 0.01 cm more or less than 8.6 cm. The zero is a significant digit because it transmits information. It is the uncertain digit because you are estimating it. The last digit given for any measurement is the uncertain digit. All nonzero digits in a measurement are significant.

Are all zeros significant?

Not all zeros are significant. For example, if you had reported the length of the strip in meters as 0.0860 m, it would still have only three significant digits. The first two zeros serve only to locate the decimal point and are not significant. The last zero, however, is the estimated digit and is significant.

How many zeros in the measurement 186 000 m are significant? There is no way to tell. The 6 may be the estimated digit, with the three zeros used to place the decimal point, or they may all have been measured. There are two ways to avoid such confusion. First, the units can be changed to move the decimal point. If the measurement were given as 186 km, it would have three significant digits, but if it were written as 186.000 km, it would have six. Second, it can be written in scientific notation: 1.86×10^5 m has three significant digits and 1.86000×10^5 m has six significant digits.

The following rules summarize how to determine the number of significant digits:

1. Nonzero digits are always significant.
2. All final zeros after the decimal point are significant.
3. Zeros between two other significant digits are always significant.
4. Zeros used solely as placeholders are not significant.

All of the following measurements have three significant digits.

245 m 18.0 g 308 km 0.00623 g

The number of significant digits in a measurement is an indication of the precision with which the measurement was taken.

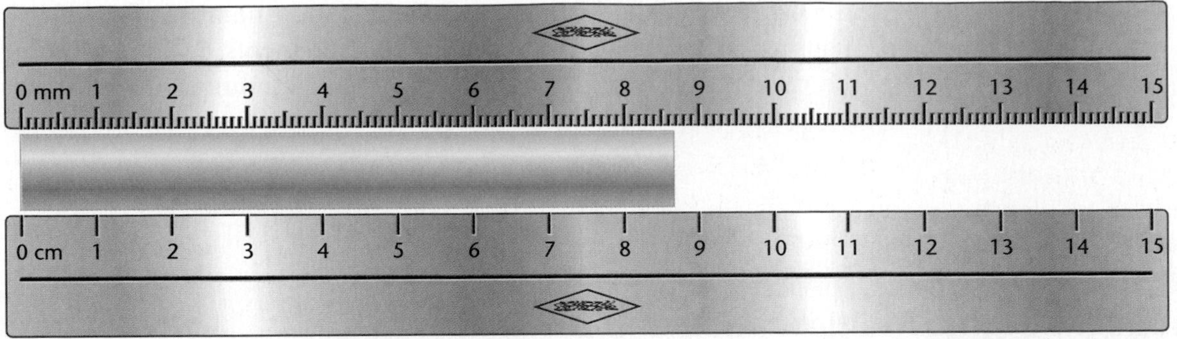

FIGURE 2–8 The accuracy and precision of any measurement depend on both the instrument used and the observer. After a calculation, keep only those digits that truly reflect the accuracy of the original measurement.

Practice Problems

State the number of significant digits in each measurement.

15. a. 2804 m
b. 2.84 km
c. 0.0029 m
d. 0.003 068 m
e. 4.6×10^5 m
f. 4.06×10^{-5} m

16. a. 75 m
b. 75.00 m
c. 0.007 060 kg
d. 1.87×10^6 mL
e. 1.008×10^8 m
f. 1.20×10^{-4} m

Arithmetic with Significant Digits

When you record the results of an experiment, be sure to record them with the correct number of significant digits. Frequently, you will need to add, subtract, multiply, or divide these measurements. When you perform any arithmetic operation, it is important to remember that the result can never be more precise than the least precise measurement.

To add or subtract measurements, first perform the operation, then round off the result to correspond to the least precise value involved.

Example Problem

Significant Digits: Addition and Subtraction

Add 24.686 m + 2.343 m + 3.21 m.

Calculate Your Answer

Strategy:

Note that 3.21 m has the least number of decimal places. Round off the result to the nearest hundredth of a meter.

Calculations:

$$\begin{array}{r} 24.686\ \text{m} \\ 2.343\ \text{m} \\ +\ 3.21\ \ \text{m} \\ \hline 30.239\ \text{m} \end{array}$$

The correct answer is 30.24 m.

A different method is used to find the correct number of significant digits when multiplying or dividing measurements. After performing the calculation, note the factor with the least number of significant digits. Round the product or quotient to this number of digits.

Example Problem

Significant Digits: Multiplication and Division

a. Multiply 3.22 cm by 2.1 cm.

b. Divide 36.5 m by 3.414 s.

Calculate Your Answer

Strategy:

a. The factor, 2.1 cm, contains two significant digits. State the product in two significant digits.

b. The less precise factor contains three significant digits. State the answer in three significant digits.

Calculations:

$$\begin{array}{r} 3.22\text{ cm} \\ \times\ 2.1\text{ cm} \\ \hline 6.762\text{ cm}^2 \end{array}$$

The correct answer is 6.8 cm^2.

$$\frac{36.5\text{ m}}{3.414\text{ s}} = 10.691\text{ m/s}$$

The correct answer is 10.7 m/s.

Practice Problems

Solve the following addition problems.

17. a. 6.201 cm, 7.4 cm, 0.68 cm, and 12.0 cm

b. 1.6 km, 1.62 m, and 1200 cm

Solve the following subtraction problems.

18. a. 8.264 g from 10.8 g

b. 0.4168 m from 475 m

Solve the following multiplication problems.

19. a. 131 cm × 2.3 cm

b. 3.2145 km × 4.23 km

c. 5.761 N × 6.20 m

Solve the following division problems.

20. a. 20.2 cm ÷ 7.41 s

b. 3.1416 cm ÷ 12.4 s

c. 13.78 g ÷ 11.3 mg

d. 18.21 g ÷ 4.4 cm^3

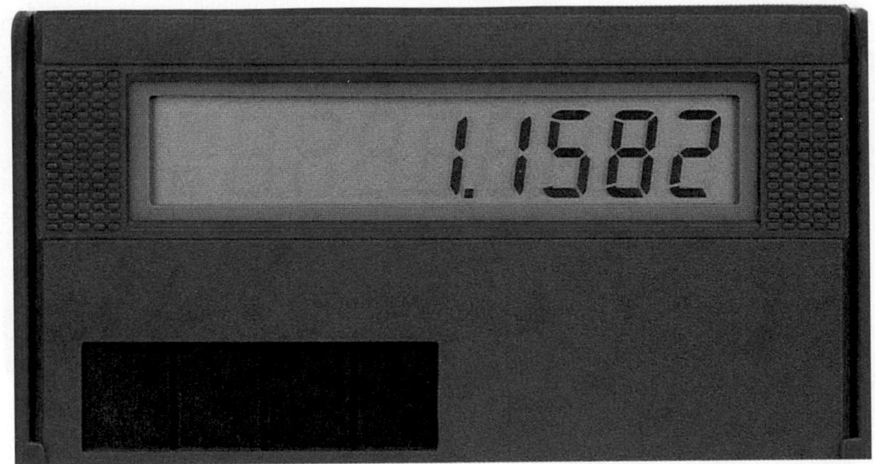

FIGURE 2–9 When using a calculator to solve problems, it is important to note that your answers cannot be more precise that the least precise measurement involved.

Some calculators display several additional, meaningless digits, as shown in **Figure 2–9;** some always display only two. Be sure to record your answer with the correct number of digits, as you have just learned in the example problems.

Note that significant digits are only considered when calculating with measurements; there is no uncertainty associated with counting. If, for example, you measure the time required for a race car to make ten counted trips around the track and want to find the average time for one trip, the measured time has an uncertainty, but the number of trips does not.

Significant digits are an important part of interpreting your work and of determining the meaning of your calculations. Be careful about significant digits when you assign them to measurements, when you do arithmetic with those measurements, and when you report the results.

2.2 Section Review

2.1 You find a micrometer in a cabinet that has been badly bent. How would it compare to a new, high-quality meterstick in precision? In accuracy?

2.2 Does parallax affect the precision of a measurement that you make? Explain.

2.3 Your friend tells you that his height is 182 cm. Explain in your own words the range of heights implied by that statement.

2.4 **Critical Thinking** Your friend states in a report that the time needed for ten laps of a race track had been measured and that the average time required to circle the 2.5-mile track was 65.421 seconds. You know that the clock used had a precision of 0.2 second. How much confidence do you have in the results of the report? Explain.

2.3 Visualizing Data

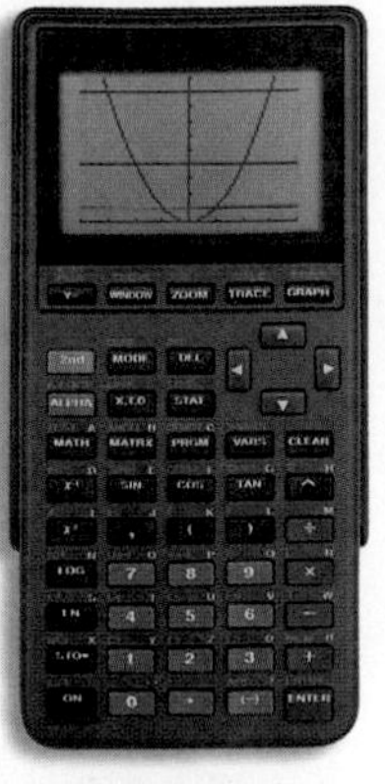

A well-designed graph is like a "picture worth a thousand words." It often can give you more information than words, columns of numbers, or equations alone. To be useful, however, a graph must be drawn properly. In this section, you will develop graphing techniques that will enable you to display data.

OBJECTIVES

- **Graph** the relationship between independent and dependent variables.
- **Recognize** linear and direct relationships and **interpret** the slope of the curve.
- **Recognize** quadratic and inverse relationships.

Graphing Data

One of the most important skills to learn in driving a car is how to stop it safely. No car can actually "stop on a dime." The faster a car is going, the farther it travels before it stops. If you studied to earn a driver's license, you probably found a table in the manual showing how far a car moves beyond the point at which the driver makes a decision to stop.

Many driving manuals also show the distance that a car travels between the time the driver decides to stop the car and the time the driver puts on the brakes. This is called the reaction distance. When the brakes are applied, the car slows down and travels the braking distance. These distances are shown in **Table 2–3** and the bar graph in **Figure 2–10a.** The total stopping distance for various speeds is the sum of the reaction distance and the braking distance. **Table 2–3** shows the English units used in driver's manuals and their SI equivalents.

TABLE 2–3

Reaction and Braking Distances Versus Speed

Original Speed		Reaction Distance		Braking Distance		Total Distance	
m/s	mph	m	ft	m	ft	m	ft
11	25	8	27	10	34	18	61
16	35	12	38	20	67	32	105
20	45	15	49	34	110	49	159
25	55	18	60	50	165	68	225
29	65	22	71	70	231	92	302

The first step in analyzing data is to look at them carefully. Which variable does the experimenter (the driver) change? In this example, it is the speed of the car. Thus, speed is the independent variable, the variable that is changed or manipulated. The independent variable is the one the experimenter can control directly. The other two variables, reaction distance and braking distance, change as a result of the change in speed. These quantities are called dependent variables, or responding variables. The value of the dependent variable depends on the independent variable.

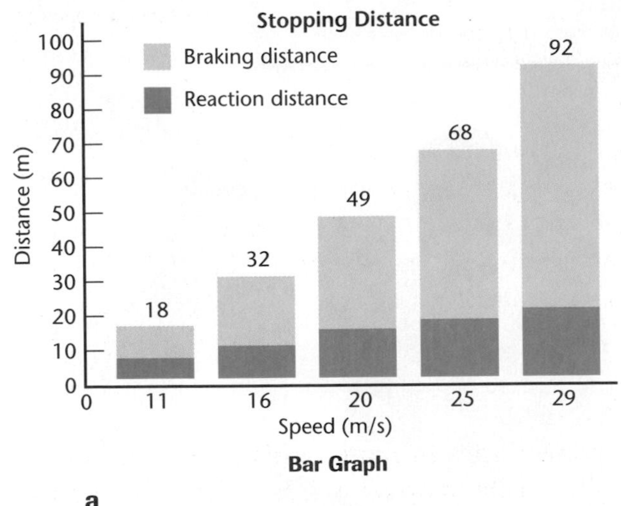

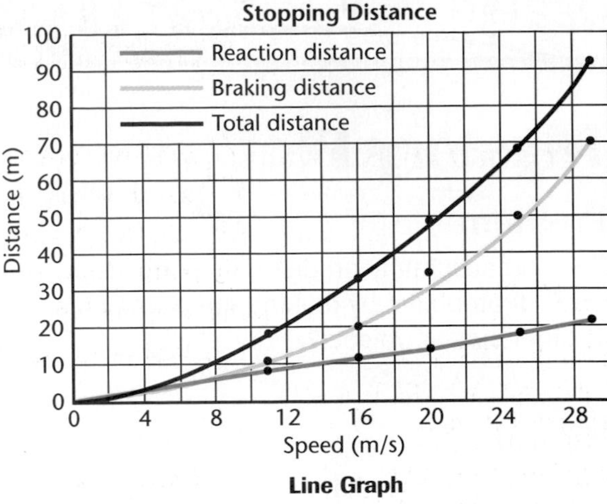

FIGURE 2–10 Graphs **(a)** and **(b)** display the same information in two different ways.

How do the distances change for a given change in the speed of the car? Notice that the reaction distance increases 3 to 4 meters for each 5-m/s increase in speed. The braking distance, however, increases by 10 m when the speed increases from 11 to 16 m/s, and by 20 m when the speed increases from 25 to 29 m/s. The way the two distances depend on speed can be seen more easily when the data are plotted on the line graph in **Figure 2-10b.** Follow the steps in the Problem Solving Strategy to create line graphs that display data from tables.

Problem Solving Strategy

Plotting Line Graphs

Use the following steps to plot line graphs from data tables.

1. Identify the independent and dependent variables in your data. The independent variable is plotted on the horizontal axis, or x-axis. The dependent variable is plotted on the vertical axis, or y-axis.
2. Determine the range of the independent variable to be plotted.
3. Decide whether the origin (0,0) is a valid data point.
4. Spread the data out as much as possible. Let each division on the graph paper stand for a convenient unit.
5. Number and label the horizontal axis.
6. Repeat steps 2–5 for the dependent variable.
7. Plot the data points on the graph.
8. Draw the "best fit" straight line or smooth curve that passes through as many data points as possible. Do not use a series of straight line segments that "connect the dots."
9. Give the graph a title that clearly tells what the graph represents.

How far around?

Use a meterstick to measure the diameter of four circular objects and a string to measure their circumferences. Record your data in a table. Graph the circumference versus the diameter.

Communicate Results Write a few sentences to summarize your graph. Write a sentence using the word that explains the meaning of the slope of your graph. Explain whether the value of the slope would be different if you had measured in different units.

Mystery Plot

Problem

Can you accurately predict the unknown mass of an object by making measurements of other similar objects?

Hypothesis

Form a hypothesis that relates the mass of an object to another measurable quantity. Describe the variables to be measured and why these measurements are necessary.

Possible Materials

4 pieces of electrical wire with lengths between 5 cm and 30 cm
3 rectangular pieces of floor tile
1 triangular piece of floor tile
metric ruler
balance
graph paper

Plan the Experiment

1. As a group, examine the pieces of floor tile and the pieces of electrical wire. Determine the quantities you want to measure. How can you assure the accuracy and precision of your measurements?
2. Identify the independent and dependent variables.
3. Which objects will be the unknown objects? Which objects will be measured? Set aside the unknowns.
4. Construct a data table that will include all your measurements and calculations.
5. **Check the Plan** Make sure your teacher has approved your final plan before you proceed with your experiment.

Analyze and Conclude

1. **Graphing Data** Make graphs of your measurements. Clearly label the axes.
2. **Analyzing Graphs** Do your graphs depict linear, quadratic, or inverse relationships? How do you know? Can you calculate the slope of each graph? Do your graphs go through the origin (0,0)? Should they?
3. **Calculating Results** Write the equations that relate your variables. Use the equations and the graphs to predict the unknown mass of wire and floor tile.
4. **Checking Your Hypothesis** Measure the unknown masses of the wire and floor tile on the balance. Do your measurements agree with the predicted values?

Apply

1. Suppose another group measures longer wires. How should the slope of your graph compare to their slope?
2. In the pharmaceutical industry, how might the weight of compressed medicine tablets be used to determine the quantity of finished tablets produced in a specific lot?

In **Figure 2–10b,** speed is plotted on the *x*-axis, and distance is plotted on the *y*-axis. Data are given for speeds between 11 and 29 m/s, so a convenient range for the *x*-axis is 0 to 30 m/s. On the *y*-axis, the maximum distance is 92 m, so a range of 0 to 100 m is used. When the speed is zero, reaction and stopping distances are both zero, so the graph includes the origin. One division equals 2 m/s on the *x*-axis and 10 m on the *y*-axis.

Linear Relationships

Look at **Figure 2–11,** the graph of reaction distance versus speed. A straight line can be drawn through all data points. That is, the dependent variable varies linearly with the independent variable; the two variables are directly proportional. There is a **linear relationship** between the two variables.

The relationship between the two variables can be written as an equation.

$$y = mx + b$$

The **slope,** *m*, is the ratio of the vertical change to the horizontal change. To find the slope, select two points, A and B, as far apart as possible on the line. These should not be data points, but points on the line. The vertical change, or rise, Δy, is the difference between the vertical values of A and B. The horizontal change, or run, Δx, is the difference between the horizontal values of A and B. The slope of the graph is then calculated in the following way using points (0,0) and (27,21) from **Figure 2–11.** Note that the units are kept with the variables.

$$m = \frac{\text{rise}}{\text{run}} = \frac{\Delta y}{\Delta x}$$

$$m = \frac{(21 - 0)\text{ m}}{(27 - 0)\text{ m/s}} = \frac{21\text{ m}}{27\text{ m/s}} = 0.78\text{ s}$$

The ***y*-intercept,** *b*, is the point at which the line crosses the *y* axis, and it is the *y* value when the value of *x* is zero. When the *y*-intercept is zero, that is, $b = 0$, the equation becomes $y = mx$. The quantity *y* varies directly with *x*. The value of *y* does not always increase with increasing *x*. If *y* gets smaller as *x* gets larger, then $\Delta y/\Delta x$ is negative.

A Graphic Display

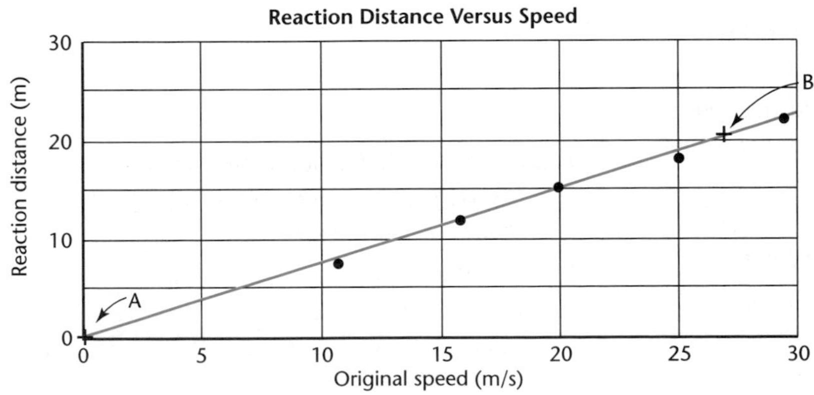

FIGURE 2–11 The graph indicates a linear relationship between reaction distance and speed.

Math Handbook

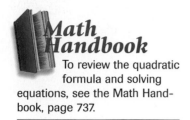

To review the quadratic formula and solving equations, see the Math Handbook, page 737.

Nonlinear Relationships

Figure 2–12a is a graph of braking distance versus speed. Note that the graph is not a straight line; the relationship is not linear. The smooth line drawn through all the data points curves upward. Sometimes, such a graph is a parabola, in which the two variables are related by a **quadratic relationship,** represented by the following equation.

$$y = ax^2 + bx + c$$

The parabolic relationship exists when one variable depends on the square of another.

A computer program or graphing calculator can easily find the values of the constants a, b, and c in this equation. In later chapters you will learn about variables that are related by this equation and learn why braking distance depends on speed in this way.

Some variables are related by the type of graph that is shown in **Figure 2–12b.** In this case, a plot has been made of the time required to travel a fixed distance as the speed is changed. When the speed is doubled, the time is reduced to one-half its original value. The relationship between speed and time is an **inverse relationship.** The graph is a hyperbola described by the general equation

$$y = \frac{a}{x} \text{ or } xy = a$$

where a is a constant. A hyperbola results when one variable depends on the inverse of the other.

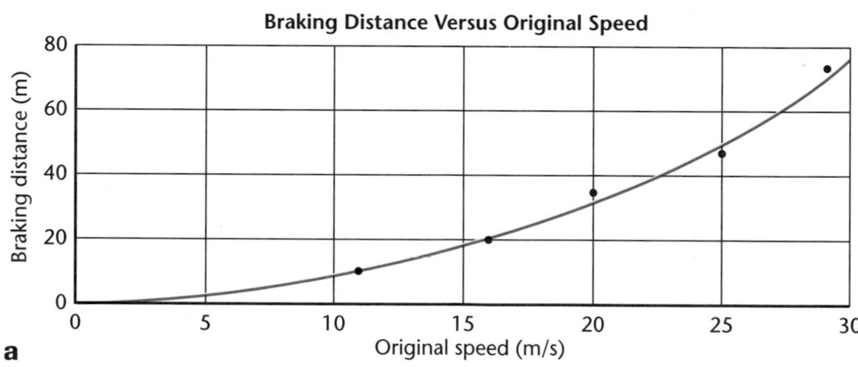

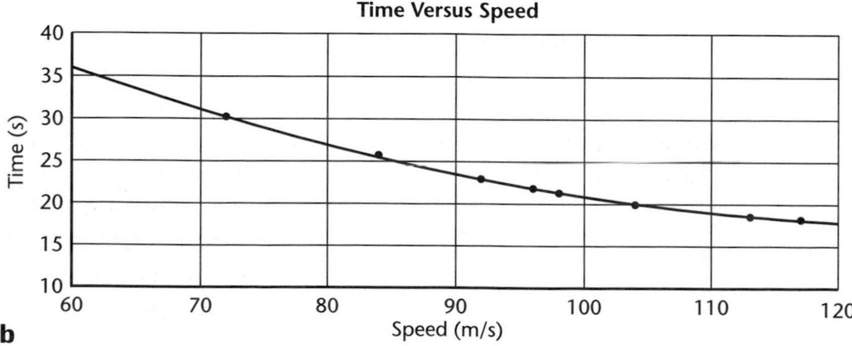

FIGURE 2–12 The graph **(a)** indicates a parabolic relationship; braking distance varies as the square of the original speed. The graph **(b)** shows the inverse relationship between the time required to travel a fixed distance and the speed.

How It Works

Electronic Calculators

A pocket calculator is a specialized computer programmed to solve arithmetic problems. The parts of a typical calculator include a power supply, a keypad for entering numbers and calculation commands, and a screen for displaying input numbers and calculation results. The brain of the calculator is a tiny silicon semiconductor chip. This chip, the calculator's processing unit, performs arithmetic operations.

1. Batteries or solar cells provide electricity.

2. Pressing a number on the keypad closes a contact between the key and the circuit board beneath it. The closed contact allows an electrical signal specific to that key to flow from the circuit board to a storage area in the calculator's processing unit.

3. The processing unit's storage area, or memory, holds all input information until the entire problem has been entered and is ready for processing.

4. With each key stroke, an electrical signal also flows from the processing unit to the screen, which displays the number.

5. When one of the function keys such as plus, minus, addition, subtraction, or square root is pressed, its unique signal is also sent to the processing unit for storage. In most calculators, this information is not displayed on the screen.

6. Pressing the equal sign sends a signal to the processing unit instructing it to perform the calculation stored in its memory. The result is sent to the screen for display.

Thinking Critically

1. What are some of the similarities and differences between pocket calculators and computers?
2. Why is it necessary to clear the memory of a calculator before beginning a new problem?

Practice Problems

21. The total distance a lab cart travels during specified lengths of time is given in the following data table.

TABLE 2–4

Time (s)	Distance (m)
1.0	0.32
2.0	0.60
3.0	0.95
4.0	1.18
5.0	1.45

a. Plot distance versus time from the values given in the table and draw the curve that best fits all points.
b. Describe the resulting curve.
c. According to the graph, what type of relationship exists between the total distance traveled by the lab cart and the time?
d. What is the slope of this graph?
e. Write an equation relating distance and time for this data.

2.3 Section Review

3.1 What would be the meaning of a nonzero y-intercept to a graph of reaction distance versus speed?

3.2 Use the graph in **Figure 2–11** to determine at what speed a car moves 10 m while the driver is reacting.

3.3 Explain in your own words the meaning of a steeper line, or greater slope, to the graph of reaction distance versus speed.

3.4 **Critical Thinking** The relationship between the circumference and the diameter of a circle is shown in **Figure 2–13.** Could a different straight line describe a different circle? What is the meaning of the slope?

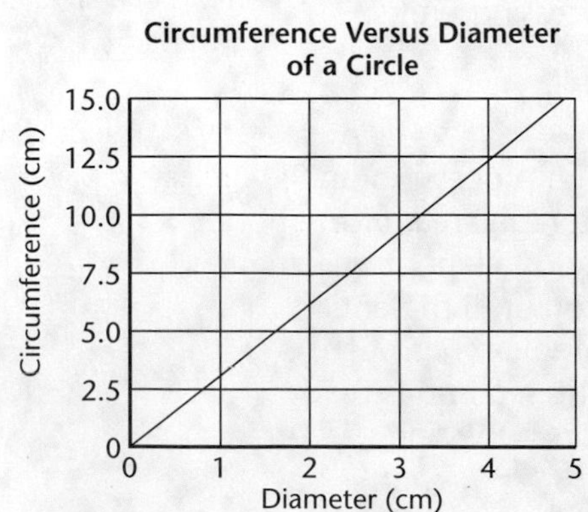

FIGURE 2–13

CHAPTER 2 REVIEW

Key Terms

2.1
- metric system
- SI
- base unit
- meter
- second
- kilogram
- derived unit
- scientific notation
- factor-label method

2.2
- precision
- accuracy
- parallax
- significant digits

2.3
- linear relationship
- slope
- *y*-intercept
- quadratic relationship
- inverse relationship

Summary

2.1 The Measures of Science

- The meter, second, and kilogram are the SI base units of length, time, and mass, respectively.
- Derived units are combinations of base units.
- Making rough estimates is a good way to start and to check the solution of a problem.
- Prefixes are used to change SI units by powers of 10.
- Very large and very small measurements are most clearly written using scientific notation.
- The method of converting one unit to another unit is called the factor-label method of unit conversion.
- To be added or subtracted, measurements written in scientific notation must be raised to the same power of 10.
- Measurements written in scientific notation need not have the same power of 10 to be multiplied or divided.

2.2 Measurement Uncertainties

- All measurements are subject to some uncertainty.
- Precision is the degree of exactness with which a quantity is measured using a given instrument.
- Accuracy is the extent to which the measured and accepted values of a quantity agree.
- The number of significant digits is limited by the precision of the measuring device.
- The last digit in a measurement is always an estimate.
- The result of any mathematical operation with measurements can never be more precise than the least precise measurement involved in the operation.

2.3 Visualizing Data

- Data are plotted in graphical form to show the relationship between two variables.
- The independent variable is the variable that the experimenter changes. It is plotted on the *x*- or horizontal axis.
- The dependent variable, which changes as a result of the changes made to the independent variable, is plotted on the *y*- or vertical axis.
- A graph in which data points lie in a straight line is a graph of a linear relationship.
- A linear relationship can be represented by the equation $y = mx + b$.
- The slope, m, of a straight-line graph is the vertical change (rise) divided by the horizontal change (run).
- The graph of a quadratic relationship is a parabolic curve. It is represented by the equation $y = ax^2 + bx + c$.
- The graph of an inverse relationship between x and y is a hyperbolic curve. It is represented by the equation $y = \frac{a}{x}$.

Reviewing Concepts

Section 2.1

1. Why is SI important?
2. List the common SI base units.
3. How are base units and derived units related?
4. You convert the speed limit of an expressway given in miles per hour into meters per second and obtain the value 1.5 m/s. Is this calculation likely to be correct? Explain.

5. Give the name for each multiple of the meter.
   a. 1/100 m **b.** 1/1000 m **c.** 1000 m
6. How may units be used to check on whether a conversion factor has been used correctly?

Section 2.2

7. What determines the precision of a measurement?
8. Explain how a measurement can be precise but not accurate.
9. How does the last digit differ from the other digits in a measurement?
10. Your lab partner recorded a measurement as 100 g.
    a. Why is it difficult to tell the number of significant digits in this measurement?
    b. How can the number of significant digits in such a number be made clear?

Section 2.3

11. How do you find the slope of a linear graph?
12. A person who has recently consumed alcohol usually has longer reaction times than a person who has not. Thus, the time between seeing a stoplight and hitting the brakes would be longer for the drinker than for the nondrinker.
    a. For a fixed speed, would the reaction distance for a driver who had consumed alcohol be longer or shorter than for a nondrinking driver?
    b. Would the slope of the graph of that reaction distance versus speed have the steeper or the more gradual slope?
13. During a laboratory experiment, the temperature of the gas in a balloon is varied and the volume of the balloon is measured. Which quantity is the independent variable? Which quantity is the dependent variable?
14. For a graph of the experiment in problem 13,
    a. What quantity is plotted on the horizontal axis?
    b. What quantity is plotted on the vertical axis?
15. A relationship between the independent variable x and the dependent variable y can be written using the equation $y = ax^2$, where a is a constant.
    a. What is the shape of the graph of this equation?
    b. If you define a quantity $z = x^2$, what would be the shape of the graph obtained by plotting y versus z?
16. Given the equation $F = mv^2/R$, what relationship exists between
    a. F and R?
    b. F and m?
    c. F and v?
17. Based on the equation in problem 16, what type of graph would be drawn for
    a. F versus R?
    b. F versus m?
    c. F versus v?

Applying Concepts

18. The density of a substance is its mass per unit volume.
    a. Give a possible metric unit for density.
    b. Is the unit for density base or derived?
19. Use **Figure 2–4** to locate the size of the following objects.
    a. The width of your thumb.
    b. The thickness of a page in this book.
    c. The height of your classroom.
    d. The distance from your home to your classroom.
20. Make a chart of sizes of objects similar to the one shown in **Figure 2–4.** Include only objects that you have measured. Some should be less than one millimeter; others should be several kilometers.
21. Make a chart similar to **Figure 2–4** of time intervals. Include intervals like the time between heartbeats, the time between presidential elections, the average lifetime of a human, the age of the United States. Find as many very short and very long examples as you can.
22. Three students use a meterstick to measure the width of a lab table. One records a measurement of 84 cm, another of 83.8 cm, and the third of 83.78 cm. Explain which answer is recorded correctly.
23. Two students measure the speed of light. One obtains $(3.001 \pm 0.001) \times 10^8$ m/s; the other obtains $(2.999 \pm 0.006) \times 10^8$ m/s.
    a. Which is more precise?
    b. Which is more accurate?

24. Why can quantities with different units never be added or subtracted but can be multiplied or divided? Give examples to support your answer.
25. Suppose you receive \$5.00 at the beginning of a week and spend \$1.00 each day for lunch. You prepare a graph of the amount you have left at the end of each day for one week. Would the slope of this graph be positive, zero, or negative? Why?
26. Data are plotted on a graph and the value on the y-axis is the same for each value of the independent variable. What is the slope? Why?
27. The graph of braking distance versus car speed is part of a parabola. Thus, we write the equation $d = av^2 + bv + c$. The distance, d, has units meters, and velocity, v, has units meter/second. How could you find the units of a, b, and c? What would they be?
28. In baseball, there is a relationship between the distance the ball is hit and the speed of the pitch. The speed of the pitch is the independent variable. Choose your own relationship. Determine which is the independent variable and which is the dependent variable. If you can, think of other possible independent variables for the same dependent variables.
29. Aristotle said that the quickness of a falling object varies inversely with the density of the medium through which it falls.
    a. According to Aristotle, would a rock fall faster in water (density 1000 kg/m^3), or in air (density 1 kg/m^3)?
    b. How fast would a rock fall in a vacuum? Based on this, why would Aristotle say that there could be no such thing as a vacuum?

Problems

Section 2.1

LEVEL 1

30. Express the following numbers in scientific notation:
    a. 5 000 000 000 000 m
    b. 0.000 000 000 166 m
    c. 2 003 000 000 m
    d. 0.000 000 103 0 m
31. Convert each of the following measurements to meters.
    a. 42.3 cm d. 0.023 mm
    b. 6.2 pm e. 214 μm
    c. 21 km f. 570 nm
32. Add or subtract as indicated.
    a. 5.80×10^9 s + 3.20×10^8 s
    b. 4.87×10^{-6} m − 1.93×10^{-6} m
    c. 3.14×10^{-5} kg + 9.36×10^{-5} kg
    d. 8.12×10^7 g − 6.20×10^6 g

LEVEL 2

33. Rank the following mass measurements from smallest to largest: 11.6 mg, 1021 μg, 0.000 006 kg, 0.31 mg.

Section 2.2

LEVEL 1

34. State the number of significant digits in each of the following measurements.
    a. 0.000 03 m c. 80.001 m
    b. 64.01 fm d. 0.720 μg
35. State the number of significant digits in each of the following measurements.
    a. 2.40×10^6 kg
    b. 6×10^8 kg
    c. 4.07×10^{16} m
36. Add or subtract as indicated.
    a. 16.2 m + 5.008 m + 13.48 m
    b. 5.006 m + 12.0077 m + 8.0084 m
    c. 78.05 cm^2 − 32.046 cm^2
    d. 15.07 kg − 12.0 kg
37. Multiply or divide as indicated.
    a. $(6.2 \times 10^{18}$ m$)(4.7 \times 10^{-10}$ m$)$
    b. $(5.6 \times 10^{-7}$ m$)/(2.8 \times 10^{-12}$ s$)$
    c. $(8.1 \times 10^{-4}$ km$)(1.6 \times 10^{-3}$ km$)$
    d. $(6.5 \times 10^5$ kg$)/(3.4 \times 10^3$ m$^3)$
38. Using a calculator, Chris obtained the following results. Give the answer to each operation using the correct number of significant digits.
    a. 5.32 mm + 2.1 mm = 7.4200000 mm
    b. 13.597 m × 3.65 m = 49.62905 m^2
    c. 83.2 kg − 12.804 kg = 70.3960000 kg
39. A rectangular floor has a length of 15.72 m and a width of 4.40 m. Calculate the area of the floor.

40. A water tank has a mass of 3.64 kg when it is empty and a mass of 51.8 kg when it is filled to a certain level. What is the mass of the water in the tank?

LEVEL 2

41. A lawn is 33.21 m long and 17.6 m wide.
   a. What length of fence must be purchased to enclose the entire lawn?
   b. What area must be covered if the lawn is to be fertilized?
42. The length of a room is 16.40 m, its width is 4.5 m, and its height is 3.26 m. What volume does the room enclose?
43. The sides of a quadrangular plot of land are 132.68 m, 48.3 m, 132.736 m, and 48.37 m. What is the perimeter of the plot?

Section 2.3

LEVEL 1

44. **Figure 2–14** shows the mass of three substances for volumes between 0 and 60 cm^3.
   a. What is the mass of 30 cm^3 of each substance?
   b. If you had 100 g of each substance, what would their volumes be?
   c. In one or two sentences, describe the meaning of the steepness of the lines in this graph.

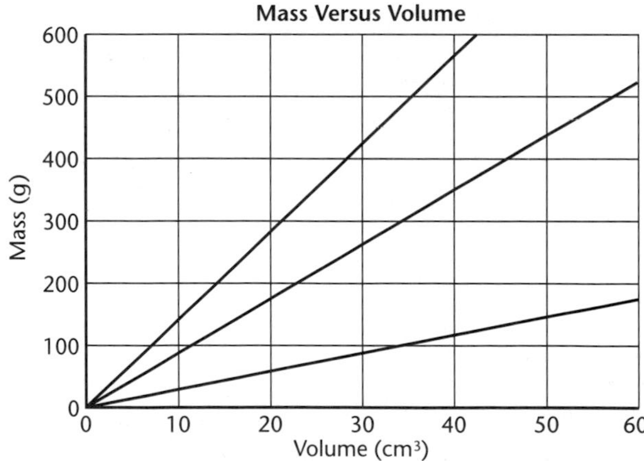

FIGURE 2–14

LEVEL 2

45. During an experiment, a student measured the mass of 10.0 cm^3 of alcohol. The student then measured the mass of 20.0 cm^3 of alcohol. In this way, the data in **Table 2–5** were collected.

TABLE 2–5

Volume (cm^3)	Mass (g)
10.0	7.9
20.0	15.8
30.0	23.7
40.0	31.6
50.0	39.6

   a. Plot the values given in the table and draw the curve that best fits all points.
   b. Describe the resulting curve.
   c. Use the graph to write an equation relating the volume to the mass of alcohol.
   d. Find the units of the slope of the graph. What is the name given to this quantity?

46. During a class demonstration, a physics instructor placed a 1.0-kg mass on a horizontal table that was nearly frictionless. The instructor then applied various horizontal forces to the mass and measured the rate at which it gained speed (was accelerated) for each force applied. The results of the experiment are shown in **Table 2–6.**

TABLE 2–6

Force (N)	Acceleration (m/s^2)
5.0	4.9
10.0	9.8
15.0	15.2
20.0	20.1
25.0	25.0
30.0	29.9

a. Plot the values given in the table and draw the curve that best fits the results.
b. Describe, in words, the relationship between force and acceleration according to the graph.
c. Write the equation relating the force and the acceleration that results from the graph.
d. Find the units of the slope of the graph.

47. The physics instructor who performed the experiment in problem 46 changed the procedure. The mass was varied while the force was kept constant. The acceleration of each mass was recorded. The results of the experiment are shown in **Table 2–7.**

TABLE 2–7

Mass (kg)	Acceleration (m/s^2)
1.0	12.0
2.0	5.9
3.0	4.1
4.0	3.0
5.0	2.5
6.0	2.0

a. Plot the values given in the table and draw the curve that best fits all points.
b. Describe the resulting curve.
c. According to the graph, what is the relationship between mass and the acceleration produced by a constant force?
d. Write the equation relating acceleration to mass given by the data in the graph.
e. Find the units of the constant in the equation.

Critical Thinking Problems

48. Find the approximate time needed for a pitched baseball to reach home plate. Report your result to one significant digit. (Use a reference source to find the distance thrown and the speed of a fastball.)

49. Have a student walk across the front of the classroom. Estimate his or her walking speed.

50. How high can you throw a ball? Find a tall building whose height you can estimate and compare the height of your throw to that of the building.

51. Use a graphing calculator or computer graphing program to graph reaction and braking distances versus original speed. Use the calculator or computer to find the slope of the reaction distance and the best quadratic fit to the braking distance.

52. If the sun suddenly ceased to shine, how long would it take Earth to become dark? You will have to look up the speed of light in a vacuum and the distance from the sun to Earth. How long would it take to become dark on the surface of Jupiter?

Going Further

Team Project Divide your class into teams. Estimate the number of students taking high school physics in the United States. Several approaches are possible. You could use the number of students taking physics in your school, the number of students in your class, and the population of your town and scale them to the population of the country. Another approach would be to estimate the proportion of the total number of students in a single grade to the population of the country; then assume that the percentage of physics students in your school holds true for all schools in which physics is offered. Compare the teams' estimates. Assume that each estimate is reported as ± 1%. Can you conclude that the estimates are in agreement? How could you determine the accuracy of your estimates?

interNET CONNECTION

Follow the link on the Glencoe Homepage at **www.glencoe.com/sec/science** to find out more about this chapter.

Burst of Motion

Sprinters, tensed at the starting block, explode into motion at the sound of the starting gun. That instant burst of motion is a key to winning the event. How would you describe the motion of a sprinter as she leaves the starting block?

CHAPTER

3 Describing Motion

The short length of the 100-m dash means that the winner must attain top speed as quickly as possible and maintain it until she crosses the finish line. Florence Griffith-Joyner needed only 10.54 s to run the course when she won the Olympic gold medal for her record-breaking performance in 1988. That record still holds. Florence Griffith-Joyner can move!

And so does almost everything else. Movement is all around you, fast and slow, in straight lines, curves, spirals, and circles. Do you ever think about what's happening when a basketball is swished through the basket, or when a football sails between the goalposts?

In this chapter, you'll begin to analyze motion in terms of displacement, velocity, and acceleration. When you understand these concepts, you can apply them in later chapters to all kinds of movement using sketches, motion diagrams, graphs, and equations. These tools will enable you to determine how fast and how far an object will move, whether the object is speeding up or slowing down, and whether it is standing still or moving at a constant speed.

WHAT YOU'LL LEARN

- You will describe motion by means of motion diagrams incorporating coordinate systems.
- You will develop descriptions of motion using vector and scalar quantities.
- You will demonstrate the first step, *Sketch the Problem,* in the strategy for solving physics problems.

WHY IT'S IMPORTANT

- Without a knowledge of velocity, time intervals, and displacement, travel by plane, train, or bus would be chaotic at best, and the landing of a space vehicle on Mars an impossibility.

interNET CONNECTION

Follow the link for this chapter on the Glencoe Homepage at **www.glencoe.com/sec/science** to find out more about motion.

3.1 Picturing Motion

OBJECTIVES

- **Draw** and **use** motion diagrams to describe motion.
- **Use** a particle model to represent a moving object.

What comes to your mind when you hear the word *motion*? A speeding automobile? A spinning ride at an amusement park? A football kicked over the crossbar of the goalpost? Or trapeze artists swinging back and forth in a regular rhythm? As you can see in **Figure 3–1,** when an object is in motion, its position changes, and that its position can change along the path of a straight line, a circle, a graceful arc, or a back-and-forth vibration.

FIGURE 3–1
An object in motion changes its position as it moves. You will learn about motion along a straight line, around a circle, along a curved arc, and along a back-and-forth path.

Motion Diagrams

A motion diagram is a powerful tool for the study of motion. You can get a good idea of what a motion diagram is by thinking about the following procedure for making a video of a student athlete training for a race. Point the camcorder in a single direction, perpendicular to the direction of the motion, and hold it still while the motion is occurring, as shown in **Figure 3–2.** The camcorder will record an image 30 times per second. Each image is called a *frame.*

FIGURE 3–2
When the race begins, the camcorder will record the position of the sprinter 30 times each second.

FIGURE 3–3 If you relate the position of the runner to the background in each frame, you will conclude that the sprinter is in motion.

Figure 3–3 shows what a series of consecutive frames might look like. Notice that the runner is in a different position in each frame, but everything in the background remains in the same position. These facts indicate that relative to the ground, only the runner is in motion.

Now imagine that you stacked the frames on top of one another as shown in **Figure 3–4.** You see more than one image of each moving object, but only a single image of all motionless objects. A series of images of a moving object that records its position after equal time intervals is called a **motion diagram.** Successive images recorded by a camcorder are at time intervals of one-thirtieth of a second. Those in **Figure 3–4** have a larger time interval.

FIGURE 3–4 This series of images, taken at regular intervals, creates a motion diagram for the student's practice run.

Some examples of motion diagrams are shown in **Figure 3–5.** In one diagram, a jogger is motionless, or at rest. In another, she is moving at a constant speed. In a third, she is speeding up, and in a fourth, she is slowing down. How can you distinguish the four situations?

In **Figure 3–3,** you saw that motionless objects in the background did not change positions. Therefore, you can associate the jogger in **Figure 3–5a** with an object at rest. Now look at the way the distance between successive positions changes in the three remaining diagrams. If the change in position gets larger, as it does in **Figure 3–5c,** the jogger is speeding up. If the change in position gets smaller, as in **Figure 3–5d,** she is slowing down. In **Figure 3–5b,** the distance between images is the same, so the jogger is moving at a constant speed.

You have just defined four concepts in the study of motion: at rest, speeding up, slowing down, and constant speed. You defined them in terms of the procedure or operation you used to identify them. For that reason, each definition is called an **operational definition.** You will find this method of defining a concept to be useful in this course.

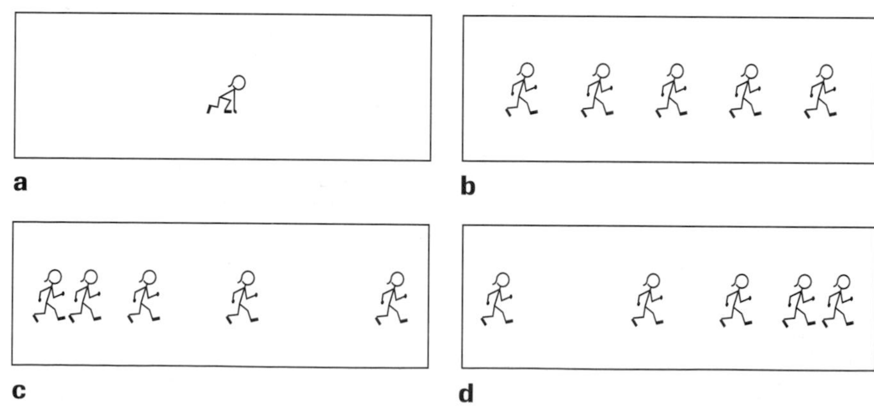

FIGURE 3–5 By noting the distance the jogger moves in equal time intervals, you can determine that the jogger in **a** is standing still, in **b** she is moving at a constant speed, in **c** she is speeding up, and in **d** she is slowing down.

HELP WANTED

AUTO MECHANIC

Our automotive technicians have vocational/technical school or military training; have successfully passed certification exams; own their own basic tools; and have excellent mechanical aptitude, reasoning ability, and knowledge of electronics and various automotive systems. They complete our educational programs and on-the-job training while advancing within the organization. For information contact:

Automotive Service Industry Association
444 North Michigan Ave.
Chicago, IL 60611

The Particle Model

Keeping track of the motion of the runner is easier if you disregard moving arms and legs and concentrate on a single point at the center of her body. In effect, you can consider all of her mass to be concentrated at that point. Replacing an object by a single point is called the **particle model.** But to use the particle model, you must make sure that the size of the object is much less than the distance it moves, and you must ignore internal motions such as the waving of the runner's arms. In a camcorder motion diagram, you could identify one central point on the runner, for example, the knot on her belt, and make measurements of distance with relation to the knot. In **Figure 3–6,** you can see that the particle model provides simplified versions of the motion diagrams in **Figure 3–5.** In the next section, you'll learn how to create and use a motion diagram that shows how much distance was covered and the time interval in which it occurred.

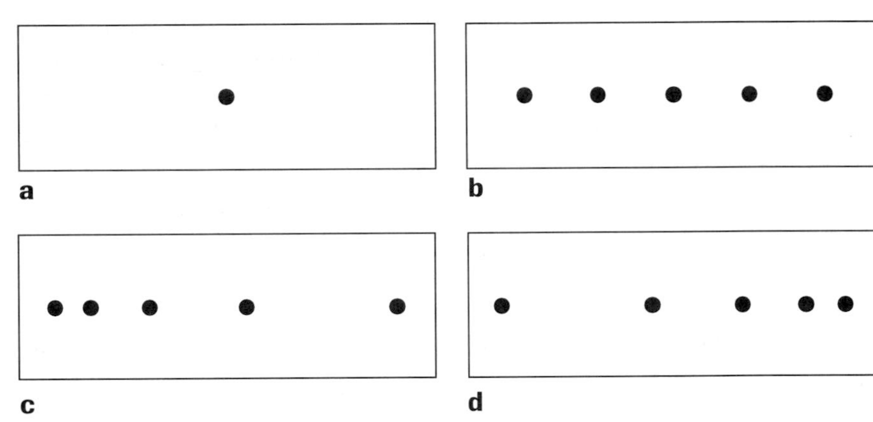

FIGURE 3–6 Using the particle model, you can draw simplified motion diagrams such as these for the jogger in **Figure 3–5.**

3.1 Section Review

1.1 Use the particle model to draw a motion diagram for a runner moving at a constant speed.

1.2 Use the particle model to draw a motion diagram for a runner starting at rest and speeding up.

1.3 Use the particle model to draw a motion diagram for a car that starts from rest, speeds up to a constant speed, and then slows to a stop.

1.4 Critical Thinking Use the particle model to draw a motion diagram for a wheel turning at a constant speed. Place the dot at the hub of the wheel. Would it make any difference if the dot were placed on the rim of the wheel? Explain.

Where and When? 3.2

Would it be possible to make measurements of distance and time from a motion diagram such as that shown in **Figure 3–7?** Before turning on the camcorder, you could place a meterstick or a measuring tape on the ground along the path of the runner. The measuring tape would tell you where the runner was in each frame. A clock within the view of the camera could tell the time. But where should you place the measuring tape? When should you start the stopwatch?

OBJECTIVES

- **Choose** coordinate systems for motion problems.
- **Differentiate** between scalar and vector quantities.
- **Define** a displacement vector and **determine** a time interval.
- **Recognize** how the chosen coordinate system affects the signs of vector quantities.

Coordinate Systems

When you decide where to put the measuring tape and when to start the stopwatch, you are defining a **coordinate system.** A coordinate system tells you where the zero point of the variable you are studying is located and the direction in which the values of the variable increase. The **origin** is the point at which the variables have the value zero. In the example of the runner, the origin, that is, the zero end of the measuring tape, can be placed at the starting line. The motion is in a straight line, thus your measuring tape should lie along that straight line. The straight line is an axis of the coordinate system. You probably would place the tape so that the meter scale increases to the right of the zero, but putting it in the opposite direction is equally correct. In **Figure 3–8,** the origin of the coordinate system is on the left.

To measure motion in two dimensions, for example, the motion of a high jumper, you need to know both the direction parallel to the ground and the height above the ground. That is, you need two axes. Normally, the horizontal direction is called the x-axis, and the vertical direction, perpendicular to the x-axis, is called the y-axis.

FIGURE 3–7 To determine time and distance, a coordinate system must be specified.

FIGURE 3–8 When the origin is at the left, the positive values of x extend horizontally to the right, and the positive values of y extend vertically upward.

F.Y.I.

The gold medal for the men's 4 × 100 m relay in the 1996 Olympics was won by Canada with a time of 37.69 s. Donovan Bailey set the pace with a time of 8.96 s in his split.

You can locate the position of a sprinter at a particular time on a motion diagram by drawing an arrow from the origin to the belt of the sprinter, as shown in **Figure 3–9a.** The arrow is called a **position vector.** The length of the position vector is proportional to the distance of the object from the origin and points from the origin to the location of the moving object at a particular time.

Is there such a thing as a negative position? If there is, what does it mean? Suppose you chose the coordinate system just described, that is, the *x*-axis extending in a positive direction to the right. A negative position would be a position to the left of the origin, as shown in **Figure 3–9b.** In the same way, a negative time would occur before the clock or stopwatch was started. Thus, both negative positions and times are possible and acceptable.

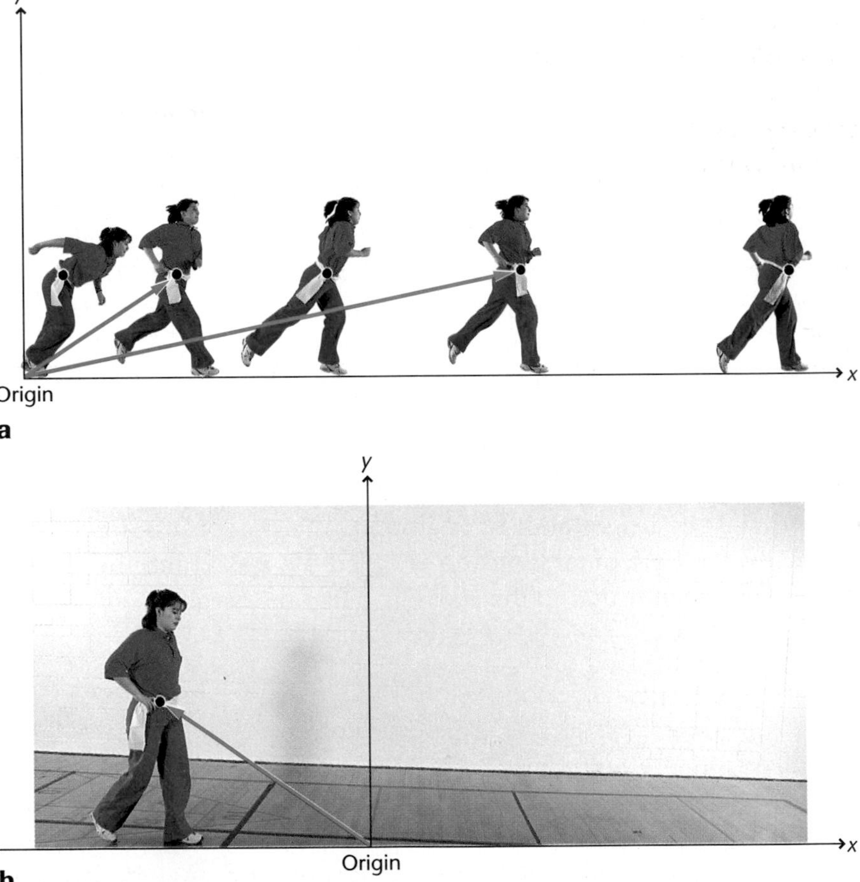

FIGURE 3–9 Two position vectors in **a**, drawn from the origin to the knot on the sprinter's belt, locate her position at two different times. The position of the sprinter in **b**, as she walks toward the starting block, is negative in this coordinate system.

Vectors and Scalars

What is the difference between the information you can obtain from the devices in **Figure 3–10a** and what you can learn from **Figure 3–10b?** In **Figure 3–10a,** you learn that 15 s have elapsed, the temperature is 25°C, and the mass of the grapes on the balance is 125.00 g. Each of these is a definite quantity easily recorded as a

a

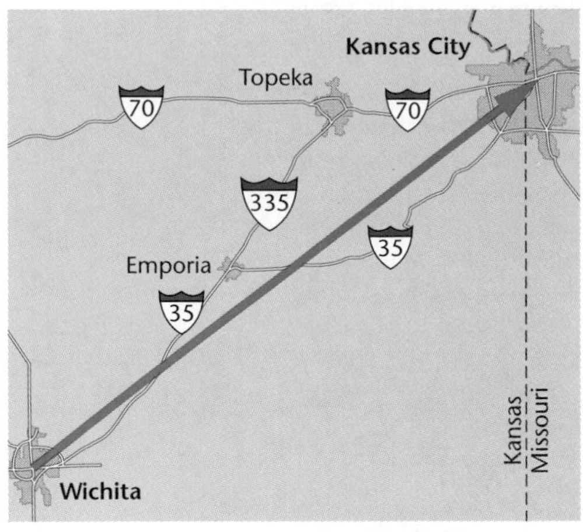

b

FIGURE 3–10 Time, temperature, and mass are scalar quantities, expressed as numbers with units. The arrow in **b** represents a vector quantity. It indicates the direction of Kansas City relative to Wichita and its length is proportional to the distance between the two cities.

number with its units. A quantity such as these that tells you only the magnitude of something is called a **scalar quantity.**

Other quantities, such as the location of one city with respect to another, require both a direction and a number with units. In **Figure 3–10b,** the length of the arrow between Wichita, Kansas, and Kansas City, Missouri, is proportional to the distance between the two cities. You can calculate the distance using the scale of miles for the map. The distance between the two cities, 192 miles, is a scalar quantity. In addition, the arrow tells you the direction of Kansas City in relation to Wichita. Kansas City is 192 miles northeast of Wichita. This information, represented by the arrow on the map, is called a **vector quantity.** A vector quantity tells you not only the magnitude of the quantity, but also its direction.

Symbols often are used to represent quantities. Scalar quantities are represented by simple letters such as m, t, and T for mass, time, and temperature, respectively. Vector quantities are often represented by a letter with an arrow above it, for example, $\vec{v}$ for velocity and $\vec{a}$ for acceleration. In this book, vectors are represented by boldface letters, for example, $\boldsymbol{v}$ represents velocity and $\boldsymbol{a}$ represents acceleration.

Time Intervals and Displacements

The motion of the runner depends upon both the scalar quantity time and the vector quantity displacement. **Displacement** defines the distance and direction between two positions. The sprinter begins at the starting line and a short time later crosses the finish line. How long did it take her to move this far? That is, what was the change in time displayed on the clock? You would find this by subtracting the time shown when she started from the time shown when she finished the race. Assign the symbol t_0 to her starting time and the symbol t_1 to her time at the finish line. The difference between t_0 and t_1 is the **time interval.** A common symbol for the time interval is Δt, where the Greek letter *delta,* Δ, is used to mean a change in a quantity. The time interval is defined mathematically as $\Delta t = t_1 - t_0$.

Color Conventions

- Displacement vectors are **green.**

FIGURE 3–11 In **a,** you can see that the sprinter ran 50 m in the time interval $t_1 - t_0$, which is 6 s. In **b,** the initial position of the sprinter is used as a reference point. The displacement vector indicates both the magnitude and direction of the sprinter's change in position during the 6-s interval.

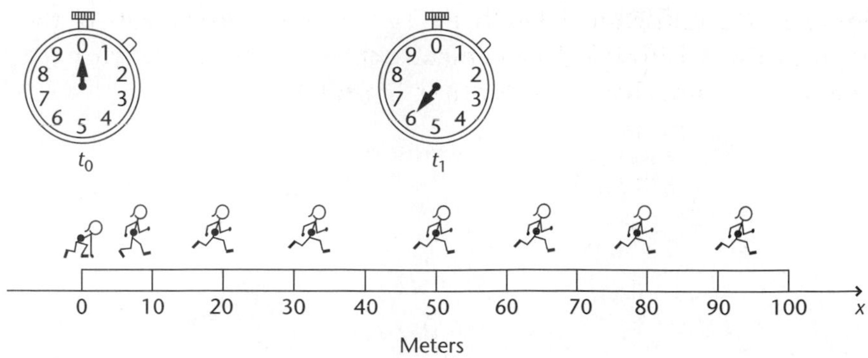

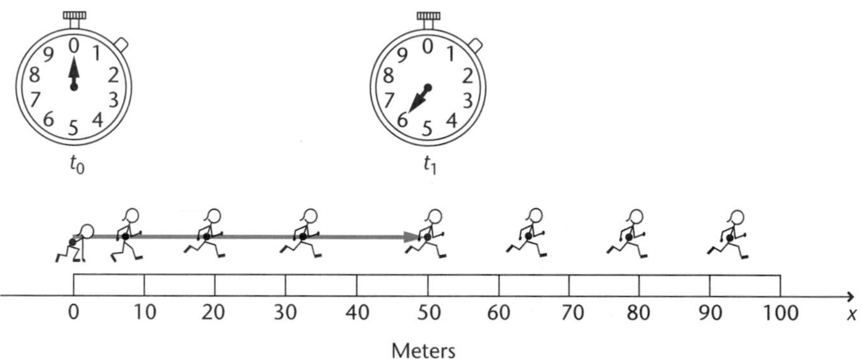

Rolling Along

Tape a 2.5- to 3-m strip of paper to the floor or other smooth, level surface. Gently roll a smooth rubber or steel ball along the paper so that it takes about 4 or 5 s to cover the distance. Now roll the ball while a recorder makes beeps every 1.0 s. Mark the paper at the position of the rolling ball every 1.0 s.

Analyze and Conclude Are the marks on the paper evenly spaced? Make a data table of position and time and use the data to plot a graph. In a few sentences, describe the graph.

Figure 3–11a shows that the time interval for the 100-m sprinter from the start to the time when she is halfway through the course is 6.0 s. What was the change in position of the sprinter as she moved from the starting block to midway in the race? The position of an object is the separation between that object and a reference point. The symbol $\boldsymbol{d}$ may be used to represent position. **Figure 3–11b** shows an arrow drawn from the runner's initial position, $\boldsymbol{d}_0$, to her position 50 m along the track, $\boldsymbol{d}_1$. This arrow is called a displacement vector and is represented by the symbol $\Delta\boldsymbol{d}$. The change in position of an object is called its displacement.

The length, or size, of the displacement vector is called the **distance** between the two positions. That is, the distance the runner moved from $\boldsymbol{d}_0$ to $\boldsymbol{d}_1$ was 50 m. Distance is a scalar quantity.

What would happen if you chose a different coordinate system, that is, if you measured the position of the runner from another location? While both position vectors would change, the displacement vector would not. You will frequently use displacement when studying the motion of an object because displacement is the same in any coordinate system. The displacement of an object that moves from position $\boldsymbol{d}_0$ to $\boldsymbol{d}_1$ is given by $\Delta\boldsymbol{d} = \boldsymbol{d}_1 - \boldsymbol{d}_0$. The displacement vector is drawn with its tail at the earlier position and its head at the later position. Note in

Figure 3–12a and **Figure 3–12b** the two different placements of the origin of the *x*-axis. The displacement, $\Delta \boldsymbol{d}$, in the time interval from 2 s to 6 s does not change, as shown in **Figure 3-12c.**

FIGURE 3–12 The displacement of the sprinter during the 4-s time interval is found by subtracting d_0 from d_1. Δd is the same in both coordinate systems.

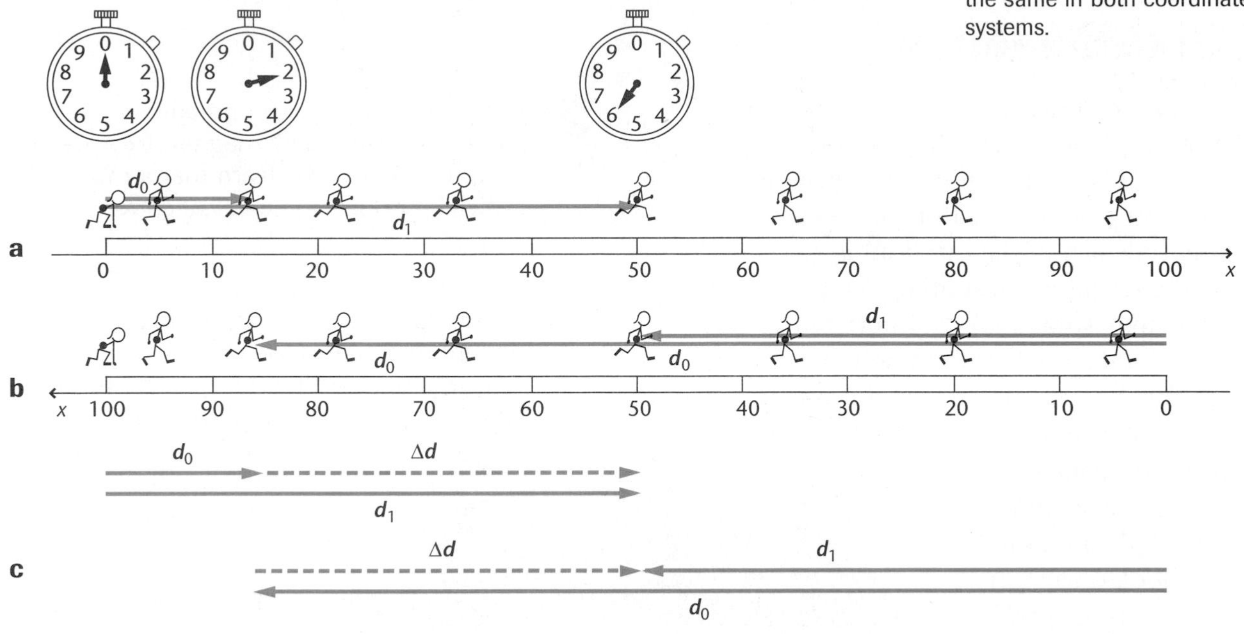

3.2 Section Review

2.1 The dots below are a motion diagram for a car speeding up. The starting point is shown. Make a copy of the motion diagram, and draw displacement vectors between each pair of dots using a green pencil.

Begin End

• • • • • •

2.2 The dots below are a motion diagram for a runner slowing to a stop at the end of a race. On a copy of the motion diagram, draw displacement vectors between each pair of dots.

Begin End

• • • • • •

2.3 The dots at the end of this item are a motion diagram for a bus that first speeds up, then moves at a constant speed, then brakes to a halt. On a copy of the diagram, draw the displacement vectors and explain where the bus was speeding up, where it was going at a constant speed, and where it was slowing down.

Begin End

• • • • • • • • • •

2.4 Critical Thinking Two students compared the position vectors they each had drawn on a motion diagram to show the position of a moving object at the same time. They found that the directions of their vectors were not the same. Explain.

How It Works

Speedometers

A speedometer is a device for measuring the speed, or rate of change in the position of an object. An automobile speedometer is related to the rotation of a gear in the transmission of the car. If all is going well and your car is not slipping on ice or experiencing some other interference, the speedometer will give an accurate reading of how fast you are moving down the road. The pathway from the transmission to the speedometer dial on your dashboard consists of four basic parts: a cable, a magnet, an aluminum ring, and a pointer.

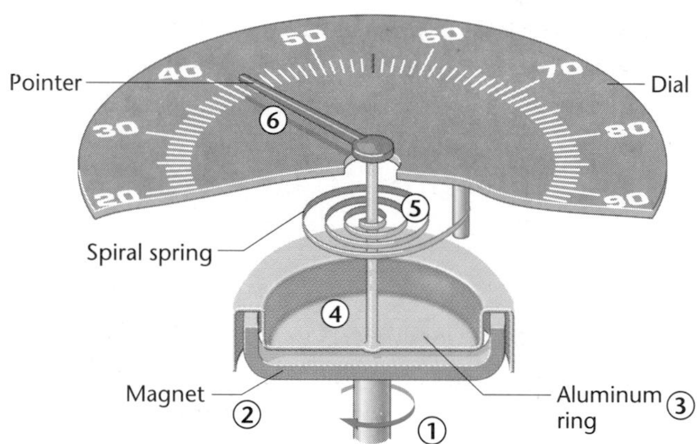

1 When a vehicle such as an automobile moves, a gear at the rear end of the transmission causes a cable to spin. The faster you go, the faster the cable spins.

2 The cable is attached to a magnet that is free to spin at the same rate as the cable. Next to the magnet is an aluminum ring.

3 Because aluminum is nonmagnetic, it is unaffected by a stationary magnet. However, in Chapter 25, you will learn that a moving magnet will produce an electrical current in metals.

4 As the magnet spins, the current in the aluminum ring causes the ring to act as a magnet. As a result, a twisting force, called torque, is applied to the ring by the spinning magnet just as if two magnets were pushin on each other.

5 The torque causes th aluminum ring to rotate. A spiral sprin maintains an opposit push against the torque from the spin ning magnet. The faster the magnet spins, the greater th torque, and thus the greater the rotation of the aluminum rin

6 The aluminum ring is connected to a pointer that rotates in front of a dial. This dial is usually graduated in both miles per hour and kilometers per hour.

Thinking Critically

1. When a car is backing up, does the pointer move? Why or why not?
2. Would the speedometer still be accurate if you put the wrong size tires on your car? Explain.

Velocity and Acceleration

3.3

You've learned how to use a motion diagram to show objects moving at different speeds. How could you measure how fast they are moving? With devices such as a meterstick and a clock, you can measure position and time. Can these two quantities be combined in some way to create a quantity that tells you the rate of motion?

OBJECTIVES

- **Define** *velocity* and *acceleration* operationally.
- **Relate** the direction and magnitude of velocity and acceleration vectors to the motion of objects.
- **Create** pictorial and physical models for solving motion problems.

Velocity

Suppose you recorded a speedy jogger and a slow walker on one motion diagram, as shown in **Figure 3–13.** From one frame to the next, you can see that the position of the jogger changes more than that of the walker. In other words, for a fixed time interval, the displacement, $\Delta \mathbf{d}$, is larger for the jogger because she is moving faster. The jogger covers a larger distance than the walker does in the same amount of time. Now, suppose that the walker and the jogger each travel 100 m. Each would need a different amount of time to go that distance. How would these time intervals compare? Certainly the time interval, Δt, would be smaller for the jogger than for the walker.

Average velocity

From these examples, you can see that both displacement, $\Delta \mathbf{d}$, and time interval, Δt, might reasonably be needed to create the quantity that tells how fast an object is moving. How could you combine them?

The ratio $\Delta \mathbf{d}/\Delta t$ has the correct properties. It is the change in position divided by the time interval during which that change took place, or $(\mathbf{d}_1 - \mathbf{d}_0)/(t_1 - t_0)$. This ratio increases when $\Delta \mathbf{d}$ increases, and it also increases when Δt gets smaller, so it agrees with the interpretation you made of the movements of the walker and runner. It is a vector in the same direction as the displacement. The ratio $\Delta \mathbf{d}/\Delta t$ is called the **average velocity, $\bar{\mathbf{v}}$.**

$$\bar{\mathbf{v}} \equiv \frac{\Delta \mathbf{d}}{\Delta t} = \frac{\mathbf{d}_1 - \mathbf{d}_0}{t_1 - t_0}$$

The symbol $\equiv$ means that the left-hand side of the equation is defined by the right-hand side.

FIGURE 3–13 Because the jogger is moving faster than the walker, the jogger's displacement is greater than the displacement of the walker in each time interval.

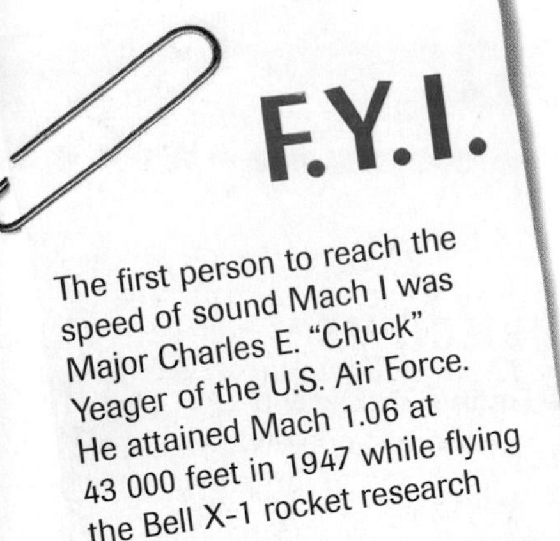

The **average speed** is the ratio of the total distance traveled to the time interval. Automobile speeds are measured in miles per hour (mph) or kilometers per hour (km/h), but in this course, the usual unit will be meters per second (m/s).

Instantaneous velocity

Why *average* velocity? A motion diagram tells you the position of a moving object at the beginning and end of a time interval. It doesn't tell you what happened within the time interval. Within a time interval, the speed of the object could have remained the same, increased, or decreased. The object may have stopped or even changed directions. All that can be determined from the motion diagram is an average velocity, which is found by dividing the total displacement by the time interval in which it took place.

What if you want to know the speed and direction of an object at a particular instant in time? The quantity you are looking for is **instantaneous velocity**. In this text, the term *velocity* will refer to instantaneous velocity, represented by the symbol $\boldsymbol{v}$.

Color Conventions

- Displacement vectors are **green.**
- Velocity vectors are **red.**

Average velocity motion diagrams

How can you show average velocity on a motion diagram? Although the average velocity vector is in the same direction as displacement, the two vectors are not measured in the same units. Nevertheless, they are proportional; when displacement is larger over a given time interval, so is average velocity. A motion diagram isn't a precise graph of average velocity, but you can indicate the direction and magnitude of the average velocity vectors on it. Use a red pencil to draw arrows proportional in length to the displacement vectors. Label them, as shown in **Figure 3–14.**

The definition of average velocity, $\overline{\boldsymbol{v}} = \Delta\boldsymbol{d}/\Delta t$, shows that you could calculate velocity from the displacement of an object, but look at the equation in a different way. Rearrange the equation $\overline{\boldsymbol{v}} = \Delta\boldsymbol{d}/\Delta t$ by multiplying both sides by Δt.

$$\Delta\boldsymbol{d} = \overline{\boldsymbol{v}}\Delta t$$

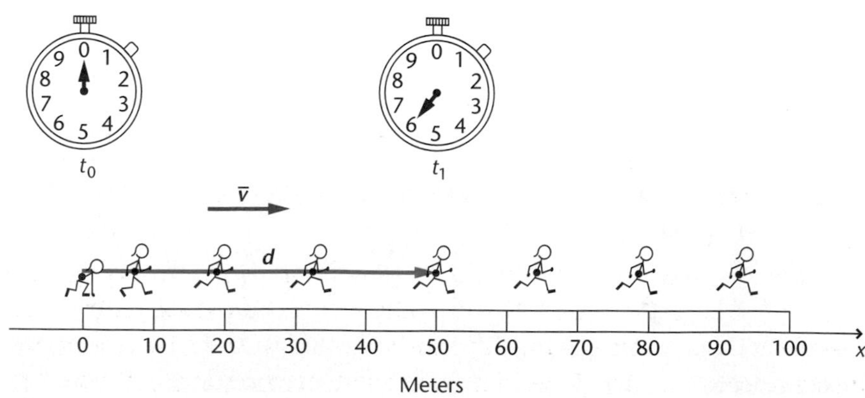

FIGURE 3–14 Average velocity vectors have the same direction as their corresponding displacement vectors. Their magnitudes are different but proportional and they have different units.

Now, write the displacement, $\Delta \boldsymbol{d}$, in terms of the two positions $\boldsymbol{d}_0$ and $\boldsymbol{d}_1$.

$$\Delta \boldsymbol{d} = \boldsymbol{d}_1 - \boldsymbol{d}_0$$

Substitute $\boldsymbol{d}_1 - \boldsymbol{d}_0$ for $\Delta \boldsymbol{d}$ in the first equation.

$$\boldsymbol{d}_1 - \boldsymbol{d}_0 = \bar{\boldsymbol{v}}\Delta t$$

Add $\boldsymbol{d}_0$ to both sides of the equation.

$$\boldsymbol{d}_1 = \boldsymbol{d}_0 + \bar{\boldsymbol{v}}\Delta t$$

This equation tells you that over the time interval Δt, the average velocity of a moving object results in a change in position equal to $\bar{\boldsymbol{v}}\Delta t$. If there were no average velocity, there would be no change in position.

The motion diagrams in **Figure 3–15** describe a long golf putt that comes to a stop at the rim of the hole. Study the diagrams to answer these questions. When is the average velocity within a time interval greatest? When is it smallest? You can see that the average velocity vector is the longest in the first time interval. There was the greatest displacement of the ball in that time interval because the average velocity was greatest. The average velocity was the least in the last time interval in which the length of the average velocity vector is shortest.

What is the direction of the average velocity vectors in **Figure 3–15?** Before answering, you must define a coordinate system. If the origin is the point at which the ball was tapped by the golf club, then the ball was moving in a positive direction and the direction of the average velocity vector is positive, as shown in **Figure 3–15a.** But suppose you chose the hole as the origin. Then the direction of the average velocity vector is negative, as shown in **Figure 3–15b.** Either choice is correct.

Swinging

Use a video recorder to capture an object swinging like a pendulum. Then attach a piece of tracing paper or other see-through material over the TV screen as you play back the video frame by frame. Use a felt marker to show the position of the center of the swinging object at every frame as it moves from one side of the screen to the opposite side.

Analyze and Conclude Does the object have a steady speed? Describe how the speed changes. Where is the object moving the fastest? Do you think that your results are true for other swinging objects? Why?

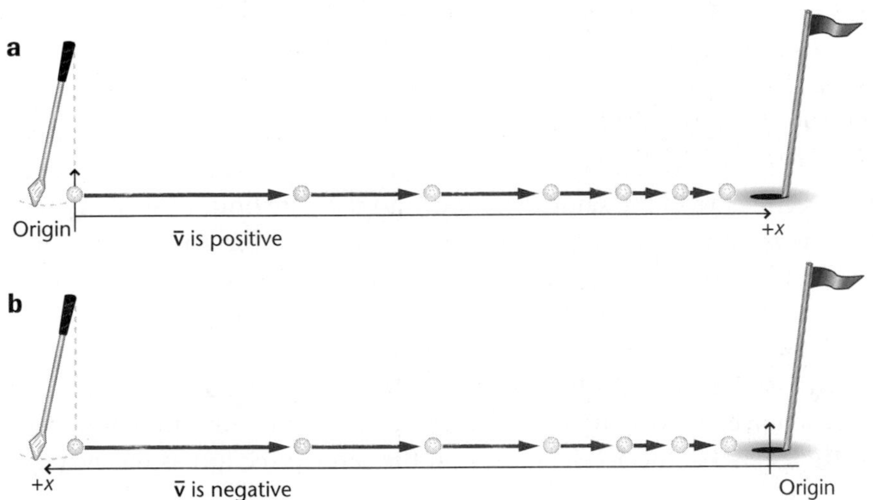

FIGURE 3–15 The sign of the average velocity depends upon the chosen coordinate system. The coordinate systems in **a** and **b** are equally correct.

Acceleration

The average velocity of the golf ball in **Figure 3–15** was changing from one time interval to the next. You can tell because the average velocity vectors in each time interval have different magnitudes. At the same time, the instantaneous velocity, or velocity, must also be changing. An object in

EARTH SCIENCE CONNECTION

The Beaufort Scale is used by meteorologists to indicate wind speeds. A wind comparable to the fastest speed run by a person is classed as 5, a strong breeze. A wind as fast as a running cheetah is classified as 11, a storm. Winds of up to 371 km/h (beyond the scale) have been registered on Mount Washington, New Hampshire.

motion whose velocity is changing is said to be accelerating. Recall that an object's velocity changes when either the magnitude or direction of the motion changes.

How can you relate the change in velocity to the time interval over which it occurs to describe acceleration? When the change in velocity is increasing or the change in velocity occurs over a shorter time interval, the acceleration is larger. The ratio $\Delta \boldsymbol{v}/\Delta t$ has the properties needed to describe acceleration.

Let $\bar{\boldsymbol{a}}$ be the **average acceleration** over the time interval Δt.

$$\bar{\boldsymbol{a}} \equiv \frac{\Delta \boldsymbol{v}}{\Delta t} = \frac{\boldsymbol{v}_1 - \boldsymbol{v}_0}{t_1 - t_0}$$

What is the unit of average acceleration? Both velocity and change in velocity are measured in meters per second, m/s, so because average acceleration is change in velocity divided by time, the unit of average acceleration is meters per second per second. The unit is abbreviated m/s^2.

Using motion diagrams to obtain average acceleration

How can you find the change in average velocity using motion diagrams? Motion diagrams indicate position and time. From position and time, you can determine average velocity. You can get a rough idea, or qualitative description, of acceleration by looking at how the average velocity changes.

Color Conventions

- Displacement vectors are **green.**
- Velocity vectors are **red.**
- Accelerations vectors are **violet.**

In a motion diagram, the average acceleration vector, $\bar{\boldsymbol{a}}$ is proportional to the change in the average velocity vector, $\Delta \bar{\boldsymbol{v}}$. You can draw the average acceleration and change in average velocity vectors the same length, but use the color violet to represent acceleration vectors.

Figure 3–16 shows a motion diagram describing a car that speeds up, then travels at a constant speed, and then slows down. The origin is at the left, so the car is moving in the positive direction. You can see that when the car is speeding up, the average velocity and average acceleration vectors are in the same direction, and they are both positive. When the car is slowing down, the average velocity vector and the average acceleration vector are in opposite directions. The average velocity is positive, but the average acceleration is negative. When the velocity is constant, the average velocity vectors are of equal length. There is no change in average velocity; therefore, the average acceleration is zero.

When average velocity is increasing, as in the first four time intervals of **Figure 3–16,** the acceleration is in the same direction as the average

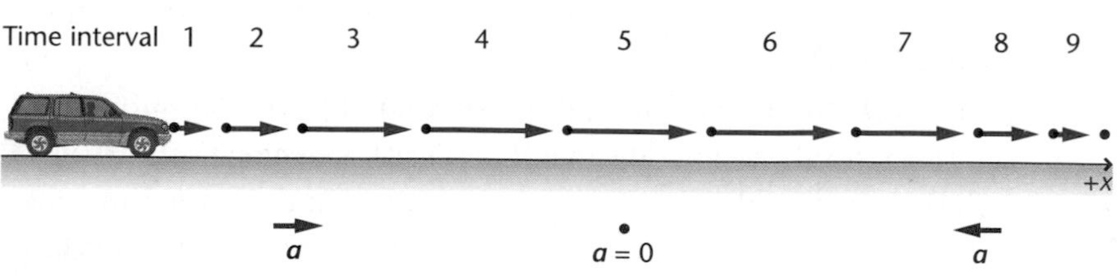

FIGURE 3–16 In this diagram, the origin is on the left. As a result, all the average velocity vectors are positive. The sign of the acceleration is determined by whether the car is speeding up or slowing down.

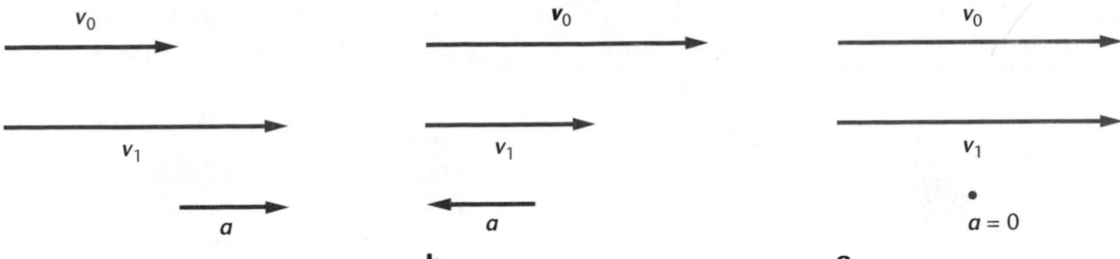

FIGURE 3–17 The direction of the acceleration is determined by whether the car is speeding up, slowing down, or traveling at constant speed.

velocity. Similarly, the vector diagram in **Figure 3–17a** represents motion that is speeding up from $\boldsymbol{v}_0$ to $\boldsymbol{v}_1$. When motion is slowing down, as in time intervals 6–9 in **Figure 3–16** and in **Figure 3–17b,** the average acceleration is in a direction opposite that of the average velocity. When average velocity is constant, as in time intervals 4–6, $\boldsymbol{v}_0$ is equal to $\boldsymbol{v}_1$ and **Figure 3–17c** shows that the acceleration is zero.

You can now describe the motion of the sprinter as she leaves the starting block in a 100-m race. Her average velocity is increasing to the right, so with the origin at the starting block, both her average velocity and average acceleration are positive. What happens to these quantities just after the sprinter crosses the finish line? The average velocity decreases but is still in a positive direction as the sprinter slows down, but slowing down means that the average acceleration is negative.

In the remainder of this chapter, you'll learn how to sketch a problem and link it with the motion diagrams you've learned to draw. In many cases throughout this book, you'll be asked to solve problems in three steps. In this chapter, however, the focus will be on the first step.

Burst of Motion

Problem Solving Strategy

Solving Problems

1. **Sketch the Problem** Carefully read the problem statement and make a mental picture of the problem situation. Decide whether the problem has more than one part. Then, sketch the situation. Establish a coordinate system and add it to your sketch. Next, reread the problem and make a list of unique symbols to represent each of the variables that are given or known. Finally, decide which quantity or quantities are unknown and give them symbols. This is called building a pictorial model. Next, create a physical model. When solving motion problems, the physical model is a motion diagram.
2. **Calculate Your Answer** Now use the physical model as a guide to the equations and graphs you will need. Use them to solve for the unknown quantity.
3. **Check Your Answer** Did you answer the question? Is the answer reasonable? This step is as important as the others, but it may be the hardest.

Design Your Own **Physics Lab**

Notion of Motion

Problem

You are to construct motion diagrams based on a steady walk and a simulated sprint.

Hypothesis

Devise a procedure for creating motion diagrams for a steady walk and a sprint.

Possible Materials

stopwatch
metersticks
10-m length of string, cord, or tape

Plan the Experiment

1. Decide on the variables to be measured and how you will measure them.
2. Decide how you will measure the distance over the course of the walk.
3. Create a data table.
4. Organize team members to perform the individual tasks of walker, sprinter, timekeeper, and recorder.
5. **Check the Plan** Make sure your teacher approves your final plan before you proceed.
6. Think about how the procedures you use for the fast sprint may differ from those you used for the steady walk, then follow steps 1–5.

Analyze and Conclude

1. **Organizing Data** Use your data to write a word description of each event.
2. **Comparing Data** Make a motion diagram for each event. Label the diagrams *Begin* and *End* to indicate the beginning and the end of the motion.
3. **Organizing Data** Draw the acceleration vectors on your motion diagram for the two events.
4. **Comparing Results** Compare the pattern of average velocity vectors for the two events. How are they different? Explain.
5. **Inferring Conclusions** Compare the acceleration vectors from the steady walk and the sprint. What can you conclude?

Apply

1. Imagine that you have a first-row seat for the 100-m world championship sprint. Write a description of the race in terms of velocity and acceleration. Include a motion diagram that would represent the race run by the winner.

Sketch the Problem

Here is a typical motion problem: *A driver, going at a constant speed of 25 m/s, sees a child suddenly run into the road. It takes the driver 0.40 s to hit the brakes. The car then slows at a rate of 8.5 m/s². What is the total distance the car moves before it stops?*

Follow **Figure 3–18** as you set up this problem. What information is given? First, the speed is constant, then the brakes are applied, so this is a two-step problem. For the first step, the constant velocity is 25 m/s, and the time interval is 0.40 s. In the second step, the initial velocity is 25 m/s; the final velocity is 0.0 m/s. The acceleration is -8.5 m/s². There are three positions in this problem—the beginning, middle, and end—$\boldsymbol{d}_1$, $\boldsymbol{d}_2$, and $\boldsymbol{d}_3$. The unknown is position $\boldsymbol{d}_3$. Use $\boldsymbol{a}_{12}$ for the acceleration between $\boldsymbol{d}_1$ and $\boldsymbol{d}_2$, and $\boldsymbol{a}_{23}$ for the acceleration between $\boldsymbol{d}_2$ and $\boldsymbol{d}_3$.

The motion diagram shows that in the first part, the acceleration is zero. In the second part, the acceleration is in the direction opposite to the velocity. In this coordinate system, the acceleration is negative.

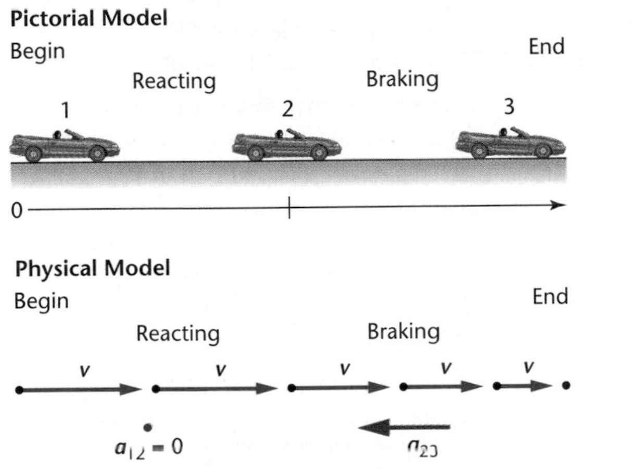

Known:
$d_1 = 0.0$ m
$v_1 = 25$ m/s
$a_{12} = 0.0$ m/s²
$t_2 = 0.40$ s
$v_2 = 25$ m/s
$v_3 = 0.0$ m/s
$a_{23} = -8.5$ m/s²

Unknown:
d_3

FIGURE 3–18 Symbols for time and velocity are subscripted to identify the position at which they are valid. The subscripts on the symbol ***a*** indicate the two positions between which each acceleration is valid.

3.3 Section Review

For the following questions, build the pictorial and physical models as shown in the preceding example. Do not solve the problems.

3.1 A dragster starting from rest accelerates at 49 m/s². How fast is it going when it has traveled 325 m?

3.2 A speeding car is traveling at a constant speed of 30 m/s when it passes a stopped police car. The police car accelerates at 7 m/s². How fast will it be going when it catches up with the speeding car?

3.3 Critical Thinking In solving a physics problem, why is it important to make a table of the given quantities and the unknown quantity, and to assign a symbol for each?

CHAPTER 3 REVIEW

Key Terms

3.1
- motion diagram
- operational definition
- particle model

3.2
- coordinate system
- origin
- position vector
- scalar quantity
- vector quantity
- displacement
- time interval
- distance

3.3
- average velocity
- average speed
- instantaneous velocity
- average acceleration

Summary

3.1 Picturing Motion

- A motion diagram shows the position of an object at successive times.
- In the particle model, the object in the motion diagram is replaced by a series of single points.
- An operational definition defines a concept in terms of the process or operation used.

3.2 Where and When?

- You can define any coordinate system you wish in describing motion, but some are more useful than others.
- While a scalar quantity has only magnitude, or size, a vector quantity has both magnitude and a direction.
- A position vector is drawn from the origin of the coordinate system to the object. A displacement vector is drawn from the position of the moving object at an earlier time to its position at a later time.
- The distance is the length or magnitude of the displacement vector.

3.3 Velocity and Acceleration

- Velocity and acceleration are defined in terms of the processes used to find them. Both are vector quantities with magnitude and direction.
- Average speed is the ratio of the total distance traveled to the time interval.
- The most important part of solving a physics problem is translating words into pictures and symbols.
- To build a pictorial model, analyze the problem, draw a sketch, choose a coordinate system, assign symbols to the known and unknown quantities, and tabulate the symbols.
- Use a motion diagram as a physical model to find the direction of the acceleration in each part of the problem.

Reviewing Concepts

Section 3.1

1. What is the purpose of drawing a motion diagram?
2. Under what circumstances is it legitimate to treat an object as a point particle?

Section 3.2

3. How does a vector quantity differ from a scalar quantity?
4. The following quantities describe location or its change: position, distance, and displacement. Which are vectors?
5. How can you use a clock to find a time interval?

Section 3.3

6. What is the difference between average velocity and average speed?
7. How are velocity and acceleration related?
8. What are the three parts of the problem solving strategy used in this book?
9. In which part of the problem solving strategy do you sketch the situation?
10. In which part of the problem solving strategy do you draw a motion diagram?

Applying Concepts

11. Test the following combinations and explain why each does not have the properties needed to describe the concept of velocity: $\Delta \boldsymbol{d} + \Delta t$, $\Delta \boldsymbol{d} - \Delta t$, $\Delta \boldsymbol{d} \times \Delta t$, $\Delta t / \Delta \boldsymbol{d}$.

12. When can a football be considered a point particle?

13. When can a football player be treated as a point particle?

14. When you enter a toll road, your toll ticket is stamped 1:00 P.M. When you leave, after traveling 55 miles, your ticket is stamped 2:00 P.M. What was your average speed in miles per hour? Could you ever have gone faster than that average speed? Explain.

15. Does a car that's slowing down always have a negative acceleration? Explain.

16. A croquet ball, after being hit by a mallet, slows down and stops. Do the velocity and acceleration of the ball have the same signs?

Problems

Create pictorial and physical models for the following problems. Do not solve the problems.

Section 3.3

LEVEL 1

17. A bike travels at a constant speed of 4.0 m/s for 5 s. How far does it go?

18. A bike accelerates from 0.0 m/s to 4.0 m/s in 4 s. What distance does it travel?

19. A student drops a ball from a window 3.5 m above the sidewalk. The ball accelerates at 9.80 m/s^2. How fast is it moving when it hits the sidewalk?

LEVEL 2

20. A bike first accelerates from 0.0 m/s to 5.0 m/s in 4.5 s, then continues at this constant speed for another 4.5 s. What is the total distance traveled by the bike?

21. A car is traveling 20 m/s when the driver sees a child standing in the road. He takes 0.8 s to react, then steps on the brakes and slows at 7.0 m/s^2. How far does the car go before it stops?

22. You throw a ball downward from a window at a speed of 2.0 m/s. The ball accelerates at 9.8 m/s^2. How fast is it moving when it hits the sidewalk 2.5 m below?

23. If you throw the ball in problem 22 up instead of down, how fast is it moving when it hits the sidewalk? **Hint:** Its acceleration is the same whether it is moving up or down.

Critical Thinking Problems

Each of the following problems involves two objects. Draw the pictorial and physical models for each. Use different symbols to represent the position, velocity, and acceleration of each object. Do not solve the problem.

24. A car is traveling 25 m/s to the east, while a truck, initially 625 m away, is moving at 20 m/s to the west along the same road. Where do they meet?

25. A truck is stopped at a stoplight. When the light turns green, it accelerates at 2.5 m/s^2. At the same instant, a car passes the truck going 15 m/s. Where and when does the truck catch up with the car?

26. A truck is traveling at 18 m/s to the north. The driver of a car, 500 m to the north and traveling south at 24 m/s, puts on the brakes and slows at 3.5 m/s^2. Where do they meet?

Going Further

Using What You Know Write a problem and make a pictorial model for each of the following motion diagrams. Be creative!

a End ● ● ● ● ● ● ● Begin

b Begin ● ● ● ● ●●● ● ● ● Stop ... End

interNET CONNECTION

Follow the link on the Glencoe Homepage at **www.glencoe.com/sec/science** to find out more about this chapter.

VECTO

displacemen

Homeward Bound

A GPS receiver told you that your home was 15.0 km at a direction of 40° north of west, but the only path led directly north. If you took that path and walked 10 km, how far and in what direction would you then have to walk in a straight line to reach your home?

CHAPTER

4 Vector Addition

Finally, you've reached your destination. The scene you had been anticipating is spread out before you. It's the reward for the long trek that has brought you here and it's yours to enjoy.

But the time comes when you need to think about the journey home. It's easy to lose track of directions in country so vast, but you know you can't get lost when you have a GPS receiver with you. These small, handheld receivers can pinpoint your location using signals from the satellites of the Global Positioning System (GPS). The satellites, located in regular orbits around the globe, have different displacements from the receiver. Thus, synchronized pulses transmitted from the satellites are received at different times. The time differentials are translated into data that provide the position of the receiver.

Recall from Chapter 3 that displacement is a vector quantity. That means it has both magnitude (distance) and direction. In this chapter, you'll learn how to represent vectors and how to combine them in order to solve problems such as finding your way home.

WHAT YOU'LL LEARN

- You will represent vector quantities graphically and algebraically.
- You will determine the sum of vectors both graphically and algebraically.

WHY IT'S IMPORTANT

- Airplane pilots would find it difficult or impossible to locate their intended airport or estimate their time of arrival without taking into account the vectors that describe both the plane's velocity with respect to the air and the velocity of the air (winds) with respect to the ground.

interNET CONNECTION

Follow the link for this chapter on the Glencoe Homepage at **www.glencoe.com/sec/science** to find out more about vectors.

4.1 Properties of Vectors

OBJECTIVES

- **Determine** graphically the sum of two or more vectors.
- **Solve** problems of relative velocity.

You've learned that vectors have both a size, or magnitude, and a direction. For some vector quantities, the magnitude is so useful that it has been given its own name. For example, the magnitude of velocity is speed, and the magnitude of displacement is distance. The magnitude of a vector is always a positive quantity; a car can't have a negative speed, that is, a speed less than zero. But, vectors can have both positive and negative directions. In order to specify the direction of a vector, it's necessary to define a coordinate system. For now, the direction of vectors will be defined by the familiar set of directions associated with a compass: north, south, east, and west and the intermediate compass points such as northeast or southwest.

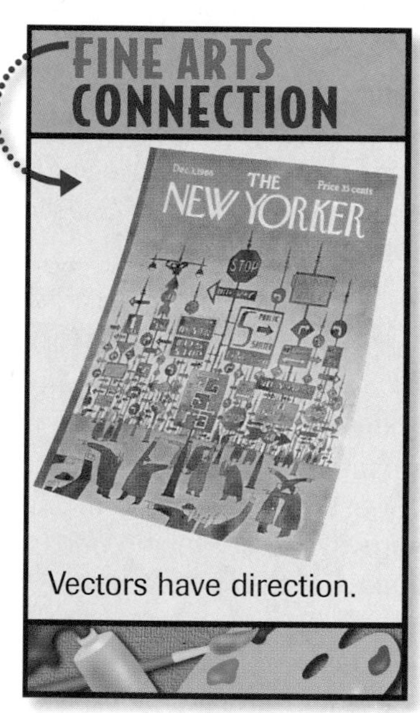

Vectors have direction.

Representing Vector Quantities

In Chapter 3, you learned that vector quantities can be represented by an arrow, or an arrow-tipped line segment. Such an arrow, having a specified length and direction, is called a **graphical representation** of a vector. You will use this representation when drawing vector diagrams. The arrow is drawn to scale so that its length represents the magnitude of the vector, and the arrow points in the specified direction of the vector.

In printed materials, an **algebraic representation** of a vector is often used. This representation is an italicized letter in boldface type. For example, a displacement can be represented by the expression $\boldsymbol{d} = 50$ km, southwest. $d = 50$ km designates only the magnitude of the vector.

Color Conventions

- Displacement vectors are **green.**
- Velocity vectors are **red.**

The resultant vector

Two displacements are equal when the two distances and directions are the same. For example, the two displacement vectors, ***A*** and ***B***, as shown in **Figure 4–1,** are equal. Even though they don't begin or end at the same point, they have the same length and direction. This property of vectors makes it possible to move vectors graphically for the purpose of adding or subtracting them. **Figure 4–1** also shows two unequal vectors, ***C*** and ***D***. Although they happen to start at the same position, they have different directions.

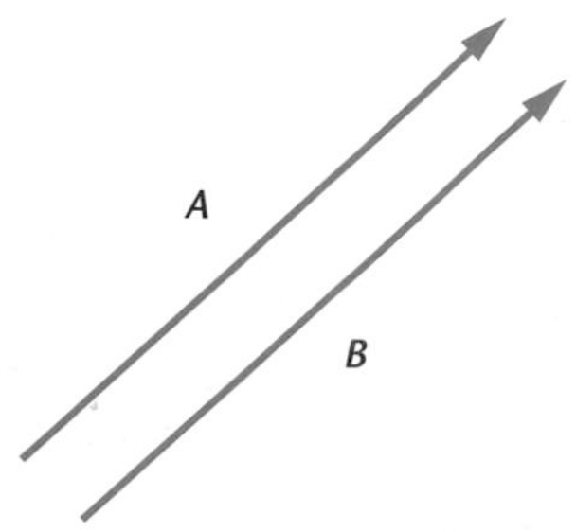

Two equal vectors

Two unequal vectors

FIGURE 4–1 Although they do not start at the same point, ***A*** and ***B*** are equal because they have the same length and direction.

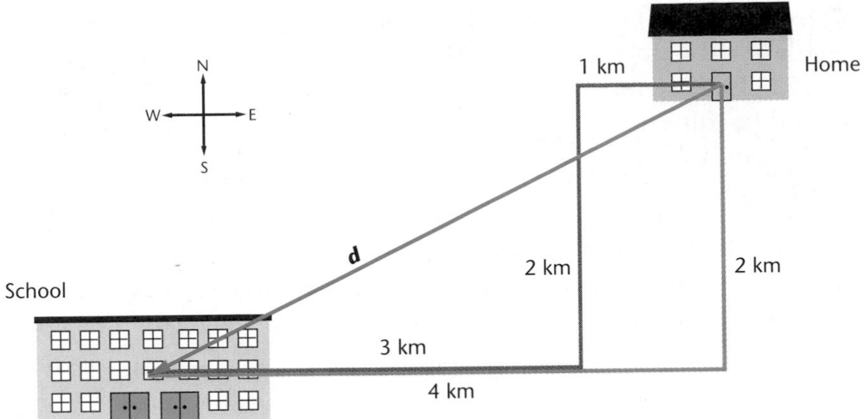

FIGURE 4–2 Your displacement from home to school is the same regardless of which route you take.

Recall that a displacement is a change in position. No matter what route you take from home to school, your displacement is the same. **Figure 4–2** shows some paths you could take. You could first walk 2 km south and then 4 km west and arrive at school, or you could travel 1 km west, then 2 km south, and then 3 km west. In each case, the displacement vector, ***d***, shown in **Figure 4–2,** is the same. This displacement vector is called a **resultant vector.** A resultant is a vector that is equal to the sum of two or more vectors. In this section, you will learn two methods of adding vectors to find the resultant vector.

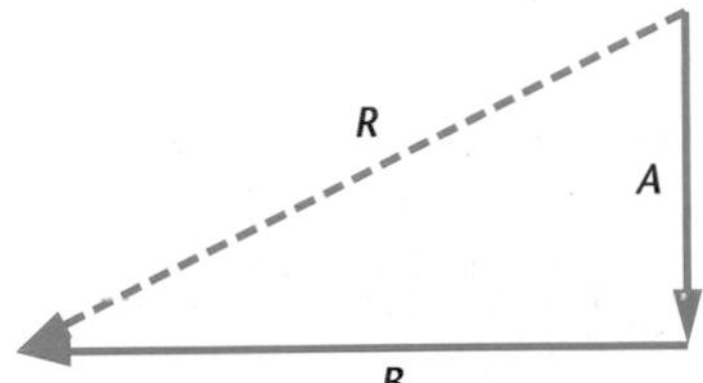

FIGURE 4–3 The length of ***R*** is proportional to the actual straight-line distance from home to school, and its direction is the direction of the displacement.

Graphical Addition of Vectors

One method for adding vectors involves manipulating their graphical representations on paper. To do so, you need a ruler to measure and draw the vectors to the correct length, and a protractor to measure the angle that establishes the direction. The length of the arrow should be proportional to the magnitude of the quantity being represented, so you must decide on a scale for your drawing. For example, you might let 1 cm on paper represent 1 km. The important thing is to choose a scale that produces a diagram of reasonable size with a vector about 5-10 cm long.

One route from home to school shown in **Figure 4–2** involves traveling 2 km south and then 4 km west. **Figure 4–3** shows how these two vectors can be added to give the resultant displacement, ***R***. First, vector ***A*** is drawn pointing directly south. Then, vector ***B*** is drawn with the tail of ***B*** at the tip of ***A*** and pointing directly west. Finally, the resultant is drawn from the tail of ***A*** to the tip of ***B***. The order of the addition can be reversed. Prove to yourself that the resultant would be the same if you drew ***B*** first and placed the tail of ***A*** at the tip of ***B***.

The magnitude of the resultant is found by measuring the length of the resultant with a ruler. To determine the direction, use a protractor to measure the number of degrees west of south the resultant is. How could you find the resultant vector of more than two vectors? **Figure 4–4** shows how to add the three vectors representing the second path you could take from home to school. Draw vector ***C***, then place the tail of ***D***

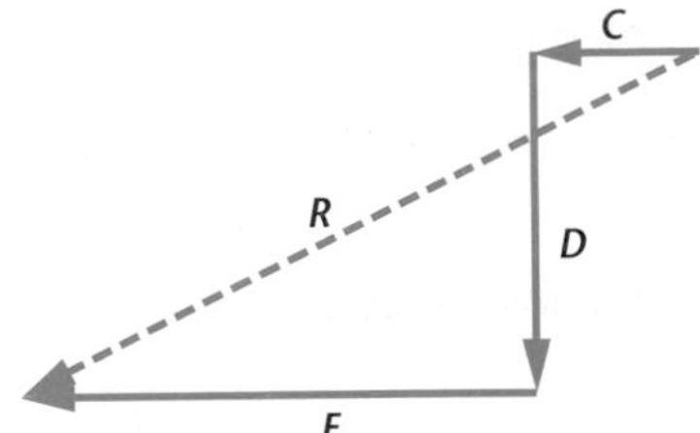

FIGURE 4–4 If you compare the displacement for route ***AB***, shown in **Figure 4–3,** with the displacement for route ***CDE***, you will find that the displacements are equal.

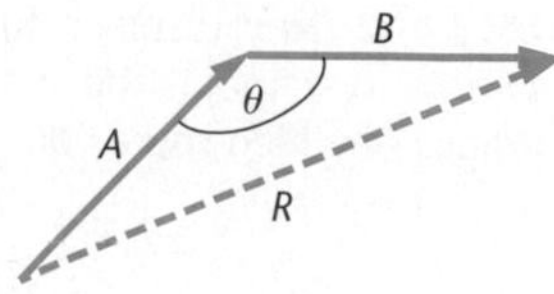

$R^2 = A^2 + B^2 - 2AB \cos \theta$

FIGURE 4–5 The Law of Cosines is used to calculate the magnitude of the resultant when the angle between the vectors is other than 90°.

at the tip of **C.** The third vector, **E,** is added in the same way. Place the tail of **E** at the tip of **D.** The resultant, **R,** is drawn from the tail of **C** to the tip of **E.** Use the ruler to measure the magnitude and the protractor to find the direction. If you measure the lengths of the resultant vectors in **Figures 4–3** and **4–4,** you will find that even though the paths that were walked are different, the resulting displacements are equal.

The magnitude of the resultant

If the two vectors to be added are at right angles, as shown in **Figure 4–3,** the magnitude can be found by using the Pythagorean theorem.

$$R^2 = A^2 + B^2$$

The magnitude of the resultant vector can be determined by calculating the square root. If the two vectors to be added are at some angle other than 90°, then you can use the Law of Cosines.

$$R^2 = A^2 + B^2 - 2AB\cos \theta$$

This equation calculates the magnitude of the resultant vector from the known magnitudes of the vectors **A** and **B** and the cosine of the angle, θ, between them. **Figure 4–5** shows the vector addition of **A** and **B.** Notice that the vectors must be placed tail to tip, and the angle θ is the angle between them.

Math Handbook

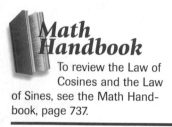

To review the Law of Cosines and the Law of Sines, see the Math Handbook, page 737.

Example Problem

Finding the Magnitude of the Sum of Two Vectors

Find the magnitude of the sum of a 15-km displacement and a 25-km displacement when the angle between them is 135°.

Sketch the Problem

- **Figure 4–5** shows the two displacement vectors, **A** and **B,** and the angle between them.

Calculate Your Answer

Known:	**Unknown:**
$A = 25$ km	$R = ?$
$B = 15$ km	
$\theta = 135°$	

Strategy:

Use the Law of Cosines to find the magnitude of the resultant vector when the angle does not equal 90°.

Calculations:

$$\begin{aligned}R^2 &= A^2 + B^2 - 2AB\cos \theta \\ &= (25\text{ km})^2 + (15\text{ km})^2 - 2(25\text{ km})(15\text{ km})\cos 135° \\ &= 625\text{ km}^2 + 225\text{ km}^2 - 750\text{ km}^2(-0.707) \\ &= 1380\text{ km}^2 \\ R &= \sqrt{1380\text{ km}^2} \\ &= 37\text{ km}\end{aligned}$$

Check Your Answer

- Is the unit correct? The unit of the answer is a length.
- Does the sign make sense? The sum should be positive.
- Is the magnitude realistic? The magnitude is in the same range as the two combined vectors but longer than either of them, as it should be because the resultant is the side opposite an obtuse angle.

Practice Problems

1. A car is driven 125 km due west, then 65 km due south. What is the magnitude of its displacement?
2. A shopper walks from the door of the mall to her car 250 m down a lane of cars, then turns 90° to the right and walks an additional 60 m. What is the magnitude of the displacement of her car from the mall door?
3. A hiker walks 4.5 km in one direction, then makes a 45° turn to the right and walks another 6.4 km. What is the magnitude of her displacement?
4. What is the magnitude of your displacement when you follow directions that tell you to walk 225 m in one direction, make a 90° turn to the left and walk 350 m, then make a 30° turn to the right and walk 125 m?

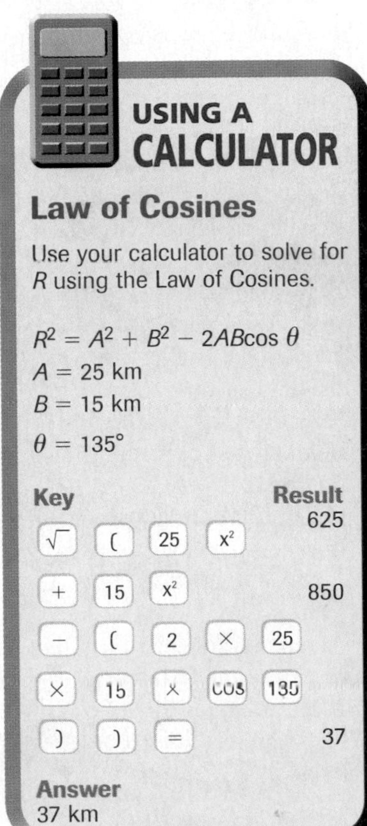

USING A
CALCULATOR
Law of Cosines
Use your calculator to solve for R using the Law of Cosines.
R² = A² + B² – 2ABcos θ
A = 25 km
B = 15 km
θ = 135°
Key
Result
√ (25 x²
625
+ 15 x²
850
– (2 × 25
× 15 × cos 135
)) =
37
Answer
37 km

Subtracting Vectors

Multiplying a vector by a scalar number changes its length but not its direction unless the scalar is negative. Then, the vector's direction is reversed. This fact can be used to subtract two vectors using the same methods you used for adding them. For example, you've learned that the difference in two velocities is defined by this equation.

$$\Delta \boldsymbol{v} = \boldsymbol{v}_2 - \boldsymbol{v}_1$$

The equation can be written as the sum of two vectors.

$$\Delta \boldsymbol{v} = \boldsymbol{v}_2 + (-\boldsymbol{v}_1)$$

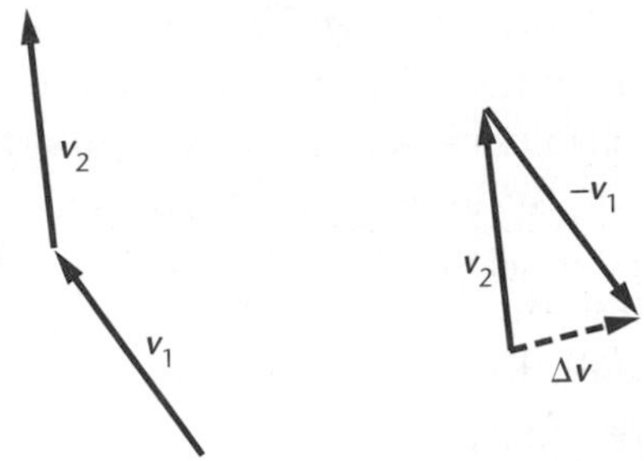

FIGURE 4–6 To subtract two vectors, reverse the direction of the second vector and then add them.

FIGURE 4–7 When a coordinate system is moving, two velocities add if both motions are in the same direction and subtract if the motions are in opposite directions.

$v_{\text{bus relative to street}}$ $v_{\text{you relative to bus}}$ $v_{\text{you relative to street}}$

$v_{\text{bus relative to street}}$ $v_{\text{you relative to bus}}$ $v_{\text{you relative to street}}$

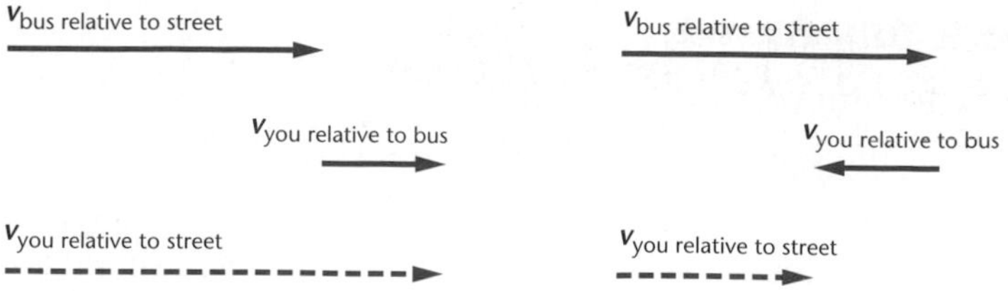

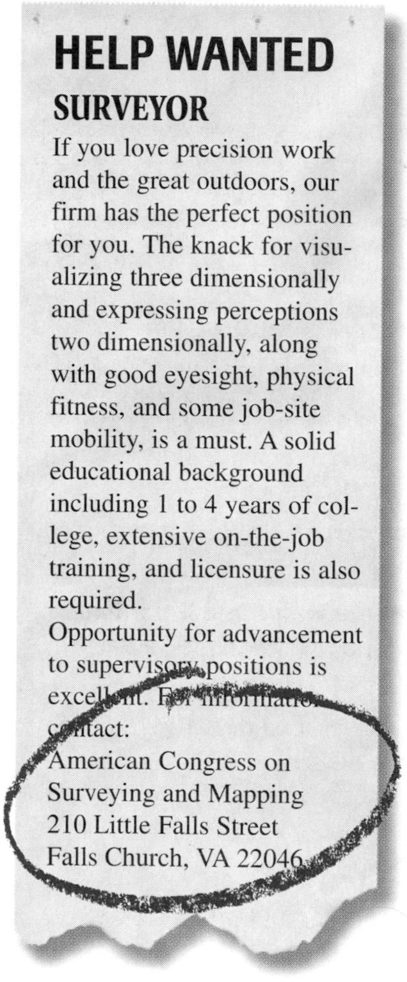

HELP WANTED

SURVEYOR

If you love precision work and the great outdoors, our firm has the perfect position for you. The knack for visualizing three dimensionally and expressing perceptions two dimensionally, along with good eyesight, physical fitness, and some job-site mobility, is a must. A solid educational background including 1 to 4 years of college, extensive on-the-job training, and licensure is also required.
Opportunity for advancement to supervisory positions is excellent. For information contact:
American Congress on Surveying and Mapping
210 Little Falls Street
Falls Church, VA 22046

If $\boldsymbol{v}_1$ is multiplied by -1, the direction of $\boldsymbol{v}_1$ is reversed as shown in **Figure 4–6.** The vector $-\boldsymbol{v}_1$ can then be added to $\boldsymbol{v}_2$ to get the resultant, which represents the difference, $\Delta \boldsymbol{v}$.

Relative Velocities: Some Applications

Graphical addition of vectors can be a useful tool when solving problems that involve relative velocity. Suppose you're in a school bus traveling at a velocity of 8 m/s in a positive direction. You walk at 3 m/s toward the front of the bus. How fast are you moving relative to the street? To solve this problem, you must translate these statements into symbols. If the bus is going 8 m/s, that means that the velocity of the bus is 8 m/s as measured in a coordinate system fixed to the street. Standing still, your velocity relative to the street is also 8 m/s but your velocity relative to the bus is zero. *Walk at 3 m/s toward the front of the bus* means that your velocity is measured relative to the bus. The question can be rephrased: Given the velocity of the bus relative to the street and your velocity relative to the bus, what is your velocity relative to the street?

A vector representation of this problem is shown in **Figure 4–7.** After looking at it and thinking about it, you'll agree that your velocity relative to the street is 11 m/s, the sum of 8 m/s and 3 m/s. Suppose you now walked at the same speed toward the rear of the bus. What would be your velocity relative to the street? **Figure 4–7** shows that because the two velocities are in opposite directions, the resultant velocity is 5 m/s, the difference between 8 m/s and 3 m/s. You can see that when the velocities are along the same line, simple addition or subtraction can be used to determine the relative velocity.

The addition of relative velocities can be extended to include motion in two dimensions. For example, airline pilots cannot expect to reach their destinations by simply aiming their planes along a compass direction. They must take into account the plane's velocity relative to the air, which is given by their airspeed indicators and their direction relative to the air. They must also consider the velocity of the wind that they must fly through relative to the ground. These two vectors must be combined, as shown in **Figure 4–8,** to obtain the velocity of the airplane relative to the ground. The resultant vector tells the pilot how fast and in what direction the plane must travel relative to the ground to reach its destination. You can add relative velocities even if they are at arbitrary angles by using a graphical method.

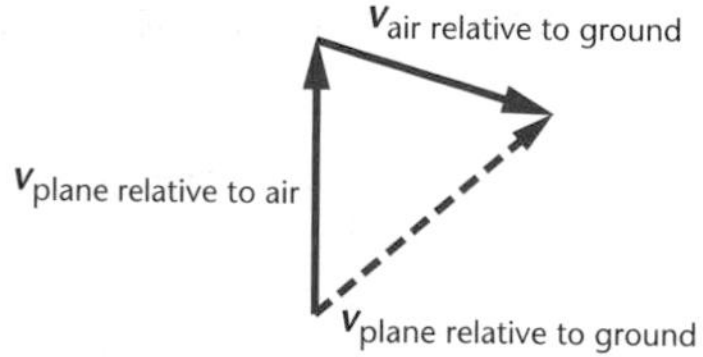

FIGURE 4–8 The plane's velocity relative to the ground can be obtained by vector addition.

Physics Lab

The Paper River

Problem

How does a boat travel on a river?

Materials

small battery-powered car (or physics bulldozer)
meterstick
protractor
stopwatch
a piece of paper, 1 m × 10 m

Procedure

1. Your car will serve as the boat. Write a brief statement to explain how the boat's speed can be determined.
2. Your boat will start with all wheels on the paper river. Measure the width of the river and predict how much time is needed for your boat to go directly across the river. Show your data and calculations.
3. Determine the time needed to cross the river when your boat is placed on the edge of the river. Make three trials and record the times.
4. Do you think it will take more or less time to cross when the river is flowing? Explain your prediction.
5. Have a student (the hydro engineer) walk slowly, at a constant speed, while pulling the river along the floor. Each group should measure the time it takes for the boat to cross the flowing river. Compare the results with your prediction.
6. Devise a method to measure the speed of the river. Have the hydro engineer pull the river at a constant speed and collect the necessary data.

Data and Observations

1. Does the boat move in the direction that it is pointing?
2. Did the motion of the water affect the time needed when the boat was pointed straight across?
3. Which had the greater speed, the river or the boat? Explain your choice.

Analyze and Conclude

1. **Calculating Results** Calculate the speed of the river.
2. **Inferring Conclusions** Using your results for the speed of the boat and the speed of the river, calculate the speed of the boat compared to the ground when the boat is headed directly downstream and directly upstream.

Apply

1. Do small propeller aircraft always move in the direction that they are pointing? Do they ever fly sideways?

Physics & Society

Assessing Risk

Nearly every decision you make involves risk. Risk is the likelihood that a decision you make will cause you, another person, or an object injury, damage, or even loss. Read the information below and assess whether you think air bags should be standard equipment in automobiles.

Air Bags—Assets or Assaults?

Since the early 1990s, nearly all cars made in the United States have had a driver's-side air bag as standard equipment. About half of these vehicles also came equipped with a front-passenger air bag located in the car's dashboard. When deployed, air bags are designed to be protective cushions between a front-seat occupant and the car's steering column or dashboard.

The National Highway Traffic Safety Administration estimates that at least 1500 drivers or passengers have been saved by air bags since the early 1990s. Through 1996, however, air bags have been responsible for the deaths of 52 people who might otherwise have survived the crash. By mid-1995, 163 000 injuries were sustained by drivers and front-seat passengers as a result of the force of air bags. Fewer than one percent were major injuries or proved to be fatal. Most of the fatalities were sustained by children or short, female adults.

Proponents of automotive air bags admit that there is a risk but believe that the number of lives saved is sufficient reason for the installation of air bags in all vehicles. Suggested design changes include sensors to assess the severity of the impact and determine the weight and location of front-seat occupants at the time of the crash. With these data, a computer could prevent air-bag deployment if the driver or passenger were in danger of being injured by the air bag. Decreasing the force with which air bags deploy also is being considered in the design of "smart" air bags.

Air-bag opponents contend that after three decades of development, a system that takes into account every possible crash scenario still doesn't exist. Many opponents feel the federal government jumped the gun by legislating the installation of air bags before determining whether the devices enhance safety. Opponents also argue that air-bag regulations are biased because they call for protecting an unbelted, 77-kg male. Some opponents propose that air bags be optional equipment, or people should have the choice of disabling air bags.

Investigating the Issue

1. **Acquiring Information** Find out more about the tests that were done that led to federal regulations on automotive air bags.
2. **Assessing Risk** Does the possible risk of injury or death due to air bags outweigh their potential to save lives? Support your answer.
3. **Thinking Critically** Would today's air bags be useful in a rear-end collision? Explain.

interNET CONNECTION

Follow the link for this chapter on the Glencoe Homepage at **www.glencoe.com/sec/science** to find out more about air bags.

Practice Problems

5. A car moving east at 45 km/h turns and travels west at 30 km/h. What are the magnitude and direction of the change in velocity?
6. You are riding in a bus moving slowly through heavy traffic at 2.0 m/s. You hurry to the front of the bus at 4.0 m/s relative to the bus. What is your speed relative to the street?
7. A motorboat heads due east at 11 m/s relative to the water across a river that flows due north at 5.0 m/s. What is the velocity of the motorboat with respect to the shore?
8. A boat is rowed directly upriver at a speed of 2.5 m/s relative to the water. Viewers on the shore find that it is moving at only 0.5 m/s relative to the shore. What is the speed of the river? Is it moving with or against the boat?
9. An airplane flies due north at 150 km/h with respect to the air. There is a wind blowing at 75 km/h to the east relative to the ground. What is the plane's speed with respect to the ground?
10. An airplane flies due west at 185 km/h with respect to the air. There is a wind blowing at 85 km/h to the northeast relative to the ground. What is the plane's speed with respect to the ground?

F.Y.I.

Vector is a term used in biology and medicine to describe any disease-carrying microorganism. In genetics, a vector is any self-replicating DNA molecule that will carry one gene from one organism to another.

4.1 Section Review

1.1 Is the distance you walk equal to the magnitude of your displacement? Give an example that supports your conclusion.

1.2 A fishing boat with a maximum speed of 3 m/s with respect to the water is in a river that is flowing at 2 m/s. What is the maximum speed of the boat with respect to the shore? The minimum speed? Give the direction of the boat, relative to the river's current, for the maximum speed and the minimum speed relative to the shore.

1.3 The order in which vectors are added doesn't matter. Mathematicians say that vector addition is commutative. Which ordinary arithmetic operations are commutative? Which are not?

1.4 **Critical Thinking** A box is moved through one displacement and then through a second displacement. The magnitudes of the two displacements are unequal. Could the displacements have directions such that the resultant displacement is zero? Suppose the box was moved through three displacements of unequal magnitude? Could the resultant displacement be zero? Support your argument with a diagram.

4.2 Components of Vectors

The graphical method of adding vectors did not require that you decide on a coordinate system. The sum, or the difference, of vectors is the same no matter what coordinate system is used. Nevertheless, as you'll find, creating and using a coordinate system allows you not only to make quantitative measurements, but also provides an alternative method of adding vectors.

OBJECTIVES

- **Establish** a coordinate system in problems involving vector quantities.
- **Use** the process of resolution of vectors to find the components of vectors.
- **Determine** algebraically the sum of two or more vectors by adding the components of the vectors.

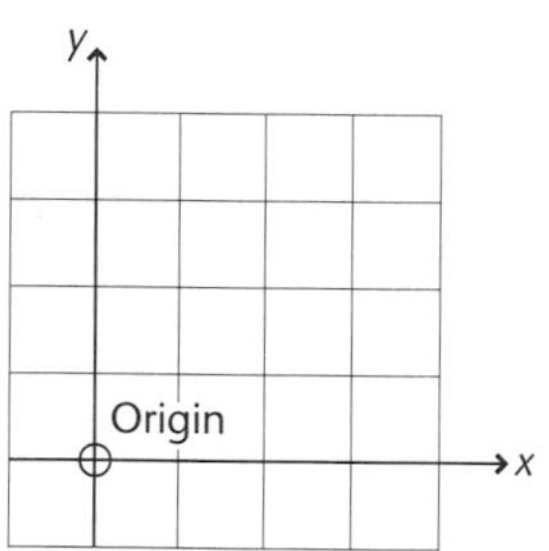

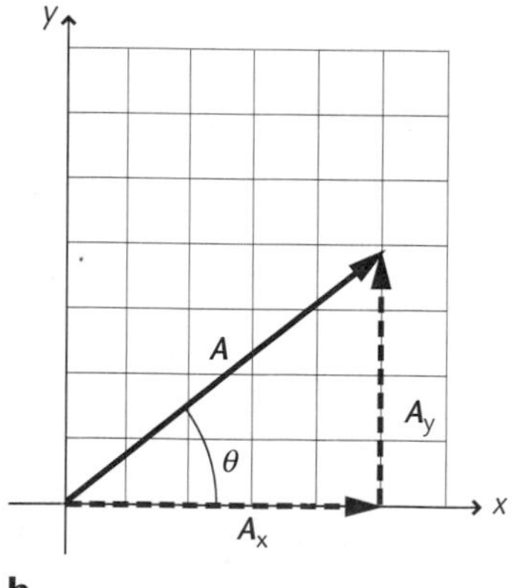

FIGURE 4–9 A coordinate system has an origin and two perpendicular axes, as in **a.** In **b,** the direction of a vector is measured counterclockwise from the *x*-axis.

Choosing a Coordinate System

Choosing a coordinate system, such as the one in **Figure 4–9a,** is similar to laying a grid drawn on a sheet of transparent plastic on top of your problem. You have to choose where to put the center of the grid (the origin) and establish the direction in which the axes point. Notice that in the coordinate system shown in **Figure 4–9a,** the *x*-axis is drawn through the origin with an arrow pointing in the positive direction. Then, the positive *y*-axis is located 90° counterclockwise from the positive *x*-axis and crosses the *x*-axis at the origin.

How do you choose the direction of the *x*-axis? There is never a single correct answer, but some choices make the problem easier to solve than others. When the motion you are describing is confined to the surface of Earth, it is often convenient to have the *x*-axis point east and the *y*-axis point north. When the motion involves an object moving through the air, the positive *x*-axis is often chosen to be horizontal and the positive *y*-axis vertical (upward). If the motion is on a hill, it's convenient to place the positive *x*-axis in the direction of the motion and the *y*-axis perpendicular to the *x*-axis.

After the coordinate system is chosen, the direction of any vector can be specified relative to those coordinates. The direction of a vector is defined as the angle that the vector makes with the *x*-axis, measured counterclockwise. In **Figure 4–9b,** the angle θ tells the direction of the vector **A.**

Components

A coordinate system allows you to expand your description of a vector. In the coordinate system shown in **Figure 4–9b,** the vector **A** is broken up or resolved into two component vectors. One, $\mathbf{A}_x$, is parallel to the *x*-axis, and the other, $\mathbf{A}_y$, is parallel to the *y*-axis. You can see that the original vector is the sum of the two component vectors.

$$\mathbf{A} = \mathbf{A}_x + \mathbf{A}_y$$

The process of breaking a vector into its components is sometimes called **vector resolution.** The magnitude and sign of component vectors are called the **components.** All algebraic calculations involve only the components of vectors, not the vectors themselves. You can find the components by using trigonometry. The components are calculated according to these equations, where the angle θ is measured counterclockwise from the positive x-axis.

$$A_x = A \cos \theta\text{; therefore, } \cos \theta = \frac{\text{adjacent side}}{\text{hypotenuse}} = \frac{A_x}{A}$$

$$A_y = A \sin \theta\text{; therefore, } \sin \theta = \frac{\text{opposite side}}{\text{hypotenuse}} = \frac{A_y}{A}$$

When the angle that a vector makes with the x-axis is larger than 90°—that is, the vector is in the second, third, or fourth quadrants—the sign of one or more components is negative, as shown in **Figure 4–10.** Although the components are scalars, they can have both positive and negative signs.

Second Quadrant	First Quadrant
$A_x < 0$, $A_y > 0$	$A_x > 0$, $A_y > 0$
$A_x < 0$, $A_y < 0$	$A_x > 0$, $A_y < 0$
Third Quadrant	Fourth Quadrant

FIGURE 4–10 The sign of a component depends upon which of the four quadrants the component is in.

Example Problem

The Components of Displacement

A bus travels 23.0 km on a straight road that is 30° north of east. What are the east and north components of its displacement?

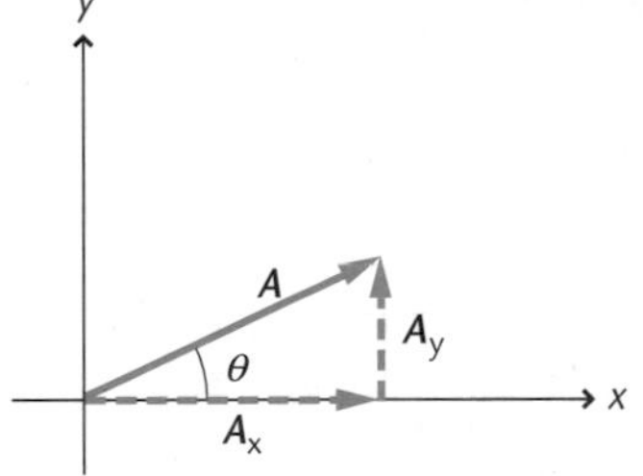

Sketch the Problem

- Draw the same sketch as in **Figure 4–9b.**
- A coordinate system is used in which the x-axis points east.
- The angle θ is measured counterclockwise from the x-axis.

Calculate Your Answer

Known:

$A = 23.0$ km

$\theta = 30°$

Strategy:

Use the trigonometric ratios to find the components.

Unknown:

$A_x = ?$

$A_y = ?$

Calculations:

$A_x = A \cos \theta$

$A_y = A \sin \theta$

$A_x = (23.0 \text{ km})\cos 30°$
$= +19.9$ km

$A_y = (23.0 \text{ km})\sin 30°$
$= +11.5$ km

Check Your Answer

- Are the units correct? The kilometer is an appropriate unit of length.
- Do the signs make sense? Both components are in the first quadrant and should be positive.
- Are the magnitudes reasonable? The magnitudes are less than the hypotenuse of the right triangle of which they are the other two sides.

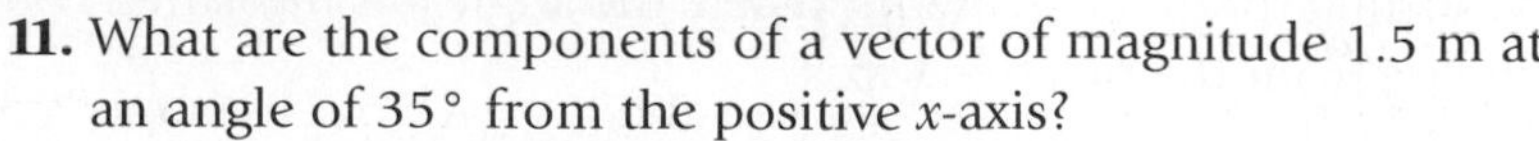

Practice Problems

11. What are the components of a vector of magnitude 1.5 m at an angle of 35° from the positive x-axis?

12. A hiker walks 14.7 km at an angle 35° south of east. Find the east and north components of this walk.

13. An airplane flies at 65 m/s in the direction 149° counterclockwise from east. What are the east and north components of the plane's velocity?

14. A golf ball, hit from the tee, travels 325 m in a direction 25° south of the east axis. What are the east and north components of its displacement?

Pocket Lab

Ladybug

You notice a ladybug moving from one corner of your textbook to the corner diagonally opposite. The trip takes the ladybug 6.0 s. Use the long side of the book as the x-axis. Find the component vectors of the ladybug's velocity, $\boldsymbol{v}_x$ and $\boldsymbol{v}_y$, and the resultant velocity ***R.***

Analyze and Conclude Does the ladybug's path from one corner to the other affect the values in your measurements or calculations? Do $\boldsymbol{v}_x + \boldsymbol{v}_y$ really add up to ***R?*** Explain.

Algebraic Addition of Vectors

Two or more vectors (***A, B, C,***. . .) may be added by first resolving each vector to its x- and y-components. The x-components are added to form the x-component of the resultant, $R_x = A_x + B_x + C_x + \ldots$ Similarly, the y-components are added to form the y-component of the resultant, $R_y = A_y + B_y + C_y + \ldots$.

The process is illustrated graphically in **Figure 4–11.** Because R_x and R_y are at a right angle (90°), the magnitude of the resultant vector can be calculated using the Pythagorean theorem.

$$R^2 = R_x^{\,2} + R_y^{\,2}$$

To find the angle or direction of the resultant, recall that the tangent of the angle that the vector makes with the x-axis is given by the following.

$$\tan\theta = \frac{R_y}{R_x}$$

You can find the angle by using the $\tan^{-1}$ key on your calculator. Note: when $\tan\theta > 0$, most calculators give the angle between 0 and 90°; when $\tan\theta < 0$, the angle is reported to be between 0 and −90°.

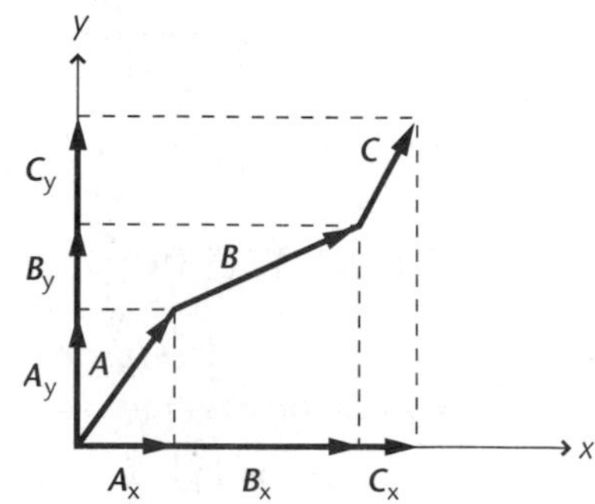

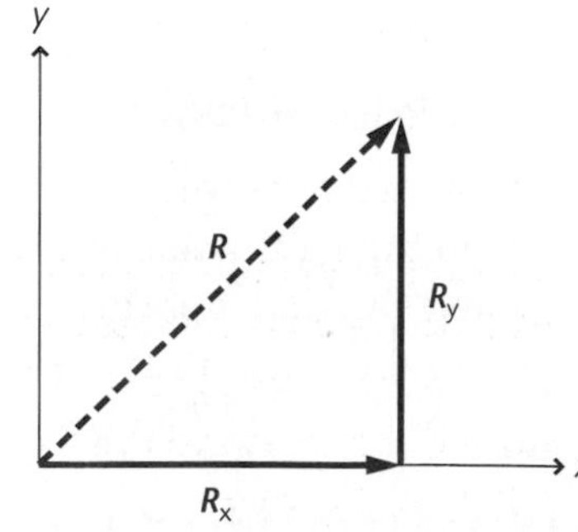

FIGURE 4–11 R_x is the sum of the x-components of ***A, B,*** and ***C.*** R_y is the sum of the y-components. The vector sum of R_x and R_y is the vector sum of ***A, B,*** and ***C.***

Example Problem

Finding Your Way Home

A GPS receiver told you that your home was 15.0 km at a direction of 40° north of west, but the only path led directly north. If you took that path and walked 10.0 km, how far, and in what direction would you then have to walk to reach your home?

Sketch the Problem

- Draw the resultant vector, **R** from your original location to home.
- Draw **A,** the known displacement.
- Draw **B,** the unknown displacement.

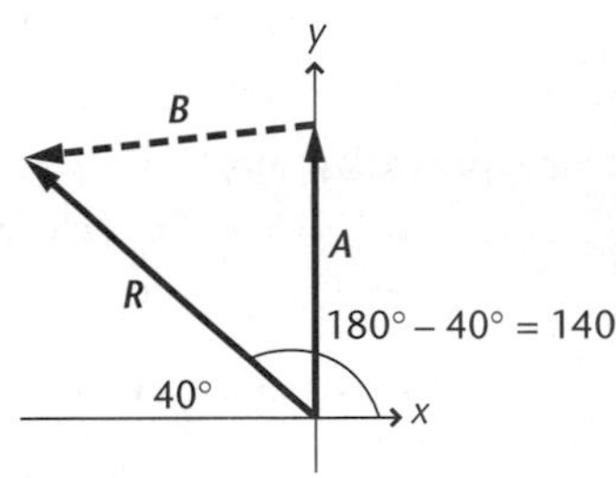

Homeward Bound

Calculate Your Answer

Known:

$\boldsymbol{A}$ = 10.0 km, due north

$\boldsymbol{R}$ = 15.0 km, 40° north of west

Unknown:

$\boldsymbol{B}$ = ?

Strategy:

Find the components of **R** and **A.**

Calculations:

$R_x = R\cos\theta$
$= (15.0\text{ km})\cos 140°$
$= -11.5\text{ km}$

$R_y = R\sin\theta$
$= (15.0\text{ km})\sin 140°$
$= +9.6\text{ km}$

$A_x = 0.0\text{ km},\ A_y = 10.0\text{ km}$

Use the components of **R** and **A** to find the components of **B.** The signs of B_x and B_y will tell you the direction of the component.

$\boldsymbol{R} = \boldsymbol{A} + \boldsymbol{B}$, so $\boldsymbol{B} = \boldsymbol{R} - \boldsymbol{A}$

$B_x = R_x - A_x = -11.5\text{ km} - 0.0\text{ km} = -11.5\text{ km}$;
This component points west.

$B_y = R_y - A_y = 9.6\text{ km} - 10.0\text{ km} = -0.4\text{ km}$;
This component points south.

Use the components of **B** to find the magnitude of **B.**

$B = \sqrt{B_x^2 + B_y^2}$
$= \sqrt{(-11.5\text{ km})^2 + (-0.4\text{ km})^2}$
$= 11.5\text{ km}$

Use the tangent to find the direction of **B.**

$$\tan\theta = \frac{B_y}{B_x} = \frac{-0.4\text{ km}}{-11.5\text{ km}} = +0.035$$

$\theta = \tan^{-1}(+0.035) = 2.0°$

Locate the tail of B at the origin of a coordinate system and draw the components $\boldsymbol{B}_x$ and $\boldsymbol{B}_y$. The direction is in the third quadrant, 2.0° south of west.

$\boldsymbol{B}$ = 11.5 km, 2.0° south of west

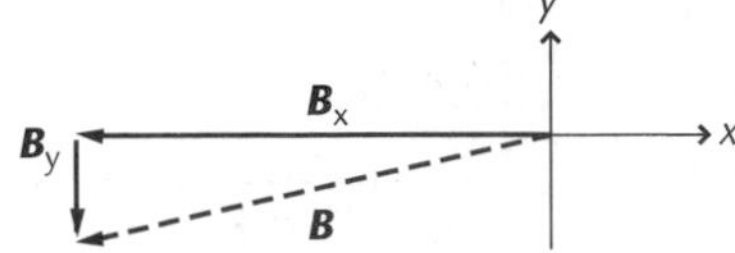

Check Your Answer

- Are the units correct? Kilometers and degrees are correct.
- Do the signs make sense? They agree with the diagram.
- Is the magnitude realistic? The length of ***B*** is reasonable because the angle between ***A*** and ***B*** is slightly less than 90°. If the angle were 90°, ***B*** would have been 11.2 km, which is close to 11.5 km. The direction of ***B*** deviates only slightly from the east-west direction.

Practice Problems

15. A powerboat heads due northwest at 13 m/s with respect to the water across a river that flows due north at 5.0 m/s. What is the velocity (both magnitude and direction) of the motorboat with respect to the shore?

16. An airplane flies due south at 175 km/h with respect to the air. There is a wind blowing at 85 km/h to the east relative to the ground. What are the plane's speed and direction with respect to the ground?

17. An airplane flies due north at 235 km/h with respect to the air. There is a wind blowing at 65 km/h to the northeast with respect to the ground. What are the plane's speed and direction with respect to the ground?

18. An airplane has a speed of 285 km/h with respect to the air. There is a wind blowing at 95 km/h at 30° north of east with respect to Earth. In which direction should the plane head in order to land at an airport due north of its present location? What would be the plane's speed with respect to the ground?

F.Y.I.

Although Oliver Heaviside was greatly respected by scientists of his day, he is almost forgotten today. His methods of describing forces by means of vectors were so successful that they were used in textbooks by other people. Unfortunately, few gave Heaviside credit for his work.

4.2 Section Review

2.1 You first walk 8.0 km north from home, then walk east until your distance from home is 10.0 km. How far east did you walk?

2.2 Could a vector ever be shorter than one of its components? Equal in length to one of its components? Explain.

2.3 In a coordinate system in which the *x*-axis is east, for what range of angles is the *x*-component positive? For what range is it negative?

2.4 **Critical Thinking** You are piloting a boat across a fast-moving river. You want to reach a pier directly opposite your starting point. Describe how you would select your heading in terms of the components of your velocity relative to the water.

CHAPTER 4 REVIEW

Key Terms

4.1

- graphical representation
- algebraic representation
- resultant vector

4.2

- vector resolution
- component

Summary

4.1 Properties of Vectors

- Vectors are quantities that have both magnitude and direction. They can be represented graphically as arrows or algebraically as symbols.
- Vectors are not changed by moving them, as long as their magnitudes (lengths) and directions are maintained.
- Vectors can be added graphically by placing the tail of one at the tip of the other and drawing the resultant from the tail of the first to the tip of the second.
- The sum of two or more vectors is the resultant vector.
- The Law of Cosines may be used to find the magnitude of the resultant of any two vectors. This simplifies to the Pythagorean theorem if the vectors are at right angles.
- Vector addition may be used to solve problems involving relative velocities.

4.2 Components of Vectors

- Placing vectors in a coordinate system that you have chosen makes it possible to decompose them into components along each of the chosen coordinate axes.
- The components of a vector are the projections of the component vectors. They are scalars and have signs, positive or negative, indicating their directions.
- Two or more vectors can be added by separately adding the *x*- and *y*-components. These components can then be used to determine the magnitude and direction of the resultant vector.

Reviewing Concepts

Section 4.1

1. Describe how you would add two vectors graphically.

2. Which of the following actions is permissible when you are graphically adding one vector to another: move the vector, rotate the vector, change the vector's length?

3. In your own words, write a clear definition of the resultant of two or more vectors. Do not tell how to find it, but tell what it represents.

4. How is the resultant displacement affected when two displacement vectors are added in a different order?

5. Explain the method you would use to subtract two vectors graphically.

6. Explain the difference between these two symbols: A and $\mathbf{A}$.

Section 4.2

7. Describe a coordinate system that would be suitable for dealing with a problem in which a ball is thrown up into the air.

8. If a coordinate system is set up such that the positive *x*-axis points in a direction 30° above the horizontal, what should be the angle between the *x*-axis and the *y*-axis? What should be the direction of the positive *y*-axis?

9. The Pythagorean theorem is usually written $c^2 = a^2 + b^2$. If this relationship is used in vector addition, what do a, b, and c represent?

10. Using a coordinate system, how is the angle or direction of a vector determined with respect to the axes of the coordinate system?

Applying Concepts

11. A vector drawn 15 mm long represents a velocity of 30 m/s. How long should you draw a vector to represent a velocity of 20 m/s?

12. A vector that is 1 cm long represents a displacement of 5 km. How many kilometers are represented by a 3-cm vector drawn to the same scale?

13. What is the largest possible displacement resulting from two displacements with magnitudes 3 m and 4 m? What is the smallest possible resultant? Draw sketches to demonstrate your answers.

14. How does the resultant displacement change as the angle between two vectors increases from 0° to 180°?

15. *A* and *B* are two sides of a right triangle. If $\tan \theta = A/B$,

a. which side of the triangle is longer if $\tan \theta$ is greater than one?

b. which side is longer if $\tan \theta$ is less than one?

c. what does it mean if $\tan \theta$ is equal to one?

16. A car has a velocity of 50 km/h in a direction 60° north of east. A coordinate system with the positive *x*-axis pointing east and a positive *y*-axis pointing north is chosen. Which component of the velocity vector is larger, *x* or *y*?

17. Under what conditions can the Pythagorean theorem, rather than the Law of Cosines, be used to find the magnitude of a resultant vector?

18. A problem involves a car moving up a hill so a coordinate system is chosen with the positive *x*-axis parallel to the surface of the hill. The problem also involves a stone that is dropped onto the car. Sketch the problem and show the components of the velocity vector of the stone.

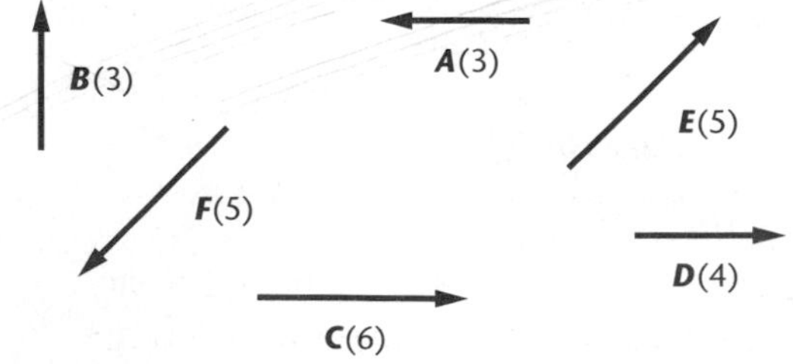

FIGURE 4–12

Problems

Section 4.1

LEVEL 1

19. A car moves 65 km due east, then 45 km due west. What is its total displacement?

20. Graphically find the sum of the following pairs of vectors whose lengths and directions are shown in **Figure 4–12.**

a. ***D*** and ***A***

b. ***C*** and ***D***

c. ***C*** and ***A***

d. ***E*** and ***F***

21. An airplane flies at 200 km/h with respect to the air. What is the velocity of the plane relative to the ground if it flies with

a. a 50-km/h tailwind?

b. a 50-km/h head wind?

22. Graphically add the following pairs of vectors as shown in **Figure 4–12.**

a. ***B*** and ***D***

b. ***D*** and ***E***

c. ***A*** and ***F***

23. Graphically add the following sets of vectors as shown in **Figure 4–12.**

a. ***A*, *C*,** and ***D***

b. ***A*, *B*,** and ***E***

c. ***B*, *D*,** and ***F***

24. Path A is 8.0 km long heading 60.0° north of east. Path B is 7.0 km long in a direction due east. Path C is 4.0 km long heading 315° counterclockwise from east.

a. Graphically add the hiker's displacements in the order **A, B, C.**

b. Graphically add the hiker's displacements in the order **C, B, A.**

c. What can you conclude about the resulting displacements?

LEVEL 2

25. A river flows toward the east. Because of your knowledge of physics, you head your boat 53° west of north and have a velocity of 6.0 m/s due north relative to the shore.

a. What is the velocity of the current?

b. What is your speed relative to the water?

Section 4.2

LEVEL 1

26. You walk 30 m south and 30 m east. Find the magnitude and direction of the resultant displacement both graphically and algebraically.
27. A ship leaves its home port expecting to travel to a port 500 km due south. Before it moves even 1 km, a severe storm blows it 100 km due east. How far is the ship from its destination? In what direction must it travel to reach its destination?
28. A descent vehicle landing on Mars has a vertical velocity toward the surface of Mars of 5.5 m/s. At the same time, it has a horizontal velocity of 3.5 m/s.
    a. At what speed does the vehicle move along its descent path?
    b. At what angle with the vertical is this path?
29. You are piloting a small plane, and you want to reach an airport 450 km due south in 3.0 hours. A wind is blowing from the west at 50 km/h. What heading and airspeed should you choose to reach your destination in time?

LEVEL 2

30. A hiker leaves camp and, using a compass, walks 4 km E, then 6 km S, 3 km E, 5 km N, 10 km W, 8 km N, and finally 3 km S. At the end of three days, the hiker is lost. By drawing a diagram, compute how far the hiker is from camp and which direction should be taken to get back to camp.
31. You row a boat perpendicular to the shore of a river that flows at 3.0 m/s. The velocity of your boat is 4.0 m/s relative to the water.
    a. What is the velocity of your boat relative to the shore?
    b. What is the component of your velocity parallel to the shore?
    c. What is the component of your velocity perpendicular to the shore?
32. A weather station releases a balloon that rises at a constant 15 m/s relative to the air, but there is a wind blowing at 6.5 m/s toward the west. What are the magnitude and direction of the velocity of the balloon?

Critical Thinking Problems

33. An airplane, moving at 375 m/s relative to the ground, fires a missile forward at a speed of 782 m/s relative to the plane. What is the speed of the missile relative to the ground?
34. A rocket in outer space that is moving at a speed of 1.25 km/s relative to an observer fires its motor. Hot gases are expelled out the rear at 2.75 km/s relative to the rocket. What is the speed of the gases relative to the observer?
35. The same rocket as in problem 34 uses small thrusters to rotate itself until its motor's exhaust is pointing at a direction of 25° from the velocity of the rocket. It now fires the motor. What is the speed of the hot gases relative to a stationary observer?

Going Further

Albert Einstein showed that the rule you learned for the addition of velocities doesn't work for objects moving near the speed of light. For example, if a rocket moving at velocity v_A releases a missile that has a velocity v_B relative to the rocket, then the velocity of the missile relative to an observer that is at rest is given by,

$$v = \frac{v_A + v_B}{1 + v_A v_B/c^2}$$ where c is the speed of light,

3.00×10^8 m/s. Einstein pointed out that this formula gives the correct values for objects moving at slow speeds as well. Test Einstein's formula with problem 34. Now, suppose a rocket moving at 11 km/s shoots a laser beam out front. What speed would an unmoving observer find for the laser light? Finally, suppose a rocket moves at a speed of $c/2$, half the speed of light, and shoots a missile forward at a speed of $c/2$ relative to the rocket. How fast would the missile be moving relative to a fixed observer?

interNET CONNECTION

Follow the link on the Glencoe Homepage at **www.glencoe.com/sec/science** to find out more about this chapter.

Free Fall

Held down by shoulder bars, you stop for a moment at the top. The supports under the car are released, and you're in free fall, dropping faster and faster for a time interval of 1.5 s. How far will you drop and how fast will you be going the instant before the car reaches the bottom?

CHAPTER

5 A Mathematical Model of Motion

For pure heart-stopping thrills, nothing beats a ride on the Demon Drop at Cedar Point Amusement Park. It's the ultimate thrill. But all amusement park rides can be thrilling, from the gentle, circling carousel and the Ferris wheel to the tilt-o-whirl and the roller coaster. Isn't the excitement of most of these rides the rapid and unexpected changes in speed and direction that you feel in your stomach? Probably the rides that you like best are those with the sharpest turns and the most precipitous drops.

It's hard to imagine that the thrill of the Demon Drop, or even the merry-go-round, could be interpreted in mathematical terms. Yet the vectors you learned about in Chapter 4, together with the concepts of velocity and acceleration, can help explain why amusement parks are fun and how such hair-raising rides as the Demon Drop and loop-de-loop roller coasters can also be safe.

WHAT YOU'LL LEARN

- You will continue your study of average and instantaneous velocity, and acceleration.
- You will use graphs and equations to solve problems involving moving objects, including freely falling objects.

WHY IT'S IMPORTANT

- The rapid pace of life today means that modern cars, planes, elevators, and other people-moving vehicles often are designed to accelerate quickly from rest to the highest speed deemed safe, and then continue at that speed until they reach their destinations.

interNET CONNECTION

Follow the link for this chapter on the Glencoe Homepage at **www.glencoe.com/sec/science** to find out more about motion.

5.1 Graphing Motion in One Dimension

OBJECTIVES

- **Interpret** graphs of position versus time for a moving object to determine the velocity of the object.
- **Describe** in words the information presented in graphs and **draw** graphs from descriptions of motion.
- **Write** equations that describe the position of an object moving at constant velocity.

You have learned how to describe motion in terms of words, sketches, and motion diagrams. In this chapter, you'll learn to represent one-dimensional motion by means of a graph of position versus time. Such a graph presents information not only about the displacement of an object, but also about its velocity.

The other tool that is useful for certain kinds of motion is an equation that describes an object's displacement versus time. You can use a position-time graph and this equation to analyze the motion of an object mathematically and to make predictions about its position, velocity, and acceleration.

Position-Time Graphs

How could you make a graph of the position of the Demon Drop at various times? Such a graph would be a position-time graph. You will learn to make a *p-t* graph for the Demon Drop, but first, consider a simpler example. A physics student uses a camcorder to record the motion of a running back as he runs straight down the football field to make a touchdown. She records one frame each second and produces the motion diagram shown in **Figure 5–1.** From the motion diagram, the physics student obtains the data in **Table 5–1.** Because she chooses the *x*-coordinate axis, the symbol *x* in this problem represents distance from his own goal line.

Notice the origin of the coordinate system in **Figure 5–1.** Time was set to zero when the running back began to move with the ball, but the origin of the *x*-axis was not chosen to be his initial position. Instead, he began 10 m from the origin.

Two graphs of the running back's motion are shown in **Figure 5–2.** In the first graph, only the recorded positions are shown. In the second graph, a curve connects each of the recorded points. These lines represent our best guess as to where the running back was in between the recorded points. You can see that this graph is not a picture of the path taken by the ball carrier as he was running; the graph is curved, but the path that he took down the field was not.

TABLE 5–1
Position versus Time

Time, t (s)	Position, x (m)
0.0	10
1.0	12
2.0	18
3.0	26
4.0	36
5.0	43
6.0	48

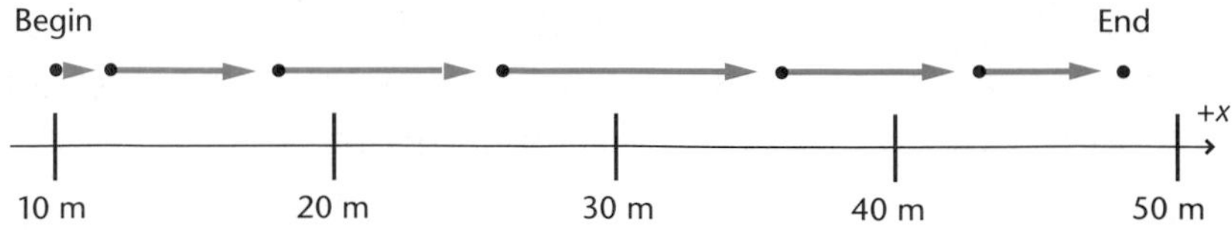

FIGURE 5–1 The football player began at 10 m and ran in the positive *x* direction.

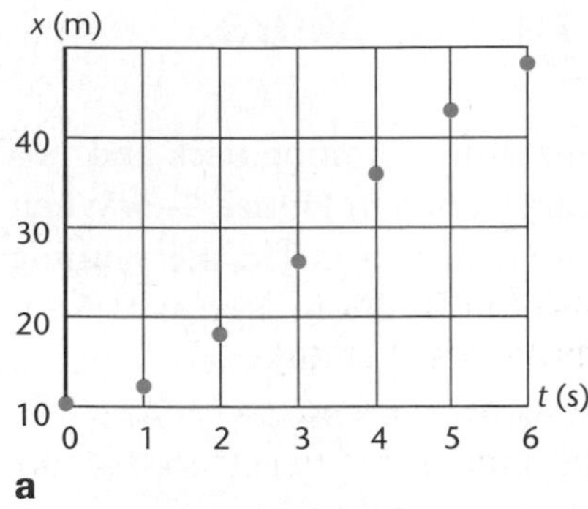

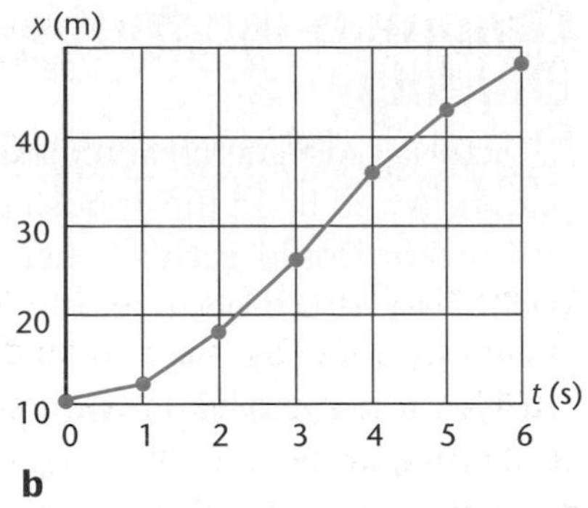

FIGURE 5–2 Only the plotted points are known. The lines joining the points are best guesses as to the position of the running back.

What is an instant of time?

How long did the running back spend at any location? Each position has been linked to a time, but how long did that time last? You could say "an instant," but how long is that? If an instant lasts for any finite amount of time, then, because the running back would be at the same position during that time, he would be at rest. But a moving object cannot be at rest; thus, an instant is not a finite period of time. This means that an instant of time lasts zero seconds. The symbol x represents the instantaneous position of the running back at a particular instant of time.

Using a Graph to Find Out *Where* and *When*

When did the running back reach the 30-m mark? Where was he 4.5 s after he started running? These questions can be answered easily with a position-time graph, as you will see in the following example problem. Note that the questions are first restated in the language of physics in terms of positions and times.

Example Problem

Data from a Position-Time Graph

When did the running back reach the 30-m mark? Where was he after 4.5 seconds?

Strategy:

Restate the questions:

Question 1: At what time was the position of the object equal to 30 m?

Question 2: What was the position of the object at 4.5 s?

To answer question 1, examine the graph to find the intersection of the curve with a horizontal line at the 30-m mark.

To answer question 2, find the intersection of the curve with a vertical line at 4.5 s (halfway between 4 s and 5 s on this graph).

The two intersections are shown on the graph.

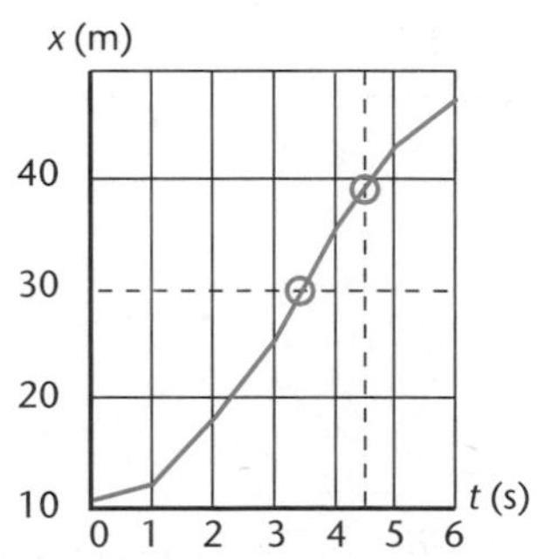

Graphing the Motion of Two or More Objects

Pictorial and graphical representations of the running back and two players, A and B, on the opposing team are shown in **Figure 5–3.** When and where would each of them have a chance to tackle the running back? First, you need to restate this question in physics terms: At what time do two objects have the same position? On a position-time graph, when do the curves representing the two objects intersect?

FIGURE 5–3 When and where can A and B tackle the running back?

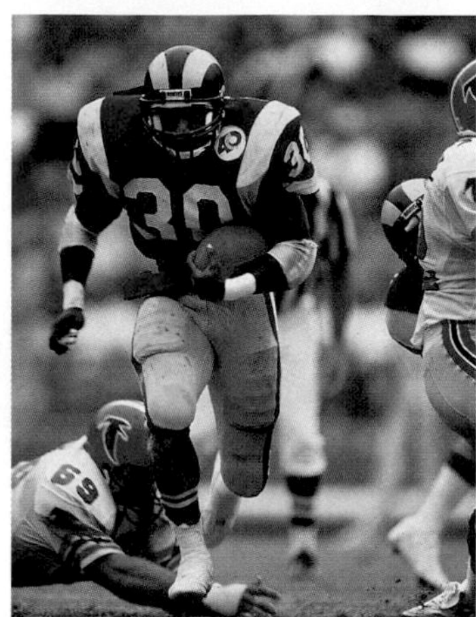

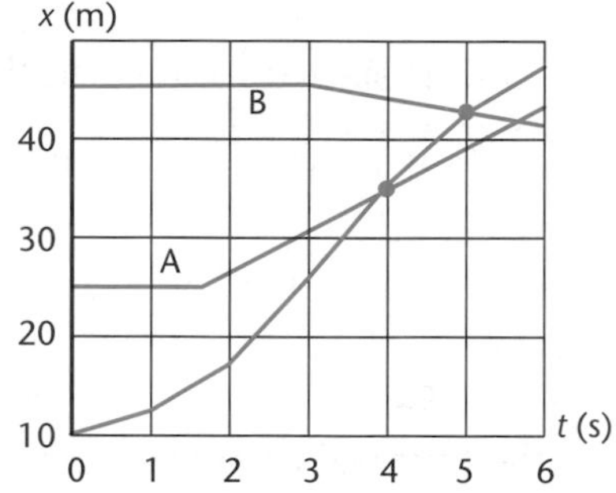

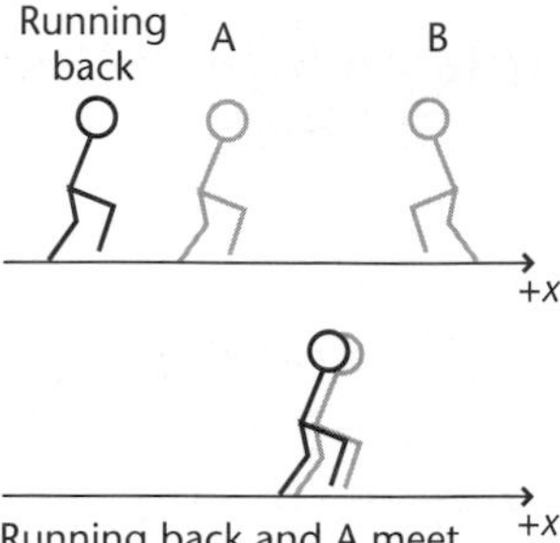

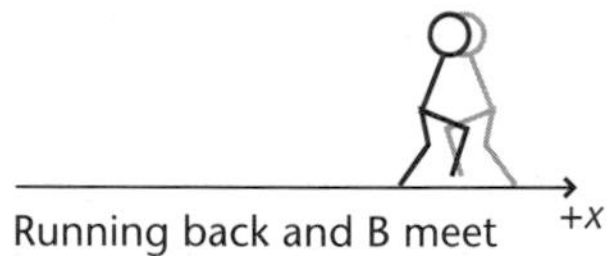

Example Problem

Interpreting Position-Time Graphs

Where and when do defenders A and B have a chance to tackle the running back?

Strategy:

In **Figure 5–3,** the intersections of the curves representing the motion of defenders A and B with the curve representing the motion of the running back are points.

Defender A intersects with running back in 4 s at about 35 m.

Defender B intersects in 5 s at about 42 m.

From Graphs to Words and Back Again

To interpret a position-time graph in words, start by finding the position of the object at $t = 0$. You have already seen that the position of the object is not always zero when $t = 0$. Then, examine the curve to see whether the position increases, remains the same, or decreases with increasing time. Motion away from the origin in a positive direction has a positive velocity, and motion in the negative direction has a negative velocity. If there is no change in position, then the velocity is zero.

Example Problem

Describing Motion from a Position-Time Graph

Describe the motion of the players in **Figure 5–3.**

Strategy:

The running back started at the 10-m mark and moved in the positive direction, that is, with a positive velocity.

Defender A started at 25 m. After waiting about 1.5 s, he also moved with positive velocity.

Defender B started at 45 m. After 3 s, he started running in the opposite direction, that is, with a negative velocity.

Practice Problems

1. Describe in words the motion of the four walkers shown by the four lines in **Figure 5–4.** Assume the positive direction is east and the origin is the corner of High Street.
2. Describe the motion of the car shown in **Figure 5–5.**
3. Answer the following questions about the car whose motion is graphed in **Figure 5–5.**
   a. When was the car 20 m west of the origin?
   b. Where was the car at 50 s?
   c. The car suddenly reversed direction. When and where did that occur?

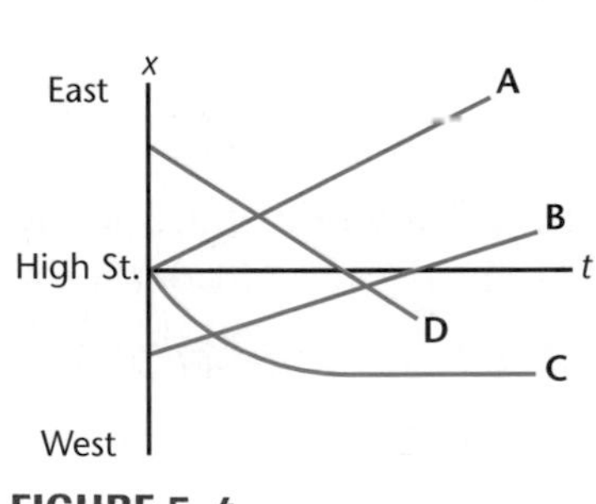

FIGURE 5–4

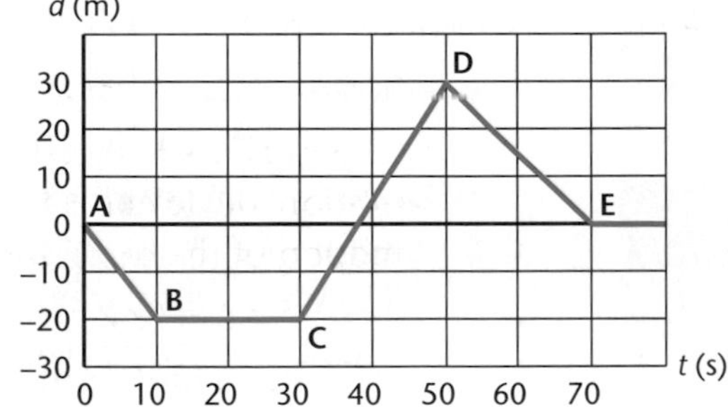

FIGURE 5–5

Uniform Motion

If an airplane travels 75 m in a straight line in the first second of its flight, 75 m in the next second, and continues in this way, then it is moving with uniform motion. **Uniform motion** means that equal displacements occur during successive equal time intervals. A motion diagram and a position-time graph can be used to describe the uniform motion of the plane.

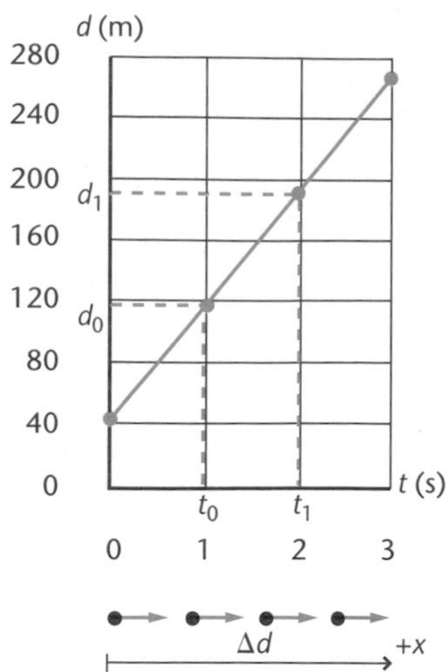

FIGURE 5–6 The regular change in position shows that this airplane is moving with uniform velocity.

The line drawn through the points representing the position of the plane each second is a straight line. Recall from Chapter 2 that one of the properties of a straight line is its slope. To find the slope, take the ratio of the vertical difference between two points on the line, the rise, to the horizontal difference between the same points, the run.

$$\text{slope} = \frac{\text{rise}}{\text{run}}$$

Figure 5–6 shows how to determine the slope using the points at 1 s and 2 s. Any set of points would produce the same result because a straight line has a constant slope.

When the line is a position-time graph, then rise = Δd and run = Δt, so the slope is the average velocity.

$$\text{slope} = \bar{v} = \frac{\Delta d}{\Delta t} = \frac{d_1 - d_0}{t_1 - t_0}$$

Thus, on a position-time graph, the slope of a straight line passing through the points on the graph at times t_0 and t_1 is the average velocity between any two times. In **Figure 5–6,** notice that when t_0 is 1 s and t_1 is 2 s, d_0 equals 115 m and d_1 equals 190 m. The average velocity of the airplane $\bar{v}$ = (190 m − 115 m)/(2 s − 1 s) = 75 m/s.

Note that the average velocity is not d/t, as you can see by calculating that ratio from the coordinates of the points on the graph. Check this for yourself using **Figure 5–6.** You will find that when t = 2 s and d = 190 m, the ratio d/t is 95 m/s, not 75 m/s, as calculated in the previous equation.

What do position-time graphs look like for different velocities? Look at the graphs of three bike riders in **Figure 5–7.** Note the initial position of the riders. Rider A has a displacement of 4.0 m in 0.4 s, so the average velocity of rider A is the following.

$$\bar{v}_A = \frac{\Delta d}{\Delta t} = \frac{2.0\text{ m} - (-2.0\text{ m})}{0.4\text{ s} - 0\text{ s}} = \frac{4.0\text{ m}}{0.4\text{ s}} = 10\text{ m/s}$$

What is the average velocity, $\bar{v}_B$, of rider B? The slope of the line is less than that for rider A, so the magnitude of the velocity of rider B should be smaller. With a displacement of 4.0 m in 0.6 s, the speed is 6.7 m/s, less than that of A, as expected.

FIGURE 5–7 You can tell in the graph below that the riders are traveling at different velocities because the three lines representing their velocities have different slopes.

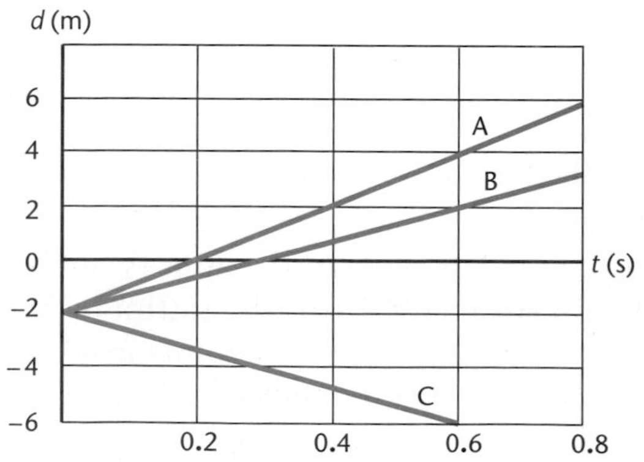

What about rider C? Although rider C moves 4.0 m in 0.6 s, her displacement, which is the final position minus the initial position, is −4.0 m. Thus, her average velocity is $\bar{v}_C = -6.7$ m/s. This tells you that rider C is moving in the negative direction. A line that slants downward and has a negative slope represents a negative average velocity. Recall that a velocity is negative if the direction of motion is opposite the direction you chose to be positive.

Practice Problems

4. For each of the position-time graphs shown in **Figure 5–8,**
   a. write a description of the motion.
   b. draw a motion diagram.
   c. rank the average velocities from largest to smallest.
5. Draw a position-time graph for a person who starts on the positive side of the origin and walks with uniform motion toward the origin. Repeat for a person who starts on the negative side of the origin and walks toward the origin.
6. Chris claims that as long as average velocity is constant, you can write $v = d/t$. Use data from the graph of the airplane's motion in **Figure 5–6** to convince Chris that this is not true.
7. Use the factor-label method to convert the units of the following average velocities.
   a. speed of a sprinter: 10 m/s into mph and km/h
   b. speed of a car: 65 mph into km/h and m/s
   c. speed of a walker: 4 mph into km/h and m/s
8. Draw a position-time graph of a person who walks one block at a moderate speed, waits a short time for a traffic light, walks the next block slowly, and then walks the final block quickly. All blocks are of equal length.

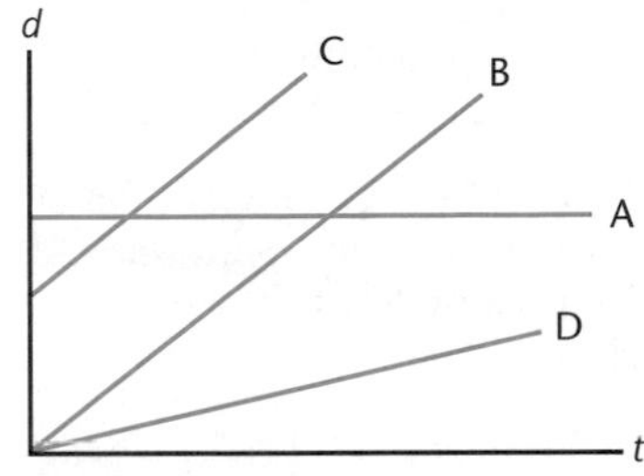

FIGURE 5–8

Uniform or Not?

Set up a U-channel on a book so that it acts as an inclined ramp, or make a channel from two metersticks taped together along the edges. Release a steel ball so that it rolls down the ramp. Using a stopwatch, measure the time it takes the ball to roll 0.40 m.

Analyze and Conclude Write a brief description of the motion of the ball. Predict how much time it would take the ball to roll 0.80 m. Explain your prediction.

Using an Equation to Find Out *Where* and *When*

Uniform motion can be represented by an algebraic equation. Recall from Chapter 3 that the average velocity was defined in this way.

$$\bar{v} = \frac{\Delta d}{\Delta t} = \frac{d_1 - d_0}{t_1 - t_0}$$

Note the absence of bold-face type. All algebra is done using the components of vectors and not the vectors themselves. Assume that you've chosen the origin of the time axis to be zero, so that $t_0 = 0$. Then, t_0 can be eliminated and the equation rearranged.

$$d_1 = d_0 + \bar{v}t_1$$

F.Y.I.

A stroboscope provides intermittent illumination of an object so that the object's motion, rotary speed, or frequency of vibration may be studied. The stroboscope causes an object to appear to slow down or stop by producing illumination in short bursts of about one microsecond duration at a frequency selected by the user.

The equation can be made more general by letting t be any value of t_1 and d be the value of the position at that time. In addition, the distinction between velocity and average velocity is not needed because you will be working with the special case of constant velocity. This means that the average velocity between any two times will be the same as the constant instantaneous velocity, $\bar{v} = v$. The symbol v represents velocity, and d_0 represents the position at $t = 0$. The following equation is then obtained for the position of an object moving at constant velocity.

$$d = d_0 + vt$$

The equation involves four quantities: the initial position, d_0, the constant velocity, v, the time, t, and the position at that time, d. If you are given three of these quantities, you can use the equation to find the fourth. When a problem is stated in words, you have to read it carefully to find out which three are given and which is unknown. When problems are given in graphical form, the slope of the curve tells you the constant velocity, and the point where the curve crosses the $t = 0$ line is the initial position. The following example problem illustrates the use of both a graph and the equation in solving a problem.

Example Problem

Finding Position from a Graph and an Equation

Write the equation that describes the motion of the airplane graphed in **Figure 5–6,** and find the position of the airplane at 2.5 s.

Calculate Your Answer

Known:

$v = 75$ m/s

$d_0 = 40$ m

$t = 2.5$ s

Unknown:

$d = ?$

Strategy:

The constant velocity is 75 m/s.

The curve intersects the $t = 0$ line at 40 m, so the initial position is 40 m.

You know the time, so the equation to use is $d = d_0 + vt$.

Calculations:

$d = d_0 + vt$

$d = 40 \text{ m} + (75 \text{ m/s})(2.5 \text{ s})$

$= 230 \text{ m}$

Check Your Answer

- Is the unit correct? m/s × s results in m.
- Does the sign make sense? It is positive, as it should be.
- Is the magnitude realistic? The result agrees with the value shown on the graph.

Practice Problems

9. Consider the motion of bike rider A in **Figure 5–7.**
   a. Write the equation that represents her motion.
   b. Where will rider A be at 1.0 s?
10. Consider the motion of bike rider C in **Figure 5–7.**
    a. Write the equation that represents her motion.
    b. When will rider C be at −10.0 m?
11. A car starts 200 m west of the town square and moves with a constant velocity of 15 m/s toward the east. Choose a coordinate system in which the *x*-axis points east and the origin is at the town square.
    a. Write the equation that represents the motion of the car.
    b. Where will the car be 10 min later?
    c. When will the car reach the town square?
12. At the same time the car in problem 11 left, a truck was 400 m east of the town square moving west at a constant velocity of 12 m/s. Use the same coordinate system as you did for problem 11.
    a. Draw a graph showing the motion of both the car and the truck.
    b. Find the time and place where the car passed the truck using both the graph and your two equations.

EARTH SCIENCE CONNECTION

Scientists use other planets' gravity to alter the path of a space probe and increase its speed. The probe is programmed to pass near a planet. The planet's gravity bends the probe's trajectory and propels it away. In 1974, *Mariner 10* swung by Venus to gain energy for three passes by Mercury. In 1992, Jupiter's gravity allowed *Ulysses* to make a hairpin turn on its way to take photos of the sun's south pole.

5.1 Section Review

1.1 You drive at constant speed toward the grocery store, but halfway there you realize you forgot your list. You quickly turn around and return home at the same speed. Describe in words the position-time graph that would represent your trip.

1.2 A car drives 3.0 km at a constant speed of 45 km/h. You use a coordinate system with its origin at the point where the car started and the direction of the car as the positive direction. Your friend uses a coordinate system with its origin at the point where the car stopped and the opposite direction as the positive direction. Would the two of you agree on the car's position? Displacement? Distance? Velocity? Speed?

1.3 Write equations for the motion of the car just described in both coordinate systems.

1.4 **Critical Thinking** A policeman clocked a driver going 20 mph over the speed limit just as the driver passed a slower car. He arrested both drivers. The judge agreed that both were guilty, saying, "If the two cars were next to each other, they must have been going the same speed." Are the judge and policeman correct? Explain with a sketch, a motion diagram, and a position-time graph.

5.2 Graphing Velocity in One Dimension

OBJECTIVES

- **Determine,** from a graph of velocity versus time, the velocity of an object at a specified time.
- **Interpret** a v-t graph to find the time at which an object has a specific velocity.
- **Calculate** the displacement of an object from the area under a v-t curve.

You've learned how to draw a position-time graph for an object moving at a constant velocity, and how to use that graph to write an equation to determine the velocity of the object. You also have learned how to use both the graph and the equation to find an object's position at a specified time, as well as the time it takes the object to reach a specific position. Now you will explore the relationship between velocity and time when velocity is not constant.

Determining Instantaneous Velocity

What does a position-time graph look like when an object is going faster and faster? **Figure 5–9a** shows a different position-time record of an airplane flight. The displacements for equal time intervals in both the motion diagram and the graph get larger and larger. This means that the average velocity during each time interval, $\bar{v} = \Delta d/\Delta t$, also gets larger and larger. The motion is certainly not uniform. The instantaneous velocity cannot equal the average velocity.

How fast was the plane going at 1.5 s? The average velocity is the slope of the straight line connecting any two points. Using **Figure 5–9a,** you can see that the average velocity between 1 s and 2 s is $\Delta d/\Delta t$, that is (10 m − 4 m)/(1 s), or 6 m/s. But the velocity could have changed within that second. To be precise, more data are needed. **Figure 5–9b** shows the slope of the line connecting 1.25 s and 1.75 s. Because this time interval is half of the 1-s time interval used previously, the average velocity is probably closer to the instantaneous velocity at 1.5 s. You could continue the process of reducing the time interval and finding the ratio of the displacement to the time interval. Each time you reduce the time interval, the ratio is likely to be closer to the quantity called the

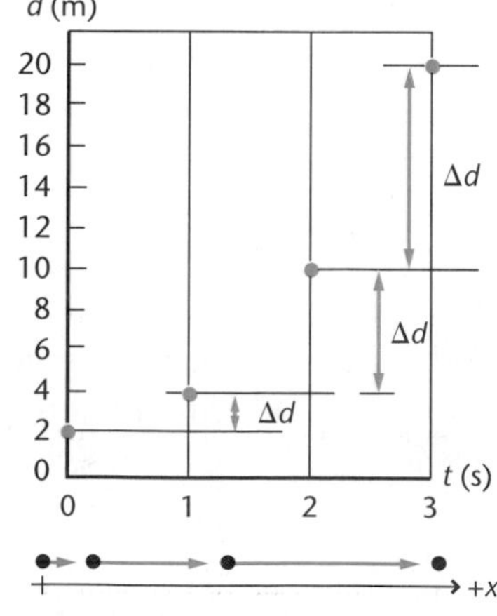

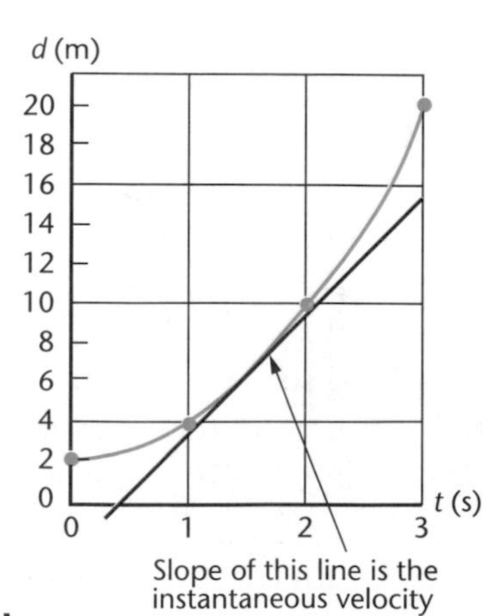

FIGURE 5–9 The displacement during equal time intervals increases as velocity increases. The graph in **b** shows that the slope of the tangent to the curve at any point is the instantaneous velocity at that time.

instantaneous velocity. Finally, you would find that the slope of the line tangent to the position-time curve at a specific time is the instantaneous velocity, v, at that time.

Velocity-Time Graphs

Now that you know what the velocity at an instant of time is and how to find it, you will be able to draw a graph of the velocity versus time of the airplane whose motion is not constant. Just as average velocity arrows on a motion diagram are drawn in red, the curves on a velocity-time (v-t) graph are also shown in red.

Motion diagrams and a velocity-time graph for the two airplanes, one with constant velocity and the other with increasing velocity, are shown in **Figure 5–10** for a 3-s time interval during their flights. Note that the velocities are given in m/s. The velocity of one airplane is increasing, while the velocity of the other airplane is constant. Which line on the graph represents the plane with uniform motion? Uniform motion, or constant velocity, is represented by a horizontal line on a v-t graph. Airplane B is traveling with a constant velocity of 75 m/s, or 270 km/h. The second line increases from 70 m/s to 82 m/s over a 3-s time interval. At 1.5 s, the velocity is 76 m/s. This is the slope of the tangent to the curve on a position-time graph at 1.5 s for the flight of airplane A.

What would the v-t graph look like if a plane were going at constant speed in the opposite direction? As long as you don't change the direction of the coordinate axis, the velocity would be negative, so the graph would be a horizontal line below the t-axis.

What can you learn from the intersection of the two lines on the graph? When the two v-t lines cross, the two airplanes have the same velocity. The planes do not necessarily have the same position, so they do not meet at this time. Velocity-time graphs give no information about position, although, as you will learn in the next section, you can use them to find displacement.

A Ball Race

Assemble an inclined ramp from a piece of U-channel or two metersticks taped together. Make a mark at 40 cm from the top and another at 80 cm from the top. If two balls are released at the same instant, one ball from the top and the other ball at 40 cm, will the balls get closer or farther apart as they roll down the ramp? Why? Try it. Now, release one ball from the top and then release another ball from the top as soon as the first ball reaches the 40-cm mark.

Analyze and Conclude Explain your observations in terms of velocities. Do the balls ever have the same velocities as they roll down the hill? Do they have the same acceleration?

To review calculating the area under a graph, see the Math Handbook, page 787.

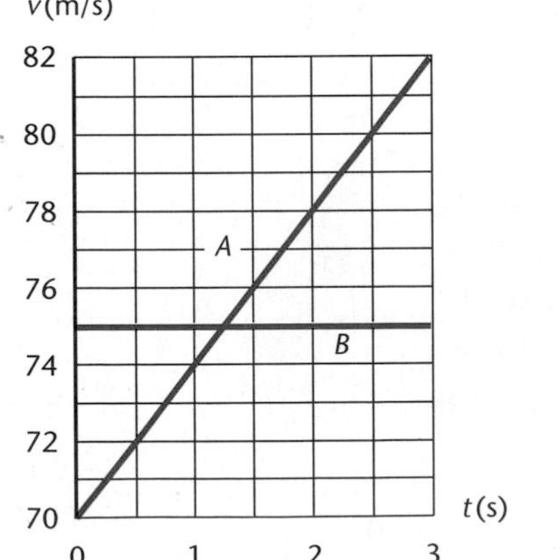

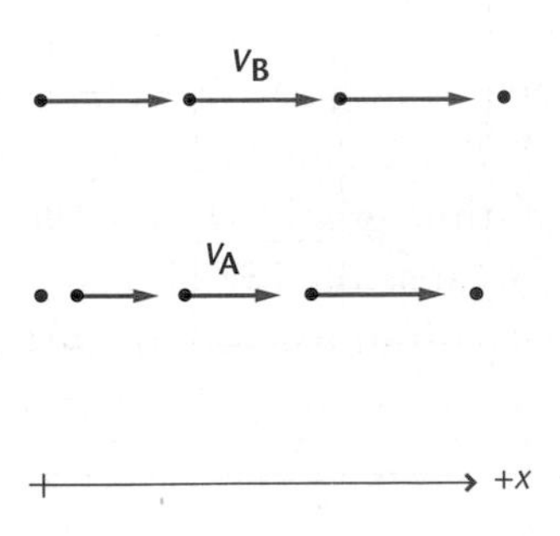

FIGURE 5–10 The lines on the graph and the motion diagrams are two different ways of representing both constant and increasing velocity.

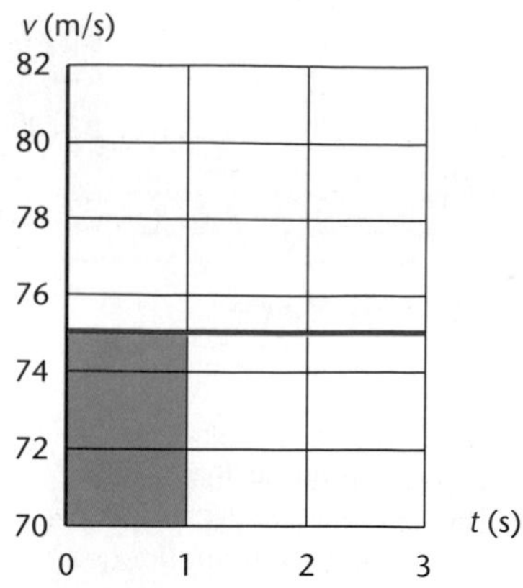

FIGURE 5–11 The displacement during a given time interval is the area under the curve of a *v-t* graph.

Displacement from a Velocity-Time Graph

For an object moving at constant velocity,

$$v = \bar{v} = \frac{\Delta d}{\Delta t}, \text{ so } \Delta d = v\Delta t.$$

As you can see in **Figure 5–11,** v is the height of the curve above the t-axis, while Δt is the width of the shaded rectangle. The area of the rectangle, then, is $v\Delta t$, or Δd. You can find the displacement of the object by determining the area under the v-t curve.

If the velocity is constant, the displacement is proportional to the width of the rectangle. Thus, if you plot the displacement versus time, you will get a straight line with slope equal to velocity. If the velocity is increasing, then the area of the rectangle increases in time, so the slope of a displacement-versus-time graph also increases.

Example Problem

Finding the Displacement of an Airplane from Its *v-t* Graph

Find the displacement of the plane in **Figure 5–11** that is moving at constant velocity after

a. 1.0 s. **b.** 2.0 s. **c.** 3.0 s.

Compare your results to the original position-time graph in **Figure 5–6.**

Calculate Your Answer

Known:

$v = 75$ m/s

$\Delta t = 1.0$ s, 2.0 s, and 3.0 s

Unknown:

Δd at 1.0 s, 2.0 s, and 3.0 s

Strategy:

The displacement is the area under the curve, or $\Delta d = v\Delta t$.

Calculations:

a. $\Delta d = v\Delta t = (75 \text{ m/s})(1.0 \text{ s}) = 75 \text{ m}$

b. $\Delta d = v\Delta t = (75 \text{ m/s})(2.0 \text{ s}) = 150 \text{ m}$

c. $\Delta d = v\Delta t = (75 \text{ m/s})(3.0 \text{ s}) = 225 \text{ m}$

Check Your Answer

- Are the units correct? m/s × s = m.
- Do the signs make sense? They are all positive, as they should be.
- Are the magnitudes realistic? The calculated positions are different from those on the position-time graph: 115 m, 190 m, and 265 m. The differences occur because the initial position of the plane at $t = 0$ is 40 m, not zero. You must add the displacement of the airplane at $t = 0$ to the value calculated at each time.

Practice Problems

13. Use **Figure 5–10** to determine the velocity of the airplane that is speeding up at
  a. 1.0 s.
  b. 2.0 s.
  c. 2.5 s.

14. Use the factor-label method to convert the speed of the airplane whose motion is graphed in **Figure 5–6** (75 m/s) to km/h.

15. Sketch the velocity-time graphs for the three bike riders in **Figure 5–7.**

16. A car is driven at a constant velocity of 25 m/s for 10.0 min. The car runs out of gas, so the driver walks in the same direction at 1.5 m/s for 20.0 min to the nearest gas station. After spending 10.0 min filling a gasoline can, the driver walks back to the car at a slower speed of 1.2 m/s. The car is then driven home at 25 m/s (in the direction opposite that of the original trip).
  a. Draw a velocity-time graph for the driver, using seconds as your time unit. You will have to calculate the distance the driver walked to the gas station in order to find the time it took the driver to walk back to the car.
  b. Draw a position-time graph for the problem using the areas under the curve of the velocity-time graph.

Pocket Lab

Bowling Ball Displacement

Take a bowling ball and three stopwatches into the hallway. Divide into three groups. Have all timers start their watches when the ball is rolled. Group 1 should stop its watch when the ball has gone 10 m, group 2 should stop its watch when the ball has rolled 20 m, and group 3 should stop its watch when the ball has rolled 30 m.

Analyze and Conclude
Record the data and calculate the average speed for each distance. Could the average speed for 30 m be used to predict the time needed to roll 100 m? Why or why not?

5.2 Section Review

2.1 What information can you obtain from a velocity-time graph?

2.2 Two joggers run at a constant velocity of 7.5 m/s toward the east. At time $t = 0$, one is 15 m east of the origin; the other is 15 m west.
  a. What would be the difference(s) in the position-time graphs of their motion?
  b. What would be the difference(s) in their velocity-time graphs?

2.3 Explain how you would use a velocity-time graph to find the time at which an object had a specified velocity.

2.4 Sketch a velocity-time graph for a car that goes 25 m/s toward the east for 100 s, then 25 m/s toward the west for another 100 s.

2.5 **Critical Thinking** If the constant velocity on a v-t graph is negative, what is the sign of the area under the curve? Explain in terms of the displacement of the object whose motion is represented on the graph.

5.3 Acceleration

OBJECTIVES

- **Determine** from the curves on a velocity-time graph both the constant and instantaneous acceleration.
- **Determine** the sign of acceleration using a *v-t* graph and a motion diagram.
- **Calculate** the velocity and the displacement of an object undergoing constant acceleration.

In Chapter 3, you learned how to use a motion diagram to get a feel for the average acceleration of an object. This method is illustrated in **Figure 5–12b** for the motion of two airplanes. Airplane A travels with non-uniform velocity, so the change in velocity, and thus the acceleration, is in the same direction as the velocity. Both velocity and acceleration have a positive sign. Airplane B travels with uniform velocity, so its acceleration is zero.

Determining Average Acceleration

Average acceleration is the rate of change of velocity between t_0 and t_1, and is represented by the following equation.

$$\bar{a} = \frac{\Delta v}{\Delta t} = \frac{v_1 - v_0}{t_1 - t_0}$$

You can find this ratio by determining the slope of the velocity-time graph in the same way you found velocity from the slope of the position-time graph. For example, **Figure 5–12a** shows that in a 1-s time interval, the velocity of plane A increases by 4 m/s. That is, $\Delta v = 4$ m/s and $\Delta t = 1$ s, so $\bar{a} = \Delta v/\Delta t = 4\ \text{m/s}^2$.

Constant and Instantaneous Acceleration

Recall that an object undergoes uniform motion, or constant velocity, if the slope of the position-time graph is constant. Does the slope of the velocity-time graph for the accelerating airplane of **Figure 5–12a** change? No, it rises by 4 m/s every second. This type of motion, which can be described by a constant slope on a velocity-time graph, is called **constant acceleration.**

What if the slope of a *v-t* graph isn't constant? You learned that you could find the instantaneous velocity by finding the slope of the tangent to the curve on the position-time graph. In the same way, you can find **instantaneous acceleration,** *a*, as the slope of the tangent to the curve on a velocity-time graph. Instantaneous acceleration is the acceleration of an object at an instant of time. Consider the velocity-time graph in the example problem.

FIGURE 5–12 Graphs and motion diagrams are useful in differentiating motion having uniform velocity and motion that is accelerated.

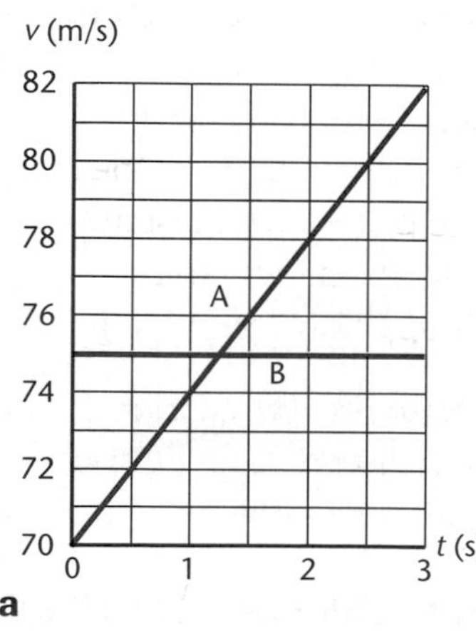

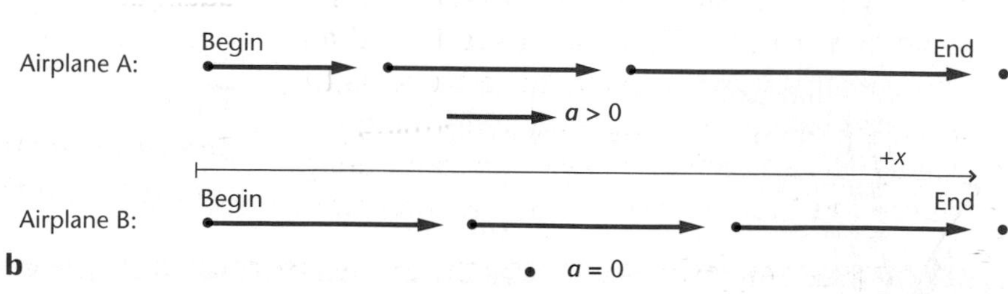

Example Problem

Determining Velocity and Acceleration from a Graph

How would you describe the sprinter's velocity and acceleration as shown on the graph?

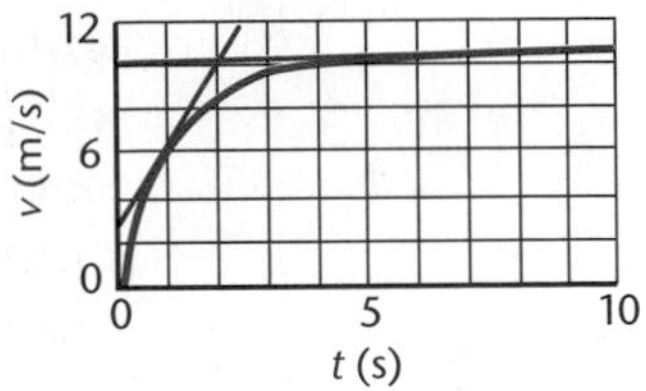

Calculate Your Answer

Strategy:

From the graph, note that the sprinter's velocity starts at zero, increases rapidly for the first few seconds, and then, after reaching about 10 m/s, remains almost constant.

Draw a tangent to the curve at two different times, $t = 1$ s and $t = 5$ s.

The slope of the lines at 1 s and 5 s is the acceleration at those times.

Calculations:

At 1 s, $a = \frac{\text{rise}}{\text{run}} = \frac{12.0 \text{ m/s} - 3.0 \text{ m/s}}{2.5 \text{ s} - 0.0 \text{ s}}$;

$a = 3.6 \text{ m/s}^2$

At 5 s, $a = \frac{10.7 \text{ m/s} - 10.0 \text{ m/s}}{10 \text{ s} - 0 \text{ s}}$

$a = 0.07 \text{ m/s}^2$

The acceleration is 3.6 m/s^2 at 1 s, and 0.07 m/s^2 at 5 s. It is larger before 1 s and smaller after 5 s. The acceleration is not constant.

Physics & Technology

The Zero Gravity Trainer

The zero-gravity trainer is an aircraft at NASA's Johnson Space Center designed for use in simulating zero-gravity conditions such as those experienced aboard spacecraft orbiting Earth. The plane is a modified Boeing 707 that mimics the free-fall environment aboard a spacecraft as it flies a series of parabola-shaped courses that take the crew from altitudes of about 8,000 m up to about 12,000 m and back down again—all in less than two minutes! These short spurts of simulated weightlessness enable the astronauts to practice eating, drinking, and performing a variety of tasks that they will carry out during future missions. Because of the rapid ascents and descents of the aptly nicknamed *Vomit Comet,* training sessions are generally limited to one or two hours.

During a run, the four-engine turbo jet accelerates from 350 knots indicated airspeed (KIA) to about 150 KIA at the top of the parabola. There, the pilot adjusts the jet's engines so that speed is constant. Then, the jet pitches over until the plane is descending again at 350 KIA. During the ascent and descent, acceleration is approximately 1.8 *g*.

Thinking Critically How are the Demon Drop amusement park ride and a ride in the zero-gravity trainer alike?

Positive and Negative Acceleration

You've considered the motion of an accelerating airplane and a sprinter and found that, in both cases, the object's velocity was positive and increasing, and the sign of the acceleration was positive. Now consider a ball being rolled up a slanted driveway. What happens? It slows down, stops briefly, then rolls back down the hill at an increasing speed. Examine the two graphs in **Figure 5–13** that represent the ball's motion and interpret them in the following example problem.

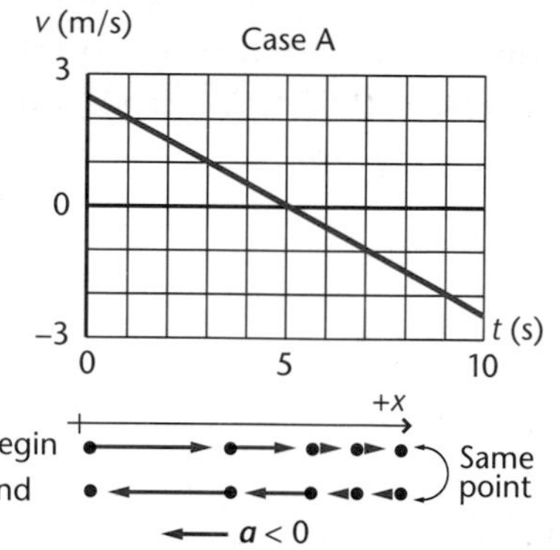

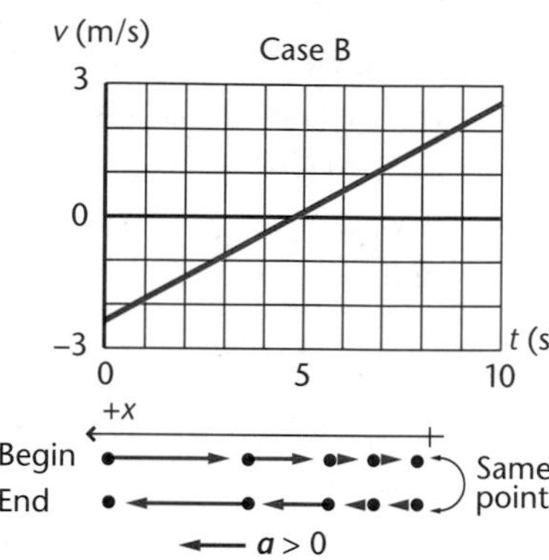

FIGURE 5–13 The sign of the acceleration depends upon the chosen coordinate system.

Example Problem

Finding the Sign of Acceleration

Describe the motion of the ball shown in **Figure 5–13.** What is the difference between the two cases? What is the sign of the ball's acceleration? What is the magnitude of the ball's acceleration?

Strategy:

In each case, the coordinate axis is parallel to the surface. In case A, the ball initially moved in the direction of the positive axis; thus, the sign of the velocity is positive. In case B, the axis was chosen in the opposite direction; thus, the sign of the velocity is negative.

In each case, the ball started with a speed of 2.5 m/s. In both cases, the ball slows down, reaches zero velocity, then speeds up in the opposite direction.

To find the ball's acceleration, use the motion diagrams. Subtract an earlier velocity vector from a later one. Whether the ball is moving uphill or downhill, the acceleration vector points to the left. In case A, the positive axis points to the right, but the acceleration vector points in the opposite direction. Therefore, the acceleration is negative. In case B, the positive axis points to the left, and the acceleration points in the same direction. The acceleration is positive.

Find the magnitude of the acceleration from the slopes of the graphs.

Calculations:

For case A, the ball slows down in the first 5 s.

$\Delta v = v_1 - v_0 = 0 \text{ m/s} - 2.5 \text{ m/s} = -2.5 \text{ m/s}$

$a = \Delta v/\Delta t$

$a = (-2.5 \text{ m/s})/(5.0 \text{ s}) = -0.50 \text{ m/s}^2$

During the next 5 s, the ball speeds up in the negative direction.

$\Delta v = -2.5 \text{ m/s} - 0 \text{ m/s} = -2.5 \text{ m/s}$

Again, $a = -0.50 \text{ m/s}^2$

Check case B for yourself. Because the axis for case B was chosen in the opposite direction, you will find that a is $+\ 0.50 \text{ m/s}^2$, both when the ball is slowing down and when it is speeding up.

Practice Problems

17. An Indy 500 race car's velocity increases from +4.0 m/s to +36 m/s over a 4.0-s time interval. What is its average acceleration?

18. The race car in problem 17 slows from +36 m/s to +15 m/s over 3.0 s. What is its average acceleration?

19. A car is coasting backwards downhill at a speed of 3.0 m/s when the driver gets the engine started. After 2.5 s, the car is moving uphill at 4.5 m/s. Assuming that uphill is the positive direction, what is the car's average acceleration?

20. A bus is moving at 25 m/s when the driver steps on the brakes and brings the bus to a stop in 3.0 s.

a. What is the average acceleration of the bus while braking?

b. If the bus took twice as long to stop, how would the acceleration compare with what you found in part **a?**

21. Look at the *v-t* graph of the toy train in **Figure 5–14.**

a. During which time interval or intervals is the speed constant?

b. During which interval or intervals is the train's acceleration positive?

c. During which time interval is its acceleration most negative?

22. Using **Figure 5–14,** find the average acceleration during the following time intervals.

a. 0 to 5 s **b.** 15 to 20 s **c.** 0 to 40 s

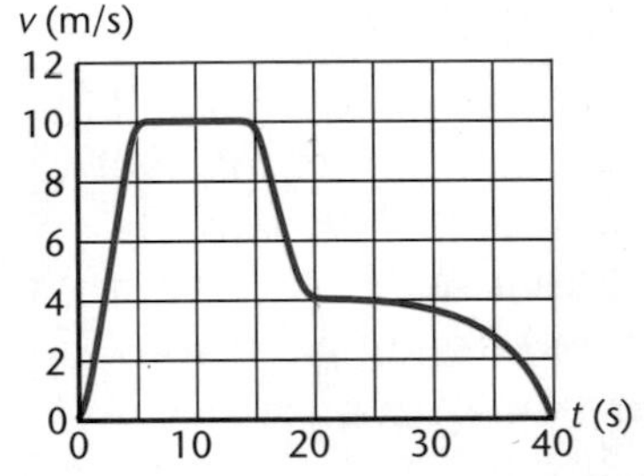

FIGURE 5–14

Acceleration when instantaneous velocity is zero

What happens to the acceleration when $v = 0$, that is, when the ball in the example problem stops and reverses direction? Consider the motion diagram that starts when the ball is still moving uphill and ends

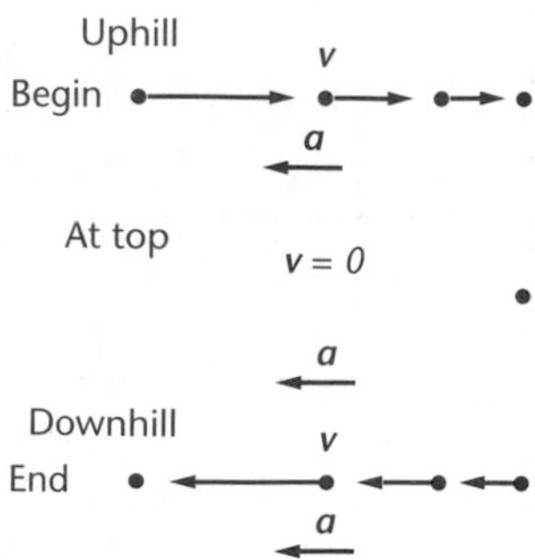

FIGURE 5–15 The acceleration of the ball is always downward.

when the ball is moving back downhill as in **Figure 5–15.** The acceleration is downhill both before and after the ball stops. What happens at the instant when the ball's instantaneous velocity is zero? Remember that the velocity is zero only *at* an instant of time, not *for* an instant. That is, the time interval over which the velocity is zero is itself zero. Thus, the acceleration points downhill as the ball reaches the top of the hill, and it continues to point downhill as the ball moves back downhill.

Calculating Velocity from Acceleration

You learned that you could use the definition of velocity to find the position of an object moving at constant velocity. In the same way, you can find the velocity of the object by rearranging the definition of average acceleration.

$$\bar{a} = \frac{\Delta v}{\Delta t} = \frac{v_1 - v_0}{t_1 - t_0}$$

Assume that $t_0 = 0$. At that time, the object has a velocity v_0. Let t be any value of t_1 and v be the value of the velocity at that time. Because only one-dimensional, straight-line motion with constant acceleration will be considered, a can be used instead of $\bar{a}$. After making these substitutions and rearranging the equation, the following equation is obtained for the velocity of an object moving at constant acceleration.

$$v = v_0 + at$$

You also can use this equation to find the time at which a constantly accelerating object has a given velocity, or, if you are given both a velocity and the time at which it occurred, you can calculate the initial velocity.

HELP WANTED

AIR-TRAFFIC CONTROLLER

If you can think fast and communicate clearly and decisively while juggling dozens of details in your mind, you might be the person we are seeking. Three years of work experience or four years of college and the ability to pass the Federal Civil Service Exam may qualify you to enter a training program. Advancement is based on job performance and completion of on-the-job training programs. Knowledge of foreign languages is a plus. For information, contact your local Job Information Center of The U. S. Office of Personnel Management.

Practice Problems

23. A golf ball rolls up a hill toward a miniature-golf hole. Assign the direction toward the hole as being positive.
  a. If the ball starts with a speed of 2.0 m/s and slows at a constant rate of 0.50 m/s^2, what is its velocity after 2.0 s?
  b. If the constant acceleration continues for 6.0 s, what will be its velocity then?
  c. Describe in words and in a motion diagram the motion of the golf ball.

24. A bus, traveling at 30 km/h, speeds up at a constant rate of 3.5 m/s^2. What velocity does it reach 6.8 s later?

25. If a car accelerates from rest at a constant 5.5 m/s^2, how long will it need to reach a velocity of 28 m/s?

26. A car slows from 22 m/s to 3.0 m/s at a constant rate of 2.1 m/s^2. How many seconds are required before the car is traveling at 3.0 m/s?

Displacement Under Constant Acceleration

You know how to use the area under the curve of a velocity-time graph to find the displacement when the velocity is constant. The same method can be used to find the displacement when the acceleration is constant. **Figure 5–16** is a graph of the motion of an object accelerating constantly from v_0 to v. If the velocity had been a constant, v_0, the displacement would have been v_0t, the area of the lightly shaded rectangle. Instead, the velocity increased from v_0 to v. Thus, the displacement is increased by the area of the triangle, $1/2(v - v_0)t$. The total displacement, then, is the sum of the two.

$$d = v_0t + 1/2(v - v_0)t$$

When the terms are combined, the following equation results.

$$d = 1/2(v + v_0)t$$

If the initial position, d_0, is not zero, then this term must be added to give the general equation for the final position.

$$d = d_0 + 1/2(v + v_0)t$$

Frequently, the velocity at time t is not known, but because $v = v_0 + at$, you can substitute $v_0 + at$ for v in the previous equation and obtain the following equation.

$$d = d_0 + 1/2(v_0 + v_0 + at)t$$

When the terms are combined, the following equation results.

$$d = d_0 + v_0t + 1/2at^2$$

Note that the third equation involves position, velocity, and time, but not acceleration. The fifth equation involves position, acceleration, and time, but not velocity. Is there an equation that relates position, velocity, and acceleration, but doesn't include time? To find that equation, start with the following equations.

$$d = d_0 + 1/2(v + v_0)t \text{ and } v = v_0 + at$$

Solve the second equation for t.

$$t = (v - v_0)/a$$

Substitute this into the equation for displacement.

$$d = d_0 + 1/2(v_0 + v)(v - v_0)/a$$

This equation can be solved for the final velocity.

$$v^2 = v_0^2 + 2a(d - d_0)$$

The four equations that have been derived for motion under constant acceleration are summarized in **Table 5–2.** One is useful for calculating velocity and three are equations for position. When solving problems involving constant acceleration, determine what information is given and what is unknown, then choose the appropriate equation. These equations, along with velocity-time and position-time graphs, provide the mathematical models you need to solve motion problems.

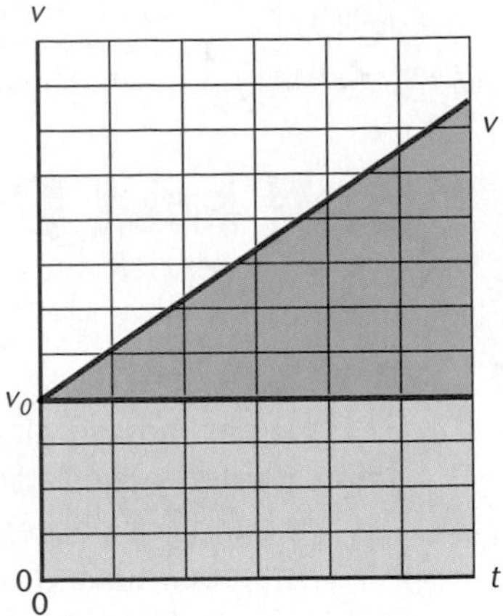

FIGURE 5–16 The area under the curve of a velocity versus time graph equals the displacement.

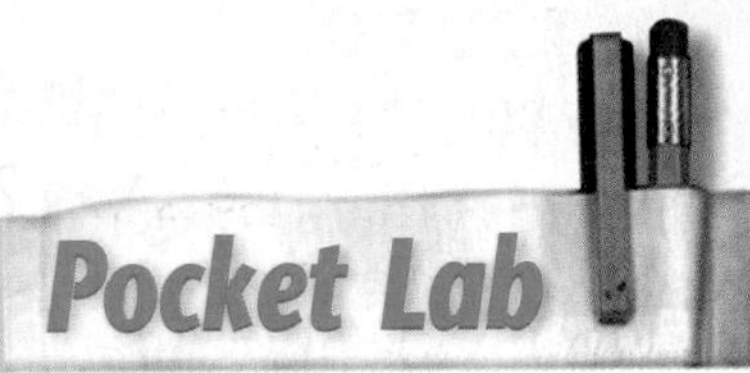

Direction of Acceleration

Tape a bubble level onto the top of a laboratory cart. Center the bubble. Observe the direction of the motion of the bubble as you pull the cart forward, move it at constant speed, and allow it to coast to a stop. Relate the motion of the bubble to the acceleration of the cart. Predict what would happen if you tie the string to the back of the cart and repeat the experiment. Try it.

Analyze and Conclude Draw motion diagrams for the cart as you moved it in the forward direction and it coasted to a stop and as you repeated the experiment in the opposite direction.

Design Your Own **Physics Lab**

Ball and Car Race

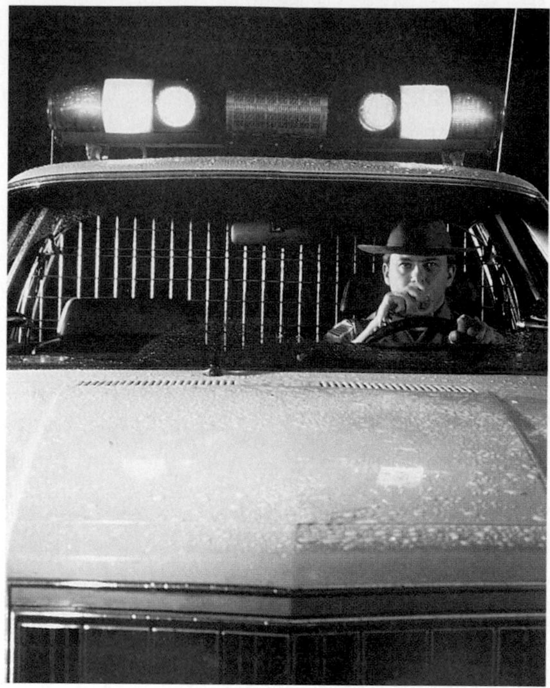

Problem

A car moving along a highway passes a parked police car with a radar detector. Just as the car passes, the police car starts to pursue, moving with a constant acceleration. The police car catches up with the car just as it leaves the jurisdiction of the policeman.

Hypothesis

Sketch the position-versus-time graphs and the velocity-versus-time graphs for this chase, then simulate the chase.

Possible Materials

battery-powered car
masking tape
wood block
90-cm-long grooved track
1-in. steel ball
stopwatch
graph paper

Plan the Experiment

1. Identify the variables in this activity.
2. Determine how you will give the ball a constant acceleration.
3. Devise a method to ensure that both objects reach the end of the track at the same time.
4. Construct a data table that will show the positions of both objects at the beginning, the halfway point, and the end of the chase.
5. **Check the Plan** Review your plan with your teacher before you begin the race.
6. Construct *p-t* and *v-t* graphs for both objects.

Analyze and Conclude

1. **Comparing and Contrasting** Compare the velocities of the cars at the beginning and at the end of the chase. Write a verbal description.
2. **Using Graphs** At any time during the chase, did the cars ever have the same velocity? If so, mark these points on the graphs.
3. **Comparing and Contrasting** Compare the average velocity of the police car to that of the car.
4. **Calculating Results** Calculate the average speed of each car.

Apply

1. Explain why it took the police car so long to catch the car after it sped by.

TABLE 5–2		
Equations of Motion for Uniform Acceleration		
Equation	**Variables**	**Initial Conditions**
$v = v_0 + at$	t v a	v_0
$d = d_0 + 1/2(v_0 + v)t$	t d v	d_0 v_0
$d = d_0 + v_0t + 1/2at^2$	t d a	d_0 v_0
$v^2 = v_0^2 + 2a(d - d_0)$	d v a	d_0 v_0

F.Y.I.

A dragster tries to obtain maximum acceleration over a quarter-mile course. The fastest time on record for the quarter mile is 4.539 s. The highest final speed on record is 322.92 mph.

Example Problem

Finding Displacement Under Constant Acceleration

In Chapter 3, you completed the first step in solving the following problem by sketching the situation and drawing the motion diagram. Now you can add the mathematical model. A car starts at rest and speeds up at 3.5 m/s^2 after the traffic light turns green. How far will it have gone when it is going 25 m/s?

Sketch the Problem

- Sketch the situation.
- Establish coordinate axes.
- Draw a motion diagram.

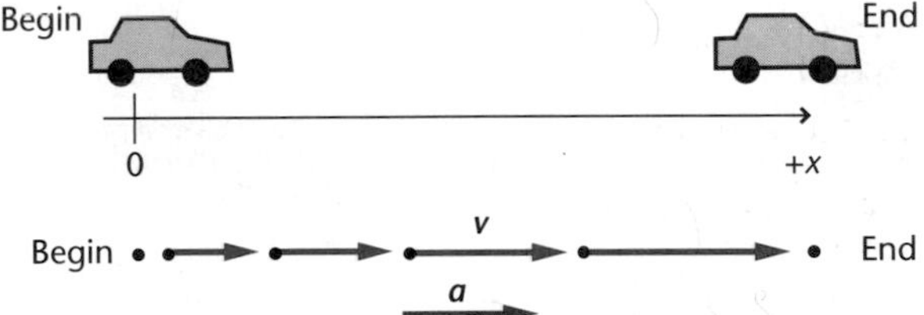

Calculate Your Answer

Known:

$d_0 - 0.0$ m $\quad v = 25$ m/s

$v_0 = 0.0$ m/s $\quad a = 3.5$ m/s^2

Unknown:

$d = ?$

Strategy:

Refer to **Table 5–2.** Use an equation containing v, a, and d.

Calculations:

$$v^2 = v_0^2 + 2a(d - d_0)$$

$$d = d_0 + (v^2 - v_0^2)/(2a)$$

$$= 0.0 \text{ m} + [(25 \text{ m/s})^2 - (0.0 \text{ m/s})^2]/(2 \cdot 3.5 \text{ m/s}^2)$$

$$= 89 \text{ m}$$

Check Your Answer

- Is the unit correct? Dividing m^2/s^2 by m/s^2 results in m, the correct unit for position.
- Does the sign make sense? It is positive, in agreement with both the pictorial and physical models.
- Is the magnitude realistic? The displacement is almost the length of a football field. It seems large, but 25 m/s is fast (about 55 mph), and the acceleration, as you will find in the next example problem, is not very great. Therefore, the result is reasonable.

Example Problem

Two-Part Motion

The driver of the car in the previous example problem, traveling at a constant 25 m/s, sees a child suddenly run into the road. It takes the driver 0.40 s to hit the brakes. As it slows, the car has a steady acceleration of 8.5 m/s^2. What's the total distance the car moves before it stops?

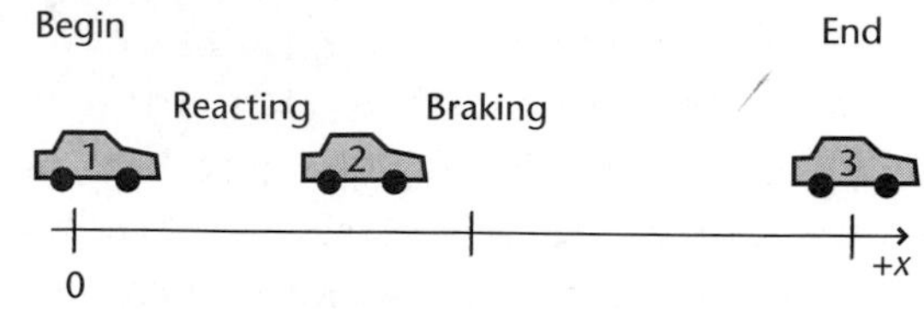

Sketch the Problem

- Label your drawing with "begin" and "end."
- Choose a coordinate system and create the motion diagram.
- Use subscripts to distinguish the three positions in the problem.

Begin v End

a_{12} a_{23}

Calculate Your Answer

Known:		**Unknown:**
$d_1 = 0.0$ m	$v_2 = 25$ m/s	$d_2 = ?$
$v_1 = 25$ m/s	$a_{23} = -8.5$ m/s^2	$d_3 = ?$
$a_{12} = 0.0$ m/s^2	$v_3 = 0.0$ m/s	
$t_2 = 0.40$ s		

Strategy:

There are two parts to the problem: the interval of reacting and the interval of braking.

Reacting: Find the distance the car travels. During this time, the velocity and time are known and the velocity is constant.

Braking: Find the distance the car moves while braking. The initial and final velocities are known. The acceleration is constant and negative, as shown in the motion diagram.

The position of the car when the brakes are applied, d_2, is the solution of the first part of the problem; it is needed to solve the second part.

Calculations:

Reacting: $d_2 = vt$

$d_2 = (25 \text{ m/s})(0.40 \text{ s})$

$= 10$ m

Braking: $v_3{}^2 = v_2{}^2 + 2a_{23}(d_3 - d_2)$

$d_3 = d_2 + (v_3{}^2 - v_2{}^2)/(2a_{23})$

$$= 10 \text{ m} + \frac{0 - (25 \text{ m/s})^2}{2(-8.5 \text{ m/s}^2)}$$

$= 47$ m

Check Your Answer

- Is the unit correct? Performing algebra on the units verifies the distance in meters.
- Do the signs make sense? Both d_3 and d_2 are positive, as they should be.
- Is the magnitude realistic? The braking distance is much smaller than it was in the previous example problem, which makes sense because the magnitude of the acceleration is larger.

Practice Problems

For all problems, sketch the situation, assign variables, create a motion diagram, and then develop a mathematical model.

27. A race car traveling at 44 m/s slows at a constant rate to a velocity of 22 m/s over 11 s. How far does it move during this time?

28. A car accelerates at a constant rate from 15 m/s to 25 m/s while it travels 125 m. How long does it take to achieve this speed?

29. A bike rider accelerates constantly to a velocity of 7.5 m/s during 4.5 s. The bike's displacement during the acceleration was 19 m. What was the initial velocity of the bike?

30. An airplane starts from rest and accelerates at a constant 3.00 m/s^2 for 30.0 s before leaving the ground.

a. How far did it move?

b. How fast was it going when it took off?

5.3 Section Review

3.1a. Give an example of an object that is slowing down but has a positive acceleration.

b. Give an example of an object that is speeding up but has a negative acceleration.

3.2a. If an object has zero acceleration, does that mean its velocity is zero? Give an example.

b. If an object has zero velocity at some instant, does that mean its acceleration is zero? Give an example.

3.3 Is km/h/s a unit of acceleration? Is this unit the same as km/s/h? Explain.

3.4 **Figure 5–17** is a strobe photo of a horizontally moving ball. What information about the photo would you need and what measurements would you make to estimate the acceleration?

3.5 If you are given a table of velocities of an object at various times, how could you find out if the acceleration was constant?

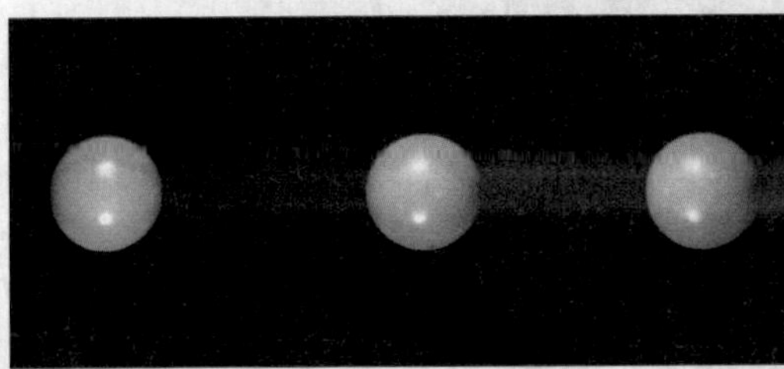

FIGURE 5–17

3.6 If you are given initial and final velocities and the constant acceleration of an object, and you are asked to find the displacement, which equation would you use?

3.7 **Critical Thinking** Describe how you could calculate the acceleration of an automobile. Specify the measuring instruments and the procedures you would use.

5.4 Free Fall

Drop a sheet of paper. Crumple it, then drop it again. Its motion is different in the two instances. So is the motion of a pebble falling through water compared with the same pebble falling through air. Do heavier objects fall faster than lighter ones? The answer depends upon whether you drop sheets of paper or rocks.

OBJECTIVES

- **Recognize** the meaning of the acceleration due to gravity.
- **Define** the magnitude of the acceleration due to gravity as a positive quantity and **determine** the sign of the acceleration relative to the chosen coordinate system.
- **Use** the motion equations to solve problems involving freely falling objects.

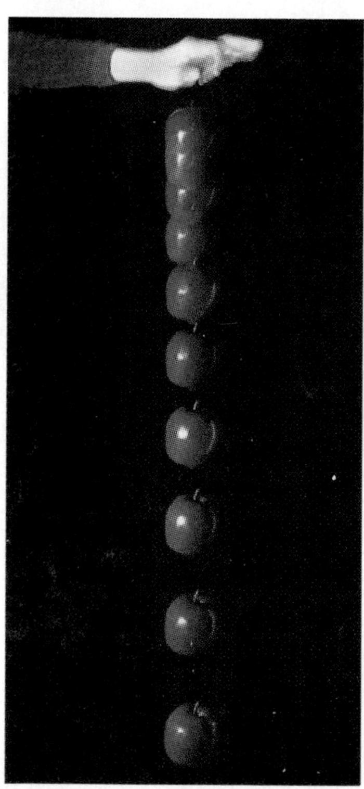

FIGURE 5–18 If the upward direction is chosen as positive, then both the velocity and the acceleration of this apple in free fall are negative.

Acceleration Due to Gravity

Galileo Galilei recognized about 400 years ago that to make progress in the study of the motion of falling bodies, the effects of air or water, the medium through which the object falls, had to be ignored. He also knew that he had no means of recording the fall of objects, so he rolled balls down inclined planes. By "diluting" gravity in this way, he could make careful measurements even with simple instruments.

Galileo found that, neglecting the effect of the air, all freely falling objects had the same acceleration. It didn't matter what they were made of, what their masses were, from how high they were dropped, or whether they were dropped or thrown. The magnitude of the acceleration of falling objects is given a special symbol, g, equal to 9.80 m/s^2. We now know that there are small variations in g at different places on Earth, and that 9.80 m/s^2 is the average value.

Note that g is a positive quantity. You will never use a negative value of g in a problem. But don't things accelerate downward, and isn't down usually the negative direction? Although this is true, remember that g is only the magnitude of the acceleration, not the acceleration itself. If upward is defined to be the positive direction, then the acceleration due to gravity is equal to $-g$. The **acceleration due to gravity** is the acceleration of an object in free fall that results from the influence of Earth's gravity. Suppose you drop a rock. One second later, its velocity is 9.80 m/s downward. One second after that, its velocity is 19.60 m/s downward. For each second that the rock is falling, its downward velocity increases by 9.80 m/s.

Look at the strobe photo of a dropped apple in **Figure 5–18.** The time interval between the photos is 1/120 s. The displacement between each pair of images increases, so the speed is increasing. If the upward direction is chosen as positive, then the velocity is becoming more and more negative.

Could this photo be of a ball thrown upward? If you again choose upward as the positive direction, then the ball leaves your hand with a positive velocity of, say, 20.0 m/s. The acceleration is downward, so a is negative. That is, $a = -g = -9.80$ m/s^2. This means that the speed of the ball becomes less and less, which is in agreement with the strobe

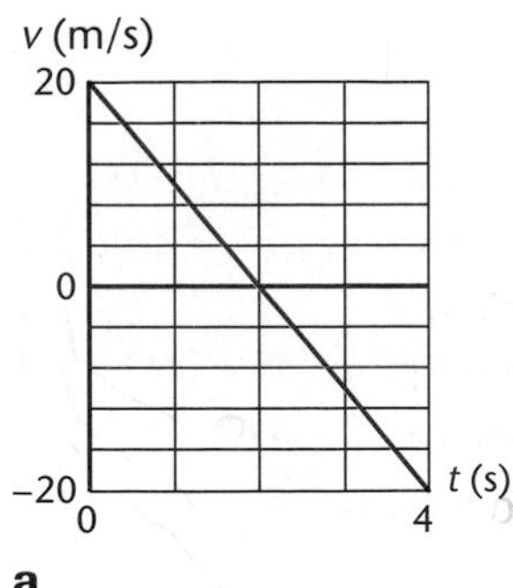

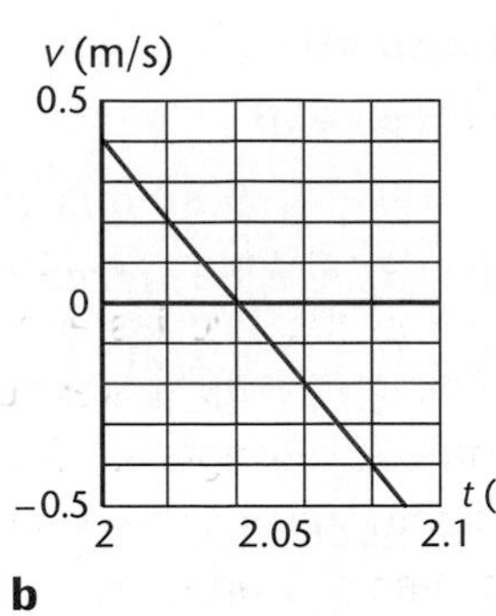

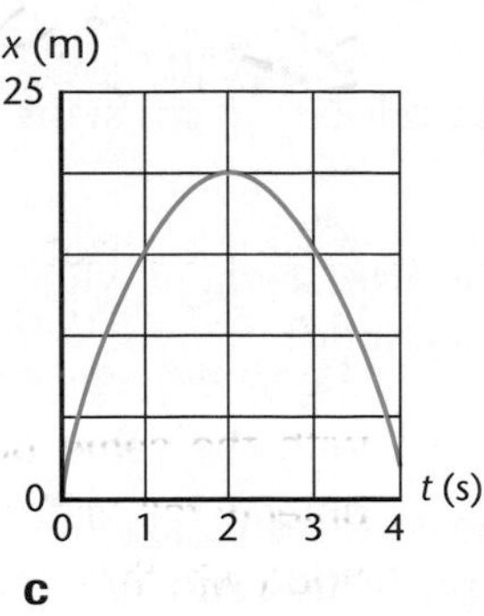

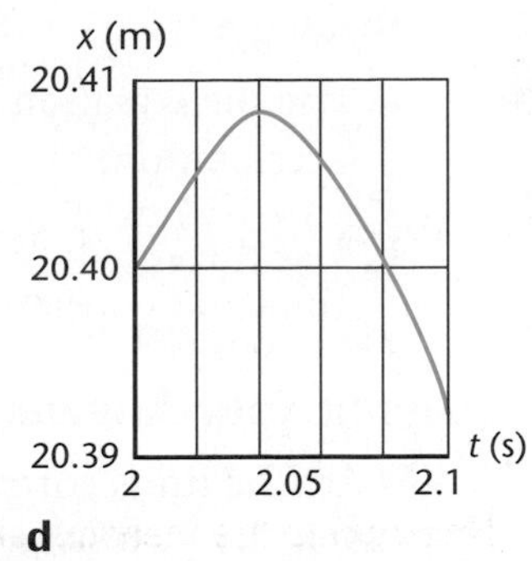

FIGURE 5–19 In a coordinate system in which the upward direction is positive, the velocity of the thrown ball decreases until it becomes zero at 2.04 s, then it increases in the negative direction.

photo. After 1 s, the ball's velocity is reduced by 9.80 m/s, so it is now traveling at 10.2 m/s. After 2 s, the velocity is 0.4 m/s, and the ball is still moving upward. After the next second, the ball's velocity, being reduced by another 9.80 m/s, is now −9.4 m/s. The ball is now moving downward. After 4 s, the velocity is −19.2 m/s, meaning that it is falling even faster. The velocity-time graph of the ball's flight is shown in **Figure 5–19a. Figure 5–19b** shows what happens at around 2 s, where the velocity changes smoothly from positive to negative. At an instant of time, near 2.04 s, the ball's velocity is zero.

The position-time graphs in **Figure 5–19c** and **d** show how the ball's height changes. The ball has its maximum height when its velocity is zero.

Example Problem

The Demon Drop

The Demon Drop ride at Cedar Point Amusement Park falls freely for 1.5 s after starting from rest.

a. What is its velocity at the end of 1.5 s?

b. How far does it fall?

Sketch the Problem

- Choose a coordinate system with a positive axis upward and the origin at the initial position of the car.
- Label "begin" and "end."
- Draw a motion diagram showing that both $\boldsymbol{a}$ and $\boldsymbol{v}$ are downward and, therefore, negative.

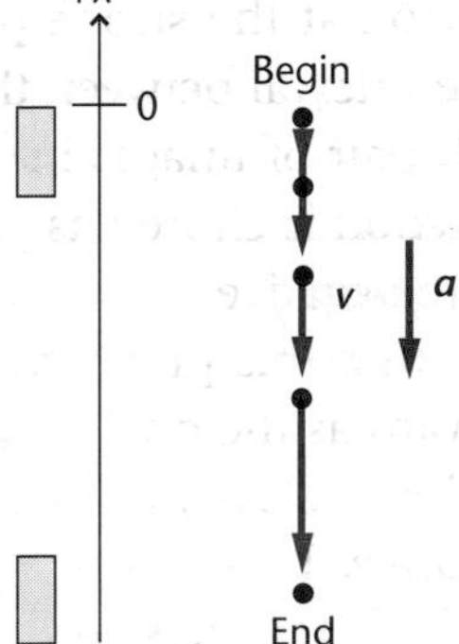

Calculate Your Answer

Known:	Unknown:
$a = -g = -9.80\ \text{m/s}^2$	$d = ?$
$d_0 = 0$	$v = ?$
$v_0 = 0$	
$t = 1.5$ s	

Free Fall

Strategy:

a. Use the equation for velocity at constant acceleration.

b. Use the equation for displacement when time and constant acceleration are known.

Calculations:

$v = v_0 + at$

$v = 0 + (-9.80\ \text{m/s}^2)(1.5\ \text{s}) = -15\ \text{m/s}$

$d = d_0 + v_0 t + 1/2at^2$

$d = 0 + 0 + 1/2(-9.80\ \text{m/s}^2)(1.5\ \text{s})^2 = -11\ \text{m}$

Check Your Answer

- Are the units correct? Performing algebra on the units verifies velocity in m/s and position in m.
- Do the signs make sense? Negative signs agree with the diagram.
- Are the magnitudes realistic? Yes, when judged by the photo at the opening of this chapter in which the car is about the height of a person, about 2 m.

Practice Problems

31. A brick is dropped from a high scaffold.
  a. What is its velocity after 4.0 s?
  b. How far does the brick fall during this time?

32. A tennis ball is thrown straight up with an initial speed of 22.5 m/s. It is caught at the same distance above ground.
  a. How high does the ball rise?
  b. How long does the ball remain in the air? (**Hint:** The time to rise equals the time to fall. Can you show this?)

33. A spaceship far from any star or planet accelerates uniformly from 65.0 m/s to 162.0 m/s in 10.0 s. How far does it move?

5.4 Section Review

4.1 Gravitational acceleration on Mars is about 1/3 that on Earth. Suppose you could throw a ball upward with the same velocity on Mars as on Earth.
  a. How would the ball's maximum height compare to that on Earth?
  b. How would its flight time compare?

4.2 **Critical Thinking** When a ball is thrown vertically upward, it continues upward until it reaches a certain position, then it falls down again. At that highest point, its velocity is instantaneously zero. Is the ball accelerating at the highest point? Devise an experiment to prove or disprove your answer.

CHAPTER 5 REVIEW

Key Terms

5.1
- uniform motion

5.3
- constant acceleration
- instantaneous acceleration

5.4
- acceleration due to gravity

Summary

5.1 Graphing Motion in One Dimension

- Position-time graphs can be used to find the velocity and position of an object, and where and when two objects meet.
- A description of motion can be obtained by interpreting graphs, and graphs can be drawn from descriptions of motion.
- Equations that describe the position of an object moving at constant velocity can be written based on word and graphical representations of problems.

5.2 Graphing Velocity in One Dimension

- Instantaneous velocity is the slope of the tangent to the curve on a position-time graph.
- Velocity-time graphs can be used to determine the velocity of an object and the time when two objects have the same velocity.
- The area under the curve on a velocity-time graph is displacement.

5.3 Acceleration

- The acceleration of an object is the slope of the curve on a velocity-time graph.
- The slope of the tangent to the curve on a *v-t* graph is the instantaneous acceleration of the object.
- Velocity-time graphs and motion diagrams can be used to find the sign of the acceleration.
- Both graphs and equations can be used to find the velocity of an object undergoing constant acceleration.
- Three different equations give the displacement of an object under constant acceleration, depending on what quantities are known.
- The mathematical model completes the solution of motion problems.
- Results obtained by solving a problem must be tested to find out whether they are reasonable.

5.4 Free Fall

- The magnitude of the acceleration due to gravity ($g = 9.80\ \text{m/s}^2$) is always a positive quantity. The sign of acceleration depends upon the choice of the coordinate system.
- Motion equations can be used to solve problems involving freely falling objects.

Reviewing Concepts

Section 5.1

1. A walker and a runner leave your front door at the same time. They move in the same direction at different constant velocities. Describe the position-time graphs of each.
2. What does the slope of the tangent to the curve on a position-time graph measure?

Section 5.2

3. If you know the positions of an object at two points along its path, and you also know the time it took to get from one point to the other, can you determine the particle's instantaneous velocity? Its average velocity? Explain.
4. What quantity is represented by the area under a velocity-time curve?
5. **Figure 5–20** shows the velocity-time graph for an automobile on a test track. Describe how the velocity changes with time.

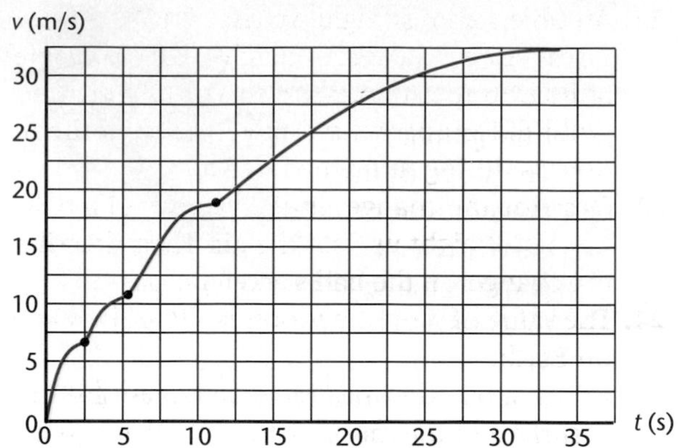

FIGURE 5–20

Section 5.3

6. What does the slope of the tangent to the curve on a velocity-time graph measure?
7. A car is traveling on an interstate highway.
   a. Can the car have a negative velocity and a positive acceleration at the same time? Explain.
   b. Can the car's velocity change signs while it is traveling with constant acceleration? Explain.
8. Can the velocity of an object change when its acceleration is constant? If so, give an example. If not, explain.
9. If the velocity-time curve is a straight line parallel to the *t*-axis, what can you say about the acceleration?
10. If you are given a table of velocities of an object at various times, how could you find out if the acceleration of the object is constant?
11. Write a summary of the equations for position, velocity, and time for an object experiencing uniformly accelerated motion.

Section 5.4

12. Explain why an aluminum ball and a steel ball of similar size and shape, dropped from the same height, reach the ground at the same time.
13. Give some examples of falling objects for which air resistance cannot be ignored.
14. Give some examples of falling objects for which air resistance can be ignored.

Applying Concepts

15. **Figure 5–20** shows the velocity-time graph of an accelerating car. The three "notches" in the curve occur where the driver changes gears.
    a. Describe the changes in velocity and acceleration of the car while in first gear.
    b. Is the acceleration just before a gear change larger or smaller than the acceleration just after the change? Explain your answer.
16. Explain how you would walk to produce each of the position-time graphs in **Figure 5–21.**

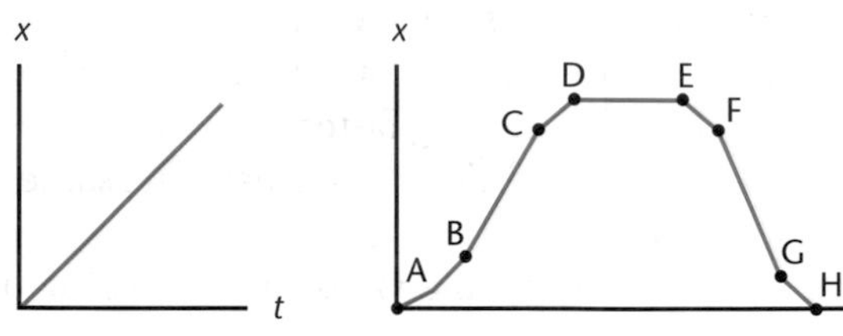

FIGURE 5–21

17. Use **Figure 5–20** to determine during what time interval the acceleration is largest and during what time interval the acceleration is smallest.
18. Solve the equation $v = v_0 + at$ for acceleration.
19. **Figure 5–22** is a position-time graph of two people running.
    a. Describe the position of runner A relative to runner B at the *y*-intercept.
    b. Which runner is faster?
    c. What occurs at point P and beyond?

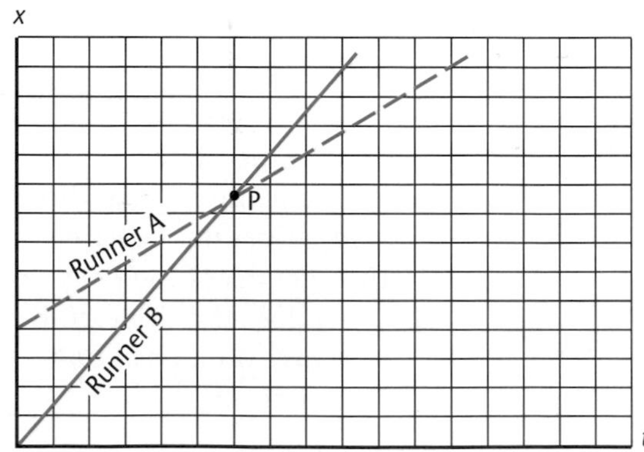

FIGURE 5–22

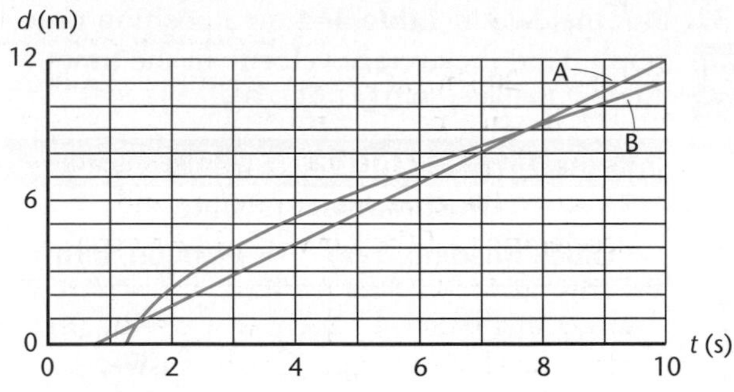

FIGURE 5–23

20. **Figure 5–23** is a position-time graph of the motion of two cars on a road.
    a. At what time(s) does one car pass the other?
    b. Which car is moving faster at 7.0 s?
    c. At what time(s) do the cars have the same velocity?
    d. Over what time interval is car B speeding up all the time?
    e. Over what time interval is car B slowing down all the time?
21. Look at **Figure 5–24.**
    a. What kind of motion is represented by **a?**
    b. What does the area under the curve represent?
    c. What kind of motion is represented by **b?**
    d. What does the area under the curve represent?

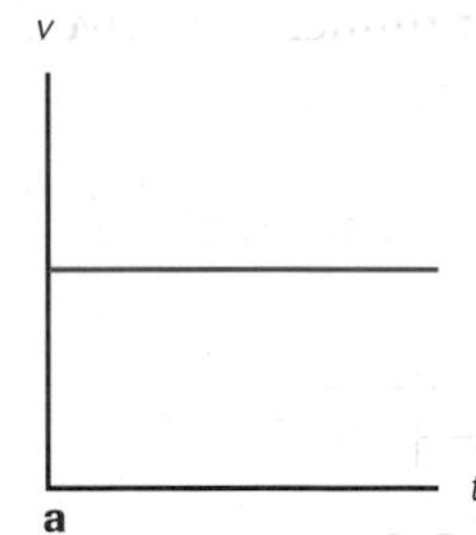

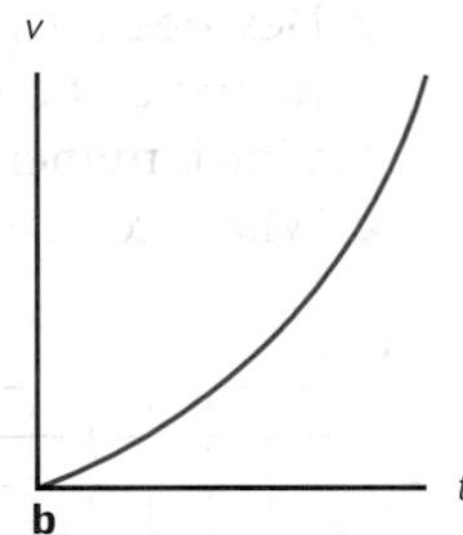

FIGURE 5–24

22. An object shot straight up rises for 7.0 s before it reaches its maximum height. A second object falling from rest takes 7.0 s to reach the ground. Compare the displacements of the two objects during this time interval.
23. Describe the changes in the velocity of a ball thrown straight up into the air. Then describe the changes in the ball's acceleration.
24. The value of g on the moon is 1/6 of its value on Earth.
    a. Will a ball dropped by an astronaut hit the surface of the moon with a smaller, equal, or larger speed than that of a ball dropped from the same height to Earth?
    b. Will it take more, less, or equal time to fall?
25. Planet Dweeb has three times the gravitational acceleration of Earth. A ball is thrown vertically upward with the same initial velocity on Earth and on Dweeb.
    a. How does the maximum height reached by the ball on Dweeb compare to the maximum height on Earth?
    b. If the ball on Dweeb were thrown with three times greater initial velocity, how would that affect your answer to **a?**
26. Rock A is dropped from a cliff; rock B is thrown upward from the same position.
    a. When they reach the ground at the bottom of the cliff, which rock has a greater velocity?
    b. Which has a greater acceleration?
    c. Which arrives first?

Problems

Section 5.1

LEVEL 1

27. Light from the sun reaches Earth in 8.3 min. The velocity of light is 3.00×10^8 m/s. How far is Earth from the sun?
28. You and a friend each drive 50 km. You travel at 90 km/h; your friend travels at 95 km/h. How long will your friend wait for you at the end of the trip?

TABLE 5–3	
Distance versus Time	
Time (s)	**Distance (m)**
0.0	0.0
1.0	2.0
2.0	8.0
3.0	18.0
4.0	32.9
5.0	50.0

29. The total distance a steel ball rolls down an incline at various times is given in **Table 5–3.**

a. Draw a position-time graph of the motion of the ball. When setting up the axes, use five divisions for each 10 m of travel on the *d*-axis. Use five divisions for 1 s of time on the *t*-axis.

b. What type of curve is the line of the graph?

c. What distance has the ball rolled at the end of 2.2 s?

30. A cyclist maintains a constant velocity of +5.0 m/s. At time $t = 0.0$, the cyclist is +250 m from point A.

a. Plot a position-time graph of the cyclist's location from point A at 10.0-s intervals for 60.0 s.

b. What is the cyclist's position from point A at 60.0 s?

c. What is the displacement from the starting position at 60.0 s?

31. From the position-time graph in **Figure 5–25,** construct a table showing the average velocity of the object during each 10-s interval over the entire 100 s.

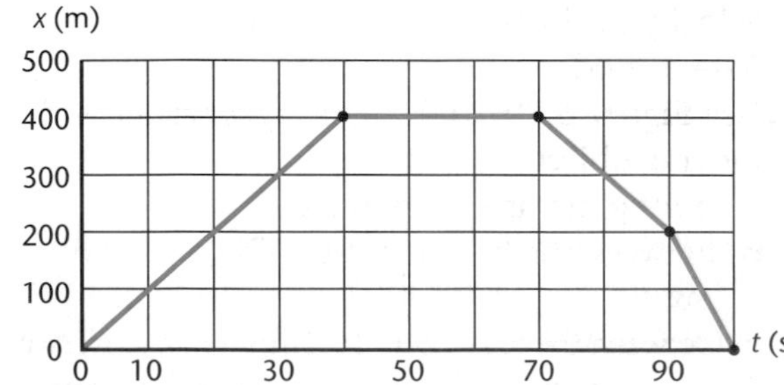

FIGURE 5–25

32. Plot the data in **Table 5–4** on a position-time graph. Find the average velocity in the time interval between 0.0 s and 5.0 s.

TABLE 5–4	
Position versus Time	
Clock Reading, *t* (s)	**Position, *d* (m)**
0.0	30
1.0	30
2.0	35
3.0	45
4.0	60
5.0	70

LEVEL 2

33. You drive a car for 2.0 h at 40 km/h, then for another 2.0 h at 60 km/h.

a. What is your average velocity?

b. Do you get the same answer if you drive 100 km at each of the two speeds?

34. Use the position-time graph in **Figure 5–25** to find how far the object travels

a. between $t = 0$ s and $t = 40$ s.

b. between $t = 40$ s and $t = 70$ s.

c. between $t = 90$ s and $t = 100$ s.

35. Do this problem on a worksheet. Both car A and car B leave school when a clock reads zero. Car A travels at a constant 75 km/h, and car B travels at a constant 85 km/h.

a. Draw a position-time graph showing the motion of both cars.

b. How far are the two cars from school when the clock reads 2.0 h? Calculate the distances using the equation for motion and show them on your graph.

c. Both cars passed a gas station 120 km from the school. When did each car pass the gas station? Calculate the times and show them on your graph.

36. Draw a position-time graph for two cars driving to the beach, which is 50 km from school. At noon Car A leaves a store 10 km closer to the beach than the school is and drives at 40 km/h. Car B starts from school at 12:30 P.M. and drives at 100 km/h. When does each car get to the beach?

37. Two cars travel along a straight road. When a stopwatch reads $t = 0.00$ h, car A is at $d_A = 48.0$ km moving at a constant 36.0 km/h. Later, when the watch reads $t = 0.50$ h, car B is at $d_B = 0.00$ km moving at 48.0 km/h. Answer the following questions, first, graphically by creating a position-time graph, and second, algebraically by writing down equations for the positions d_A and d_B as a function of the stopwatch time, t.
   a. What will the watch read when car B passes car A?
   b. At what position will car B pass car A?
   c. When the cars pass, how long will it have been since car A was at the reference point?

38. A car is moving down a street at 55 km/h. A child suddenly runs into the street. If it takes the driver 0.75 s to react and apply the brakes, how many meters will the car have moved before it begins to slow down?

Section 5.2

LEVEL 1

39. Refer to **Figure 5–23** to find the instantaneous speed for
   a. car B at 2.0 s.
   b. car B at 9.0 s.
   c. car A at 2.0 s.

40. Refer to **Figure 5–26** to find the distance the moving object travels between
   a. $t = 0$ s and $t = 5$ s.
   b. $t = 5$ s and $t = 10$ s.
   c. $t = 10$ s and $t = 15$ s.
   d. $t = 0$ s and $t = 25$ s.

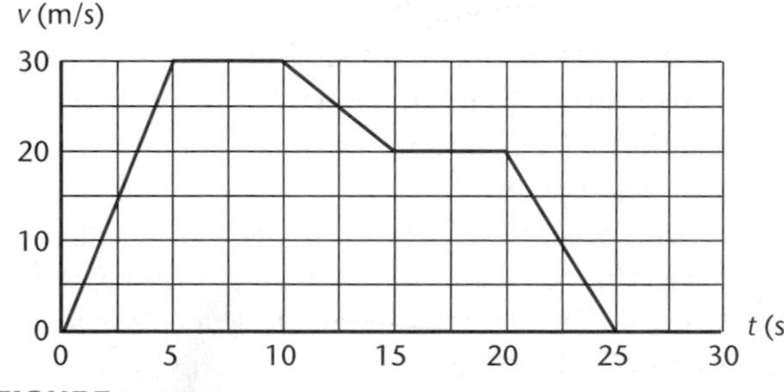

FIGURE 5–26

41. Find the instantaneous speed of the car in **Figure 5–20** at 15 s.

42. You ride your bike for 1.5 h at an average velocity of 10 km/h, then for 30 min at 15 km/h. What is your average velocity?

LEVEL 2

43. Plot a velocity-time graph using the information in **Table 5–5,** then answer the questions.
   a. During what time interval is the object speeding up? Slowing down?
   b. At what time does the object reverse direction?
   c. How does the average acceleration of the object in the interval between 0 s and 2 s differ from the average acceleration in the interval between 7 s and 12 s?

TABLE 5–5
Velocity versus Time

Time (s)	Velocity (m/s)	Time (s)	Velocity (m/s)
0.0	4.0	7.0	12.0
1.0	8.0	8.0	8.0
2.0	12.0	9.0	4.0
3.0	14.0	10.0	0.0
4.0	16.0	11.0	−4.0
5.0	16.0	12.0	−8.0
6.0	14.0		

Section 5.3

LEVEL 1

44. Find the uniform acceleration that causes a car's velocity to change from 32 m/s to 96 m/s in an 8.0-s period.

45. Use **Figure 5–26** to find the acceleration of the moving object
   a. during the first 5 s of travel.
   b. between the fifth and the tenth second of travel.
   c. between the tenth and the 15th second of travel.
   d. between the 20th and 25th second of travel.

46. A car with a velocity of 22 m/s is accelerated uniformly at the rate of 1.6 m/s^2 for 6.8 s. What is its final velocity?

47. A supersonic jet flying at 145 m/s is accelerated uniformly at the rate of 23.1 m/s^2 for 20.0 s.
a. What is its final velocity?
b. The speed of sound in air is 331 m/s. How many times the speed of sound is the plane's final speed?

48. Determine the final velocity of a proton that has an initial velocity of 2.35×10^5 m/s, and then is accelerated uniformly in an electric field at the rate of -1.10×10^{12} m/s^2 for 1.50×10^{-7} s.

49. Determine the displacement of a plane that is uniformly accelerated from 66 m/s to 88 m/s in 12 s.

50. How far does a plane fly in 15 s while its velocity is changing from 145 m/s to 75 m/s at a uniform rate of acceleration?

51. A car moves at 12 m/s and coasts up a hill with a uniform acceleration of -1.6 m/s^2.
a. How far has it traveled after 6.0 s?
b. How far has it gone after 9.0 s?

52. A plane travels 5.0×10^2 m while being accelerated uniformly from rest at the rate of 5.0 m/s^2. What final velocity does it attain?

53. A race car can be slowed with a constant acceleration of -11 m/s^2.
a. If the car is going 55 m/s, how many meters will it take to stop?
b. How many meters will it take to stop a car going twice as fast?

54. An engineer must design a runway to accommodate airplanes that must reach a ground velocity of 61 m/s before they can take off. These planes are capable of being accelerated uniformly at the rate of 2.5 m/s^2.
a. How long will it take the planes to reach takeoff speed?
b. What must be the minimum length of the runway?

55. Engineers are developing new types of guns that might someday be used to launch satellites as if they were bullets. One such gun can give a small object a velocity of 3.5 km/s, moving it through only 2.0 cm.
a. What acceleration does the gun give this object?
b. Over what time interval does the acceleration take place?

56. Highway safety engineers build soft barriers so that cars hitting them will slow down at a safe rate. A person wearing a seat belt can withstand an acceleration of -300 m/s^2. How thick should barriers be to safely stop a car that hits a barrier at 110 km/h?

57. A baseball pitcher throws a fastball at a speed of 44 m/s. The acceleration occurs as the pitcher holds the ball in his hand and moves it through an almost straight-line distance of 3.5 m. Calculate the acceleration, assuming it is uniform. Compare this acceleration to the acceleration due to gravity, 9.80 m/s^2.

LEVEL 2

58. Rocket-powered sleds are used to test the responses of humans to acceleration. Starting from rest, one sled can reach a speed of 444 m/s in 1.80 s and can be brought to a stop again in 2.15 s.
a. Calculate the acceleration of the sled when starting, and compare it to the magnitude of the acceleration due to gravity, 9.80 m/s^2.
b. Find the acceleration of the sled when braking and compare it to the magnitude of the acceleration due to gravity.

59. Draw a velocity-time graph for each of the graphs in **Figure 5–27.**

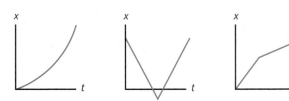

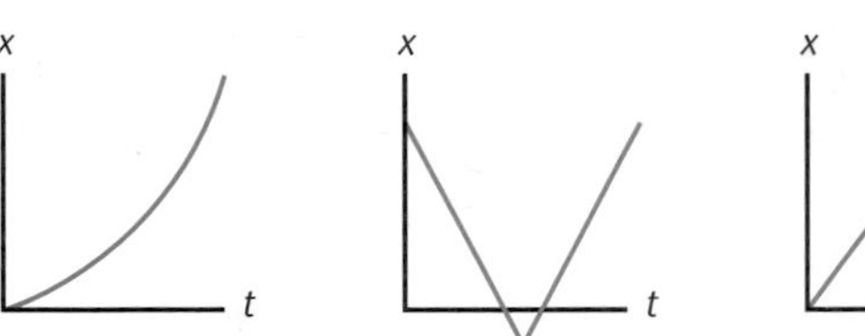

FIGURE 5–27

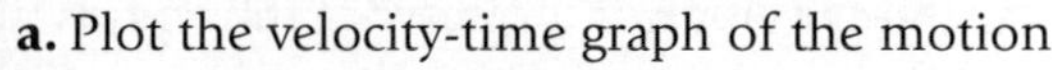

60. The velocity of an automobile changes over an 8.0-s time period as shown in **Table 5–6.**
   a. Plot the velocity-time graph of the motion.
   b. Determine the displacement of the car during the first 2.0 s.
   c. What displacement does the car have during the first 4.0 s?
   d. What displacement does the car have during the entire 8.0 s?
   e. Find the slope of the line between $t = 0.0$ s and $t = 4.0$ s. What does this slope represent?
   f. Find the slope of the line between $t = 5.0$ s and $t = 7.0$ s. What does this slope indicate?

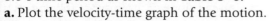

TABLE 5–6
Velocity versus Time

Time (s)	Velocity (m/s)	Time (s)	Velocity (m/s)
0.0	0.0	5.0	20.0
1.0	4.0	6.0	20.0
2.0	8.0	7.0	20.0
3.0	12.0	8.0	20.0
4.0	16.0		

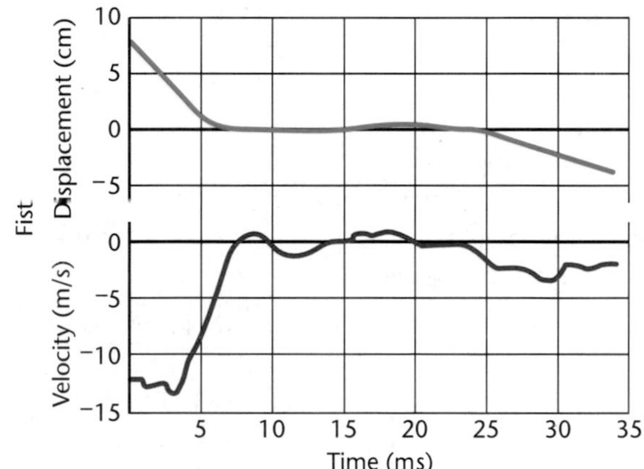

FIGURE 5-28

61. **Figure 5–28** shows the position-time and velocity-time graphs of a karate expert's fist as it breaks a wooden board.
   a. Use the velocity-time graph to describe the motion of the expert's fist during the first 10 ms.
   b. Estimate the slope of the velocity-time graph to determine the acceleration of the fist when it suddenly stops.
   c. Express the acceleration as a multiple of the gravitational acceleration, $g = 9.80$ m/s^2.
   d. Determine the area under the velocity-time curve to find the displacement of the fist in the first 6 ms. Compare this with the position-time graph.

62. The driver of a car going 90.0 km/h suddenly sees the lights of a barrier 40.0 m ahead. It takes the driver 0.75 s to apply the brakes, and the average acceleration during braking is -10.0 m/s^2.
   a. Determine whether the car hits the barrier.
   b. What is the maximum speed at which the car could be moving and not hit the barrier 40.0 m ahead? Assume that the acceleration rate doesn't change.

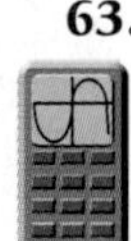

63. The data in **Table 5–7,** taken from a driver's handbook, show the distance a car travels when it brakes to a halt from a specific initial velocity.

TABLE 5–7
Initial Velocity versus Braking Distance

Initial Velocity (m/s)	Braking Distance (m)
11	10
15	20
20	34
25	50
29	70

   a. Plot the braking distance versus the initial velocity. Describe the shape of the curve.
   b. Plot the braking distance versus the square of the initial velocity. Describe the shape of the curve.
   c. Calculate the slope of your graph from part **b.** Find the value and units of the quantity 1/slope.
   d. Does this curve agree with the equation $v_0^2 = -2ad$? What is the value of a?

64. As a traffic light turns green, a waiting car starts with a constant acceleration of 6.0 m/s^2. At the instant the car begins to accelerate, a truck with a constant velocity of 21 m/s passes in the next lane.

a. How far will the car travel before it overtakes the truck?
b. How fast will the car be traveling when it overtakes the truck?

65. Use the information given in problem 64.
a. Draw velocity-time and position-time graphs for the car and truck.
b. Do the graphs confirm the answer you calculated for problem 64?

Section 5.4

LEVEL 1

66. An astronaut drops a feather from 1.2 m above the surface of the moon. If the acceleration of gravity on the moon is 1.62 m/s^2 downward, how long does it take the feather to hit the moon's surface?

67. A stone falls freely from rest for 8.0 s.
a. Calculate the stone's velocity after 8.0 s.
b. What is the stone's displacement during this time?

68. A student drops a penny from the top of a tower and decides that she will establish a coordinate system in which the direction of the penny's motion is positive. What is the sign of the acceleration of the penny?

69. A bag is dropped from a hovering helicopter. When the bag has fallen 2.0 s,
a. what is the bag's velocity?
b. how far has the bag fallen?

70. A weather balloon is floating at a constant height above Earth when it releases a pack of instruments.
a. If the pack hits the ground with a velocity of −73.5 m/s, how far did the pack fall?
b. How long did it take for the pack to fall?

71. During a baseball game, a batter hits a high pop-up. If the ball remains in the air for 6.0 s, how high does it rise? **Hint:** Calculate the height using the second half of the trajectory.

LEVEL 2

72. **Table 5–8** gives the positions and velocities of a ball at the end of each second for the first 5.0 s of free fall from rest.
a. Use the data to plot a velocity-time graph.
b. Use the data in the table to plot a position-time graph.
c. Find the slope of the curve at the end of 2.0 s and 4.0 s on the position-time graph. Do the values agree with the table of velocity?
d. Use the data in the table to plot a position-versus-time-squared graph. What type of curve is obtained?
e. Find the slope of the line at any point. Explain the significance of the value.
f. Does this curve agree with the equation $d = 1/2\ gt^2$?

TABLE 5–8
Position and Velocity in Free Fall

Time (s)	Position (m)	Velocity (m/s)
0.0	0.0	0.0
1.0	−4.9	−9.8
2.0	−19.6	−19.6
3.0	−44.1	−29.4
4.0	−78.4	−39.2
5.0	−122.5	−49.0

73. The same helicopter in problem 69 is rising at 5.0 m/s when the bag is dropped. After 2.0 s,
a. what is the bag's velocity?
b. how far has the bag fallen?
c. how far below the helicopter is the bag?

74. The helicopter in problems 69 and 73 now descends at 5.0 m/s as the bag is released. After 2.0 s,
a. what is the bag's velocity?
b. how far has the bag fallen?
c. how far below the helicopter is the bag?

75. What is common to the answers to problems 69, 73, and 74?

76. A tennis ball is dropped from 1.20 m above the ground. It rebounds to a height of 1.00 m.
a. With what velocity does it hit the ground?
b. With what velocity does it leave the ground?
c. If the tennis ball were in contact with the ground for 0.010 s, find its acceleration while touching the ground. Compare the acceleration to g.

Critical Thinking Problems

77. An express train, traveling at 36.0 m/s, is accidentally sidetracked onto a local train track. The express engineer spots a local train exactly 1.00×10^2 m ahead on the same track and traveling in the same direction. The local engineer is unaware of the situation. The express engineer jams on the brakes and slows the express at a constant rate of 3.00 m/s^2. If the speed of the local train is 11.0 m/s, will the express train be able to stop in time or will there be a collision? To solve this problem, take the position of the express train when it first sights the local train as a point of origin. Next, keeping in mind that the local train has exactly a 1.00×10^2 m lead, calculate how far each train is from the origin at the end of the 12.0 s it would take the express train to stop.

a. On the basis of your calculations, would you conclude that a collision will occur?

b. The calculations you made do not allow for the possibility that a collision might take place before the end of the 12 s required for the express train to come to a halt. To check this, take the position of the express train when it first sights the local train as the point of origin and calculate the position of each train at the end of each second after sighting. Make a table showing the distance of each train from the origin at the end of each second. Plot these positions on the same graph and draw two lines. Use your graph to check your answer to part **a.**

78. Which has the greater acceleration: a car that increases its speed from 50 to 60 km/h, or a bike that goes from 0 to 10 km/h in the same time? Explain.

79. You plan a car trip on which you want to average 90 km/h. You cover the first half of the distance at an average speed of only 48 km/h. What must your average speed be in the second half of the trip to meet your goal? Is this reasonable? Note that the velocities are based on half the distance, not half the time.

Going Further

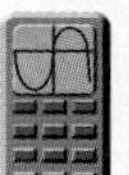

Applying Calculators Members of a physics class stood 25 m apart and used stopwatches to measure the time a car driving down the highway passed each person. The data they compiled are shown in **Table 5–9.**

TABLE 5–9
Position versus Time

Time (s)	Position (m)	Time (s)	Position (m)
0.0	0.0	5.9	125.0
1.3	25.0	7.0	150.0
2.7	50.0	8.6	175.0
3.6	75.0	10.3	200.0
5.1	100.0		

Use a graphing calculator to fit a line to a position-time graph of the data and to plot this line. Be sure to set the display range of the graph so that all the data fit on it. Find the slope of the line. What was the speed of the car?

***inter*NET CONNECTION**

Follow the link on the Glencoe Homepage at **www.glencoe.com/sec/science** to find out more about this chapter.

Going Down

You might expect a skydiver to plummet to Earth in a rapid, uncontrolled descent. Yet a group of skydivers can perform beautiful maneuvers as they drop toward Earth at high speeds. How do skydivers control their velocities?

CHAPTER

6 Forces

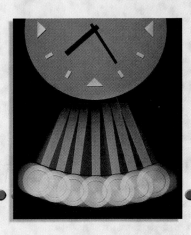

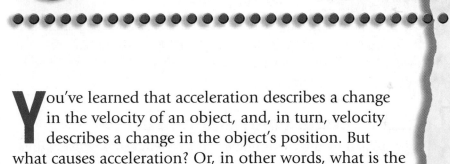

You've learned that acceleration describes a change in the velocity of an object, and, in turn, velocity describes a change in the object's position. But what causes acceleration? Or, in other words, what is the cause of motion?

The answer was given by Sir Isaac Newton more than 300 years ago. He explained the way in which forces influence motion. Newton summed it up in three clear and concise laws. They explain why sky divers begin to fall rapidly downward when they leap from a plane but, after a time, continue their descent at a steady but still rapid speed. Sky divers understand, and you will, too, how changing the position of their arms and legs changes their motion, allowing them the freedom to perform extraordinary feats in the sky.

Many experiments show that real objects move the way that Newton said they do, but Newton's physics sometimes seems contrary to common sense. To fully understand the laws that govern the motion of every moving object you can see around you, you'll have to do experiments and observe the evidence yourself.

WHAT YOU'LL LEARN

- You will use Newton's laws of motion to solve motion problems.
- You will determine the magnitude and direction of the net force that determines the motion of an object.

WHY IT'S IMPORTANT

- The extent to which your environment can withstand the forces of nature depends upon how well the magnitudes of those forces can be predicted and upon the building of homes, bridges, and other structures that are capable of withstanding those forces.

*inter*NET CONNECTION

Follow the link for this chapter on the Glencoe Homepage at **www.glencoe.com/sec/science** to find out more about forces.

6.1 Force and Motion

Push your book slowly across your desk. Then fasten a string to the book and pull it. By pushing or pulling the book, you are exerting a force on the book. An object that experiences a push or a pull has a **force** exerted on it. Notice that it is the object that is considered. The object is called the **system.** The world around the object that exerts forces on it is called the **environment.**

You can push hard or gently, left or right, so force has both magnitude and direction. Force is a vector quantity. The symbol $\boldsymbol{F}$ is used to represent the force vector, and F represents its magnitude, or size. How many different kinds of forces can you identify?

OBJECTIVES

- **Define** a force and **differentiate** between contact forces and long-range forces.
- **Recognize** the significance of Newton's second law of motion and use it to solve motion problems.
- **Explain** the meaning of Newton's first law and **describe** an object in equilibrium.

Color Convention

- Force vectors are **blue.**
- Position vectors are **green.**
- Velocity vectors are **red.**
- Acceleration vectors are **violet.**

Contact Versus Long-Range Forces

Forces exerted by the environment on a system can be divided into two types. The first type is a contact force. A **contact force** acts on an object only by touching it. Either the desk or your hands are probably touching your physics book right now, exerting a contact force on it. Your hand and the desk exert forces only when they touch the book.

The second kind of force is a **long-range force.** A long-range force is exerted without contact. If you have ever played with magnets, you know that they exert forces without touching.

Suppose that you are holding a ball in your hand. The ball has a contact force exerted on it, the force of your hand. Now, suppose that you let go of the ball. Although nothing is touching the ball, it moves because there is a long-range force, the **force of gravity,** acting on the ball. The force of gravity is an attractive force that exists between all objects. In the first half of this book, the only long-range force that will be considered is the force of gravity.

Forces have agents

Each force has a specific, identifiable, immediate cause called the **agent.** You should be able to name the agent of each force, for example, the force of the desk or your hand on your book. The agent can be animate, such as a person, or inanimate, such as a desk, floor, or a magnet. The agent for the force of gravity is Earth's mass. If you can't name an agent, the force doesn't exist!

The first step in solving any problem is to create a pictorial model. To represent the forces on a book as it rests on a table, sketch the situation, as shown in **Figure 6–1.** Circle the system and identify every place where the system touches the environment. It is at these places that contact forces are exerted. Identify the contact forces. Then identify any long-range forces on the system.

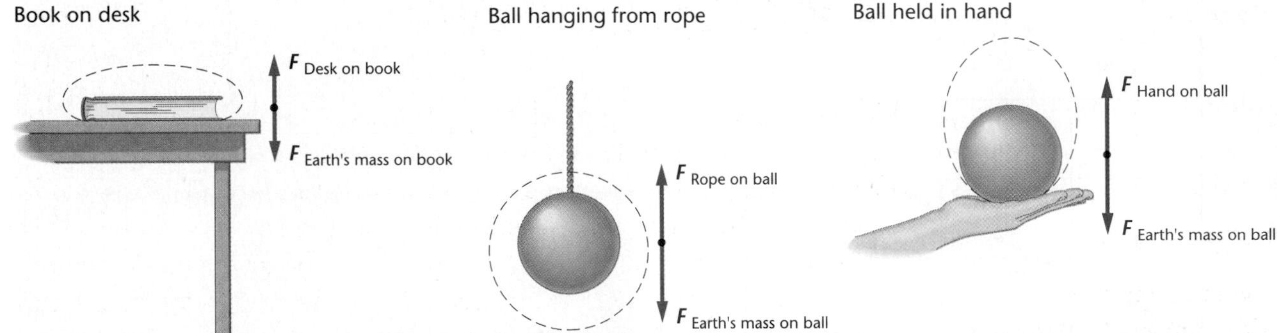

Figure 6–1 To analyze the forces on an object, sketch the situation, circle the system, and identify the agents and the directions of all the forces. Earth's mass is the agent for the force of gravity.

Next, replace the object by a dot, that is, use the particle model. Each force is represented as a blue arrow that points in the correct direction. The length of the arrow is proportional to the size of the force. The tail of the force vector is always on the particle, even when the force is a push. Finally, label the force. For now, use the symbol $\boldsymbol{F}$ with a subscript label to identify both the agent and the object on which the force is exerted. **Figure 6–1** shows pictorial models of the three situations.

Practice Problems

1. Draw pictorial models for the following situations. Circle each system. Draw the forces exerted on the system. Name the agent for each force acting on each system.
   a. a book held in your hand
   b. a book pushed across the desk by your hand
   c. a book pulled across the desk by a string
   d. a book on a desk when your hand is pushing down on it
   e. a ball just after the string that was holding it broke

Pocket Lab

How far is forever?

Galileo proposed that if a perfectly smooth ball were rolled on a perfectly smooth surface in a vacuum, it would roll forever at a steady speed. Use a stopwatch with a lap or split timer and measure the time it takes a ball to roll the first meter and then the total time it takes to roll 2.0 m.

Analyze and Conclude Make a motion diagram to describe the motion and a free-body diagram showing the forces acting on the ball. Indicate the direction of the net force, $\boldsymbol{F}_{net}$, and the acceleration.

Newton's Second Law of Motion

How does an object move when one or more forces are exerted on it? The only way to find out is by doing experiments. Experiments are easier when the influences of gravity and friction can be avoided or minimized. A good way to begin is by studying horizontal forces because gravity does not act in the horizontal direction, and friction can be minimized by doing the experiments either on ice or with carts with low-friction wheels.

How can you exert a controlled force? A stretched rubber band exerts a pulling force; the farther you stretch it, the larger the force. If you always stretch the rubber band the same amount, you always have the same force.

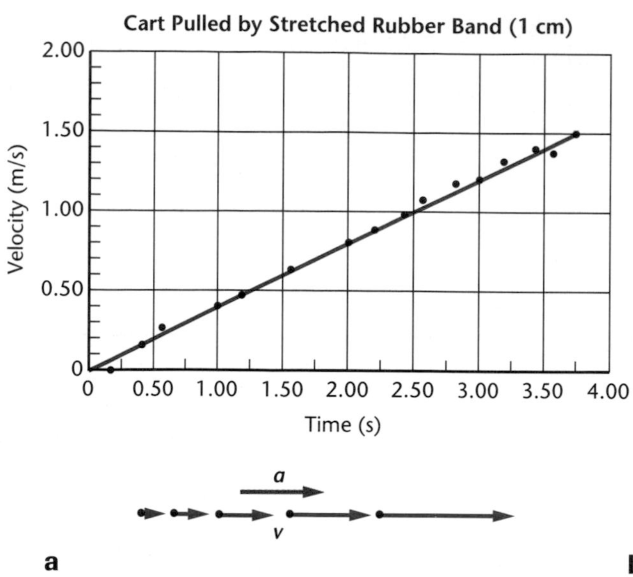

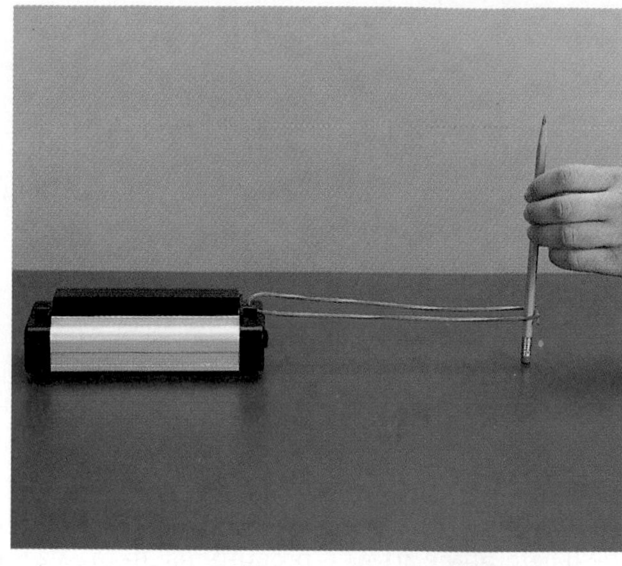

Figure 6–2 The constant slope of the line indicates that the acceleration of the cart is constant. The cart used in this experiment is shown in **b.** It is designed to minimize friction.

The graph in **Figure 6–2a** shows some typical data taken when a rubber band, stretched a constant 1 cm, was used to pull the low-friction cart shown in **Figure 6–2b.** Notice that the velocity-time graph is linear so the cart's acceleration is constant. You can determine the acceleration by calculating the slope of the line. What is it?

How does acceleration depend upon the force?

You could repeat the experiment, this time with the rubber band stretched to a constant 2 cm, and then repeat it again with the rubber band stretched longer and longer. For each experiment, you could determine the acceleration from a velocity-time graph like the one in **Figure 6–2,** and then plot the accelerations for all the trials, as shown in **Figure 6–3a.** Note that this is a force-acceleration graph, and that the acceleration, a, and force, F, are proportional. The larger the force, the greater the acceleration. A linear relationship that goes through the origin, is represented by the equation $F = ka$, where k is the slope of the line.

F.Y.I.

Small insects have very little mass, but the ratio of the surface areas of their bodies to their mass (surface-to-mass ratio) is large. When they are in free fall, their bodies act like parachutes quickly reaching a terminal velocity of only a few cm/s. An ant falling from a 50-story building will walk away unharmed after hitting the sidewalk.

How does acceleration depend upon the object?

This experiment shows that the acceleration of an object is proportional to the net force exerted on it. What happens if the object changes? Suppose that a second cart is placed on top of the first, and then a third cart is added. The rubber band would be pulling two carts, then three. A plot of the force versus the acceleration for one, two, and three carts, is shown in **Figure 6–3b.** The graph shows that for an equal force, the acceleration of two carts is 1/2 the acceleration of one, and the acceleration of three carts is 1/3 the acceleration of one. This means that as the number of carts is increased, a greater force is needed to produce the same acceleration. The slopes of the lines in **Figure 6–3b** depend upon the number of carts, or upon mass. If the mass is defined as the slope of the F-a graph, then, $m = F/a$, or $F = ma$.

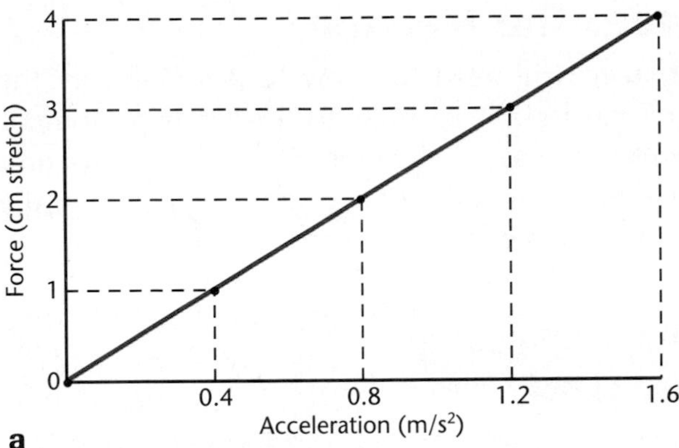

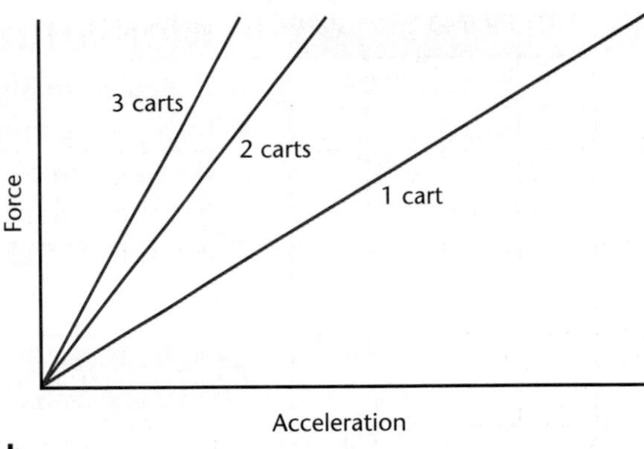

Figure 6–3 The graph in **a** shows that as the force increases, so does the acceleration. In **b**, you can see that the slope of the force-acceleration graph depends upon the number of carts.

Combining forces

What if two or more rubber bands exert forces on a cart? They could act in the same direction, in opposite directions, or in directions at an angle to one another. In **Figure 6–4,** the carts are represented by dots, and the forces operating on each dot (cart) are drawn in the direction of the force with their tails on the dot. This is called a **free-body diagram.**

Because forces are vectors, the total force on an object is the vector sum of all forces exerted on the object. You have learned how to add vectors and find the resultant as shown in **Figure 6–4.** The vector sum of two or more forces on an object is called the **net force.** Experiments show that the acceleration of an object is proportional to the net force exerted on the object and inversely proportional to the mass of the object being accelerated. This is a statement of **Newton's second law,** which can be written as an equation.

$$\boldsymbol{a} = \frac{\boldsymbol{F}_{\text{net}}}{m}$$

Here is a strategy for finding how the motion of an object depends on the forces exerted on the object. First, identify all the forces on the object. Draw a free-body diagram showing the direction and relative magnitude of each force acting on the system. Then, add the force vectors to find the net force. Next, use Newton's second law to calculate the acceleration. Finally, use kinematics to find the velocity and position of the object. You learned about kinematics in Chapters 3, 4, and 5 when you studied the motion of objects without regard for the causes of motion. You now know that an unbalanced force is the cause of a change in velocity.

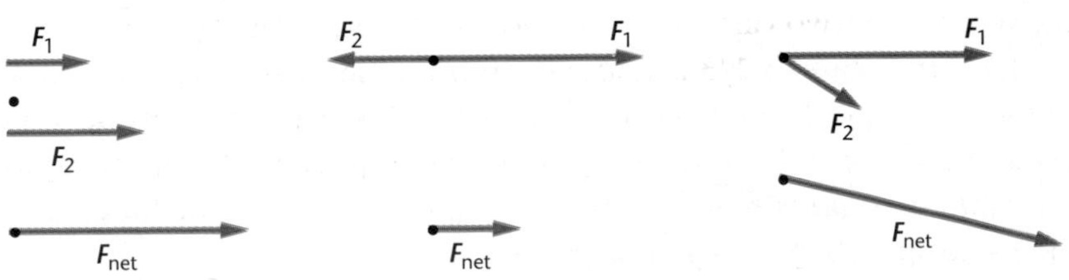

Figure 6–4 The net force is the vector sum of $\boldsymbol{F}_1$ and $\boldsymbol{F}_2$.

TABLE 6–1	
Description	***F* (N)**
Force of gravity on coin (nickel)	0.05
Force of gravity on 1 lb sugar	4.5
Force of gravity on 150-lb person	668
Force accelerating a car	3 000
Force of a rocket motor	5 000 000

HELP WANTED

PHYSICS TEACHER

To qualify, you must enjoy science and math and be skilled in explaining technical concepts clearly. You must be able to tailor your explanations and demonstrations to individual student's needs. The position requires a bachelor's degree in physics, chemistry, biology, math, or a related area. You may be required to teach any of these courses in addition to physics. Knowledge of computers is vital. A master's degree in education or administration is a plus. For more information, contact:

American Association of Physics Teachers
One Physics Ellipse
College Park, MD 20740

Measuring Force: The Newton

Before trying the strategy, you need to know how to measure the force. One unit of force causes a 1-kg mass to accelerate at 1 m/s^2. Because force is equal to mass times acceleration, $F = ma$, one force unit has the dimensions 1 $kg \cdot m/s^2$. The unit of force, in the SI system, is the newton, N. **Table 6–1** shows some typical forces.

Practice Problems

2. Two horizontal forces, 225 N and 165 N, are exerted in the same direction on a crate. Find the net horizontal force on the crate.
3. If the same two forces are exerted in opposite directions, what is the net horizontal force on the crate? Be sure to indicate the direction of the net force.
4. The 225-N force is exerted on the crate toward the north and the 165-N force is exerted toward the east. Find the magnitude and direction of the net force.
5. Your hand exerts a 6.5-N upward force on a pound of sugar. Considering the force of gravity on the sugar, what is the net force on the sugar? Give the magnitude and direction.
6. Calculate the force you exert as you stand on the floor (1 lb = 0.454 kg). Is the force the same if you lie on the floor?

Newton's First Law of Motion

What is the motion of an object with no net force on it? Think of a ball rolling on a surface. How long will the ball continue to roll? That depends on the quality of the surface. If you roll it on thick carpet or soft sand, it will quickly come to rest. If you roll it on a surface that is hard and smooth, such as a bowling alley, the ball will roll for a long time with little change in velocity. You could imagine that if all friction were eliminated, the ball might roll at the same velocity forever. Galileo did many experiments on the motion of balls on very smooth surfaces. He concluded that in the ideal case, horizontal motion was eternal: it would never stop. Galileo was the first to recognize that the general principles of motion could be found only by extrapolating experimental results to the ideal case, in which there is no friction or other drag force.

Newton generalized Galileo's results to motion in any direction. He stated, "An object that is at rest will remain at rest or an object that is moving will continue to move in a straight line with constant speed, if and only if the net force acting on that object is zero." This statement is called **Newton's first law** of motion.

Inertia

Newton's first law is often called the law of inertia. **Inertia** is the tendency of an object to resist change. If an object is at rest, it tends to remain at rest. If it is moving at a constant velocity, it tends to continue moving at that velocity.

Equilibrium

If the net force on an object is zero, then the object is in equilibrium. An object is in **equilibrium** if it is at rest or if it is moving at constant velocity. Note that being at rest is just a special case of constant velocity. Newton's first law identifies a net force as something that disturbs a state of equilibrium. That means that a net force changes the velocity of an object. Thus, change in velocity, or acceleration, is the result of a net force acting on an object.

The physical model: Free-body diagrams

Because the net force on an object causes the acceleration of the object, it is important to know how to find the net force. The net force is the sum of all the forces on an object. **Table 6-2** will help you identify some common types of forces.

Tug-of-War Challenge

In a tug-of-war, predict how the force you exert on your end of the rope compares to the force your opponent exerts if you pull and your opponent just holds the end of the rope. Predict how the forces compare if the rope moves in your direction. Try it.

Analyze and Conclude What did you notice about the forces? What happened when you started to move your opponent in your direction?

TABLE 6–2

Some Types of Forces

Force	Symbol	Definition	Direction
Friction	$\boldsymbol{F}_{\mathrm{f}}$	The contact force that acts to oppose sliding motion between surfaces	Parallel to the surface and opposite the direction of sliding
Normal	$\boldsymbol{F}_{\mathrm{N}}$	The contact force exerted by a surface on an object	Perpendicular to and away from the surface
Spring	$\boldsymbol{F}_{\mathrm{sp}}$	A restoring force, that is, the push or pull a spring exerts on an object	Opposite the displacement of the object at the end of the spring
Tension	$\boldsymbol{F}_{\mathrm{T}}$	The pull exerted by a string, rope, or cable when attached to a body and pulled taut	Away from the object and parallel to the string, rope, or cable at the point of attachment
Thrust	$\boldsymbol{F}_{\mathrm{thrust}}$	A general term for the forces that move objects such as rockets, planes, cars, and people	In the same direction as the acceleration of the object barring any resistive forces
Weight	$\boldsymbol{F}_{\mathrm{g}}$	A long-range force due to gravitational attraction between two objects, generally Earth and an object	Straight down toward the center of Earth

Example Problem

Constructing a Free-Body Diagram

A rope is lifting a heavy bucket. The speed of the bucket is increasing. How can the forces on the bucket be related to the change in speed?

Sketch the Problem

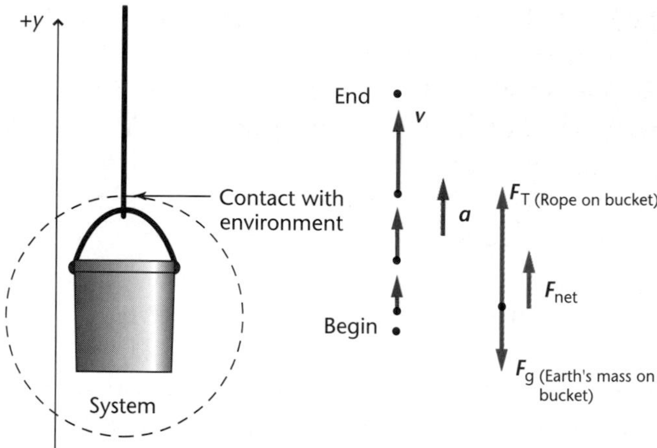

- Choose a coordinate system defining the positive direction of the velocity.
- Locate every point at which the environment touches the system.
- Draw a motion diagram including the velocity and acceleration. The bucket is moving upward, so the direction of $\boldsymbol{v}$ is upward. The speed is increasing so the direction of $\boldsymbol{a}$ is upward. Indicate "begin" and "end."
- Draw the free-body diagram. Replace the bucket by a dot and draw arrows to represent $\boldsymbol{F}_{\text{T (rope on bucket)}}$ and $\boldsymbol{F}_{\text{g (Earth's mass on bucket)}}$.

Check Your Answer

- Velocity is increasing in the upward direction, so acceleration is upward.
- According to Newton's second law, $\boldsymbol{F}_{\text{net}}$ and $\boldsymbol{a}$ are in the same direction.
- Therefore, vector addition of the positive $\boldsymbol{F}_{\text{T}}$ and the negative $\boldsymbol{F}_{\text{g}}$ results in a positive $\boldsymbol{F}_{\text{net}}$.
- Draw an arrow showing $\boldsymbol{F}_{\text{net}}$.

Practice Problems

For each problem, draw a motion diagram and a free-body diagram labeling all forces with their agents and indicating the direction of the acceleration and the net force. Draw arrows the appropriate lengths.

7. A skydiver falls downward through the air at constant velocity (air drag is important).
8. A cable pulls a crate at constant speed across a horizontal surface (there is friction).
9. A rope lifts a bucket upward at constant speed (ignore air drag).
10. A rope lowers a bucket at constant speed (ignore air drag).
11. A rocket blasts off and its vertical velocity increases with time (ignore air drag).

Common misconceptions about forces

The world is dominated by friction, and so Newton's ideal, friction-free world is not easy to visualize. In addition, many terms used in physics have everyday meanings that are different from those understood in physics. Here are some examples of common, but mistaken ideas about forces.

- **When a ball has been thrown, the force of the hand that threw it remains on it.** No, the force of the hand is a contact force; therefore, once contact is broken, the force is no longer exerted.
- **A force is needed to keep an object moving.** If there is no net force, then the object keeps moving with unchanged velocity. If friction is a factor, then there is a net force and the object's velocity will change.
- **Inertia is a force.** Inertia is the tendency of an object to resist changing its velocity. Forces are exerted on objects by the environment; they are not properties of objects.
- **Air does not exert a force.** Air exerts a huge force, but because it is balanced on all sides, it usually exerts no net force unless an object is moving. You can experience this force only if you remove the air from one side. For example, when you stick a suction cup on a wall or table, you remove air from one side. The suction cup is difficult to remove because of the large unbalanced force of the air on the other side.
- **The quantity *ma* is a force.** The equals sign in $F = ma$ does not define ma as a force. Rather, the equal sign means that experiments have shown that the two sides of the equation are equal.

6.1 Section Review

1.1 Identify each of the following as either **a, b,** or **c:** weight, mass, inertia, the push of a hand, thrust, tension, friction, air drag, spring force, acceleration, and mass times acceleration.

a. a contact force

b. a long-range force

c. not a force

1.2 Can you feel the inertia of a pencil? Of a book? If you can, describe how.

1.3 If you push a book in the forward direction, does that mean its velocity has to be forward?

1.4 Draw a free-body diagram of a water bucket being lifted by a rope at a decreasing speed. Label all forces with their agents and make the arrows the correct lengths.

1.5 **Critical Thinking** A force of 1 N is the only force exerted on a block, and the acceleration of the block is measured. When the same force is the only force exerted on a second block, the acceleration is three times as large. What can you conclude about the masses of the two blocks?

6.2 Using Newton's Laws

Newton's second law describes the connection between the net force exerted on an object and its acceleration. The second law identifies the cause of a change in velocity and the resulting displacement. Newton called this *a law of nature* because he thought it held true for all motions. Early in the twentieth century, more than 200 years after Newton's time, physicists discovered that the second law is not true for velocities close to the speed of light, nor for objects the size of atoms. Nevertheless, all of our everyday experiences are governed by this physical law which was formulated over 300 years ago.

OBJECTIVES

- **Describe** how the weight and the mass of an object are related.
- **Differentiate** between the gravitational force weight and what is experienced as apparent weight.
- **Define** the friction force and **distinguish** between static and kinetic friction.
- **Describe** simple harmonic motion and **explain** how the acceleration due to gravity influences such motion.

Using Newton's Second Law

Aristotle's followers believed that the heavier an object is, the faster it falls. Test this idea yourself. Drop a feather and a coin. Doesn't the coin fall faster? You can see for yourself that the evidence seems to be in Aristotle's favor. But Galileo knew that if he was to understand the nature of the force that causes an object to fall, he had to simulate an idealized world in which there is no air drag.

Mass and weight

While there is no evidence that Galileo actually dropped two balls from the Leaning Tower of Pisa to test his ideas, he did describe the following thought experiment. Two cannon balls of equal weight, dropped side by side, should fall at an equal rate. But what happens if the cannon balls are tied together? According to Aristotle, they should fall twice as fast. But Galileo hypothesized that all objects, no matter what their weight, gain speed at the same rate, which means that they have the same downward acceleration. This hypothesis has been tested and found to be true.

What is the weight force, $\boldsymbol{F}_g$, exerted on an object of mass m? Galileo's hypothesis and Newton's second law can answer this question. Consider the pictorial and physical models in **Figure 6–5,** which show a falling ball in midair. Because it is touching nothing and air resistance can be neglected, there are no contact forces on it, only $\boldsymbol{F}_g$. The ball's acceleration is $\boldsymbol{g}$. Newton's second law then becomes $\boldsymbol{F}_g = m\boldsymbol{g}$. Both the force and the acceleration are downward. The magnitude of an object's weight is equal to its mass times the acceleration it would have if it were falling freely.

This result is true on Earth, as well as on any other planet, although the magnitude of g will be different on other planets. Future astronauts will find that their weights vary from planet to planet, but their masses will not change.

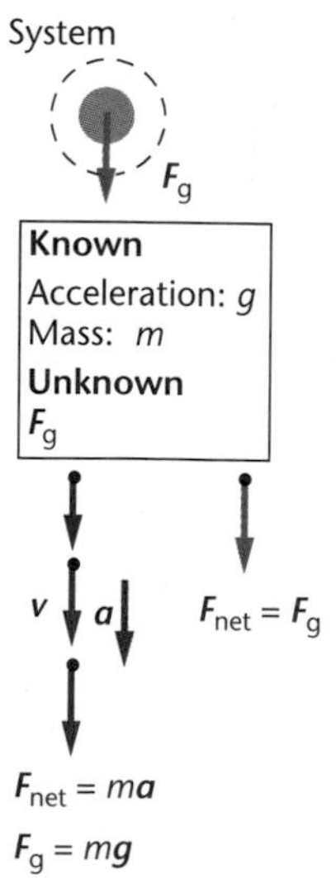

FIGURE 6–5 The net force on the ball is the weight force, $\boldsymbol{F}_g$.

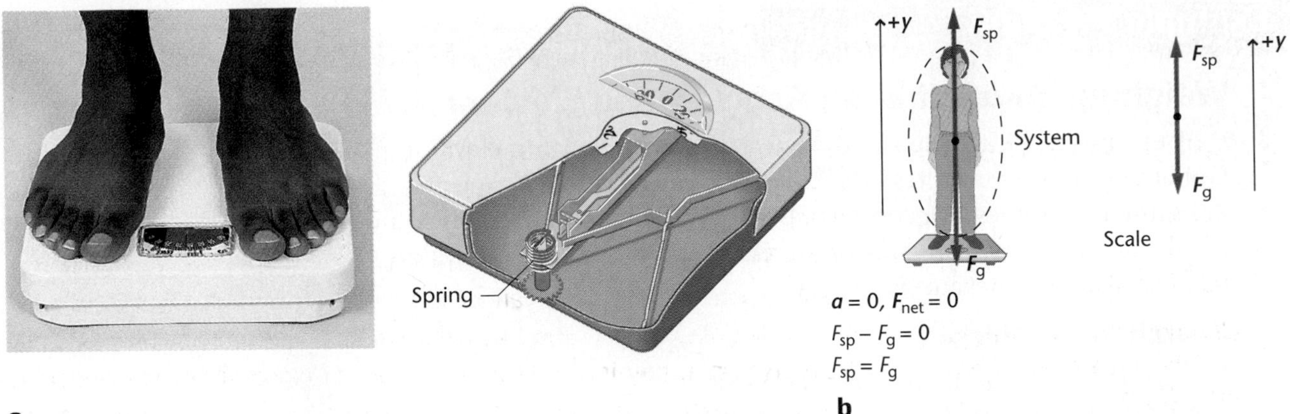

FIGURE 6–6 The upward force of the spring in the scale is equal to your weight, F_g, when you step on the bathroom scale. The sketch and free-body diagram in **b** show that the system is in equilibrium, so $F_g = F_{sp}$.

Scales

Figure 6–6a asks this question: "What is being measured, mass or weight?" A bathroom scale contains springs. When you step on the scale, the scale exerts an upward force on you. The pictorial and physical models in **Figure 6–6b** show that, because you are not accelerating, the net force is zero. Therefore, the magnitude of the spring force, F_{sp}, is equal to your weight, F_g. A spring scale, therefore, measures weight, not mass. If you were on a different planet, the compression of the spring would be different, and consequently, the scale's reading would be different.

Problem Solving Strategy

Force and Motion

When using Newton's laws to solve force and motion problems, use the following strategy.

1. Read the problem carefully. Visualize the situation and create the pictorial model with a sketch.
2. Circle the system and choose a coordinate system.
3. Decide which quantities are known and which quantity you need to find. Assign symbols to the known and unknown quantities.
4. Create the physical model, which includes a motion diagram showing the direction of the acceleration, and a free-body diagram, which includes the net force.
5. To calculate your answer, use Newton's laws to link acceleration and net force.
6. Rearrange the equation to solve for the unknown quantity, $\boldsymbol{a}$ or $\boldsymbol{F}_{net}$. Newton's second law involves vectors, so the equation must be solved separately in the x and y directions.
7. Substitute the known quantities with their units in the equation and solve.
8. Check your results to see if they are reasonable.

Example Problem

Weighing Yourself in an Accelerating Elevator

Your mass is 75 kg. You stand on a bathroom scale in an elevator. Going up! Starting from rest, the elevator accelerates at 2.0 m/s^2 for 2.0 s, then continues at a constant speed. What is the scale reading during the acceleration? Is it larger than, equal to, or less than the scale reading when the elevator is at rest?

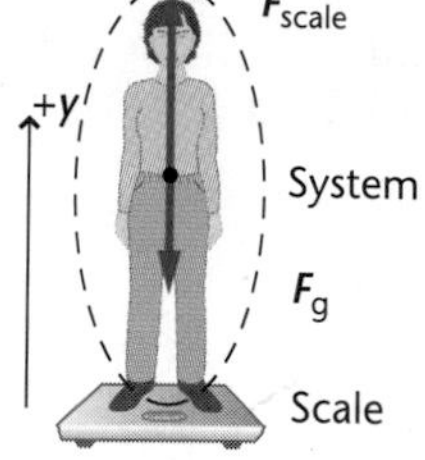

Sketch the Problem

- Sketch the situation as in **Figure 6–6b.**
- Draw the motion diagram. Label ***v*** and ***a***.
- Choose a coordinate system with the positive direction up.
- The net force is in the same direction as the acceleration, so the upward force is greater than the downward force.

Calculate Your Answer

Known:

$m = 75$ kg

$a = +2.0\ m/s^2$

$t = 2.0$ s

Unknown:

$F_{scale} = ?$

Strategy:

F_{net} is the sum of the positive force of the scale on you, F_{scale}, and the negative weight force, $F_{net} = F_{scale} - F_g$.

Solve for F_{scale} and substitute ma for F_{net} and mg for F_g.

Calculations:

$F_{scale} = ma + mg$

$F_{scale} = m(a + g)$

$= (75\ kg)(2.0\ m/s^2 + 9.80\ m/s^2)$

$F_{scale} = 890$ N

Check Your Answer

- Are the units correct? $kg{\cdot}m/s^2$ is the force unit, N.
- Does the sign make sense? The positive sign agrees with the diagram.
- Is the magnitude realistic? F_{scale} is larger than it would be at rest when F_{scale} would be 7.4×10^2 N, so the magnitude is reasonable.

Example Problem

Lifting a Bucket

A 50-kg bucket is being lifted by a rope. The rope is guaranteed not to break if the tension is 500 N or less. The bucket started at rest, and after being lifted 3.0 m, it is moving at 3.0 m/s. Assuming that the acceleration is constant, is the rope in danger of breaking?

Sketch the Problem

- Draw the situation; identify the forces on the system.

- Establish a coordinate system with a positive axis up.
- Draw a motion diagram including $\boldsymbol{v}$ and $\boldsymbol{a}$.
- Draw the free-body diagram. Position the force vectors with their tails on the dot.

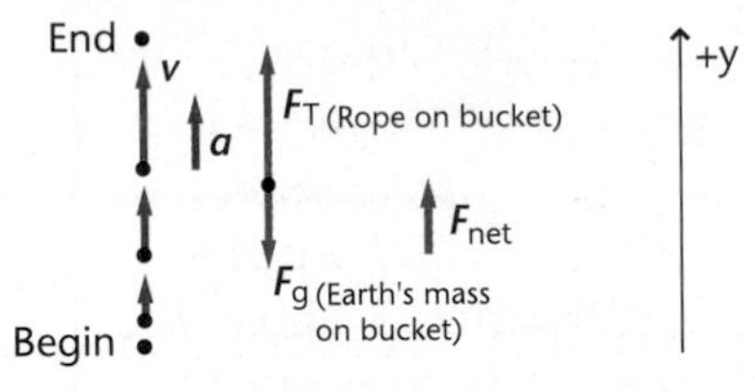

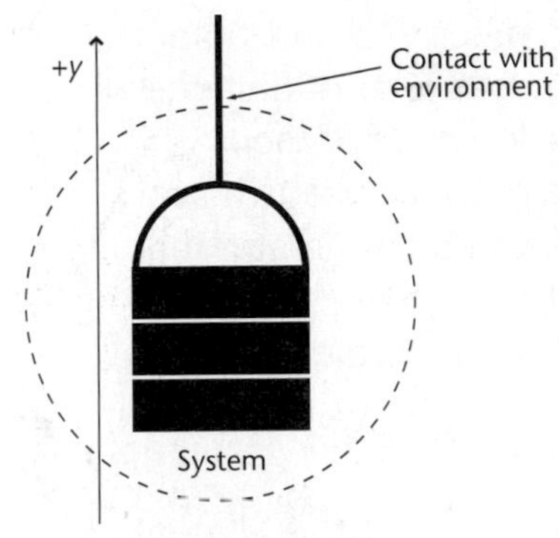

Calculate Your Answer

Known:

$m = 50$ kg $\quad v = 3.0$ m/s

$v_0 = 0.0$ m/s $\quad d = 3.0$ m

Unknown:

$F_T = ?$

Strategy:

The net force is the vector sum of $\boldsymbol{F}_T$ (positive) and $\boldsymbol{F}_g$ (negative). $F_{net} = F_T - F_g$.

Rearrange the equation: $F_T = F_{net} + F_g = ma + mg$

Because v_0 is zero, $a = v^2/2d$

Calculations:

$$F_T = m(a + g) = m(v^2/2d + g)$$

$$F_T = (50 \text{ kg})\left(\frac{9.0 \text{ m}^2/\text{s}^2}{2(3.0 \text{ m})} + 9.80 \text{ m/s}^2\right)$$

$F_T = 570$ N; the rope is in danger of breaking because the tension exceeds 500 N.

Check Your Answer

- Are the units correct? Performing algebra on the units verifies kg·m/s^2 which is N.
- Does the sign make sense? The upward force should be positive.
- Is the magnitude realistic? Yes, the magnitude is a little larger than 490 N, which is the weight of the bucket.

Practice Problems

12. On Earth, a scale shows that you weigh 585 N.
  a. What is your mass?
  b. What would the scale read on the moon ($g = 1.60$ m/s^2)?

13. Use the results from the first example problem to answer these questions about a scale in an elevator on Earth. What force would the scale exert when
  a. the elevator moves up at a constant speed?
  b. it slows at 2.0 m/s^2 while moving upward?
  c. it speeds up at 2.0 m/s^2 while moving downward?
  d. it moves downward at a constant speed?
  e. it slows to a stop at a constant magnitude of acceleration?

Pocket Lab

Friction Depends On What?

Find out! Tape a 0.5-kg mass to a 10-cm × 10-cm piece of cardboard. Tie one end of a string to the mass and the other end to a spring scale. Pull until the mass begins to move. Record the maximum force before the mass began to slide as the static force of friction. Repeat for a 1.0-kg mass. Repeat with the two masses on a 10-cm × 20-cm piece of cardboard.

Analyze and Conclude Describe your results. Does the force of friction depend on the mass? Does the force of friction depend on the surface area?

EARTH SCIENCE CONNECTION

The San Andreas Fault in California is a series of fractures in Earth's crust. Forces in Earth's interior cause the rocks to slide past each other in a horizontal direction. At first, the forces of friction between the two surfaces are greater than the forces that cause the slide, so the rocks stretch and twist. Eventually, forces within the rocks become greater than the forces of friction, and, much like the release of a stretched rubber band, the rocks snap back in place. This movement, with the resulting release of tremendous amounts of energy, is an earthquake.

Apparent weight

What is weight? What does a bathroom scale measure? The weight force is defined as $F_g = mg$, so F_g changes when g varies. On or near the surface of Earth, however, g is approximately constant. If a bathroom scale supports you—it provides the only upward force on you—then it reads your weight. But, suppose you stood with one foot on the scale and one foot off? Or what if a friend pushed down on your shoulders or pushed up on your elbows? Then there would be other contact forces on you, and the scale would not read your weight.

What happens if you are standing on a scale in an elevator? As long as the elevator is in equilibrium, that is, at rest or moving at constant speed, the scale reads your weight. But if the elevator accelerates upward, then the scale reads a larger force. What does it feel like to be in an elevator like this? You feel heavier; the floor presses harder on your feet. On the other hand, if the acceleration is downward, then you feel lighter, and the scale reads less. The force exerted by the scale is called the **apparent weight.**

Imagine that the cable holding the elevator breaks. The scale with you on it would accelerate with $a = -g$. According to the solution to the first example problem, the scale would read zero! Your apparent weight would be zero. That is, you would be weightless. However, **weightlessness** doesn't mean your weight is zero, but that there are no contact forces pushing up on you. Weightlessness means that your apparent weight is zero.

The Friction Force

Push your hand across your desktop and feel the force called friction opposing the motion. Friction is often minimized in solving force and motion problems, but in the real world, friction is everywhere. You need it to both start and stop a bike and a car. If you've ever walked on ice, you know how important friction is. Friction lets a pencil make a mark on paper and an eraser fix mistakes.

Static and kinetic friction

Think about friction as you push a heavy crate across the floor. You give the crate a push, but it doesn't move. Newton's laws tell you it should move unless there is a second horizontal force on the crate, opposite in direction to your force, and equal in size. That force is called the **static friction force.** It is exerted on one surface by the other when there is no relative motion between the two surfaces. You can push harder and harder, as shown in **Figure 6–7,** but if the crate still doesn't move, the friction force also must be getting larger. The static friction force acts in response to other forces. Finally, when your push gets hard enough, the crate begins to move. Evidently, the static friction force can grow only so large.

The crate may be moving, but friction is still acting because if you stop pushing, the crate slows. The force that is acting is called the kinetic friction force. The **kinetic friction force** is the force exerted on one surface by the other when the surfaces are in relative motion.

a

A model for friction forces

Although friction forces are complicated, a simplified model can be used to find solutions close to those found by experiments. The model assumes that friction depends on the surfaces in contact, but not on the area of the surfaces nor the speed of their relative motion. In the model, the magnitude of the friction force is proportional to the magnitude of the force pushing one surface against the other. That force, perpendicular to the surface, is the normal force, F_N.

$$F_{f,\ kinetic} = \mu_k F_N$$

In this equation, μ_k is a proportionality constant called the kinetic coefficient of friction.

The static friction force is related to the normal force by this expression.

$$0 \leq F_{f,\ static} \leq \mu_s F_N$$

where μ_s is the static coefficient of friction. The equation tells you that the static friction force can vary from zero to $\mu_s F_N$ where $\mu_s F_N$ is the maximum static friction force that must be balanced before motion can begin. In **Figure 6-7c,** the static friction force has just been balanced the instant before the box begins to move.

b

c

FIGURE 6–7
There is a limit to the ability of the static friction force to match the applied force.

Note that the preceding equations involve the magnitudes of the forces only. The forces themselves, $\mathbf{F_f}$ and $\mathbf{F_N}$, are at right angles to each other. **Table 6–3** shows coefficients of friction for various surfaces. You will need to use these in solving problems. Although all the listed coefficients are less than 1, this doesn't mean that the coefficient of friction must be less than 1. Coefficients as large as 5.0 are experienced in drag racing.

TABLE 6–3
Typical Coefficients of Friction

Surface	μ_s	μ_k
Rubber on concrete	0.80	0.65
Rubber on wet concrete	0.60	0.40
Wood on wood	0.50	0.20
Steel on steel (dry)	0.78	0.58
Steel on steel (with oil)	0.15	0.06
Teflon on steel	0.04	0.04

Example Problem

Balanced Friction Forces

You push a 25-kg wooden box across a wooden floor at a constant speed of 1.0 m/s. How much force do you exert on the box?

Sketch the Problem

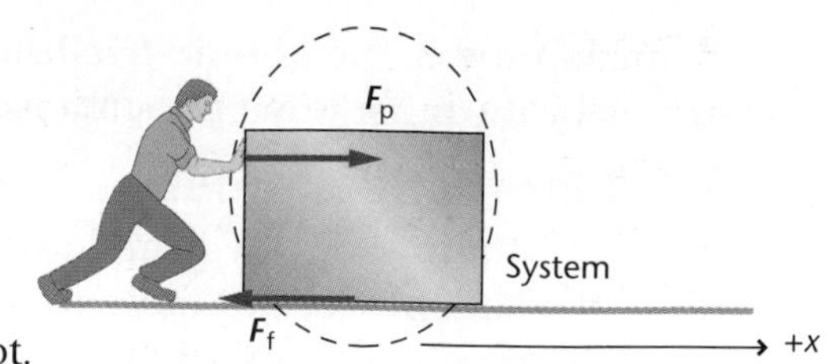

- Identify the forces and establish a coordinate system.
- Draw a motion diagram indicating constant $\boldsymbol{v}$ and $\boldsymbol{a}$ = zero.
- Draw the free-body diagram with the tails of the four forces ($\boldsymbol{F}_f$, $\boldsymbol{F}_N$, $\boldsymbol{F}_g$, and $\boldsymbol{F}_p$, your pushing force) on the dot.

Calculate Your Answer

Known:

$m = 25$ kg

$v = 1.0$ m/s

$a = 0.0$ m/s^2

Unknown:

$F_p = ?$

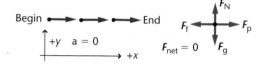

Strategy:

y-direction: Because there is no acceleration, $F_N = F_g = mg$.

x-direction: Because v is constant, there is no acceleration. Therefore, the pushing force, $F_p = F_f = \mu_k mg$.

Calculations:

$F_p = \mu_k mg$

$F_p = (0.20)(25 \text{ kg})(9.80 \text{ m/s}^2)$
$= 49$ N

Check Your Answer

- Are the units correct? Performing algebra on the units verifies that force = kg·m/s^2 or N.
- Does the sign make sense? The positive sign agrees with the sketch.
- Is the magnitude realistic? It is a reasonable force for moving a 25-kg box.

Example Problem

Unbalanced Friction Forces

If the force you exert on the box is doubled, what is the resulting acceleration of the box?

Sketch the Problem

- The sketch is the same as in the preceding example problem.
- Draw a motion diagram showing increasing $\boldsymbol{v}$ and the direction of $\boldsymbol{a}$.
- Draw the free-body diagram with doubled pushing force, $\boldsymbol{F}_p$.

Calculate Your Answer

Known:

$m = 25$ kg

$v = 1.0$ m/s

$\mu_k = 0.20$

$F_p = 2(49 \text{ N}) = 98$ N

Unknown:

$a = ?$

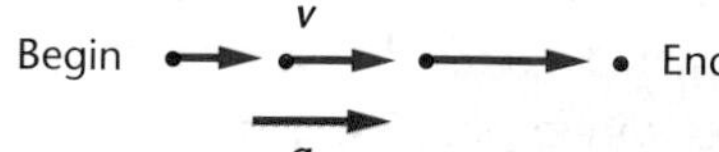

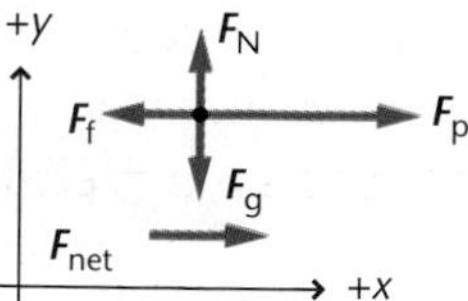

Strategy:

Friction force is the same; it is independent of speed.

There is a net horizontal force; the crate accelerates.

Apply Newton's laws separately in two directions.

Calculations:

y-direction: $F_N = F_g = mg;\ F_N = mg$

x-direction: $F_p - F_f = ma$

$F_f = \mu_k F_N = \mu_k mg$

$$a = \frac{F_{net}}{m} = \frac{F_p - \mu_k mg}{m} = \frac{F_p}{m} - \mu_k g$$

$$a = \frac{98\ \text{N}}{25\ \text{kg}} - (0.20)(9.80\ \text{m/s}^2)$$

$$a = 2.0\ \text{m/s}^2$$

Check Your Answer

- Are the units correct? Performing algebra on units verifies that a is in m/s^2.
- Does the sign make sense? For the chosen coordinate system, the sign should be positive.
- Is the magnitude realistic? In the calculation of a, if the force were cut in two, a would be zero as in the preceding example problem.

Practice Problems

14. A boy exerts a 36-N horizontal force as he pulls a 52-N sled across a cement sidewalk at constant speed. What is the coefficient of kinetic friction between the sidewalk and the metal sled runners? Ignore air resistance.
15. Suppose the sled runs on packed snow. The coefficient of friction is now only 0.12. If a person weighing 650 N sits on the sled, what force is needed to pull the sled across the snow at constant speed?
16. Consider the doubled force pushing the crate in the example problem *Unbalanced Friction Forces*. How long would it take for the velocity of the crate to double to 2.0 m/s?

Upside Down Parachute

How long does it take for a falling object to reach a terminal velocity? How fast is the terminal velocity? Does the terminal velocity depend on the mass? Find out.

Use coffee filters, a meterstick, a stopwatch, and your creativity to answer each question.

Analyze and Conclude
Describe your procedures, results, and conclusions to the class.

Causes of friction

All surfaces, even those that appear to be smooth, are rough at a microscopic level as shown in **Figure 6–8**. When two surfaces touch, the high points on each are in contact and temporarily bond. When you try to move one of the pieces, you must break the bonds. This is the origin of static friction. As the surfaces move past each other, the electrostatic forces that caused the bonds continue to create an attraction between the high points on the moving surfaces and this results in the weaker kinetic friction. The details of this process are still unknown and are the subject of research in both physics and engineering.

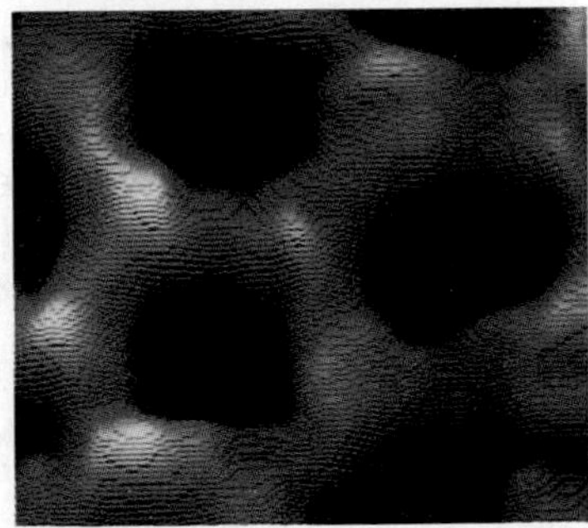

FIGURE 6–8 This photograph of a graphite crystal, magnified by a scanning tunneling microscope, reveals the surface irregularities of the crystal at the atomic level.

Air drag and terminal velocity

When an object moves through air or any other fluid, the fluid exerts a frictionlike force on the moving object. Unlike the friction between surfaces, however, this force depends upon the speed of the motion, becoming larger as the speed increases. It also depends upon the size and shape of the object and the density and kind of fluid.

If you drop a table tennis ball from a tower, it has very little velocity at the start, and thus only a small drag force. The downward force of gravity is much stronger than the upward drag force, so there is a downward acceleration. As the ball's velocity increases, so does the drag force. Soon, the drag force equals the force of gravity. With no net force, there is no acceleration. The velocity of the ball becomes constant. The constant velocity that is reached when the drag force equals the force of gravity is called the **terminal velocity.**

The terminal velocity of table tennis ball in air is 9 m/s. A basketball has a terminal velocity of 20 m/s; the terminal velocity of a baseball is 42 m/s. Skiers increase their terminal velocities by decreasing the drag force. They hold their bodies in an egg shape and wear smooth clothing and streamlined helmets. How do sky divers control their velocities? By changing body orientation and shape, sky divers can both increase and decrease their terminal velocity so that they can perform maneuvers in the air. A horizontal spread-eagle shape gives the slowest terminal velocity, about 60 m/s. When the parachute opens, the sky diver becomes part of a very large object with a correspondingly large drag force and a terminal velocity of about 5 m/s.

Periodic Motion

A playground swing, moving back and forth over the same path, is one example of vibrational motion. Other examples are a pendulum, a metal block bobbing up and down on a spring, and a vibrating guitar string, as shown in **Figure 6–9.**

FIGURE 6–9 A plucked guitar string continues to move rapidly back and forth in simple harmonic motion.

In each example, the object has one position in which the net force on it is zero. At that position, the object is in equilibrium. Whenever the object is pulled away from its equilibrium position, the net force on the system becomes nonzero and pulls it back toward equilibrium. If the force that restores the object to its equilibrium position is directly proportional to the displacement of the object, the motion that results is called **simple harmonic motion**.

Two quantities describe simple harmonic motion. One is the period, represented by the symbol T. The **period** is the time needed to repeat one complete cycle of motion. The other quantity, called the **amplitude** of the motion, is the maximum distance the object moves from equilibrium.

The mass on a spring

How do you describe the simple harmonic motion of objects? **Figure 6–10a** shows a block hanging on a spring. Two forces are exerted on the block. The weight force is a constant downward force, $\mathbf{F}_g$. The upward force of the spring is directly proportional to the amount the spring is stretched. A spring that acts this way is said to obey Hooke's law.

How does the net force depend upon position? When a block hangs on a spring, the spring stretches until its force balances the object's weight as shown in **Figure 6–10a.** The block is then in its equilibrium position. If you pull the block down, as in **Figure 6–10b,** the spring force increases, producing a net force upward. When you let go of the block, it accelerates upward, as in **Figure 6–10c.** But as the spring stretch is reduced, the upward force decreases. In **Figure 6–10d,** the upward force of the spring and the object's weight are equal; there is no acceleration. But with no net force, the block's inertia causes it to continue its upward motion above the equilibrium position. In **Figure 6–10e,** the net force is in the direction opposite the displacement of the block and is directly proportional to the displacement, so the motion is simple harmonic. The block returns to the equilibrium position, as in **Figure 6–10f.**

Again, at this position, the net force is zero and so is the acceleration. Does the block stop? No, it would take a net upward force to slow the block, and that doesn't exist until the block falls below the equilibrium position. When it comes to the position at which it was released, the net force and acceleration are at their maximum in the upward direction. The block moves up and continues to move in this vibratory manner. The period of oscillation, T, depends upon the mass of the block and the strength of the spring, but not on the amplitude of the motion.

The pendulum

The swing of a pendulum also demonstrates simple harmonic motion. A simple pendulum consists of a massive object, called the bob, suspended by a string or rod of length l. After the bob is pulled to one side and released, it swings back and forth, as shown in **Figure 6–11.** The string or rod exerts a tension force, F_T, and gravity exerts the weight force, F_g, on the bob. The vector sum of the two forces produces the net force, shown at three positions in **Figure 6–11.** You can see that the net force is restoring, that is, it is opposite the direction of the displacement of the bob. For small angles (under about 15°) the force is linear to the displacement, so the motion is simple harmonic.

The period of a pendulum of length l is given by the following equation.

$$T = 2\pi\sqrt{\frac{l}{g}}$$

Notice that the period depends only upon the length of the pendulum and the acceleration due to gravity, not on the mass of the bob or the amplitude of oscillation. One application of the pendulum is to measure g, which can vary slightly at different locations on Earth.

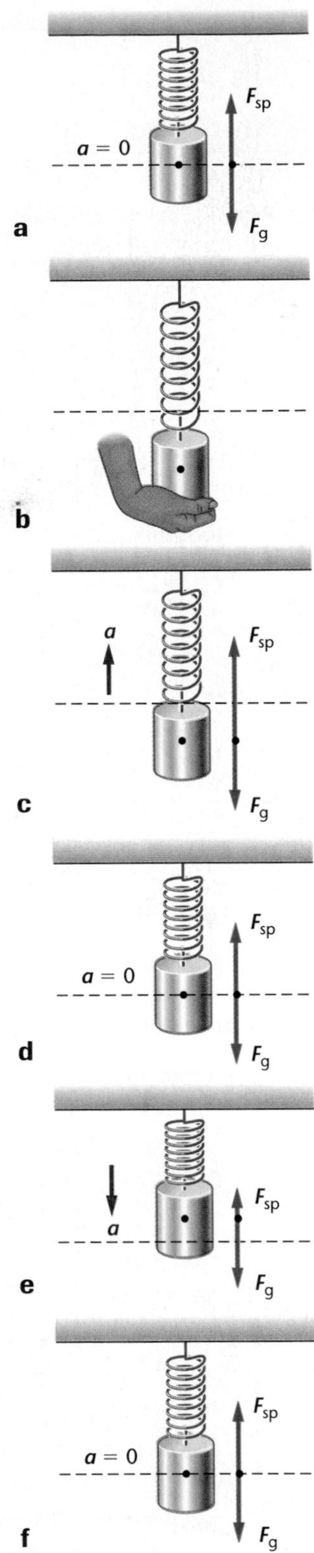

FIGURE 6-10
Simple harmonic motion is demonstrated by the vibration of a block hanging on a spring.

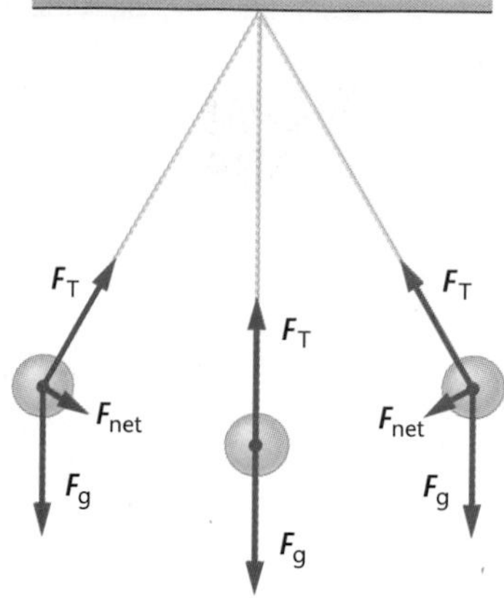

FIGURE 6–11 F_{net}, the vector sum of F_T and F_g, is the restoring force of the pendulum.

Practice Problems

17. What is the length of a pendulum with a period of 1.00 s?

18. Would it be practical to make a pendulum with a period of 10.0 s? Calculate the length and explain.

19. On a planet with an unknown value of g, the period of a 0.65-m -long pendulum is 2.8 s. What is g for this planet?

Resonance

To get a playground swing going, you "pump" it by leaning back and pulling the chains at the same point in each swing, or your friend gives you repeated pushes at just the right times. When small forces are applied at regular intervals to a vibrating or oscillating object, the amplitude of the vibration increases. Such an increase in amplitude is called **mechanical resonance.** The time interval between applications of the force is equal to the period of oscillation. Other familiar examples of resonance include rocking a car to free it from a snow bank and jumping rhythmically on a trampoline or diving board. The large amplitude oscillations caused by resonance can create stresses. Audiences in theater balconies, for example, have damaged the structures by jumping up and down with a period equal to the natural oscillation period of the balcony.

6.2 Section Review

2.1 Compare the force needed to hold a 10.0-kg rock on Earth and on the moon. (The acceleration due to gravity on the moon is 1.62 m/s^2.) Then compare the force needed to throw the same rock horizontally at the same speed in the two locations.

2.2 You take a ride in a fast elevator to the top of a tall building and ride back down while standing on a bathroom scale. During which parts of the ride will your apparent and real weights be the same? During which parts will your apparent weight be less than your real weight? More than your real weight?

2.3 A box is in the back of a pickup truck when the truck accelerates forward. What force accelerates the box? Under what circumstances could the box slide? In which direction?

2.4 A skydiver falling at constant speed in the spread-eagle position opens the parachute. Is the skydiver accelerated? In which direction? Explain your answer using Newton's laws.

2.5 **Critical Thinking** The speed of a pendulum bob is largest when it is directly below the support. Give two ways you could increase this speed.

Physics Lab

The Elevator Ride

Problem

Why do you feel heavier or lighter when riding in an elevator?

Materials

1-kg mass
20-N spring scale
10 cm masking tape

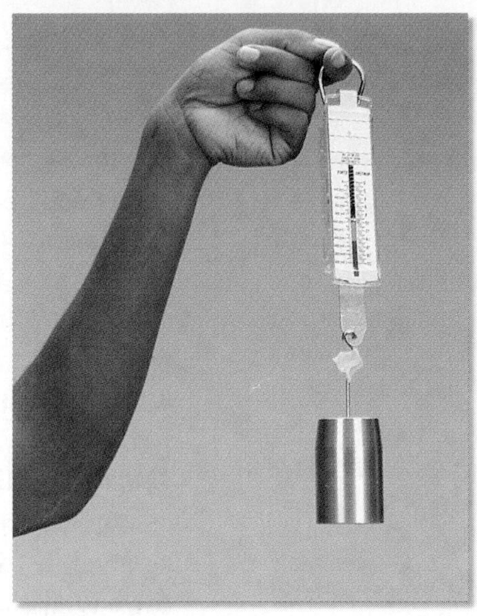

Procedure

1. Imagine that you take an upward elevator ride. Write a few sentences describing when you feel normal, heavier than normal, and lighter than normal. Repeat for a downward elevator ride.
2. Hold the 1-kg mass in your hand and give it an upward elevator ride. Describe when the mass feels normal, heavier than normal, and lighter than normal.
3. Hold the mass in your hand and give it a downward elevator ride. Describe when the mass feels normal, heavier than normal, and lighter than normal.
4. Securely tape the mass to the hook on the spring scale. **Caution:** *A falling mass can cause serious damage to feet or toes.*
5. Start with the mass just above the floor and take it on an upward and then a downward elevator ride.

Data and Observations

1. Watch the spring scale and record the readings for different parts of the ride.

Analyze and Conclude

1. **Interpreting Data** Identify those places in the ride when the spring scale records a normal value for the mass. Describe the motion of the mass. Are the forces balanced or unbalanced?
2. **Interpreting Data** Identify those places in the ride when the spring scale records a heavier value. Which direction is the F_{net}? Which direction is the acceleration?
3. **Interpreting Data** Identify those places in the ride when the spring scale records a lighter value. Which direction is the F_{net}? Which direction is the acceleration?

Apply

1. Do you feel heavier or lighter when riding on an escalator? Explain your answer in terms of the motion and the forces.
2. Identify the places on a roller coaster where you feel heavier or lighter. Explain your answer in terms of the motion and the forces.

6.3 Interaction Forces

You have explored the acceleration given an object when a net force acts on it, $\boldsymbol{a} = \boldsymbol{F}_{net}/m$. You know that forces are exerted on objects by agents, and that forces can be either contact or long-range. But what causes the force? If a rope pulls on a block, something or someone has to pull the rope. If you pull a rope, you feel the rope pulling you. Which is the object? Which is the agent? Long-range forces are similar. If you play with two magnets you feel each magnet pushing or pulling the other. Forces are the pushing or pulling of two objects on each other.

OBJECTIVES

- **Explain** the meaning of interaction pairs of forces and how they are related by Newton's third law.
- **List** the four fundamental forces and **illustrate** the environment in which each can be observed.
- **Explain** the tension in ropes and strings in terms of Newton's third law.

Identifying Interaction Forces

When a fast-moving baseball is caught, the motion of the ball is stopped. That requires a force, a force the catcher exerts on the ball. But the ball also exerts a force on the catcher, a force that can be felt. How do those forces compare? You've probably heard the answer for every action there is an equal and opposite reaction. But what is an action, what is a reaction, and why are they equal?

Systems and the environment

You have already studied the situation diagrammed in **Figure 6–12a.** A system, whose motion you want to study, is isolated, and agents that exert forces on the system are identified. Now, consider the two systems whose motion you want to study, illustrated in **Figure 6–12b.** They are interacting with each other as well as with other agents. Recall that the environment consists of all the other systems whose motion is not being studied.

Now look at the interaction of the catcher's hand with a baseball illustrated in **Figure 6–13.** The ball is one system, the catcher's hand is the other. What forces act on each of the two systems? The weight forces on the ball and the hand, and the force of the arm on the hand, are considered external forces. The two forces, $\boldsymbol{F}_{\text{hand on ball}}$ and $\boldsymbol{F}_{\text{ball on hand}}$, are the forces of interaction between the ball and the hand. Notice the symmetry in the subscripts: hand on ball and ball on hand, or more generally, A on B and B on A.

The forces $\boldsymbol{F}_{\text{A on B}}$ and $\boldsymbol{F}_{\text{B on A}}$ are sometimes called action-reaction pairs of forces. This suggests that one causes the other, but this is not true. The force of the hand on the ball doesn't cause the ball to exert a

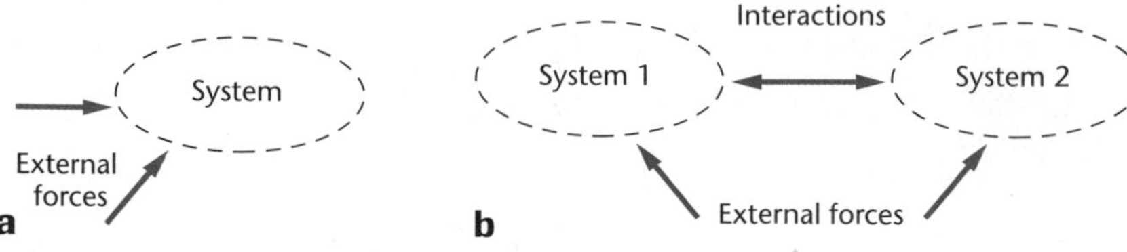

FIGURE 6–12 In **a**, external forces act on the system that is isolated for study. In **b**, two isolated systems are acted on by external forces and they also interact with each other.

force on the hand. The two forces either exist together or not at all. What about the directions and magnitudes of the forces?

Newton's Third Law

According to Newton, an **interaction pair** is two forces that are in opposite directions and have equal magnitude. The force of the catcher's hand on the ball is equal in magnitude and opposite in direction to the force of the ball on the catcher's hand. This is summarized in **Newton's third law** of motion, which states that all forces come in pairs. The two forces in the pair act on different objects and are equal in magnitude and opposite in direction: $\boldsymbol{F}_{\text{A on B}} = -\boldsymbol{F}_{\text{B on A}}$.

To illustrate Newton's third law, consider how a car accelerates. First, treat the car as a system, as in **Figure 6–14.** The car touches the road, so the road exerts contact forces on the car. There is the upward normal force and the forward friction force in the direction of the acceleration. There is also the downward, long-range weight force on the car. To keep the picture simple, ignore forces on the rear car tires.

But all of these forces are part of force pairs. If the road exerts forces on the car, then the car must exert equal and opposite forces on the road. Thus, the car exerts a downward normal force, and a backward friction force on the road. Finally, the car exerts an upward force on Earth. In **Figure 6–14,** dashed lines connect the three pairs of forces.

These forces can be confusing. You accelerate a car by pressing on the "accelerator." Through a variety of gears and rods, the engine turns the wheels. The wheels exert a backwards force on the road. But it is not this force that accelerates the car. First, it is in the wrong direction. Second, it is exerted on the road, not the car. It is the forward force of the road on the car that propels the car forward. If it weren't for the interaction between the car and the road, the car wouldn't move. If you've ever tried to accelerate a car on ice or on loose sand, where the frictional interaction is reduced, you can appreciate the importance of Newton's third law.

Thus, there is a backward force on the road and an upward force on Earth. As a result of this force, does Earth accelerate upward? Consider the simpler case of the interaction of a ball and Earth in the next example problem using the following problem solving strategy.

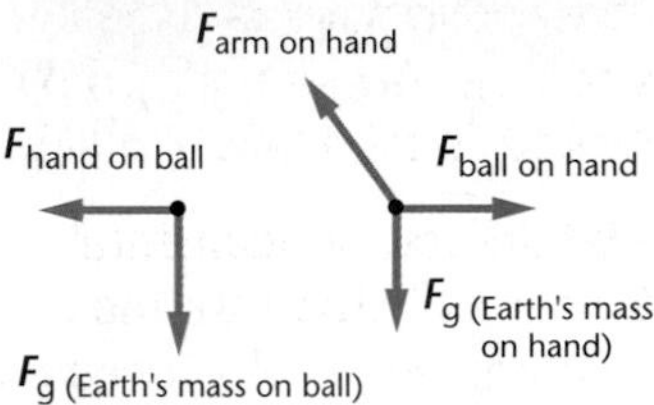

FIGURE 6-13 In addition to the external forces on the two systems, there are forces of interaction between the hand and the ball.

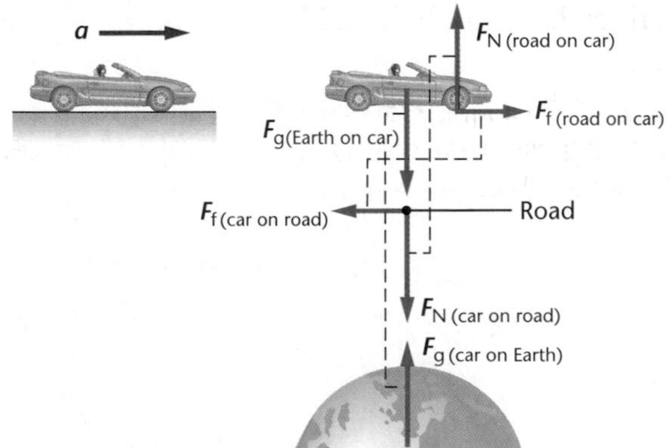

FIGURE 6–14 In the diagram, you can identify three interaction pairs.

Problem Solving Strategy

Interaction Pairs

1. Separate the system or systems from the environment.
2. Draw a pictorial model with coordinate systems for each system and a physical model which includes free-body diagrams for each system.
3. Connect interaction pairs by dashed lines.
4. To calculate your answer, use Newton's second law to relate the net force and acceleration for each system.
5. Newton's third law equates the magnitudes of the force pairs and gives the relative directions.
6. Solve the problem and check the units, signs, and magnitudes for reasonableness.

Example Problem

Earth's Acceleration

When a softball with a mass of 0.18 kg is dropped, its acceleration toward Earth is equal to g, the acceleration due to gravity. What is the force on Earth due to the ball, and what is Earth's acceleration? Earth's mass is 6.0×10^{24} kg.

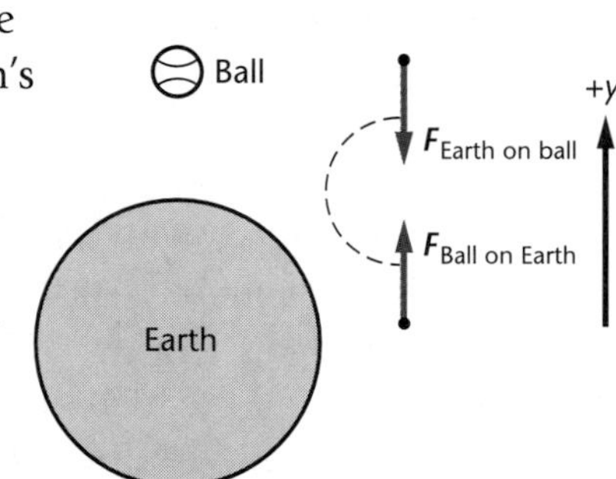

Sketch the Problem

- Draw the forces on the two systems, ball and Earth, and connect them by dotted lines as an interactive pair.

Calculate Your Answer

Known:	**Unknown:**
$m_{\text{ball}} = 0.18$ kg	$F_{\text{Earth on ball}} = ?$
$m_{\text{Earth}} = 6.0 \times 10^{24}$ kg	$a_{\text{Earth}} = ?$
$g = 9.80\ \text{m/s}^2$	

Strategy:	**Calculations:**	
Use Newton's second law to find the weight of the ball.	$F_{(\text{Earth on ball})}$	$= m_{\text{ball}} \times g$ $= (0.18\ \text{kg})(-9.80\ \text{m/s}^2)$ $= -1.8\ \text{N}$
Use Newton's third law to find $F_{(\text{ball on Earth})}$.	$F_{(\text{Earth on ball})}$	$= -F_{\text{Earth on ball}}$ $= -(-1.8\ \text{N}) = +\ 1.8\ \text{N}$
Use Newton's second law to find a_{Earth}.	a_{Earth}	$= \dfrac{F_{\text{net}}}{m_{\text{Earth}}} = \dfrac{1.8\ \text{N}}{6.0 \times 10^{24}\ \text{kg}}$ $= 2.9 \times 10^{-25}\ \text{m/s}^2$

Check Your Answer

- Are the units correct? Performing algebra on the units verifies force in N and acceleration in m/s^2.
- Do the signs make sense? Yes, force and acceleration should be positive for the directions of the force vectors in the diagram.
- Is the magnitude realistic? Because of Earth's large mass, the acceleration should be small.

The acceleration is such a small number that there is no question that, when doing problems involving falling objects, Earth can be treated as part of the environment rather than as a second system.

Practice Problems

20. You lift a bowling ball with your hand, accelerating it upward. What are the forces on the ball? What are the other parts of the action-reaction pairs? On what objects are they exerted?

21. A car brakes to a halt. What forces act on the car? What are the other parts of the action-reaction pairs? On what objects are they exerted?

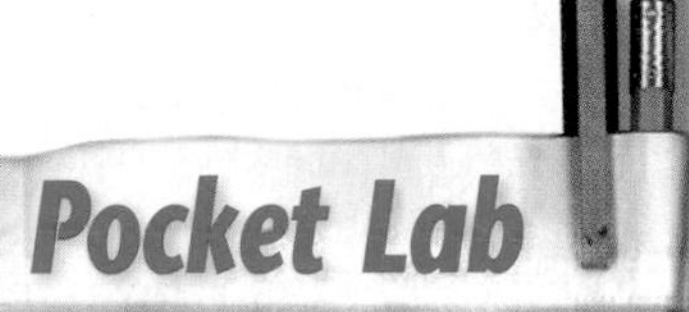

Stopping Forces

Tie two 1-m long strings to the backs of two lab carts and attach 0.2 kg masses to the other ends. Hang the masses over the end of a lab table so that the masses are just above the floor. Add mass to one of the carts so that its mass is about twice its original mass. Predict how the motion of the carts might be different when you push them at the same speed and then let them coast. Try it. Predict how you could change the mass on one of the strings so that the motion of the carts would be the same when given the same initial speed. Test your prediction.

Analyze and Conclude Describe your observations in words and in a motion diagram. Explain your results in terms of inertia, force, mass, and acceleration.

The four fundamental forces

You have investigated several contact interactions and one long-range interaction. Are they all different, or are they the result of a single, fundamental force? At this time, physicists recognize four fundamental forces. One is the gravitational interaction. All objects attract one another through the gravitational interaction, which is an attractive force due to the masses of the objects. You'll learn more about this in Chapter 8. Magnetic forces and the electric forces, such as those that cause static cling, are part of the electromagnetic interaction that you will learn more about in later chapters. The electromagnetic interaction is a force that holds atoms and molecules together, so it is actually responsible for all the contact forces. Two more fundamental interactions occur within the nucleus of the atom. The strong nuclear interaction acts between the protons and neutrons that hold the nucleus together. The weak nuclear interaction makes itself known in some kinds of radioactive decay.

Physicists have long searched for ways in which these interactions might be related. The unification of electric and magnetic interactions was a triumph of nineteenth-century physics. In the 1970s, physicists showed that the electromagnetic and weak interactions were part of a single electro-weak interaction. The ultimate goal is to show that at some level, all four interactions are really one.

How It Works

Piano

The forerunner of the modern piano was invented in the early 1700s by an Italian instrument maker named Bartolommeo Cristofori. Cristofori's instrument consisted of a keyboard and strings that were struck with hammers. The *gravicembalo col piano e forte,* as the instrument was called, was unique in that it had the ability to vary the loudness of its tone, which, of course, depended on the force exerted by the player's fingers.

1. **Keyboard** On nearly all pianos, the keyboard consists of 36 black keys and 52 white keys. The force from the pianist's fingers causes the keys to move a set of felt-covered hammers in the piano's action.

2. **Action** Attached to the keyboard is a complex mechanism called the action, which consists of thousands of wooden parts. Various pieces of the action make it possible for the hammers to strike any of the over 220 strings in a typical piano.

3. **Soundboard** A piano's soundboard is a thin sheet of wood that amplifies the sound created by the vibrating strings.

4. **Strings** The strings of most pianos are made of steel and vary in length from about 15 to 200 cm. The pitch of a tone depends primarily upon the length of the strings, the longest being lowest in pitch. But pitch also depends on the thickness of the strings and their tension.

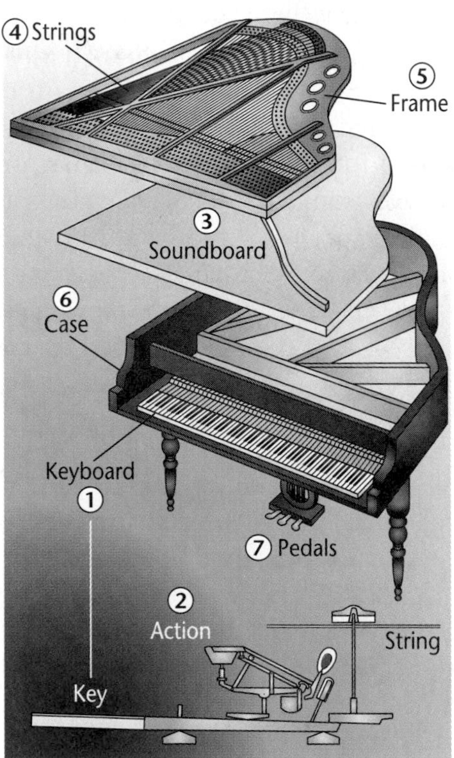

5. **Frame** To sustain the tremendous tension of the hundreds of taut strings, the frame of a piano is made of cast iron.

6. **Case** A piano's case houses the instrument's strings, action, frame, and soundboard and is made of a hard wood.

7. **Pedals** Most pianos have three pedals. Force from the pianist's foot moves the pedals, which varies the quality of the piano's tones.

Thinking Critically

1. The bridge is the wooden ridge that runs diagonally across the soundboard. What do you think is the function of this bridge?
2. The damper pedal lifts all of the dampers in the piano's action. In terms of force and motion, what happens to the strings when the damper pedal is depressed?

Forces of Ropes and Strings

You have already dealt with problems involving the force called tension which is exerted by strings or ropes. For example, **Figure 6–15** shows a bucket hanging from a rope attached to the ceiling. If the rope breaks, the bucket will fall, so before it breaks, there must be forces holding the rope together. The force that the top part of the rope exerts on the bottom part is $\boldsymbol{F}_{\text{T(top on bottom)}}$. Newton's third law states that this force must be part of an interaction pair. The other member of the pair is the force the bottom part exerts on the top, $\boldsymbol{F}_{\text{T(bottom on top)}}$. These forces, equal in magnitude but opposite in direction, are shown in **Figure 6–15.**

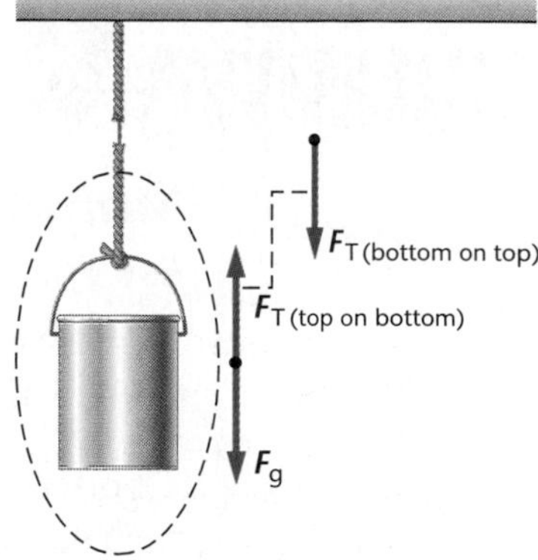

FIGURE 6–15 The tension in the rope is equal to the weight of all objects hanging from it.

The origin of the tension forces holding the rope together are the electromagnetic forces between the molecules and atoms of the rope. At any point in the rope, the tension forces are pulling equally in both directions. But the bucket is in equilibrium, so according to Newton's second law, the net force on the bucket is zero. That is, $F_{\text{T(top on bottom)}} - F_g = 0$, or $F_{\text{T(top on bottom)}} = F_g$. Thus, the tension in the rope is the weight of all objects below it.

Tension forces are also at work in a tug-of-war. If team A on the left is exerting a force of 500 N and the rope doesn't move, then team B on the right must also be pulling with a force of 500 N.

But, what is the tension in the rope? If each team pulls with 500 N of force, is the tension 1000 N? To decide, think of the rope about to break into two pieces. The left-hand end isn't moving, so the net force on it is zero. That is, $F_{\text{T(A on rope)}} = F_{\text{T(right on left)}} = 500$ N. Similarly, $F_{\text{T(B on rope)}} = F_{\text{T(left on right)}} = 500$ N. But the two tensions, $F_{\text{T(right on left)}}$ and $F_{\text{T(left on right)}}$, are an interaction pair, so they are equal and opposite. Thus, the tension in the rope equals the force with which each team pulls, or 500 N.

6.3 Section Review

3.1 You hold a book in your hand, motionless in the air. Identify the forces on the book.

For each force, identify the other force that makes up the interaction pair.

3.2 You now lower the book at increasing speed. Do any of the forces on the book change? Explain. Do their interaction pair partners change? Explain.

3.3 **Critical Thinking** Suppose a curtain prevented each tug-of-war team from seeing its opposing team. One team ties its end of the rope to a tree. If the opposing team pulls with a 500-N force, what is the tension in the rope? Explain.

CHAPTER 6 REVIEW

Key Terms

6.1
- force
- system
- environment
- contact force
- long-range force
- force of gravity
- agent
- free-body diagram
- net force
- Newton's second law
- Newton's first law
- inertia
- equilibrium

6.2
- apparent weight
- weightlessness
- static friction force
- kinetic friction force
- terminal velocity
- simple harmonic motion
- period
- amplitude
- mechanical resonance

6.3
- interaction pair
- Newton's third law

Summary

6.1 Force and Motion

- An object that experiences a push or a pull has a force exerted on it.
- Forces are vector quantities, having both direction and magnitude.
- Forces may be divided into contact and long-range forces.
- Newton's second law states that the acceleration of a system equals the net force on it divided by its mass.
- Newton's first law states that if, and only if, an object has no net force on it, then its velocity will not change.
- The inertia of an object is its resistance to changing velocity.

6.2 Using Newton's Laws

- The weight of an object depends upon the acceleration due to gravity and the mass of the object.
- An object's apparent weight is what is sensed as a result of contact forces on it.
- The friction force acts when two surfaces touch.
- The friction force is proportional to the force pushing the surfaces together.
- An object undergoes simple harmonic motion if the net restoring force on it is directly proportional to the object's displacement.
- Mechanical resonance can greatly increase the amplitude of simple harmonic motion when a small, periodic force acts on an oscillating object at its natural frequency.

6.3 Interaction Forces

- All forces result from interactions between objects.
- Newton's third law states that the two forces that make up an interaction pair of forces are equal in magnitude but opposite in direction and act on different objects.
- Although there are many different forces, they are all forms of the four fundamental forces.

Reviewing Concepts

Section 6.1

1. A physics book is motionless on the top of a table. If you give it a hard push with your hand, it slides across the table and slowly comes to a stop. Use Newton's laws of motion to answer the following questions.
   a. Why does the book remain motionless before the force of the hand is applied?
   b. Why does the book begin to move when your hand pushes hard enough on it?
   c. Why does the book eventually come to a stop?
   d. Under what conditions would the book remain in motion at constant speed?
2. Why do you have to push harder on the pedals of a single-speed bicycle to start it moving than to keep it moving at a constant velocity?
3. Suppose the acceleration of an object is zero. Does this mean that there are no forces acting on it? Give an example supporting your answer.
4. When a basketball player dribbles a ball, it falls to the floor and bounces up. Is a force required to make it bounce? Why? If a force is needed, what is the agent involved?

Section 6.2

5. Before a sky diver opens his parachute, he may be falling at a velocity higher than the terminal velocity he will have after the parachute opens.
   a. Describe what happens to his velocity as he opens the parachute.
   b. Describe his velocity from after his parachute has been open for a time until he is about to land.
6. What is the difference between the period and the amplitude of a pendulum?
7. When an object is vibrating on a spring and passes through the equilibrium position, there is no net force on it. Why is the velocity not zero at this point? What quantity is zero?

Section 6.3

8. A rock is dropped from a bridge into a valley. Earth pulls on the rock and accelerates it downward. According to Newton's third law, the rock must also be pulling on Earth, yet Earth doesn't seem to accelerate. Explain.
9. All forces can be divided into just four fundamental kinds. Name the fundamental force that best describes the following.
   a. holds the nucleus together
   b. holds molecules together
   c. holds the solar system together

Applying Concepts

10. If you are in a car that is struck from behind, you can receive a serious neck injury called whiplash.
    a. Using Newton's laws of motion, explain what happens to cause the injury.
    b. How does a headrest reduce whiplash?
11. Should astronauts choose pencils with hard or soft lead for making notes in space? Explain.
12. If you find a pendulum clock running slightly fast, how can you adjust it to keep better time?
13. Dragsters often set their tires on fire to soften the rubber and increase the coefficient of friction, μ, to nearly 5.0. What role does friction play in accelerating the dragster?
14. What is the meaning of a coefficient of friction that is greater than 1? How would you measure it?
15. Using the model of friction described in this book, would the friction between the tire and the road be increased by a wide rather than a narrow tire? Explain.
16. From the top of a tall building, you drop two table tennis balls, one filled with air and the other with water. Both experience air resistance as they fall. Which ball reaches terminal velocity first? Do both hit the ground at the same time?
17. It is often said that 1 kg equals 2.2 lb. What does this statement mean? What would be the proper way of making the comparison?
18. Which of the four fundamental forces makes paint cling to a wall? Which force makes adhesive sticky? Which force makes wax stick to a car?
19. According to legend, a horse learned Newton's laws. When the horse was told to pull a cart, it refused, saying that if it pulled the cart forward, according to Newton's third law there would be an equal force backwards. Thus, there would be balanced forces, and, according to Newton's second law, the cart wouldn't accelerate. How would you reason with this horse?

Problems

Section 6.1

LEVEL 1

20. A 873-kg (1930 lb) dragster, starting from rest, attains a speed of 26.3 m/s (58.9 mph) in 0.59 s.
    a. Find the average acceleration of the dragster during this time interval.
    b. What is the magnitude of the average net force on the dragster during this time?
    c. Assume that the driver has a mass of 68 kg. What horizontal force does the seat exert on the driver?
21. The dragster in problem 20 completed the 402.3 m (0.2500 mile) run in 4.936 s. If the car had a constant acceleration, what would be its acceleration and final velocity?
22. After a day of testing race cars, you decide to take your own 1550-kg car onto the test track. While moving down the track at 10.0 m/s, you

uniformly accelerate to 30.0 m/s in 10 s. What is the average net force that you have applied to the car during the 10-s interval?

23. A 65-kg swimmer jumps off a 10.0-m tower.

a. Find the swimmer's velocity on hitting the water.

b. The swimmer comes to a stop 2.0 m below the surface. Find the net force exerted by the water.

LEVEL 2

24. The dragster in problem 21 crossed the finish line going 126.6 m/s (283.1 mph). Does the assumption of constant acceleration hold true? What other piece of evidence could you use to see if the acceleration is constant?

25. A race car has a mass of 710 kg. It starts from rest and travels 40.0 m in 3.0 s. The car is uniformly accelerated during the entire time. What net force is exerted on it?

Section 6.2

LEVEL 1

26. What is your weight in newtons?

27. Your new motorcycle weighs 2450 N. What is its mass in kg?

28. A pendulum has a length of 0.67 m.

a. Find its period.

b. How long would the pendulum have to be to double the period?

29. You place a 7.50-kg television set on a spring scale. If the scale reads 78.4 N, what is the acceleration due to gravity at that location?

30. If you use a horizontal force of 30.0 N to slide a 12.0-kg wooden crate across a floor at a constant velocity, what is the coefficient of kinetic friction between the crate and the floor?

31. A 4500-kg helicopter accelerates upward at 2.0 m/s^2. What lift force is exerted by the air on the propellers?

32. The maximum force a grocery sack can withstand and not rip is 250 N. If 20.0 kg of groceries are lifted from the floor to the table with an acceleration of 5.0 m/s^2, will the sack hold?

33. A force of 40.0 N accelerates a 5.0-kg block at 6.0 m/s^2 along a horizontal surface.

a. How large is the frictional force?

b. What is the coefficient of friction?

34. A 225-kg crate is pushed horizontally with a force of 710 N. If the coefficient of friction is 0.20, calculate the acceleration of the crate.

LEVEL 2

35. You are driving a 2500.0-kg car at a constant speed of 14.0 m/s along an icy, but straight, level road. As you approach an intersection, the traffic light turns red. You slam on the brakes. Your wheels lock, the tires begin skidding, and the car slides to a halt in a distance of 25.0 m. What is the coefficient of kinetic friction between your tires and the icy road?

36. A student stands on a bathroom scale in an elevator at rest on the 64th floor of a building. The scale reads 836 N.

a. As the elevator moves up, the scale reading increases to 936 N, then decreases back to 836 N. Find the acceleration of the elevator.

b. As the elevator approaches the 74th floor, the scale reading drops to 782 N. What is the acceleration of the elevator?

c. Using your results from parts a and b, explain which change in velocity, starting or stopping, would take the longer time.

d. What changes would you expect in the scale readings on the ride back down?

37. A sled of mass 50.0 kg is pulled along flat, snow-covered ground. The static friction coefficient is 0.30, and the kinetic friction coefficient is 0.10.

a. What does the sled weigh?

b. What force will be needed to start the sled moving?

c. What force is needed to keep the sled moving at a constant velocity?

d. Once moving, what total force must be applied to the sled to accelerate it at 3.0 m/s^2?

38. The instruments attached to a weather balloon have a mass of 5.0 kg. The balloon is released and exerts an upward force of 98 N on the instruments.

a. What is the acceleration of the balloon and instruments?
b. After the balloon has accelerated for 10 s, the instruments are released. What is the velocity of the instruments at the moment of their release?
c. What net force acts on the instruments after their release?
d. When does the direction of their velocity first become downward?

Section 6.3

LEVEL 1

39. A 65-kg boy and a 45-kg girl use an elastic rope while engaged in a tug-of-war on an icy, frictionless surface. If the acceleration of the girl toward the boy is 3.0 m/s^2, find the magnitude of the acceleration of the boy toward the girl.
40. As a baseball is being caught, its speed goes from 30.0 m/s to 0.0 m/s in about 0.0050 s. The mass of the baseball is 0.145 kg.
    a. What are the baseball's acceleration?
    b. What are the magnitude and direction of the force acting on it?
    c. What is the magnitude and direction of the force acting on the player who caught it?

LEVEL 2

41. A 2.0-kg mass (m_A) and a 3.0-kg mass (m_B) are attached to a lightweight cord that passes over a frictionless pulley, as shown in **Figure 6–16.**

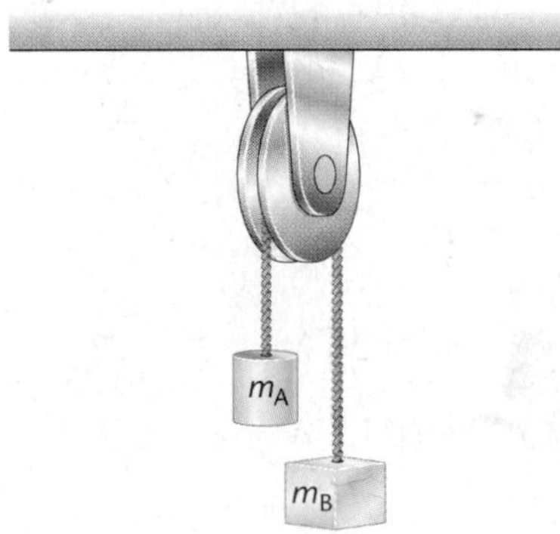

FIGURE 6–16

The hanging masses are free to move. Choose coordinate systems for the two masses with the positive direction up for m_A and down for m_B.
    a. Create a pictorial model.
    b. Create a physical model with motion and free-body diagrams.
    c. Find the acceleration of the smaller mass.
42. Suppose the masses in problem 41 are now 1.00 kg and 4.00 kg. Find the acceleration of the larger mass.
43. Replace the 1.00-kg mass in problem 42 with a 2.00-kg mass. Find the acceleration of the smaller mass.

Critical Thinking Problems

44. The force exerted on a 0.145-kg baseball by a bat changes from 0.0 N to 1.0 × 10^4 N over 0.0010 s, then drops back to zero in the same amount of time. The baseball was going toward the bat at 25 m/s.
    a. Draw a graph of force versus time. What is the average force exerted on the ball by the bat?
    b. What is the acceleration of the ball?
    c. What is the final velocity of the ball, assuming that it reverses direction?

Going Further

Team Project Using the example problems in this chapter as models, write an example problem to solve the following problem. Include Sketch the Problem, Calculate Your Answer (with a complete strategy), and Check Your Answer.

A driver of a 975-kg car, traveling 25 m/s, puts on the brakes. What is the shortest distance it will take for the car to stop? Assume that the road is concrete and that the frictional force of the road on the tires is constant. Assume that the tires don't slip.

interNET CONNECTION

Follow the link on the Glencoe Homepage at **www.glencoe.com/sec/science** to find out more about this chapter.

Sand Blast

More than just the golf ball was put into motion by Lee Trevino's swing. The upward swing of the golf club propels the sand along an upwardly curving path. How would you describe the motion of the sand?

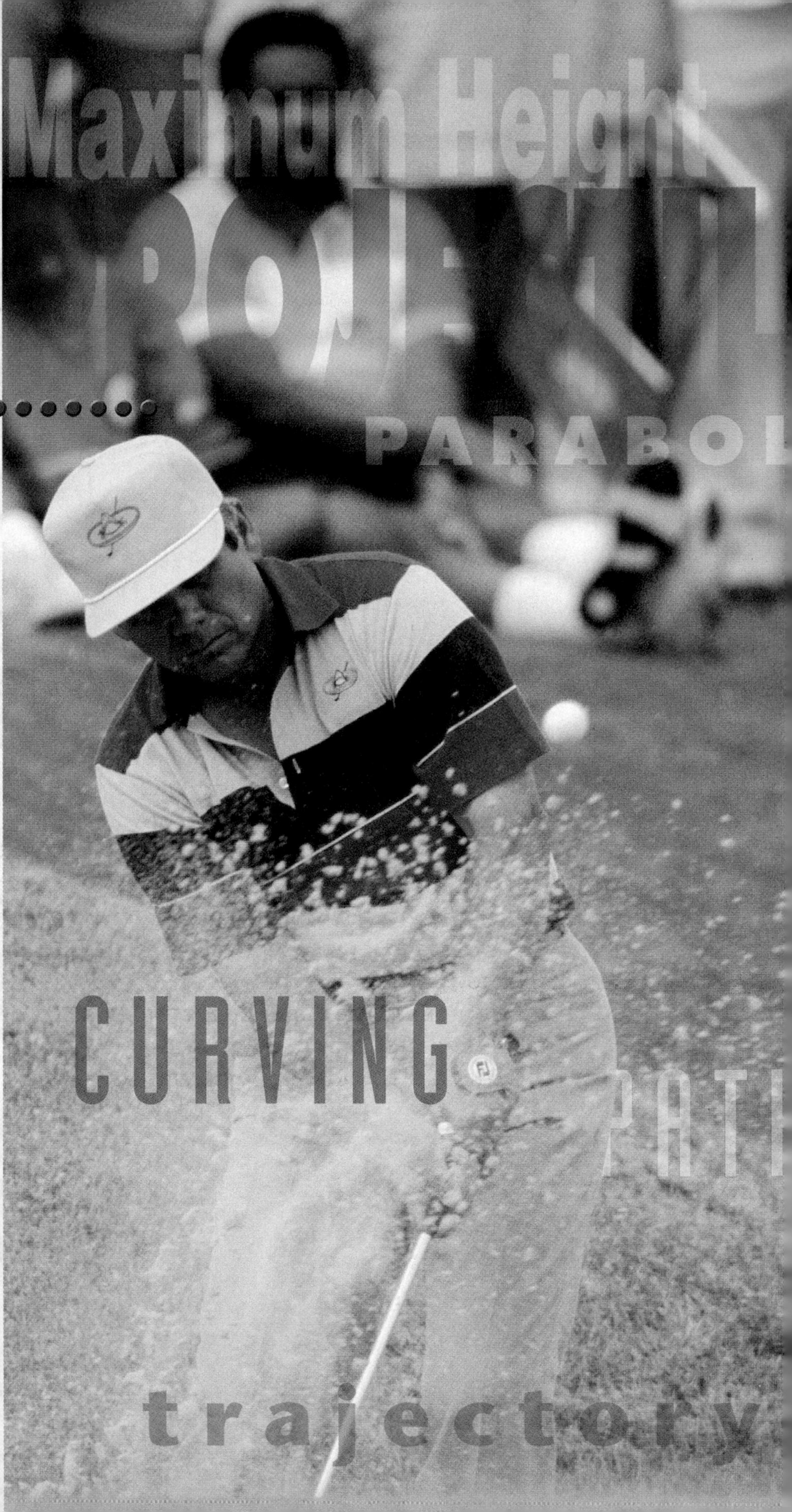

CHAPTER

7 Forces and Motion in Two Dimensions

If you could watch the movement of a golf ball as it leaves a sand trap, you would see that it follows a path similar to that of the sand. All kinds of objects move through the air along such a path. The flights of baseballs and basketballs, arrows, bullets, and rockets follow similar courses. You may be familiar with this curve, called a parabola, from your math class.

Can Newton's laws of motion describe the motion of the sand and the golf ball? They move not only in a horizontal direction, but vertically as well, so the problem becomes more complex. With your knowledge of vectors and Newton's laws, however, you will soon be able to predict how high the golf ball will rise above the ground, where the ball will land, how long it will remain in the air, and how fast it will be moving the instant before it hits the ground. The same equations you used for solving motion problems in one dimension can be applied again to the solution of problems in two dimensions.

WHAT YOU'LL LEARN

- You will use Newton's laws and your knowledge of vectors to analyze motion in two dimensions.
- You will solve problems dealing with projectile and circular motion, and demonstrate your understanding of acceleration and torque.

WHY IT'S IMPORTANT

- The worldwide space program depends fundamentally on the application of Newton's laws to the launching of space vehicles and their guidance into stable orbits.

interNET CONNECTION

Follow the link for this chapter on the Glencoe Homepage at **www.glencoe.com/sec/science** to find out more about forces in two dimensions.

7.1 Forces in Two Dimensions

OBJECTIVES

- **Determine** the force that produces equilibrium when three forces act on an object.
- **Analyze** the motion of an object on an inclined plane with and without friction.

You already know one example of forces in two dimensions. When friction acts between two surfaces, you must take into account both the friction force that is parallel to the surface, and the normal force perpendicular to it. So far, you have considered only motion along the surface. Now you will use your skill in adding vectors to analyze two situations in which the forces on an object are at angles other than 90°.

Equilibrium and the Equilibrant

An object is in equilibrium when the net force on it is zero. When in equilibrium, an object is motionless or moves with constant velocity. According to Newton's laws, the object will not be accelerated because there is no net force on it. You have already added two force vectors to find that the net force is zero. Equilibrium also occurs when the resultant of three or more forces equals a net force of zero.

Figure 7-1a shows three forces exerted on a point object. What is the sum of ***A, B,*** and ***C,*** or what is the net force on the object? Remember that vectors may be moved if you don't change their direction (angle) or length. **Figure 7-1b** shows the addition of the three forces, ***A, B,*** and ***C.*** Note that the three vectors form a closed triangle. There is no net force so the sum is zero and the object is in equilibrium.

Suppose two forces are exerted on an object and the sum is not zero. How could you find a third force that, when added to the other two, would add up to zero? Such a force, one that produces equilibrium, is called the **equilibrant.**

To find the equilibrant, first find the sum of the two forces exerted on the object. This sum is the resultant force, ***R,*** the single force that would produce the same effect as the two individual forces added together. The equilibrant is thus a force with a magnitude equal to the resultant, but in the opposite direction. **Figure 7-2** illustrates this procedure for two vectors, but any number of vectors could be used.

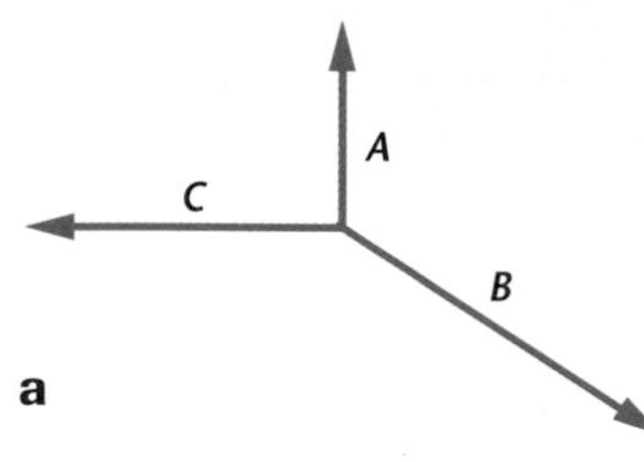

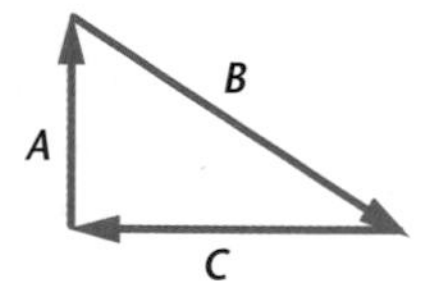

FIGURE 7-1 An object is in equilibrium when all the forces on it add up to zero.

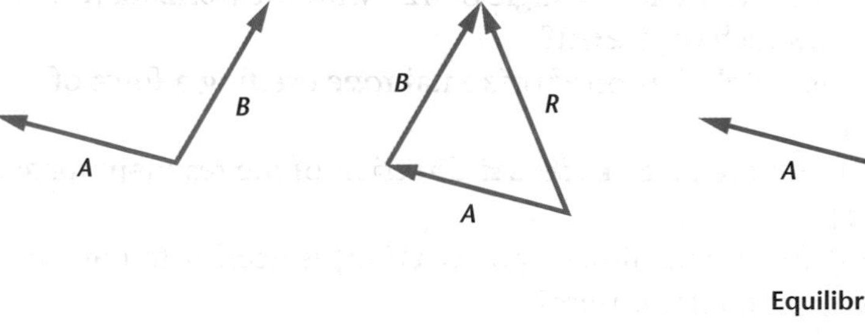

FIGURE 7-2 The equilibrant is the same magnitude as the resultant but opposite in direction.

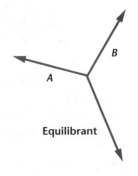

Example Problem

Creating Equilibrium

A 168-N sign is supported in a motionless position by two ropes that each make 22.5° angles with the horizontal. What is the tension in the ropes?

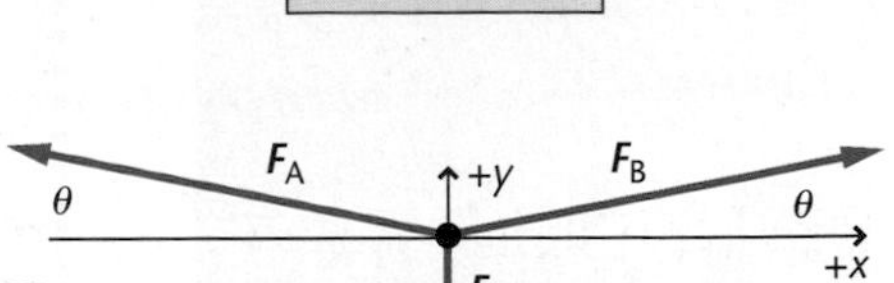

Sketch the Problem

- Draw the ropes at equal angles and establish a coordinate system.
- Draw the free-body diagram with the dot at the origin.

Calculate Your Answer

Known:

$\theta = 22.5°$

$F_g = 168$ N

Unknown:

$F_A = ?$

$F_B = ?$

Strategy:

The sum of the two rope forces and the downward weight force is zero. Write equations for equilibrium in the *x*-direction and in the *y*-direction.

Calculations:

$F_{net,x} = 0$, thus $-F_{Ax} + F_{Bx} = 0$

$-F_A\cos\theta + F_B\cos\theta = 0$

so, $F_A = F_B$

$F_{net,y} = 0$, thus $F_{Ay} + F_{By} - F_g = 0$

$F_A \sin\theta + F_B \sin\theta - F_g = 0$

$2F_A \sin\theta = F_g$

$$F_A = \frac{F_g}{2 \sin 22.5°} = \frac{168 \text{ N}}{2 \times 0.383}$$

$F_A = 2.20 \times 10^2$ N

Check Your Answer

- Is the unit correct? N is the only unit in the calculation.
- Do the signs make sense? Yes, the tension forces are in the positive *y*-direction.
- Is the magnitude realistic? It is greater than the weight of the sign, which is reasonable, because only the small vertical components of $\mathbf{F}_A$ and $\mathbf{F}_B$ are available to balance the sign's weight.

Practice Problems

1. The sign from the preceding example problem is now hung by ropes that each make an angle of 42° with the horizontal. What force does each rope exert?
2. An 8.0-N weight has one horizontal rope exerting a force of 6.0 N on it.
   a. What are the magnitude and direction of the resultant force on the weight?
   b. What force (magnitude and direction) is needed to put the weight into equilibrium?

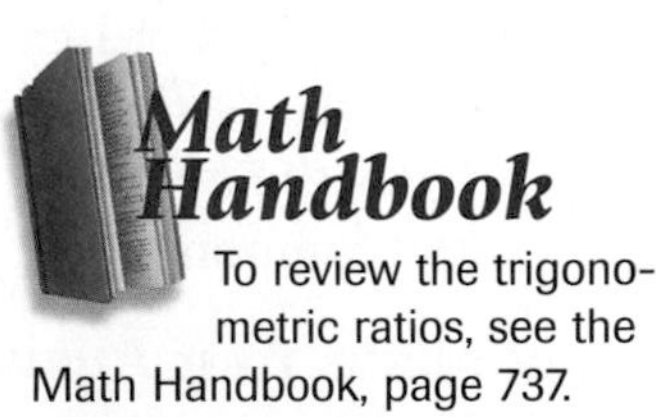

To review the trigonometric ratios, see the Math Handbook, page 737.

FIGURE 7–3 If the *x*-axis is chosen to be parallel to the road, $\boldsymbol{F}_f$ and $\boldsymbol{F}_N$ are parallel to the *x*- and *y*-axes respectively, but $\boldsymbol{F}_g$ points in the direction of the center of Earth as shown.

3. Two ropes pull on a ring. One exerts a 62-N force at 30.0°, the other a 62-N force at 60.0°
   a. What is the net force on the ring?
   b. What are the magnitude and direction of the force that would cause the ring to be in equilibrium?
4. Two forces are exerted on an object. A 36-N force acts at 225° and a 48-N force acts at 315°. What are the magnitude and direction of the equilibrant?

Motion Along an Inclined Plane

The gravitational force is directed toward the center of Earth, in the downward direction. But if a vehicle such as the one in **Figure 7–3** is on a hill, there is a normal force perpendicular to the hill, and the forces of friction that will either speed up or slow down the car are parallel to the hill. What strategy should you use to find the net force that causes the car to accelerate? The most important decision to be made is what coordinate system to use.

Because the direction of the vehicle's velocity and acceleration will be parallel to the hill, one axis, usually the *x*-axis, should be in that direction. The *y*-axis is, as usual, perpendicular to the *x*-axis and perpendicular, or normal, to the surface of the hill.

For such a coordinate system, the normal and friction forces are both in the direction of a coordinate axis, but the weight is not. In most problems, you'll have to find the *x*- and *y*-components of this force.

Example Problem

Components of Weight for an Object on an Incline

A trunk weighing 562 N is resting on a plane inclined 30.0° above the horizontal. Find the components of the weight force parallel and perpendicular to the plane.

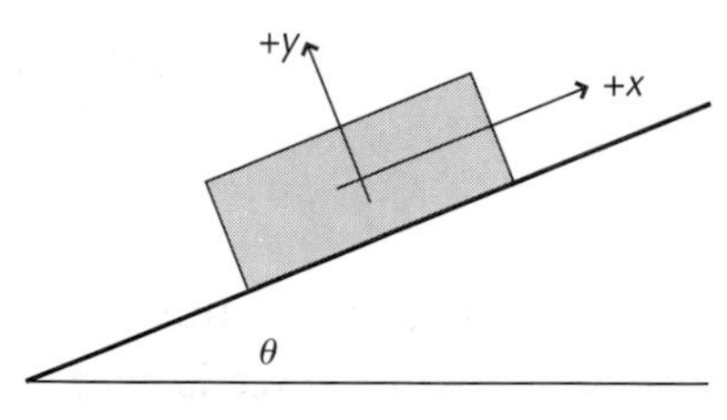

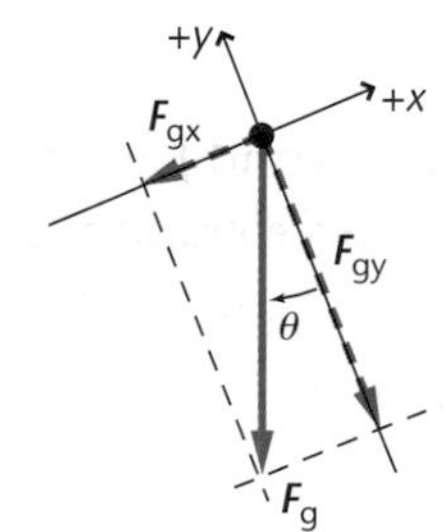

Sketch the Problem

- Include a coordinate system with the positive *x*-axis pointing uphill.
- Draw the free-body diagram showing $\boldsymbol{F}_g$, the components $\boldsymbol{F}_{gx}$ and $\boldsymbol{F}_{gy}$, and the angle θ.

Calculate Your Answer

Known:	**Unknown:**
$F_g = 562$ N	$F_{gx} = ?$
$\theta = 30.0°$	$F_{gy} = ?$

Strategy:

F_{gx} and F_{gy} are negative because they point in directions opposite to the positive axes.

Vector components are scalars, but they have signs indicating their direction relative to the axes.

Calculations:

$F_{gx} = -F_g \sin\theta$

$F_{gx} = -(562\ \text{N})\sin 30.0° = -281\ \text{N}$

$F_{gy} = -F_g \cos\theta$

$F_{gy} = -(562\ \text{N})\cos 30.0° = -487\ \text{N}$

Check Your Answer

- Are the units correct? Only newtons appears in the calculations.
- Do the signs make sense? Yes, the components point in directions opposite to the positive axes.
- Are the magnitudes realistic? The values are less than F_g.

Example Problem

Skiing Downhill

A 62-kg person on skis is going down a hill sloped at 37°. The coefficient of kinetic friction between the skis and the snow is 0.15. How fast is the skier going 5.0 s after starting from rest?

Sketch the Problem

- Circle the system and identify points of contact.
- Establish a coordinate system.
- Draw a free-body diagram.
- Draw a motion diagram showing increasing $\boldsymbol{v}$, and both $\boldsymbol{a}$ and $\boldsymbol{F}_{net}$ in the $+x$ direction.

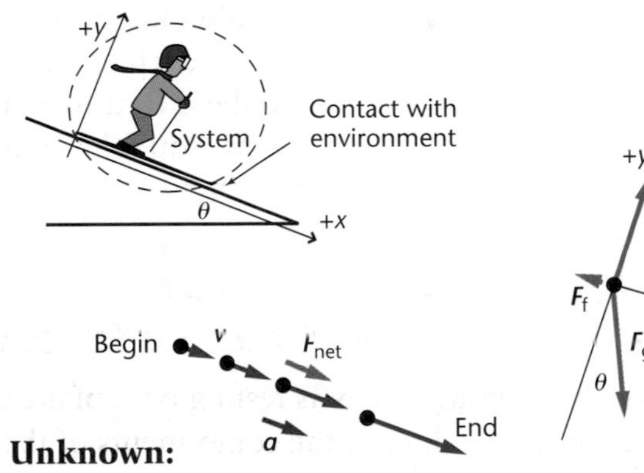

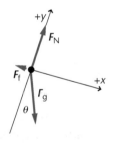

Calculate Your Answer

Known:

$m = 62$ kg $\quad \mu_k = 0.15 \quad t = 5.0$ s

$\theta = 37° \quad v_0 = 0.0$ m/s

Unknown:

$a = ?$

$v = ?$

Strategy:

There is no acceleration in the y-direction, so the net force is zero. Solve for F_N.

Apply Newton's second law of motion to relate acceleration to the downhill force. Solve for a by substituting $\mu_k F_N$ for F_f.

Calculations:

y-direction:

$F_{net,y} = ma_y = 0$

$F_N - F_{gy} = 0$

$F_N = F_{gy} = mg\cos\theta$

x-direction:

$F_{net,x} = ma_x = ma$

$F_{gx} - F_f = ma$

$ma = mg\sin\theta - \mu_k F_N$

$$ma = mg\sin\theta - \mu_k mg\cos\theta$$

$$a = g(\sin\theta - \mu_k\cos\theta)$$

$$a = 9.80\ \text{m/s}^2(\sin 37° - 0.15\cos 37°) = 4.7\ \text{m/s}^2$$

Use velocity-acceleration relation to find speed.

$$v = v_0 + at$$

$$v = 0 + (4.7\ \text{m/s}^2)\ (5.0\text{s}) = 24\ \text{m/s}$$

Check Your Answer

- Are the units correct? Performing algebra on the units verifies that v is in m/s and a is in m/s^2.
- Do the signs make sense? Yes, because v and a are both in the $+x$ direction.
- Are the magnitudes reasonable? The velocity is fast, over 50 mph, but 37° is a steep incline, and the friction with snow is not large.

Practice Problems

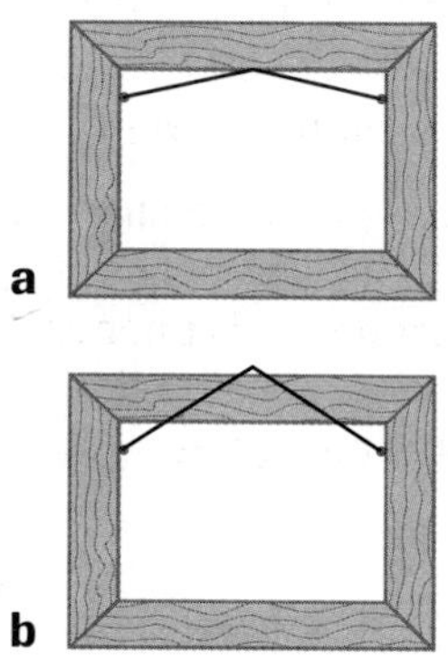

FIGURE 7-4

5. Consider the trunk on the incline in the Example Problem.
   a. Calculate the magnitude of the acceleration.
   b. After 4.00 s, how fast would the trunk be moving?
6. For the Example Problem *Skiing Downhill,* find the x- and y-components of the weight of the skier going downhill.
7. If the skier were on a 30° downhill slope, what would be the magnitude of the acceleration?
8. After the skier on the 37° hill had been moving for 5.0 s, the friction of the snow suddenly increased making the net force on the skier zero. What is the new coefficient of friction? How fast would the skier now be going after skiing for 5.0 s?

7.1 Section Review

1.1 You are to hang a painting using two lengths of wire. The wires will break if the force on them is too great. Should the painting look like **Figure 7–4a** or **b**? Explain.

1.2 One way to get a car unstuck is to tie one end of a strong rope to the car and the other end to a tree. Then push the rope at its midpoint at right angles to the rope. Draw a free-body diagram and explain why even a small force on the rope can exert a large force on the car.

1.3 The skier in the Example Problem finishes the downhill run, turns, and continues to slide uphill for a time. Draw the free-body diagram for the uphill slide. In which direction is the net force?

1.4 **Critical Thinking** Can the coefficient of friction ever have a value such that a skier could slide uphill at a constant velocity? Explain.

Projectile Motion 7.2

A projectile can be a football, a bullet, or a drop of water. No matter what the object is, after a **projectile** has been given an initial thrust, ignoring air resistance, it moves through the air only under the force of gravity. Its path through space is called its **trajectory.** If you know the force of the initial thrust on a projectile, you can figure out its trajectory.

OBJECTIVES

- **Recognize** that the vertical and horizontal motions of a projectile are independent.
- **Relate** the height, time in the air, and initial vertical velocity of a projectile using its vertical motion, then **determine** the range.
- **Explain** how the shape of the trajectory of a moving object depends upon the frame of reference from which it is observed.

Independence of Motion in Two Dimensions

After a golf ball leaves the golf club, what forces are exerted on the ball? If you ignore air resistance, there are no other contact forces on the golf ball. There is only the long-range force of gravity in the downward direction. How does this affect the ball's motion?

Figure 7–5 shows the trajectories of two golf balls. One was dropped, and the other was given an initial horizontal velocity of 2.0 m/s. What is similar about the two paths?

Look at the vertical positions of the balls. At each flash, the heights of the two balls are the same. Because the change in vertical position is the same for both balls, their average vertical velocities during each interval are the same. The increasingly large distances traveled vertically by the two balls, from one time interval to the next, show that the balls are accelerated downward by the force of gravity. Notice that the horizontal motion of the launched ball doesn't affect its vertical motion. A projectile launched horizontally has no initial vertical velocity. Therefore, its vertical motion is like that of a dropped object.

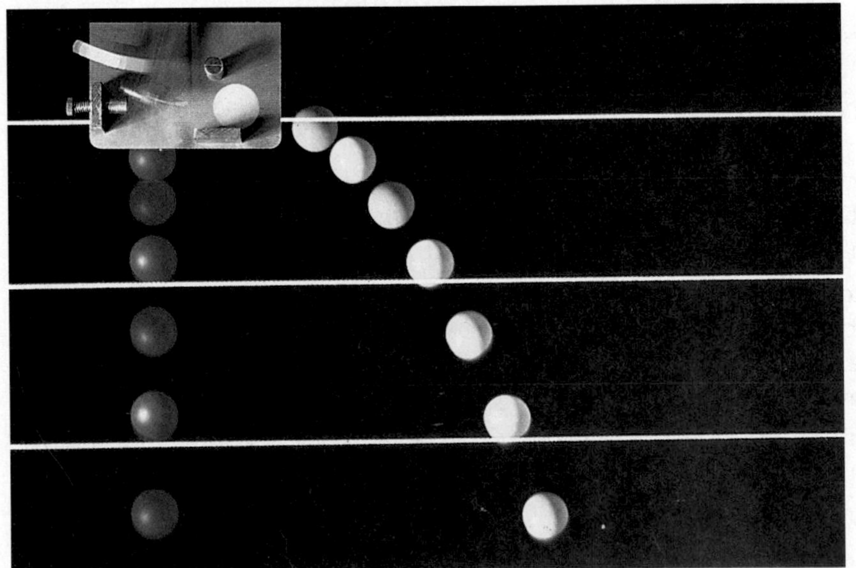

FIGURE 7-5 The ball on the right was given a horizontal velocity; the ball on the left was dropped. The balls were photographed using a strobe light that flashed 30 times each second. Note that the vertical positions of the two balls are the same at each flash of the strobe light.

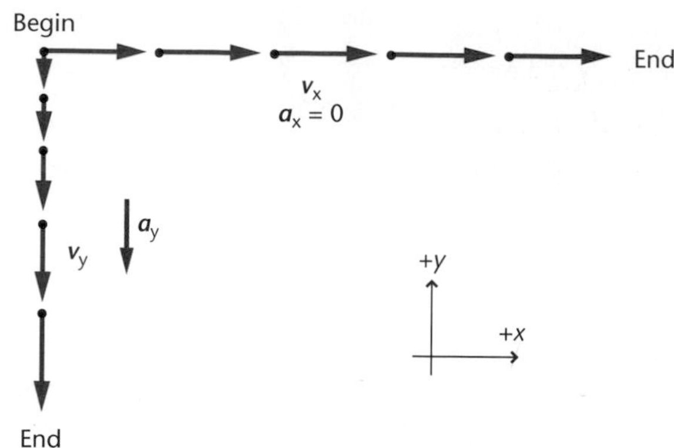

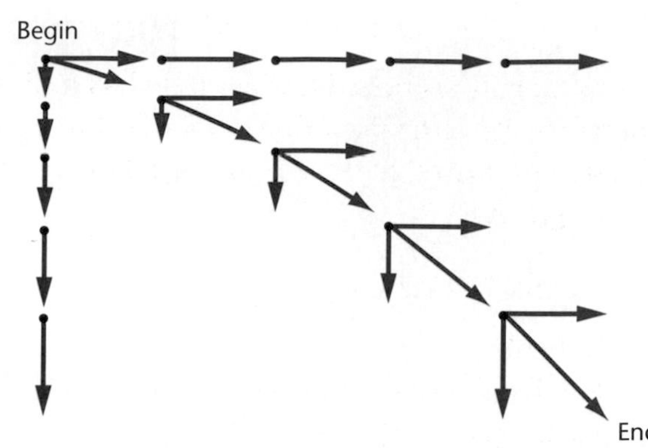

FIGURE 7-6 When the horizontal and vertical components of the ball's velocity are combined in **b,** the resultant vectors are tangent to a parabola.

Separate motion diagrams for the horizontal and vertical motions are shown in **Figure 7–6a.** The vertical motion diagram represents the motion of the dropped ball. The horizontal motion diagram shows the constant velocity in the *x*-direction of the launched ball.

In **Figure 7–6b,** the horizontal and vertical components are added to form the velocity vector for the projectile. You can see how the combination of constant horizontal velocity and uniform vertical acceleration produces a trajectory that has the shape of the mathematical curve called the parabola.

Problem Solving Strategy

Projectile Motion

1. Motion in two dimensions can be solved by breaking the problem into two interconnected one-dimensional problems. For instance, projectile motion can be divided into a vertical motion problem and a horizontal motion problem.
2. The vertical motion of a projectile is exactly that of an object dropped or thrown straight up or down. A gravitational force acts on the object accelerating it by an amount **g**. Review Section 5.4 on Free Fall to refresh your problem solving skills for vertical motion.
3. Analyzing the horizontal motion of a projectile is the same as solving a constant velocity problem. A projectile has no thrust force and air drag is neglected, consequently there are no forces acting in the horizontal direction and thus, no acceleration, $a = 0$. To solve, use the same methods you learned in Section 5.1, Uniform Motion.
4. Vertical motion and horizontal motion are connected through the variable time. The time from the launch of the projectile to the time it hits the target is the same for vertical motion and for horizontal motion. Therefore, solving for time in one of the dimensions, vertical or horizontal, automatically gives you the time for the other dimension.

Pocket Lab

Over the Edge

Obtain two balls, one twice the mass of the other. Predict which ball will hit the floor first when you roll them over the surface of a table with the same speed and let them roll off. Predict which ball will hit the floor farther from the table. Explain your predictions.

Analyze and Conclude Does the mass of the ball affect its motion? Is mass a factor in any of the equations for projectile motion?

Projectiles Launched Horizontally

A projectile launched horizontally has no initial vertical velocity. Therefore, its vertical motion is identical to that of a dropped object. The downward velocity increases regularly because of the acceleration due to gravity.

Example Problem

A Projectile Launched Horizontally

A stone is thrown horizontally at 15 m/s from the top of a cliff 44 m high.

a. How far from the base of the cliff does the stone hit the ground?

b. How fast is it moving the instant before it hits the ground?

Sketch the Problem

- Establish a coordinate system with the launch point labeled "begin"at the origin.
- The point to be labeled "end" is at $y = -44$ m; x is unknown.
- Draw a motion diagram for the trajectory showing the downward acceleration and net force.

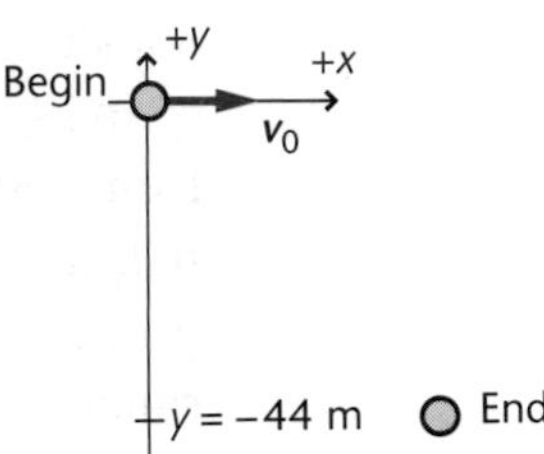

Calculate Your Answer

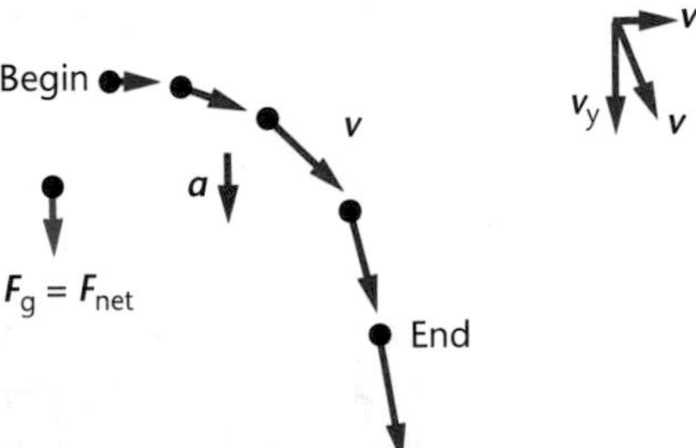

Known:

$x_0 = 0$

$v_{x0} = 15$ m/s

$y_0 = 0$

$v_{y0} = 0$

$a = -g$

Unknown:

x when $y = -44$ m

v at that time

Strategy:

a. Use the equation for the y-position to get and solve an equation for the time the stone is in the air.

b. Velocity is a vector quantity; find the two components, then the magnitude, or speed. Use the Pythagorean relationship to find v.

Calculations:

y-direction:

$v_y = -gt$

$y = y_0 - 1/2gt^2$

$$t = \sqrt{\frac{-2(y - y_0)}{g}} = \sqrt{\frac{-2y}{g}}$$

$$\sqrt{\frac{-2(-44\text{ m})}{-9.80\text{ m/s}^2}} = 3.0\text{ s}$$

x-direction:

$x = x_0 + v_{x0}t$

$x = (15\text{ m/s})(3.0\text{ s}) = 45$ m from the base

$v_y = -gt$

$v_y = -(9.80\text{ m/s}^2)(3.0\text{ s}) = -29\text{ m/s}$

$v = \sqrt{v_x^2 + v_y^2}$

$v = \sqrt{(15\text{ m/s})^2 + (-29\text{ m/s})^2} = 33\text{ m/s}$

Check Your Answer

- Are the units correct? Performing algebra on the units verifies that x is in m and v is in m/s.
- Do the signs make sense? Both x and v should be positive.
- Are the magnitudes realistic? The projectile is in the air 3.0 s. The horizontal distance is about the same magnitude as the vertical distance. The final velocity is larger than the initial horizontal velocity but of the same order of magnitude.

Practice Problems

9. A stone is thrown horizontally at a speed of 5.0 m/s from the top of a cliff 78.4 m high.

a. How long does it take the stone to reach the bottom of the cliff?

b. How far from the base of the cliff does the stone hit the ground?

c. What are the horizontal and vertical components of the stone's velocity just before it hits the ground?

10. How would the three answers to problem 9 change if

a. the stone were thrown with twice the horizontal speed?

b. the stone were thrown with the same speed, but the cliff were twice as high?

11. A steel ball rolls with constant velocity across a tabletop 0.950 m high. It rolls off and hits the ground 0.352 m from the edge of the table. How fast was the ball rolling?

BIOLOGY CONNECTION

Have you ever watched a frog jump? The launch angle of a frog's jump is approximately 45°. Jumping at this angle is innate behavior that helps the frog cover maximum distance on flat ground.

Sand Blast

Projectiles Launched at an Angle

When a projectile is launched at an angle, the initial velocity has a vertical component as well as a horizontal component. If the object is launched upward, then it rises with slowing speed, reaches the top of its path, and descends with increasing speed. This is what happens to the sand in the photo at the beginning of this chapter. **Figure 7–7a** shows the separate vertical and horizontal motion diagrams for the trajectory. The coordinate system is chosen with $+x$ horizontal and $+y$ vertical. Note the symmetry. At each point in the vertical direction, the velocity of the object as it is moving up has the same magnitude as when it is moving down, but the directions of the two velocities are opposite.

Figure 7–7b defines two quantities associated with the trajectory. One is the **maximum height,** which is the height of the projectile when the vertical velocity is zero and the projectile has only its horizontal velocity component. The other quantity depicted is the **range,** R, which is the horizontal distance the projectile travels. Not shown is the **flight time,** which is the time the projectile is in the air. In the game of football, flight time is usually called hang time.

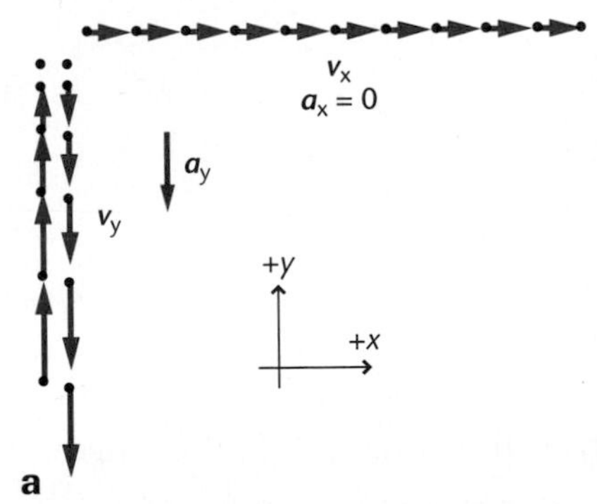

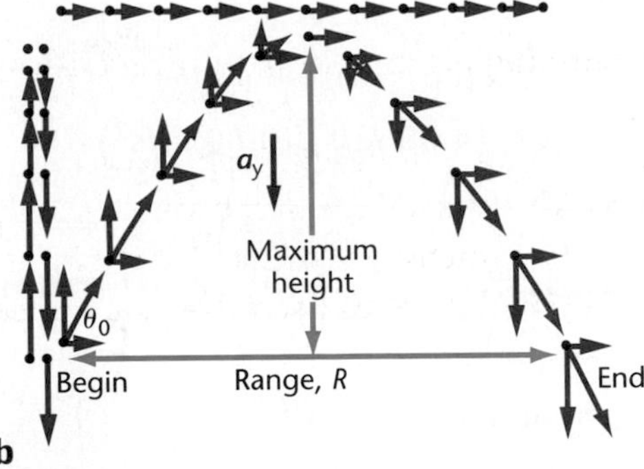

FIGURE 7–7 The vector sum of $\boldsymbol{v}_x$ and $\boldsymbol{v}_y$, at each position, points in the direction of the flight.

Example Problem

The Flight of a Ball

The ball in the strobe photo was launched with an initial velocity of 4.47 m/s at an angle of 66° above the horizontal.

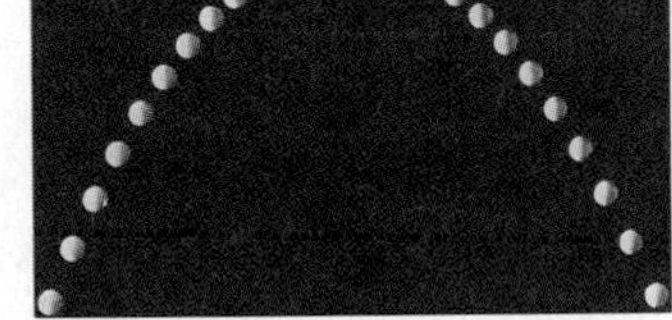

a. What was the maximum height the ball attained?

b. How long did it take the ball to return to the launching height?

c. What was its range?

Sketch the Problem

- Establish a coordinate system. One choice for the initial position of the ball is at the origin.
- Show the positions of the ball at maximum height and at the end of the flight.
- Draw a motion diagram showing the $\boldsymbol{v}$, $\boldsymbol{a}$, and $\boldsymbol{F}_{net}$.

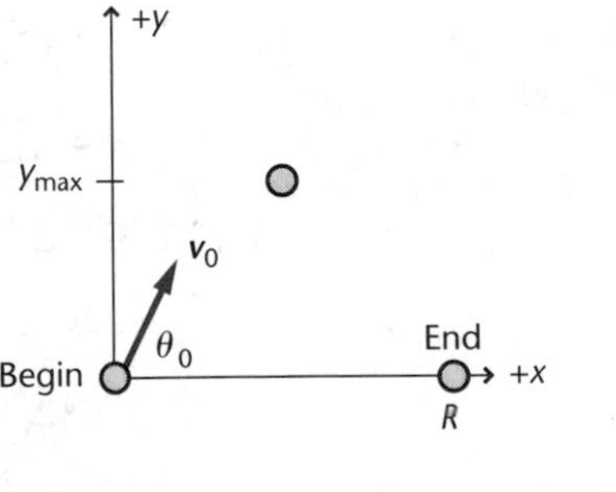

Calculate Your Answer

Known:	Unknown:
$x_0 = 0$	y, when $v_y = 0$
$y_0 = 0$	$t = ?$
$v_0 = 4.47$ m/s	x, when $y = 0$
$\theta_0 = 66°$	
$a = -g$	

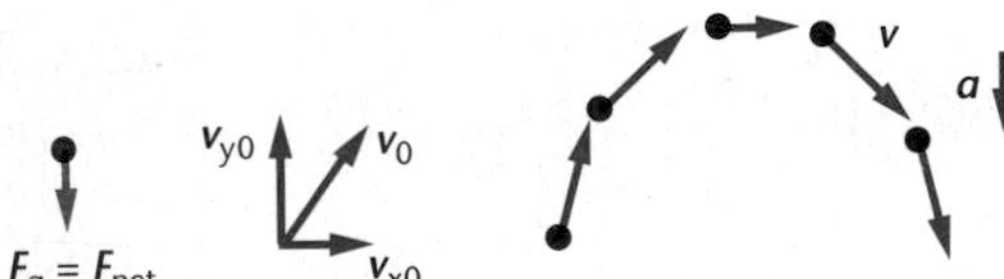

Strategy:

a. Write the equations for the initial velocity components, the velocity components at time t, and the position in both directions. The vertical velocity is zero when the ball reaches maximum height. Solve the velocity equation for the time of maximum height. Substitute this time into the vertical-position equation to find the height.

b. Solve the vertical-position equation for the time of the end of the flight, when $y = 0$.

c. Substitute that time into the equation for horizontal distance to get the range.

Calculations:

y-direction:

$v_{y0} = v_0 \sin \theta_0$

$v_{y0} = (4.47 \text{ m/s}) \sin 66°$

$v_{y0} = 4.08 \text{ m/s}$

$v_y = v_{y0} - gt$

$y = y_0 + v_{y0}t - 1/2gt^2$

x-direction:

$v_{x0} = v_0 \cos \theta_0$

$v_x = v_{x0}$

$x = x_0 + v_{x0}t$

a. When $v_y = 0$, $t = v_{y0}/g$

$t = (4.08 \text{ m/s})/(9.80 \text{ m/s}^2)$

$t = 0.42 \text{ s}$

$y_{max} = v_{y0}t - 1/2gt^2$

$y_{max} = (4.08 \text{ m/s})(0.42 \text{ s}) - 1/2(9.80 \text{ m/s}^2)(0.42 \text{ s})^2 = 0.85 \text{ m}$

b. At landing, $y = 0$

$0 = 0 + v_{y0}t - 1/2gt^2$

$t = 2v_{y0}/g$

$= 2(4.08 \text{ m/s})/(9.80 \text{ m/s}^2)$

$= 0.83 \text{ s}$

c. At this time, $x = R$, the range

$R = v_{x0}t$

$= (4.47 \text{ m/s})(\cos 66°)(0.83 \text{ s})$

$= 1.5 \text{ m}$

Check Your Answer

- Are the units correct? Performing algebra on the units verifies that time is in s, velocity is in m/s, and distance is in m.
- Do the signs make sense? All should be positive.
- Are the magnitudes realistic? Compare them with those in the photo. The calculated flight time is 0.83 s. At 30 flashes/s, this would be 25 flashes, and 25 are visible. The scale of the photo is unknown, as it is, but the ratio of the maximum height to range is (0.85 m)/(1.5 m), or 0.57/1, in the photo.

Practice Problems

12. A player kicks a football from ground level with an initial velocity of 27.0 m/s, 30.0° above the horizontal, as shown in **Figure 7–8.** Find the ball's hang time, range, and maximum height. Assume air resistance is negligible.

13. The player then kicks the ball with the same speed, but at 60.0° from the horizontal. What is the ball's hang time, range, and maximum height?

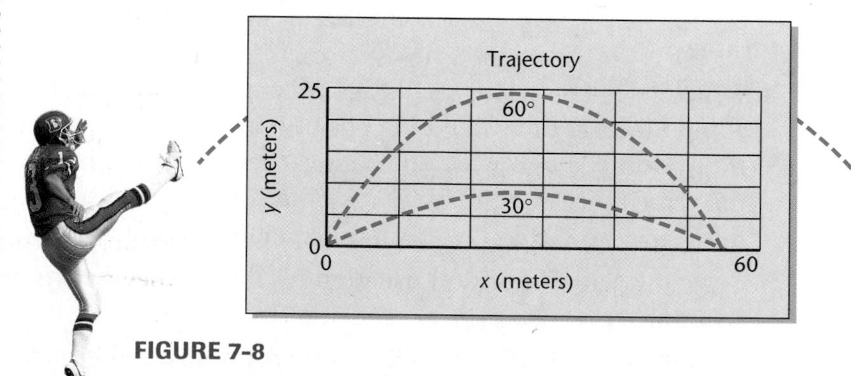

FIGURE 7-8

Trajectories Depend upon the Frame of Reference

Suppose you toss a ball up and catch it while riding in a bus. To you, the ball would seem to go straight up and down. But what would an observer on the sidewalk see? The observer would see the ball leave your hand, rise up, and return to your hand, but because the bus would be moving, your hand also would be moving. The bus, your hand, and the ball would all have the same horizontal velocity. Thus, the trajectory of the ball would be similar to that of the ball in the previous Example Problem. Although you and the observer would disagree on the horizontal motion of the ball, you would agree on the vertical motion. You would both find the vertical velocity, displacement, and time in the air to be the same.

Effects of Air Resistance

The force of air, or air resistance, has been ignored in the analysis of the motion of a projectile, but that doesn't mean that air resistance is unimportant. It's true that for some projectiles, the effect is very small. But for others, the effects are large and very complex. For example, the shape and pattern of dimples on a golf ball have been carefully designed to maximize its range. In baseball, the spin of the ball creates forces that can deflect the ball up, down, or to either side. If the spin is very slow, as in a knuckleball, the interaction of the laces with the air results in a very unpredictable trajectory. Rings, disks, and boomerangs generate enough upward force, or lift, from the air that they seem to float through the air.

Where the Ball Bounces

Place a golf ball in your hand and extend your arm sideways so that the ball is at shoulder height. Drop the ball and have a lab partner start a stopwatch when the ball strikes the floor and stop it the next time the ball strikes the floor. Predict where the ball will hit when you walk at a steady speed and drop the ball. Would the ball take the same time to bounce? Try it.

Analyze and Conclude Where does the ball hit? Does it take more time?

7.2 Section Review

2.1 Two baseballs are pitched horizontally from the same height but at different speeds. The faster ball crosses home plate within the strike zone, but the slower ball is below the batter's knees. Why does the faster ball not fall as far as the slower one?

2.2 An ice cube slides without friction across a table at constant velocity. It slides off and lands on the floor. Draw free-body diagrams of the cube at two points while it is on the table and at two points when it is in the air.

2.3 For the same ice cube, draw motion diagrams showing the velocity and acceleration of the ice cube both when it is on the table and in the air.

2.4 **Critical Thinking** Suppose an object is thrown with the same initial velocity and direction on Earth and on the moon, where g is 1/6 as large as it is on Earth. Will the following quantities change? If so, will they become larger or smaller?

a. v_x
b. time of flight
c. maximum height
d. range

Design Your Own **Physics Lab**

The Softball Throw

Problem

What advice can you give the center fielder on your softball team on how to throw the ball to the catcher at home plate so that it gets there before the runner?

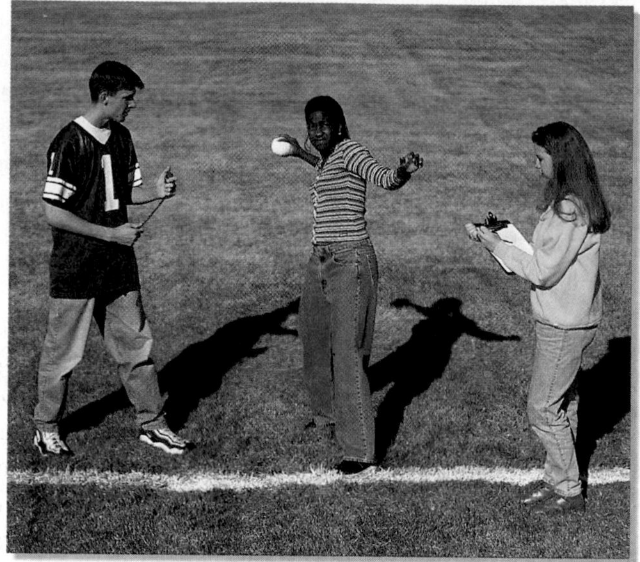

Hypothesis

Formulate a hypothesis using what you know about the horizontal and vertical motion of a projectile to advise the center fielder about how to throw the ball. Consider the factors that affect the time it will take for the ball to arrive at home plate.

Possible Materials

stopwatch
softball
football field or large open area with premeasured distances

Plan the Experiment

1. As a group, determine the variable(s) you want to measure. How do horizontal and vertical velocity affect the range?
2. Who will time the throws? How will you determine the range? Will the range be a constant or a variable?
3. Construct a data table for recording data from all the trial throws and the calculations.
4. **Check the Plan** Make sure your teacher approves your final plan before you proceed.

Analyze and Conclude

1. **Calculating Results** Determine the initial values for v_x and v_y for each trial. Use the Pythagorean theorem to find the value of the initial velocity, v_0, for each throw.
2. **Analyzing Data** Was the range of each person's throw about the same? Did the initial velocity of the throws vary?
3. **Analyzing Data** How did the angle at which the ball was thrown affect the range? The time?
4. **Checking Your Hypothesis** Should the center fielder throw the ball to the catcher at home plate with a larger v_x or v_y?

Apply

1. Why might a kickoff in a football game be made at a different angle than a punt?

Circular Motion 7.3

Can an object be accelerated if its speed remains constant? Yes, because velocity is a vector quantity; just as a change in speed means that there is a change in velocity, so too does a change in direction indicate a change in velocity. Consider an object moving in a circle at constant speed. **Figure 7–9** shows a person riding on a merry-go-round moving at a steady speed. That person is in **uniform circular motion.** So is a sock among the clothes spinning in a washing machine. Uniform circular motion is the movement of an object or point mass at constant speed around a circle with a fixed radius.

OBJECTIVES

- **Explain** the acceleration of an object moving in a circle at constant speed.
- **Describe** how centripetal acceleration depends upon the object's speed and the radius of the circle.
- **Recognize** the direction of the force that causes centripetal acceleration.
- **Explain** how the rate of circular motion is changed by exerting torque on it.

Describing Circular Motion

An object's position relative to the center of the circle is given by the position vector $\boldsymbol{r}$, shown in **Figure 7–10a.** As the object moves around the circle, the length of the position vector doesn't change, but its direction does. To find the object's velocity, you need to find its displacement vector over a time interval. The change in position, or the object's displacement, is represented by $\Delta \boldsymbol{r}$. **Figure 7–10b** shows two position vectors, $\boldsymbol{r}_1$ at the beginning of a time interval, and $\boldsymbol{r}_2$ at the end of the time interval. In the vector diagram, $\boldsymbol{r}_1$ and $\boldsymbol{r}_2$ are subtracted to give the resultant $\Delta \boldsymbol{r}$, the displacement during the time interval. Recall that a moving object's average velocity is $\Delta \boldsymbol{d}/\Delta t$, so for an object in circular motion $\bar{\boldsymbol{v}} = \Delta \boldsymbol{r}/\Delta t$. The velocity vector has the same direction as the displacement but a different length. You can see in **Figure 7–10a** that the velocity is at right angles to the position vector and tangent to its circular path. As the velocity vector moves around the circle, its direction changes but its length remains the same.

What is the direction of the object's acceleration? **Figure 7–11a** shows the velocity vectors $\boldsymbol{v}_1$ and $\boldsymbol{v}_2$ at the beginning and end of a time interval. The difference in the two vectors, $\Delta \boldsymbol{v}$, is found by subtracting the vectors, as shown in **Figure 7–11b.** The acceleration, $\boldsymbol{a} = \Delta \boldsymbol{v}/\Delta t$, is in the same direction as $\Delta \boldsymbol{v}$, that is, toward the center of the circle. As the

FIGURE 7-9 The rider is in uniform circular motion.

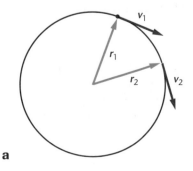

a

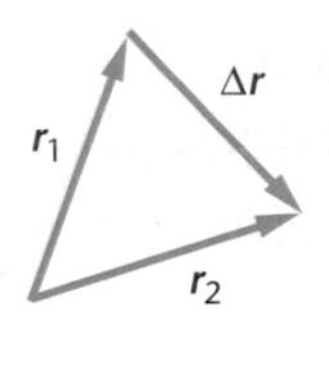

b

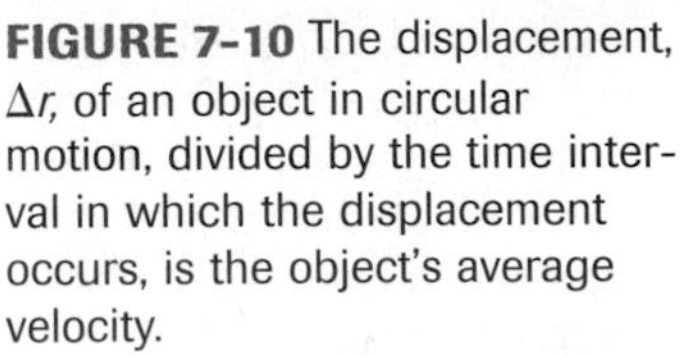

FIGURE 7-10 The displacement, Δr, of an object in circular motion, divided by the time interval in which the displacement occurs, is the object's average velocity.

FIGURE 7-11 The direction of the change in velocity is toward the center of the circle and so the acceleration vector also points to the center of the circle.

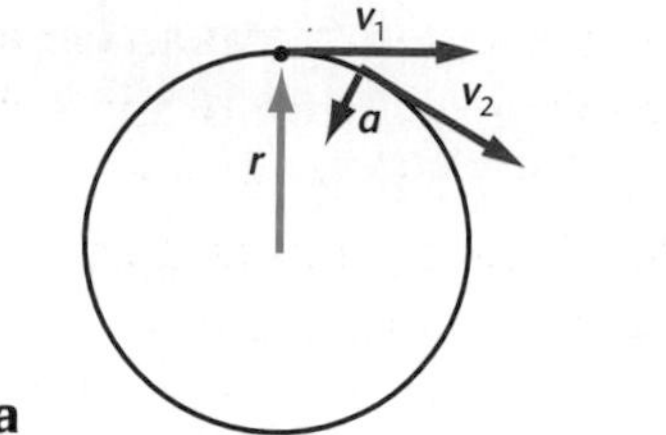

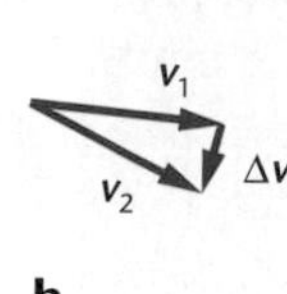

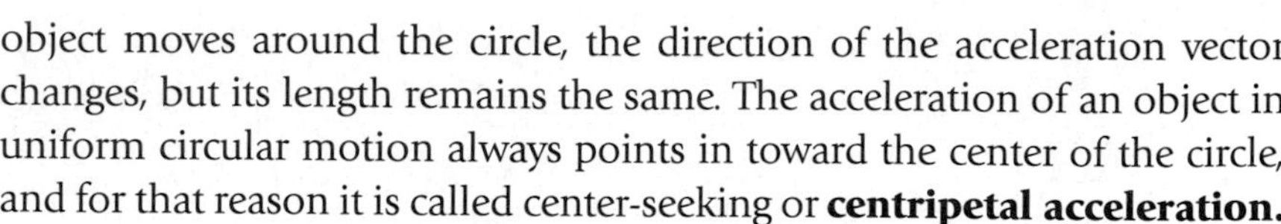

object moves around the circle, the direction of the acceleration vector changes, but its length remains the same. The acceleration of an object in uniform circular motion always points in toward the center of the circle, and for that reason it is called center-seeking or **centripetal acceleration.**

Pocket Lab

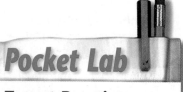

Target Practice

Tie a 1.0-m length of string onto a one-hole rubber stopper. *Note:* Everyone in the classroom should be wearing goggles. Swing the stopper around your head in a horizontal circle. Release the string from your hand when the string is lined up with a spot on the wall. Repeat the experiment until the stopper flies toward the spot on the wall.

Analyze and Conclude Did the stopper travel toward the spot on the wall? What does this indicate about the direction of the velocity compared to the orientation of the string?

Centripetal Acceleration

What is the magnitude of the centripetal acceleration? Compare the triangle made from the position vectors in **Figure 7–10b** with the triangle made by the velocity vectors in **Figure 7–11b.** The angle between $\boldsymbol{r}_1$ and $\boldsymbol{r}_2$ is the same as that between $\boldsymbol{v}_1$ and $\boldsymbol{v}_2$. Therefore, the two triangles formed by subtracting the two sets of vectors are similar triangles, and the ratios of the lengths of two corresponding sides are equal. Thus, $\Delta r/r = \Delta v/v$. The equation is not changed if both sides are divided by Δt.

$$\frac{\Delta r}{r\Delta t} = \frac{\Delta v}{v\Delta t}$$

But $v = \Delta r/\Delta t$ and $a = \Delta v/\Delta t$. Substituting these expressions, the following equation is obtained.

$$\frac{v}{r} = \frac{a}{v}$$

Solve this equation for the acceleration and give it the special symbol a_c for centripetal acceleration.

$$a_c = \frac{v^2}{r}$$

Centripetal acceleration always points toward the center of the circular motion.

How can you measure the speed of an object moving in a circle? One way is to measure its period, T, the time needed for the object to make a complete revolution. During this time, it travels a distance equal to the circumference of the circle, $2\pi r$. The object's speed, then, is represented by $v = 2\pi r/T$.

If this expression is substituted for v in the equation for centripetal acceleration, the following equation is obtained.

$$a_c = \frac{(2\pi r/T)^2}{r} = \frac{4\pi^2 r}{T^2}$$

FIGURE 7-12 When the thrower lets go, the hammer moves in a straight line tangent to the point of release.

What causes an object to have a centripetal acceleration? There must be a net force on the object in the direction of the acceleration, toward the center of the circle. For Earth circling the sun, the force is the sun's gravitational force on Earth. When a hammer thrower swings the hammer, as in **Figure 7-12,** the force is the tension in the chain attached to the massive ball. When a car turns around a bend, the inward force is the frictional force of the road on the tires. Sometimes, the necessary net force that causes centripetal acceleration is called a **centripetal force.**

This, however, can be misleading. To understand centripetal acceleration, you must identify the agent of the contact or long-range force that causes the acceleration. Then you can write Newton's second law for the component in the direction of the acceleration in the following way.

$$F_{net} = ma_c$$

$$F_{net} = \frac{mv^2}{r}$$

$$F_{net} = m\left(\frac{4\pi^2 r}{T^2}\right)$$

When solving circular motion problems, choose a coordinate system in the usual way, with one axis in the direction of the acceleration. But remember that for circular motion, the direction of the acceleration is always toward the center of the circle. Rather than labeling this axis *x* or *y*, call it *c*, for centripetal. The other axis, which, as always, must be perpendicular to the first, is in the direction of the velocity, tangent to the circle. It is labeled *tang* for tangential. The next Example Problem shows the labeled coordinate axes.

In the case of the hammer thrower, the purpose of circular motion is to give the hammer great speed. In what direction does the ball fly when the thrower releases the chain? Once the contact force of the chain is gone, there is no force accelerating the ball toward the center of a circle, so the hammer flies off in the direction of its velocity, which is tangent to the circle. After release, only gravitational force acts on the ball, and it moves like any other projectile.

HELP WANTED

CIVIL ENGINEER

Federal agencies need civil engineers to fill positions as inspectors of roads, bridges, airports, and water/waste systems. An analytical mind, a bachelor's degree in civil engineering, and excellent communication skills (oral and written) are required.

Apply your generalized training or any specialized experience you may have to a position that can be an important step in your professional career. For information contact:

American Society of Civil Engineers
345 East 47th Street
New York, NY 10017

Example Problem

Uniform Circular Motion

A 13-g rubber stopper is attached to a 0.93-m string. The stopper is swung in a horizontal circle, making one revolution in 1.18 s. Find the tension force exerted by the string on the stopper.

Sketch the Problem

- In your sketch, include the radius and the direction of motion.
- Establish a coordinate system labeled *tang* and *c*. Show that the directions of $\boldsymbol{a}$ and $\boldsymbol{F}_T$ are parallel to *c*.

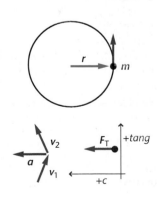

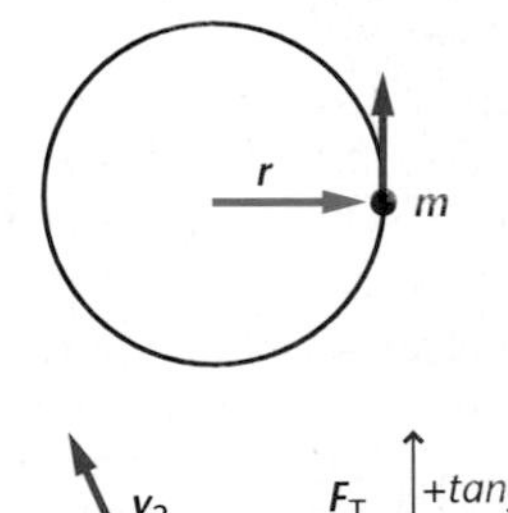

Calculate Your Answer

Known:	Unknown:	Calculations:
$m = 13$ g	$F_T = ?$	$a_c = 4\pi^2 r/T^2$
$r = 0.93$ m		$a_c = 4(3.14)^2(0.93 \text{ m})/(1.18 \text{ s})^2 = 26 \text{ m/s}^2$
$T = 1.18$ s		$F_T = ma = (0.013 \text{ kg})(26 \text{ m/s}^2) = 0.34$ N

Check Your Answer

- Are the units correct? Performing algebra on the units verifies that a is in m/s^2 and F is in N.
- Do the signs make sense? The signs should all be positive.
- Are the magnitudes realistic? The force is almost three times the weight of the stopper, but the acceleration is almost three times that of gravity, so the answer is reasonable.

Practice Problems

14. Consider the following changes to the Example Problem.

a. The mass is doubled, but all other quantities remain the same. What would be the effect on the velocity, acceleration, and force?

b. The radius is doubled, but all other quantities remain the same. What would be the effect on the velocity, acceleration, and force?

c. The period of revolution is half as large, but all other quantities remain the same. What would be the effect on the velocity, acceleration, and force?

15. A runner moving at a speed of 8.8 m/s rounds a bend with a radius of 25 m.

a. What is the centripetal acceleration of the runner?

b. What agent exerts the force on the runner?

16. Racing on a flat track, a car going 32 m/s rounds a curve 56 m in radius.

a. What is the car's centripetal acceleration?

b. What minimum coefficient of static friction between the tires and road would be needed for the car to round the curve without slipping?

Falling Sideways

Will a ball dropped straight down hit the floor before or after a ball that is tossed directly sideways at the same instant? Try it. You may need to repeat the experiment several times before you are sure of your results. Toss the ball sideways and not up or down.

Analyze and Conclude Compare the downward force on each ball. Compare the distance that each ball falls in the vertical direction.

A Nonexistent Force

If a car in which you are riding stops suddenly, you will be thrown forward into your seat belt. Is there a forward force on you? No, because according to Newton's first law, you will continue moving with the same velocity unless there is a net force acting on you. The seat belt applies the force that accelerates you to a stop. Similarly, if a car makes a sharp

Physics & Technology

Looping Roller Coasters

How do roller coaster cars stay on the tracks when they are upside down? The answer involves the speed of the cars, the shape of the loop, and the laws of physics that govern circular motion.

Roller coaster cars are always trying to move in a straight line, but they are prevented from doing so by the tracks which force them along a curving path. Wheels and tracks will remain in contact as long as the forward motion of the cars is great enough, and the curvature of the tracks is tight enough.

The curving tracks and the forward motion of the cars combine to create centripetal acceleration directed toward the center of the curving path. The magnitude of the acceleration is inversely proportional to the radius of the loop. The smaller the radius, the greater the acceleration. Forces associated with centripetal acceleration are measured in units of g. The greater the g force experienced by a roller coaster rider, the heavier the rider feels. The smaller the g force, the lighter the rider feels. Most of the thrills of roller coaster riding result from constantly changing g forces.

Most roller coaster loops are shaped like a teardrop. The upper arc of the loop has a smaller radius of curvature than the lower arc and so the acceleration at the top is greater than at the bottom. The higher acceleration at the top helps maintain contact between the wheels and the track. If the same rate of acceleration were maintained everywhere in the loop, the riders would experience higher g forces than most people would find comfortable.

Thinking Critically Which of Newton's laws of motion explains why the roller coaster car wheels and the tracks stay in contact at the top of the loop? Explain.

left turn, a passenger on the right side may be thrown against the right door. Is there an outward force on the passenger? **Figure 7–13** shows such a car turning to the left as viewed from above. A passenger would continue to move straight ahead if it were not for the force of the door acting in the direction of the acceleration, that is, toward the center of the circle. So there is no outward force on the passenger. The so-called centrifugal, or outward force, is a fictitious, nonexistent force. Newton's laws, which are used in nonaccelerating frames of reference, can explain motion in both straight lines and circles.

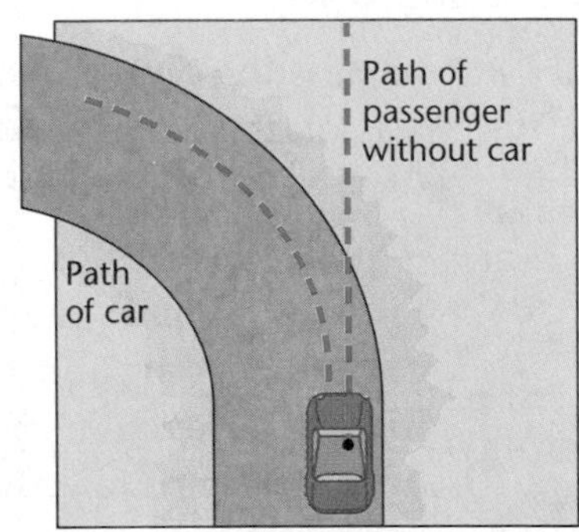

FIGURE 7-13 The passenger would move forward in a straight line if the car did not exert an inward force.

Changing Circular Motion: Torque

In relation to uniform circular motion, you have considered objects such as a person on a merry-go-round and a sock spinning in a washing machine. These can be considered point masses. Now, consider rigid rotating objects. A **rigid rotating object** is a mass that rotates around its own axis. For example, the merry-go-round itself is a rotating object turning on a central axis. A spinning washing machine tub and a revolving

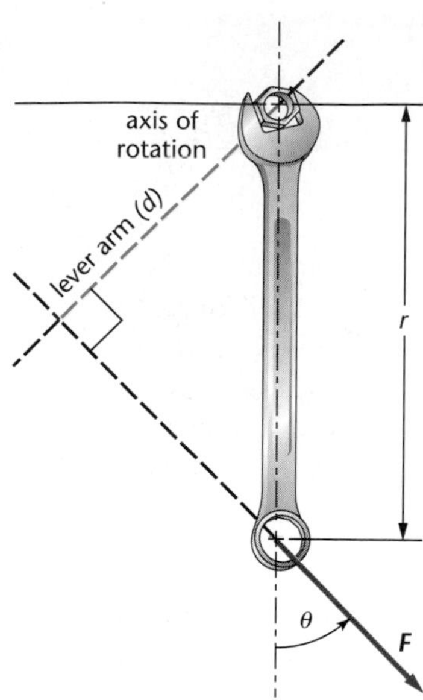

FIGURE 7-14 Torque is the product of the lever arm and the applied force.

door are rotating objects. An ordinary door is also a rigid rotating object, although it usually rotates only through a portion of a circle.

How do you make a door rotate about its axis of rotation, which is its hinges? You exert a force. But where? Pushing on the hinges has little effect, but pushing as far from them as possible starts the door rotating easily. In what direction should you push? Perpendicular to the door is effective; pushing toward the hinges is not.

To open the door most easily, you push at a distance from the hinges (axis of rotation) and in a direction perpendicular to the door. This information about distance and direction is combined in one concept called the **lever arm.** The lever arm in **Figure 7–14** is defined as the perpendicular distance from the axis of rotation to a line along which the force acts. The product of the force and the lever arm is called **torque.** The greater the torque, the greater the change in rotational motion. Thus, torque plays the role of force for rotational motion.

Torque can stop, start, or change the direction of rotation. To stop the door from opening, or to close it, you exert a force in the opposite direction. To start the lug nuts moving when you are changing a tire, you use a lug wrench to apply torque. Sometimes additional length is added to these wrenches to increase the torque.

A seesaw is another example of torque. If a seesaw is balanced, there is no net torque. How, then, do two children, one small, the other large, manage to balance? Each child must exert a torque of the same magnitude but opposite in direction. Because torque is the product of the lever arm, *d*, and the weight of a child, *mg*, the smaller child must sit farther from the axis of rotation, or the pivot point. The seesaw will balance when $m_A g d_A = m_B g d_B$. This concept is the basis for the design of triple beam balances which you may have used in your science courses.

7.3 Section Review

3.1 What is the direction of the force that acts on the clothes in the spin cycle of a washing machine? What exerts the force?

3.2 You are sitting on the back seat of a car that is going around a curve to the right. Sketch motion and free-body diagrams to answer the following questions.

a. What is the direction of your acceleration?

b. What is the direction of the net force acting on you?

c. What exerts that force?

3.3 **Critical Thinking** Thanks to Earth's daily rotation, you always move with uniform circular motion. What supplies the force that accelerates you? How does this motion affect your apparent weight?

CHAPTER 7 REVIEW

Key Terms

7.1
- equilibrant

7.2
- projectile
- trajectory
- maximum height
- range
- flight time

7.3
- uniform circular motion
- centripetal acceleration
- centripetal force
- rigid rotating object
- lever arm
- torque

Summary

7.1 Forces in Two Dimensions

- The force that must be exerted on an object in order to put it in equilibrium is called the equilibrant.
- The equilibrant is found by finding the sum of all forces on an object, then applying a force with the same magnitude but opposite direction.
- An object on an inclined plane has a component of the force of gravity in a direction parallel to the plane; the component can accelerate the object down the plane.

7.2 Projectile Motion

- The vertical and horizontal motions of a projectile are independent.
- Projectile problems are solved by first using the vertical motion to relate height, time in the air, and initial vertical velocity. Then the range, the distance traveled horizontally, is found.
- The range of a projectile depends upon the acceleration due to gravity and upon both components of the initial velocity.

7.3 Circular Motion

- An object moving in a circle at constant speed is accelerating toward the center of the circle (centripetal acceleration).
- Centripetal acceleration depends directly on the square of the object's speed and inversely on the radius of the circle.
- A force must be exerted in the centripetal direction to cause that acceleration.
- The torque that changes the velocity of circular motion is proportional to the force applied and the lever arm.

Reviewing Concepts

Section 7.1

1. Explain how you would set up a coordinate system for motion on a hill.
2. If your textbook is in equilibrium, what can you say about the forces acting on it?
3. Can an object in equilibrium be moving? Explain.
4. What is the sum of three vectors that, when placed tip to tail, form a triangle? If these vectors represent forces on an object, what does this imply about the object?
5. You are asked to analyze the motion of a book placed on a sloping table.
   a. Describe the best coordinate system for analyzing the motion.
   b. How are the components of the weight of the book related to the angle of the table?
6. For the book on the sloping table, describe what happens to the component of the weight force along the table and the friction force on the book as you increase the angle the table makes with the horizontal.
   a. Which components of force(s) increase when the angle increases?
   b. Which components of force(s) decrease?

Section 7.2

7. Consider the trajectory of the ball shown in **Figure 7–15.**
   a. Where is the magnitude of the vertical-velocity component greatest?
   b. Where is the magnitude of the horizontal-velocity component largest?

c. Where is the vertical velocity smallest?
d. Where is the acceleration smallest?

FIGURE 7-15

8. A student is playing with a radio-controlled race car on the balcony of a sixth-floor apartment. An accidental turn sends the car through the railing and over the edge of the balcony. Does the time it takes the car to fall depend upon the speed it had when it left the balcony?
9. An airplane pilot flying at constant velocity and altitude drops a heavy crate. Ignoring air resistance, where will the plane be relative to the crate when the crate hits the ground?

Section 7.3

10. Can you go around a curve
    a. with zero acceleration? Explain
    b. with constant acceleration? Explain.
11. To obtain uniform circular motion, how must the net force depend on the speed of the moving object?
12. If you whirl a yo-yo about your head in a horizontal circle, in what direction must a force act on the yo-yo? What exerts the force?
13. In general, a long-handled wrench removes a stuck bolt more easily than a short-handled wrench does. Explain.

Applying Concepts

14. If you are pushing a lawnmower across the grass, can you increase the horizontal component of the force you exert on the mower without increasing the magnitude of the force? Explain.
15. The transmitting tower of a TV station is held upright by guy wires that extend from the top of the tower to the ground. The force along the guy wires can be resolved into two perpendicular components. Which one is larger?
16. When stretching a tennis net between two posts, it is relatively easy to pull one end of the net hard enough to remove most of the slack, but you need a winch to take the last slack out of the net to make the top almost completely horizontal. Why is this true?
17. The weight of a book on an inclined plane can be resolved into two vector components, one along the plane, the other perpendicular to it.
    a. At what angle are the components equal?
    b. At what angle is the parallel component equal to zero?
    c. At what angle is the parallel component equal to the weight?
18. A student puts two objects on a physics book and carefully tilts the cover. At a small angle, object 1 starts to slide. At a large angle, object 2 begins to slide. Which has the greater coefficient of static friction?
19. A batter hits a pop-up straight up over home plate at an initial velocity of 20 m/s. The ball is caught by the catcher at the same height that it was hit. At what velocity does the ball land in the catcher's mitt? Neglect air resistance.
20. In baseball, a fastball takes about 1/2 s to reach the plate. Assuming that such a pitch is thrown horizontally, compare the distance the ball falls in the first 1/4 s with the distance it falls in the second 1/4 s.
21. You throw a rock horizontally. In a second throw, you gave it even more speed.
    a. How would the time it took to hit the ground be affected? Neglect air resistance.
    b. How would the increased speed affect the distance from the edge of the cliff to where the stone hit the ground?
22. A zoologist standing on a cliff aims a tranquilizer gun at a monkey hanging from a distant tree branch. The barrel of the gun is horizontal. Just as the zoologist pulls the trigger, the monkey lets go and begins to fall. Will the dart hit the monkey? Neglect air resistance.
23. A quarterback threw a football at 24 m/s at a 45° angle. If it took the ball 3.0 s to reach the top of its path, how long was it in the air?
24. You are working on improving your performance in the long jump and believe that the

information in this chapter can help. Does the height you reach make any difference? What does influence the length of your jump?

25. Imagine that you are sitting in a car tossing a ball straight up into the air.
  a. If the car is moving at constant velocity, will the ball land in front of, behind, or in your hand?
  b. If the car rounds a curve at constant speed, where will the ball land?

26. You swing one yo-yo around your head in a horizontal circle, then you swing another one with twice the mass, but you don't change the length of the string or the period. How do the tensions in the strings differ?

27. The curves on a race track are banked to make it easier for cars to go around the curves at high speed. Draw a free-body diagram of a car on a banked curve. From the motion diagram, find the direction of the acceleration.
  a. What exerts the force in the direction of the acceleration?
  b. Can you have such a force without friction?

28. Which is easier for turning a stuck screw, a screwdriver with a large diameter or one with a long handle?

29. Some doors have a doorknob in the center rather than close to the edge. Do these doors require more or less force to produce the same torque as a standard door of the same width and mass?

Problems

Section 7.1

LEVEL 1

30. An object in equilibrium has three forces exerted on it. A 33-N force acts at 90° from the x-axis and a 44-N force acts at 60°. What are the magnitude and direction of the third force?

31. A street lamp weighs 150 N. It is supported by two wires that form an angle of 120° with each other. The tensions in the wires are equal.
  a. What is the tension in each wire?
  b. If the angle between the wires is reduced to 90.0°, what is the tension in each wire?

32. A 215-N box is placed on an inclined plane that makes a 35.0° angle with the horizontal. Find the component of the weight force parallel to the plane's surface.

LEVEL 2

33. Five forces act on an object: (1) 60 N at 90°, (2) 40 N at 0°, (3) 80 N at 270°, (4) 40 N at 180°, and (5) 50 N at 60° What are the magnitude and direction of a sixth force that would produce equilibrium?

34. Joe wishes to hang a sign weighing 750 N so that cable A attached to the store makes a 30.0° angle, as shown in **Figure 7–16.** Cable B is horizontal and attached to an adjoining building. What is the tension in cable B?

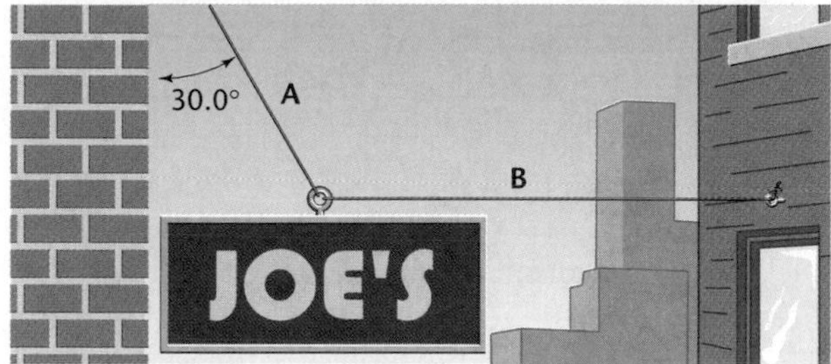

FIGURE 7-16

35. You pull your 18-kg suitcase at constant speed on a horizontal floor by exerting a 43-N force on the handle, which makes an angle θ with the horizontal. The force of friction on the suitcase is 27 N.
  a. What angle does the handle make with the horizontal?
  b. What is the normal force on the suitcase?
  c. What is the coefficient of friction?

36. You push a 325-N trunk up a 20.0° inclined plane at a constant velocity by exerting a 211-N force parallel to the plane's surface.
  a. What is the component of the trunk's weight parallel to the plane?
  b. What is the sum of all forces parallel to the plane's surface?
  c. What are the magnitude and direction of the friction force?
  d. What is the coefficient of friction?

37. What force must be exerted on the trunk in problem 36 so that it would slide down the plane with a constant velocity? In which direction should the force be exerted?

38. A 2.5-kg block slides down a 25° inclined plane with constant acceleration. The block starts from rest at the top. At the bottom, its velocity is 0.65 m/s. The incline is 1.6 m long.
 a. What is the acceleration of the block?
 b. What is the coefficient of friction?
 c. Does the result of either **a** or **b** depend on the mass of the block?

Section 7.2

LEVEL 1

39. You accidentally throw your car keys horizontally at 8.0 m/s from a cliff 64 m high. How far from the base of the cliff should you look for the keys?

40. A toy car runs off the edge of a table that is 1.225 m high. If the car lands 0.400 m from the base of the table,
 a. how long did it take the car to fall?
 b. how fast was the car going on the table?

41. You take a running leap off a high-diving platform. You were running at 2.8 m/s and hit the water 2.6 s later. How high was the platform, and how far from the edge of the platform did you hit the water? Neglect air resistance.

42. An arrow is shot at 30.0° above the horizontal. Its velocity is 49 m/s and it hits the target.
 a. What is the maximum height the arrow will attain?
 b. The target is at the height from which the arrow was shot. How far away is it?

43. A pitched ball is hit by a batter at a 45° angle and just clears the outfield fence, 98 m away. Assume that the fence is at the same height as the pitch and find the velocity of the ball when it left the bat. Neglect air resistance.

LEVEL 2

44. The two baseballs in **Figure 7–17** were hit with the same speed, 25 m/s. Draw separate graphs of y versus t and x versus t for each ball.

45. An airplane traveling 1001 m above the ocean at 125 km/h is to drop a box of supplies to shipwrecked victims below.

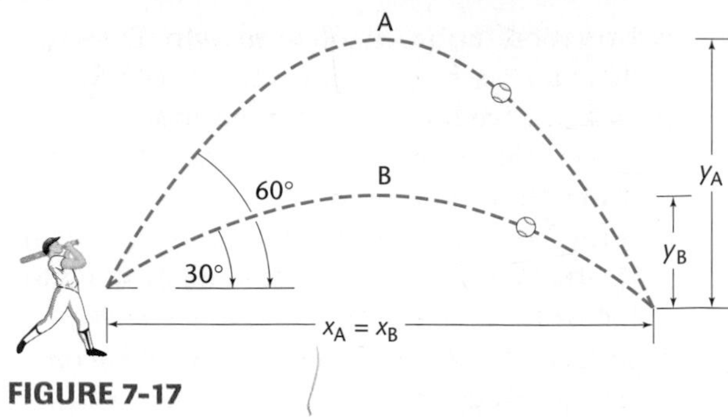

FIGURE 7-17

 a. How many seconds before being directly overhead should the box be dropped?
 b. What is the horizontal distance between the plane and the victims when the box is dropped?

46. Divers in Acapulco dive from a cliff that is 61 m high. If the rocks below the cliff extend outward for 23 m, what is the minimum horizontal velocity a diver must have to clear the rocks?

47. A dart player throws a dart horizontally at a speed of 12.4 m/s. The dart hits the board 0.32 m below the height from which it was thrown. How far away is the player from the board?

48. A basketball player tries to make a half-court jump shot, releasing the ball at the height of the basket. Assuming that the ball is launched at 51.0°, 14.0 m from the basket, what speed must the player give the ball?

Section 7.3

LEVEL 1

49. A 615-kg racing car completes one lap in 14.3 s around a circular track with a radius of 50.0 m. The car moves at constant speed.
 a. What is the acceleration of the car?
 b. What force must the track exert on the tires to produce this acceleration?

50. An athlete whirls in a 7.00-kg hammer tied to the end of a 1.3-m chain in a horizontal circle. The hammer makes one revolution in 1.0 s.
 a. What is the centripetal acceleration of the hammer?
 b. What is the tension in the chain?

51. A coin is placed on a vinyl stereo record making 33 1/3 revolutions per minute.

a. In what direction is the acceleration of the coin?
b. Find the magnitude of the acceleration when the coin is placed 5.0, 10, and 15 cm from the center of the record.
c. What force accelerates the coin?
d. In which of the three radii listed in **b** would the coin be most likely to fly off? Why?

52. According to the *Guinness Book of World Records* (1990) the highest rotary speed ever attained was 2010 m/s (4500 mph). The rotating rod was 15.3 cm (6 in.) long. Assume that the speed quoted is that of the end of the rod.
a. What is the centripetal acceleration of the end of the rod?
b. If you were to attach a 1.0-g object to the end of the rod, what force would be needed to hold it on the rod?

53. Early skeptics of the idea of a rotating Earth said that the fast spin of Earth would throw people at the equator into space. The radius of Earth is about 6400 km. Show why this objection is wrong by calculating
a. the speed of a 97-kg person at the equator.
b. the force needed to accelerate the person in the circle.
c. the weight of the person.
d. the normal force of Earth on the person, that is, the person's apparent weight.

LEVEL 2

54. The carnival ride shown in **Figure 7–18** has a 2.0-m radius and rotates once each 0.90 s.
a. Find the speed of a rider.
b. Find the centripetal acceleration of a rider.

FIGURE 7-18

c. What produces this acceleration?
d. When the floor drops down, riders are held up by friction. Draw motion and free-body diagrams of the situation.
e. What coefficient of static friction is needed to keep the riders from slipping?

55. Friction provides the force needed for a car to travel around a flat, circular race track. What is the maximum speed at which a car can safely travel if the radius of the track is 80.0 m and the coefficient of friction is 0.40?

Critical Thinking Problems

56. A 3-point jump shot is released 2.2 m above the ground, 6.02 m from the basket, which is 3.05 m high. For launch angles of 30° and 60°, find the speed needed to make the basket.

57. For which angle in problem 56 is it more important that the player get the speed right? To explore this question, vary the speed at each angle by 5% and find the change in the range of the throw.

Going Further

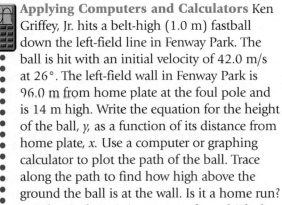

Applying Computers and Calculators Ken Griffey, Jr. hits a belt-high (1.0 m) fastball down the left-field line in Fenway Park. The ball is hit with an initial velocity of 42.0 m/s at 26°. The left-field wall in Fenway Park is 96.0 m from home plate at the foul pole and is 14 m high. Write the equation for the height of the ball, *y*, as a function of its distance from home plate, *x*. Use a computer or graphing calculator to plot the path of the ball. Trace along the path to find how high above the ground the ball is at the wall. Is it a home run?
a. What is the minimum speed at which the ball could be hit and clear the wall?
b. If the initial velocity of a ball is 42.0 m/s, for what range of angles will the ball go over the wall?

interNET CONNECTION

Follow the link on the Glencoe Homepage at **www.glencoe.com/sec/science** to find out more about this chapter.

Kepler's planetar laws of motio

gravitational force

What Goes Up, Must Come Down

Pathfinder **ferried its rover, *Sojourner,* across millions of kilometers of space to land on Mars. Why were NASA scientists certain that the law of gravitation that's valid on Earth works everywhere in the solar system?**

CHAPTER

8 Universal Gravitation

Why do objects fall toward Earth? Ancient Greek scientists believed that objects simply either rose or fell according to their nature; such things as hot air and smoke rose, while others, such as rocks and shoes, fell. The Greeks gave the names *levity,* meaning lightweight, and *gravity,* meaning heavy, to these properties. If you ask a friend why things fall, he or she will probably say, "because of gravity." But, how does the name *gravity* explain why objects fall to Earth?

Almost 400 years ago, Galileo wrote in response to a statement that "gravity" is why stones fall downward,

> *What I am asking you for is not the name of the thing, but its essence, of which essence you know not a bit more than you know about the essence of whatever moves the stars around . . . we do not really understand what principle or what force it is that moves stones downward.*

During the twentieth century, Albert Einstein gave a much different and deeper description of the gravitational attraction. Today, however, we still know only how things fall, not why.

WHAT YOU'LL LEARN

- You will learn the nature of the gravitational force.
- You will relate Kepler's laws of planetary motion to Newton's laws of motion.
- You will describe the orbits of planets and satellites using the law of universal gravitation.

WHY IT'S IMPORTANT

- Without a knowledge of universal gravitation, space travel and an understanding of planetary motion would be impossible.

interNET CONNECTION

Follow the link for this chapter on the Glencoe Homepage at **www.glencoe.com/sec/science** to find out more about universal gravitation.

8.1 Motion in the Heavens and on Earth

OBJECTIVES

- **Relate** Kepler's laws of planetary motion to Newton's law of universal gravitation.
- **Calculate** the periods and speeds of orbiting objects.
- **Describe** the method Cavendish used to measure G and the results of knowing G.

We know how objects move on Earth. We can describe and even calculate projectile motion. Early humans could not do that, but they did notice that the motions of stars and other bodies in the heavens were quite different. Stars moved in regular paths. Planets—or wanderers, as they were called—moved through the sky in much more complicated paths. Comets were even more erratic. These mysterious bodies spouting bright tails appeared without warning. Because of the work of Galileo, Kepler, Newton, and others, we now know that all of these objects follow the same laws that govern the motion of golf balls and other objects here on Earth.

Observed Motion

As a boy of 14 in Denmark, Tycho Brahe (1546–1601) observed an eclipse of the sun on August 21, 1560, and vowed to become an astronomer. In 1563, he observed two planets in conjunction, that is, located at the same point in the sky. The date of that event as predicted by all the books of that period was off by two days, so Brahe decided to dedicate his life to making accurate predictions of astronomical events.

Brahe studied astronomy as he traveled throughout Europe for five years. In 1576, he persuaded King Frederick II of Denmark to give him the island of Hven as the site for the finest observatory of its time. Using huge instruments like those shown in **Figure 8–1,** Brahe spent the next 20 years carefully recording the exact positions of the planets and stars.

FIGURE 8–1 Among the huge astronomical instruments that Tycho Brahe had constructed to use at Hven **(a)** were an astrolabe **(b)** and a sextant **(c).**

a

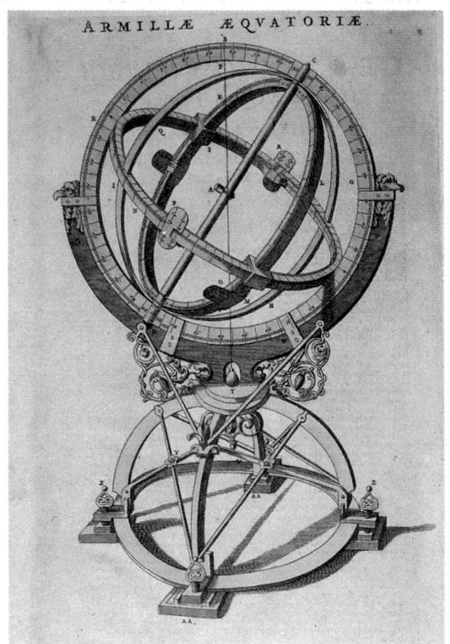

b

c

Kepler's laws

In 1597, after falling out of favor with the new Danish king, Brahe moved to Prague. There, he became the astronomer to the court of Emperor Rudolph of Bohemia where, in 1600, a 29-year-old German named Johannes Kepler (1571–1630) became one of his assistants. Although Brahe still believed strongly that Earth was the center of the universe, Kepler wanted to use a sun-centered system to explain Brahe's precise data. He was convinced that geometry and mathematics could be used to explain the number, distance, and motion of the planets. By doing a careful mathematical analysis of Brahe's data, Kepler discovered three mathematical laws that describe the behavior of every planet and satellite. **Kepler's laws of planetary motion** can be stated as follows.

1. The paths of the planets are ellipses, with the sun at one focus.

2. An imaginary line from the sun to a planet sweeps out equal areas in equal time intervals. Thus, planets move faster when they are closer to the sun and slower when they are farther away from the sun, as illustrated in **Figure 8–2.**

3. The square of the ratio of the periods of any two planets revolving about the sun is equal to the cube of the ratio of their average distances from the sun. Thus, if T_A and T_B are the planets' periods, and r_A and r_B are their average distances from the sun, the following is true.

$$\left(\frac{T_A}{T_B}\right)^2 = \left(\frac{r_A}{r_B}\right)^3$$

Note that the first two laws apply to each planet, moon, or satellite individually. The third law, however, relates the motion of several satellites about a single body. For example, it can be used to compare the distances and periods of the planets about the sun. It also can be used to compare distances and periods of the moon and artificial satellites orbiting around Earth. **Table 8–1** on the next page shows some of these data.

Strange Orbit

Does the moon affect the motion of Earth in its orbit around the sun? Make your prediction. Then, build the following model planet and moon system. Push a small ball of clay onto the end of a drinking straw to represent the moon. Push a larger ball of clay, representing the planet, onto the opposite end. Tape a piece of string to the balance point on the straw so that the straw will stay parallel to the table when it is lifted. Give the moon a gentle push so that it moves in a slow circle.

Analyze and Conclude Does the planet move in response to the motion of the moon? What effect would a more massive moon have on the planet? What might you conclude about Earth's motion?

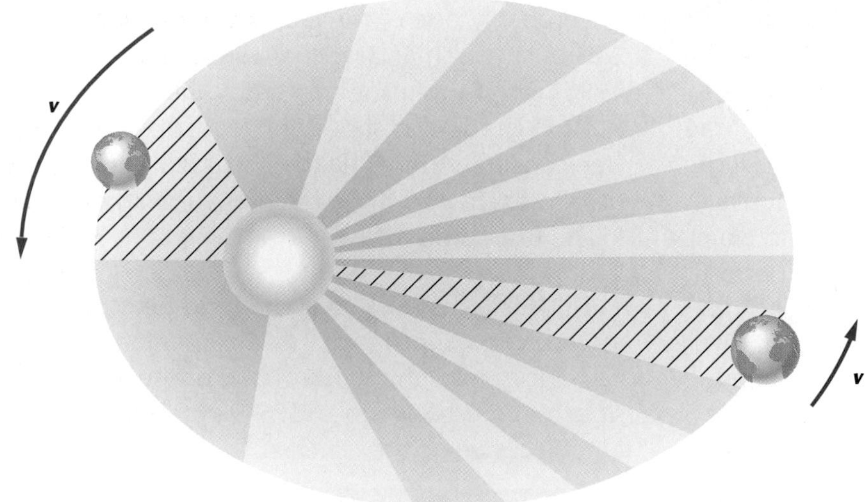

FIGURE 8–2 An imaginary line from Earth to the sun sweeps out equal areas each second, whether Earth is close to or far from the sun.

TABLE 8–1

Planetary Data

Name	Average Radius (m)	Mass (kg)	Mean Distance from Sun (m)
Sun	696×10^{6}	1.99×10^{30}	–
Mercury	2.44×10^{6}	3.30×10^{23}	5.79×10^{10}
Venus	6.05×10^{6}	4.87×10^{24}	1.08×10^{11}
Earth	6.38×10^{6}	5.97×10^{24}	1.50×10^{11}
Mars	3.40×10^{6}	6.42×10^{23}	2.28×10^{11}
Jupiter	71.5×10^{6}	1.90×10^{27}	7.78×10^{11}
Saturn	60.3×10^{6}	5.69×10^{26}	1.43×10^{12}
Uranus	25.6×10^{6}	8.66×10^{25}	2.87×10^{12}
Neptune	24.8×10^{6}	1.03×10^{26}	4.50×10^{12}
Pluto	1.15×10^{6}	1.5×10^{22}	5.91×10^{12}

Physics & Technology

Global Positioning Systems

Do you find it difficult to navigate around town? If so, then a global positioning system, or GPS, might be just what you need. A GPS consists of two parts: a system of transmitters and a receiver. Transmitters aboard two dozen Earth-orbiting satellites send out radio signals that give the exact time and the location of the satellites when the signals are sent. A handheld GPS receiver, which is about the size of a pocket calculator, determines the time it takes the signals to arrive from the satellites. The receiver then calculates the distance to each satellite. When the distances to at least four different satellites are known, the exact location of the person or object using the receiver can be calculated by triangulation. Then, the latitude, longitude, and altitude—all within about 16 m—of the person or object are displayed on the receiver. And because each satellite carries one or more clocks, which are set to agree with the atomic clocks on Earth, the user of a GPS can tell the time to within a few billionths of a second!

Initially, global positioning systems were used primarily by the U.S. Department of Defense. Today, however, for a few hundred dollars, a person can purchase a GPS receiver that will allow him or her to hike the Rockies or sail the Pacific without getting lost. Geologists who use a GPS are able to measure the rates at which Earth's landmasses are moving. Biologists are trying to use a GPS to track grizzly bears in Yellowstone National Park.

Probably one of the best-known uses of a GPS by a search-and-rescue team occurred in 1995, when the plane piloted by Captain Scott O'Grady was shot down over Bosnia-Herzegovina. After O'Grady had spent four days in enemy territory, rescuers were able to locate him using a GPS.

Thinking Critically In addition to determining an object's location, a GPS can determine its velocity to within 0.03 m/s. Propose a possible method by which a GPS can determine an object's velocity.

Physics Lab

The Orbit

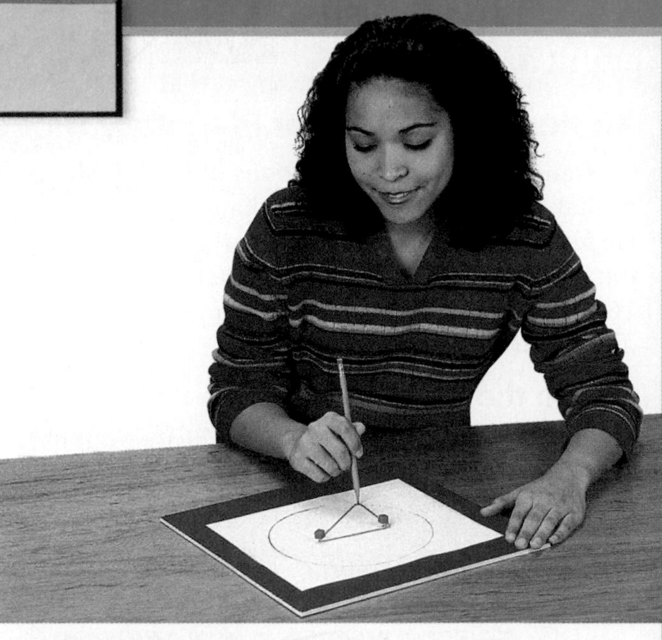

Problem

How does the gravitational force vary at different points of an elliptical orbit?

Materials

2 pushpins
21-cm × 28-cm piece of cardboard or corkboard
sheet of paper
30-cm piece of string or thread
pencil
metric ruler

Procedure

1. Place the paper on top of the cardboard. Push the pushpins into the paper and cardboard so that they are between 7 and 10 cm apart.
2. Make a loop with the string. Place the loop over the two pushpins. Keep the loop tight as you draw the ellipse, as shown.
3. Remove the pins and string. Draw a small star centered at one of the pinholes.
4. Draw the position of a planet in the orbit where it is farthest from the star. Measure and record the distance from this position to the center of the star.
5. Draw a 1-cm-long force vector from this planet directly toward the star. Label this vector 1.0 ***F***.
6. Draw the position of a planet when it is nearest the star. Measure and record the distance from this position to the star's center.

Data and Observations	
Farthest Distance	
Nearest Distance	

Analyze and Conclude

1. **Calculating Results** Calculate the amount of force on the planet at the closest distance. Gravity is an inverse square force. If the planet is 0.45 times as far as the closest distance, the force is $1/(0.45)^2$ as much, or 4.9 ***F***. **Hint:** The force will be more than 1.0 ***F***.
2. **Diagramming Results** Draw the force vector, using the correct length and direction, for this position and at two other positions in the orbit. Use the scale 1.0 ***F*** : 1.0 cm.

Apply

1. Draw a velocity vector at each planet position to show the direction of motion. Assume that the planet moves in a clockwise pattern on the ellipse. Where does the planet move fastest?
2. Look at the direction of the velocity vectors and the direction of the force vectors at each position of the planet. Where does the planet gain and lose speed? Why?

Example Problem

Kepler's Third Law of Planetary Motion

Galileo discovered the moons of Jupiter. He could measure their orbital sizes only by using the diameter of Jupiter as a unit of measure. He found that Io, which had a period of 1.8 days, was 4.2 units from the center of Jupiter. Callisto, Jupiter's fourth moon, had a period of 16.7 days. Using the same units that Galileo used, predict Callisto's distance from Jupiter.

Sketch the Problem

- Sketch the orbits of Io and Callisto, noting that a longer period implies a larger orbit.
- Label radii and periods.

Calculate Your Answer

Known:

$T_C = 16.7$ days

$T_I = 1.8$ days

$r_I = 4.2$ units

Unknown:

$r_C = ?$

Strategy:

Start with Kepler's third law.

Calculations:

$$\left(\frac{T_C}{T_I}\right)^2 = \left(\frac{r_C}{r_I}\right)^3 \text{ or } r_C^{\,3} = r_I^{\,3}\left(\frac{T_C}{T_I}\right)^2$$

Rearrange to isolate the unknown r_C.

$$r_C^{\,3} = (4.2 \text{ units})^3\left(\frac{16.7 \text{ days}}{1.8 \text{ days}}\right)^2$$

$$= 6.4 \times 10^3 \text{ units}^3$$

$$r_C = (6.4 \times 10^3 \text{ units}^3)^{1/3} = 19 \text{ units}$$

Check Your Answer

- Are the units correct? Work algebra on the units to ensure that your answer is in Galileo's units.
- Do the signs make sense? All quantities are positive. Radius and period are never negative.
- Is the magnitude realistic? Expect a larger radius because the period is larger.

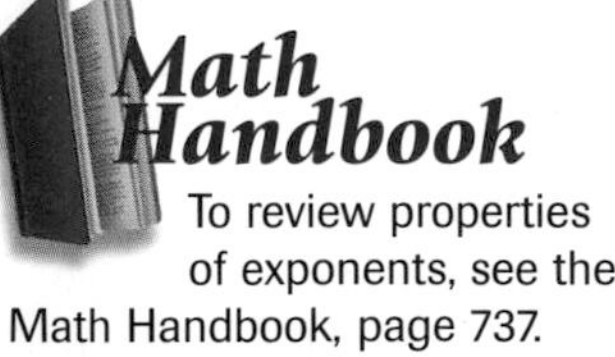

To review properties of exponents, see the Math Handbook, page 737.

Practice Problems

1. An asteroid revolves around the sun with a mean (average) orbital radius twice that of Earth's. Predict the period of the asteroid in Earth years.

2. From **Table 8–1,** you can calculate that, on the average, Mars is 1.52 times as far from the sun as Earth is. Predict the time required for Mars to circle the sun in Earth days.
3. The moon has a period of 27.3 days and has a mean distance of 3.90×10^5 km from the center of Earth. Find the period of a satellite that is in orbit 6.70×10^3 km from the center of Earth.
4. Using the data on the period and radius of revolution of the moon in problem 3, predict what the mean distance from Earth's center would be for an artificial satellite that has a period of 1.00 day.

USING A CALCULATOR

Cube Root

When you use Kepler's third law of motion to find the radius of the orbit of a planet or satellite, first solve for the cube of the radius, then take the cube root. This is easier to do if your calculator has a cube-root key, $\sqrt[3]{x}$. If your calculator has the key y^x or x^y , you also can find the cube root using this key. Check the instructions of your calculator, but you usually enter the cube of the radius, press the y^x key, then enter 0.3333333 and press = .

Universal Gravitation

In 1666, some 45 years after Kepler did his work, 24-year-old Isaac Newton was living at home in rural England because an epidemic of the black plague had closed all the schools. Newton had used mathematical arguments to show that if the path of a planet were an ellipse, which was in agreement with Kepler's first law of planetary motion, then the magnitude of the force, *F,* on the planet resulting from the sun must vary inversely with the square of the distance between the center of the planet and the center of the sun.

$$F \propto \frac{1}{d^2}$$

The symbol $\propto$ means *is proportional to,* and *d* is the distance between the centers of the two bodies. Newton also showed that the force acted in the direction of the line connecting the centers of the two bodies. But was the force that acted between the planet and the sun the same force that caused objects to fall to Earth?

Newton later wrote that the sight of a falling apple made him think about the problem of the motion of the planets. He recognized that the apple fell straight down because Earth attracted it. He wondered whether this force might extend beyond the trees to the clouds, to the moon, and even beyond. Could gravity be the force that also attracts the planets to the sun? Newton hypothesized that the force on the apple must be proportional to its mass. In addition, according to his own third law of motion, the apple also would attract Earth. Thus, the force of attraction also must be proportional to the mass of Earth. This attractive force that exists between all objects is known as **gravitational force.**

Newton was so confident that the laws governing motion on Earth would work anywhere in the universe that he assumed that the same force of attraction would act between any two masses, m_A and m_B. He proposed his **law of universal gravitation,** which is represented by the following equation.

$$F = G\frac{m_A m_B}{d^2}$$

What Goes Up, Must Come Down

In the equation, d is the distance between the centers of the masses, and G is a universal constant—one that is the same everywhere. According to Newton's equation, if the mass of a planet near the sun were doubled, the force of attraction would be doubled. Similarly, if the planet were near a star having twice the mass of the sun, the force between the two bodies would be twice as great. In addition, if the planet were twice the distance from the sun, the gravitational force would be only one quarter as strong. **Figure 8–3** illustrates these relationships pictorially, and **Figure 8–4** illustrates them graphically. Because the force depends on $1/d^2$, it is called an inverse square law.

Using Newton's Law of Universal Gravitation

Newton was able to state his law of universal gravitation in terms that applied to the motion of the planets about the sun. This agreed with Kepler's third law of planetary motion and provided confirmation that Newton's law fit the best observations of the day.

You can use the symbol m_p for the mass of a planet, m_s for the mass of the sun, and r for the radius of the planet's orbit. Then, Newton's second law of motion, $F = ma$, can be stated as $F = m_p a_c$, where F is the gravitational force, m_p is the mass, and a_c is the centripetal acceleration of the planet. For the sake of simplicity, assume circular orbits. Recall from your study of uniform circular motion in Chapter 7 that, for a circular orbit, $a_c = 4\pi^2 r/T^2$. This means that $F = m_p a_c$ may now be written as

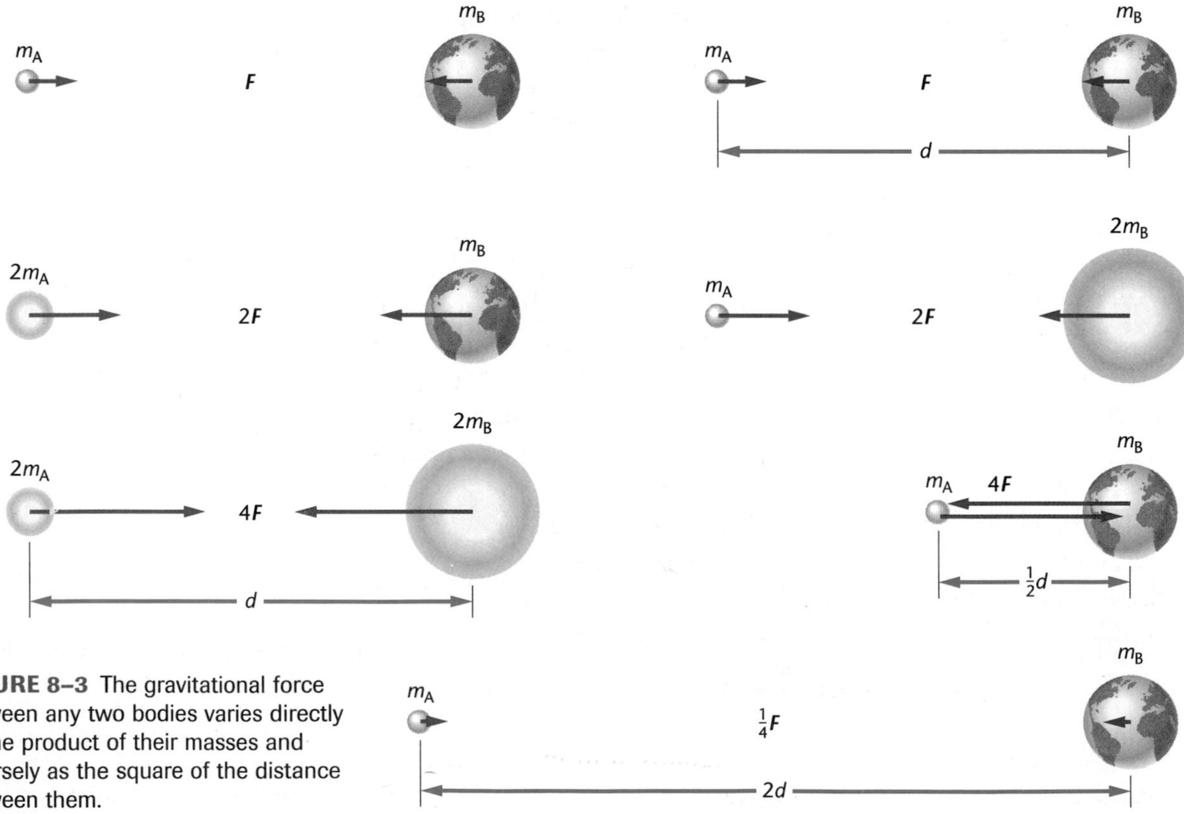

FIGURE 8–3 The gravitational force between any two bodies varies directly as the product of their masses and inversely as the square of the distance between them.

$F = m_p 4\pi^2 r/T^2$. If you set the right side of this equation equal to the right side of Newton's law of universal gravitation, you arrive at the following result.

$$G\frac{m_s m_p}{r^2} = \frac{m_p 4\pi^2 r}{T^2}$$

In this equation, T is the time required for the planet to make one complete revolution about the sun. The equation can be rearranged into the following form.

$$T^2 = \left(\frac{4\pi^2}{Gm_s}\right)r^3$$

This equation is Kepler's third law of planetary motion—the square of the period is proportional to the cube of the distance that separates the masses. The proportionality constant, $4\pi^2/Gm_s$, depends only on the mass of the sun and Newton's universal gravitational constant, G. It does not depend on any property of the planet. Thus, Newton's law of universal gravitation leads to Kepler's third law. In the derivation of this equation, it is assumed that the orbits of the planets are circles. Newton found the same result for elliptical orbits.

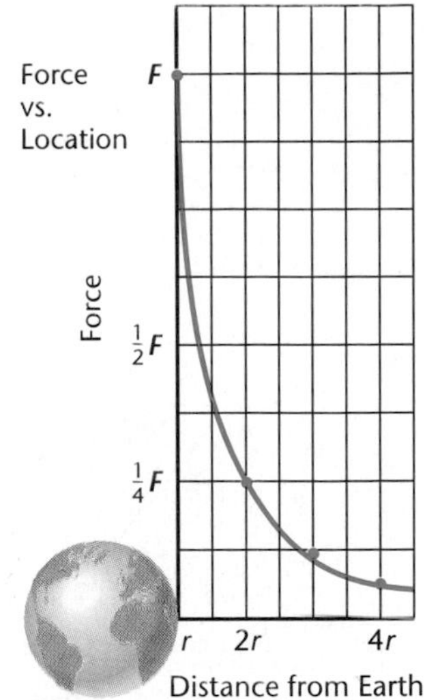

FIGURE 8–4 The change in gravitational force with distance follows the inverse square law.

Weighing Earth

How large is the constant G? As you know, the force of gravitational attraction between two objects on Earth is relatively small. You can't feel the slightest attraction even between two massive bowling balls. In fact, it took 100 years from the time of Newton's work before an apparatus that was sensitive enough to measure the force was developed. In 1798, Englishman Henry Cavendish (1731–1810) used equipment similar to the apparatus sketched in **Figure 8–5** to measure the gravitational force between two objects. Rod A, about 20 cm long, had a small lead ball, B, attached to each end. The rod was suspended by a thin wire, C, so that it could rotate. Cavendish measured the force on the balls that was needed to rotate the rod through given angles by the twisting of the wire. Then he placed a large lead ball, D, close to each of the two small balls. The position of the large balls was fixed. The force of attraction between the large and the small balls caused the rod to rotate. It stopped rotating only when the force required to twist the wire equaled the gravitational forces between the balls. By measuring the angle through which the rod turned, Cavendish was able to calculate the attractive force between the masses. He measured the masses of the balls and the distance between their centers. Substituting these values for force, mass, and distance into Newton's law, he found an experimental value for G. Newton's law of universal gravitation is stated as follows.

$$F = G\frac{m_A m_B}{d^2}$$

When m_A and m_B are measured in kilograms, d in meters, and F in newtons, then $G = 6.67 \times 10^{-11}$ N·m²/kg².

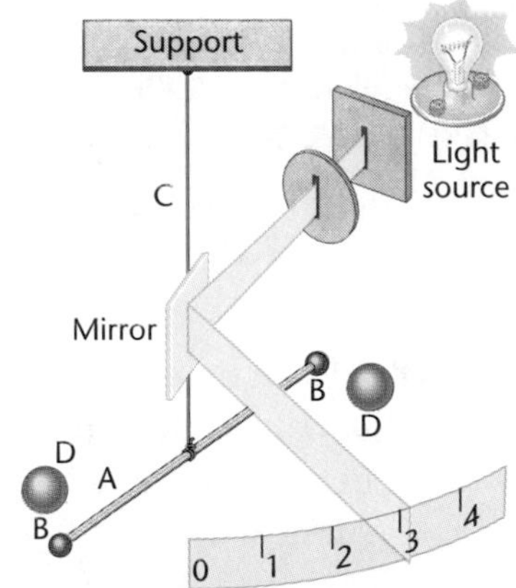

FIGURE 8–5 Cavendish verified the existence of gravitational forces between masses by measuring, with the help of a mirror and light source, the amount of twist in the suspending wire.

Now that you know the value of G, you can use Newton's law to find the gravitational force between two objects. For example, the attractive gravitational force between two bowling balls, each of mass 7.26 kg, with their centers separated by 0.30 m, is represented as follows.

$$F_g = \frac{(6.67 \times 10^{-11} \text{N}\cdot\text{m}^2/\text{kg}^2)(7.26 \text{ kg})(7.26 \text{ kg})}{(0.30 \text{ m})^2} = 3.9 \times 10^{-8} \text{ N}$$

Cavendish's experiment is often called "weighing the earth." You know that on Earth's surface, the weight of an object of mass m is a measure of Earth's gravitational attraction: $F_g = mg$. According to Newton, however, the following is true.

$$F_g = \frac{Gm_E m}{r^2}, \text{ so } g = \frac{Gm_E}{r^2}$$

Because Cavendish measured the constant G, this equation can be rearranged.

$$m_E = \frac{gr^2}{G}$$

Using 6.38×10^6 m as the radius of Earth, 9.80 m/s^2 as gravitational acceleration, and $G = 6.67 \times 10^{-11}$ N·m^2/kg^2, the following result is obtained.

$$m_E = \frac{(9.80 \text{ m/s}^2)(6.38 \times 10^6 \text{ m})^2}{6.67 \times 10^{-11} \text{ N}\cdot\text{m}^2/\text{kg}^2} = 5.98 \times 10^{24} \text{ kg}$$

When you compare the mass of Earth to that of a bowling ball, you can see why the gravitational attraction between everyday objects is not easily observed.

8.1 Section Review

1.1 Earth is attracted to the sun by the force of gravity. Why doesn't Earth fall into the sun? Explain.

1.2 If Earth began to shrink but its mass remained the same, what would happen to the value of g on Earth's surface?

1.3 Cavendish did his experiment using lead balls. Suppose he had used equal masses of copper instead. Would his value of G be the same or different? Explain.

1.4 Critical Thinking An astronaut can pick up a rock with less effort on the moon than on Earth.

a. How will the weaker gravitational force on the moon's surface affect the path of the rock if the astronaut throws it horizontally?

b. If the rock drops on the astronaut's toe, will it hurt more or less than it would on Earth? Explain.

Using the Law of Universal Gravitation

8.2

The planet Uranus was discovered in 1741. By 1830, it was clear that Newton's law of gravitation didn't correctly predict its orbit. This fact puzzled astronomers. Then, two astronomers proposed that Uranus was being attracted not only by the sun but also by an unknown planet, not yet discovered. They calculated the orbit of such a planet in 1845 and, one year later, astronomers at the Berlin Observatory began to search for it. During the first evening of their search, they found the giant planet now called Neptune.

OBJECTIVES

- **Solve** problems involving orbital speed and period.
- **Relate** weightlessness to objects in free fall.
- **Describe** gravitational fields.
- **Distinguish** between inertial mass and gravitational mass.
- **Contrast** Newton's and Einstein's views about gravitation.

Motion of Planets and Satellites

Newton used a drawing similar to the one shown in **Figure 8–6** to illustrate a thought experiment on the motion of satellites. Imagine a cannon, perched high atop a mountain, firing a cannonball horizontally with a given horizontal speed. The cannonball is a projectile, and its motion has both vertical and horizontal components. Like all projectiles on Earth, it follows a parabolic trajectory. During its first second of flight, the ball falls 4.9 m. If its horizontal speed were increased, it would travel farther across the surface of Earth, but it would still fall 4.9 m in the first second of flight. Because the surface of Earth is curved, it is possible for a cannonball with just the right horizontal speed to fall 4.9 m at a point where Earth's surface has curved 4.9 m away from the horizontal. This means that, after one second, the cannonball is at the same height above Earth as it was initially. The curvature of the projectile will continue to just match the curvature of Earth, so that the cannonball never gets any closer or farther away from Earth's curved surface. When this happens, the ball is said to be in orbit.

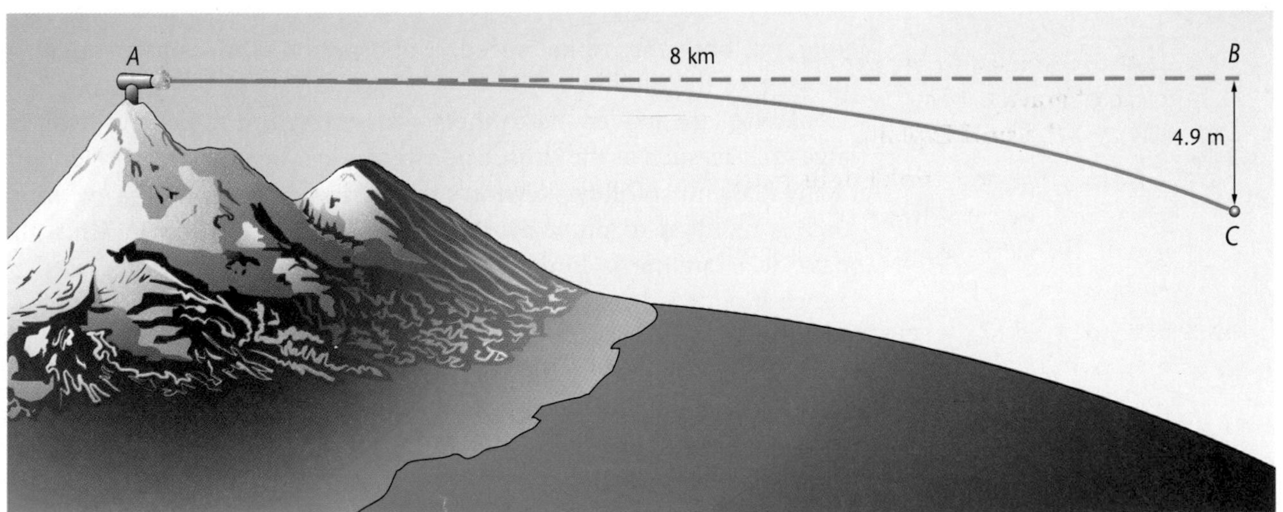

FIGURE 8–6 If the cannonball travels 8 km horizontally in 1 s, it will fall the same distance toward Earth as Earth curves away from the cannonball.

HELP WANTED

PILOT

We need "safety-first" people who can make quick but correct decisions, even under occasionally adverse conditions.
A current FAA commercial license with instrument rating and the ability to continue to pass strict physical, vision, hearing, flying, and rules exams are a must. Graduates of accredited flying schools or colleges or those with military flying experience will be considered equally. For information contact:
Air Pilots Association
1625 Massachusetts Ave. NW
Washington, DC 20036

Newton's drawing shows that Earth curves away from a line tangent to its surface at a rate of 4.9 m for every 8 km. That is, the altitude of the line tangent to Earth at A will be 4.9 m above Earth at B. If the cannonball were given just enough horizontal speed to travel from A to B in one second, it would also fall 4.9 m and arrive at C. The altitude of the ball in relation to Earth's surface would not have changed. The cannonball would fall toward Earth at the same rate that Earth's surface curves away. An object at Earth's surface with a horizontal speed of 8 km/s will keep the same altitude and circle Earth as an artificial satellite.

Newton's thought experiment ignored air resistance. The mountain would have had to be more than 150 km above Earth's surface to be above most of the atmosphere. A satellite at or above this altitude encounters little air resistance and can orbit Earth for a long time.

A satellite in an orbit that is always the same height above Earth moves with uniform circular motion. Recall from Chapter 7 that its centripetal acceleration is given by $a_c = v^2/r$. Newton's second law, $F = ma$, can be rewritten as $F = mv^2/r$. Combining this with Newton's inverse square law produces the following equation.

$$\frac{Gm_E m}{r^2} = \frac{mv^2}{r}$$

Solving this for the speed of an object in circular orbit, v, yields the following.

$$v = \sqrt{\frac{Gm_E}{r}}$$

By using Newton's law of universal gravitation, you saw that the time, T, for a satellite to circle Earth is given by the following.

$$T = 2\pi\sqrt{\frac{r^3}{Gm_E}}$$

Note that both the orbital speed, v, and period, T, are independent of the mass of the satellite.

Satellites are accelerated to the speeds they need to achieve orbit by large rockets, such as the shuttle booster rocket. Because the acceleration of any mass must follow Newton's second law of motion, $F = ma$, more force is required to put a more massive satellite into orbit. Thus, the mass of a satellite is limited by the capability of the rocket used to launch it.

These equations for the speed and period of a satellite can be used for any body in orbit about another. The mass of the central body would replace m_E in the equations, and r would be the distance between the centers of the sun and the orbiting body. If the mass of the central body is much greater than the mass of the orbiting body, then r is equal to the distance between the central body and the orbiting body.

Example Problem

Finding the Speed of a Satellite

A satellite orbits Earth 225 km above its surface. What is its speed in orbit and its period?

Sketch the Problem

- Draw Earth, showing the height of the satellite's orbit.

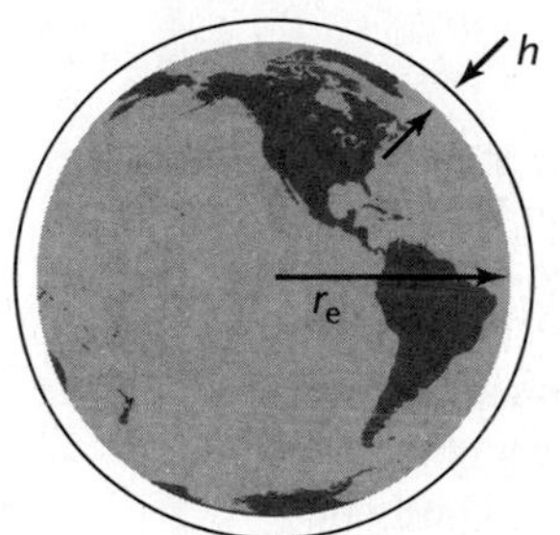

Calculate Your Answer

Known:

$h = 2.25 \times 10^5$ m

$r_E = 6.38 \times 10^6$ m

$m_E = 5.97 \times 10^{24}$ kg

$G = 6.67 \times 10^{-11}$ N·m²/kg²

Unknown:

$v = ?$

$T = ?$

Strategy:

Determine the radius of the satellite's orbit by adding the height to Earth's radius.

Calculations:

$r = h + r_E$

$r = 2.25 \times 10^5 \text{ m} + 6.38 \times 10^6 \text{ m} = 6.61 \times 10^6 \text{ m}$

Use the velocity equation.

$$v = \sqrt{\frac{Gm_E}{r}}$$

$$v = \sqrt{\frac{(6.67 \times 10^{-11} \text{ N·m}^2/\text{kg}^2)(5.97 \times 10^{24} \text{ kg})}{6.61 \times 10^6 \text{ m}}}$$

$$= 7.76 \times 10^3 \text{ m/s}$$

Use the definition of velocity to find the orbital period.

$$v = \frac{d}{t} = \frac{2\pi r}{T}$$

Rearrange and solve for T.

$$T = \frac{2\pi r}{v}$$

$$T = \frac{2(3.14)(6.61 \times 10^6 \text{ m})}{7.76 \times 10^3 \text{ m/s}} = 5350 \text{ s} = 89.2 \text{ min} \approx 1.5 \text{ h}$$

Check Your Answer

- Are the units correct? Be sure that v is in m/s and T is in s.
- Do the signs make sense? Orbital speed and period are always positive.
- Is the magnitude realistic? The speed is close to the 8 km/s obtained in Newton's thought experiment. The period, about 1 1/2 hours, is typical of low Earth orbits.

Pocket Lab

Weight in a Free Fall

Tie a string to the top of a spring scale. Hang a 1.0-kg mass on the spring scale. Hold the scale in your hand.

Analyze and Conclude
Observe the weight of the mass. What will the reading be when the string is released (as the mass and scale are falling)? Why?

Practice Problems

Assume a circular orbit for all calculations.

5. Use Newton's thought experiment on the motion of satellites to solve the following.
   a. Calculate the speed that a satellite shot from the cannon must have in order to orbit Earth 150 km above its surface.
   b. How long, in seconds and minutes, would it take for the satellite to complete one orbit and return to the cannon?

6. Use the data for Mercury in **Table 8–1** to find
   a. the speed of a satellite in orbit 265 km above Mercury's surface.
   b. the period of the satellite.

7. Find the speeds with which Mercury and Saturn move around the sun. Does it make sense that Mercury is named after a speedy messenger of the gods, whereas Saturn is named after the father of Jupiter?

8. The sun is considered to be a satellite of our galaxy, the Milky Way. The sun revolves around the center of the galaxy with a radius of 2.2×10^{20} m. The period of one revolution is 2.5×10^{8} years.
   a. Find the mass of the galaxy.
   b. Assuming that the average star in the galaxy has the same mass as the sun, find the number of stars.
   c. Find the speed with which the sun moves around the center of the galaxy.

Weight and Weightlessness

The acceleration of objects due to Earth's gravitation can be found by using Newton's law of universal gravitation and second law of motion. For a free-falling object,

$$F = \frac{Gm_{\text{E}}m}{d^2} = ma, \text{ so } a = \frac{Gm_{\text{E}}}{d^2}.$$

On Earth's surface, $d = r_{\text{E}}$, and the following equation can be written.

$$g = \frac{Gm_{\text{E}}}{{r_{\text{E}}}^2}$$

Thus,

$$a = g\left(\frac{r_{\text{E}}}{d}\right)^2.$$

As you move farther from Earth's center, that is, as d becomes larger, the acceleration due to gravity is reduced according to this inverse square relationship.

You have probably seen photos similar to the one in **Figure 8–7,** in which astronauts on a space shuttle are working and relaxing in what is often called "zero-g" or "weightlessness." The shuttle orbits Earth about 400 km above its surface. At that distance, $g = 8.7\ \text{m/s}^2$, only slightly less than on Earth's surface. Thus, Earth's gravitational force is certainly not zero in the shuttle. In fact, gravity causes the shuttle to circle Earth. Why, then, do the astronauts appear to have no weight? Just as with Newton's cannonball, the shuttle and everything in it are falling freely toward Earth as they orbit around it.

Astronauts have weight because the gravitational force is exerted on them, but do they have any apparent weight? Remember that you sense weight when something such as the floor or your chair exerts a force on you. But if you, your chair, and the floor are all accelerating toward Earth together, then no contact forces are exerted on you. Your apparent weight is zero. You are experiencing weightlessness.

Pocket Lab

Water, Water, Everywhere

This activity is best done outdoors. Use a pencil to poke a hole in the bottom and side of a cup. Hold your fingers over the two holes as you pour colored water into the cup until it is 2/3 full. Predict what will happen as the cup is allowed to fall. Drop the cup and watch closely.

Analyze and Conclude What happened? Why?

The Gravitational Field

You may recall from Chapter 6 that many common forces are contact forces. Friction is exerted where two objects touch; the floor and your chair or desk push on you. Gravity is different. It acts on an apple falling from a tree and on the moon in orbit; it even acts on you in midair. In other words, gravity acts over a distance. It acts between bodies that are not touching or even close to one another. Newton himself was uneasy with this idea. He wondered how the sun could exert a force on planet Earth, which was hundreds of millions of kilometers away.

The answer to the puzzle arose from a study of magnetism. In the nineteenth century, Michael Faraday developed the concept of the field to explain how a magnet attracts objects. Later, the field concept was applied to gravity. It was proposed that anything with a mass is surrounded by a gravitational field. It is this gravitational field that interacts with objects, resulting in a force of attraction. The field acts on a body at the location of that body.

FIGURE 8–7 Astronauts train on Earth in a diving aircraft to practice procedures they will perform in space. They experience about 20 seconds of weightlessness during each dive. In space, however, they would experience weightlessness as long as they were in orbit. A portion of the 1995 class of astronaut candidates poses here for their class picture.

LITERATURE CONNECTION

Jules Verne (1828–1905) was a French science fiction novelist. He wrote about airplanes, submarines, guided missiles, and space satellites years before their invention, and he accurately predicted their uses. His use of detail and knowledge of geography add realism to his stories as his readers travel over, through, and around Earth. Perhaps you have read some of his exciting works: *Twenty Thousand Leagues Under the Sea, Around the World in Eighty Days, A Journey to the Center of the Earth, From the Earth to the Moon,* and *Around the Moon.* Check out Verne's perspective on adventure and science. Why do you think his books remain popular today?

To find the strength of a gravitational field, you can place a small body of mass m in the field and measure the force on the body. The gravitational field, $\mathbf{g}$, is defined as the force divided by the mass.

$$\mathbf{g} = \frac{\mathbf{F}}{m}$$

Gravitational fields are often measured in newtons per kilogram. The direction of $\mathbf{g}$ is in the direction of the force. Recall that $\mathbf{g}$ is also called acceleration due to gravity.

On Earth's surface, the strength of the gravitational field is 9.80 N/kg, and its direction is toward Earth's center. From Newton's law of universal gravitation, you know that the gravitational field is independent of the size of an object's mass. The field can be represented by a vector of length g pointing toward the center of the object producing the field being measured. You can picture the gravitational field of Earth as a collection of vectors surrounding Earth and pointing toward it, as shown in **Figure 8–8.** The strength of the field varies inversely with the square of the distance from the center of Earth.

Two Kinds of Mass

When the concept of mass was first introduced in Chapter 6, it was defined as the slope of a graph of force versus acceleration; that is, the ratio of the net force exerted on an object and its acceleration. This kind of mass is related to the inertia of an object and is called the **inertial mass.** The inertial mass of an object is measured by applying a force to the object and measuring its acceleration.

$$m_{\text{inertial}} = \frac{F_{\text{net}}}{a}$$

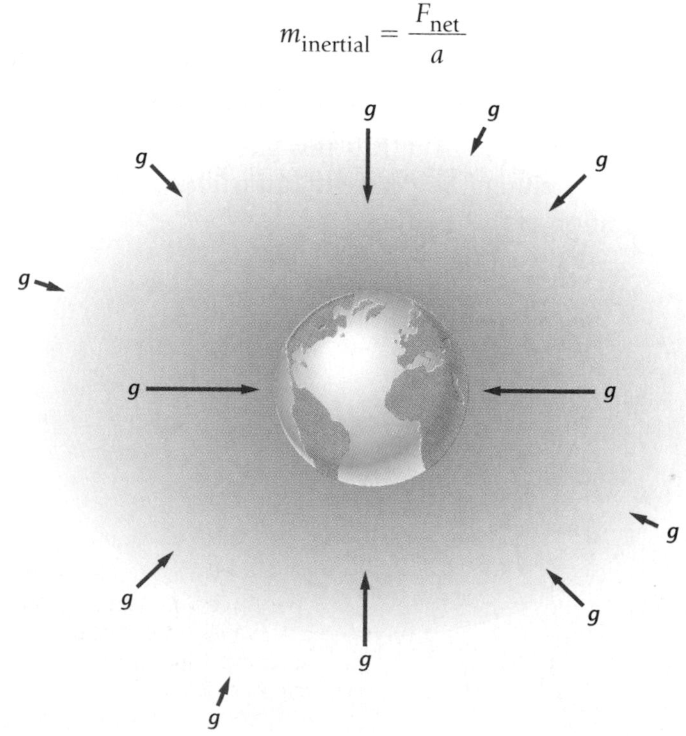

FIGURE 8–8 Vectors can be used to show Earth's gravitational field.

Newton's law of universal gravitation, $F = Gm_Am_B/d^2$, also involves mass, but it is a different kind of mass. Mass as used in the law of gravitation determines the size of the gravitational attraction between two objects. This kind of mass is called **gravitational mass.** It can be measured using a simple balance, such as the one shown in **Figure 8–9.** If you measure the attractive force exerted on an object by another object of mass m, at a distance r, then you can define the gravitational mass in the following way.

$$m_{\text{gravitational}} = \frac{r^2F_{\text{grav}}}{Gm}$$

FIGURE 8–9 The platform balance shown here allows you to compare an unknown mass to a known mass. Using an inertial balance, you can calculate the mass from the back-and-forth motion of the mass.

How different are these two kinds of masses? Suppose you have a block of ice in the back of a pickup truck. If you accelerate the truck forward, the ice will slide backwards relative to the bed of the truck. This is a result of its inertial mass—its resistance to acceleration. Now suppose the truck climbs a steep hill at a constant speed. The ice will again slide backwards. But this time, it moves as a result of its gravitational mass. The ice is being attracted downward toward Earth. Newton made the claim that these two masses are identical. This hypothesis is called the principle of equivalence. It has been tested very carefully in many experiments. If any difference exists between the two kinds of mass, it is less than one part in 100 billion. But why should the two masses be equivalent? Albert Einstein (1879–1955) was intrigued by this equivalence and made it a central point in the treatment of gravity in his general theory of relativity.

Einstein's Theory of Gravity

Newton's law of universal gravitation allows us to calculate the force that exists between two bodies because of their masses. The concept of a gravitational field allows us to picture the way gravity acts on bodies far away. However, neither explains the origin of gravity.

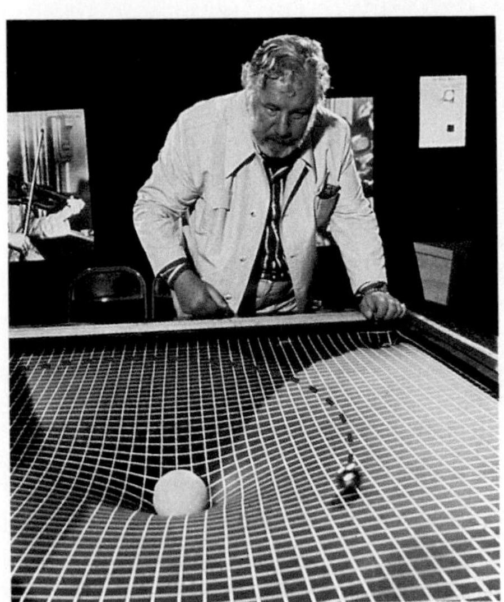

FIGURE 8–10 Matter causes space to curve just as a mass on a rubber sheet curves the sheet around it. Moving bodies near the mass follow the curvature of space, as indicated by the dotted line.

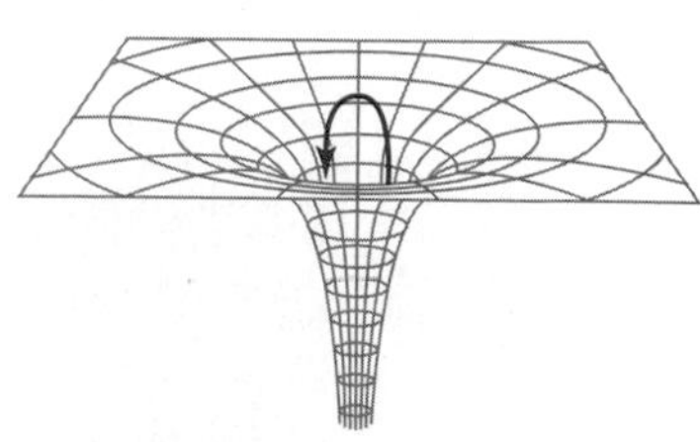

FIGURE 8–11 A black hole is so massive and of such unimaginable density that light leaving it will be bent back to it.

To do this, Einstein proposed that gravity is not a force, but an effect of space itself. According to Einstein, a mass changes the space around it. Mass causes space to be curved, and other bodies are accelerated because of the way they follow this curved space.

One way to picture how space is affected by mass is to compare space to a large, two-dimensional rubber sheet, as shown in **Figure 8–10** on the previous page. The yellow ball on the sheet represents a massive object. It forms an indentation. A marble rolling across the sheet simulates the motion of an object in space. If the marble moves near the sagging region of the sheet, it will be accelerated. In the same way, Earth and the sun are attracted to one another because of the way space is distorted by the two bodies.

Einstein's theory, called the general theory of relativity, makes many predictions about how massive objects affect one another. In every test conducted to date, Einstein's theory has been shown to give the correct results.

One of the most interesting predictions to come out of Einstein's theory is the deflection of light by massive objects. In 1919, during an eclipse of the sun, astronomers found that light from distant stars that passed near the sun was deflected in agreement with Einstein's predictions. Astronomers have seen light from a distant, bright galaxy bend as it passed by a closer, dark galaxy. The result is two or more images of the bright galaxy. Another result of general relativity is the effect on light of very massive objects. If an object is massive and dense enough, light leaving it will be totally bent back to the object, as **Figure 8–11** shows. No light ever escapes the object. Such an object, called a black hole, is believed to have been identified as a result of its effect on nearby stars.

While Einstein's theory provides very accurate predictions of gravity's effects, it still is not yet complete. It does not explain how masses curve space. Physicists are working to understand the true nature of gravity.

8.2 Section Review

2.1 What is the strength of the gravitational field on the surface of the moon?

2.2 Two satellites are in circular orbits about Earth. One is 150 km above the surface, the other 160 km.

a. Which satellite has the larger orbital period?

b. Which one has the greater speed?

2.3 What is g? Explain in your own words.

2.4 **Critical Thinking** It is easier to launch a satellite from Earth into an orbit that circles eastward than it is to launch one that circles westward. Explain.

CHAPTER 8 REVIEW

Key Terms

8.1
- Kepler's laws of planetary motion
- gravitational force
- law of universal gravitation

8.2
- inertial mass
- gravitational mass

Summary

8.1 Motion in the Heavens and on Earth

- Kepler's three laws of planetary motion state that planets move in elliptical orbits, that they sweep out equal areas in equal times, and that the square of the ratio of the periods of any two planets is equal to the cube of the ratio of their distances from the sun.
- Newton's law of universal gravitation states that the gravitational force between any two bodies is directly proportional to the product of their masses and inversely proportional to the square of the distance between their centers. The force is attractive and along a line connecting their centers.
- The mass of the sun can be found from the period and radius of a planet's orbit. The mass of the planet can be found only if it has a satellite orbiting it.
- Cavendish was the first to measure the gravitational attraction between two bodies on Earth.

8.2 Using the Law of Universal Gravitation

- A satellite in a circular orbit accelerates toward Earth at a rate equal to the acceleration of gravity at its orbital radius.
- All bodies have gravitational fields surrounding them that can be represented by a collection of vectors representing the force per unit mass at all locations.
- Gravitational mass and inertial mass are two essentially different concepts. The gravitational and inertial masses of a body, however, are numerically equal.
- Einstein's theory of gravity describes gravitational attraction as a property of space itself.

Reviewing Concepts

Section 8.1

1. In 1609, Galileo looked through his telescope at Jupiter and saw four moons. The name of one of the moons is Io. Restate Kepler's first law for Io and Jupiter.
2. Earth moves more slowly in its orbit during summer in the northern hemisphere than during winter. Is it closer to the sun in summer or in winter?
3. Is the area swept out per unit time by Earth moving around the sun equal to the area swept out per unit time by Mars moving around the sun?
4. Why did Newton think that a force must act on the moon?
5. The force of gravity acting on an object near Earth's surface is proportional to the mass of the object. Why does a heavy object not fall faster than a light object?
6. What information do you need to find the mass of Jupiter using Newton's version of Kepler's third law?
7. The mass of Pluto was not known until a satellite of the planet was discovered. Why?
8. How did Cavendish demonstrate that a gravitational force of attraction exists between two small bodies?

Section 8.2

9. What provides the force that causes the centripetal acceleration of a satellite in orbit?
10. How do you answer the question, "What keeps a satellite up?"

11. A satellite is going around Earth. On which of the following does the speed depend?
   a. mass of the satellite
   b. distance from Earth
   c. mass of Earth
12. Chairs in an orbiting spacecraft are weightless. If you were on board and you were barefoot, would you stub your toe if you kicked a chair? Explain.
13. During space flight, astronauts often refer to forces as multiples of the force of gravity on Earth's surface. What does a force of 5 g mean to an astronaut?
14. Show that the dimensions of g in the equation $g = F/m$ are m/s^2.
15. Newton assumed that the gravitational force acts directly between Earth and the moon. How does Einstein's view of the attractive force between the two bodies differ from the view of Newton?

Applying Concepts

16. Tell whether each of the orbits shown in **Figure 8–12** is a possible orbit for a planet.

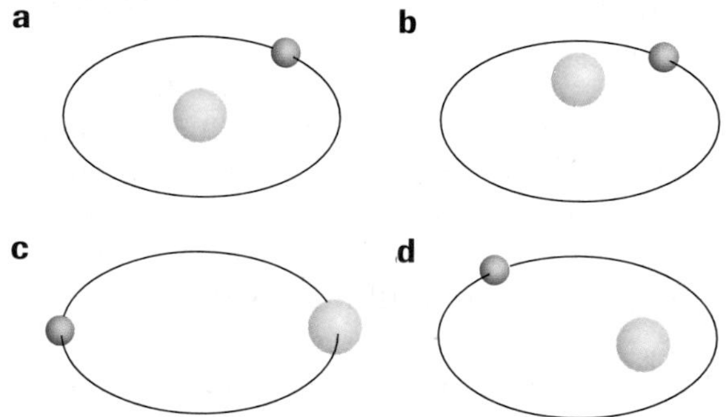

FIGURE 8–12

17. What happens to the gravitational force between two masses when the distance between the masses is doubled?
18. The moon and Earth are attracted to each other by gravitational force. Does the more massive Earth attract the moon with a greater force than the moon attracts Earth? Explain.
19. According to Newton's version of Kepler's third law, how does the ratio $\left(\frac{T^2}{r^3}\right)$ change if the mass of the sun is doubled?
20. If Earth were twice as massive but remained the same size, what would happen to the value of G?
21. Examine the equation relating the speed of an orbiting satellite and the distance from the center of Earth.
   a. Does a satellite with a large or small orbital radius have the greater velocity?
   b. When a satellite is too close to Earth, it can move into the atmosphere where there is air drag. As a result, its orbit gets smaller. Does its speed increase or decrease?
22. If a space shuttle goes into a higher orbit, what happens to the shuttle's period?
23. Mars has about one-ninth the mass of Earth. Satellite M orbits Mars with the same orbital radius as satellite E, which orbits Earth. Which satellite has a smaller period?
24. A satellite is one Earth radius above the surface of Earth. How does the acceleration due to gravity at that location compare to acceleration at the surface of Earth?
25. If Earth were twice as massive but remained the same size, what would happen to the value of g?
26. Jupiter has about 300 times the mass of Earth and about ten times Earth's radius. Estimate the size of g on the surface of Jupiter.
27. If a mass in Earth's gravitational field is doubled, what will happen to the force exerted by the field upon the mass?
28. Suppose that yesterday you had a mass of 50.0 kg. This morning you stepped on a scale and found that you had gained weight.
   a. What happened, if anything, to your mass?
   b. What happened, if anything, to the ratio of your weight to your mass?
29. As an astronaut in an orbiting space shuttle, how would you go about "dropping" an object down to Earth?

30. The weather pictures you see every day on TV come from a spacecraft in a stationary position relative to the surface of Earth, 35 700 km above Earth's equator. Explain how it can stay exactly in position day after day. What would happen if it were closer? Farther out? **Hint:** Draw a pictorial model.

Problems

Section 8.1

LEVEL 1

Use $G = 6.67 \times 10^{-11}\ \text{N·m}^2/\text{kg}^2$.

31. Jupiter is 5.2 times farther from the sun than Earth is. Find Jupiter's orbital period in Earth years.

32. An apparatus like the one Cavendish used to find G has a large lead ball that is 5.9 kg in mass and a small one that is 0.047 kg. Their centers are separated by 0.055 m. Find the force of attraction between them.

33. Use the data in **Table 8–1** to compute the gravitational force that the sun exerts on Jupiter.

34. Tom has a mass of 70.0 kg and Sally has a mass of 50.0 kg. Tom and Sally are standing 20.0 m apart on the dance floor. Sally looks up and sees Tom. She feels an attraction. If the attraction is gravitational, find its size. Assume that both Tom and Sally can be replaced by spherical masses.

35. Two balls have their centers 2.0 m apart. One ball has a mass of 8.0 kg. The other has a mass of 6.0 kg. What is the gravitational force between them?

36. Two bowling balls each have a mass of 6.8 kg. They are located next to each other with their centers 21.8 cm apart. What gravitational force do they exert on each other?

37. Assume that you have a mass of 50.0 kg and Earth has a mass of 5.97×10^{24} kg. The radius of Earth is 6.38×10^6 m.

a. What is the force of gravitational attraction between you and Earth?

b. What is your weight?

38. The gravitational force between two electrons 1.00 m apart is 5.42×10^{-71} N. Find the mass of an electron.

39. A 1.0-kg mass weighs 9.8 N on Earth's surface, and the radius of Earth is roughly 6.4×10^6 m.

a. Calculate the mass of Earth.

b. Calculate the average density of Earth.

40. Use the information for Earth in **Table 8–1** to calculate the mass of the sun, using Newton's version of Kepler's third law.

LEVEL 2

41. Uranus requires 84 years to circle the sun. Find Uranus's orbital radius as a multiple of Earth's orbital radius.

42. Venus has a period of revolution of 225 Earth days. Find the distance between the sun and Venus as a multiple of Earth's orbital radius.

43. If a small planet were located 8.0 times as far from the sun as Earth is, how many years would it take the planet to orbit the sun?

44. A satellite is placed in an orbit with a radius that is half the radius of the moon's orbit. Find its period in units of the period of the moon.

45. Two spherical balls are placed so that their centers are 2.6 m apart. The force between the two balls is 2.75×10^{-12} N. What is the mass of each ball if one ball is twice the mass of the other ball?

46. The moon is 3.9×10^5 km from Earth's center and 1.5×10^8 km from the sun's center. If the masses of the moon, Earth, and the sun are 7.3×10^{22} kg, 6.0×10^{24} kg, and 2.0×10^{30} kg, respectively, find the ratio of the gravitational forces exerted by Earth and the sun on the moon.

47. A force of 40.0 N is required to pull a 10.0-kg wooden block at a constant velocity across a smooth glass surface on Earth. What force would be required to pull the same wooden block across the same glass surface on the planet Jupiter?

48. Mimas, one of Saturn's moons, has an orbital radius of 1.87×10^8 m and an orbital period of about 23 h. Use Newton's version of Kepler's third law and these data to find Saturn's mass.

49. Use Newton's version of Kepler's third law to find the mass of Earth. The moon is 3.9×10^8 m away from Earth, and the moon has a period of 27.33 days. Compare this mass to the mass found in problem 39.

Section 8.2

LEVEL 1

50. A geosynchronous satellite is one that appears to remain over one spot on Earth. Assume that a geosynchronous satellite has an orbital radius of 4.23×10^7 m.
   a. Calculate its speed in orbit.
   b. Calculate its period.

51. The asteroid Ceres has a mass of 7×10^{20} kg and a radius of 500 km.
   a. What is g on the surface?
   b. How much would an 85-kg astronaut weigh on Ceres?

52. A 1.25-kg book in space has a weight of 8.35 N. What is the value of the gravitational field at that location?

53. The moon's mass is 7.34×10^{22} kg, and it is 3.8×10^8 m away from Earth. Earth's mass can be found in **Table 8–1.**
   a. Calculate the gravitational force of attraction between Earth and the moon.
   b. Find Earth's gravitational field at the moon.

54. Earth's gravitational field is 7.83 N/kg at the altitude of the space shuttle. What is the size of the force of attraction between a student with a mass of 45.0 kg and Earth?

LEVEL 2

55. On July 19, 1969, *Apollo11's* orbit around the moon was adjusted to an average orbit of 111 km. The radius of the moon is 1785 km, and the mass of the moon is 7.3×10^{22} kg.
   a. How many minutes did *Apollo 11* take to orbit the moon once?
   b. At what velocity did it orbit the moon?

56. The radius of Earth is about 6.38×10^3 km. A 7.20×10^3-N spacecraft travels away from Earth. What is the weight of the spacecraft at the following distances from Earth's surface?
   a. 6.38×10^3 km
   b. 1.28×10^4 km

57. How high does a rocket have to go above Earth's surface before its weight is half what it would be on Earth?

58. The following formula represents the period of a pendulum, T.

$$T = 2\pi\sqrt{\frac{l}{g}}$$

   a. What would be the period of a 2.0-m-long pendulum on the moon's surface? The moon's mass is 7.34×10^{22} kg, and its radius is 1.74×10^6 m.
   b. What is the period of this pendulum on Earth?

Critical Thinking Problems

59. Some people say that the tides on Earth are caused by the pull of the moon. Is this statement true?
   a. Determine the forces that the moon and the sun exert on a mass, m, of water on Earth. Your answer will be in terms of m with units of N.
   b. Which celestial body, the sun or the moon, has a greater pull on the waters of Earth?
   c. Determine the difference in force exerted by the moon on the water at the near surface and the water at the far surface (on the opposite side of Earth), as illustrated in **Figure 8–13.** Again, your answer will be in terms of m with units of N.

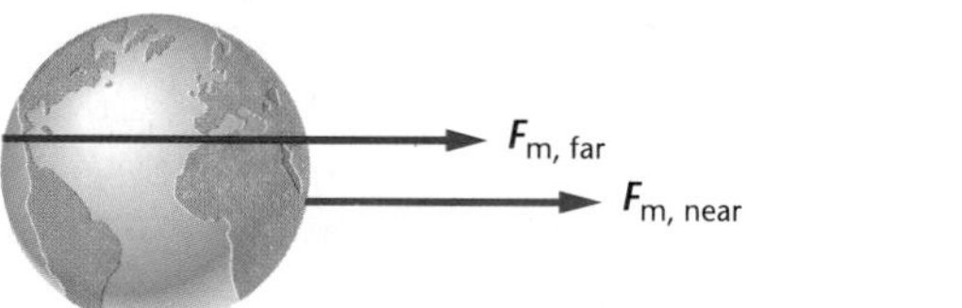

FIGURE 8–13

   d. Determine the difference in force exerted by the sun on water at the near surface and water at the far surface (on the opposite side of Earth).
   e. Which celestial body has a greater difference in pull from one side of Earth to the other?
   f. Why is the statement that the tides are due to the pull of the moon misleading? Make a correct statement to explain how the moon causes tides on Earth.

60. Graphing Calculator Use Newton's law of universal gravitation to find an equation where x is equal to an object's distance from Earth's center, and y is its acceleration due to gravity. Use a graphing calculator to graph this equation, using 6400-6600 km as the range for x and 9-10 m/s^2 as the range for y. The equation should be of the form $y = c(1/x^2)$. Trace along this graph and find y

a. at sea level, 6400 km.

b. on top of Mt. Everest, 6410 km.

c. in a typical satellite orbit, 6500 km.

d. in a much higher orbit, 6600 km.

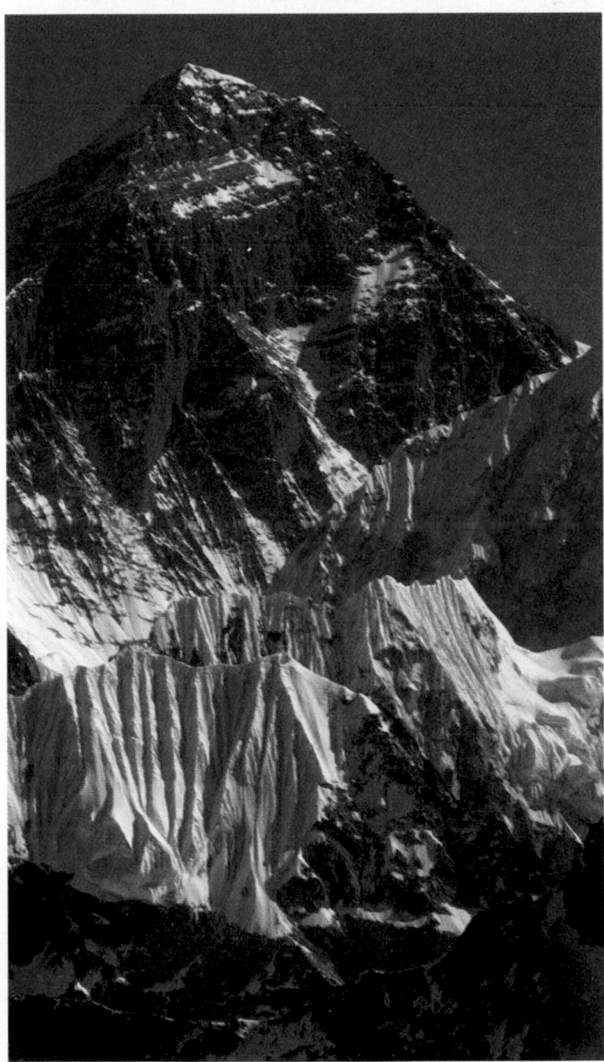

Mt. Everest

Going Further

Team Project Design a set of sports competitions to be held in the human base camp on Mars. Assume that Martian explorers would live in a dome filled with an atmosphere at normal Earth pressure and temperature, and that they would wear suits to keep them warm and provide air to breathe when they were outside the dome.

You will need to determine how each sports event would be affected by the Martian gravity and, if the event were to be held outside, how it would be affected by the extremely thin, dry, oxygen-free atmosphere. Consider, for example, how high a bar the high jumpers could clear. How would a discus, shot put, or javelin event have to be adjusted? If there were a pool under the dome, would the swimming and diving events have to be designed in a different way from those on Earth? Could you invent an event that would work only on Mars and not on Earth?

Each team should decide on the new rules for a set of events in one area and create a poster presentation of its designs. The links to physics should be highlighted.

*inter***NET**
CONNECTION

Follow the link on the Glencoe Homepage at **www.glencoe.com/sec/science** to find out more about this chapter.

How Safe?

Many of today's cars have air bags. In a head-on crash of two cars equipped with air bags, both drivers walked away uninjured. How does an air bag help to reduce the injury to a person in an automobile accident?

CHAPTER

9 Momentum and Its Conservation

In the crash of a car moving at high speed, the passengers are brought to a stop so quickly that they often are injured. So far in your study of physics, you have examined the causes of change, which is the part of physics called dynamics. You found that position is changed by velocity, that velocity is changed by acceleration, and that acceleration is caused by a net force. In most real-life situations, such as a car crash, the forces and accelerations change so rapidly that it would be nearly impossible to study them without a sophisticated computer.

However, you can learn more about forces by studying the properties of interacting bodies. You will examine some of the properties of objects before and after an interaction takes place, and you will discover how these properties are affected. You especially will look for properties that remain constant. Properties that remain constant can be described as being conserved.

WHAT YOU'LL LEARN

- You will describe momentum and impulse and apply them to the interaction of objects.
- You will relate Newton's third law of motion to conservation of momentum.

WHY IT'S IMPORTANT

- You will be able to explain how air bags can help reduce injuries and save lives in a car crash.
- You will understand how conservation of momentum explains the propulsion of rockets.

interNET CONNECTION

Follow the link for this chapter on the Glencoe Homepage at **www.glencoe.com/sec/science** to find out more about momentum and its conservation.

9.1 Impulse and Momentum

The word *momentum* is used often in everyday speech. For example, a winning sports team is said to have momentum. In physics, however, momentum has its own definition. Newton wrote his three laws of motion in terms of momentum, which he called the quantity of motion.

OBJECTIVES

- **Compare** the system before and after an event in momentum problems.
- **Define** the momentum of an object.
- **Determine** the impulse given to an object.
- **Recognize** that impulse equals the change in momentum of an object.

Impulse and Momentum

A service ace in tennis is an exciting shot. The server lobs the ball overhead and swings the racket through a smooth arc to meet the ball. The ball explodes away from the racket at high speed. The first step in analyzing this interaction is to define "before," "during," and "after" and to sketch them as shown in **Figure 9–1.**

You can simplify the collision between the ball and the racket by assuming that all motion is in the horizontal direction. Before the hit, the ball is moving slowly. During the hit, the ball is squashed against the racket. After the hit, the ball moves at a higher velocity and the racket continues in its path, but at a slower velocity.

How is velocity affected by force?

How are the velocities of the ball before and after the collision related to the force acting on it? According to Newton's first law of motion, if no net force acts on a body, its velocity is constant. Newton's second law of motion describes how the velocity of a body is changed by a net force acting on it.

Color Convention

- Displacement vectors are **green.**
- Velocity vectors are **red.**
- Acceleration vectors are **violet.**
- Force vectors are **blue.**
- Momentum and impulse vectors are **orange.**

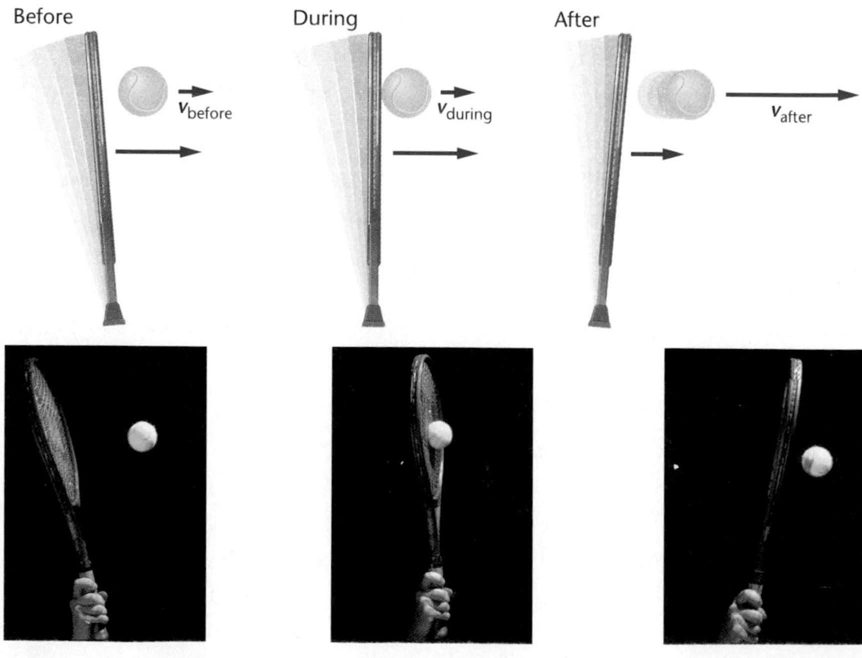

FIGURE 9–1 The motions of a tennis racket and ball are shown before, during, and after their interaction.

The change in velocity of the ball must have been caused by the force exerted by the racket on the ball. The force changes over time, as shown in **Figure 9–2.** Just after contact is made, the ball is squeezed, the racket strings are stretched, and the force increases. After the force reaches a maximum, the ball recovers its shape and snaps away from the strings of the racket. The force rapidly returns to zero. The maximum force is more than 1000 times greater than the weight of the ball! The whole event takes place within only a few thousandths of a second.

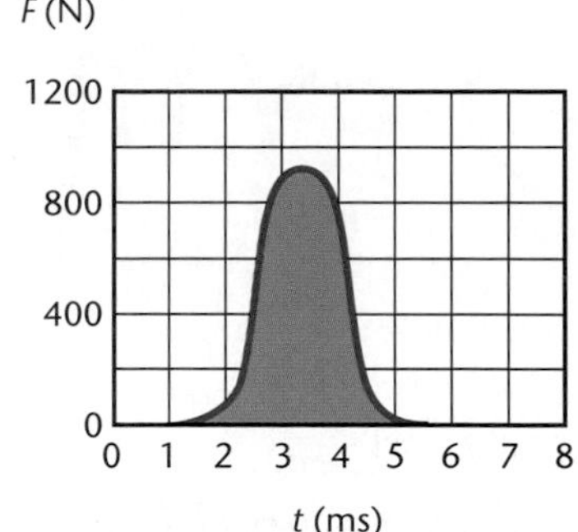

FIGURE 9–2 The force acting on a tennis ball increases, then rapidly decreases during a hit, as shown in this force-time graph.

Relating impulse and momentum

Newton's second law of motion can help explain how the momentum of an object is changed by a net force acting on it. Newton's second law of motion, $\mathbf{F}=m\mathbf{a}$, can be rewritten by using the definition of acceleration as "the change in velocity divided by the time interval."

$$\mathbf{F} = m\mathbf{a} = m\frac{\Delta \mathbf{v}}{\Delta t}$$

Multiplying both sides of the equation by the time interval, Δt, results in the following equation.

$$\mathbf{F}\Delta t = m\Delta \mathbf{v}$$

The left-hand side, $\mathbf{F}\Delta t$, is the product of the average force and the time interval over which it acts. This product is called the **impulse,** and its unit of measurement is the newton-second (N·s). The magnitude of an impulse is found by determining the area under the curve of a force-time graph, such as the one shown in **Figure 9–2.**

The right-hand side of the equation, $m\Delta \mathbf{v}$, shows the change in velocity, $\Delta \mathbf{v} = \mathbf{v}_2 - \mathbf{v}_1$, which also can be stated as $m\mathbf{v}_2 - m\mathbf{v}_1$. The product of mass and velocity of an object such as a tennis ball is defined as the **linear momentum** (plural: momenta) of the object. The symbol for momentum is $\mathbf{p}$. Thus, $\mathbf{p} = m\mathbf{v}$. The right-hand side of the equation can be written $\mathbf{p}_2 - \mathbf{p}_1$, which expresses the change in momentum of the tennis ball. Thus, the impulse on an object is equal to the change in its momentum.

$$\mathbf{F}\Delta t = \mathbf{p}_2 - \mathbf{p}_1$$

This equation is called the **impulse-momentum theorem.** The impulse on an object is equal to the change in momentum that it causes. If the force is constant, the impulse is simply the product of the force times the time interval over which it acts. Generally, the force is not constant, and the impulse is found by using an average force times the time interval, or by finding the area under the curve on a force-time graph.

F.Y.I.

Each time a runner's foot strikes the ground, it must absorb the force of two to four times the runner's weight. The goal of athletic shoe design is to reduce the stress on the foot. By using materials that lengthen the time of impact on the foot, the force of the impact on the foot is reduced.

Using the impulse-momentum theorem

What is the change in momentum of the tennis ball? From the impulse-momentum theorem, you know that the change in momentum is equal to the impulse. The impulse on the tennis ball can be calculated by using the force-time graph. In **Figure 9–2,** the area under the curve is approximately 1.4 N·s. Therefore, the change in momentum of the

F.Y.I.

A fully loaded supertanker traveling at a normal speed of 16 knots, about 18 miles/hour, will require at least 20 minutes to stop. The supertanker would collide with an immobile object at least 3 miles away.

ball is also 1.4 N·s. Because one newton-second is equal to one kg·m/s, the momentum gained by the ball is 1.4 kg·m/s.

What is the momentum of the ball after the hit? Rearrange the impulse-momentum theorem to answer this question.

$$\boldsymbol{p}_2 = \boldsymbol{F}\Delta t + \boldsymbol{p}_1$$

You can see now that the ball's final momentum is the sum of the initial momentum and the impulse. If the tennis ball was at rest before it was hit, its final momentum is equal to the impulse, 1.4 kg·m/s.

$$p_2 = mv = 1.4 \text{ kg·m/s}$$

If the ball has a mass of 0.060 kg, then its velocity will be 23 m/s.

$$v = \frac{p_2}{m} = \frac{1.4 \text{ kg·m/s}}{0.060 \text{ kg}} = 23 \text{ m/s}$$

Physics & Technology

High-Tech Tennis Rackets

Strings along the outer edges of a tennis racket are less flexible than the strings at the center. The more flexible area at the center of a racket is known as the "sweet spot." Striking a tennis ball near the edge of the racket imparts greater momentum to the ball, but the shock of the impact is transferred to the player's arm. Hitting a ball at the sweet spot imparts less momentum, but the strings absorb more of the shock of impact, thereby increasing the player's control and reducing the risk of injury.

Sports enthusiasts are always willing to try new technologies that could help improve their game. One of the goals of tennis equipment manufacturers is to design a racket with a larger sweet spot, so that players who don't always hit the ball with the center of the racket won't suffer arm injuries. Using information developed during research on how best to connect platforms in space, NASA researchers discovered that using strings that are thicker in the center and thinner near the edges of the racket enlarges the sweet spot. But racket makers found that implementing this idea is too complicated for practical use. The NASA researchers then developed a way to chemically treat strings so that they become more flexible as they are stretched tighter. In the manufacture of tennis rackets, these new strings enlarge the sweet spot.

The sweet spot can also be enlarged by widening the upper portion of the racket frame or using a thinner gauge string, more flexible string material, or less string tension.

Thinking Critically Why does striking a tennis ball with taut strings at the edge of the racket impart more speed to the ball than striking it at the sweet spot?

Because velocity is a vector, so is momentum. Similarly, because force is a vector, so is impulse. This means that signs are important for motion in one dimension. If you choose the positive direction to be to the right, then negative velocities, momenta, and impulses will be directed to the left.

Using the impulse-momentum theorem to save lives

A large change in momentum occurs only when there is a large impulse. A large impulse, however, can result either from a large force acting over a short period of time, or from a smaller force acting over a longer period of time.

What happens to the driver when a crash suddenly stops a car? An impulse is needed to bring the driver's momentum to zero. The steering wheel can exert a large force during a short period of time. An air bag reduces the force exerted on the driver by greatly increasing the length of the time the force is exerted. If you refer back to the equation

$$\mathbf{F} = m\frac{\Delta \mathbf{v}}{\Delta t}$$

$\Delta \mathbf{v}$ is the same with or without the air bag. However, the air bag reduces $\mathbf{F}$ by increasing Δt. The product of the average force and the time interval of the crash would be the same for both kinds of crashes. Remember that mass has not changed and the change in velocity will not be any different regardless of the time needed to stop.

How Safe?

Example Problem

Stopping a Vehicle

A 2200-kg sport utility vehicle (SUV) traveling at 94 km/h (26 m/s) can be stopped in 21 s by gently applying the brakes, in 5.5 s in a panic stop, or in 0.22 s if it hits a concrete wall. What average force is exerted on the SUV in each of these stops?

Sketch the Problem

- Sketch the system before and after the event.
- Show the SUV coming to rest. Label the velocity vectors.
- Include a coordinate axis to select the positive direction.
- Draw a vector diagram for momentum and impulse.

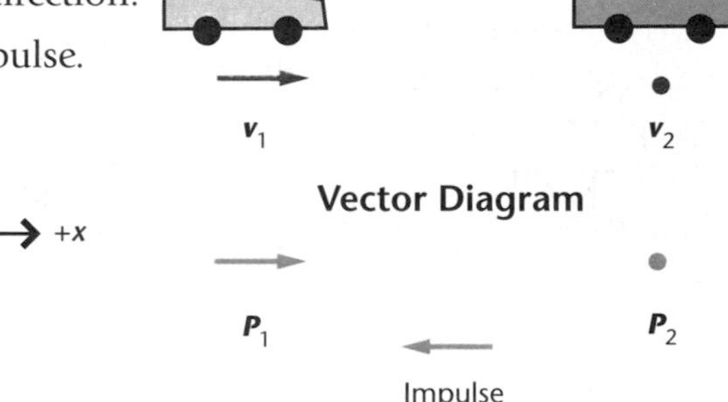

Calculate Your Answer

Known:	Unknown:
$m = 2200$ kg	$F = ?$
$v_1 = 26$ m/s	
$v_2 = 0$ m/s	
Δt: 21 s, 5.5 s, 0.22 s	

Strategy:

Determine the momentum before, p_1, and after, p_2, the crash.

Calculations:

$p_1 = mv_1 = (2200 \text{ kg})(26 \text{ m/s})$
$= 5.7 \times 10^4 \text{ kg·m/s}$

$p_2 = mv_2 = 0$

Apply the impulse-momentum theorem to obtain the force needed to stop the SUV.

$F\Delta t = p_2 - p_1$

$F\Delta t = -5.7 \times 10^4 \text{ kg·m/s}$

$F = (-5.7 \times 10^4 \text{ kg·m/s})/\Delta t$

For gentle braking	$F = -2.7 \times 10^3 \text{ N} = -2700 \text{ N}$
For panic braking	$F = -1.0 \times 10^4 \text{ N} = -10\ 000 \text{ N}$
When hitting the wall	$F = -2.6 \times 10^5 \text{ N} = -260\ 000 \text{ N}$

Check Your Answer

- Are the units correct? Force is measured in newtons.
- Is the magnitude realistic? People weigh hundreds of newtons, so you would expect that the force to stop a car would be in the thousands of newtons. The impulse is the same for all three stops. So, as the stopping time is shortened by a factor of ten, the force is increased by a factor of ten.
- Does the direction make sense? Force is negative; it pushes back against the motion of the car.

Practice Problems

1. A compact car, mass 725 kg, is moving at 100 km/h toward the east. Sketch the moving car.

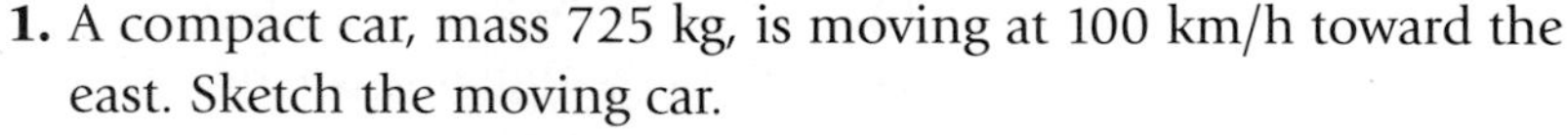

   a. Find the magnitude and direction of its momentum. Draw an arrow on your picture showing the momentum.

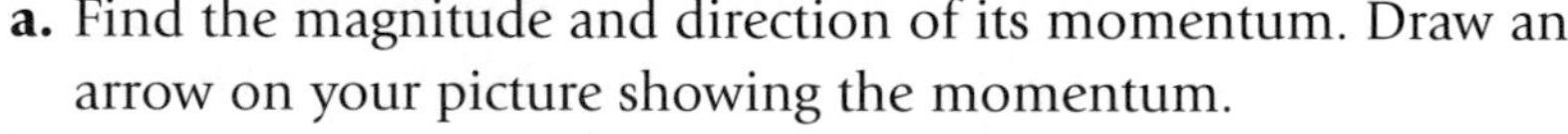

   b. A second car, mass 2175 kg, has the same momentum. What is its velocity?

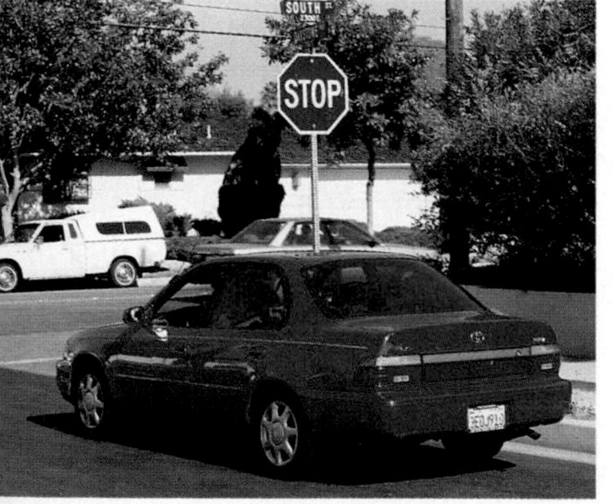

2. The driver of the compact car suddenly applies the brakes hard for 2.0 s. As a result, an average force of 5.0×10^3 N is exerted on the car to slow it. Sketch the situation.
   a. What is the change in momentum, that is the magnitude and direction of the impulse, on the car?
   b. Complete the "before" and "after" diagrams, and determine the new momentum of the car.
   c. What is the velocity of the car now?
3. A 7.0-kg bowling ball is rolling down the alley with a velocity of 2.0 m/s. For each impulse, **a** and **b,** as shown in **Figure 9–3,** find the resulting speed and direction of motion of the bowling ball.

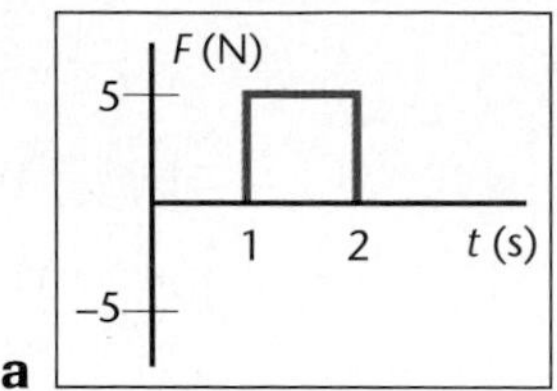

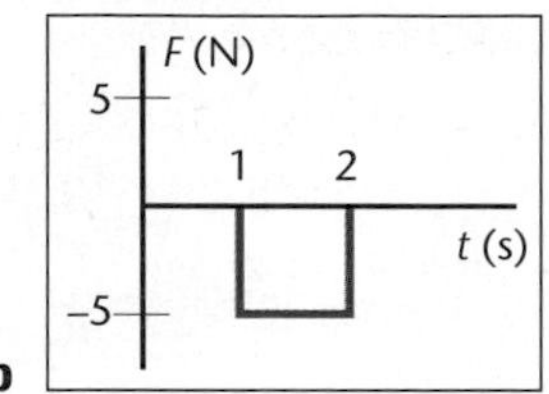

FIGURE 9–3

4. The driver accelerates a 240.0 kg snowmobile, which results in a force being exerted that speeds the snowmobile up from 6.00 m/s to 28.0 m/s over a time interval of 60.0 s.
   a. Sketch the event, showing the initial and final situations.
   b. What is the snowmobile's change in momentum? What is the impulse on the snowmobile?
   c. What is the magnitude of the average force that is exerted on the snowmobile?
5. A 0.144-kg baseball is pitched horizontally at 38.0 m/s. After it is hit by the bat, it moves at the same speed, but in the opposite direction.
   a. Draw arrows showing the ball's momentum before and after it hits the bat.
   b. What was the change in momentum of the ball?
   c. What was the impulse delivered by the bat?
   d. If the bat and ball were in contact for 0.80 ms, what was the average force the bat exerted on the ball?
6. A 60-kg person was in the car that hit the concrete wall in the example problem. The velocity of the person equals that of the car both before and after the crash, and the velocity changes in 0.20 s. Sketch the problem.
   a. What is the average force exerted on the person?
   b. Some people think that they can stop themselves rushing forward by putting their hands on the dashboard. Find the mass of the object that has a weight equal to the force you just calculated. Could you lift such a mass? Are you strong enough to stop yourself with your arms?

Angular Momentum

As you have seen in Chapter 7, if an object rotates, its speed changes only if torque is applied to it. This is a statement of Newton's law for rotating objects. The quantity of angular motion that is used with rotating objects is called angular momentum. **Angular momentum** is the quantity of motion used with objects rotating about a fixed axis. Just as the linear momentum of an object changes when force acts on the object, the angular momentum of an object changes when torque acts on the object.

Pocket Lab

Cart Momentum

Attach a spring scale to a laboratory cart. First, pull the cart for 1.0 s while exerting 1.0 N of force. Next, pull the cart for 2.0 s while exerting about 0.50 N of force. Predict which trial will give the cart more acceleration. Explain. Predict which trial will give the cart more velocity. Explain. Then try it.

Recognizing Cause and Effect Which factor, F or Δt, seems to be more important in changing the velocity of the cart?

FINE ARTS CONNECTION

Alexander Calder (1898–1976), one of the first American sculptors to gain international recognition, received a degree in mechanical engineering in 1919. His training remained evident in his sculptures. Calder suspended abstract metal shapes from wires and delicately balanced them so they could be set into motion when slight currents of air provided the impulse. Alexander Calder created the mobile, a work of art in which movement is beautiful.

FIGURE 9–4 This hurricane was photographed from space. The huge, rotating mass of air possesses a large angular momentum.

Linear momentum is a product of an object's mass and velocity $\boldsymbol{p} = m\boldsymbol{v}$. Angular momentum is a product of the object's mass, displacement from the center of rotation, and the component of velocity perpendicular to that displacement, as illustrated by **Figure 9–4.** If angular momentum is constant and the distance to the center of rotation decreases, then velocity increases. For example, the torque on the planets orbiting the sun is zero because the gravitational force is directly toward the sun. Therefore, each planet's angular momentum is constant. Thus, when a planet's distance from the sun becomes smaller, the planet moves faster. This is an explanation of Kepler's second law of planetary motion based on Newton's laws of motion.

9.1 Section Review

1.1 Is the momentum of a car traveling south different from that of the same car when it travels north at the same speed? Draw the momentum vectors to support your answer.

1.2 A basketball is dribbled. If its speed while going toward the floor is the same as it is when it rises from the floor, is the ball's change in momentum equal to zero when it hits the floor? If not, in which direction is the change in momentum? Draw the ball's momentum vectors before and after it hits the floor.

1.3 Which has more momentum, a supertanker tied to a dock or a raindrop falling?

1.4 If you jump off a table, you let your legs bend at the knees as your feet hit the floor. Explain why you do this in terms of the physics concepts introduced in this chapter.

1.5 **Critical Thinking** An archer shoots arrows at a target. Some arrows stick in the target, while others bounce off. Assuming that their masses and velocities are the same, which arrows give a bigger impulse to the target? **Hint**: Draw a diagram to show the momentum of the arrows before and after hitting the target for the two cases.

The Conservation of Momentum 9.2

You have seen how a force applied during a time interval changes the momentum of a tennis ball. But in the discussion of Newton's third law of motion, you learned that forces are the result of interactions between objects moving in opposite directions. The force of a tennis racket on the ball is accompanied by an equal and opposite force of the ball on the racket. Is the momentum of the racket, therefore, also changed?

OBJECTIVES

- **Relate** Newton's third law of motion to conservation of momentum in collisions and explosions.
- **Recognize** the conditions under which the momentum of a system is conserved.
- **Apply** conservation of momentum to explain the propulsion of rockets.
- **Solve** conservation of momentum problems in two dimensions by using vector analysis.

Two-Particle Collisions

Although it would be simple to consider the tennis racket as a single object, the racket, the hand of the player, and the ground on which the player is standing are all objects that interact when the tennis player hits the ball. To begin your study of interactions in collisions, examine the much simpler system, shown in **Figure 9–5.**

During the collision of two balls, each briefly exerts a force on the other. Despite the differences in sizes and velocities of the balls, the forces they exert on each other are equal and opposite, according to Newton's third law of motion. These forces are represented by the following equation.

$$\boldsymbol{F}_{\text{B on A}} = -\boldsymbol{F}_{\text{A on B}}$$

Because the time intervals over which the forces are exerted are the same, how do the impulses received by both balls compare? They must be equal in magnitude but opposite in direction. How do the momenta of the balls compare after the collision?

According to the impulse-momentum theorem, the final momentum is equal to the initial momentum plus the impulse. Compare the momenta of the two balls.

For ball A: $\boldsymbol{p}_{A2} = \boldsymbol{F}_{\text{B on A}}\Delta t + \boldsymbol{p}_{A1}$

For ball B: $\boldsymbol{p}_{B2} = \boldsymbol{F}_{\text{A on B}}\Delta t + \boldsymbol{p}_{B1}$

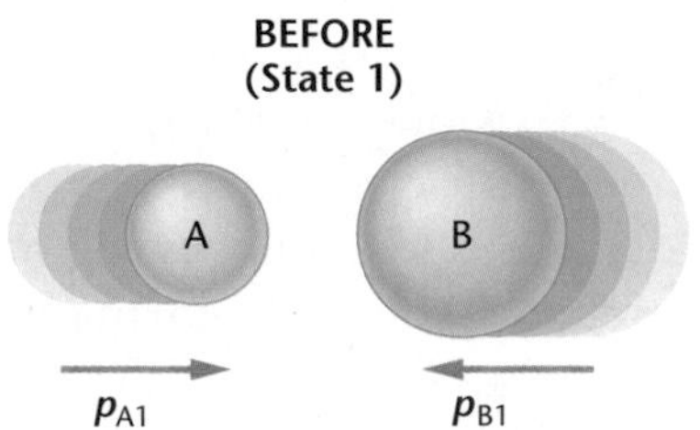

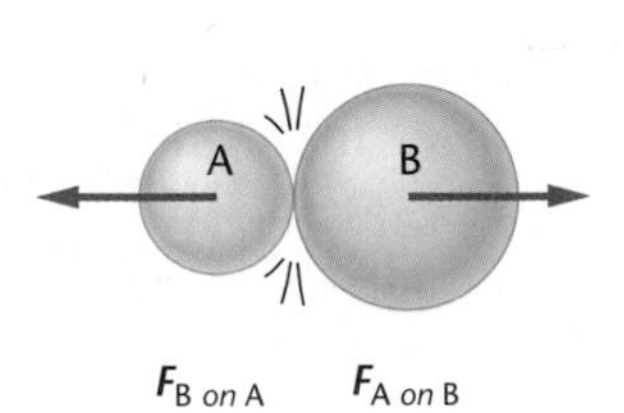

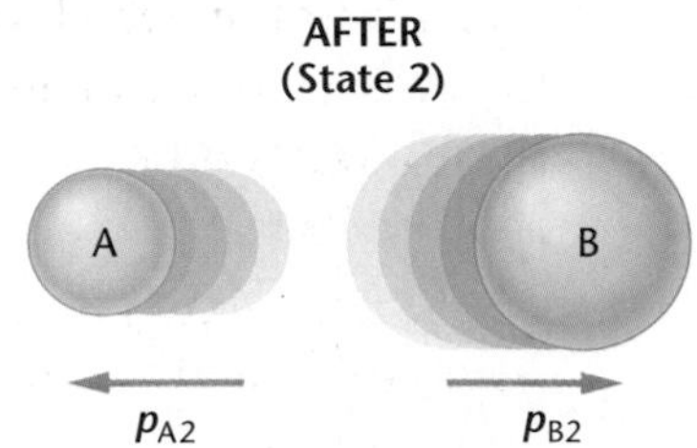

FIGURE 9–5 When two balls collide, they exert forces on each other, changing their momenta.

Pocket Lab

Skateboard Fun

Have two students sit facing each other on skateboards approximately 3 to 5 meters apart. Place a rope in their hands. Predict what will happen when one student pulls on the rope while the other just holds his or her end. Explain your prediction. Which person is exerting more force on the rope? Compare the amount of time that the force is acting on each person. Which person will have a greater change in momentum? Explain. Then try it. Describe what really happened.

Design An Experiment Can you devise a method to pull only one student to the other so that the other student doesn't move?

Use the result of Newton's third law of motion $-\boldsymbol{F}_{\text{A on B}} = \boldsymbol{F}_{\text{B on A}}$.

$$\boldsymbol{p}_{A2} = -\boldsymbol{F}_{\text{A on B}}\Delta t + \boldsymbol{p}_{A1}$$

Add the momenta of the two balls.

$$\begin{aligned} \boldsymbol{p}_{A2} &= -\boldsymbol{F}_{\text{A on B}}\Delta t + \boldsymbol{p}_{A1} \\ \boldsymbol{p}_{B2} &= \boldsymbol{F}_{\text{A on B}}\Delta t + \boldsymbol{p}_{B1} \\ \hline \boldsymbol{p}_{A2} + \boldsymbol{p}_{B2} &= \boldsymbol{p}_{A1} + \boldsymbol{p}_{B1} \end{aligned}$$

This shows that the sum of the momenta of the balls is the same before and after the collision. That is, the momentum gained by ball 2 is equal to the momentum lost by ball 1. If the system is defined as the two balls, the momentum of the system is constant. For the system, momentum is conserved.

Momentum in a Closed System

Under what conditions is the momentum of the system of two balls conserved? The first and most obvious condition is that at all times only two balls collide. No balls are lost, and none are gained. A system that doesn't gain or lose mass is said to be a **closed system.** All the forces within a closed system are **internal forces.** The second condition required to conserve momentum of the system is that the only forces involved are internal forces. All the forces outside the system are **external forces.** When the net external force on a closed system is zero, it is described as an **isolated system.** No system on Earth can be said to be absolutely closed and isolated. That is, there will always be some interaction between a system and its environment. Often, these interactions are small and can be ignored when solving physics problems.

Systems can contain any number of objects, and the objects can stick together or come apart in the collision. Under these conditions, the **law of conservation of momentum** states that the momentum of any closed system with no net external force does not change. This law will enable you to make a connection between conditions before and after an interaction without knowing any of the details of the interaction.

A flask filled with gas and closed with a stopper, as shown in **Figure 9–6,** is a system with many particles. The gas molecules are in constant, random motion at all temperatures above absolute zero, and they are constantly colliding with each other and with the walls of the flask. The momenta of the particles are changing with every collision. In a two-particle collision, the momentum gained by one particle is equal to that lost by the other. Momentum is also conserved in collisions between particles and the flask wall. Although the wall's velocity might change very slightly in each collision, there are as many momenta of particles to the right as to the left, and as many up as down, so the net change in the momentum of the flask is zero. The total momentum of the system doesn't change; it is conserved.

FIGURE 9–6 The total momentum of a closed, isolated system is constant.

Example Problem

Car Collisions

A 2275-kg car going 28 m/s rear-ends an 875-kg compact car going 16 m/s on ice in the same direction. The two cars stick together. How fast does the wreckage move immediately after the collision?

Sketch the Problem

- Establish a coordinate axis.
- Show the before and after states.
- Label car A and car B and include velocities.
- Draw a vector diagram for the momentum.
- The length of the arrow representing the momentum after the collision equals the sum of the lengths of the arrows for the momenta before the collision.

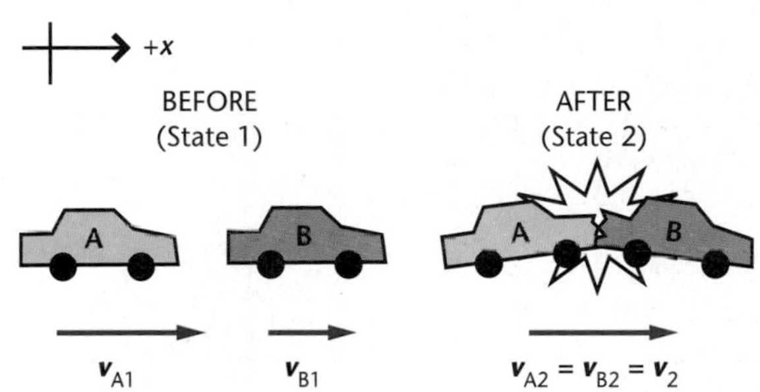

Calculate Your Answer

Vector Diagram

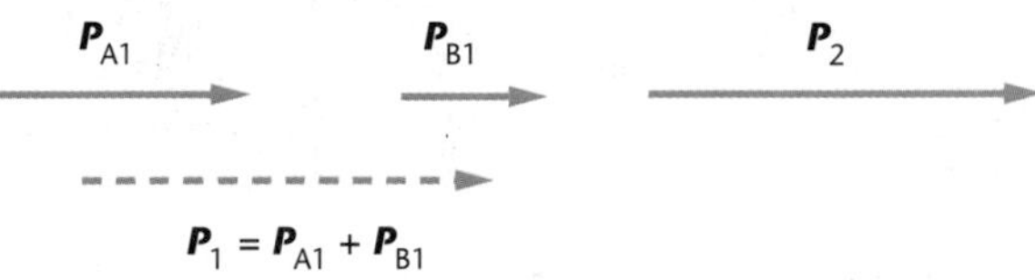

Known:

$m_A = 2275$ kg
$v_{A1} = 28$ m/s
$m_B = 875$ kg
$v_{B1} = 16$ m/s

Unknown:

$v_2 = ?$

Strategy:

The law of conservation of momentum can be used because the ice makes total external force on the cars nearly zero.

Because the two cars stick together, their velocities after the collision, denoted as v_2, are equal.

Calculations:

$$p_1 = p_2$$
$$p_{A1} + p_{B1} = p_{A2} + p_{B2}$$
$$m_A v_{A1} + m_B v_{B1} = m_A v_{A2} + m_B v_{B2}$$

$$v_{A2} = v_{B2} = v_2$$
$$m_A v_{A1} + m_B v_{B1} = (m_A + m_B)\, v_2$$
$$v_2 = \frac{m_A v_{A1} + m_B v_{B1}}{m_A + m_B}$$

$$v_2 = \frac{(2275 \text{ kg})(28 \text{ m/s}) + (875 \text{ kg})(16 \text{ m/s})}{2275 \text{ kg} + 875 \text{ kg}}$$
$$v_2 = 25 \text{ m/s}$$

Check Your Answer

- Are the units correct? The correct unit for speed is m/s.
- Does the direction make sense? All the initial speeds are in the positive direction. You would, therefore, expect v_2 to be positive.
- Is the magnitude realistic? The magnitude of v_2 is between the initial speeds of the two cars, so it is reasonable.

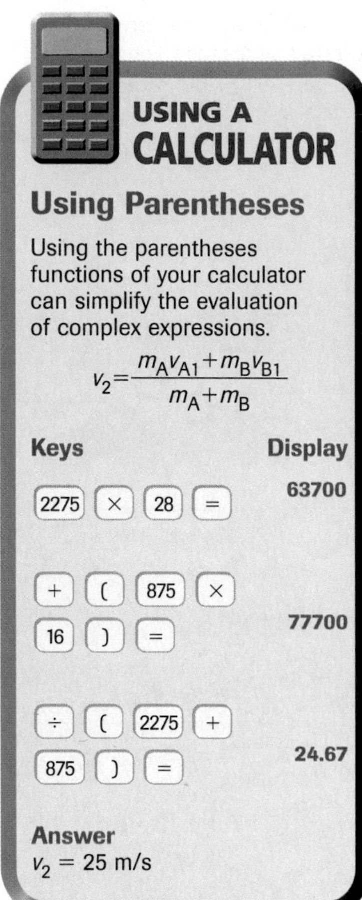

Practice Problems

7. Two freight cars, each with a mass of 3.0×10^5 kg, collide. One was initially moving at 2.2 m/s; the other was at rest. They stick together. What is their final speed?
8. A 0.105-kg hockey puck moving at 24 m/s is caught and held by a 75-kg goalie at rest. With what speed does the goalie slide on the ice?
9. A 35.0-g bullet strikes a 5.0-kg stationary wooden block and embeds itself in the block. The block and bullet fly off together at 8.6 m/s. What was the original speed of the bullet?
10. A 35.0-g bullet moving at 475 m/s strikes a 2.5-kg wooden block that is at rest. The bullet passes through the block, leaving at 275 m/s. How fast is the block moving when the bullet leaves?
11. Glider A, with a mass of 0.355 kg, moves along a frictionless air track with a velocity of 0.095 m/s, as in **Figure 9–7.** It collides with glider B, with a mass of 0.710 kg and a speed of 0.045 m/s in the same direction. After the collision, glider A continues in the same direction at 0.035 m/s. What is the speed of glider B?

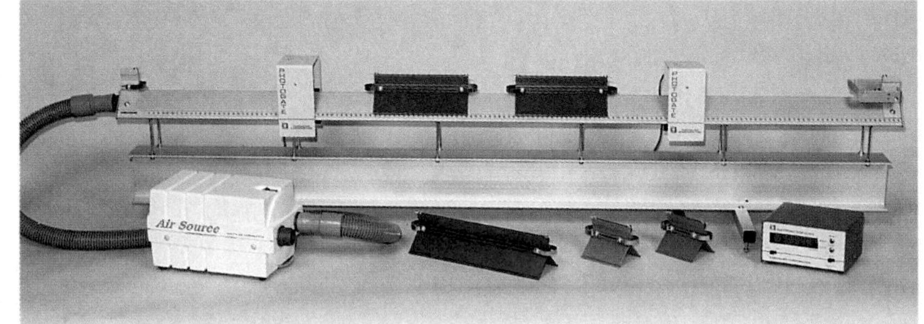

FIGURE 9–7

12. A 0.50-kg ball traveling at 6.0 m/s collides head-on with a 1.00-kg ball moving in the opposite direction at a speed of 12.0 m/s. The 0.50-kg ball bounces backward at 14 m/s after the collision. Find the speed of the second ball after the collision.

Explosions

You have seen how important it is to define each system carefully. The momentum of the tennis ball changed when the external force of the racket was exerted on it. The tennis ball was not an isolated system. On the other hand, the total momentum of the two colliding balls within the isolated system didn't change because all forces were between objects within the system.

Can you find the final velocities of the two in-line skaters in **Figure 9–8?** Assume that they are skating on such a smooth surface that there are no external forces. They both start at rest one behind the other.

FIGURE 9–8 The internal forces exerted by these in-line skaters cannot change the total momentum of the system.

Skater A gives skater B a push. Now both skaters are moving, making this situation similar to that of an explosion. Because the push was an internal force, you can use the law of conservation of momentum to find the skaters' relative velocities. The total momentum of the system was zero before the push. Therefore, it also must be zero after the push.

	BEFORE (State 1)		AFTER (State 2)
	$\boldsymbol{p}_{A1} + \boldsymbol{p}_{B1}$	$=$	$\boldsymbol{p}_{A2} + \boldsymbol{p}_{B2}$
	0	$=$	$\boldsymbol{p}_{A2} + \boldsymbol{p}_{B2}$
or:	$\boldsymbol{p}_{A2}$	$=$	$-\boldsymbol{p}_{B2}$
	$m_A\boldsymbol{v}_{A2}$	$=$	$-m_B\boldsymbol{v}_{B2}$

The momenta of the skaters after the push are equal in magnitude but opposite in direction. The backward motion of skater A is an example of recoil. Are the skaters' velocities equal and opposite? Solve the last equation for the velocity of skater A.

$$\boldsymbol{v}_{A2} = -\left(\frac{m_B}{m_A}\right)\boldsymbol{v}_{B2}$$

The velocities depend on the skaters' relative masses. If skater A has a mass of 45.0 kg and skater B's mass is 60.0 kg, then the ratio of their velocities will be 60/45 or 1.33. The less massive skater moves at the greater velocity. But, without more information about how hard they pushed, you can't find the velocity of each skater.

Explosions in space

How does a rocket in space change its velocity? The rocket carries both fuel and oxidizer. They are combined chemically in the rocket motor, and the resulting hot gases leave the exhaust nozzle at high speed. If the rocket and chemicals are the system, then the system is closed. The forces that expel the gases are internal forces, so the system is also isolated. Therefore, the law of conservation of momentum can be applied to this situation. The movement of an astronaut in space can be used to demonstrate an isolated system.

F.Y.I.

Forensic investigations frequently involve the study of momentum. Careful analysis of skid marks, bullet tracks, wounds, and cracks in fragile materials can indicate the initial velocities of moving objects and provide important evidence about crimes.

Example Problem

Recoil of an Astronaut

An astronaut at rest in space fires a thruster pistol that expels 35 g of hot gas at 875 m/s. The combined mass of the astronaut and pistol is 84 kg. How fast and in what direction is the astronaut moving after firing the pistol?

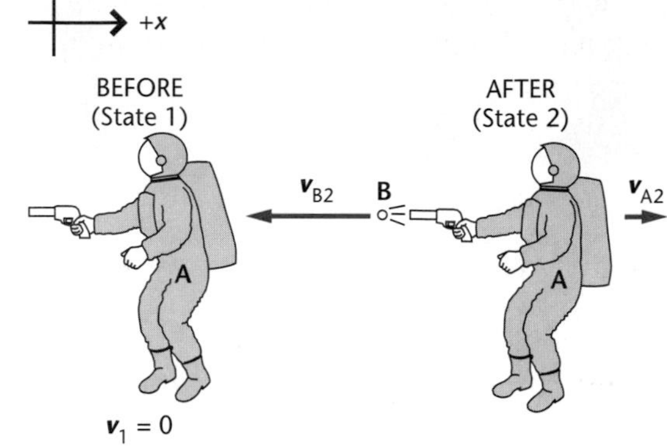

Sketch the Problem

- Establish a coordinate axis.
- Show the before and after conditions.
- Label the astronaut A and the expelled gas B, and include their velocities.
- Draw a vector diagram including all momenta.

Calculate Your Answer

Known:	**Unknown:**
$m_A = 84$ kg	$v_{A2} = ?$
$m_B = 0.035$ kg	
$v_{A1} = v_{B1} = 0$ m/s	
$v_{B2} = -875$ m/s	

Strategy:

Before the firing, both parts of the system are at rest, thus, initial momentum is zero.

The momentum of the astronaut is equal in magnitude but opposite in direction to that of the gas leaving the pistol.

Use the law of conservation of momentum to find p_2 and solve for the final velocity of the astronaut, v_{A2}

Calculations:

$$p_1 = p_{A1} + p_{B1} = 0$$

$$p_{A1} + p_{B1} = p_{A2} + p_{B2}$$
$$0 = p_{A2} + p_{B2}$$
$$p_{A2} = -p_{B2}$$
$$m_A v_{A2} = -m_B v_{B2} \text{ or } v_{A2} = \left(\frac{-m_B v_{B2}}{m_A}\right)$$

$$v_{A2} = \frac{-(0.035 \text{ kg})(-875 \text{ m/s})}{84 \text{ kg}} = +0.36 \text{ m/s}$$

Check Your Answer

- Are the units correct? The correct unit for velocity is m/s.
- Do the direction and magnitude make sense? The astronaut's mass is much larger than that of the gas. So the velocity of the astronaut is much less than that of the expelled gas, and opposite in direction.

Have you ever wondered how a rocket can accelerate in space? In this example, you see that the astronaut didn't push on anything external. According to Newton's third law, the pistol pushes the gases out, and the gases in turn push on the pistol and the astronaut. All the system's forces are internal.

Physics Lab

The Explosion

Problem

How do the forces and changes in momenta acting on different masses compare during an explosion?

Materials

two laboratory carts
(one with a spring mechanism)
two C-clamps
two blocks of wood
20-N spring balance
0.50-kg mass
stopwatch
masking tape

Procedure

1. Securely tape the 0.50-kg mass to cart 2 and then use the balance to determine the mass of each cart.
2. Arrange the equipment as shown in the diagram.
3. Predict the starting position so that the carts will hit the blocks at the same instant when the spring mechanism is released.
4. Place pieces of tape on the table at the front of the carts to mark starting positions.
5. Depress the mechanism to release the spring and explode the carts.
6. Notice which cart hits the block first.
7. Adjust the starting position until the carts hit at the same instant. Remember to move the tapes.

Data and Observations

1. Which cart moved farther?
2. Which cart moved faster? Explain.

Analyze and Conclude

1. **Estimating** Estimate the velocity of each cart. Which was greater?
2. **Comparing** Compare the change in momentum of each cart.
3. **Applying** Suppose that the spring pushed on cart 1 for 0.05 s. How long did cart 2 push on the spring? Explain.
4. **Comparing** Using $\boldsymbol{F}\Delta t = m\Delta \boldsymbol{v}$, which cart had the greater impulse?

Apply

1. Explain why a target shooter might prefer to shoot a more massive gun.

HELP WANTED

PARAMEDIC TRAINEE

Trainee position open to Emergency Medical Technicians certified in basic life support. Will complete up to 2000 hours of advanced training and work with patients in hospital and field settings. Must be skilled in communicating with both medical staff and injured, frightened patients. Position requires strength, endurance, coordination, high tolerance for stress, and excellent driving skills. Must also pass state exam. For information contact:

American Ambulance Association
3800 Auburn Blvd.
Suite C
Sacramento, CA 95821 or
National Registry of Emergency Medical Technicians
Box 29233
6610 Busch Blvd.
Columbus, OH 43229

Practice Problems

13. A 4.00-kg model rocket is launched, shooting 50.0 g of burned fuel from its exhaust at a speed of 625 m/s. What is the velocity of the rocket after the fuel has burned? **Hint:** Ignore the external forces of gravity and air resistance.
14. A thread holds two carts together, as shown in **Figure 9–9.** After the thread is burned, a compressed spring pushes the carts apart, giving the 1.5-kg cart a speed of 27 cm/s to the left. What is the velocity of the 4.5-kg cart?

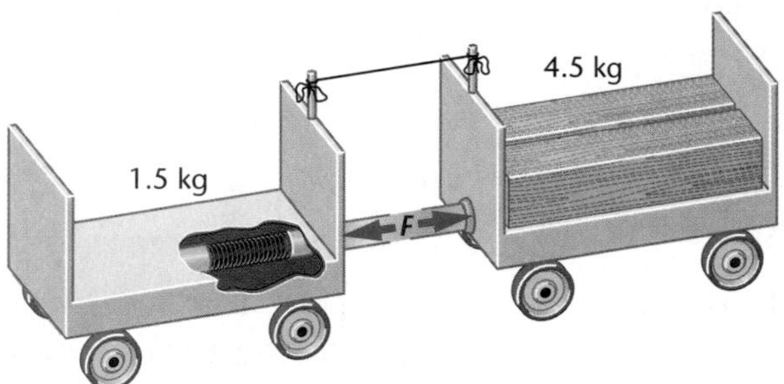

FIGURE 9–9

15. Two campers dock a canoe. One camper has a mass of 80.0 kg and moves forward at 4.0 m/s as she leaves the boat to step onto the dock. With what speed and direction do the canoe and the other camper move if their combined mass is 115 kg?
16. A colonial gunner sets up his 225-kg cannon at the edge of the flat top of a high tower. It shoots a 4.5-kg cannonball horizontally. The ball hits the ground 215 m from the base of the tower. The cannon also moves on frictionless wheels, falls off the back of the tower, and lands on the ground.
    a. What is the horizontal distance of the cannon's landing, measured from the base of the back of the tower?
    b. Why don't you need to know the width of the tower?

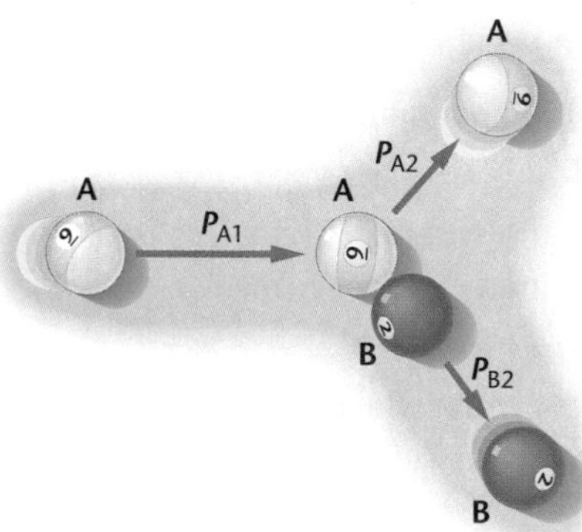

FIGURE 9–10 The law of conservation of momentum holds for all isolated, closed systems, regardless of the directions of objects before and after collision.

Two-Dimensional Collisions

Up until now, you have looked at momentum in one dimension only. The law of conservation of momentum holds for all closed systems with no external forces. It is valid regardless of the directions of the particles before or after they interact.

Now you will look at momentum in two dimensions. **Figure 9–10** shows the result of billiard ball A striking stationary ball B. Consider the two balls to be the system. The original momentum of the moving ball is $\boldsymbol{p}_{A1}$; the momentum of the stationary ball is zero. Therefore, the momentum of the system before the collision is equal to $\boldsymbol{p}_{A1}$.

After the collision, both balls are moving and have momenta.

According to the law of conservation of momentum, the initial momentum equals the vector sum of the final momenta, so $\boldsymbol{p}_{A1} = \boldsymbol{p}_{A2} + \boldsymbol{p}_{B2}$.

The equality of the momentum before and after the collision also means that the sum of the components of the vectors before and after the collision must be equal. If you define the *x*-axis to be in the direction of the initial momentum, the *y*-component of the initial momentum is zero. Therefore, the sum of the final *y*-components must be zero.

$$p_{A2y} + p_{B2y} = 0$$

They are equal in magnitude but have opposite signs. The sum of the horizontal components is also equal.

$$p_{A1} = p_{A2x} + p_{B2x}$$

Example Problem

Two-Dimensional Collision

A 2.00-kg ball, A, is moving at a speed of 5.00 m/s. It collides with a stationary ball, B, of the same mass. After the collision, A moves off in a direction 30.0° to the left of its original direction. Ball B moves off in a direction 90.0° to the right of ball A's final direction. How fast are they moving after the collision?

BEFORE (State 1)
A $\boldsymbol{v}_{A1}$ B $\boldsymbol{v}_{B1} = 0$
+y +x

AFTER (State 2)
A $\boldsymbol{v}_{A2}$
B $\boldsymbol{v}_{B2}$

Sketch the Problem

- Sketch the before and after states.
- Establish the coordinate axis with the *x*-axis in the original direction of ball A.
- Draw a momentum vector diagram. Note that p_{A2} and p_{B2} form a 90° angle.

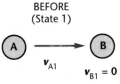

Calculate Your Answer

Known:

$m_A = m_B = 2.00$ kg
$v_{A1} = 5.00$ m/s
$v_{B1} = 0$

Unknowns:

$v_{A2} = ?$
$v_{B2} = ?$

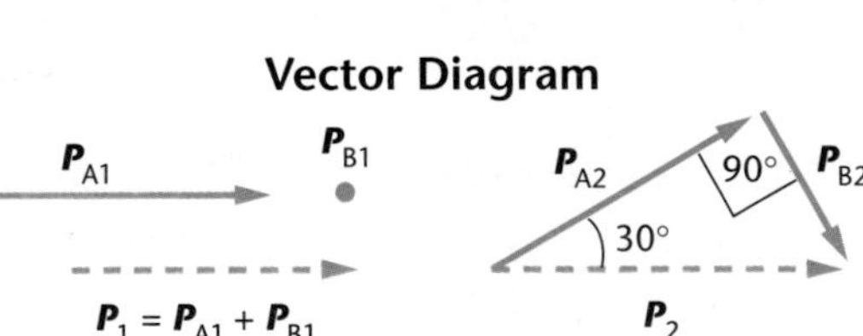

Strategy:

Determine the initial momenta.

Calculations:

$p_{A1} = m_A v_{A1} = (2.00 \text{ kg})(5.00 \text{ m/s}) = 10.0 \text{ kg·m/s}$
$p_{B1} = 0$
$p_1 = p_{A1} + p_{B1} = 10.0 \text{ kg·m/s}$

Use conservation of momentum to find p_2.

$p_2 = p_1 = 10.0 \text{ kg·m/s}$

Use the diagram to set up equations for p_{A2} and p_{B2}.

$p_{A2} = p_2 \cos 30.0°$
$= (10.0 \text{ kg·m/s})(0.866)$
$= 8.66 \text{ kg·m/s}$

$p_{B2} = p_2 \sin 30.0°$
$= (10.0 \text{ kg·m/s})(0.500)$
$= 5.00 \text{ kg·m/s}$

Determine the final speeds:

$$v_{A2} = \frac{p_{A2}}{m_A} = \frac{8.66 \text{ kg·m/s}}{2.00 \text{ kg}} = 4.33 \text{ m/s}$$

$$v_{B2} = \frac{p_{B2}}{m_B} = \frac{5.00 \text{ kg·m/s}}{2.00 \text{ kg}} = 2.50 \text{ m/s}$$

Check Your Answer

- Are the units correct? The correct unit for speed is m/s.
- Does the sign make sense? Answers are both positive and at the appropriate angles.
- Is the magnitude realistic? In this system in which two equal masses collide, v_{A2} and v_{B2} must be smaller than v_{A1}.

Practice Problems

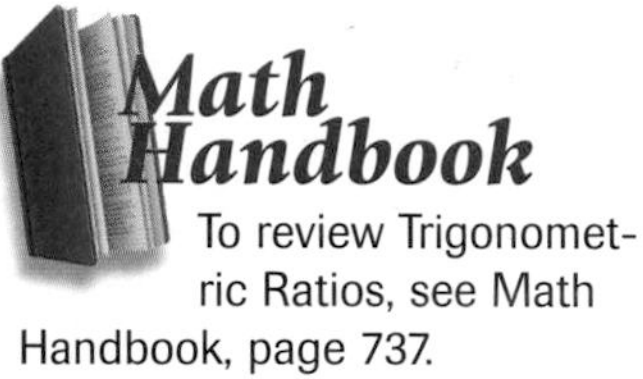

To review Trigonometric Ratios, see Math Handbook, page 737.

17. A 1325-kg car moving north at 27.0 m/s collides with a 2165-kg car moving east at 17.0 m/s. They stick together. In what direction and with what speed do they move after the collision?

18. A stationary billiard ball, mass 0.17 kg, is struck by an identical ball moving at 4.0 m/s. After the collision, the second ball moves off at 60° to the left of its original direction. The stationary ball moves off at 30° to the right of the moving ball's original direction. What is the velocity of each ball after the collision?

9.2 Section Review

2.1 Two soccer players come from opposite directions and collide when trying to head the ball. They come to rest in midair and fall to the ground. What can you say about their initial momenta?

2.2 During a tennis serve, the racket continues forward after hitting the ball. Is momentum conserved in the collision? Explain, making sure you are careful in defining the system.

2.3 A pole-vaulter runs toward the launch point with horizontal momentum. Where does the vertical momentum come from as the athlete vaults over the crossbar?

2.4 **Critical Thinking** You catch a heavy ball while you are standing on a skateboard, and roll backward. If you were standing on the ground, however, you would be able to avoid moving. Explain both results using the law of conservation of momentum. Explain the system you use in each case.

CHAPTER 9 REVIEW

ey Terms

ıpulse

ıear
omentum

ıpulse-
omentum
eorem

ıgular
omentum

osed system

ternal force

xternal force

olated system

w of
nservation of
omentum

Summary

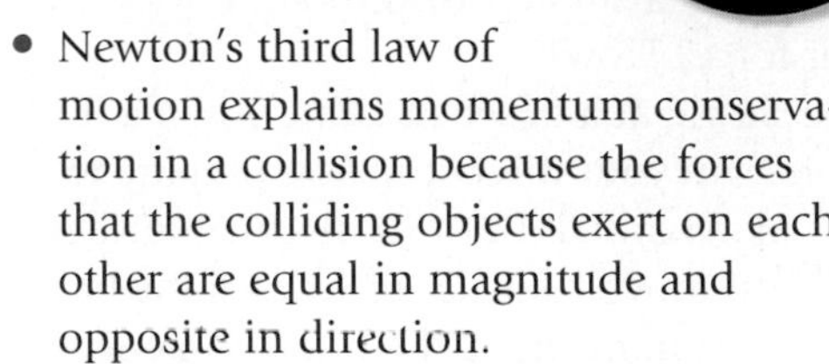

9.1 Impulse and Momentum

- When doing a momentum problem, first examine the system before and after the event.
- The momentum of an object is the product of its mass and velocity and is a vector quantity.
- The impulse given an object is the average net force exerted on the object multiplied by the time interval over which the force acts.
- The impulse given an object is equal to the change in momentum of the object.

9.2 The Conservation of Momentum

- Newton's third law of motion explains momentum conservation in a collision because the forces that the colliding objects exert on each other are equal in magnitude and opposite in direction.
- The momentum is conserved in a closed, isolated system.
- Conservation of momentum is used to explain the propulsion of rockets.
- Vector analysis is used to solve momentum-conservation problems in two dimensions.

Reviewing Concepts

Section 9.1

1. Can a bullet have the same momentum as a truck? Explain.

2. A pitcher throws a fastball to the catcher. Assuming that the speed of the ball doesn't change in flight,
   a. which player exerts the larger impulse on the ball?
   b. which player exerts the larger force on the ball?

3. Newton's second law of motion states that if no net force is exerted on a system, no acceleration is possible. Does it follow that no change in momentum can occur?

4. Why are cars made with bumpers that can be pushed in during a crash?

Section 9.2

5. What is meant by an isolated system?

6. A spacecraft in outer space increases its velocity by firing its rockets. How can hot gases escaping from its rocket engine change the velocity of the craft when there is nothing in space for the gases to push against?

7. The cue ball travels across the pool table and collides with the stationary eight-ball. The two balls have equal mass. After the collision, the cue ball is at rest. What must be true regarding the speed of the eight-ball?

8. Consider a ball falling toward Earth.
   a. Why is the momentum of the ball not conserved?
   b. In what system that includes the falling ball is the momentum conserved?

9. A falling ball hits the floor. Just before it hits, the momentum is in the downward direction, and the momentum is in the upward direction after it hits.
   a. The bounce is a collision, so why isn't the momentum of the ball conserved?
   b. In what system is it conserved?

10. Only an external force can change the momentum of a system. Explain how the internal force of a car's brakes brings the car to a stop.

Applying Concepts

11. Explain the concept of impulse using physical ideas rather than mathematics.

12. Is it possible for an object to obtain a larger impulse from a smaller force than it does from a larger force? How?

13. You are sitting at a baseball game when a foul ball comes in your direction. You prepare to catch it barehanded. In order to catch it safely, should you move your hands toward the ball, hold them still, or move them in the same direction as the moving ball? Explain.

14. A 0.11-g bullet leaves a pistol at 323 m/s, while a similar bullet leaves a rifle at 396 m/s. Explain the difference in exit speeds of the two bullets assuming that the forces exerted on the bullets by the expanding gases have the same magnitude.

15. An object initially at rest experiences the impulses described by the graph in **Figure 9–11.** Describe the object's motion after impulses A, B, and C.

FIGURE 9–11

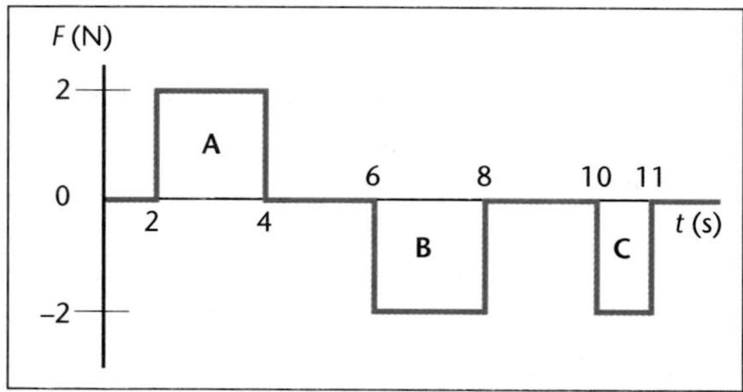

16. During a space walk, the tether connecting an astronaut to the spaceship breaks. Using a gas pistol, the astronaut manages to get back to the ship. Explain why this method was effective, using the language of the impulse-momentum theorem and a diagram.

17. As a tennis ball bounces off a wall, its momentum is reversed. Explain this action in terms of the law of conservation of momentum, defining the system and using a diagram.

18. You command *Spaceship Zero,* which is moving through interplanetary space at high speed. How could you slow your ship by applying the law of conservation of momentum?

19. Two trucks that appear to be identical collide on an icy road. One was originally at rest. The trucks stick together and move off at more than half the original speed of the moving truck. What can you conclude about the contents of the two trucks?

20. Explain, in terms of impulse and momentum, why it is advisable to place the butt of a rifle against your shoulder when first learning to shoot.

21. Two bullets of equal mass are shot at equal speeds at blocks of wood on a smooth ice rink. One bullet, which is made of rubber, bounces off the wood. The other bullet, made of aluminum, burrows into the wood. In which case does the wood move faster? Explain.

Problems

Section 9.1

LEVEL 1

22. Your brother's mass is 35.6 kg, and he has a 1.3-kg skateboard. What is the combined momentum of your brother and his skateboard if they are going 9.50 m/s?

23. A hockey player makes a slap shot, exerting a constant force of 30.0 N on the hockey puck for 0.16 s. What is the magnitude of the impulse given to the puck?

24. A hockey puck has a mass of 0.115 kg and is at rest. A hockey player makes a shot, exerting a constant force of 30.0 N on the puck for 0.16 s. With what speed does it head toward the goal?

25. Before a collision, a 25-kg object is moving at +12 m/s. Find the impulse that acted on the object if, after the collision, it moves at:
a. +8.0 m/s. **b.** −8.0 m/s.

26. A constant force of 6.00 N acts on a 3.00-kg object for 10.0 s. What are the changes in the object's momentum and velocity?

27. The velocity of a 625-kg auto is changed from 10.0 m/s to 44.0 m/s in 68.0 s by an external, constant force.
a. What is the resulting change in momentum of the car?
b. What is the magnitude of the force?

28. An 845-kg dragster accelerates from rest to 100 km/h in 0.90 seconds.
    a. What is the change in momentum of the car?
    b. What is the average force exerted on the car?
    c. What exerts that force?

LEVEL 2

29. A 0.150-kg ball, moving in the positive direction at 12 m/s, is acted on by the impulse shown in the graph in **Figure 9–12**. What is the ball's speed at 4.0 s?

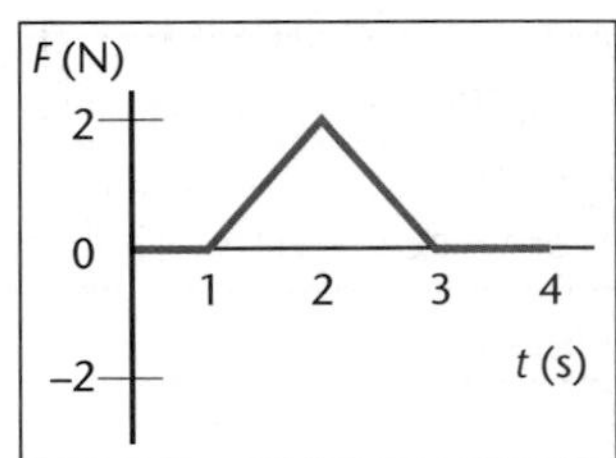

FIGURE 9–12

30. Small rockets are used to make tiny adjustments in the speed of satellites. One such rocket has a thrust of 35 N. If it is fired to change the velocity of a 72 000-kg spacecraft by 63 cm/s, how long should it be fired?
31. A car moving at 10 m/s crashes into a barrier and stops in 0.050 s. There is a 20-kg child in the car. Assume that the child's velocity is changed by the same amount as the car's in the same time period.
    a. What is the impulse needed to stop the child?
    b. What is the average force on the child?
    c. What is the approximate mass of an object whose weight equals the force in part **b**?
    d. Could you lift such a weight with your arm?
    e. Why is it advisable to use a proper infant restraint rather than hold a child on your lap?
32. An animal-rescue plane flying due east at 36.0 m/s drops a bale of hay from an altitude of 60.0 m. If the bale of hay weighs 175 N, what is the momentum of the bale the moment before it strikes the ground? Give both magnitude and direction.
33. A 60-kg dancer leaps 0.32 m high.
    a. With what momentum does the dancer reach the ground?
    b. What impulse is needed to stop the dancer?
    c. As the dancer lands, his knees bend, lengthening the stopping time to 0.050 s. Find the average force exerted on the dancer's body.
    d. Compare the stopping force to the dancer's weight.

Section 9.2

LEVEL 1

34. A 95-kg fullback, running at 8.2 m/s, collides in midair with a 128-kg defensive tackle moving in the opposite direction. Both players end up with zero speed.
    a. Identify "before" and "after" and make a diagram of the situations.
    b. What was the fullback's momentum before the collision?
    c. What was the change in the fullback's momentum?
    d. What was the change in the tackle's momentum?
    e. What was the tackle's original momentum?
    f. How fast was the tackle moving originally?
35. Marble A, mass 5.0 g, moves at a speed of 20.0 cm/s. It collides with a second marble, B, mass 10.0 g, moving at 10.0 cm/s in the same direction. After the collision, marble A continues with a speed of 8.0 cm/s in the same direction.
    a. Sketch the situation, identify the system, define "before" and "after," and assign a coordinate axis.
    b. Calculate the marbles' momenta before the collision.
    c. Calculate the momentum of marble A after the collision.
    d. Calculate the momentum of marble B after the collision.
    e. What is the speed of marble B after the collision?
36. A 2575-kg van runs into the back of a 825-kg compact car at rest. They move off together at 8.5 m/s. Assuming the friction with the road can be negligible, find the initial speed of the van.
37. A 0.115-kg hockey puck, moving at 35.0 m/s, strikes a 0.265-kg octopus thrown onto the ice by a hockey fan. The puck and octopus slide off together. Find their velocity.

38. A 50-kg woman, riding on a 10-kg cart, is moving east at 5.0 m/s. The woman jumps off the front of the cart and hits the ground at 7.0 m/s eastward, relative to the ground.
  a. Sketch the situation, identifying "before" and "after," and assigning a coordinate axis.
  b. Find the velocity of the cart after the woman jumps off.

39. Two students on roller skates stand face-to-face, then push each other away. One student has a mass of 90 kg; the other has a mass of 60 kg.
  a. Sketch the situation, identifying "before" and "after," and assigning a coordinate axis.
  b. Find the ratio of the students' velocities just after their hands lose contact.
  c. Which student has the greater speed?
  d. Which student pushed harder?

40. A 0.200-kg plastic ball moves with a velocity of 0.30 m/s. It collides with a second plastic ball of mass 0.100 kg, which is moving along the same line at a speed of 0.10 m/s. After the collision, both balls continue moving in the same, original direction, and the speed of the 0.100-kg ball is 0.26 m/s. What is the new velocity of the first ball?

LEVEL 2

41. A 92-kg fullback, running at 5.0 m/s, attempts to dive directly across the goal line for a touchdown. Just as he reaches the line, he is met head-on in midair by two 75-kg linebackers both moving in the direction opposite the fullback. One is moving at 2.0 m/s, the other at 4.0 m/s. They all become entangled as one mass.
  a. Sketch the situation, identifying "before" and "after."
  b. What is their velocity after the collision?
  c. Does the fullback score?

42. A 5.00-g bullet is fired with a velocity of 100.0 m/s toward a 10.00-kg stationary solid block resting on a frictionless surface.
  a. What is the change in momentum of the bullet if it is embedded in the block?
  b. What is the change in momentum of the bullet if it ricochets in the opposite direction with a speed of 99 m/s?
  c. In which case does the block end up with a greater speed?

43. The diagrams in **Figure 9–13** show a brick weighing 24.5 N being released from rest on a 1.00-m frictionless plane, inclined at an angle of 30.0°. The brick slides down the incline and strikes a second brick weighing 36.8 N.

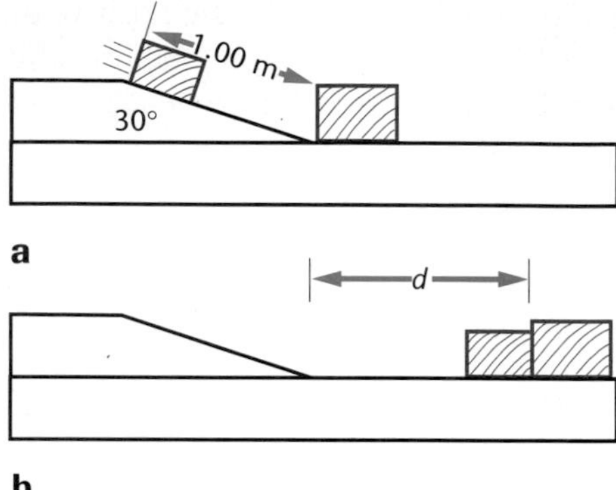

FIGURE 9-13

  a. Calculate the speed of the first brick at the bottom of the incline.
  b. If the two bricks stick together, with what initial speed will they move along?
  c. If the force of friction acting on the two bricks is 5.0 N, how much time will elapse before the bricks come to rest?
  d. How far will the two bricks slide before coming to rest?

44. Ball A, rolling west at 3.0 m/s, has a mass of 1.0 kg. Ball B has a mass of 2.0 kg and is stationary. After colliding with ball B, ball A moves south at 2.0 m/s.
  a. Sketch the system, showing the velocities and momenta before and after the collision.
  b. Calculate the momentum and velocity of ball B after the collision.

45. A space probe with a mass of 7.600×10^3 kg is traveling through space at 125 m/s. Mission control decides that a course correction of 30.0° is needed and instructs the probe to fire rockets perpendicular to its present direction of motion. If the gas expelled by the rockets has a speed of 3.200 km/s, what mass of gas should be released?

46. The diagram in **Figure 9–14,** which is drawn to scale, shows two balls during a collision. The balls enter from the left, collide, and then bounce away. The heavier ball at the bottom of the diagram has a mass of 0.600 kg, and the other has a mass of 0.400 kg. Using a vector diagram, determine whether momentum is conserved in this collision. What could explain any difference in the momentum of the system before and after the collision?

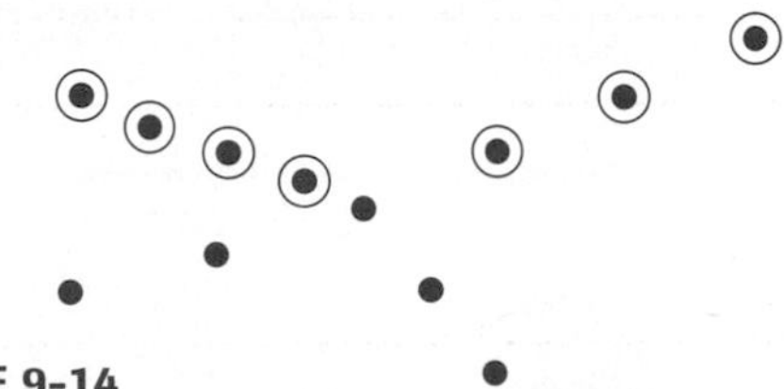

FIGURE 9-14

Critical Thinking Problems

47. A compact car, mass 875 kg, moving south at 15 m/s, is struck by a full-sized car, mass 1584 kg, moving east at 12 m/s. The two cars stick together, and momentum is conserved in the collision.

a. Sketch the situation, assigning coordinate axes and identifying "before" and "after."

b. Find the direction and speed of the wreck immediately after the collision, remembering that momentum is a vector quantity.

c. The wreck skids along the ground and comes to a stop. The coefficient of kinetic friction while the wreck is skidding is 0.55. Assume that the acceleration is constant. How far does the wreck skid?

48. Your friend has been in a car accident and wants your help. She was driving her 1265-kg car north on Oak Street when she was hit by a 925-kg compact car going west on Maple Street. The cars stuck together and slid 23.1 m at 42° north of west. The speed limit on both streets is 50 mph (22 m/s). Your friend claims that she wasn't speeding, but that the other car was. Can you support her case in court? Assume that momentum was conserved during the collision and that acceleration was constant during the skid. The coefficient of kinetic friction between the tires and the pavement is 0.65.

Going Further

Team Project How can you survive a car crash? Work with a team to design a model for testing automobile safety devices. Your car can be a dynamics cart or other device with low-friction wheels. Make a seat out of wood that you securely mount on the car. Use clay to model a person. For a dashboard, use a piece of metal fastened to the front of the cart. Allow the car to roll down a ramp and collide with a block at the bottom of the ramp. Devise a testing procedure so that the car starts from the same distance up the ramp and comes to rest at the same place in every test. First, crash the car with no protection for the person. Examine the clay and describe the damage done. Then, design a padded dash by using a piece of rubber tubing. Use a piece of string or ribbon to make a lap and shoulder belt. Fasten the belt to the seat. Finally, model an airbag by placing a small, partially inflated balloon between the passenger and the dashboard.

Summarize your experiments, including an explanation of the forces placed on the passenger in terms of the change in momentum, the impulse, the average force, and the time interval over which the impulse occurred.

***inter*NET CONNECTION**

Follow the link on the Glencoe Homepage at **www.glencoe.com/sec/science** to find out more about this chapter.

A Not-So-Simple Machine

How does a multispeed bicycle let a cyclist ride over any kind of terrain with the least effort?

CHAPTER

10 Energy, Work, and Simple Machines

What is energy? Energy is needed to make cars run, to heat or cool our homes, and to make computers work. Solar energy is required for crops and forests to grow. The energy stored in food gives you the energy needed to play sports or walk to the store. Note, however, all these statements indicate that having energy enables something to perform an action, rather than directly saying what energy is. It is hard to give a good definition of energy without examples of how energy is used and the changes that result.

In this chapter, you'll concentrate on one method of changing the energy of a system, work. For thousands of years, doing work has been of vital concern to the human race. However, the forces that the human body can exert are limited by physical strength and body design. Consequently, humans have developed machines that increase the amount of force the human body can produce. A mountain bike is a machine that uses sprockets and a chain to transfer the force of the legs to a force exerted by the rear wheel. Different combinations of sprockets are used to match the forces of the leg to the task of riding at high speed on level ground or while climbing a steep hill. In this chapter, you'll investigate how machines make doing work easier.

WHAT YOU'LL LEARN

- You will recognize that work and power describe how energy moves through the environment.
- You will relate force to work and explain how machines make work easier by changing forces.

WHY IT'S IMPORTANT

- A little mental effort in identifying the right machine for a task can save you much physical effort. From opening a can of paint to releasing a car stuck in the mud to sharpening a pencil, machines are a part of everyday life.

interNET CONNECTION

Follow the link for this chapter on the Glencoe Homepage at **www.glencoe.com/sec/science** to find out more about work, energy, and machines.

10.1 Energy and Work

OBJECTIVES

- **Describe** the relationship between work and energy.
- **Display** an ability to calculate work done by a force.
- **Identify** the force that does work.
- **Differentiate** between work and power and correctly **calculate** power used.

If you had a job moving boxes around a warehouse, you would know something about work and energy. It's not easy to lift the boxes onto a truck, slide them across a rough floor, or get them moving quickly on a roller belt, as in **Figure 10–1.** You have probably thought on more than one occasion that physics is hard work and that you expend a lot of energy solving problems. Your meaning of the words work and energy is different from their meaning in physics. You need to be more specific about the meaning of work and energy to communicate effectively in science.

FIGURE 10–1 In physics, work is done only when a force causes an object to move.

F.Y.I.

In physics, work and energy have precise meanings, which must not be confused with their everyday meanings. Robert Oppenheimer wrote, "Often the very fact that the words of science are the same as those of our common life and tongue can be more misleading than enlightening."

Energy

When describing an object, you might say that it is blue, it is 2 m tall, and it can produce a change. This property, the ability to produce change in itself or the environment, is called **energy.** The energy of an object can take many forms, including thermal energy, chemical energy, and energy of motion. For example, the position of a moving object is changing over time; this change in position indicates that the object has energy. The energy of an object resulting from motion is called **kinetic energy.** To describe kinetic energy mathematically, you need to use motion equations and Newton's second law of motion, $F = ma$.

Energy of motion

Start with an object of mass m, moving at speed v_0. Now apply a force, F, to the object to accelerate it to a new speed, v_1. In Chapter 5 you learned the motion equation that describes this situation.

$$v_1^2 = v_0^2 + 2ad$$

To see how energy is expressed in this relationship, you need to do some rearranging. First add a negative v_0^2 to both sides.

$$v_1^2 - v_0^2 = 2ad$$

Using Newton's second law of motion, substitute F/m for a.

$$v_1^2 - v_0^2 = 2Fd/m$$

And finally, multiply both sides of the equation by 1/2 *m*.

$$1/2mv_1^2 - 1/2mv_0^2 = Fd$$

On the left-hand side are the terms that describe the energy of the system. This energy results from motion and is represented by the symbol *K*, for kinetic.

$$K = 1/2mv^2$$

Because mass and velocity are both properties of the system, kinetic energy describes a property of the system. In contrast, the right-hand side of the equation refers to the environment: a force exerted and the resulting displacement. Thus, some agent in the environment changed a property of the system. The process of changing the energy of the system is called **work,** and it is represented by the symbol *W*.

$$W = Fd$$

Substituting *K* and *W* into the equation, you obtain $K_1 - K_0 = W$. The left-hand side is simply the difference or change in kinetic energy and can be expressed by using a delta.

$$\Delta K = W$$

In words, this equation says that when work is done on an object, a change in kinetic energy results. This hypothesis, $\Delta K = W$, has been tested experimentally many times and has always been found to be correct. It is called the **work-energy theorem.** This relationship between doing work and a resulting change in energy was established by the nineteenth-century physicist James Prescott Joule. To honor his work, a unit of energy is called a **joule.** For example, if a 2-kg object moves at 1 m/s, it has a kinetic energy of $1\text{kg}\cdot\text{m}^2/\text{s}^2$ or 1 J.

Working Out

Attach a spring scale to a 1.0-kg mass with a string. Pull the mass along the table at a slow, steady speed while keeping the scale parallel to the tabletop. Note the reading on the spring scale.

Analyze and Conclude What are the physical factors that determine the amount of force? How much work is done in moving the mass 1.0 m? Predict the force and the work when a 2.0-kg mass is pulled along the table. Try it. Was your prediction accurate?

Work

While the change in kinetic energy describes the change in a property of an object, the term *Fd*, describes something done to the object. An agent in the environment exerted a force *F* that displaced the object an amount *d*. The work done on an object by external forces changes the amount of energy the object has.

Energy transfer

Remember when you studied Newton's Laws of motion and momentum, that a system was the object of interest, and the environment was everything else. For example, the one system might be a box in the warehouse and the environment is you, gravity, and anything else external to the box. Through the process of doing work, energy can move between the environment and the system, as diagrammed in **Figure 10–2.**

Notice that the direction of energy transfer can go both ways. If the environment does work on the system, then *W* is positive and the energy of the system increases. If, however, the system does work on the environment, then *W* is negative, and the energy of the system decreases.

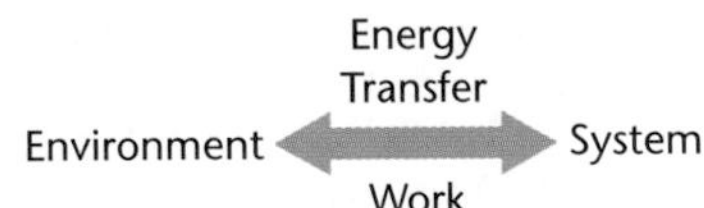

FIGURE 10–2 Work transfers energy between an environment and a system. Energy transfers can go either direction.

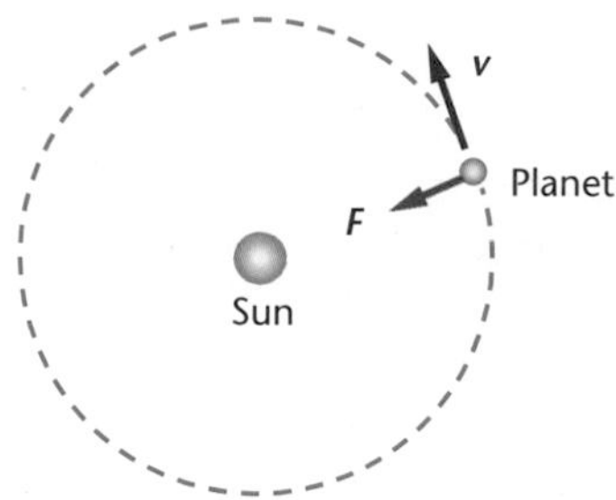

FIGURE 10–3 If a planet is in a circular orbit, then the force is perpendicular to the direction of motion. Consequently, the gravitational force does no work on the planet.

Calculating work

The equation for work is $W = Fd$, however this equation holds only for constant forces exerted in the direction of the motion. What happens if the force is exerted perpendicular to the direction of motion? An everyday example is the motion of a planet around the sun, as diagramed in **Figure 10–3.** If the orbit is circular, then the force is always perpendicular to the direction of motion. Remember from Chapter 7 that a perpendicular force does not change the speed of an object, only its direction. Consequently, the speed of the planet doesn't change. Therefore, its kinetic energy is also constant. Using the equation $\Delta K = W$, you see for constant K that $\Delta K = 0$ and thus $W = 0$. This means that if $\mathbf{F}$ and $\mathbf{d}$ are at right angles, then $W = 0$.

Because the work done on an object equals the change in energy, work is also measured in joules. A joule of work is done when a force of one newton acts on an object over a displacement of one meter. An apple weighs about one newton. Thus, when you lift an apple a distance of one meter, you do one joule of work on it.

Example Problem

Calculating Work

A 105-g hockey puck is sliding across the ice. A player exerts a constant 4.5-N force over a distance of 0.15 m. How much work does the player do on the puck? What is the change in the puck's energy?

Sketch the Problem

- Establish a coordinate axis.
- Show the hockey puck with initial conditions.
- Draw a vector diagram.

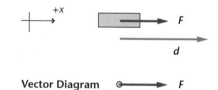

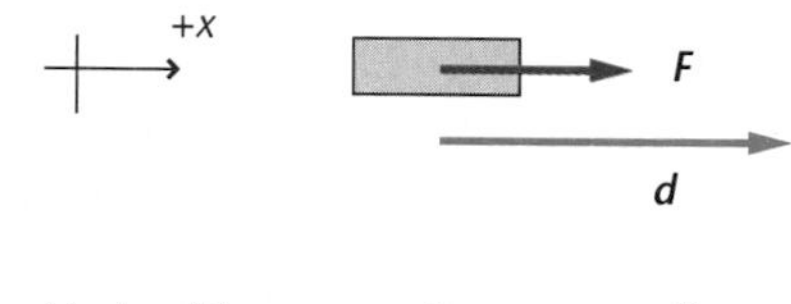

Calculate Your Answer

Known:

$m = 105$ g

$F = 4.5$ N

$d = 0.15$ m

Unknown:

$W = ?$

$\Delta K = ?$

Strategy:

Use the basic equation for work when a constant force is exerted in same direction as displacement.

Use the work-energy theorem to determine the change in energy of the system.

Calculations:

$W = Fd$

$W = (4.5 \text{ N})(0.15 \text{ m})$

$W = 0.68 \text{ N·m} = 0.68 \text{ J}$

$\Delta K = W = 0.68 \text{ J}$

Check Your Answer

- Are the units correct? Work is measured in joules, J = N·m.
- Does the sign make sense? The player does work on the puck, which agrees with a positive sign for work.
- A magnitude of about 1 J fits with the quantities given.

Practice Problems

1. A student lifts a box of books that weighs 185 N. The box is lifted 0.800 m. How much work does the student do on the box?
2. Two students together exert a force of 825 N in pushing a car 35 m.
   a. How much work do they do on the car?
   b. If the force were doubled, how much work would they do pushing the car the same distance?
3. A 0.180-kg ball falls 2.5 m. How much work does the force of gravity do on the ball?
4. A forklift raises a box 1.2 m doing 7.0 kJ of work on it. What is the mass of the box?
5. You and a friend each carry identical boxes to a room one floor above you and down the hall. You choose to carry it first up the stairs, then down the hall. Your friend carries it down the hall, then up another stairwell. Who does more work?

Constant force at an angle

You've learned that a force exerted in the direction of motion does an amount of work given by $W = Fd$. A force exerted perpendicular to the motion does no work. What work does a force exerted at an angle do? For example, what work does the person pushing the lawn mower in **Figure 10–4a** do? You know that any force can be replaced by its components. The 125-N force, ***F***, exerted in the direction of the handle has two components. If we choose the coordinate system shown in **Figure 10–4b,** then the magnitude of the horizontal component, F_x, is related to the magnitude of the force, F, by a cosine function: $\cos 25.0° = F_x/F$. By solving for F_x, you obtain $F_x = F \cos 25.0° = 113$ N. Using the same method, the vertical component is $F_y = -F \sin 25.0° = -52.8$ N, where the negative sign shows that the force is down. Because the displacement is in the x direction, only the x-component does work. The y-component does no work. The work you do when you exert a force at an angle to the motion is equal to the component of the force in the direction of the displacement times the distance moved.

An Inclined Mass

Attach a spring scale to a 1.0-kg mass with a string. Increase the angle between the string and the tabletop, for example, to 30°. Try to keep the angle constant as you pull the 1.0-kg mass along the table at a slow, steady speed. Note the reading on the scale.

Analyze and Conclude How much force is in the direction of motion? How much work is done when the 1.0-kg mass moves 1.0 m? How does the work compare to the previous value?

a

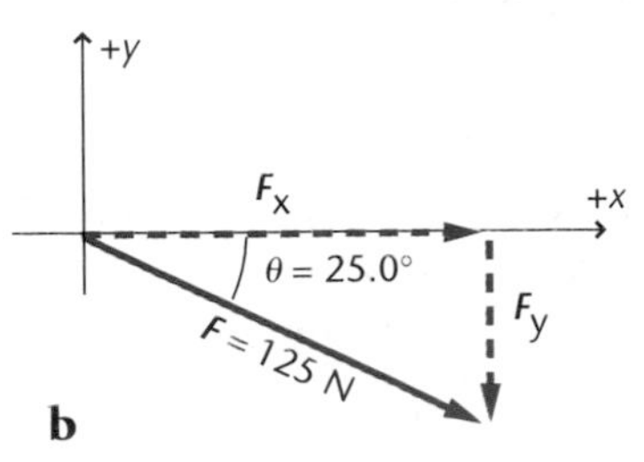

b

FIGURE 10–4 If a force is applied to the mower at an angle, the net force doing the work is the component that acts in the direction of the displacement.

The magnitude of the component force acting in the direction of displacement is found by multiplying the magnitude of force **F** by the cosine of the angle between **F** and the direction of the displacement, $F_x = F \cos \theta$. Thus, the work done is represented the following way.

$$W = F \cos \theta \, d$$

$$= Fd \cos \theta$$

Other agents exert forces on the lawn mower. Which of these agents do work? Earth's gravity acts downward, the ground exerts a normal force upward, and friction exerts a horizontal force opposite the motion. The upward and downward forces are perpendicular to the motion and do no work. For these forces, $\theta = 90°$, which makes $\cos \theta = 0$, and thus, no work is done.

The work done by friction acts at an angle of $180°$. Because $\cos 180° = -1$, the work done by friction is negative. Negative work done by a force in the environment reduces the energy of the system. If the person in **Figure 10–4a** were to stop pushing, the mower would quickly stop moving; its energy of motion would be reduced. Positive work done by a force increases the energy; negative work decreases it.

Problem Solving Strategy

Work Problems

1. Sketch the system and show the force that is doing the work.
2. Diagram the vectors of the system.
3. Find the angle, θ, between each force and displacement.
4. Calculate the work done by each force using $W = Fd \cos \theta$.
5. Check the sign of the work using the direction of energy transfer. If the energy of the system has increased, the work done by that force is positive. If the energy has decreased, then the work done is negative.

Example Problem

Force and Displacement at an Angle

A sailor pulls a boat 30.0 m along a dock using a rope that makes a 25.0° angle with the horizontal. How much work does the sailor do on the boat if he exerts a force of 255 N on the rope?

Sketch the Problem

- Establish coordinate axes.
- Show the boat with initial conditions.
- Draw a vector diagram showing the force and its component in the direction of the displacement.

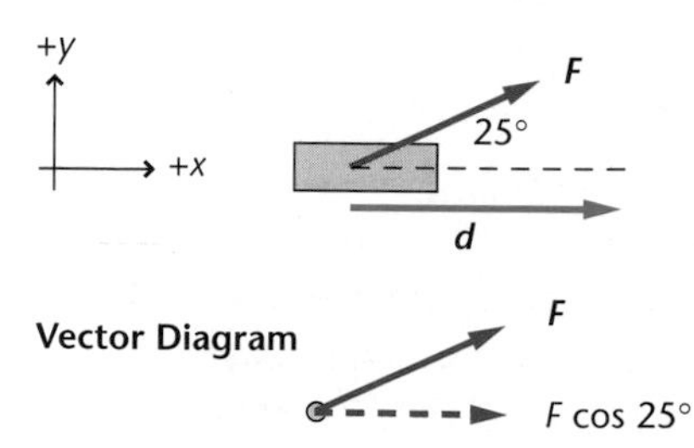

Calculate Your Answer

Known:

$F = 255$ N

$d = 30.0$ m

$\theta = 25.0°$

Unknown:

$W = ?$

Strategy:

Use the equation for work when there is an angle between force and displacement.

Calculations:

$W = Fd \cos \theta$

$W = (255\text{ N})(30.0\text{ m})(\cos 25.0°)$

$W = 6.93 \times 10^3$ J

Check Your Answer

- Are the units correct? N·m = J, and work is measured in joules.
- Does the sign make sense? The sailor does work on the boat, which agrees with a positive sign for work.
- Is the magnitude realistic? Magnitude of about 7000 J fits with the quantities given.

Practice Problems

6. How much work does the force of gravity do when a 25-N object falls a distance of 3.5 m?

7. An airplane passenger carries a 215-N suitcase up the stairs, a displacement of 4.20 m vertically and 4.60 m horizontally.

a. How much work does the passenger do?

b. The same passenger carries the same suitcase back down the same stairs. How much work does the passenger do now?

8. A rope is used to pull a metal box 15.0 m across the floor. The rope is held at an angle of 46.0° with the floor and a force of 628 N is used. How much work does the force on the rope do?

Finding work done when forces change

A graph of the force versus the displacement lets you determine the work done by a force. This graphical method can be used to solve problems for which the force is changing. **Figure 10–5** shows how to find the work done by a constant force of 20 N that is exerted lifting an object 1.5 m. The work done by this constant force is represented by $W = Fd = (20\text{ N})(1.5\text{ m}) = 30$ J. The shaded area under the curve is equal to 20 N × 1.5 m, or 30 J. The area under the curve of a force-displacement graph is equal to the work done by that force. The area is the work done even if the force changes. **Figure 10–5** shows the force exerted by a spring, which varies linearly from 0 to 20 N as it is compressed 1.5 m. The work done by the force that compressed the spring is the area under the curve, which is the area of a triangle, 1/2 (base)(altitude), or $W = 1/2(20\text{ N})(1.5\text{ m}) = 15$ J.

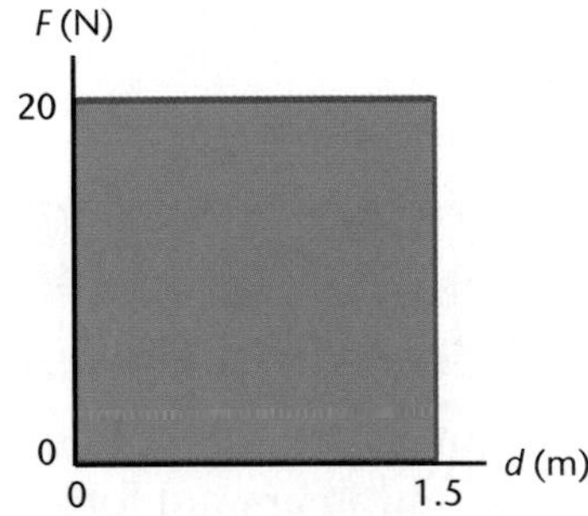

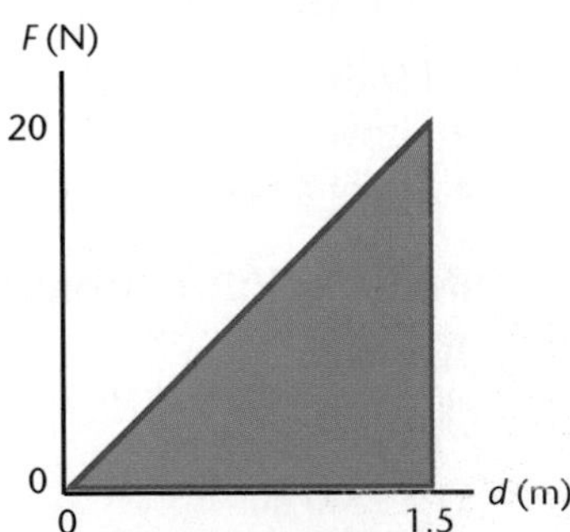

FIGURE 10–5 Work can be obtained graphically by finding the area under a force-displacement curve.

FIGURE 10–6 These students are expending energy at different rates while climbing the stairs.

Power

Until now, none of the discussions of work has mentioned the time it takes to move an object. The work done by a person lifting a box of books is the same whether the box is lifted onto a shelf in 2 seconds or each book is lifted separately, so that it takes 20 minutes to put them all on the shelf. Although the work done is the same, the power is different. **Power** is the rate of doing work. That is, power is the rate at which energy is transferred. Consider the three students in **Figure 10–6,** the girl hurrying up the stairs is more powerful than the boy walking. Even though the same work is accomplished, the girl accomplishes it in less time. In the case of the two students walking up the stairs, both accomplish work in the same amount of time. However, the girl carrying books does more work and, consequently, her power is greater. To calculate power, use the following formula.

$$P = \frac{W}{t}$$

Power is measured in watts (W). One **watt** is one joule of energy transferred in one second. A machine that does work at a rate of one joule per second has a power of one watt. A watt is a relatively small unit of power. For example, a glass of water weighs about 2 N. If you lift it 0.5 m to your mouth, you do one joule of work. If you lift the glass in one second, you are doing work at the rate of one watt. Because a watt is such a small unit, power is often measured in kilowatts (kW). A kilowatt is 1000 watts.

Example Problem

Calculating Power

An electric motor lifts an elevator 9.00 m in 15.0 s by exerting an upward force of 1.20×10^4 N. What power does the motor produce in watts and kilowatts?

Sketch the Problem

- Establish a coordinate axis, up being positive.
- Show the elevator with initial conditions.
- Draw a vector diagram for the force.

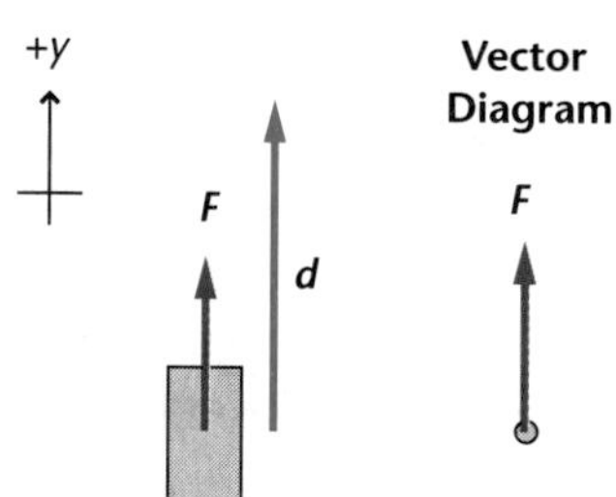

Calculate Your Answer

Known:

$d = 9.00$ m

$t = 15.0$ s

$F = 1.20 \times 10^4$ N

Unknown:

$P = ?$

Strategy:

Use work and time to find power.

Calculations:

$W = Fd$ and $P = \frac{W}{t}$, so

$$P = \frac{Fd}{t}$$

$$P = \frac{(1.20 \times 10^4\ \text{N})(9.00\ \text{m})}{15.0\ \text{s}} = 7.20\ \text{kW}$$

Check Your Answer

- Are the units correct? Check algebra on units to ensure that power is measured in watts.
- Does the sign make sense? Positive sign agrees with the upward direction of force.
- Is the magnitude realistic? Lifting an elevator requires a high power. 7200 watts is about right.

F.Y.I.

A bicyclist in the *Tour de France* rides at about 20 mph for more than six hours a day. The power output of the racer is about one kilowatt. One quarter of that power goes into moving the bike against the resistance of the air, gears, and tires. Three quarters of the power is used to cool the racer's body.

Practice Problems

9. A box that weighs 575 N is lifted a distance of 20.0 m straight up by a cable attached to a motor. The job is done in 10.0 s. What power is developed by the motor in watts and kilowatts?
10. A rock climber wears a 7.5-kg knapsack while scaling a cliff. After 30 min, the climber is 8.2 m above the starting point.
    a. How much work does the climber do on the knapsack?
    b. If the climber weighs 645 N, how much work does she do lifting herself and the knapsack?
    c. What is the average power developed by the climber?
11. An electric motor develops 65 kW of power as it lifts a loaded elevator 17.5 m in 35 s. How much force does the motor exert?
12. Your car has stalled and you need to push it. You notice as the car gets going that you need less and less force to keep it going. Suppose that for the first 15 m your force decreased at a constant rate from 210 N to 40 N. How much work did you do on the car? Draw a force-displacement graph to represent the work done during this period.

10.1 Section Review

1.1 Explain in words, without the use of a formula, what work is.

1.2 When a bowling ball rolls down a level alley, does Earth's gravity do any work on the ball? Explain.

1.3 Does the work required to lift a book to a high shelf depend on how fast you raise it? Does the power required for the lift depend on how fast you raise the book? Explain.

1.4 **Critical Thinking** If three objects exert forces on a body, can they all do work at the same time? Explain.

Design Your Own **Physics Lab**

Your Power

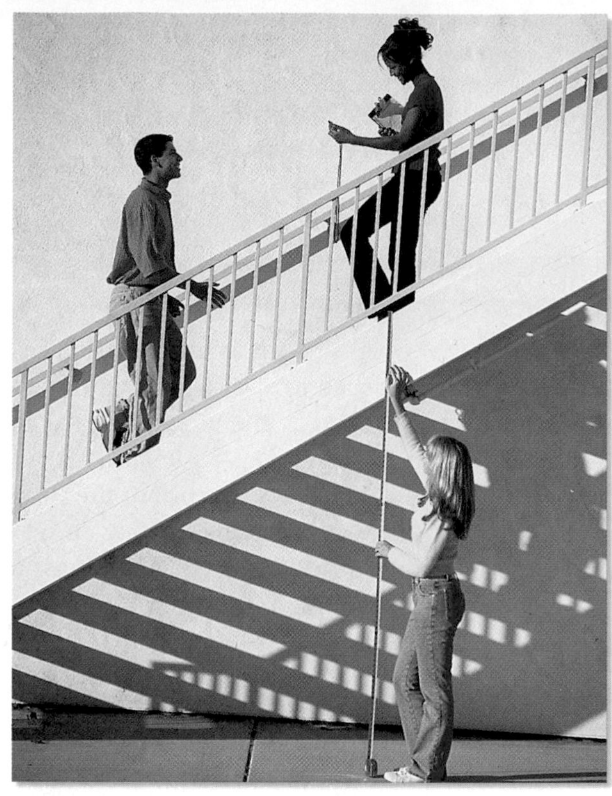

Problem

Can you estimate the power that you generate as you climb stairs? Climbing stairs requires energy. As you move your weight through a distance, you accomplish work. The rate at which you do this work is power.

Hypothesis

Form a hypothesis that relates estimating power to measurable quantities. Predict the difficulties you may encounter as you are trying to solve the problem.

Possible Materials

Determine which variables you will measure and then plan a procedure for taking measurements. Tell your teacher what materials you would like to use to accomplish your plan.

Plan the Experiment

In your group, develop a plan to measure your power as you climb stairs. Be prepared to present your plan, your data, your calculations, and your results to the rest of the class. Take measurements for at least two students.

1. Identify the dependent and independent variables.
2. Describe your procedures.
3. Set up data tables.
4. Write any equations that you will need for the calculations.
5. **Check the Plan** Show your teacher your plan before you leave the room to start the experiment.

Analyze and Conclude

1. **Calculating Results** Show your calculations for the power rating of each climber.
2. **Comparing Results** Did each climber have the same power rating?
3. **Analyzing Data** Explain how your power could be increased.
4. **Making Inferences** Explain why the fastest climber might not have the highest power rating. Explain why the largest climber might not have the highest power rating.

Apply

1. Your local electric company charges you about 11 cents for a kilowatt-hour of energy. At this rate, how much money could you earn by climbing stairs continuously for one hour? Show your calculations.

Machines 10.2

Everyone uses some machines every day. Some are simple tools, such as bottle openers and screwdrivers; others are complex, such as bicycles and automobiles. Machines, whether powered by engines or people, make tasks easier. A **machine** eases the load by changing either the magnitude or the direction of a force as it transmits energy to the task.

OBJECTIVES

- **Demonstrate** knowledge of why simple machines are useful.
- **Communicate** an understanding of mechanical advantage in ideal and real machines.
- **Analyze** compound machines and **describe** them in terms of simple machines.
- **Calculate** efficiencies for simple and compound machines.

Simple and Compound Machines

Consider the bottle opener in **Figure 10–7.** When you use the opener, you lift the handle, thereby doing work on the opener. The opener lifts the cap, doing work on it. The work you do is called the input work, W_i. The work the machine does is called the output work, W_o.

Work, as you recall, is the transfer of energy by mechanical means. You put work into a machine, in this case, the bottle opener. That is, you transfer energy to the opener. The opener, in turn, does work on the cap, transferring energy to it. The opener is not a source of energy, so the cap cannot receive more energy than you put into the opener. Thus, the output work can never be greater than the input work. The machine simply aids in the transfer of energy from you to the bottle cap.

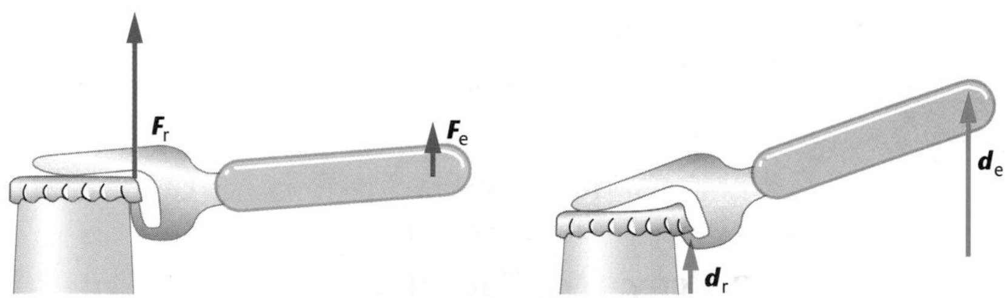

FIGURE 10–7 A bottle opener is an example of a simple machine. It makes opening a bottle easier, but not less work than it would be otherwise.

Mechanical advantage

The force you exert on a machine is called the **effort force,** F_e. The force exerted by the machine is called the **resistance force,** F_r. **Figure 10–8** shows a typical pulley setup, where F_e is the downward force exerted by the man and the F_r is the upward force exerted by the rope. The ratio of resistance force to effort force, F_r/F_e, is called the **mechanical advantage** (MA) of the machine.

$$MA = \frac{F_r}{F_e}$$

Many machines, such as the bottle opener, have a mechanical advantage greater than one. When the mechanical advantage is greater than one, the machine increases the force you apply. In the case of the pulley system in **Figure 10–8,** the forces F_e and F_r are equal, consequently MA is 1. So what is the advantage? The usefulness of this pulley arrangement is not that the effort force is lessened, but that the direction is changed; now the direction of effort is in the same direction as displacement.

FIGURE 10–8 The pulley system makes work easier not by increasing the force the man can apply to the resistance, but by allowing him to apply the force parallel to displacement.

HELP WANTED

CHIROPRACTOR

Do you like to work on "the human machine?" A group practice in a small city desires a new partner to expand services. A belief in the holistic (nondrug, nonsurgical) approach to medicine and a positive, professional demeanor are required for one to be a successful chiropractor. Completion of an accredited, six-year Chiropractic College program with a Doctor of Chiropractic (D.C.) degree and a state license are required. For information contact:

International Chiropractors Association
1901 L Street, N.W. Suite 800
Washington, DC 20036

You can write the mechanical advantage of a machine in another way using the definition of work. The input work is the product of the effort force you exert, F_e, and the distance your hand moved, d_e. In the same way, the output work is the product of the resistance force, F_r, and the displacement of the object, d_r. A machine can increase force, but it cannot increase energy. An ideal machine transfers all the energy, so the output work equals the input work.

$$W_o = W_i$$

or

$$F_r d_r = F_e d_e$$

This equation can be rewritten $F_r/F_e = d_e/d_r$. We know that the mechanical advantage is given by $MA = F_r/F_e$. For an ideal machine, $MA = d_e/d_r$. Because this equation is characteristic of an ideal machine, the mechanical advantage is called the **ideal mechanical advantage,** *IMA*.

$$IMA = \frac{d_e}{d_r}$$

Note that you measure distances moved to calculate the ideal mechanical advantage, *IMA*, but you measure the forces exerted to find the actual mechanical advantage, *MA*.

Efficiency

In a real machine, not all of the input work is available as output work. Some of the energy transferred by the work may be "lost" to thermal energy. Any energy removed from the system means less output work from the machine. Consequently, the machine is less efficient at accomplishing the task.

The **efficiency** of a machine is defined as the ratio of output work to input work.

$$\text{efficiency} = \frac{W_o}{W_i} \times 100\%$$

An ideal machine has equal output and input work, $W_o/W_i = 1$, and its efficiency is 100%. All real machines have efficiencies less than 100%. We can express the efficiency in terms of the mechanical advantage and ideal mechanical advantage.

$$\text{efficiency} = \frac{F_r/F_e}{d_e/d_r} \times 100\%$$

$$\text{efficiency} = \frac{MA}{IMA} \times 100\%$$

The *IMA* of most machines is fixed by the machine's design. An efficient machine has an *MA* almost equal to its *IMA*. A less efficient machine has a small *MA* relative to its *IMA*. Lower efficiency means that a greater effort force is needed to exert the same resistance force as a comparable machine of higher efficiency.

F.Y.I.

Efficiency Statistic: Most automobiles get about 28% more miles per gallon of fuel at 50 miles per hour than at 70 miles per hour.

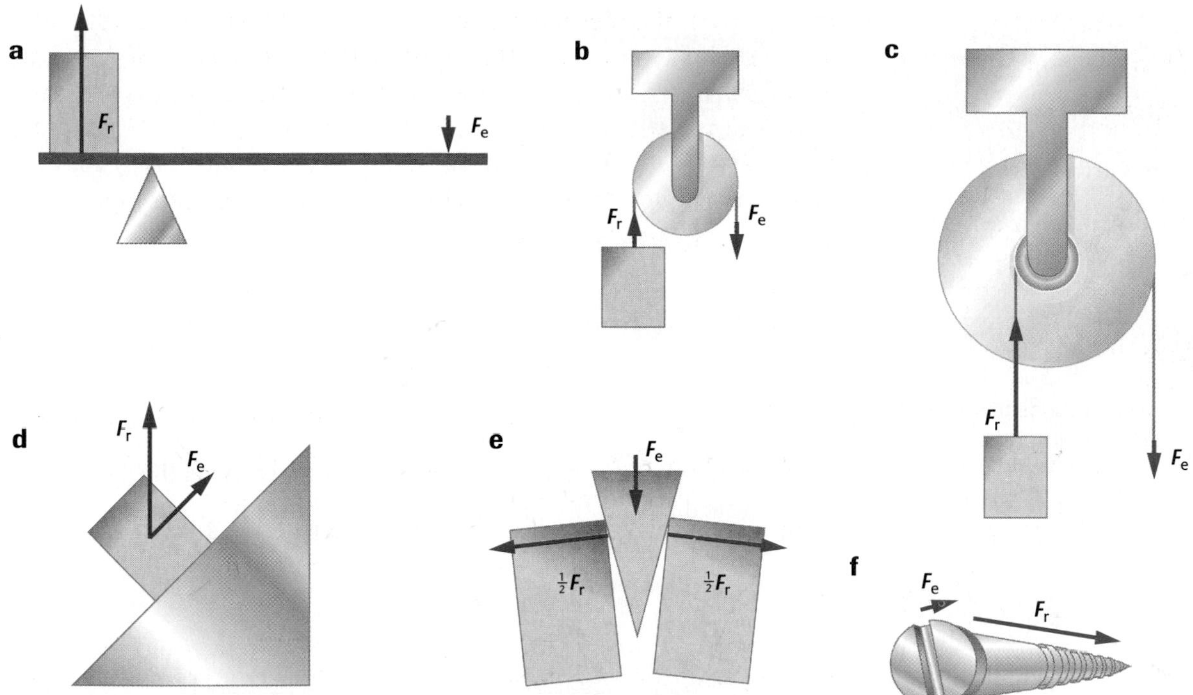

FIGURE 10–9 The simple machines pictured are the lever **(a)**; pulley **(b)**; wheel and axle **(c)**; inclined plane **(d)**; wedge **(e)**; and screw **(f)**.

Simple machines

Most machines, no matter how complex, are combinations of one or more of the six simple machines shown in **Figure 10–9.** They are the lever, pulley, wheel and axle, inclined plane, wedge, and screw. Gears, one of the simple machines used in a bicycle, are really a form of the wheel and axle. The *IMA* of all machines is the ratio of distances moved. **Figure 10–10** shows that for levers and wheel and axles this ratio can be replaced by the ratio of the distance between the place where the force is applied and the pivot point. A common version of the wheel and axle is a pair of gears on a rotating shaft. The *IMA* is the ratio of the radii of the two gears.

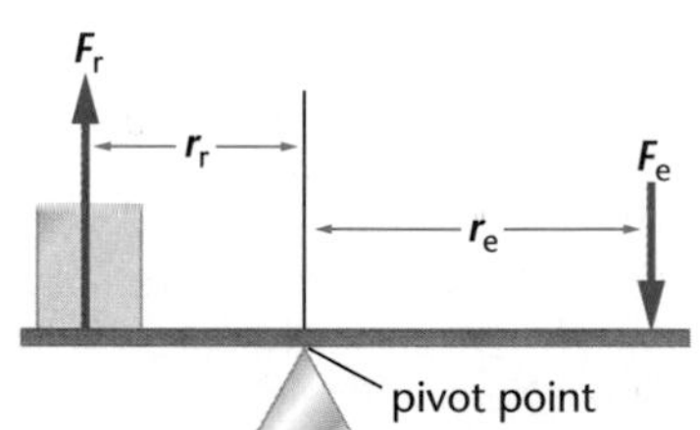

Compound machines

A **compound machine** consists of two or more simple machines linked so that the resistance force of one machine becomes the effort force of the second. For example, in the bicycle, the pedal and front sprocket (or gear) act like a wheel and axle. The effort force is the force you exert on the pedal, $F_{\text{on pedal}}$. The resistance is the force the front sprocket exerts on the chain, $F_{\text{on chain}}$, as illustrated in **Figure 10–11.**

The chain exerts an effort force on the rear wheel sprocket, $F_{\text{by chain}}$, equal to the force exerted on the chain. This sprocket and the rear wheel act like another wheel and axle. The resistance force is the force the wheel exerts on the road, $F_{\text{on road}}$. According to Newton's third law of motion, the ground exerts an equal forward force on the wheel. This force accelerates the bicycle forward.

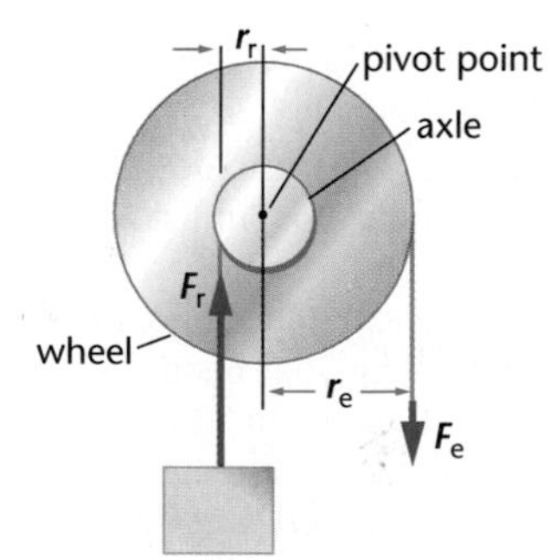

FIGURE 10–10 For levers and wheel and axles, the *IMA* is r_e/r_r.

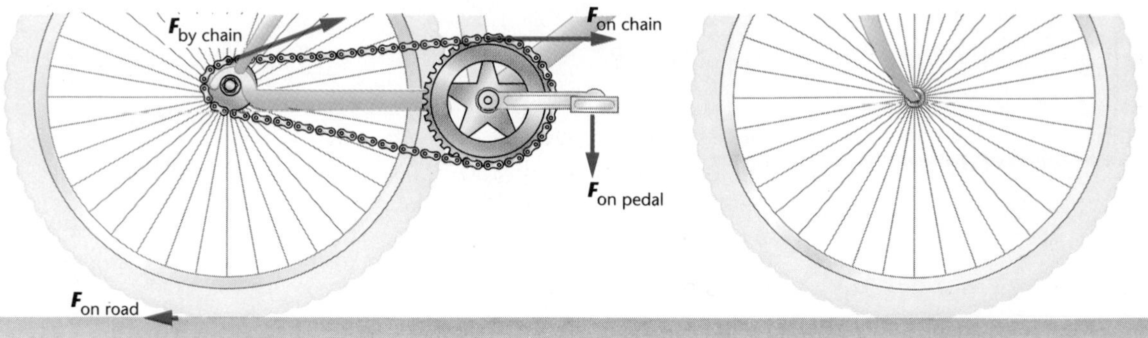

FIGURE 10–11 A series of simple machines combines to transmit the force the rider exerts on the pedal to the road.

The mechanical advantage of a compound machine is the product of the mechanical advantages of the simple machines it is made up of. For example, **Figure 10–11** illustrates the case of the bicycle.

$$MA = MA_{\text{machine 1}} \times MA_{\text{machine 2}}$$

$$MA = \frac{F_{\text{on chain}}}{F_{\text{on pedal}}} \times \frac{F_{\text{on road}}}{F_{\text{by chain}}} = \frac{F_{\text{on road}}}{F_{\text{on pedal}}}$$

The *IMA* of each wheel and axle machine is the ratio of the distances moved. For the pedal sprocket,

$$IMA = \frac{\textit{pedal radius}}{\textit{front sprocket radius}}.$$

For the rear wheel,

$$IMA = \frac{\textit{rear sprocket radius}}{\textit{wheel radius}}.$$

For the bicycle, then,

$$IMA = \frac{\textit{pedal radius}}{\textit{front sprocket radius}} \times \frac{\textit{rear sprocket radius}}{\textit{wheel radius}}$$

$$= \frac{\textit{rear sprocket radius}}{\textit{front sprocket radius}} \times \frac{\textit{pedal radius}}{\textit{wheel radius}}.$$

Because both sprockets use the same chain and have teeth of the same size, you can simply count the number of teeth on the gears and find that

$$IMA = \frac{\textit{teeth on rear sprocket}}{\textit{teeth on front sprocket}} \times \frac{\textit{pedal arm length}}{\textit{wheel radius}}.$$

Shifting gears on your bicycle is a way of adjusting the ratio of sprocket radii to obtain the desired *IMA*. *MA* depends on forces. You know that if the pedal is at the top or bottom of its circle, no matter how much downward force you exert, the pedals will not turn. The force of your foot is most effective when the force is exerted perpendicular to the arm of the pedal. Whenever a force on a pedal is specified, you should assume that it is applied perpendicular to the arm.

Pocket Lab

Wheel and Axle

The gear mechanism on your bicycle multiplies the distance that you travel. What does it do to the force? Try this activity to find out. Mount a wheel and axle on a solid support rod. Wrap a string clockwise around the small diameter wheel and a different string counterclockwise around the large diameter wheel. Hang a 500-gram mass from the end of the string on the larger wheel. Pull the string down so that the mass is lifted by about 10 cm.

Analyze and Conclude What did you notice about the force on the string in your hand? What did you notice about the distance that your hand needed to move to lift the mass? Explain the results in terms of the work done on both strings.

Example Problem

Bicycle Wheel

You are studying the rear wheel on your bicycle. It has a radius of 35.6 cm and has a gear with radius of 4.00 cm. When the chain is pulled with a force of 155 N, the wheel rim moves 14.0 cm. The efficiency of this part of the bicycle is 95.0 %.

a. What is the *IMA* of the wheel and gear?

b. What is the *MA* of the wheel and gear?

c. What force does the scale register?

d. How far was the chain pulled to move the rim that amount?

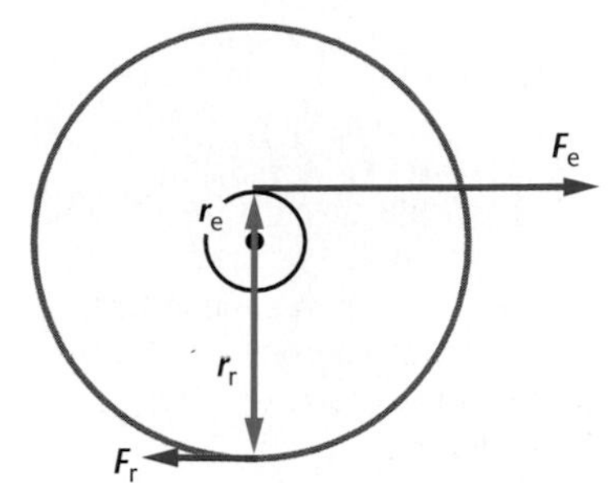

Sketch the Problem

- Diagram the wheel and axle.
- Add the force vectors and distance vectors.

Calculate Your Answer

Known:

$r_e = 4.00$ cm

$r_r = 35.6$ cm

$F_e = 155$ N

$e = 95.0\%$

$d_r = 14.0$ cm

Unknown:

$IMA = ?$

$MA = ?$

$F_r = ?$

$d_e = ?$

Strategy:

a. For a wheel and axle machine, *IMA* is represented by the ratio of radii.

Calculations:

$$IMA = \frac{r_e}{r_r}$$

$$IMA = \frac{4.00 \text{ cm}}{35.6 \text{ cm}} = 0.112$$

b. Use efficiency ratio to obtain *MA*.

$$e = \frac{MA}{IMA} \times 100\%$$

$$MA = (e/100\%) \times IMA$$

$$MA = (95\%/100\%) \times 0.112 = 0.107$$

c. Use *MA* equation to find force.

$$F_r = (MA)(F_e)$$

$$F_r = (0.107)(155 \text{ N}) = 16.5 \text{ N}$$

d. Use *IMA* equation to find distance.

$$d_e = (IMA)(d_r)$$

$$d_e = (0.112)(14.0 \text{ cm}) = 1.57 \text{ cm}$$

Check Your Answer

- Are the units correct? Perform the algebra with the units to confirm that the answer's units are correct.
- Does the sign make sense? All answers should be positive.
- Is the magnitude realistic? **a.** Expect a low *IMA* for a bicycle because you want to trade greater F_e for a greater d_r. **b.** *MA* is always smaller than *IMA*. **c.** Expect low F_r because *MA* is low. **d.** Expect d_e to be very small: Small distance of the axle results in a large distance of the wheel.

A Not-So-Simple Machine

On a multigear bike, the rider can change the mechanical advantage of the machine by choosing the size of one or both sprockets. When accelerating or climbing a hill, the rider increases the ideal mechanical advantage to increase the force the wheel exerts on the road. Looking at the *IMA* equation on page 236, to increase the *IMA*, the rider needs to make the rear sprocket radius large compared to the front sprocket radius.

On the other hand, when the rider is going at high speed on a level road, less force is needed, and the rider decreases the ideal mechanical advantage to reduce the distance the pedals must move for each revolution of the wheel.

Practice Problems

13. A sledgehammer is used to drive a wedge into a log to split it. When the wedge is driven 0.20 m into the log, the log is separated a distance of 5.0 cm. A force of 1.9×10^4 N is needed to split the log, and the sledgehammer exerts a force of 9.8×10^3 N.
- **a.** What is the *IMA* of the wedge?
- **b.** Find the *MA* of the wedge.
- **c.** Calculate the efficiency of the wedge as a machine.

14. A worker uses a pulley system to raise a 24.0 kg carton 16.5 m. A force of 129 N is exerted and the rope is pulled 33.0 m.
- **a.** What is the mechanical advantage of the pulley system?
- **b.** What is the efficiency of the system?

15. A boy exerts a force of 225 N on a lever to raise a 1.25×10^3-N rock a distance of 13 cm. If the efficiency of the lever is 88.7%, how far did the boy move his end of the lever?

16. If the gear radius in the bicycle in the Example Problem is doubled, while the force exerted on the chain and the distance the wheel rim moves remain the same, what quantities change, and by how much?

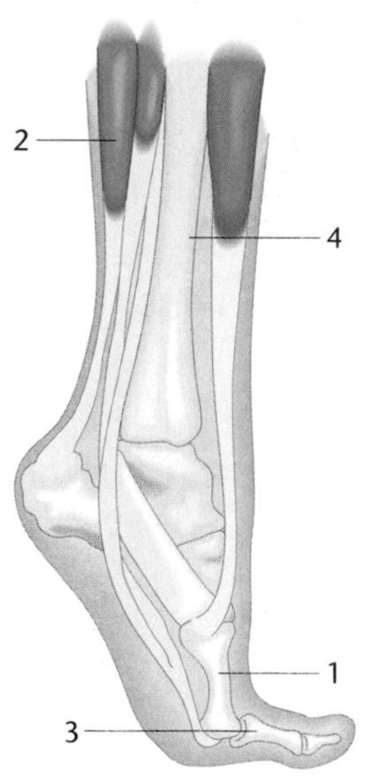

FIGURE 10–12 The human walking machine.

The Human Walking Machine

Movement of the human body is explained by the same principles of force and work that describe all motion. Simple machines, in the form of levers, give us the ability to walk and run. Lever systems of the body are complex, but each system has four basic parts: (1) a rigid bar (bone), (2) a source of force (muscle contraction), (3) a fulcrum or pivot (movable joints between bones), and (4) a resistance (the weight of the body or an object being lifted or moved), as shown in **Figure 10–12.** Lever systems of the body are not very efficient, and mechanical advantages are low. This is why walking and jogging require energy (burn calories) and help individuals lose weight.

When a person walks, the hip acts as a fulcrum and moves through the arc of a circle centered on the foot. The center of mass of the body moves as a resistance around the fulcrum in the same arc. The length of the radius of the circle is the length of the lever formed by the bones of the leg. Athletes in walking races increase their velocity by swinging their hips upward to increase this radius.

A tall person has lever systems with less mechanical advantage than a short person does. Although tall people can usually walk faster than short people can, a tall person must apply a greater force to move the longer lever formed by the leg bones. Walking races are usually 20 or 50 km long. Because of the inefficiency of their lever systems and the length of a walking race, very tall people rarely have the stamina to win.

10.2 Section Review

2.1 Many hand tools are simple machines. Classify the tools below as levers, wheel and axles, inclined planes, wedges, or pulleys.

a. screwdriver

b. pliers

c. chisel

d. nail puller

e. wrench

2.2 If you increase the efficiency of a simple machine, does the

a. *MA* increase, decrease, or remain the same?

b. *IMA* increase, decrease, or remain the same?

2.3 A worker exerts a force of 20 N on a machine with $IMA = 2.0$, moving it 10 cm.

a. Draw a graph of the force as a function of distance. Shade in the area representing the work done by this force and calculate the amount of work done.

b. On the same graph, draw the force supplied by the machine as a function of resistance distance. Shade in the area representing the work done by the machine. Calculate this work and compare to your answer above.

2.4 **Critical Thinking** The mechanical advantage of a multigear bike is changed by moving the chain to a suitable back sprocket.

a. To start out, you must accelerate the bike, so you want to have the bike exert the greatest possible force. Should you choose a small or large sprocket?

b. As you reach your traveling speed, you want to rotate the pedals as few times as possible. Should you choose a small or large sprocket?

c. Many bikes also let you choose the size of the front sprocket. If you want even more force to accelerate while climbing a hill, would you move to a larger or smaller front sprocket?

How It Works

Zippers

In 1893, the first known zipper, then known as a clasp locker, was exhibited at the World's Columbian Exposition in Chicago by its inventor, Whitcomb Judson. The clasp locker was a series of hooks and eyes that opened and closed when a slide clasp passed over them. It wasn't until two decades later that the modern zipper, often an integral part of garments, handbags, tents, backpacks, boots, and many other items, had evolved. Most of today's zippers consist of two basic parts: teeth and a slide. Some sort of tab or pull is usually attached to the slide.

1. When a zipper is opened, a wedge in the upper portion of the slide forces the teeth apart.

2. As a zipper closes, wedges on either side of the lower part of the slide force the teeth into sockets, one after the other. In some plastic zippers, spirals mesh as the zipper slide passes over them.

3. Most zippers consis of a slide and two rows of interlocking teeth. In some zippers, intermeshing spirals replace the teeth.

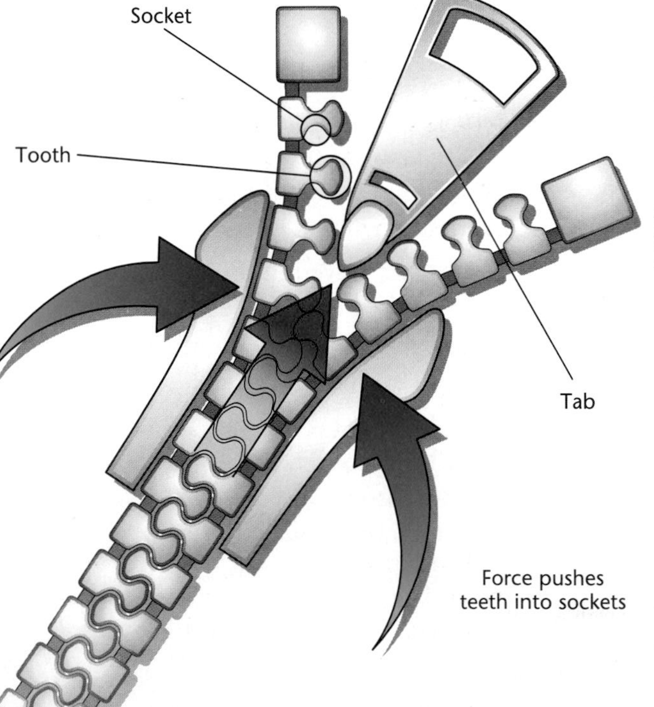

Thinking Critically

1. Classify the zipper as either a simple or compound machine. Justify your answer.
2. Look closely at a zipper slide. What prevents a zipper from opening unintentionally?
3. What role does the ta on a zipper's slide pl

CHAPTER 10 REVIEW

Key Terms

10.1
- energy
- kinetic energy
- work
- work-energy theorem
- joule
- power
- watt

10.2
- machine
- effort force
- resistance force
- mechanical advantage
- ideal mechanical advantage
- efficiency
- compound machine

Summary

10.1 Energy and Work

- Work is the transfer of energy by means of forces. The work done on the system is equal to the change in energy of the system.
- Work is the product of the force exerted on an object and the distance the object moves in the direction of the force.
- The area under the force-displacement graph is work.
- Power is the rate of doing work. That is, power is the rate at which energy is transferred.

10.2 Machines

- Machines, whether powered by engines or humans, do not change work, but make it easier.
- A machine eases the load either by changing the magnitude or the direction of the force exerted to do work.
- The mechanical advantage, *MA*, is the ratio of resistance force to effort force.
- The ideal mechanical advantage, *IMA*, is the ratio of the distances. In all real machines, *MA* is less than *IMA*.

Reviewing Concepts

Section 10.1

1. In what units is work measured?
2. A satellite revolves around Earth in a circular orbit. Does Earth's gravity do any work on the satellite?
3. An object slides at constant speed on a frictionless surface. What forces act on the object? What work is done by each force?
4. Define work and power.
5. What is a watt equivalent to in terms of kg, m, and s?

Section 10.2

6. Is it possible to get more work out of a machine than you put in?
7. How are the pedals of a bicycle a simple machine?

Applying Concepts

8. Which requires more work, carrying a 420-N knapsack up a 200-m hill or carrying a 210–N knapsack up a 400-m hill? Why?
9. You slowly lift a box of books from the floor and put it on a table. Earth's gravity exerts a force, magnitude mg, downward, and you exert a force, magnitude mg, upward. The two forces have equal magnitudes and opposite directions. It appears that no work is done, but you know you did work. Explain what work is done.
10. Grace has an after-school job carrying cartons of new copy paper up a flight of stairs, and then carrying recycled paper back down the stairs. The mass of the paper does not change. Grace's physics teacher suggests that Grace does no work all day, so she should not be paid. In what sense is the physics teacher correct? What arrangement of payments might Grace make to ensure compensation?
11. Grace now carries the copy paper boxes down a level, 15-m-long hall. Is Grace working now? Explain.
12. Two people of the same mass climb the same flight of stairs. The first person climbs the stairs in 25 s; the second person does so in 35 s.
    - **a.** Which person does more work? Explain your answer
    - **b.** Which person produces more power? Explain your answer.

13. Show that power delivered can be written as $P = Fv$.
14. Guy has to get a piano onto a 2.0-m-high platform. He can use a 3.0-m-long frictionless ramp or a 4.0-m-long frictionless ramp. Which ramp will Guy use if he wants to do the least amount of work?
15. How could you increase the ideal mechanical advantage of a machine?
16. A claw hammer is used to pull a nail from a piece of wood. How can you place your hand on the handle and locate the nail in the claw to make the effort force as small as possible?
17. How could you increase the mechanical advantage of a wedge without changing the ideal mechanical advantage?

Problems

Section 10.1

LEVEL 1

18. Lee pushes a 20-kg mass 10 m across a floor with a horizontal force of 80 N. Calculate the amount of work Lee does.
19. The third floor of a house is 8 m above street level. How much work is needed to move a 150-kg refrigerator to the third floor?
20. Stan does 176 J of work lifting himself 0.300 m. What is Stan's mass?
21. Mike pulls a 4.5-kg sled across level snow with a force of 225 N along a rope that is 35.0° above the horizontal. If the sled moves a distance of 65.3 m, how much work does Mike do?
22. Sau-Lan has a mass of 52 kg. She rides the up escalator at Ocean Park in Hong Kong. This is the world's longest escalator, with a length of 227 m and an average inclination of 31°. How much work does the escalator do on Sau-Lan?
23. Chris carries a carton of milk, weight 10 N, along a level hall to the kitchen, a distance of 3.5 m. How much work does Chris do?
24. A student librarian picks up a 2.2-kg book from the floor to a height of 1.25 m. He carries the book 8.0 m to the stacks and places the book on a shelf that is 0.35 m above the floor. How much work does he do on the book?
25. Brutus, a champion weightlifter, raises 240 kg of weights a distance of 2.35 m.
    a. How much work is done by Brutus lifting the weights?
    b. How much work is done by Brutus holding the weights above his head?
    c. How much work is done by Brutus lowering them back to the ground?
    d. Does Brutus do work if he lets go of the weights and they fall back to the ground?
    e. If Brutus completes the lift in 2.5 s, how much power is developed?
26. A force of 300.0 N is used to push a 145-kg mass 30.0 m horizontally in 3.00 s.
    a. Calculate the work done on the mass.
    b. Calculate the power developed.
27. Robin pushes a wheelbarrow by exerting a 145-N force horizontally. Robin moves it 60.0 m at a constant speed for 25.0 s.
    a. What power does Robin develop?
    b. If Robin moves the wheelbarrow twice as fast, how much power is developed?
28. A horizontal force of 805 N is needed to drag a crate across a horizontal floor with a constant speed. You drag the crate using a rope held at an angle of 32°.
    a. What force do you exert on the rope?
    b. How much work do you do on the crate when moving it 22 m?
    c. If you complete the job in 8.0 s, what power is developed?
29. Wayne pulls a 305-N sled along a snowy path using a rope that makes a 45.0° angle with the ground. Wayne pulls with a force of 42.3 N. The sled moves 16 m in 3.0 s. What power does Wayne produce?
30. A lawn roller is pushed across a lawn by a force of 115 N along the direction of the handle, which is 22.5° above the horizontal. If you develop 64.6 W of power for 90.0 s, what distance is the roller pushed?

LEVEL 2

31. A crane lifts a 3.50×10^3-N bucket containing 1.15 m^3 of soil (density = 2.00×10^3 kg/m^3) to a height of 7.50 m. Calculate the work the crane performs. Disregard the weight of the cable.

32. In **Figure 10–13,** the magnitude of the force necessary to stretch a spring is plotted against the distance the spring is stretched.
    a. Calculate the slope of the graph and show that $F = kd$, where $k = 25$ N/m.
    b. Find the amount of work done in stretching the spring from 0.00 m to 0.20 m by calculating the area under the curve from 0.00 m to 0.20 m.
    c. Show that the answer to part b can be calculated using the formula $W = 1/2kd^2$, where W is the work, $k = 25$ N/m (the slope of the graph), and d is the distance the spring is stretched (0.20 m).
33. The graph in **Figure 10–13** shows the force needed to stretch a spring. Find the work needed to stretch it from 0.12 m to 0.28 m.

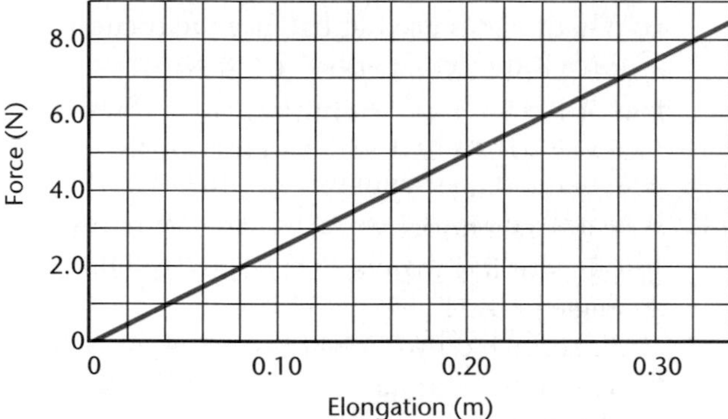

FIGURE 10–13

34. John pushes a crate across the floor of a factory with a horizontal force. The roughness of the floor changes, and John must exert a force of 20 N for 5 m, then 35 N for 12 m, and then 10 N for 8 m.
    a. Draw a graph of force as a function of distance.
    b. Find the work John does pushing the crate.
35. Sally expends 11 400 J of energy to drag a wooden crate 25.0 m across a floor with a constant speed. The rope makes an angle of 48.0° with the horizontal.
    a. How much force does the rope exert on the crate?
    b. What is the force of friction acting on the crate to impede its motion?
    c. What work is done by the floor through the force of friction between the floor and the crate?
36. An 845-N sled is pulled a distance of 185 m. The task requires 1.20×10^4 J of work and is done by pulling on a rope with a force of 125 N. At what angle is the rope held?
37. You slide a crate up a ramp at an angle of 30.0° by exerting a 225-N force parallel to the ramp. The crate moves at constant speed. The coefficient of friction is 0.28. How much work have you done on the crate when it is raised a vertical distance of 1.15 m?
38. A 4.2-kN piano is to be slid up a 3.5-m frictionless plank at a constant speed. The plank makes an angle of 30.0° with the horizontal. Calculate the work done by the person sliding the piano up the plank.
39. Rico slides a 60-kg crate up an inclined ramp 2.0-m long onto a platform 1.0 m above floor level. A 400-N force, parallel to the ramp, is needed to slide the crate up the ramp at a constant speed.
    a. How much work does Rico do in sliding the crate up the ramp?
    b. How much work would be done if Rico simply lifted the crate straight up from the floor to the platform?
40. A worker pushes a crate weighing 93 N up an inclined plane. The worker pushes the crate horizontally, parallel to the ground, as illustrated in **Figure 10–14.**
    a. The worker exerts a force of 85 N. How much work does he do?
    b. How much work is done by gravity? (Be careful with the signs you use.)
    c. The coefficient of friction is $\mu = 0.20$. How much work is done by friction? (Be careful with the signs you use.)

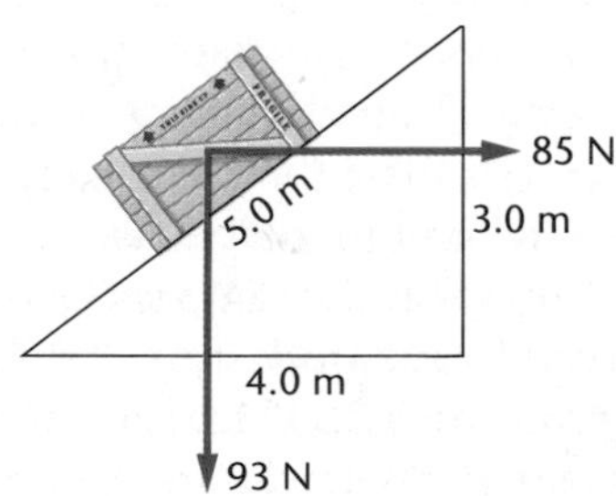

FIGURE 10–14

41. The graph in **Figure 10–15** shows the force and displacement of an object being pulled.
   a. Calculate the work done to pull the object 7.0 m.
   b. Calculate the power developed if the work were done in 2.0 s.

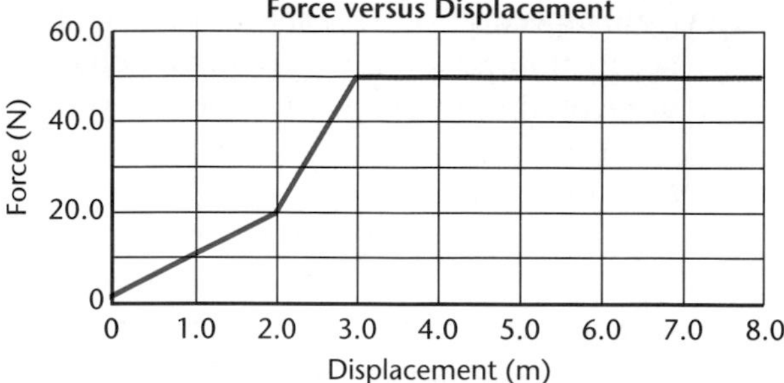

FIGURE 10–15

42. In 35.0 s, a pump delivers 0.550 m^3 of oil into barrels on a platform 25.0 m above the pump intake pipe. The density of the oil is 0.820 g/cm^3.
   a. Calculate the work done by the pump.
   b. Calculate the power produced by the pump.
43. A 12.0-m-long conveyor belt, inclined at 30.0°, is used to transport bundles of newspapers from the mailroom up to the cargo bay to be loaded on to delivery trucks. Each newspaper has a mass of 1.0 kg, and there are 25 newspapers per bundle. Determine the power of the conveyor if it delivers 15 bundles per minute.
44. An engine moves a boat through the water at a constant speed of 15 m/s. The engine must exert a force of 6.0×10^3 N to balance the force that water exerts against the hull. What power does the engine develop?
45. A 188-W motor will lift a load at the rate (speed) of 6.50 cm/s. How great a load can the motor lift at this rate?
46. A car is driven at a constant speed of 76 km/h down a road. The car's engine delivers 48 kW of power. Calculate the average force that is resisting the motion of the car.
47. Two cars travel at the same speed of 105 km/h. One of the cars, a sleek sports car, has a motor that delivers only 35 kW of power at this speed. The other car's motor needs to produce 65 kW to move the car at this speed. Forces exerted from air resistance cause the difference.
   a. For each car, list the external horizontal forces exerted on it, and give the cause of each force. Compare their magnitudes.
   b. According to Newton's third law of motion, each car exerts forces. What are the directions of these forces?
   c. Calculate the magnitude of the forward force exerted by each car on the air.
   d. The car engines did work. Where did the energy they transferred come from?

Section 10.2

LEVEL 1

48. Stan raises a 1200-N piano a distance of 5.00 m using a set of pulleys. Stan pulls in 20.0 m of rope.
   a. How much effort force would Stan apply if this were an ideal machine?
   b. What force is used to balance the friction force if the actual effort is 340 N?
   c. What is the work output?
   d. What is the input work?
   e. What is the mechanical advantage?
49. A mover's dolly is used to transport a refrigerator up a ramp into a house. The refrigerator has a mass of 115 kg. The ramp is 2.10 m long and rises 0.850 m. The mover pulls the dolly with a force of 496 N up the ramp. The dolly and ramp constitute a machine.
   a. What work does the mover do?
   b. What is the work done on the refrigerator by the machine?
   c. What is the efficiency of the machine?
50. A pulley system lifts a 1345-N weight a distance of 0.975 m. Paul pulls the rope a distance of 3.90 m, exerting a force of 375 N.
   a. What is the ideal mechanical advantage of the system?
   b. What is the mechanical advantage?
   c. How efficient is the system?
51. Because there is very little friction, the lever is an extremely efficient simple machine. Using a 90.0% efficient lever, what input work is required to lift an 18.0-kg mass through a distance of 50 cm?
52. What work is required to lift a 215-kg mass a distance of 5.65 m using a machine that is 72.5% efficient?

LEVEL 2

53. The ramp in **Figure 10–16** is 18 m long and 4.5 m high.

a. What force parallel to the ramp (F_A) is required to slide a 25-kg box at constant speed to the top of the ramp if friction is disregarded?

b. What is the *IMA* of the ramp?

c. What are the real *MA* and the efficiency of the ramp if a parallel force of 75 N is actually required?

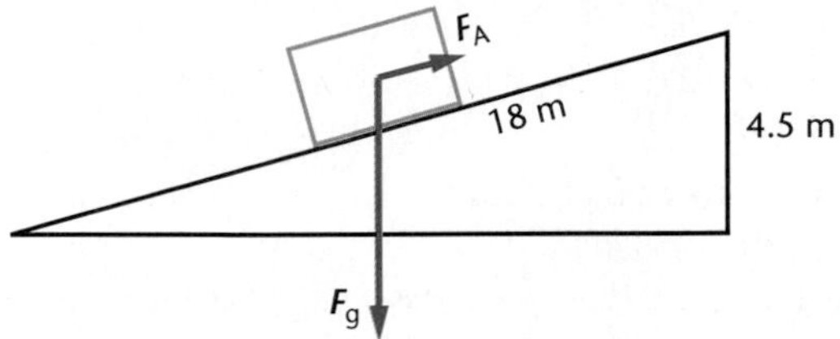

FIGURE 10–16

54. A motor having an efficiency of 88% operates a crane having an efficiency of 42%. With what constant speed does the crane lift a 410-kg crate of machine parts if the power supplied to the motor is 5.5 kW?

55. A compound machine is constructed by attaching a lever to a pulley system. Consider an ideal compound machine consisting of a lever with an *IMA* of 3.0 and a pulley system with an *IMA* of 2.0.

a. Show that the *IMA* of this compound machine is 6.0.

b. If the compound machine is 60.0% efficient, how much effort must be applied to the lever to lift a 540-N box?

c. If you move the effort side of the lever 12.0 cm, how far is the box lifted?

Critical Thinking Problems

56. A sprinter, mass 75 kg, runs the 50-meter dash in 8.50 s. Assume that the sprinter's acceleration is constant throughout the race.

a. What is the average power of the sprinter over the 50.0 m?

b. What is the maximum power generated by the sprinter?

c. Make a quantitative graph of power versus time for the entire race.

57. A sprinter in problem 56 runs the 50-meter dash in the same time, 8.50 s. However, this time the sprinter accelerates in the first second and runs the rest of the race at a constant velocity.

a. Calculate the average power produced for that first second.

b. What is the maximum power the sprinter now generates?

Going Further

Task Analysis You work at a store carrying boxes to a storage loft, 12 m above the ground. You have 30 boxes with a total mass of 150 kg that must be moved as quickly as possible, so you consider carrying more than one up at a time. If you try to move too many at once, you know you'll go very slowly, resting often. If you carry only one, most of the energy will go into raising your own body. The power (in watts) that your body can develop over a long time depends on the mass you carry as shown in **Figure 10–17.** This is an example of a power curve that applies to machines as well as people. Find the number of boxes to carry on each trip that would minimize the time required. What time would you spend doing the job? (Ignore the time needed to go back down the stairs, lift and lower each box, etc.)

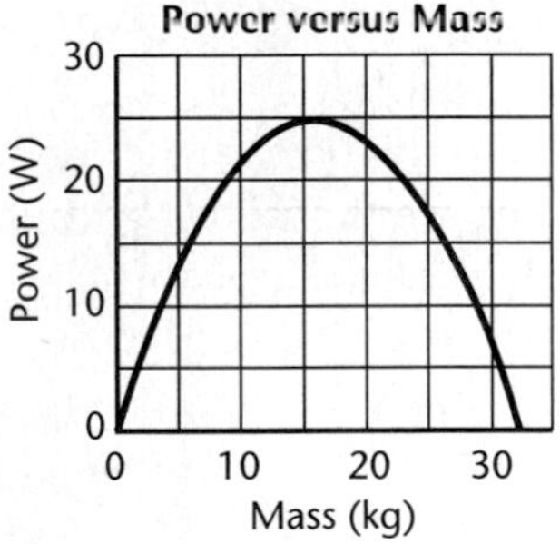

FIGURE 10–17

*inter*NET CONNECTION

Follow the link on the Glencoe Homepage at **www.glencoe.com/sec/science** to find out more about this chapter.

Whoosh!

The roller coaster slowly climbs the first hill. Then, hang onto your hat! Down, down, down it flies until it reaches the bottom and begins to climb the next hill. Must the first hill of a roller coaster be the highest one?

CHAPTER

11 Energy

In everyday speech, the word *energy* is used in many ways. A child who runs and plays long after adults are tired is said to be full of energy. Companies that supply your home with electricity, natural gas, or heating fuel and your car with gasoline often are called energy companies. When these resources become more scarce and more expensive, the media report stories of an energy crisis.

Physicists use the term *energy* in a much more precise way. You began your study of energy in the last chapter. Now you will build on that knowledge.

In this chapter, you'll investigate a variety of forms of energy that objects can have. You'll learn about ways in which energy is transferred from one form to another and about methods of keeping track of all those changes. And, you'll also be able to determine whether the first hill in a roller coaster must always be the highest one.

WHAT YOU'LL LEARN

- You will learn that energy is simply the ability to do something.
- You will learn that the total amount of energy in a closed system never changes; energy just changes form.

WHY IT'S IMPORTANT

- How energy changes form, and how much energy remains in a useful form, can influence how much you accomplish in an hour, in a day, and in your lifetime.

interNET CONNECTION

Follow the link for this chapter on the Glencoe Homepage at **www.glencoe.com/sec/science** to find out more about energy.

11.1 The Many Forms of Energy

OBJECTIVES

- **Use a model** to relate work and energy.
- **Calculate** the kinetic energy of a moving object.
- **Determine** how to find the gravitational potential energy of a system.
- **Identify** ways in which elastic potential energy is stored in a system.

In Chapter 10, you were introduced to the work-energy theorem. You learned that when work is done on a system, the energy of that system increases. On the other hand, if the system does work, then the energy of the system decreases. Those are pretty abstract ideas. Let's make them more like real life and develop a graphic model to give you a picture of what is going on.

A Model of the Work-Energy Theorem

If you have a job, the amount of money you have increases every time you are paid. This process can be represented with a bar graph, as shown in **Figure 11–1a.** The bar representing the money you have before you are paid is shorter than the bar representing the amount you have after you are paid. The difference in height of the two bars, the cash flow, is equal to your pay.

Now, what happens when you spend money? The total amount of money you have decreases. As shown in **Figure 11–1b,** the bar that represents the amount of money you had before you bought that new CD is higher than the bar that stands for the amount remaining after your shopping trip. The difference is the cost of the CD. Cash flow is shown as a bar below the axis, representing a negative number.

Throwing a ball

But what does this have to do with work and energy? In Chapter 10, you read that when you exert a constant force, ***F,*** on an object through a distance, ***d,*** in the direction of the force, you do an amount of work, represented by $W = Fd$, on that object. The work is positive, and the energy of the object increases by an amount equal to W. Suppose the object is a ball, and you exert a force to throw the ball. As a result of the

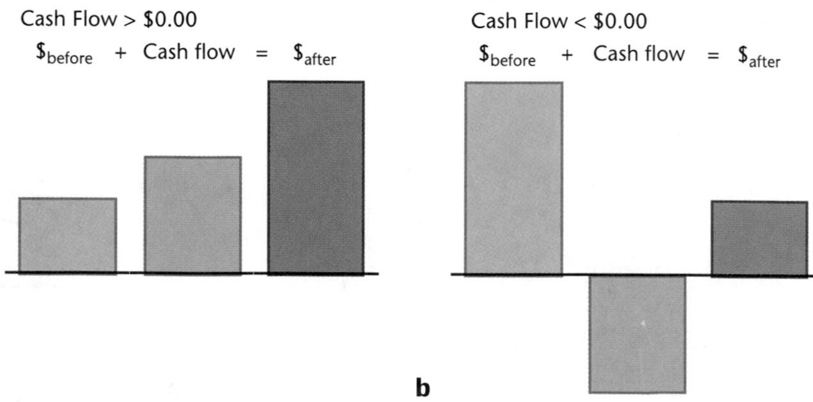

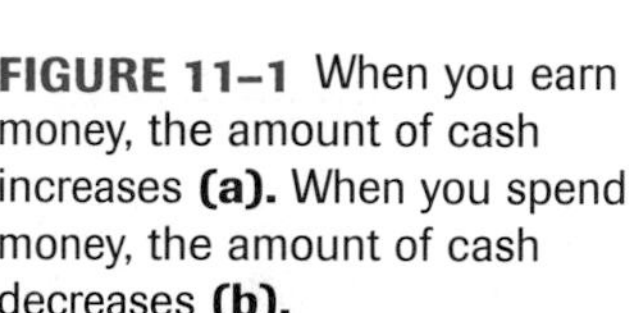

FIGURE 11–1 When you earn money, the amount of cash increases **(a).** When you spend money, the amount of cash decreases **(b).**

force you apply, the ball gains an energy of motion that is referred to as kinetic energy. This process is shown in **Figure 11–2a.** You can again use a bar graph to represent the process. This time, the height of the bar represents an amount of work, or energy, measured in joules. The kinetic energy after the work is done is equal to the sum of the initial kinetic energy plus the work done on the ball.

Catching a ball

What happens when you catch a ball? Before hitting your hands or glove, the ball is moving, so it has kinetic energy. In catching it, you exert a force on the ball in the direction opposite to its motion. Therefore, you do negative work on it, causing it to stop. Now that the ball isn't moving, it has no kinetic energy. This process and the bar graph that represents it are shown in **Figure 11–2b.** The initial kinetic energy of the ball is positive: kinetic energy is always positive. The work done on the ball is negative, so the bar graph representing the work is below the axis. The final kinetic energy is zero because the ball has stopped. Again, kinetic energy after the ball has stopped is equal to the sum of the initial kinetic energy plus the work done on the ball.

Kinetic Energy

As you learned in Chapter 10, the kinetic energy of an object is represented by the equation $K = 1/2mv^2$, where m is the mass of the object and v is its velocity. The kinetic energy is proportional to the object's mass. Thus, a 7.26-kg shot thrown through the air has much more kinetic energy than a 148-g baseball with the same velocity. The kinetic energy of an object is also proportional to the square of the velocity of the object. A car speeding at 30 m/s has four times the kinetic energy of the same car moving at 15 m/s. Kinetic energy, like work, is measured in joules. **Table 11–1** on the next page shows the kinetic energies of some typical moving objects.

Energy in Coins

Does your car require more or less stopping distance when it is loaded with passengers than when you are driving alone? A short activity will help you to answer this question. Lay a ruler on a smooth table. Place two quarters against the edge of the ruler. Momentarily push the two quarters at the same speed across the table, and then stop the ruler. The two quarters should slide the same distance before stopping. Now tape another coin on top of one quarter to increase its mass. Again push the coins with the ruler.

Analyze and Conclude Does the stopping distance depend upon the mass? Explain.

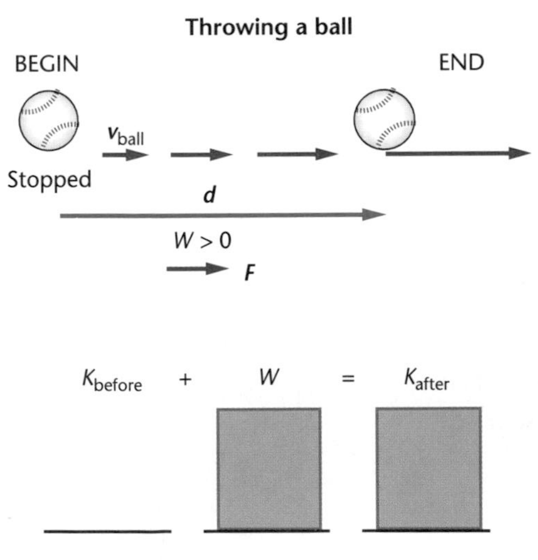

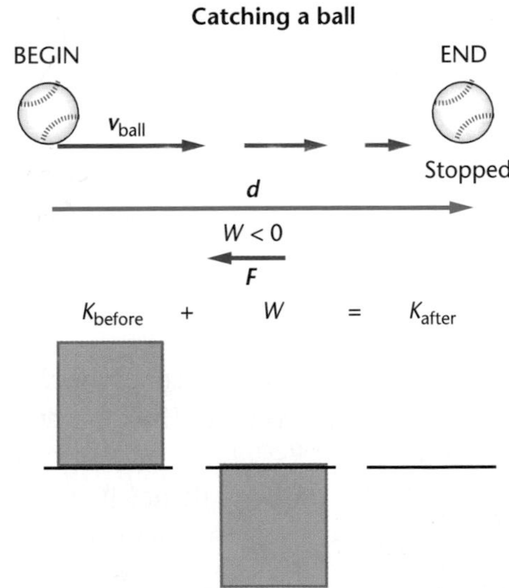

FIGURE 11–2 When you throw a ball, you do work on it. The ball gains kinetic energy as a result **(a).** When you catch a ball, it stops because you exert a force on it in the opposite direction from its motion **(b).**

TABLE 11–1
Typical Kinetic Energy

Object	Mass	Kinetic Energy (J)
Aircraft carrier	91 400 tons at 30 knots	9.9×10^9
Orbiting satellite	100 kg at 7.8 km/s	3.0×10^9
Trailer truck	5700 kg at 100 km/h	2.2×10^6
Compact car	750 kg at 100 km/h	2.9×10^5
Football linebacker	110 kg at 9.0 m/s	4.5×10^3
Pitched baseball	148 g at 45 m/s	1.5×10^2
Falling nickel	5 g from 50-m height	2.5
Bumblebee	2 g at 2 m/s	4×10^{-3}
Snail	5 g at 0.05 km/h	4.5×10^{-7}

Example Problem

Kinetic Energy and Work

An 875-kg compact car speeds up from 22.0 to 44.0 m/s while passing another car. What were its initial and final energies, and how much work was done on the car to increase its speed?

Sketch the Problem

- Sketch the initial and final conditions.
- Make a bar graph.

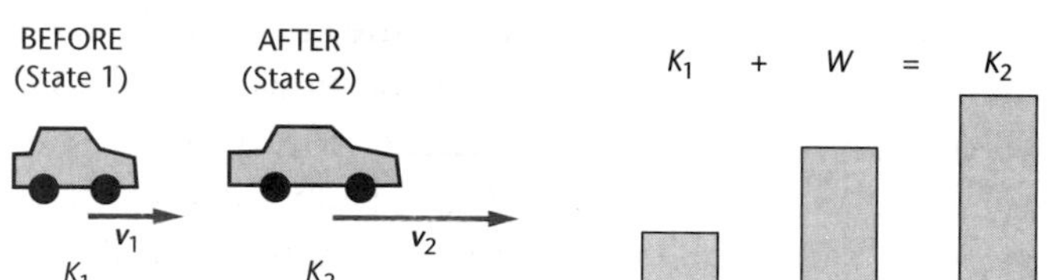

Calculate Your Answer

Known:

$m = 875$ kg

$v_1 = 22.0$ m/s

$v_2 = 44.0$ m/s

Unknown:

$K_1 = ?$

$K_2 = ?$

$W = ?$

Strategy:

Use the initial and final speeds of the car to calculate kinetic energy.

The work done equals the change in kinetic energy.

Calculations:

$K = 1/2\ mv^2$

$K_1 = 1/2(875\text{ kg})(22.0\text{ m/s})^2 = 2.12 \times 10^5\text{ J}$

$K_2 = 1/2(875\text{ kg})(44.0\text{ m/s})^2 = 8.47 \times 10^5\text{ J}$

$W = K_2 - K_1 = (8.47 - 2.12) \times 10^5\text{ J}$
$= 6.35 \times 10^5\text{ J}$

Check Your Answer

- Do the signs make sense? K should be positive. $W > 0$ as K increases.
- Are the units correct? The answer should be in J, which equals $kg{\cdot}m^2/s^2$.
- Is the magnitude realistic? It should be similar to the listing in **Table 11–1** for a compact car.

Practice Problems

1. **Consider** the compact car in the Example Problem.
   a. Write 22 m/s and 44 m/s in km/h.
   b. How much work is done in slowing the car to 22.0 m/s?
   c. How much work is done in bringing it to rest?
   d. Assume that the force that does the work slowing the car is constant. Find the ratio of the distance needed to slow the car from 44.0 m/s to 22.0 m/s to the distance needed to slow it from 22.0 m/s to rest.
2. A rifle can shoot a 4.20-g bullet at a speed of 965 m/s.
   a. Draw work-energy bar graphs and free-body diagrams for all parts of this problem.
   b. Find the kinetic energy of the bullet as it leaves the rifle.
   c. What work is done on the bullet if it starts from rest?
   d. If the work is done over a distance of 0.75 m, what is the average force on the bullet?
   e. If the bullet comes to rest by penetrating 1.5 cm into metal, what is the magnitude and direction of the force the metal exerts? Again, assume that the force is constant.
3. A comet with a mass of 7.85×10^{11} kg strikes Earth at a speed of 25.0 km/s.
   a. Find the kinetic energy of the comet in joules.
   b. Compare the work that is done by Earth in stopping the comet to the 4.2×10^{15} J of energy that were released by the largest nuclear weapon ever built. Such a comet collision has been suggested as having caused the extinction of the dinosaurs.
4. **Table 11–1** shows that 2.2×10^6 J of work are needed to accelerate a 5700-kg trailer truck to 100 km/h.
   a. What would be the truck's speed if half as much work were done on it?
   b. What would be the truck's speed if twice as much work were done on it?

F.Y.I.

The matter in one pound of anything, when it is completely converted into energy according to $E_0 = mc^2$, will produce 4.086×10^{16} J (11 400 000 000 kWh) of energy.

Stored Energy

Imagine a group of boulders high on a hill. Objects such as these boulders that have been lifted up against the force of gravity have stored, or potential, energy. Now imagine a rock slide in which the boulders are shaken loose. They fall, picking up speed as their potential energy is converted to kinetic energy. In the same way, a small spring-loaded toy such as the one pictured in **Figure 11–3** has energy stored in a compressed spring. While both of these examples represent energy stored by mechanical means, there are many other means of storing energy. Automobiles, for example, carry their energy stored in the form

FIGURE 11–3 This spring-loaded Jack-in-the-box had energy stored in its compressed spring. When the spring was released, the energy was converted to kinetic energy.

F.Y.I.

The gravitational potential energy stored in a block of stone when it was lifted to the top of a pyramid 2500 years ago is still there undiminished. Making allowances for erosion, if the stone were dropped today, it would do as much work as when it was first put into place.

of chemical energy in the gasoline tank. The conversion of energy from one form to another is the focus of many nations' industries and is an integral part of modern life.

How does the money model that was discussed earlier illustrate the transformation of energy from one form to another? Money, too, can come in different forms. You can think of money in the bank as stored money, or potential spending money. Depositing money in your bank account or getting it out with an ATM card doesn't change the total amount of money you have; it just converts it from one form to another. The height of the bar graph in **Figure 11–4** represents the amount of money in each form. In the same way, you can use a bar graph to represent the amount of energy in various forms that a system has.

Gravitational Potential Energy

Look at the balls being juggled in **Figure 11–5.** If you consider the system to be only one ball, then it has several external forces exerted on it. The force of the juggler's hand does work, giving the ball its original kinetic energy. After the ball leaves his hand, only the force of gravity acts on it. How much work does gravity do on the ball as its height changes?

Let h represent the ball's height measured from the juggler's hand. On the way up, its displacement is up, but the force on the ball, $\mathbf{F}_g$, is downward, so the work done by gravity, $W_{\text{by gravity on ball}} = -mgh$, is negative. On the way back down, the force and displacement are in the same direction, so the work done by gravity, $W_{\text{by gravity on ball}} = mgh$, is positive. The magnitude of the work is the same; only the sign changes. Thus, while the ball is moving upward, gravity does negative work, slowing the ball to a stop. On the way back down, gravity does positive work, increasing the ball's speed and thereby increasing its kinetic energy. The ball recovers all of the kinetic energy it originally had when it returns to the height at which it left the juggler's hand. It is as if the ball's kinetic energy is stored in another form as the ball rises and is returned to kinetic energy as the ball falls. The notion that energy may take different forms will be discussed in more detail in a later section.

If you choose a system to consist of an object plus Earth, then the gravitational attraction between the object and Earth is an interaction force between members of the system. If the object moves away from Earth, energy is stored in the system as a result of the gravitational interaction between the object and Earth. This stored energy is called the **gravitational potential energy** and is represented by the symbol U_g. In the example of the juggler, where gravity was an external force, the change in the potential energy of the object is then the negative of the work done. Thus, the gravitational potential energy is represented by

$$U_g = mgh,$$

where m is the mass of the object, g is the acceleration resulting from gravity, and h is the distance the object has risen above the position it

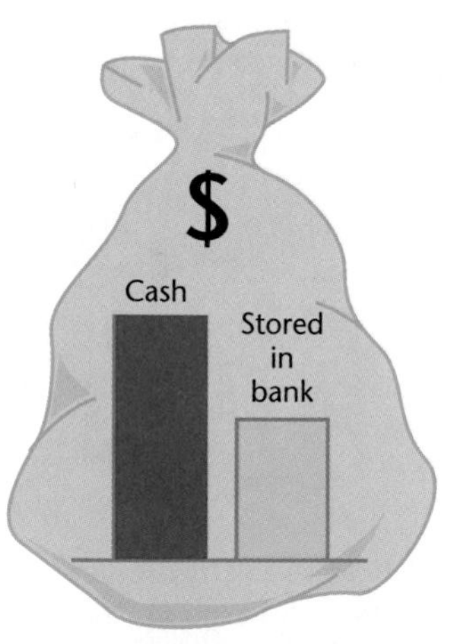

Total $ = Cash + Money in bank

FIGURE 11–4 Money in the bank and cash in your pocket are different forms of the same thing.

had when U_g was defined to be zero. **Figure 11–6** shows the energy of a system consisting of one of the juggler's balls plus Earth. The energy in the system exists in two forms: kinetic energy and gravitational potential energy. At the beginning of the ball's flight, the energy is all in the form of kinetic energy. On the way up, as the ball slows, energy is changed from kinetic to potential. At the top of the ball's flight, when the ball instantaneously comes to rest, the energy is all potential. On the way back down, potential energy is changed back into kinetic energy. The sum of kinetic and potential energy is constant because no work is done on the system by any force external to the system.

FIGURE 11–5 Kinetic and potential energy are constantly being exchanged when juggling.

Choosing a reference level

In the example of juggling a ball, height was measured from the place at which the ball left the juggler's hand. At that height, $h = 0$, the gravitational potential energy was defined to be zero. This choice of a **reference level,** at which the potential energy is defined to be zero, is arbitrary; the reference level may be taken as any position that is convenient for solving a given problem.

If the highest point the ball reached had been chosen as zero, as illustrated in **Figure 11–6b,** then the potential energy of the system would have been zero there and negative at the beginning and end of the flight. Although the total energy in the system would have been different, it would not have changed at any point during the flight. Only *changes* in energy, both kinetic and potential, can be measured. Because the initial energy of a real physical system is not known, the total energy of the system cannot be determined.

FIGURE 11–6 The energy of a ball is converted from one form to another in various stages of its flight **(a).** Note that the choice of a reference level is arbitrary but that the total energy remains constant **(b).**

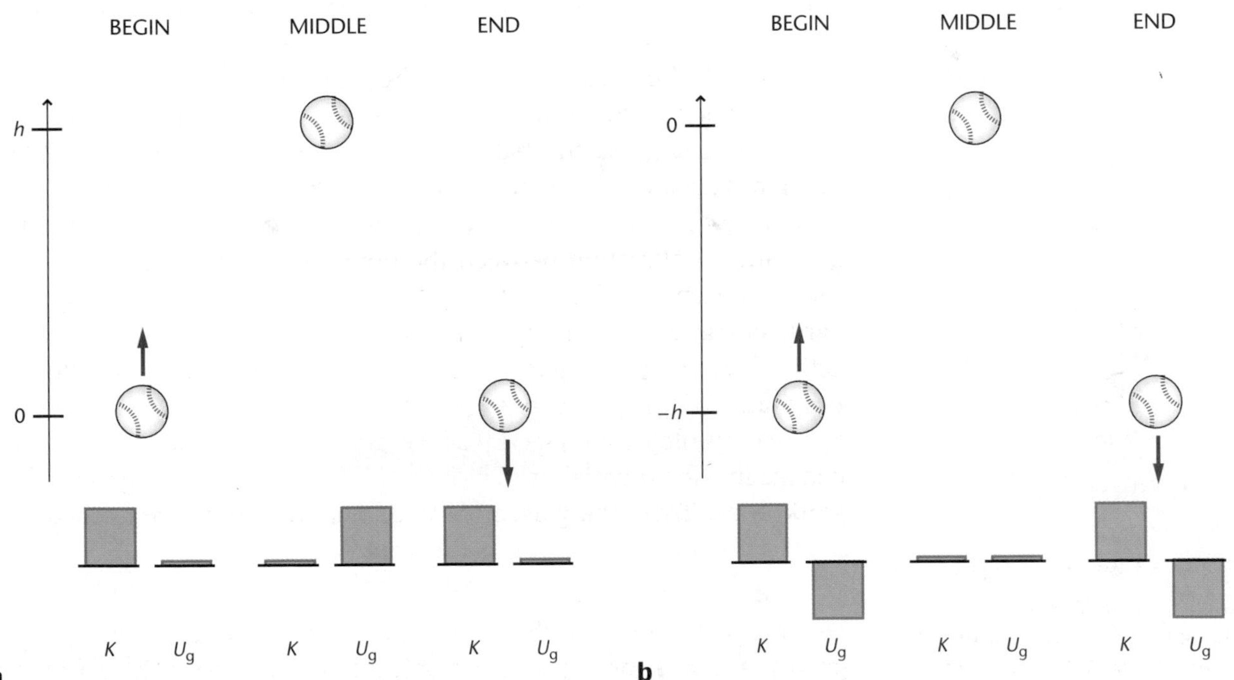

Example Problem

Gravitational Potential Energy

You lift a 2.00-kg textbook from the floor to a shelf 2.10 m above the floor.

a. What is the book's gravitational potential energy relative to the floor?

b. What is its gravitational potential energy relative to your head, assuming that you're 1.65 m tall?

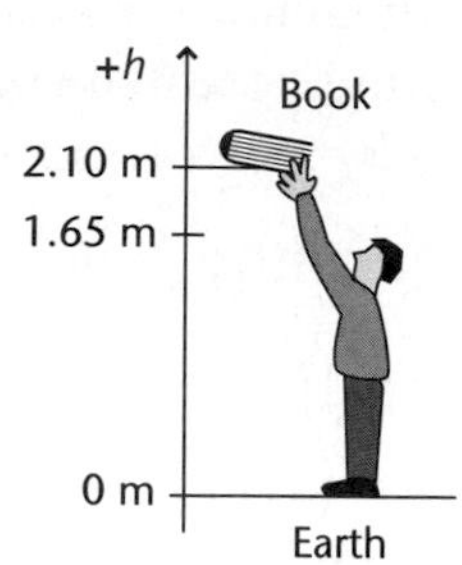

Sketch the Problem

- Sketch the initial and final conditions.
- Choose a reference level.
- Make a bar graph.

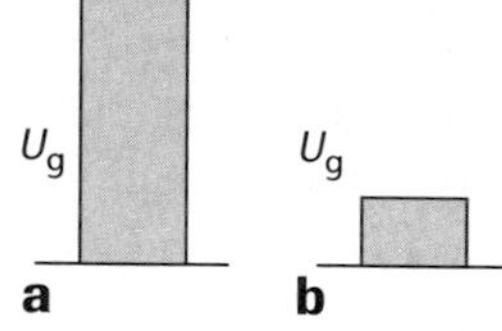

Calculate Your Answer

Known:

$m = 2.00$ kg

$h_{shelf} = 2.10$ m

$h_{head} = 1.65$ m

$g = 9.80\ m/s^2$

Unknown:

U_g (relative to floor) = ?

U_g (relative to head) = ?

Strategy:

In both **a** and **b,** the reference level is below the book, so the gravitational potential energy of the system is positive.

Calculations:

$U_g = mgh$

a. $h = 2.10\ m - 0.00\ m = 2.10\ m$

$U_g = (2.00\ kg)(9.80\ m/s^2)(2.10\ m) = 41.2\ J$

b. $h = 2.10\ m - 1.65\ m = 0.45\ m$

$U_g = (2.00\ kg)(9.80\ m/s^2)(0.45\ m) = 8.82\ J$

Check Your Answer

- Are the units correct? Energy is in $kg{\cdot}m^2/s^2 = J$.
- Do the signs make sense? Both are positive because the book is above the reference level.
- Is the magnitude realistic? The energy relative to your head is less than 1/4 of the energy relative to the floor, which is similar to the ratio of the distances.

Practice Problems

5. For the preceding Example Problem, select the shelf as the reference level. The system is the book plus Earth.

a. What is the gravitational potential energy of the book at the top of your head?

b. What is the gravitational potential energy of the book at the floor?

6. A 90-kg rock climber first climbs 45 m up to the top of a quarry, then descends 85 m from the top to the bottom of the quarry. If the initial height is the reference level, find the potential energy of the system (the climber plus Earth) at the top and the bottom. Draw bar graphs for both situations.
7. A 50.0-kg shell is shot from a cannon at Earth's surface to a height of 425 m. The system is the shell plus Earth, and the reference level is Earth's surface.
   a. What is the gravitational potential energy of the system when the shell is at this height?
   b. What is the change in the potential energy when the shell falls to a height of 225 m?
8. A 7.26-kg bowling ball hangs from the end of a 2.5-m rope. The ball is pulled back until the rope makes a 45° angle with the vertical.
   a. What is the gravitational potential energy of the system?
   b. What system and what reference level did you use in your calculation?

HELP WANTED

LOCK-AND-DAM CONSTRUCTION ENGINEERS

Wanted: civil, mechanical, and geotechnical engineers to design a new dam and supervise its construction. Will work with geologists to analyze the underlying soil and rock for seepage or earthquake activity, and with biologists to consider the environmental impact of the dam. College degree in engineering required. For more information, contact:
U.S. Army Corps of Engineers
Environmental Division
Directorate of Military Programs
Washington, DC 20314-1000

Elastic Potential Energy

When the string on the bow shown in **Figure 11–7** is pulled, work is done on the bow, storing energy in it. If you choose the system to be the bow, the arrow, and Earth, then you increase the energy of the system. When the string and arrow are released, energy is changed into kinetic energy. The energy stored in the pulled string is called **elastic potential energy.** Elastic potential energy is often stored in rubber balls, rubber bands, slingshots, and trampolines.

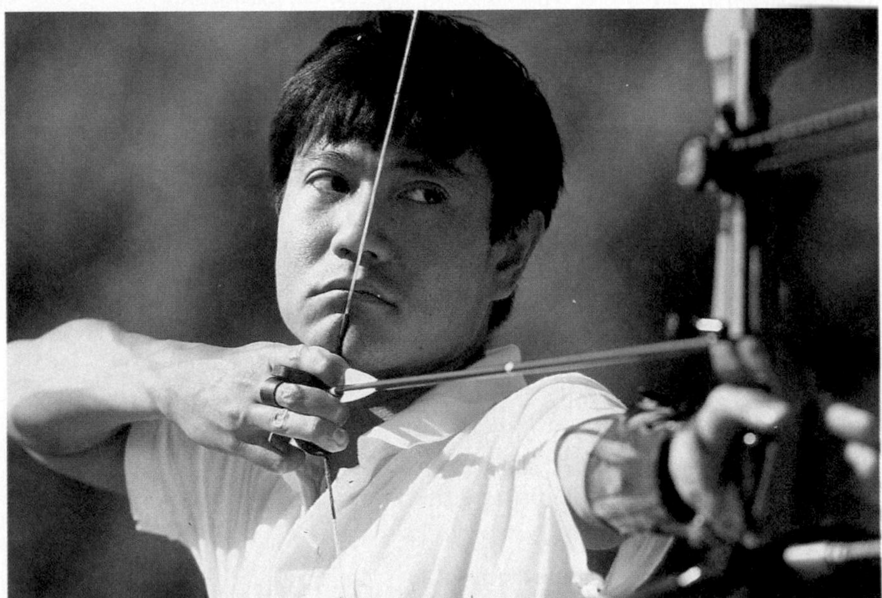

FIGURE 11–7 Elastic potential energy is stored in the string of this bow. Before the string is released, the energy is all potential. As the string is released, the energy is transferred to the arrow as kinetic energy.

Energy can also be stored in the bending of an object. When stiff metal or bamboo poles were used in pole-vaulting, the poles did not bend easily. Little work was done on the poles and consequently, the poles did not store much potential energy. Since the flexible fiberglass pole was introduced, record heights have soared. The pole-vaulter runs with the flexible pole, then plants its end into a socket in the ground. Work is done to bend the pole as the kinetic energy of the runner is converted to elastic potential energy. Then, as the pole straightens, the elastic potential energy is converted to gravitational potential energy and kinetic energy as the vaulter is lifted as high as 6 m above the ground. The increased capacity for the pole to store energy is reflected in additional height.

11.1 Section Review

1.1 In each of the following situations, the system consists of a ball and Earth. Draw bar graphs that describe the work done and changes in energy forms.

a. You throw a ball horizontally.

b. The horizontally thrown ball is caught in a mitt.

c. You throw a ball vertically, and it comes to rest at the top of its flight.

d. The ball falls back to Earth, where you catch it.

1.2 You use a toy dart gun to shoot the dart straight up. The system is the gun, the dart, and Earth. Draw bar graphs that describe the energy forms when

a. you have pushed the dart into the gun barrel, thus compressing the spring.

b. the spring expands and the dart leaves the gun barrel after you pull the trigger.

c. the dart reaches the top of its flight.

1.3 You use an air hose to exert a constant horizontal force on a puck on a frictionless air table. You keep the hose aimed at the puck so that the force on it is constant as the puck moves a fixed distance.

a. Explain what happens in terms of work and energy. Draw bar graphs as part of your explanation.

b. You now repeat the experiment. Everything is the same except that the puck has half the mass. What will be the same? What will be different? In what way?

1.4 **Critical Thinking** You now modify the experiment slightly by applying the constant force for a fixed amount of time.

a. Explain what happens in terms of impulse and momentum.

b. What happens now when the mass of the puck is reduced?

c. Compare the kinetic energies of the two pucks.

Design Your Own **Physics Lab**

Down the Ramp

Problem

What factors affect the speed of a cart at the bottom of a ramp? Along the floor?

Hypothesis

Form a hypothesis that relates the speed or energy of the cart at the bottom of the ramp to the mass of the cart on the ramp.

Possible Materials 

cart
0.50-kg mass
1.0-kg mass
board to be used as a ramp
stopwatch
meterstick
masking tape

Plan the Experiment

1. Your lab group should develop a plan to answer the questions stated in the problem. How should you structure your investigation? How many trials do you need for each setup? Be prepared to present and defend your plan, data, and results to the class.
2. Identify the independent and dependent variables. Which will you keep constant?
3. Describe your procedures.
4. Describe the energy changes as the cart rolls down the ramp and onto the floor.
5. Construct data tables that will show the measurements that you make.
6. **Check the Plan** Make sure your teacher has approved your plan before you proceed with your experiment.

Analyze and Conclude

1. **Checking Your Hypothesis** Did the speed at the bottom of the ramp depend on the mass of the cart? Does twice the mass have twice the speed? Does three times the mass go three times as fast?
2. **Calculating Results** List and explain the equations that you used for your energy calculations. What do the equations suggest about the speed at the bottom when the mass is changed?
3. **Comparing and Contrasting** Compare the gravitational potential energy of the cart at the starting position to the kinetic energy of the cart along the floor. What is your conclusion?
4. **Thinking Critically** Suppose one lab group finds that the cart has 30% more kinetic energy along the floor than the starting gravitational potential energy. What would you tell the group?

Apply

1. A Soap Box Derby is a contest in which riders coast down a long hill. Does the mass of the cart have a significant effect on the results? What other factors may be more important in winning the race?

11.2 Conservation of Energy

You have read that when you consider a ball and Earth as a system, the sum of gravitational potential energy and kinetic energy in that system is constant. As the height of the ball changes, energy is converted from kinetic to potential energy and back again, but the total amount of energy stays the same. Conservation of energy, unlike conservation of momentum, does not follow directly from Newton's laws. It is a separate fact of nature. But this fact wasn't easy to discover. Rather, the law of conservation of energy was discovered in the middle of the 1800s, more than 150 years after Newton's work had been published. Since that time, the law has survived another 150 years of questioning and probing by scientists in many fields.

OBJECTIVES

- **Solve** problems using the law of conservation of energy.
- **Analyze** collisions to find the change in kinetic energy.

Choosing a System

If you have ever observed energy in action around you, you may have noticed that energy doesn't always seem to be conserved. The kinetic energy of a rolling soccer ball is soon gone. Even on smooth ice, a hockey puck eventually stops moving. The swings of a pendulum soon die away. To get an idea of what is happening, let's go back to the money model.

Suppose you had a total of $100 in cash and in the bank. One day, you counted your money and, despite neither earning nor spending anything, you were 50¢ short. Would you assume that the money just disappeared? Probably not. More likely, you'd go hunting through your purse or pants pockets trying to locate the lost change. If you couldn't find it there, you might look under the couch cushions or even in the dryer. In other words, rather than giving up on the conservation of money, you would seek new and different places where it might be.

Scientists have done the same with the conservation of energy. Rather than concluding that energy is not conserved, they have discovered new forms into which energy can be converted. They have concluded that the **law of conservation of energy** is a description of nature. That is, as long as the system under investigation is closed so that objects do not move in and out, and as long as the system is isolated from external forces, then energy can only change form. The total amount of energy is constant. In other words, energy can be neither created nor destroyed. In a closed, isolated system, energy is conserved.

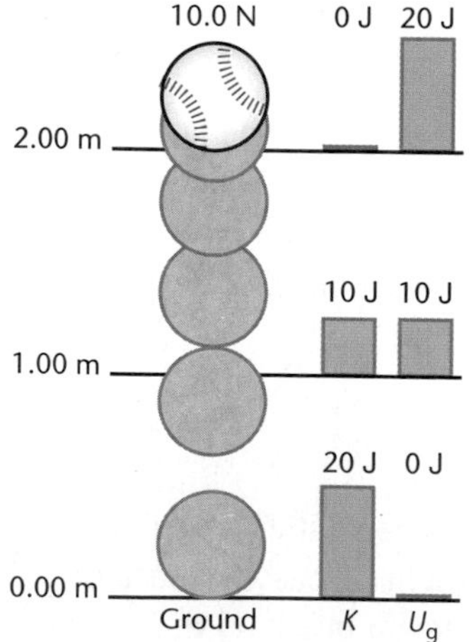

FIGURE 11–8 The decrease in potential energy is equal to the increase in kinetic energy.

Conservation of mechanical energy

Although there are many forms of energy, you will be concerned only with kinetic and gravitational potential energy while investigating motion in this book. The sum of these forms of energy is referred to as **mechanical energy.** In any given system, *if no other forms of energy are present,*

$$E = K + U_g.$$

Imagine a system consisting of a ball and Earth, as shown in **Figure 11–8.** Suppose the ball has a weight of 10.0 N and will be released 2.00 m above the ground, which you define to be the reference level. Because the ball isn't yet moving, it has no kinetic energy. Its potential energy is represented by the following.

$$U_g = mgh$$

$$U_g = (10.0\ \text{N})(2.00\ \text{m}) = 20.0\ \text{J}$$

The ball's total mechanical energy is therefore 20.0 J. As the ball falls, it loses potential energy and gains kinetic energy. When it is 1.00 m above Earth's surface, its potential energy is

$$U_g = mgh.$$

$$U_g = (10.0\ \text{N})(1.00\ \text{m}) = 10.0\ \text{J}$$

What is the ball's kinetic energy when it is at a height of 1.00 m? The system, which consists of the ball and Earth, is closed and, with no external forces acting upon it, is isolated. Thus, its total energy remains constant at 20.0 J.

$$E = K + U_g,\ \text{so}\ K = E - U_g$$

$$K = 20.0\ \text{J} - 10.0\ \text{J} = 10.0\ \text{J}$$

When the ball reaches the ground, its potential energy is zero, and its kinetic energy is now the full 20.0 J. The equation that describes conservation of mechanical energy is $E_{before} = E_{after}$, which may be rewritten as follows.

$$K_{before} + U_{g\ before} = K_{after} + U_{g\ after}$$

What happens if the ball doesn't fall down, but rolls down a ramp, as shown in **Figure 11–9?** Without friction, there are still no net external forces, so the system is still closed and isolated. The ball still moves down a vertical distance of 2.00 m, so its loss of potential energy is still 20.0 J. Therefore, it still gains 20.0 J of kinetic energy. The path it takes doesn't matter.

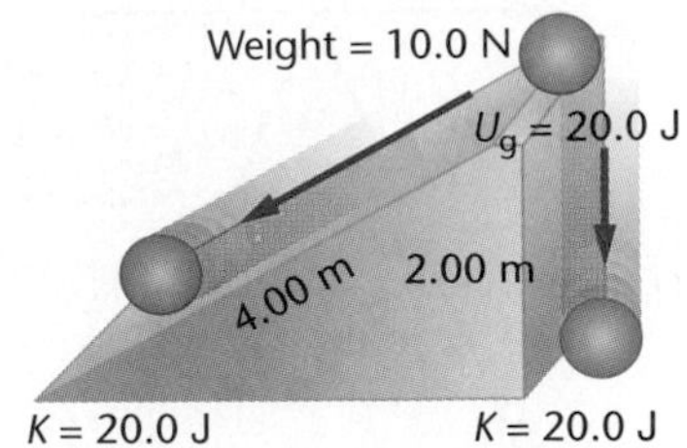

FIGURE 11–9 The path an object follows in reaching the ground does not affect the final kinetic energy of the object.

In the case of a roller coaster that is nearly at rest at the top of the first hill, the total mechanical energy in the system is the coaster's gravitational potential energy at this point. Suppose some other hill along the track were higher than the first one. The roller coaster could not climb such a hill because doing so would require more mechanical energy than is in the system.

The simple oscillation of a pendulum also demonstrates the conservation of energy. The system is the pendulum bob plus Earth. Usually, the reference level is chosen to be the equilibrium position of the bob. Work is done by an external force in pulling the bob to one side. This gives the system mechanical energy. At the instant the bob is released, all the energy is potential; but as it swings down, the energy is converted to kinetic energy. When the bob is at the equilibrium point, its gravitational potential energy is zero, and its kinetic energy is equal

Whoosh!

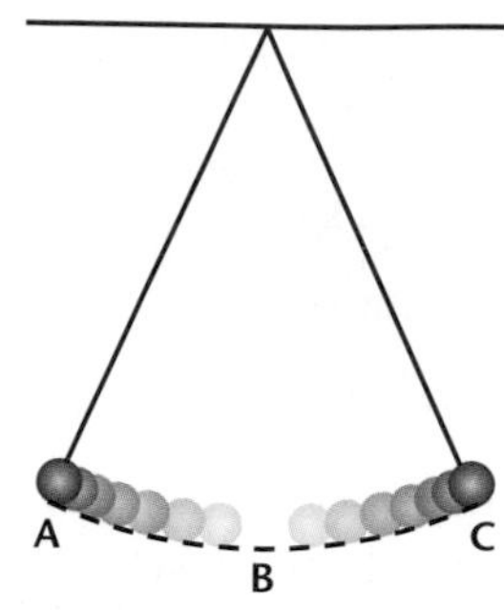

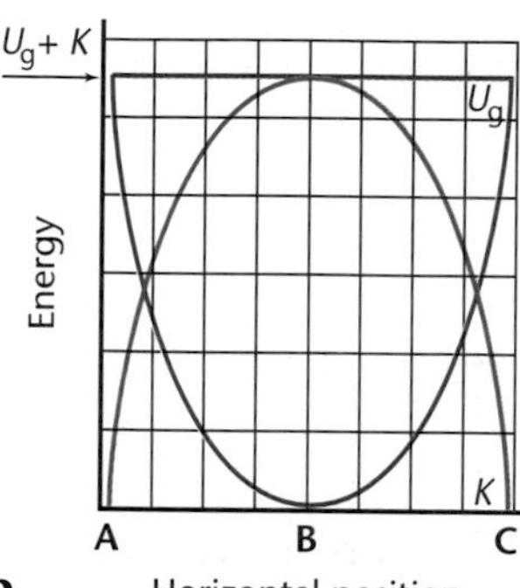

FIGURE 11–10 For the simple harmonic motion of a pendulum bob **(a)**, the mechanical energy—the sum of the potential and kinetic energies—is a constant **(b)**.

to the total mechanical energy in the system. **Figure 11–10** shows a graph of the changing potential and kinetic energy during one-half period of oscillation. Note that the mechanical energy is constant.

Loss of mechanical energy

As you know, the oscillations of a pendulum eventually die away, a bouncing ball comes to rest, and the heights of roller-coaster hills get lower and lower. Where does the mechanical energy in such systems go? Any object moving through the air experiences the forces of air resistance. In a roller coaster, there are frictional forces in the wheels. When a ball bounces, the elastic potential energy stored in the deformed ball is not all converted back into kinetic energy after the bounce. In each of these cases, some of the original mechanical energy is converted into thermal energy within members of the system. As a result of the increased thermal energy, the temperature of the objects in the system will rise slightly. You will study this form of energy in Chapter 12.

Albert Einstein recognized yet another form of potential energy, mass itself. He said that mass, by its very nature, is energy. This energy, E_0, called the rest energy, is represented by the famous formula $E_0 = mc^2$. According to this equation, stretching a spring or bending a vaulting pole causes the spring or pole to gain mass. Likewise, a hot potato weighs more than a cold one. In these cases, the change in mass is too small to be detected. When forces within the nucleus of an atom are involved, however, the energy released into other forms, such as kinetic energy, by changes in mass can be relatively large.

Problem Solving Strategy

Conservation of Energy

When solving conservation of energy problems, use an orderly procedure.

1. Carefully identify the system. Make sure it is closed; that is, no objects enter or leave.
2. Identify the initial and final states of the system.
3. Is the system isolated?
   a. If there are no external forces acting on the system, then the total energy of the system is constant: $E_{before} = E_{after}$.
   b. If there are external forces, then $E_{before} + W = E_{after}$.
4. Identify the forms of energy in the system.
   a. If mechanical energy is conserved, then there is no thermal energy change. If it is not conserved, then the final thermal energy is larger than the initial thermal energy.
   b. Decide on the reference level for potential energy, and draw bar graphs showing initial and final energy.

Example Problem

Conservation of Mechanical Energy

A large chunk of ice with mass 15.0 kg falls from a roof 8.00 m above the ground.

a. Ignoring air resistance, find the kinetic energy of the ice when it reaches the ground.

b. What is the speed of the ice when it reaches the ground?

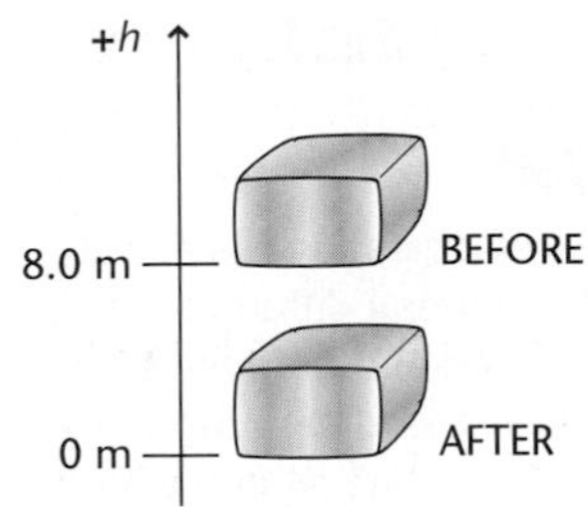

Sketch the Problem

- Sketch the initial and final conditions.
- Choose a reference level.
- Draw a bar graph.

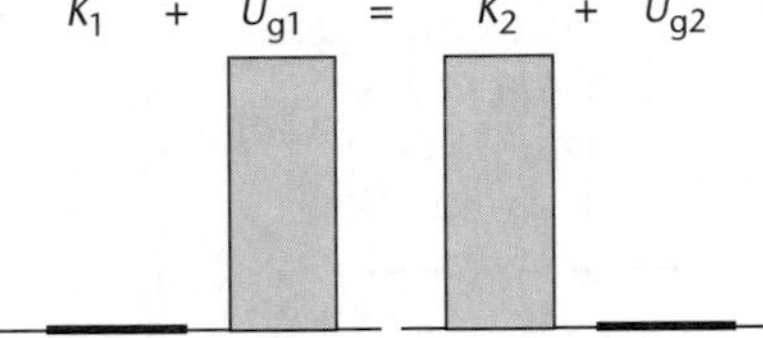

Calculate Your Answer

Known:

$m = 15.0$ kg $K_1 = 0$

$h = 8.00$ m $U_{g2} = 0$

$g = 9.80$ m/s^2

Unknown:

$U_{g1} = ?$

$K_2 = ?$

$v_2 = ?$

Strategy:

The system is the ice chunk plus Earth. With no external forces, it is isolated, so total energy is conserved. Falling does not significantly increase the thermal energy of the system, so mechanical energy is conserved and exists in two forms, kinetic and gravitational potential. Because the ice starts at rest and ends at the reference level, initial kinetic and final potential energies are zero. Thus, the initial potential energy equals the final kinetic energy.

Calculations:

a. $U_{g1} = mgh$

$= (15.0 \text{ kg})(9.80 \text{ m/s}^2)(8.00 \text{ m})$

$= 1.18 \times 10^3 \text{ J}$

$K_2 = U_{g1} = 1.18 \times 10^3 \text{ J}$

b. $K_2 = 1/2\, m{v_2}^2$

$$v_2 = \sqrt{\frac{2K_2}{m}} = \sqrt{\frac{2(1.18 \times 10^3 \text{ J})}{15.0 \text{ kg}}}$$

$= 12.5$ m/s

Check Your Answer

- Are your units correct? Velocity is in m/s and energy is in kg·m^2/s^2 = J.
- Do the signs make sense? K and speed are always positive.
- Is the magnitude realistic? Check with ${v_2}^2 = {v_1}^2 + 2ad$ from Newton's laws. $v_2 = \sqrt{2gh} = \sqrt{2(9.80 \text{ m/s}^2)(8.00 \text{ m})} = 12.5$ m/s

Practice Problems

9. A bike rider approaches a hill at a speed of 8.5 m/s. The mass of the bike and rider together is 85 kg.

a. Identify a suitable system.

Pocket Lab

Energy Exchange

Wear goggles for this activity. Select several different-sized steel balls and determine their masses. Stand a spring-loaded laboratory cart on end with the spring mechanism pointing upward. Place a ball on top of the spring mechanism. Press down on the ball to compress the spring until the ball is touching the cart. Quickly release the ball so that the spring shoots it upward. Repeat several times, and measure the average height. Predict how high the other sizes of steel balls should go. Try it. Record the values in a data table.

Analyze and Conclude Classify the balls in order of height attained. What conclusions can you reach?

b. Find the initial kinetic energy of the system.

c. The rider coasts up the hill. Assuming that there is no friction, at what height will the bike come to rest?

d. Does your answer to **c** depend on the mass of the system? Explain.

10. Tarzan, mass 85 kg, swings down on the end of a 20-m vine from a tree limb 4.0 m above the ground. Sketch the situation.

a. How fast is Tarzan moving when he reaches the ground?

b. Does your answer to **a** depend on Tarzan's mass?

c. Does your answer to **a** depend on the length of the vine?

11. A skier starts from rest at the top of a 45-m hill, skis down a 30° incline into a valley, and continues up a 40-m-high hill. Both hill heights are measured from the valley floor. Assume that you can neglect friction and the effect of ski poles.

a. How fast is the skier moving at the bottom of the valley?

b. What is the skier's speed at the top of the next hill?

c. Does your answer to **a** or **b** depend on the angles of the hills?

12. Suppose, in the case of problem 9, that the bike rider pedaled up the hill and never came to a stop.

a. In what system is energy conserved?

b. From what form of energy did the bike gain mechanical energy?

Analyzing Collisions

In a system containing two objects, collisions sometimes occur. How can you analyze a collision between two objects? The details of a collision can be very complex, so you should find the motion of the objects just before and just after the collision. What conservation laws can be used to analyze such a system? If there are no external forces—that is, the system is isolated—then momentum is conserved. Energy also is conserved, but the potential energy or thermal energy could decrease,

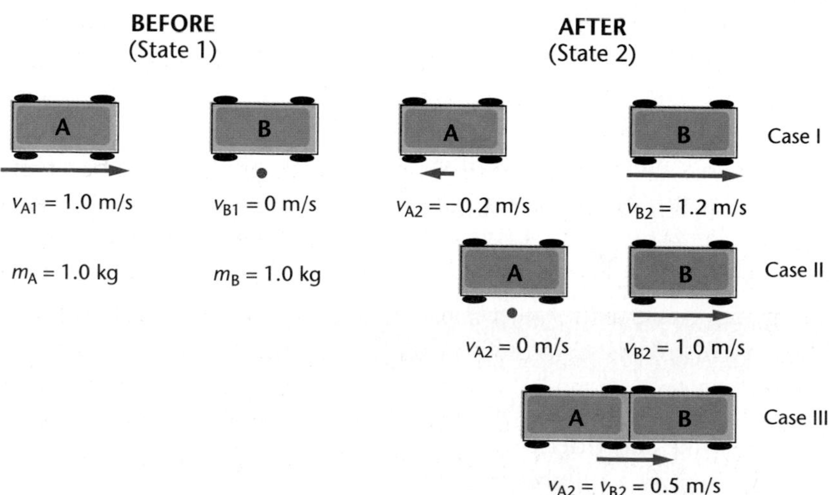

FIGURE 11–11 The energy considerations in these three kinds of collisions between a moving object and a stationary object are different. In Case I, the two objects move apart in opposite directions. In Case II, the moving object comes to rest, and the stationary object begins to move. In Case III, the two objects stick together and move as one.

remain the same, or increase. Therefore, you cannot tell whether or not kinetic energy is conserved. **Figure 11–11** shows three different kinds of collisions, labeled Cases I, II, and III. Let's check the laws of conservation of momentum and energy using the data given for these three cases.

Is momentum conserved in each kind of collision? The momentum of the system before and after the collision is represented as

$$p_1 = p_{A1} + p_{B1} = (1.0\ \text{kg})(1.0\ \text{m/s}) + (1.0\ \text{kg})(0\ \text{m/s}) = 1.0\ \text{kg·m/s}$$

$$p_2 = p_{A2} + p_{B2} = (1.0\ \text{kg})(-0.2\ \text{m/s}) + (1.0\ \text{kg})(1.2\ \text{m/s}) = 1.0\ \text{kg·m/s}.$$

Thus, in Case I, the momentum is conserved. You can check to see if the momentum is conserved in the other two cases.

Is kinetic energy conserved in each of these cases? The kinetic energy of the system before and after the collision is represented as

$$K_{A1} + K_{B1} = 1/2(1.0\ \text{kg})(1.0\ \text{m/s})^2 + 1/2(1.0\ \text{kg})(0\ \text{m/s})^2 = 0.50\ \text{J}$$

$$K_{A2} + K_{B2} = 1/2(1.0\ \text{kg})(-0.2\ \text{m/s})^2 + 1/2(1.0\ \text{kg})(1.2\ \text{m/s})^2 = 0.74\ \text{J}.$$

The kinetic energy increased in Case I. How could this be? If energy is conserved, then one or more other forms of energy of the system must have decreased. Perhaps when the two carts collided, a compressed spring was released, adding kinetic energy to the system. This kind of collision sometimes is called super-elastic or explosive.

After the collision in Case II, the kinetic energy is represented as

$$K_{A2} + K_{B2} = 1/2(1.0\ \text{kg})(0\ \text{m/s})^2 + 1/2(1.0\ \text{kg})(1.0\ \text{m/s})^2 = 0.50\ \text{J}.$$

Kinetic energy remained the same. This kind of collision, in which the kinetic energy doesn't change, is called an **elastic collision.** Collisions between hard, elastic objects, such as those made of steel, glass, or hard plastic, are often nearly elastic.

After the collision in Case III, the kinetic energy is represented as

$$K_{A2} + K_{B2} = 1/2(1.0\ \text{kg})(0.5\ \text{m/s})^2 + 1/2(1.0\ \text{kg})(0.5\ \text{m/s})^2 = 0.25\ \text{J}.$$

Kinetic energy decreased; some kinetic energy was converted to thermal energy. This kind of collision, in which kinetic energy decreases, is called an **inelastic collision.** Objects made of soft, sticky material, such as clay, act in this way. When the two carts in Case III stuck together, they lost kinetic energy. Collisions are inelastic whenever kinetic energy is converted to thermal energy.

The three kinds of collisions can be represented using bar graphs, such as those shown in **Figure 11–12.** Although the kinetic energy before and after the collisions can be calculated, only the change in other forms of energy can be found. In automobile collisions, kinetic energy is transferred into other forms of energy such as heat and sound. These forms of energy are the result of bent metal, shattered glass, and cracked plastics. Unfortunately, you can't simply put the heat and sound back into the car and make it new again. Although you can recover the energy you put into gravitational potential energy, that is not the case for many other forms of energy.

EARTH SCIENCE CONNECTION

Fossil fuels—coal, oil, and natural gas—now supply about 86% of the energy consumed by industrial nations. Coal was formed when prehistoric swamp plants decayed; oil and natural gas were formed by the decay of tiny marine organisms. Millions of years ago, these living things were buried under tons of sediment and subjected to great pressure and heat. The energy that was trapped in the decayed organisms so long ago is released as heat and light when the fuels are burned.

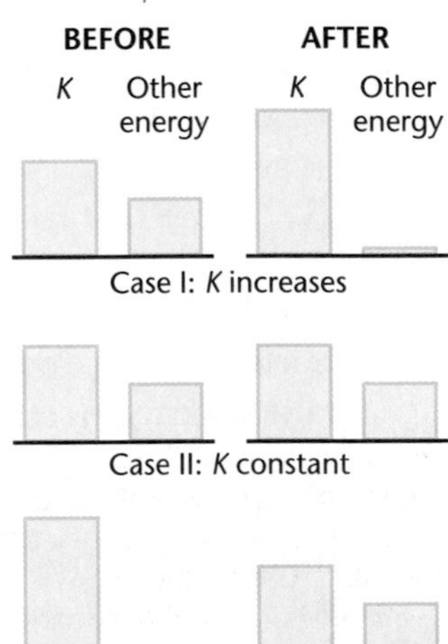

FIGURE 11–12 Bar graphs can be drawn to represent the three kinds of collisions.

Example Problem

Kinetic Energy Loss in a Collision

In an accident on a slippery road, a compact car with mass 575 kg, moving at 15.0 m/s, smashes into the rear end of a car with mass 1575 kg moving at 5.00 m/s in the same direction.

a. What is the final velocity if the wrecked cars lock together?

b. How much kinetic energy was lost in the collision?

c. What fraction of the original kinetic energy was lost?

BEFORE (State 1) AFTER (State 2)

A B A B

v_{A1} v_{B1} v_2

p_{A1} p_{B1} $p_{A2} + p_{B2}$

Sketch the Problem

- Sketch the initial and final conditions.
- Sketch the momentum diagram.

Calculate Your Answer

Known:

$m_A = 575$ kg

$v_{A1} = 15.0$ m/s

$m_B = 1575$ kg

$v_{B1} = 5.00$ m/s

$v_{A2} = v_{B2} = v_2$

Unknown:

$v_2 = ?$

$\Delta K = K_2 - K_1 = ?$

Fraction of K_1 lost, $\Delta K/K_1 = ?$

Strategy:

a. Use the conservation of momentum equation to find the final velocity.

Calculations:

$$p_{A1} + p_{B1} = p_{A2} + p_{B2}$$

$$m_A v_{A1} + m_B v_{B1} = (m_A + m_B)v_2$$

$$v_2 = \frac{m_A v_{A1} + m_B v_{B1}}{m_A + m_B}$$

$$= \frac{(575 \text{ kg})(15.0 \text{ m/s}) + (1575 \text{ kg})(5.00 \text{ m/s})}{575 \text{ kg} + 1575 \text{ kg}} = 7.67 \text{ m/s}$$

b. Determine the change in kinetic energy of the system.

$$\Delta K = K_2 - K_1$$

$$K_2 = 1/2(m_A + m_B)v_2^2 = 1/2(2150 \text{ kg})(7.67 \text{ m/s})^2 = 6.33 \times 10^4 \text{ J}$$

$$K_1 = 1/2 m_A v_{A1}^2 + 1/2 m_B v_{B1}^2$$

$$= 1/2(575 \text{ kg})(15.0 \text{ m/s})^2 + 1/2(1575 \text{ kg})(5.00 \text{ m/s})^2$$

$$= 8.44 \times 10^4 \text{ J}$$

c. Use ΔK to calculate the fraction of the original kinetic energy lost.

$$\Delta K = 6.33 \times 10^4 \text{ J} - 8.44 \times 10^4 \text{ J} = -2.11 \times 10^4 \text{ J}$$

$$\Delta K/K_1 = (-2.11 \times 10^4 \text{ J})/(8.44 \times 10^4 \text{ J}) = -0.25$$

25% was lost.

Check Your Answer

- Are the units correct? Velocity is in m/s; energy is in J.
- Does the sign make sense? Velocity is positive, consistent with the original velocities.
- Is the magnitude realistic? v_2 is faster than v_{B1}. Energies are typical of moving cars.

Practice Problems

13. A 2.00-g bullet, moving at 538 m/s, strikes a 0.250-kg piece of wood at rest on a frictionless table. The bullet sticks in the wood, and the combined mass moves slowly down the table.

a. Draw energy bar graphs and momentum vectors for the collision.
b. Find the speed of the system after the collision.
c. Find the kinetic energy of the system before the collision.
d. Find the kinetic energy of the system after the collision.
e. What percentage of the system's original kinetic energy was lost?

14. An 8.00-g bullet is fired horizontally into a 9.00-kg block of wood on an air table and is embedded in it. After the collision, the block and bullet slide along the frictionless surface together with a speed of 10.0 cm/s. What was the initial speed of the bullet?

15. Bullets bounce from Superman's chest. Suppose that Superman, with mass 104 kg, while not moving, is struck by a 4.20-g bullet moving with a speed of 835 m/s. The bullet drops straight down with no horizontal velocity. How fast was Superman moving after the collision if his superfeet are frictionless?

16. A 0.73-kg magnetic target is suspended on a string. A 0.025-kg magnetic dart, shot horizontally, strikes it head-on. The dart and the target together, acting like a pendulum, swing up 12 cm above the initial level before instantaneously coming to rest.

a. Sketch the situation and decide on the system.
b. This is a two-part problem. Decide what is conserved in each part and explain your decision.
c. What was the initial velocity of the dart?

11.2 Section Review

2.1 Is "spaceship Earth" a closed, isolated system? Support your answer.

2.2 A surfer crouches and moves high up into the curl of a wave. How is she using conservation of energy?

2.3 A child jumps on a trampoline. Draw bar graphs to show what form energy is in when

a. the child is at the highest point.
b. the child is at the lowest point.
c. the child's feet are just touching the trampoline.

2.4 Critical Thinking A ball drops 20 m. When it has fallen half the distance, or 10 m, half of its energy is potential and half is kinetic. When the ball has fallen instead for half the amount of time it takes to fall, will more, less, or exactly half of its energy be potential energy?

Physics & Society

Energy from Tides

Tides are the periodic rise and fall of the surface of a body of water. They result from the gravitational attraction among the sun, Earth, and its moon. When these bodies form a straight line in relation to each other, the effects of gravity are additive, and tidal ranges—the difference between high and low tides—are greatest. In regions where tidal ranges are at least 10 m, people are trying to harness the energy available in the tides to produce electricity.

Tidal power plants are built across estuaries, which are flooded river valleys that empty directly into an ocean. As an incoming tide arrives, the floodgates of a barrier near the mouth of the river are opened to capture the water. Then, as the tide ebbs, or recedes, the water exits the dam through turbines to generate electricity.

After three successful decades of operation, the La Rance tidal power plant in northern France produces 240 megawatts (MW) of electricity. It provides about 90% of the energy needs for this region of France. Eight trial tidal plants in China generate more than 6 MW of electricity. The newest plant in Jiangsha provides residents in the Zhejiang Province with 3.2 MW of power. The Jiangsha station is unlike other tidal plants in that it produces electricity both when the tide flows and when it ebbs.

Tidal Energy—At What Cost?

For all energy resources, certain factors such as cost, environmental impact, availability, and wastes produced must be evaluated to determine whether the resource is feasible. Tidal power is relatively cheap. Electricity at La Rance is generated at about 3.7 cents per kilowatt hour (kWh). French nuclear plants provide the same power at about 3.8 cents per kWh. Only hydroelectric power plants are able to beat the cost of the tidal plants by producing electricity at about 3.2 cents per kWh.

Energy produced by tides is nonpolluting. Moreover, unlike nuclear resources, tidal power requires no hazardous fuel to produce the energy. Nor does tidal power require disposal of dangerous substances.

Like all methods of producing energy, however, tidal power does have its problems. The environmental impact can include changes in water temperatures that can be detrimental to marine plants and animals in tidal zones, and a decrease in the number of marine organisms and migrating birds to an area as a result of changes in tidal levels.

Investigating the Issue

1. **Acquiring Information** Find out more about how a tidal power plant produces electricity. Display your findings with sketches that detail how a plant operates.
2. **Debating the Issue** Research the cost, environmental impact, availability, and wastes produced for each of the following resources: nuclear energy, hydroelectric power, solar energy, and fossil fuels. Now debate whether or not you think tidal power plants are a plausible energy alternative.

*inter***NET**
CONNECTION

Follow the link for this chapter on the Glencoe Homepage at **www.glencoe.com/sec/science** to find out more about tidal energy.

CHAPTER 11 REVIEW

ey Terms

.1

avitational
otential energy

ference level

astic potential
ergy

.2

w of
nservation of
ergy

echanical
ergy

astic collision

elastic
llision

Summary

11.1 The Many Forms of Energy

- The kinetic energy of an object is proportional to its mass and the square of its velocity.
- When Earth is included in a system, the work done by gravity is replaced by gravitational potential energy.
- The gravitational potential energy of an object depends on the object's weight and its distance from Earth's surface.
- The sum of kinetic and potential energy is called mechanical energy.
- Elastic potential energy may be stored in an object as a result of its change in shape.

11.2 Conservation of Energy

- The total energy of a closed, isolated system is constant. Within the system, energy can change form, but the total amount of energy doesn't change.
- Momentum is conserved in collisions if the external force is zero. The mechanical energy may be unchanged or decreased by the collision, depending on whether the collision is elastic or inelastic.

Reviewing Concepts

Unless otherwise directed, assume that air resistance is negligible.

Section 11.1

1. Explain how work and a change in the form of energy are related.
2. What form of energy does a wound watch spring have? What form of energy does a running mechanical watch use? When a watch runs down, what has happened to the energy?
3. Explain how energy and force are related.
4. A ball is dropped from the top of a building. You choose the top to be the reference level, while your friend chooses the bottom. Do the two of you agree on
   a. the ball's potential energy at any point?
   b. the change in the ball's potential energy as a result of the fall?
   c. the kinetic energy of the ball at any point?
5. Can the kinetic energy of a baseball ever be negative? Explain without using a formula.
6. Can the gravitational potential energy of a baseball ever be negative? Explain without using a formula.
7. If a baseball's velocity is increased to three times its original velocity, by what factor does its kinetic energy increase?
8. Can a baseball have kinetic energy and gravitational potential energy at the same time?
9. One athlete lifts a barbell three times as high as another. What is the ratio of changes in potential energy in the two lifts?

Section 11.2

10. What energy transformations take place when an athlete is pole-vaulting?
11. The sport of pole-vaulting was drastically changed when the stiff, wooden pole was replaced by the flexible, fiberglass pole. Explain why.
12. You throw a clay ball at a hockey puck on ice. The ball sticks, and the puck moves slowly.
    a. Is momentum conserved in the collision? Explain.

b. Is kinetic energy conserved? Explain.

13. Draw energy bar graphs for the following processes.
    a. An ice cube, initially at rest, slides down a frictionless slope.
    b. An ice cube, initially moving, slides up a frictionless slope and instantaneously comes to rest.
14. An earthquake can release energy to devastate a city. Where does this energy reside moments before the earthquake takes place?
15. A rubber ball is dropped from a height of 1.2 m. After striking the floor, the ball bounces to a height of 0.8 m.
    a. Has the energy of the ball changed as a result of the collision with the floor? Explain.
    b. How high a bounce would you observe if the collision were completely elastic?
    c. How high a bounce would you observe if the collision were completely inelastic?
16. A speeding car puts on its brakes and comes to a stop. The system includes the car, but not the road. Apply the work-energy theorem when
    a. the car's wheels do not skid.
    b. the brakes lock and the car wheels skid.

Applying Concepts

17. A compact car and a trailer truck are both traveling at the same velocity. Which has more kinetic energy?
18. Carmen and Lisa have identical compact cars. Carmen drives on the freeway with a greater speed than Lisa does. Which car has more kinetic energy?
19. Carmen and Lisa have identical compact cars. Carmen is northbound on the freeway, while Lisa is traveling south at the same speed. Which car has more kinetic energy?
20. During a process, positive work is done on a system, and the potential energy decreases. Can you determine anything about the change in kinetic energy of the system? Explain.
21. During a process, positive work is done on a system, and the potential energy increases. Can you tell whether the kinetic energy increased, decreased, or remained the same? Explain.
22. Two bodies of unequal mass have the same kinetic energy and are moving in the same direction. If the same retarding force is exerted on each body, how will the stopping distances of the bodies compare?
23. Roads seldom go straight up a mountain; they wind around and go up gradually. Explain.
24. You swing a 625-g mass on the end of a 0.75-m string around your head in a nearly horizontal circle at constant speed.
    a. How much work is done on the mass by the tension of the string in one revolution?
    b. Is your answer to **a** in agreement with the work-energy theorem? Explain.
25. Give specific examples that illustrate the following processes.
    a. Work is done on a system, increasing kinetic energy with no change in potential energy.
    b. Potential energy is changed to kinetic energy with no work done on the system.
    c. Work is done on a system, increasing potential energy with no change in kinetic energy.
    d. Kinetic energy is reduced but potential energy is unchanged. Work is done by the system.
26. You have been asked to make a roller coaster more exciting. The owners want the speed at the bottom of the first hill doubled. How much higher must the first hill be built?
27. If you drop a tennis ball onto a concrete floor, it will bounce back farther than it will if you drop it onto a rug. Where does the lost mechanical energy go when the ball strikes the rug?
28. Most Earth satellites follow an elliptical path rather than a circular path around Earth. The value of U_g increases when the satellite moves farther from Earth. According to the law of conservation of energy, does a satellite have its greatest speed when it is closest to or farthest from Earth?
29. In mountainous areas, road designers build escape ramps to help trucks with failed brakes stop. These escape ramps are usually roads made of loose gravel that go uphill. Describe changes in energy forms when a fast-moving truck uses one of these escape ramps.

30. If two identical bowling balls are raised to the same height, one on Earth, the other on Mars, which has the larger potential energy increase?
31. What will be the largest possible kinetic energy of an arrow shot from a bow that has been pulled back so that it stores 50 J of elastic potential energy?
32. Two pendulums swing side-by-side. At the bottom of the swing, the speed of one is twice the speed of the other. Compare the heights of the bobs at the end of their swings.
33. In a baseball game, two pop-ups are hit in succession. The second rises twice as high as the first. Compare the speeds of the two balls when they leave the bat.
34. Two identical balls are thrown from the top of a cliff, each with the same speed. One is thrown straight up, the other straight down. How do the kinetic energies and speeds of the balls compare as they strike the ground?
35. A ball is dropped from the top of a tall building and reaches terminal velocity as it falls. Will the decrease in potential energy of the ball equal the increase in its kinetic energy? Explain.

Problems

Unless otherwise directed, assume that air resistance is negligible. Draw energy bar graphs to solve the problems.

Section 11.1

LEVEL 1

36. A 1600-kg car travels at a speed of 12.5 m/s. What is its kinetic energy?
37. A racing car has a mass of 1525 kg. What is its kinetic energy if it has a speed of 108 km/h?
38. Toni has a mass of 45 kg and is moving with a speed of 10.0 m/s.
    a. Find Toni's kinetic energy.
    b. Toni's speed changes to 5.0 m/s. Now what is her kinetic energy?
    c. What is the ratio of the kinetic energies in **a** and **b?** Explain the ratio.
39. Shawn and his bike have a total mass of 45.0 kg. Shawn rides his bike 1.80 km in 10.0 min at a constant velocity. What is Shawn's kinetic energy?
40. Ellen and Angela each has a mass of 45 kg, and they are moving together with a speed of 10.0 m/s.
    a. What is their combined kinetic energy?
    b. What is the ratio of their combined mass to Ellen's mass?
    c. What is the ratio of their combined kinetic energy to Ellen's kinetic energy? Explain.
41. In the 1950s, an experimental train that had a mass of 2.50×10^4 kg was powered across a level track by a jet engine that produced a thrust of 5.00×10^5 N for a distance of 509 m.
    a. Find the work done on the train.
    b. Find the change in kinetic energy.
    c. Find the final kinetic energy of the train if it started from rest.
    d. Find the final speed of the train if there were no friction.
42. A 14 700-N car is traveling at 25 m/s. The brakes are applied suddenly, and the car slides to a stop. The average braking force between the tires and the road is 7100 N. How far will the car slide once the brakes are applied?
43. A 15.0-kg cart is moving with a velocity of 7.50 m/s down a level hallway. A constant force of -10.0 N acts on the cart, and its velocity becomes 3.20 m/s.
    a. What is the change in kinetic energy of the cart?
    b. How much work was done on the cart?
    c. How far did the cart move while the force acted?
44. How much potential energy does Tim, with mass 60.0 kg, gain when he climbs a gymnasium rope a distance of 3.5 m?
45. A 6.4-kg bowling ball is lifted 2.1 m into a storage rack. Calculate the increase in the ball's potential energy.
46. Mary weighs 505 N. She walks down a flight of stairs to a level 5.50 m below her starting point. What is the change in Mary's potential energy?
47. A weight lifter raises a 180-kg barbell to a height of 1.95 m. What is the increase in the potential energy of the barbell?
48. A 10.0-kg test rocket is fired vertically from Cape Canaveral. Its fuel gives it a kinetic energy of 1960 J by the time the rocket engine burns

all of the fuel. What additional height will the rocket rise?

49. Antwan raised a 12.0-N physics book from a table 75 cm above the floor to a shelf 2.15 m above the floor. What was the change in the potential energy of the system?

50. A hallway display of energy is constructed in which several people pull on a rope that lifts a block 1.00 m. The display indicates that 1.00 J of work is done. What is the mass of the block?

LEVEL 2

51. It is not uncommon during the service of a professional tennis player for the racket to exert an average force of 150.0 N on the ball. If the ball has a mass of 0.060 kg and is in contact with the strings of the racket for 0.030 s, what is the kinetic energy of the ball as it leaves the racket? Assume that the ball starts from rest.

52. Pam, wearing a rocket pack, stands on frictionless ice. She has a mass of 45 kg. The rocket supplies a constant force for 22.0 m, and Pam acquires a speed of 62.0 m/s.
 a. What is the magnitude of the force?
 b. What is Pam's final kinetic energy?

53. A 2.00×10^3-kg car has a speed of 12.0 m/s. The car then hits a tree. The tree doesn't move, and the car comes to rest.
 a. Find the change in kinetic energy of the car.
 b. Find the amount of work done in pushing in the front of the car.
 c. Find the size of the force that pushed in the front of the car by 50.0 cm.

54. A constant net force of 410 N is applied upward to a stone that weighs 32 N. The upward force is applied through a distance of 2.0 m, and the stone is then released. To what height, from the point of release, will the stone rise?

Section 11.2

LEVEL 1

55. A 98-N sack of grain is hoisted to a storage room 50 m above the ground floor of a grain elevator.
 a. How much work was required?
 b. What is the increase in potential energy of the sack of grain at this height?
 c. The rope being used to lift the sack of grain breaks just as the sack reaches the storage room. What kinetic energy does the sack have just before it strikes the ground floor?

56. A 20-kg rock is on the edge of a 100-m cliff.
 a. What potential energy does the rock possess relative to the base of the cliff?
 b. The rock falls from the cliff. What is its kinetic energy just before it strikes the ground?
 c. What speed does the rock have as it strikes the ground?

57. An archer puts a 0.30-kg arrow to the bowstring. An average force of 201 N is exerted to draw the string back 1.3 m.
 a. Assuming that all the energy goes into the arrow, with what speed does the arrow leave the bow?
 b. If the arrow is shot straight up, how high does it rise?

58. A 2.0-kg rock initially at rest loses 407 J of potential energy while falling to the ground.
 a. Calculate the kinetic energy that the rock gains while falling.
 b. What is the rock's speed just before it strikes the ground?

59. A physics book of unknown mass is dropped 4.50 m. What speed does the book have just before it hits the ground?

60. A 30.0-kg gun is resting on a frictionless surface. The gun fires a 50.0-g bullet with a muzzle velocity of 310.0 m/s.
 a. Calculate the momenta of the bullet and the gun after the gun is fired.
 b. Calculate the kinetic energy of both the bullet and the gun just after firing.

61. A railroad car with a mass of 5.0×10^5 kg collides with a stationary railroad car of equal mass. After the collision, the two cars lock together and move off at 4.0 m/s.
 a. Before the collision, the first railroad car was moving at 8.0 m/s. What was its momentum?

b. What was the total momentum of the two cars after the collision?

c. What were the kinetic energies of the two cars before and after the collision?

d. Account for the loss of kinetic energy.

62. From what height would a compact car have to be dropped to have the same kinetic energy that it has when being driven at 100 km/h?

63. A steel ball has a mass of 4.0 kg and rolls along a smooth, level surface at 62 m/s.

a. Find its kinetic energy.

b. At first, the ball was at rest on the surface. A constant force acted on it through a distance of 22 m to give it the speed of 62 m/s. What was the magnitude of the force?

LEVEL 2

64. Kelli weighs 420 N, and she is sitting on a playground swing seat that hangs 0.40 m above the ground. Tom pulls the swing back and releases it when the seat is 1.00 m above the ground.

a. How fast is Kelli moving when the swing passes through its lowest position?

b. If Kelli moves through the lowest point at 2.0 m/s, how much work was done on the swing by friction?

65. Justin throws a 10.0-g ball straight down from a height of 2.0 m. The ball strikes the floor at a speed of 7.5 m/s. What was the initial speed of the ball?

66. Megan's mass is 28 kg. She climbs the 4.8-m ladder of a slide and reaches a velocity of 3.2 m/s at the bottom of the slide. How much work was done by friction on Megan?

67. A person weighing 635 N climbs up a ladder to a height of 5.0 m. Use the person and Earth as the system.

a. Draw energy bar graphs of the system before the person starts to climb the ladder and after the person stops at the top. Has the mechanical energy changed? If so, by how much?

b. Where did this energy come from?

Critical Thinking Problems

68. A golf ball with mass 0.046 kg rests on a tee. It is struck by a golf club with an effective mass of 0.220 kg and a speed of 44 m/s. Assuming that the collision is elastic, find the speed of the ball when it leaves the tee.

69. In a perfectly elastic collision, both momentum and mechanical energy are conserved. Two balls with masses m_A and m_B are moving toward each other with speeds v_A and v_B, respectively. Solve the appropriate equations to find the speeds of the two balls after the collision.

70. A 25-g ball is fired with an initial speed v_1 toward a 125-g ball that is hanging motionless from a 1.25-m string. The balls have a perfectly elastic collision. As a result, the 125-g ball swings out until the string makes an angle of 37° with the vertical. What was v_1?

Going Further

Critical Thinking One of the important concepts in golf, tennis, baseball, and other sports is follow-through. How can applying a force to a ball for the longest possible time affect the speed of the ball?

***inter*NET CONNECTION**

Follow the link on the Glencoe Homepage at **www.glencoe.com/sec/science** to find out more about this chapter.

Perpetual Motion?

The engine converts the chemical energy stored in the fuel and oxygen into kinetic energy. Could we ever invent an engine that converts all the energy into useful energy of motion?

CHAPTER

12 Thermal Energy

In Chapter 11, you learned that one of the forms of energy is stored energy. If stored energy is converted into mechanical energy, an object, such as this racing car, can gain kinetic energy. However, it takes more than the elastic potential energy of a compressed spring to accelerate this car. It takes the energy stored in the fuel and the oxygen in the air to get it to move faster than 350 km per hour in a few seconds.

In Chapter 10, you learned that work can increase the energy of a system. In this chapter, you'll learn about another way to transfer energy, and how it has changed our society and our individual lives. The steam engine, the first invention that produced mechanical energy from fuel, transformed the United States from a society of farms to one with many factories in the 1800s. The gasoline engine, invented in 1876 by Nikolaus Otto in Germany, has changed the way you travel, the way you work, and the way you live.

WHAT YOU'LL LEARN

- You will define temperature.
- You will calculate heat transfer.
- You will distinguish heat from work.

WHY IT'S IMPORTANT

- Thermal energy provides the energy to keep you warm, to prepare and preserve food and to manufacture many of the objects you use on a daily basis.

*inter*NET CONNECTION

Follow the link for this chapter on the Glencoe Homepage at **www.glencoe.com/sec/science** to find out more about thermal energy.

12.1 Temperature and Thermal Energy

OBJECTIVES

- **Describe** the nature of thermal energy.
- **Define** temperature and **distinguish** it from thermal energy.
- **Use** the Celsius and Kelvin temperature scales and **convert** one to the other.
- **Define** specific heat and **calculate** heat transfer.

Europe went through a "Little Ice Age" in the 1600s and 1700s, when temperatures were lower than any other period during the previous one thousand years. Keeping warm was vitally important, and many people devoted themselves to the study of heat. One result was the invention of machines that used the energy produced by burning fuel to do useful work. These machines freed society from its dependence on the energy provided solely by people and animals. As inventors tried to make these machines more powerful and more efficient, they developed the science of **thermodynamics,** the study of heat.

What makes a hot body hot?

Internal combustion engines require very high temperatures to operate. These high temperatures are usually produced by burning fuel. Although the effects of fire have been known since ancient times, only in the eighteenth century did scientists begin to understand how a hot body differs from a cold body. They proposed that when a body is heated, an invisible fluid called "caloric" is added to the body. Hot bodies contain more caloric than cold bodies. The caloric theory could explain observations such as the expansion of objects when heated, but it could not explain why hands get warm when they are rubbed together.

In the mid-nineteenth century, scientists developed a new theory to replace the caloric theory. This theory is based on the assumption that matter is made up of many tiny particles that are always in motion. In a hot body, the particles move faster, and thus have greater kinetic energy than particles in a cooler body. This theory is called the **kinetic-molecular theory.**

The model of a solid shown in **Figure 12–1** can help you understand the kinetic-molecular theory. This model is a solid made up of tiny spherical particles held together by massless springs. The springs represent the electromagnetic forces that bind the solid together. The particles vibrate back and forth and thus have kinetic energy. The vibrations compress and extend the springs, so the solid has potential energy as well. The overall energy of motion of the particles that make up an object is called the **thermal energy** of that object.

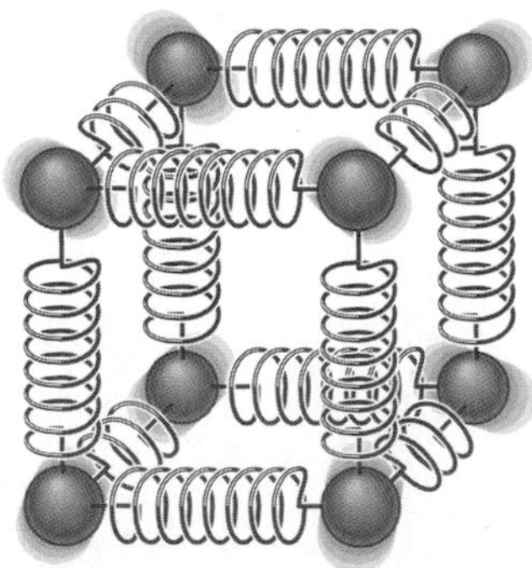

FIGURE 12–1 Molecules of a solid behave in some ways as if they were held together by springs.

Thermal Energy and Temperature

According to the kinetic-molecular theory, a hot body has more thermal energy than a similar cold body, shown in **Figure 12–2.** This means that, as a whole, the particles in a hot body have greater thermal energy than the particles in a cold body. It does not mean that all the particles in a body have exactly the same energy. The particles have a range of energies, some high, others low. It is the *average* energy of particles in a hot body that is higher than that of particles in a cold body. To help you understand this, consider the heights of students in a twelfth-grade class. The heights vary, but you can calculate the average height. This average is likely to be larger than the average height of students in a ninth-grade class, even though some ninth-graders might be taller than some twelfth-graders.

How can you determine the hotness of an object? "Hotness" is a property of an object called its **temperature,** and is measured on a definite scale. Consider two objects. In the one you call hotter, the particles are moving faster. That is, they have a greater average kinetic energy. Because the temperature is a property of matter, the temperature does not depend on the number of particles in the object. Temperature only depends on the average kinetic energy of the particles in the object. To illustrate this, consider two blocks of steel. The first block has a mass of 1 kg and the second block has twice the mass of the first at 2 kg. If the 1 kg mass of steel is at the same temperature as a 2 kg mass of steel, the average kinetic energy of the particles in the both blocks is the same. But the 2 kg mass of steel has twice the mass and the total amount of kinetic energy of particles in the 2 kg mass is twice the amount in the 1 kg mass. Total kinetic energy is divided over all the particles in an object to get the average. Therefore, the thermal energy in an object is proportional to the number of particles in it, but its temperature is not, as is shown in **Figure 12–3.**

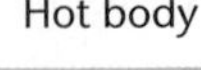

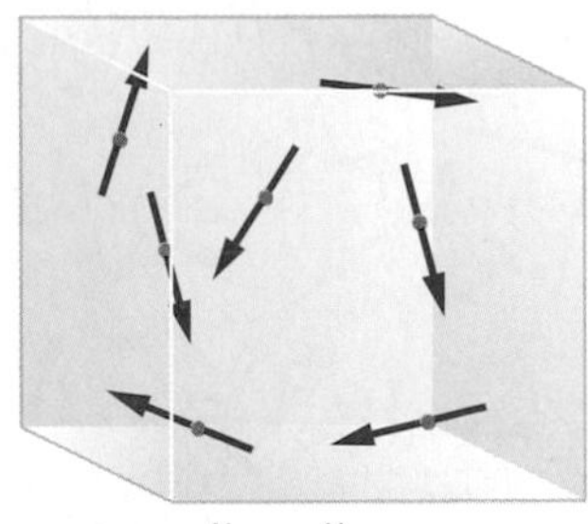

FIGURE 12–2 Particles in a hot body have greater kinetic and potential energies than particles in a cold body.

Equilibrium and Thermometry

How do you measure your temperature? If you suspect that you have a fever, you may place a thermometer in your mouth and wait two or three minutes. The thermometer then provides a measure of the temperature of your body.

You are probably less familiar with the microscopic process involved in measuring temperature. Your body is hot compared to the thermometer, which means that the particles in your body have greater thermal energy and are moving faster. The thermometer is made of a glass tube. When the cold glass touches your hotter body, the faster-moving particles in your skin collide with the slower-moving particles in the glass. Energy is transferred from your skin to the glass particles by the process of **conduction,** the transfer of kinetic energy when particles collide. The thermal energy of the particles that make up the thermometer increases and, at the same time, the thermal energy of the particles in your skin decreases. As the particles in the glass become more energetic, they begin to transfer energy back to the particles in your body. At some point, the rate of transfer of energy back and forth between the glass and your body will become equal and both will be at the same temperature. Your body and the thermometer are then in **thermal equilibrium.** That is, the rate at which energy flows from your body to the glass is equal to the rate at which energy

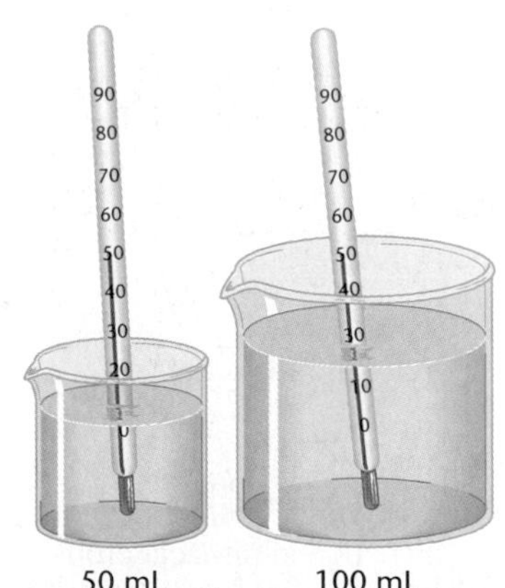

FIGURE 12–3 Temperature does not depend on the number of particles in a body.

Before thermal equilibrium

Hot body (A) Cold body (B)

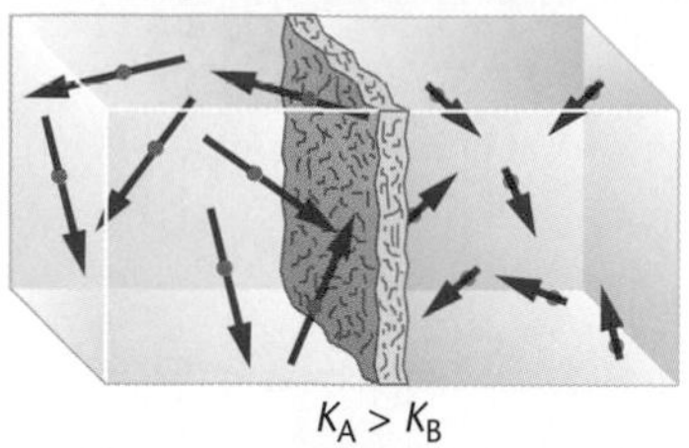

After thermal equilibrium

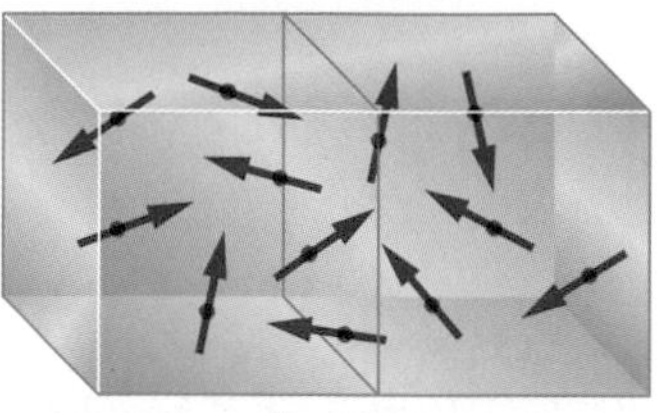

$K_A = K_B$

FIGURE 12–4 Thermal energy is transferred from a hot body to a cold body. When thermal equilibrium is reached, the transfer of energy between bodies is equal.

flows from the glass to your body. Objects that are in thermal equilibrium are at the same temperature, as is shown in **Figure 12–4.**

A **thermometer** is a device that measures temperature. It is placed in contact with an object and allowed to come to thermal equilibrium with that object. The operation of a thermometer depends on some property, such as volume, that changes with temperature. Many household thermometers contain colored alcohol that expands when heated and rises in a narrow tube. The hotter the thermometer, the more the alcohol expands and the higher it rises. Mercury is another liquid commonly used in thermometers. In liquid crystal thermometers, such as the one shown in **Figure 12–5,** a set of different kinds of liquid crystals is used. Each crystal's molecules rearrange at a specific temperature which causes the color of the crystal to change, and creates an instrument that indicates the temperature by color.

Temperature Scales: Celsius and Kelvin

Temperature scales were developed by scientists to allow them to compare their temperature measurements with those of other scientists. A scale based on the properties of water was devised in 1741 by the Swedish astronomer and physicist Anders Celsius. On this scale, now called the Celsius scale, the freezing point of pure water is 0 degrees (0°C). The boiling point of pure water at sea level is 100 degrees (100°C). On the Celsius scale, the average temperature of the human body is 37°C. **Figure 12–6** shows representative temperatures on the three most common scales: Fahrenheit, Celsius, and Kelvin.

The wide range of temperatures in the universe is shown in **Figure 12–7.** Temperatures do not appear to have an upper limit. The interior of the sun is at least 1.5×10^7°C and other stars are even hotter. Temperatures do, however, have a lower limit. Generally, materials contract as they cool. If you cool an "ideal" gas, one in which the particles occupy a tremendously large volume compared to their own size and which don't interact, it contracts in such a way that it occupies a volume that is only the size of the molecules at −273.15°C. At this temperature, all the thermal energy that can has been removed from the gas. It is impossible to reduce the temperature any further. Therefore, there can be no temperature lower than −273.15°C. This is called **absolute zero,** and is usually rounded to −273°C.

The Kelvin temperature scale is based on absolute zero. Absolute zero is the zero point of the Kelvin scale. On the Kelvin scale, the freezing point of water (0°C) is 273 K and the boiling point of water is 373 K.

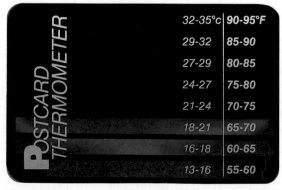

FIGURE 12–5 Thermometers use a change in physical properties to measure temperature. A liquid crystal thermometer changes color with temperature change.

Each interval on this scale, called a **kelvin,** is equal to the size of one Celsius degree. Thus, $T_C + 273 = T_K$. Very cold temperatures are reached by liquefying gases. Helium liquefies at 4.2 K, or −269 °C. Even colder temperatures can be reached by making use of special properties of solids, helium isotopes, or atoms and lasers. Using these techniques, physicists have reached temperatures as low as 2.0×10^{-9} K.

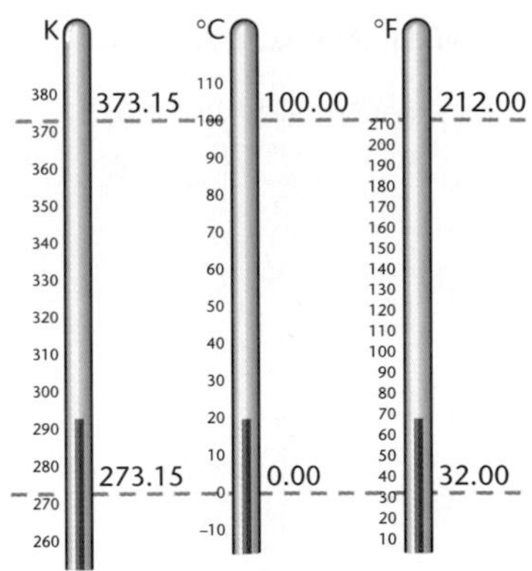

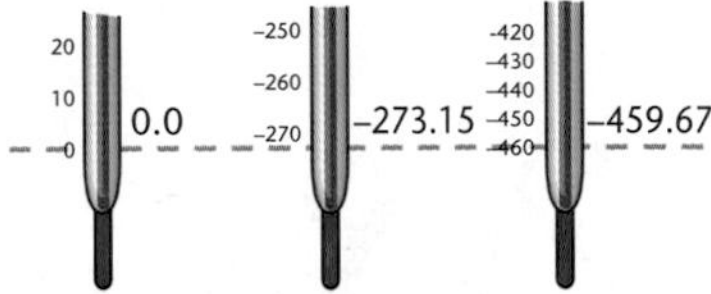

FIGURE 12–6 The three most common temperature scales are Kelvin, Celsius, and Fahrenheit.

Example Problem

Temperature Conversion

Convert 25 °C to kelvins.

Calculate Your Answer

Known:	**Unknown:**
Celsius temperature = 25 °C	$T_K = ?$
Strategy:	**Calculations:**
Change Celsius temperatures to Kelvin by using the relationship $T_K = T_C + 273$.	$T_K = T_C + 273$ $= 25 + 273$ $= 298$ K

Check Your Answer

- Are your units correct? Temperature is measured in kelvins.
- Does the magnitude make sense? Temperatures on the Kelvin scale are all larger than those on the Celsius scale.

FIGURE 12–7 There is an extremely wide range of temperatures throughout the universe. Note the scale has been expanded in areas of particular interest.

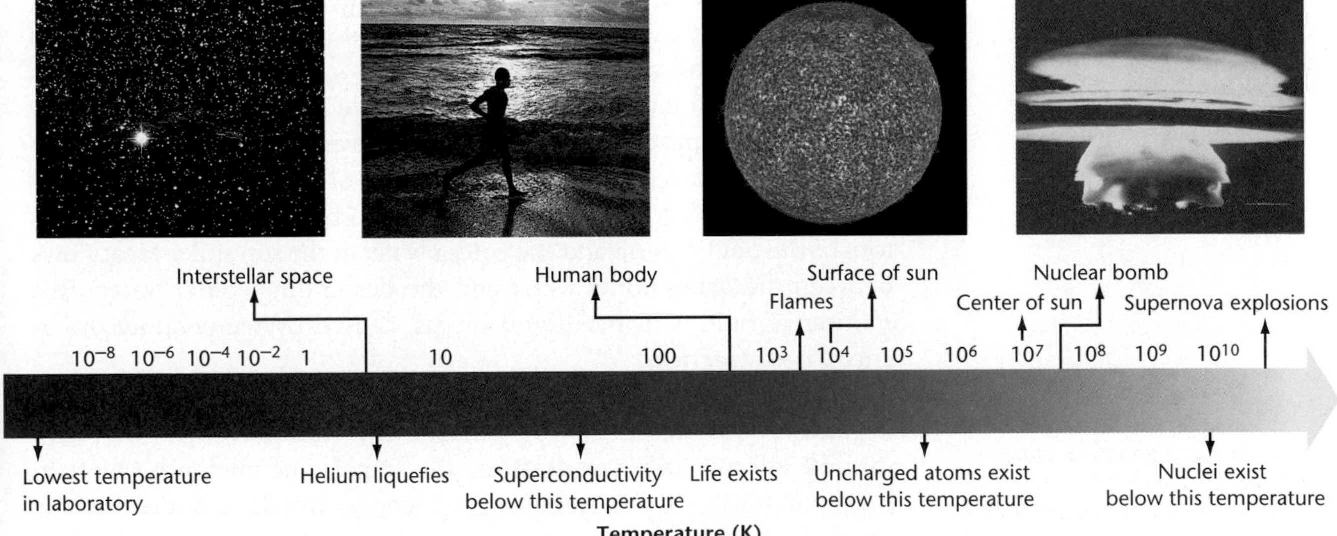

Practice Problems

1. Make the following conversions.
   a. 0°C to kelvins
   b. 0 K to degrees Celsius
   c. 273°C to kelvins
   d. 273 K to degrees Celsius
2. Convert the following Celsius temperatures to Kelvin temperatures.
   a. 27°C
   b. 150°C
   c. 560°C
   d. −50°C
   e. −184°C
   f. −300°C
3. Convert the following Kelvin temperatures to Celsius temperatures.
   a. 110 K
   b. 70 K
   c. 22 K
   d. 402 K
   e. 323 K
   f. 212 K
4. Find the Celsius and Kelvin temperatures for the following.
   a. room temperature
   b. refrigerator temperature
   c. typical hot summer day
   d. typical winter night

F.Y.I.

Why are degrees used to measure temperature? Fahrenheit set up his temperature scale so that there were 180 degrees separating the temperature where water freezes to where water boils. This was because there are 180 degrees in a straight angle.

Heat and Thermal Energy

One way to increase the temperature of an object is to place it in contact with a hotter object. The thermal energy of the hotter object is decreased, and the thermal energy of the cooler object is increased. Energy always flows from the hotter object to the cooler object. Energy never flows from a colder object to a hotter object. **Heat** is the energy that flows between two objects as a result of a difference in temperature. The symbol Q is used to represent the amount of heat. If Q has a negative value, heat has left the object; if Q has a positive value, heat has been absorbed by the object. Heat, like other forms of energy, is measured in joules.

Thermal energy transfer

You have already learned one way that heat flows from a warmer body to a colder one. If you place one end of a metal rod in a flame, it becomes hot. The other end also becomes warm very quickly. Heat is conducted because the particles in the rod are in direct contact.

A second means of thermal transfer involves particles that are not in direct contact. Have you ever looked in a pot of water just about to boil? You can see motion of water, as water heated by conduction at the bottom of the pot flows up and the colder water at the top sinks. Heat flows between the rising hotter water and the descending colder water. This motion of fluid, whether liquid or gas, caused by temperature differences, is **convection.**

The third method of thermal transfer, unlike the first two, does not depend on the presence of matter. The sun warms us from over 150 million kilometers via **radiation,** the transfer of energy by electromagnetic waves. These waves carry the energy from the hot sun to the much cooler Earth.

Specific heat

When heat flows into an object, its thermal energy increases, and so does its temperature. The amount of the increase depends on the size of the object. It also depends on the material from which the object is made. The **specific heat** of a material is the amount of energy that must be added to the material to raise the temperature of a unit mass one temperature unit. In SI units, specific heat, represented by C (not to be confused with °C), is measured in J/kg·K. **Table 12–1** provides values of specific heat for some common substances. For example, 903 J must be added to one kilogram of aluminum to raise the temperature one kelvin. The specific heat of aluminum is 903 J/kg·K.

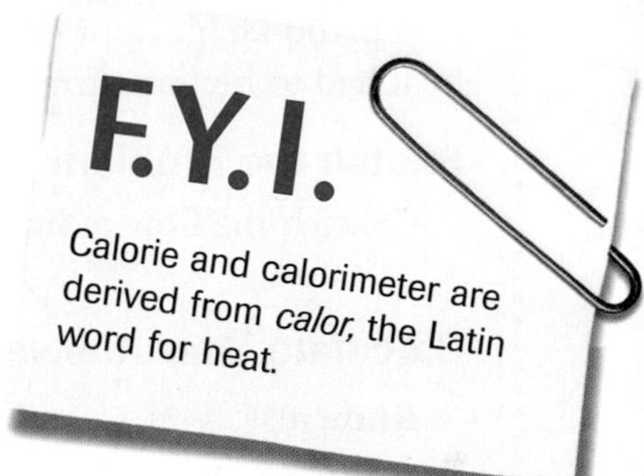

F.Y.I.
Calorie and calorimeter are derived from *calor,* the Latin word for heat.

Note that water has a high specific heat compared to those of other substances, even ice and steam. One kilogram of water requires the addition of 4180 J of energy to increase its temperature by one kelvin. The same mass of copper requires only 385 J to increase its temperature by one kelvin. The 4180 J of energy needed to raise the temperature of one kilogram of water by one kelvin would increase the temperature of the same mass of copper by 11 K. The high specific heat of water is the reason water is used in car radiators to remove thermal energy from the engine block.

The heat gained or lost by an object as its temperature changes depends on the mass, the change in temperature, and the specific heat of the substance. The amount of heat transferred can be determined using the following equation.

$$Q = mC\Delta T = mC(T_{\text{final}} - T_{\text{initial}})$$

where Q is the heat gained or lost, m is the mass of the object, C is the specific heat of the substance, and ΔT is the change in temperature. When the temperature of 10.0 kg of water is increased by 5.0 K, the heat absorbed, Q, is

$$Q = (10.0 \text{ kg})(4180 \text{ J/kg·K})(5.0 \text{ K}) = 2.1 \times 10^5 \text{ J}.$$

Because one Celsius degree is equal in magnitude to one kelvin, temperature changes can be measured in either kelvins or Celsius degrees.

TABLE 12–1

Specific Heat of Common Substances

Material	Specific heat J/kg • K	Material	Specific heat J/kg • K
aluminum	903	lead	130
brass	376	methanol	2450
carbon	710	silver	235
copper	385	steam	2020
glass	664	water	4180
ice	2060	zinc	388
iron	450		

Example Problem

Heat Transfer

A 0.400-kg block of iron is heated from 295K to 325K. How much heat had to be transferred to the iron?

Sketch the Problem

- Sketch the flow of heat into the block of iron

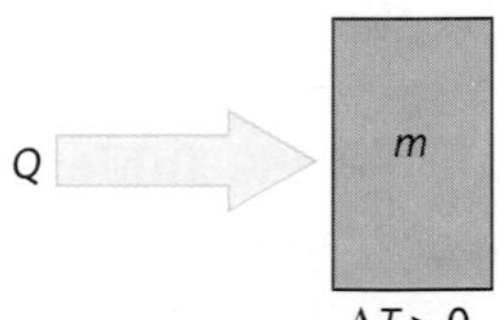

Calculate Your Answer

Known:

$m = 0.400$ kg

$C = 450$ J/kg·K

$T_i = 295$ K

$T_f = 325$ K

Unknown:

$Q = ?$

Strategy:

The heat transferred is a product of the mass, specific heat, and the temperature change.

Calculations:

$$Q = mC(T_f - T_i)$$
$$= (0.400 \text{ kg})(450 \text{ J/kg·K})(325 - 295 \text{ K})$$
$$= 5.4 \times 10^3 \text{ J}$$

Check Your Answer

- Are your units correct? Heat is measured in joules.
- Does the sign make sense? Temperature increased so Q is positive.
- Is the magnitude realistic? Magnitudes of thousands of joules are typical of solids with masses around 1 kg and temperature changes of tens of kelvins.

Practice Problems

5. How much heat is absorbed by 60.0 g of copper when its temperature is raised from 20.0°C to 80.0°C?
6. The cooling system of a car engine contains 20.0 L of water (1 L of water has a mass of 1 kg).
   a. What is the change in the temperature of the water if the engine operates until 836.0 kJ of heat are added?
   b. Suppose it is winter and the system is filled with methanol. The density of methanol is 0.80 g/cm^3. What would be the increase in temperature of the methanol if it absorbed 836.0 kJ of heat?
   c. Which is the better coolant, water or methanol? Explain.

F.Y.I.

Evaporation of perspiration from the skin is an effective way of cooling your body. Over two million joules of thermal energy are carried away for each liter of liquid evaporated.

Physics Lab

Heating Up

Problem

How does a constant supply of thermal energy affect the temperature of water?

Materials

hot plate (or Bunsen burner)
250-mL Pyrex beaker
water
thermometer
stopwatch
goggles
apron

Data and Observations	
Time	Temperature

Procedure

1. Turn your hot plate to a medium setting (or as recommended by your teacher). Allow a few minutes for the plate to heat up. Wear goggles.
2. Pour 150 mL of room temperature water into the 250-mL beaker.
3. Make a data and observation table.
4. Record the initial temperature of the water. The thermometer must not touch the bottom or sides of the beaker.
5. Place the beaker on the hot plate and record the temperature every 1.0 minute. Carefully stir the water before taking a temperature reading.
6. Record the time when the water starts to boil. Continue recording the temperature for an additional 4.0 minutes.
7. Carefully remove the beaker from the hot plate. Record the temperature of the remaining water.

Analyze and Conclude

1. **Analyzing Data** Make a graph of temperature (vertical axis) versus time (horizontal axis).
2. **Interpreting Graphs** What is the slope of the graph for the first 3.0 minutes? Be sure to include units.
3. **Relating Concepts** What is the thermal energy given to the water in the first 3.0 minutes. **Hint:** $Q = mC\Delta T$.
4. **Making Predictions** Use a dotted line on the same graph to predict what the graph would look like if the same procedure was followed with only half as much water.

Apply

1. Would you expect that the hot plate transferred energy to the water at a steady rate?
2. Where is the energy going when the water is boiling?

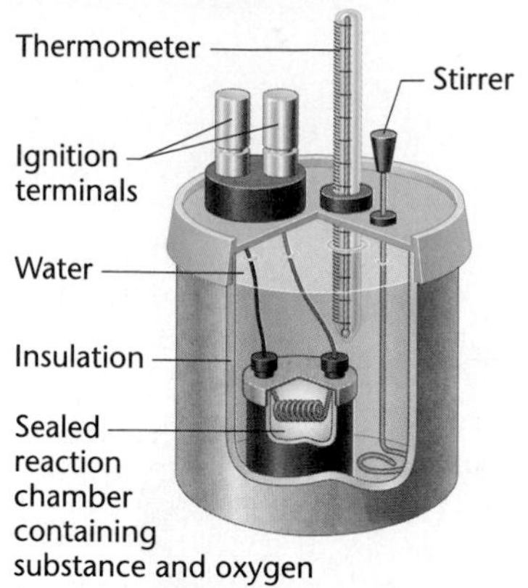

FIGURE 12–8 A calorimeter provides a closed, isolated system in which to measure energy transfer.

Calorimetry: Measuring Specific Heat

A **calorimeter,** shown in **Figure 12–8,** is a device used to measure changes in thermal energy. A calorimeter is carefully insulated so that heat transfer is very small. A measured mass of a substance is placed in the calorimeter and heated to a high temperature. The calorimeter contains a known mass of cold water at a measured temperature. The heat released by the substance is transferred to the cooler water. From the resulting increase in water temperature, the change in thermal energy of the substance is calculated.

The operation of a calorimeter depends on the conservation of energy in isolated, closed systems. Energy can neither enter nor leave an isolated system. As a result of the isolation, if the energy of one part of the system increases, the energy of another part must decrease by the same amount. Consider a system composed of two blocks of metal, block A and block B, as in **Figure 12–9a.** The total energy of the system is constant.

$$E_A + E_B = constant$$

Suppose that the two blocks are initially separated but can be placed in contact. If the thermal energy of block A changes by an amount ΔE_A, then the change in thermal energy of block B, ΔE_B, must be related by the following equation.

$$\Delta E_A + \Delta E_B = 0$$

Which means that the following relationship is true.

$$\Delta E_A = -\Delta E_B$$

The change in energy of one block is positive, while the change in energy of the other block is negative. If the thermal energy change is positive, the temperature of that block rises. If the change is negative, the temperature falls.

Assume that the initial temperatures of the two blocks are different. When the blocks are brought together, heat flows from the hotter block to the colder block, as shown in **Figure 12–9b.** The flow continues until the blocks are in thermal equilibrium. The blocks then have the same temperature.

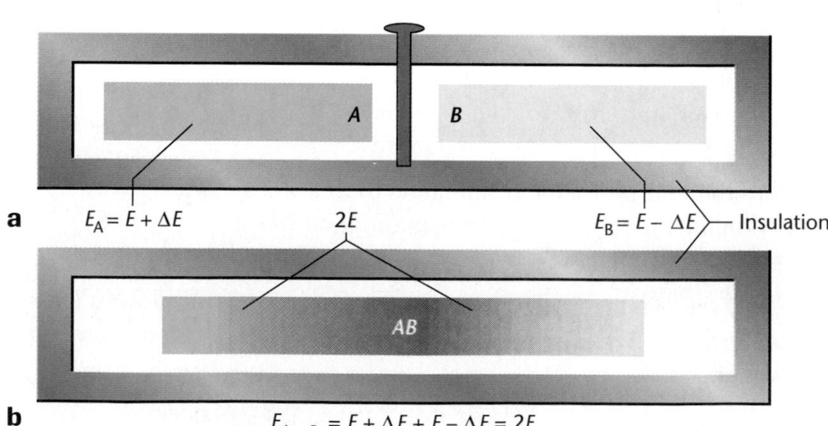

FIGURE 12–9 The total energy for this system is constant.

In a calorimeter, the change in thermal energy is equal to the heat transferred because no work is done. Therefore, the change in energy can be expressed by the following equation.

$$\Delta E = Q = mC\Delta T$$

The increase in thermal energy of block A is equal to the decrease in thermal energy of block B. Thus, the following relationship is true.

$$m_A C_A \Delta T_A + m_B C_B \Delta T_B = 0$$

The change in temperature is the difference between the final and initial temperatures, that is, $\Delta T = T_f - T_i$. If the temperature of a block increases, $T_f > T_i$, and ΔT is positive. If the temperature of the block decreases, $T_f < T_i$, and ΔT is negative. The final temperatures of the two blocks are equal. The equation for the transfer of energy is

$$m_A C_A (T_f - T_{Ai}) + m_B C_B (T_f - T_{Bi}) = 0.$$

To solve for T_f, expand the equation.

$$m_A C_A T_f - m_A C_A T_{Ai} + m_B C_B T_f - m_B C_B T_{Bi} = 0$$

$$T_f (m_A C_A + m_B C_B) = m_A C_A T_{Ai} + m_B C_B T_{Bi}.$$

$$T_f = \frac{m_A C_A T_{Ai} + m_B C_B T_{Bi}}{m_A C_A + m_B C_B}$$

F.Y.I.

For many years, thermal energy was measured in calories. Calories are not part of the SI system of measurements, so today, thermal energy is measured in joules. One calorie is equal to 4.18 joules.

Example Problem

Heat Transfer in a Calorimeter

A calorimeter contains 0.50 kg of water at 15°C. A 0.040-kg block of zinc at 115°C is placed in the water. What is the final temperature of the system?

Sketch the Problem

- Let zinc be sample A and water be sample B.
- Sketch the transfer of heat from hotter zinc to cooler water.

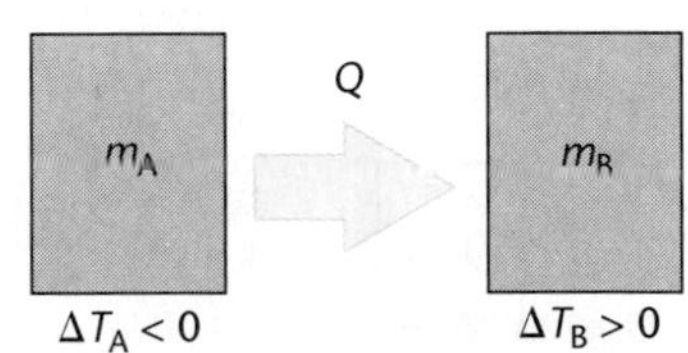

Calculate Your Answer

Known:

Zinc $m_A = 0.040$ kg, $C_A = 388$ J/kg·°C, $T_{Ai} = 115$°C

Water $m_B = 0.50$ kg, $C_B = 4180$ J/kg·°C, $T_{Bi} = 15$°C

Unknown:

$T_f = ?$

Strategy: Determine final temperature using $T_f = \dfrac{m_A C_A T_{Ai} + m_B C_B T_{Bi}}{m_A C_A + m_B C_B}$.

Calculations: $$T_f = \frac{(0.040\text{ kg})(388\text{ J/kg·°C})(115\text{°C}) + (0.50\text{ kg})(4180\text{ J/kg·°C})(15\text{°C})}{(0.040\text{ kg})(388\text{ J/kg·°C}) + (0.50\text{ kg})(4180\text{ J/kg·°C})}$$

$$= \frac{(1.78 \times 10^3 + 3.14 \times 10^4)\text{J}}{(15.5 + 2.09 \times 10^3)\text{J/°C}} = 16\text{°C}$$

Check Your Answer

- Are the units correct? Temperature is measured in °C.
- Is the magnitude realistic? The answer is between the initial temperatures of the two samples and closer to water.

Practice Problems

7. A 2.00×10^2-g sample of water at 80.0°C is mixed with 2.00×10^2 g of water at 10.0°C. Assume no heat loss to the surroundings. What is the final temperature of the mixture?
8. A 4.00×10^2-g sample of methanol at 16.0°C is mixed with 4.00×10^2 g of water at 85.0°C. Assume that there is no heat loss to the surroundings. What is the final temperature of the mixture?
9. A 1.00×10^2-g brass block at 90.0°C is placed in a plastic foam cup containing 2.00×10^2 g of water at 20.0°C. No heat is lost to the cup or the surroundings. Find the final temperature of the mixture.
10. A 1.00×10^2-g aluminum block at 100.0°C is placed in 1.00×10^2 g of water at 10.0°C. The final temperature of the mixture is 25.0°C. What is the specific heat of the aluminum?

Melting

Label two foam cups A and B. Measure 75 mL of room-temperature water into each of the two cups. Add an ice cube to cup A. Add ice water to cup B until the water levels are equal. Measure the temperature of each cup at one minute intervals until the ice has melted.

Analyze and Conclude Do the samples reach the same final temperature? Why?

12.1 Section Review

1.1 Could the thermal energy of a bowl of hot water equal that of a bowl of cold water? Explain.

1.2 On cold winter nights before central heating existed, people often placed hot water bottles in their beds. Why would this be more efficient than warmed bricks?

1.3 If you take a spoon out of a cup of hot coffee and put it in your mouth, you are not likely to burn your tongue. But, you could very easily burn your tongue if you put the hot coffee in your mouth. Why?

1.4 **Critical Thinking** You use an aluminum cup instead of a plastic foam cup as a calorimeter, allowing heat to flow between the water and the environment. You measure the specific heat of a sample by putting the hot object into room temperature water. How might your experiment be affected? Would your result be too large or too small?

Change of State and Laws of Thermodynamics

12.2

OBJECTIVES

- **Define** heats of fusion and vaporization.
- **State** the first and second laws of thermodynamics.
- **Define** heat engine, refrigerator, and heat pump.
- **Define** entropy.

If you rub your hands together, you exert a force and move your hands over a distance. You do work against friction. Your hands start and end at rest, so there is no net change in kinetic energy. They remain the same distance above Earth, so there is no change in potential energy. Yet, if the law of conservation of energy is true, then the energy transferred by the work you did must have gone somewhere. You notice that your hands feel warm; their temperature has increased. The energy to do the work against friction has changed form to thermal energy.

Change of State

The three most common states of matter are solids, liquids, and gases, as shown in **Figure 12–10.** As the temperature of a solid is raised, it will usually change to a liquid. At even higher temperatures, it will become a gas. How can we explain these changes? Consider a material in a solid state. Your simplified model of the solid consists of tiny particles bonded together by massless springs. These massless springs represent the electromagnetic forces between the particles. When the thermal energy of a solid is increased, the motion of the particles is increased and the temperature increases.

FIGURE 12–10 The three states of water are represented in this photograph. The gaseous state, water vapor, is dispersed in the air and is invisible until it condenses.

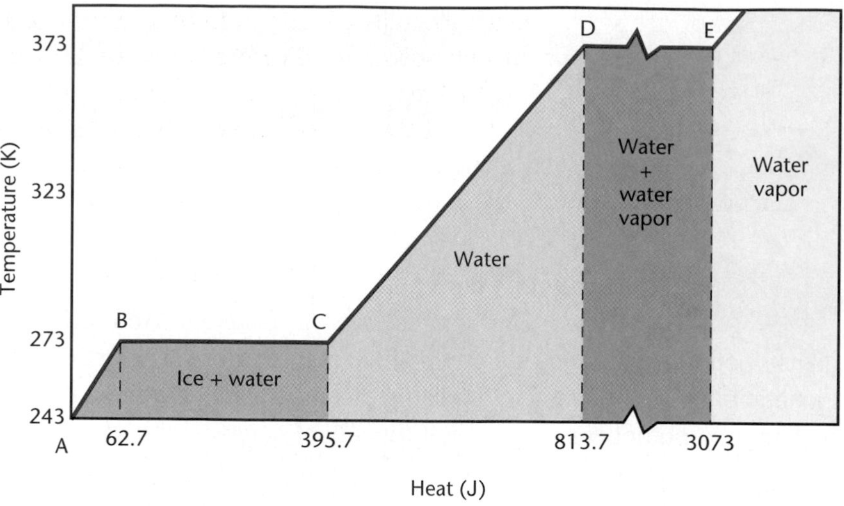

FIGURE 12–11 A plot of temperature versus heat added when 1 g of ice is initially converted to steam.

HELP WANTED

HVAC TECHNICIAN

Knowledge of thermodynamics; ability to read blueprints, specifications and manuals; familiarity with current products, procedures, tools, and test equipment; and ability to work hard in sometimes physically demanding conditions make you an ideal candidate for this position. Knowledge of all kinds of heating, ventilation, and air conditioning systems is required. Work your way up into other positions based on your work performance and completion of further training. For information contact:
Association Builders and Contractors
729 15th Street, NW
Washington, DC 20005

Figure 12–11 diagrams this process throughout all the changes of state as thermal energy is added to 1.0 g of H_2O starting at an initial temperature of 243 K and continuing until the temperature is 473 K. Between points A and B, the ice is warmed to 273 K. At some point, the added thermal energy causes the particles to move rapidly enough that their motion overcomes the forces holding the particles together in a fixed location. The particles are still touching, but they have more freedom of movement. Eventually, the particles become free enough to slide past each other. At this point, the substance has changed from a solid to a liquid. The temperature at which this change occurs is the **melting point.** When a substance is melting, all of the added thermal energy goes to overcome the forces holding the particles together in the solid state. None of the added thermal energy increases the kinetic energy of the particles. This can be observed between points B and C in the diagram where the added thermal energy melts the ice at a constant 273 K. Since the kinetic energy of the particles does not increase, the temperature does not increase here.

Once the solid is completely melted, there are no more forces holding the particles in the solid state and the added thermal energy again increases the motion of the particles, and the temperature rises. In the diagram, this is between points C and D. As the temperature increases, some particles in the liquid acquire enough energy to break free from other particles. At a specific temperature, known as the **boiling point,** further addition of energy causes another change of state where all the added thermal energy converts the material from the liquid state to the gas state. The motion of the particles does not increase so the temperature is not raised during this transition. In the diagram, this transition is represented between points D and E. After the material is entirely converted to the gas, any added thermal energy again increases the motion of the particles and the temperature rises. Above point E , steam is heated to a temperature of 473 K.

The amount of energy needed to melt one kilogram of a substance is called the **heat of fusion** of that substance. For example, the heat of fusion of ice is 3.34×10^5 J/kg. If 1 kg of ice at its melting point, 273 K, absorbs 3.34×10^5 J, the ice becomes 1 kg of water at the same temperature, 273 K. The added energy causes a change in state but not in temperature. The horizontal distance from point B to point C represents the heat of fusion in **Figure 12–11.**

At normal atmospheric pressure, water boils at 373 K. The thermal energy needed to vaporize one kilogram of a liquid is called the **heat of vaporization.** For water, the heat of vaporization is 2.26×10^6 J/kg. Each substance has a characteristic heat of vaporization. The distance from point D to point E represents the heat of vaporization.

Between points A and B in the diagram, there is a definite slope to the line as the temperature is raised. This slope represents the specific heat of the ice. The slope between points C and D represents the specific heat of water. And the slope above point E represents the specific heat of steam. Note that the slope for water is less than that of ice or steam. Water has a greater specific heat than does ice or steam.

The heat, Q, required to melt a solid of mass m is given by the equation,

$$Q = mH_f$$

where H_f is the heat of fusion. Similarly, the heat, Q, required to vaporize a mass, m, of liquid is given by the equation,

$$Q = mH_v$$

where H_v is the heat of vaporization. The values of some heats of fusion, H_f, and heats of vaporization, H_v, can be found in **Table 12–2.**

When a liquid freezes, an amount of heat $Q = -mH_f$ must be removed from the liquid to turn it into a solid. The negative sign indicates the heat is transferred from the sample to the environment. In the same way, when a vapor condenses to a liquid, an amount of heat, $Q = -mH_v$, must be removed.

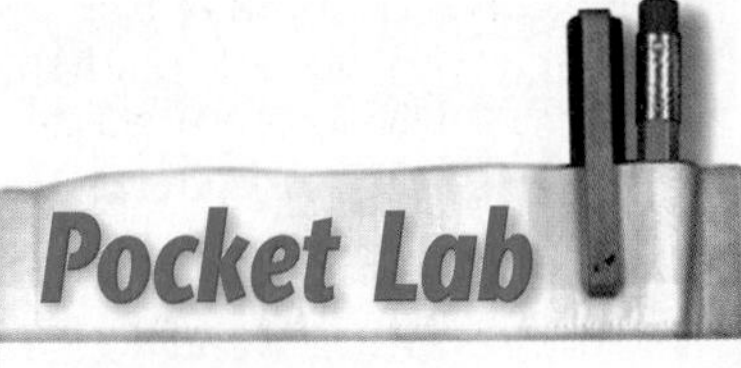

Cool Times

Place a 100-mL beaker in a 250-mL beaker. Put a thermometer in each beaker. Fill the small beaker with hot, colored water. Determine the temperature of the colored water. Slowly pour tap water into the large beaker until the water is at the same height in both beakers. Record the temperature in the large beaker. Record the temperature in both beakers every minute for five minutes. Plot your data for both beakers on a graph of temperature versus time. Measure and record the mass of water in each beaker.

Calculate and Conclude
Predict the final temperature. Describe each curve.

TABLE 12–2

Heats of Fusion and Vaporization of Some Common Substances

Material	Heat of fusion H_f (J/kg)	Heat of vaporization H_v (J/kg)
copper	2.05×10^5	5.07×10^6
mercury	1.15×10^4	2.72×10^5
gold	6.30×10^4	1.64×10^6
methanol	1.09×10^5	8.78×10^5
iron	2.66×10^5	6.29×10^6
silver	1.04×10^5	2.36×10^6
lead	2.04×10^4	8.64×10^5
water (ice)	3.34×10^5	2.26×10^6

Example Problem

Melting a Solid and Warming the Resulting Liquid

You are asked to melt 0.100 kg of ice at its melting point and warm the resulting water to 20.0°C. How much heat is needed?

Sketch the Problem

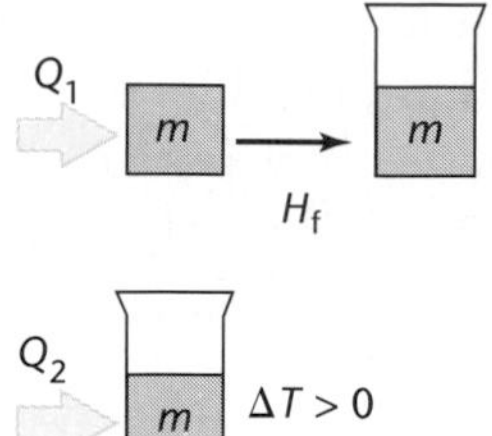

- Sketch the relationship between heat and water in its solid and liquid states.
- Sketch the transfer of heat as the temperature of the water increased.

Calculate Your Answer

Known:

$m = 0.100$ kg

$H_f = 3.34 \times 10^5$ J/kg

$T_i = 0.0°C$

$T_f = 20.0°C$

$C = 4180$ J/kg·°C

Unknown:

$Q_1 + Q_2 = ?$

Strategy:

Calculate heat needed to melt ice.

Calculations:

$$Q_1 = mH_f = (0.100 \text{ kg})(3.34 \times 10^5 \text{ J/kg}) = 3.34 \times 10^4 \text{ J} = 33.4 \text{ kJ}$$

Calculate the temperature change.

$$\Delta T = T_f - T_i = 20.0°C - 0.0°C = 20.0°C$$

Calculate heat needed to raise water temperature.

$$Q_2 = mC\Delta T = (0.100 \text{ kg})(4180 \text{ J/kg·°C})(20.0°C) = 8.36 \times 10^3 \text{ J} = 8.36 \text{ kJ}$$

Add all heats together to get total heat needed.

$$Q_1 + Q_2 = 33.4 \text{ kJ} + 8.36 \text{ kJ} = 41.8 \text{ kJ}$$

Check Your Answer

- Are the units correct? Energy units are in joules.
- Does the sign make sense? Q is positive when heat is absorbed.
- Is the magnitude realistic? The amount of heat needed to melt the ice is about four times greater than the heat needed to increase the water temperature. It takes more energy to overcome the forces holding the particles in the solid state than to raise the temperature of water.

Practice Problems

11. How much heat is absorbed by 1.00×10^2 g of ice at $-20.0°C$ to become water at $0.0°C$?
12. A 2.00×10^2-g sample of water at $60.0°C$ is heated to steam at $140.0°C$. How much heat is absorbed?
13. How much heat is needed to change 3.00×10^2 g of ice at $-30.0°C$ to steam at $130.0°C$?
14. A 175-g lump of molten lead at its melting point, $327°C$, is dropped into 55 g of water at $20.0°C$.
    a. What is the temperature of the water when the lead becomes solid?
    b. When the lead and water are in thermal equilibrium, what is the temperature?

CHEMISTRY CONNECTION

Hot packs keep your hands warm, and cold packs keep your soda chilled. In both packs, a thin membrane separates water from a chamber of salt. In a hot pack, the salt might be calcium chloride; in a cold pack, ammonium nitrate. Squeezing the pack breaks the membrane. As the salt dissolves in the water, the solution either releases or absorbs energy. What form does this energy take?

The First Law of Thermodynamics

There are additional means of changing the amount of thermal energy in a system. If you use a hand pump to inflate a bicycle tire, the air and pump become warm. The mechanical energy in the moving piston is converted into thermal energy of the gas. Other forms of energy—light, sound, and electric—can be changed into thermal energy. Some examples include a toaster, which converts electric energy into heat to cook bread, and the sun, which you have already learned warms Earth with light from a distance of over 150 million kilometers.

Thermal energy can be increased either by adding heat or by doing work on a system. Thus, the total increase in the thermal energy of a system is the sum of the work done on it and the heat added to it. This fact is called the **first law of thermodynamics.** Recall that thermodynamics is the study of the changes in thermal properties of matter. The first law of thermodynamics is merely a restatement of the law of conservation of energy, which states that energy is neither created nor destroyed but can be changed into other forms.

The conversion of mechanical energy to thermal energy, as when you rub your hands together, is easy. The reverse process, conversion of thermal to mechanical energy, is more difficult. A device able to convert thermal energy to mechanical energy continuously is called a **heat engine.**

Heat engine

A heat engine requires a high temperature source from which thermal energy can be removed, a low temperature receptacle, called a sink, into which thermal energy can be delivered, and a way to convert the thermal energy into work. A diagram is presented in **Figure 12–12.** An automobile internal combustion engine is one example of a heat engine and

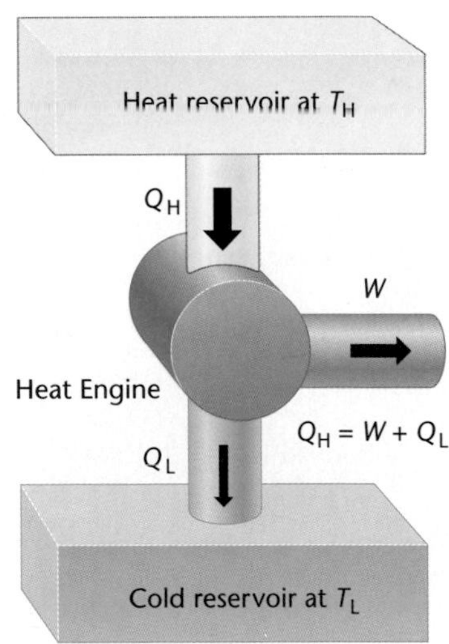

FIGURE 12–12 This diagram represents heat at high temperature transformed into mechanical energy and low-temperature waste heat.

FIGURE 12–13 The heat produced by burning gasoline causes the gases produced to expand and exert force and do work on the piston.

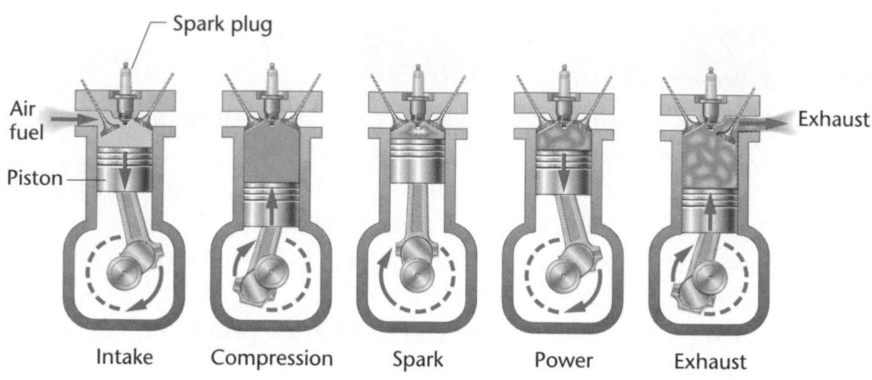

is shown in **Figure 12–13.** In the automobile, a mixture of air and gasoline vapor is ignited, producing a high-temperature flame. Heat flows from the flame to the air in the cylinder. The hot air expands and pushes on a piston, changing thermal energy into mechanical energy. In order to obtain continuous mechanical energy, the engine must be returned to its starting condition. The heated air is expelled and replaced by new air, and the piston is returned to the top of the cylinder. The entire cycle is repeated many times each minute. The thermal energy from the gasoline is converted into mechanical energy that eventually results in the movement of the car.

Perpetual Motion?

Not all the thermal energy from the high-temperature flame is converted into mechanical energy. The exhaust gases and the engine parts become hot. The exhaust comes in contact with outside air, transferring heat to it, and raising its temperature. Heat flow from the hot engine is transferred to a radiator. Outside air passes through the radiator and the air temperature is raised.

This energy transferred out of the engine is called waste heat, that is, heat that cannot be converted into work. The overall change in total energy of the car-air system is zero. In a car engine, the thermal energy in the flame equals the sum of the mechanical energy produced and the waste heat expelled. All heat engines generate waste heat. Because of the waste heat, no engine can ever convert all energy into useful motion or work.

Refrigerators and heat pumps

Heat flows spontaneously from a warm body to a cold body. It is possible to remove thermal energy from a colder body and add it to a warmer body. A refrigerator is a common example of a device that accomplishes this transfer with the use of mechanical energy. Electric energy runs a motor that does work on a gas and compresses it. As gas that has drawn heat from the interior of the refrigerator passes from the compressor through the condenser coils on the outside of the refrigerator, the gas cools into a liquid. Thermal energy is transferred into the air in the room. The liquid reenters the interior, vaporizes, and absorbs thermal energy from its surroundings. The resulting gas returns to the compressor and repeats the process. The overall change in the thermal

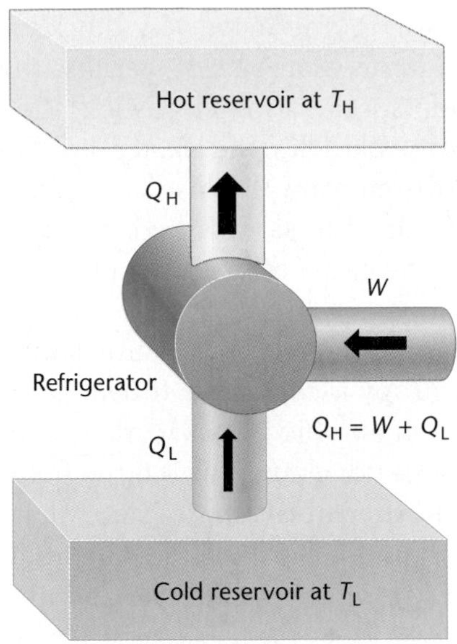

FIGURE 12–14 A refrigerator absorbs heat Q_L from the cold reservoir and gives off heat Q_H to the hot reservoir. Work, W, is done on the refrigerator.

energy of the gas is zero. Thus, according to the first law of thermodynamics, the sum of the heat removed from the refrigerator contents and the work done by the motor is equal to the heat expelled to the outside air at a higher temperature, as is shown in **Figure 12–14.** A heat pump is a refrigerator that can be run in two directions. In summer, heat is removed from the house, and thus cooling the house. The heat is expelled into the warmer air outside. In winter, heat is removed from the cold outside air and transferred into the warmer house. In both cases, mechanical energy is required to transfer heat from a cold object to a warmer one.

The Second Law of Thermodynamics

Many processes that do not violate the first law of thermodynamics have never been observed to occur spontaneously. Three such processes are presented in **Figure 12–15.** For example, the first law of thermodynamics does not prohibit heat flowing from a cold body to a hot body. Still, when a hot body has been placed in contact with a cold body, the hot body has never been observed to become hotter and the cold body colder. Heat flows spontaneously from hot to cold bodies. Another example, heat engines could convert thermal energy completely into mechanical energy with no waste heat and the first law of thermodynamics would still be obeyed. Yet waste heat is always generated, and randomly distributed particles of a gas are not observed to spontaneously arrange themselves in specific ordered patterns.

In the nineteenth century, the French engineer Sadi Carnot studied the ability of engines to convert thermal energy into mechanical energy. He developed a logical proof that even an ideal engine would generate some waste heat. Carnot's result is best described in terms of a quantity called **entropy,** which is a measure of the disorder in a system. Consider what happens when heat is added to an object. The particles

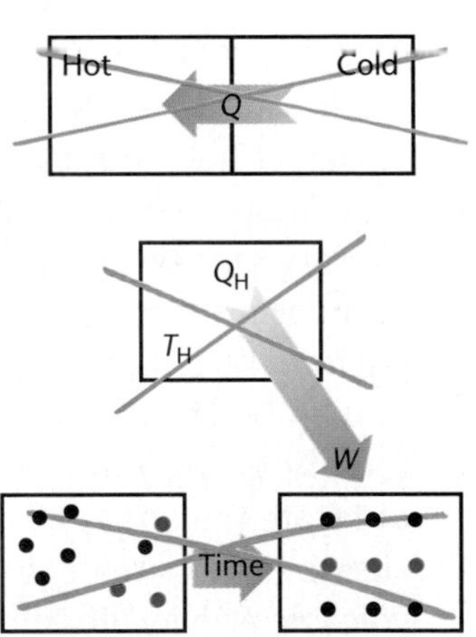

FIGURE 12–15 These three processes are not forbidden by the first law of thermodynamics, yet they do not occur spontaneously.

move in a random way. Some move very quickly, others move slowly, and many move at intermediate speeds. Because the particles are not all traveling at the same speed, disorder ensues. The greater the range of speeds exhibited by the particles, the greater the disorder. The greater the disorder, the larger the entropy. While it is theoretically possible that all the particles could have the same speed, the random collisions and energy exchanges of the particles make the probability of this extremely unlikely.

Entropy, like thermal energy, is contained in an object. If heat is added to a body, entropy is increased. If heat is removed from a body, entropy is decreased. If an object does work with no change in temperature, the entropy does not change, as long as friction is ignored.

The **second law of thermodynamics** states that natural processes go in a direction that maintains or increases the total entropy of the universe. That is, all things will become more and more disordered unless

Physics & Technology

Infrared Detectors

Infrared radiation (IR) is electromagnetic energy that radiates through space. Any object with a temperature above absolute zero gives off infrared energy. The higher the object's temperature, the more IR it emits. Instruments capable of detecting IR can measure the temperature of very cold or very distant objects. They can also be used to "see" objects in total darkness.

Rescue teams can use night-vision goggles to detect the infrared glow given off by people or animals, making them visible in the dark. IR security sensors trigger an alarm when an intruder is detected, and IR-sensitive medical equipment can produce thermal maps of the human body to help locate tumors or circulatory problems. Airborne IR cameras are used to locate the edges of wildfires obscured from view by thick smoke, or to study underground volcanic activity. IR detectors aboard weather satellites map cloud patterns by measuring temperature differences between Earth's surface and high-altitude clouds.

Perhaps the most far-reaching applications for IR detectors are in the fields of astronomy and cosmology. Highly sensitive detectors measure such tiny amounts of IR that they can be used to view objects in the universe that are not hot enough to emit visible light. These instruments can produce images of faraway stars and galaxies, search for planets orbiting nearby stars, and survey the most distant regions of the sky for clues about the origins of the universe. Because Earth's atmosphere absorbs almost all the IR, these studies are usually conducted using arrays of IR detectors aboard orbiting satellites. The detectors are thin layers of electrically conductive materials that absorb IR. This absorption causes changes in conductivity that are monitored with computers.

Thinking Critically Highly sensitive IR detectors used by astronomers are supercooled to extremely low temperatures, sometimes as low as 3 or 4 K. Why is this necessary?

FIGURE 12–16 The spontaneous mixing of the food coloring and water is an example of the second law of thermodynamics.

some action is taken to keep them ordered. Entropy increase and the second law of thermodynamics can be thought of as statements of the probability of events happening. **Figure 12–16** illustrates an increase in entropy as food-coloring molecules, originally separate from the clear water, are thoroughly mixed with the water molecules over time. **Figure 12–17** shows a familiar example of the second law of thermodynamics that most teenagers can readily recognize.

The second law of thermodynamics predicts that heat flows spontaneously only from a hot body to a cold body. Consider a hot iron bar and a cold cup of water. On the average, the particles in the iron will be moving very fast, whereas the particles in the water move more slowly. When you plunge the bar into the water and allow thermal equilibrium to be reached, the average kinetic energy of the particles in the iron and the water will be the same. More particles now have a more random motion than they did at the beginning of your test. This final state is less ordered than the initial state. No longer are the fast particles confined solely to the iron and the slower particles confined to the water. All speeds are evenly distributed. The entropy of the final state is larger than that of the initial state.

You take for granted many daily events that occur spontaneously, or naturally, in one direction, but that would shock you if they happened in reverse. You are not surprised when a metal spoon, heated at one end,

FIGURE 12–17 If no work is done on a system, entropy spontaneously reaches a maximum.

Drip, Drip, Drip

Measure equal amounts of very hot and very cold water into two clear glasses (or beakers).

Hypothesize and Test Predict what will happen if you simultaneously put one drop of food coloring in each glass. Try it. What happened? Why? Was the mixing symmetric?

soon becomes uniformly hot, or when smoke from a too-hot frying pan diffuses throughout the kitchen. Consider your reaction, however, if a spoon lying on a table suddenly, on its own, became red hot at one end and icy cold at the other, or if all the smoke from the frying pan collected in a 9-cm^3 cube in the exact center of the kitchen. Neither of the reverse processes violates the first law of thermodynamics. These are simply examples of the countless events that do not occur because their processes would violate the second law of thermodynamics.

The second law of thermodynamics and increase in entropy also give new meaning to what has been commonly called the "energy crisis," the continued use of limited resources of fossil fuels, such as natural gas and petroleum. When you use a resource such as natural gas to heat your home, you do not use up the energy in the gas. The internal chemical energy contained in the molecules of the gas is converted into thermal energy of the flame, which is then transferred to thermal energy in the air of your home. Even if this warm air leaks to the outside, the energy is not lost. Energy has not been used up. The entropy, however, has been increased. The chemical structure of natural gas is very ordered. In contrast, the random motion of the warmed air is very disordered. While it is mathematically possible for the original order to be reestablished, the probability of this occurring is essentially zero. For this reason, entropy is often used as a measure of the unavailability of useful energy. The energy in the warmed air in a house is not as available to do mechanical work or to transfer heat to other bodies as the original gas molecules. The lack of usable energy is really a surplus of entropy.

12.2 Section Review

2.1 Old-fashioned heating systems sent steam into radiators in each room. In a radiator, the steam condenses back to water. How does this process heat the room?

2.2 James Joule carefully measured the difference in temperature of water at the top and bottom of a waterfall. Why would he have expected a difference?

2.3 A man uses a 320-kg hammer moving at 5 m/s to smash a 3-kg block of lead against a 450-kg rock. He finds that the temperature of the lead increased by 5°C. Explain how this happens.

2.4 **Critical Thinking** A new deck of cards has all the suits (clubs, diamonds, hearts, and spades) in order, and the cards are ordered by number within the suits. If you shuffle the cards many times, are you likely to return the cards to the original order? Of what physical law is this an example?

CHAPTER 12 REVIEW

Key Terms

12.1

- thermodynamics
- kinetic-molecular theory
- thermal energy
- temperature
- conduction
- thermal equilibrium
- thermometer
- absolute zero
- kelvin
- heat
- convection
- radiation
- specific heat
- calorimeter

12.2

- melting point
- boiling point
- heat of fusion
- heat of vaporization
- first law of thermodynamics
- heat engine
- entropy
- second law of thermodynamics

Summary

12.1 Temperature and Thermal Energy

- The temperature of a gas is proportional to the average kinetic energy of its particles.
- Thermal energy is a measure of the internal motion of the particles.
- Thermometers reach thermal equilibrium with the objects they contact, then their temperature-dependent property is measured.
- The Celsius and Kelvin temperature scales are used in scientific work. The magnitude of one kelvin is equal to the magnitude of one Celsius degree.
- At absolute zero, no more thermal energy can be removed from a substance.
- Heat is energy transferred because of a difference in temperature. Heat flows naturally from a hot to a cold body.
- Specific heat is the quantity of heat required to raise the temperature of one kilogram of a substance by one kelvin.
- In a closed, isolated system, heat may flow and change the thermal energy of parts of the system, but the total energy of the system is constant.

12.2 Change of State and Laws of Thermodynamics

- The heat of fusion is the quantity of heat needed to change one kilogram of a substance from its solid to liquid state at its melting point.
- The heat of vaporization is the quantity of heat needed to change one kilogram of a substance from its liquid to gaseous state at its boiling point.
- Heat transferred during a change of state doesn't change the temperature.
- The first law of thermodynamics states that the total increase in energy of a system is the sum of the heat added to it and the work done on it.
- A heat engine continuously converts thermal energy to mechanical energy.
- A heat pump or refrigerator uses mechanical energy to transfer heat from a region of lower temperature to one of higher temperature.
- Entropy is a measure of the disorder of a system.
- The entropy of the universe always increases, even if the entropy of a system may decrease because of some action taken to increase its order.

Reviewing Concepts

Section 12.1

1. Explain the difference between the mechanical energy of a ball, its thermal energy, and its temperature. Give an example.
2. Can temperature be assigned to a vacuum? Explain.
3. Do all of the molecules or atoms in a liquid have the same speed?
4. Your teacher just told your class that the temperature of the core of the sun is 1.5×10^7 degrees.
   a. Sally asks whether this is the Kelvin or Celsius scale. What will be the teacher's answer?
   b. Would it matter whether you use the Celsius or Fahrenheit scale?
5. Is your body a good judge of temperature? On a cold winter day, a metal doorknob feels much colder to your hand than the wooden door. Explain why this is true.
6. Do we ever measure heat transfer directly? Explain.

7. How does heat flow when a warmer object is in contact with a colder object? Do the two have the same temperature changes?

Section 12.2

8. Can you add thermal energy to an object without increasing its temperature? Explain.
9. When wax freezes, is energy absorbed or released by the wax?
10. Why does water in a canteen stay cooler if it has a canvas cover that is kept wet?
11. Which process occurs at the coils of a running air conditioner inside the house, vaporization or condensation? Explain.
12. Which situation has more entropy, an unbroken egg or a scrambled egg?

Applying Concepts

13. Sally is cooking pasta in a pot of boiling water. Will the pasta cook faster if the water is boiling vigorously or if it is boiling gently?
14. On the following pairs of scales, is there one temperature on each scale that has the same value? $T_F = 9/5\ T_C + 32$
    a. Celsius and Fahrenheit
    b. Kelvin and Fahrenheit
    c. Celsius and Kelvin
15. Which liquid would an ice cube cool faster, water or methanol? Explain.
16. Explain why the high specific heat of water makes it desirable for use in hot-water heating systems.
17. Equal masses of aluminum and lead are heated to the same temperature. The pieces of metal are placed on a block of ice. Which metal melts more ice? Explain.
18. Why do easily vaporized liquids, such as acetone and methanol, feel cool to the skin?
19. Explain why fruit growers spray their trees with water, when frost is expected, to protect the fruit from freezing.
20. Two blocks of lead have the same temperature. Block A has twice the mass of block B. They are dropped into identical cups of water of equal temperature. Will the two cups of water have equal temperatures after equilibrium is achieved? Explain.

Problems

Section 12.1

LEVEL 1

21. Liquid nitrogen boils at 77K. Find this temperature in degrees Celsius.
22. The melting point of hydrogen is $-259.14°C$. Find this temperature in kelvins.
23. How much heat is needed to raise the temperature of 50.0 g of water from 4.5°C to 83.0°C?
24. A 5.00×10^2-g block of metal absorbs 5016 J of heat when its temperature changes from 20.0°C to 30.0°C. Calculate the specific heat of the metal.
25. A 4.00×10^2-g glass coffee cup is at room temperature, 20.0°C. It is then plunged into hot dishwater, 80.0°C. If the temperature of the cup reaches that of the dishwater, how much heat does the cup absorb? Assume the mass of the dishwater is large enough so its temperature doesn't change appreciably.
26. A 1.00×10^2-g mass of tungsten at 100.0°C is placed in 2.00×10^2g of water at 20.0°C. The mixture reaches equilibrium at 21.6°C. Calculate the specific heat of tungsten.
27. A 6.0×10^2-g sample of water at 90.0°C is mixed with 4.00×10^2g of water at 22.0°C. Assume no heat loss to the surroundings. What is the final temperature of the mixture?
28. A 10.0-kg piece of zinc at 71.0°C is placed in a container of water. The water has a mass of 20.0 kg and has a temperature of 10.0°C before the zinc is added. What is the final temperature of the water and zinc?

LEVEL 2

29. To get a feeling for the amount of energy needed to heat water, recall from **Table 11–1** that the kinetic energy of a compact car moving at 100 km/h is 2.9×10^5 J. What volume of water (in liters) would 2.9×10^5 J of energy warm from room temperature (20°C) to boiling (100°C)?
30. A 3.00×10^2-W electric immersion heater is used to heat a cup of water. The cup is made of glass and its mass is 3.00×10^2 g. It contains

250 g of water at 15°C. How much time is needed to bring the water to the boiling point? Assume that the temperature of the cup is the same as the temperature of the water at all times and that no heat is lost to the air.

31. A 2.50×10^2-kg cast-iron car engine contains water as a coolant. Suppose the engine's temperature is 35.0°C when it is shut off. The air temperature is 10.0°C. The heat given off by the engine and water in it as they cool to air temperature is 4.4×10^6 J. What mass of water is used to cool the engine?

Section 12.2

LEVEL 1

32. Years ago, a block of ice with a mass of about 20.0 kg was used daily in a home icebox. The temperature of the ice was 0.0°C when delivered. As it melted, how much heat did a block of ice that size absorb?

33. A 40.0-g sample of chloroform is condensed from a vapor at 61.6°C to a liquid at 61.6°C. It liberates 9870 J of heat. What is the heat of vaporization of chloroform?

34. A 750-kg car moving at 23 m/s brakes to a stop. The brakes contain about 15 kg of iron which absorbs the energy. What is the increase in temperature of the brakes?

LEVEL 2

35. How much heat is added to 10.0 g of ice at −20.0°C to convert it to steam at 120.0°C?

36. A 4.2-g lead bullet moving at 275 m/s strikes a steel plate and stops. If all its kinetic energy is converted to thermal energy and none leaves the bullet, what is its temperature change?

37. A soft drink from Australia is labeled "Low Joule Cola." The label says "100 mL yields 1.7 kJ." The can contains 375 mL. Sally drinks the cola and then wants to offset this input of food energy by climbing stairs. How high would Sally have to climb if she has a mass of 65.0 kg?

Critical Thinking Problems

38. Your mother demands that you clean your room. You know that reducing the disorder of your room will reduce its entropy. But, you also know that the entropy of the universe cannot be decreased. Describe processes that cause an increase in entropy as you obey your mother.

Going Further

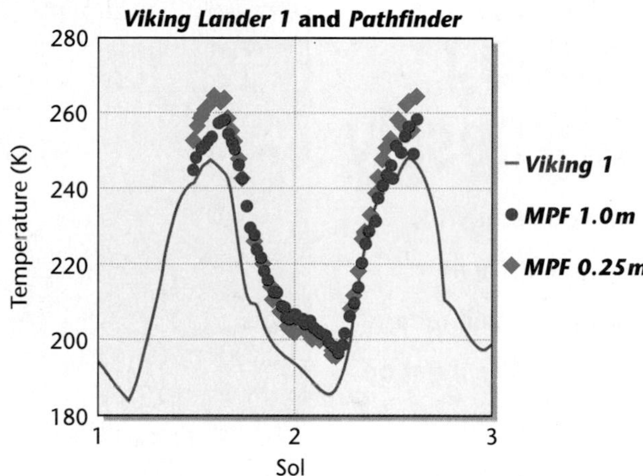

Interpreting Graphs The daily cycle of temperatures for the first two robot explorers of Mars is shown plotting the temperature versus time of the solar day starting at midnight. The points labeled "MPF" were recorded by the *Mars Pathfinder* in 1997 at various distances above the ground. Interpret this graph by finding the lowest night temperatures and highest day temperatures recorded by each planetary probe on both the Celsius and Fahrenheit scales.

interNET CONNECTION

Follow the link on the Glencoe Homepage at **www.glencoe.com/sec/science** to find out more about this chapter.

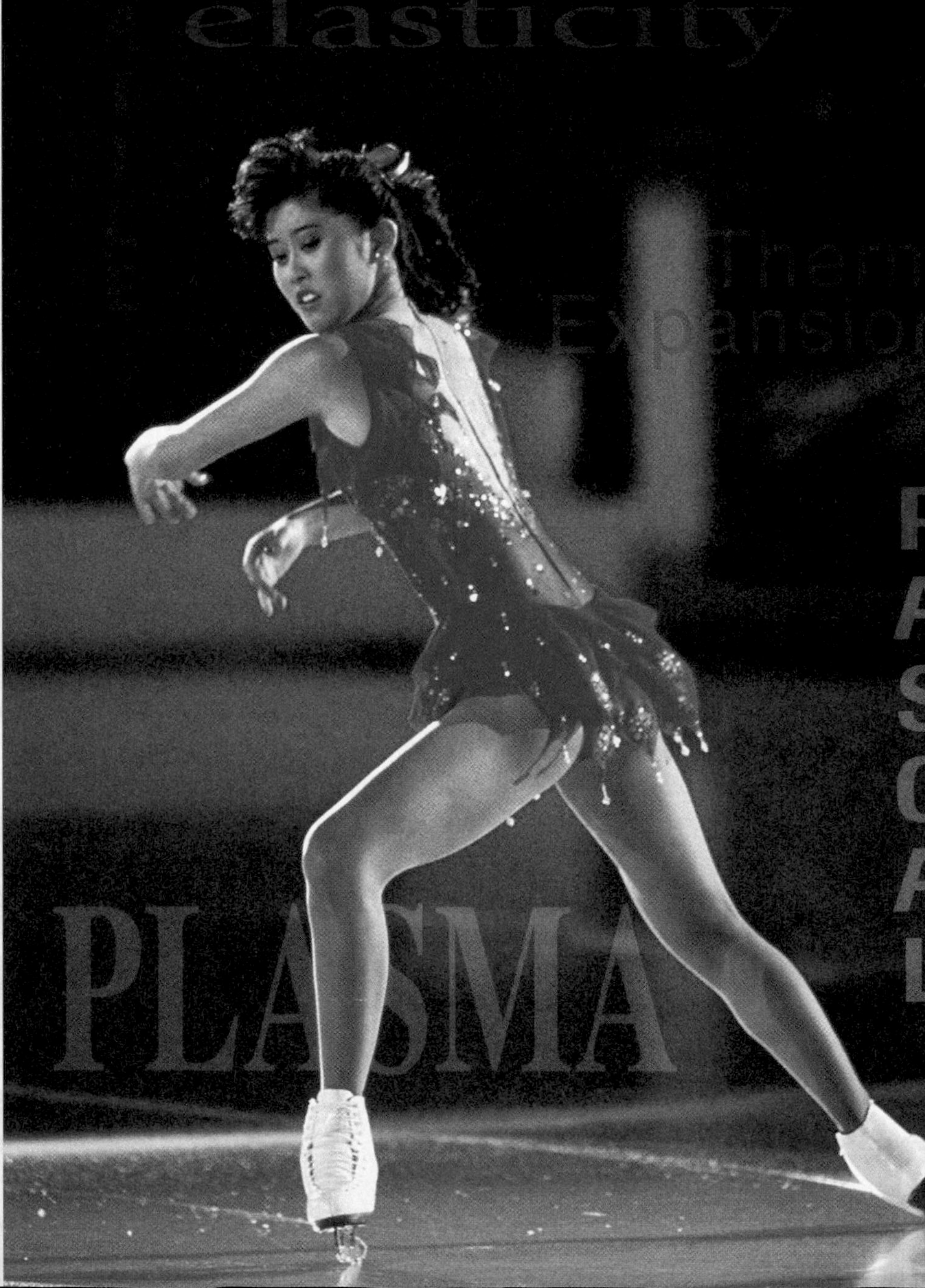

Skating Effortlessly

From your study of Newton's laws of motion, you will recognize that there must be very little friction between the skater's blades and the ice. What properties of ice and water make this possible?

CHAPTER

13 States of Matter

You are already familiar with the three common states of matter: solid, liquid, and gas. You breathe air, a gas; you drink and swim in water, a liquid; and you depend on solid objects for transport and shelter. A less-understood state of matter is the plasma state. At this moment, you are probably reading with the help of this fourth state of matter. Whether your reading light is from a fluorescent lamp or the sun, the light's source is plasma. This somewhat mysterious state of matter is the most common state in the universe.

In this chapter, your exploration of matter will go far beyond most everyday, casual observations. You will explore how liquids and gases create pressure, how it is possible for heavy ships to float on water, and how airplanes fly. You will also find out what causes some solids to be elastic, and learn that the expansion and contraction of matter that occurs when a solid's temperature changes can be important to your everyday life.

WHAT YOU'LL LEARN

- You will describe Pascal's, Archimedes', and Bernoulli's principles and relate them to everyday applications.
- You will explain how temperature changes cause the expansion and contraction of solids.

WHY IT'S IMPORTANT

- You will be able to explain how brakes stop a car.
- With an understanding of Archimedes' and Bernoulli's principles, it is possible to design larges ships, submarines, and aircraft.
- Thermal expansion allows you to control temperatures with thermostats.

interNET CONNECTION

Follow the link for this chapter on the Glencoe Homepage at **www.glencoe.com/sec/science** to find out more about the states of matter.

13.1 The Fluid States

If you think further about the air you breathe, a gas, and the water you drink, a liquid, you will realize they have some things in common. Both air and water flow and have indefinite shapes—unlike concrete, which is a solid. Because of their common characteristics, gases and liquids are called fluids.

OBJECTIVES

- **Describe** how fluids create pressure and **relate** Pascal's principle to some everyday occurrences.
- **Apply** Archimedes' and Bernoulli's principles.
- **Explain** how forces within liquids cause surface tension and capillary action, and **relate** the kinetic model to evaporation and condensation.

Properties of Fluids

How does a fluid differ from a solid? Suppose you put an ice cube in a glass. The ice cube has a certain mass and shape, and neither of these quantities depends on the size or shape of the glass. What happens, however, when the ice melts? Its mass remains the same, but its shape changes. The water flows to take the shape of its container and forms a definite upper surface, as in **Figure 13–1.** If you boiled the water, it would change into a gas, water vapor, and it also would flow and expand to fill whatever container it was in. But the water vapor would not have any definite surface. Both liquids and gases are fluids. **Fluids** are materials that flow and have no definite shape of their own.

FIGURE 13–1 The ice cube, a solid, has a definite shape. But water, a fluid, takes the shape of its container.

Pressure and fluids

Can Newton's laws of motion and the laws of conservation of momentum and energy be applied to fluids? In most cases, the answer is yes. But in some cases, the mathematics is so complicated that even the most advanced computers can't reach definite solutions. By studying the concept of pressure, you can learn more about the motion of fluids. Applying force to a surface is **pressure**. Pressure, which can be represented by the following equation, is the force on a surface divided by the area of the surface.

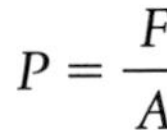

$$P = \frac{F}{A}$$

The force (F) on the surface is assumed to be perpendicular to the surface area (A).

If you stand on ice, the ice exerts on your body an upward normal force that has the same magnitude as your weight. The upward force is spread over the area of your body that touches the ice, which is the soles of your feet. **Figure 13–2** helps illustrate the relationships between force, area, and pressure.

Pressure (P) is a scalar quantity. In the SI system, the unit of pressure is the **pascal** (Pa), which is one newton per square meter. Because the pascal is a small unit, the kilopascal (kPa), equal to 1000 Pa, is more commonly used. **Table 13–1** shows how pressures vary in different situations, and **Figure 13–3** shows a few commonly used instruments for measuring pressure.

How does matter exert pressure?

Imagine that you are standing on the surface of a frozen lake. Your feet exert forces on the ice, a solid made up of vibrating water molecules. The forces that hold the water molecules apart cause the ice to exert upward forces on your feet that equal your weight.

What if the ice melts? Most of the bonds between the water molecules are broken and although the molecules continue to vibrate and remain close to each other they can also slide past one another. If the lake is deep enough, the water will eventually surround your entire body. The collisions of the moving water molecules continue to exert forces on your body.

FIGURE 13–2 Pressure is the force exerted on a unit area of a surface. Similar forces may produce vastly different pressures; for example, the pressure on the ground under the high heel of a woman's shoe is far greater than that under an elephant's foot.

TABLE 13–1

Some Typical Pressures

Location	Pressure (Pascals)
The center of the sun	2×10^{16}
The center of Earth	4×10^{11}
The deepest ocean trench	1.1×10^{8}
An automobile tire	2×10^{5}
Standard atmosphere	1.01325×10^{5}
Blood pressure	1.6×10^{4}
Air pressure on top of Mt Everest	4×10^{4}
Atmospheric pressure on Mars	7×10^{2}
The best vacuum	1×10^{-12}

Gas particles and pressure

The force exerted by a gas can be understood by using the kinetic-molecular theory of gases. According to this theory, gases are made up of very small particles, the same particles that make up solids and liquids. But the particles are now widely separated, in constant, random motion at high speed, and making elastic collisions with each other. When gas particles hit a surface, they rebound without losing kinetic energy. The forces exerted by these collisions result in gas pressure on the surface.

Atmospheric pressure

On every square centimeter of Earth's surface at sea level, the atmosphere exerts a force of approximately 10 N, about the weight of a 1-kg object. The pressure of Earth's atmosphere on your body is so well balanced by your body's outward forces that you seldom notice it. You become aware of this pressure only when your ears "pop" as the result of pressure changes when you ride an elevator in a tall building, drive up a steep mountain road, or fly in an airplane. Atmospheric pressure is about 10 N divided by the area of 1 cm^2, or 10^{-4} m^2, which is about 1.0×10^5 N/m^2, or 100 kPa.

a

b

FIGURE 13–3 A tire gauge measures tire air pressure **(a)**, and an aircraft altimeter uses pressure to indicate altitude **(b)**.

Example Problem

Calculating Pressure

A woman weighs 495 N and wears shoes that touch the ground over an area of 412 cm^2.

a. What is the average pressure in kPa that her shoes exert on the ground?
b. How does the pressure change when she stands on only one foot?
c. What is the pressure if she puts all her weight on the heel of one shoe with the area of the high heel of 2.0 cm^2?

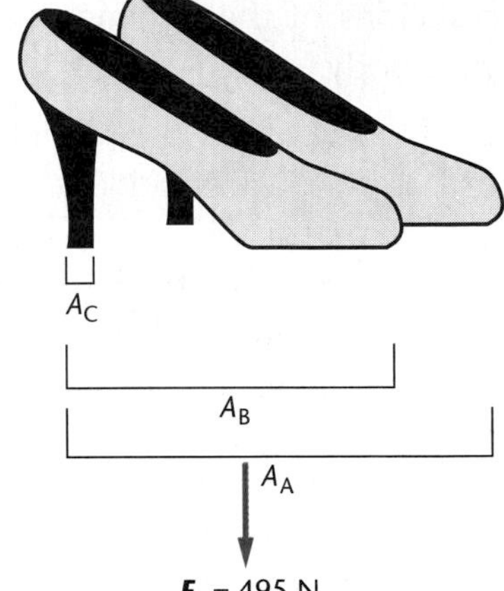

Sketch the Problem

- Sketch the shoes, labeling A_A, A_B, and A_C.
- Show the vector for the force the woman exerts on the ground, and indicate the area on which the force is exerted.

Calculate Your Answer

Known:

$F = 495$ N

$A_A = 412\ cm^2$

$A_B = 206\ cm^2$

$A_C = 2.0\ cm^2$

Unknown:

$P_A, P_B, P_C = ?$

Strategy:

Convert area to the correct units, m^2.

Find each pressure by dividing force by each area.

$$P = \frac{F}{A}$$

Calculations:

a. $$A_A = \frac{412\ cm^2 \times (1\ m)^2}{(100\ cm)^2} = 0.0412\ m^2$$

$$P_A = \frac{495\ N}{0.0412\ m^2}$$

$$= \frac{1.20 \times 10^4\ N/m^2 \times 1\ kPa}{1000 N/m^2} = 12.0\ kPa$$

b. $A_B = 0.0206\ m^2$

$$P_B = \frac{495\ N}{0.0206\ m^2}$$

$$= 24.0\ kPa$$

c. $A_C = 0.00020\ m^2$

$$P_C = \frac{495\ N}{0.00020\ m^2}$$

$$= 2500\ kPa$$

Check Your Answer

- Are your units correct? The units for pressure should be Pa, and $1\ N/m^2 = 1$ Pa.
- You can see that keeping the force the same but reducing the area increases the pressure.

Practice Problems

1. The atmospheric pressure at sea level is about 1.0×10^5 Pa. What is the force at sea level that air exerts on the top of a typical office desk, 152 cm long and 76 cm wide?
2. A car tire makes contact with the ground on a rectangular area of 12 cm by 18 cm. The car's mass is 925 kg. What pressure does the car exert on the ground?
3. A lead brick, $5.0 \times 10.0 \times 20.0$ cm, rests on the ground on its smallest face. What pressure does it exert on the ground? (Lead has a density of 11.8 g/cm^3.)
4. In a tornado, the pressure can be 15% below normal atmospheric pressure. Sometimes a tornado can move so quickly that this pressure drop can occur in one second. Suppose a tornado suddenly occurred outside your front door, which is 182 cm high and 91 cm wide. What net force would be exerted on the door? In what direction would the force be exerted?

Foot Pressure

How much pressure do you exert when standing on the ground with one foot? Is it more or less than air pressure? Estimate your weight in newtons. **Hint:** 500 N = 110 lb. Stand on a piece of paper and have a lab partner trace the outline of your foot. Draw a rectangle that has about the same area as the outline.

Using SI Measurement Calculate the area of your rectangle in square meters, and use the definition of pressure to estimate your pressure. $P = F/A$.

Fluids at Rest

The fluid most familiar to you is probably liquid water. Other fluids include honey, oil, tar, and air. Think about how these fluids are alike, and how they differ. Every fluid has its own properties, such as sticky or watery. To make it easier to compare the behavior of fluids, you can use an ideal fluid as a model. An ideal fluid has no internal friction among its particles. For the following discussion, imagine that you are dealing with an ideal fluid.

Pascal's principle

You are now familiar with how the atmosphere exerts pressure. If you have ever dived deep into a swimming pool or lake, you know that water also exerts pressure. Your body is sensitive to water pressure. You probably noticed that the pressure you felt on your ears did not depend on whether your head was upright or tilted. If your body is vertical or horizontal, the pressure is nearly the same on all parts of your body.

Blaise Pascal (1623–1662), a French physician, noted that the shape of a container has no effect on the pressure of the fluid it contains at any given depth. He discovered that any change in pressure applied at any point on a confined fluid is transmitted undiminished throughout the fluid. This fact became known as **Pascal's principle.** Every time you squeeze a tube of toothpaste, you demonstrate Pascal's principle. The pressure your fingers exert at the bottom of the tube is transmitted through the toothpaste, forcing the paste out at the top.

When fluids are used in machines, such as hydraulic lifts, to multiply forces, Pascal's principle is being applied. In a hydraulic system, a fluid is confined to two connecting chambers, as shown in

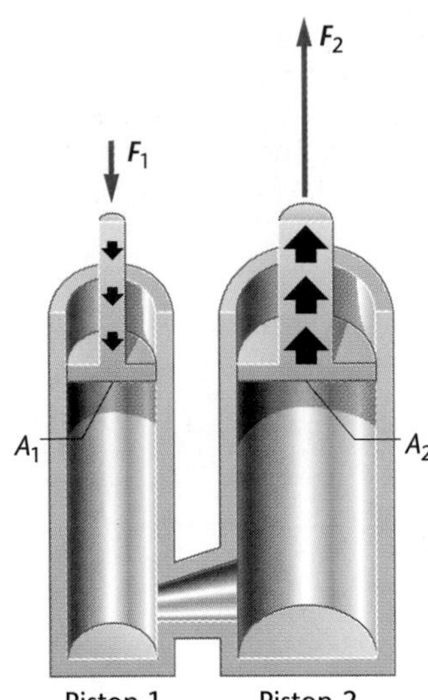

FIGURE 13–4 The pressure exerted by the force of the small piston is transmitted throughout the fluid and results in a multiplied force on the larger piston.

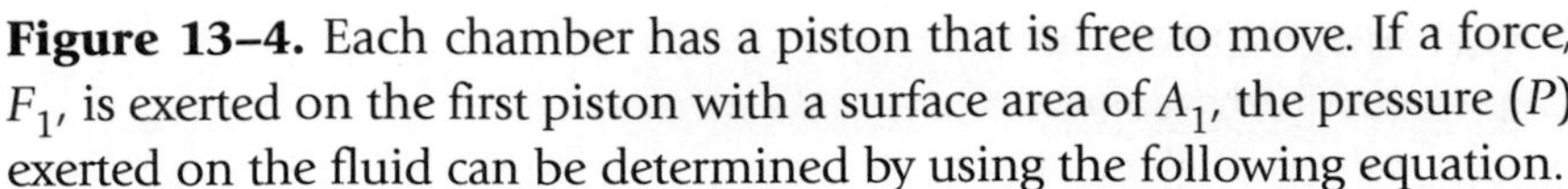

Figure 13–4. Each chamber has a piston that is free to move. If a force, F_1, is exerted on the first piston with a surface area of A_1, the pressure (P) exerted on the fluid can be determined by using the following equation.

$$P_1 = \frac{F_1}{A_1}$$

According to Pascal's principle, pressure is transmitted without change throughout a fluid, and therefore, the pressure exerted by the fluid on the second piston, with a surface area A_2, can also be determined.

$$P_2 = \frac{F_2}{A_2}$$

Because the pressure P_2 is equal in value to P_1, you can determine the force exerted by the hydraulic lift. Because $F_1/A_1 = F_2/A_2$, then the force exerted by the second piston is shown by the following equation.

$$F_2 = \frac{F_1 A_2}{A_1}$$

Practice Problems

5. Dentists' chairs are examples of hydraulic-lift systems. If a chair weighs 1600 N and rests on a piston with a cross-sectional area of 1440 cm^2, what force must be applied to the smaller piston with a cross-sectional area of 72 cm^2 to lift the chair?

Swimming Under Pressure

While you are swimming, you can observe another property of fluids. You feel the pressure of the water as you dive deeper. Recall that the pressure in a fluid is the same in all directions, and that pressure is the force on a surface divided by its area. The downward pressure of the water is equal to the weight, F_g, of the column of water above its surface area, A.

$$P = \frac{F_g}{A}$$

You can find the pressure of the water above you by applying this equation. The weight of the column of water is $F_g = mg$, and the mass is equal to the density, ρ, of the water times its volume, $m = \rho V$. You also know that the volume of the water is the area of the column times its height, $V = Ah$. Therefore, $F_g = \rho Ahg$. The pressure can now be determined.

$$P = \frac{F_g}{A} = \frac{\rho Ahg}{A} = \rho hg$$

The pressure of the water on a body depends on the density of the fluid, its depth, and g. As noted by Pascal, the shape of the container, such as those shown in **Figure 13–5**, has no effect on pressure.

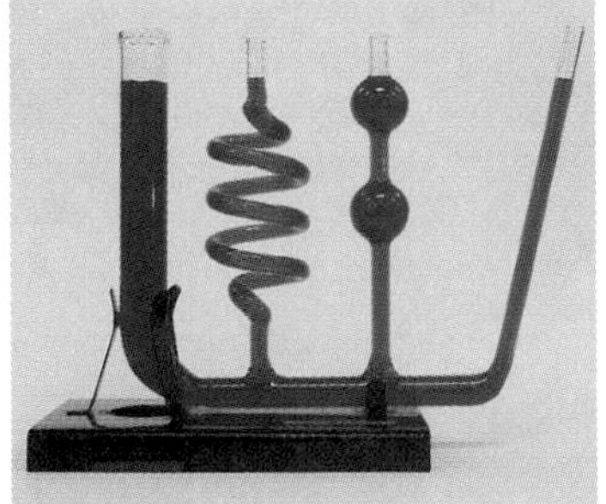

FIGURE 13–5 Equilibrium tubes show that a container's shape has no effect on pressure.

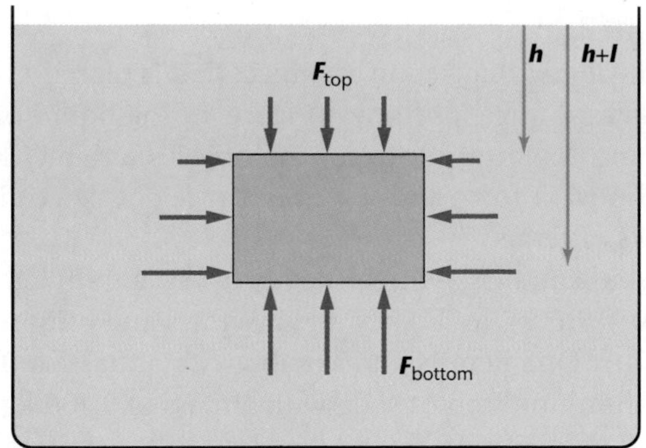

FIGURE 13–6 A fluid exerts a greater upward force on the bottom of an immersed object than the downward force on the top of the object. The net upward force is called the buoyant force.

Buoyancy

The increase of pressure with depth has an important consequence. It allows you to swim. That is, it creates an upward force, called the **buoyant force,** on all objects. By comparing the buoyant force on an object with its weight, you will know if the object will sink or float.

Suppose that a box is immersed in water. It has a height l, and its top and bottom each have surface area A. Its volume, then, is $V = lA$. Water pressure exerts forces on all sides, as shown in **Figure 13–6.** Will the box sink or float? As you know, the pressure on the box depends on its depth, h. To find out if the box will float in the water, you will need to determine how the pressure on the top of the box compares with the pressure from below the box. Compare these two equations.

$$F_{\text{top}} = P_{\text{top}}A = \rho hgA$$

$$F_{\text{bottom}} = P_{\text{bottom}}A = \rho(l + h)gA$$

On the four vertical sides, the forces are equal in all directions, so there is no net horizontal force. The upward force on the bottom is larger than the downward force on the top, so there is a net upward force. The buoyant force can now be determined.

$$F_{\text{buoyant}} = F_{\text{bottom}} - F_{\text{top}}$$

$$= \rho(l + h)gA - \rho hgA$$

$$= \rho lgA = \rho Vg$$

This shows that the net upward force is proportional to the volume of the box. This volume is equal to the volume of the fluid that was displaced, or pushed out of the way, by the box. Therefore, the magnitude of the buoyant force, ρVg, is equal to the weight of the fluid displaced by the object. This relationship was discovered in 212 B.C. by the Greek scientist Archimedes. **Archimedes' principle** states that an object immersed in a fluid has an upward force on it equal to the weight of the fluid displaced by the object. The force does not depend on the weight of the object, only on the weight of the displaced fluid.

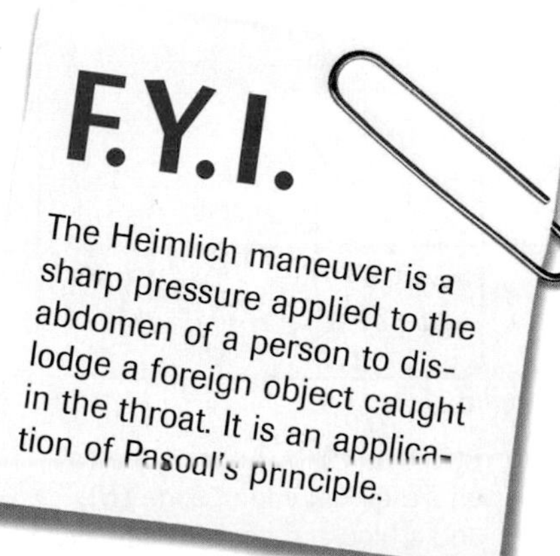

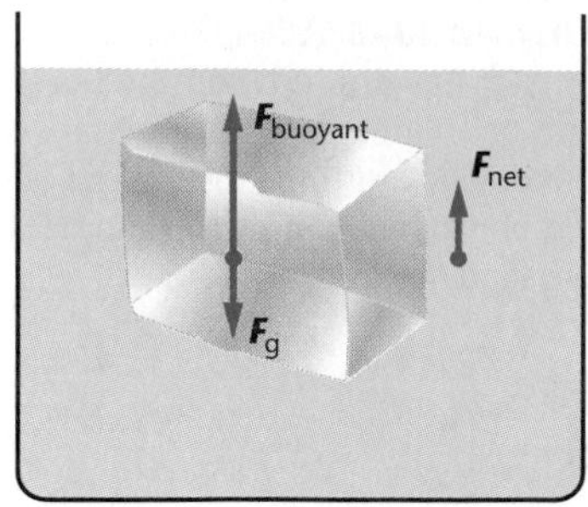

a

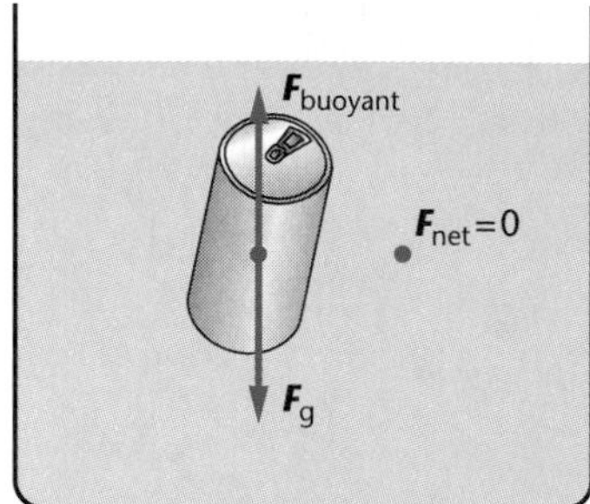

b

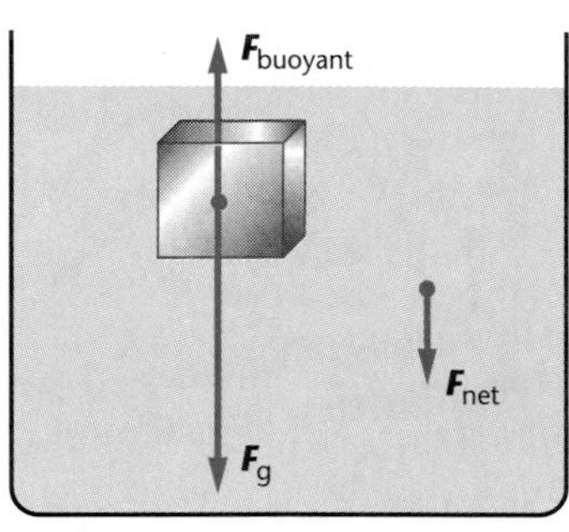

c

FIGURE 13–7 An ice cube **(a)**, an aluminum can of soda **(b)**, and a block of steel **(c)** all have the same volume, displace the same amount of water, and experience the same buoyant force. However, because their weights are different, the net forces on the three objects are also different.

Sink or float?

What are the forces that act on an object that is placed in a fluid? One force is the upward force, or buoyant force, of the liquid on the object. There is also the downward force of the object's weight. The difference between the buoyant force and the object's weight determines whether an object sinks or floats.

Suppose you submerge three items in a tank filled with water ($\rho_{water} = 1.00 \times 10^3$ kg/m^3). They all have the same volume, 100 cm^3 or 1.00×10^{-4} m^3. One item is an ice cube with a mass of 0.090 kg. The second is an aluminum can of soda with a mass of 0.100 kg. The third is a block of steel with a mass of 0.90 kg. How will each item move in the water? The upward force on all three objects, as shown in **Figure 13–7,** is the same, because all displace the same weight of water. The upward force can be calculated as shown by the following equation.

$$F_{buoyant} = \rho_{water}Vg$$

$$= (1.00 \times 10^3 \text{ kg/m}^3)(1.00 \times 10^{-4} \text{ m}^3)(9.80 \text{ m/s}^2)$$

$$= 0.980 \text{ N}.$$

The weight of the ice cube ($F_g = mg$) is 0.88 N, so there is a net upward force, and the ice cube will rise. When it reaches the surface there will still be a net upward force that will lift part of the ice cube out of the water. As a result, less water will be displaced, and the upward force will be reduced. The ice cube will float with just enough volume in the water so that the weight of water displaced equals the weight of the ice cube. An object will float because its density is less than the density of the fluid.

The weight of the soda can is 0.98 N, the same as the weight of the water displaced. There is, therefore, no net force, and the can will remain wherever it is placed in the water. It has neutral buoyancy. Objects at neutral buoyancy are described as being weightless; this property is similar to that experienced by astronauts in orbit. This environment is simulated for astronauts when they train in swimming pools.

The weight of the block of steel is 8.8 N. It has a net downward force, so it will sink to the bottom of the tank. The net downward force, its apparent weight, is less than its real weight. All objects in a liquid, even those that sink, have an apparent weight that is less than when the object is in air. The apparent weight can be expressed by the equation $F_{apparent} = F_g - F_{buoyant}$.

As a result of the buoyant force of Archimedes' principle, ships can be made of steel and still float as long as the hull is hollow and large enough so that the density of the ship is less than the density of water. You may have noticed that a ship loaded with cargo rides much lower in the water than a ship with an empty cargo hold.

Submarines take advantage of Archimedes' principle as water is pumped into or out of a number of different chambers to change the submarine's net vertical force, causing it to rise or sink. Archimedes' principle also explains the buoyancy of fishes that have air bladders. The density of a fish is about the same as that of water. Fishes that have an air bladder can expand or contract the bladder. By expanding its air bladder, a fish can move upward in the water. The expansion displaces more water and increases the buoyant force. The fish moves downward by contracting the volume of the air bladder.

Example Problem

Archimedes' Principle

A cubic decimeter, 1.00×10^{-3} m^3, of aluminum is submerged in water. The density of aluminum is 2.70×10^3 kg/m^3.

a. What is the magnitude of the buoyant force acting on the metal?

b. What is the apparent weight of the block?

Sketch the Problem

- Sketch the cubic decimeter of aluminum immersed in water.
- Show the upward buoyant force and the downward force due to gravity acting on the aluminum.

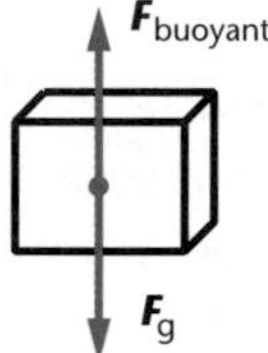

Calculate Your Answer

Known:

$V = 1.00 \times 10^{-3}$ m^3

$\rho_{aluminum} = 2.70 \times 10^3$ kg/m^3

$\rho_{water} = 1.00 \times 10^3$ kg/m^3

Unknown:

$F_{buoyant} = ?$

$F_{apparent} = ?$

Strategy:

a. Calculate the buoyant force on the aluminum block.

b. The aluminum's apparent weight equals its weight minus the upward buoyant force.

Calculations:

$F_{buoyant} = \rho_{water} V g$

$= (1.00 \times 10^3 \text{ kg/m}^3)(1.00 \times 10^{-3} \text{ m}^3)(9.80 \text{ m/s}^2)$

$= 9.80$ N

$F_g = mg = \rho_{aluminum} V g$

$= (2.70 \times 10^3 \text{ kg/m}^3)(1.00 \times 10^{-3} \text{ m}^3)(9.80 \text{ m/s}^2)$

$= 26.5$ N

$F_{apparent} = F_g - F_{buoyant}$

$= 26.5 \text{ N} - 9.80 \text{ N} = 16.7$ N

Check Your Answer

- Are your units correct? The forces and apparent weight are in newtons, as expected.
- Is the magnitude realistic? The buoyant force is about one third the weight of the aluminum, a sensible answer because the density of water is about one third that of aluminum.

Physics Lab

Float or Sink?

Problem

How can you measure the buoyancy of objects?

Materials

beaker
water
film canister with lid
25 pennies
250-g spring scale
pan balance

Procedure

1. Measure and calculate the volume of a film canister. Record the volume in a data table like the one shown.
2. Fill the canister with water. Find the mass of the filled canister on the pan balance. Record the value in your data table.
3. Empty the canister of water.
4. Place a few pennies in the canister and put the top on tightly. Find its mass and record the value in your data table.
5. Put the capped film canister into a beaker of water to see if it floats.
6. If it floats, estimate the percentage that is under water. Record this amount in your data table.
7. If it sinks, use the spring scale to measure the apparent weight while it is under water (but not touching the bottom). Record this value in your data table.
8. Repeat steps 4 through 7 using different numbers of pennies for each trial.

Data and Observations		
Volume of canister = ____cm^3		
Mass of canister with water = ____g		
Floaters		
Mass with pennies	% below water	Density
Sinkers		
Mass with pennies	Apparent weight	Density

9. Calculate the density for each trial in g/cm^3.

Analyze and Conclude

1. **Recognizing Spatial Relationships** Look closely at the mass of the floaters and the percentages below the water. What seems to be the rule?
2. **Comparing and Contrasting** Look closely at the sinkers. How much lighter are the canisters when weighed underwater?

Apply

1. Explain why a steel-hulled boat can float, even though it is quite massive.
2. Icebergs float in salt water (density 1.03 g/cm^3) with 1/9 of their volume above water. What is the density of an iceberg?

Practice Problems

6. A girl is floating in a freshwater lake with her head just above the water. If she weighs 600 N, what is the volume of the submerged part of her body?
7. What is the tension in a wire supporting a 1250-N camera submerged in water? The volume of the camera is 8.3×10^{-2} m^3.

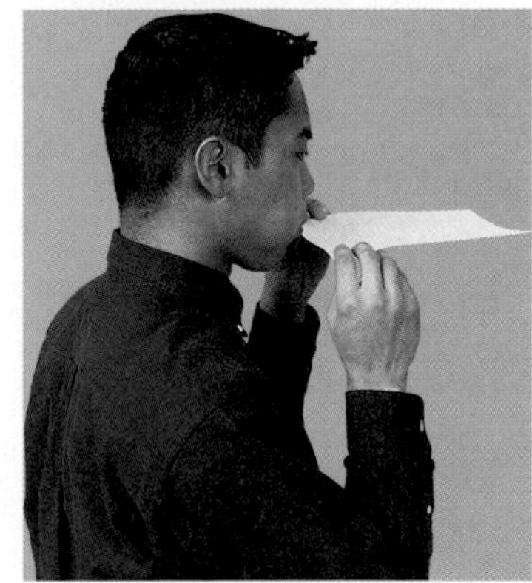

FIGURE 13–8 Blowing across the surface of a sheet of paper demonstrates Bernoulli's principle.

Fluids in Motion

To see an effect of moving fluids, try the experiment in **Figure 13–8.** Hold a strip of notebook paper just under your lower lip. Now blow hard across the top surface. The strip of paper will rise. This is because the pressure on top of the paper, where air is flowing rapidly is lower than the pressure beneath it, where air is not in motion.

The relationship between the velocity and pressure exerted by a moving fluid is described by **Bernoulli's principle,** named for the Swiss scientist Daniel Bernoulli (1700–1782). Bernoulli's principle states that as the velocity of a fluid increases, the pressure exerted by that fluid decreases.

When a fluid flows through a constriction, its velocity increases. You may have seen this happen as the water flow in a stream speeds up as it passes through narrowed sections. Consider a horizontal pipe completely filled with a smoothly flowing ideal fluid. If a certain mass of the fluid enters one end of the pipe, then an equal mass must come out the other end. Now consider a section of pipe with a cross section that becomes narrower, as shown in **Figure 13–9.** To keep the same mass of fluid moving through the narrow section in a fixed amount of time, the velocity of the fluid must increase. If the velocity increases, so does the kinetic energy. This means that net work must be done on the fluid. The net work is the difference between the work done on the mass of fluid to move it into the pipe and the work done by the fluid pushing the same mass out of the pipe. The work is proportional to the force on the fluid, which, in turn, depends on the pressure. If the net work is positive, the pressure at the input end of the section, where the velocity is lower, must be larger than the pressure at the output end, where the velocity is higher.

Most aircraft get part of their lift by taking advantage of Bernoulli's principle. Airplane wings are airfoils, devices designed to produce lift when moving through a fluid. The curvature of the top surface of a wing is greater than that of the bottom. As the wing travels through the air, the air moving over the top surface travels farther, and therefore must go faster than air moving past the bottom surface. The decreased air pressure created on the top surface results in a net upward pressure that produces an upward force on the wings, or lift, which helps to hold the

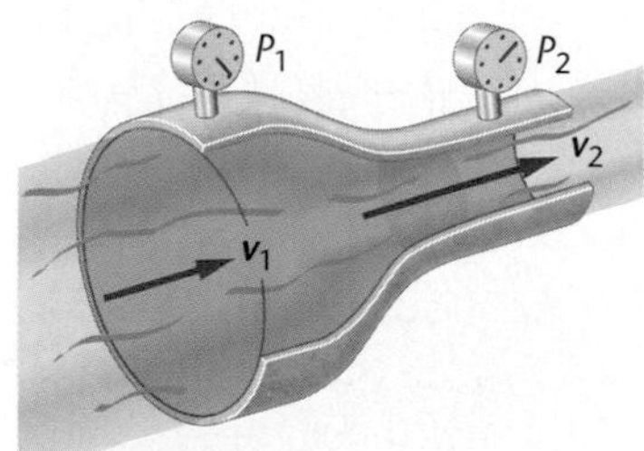

FIGURE 13–9 The pressure P_1 is greater than P_2 because v_1 is less than v_2.

FIGURE 13–10 The smooth streamlines shown in this computer simulation indicate smooth air flow around a jet fighter.

airplane aloft. Race cars use airfoils with a greater curvature on the bottom surface. The airfoils, called spoilers, produce a net downward pressure that helps to hold the rear wheels of the cars on the road at high speeds.

Streamlines

Automobile and aircraft manufacturers spend a great deal of time and money testing new designs in wind tunnels to ensure the greatest efficiency of movement through air, a fluid. The flow of fluids around objects is represented by streamlines, as shown in **Figure 13–10.** Objects require less energy to move through a smooth streamline flow.

Streamlines can best be illustrated by a simple demonstration. Imagine carefully squeezing tiny drops of food coloring into a smoothly flowing fluid. If the colored lines that form stay thin and well defined, the flow is said to be streamlined. Notice that if the flow narrows, the streamlines move closer together. Closely spaced streamlines indicate greater velocity, and therefore reduced pressure. If streamlines swirl and become diffused, the flow of the fluid is said to be turbulent. Bernoulli's principle does not apply to turbulent flow.

Forces Within Liquids

The liquids considered thus far have been ideal liquids, in which the particles are totally free to slide over and around one another. In real liquids, however, particles exert electromagnetic forces of attraction on each other. These forces, called **cohesive forces,** affect the behavior of liquids.

Surface tension

FIGURE 13–11 Rainwater beads up on a freshly waxed automobile because water drops have surface tension.

Have you ever noticed that dewdrops on spiderwebs and falling drops of oil are nearly spherical? What happens when it rains just after you have washed and waxed your car? The water drops bead up into rounded shapes, as shown in **Figure 13–11.** All of these phenomena are examples of surface tension. **Surface tension** is a result of the cohesive forces among the particles of a liquid. It is the tendency of the surface of a liquid to contract to the smallest possible area.

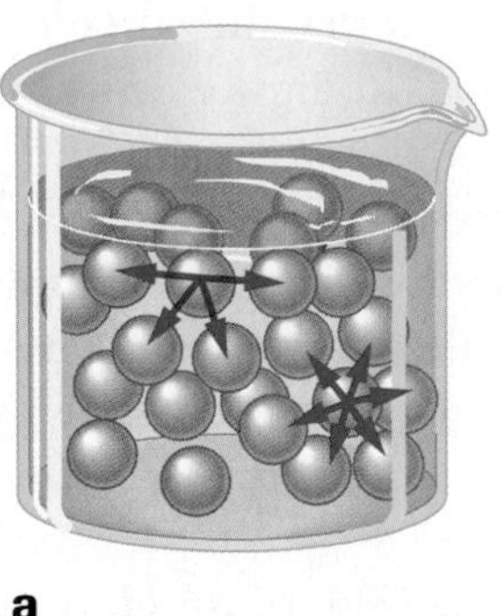

a

b

FIGURE 13–12 Molecules in the interior of a liquid are attracted in all directions **(a).** A water strider can walk on water because molecules at the surface have a net inward attraction that results in surface tension **(b).**

Notice that beneath the surface of the liquid shown in **Figure 13–12,** each particle of the liquid is attracted equally in all directions by neighboring particles, and even to the particles of the wall of the container. As a result, no net force acts on any of the particles beneath the surface. At the surface, however, the particles are attracted downward and to the sides, but not upward. There is a net downward force, which acts on the top layers and causes the surface layer to be slightly compressed. The surface layer acts like a tightly stretched rubber sheet or a film that is strong enough to support the weight of very light objects, such as water bugs. The surface tension of water also can support a steel razor blade, even though the density of steel is nine times greater than that of water. Try it!

Why does surface tension produce spherical drops? The force pulling the surface particles into the liquid causes the surface to become as small as possible. The shape that has the least surface for a given volume is a sphere. The higher the surface tension of the liquid, the more resistant the liquid is to having its surface broken. Liquid mercury has much stronger cohesive forces among its particles than water does. Mercury forms spherical drops, even when placed on a smooth surface. On the other hand, liquids such as alcohol and ether have weaker cohesive forces among their particles. A drop of either of these liquids flattens out on a smooth surface.

Capillary action

A force similar to cohesion is adhesion. **Adhesion** is the attractive force that acts between particles of different substances. Like cohesive forces, adhesive forces are electromagnetic in nature. If a piece of glass tubing with a small inner diameter is placed in water, the water rises inside the tube. The water rises because the adhesive forces between glass and water molecules are stronger than the cohesive forces between water molecules. This phenomenon is called **capillary action.** The water continues to rise until the weight of the water lifted balances the total adhesive force between the glass and water molecules. If the radius of the tube increases, the volume, and therefore the weight, of the water increases proportionally faster than does the surface area of the tube. For this reason, water is lifted higher in a narrow tube than in one that is wider.

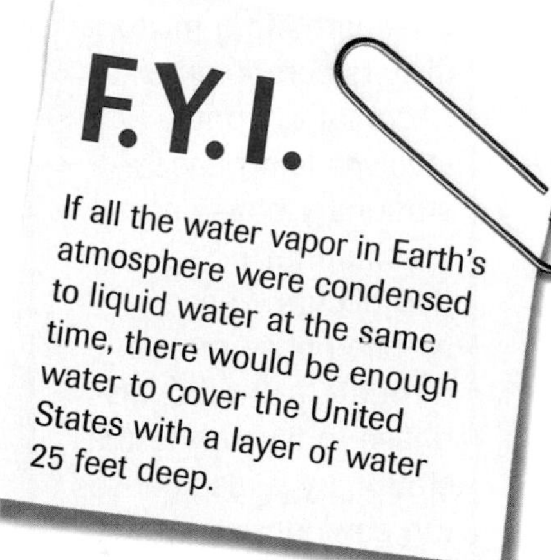

F.Y.I.
If all the water vapor in Earth's atmosphere were condensed to liquid water at the same time, there would be enough water to cover the United States with a layer of water 25 feet deep.

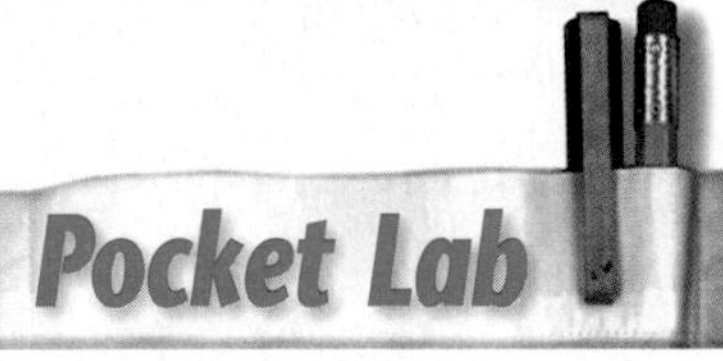

Floating?

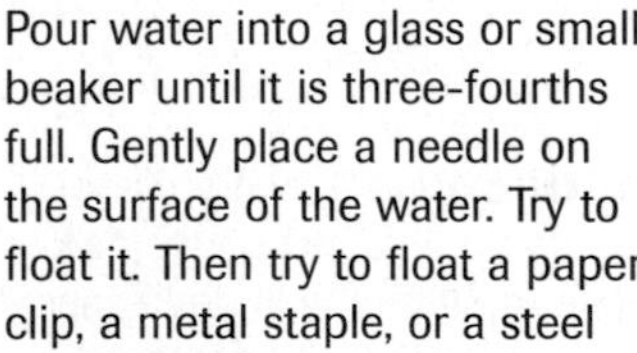

Pour water into a glass or small beaker until it is three-fourths full. Gently place a needle on the surface of the water. Try to float it. Then try to float a paper clip, a metal staple, or a steel razor blade.

Relate Cause and Effect Explain your results.

CHEMISTRY CONNECTION

Most types of matter you encounter every day are mixtures of two or more components. Chemists must often determine the identity of the components. Chromatography is a method of separating a mixture into its components. Chromatography involves two phases—a stationary phase and a mobile phase. Separation occurs when the mobile phase moves through the stationary phase. Adhesion to the stationary phase causes the components of the mixture to move with the mobile phase at different rates and eventually separate. Once separated, the components can be identified.

When a glass tube is placed in a beaker of water, the surface of the water climbs the outside of the tube, as shown in **Figure 13–13a.** This is because the adhesive forces between the glass molecules and water molecules are greater than the cohesive force between the water molecules. In contrast, the cohesive forces between the mercury molecules are greater than the adhesive forces between the mercury and glass molecules, and so the liquid does not climb the tube. These forces also cause the center of the mercury's surface to depress as shown in **Figure 13–13b.**

Molten wax rises in the wick of a candle because of capillary action. Paint moves up through the bristles of a brush for the same reason. It is also capillary action that causes water to move up through the soil and into the roots of plants.

Evaporation and condensation

What happens to a puddle of water on a hot, dry day? After a few hours the water is gone. Why? The particles in a liquid move at random speeds. The temperature of a liquid is dependent on the average kinetic energy of its particles. Some are moving rapidly; others are moving slowly. Suppose a fast-moving particle is near the surface of the liquid. If it can break through the surface layers, it will escape from the liquid. Because there is a net downward cohesive force at the surface, only the more energetic particles escape. The escape of particles from a liquid is called **evaporation.**

Evaporation has a cooling effect. On a hot day your body perspires. The evaporation of your sweat cools you down. In a puddle of water, evaporation causes the remaining liquid to cool down. Each time a particle with higher than average kinetic energy escapes from the liquid, the average kinetic energy of the remaining particles decreases. A decrease in kinetic energy is a decrease in temperature. You can test this cooling effect by pouring a little rubbing alcohol into the palm of your hand. Alcohol molecules evaporate easily because they have weak cohesive forces. The cooling effect is quite noticeable. A liquid that evaporates quickly is called a **volatile** liquid.

a

b

FIGURE 13–13 Water climbs the wall of this glass tube **(a),** while the mercury is depressed in the tube **(b).** The forces of attraction between mercury atoms are stronger than any adhesive forces between the mercury and the glass.

Have you ever wondered why humid days feel warmer than dry days at the same temperature? On a day that is humid, the water vapor content of the air is high. High humidity reduces the evaporation of perspiration from the skin, which is the body's primary mechanism for regulating body temperature.

Particles of liquid that have evaporated into the air can also return to the liquid phase if the kinetic energy or temperature decreases in a process called **condensation.** What happens if you bring a cold glass into a hot, humid area? The outside of the glass soon becomes coated with condensed water. Water molecules moving randomly in the air surrounding the glass strike the cold surface, and if they lose enough energy, the cohesive forces become strong enough to prevent their escape.

FIGURE 13–14 Warm moist surface air rises until it reaches a height where the temperature is at the point at which water vapor condenses and forms clouds.

The air above any body of water contains evaporated water vapor, which is water in the form of gas, as shown in **Figure 13–14.** If the temperature is reduced, the water vapor condenses around tiny dust particles in the air, producing droplets only 0.01 mm in diameter. A cloud of these droplets is called fog. Fog often forms when moist air is chilled by the cold ground. Fog also can be formed in your home. When a carbonated drink is opened, the sudden decrease in pressure causes the temperature of the liquid to drop, which condenses the surrounding water vapor.

13.1 Section Review

1.1 You have two boxes. One is 20 cm by 20 cm by 20 cm. The other is 20 cm by 20 cm by 40 cm.

a. How does the pressure of the air on the outside of the two boxes compare?

b. How does the magnitude of the total force of the air on the two boxes compare?

1.2 Does a full soft-drink can float or sink in water? Try it. Does it matter whether the drink is diet or not? All drink cans contain the same volume of liquid, 354 mL, and displace the same volume. What is the difference between a can that sinks and one that floats?

1.3 When a baby had a high fever, some doctors used to suggest gently sponging off the baby with rubbing alcohol. Why would this help?

1.4 **Critical Thinking** It was a hot, humid day. Beth sat on the patio with a glass of cold water. The outside of the glass was coated with water. Her younger sister, Jo, suggested that the water had leaked through the glass from the inside to the outside. Suggest an experiment that Beth could do to show Jo where the water came from.

13.2 The Solid State

How does a liquid differ from a solid? Solids are stiff. You can push them. Liquids are soft. Can you rest your finger on water? No, it goes right in. Of course, if you belly flop into a swimming pool, you recognize that a liquid may feel solid.

OBJECTIVES

- **Compare** solids, liquids, gases, and plasmas at a microscopic level, and **relate** their properties to their structures.
- **Explain** why solids expand and contract when the temperature changes.
- **Calculate** the expansion of solids and discuss the problems caused by expansion.

Solid Bodies

Under certain laboratory conditions, solids and liquids are not easily distinguished, as suggested by the computer images in **Figure 13–15.** Researchers have made clusters containing only a few dozen atoms. These clusters have properties of both liquids and solids at the same temperature. Studies of this strange state of matter may help scientists to invent new and useful materials in the future.

When the temperature of a liquid is lowered, the average kinetic energy of the particles is lowered. As the particles slow down, the cohesive forces become more effective, and the particles are no longer able to slide over one another. The particles become frozen into a fixed pattern called a **crystal lattice,** shown in **Figure 13–16.** Although the forces hold the particles in place, the particles in a crystalline solid do not stop moving completely. Rather, they vibrate around their fixed positions in the crystal lattice. In some solid materials, such as butter and glass, the particles do not form a fixed crystalline pattern. These substances have no regular crystal structure but have definite volume and shape, so they are called **amorphous solids.** Amorphous solids are also classified as viscous, or slowly flowing, liquids.

The effects of freezing

As a liquid freezes, its particles usually fit more closely together than in the liquid state. Solids usually are more dense than liquids. However, water is an exception because it is most dense at 4°C. As water freezes, the cohesive forces between particles decrease and the particles take up more space. At 0°C, ice has a lower density than liquid water does, which is why it floats.

a

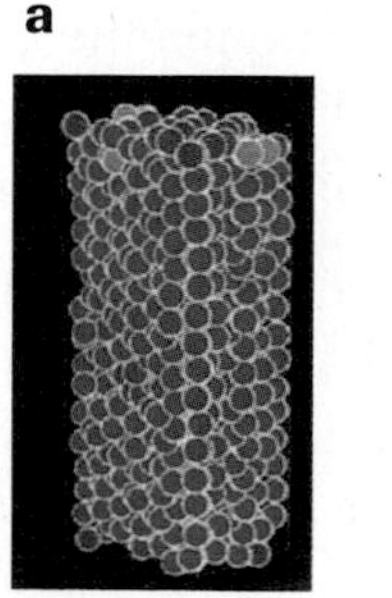

b

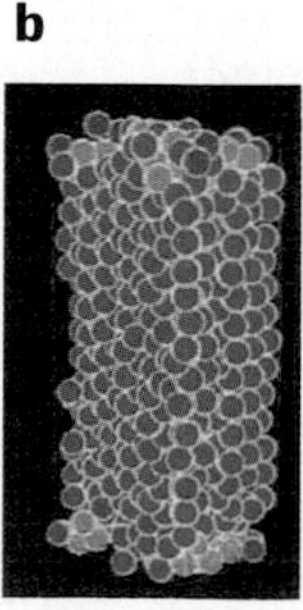

c

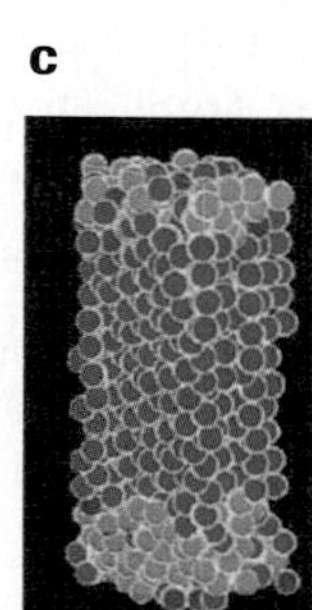

d

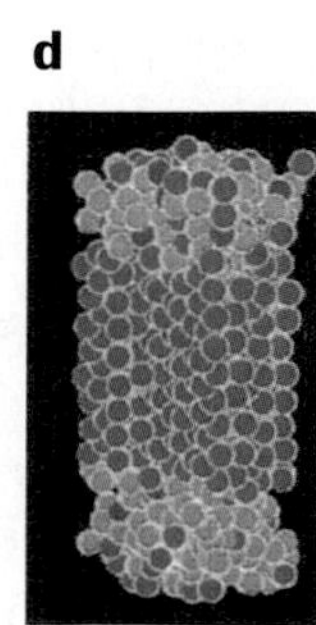

FIGURE 13–15 The melting of a solid is represented by this computer model. The green and blue spheres represent the liquid phase, and the red spheres represent the solid phase. Notice how similar they are.

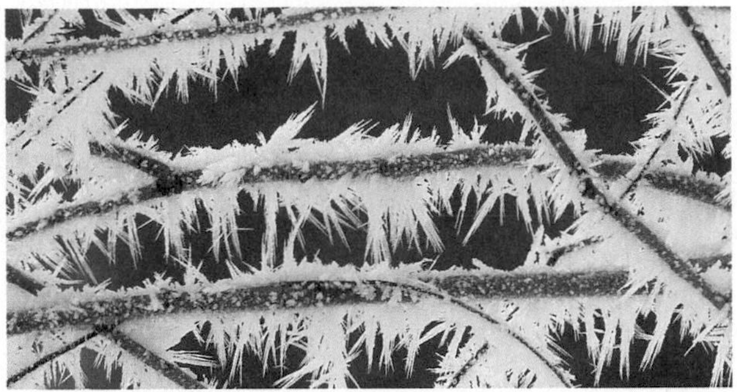

a

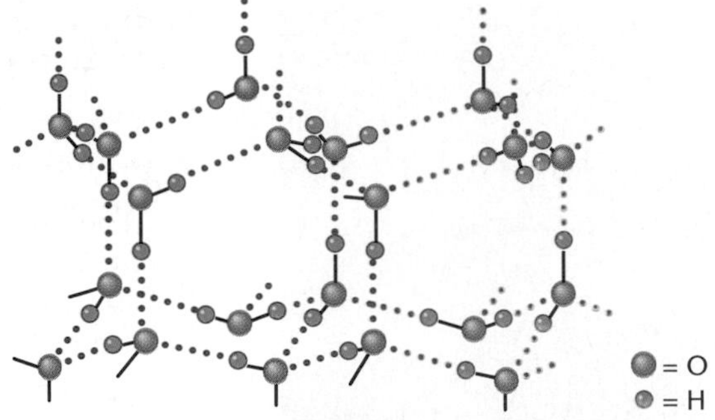

b

FIGURE 13–16 Ice, the solid form of water, has a larger volume than an equal mass of its liquid form **(a).** The crystalline structure of ice is in the form of a lattice **(b).**

For most liquids, an increase in the pressure on the surface of the liquid increases its freezing point. Water is an exception. Because water expands as it freezes, an increase in pressure forces the molecules closer together and increases the cohesive forces among them. The freezing point of water is lowered very slightly.

It has been hypothesized that the drop in water's freezing point caused by the pressure of an ice skater's blades may produce a thin film of liquid between the ice and the blades, which makes them move so effortlessly. Some calculations of the pressure caused by even the sharpest blade show that the ice would still be too cold to melt. But, more recent measurements have shown that the friction between the blade and the ice generates enough thermal energy to melt the ice and create a thin layer of water. If you've ever tried to walk on ice covered with water, you know how slippery it is. This explanation is supported by measurements of the spray of ice particles, such as those in the photo of the ice skater at the beginning of the chapter, which are considerably warmer than the ice itself. The same process of melting occurs during snow skiing.

Skating Effortlessly

Elasticity of solids

External forces applied to a solid object may twist or bend it out of shape. The ability of an object to return to its original form when the external forces are removed is called the **elasticity** of the solid. If too much deformation occurs, the object will not return to its original shape—its elastic limit has been exceeded. Elasticity depends on the electromagnetic forces that hold the particles of a substance together. Malleability and ductility are two properties that depend on the structure and elasticity of a substance. Gold is a malleable metal because it can be flattened and shaped by hammering into thin sheets. Copper is a ductile metal because it can be pulled into thin strands of wire and used in electric circuits.

Thermal Expansion of Matter

Temperature changes cause materials in both solid and fluid states, to expand when heated and to contract when cooled. This property, known as **thermal expansion,** has many useful applications such as using a thermometer to monitor your body temperature as the mercury expands up a glass tube. When heated, all forms of matter—solid,

FIGURE 13–17 The extreme temperatures of a July day caused these railroad tracks to buckle.

liquid, or gas—generally become less dense and expand to fill more space. When the air near the floor of a room is warmed, gravity pulls the denser, colder ceiling air down, which pushes the warmer air upward. This motion results in the circulation of air within the room, called a convection current.

Expansion allows a liquid to be heated rapidly. When a container of liquid is heated from the bottom, convection currents form, just as they do in air. Cold, more dense liquid sinks to the bottom where it is warmed and then pushed up by the continuous flow of cooler liquid from the top. Thus, all the liquid is able to come in contact with the hot surface.

Thermal expansion in solids

The expansion of concrete and steel in highway bridges means that the structures are longer in the summer than in the winter. Temperature extremes must be considered when bridges are designed. Gaps, called expansion joints, are built in to allow for seasonal changes in length. High temperatures can damage railroad tracks that are laid without expansion joints, as shown in **Figure 13–17.**

Physics & Technology

Aerogels

A gel is a mixture consisting of a network of solid particles surrounded by liquid. Gelatin and butter are examples of gels. Removing the liquid in a gel without disturbing its network of solids results in an aerogel—the lightest of all known solid materials. An aerogel has the same shape as the liquid-solid gel it is made from, but it is riddled with microscopic pores filled with air. Some aerogels are 95 percent air. A piece the size of an average adult person weighs less than a pound. Aerogels also have an extremely large surface area. If you could unfold a one-inch cube of aerogel and lay it out flat, it could cover two basketball courts. These characteristics make aerogels amazingly useful.

Aerogels are excellent thermal insulators. Because they contain so much air, they block heat transfer very effectively. A small piece can protect a flower from the heat of an open flame. Aerogels have been experimented with for many years for use in heating and cooling applications. Researchers are now developing ways to make very thin aerogel films for use as electronic insulators. Electric charge can sometimes leak out of computer components. Insulators that prevent this leakage could help increase the efficiency and reduce the size of all kinds of electronic devices.

Most aerogels are made from the mineral silica, which is the same material used to make glass. Aerogels produced in the microgravity of Earth's orbit are clear. Clear aerogels would make very energy-efficient windows, but those made on Earth are hazy and have a bluish color. NASA scientists are now working on a method for making silica aerogels in space.

Thinking Critically What do you think would be the most challenging step in producing an aerogel from a liquid-solid gel?

Some materials, such as Pyrex glass for cooking and laboratory glassware, are designed to have the least possible thermal expansion. Blocks used as standard lengths in machine shops are often made of Invar, a metal alloy that does not expand significantly when heated. Large telescope mirrors are made of a ceramic material called Zerodur, designed to have a coefficient of thermal expansion that is essentially zero.

The expansion of heated solids can be explained in terms of the kinetic theory. One model pictures a solid as a collection of particles connected by springs. The springs represent the forces that attract the particles to each other. When the particles get too close, the springs push them apart. If a solid did not have these forces of repulsion, it would collapse into a tiny sphere. When a solid is heated, the kinetic energy of the particles increases, and they vibrate rapidly and move farther apart. When the particles move farther apart, the attractive forces between particles become weaker. As a result, when the particles vibrate more violently with increased temperature, their average separation increases and the solid expands. The coefficients of thermal expansion for a variety of materials are given in **Table 13–2.**

Jumpers

Put on a pair of safety goggles. Examine the jumping disk. Notice that it is slightly curved. Now rub the disk for several seconds until it becomes curved in the other direction. Place the disk on a flat, level surface and stand back.

Make a Hypothesis Suggest a hypothesis that might explain the jumping. Suggest a method to test your hypothesis.

TABLE 13–2

Coefficients of Thermal Expansion at 20°C

Material	Coefficient of linear expansion, $\alpha(°C)^{-1}$	Coefficient of volume expansion, $\beta(°C)^{-1}$
Solids		
Aluminum	25×10^{-6}	75×10^{-6}
Iron, steel	12×10^{-6}	35×10^{-6}
Glass (soft)	9×10^{-6}	27×10^{-6}
Glass (Pyrex)	3×10^{-6}	9×10^{-6}
Concrete	12×10^{-6}	36×10^{-6}
Copper	16×10^{-6}	48×10^{-6}
Liquids		
Methanol		1100×10^{-6}
Gasoline		950×10^{-6}
Mercury		180×10^{-6}
Water		210×10^{-6}
Gases		
Air (and most other gases)		3400×10^{-6}

The change in length of a solid is proportional to the change in temperature, as shown in **Figure 13–18.** A solid will expand twice as much if the temperature is increased by 20°C than if it is increased by 10°C. The change is also proportional to its length. A 2-meter bar will expand twice as much as a 1-meter bar. The length, L, of a solid at temperature T can be found with the following equation.

$$L_2 = L_1 + \alpha L_1(T_2 - T_1)$$

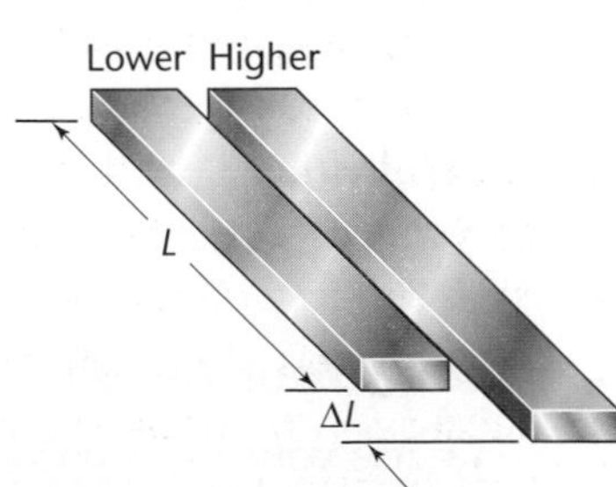

FIGURE 13–18 The change in length of a material is proportional to the original length and the change in temperature.

L_1 is the length at temperature T_1, and the alpha, α, is called the **coefficient of linear expansion.** Using simple algebra, α can be defined by the following equations.

$$L_2 - L_1 = \alpha L_1(T_2 - T_1)$$

$$\Delta L = \alpha L_1 \Delta T$$

$$\Delta L/L_1 = \alpha \Delta T$$

$$\alpha = \Delta L/L_1 \Delta T$$

Therefore, the unit for the coefficient of linear expansion is 1/°C or $(°C)^{-1}$. The **coefficient of volume expansion,** β, is approximately three times the coefficient of linear expansion because volume expansion is three dimensional. The equation to determine volume expansion is $\Delta V = \beta V \Delta T$.

Example Problem

Linear Expansion

A metal bar is 2.60 m long at room temperature, 21°C. The bar is put into an oven and heated to a temperature of 93°C. It is then measured and found to be 3.4 mm longer. What is the coefficient of linear expansion of this material?

Sketch the Problem

- Sketch the bar, which is 3.4 mm longer at 93°C than at 21°C.
- Identify the initial length of the bar, L_1, and the change in length, ΔL.

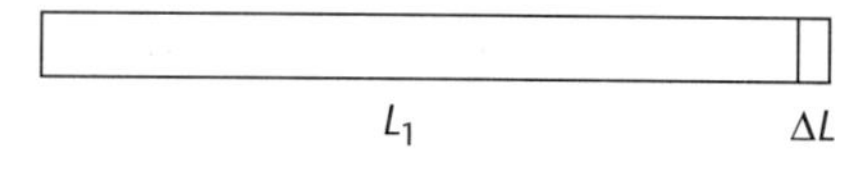

Calculate Your Answer

Known:

$L_1 = 2.60$ m

$\Delta L = 3.4 \times 10^{-3}$ m

$T_1 = 21°C$

$T_2 = 93°C$

Unknown:

$\alpha = ?$

Strategy:

Calculate the coefficient of linear expansion using the known length, change in length, and temperature change.

Calculations:

$$\Delta L = \alpha L_1(T_2 - T_1)$$

$$\alpha = \frac{\Delta L}{L_1(T_2 - T_1)}$$

$$= \frac{3.4 \times 10^{-3} \text{m}}{(2.60 \text{ m})(93°C - 21°C)}$$

$$= 1.8 \times 10^{-5} \text{ °C}^{-1}$$

Check Your Answer

- Are your units correct? The unit is correct, $°C^{-1}$.
- Is the magnitude realistic? The magnitude of the coefficient is close to the accepted value for copper.

Practice Problems

8. A piece of aluminum house siding is 3.66 m long on a cold winter day of $-28°C$. How much longer is it on a very hot summer day at $39°C$?
9. A piece of steel is 11.5 m long at $22°C$. It is heated to $1221°C$, close to its melting temperature. How long is it?
10. An aluminum soft drink can, with a capacity of 354 mL is filled to the brim with water and put in a refrigerator set at $4.4°C$. The can of water is later taken from the refrigerator and allowed to reach the temperature outside, which is $34.5°C$.
    a. What will be the volume of the liquid?
    b. What will be the volume of the can?
    Hint: The can will expand as much as a block of metal the same size.
    c. How much liquid will spill?
11. A tank truck takes on a load of 45 725 liters of gasoline in Houston at $32.0°C$. The coefficient of volume expansion, β, for gasoline is $950 \times 10^{-6}(°C)^{-1}$. The truck delivers its load in Omaha, where the temperature is $-18.0°C$.
    a. How many liters of gasoline does the truck deliver?
    b. What happened to the gasoline?

HELP WANTED

MATERIALS ENGINEER

A leading corporation that develops new materials and new uses for old ones seeks an engineer to join our research and development team. A creative yet analytical mind, a bachelor's degree in materials, metallurgical, or ceramics engineering, and the ability to work both independently and on teams are required.

Our professionals work with the finest resources in the industry. Your commitment to lifelong education and to outstanding performance will allow you to move into positions of increasing responsibility. For more information, contact:

ASU International
Metals Park, OH 44073, or
National Institute of
Ceramic Engineers
757 Brooksedge Plaza Dr.
Westerville, OH 43081

Thermal expansion in liquids

Most liquids also expand when heated. A good model for all liquids does not exist, but it is useful to think of a liquid as a ground-up solid. Groups of two, three, or more particles move together as if they were tiny pieces of a solid. When a liquid is heated, particle motion causes these groups to expand in the same way that particles in a solid are pushed apart. The spaces between groups increase. As a result, the whole liquid expands. With an equal change in temperature, liquids expand considerably more than solids. Gases expand even more.

You have learned that water is most dense at $4°C$. When water is heated from $0°C$ to $4°C$, instead of the expected expansion, water contracts as the cohesive forces increase and ice crystals collapse. The result is that the liquid form of water has a smaller volume than an equal mass of its solid form. The forces between water molecules are strong, and the crystals that make up ice have a very open structure. Even when ice melts, tiny crystals remain. These remaining crystals are melting, and the volume decreases. However, once the temperature of water moves above $4°C$, the volume increases because of greater molecular motion. The practical result is that ice floats and lakes, rivers, and other bodies of water freeze from the top down.

a

b

FIGURE 13–19 The properties of a bimetallic strip cause it to bend when heated **(a).** In this thermostat, a coiled bimetallic strip controls the flow of mercury for opening and closing electrical switches **(b)**.

Importance of thermal expansion

Different materials expand at different rates, as indicated by the different coefficients of expansion given in **Table 13–2.** Engineers must consider these different expansion rates when designing structures. Steel bars are often used to reinforce concrete, and therefore, the steel and concrete must have the same expansion coefficient. Otherwise, the structure may crack on a hot day. Similarly, a dentist must use filling materials that expand and contract at the same rate as tooth enamel.

Different rates of expansion are sometimes useful. Engineers have taken advantage of these differences to construct a useful device called a bimetallic strip, which is used in thermostats. A bimetallic strip consists of two strips of different metals welded or riveted together. Usually, one strip is brass and the other is iron. When heated, brass expands more than iron does. Thus, when the bimetallic strip of brass and iron is heated, the brass part of the strip becomes longer than the iron part. The bimetallic strip bends with the brass on the outside of the curve. If the bimetallic strip is cooled, it bends in the opposite direction. The brass is then on the inside of the curve.

In a home thermostat shown in **Figure 13–19**, the bimetallic strip is installed so that it bends toward an electric contact as the room cools. When the room cools below the setting on the thermostat, the bimetallic strip bends enough to make electric contact with the switch, which turns on the heater. As the room warms, the bimetallic strip bends in the other direction. The electric circuit is broken, and the heater is switched off.

Plasma

If you heat a solid, it melts to form a liquid. Further heating results in a gas. What happens if you increase the temperature still further? Collisions between the particles become violent enough to tear the particles apart. Electrons are pulled off the atoms, producing positively-charged ions. The gaslike state of negatively-charged electrons and positively-charged ions is called **plasma.** Plasma is another fluid state of matter.

The plasma state may seem to be uncommon, however, most of the matter in the universe is plasma. Stars consist mostly of plasma. Much of the matter between the stars and galaxies consists of energetic hydrogen that has no electrons. This hydrogen is in a plasma state. The primary difference between a gas and plasma is that plasma can conduct electricity, whereas a gas cannot. A lightning bolt is in the plasma state. Neon signs, such as the one in **Figure 13–20,** fluorescent bulbs, and sodium vapor lamps contain glowing plasmas.

FIGURE 13–20 The spectacular lighting effects in "neon" signs are caused by luminous plasmas formed in the glass tubing.

13.2 Section Review

2.1 Starting at 0°C, how will the density of water change if it is heated to 4°C? To 8°C?

2.2 You are installing a new aluminum screen door on a hot day. The door frame is concrete. You want the door to fit well on a cold winter day. Should you make the door fit tightly in the frame or leave extra room?

2.3 Why could candle wax be considered a solid? Why might it also be a viscous liquid?

2.4 Critical Thinking If you heat an iron ring with a small gap in it, as in **Figure 13–21,** will the gap become wider or narrower?

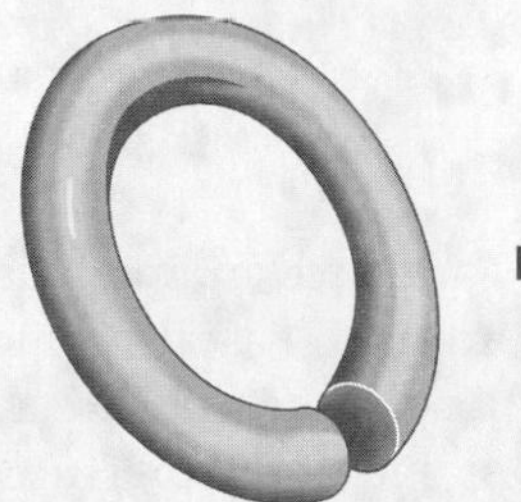

FIGURE 13–21

CHAPTER 13 REVIEW

Key Terms

13.1

- fluid
- pressure
- pascal
- Pascal's principle
- buoyant force
- Archimedes' principle
- Bernoulli's principle
- cohesive force
- surface tension
- adhesion
- capillary action
- evaporation
- volatile
- condensation

13.2

- crystal lattice
- amorphous solids
- elasticity
- thermal expansion
- coefficient of linear expansion
- coefficient of volume expansion
- plasma

Summary

13.1 The Fluid States

- Solids have fixed volumes, definite surfaces, and shapes. A liquid has a fixed volume and a definite surface, but takes the shape of its container. The volume of a gas expands to fill its container. Plasma is similar to a gas but is electrically charged, and thus other forces can contain it.
- Pressure is the force divided by the area on which it is exerted. The SI unit of pressure is the pascal, Pa.
- The constantly moving particles that make up a fluid exert pressure as they collide with all surfaces in contact with the fluid.
- According to Pascal's principle, an applied pressure is transmitted undiminished throughout a fluid.
- The buoyant force is an upward force exerted on an object immersed in a fluid.
- According to Archimedes' principle, the buoyant force on an object immersed in a fluid is equal to the weight of the fluid displaced by that object.
- Bernoulli's principle states that the pressure exerted by a fluid decreases as its velocity increases.
- Cohesive forces are the attractive forces that like particles exert on one another.
- Adhesive forces are the attractive forces that particles of different substances exert on one another.
- Evaporation, the process in which a liquid becomes a gas, occurs when the most energetic particles in a liquid have enough energy to escape into the gas phase.
- In condensation, the least energetic particles in a gas phase bind to each other and form or add to the liquid phase. Evaporation cools the remaining liquid; condensation warms the remaining gas.

13.2 The Solid State

- A crystalline solid has a regular pattern of particles. An amorphous solid has an irregular pattern of particles.
- As a liquid solidifies, its particles become frozen into a fixed pattern.
- The elasticity of a solid is its ability to return to its original form when external forces are removed.
- When the temperature of a solid or liquid is increased, the kinetic energy of its particles increases and it generally increases in size, or expands. The expansion is proportional to the temperature change and original size, and depends on the material.
- Plasma is a gaslike state of matter made up of positive or negative particles or a mixture of them.

Reviewing Concepts

Section 13.1

1. How are force and pressure different?
2. According to Pascal's principle, what happens to the pressure at the top of a container if the pressure at the bottom is increased?
3. How does the water pressure one meter below the surface of a small pond compare with the water pressure the same distance below the surface of a lake?
4. Does Archimedes' principle apply to an object inside a flask that is inside a spaceship in orbit?

5. A river narrows as it enters a gorge. As the water speeds up, what happens to the water pressure?
6. A gas is placed in a sealed container and some liquid is placed in a container the same size. They both have definite volume. How do the gas and the liquid differ?
7. A razor blade, which has a density greater than that of water, can be made to float on water. What procedures must you follow for this to happen? Explain.
8. In terms of adhesion and cohesion, explain why alcohol clings to the surface of a glass rod and mercury does not.
9. A frozen lake melts in the spring. What effect does it have on the temperature of the air above it?
10. Canteens used by hikers are often covered with a canvas bag. If you wet the bag, the water in the canteen will be cooled. Explain.
11. Why does high humidity make a hot day even more uncomfortable?

Section 13.2

12. How does the arrangement of atoms in a crystalline substance differ from that in an amorphous substance?
13. Can a spring be considered elastic?
14. Does the coefficient of linear expansion depend on the unit of length used? Explain.
15. In what way are gases and plasmas similar? In what way are they different?
16. Some of the mercury in a fluorescent lamp is in the gaseous form; some is in the form of plasma. How can you distinguish between the two?

Applying Concepts

17. A rectangular box with its largest surface resting on a table is rotated so that its smallest surface is now on the table. Has the pressure on the table increased, decreased, or remained the same?
18. Show that a pascal is equivalent to $kg/m{\cdot}s^2$.
19. Is there more pressure at the bottom of a bathtub of water 30 cm deep or at the bottom of a pitcher of water 35 cm deep? Explain.
20. Compared to an identical empty ship, would a ship filled with table-tennis balls sink deeper into the water or rise in the water? Explain.
21. Which has a greater weight, a liter of ice or a liter of water?
22. Drops of mercury, water, and acetone are placed on a smooth, flat surface, as shown in **Figure 13–22.** The mercury drop is almost a perfect sphere. The water drop is a flattened sphere. The acetone, however, spreads out over the surface. What do these observations tell you about the cohesive forces in mercury, water, and acetone?

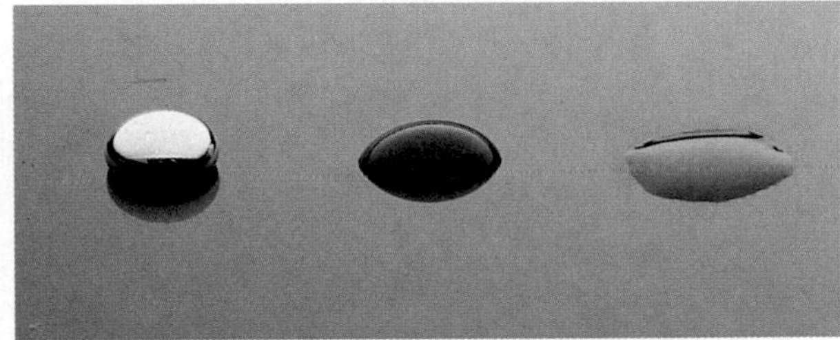

FIGURE 13–22

23. Alcohol evaporates more quickly than water does at the same temperature. What does this observation allow you to conclude about the properties of the particles in the two liquids?
24. Based on the observation in Question 22, which liquid would vaporize easier? Which would have the lower boiling point? Explain.
25. Suppose you use a punch to make a circular hole in aluminum foil. If you heat the foil, will the size of the hole decrease or increase? Explain. **Hint:** Pretend that you put the circle you punched out back into the hole. What happens when you heat the foil now?
26. Equal volumes of water are heated in two narrow tubes that are identical except that tube A is made of soft glass and tube B is made of Pyrex glass. As the temperature increases, the water level rises higher in tube B than in tube A. Give a possible explanation. Why are many cooking utensils made from Pyrex glass?
27. A platinum wire can be easily sealed in a glass tube, but a copper wire does not form a tight seal with the glass. Explain.

28. Often before a thunderstorm, when the humidity is high, someone will say, "The air is very heavy today." Is this statement correct? Describe a possible origin for the statement.

29. Five objects with the following densities are put into a tank of water:

a. 0.85 g/cm^3 **d.** 1.15 g/cm^3
b. 0.95 g/cm^3 **e.** 1.25 g/cm^3
c. 1.05 g/cm^3

The density of water is 1.00 g/cm^3. The diagram in **Figure 13–23** shows six possible positions of these objects. Select a position, 1 to 6, for each of the five objects. Not all positions need be selected.

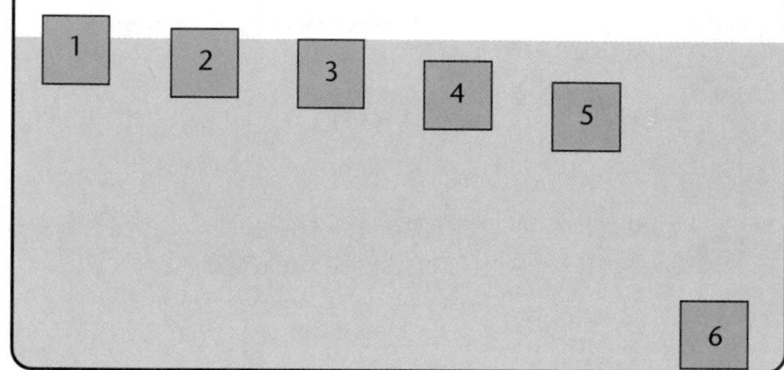

FIGURE 13–23

Problems

Section 13.1

LEVEL 1

30. A 0.75-kg physics book with dimensions of 24 cm by 20 cm is on a table.
a. What force does the book apply to the table?
b. What pressure does the book apply?

31. A reservoir behind a dam is 15 m deep. What is the pressure of the water in the following situations?
a. at the base of the dam
b. 5.0 m from the top of the dam

32. A 75-kg solid cylinder, 2.5 m long and with an end radius of 5.0 cm, stands on one end. How much pressure does it exert?

33. A test tube standing vertically in a test-tube rack contains 2.5 cm of oil ($\rho = 0.81$ g/cm^3) and 6.5 cm of water. What is the pressure on the bottom of the test tube?

34. A metal object is suspended from a spring scale. The scale reads 920 N when the object is suspended in air, and 750 N when the object is completely submerged in water.
a. Find the volume of the object.
b. Find the density of the metal.

35. During an ecology experiment, an aquarium half filled with water is placed on a scale. The scale reads 195 N.
a. A rock weighing 8 N is added to the aquarium. If the rock sinks to the bottom of the aquarium, what will the scale read?
b. The rock is removed from the aquarium, and the amount of water is adjusted until the scale again reads 195 N. A fish weighing 2 N is added to the aquarium. What is the scale reading with the fish in the aquarium?

36. What is the size of the buoyant force that acts on a floating ball that normally weighs 5.0 N?

37. What is the apparent weight of a rock submerged in water if the rock weighs 54 N in air and has a volume of 2.3×10^{-3} m^3?

38. If a rock weighing 54 N is submerged in a liquid with a density exactly twice that of water, what will be its new apparent weight reading in the liquid?

39. A 1.0-L container completely filled with mercury has a weight of 133.3 N. If the container is submerged in water, what is the buoyant force acting on it? Explain.

40. What is the maximum weight that a balloon filled with 1.00 m^3 of helium can lift in air? Assume that the density of air is 1.20 kg/m^3 and that of helium is 0.177 kg/m^3. Neglect the mass of the balloon.

LEVEL 2

41. A hydraulic jack used to lift cars is called a three-ton jack. The large piston is 22 mm in diameter, the small one 6.3 mm. Assume that a force of 3 tons is 3.0×10^4 N.
a. What force must be exerted on the small piston to lift the 3-ton weight?
b. Most jacks use a lever to reduce the force needed on the small piston. If the resistance arm is 3.0 cm, how long is the effort arm of an ideal lever to reduce the force to 100 N?

42. In a machine shop, a hydraulic lift is used to raise heavy equipment for repairs. The system has a small piston with a cross-sectional area of 7.0×10^{-2} m^2 and a large piston with a cross-sectional area of 2.1×10^{-1} m^2. An engine weighing 2.7×10^3 N rests on the large piston.
   a. What force must be applied to the small piston in order to lift the engine?
   b. If the engine rises 0.20 m, how far does the smaller piston move?

43. What is the acceleration of a small metal sphere as it falls through water? The sphere weighs 2.8×10^{-1} N in air and has a volume of 13 cm^3.

Section 13.2

LEVEL 1

44. What is the change in length of a 2.00-m copper pipe if its temperature is raised from 23°C to 978°C?

45. Bridge builders often use rivets that are larger than the rivet hole to make the joint tighter. The rivet is cooled before it is put into the hole. A builder drills a hole 1.2230 cm in diameter for a steel rivet 1.2250 cm in diameter. To what temperature must the rivet be cooled if it is to fit into the rivet hole that is at 20°C?

LEVEL 2

46. A steel tank is built to hold methanol. The tank is 2.000 m in diameter and 5.000 m high. It is completely filled with methanol at 10.0°C. If the temperature rises to 40.0°C, how much methanol (in liters) will flow out of the tank? Remember that both the tank and the methanol expand as the temperature rises.

47. An aluminum sphere is heated from 11°C to 580°C. If the volume of the sphere is 1.78 cm^3 at 11°C, what is the increase in volume of the sphere at 580°C?

48. The volume of a copper sphere is 2.56 cm^3 after being heated from 12°C to 984°C. What was the volume of the copper sphere at 12°C?

Critical Thinking Problems

49. Persons confined to bed are less likely to develop bedsores if they use a water bed rather than an ordinary mattress. Explain.

50. Hot air balloons contain a fixed volume of gas. When the gas is heated, it expands, and pushes some gas out at the lower open end. As a result, the mass of the gas in the balloon is reduced. Why would the air in a balloon have to be hotter to lift the same number of people above Vail, Colorado, which has an altitude of 2400 m, than above the tidewater flats of Virginia, at an altitude of 6 m?

Going Further

Team Project The braking systems of cars are combinations of many simple machines, one of which applies Pascal's principle. As a team, analyze the system for a real car with the goal of constructing a physical model that predicts the force on the brake pedal needed to stop the car from a given speed in a certain distance. You may need to find answers to the following questions: Does the brake pedal function as a lever when exerting a force on the master cylinder? What is the ratio of areas of the master and wheel cylinders? For disc brakes, what is the coefficient of friction between the brake pads and rotor? For drum brakes, what is the mechanical advantage (*MA*) of the lever that converts the force exerted by the wheel piston to the normal force of the pad on the drum? For all kinds of brakes, how do the brake and tire act as a wheel-and-axle machine? What force is needed to slow the car from the chosen speed in the chosen distance? You may find other questions that require answers. Devise a means of testing your model that is both realistic and safe.

***inter*NET CONNECTION**

Follow the link on the Glencoe Homepage at **www.glencoe.com/sec/science** to find out more about this chapter.

Surf's Up

From where does the surfer's kinetic energy come? Trace the energy source back as far as you can.

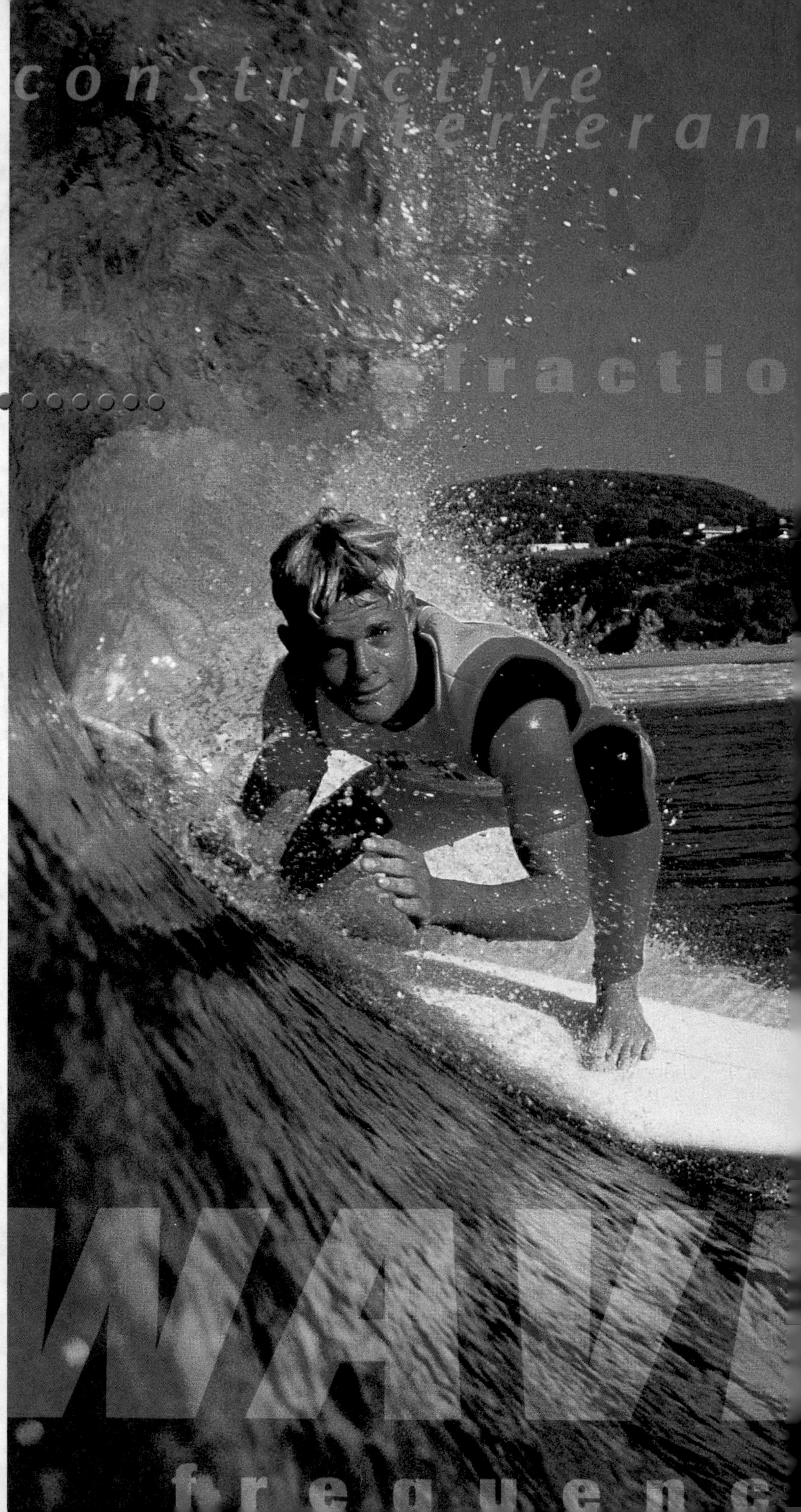

CHAPTER

14 Waves and Energy Transfer

Have you ever ridden a wave on a surfboard at the ocean, or have you ever caught a wave with your body in a wave pool? As the wave pushes you to the shore, you gain speed and stay just ahead of the breaking surf. Your body's kinetic energy increases. But where does this extra energy come from? If you can, you will try to do as this surfer does, and ride almost parallel to the wave at a very high speed. Surfing, however, can be dangerous. Unless you are skilled, the energy carried by the wave can cause you to wipe out and be thrown into the water or sand.

Catching a wave allows you to come rapidly to shore. Although the movement of the wave may make it seem that the water is moving toward the shore, the water is mostly moving up and down as the wave carries its energy toward the shore. Light and sound waves show similar patterns of behavior, as you will study in the next few chapters.

WHAT YOU'LL LEARN

- You will determine how waves transfer energy.
- You will describe wave reflection and discuss its practical significance.

WHY IT'S IMPORTANT

- Waves enable the sun's energy to reach Earth and make possible all communication through sound.
- Because light waves can be reflected, you are able to see the world around you and even read these very words.
- Knowledge of the behavior of waves is essential to the designing of bridges and many other structures.

*inter*NET CONNECTION

Follow the link for this chapter on the Glencoe Homepage at **www.glencoe.com/sec/science** to find out more about waves and energy transfer.

14.1 Wave Properties

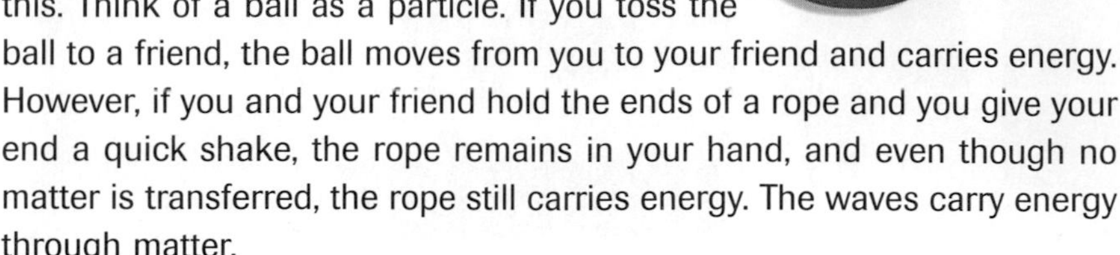

Both particles and waves carry energy, but there is an important difference in how they do this. Think of a ball as a particle. If you toss the ball to a friend, the ball moves from you to your friend and carries energy. However, if you and your friend hold the ends of a rope and you give your end a quick shake, the rope remains in your hand, and even though no matter is transferred, the rope still carries energy. The waves carry energy through matter.

OBJECTIVES

- **Identify** how waves transfer energy without transferring matter.
- **Contrast** transverse and longitudinal waves.
- **Relate** wave speed, wavelength, and frequency.

Mechanical Waves

You have learned how Newton's laws of motion and conservation of energy principles govern the behavior of particles. These laws also govern the motion of waves. There are many kinds of waves. All kinds of waves transmit energy, including the waves you cannot see, such as the sound waves you create when you speak and the light waves that reflect from the leaves on the trees.

Transverse waves

A **wave** is a rhythmic disturbance that carries energy through matter or space. Water waves, sound waves, and the waves that travel down a rope or spring are types of mechanical waves. Mechanical waves require a medium. Water, air, ropes, or springs are the materials that carry the energy of mechanical waves. Other kinds of waves, including electromagnetic waves and matter waves, will be described in later chapters. Because many of these waves cannot be directly observed, mechanical waves can serve as models for their study.

The two disturbances that go down the rope shown in **Figure 14–1** are called wave pulses. A **wave pulse** is a single bump or disturbance that travels through a medium. If the person continues to move the rope up and down, a **continuous wave** is generated. Notice that the rope is disturbed in the vertical direction, but the pulse travels horizontally. This wave motion is called a transverse wave. A **transverse wave** is a wave that vibrates perpendicular to the direction of wave motion.

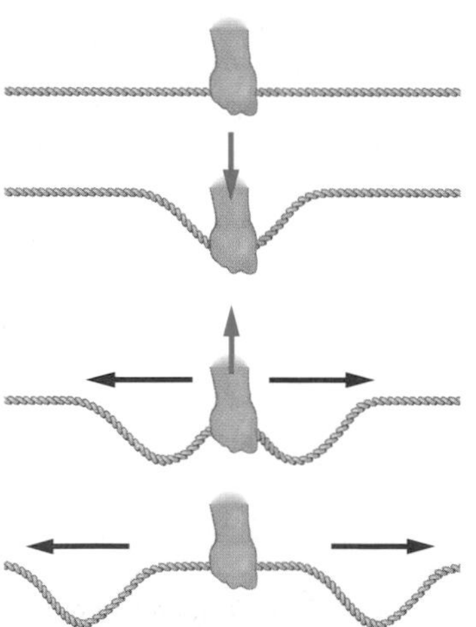

FIGURE 14–1 A quick shake of a rope sends out wave pulses in both directions.

Longitudinal and surface waves

In a coiled spring such as a Slinky toy, you can create a wave pulse in a different way. If you squeeze together several turns of the coiled spring and then suddenly release them, pulses of closely spaced turns will move away in both directions, as in **Figure 14–2.** In this case, the disturbance is in the same direction as, or parallel to, the direction of wave motion. Such a wave is called a **longitudinal wave.** Sound waves are longitudinal waves. Fluids such as liquids and gases usually transmit only longitudinal waves.

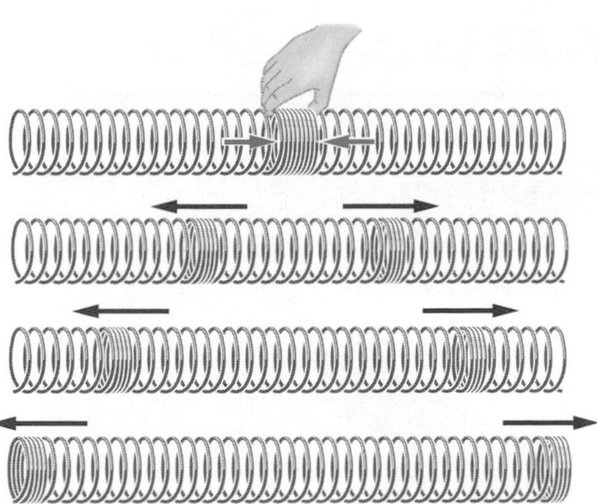

FIGURE 14–2 The squeeze and release of a coiled spring toy sends out wave pulses in both directions.

Although waves deep in a lake or ocean are longitudinal, at the surface of the water, the particles move in a direction that is both parallel and perpendicular to the direction of wave motion, as shown in **Figure 14–3.** These are **surface waves,** which have characteristics of both transverse and longitudinal waves. The energy of water waves usually comes from storms far away. The energy of the storms initially came from the heating of Earth by solar energy. This energy, in turn, was carried to Earth by transverse electromagnetic waves.

Surf's Up

Measuring a Wave

There are many ways to describe or measure a wave. Some methods depend on how the wave is produced, whereas others depend on the medium through which the wave travels.

a

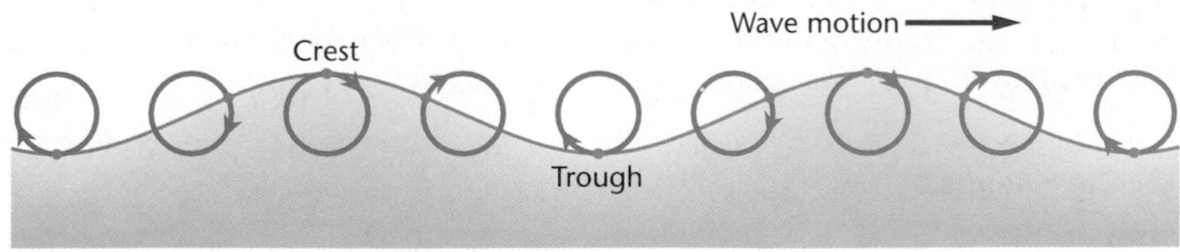

b

FIGURE 14–3 Surface waves have properties of both longitudinal and transverse waves **(a)**. The paths of the particles are circular **(b)**.

Design Your Own **Physics Lab**

Waves on a Coiled Spring

Problem

How can you model the properties of transverse waves?

Hypothesis

A coiled spring toy can be used to model transverse waves and to investigate wave properties such as speed, frequency, amplitude, and wavelength.

Possible Materials

a long coiled spring toy, such as a Slinky
stopwatch
meterstick

Plan the Experiment

1. Work in pairs or groups, and clear a path of about 6 meters for this activity.
2. One member of the team should grip the Slinky firmly with one hand. Another member of the team should stretch the spring to the length suggested by your teacher. Team members should take turns holding the end of the spring. **CAUTION:** *Coiled springs easily get out of control. Do not allow them to get tangled or overstretched.*
3. The second team member should then make a quick sideways snap of the wrist to produce transverse wave pulses. Other team members can assist in measuring, timing, and recording data. It is easier to see the motion from one end of the Slinky, rather than from the side.
4. Design and perform experiments to answer the following questions.
5. **Check the Plan** Make sure your teacher has approved your final plan before you proceed with your experiment.

Analyze and Conclude

1. **Interpreting Data** What happens to the amplitude of the transverse wave as it travels?
2. **Recognizing Cause and Effect** Does the transverse wave's speed depend upon its amplitude?
3. **Observing and Interpreting** If you put two quick transverse wave pulses into the spring and consider the wavelength to be the distance between the pulses, does the wavelength change as the pulses move?
4. **Applying** How can you decrease the wavelength of a transverse wave?
5. **Interpreting** As transverse wave pulses travel back and forth on the spring, do they bounce off each other or pass through each other?

Apply

1. How do the speeds of high frequency (short wavelength) transverse waves compare with the speeds of low frequency (long wavelength) transverse waves?

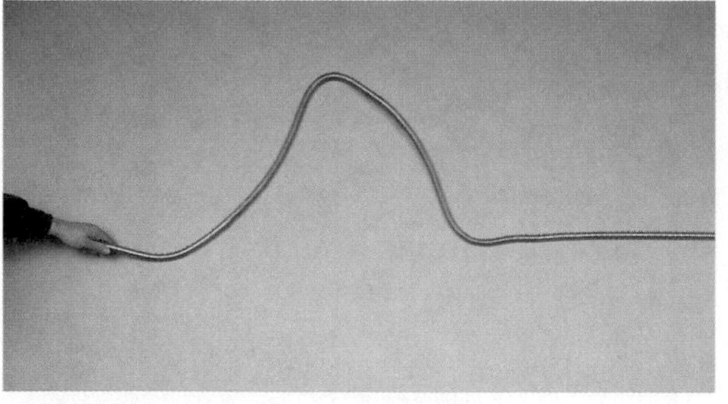

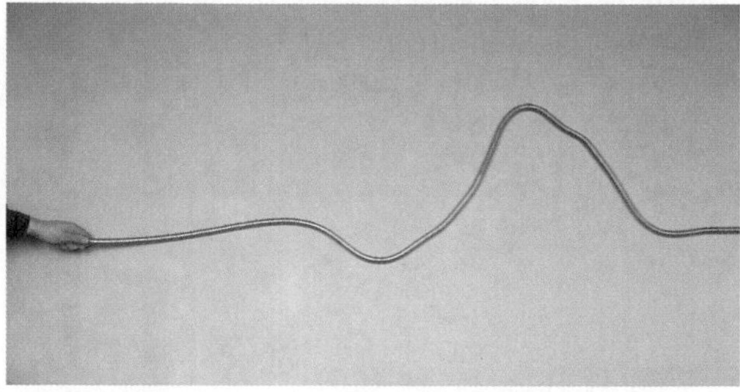

FIGURE 14–4 These two photographs were taken 0.20 s apart. During that time, the crest moved 0.80 m. The velocity of the wave is 4.0 m/s.

Speed and amplitude

How fast does a wave move? The speed of the pulse shown in **Figure 14–4** can be found in the same way in which you would determine the speed of a moving car. First, you measure the displacement of the wave peak, Δd; then you divide this by the time interval, Δt, to find the speed, as shown by $v = \bar{v} = \Delta d/\Delta t$. The speed of a continuous wave, can be found the same way. For most mechanical waves, both transverse and longitudinal, the speed depends only on the medium through which the waves move.

How does the pulse generated by gently shaking a rope differ from the pulse produced by a violent shake? The difference is similar to the difference between a ripple in a pond and a tidal wave. They have different amplitudes. The amplitude of a wave is its maximum displacement from its position of rest, or equilibrium. Two similar waves having different amplitudes are shown in **Figure 14–5.** A wave's amplitude depends on how the wave is generated, but not on its speed. More work has to be done to generate a wave with a larger amplitude. For example, strong winds produce larger water waves than those formed by gentle breezes. Waves with larger amplitudes transfer more energy. Thus, although a small wave might move sand on a beach a few centimeters, a giant wave can uproot and move a tree. For waves that move at the same speed, the rate at which energy is transferred is proportional to the square of the

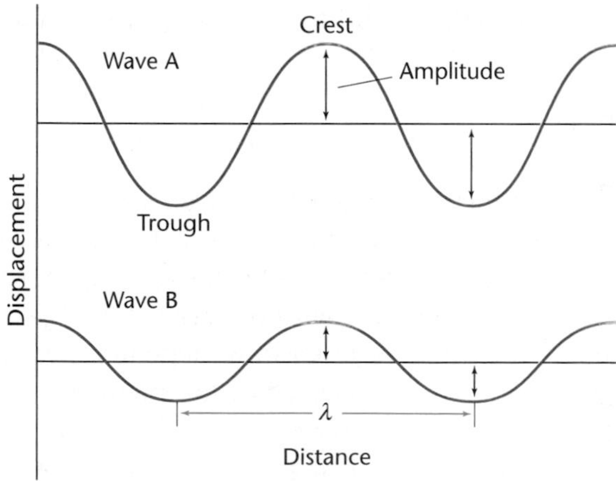

FIGURE 14–5 The amplitude of wave A is larger than that of wave B.

FIGURE 14–6 One end of a string, with a piece of tape at point P, is attached to a vibrating blade. Note the change in position of point P over time.

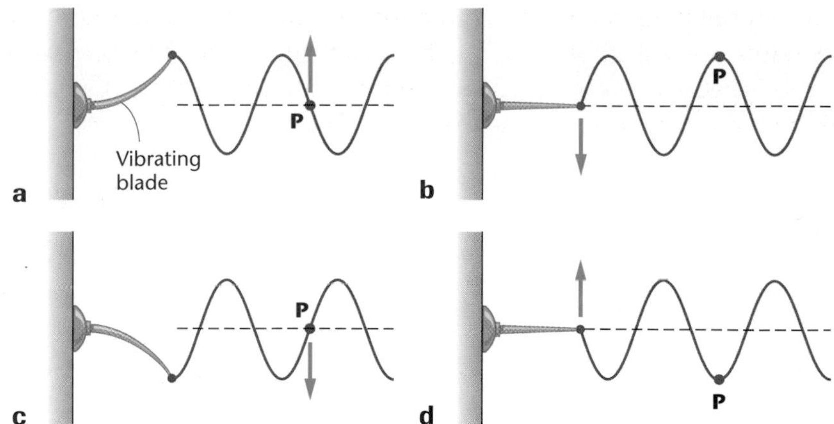

HELP WANTED

GEOLOGIST

Major oil company seeks geologist or geophysicist to work on a team for a major exploration effort aimed at discovering new and better oil sources.

The ideal candidate has a master's degree, with a specialty in areas relating to our industry, and a drive to succeed. He or she will persist when the average person says "It can't be done." He or she must be able to "think big," communicate well, and have the stamina and desire to do field work around the world. For information contact:

American Geological Institute
4220 King Street
Alexandria, VA 22302

amplitude. Thus, doubling the amplitude of a wave increases the amount of energy it transfers each second by a factor of four.

Wavelength

Rather than focusing on one point on a wave, imagine taking a snapshot of a wave, so that you can see the whole wave at one instant in time. **Figure 14–5** shows the low points, or **troughs,** and the high points, or **crests,** of a wave. The shortest distance between points where the wave pattern repeats itself is called the **wavelength.** Crests are spaced by one wavelength. Each trough is also one wavelength from the next. The Greek letter lambda, λ, represents wavelength.

Period and frequency

Although wave speed and amplitude can describe both pulses and continuous waves, period (T) and frequency (f) apply only to continuous waves. You learned in Chapter 6 that the period of a simple harmonic oscillator, such as a pendulum, is the time it takes for the motion of the oscillator to repeat itself. Such an oscillator is usually the source, or cause, of a continuous wave. The period of a wave is equal to the period of the source. In **Figure 14–6,** the period, T, equals 0.04 s, which is the time it takes the source to return to the same point in its oscillation. The same time is taken by point P, a point on the rope, to return to its initial position.

The **frequency** of a wave, f, is the number of complete oscillations it makes each second. Frequency is measured in hertz. One hertz (Hz) is one oscillation per second. The frequency (f) and period (T) of a wave are related by the following equation.

$$f = \frac{1}{T}$$

Both the period and the frequency of a wave depend only on its source. They do not depend on the wave's speed or the medium.

Although you can measure a wavelength directly, the wavelength depends on both the frequency of the oscillator and the speed of the wave. In the time interval of one period, a wave moves one wavelength. Therefore, the speed of a wave is the wavelength divided by the period, $v = \lambda/T$. Because the frequency is more easily found than the period, this equation is more often written as $v = \lambda f$.

F.Y.I.

Because radio waves travel at 3.00×10^8 m/s and sound waves are slower, 3.4×10^2 m/s, a broadcast voice can be heard sooner 10 000 miles away than it can be heard at the back of the room in which it originated.

Example Problem

Speed of a Wave

A sound wave has a frequency of 262 Hz and a wavelength measured at 1.29 m.

a. What is the speed of the wave?

b. How long will it take the wave to travel the length of a football field, 91.4 m?

c. What is the period of the wave?

Sketch the Problem

- Draw a model of one wavelength.
- Diagram a velocity vector.

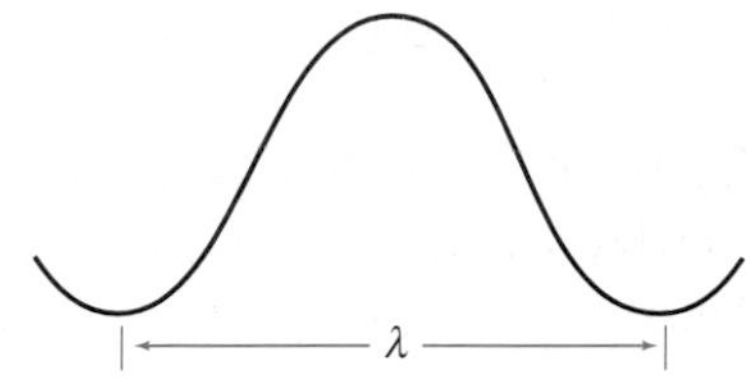

$\xrightarrow{v}$

Calculate Your Answer

Known:	**Unknown:**
$f = 262$ Hz	$v = ?$
$\lambda = 1.29$ m	$t = ?$
$d = 91.4$ m	$T = ?$

Strategy:	**Calculations:**
a. Find the speed of sound from the frequency and wavelength.	$v = \lambda f$ $= (1.29 \text{ m})(262 \text{ Hz}) = 338 \text{ m/s}$
b. Find the time required from speed and distance.	$v = \frac{d}{t}$, so $t = \frac{d}{v}$ $t = \frac{91.4 \text{ m}}{338 \text{ m/s}} = 0.270 \text{ s}$
c. Find the period from the frequency.	$T = \frac{1}{f} = \frac{1}{262 \text{ Hz}} = 0.00382 \text{ s}$

Check Your Answer

- Hz has the units s^{-1} and so m·Hz equals m/s, which is correct.
- Are the magnitudes realistic? A typical sound wave travels approximately 343 m/s. You can notice the delay in sound across a football field, so a few tenths of a second is reasonable.

Physics & Society

Predicting Earthquakes

Earthquakes are produced by the sudden motion of rock masses within Earth's crust. Although rocks can bend and twist to some extent, they break if they are exposed to forces that exceed their strength. The friction and crushing motions of breaking rock create seismic waves of energy that radiate outward. These seismic waves cause the shaking and trembling called an earthquake. The breaks in the rock are known as earthquake faults.

The subsurface region where the rock ruptures is called the focus of an earthquake. The point on the surface directly above the focus is known as the epicenter. Several types of seismic waves travel outward from the focus. Waves that travel along the surface are called surface waves. They have characteristics of both transverse and longitudinal waves and cause most major earthquake damage to homes and cities.

How much damage?

Many earthquakes are so small that they are hardly felt. But catastrophic earthquakes have been recorded in many regions of the world. One example is the powerful earthquake that hit the densely populated city of Kobe, Japan in 1995. Thousands of people were killed or injured and damage estimates reached $500 billion.

Can earthquakes be predicted?

Most geologists agree that regions along portions of California's famous San Andreas Fault will experience a major earthquake before the middle of the 21st century. The broad time frame of a prediction such as this doesn't enable people to evacuate an area in anticipation of a quake on any particular date.

Efforts are under way to develop more precise earthquake predictions. Scientists constantly use seismographs to monitor major faults such as the San Andreas Fault for the smallest Earth tremors. A seismograph records the magnitude of seismic waves by suspending a pen on a pendulum over a paper-covered drum. The stronger the motion, the larger the arc of the pen's motion on the drum.

Lasers and creep meters measure differences in land movement on the two sides of a fault. A creep meter consists of wires stretched across a fault, and a laser beam is timed as it returns to its source. Scientists monitor radon and hydrogen concentrations in groundwater to determine how they change prior to an earthquake. Antennae monitor changes in electromagnetic waves coming from deep beneath Earth's surface.

Investigating the Issue

1. **Acquiring Information** Use your library skills to find out more about the Parkfield experiment near the San Andreas Fault in California. What kinds of instruments are being used by seismologists to monitor the fault? Have any earthquakes predicted for Parkfield actually occurred?
2. **Formulating Models** Find out what kinds of structures in Kobe, Japan suffered the most damage in the 1995 earthquake. How do scientists plan to make these structures more earthquake-resistant?

*inter***NET**
CONNECTION

Follow the link for this chapter on the Glencoe Homepage at **www.glencoe.com/sec/science** to find out more about earthquakes.

Practice Problems

1. A sound wave produced by a clock chime is heard 515 m away 1.50 s later.
   a. What is the speed of sound of the clock's chime in air?
   b. The sound wave has a frequency of 436 Hz. What is its period?
   c. What is its wavelength?
2. A hiker shouts toward a vertical cliff 685 m away. The echo is heard 4.00 s later.
   a. What is the speed of sound of the hiker's voice in air?
   b. The wavelength of the sound is 0.750 m. What is its frequency?
   c. What is the period of the wave?
3. If you want to increase the wavelength of waves in a rope, should you shake it at a higher or lower frequency?
4. What is the speed of a periodic wave disturbance that has a frequency of 2.50 Hz and a wavelength of 0.600 m?
5. The speed of a transverse wave in a string is 15.0 m/s. If a source produces a disturbance that has a frequency of 5.00 Hz, what is its wavelength?
6. Five pulses are generated every 0.100 s in a tank of water. What is the speed of propagation of the wave if the wavelength of the surface wave is 1.20 cm?
7. A periodic longitudinal wave that has a frequency of 20.0 Hz travels along a coil spring. If the distance between successive compressions is 0.400 m, what is the speed of the wave?

EARTH SCIENCE CONNECTION

An earthquake produces both transverse and longitudinal waves that travel through Earth. Geologists studying these waves with seismographs found that only the longitudinal waves could pass through Earth's core. Only longitudinal waves can move through a liquid or a gas. From this observation, what can be deduced about the nature of Earth's core?

14.1 Section Review

1.1 Suppose you and your lab partner are asked to measure the speed of a transverse wave in a giant, coiled spring toy. How could you do it? List the equipment you would need.

1.2 You are creating transverse waves in a rope by shaking your hand from side to side. Without changing the distance your hand moves, you begin to shake it faster and faster. What happens to the amplitude, frequency, period, and velocity of the wave?

1.3 If you pull on one end of a coiled spring toy, does the pulse reach the other end instantaneously? What if you pull on a rope? What if you hit the end of a metal rod? Compare the responses of these three materials.

1.4 **Critical Thinking** If a raindrop falls into a pool, small-amplitude waves result. If a swimmer jumps into a pool, a large-amplitude wave is produced. Why doesn't the heavy rain in a thunderstorm produce large waves?

14.2 Wave Behavior

When a wave encounters the boundary of the medium in which it is traveling, it sometimes reflects back into the medium. In other instances, some or all of the wave passes through the boundary into another medium, often changing direction at the boundary. In addition, many properties of wave behavior result from the fact that two or more waves can exist in the same medium at the same time—quite unlike particles, which consist of matter and take up space.

OBJECTIVES

- **Relate** a wave's speed to the medium in which the wave travels.
- **Describe** how waves are reflected and refracted at boundaries between media, and **explain** how waves diffract.
- **Apply** the principle of superposition to the phenomenon of interference.

Waves at Boundaries

Recall that the speed of a mechanical wave depends only on the properties of the medium it passes through, not on the wave's amplitude or frequency. For water waves, the depth of the water affects wave speed. For sound waves in air, the temperature affects wave speed. For waves on a spring, the speed depends upon the spring's rigidity and its mass per unit length.

Examine what happens when a wave moves across a boundary from one medium into another, as in two springs of different thicknesses joined end to end. **Figure 14–7** shows a wave pulse moving from a large spring into a smaller one. The wave that strikes the boundary is called the **incident wave.** One pulse from the larger spring continues in the smaller spring, at the speed of waves on the smaller spring. Note that this transmitted wave pulse remains upward.

Some of the energy of the incident wave's pulse is reflected backward in the larger spring. This returning wave is called the **reflected wave.** Whether or not the reflected wave is upward (erect) or downward (inverted) depends on the comparative thicknesses of the two springs. If waves in the smaller spring have a higher speed because the spring is heavier or stiffer, then the reflected wave will be inverted.

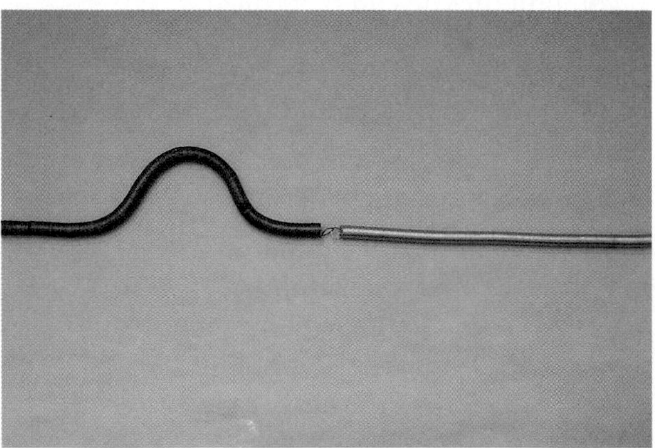

a

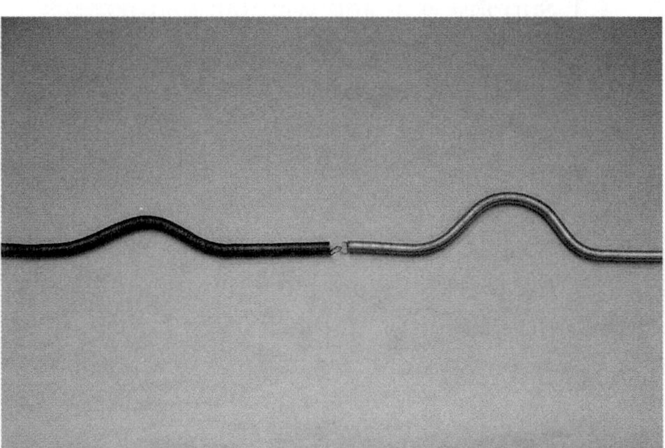

b

FIGURE 14–7 The junction of the two springs is a boundary between two media. A pulse reaching the boundary **(a)** is partially reflected and partially transmitted **(b).**

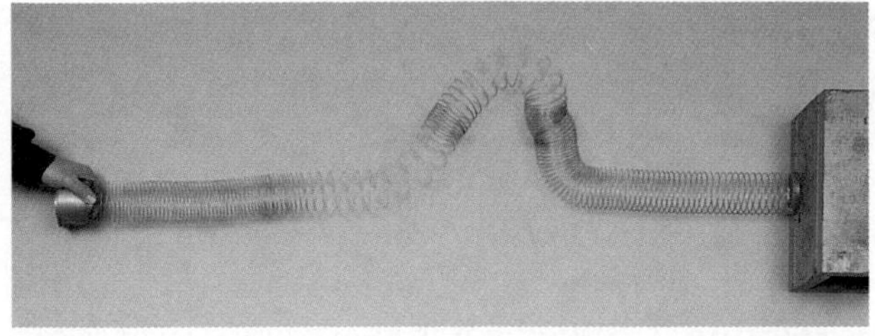
a

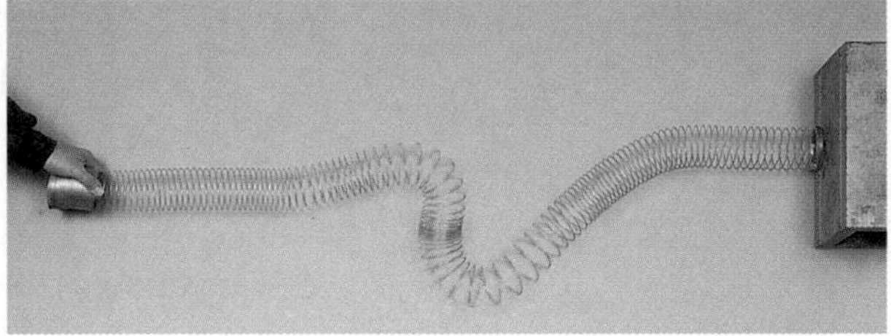
b

FIGURE 14–8 A pulse approaches a rigid wall **(a)** and is reflected back **(b).** Note that the amplitude of the reflected pulse is nearly equal to the amplitude of the incident pulse, but it is inverted and appears as a downward curve.

What happens if the boundary is a wall rather than another spring? When a wave pulse is sent down a spring connected to a rigid wall, the energy transmitted is reflected back from the wall, as shown in **Figure 14–8.** The wall is the boundary of a new medium through which the wave attempts to pass. The pulse is reflected from the wall with almost exactly the same amplitude as the pulse of the incident wave. Thus, almost all the wave's energy is reflected back. Very little energy is transmitted into the wall. Note also that the pulse is inverted.

Practice Problems

8. A pulse is sent along a spring. The spring is attached to a lightweight thread that is tied to a wall, as shown in **Figure 14–9.**
   a. What happens when the pulse reaches point A?
   b. Is the pulse reflected from point A erect or inverted?
   c. What happens when the transmitted pulse reaches point B?
   d. Is the pulse reflected from point B erect or inverted?

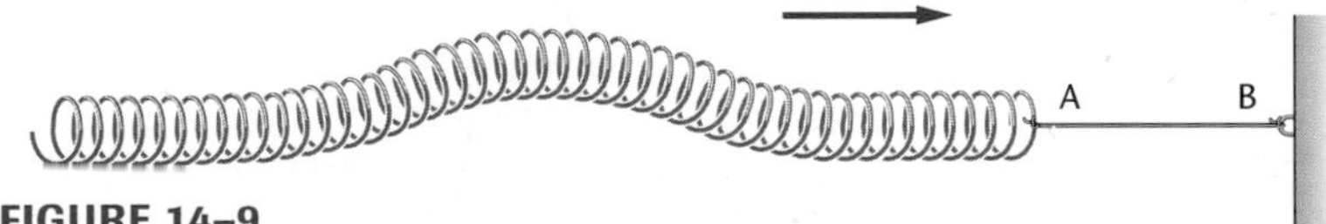

FIGURE 14–9

9. A long spring runs across the floor of a room and out the door. A pulse is sent along the spring. After a few seconds, an inverted pulse returns. Is the spring attached to the wall in the next room or is it lying loose on the floor?
10. A pulse is sent along a thin rope that is attached to a thick rope, which is tied to a wall, as shown in **Figure 14–10.**
    a. What happens when the pulse reaches point A? Point B?
    b. Is the pulse reflected from point A displaced in the same direction as the incident pulse, or is it inverted? What about the pulse reflected from point B?

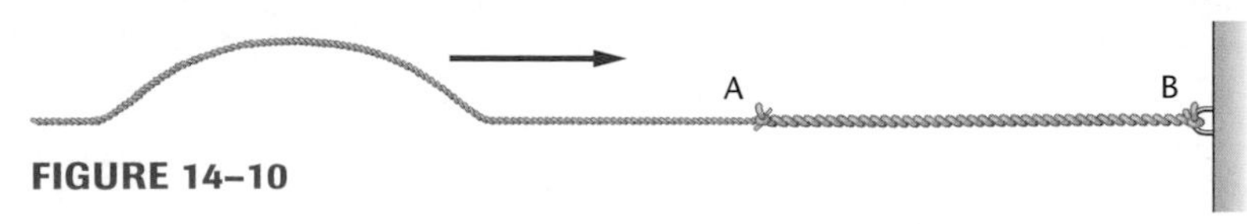

FIGURE 14–10

Pocket Lab

Wave Reflections

Waves lose amplitude and transfer energy when they reflect from a barrier. What happens to the speed of the waves? Use a wave tank with a projection system. Half-fill the tank with water. Dip your finger into the water near one end of the tank and notice how fast the wave that you make moves to the other end.

Analyze Does the wave slow down as it travels? Use a stopwatch to measure the time for a wave to cover two lengths, then four lengths, of the wave tank.

Superposition of Waves

Suppose a pulse traveling down a spring meets a reflected pulse coming back. In this case, two waves exist in the same place in the medium at the same time. Each wave affects the medium independently. The displacement of a medium caused by two or more waves is the algebraic sum of the displacements caused by the individual waves. This is called the **principle of superposition.** In other words, two or more waves can combine to form a new wave. If the waves are in opposite directions, they can cancel or form a new wave of less or greater amplitude. The result of the superposition of two or more waves is called **interference.**

Wave interference

Wave interference can be either constructive or destructive. When two pulses with equal but opposite amplitudes meet, the displacement of the medium at each point in the overlap region is reduced. The superposition of waves with equal but opposite amplitudes causes **destructive interference,** as shown in **Figure 14–11a.** When the pulses meet and are in the same location, the displacement is zero. Point N, which doesn't move at all, is called a **node.** The pulses continue to move and eventually resume their original form. As in constructive interference, the waves pass through each other unchanged.

Constructive interference occurs when the wave displacements are in the same direction. The result is a wave that has an amplitude larger than any of the individual waves. **Figure 14–11b** shows the constructive interference of two equal pulses. A larger pulse appears at point A when the two waves meet. Point A has the largest displacement and is called the **antinode.** The two pulses pass through each other without

FIGURE 14–11 When two equal pulses meet there is a point, called the node, (N), where the medium remains undisturbed **(a)**. Constructive interference results in maximum displacement at the antinode, (A), **(b)**. If the opposite pulses have unequal amplitudes, cancellation is incomplete **(c)**.

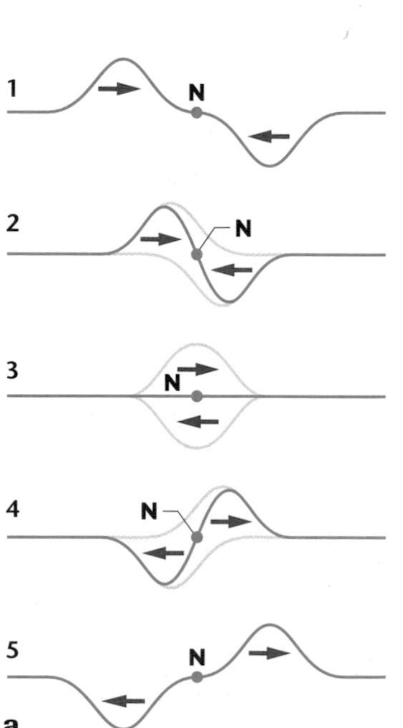

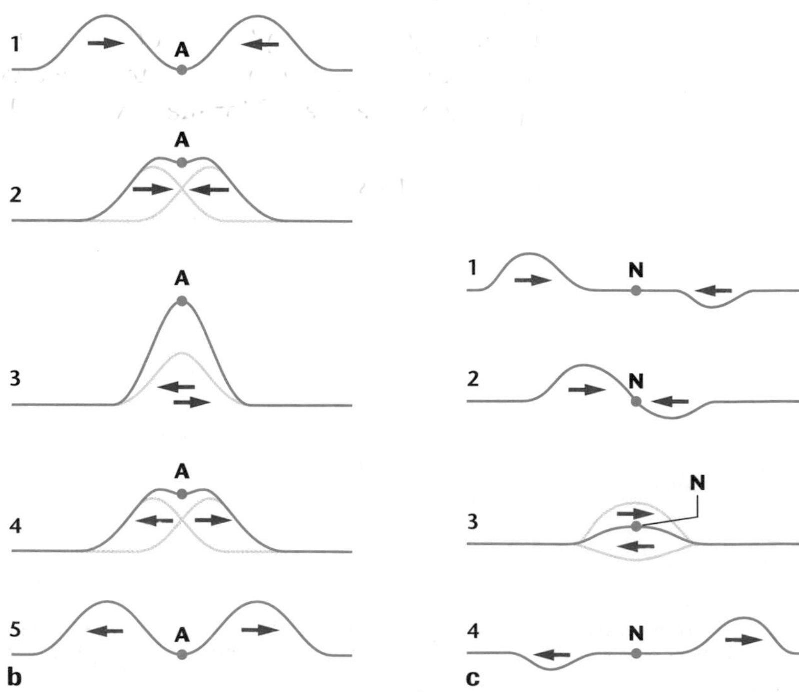

changing their shapes or sizes. If the pulses have opposite and unequal amplitudes, the resultant pulse at the overlap is the algebraic sum of the two pulses, as shown in **Figure 14–11c.**

Continuous waves

You have read how wave pulses move through each other and are reflected from boundaries. What happens when continuous waves meet a boundary? **Figure 14–12a** shows a continuous wave moving from a region with higher speed to a region with lower speed. On the left-hand side of the boundary, the amplitude of the reflected wave has been added to that of the incident wave because some of the wave energy has been reflected.

Recall that velocity of a wave is the product of wavelength and frequency, $v = \lambda f$. Thus, the transmitted wave on the other side of the boundary has a shorter wavelength because of its slower speed. The transmitted wave also has a smaller amplitude because less wave energy is available. Notice in **Figure 14–12b** how the relative amplitudes and wavelengths change when a continuous wave moves from a region with lower speed to one with higher speed.

Standing waves

You can use the concept of superimposed waves to control the frequency and formation of waves. If you attach one end of a rope or coiled spring to a fixed point such as a doorknob, and then start to vibrate the other end, the wave leaves your hand, is reflected at the fixed end, is inverted, and returns to your hand. When it reaches your hand, it is inverted and reflected again. Thus, when the wave leaves your hand the second time, its displacement is in the same direction it was when it left your hand the first time.

But what if you want to magnify the amplitude of the wave you create? Suppose you adjust the motion of your hand so that the period of the rope's vibration equals the time needed for the wave to make one

Pocket Lab

Wave Interaction

What happens to the waves coming from different directions when they meet? Do they slow down, bounce off each other, or go through each other?

Design an Experiment

Use a coiled spring toy to test these questions. Record your procedures and observations.

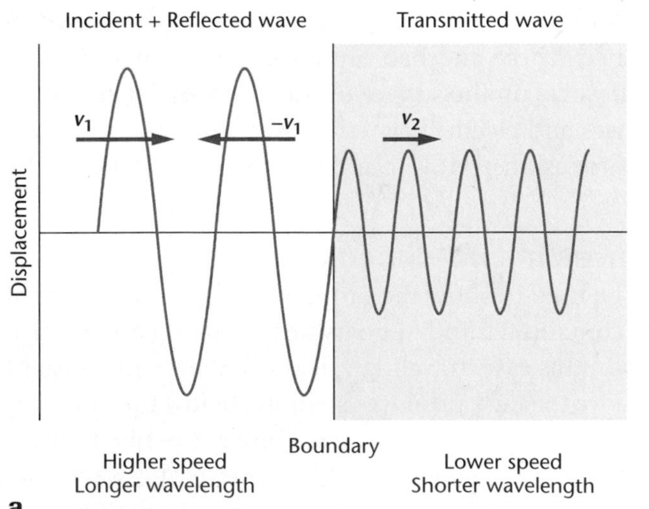

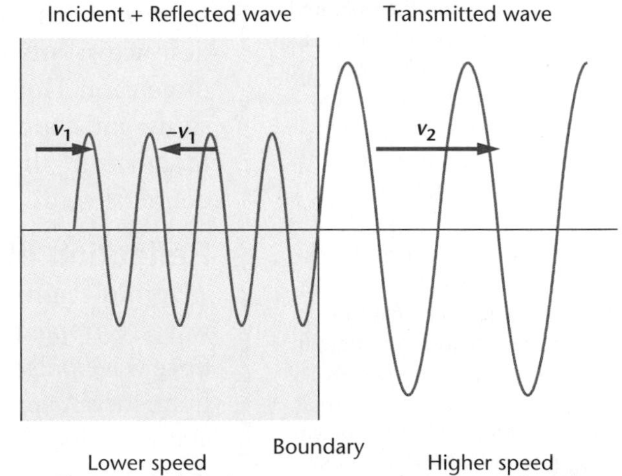

FIGURE 14–12 In each of the two examples below, continuous waves pass from the medium on the left to the medium on the right. Wave speed depends upon the medium. For example, light waves travel faster in air than they do in water.

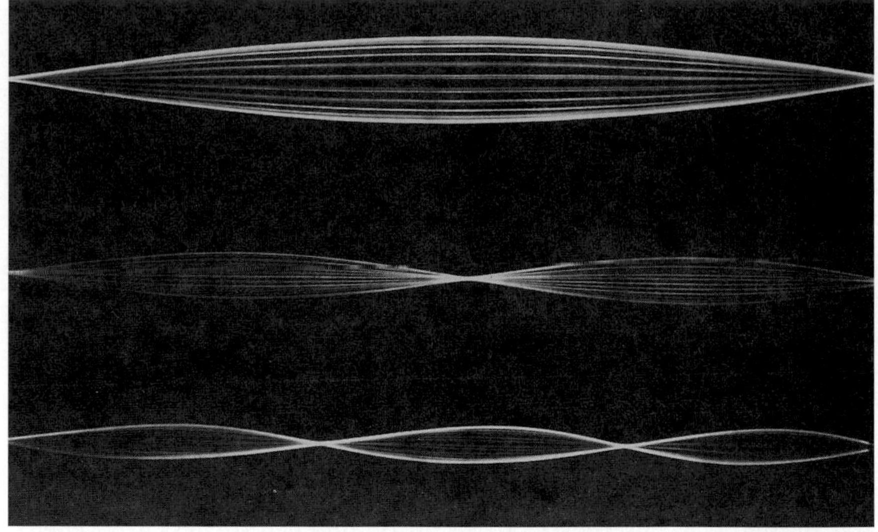

FIGURE 14–13 Interference produces standing waves in a rope. As the frequency of vibration is increased, as shown from top to bottom, the numbers of nodes and antinodes increase.

round-trip from your hand to the door and back. Then, the displacement given by your hand to the rope each time will add to the displacement of the reflected wave. As a result, the oscillation of the rope in one segment will be much larger than the motion of your hand, as expected from your knowledge of constructive interference. This large-amplitude oscillation is an example of mechanical resonance. The nodes are at the ends of the rope and an antinode is in the middle, as shown in **Figure 14–13 top.** Thus, the wave appears to be standing still and is called a **standing wave.** If you double the frequency of vibration, you can produce one more node and one more antinode in the rope. Now it appears to vibrate in two segments, as in **Figure 14–13 center.** Further increases in frequency produce even more nodes and antinodes, as shown in **Figure 14–13 bottom.**

Bent Out of Shape

What happens to a water wave's speed when the depth of the water changes? How does a change in speed affect the shape of the waves? Try this activity to find out. Use a wave tank with a projection system. Adjust the tank so that the water is shallow on one side and deep on the opposite side. Dip your finger or pencil eraser into the middle of the tank, and gently tap the water to make a circular wave. Watch closely. Do the waves hit the sides of the tank at the same time? What happens to the wavelength in different directions?

Modeling Make a sketch of the shape of the waves. Label the deep and shallow ends on your drawing. Describe the relationship between the depth of the water and the wave's speed (inverse or direct). What did you notice about the wavelength in different directions?

Waves in Two Dimensions

You have studied waves on a rope or spring reflecting from a rigid support, where the amplitude of the wave is forced to be zero by destructive interference. These mechanical waves move in only one dimension. However, waves on the surface of water move in two dimensions, and sound waves and electromagnetic waves will later be shown to move in three dimensions. How can two-dimensional waves be demonstrated?

Reflection of waves in two dimensions

A ripple tank can be used to show the properties of two-dimensional waves. A ripple tank contains a thin layer of water. Vibrating boards produce wave pulses or, in this case, traveling waves of water with constant frequency. A lamp above the tank produces shadows below the tank that show the locations of the crests of the waves. **Figure 14–14a** shows a wave pulse traveling toward a rigid barrier that reflects the wave. The incident wave moves upward. The reflected wave moves to the right.

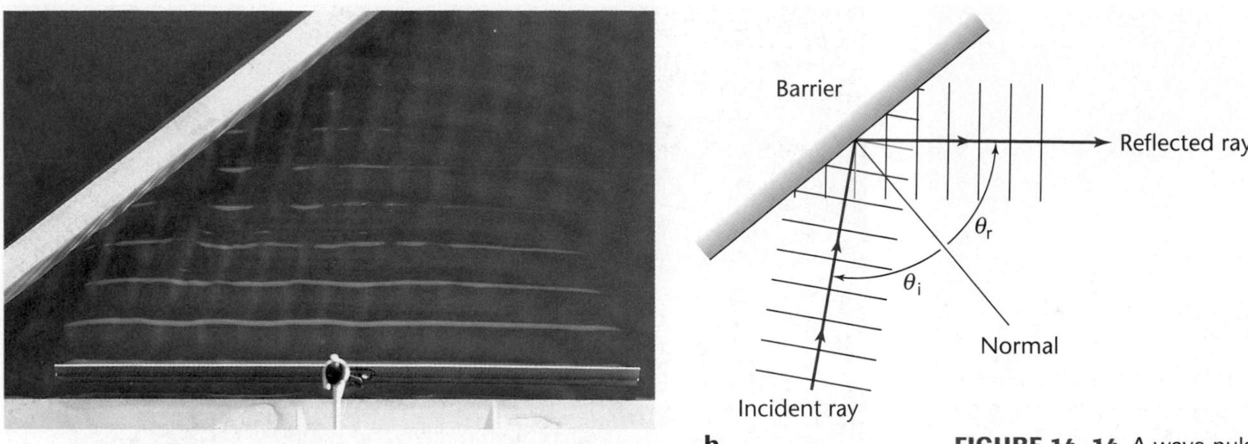

FIGURE 14–14 A wave pulse in a ripple tank is reflected by a barrier **(a).** The ray diagram models the wave in time sequence as it approaches the barrier and then is reflected to the right **(b).**

The direction of waves moving in two dimensions can be modeled by ray diagrams. Ray diagrams model the movement of waves. A ray is a line drawn at a right angles to the crests of the waves. **Figure 14–14b** shows the ray diagram for the wave in the ripple tank. The ray representing the incident ray is the arrow pointing upward. The ray representing the reflected ray points to the right.

The direction of the barrier is also shown by a line, which is drawn at a right angle to the barrier. This line is called the normal. The angle between the incident ray and the normal is called the angle of incidence. The angle between the normal and the reflected ray is called the angle of reflection. The **law of reflection** states that the angle of incidence is equal to the angle of reflection.

Refraction of waves in two dimensions

A ripple tank also can be used to model the behavior of waves as they move from one medium into another. **Figure 14–15a** shows a glass plate placed in a ripple tank. The water above the plate is shallower than the water in the rest of the tank, and the water there acts like a different medium. As the waves move from deep to shallow water, their wavelength decreases, and the direction of the waves changes. Because the waves in the shallow water are generated by the waves in deep water, their frequency is not changed. Based on the equation $v = \lambda f$, the decrease in the wavelength of the waves means that the velocity is lower in the shallower water. This is similar to what happens to a sound wave in the air that collides with and then travels through another medium such as a wall.

This same phenomenon can be seen at the coast when the land gently slopes down into the sea. In **Figure 14–15b,** the waves approach the shore. The ray direction is not parallel to the normal. Not only does the wavelength decrease over the shallower bottom, but also the direction of the waves changes. The change in the direction of waves at the boundary between two different media is known as **refraction.**

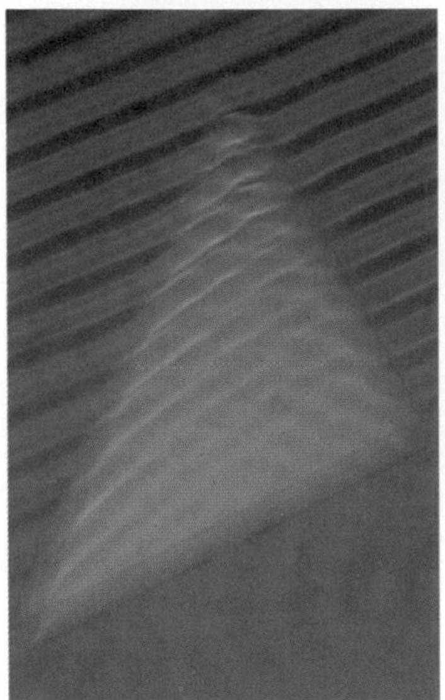

a

b

FIGURE 14–15 As the water waves move over a shallower region of the ripple tank where a triangular glass plate is placed, they slow down and their wavelength decreases **(a).** When waves approach the shore, they are refracted by the change in depth of the water **(b).**

Diffraction and Interference of Waves

If particles are thrown at a barrier with holes in it, the particles will either reflect off the barrier or pass straight through the holes. When waves encounter a small hole in a barrier, however, they do not pass straight through. Rather, they bend around the edges of the barrier, forming circular waves that radiate out, as shown in **Figure 14–16.** The spreading of waves around the edge of a barrier, such as a small barrier coral reef, is called **diffraction.** Diffraction also occurs when waves meet a small obstacle. They can bend around the obstacle, producing waves behind it. The smaller the wavelength in comparison to the size of the obstacle, the less the diffraction.

If a barrier has two closely spaced holes, the waves are diffracted by each hole and form circular waves, as shown in **Figure 14–17a.** These two sets of circular waves interfere with each other. There are regions of constructive interference where the resulting waves are large, and bands of destructive interference where the water remains almost undisturbed. Constructive interference occurs where two crests or two troughs of the circular waves meet. The antinodes formed lie on antinodal lines that

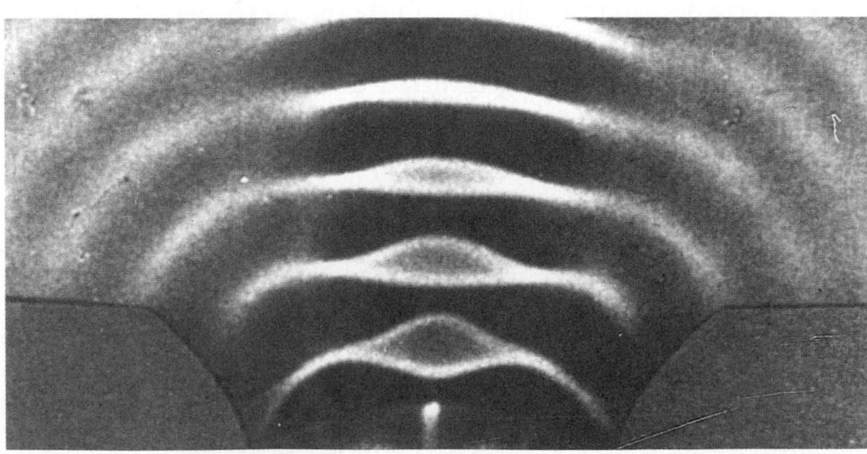

FIGURE 14–16 Waves that bend around the edges of barriers demonstrate diffraction.

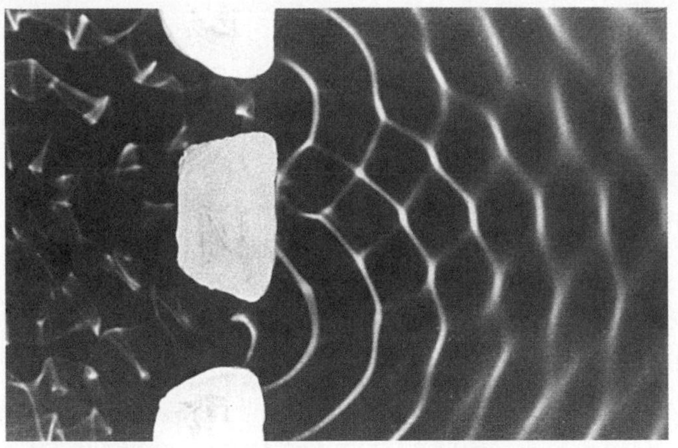

a

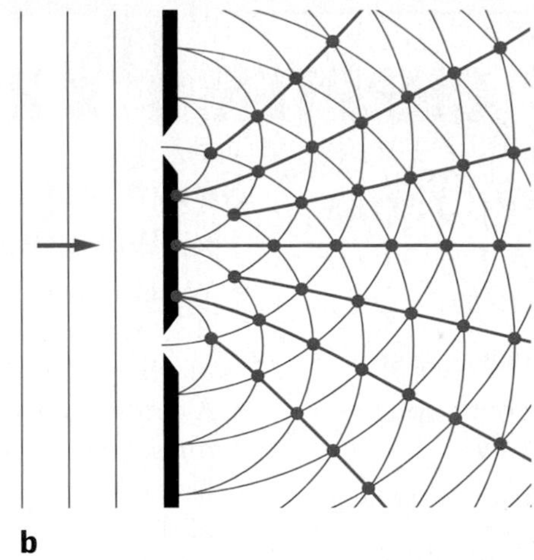

b

FIGURE 14–17 Waves are diffracted at two openings in the barrier. At each opening, circular waves are formed that interfere with each other. Points of constructive interference appear as bright spots in the photograph **(a).** Lines of constructive interference, called antinodal lines, occur where crest meets crest **(b).**

radiate outward from the barrier, **Figure 14–17b.** Between these antinodal lines are areas where a crest of one wave meets a trough from another wave. Destructive interference produces nodes where the water is undisturbed. The lines of nodes, or nodal lines, lie between adjacent antinodal lines. Thus, it is interference of the water waves that produces the series of light and dark lines observed in **Figure 14–17a**.

14.2 Section Review

2.1 Which of the following wave characteristics remain unchanged when a wave crosses a boundary into a different medium: frequency, amplitude, wavelength, velocity, or direction?

2.2 A rope vibrates with the two waves shown in **Figure 14–18.** Sketch the resulting wave.

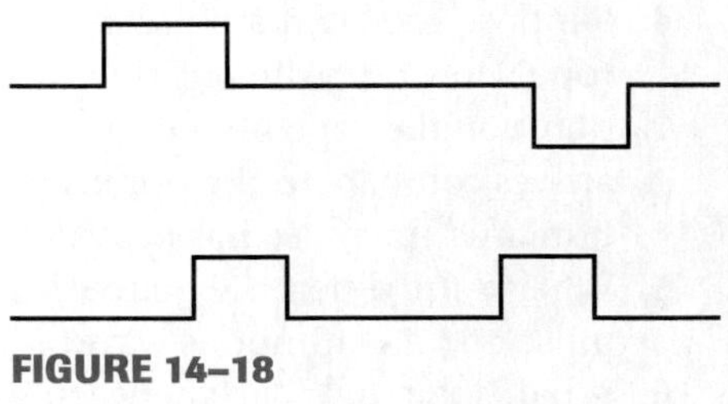

FIGURE 14–18

2.3 Would you expect high-frequency or low-frequency sound waves to be more diffracted when they pass through an open door?

2.4 Critical Thinking As another way to understand wave reflection, cover the right-hand side of each drawing in **Figure 14–11a** with a piece of paper. The edge of the paper should be at point N, the node. Now, concentrate on the resultant wave, shown in blue. Note that it acts like a wave reflected from a boundary. Is the boundary a rigid wall or open ended? Repeat this exercise for **Figure 14–11b.**

CHAPTER 14 REVIEW

Key Terms

14.1

- wave
- wave pulse
- continuous wave
- transverse wave
- longitudinal wave
- surface wave
- trough
- crest
- wavelength
- frequency

14.2

- incident wave
- reflected wave
- principle of superposition
- interference
- destructive interference
- node
- constructive interference
- antinode
- standing wave
- law of reflection
- refraction
- diffraction

Summary

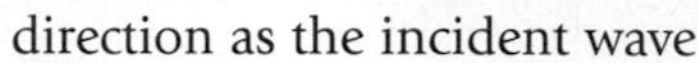

14.1 Wave Properties

- Waves transfer energy without transferring matter.
- Mechanical waves require a medium.
- A continuous wave is a regularly repeating sequence of wave pulses.
- In transverse waves, the displacement of the medium is perpendicular to the direction of wave motion. In longitudinal waves, the displacement is parallel to the wave direction. In surface waves, matter is displaced in both directions.
- The wave source determines the frequency of the wave, f, which is the number of vibrations per second.
- The wavelength of a wave, λ, is the shortest distance between points where the wave pattern repeats itself.
- The medium determines wave speed, which can be calculated for continuous waves using the equation $v = \lambda f$.

14.2 Wave Behavior

- When a wave crosses a boundary between two media, it is partially transmitted and partially reflected, depending on how much the wave velocities in the two media differ.
- When a wave moves to a medium with higher wave speed, the reflected wave is inverted. When moving to a medium with lower wave speed, the displacement of the reflected wave is in the same direction as the incident wave.
- The principle of superposition states that the displacement of a medium resulting from two or more waves is the algebraic sum of the displacements of the individual waves.
- Interference occurs when two or more waves move through a medium at the same time.
- Destructive interference results in decreased wave displacement with its least amplitude at the node.
- Constructive interference results in increased wave displacement with its greatest amplitude at the antinode.
- A standing wave has stationary nodes and antinodes.
- When two-dimensional waves are reflected from boundaries, the angles of incidence and reflection are equal.
- The change in direction of waves at the boundary between two different media is called refraction.
- The spreading of waves around a barrier is called diffraction.

Reviewing Concepts

Section 14.1

1. How many general methods of energy transfer are there? Give two examples of each.
2. What is the primary difference between a mechanical wave and an electromagnetic wave?
3. What are the differences among transverse, longitudinal, and surface waves?
4. Suppose you send a pulse along a rope. How does the position of a point on the rope before the pulse arrives compare to the point's position after the pulse has passed?
5. What is the difference between a wave pulse and a continuous wave?
6. What is the difference between wave frequency and wave velocity?

7. Suppose you produce a transverse wave by shaking one end of a spring from side to side. How does the frequency of your hand compare with the frequency of the wave?
8. Waves are sent along a spring of fixed length.
   a. Can the speed of the waves in the spring be changed? Explain.
   b. Can the frequency of a wave in the spring be changed? Explain.
9. What is the difference between the speed of a transverse wave pulse down a spring and the motion of a point on the spring?
10. Suppose you are lying on a raft in a wave pool. Describe, in terms of the waves you are riding, each of the following: amplitude, period, wavelength, speed, and frequency.
11. What is the amplitude of a wave and what does it represent?
12. What is the relationship between the amplitude of a wave and the energy it carries?

Section 14.2

13. When a wave reaches the boundary of a new medium, part of the wave is reflected and part is transmitted. What determines the amount of reflection?
14. A pulse reaches the boundary of a medium in which the speed of the pulse becomes higher. Is the reflection of the pulse the same as for the incident pulse or is it inverted?
15. A pulse reaches the boundary of a medium in which the speed is lower than the speed of the medium from which it came. Is the reflected pulse erect or inverted?
16. When a wave crosses a boundary between a thin and a thick rope, its wavelength and speed change, but its frequency does not. Explain why the frequency is constant.
17. When two waves interfere, is there a loss of energy in the system? Explain.
18. What happens to a spring at the nodes of a standing wave?
19. A metal plate is held fixed in the center and sprinkled with sugar. With a violin bow, the plate is stroked along one edge and made to vibrate. The sugar begins to collect in certain areas and move away from others. Describe these regions in terms of standing waves.
20. If a string is vibrating in four parts, there are points where it can be touched without disturbing its motion. Explain. How many of these points exist?
21. How does a spring pulse reflected from a rigid wall differ from the incident pulse?
22. Is interference a property of only some types of waves or all types of waves?

Applying Concepts

23. Suppose you hold a 1-m metal bar in your hand and hit its end with a hammer, first, in a direction parallel to its length, second, in a direction at right angles to its length. Describe the waves you produce in the two cases.
24. Suppose you repeatedly dip your finger into a sink full of water to make circular waves. What happens to the wavelength as you move your finger faster?
25. What happens to the period of a wave as the frequency increases?
26. What happens to the wavelength of a wave as the frequency increases?
27. Suppose you make a single pulse on a stretched spring. How much energy is required to make a pulse with twice the amplitude?
28. Sonar is the detection of sound waves reflected off boundaries in water. A region of warm water in a cold lake can produce a reflection, as can the bottom of the lake. Which would you expect to produce the stronger echo? Explain.
29. You can make water slosh back and forth in a shallow pan only if you shake the pan with the correct frequency. Explain.
30. AM-radio signals have wavelengths between 600 m and 200 m, whereas FM signals have wavelengths of about 3 m, **Figure 14–19**. Explain why AM signals can often be heard behind hills whereas FM signals cannot.

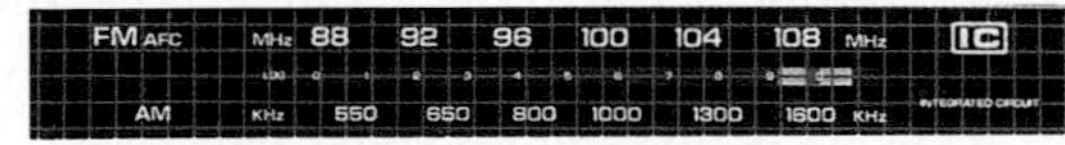

FIGURE 14–19

31. In each of the four waves in **Figure 14–20,** the pulse on the left is the original pulse moving toward the right. The center pulse is a reflected pulse; the pulse on the right is a transmitted pulse. Describe the boundaries at A, B, C, and D.

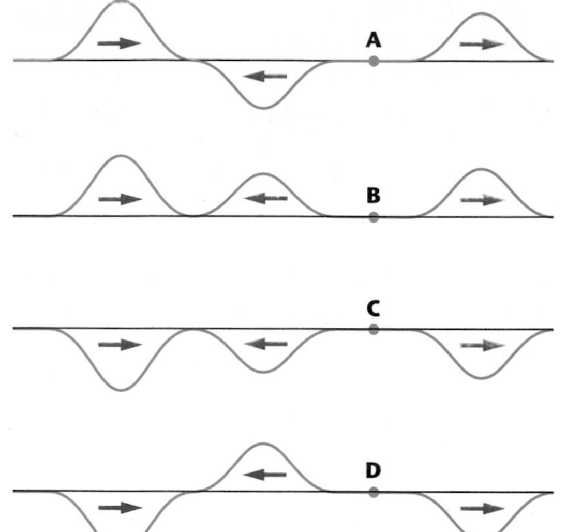

FIGURE 14–20

Problems

Section 14.1

LEVEL 1

32. The Sears Building in Chicago sways back and forth in the wind with a frequency of about 0.10 Hz. What is its period of vibration?
33. An ocean wave has a length of 10.0 m. A wave passes a fixed location every 2.0 s. What is the speed of the wave?
34. Water waves in a shallow dish are 6.0 cm long. At one point, the water oscillates up and down at a rate of 4.8 oscillations per second.
    a. What is the speed of the water waves?
    b. What is the period of the water waves?
35. Water waves in a lake travel 4.4 m in 1.8 s. The period of oscillation is 1.2 s.
    a. What is the speed of the water waves?
    b. What is their wavelength?
36. The frequency of yellow light is 5.0×10^{14} Hz. Find the wavelength of yellow light. The speed of light is 300 000 km/s.
37. AM-radio signals are broadcast at frequencies between 550 kHz and 1600 kHz (kilohertz) and travel 3.0×10^8 m/s.
    a. What is the range of wavelengths for these signals?
    b. FM frequencies range between 88 MHz and 108 MHz (megahertz) and travel at the same speed. What is the range of FM wavelengths?
38. A sonar signal of frequency 1.00×10^6 Hz has a wavelength of 1.50 mm in water.
    a. What is the speed of the signal in water?
    b. What is its period in water?
    c. What is its period in air?
39. A sound wave of wavelength 0.70 m and velocity 330 m/s is produced for 0.50 s.
    a. What is the frequency of the wave?
    b. How many complete waves are emitted in this time interval?
    c. After 0.50 s, how far is the front of the wave from the source of the sound?
40. The speed of sound in water is 1498 m/s. A sonar signal is sent straight down from a ship at a point just below the water surface, and 1.80 s later the reflected signal is detected. How deep is the ocean beneath the ship?
41. The time needed for a water wave to change from the equilibrium level to the crest is 0.18 s.
    a. What fraction of a wavelength is this?
    b. What is the period of the wave?
    c. What is the frequency of the wave?

LEVEL 2

42. Pepe and Alfredo are resting on an offshore raft after a swim. They estimate that 3.0 m separates a trough and an adjacent crest of surface waves on the lake. They count 14 crests that pass by the raft in 20.0 s. Calculate how fast the waves are moving.
43. The velocity of the transverse waves produced by an earthquake is 8.9 km/s, and that of the longitudinal waves is 5.1 km/s. A seismograph records the arrival of the transverse waves 73 s before the arrival of the longitudinal waves. How far away was the earthquake?
44. The velocity of a wave on a string depends on how hard the string is stretched, and on the

mass per unit length of the string. If F_T is the tension in the string, and μ is the mass/unit length, then the velocity, v, can be determined.

$$v = \sqrt{\frac{F_T}{\mu}}$$

A piece of string 5.30 m long has a mass of 15.0 g. What must the tension in the string be to make the wavelength of a 125-Hz wave 120.0 cm?

Section 14.2

LEVEL 1

45. Sketch the result for each of the three cases shown in **Figure 14–21,** when centers of the two wave pulses lie on the dashed line so that the pulses exactly overlap.

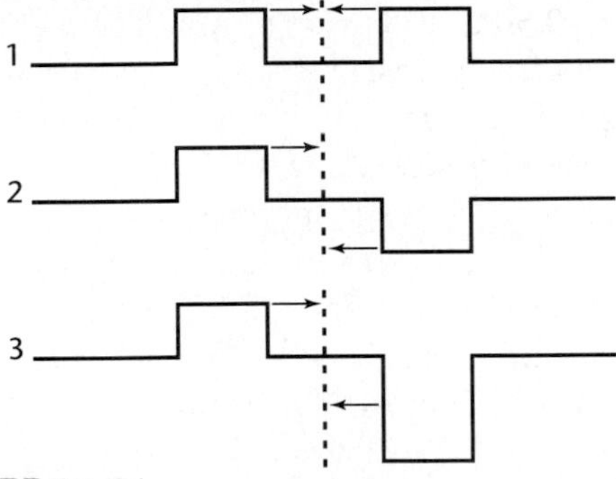

FIGURE 14–21

46. If you slosh the water back and forth in a bathtub at the correct frequency, the water rises first at one end and then at the other. Suppose you can make a standing wave in a 150-cm-long tub with a frequency of 0.30 Hz. What is the velocity of the water wave?

LEVEL 2

47. The wave speed in a guitar string is 265 m/s. The length of the string is 63 cm. You pluck the center of the string by pulling it up and letting go. Pulses move in both directions and are reflected off the ends of the string.

a. How long does it take for the pulse to move to the string end and return to the center?

b. When the pulses return, is the string above or below its resting location?

c. If you plucked the string 15 cm from one end of the string, where would the two pulses meet?

Critical Thinking Problems

48. Gravel roads often develop regularly spaced ridges that are perpendicular to the road. This effect, called washboarding occurs because most cars travel at about the same speed and the springs that connect the wheels to the cars oscillate at about the same frequency. If the ridges are 1.5 m apart and cars travel at about 5 m/s, what is the frequency of the springs' oscillation?

Going Further

Applying Calculators or Computers Use a graphing calculator or computer program to plot the following equation that describes a snapshot of a wave at a fixed time: $y = A \sin (2\pi x/\lambda)$, where y is the displacement, λ the wavelength, x the distance along the wave, and A the amplitude. Evaluate this equation for radians, not degrees. Start with $A = 10$, $\lambda = 6.28$, and let x vary from 0 to 1. Repeat plotting for shorter and longer wavelengths and larger and smaller amplitudes. Display your printed graphs that describe the wave that each set of data represents.

***inter*NET CONNECTION**

Follow the link on the Glencoe Homepage at **www.glencoe.com/sec/science** to find out more about this chapter.

Bat Music

How can a bat determine how far away an insect is? How can it tell when it is getting closer to the insect and how large the insect is?

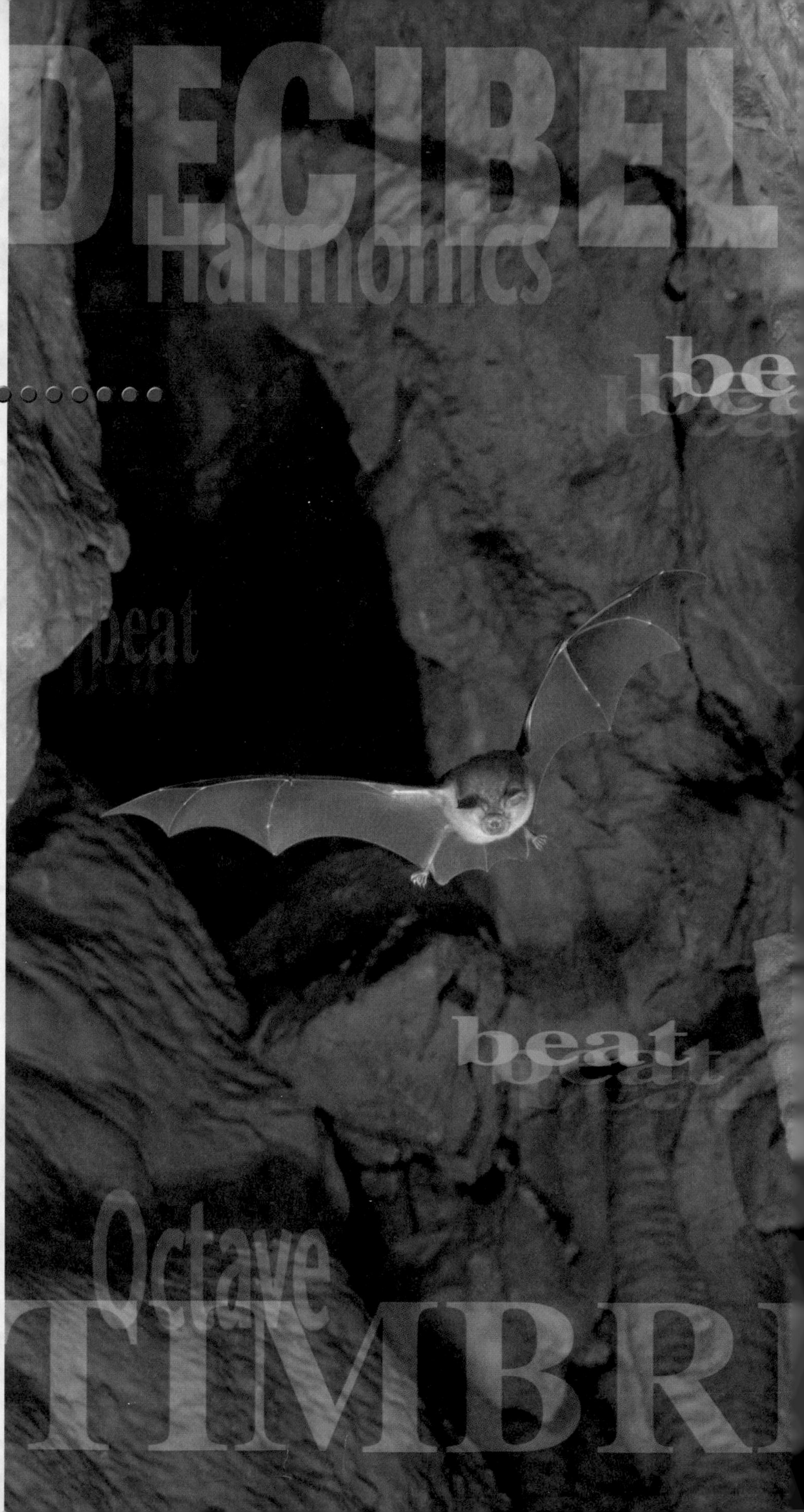

CHAPTER

15 Sound

Sound and music are important to the human experience. Primitive people made sounds not only with their voices, but also with drums, rattles, and whistles. Recently a flute has been found that was made from bear bone over 43 000 years ago. Most likely the flute was played for pleasure and not for survival. For many animals, however, sound is a matter of life or death. Animals use sound to attract mates or to warn of approaching predators. Some animals even use sound to hunt. Bats, in particular, hunt flying insects by emitting pulses of very high-frequency sound. Bats can determine how far away an insect is, whether they are getting closer to the insect, and how large the insect is. In addition, because bats live in large colonies, they must separate the echoes of their own sounds from the sounds of hundreds of other bats. The methods they use are truly amazing and are based on a few physical principles of sound.

WHAT YOU'LL LEARN

- You will describe sound in terms of wave phenomena.
- You will discover the principles behind what makes a sound either music or noise.

WHY IT'S IMPORTANT

- Human communication relies on cords vibrating in throats to send waves through gas, liquids, and solids that end up as electrical impulses in listeners' brains.

interNET CONNECTION

Follow the link for this chapter on the Glencoe Homepage at **www.glencoe.com/sec/science** to find out more about sound.

15.1 Properties of Sound

In the last chapter, you learned how to describe a wave: its speed, its periodicity, its amplitude. You also figured out how waves interact with each other and with matter. If you were told that sound is a type of wave, then you could immediately start describing its properties and interactions. However, the first question you need to answer is just what type of wave is sound?

OBJECTIVES

- **Demonstrate** knowledge of the nature of sound waves and the properties sound shares with other waves.
- **Solve** problems relating the frequency, wavelength, and velocity of sound.
- **Relate** the physical properties of sound waves to the way we perceive sound.
- **Define** the Doppler shift and **identify** some of its applications.

Sound Waves

How is sound produced? Put your fingers against your throat as you hum or speak. Can you feel the vibrations? Have you ever put your hand on the loudspeaker of a boom box? **Figure 15–1** shows a vibrating piston that represents your vocal cords, a loudspeaker, or any other sound source. As it moves back and forth, the piston strikes the molecules in the air. When the piston moves forward, air molecules are driven forward; that is, the air molecules bounce off the piston with a greater velocity. When the piston is at rest, air molecules bounce off the piston with little or no change in velocity. When the piston moves backward, air molecules bounce off the piston with a smaller velocity.

The result of these velocity changes is that the forward motion of the piston produces a region where the air pressure is slightly higher than average. The backward motion produces slightly below-average pressure. Collisions among the air molecules cause the pressure variations to move away from the piston. You can see how this happens by looking at the small red arrows that represent the velocities of the air molecules in **Figure 15–1.** The molecules converge just after the passing of a low-pressure region and the molecules diverge just after the passing of a high-pressure region. If you were to focus at one spot, you would see the value of the air pressure rise and fall, not unlike the behavior of a pendulum. In this way, the pressure variation is transmitted through matter.

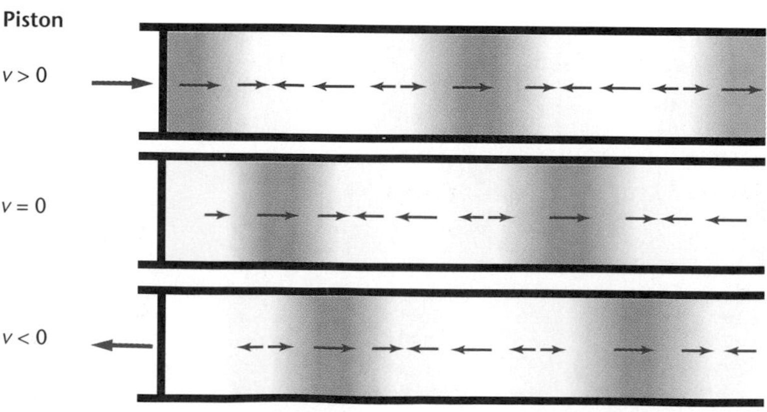

FIGURE 15–1 A vibrating piston produces sound waves. The dark areas represent regions of higher pressure; the light areas regions of lower pressure. Note that these regions move forward in time.

Describing sound

A sound wave is simply a pressure variation that is transmitted through matter. Sound waves move through air because a vibrating source produces regular oscillations in air pressure. The air molecules collide, transmitting the pressure variations away from the source of the sound. The pressure of the air varies, or oscillates, about an average value, the mean air pressure, as shown in **Figure 15–2.** The frequency of the wave is the number of oscillations in pressure each second. The wavelength is the distance between successive regions of high or low pressure. Because the motion of the air molecules is parallel to the direction of motion of the wave, sound is a longitudinal wave.

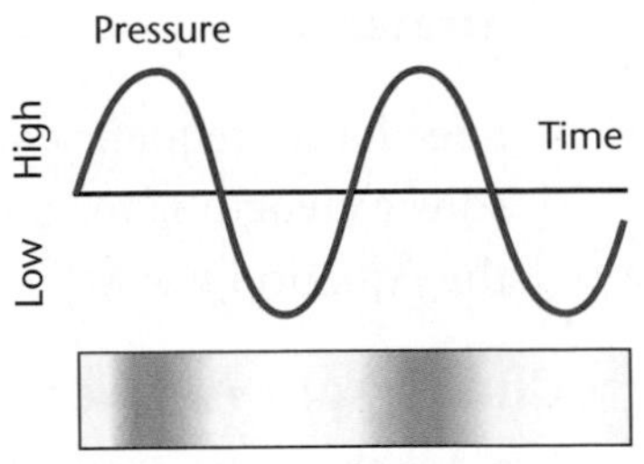

FIGURE 15–2 A graphic representation of the change in pressure over time in a sound wave is shown. Dark areas indicate high pressure; yellow areas indicate average pressure; white areas indicate low pressure.

The speed of a sound wave in air depends on the temperature of the air. Sound waves move through air at sea level at a speed of 343 m/s at room temperature (20°C). Sound can also travel through liquids and solids. In general, the speed of sound is greater in solids and liquids than in gases. Sound cannot travel through a vacuum because there are no particles to move and collide.

Sound waves share the general properties of other waves. They reflect off hard objects, such as the walls of a room. Reflected sound waves are called echoes. The time required for an echo to return to the source of the sound can be used to find the distance between the source and the reflective object. This principle is used by bats, by some cameras, and by ships that employ sonar. Sound waves also can be diffracted, spreading outward after passing through narrow openings. Two sound waves can interfere, causing "dead spots" at nodes where little sound can be heard. And as you learned in Chapter 14, the frequency and wavelength of a wave are related to the speed of the wave by the equation $v = \lambda f$.

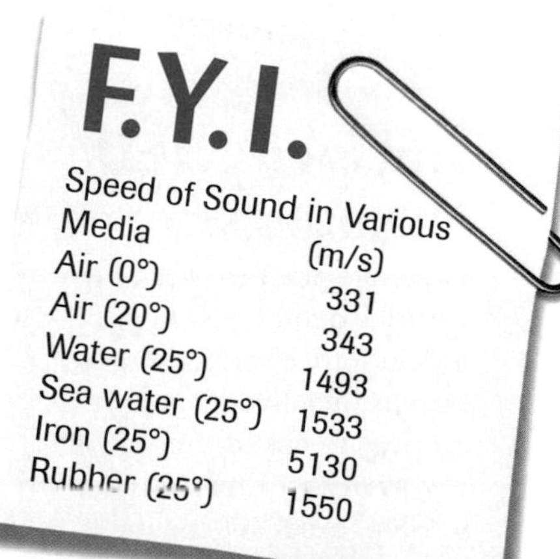

F.Y.I.

Speed of Sound in Various Media

Media	(m/s)
Air (0°)	331
Air (20°)	343
Water (25°)	1493
Sea water (25°)	1533
Iron (25°)	5130
Rubber (25°)	1550

Example Problem

Finding the Wavelength of Sound

A tuning fork produces a sound wave in air with frequency of 261.6 Hz. At room temperature the speed of sound is 343 m/s. What is the wavelength?

Sketch the Problem

- Draw one wavelength using the sine wave model.
- Label the wavelength.

Calculate Your Answer

Known:	**Unknown:**
$f = 261.6$ Hz	$\lambda = ?$
$v = 343$ m/s	

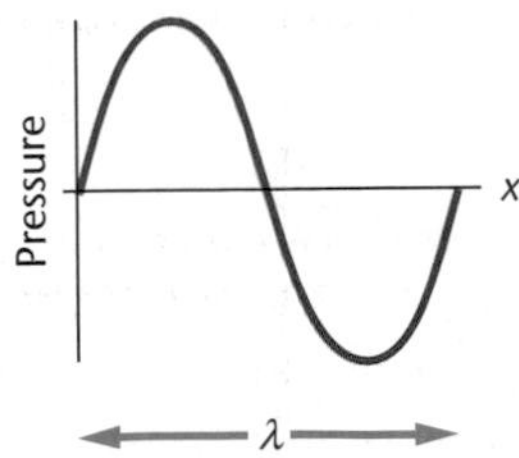

Strategy:

Speed and frequency are known, so wavelength is found by rearranging the equation $v = \lambda f$.

Calculations:

$$v = \lambda f, \text{ so } \lambda = \frac{v}{f}$$

$$\lambda = \frac{343 \text{ m/s}}{261.6 \text{ Hz}} = 1.31 \text{ m}$$

Check Your Answer

- Are the units correct? Work the algebra on the units to verify that the answer is in meters.
- Does the sign make sense? Positive: v, λ, f are always positive.
- Is the magnitude realistic? Audible sound waves in air have wavelengths from 17 mm to 17 m. The answer falls within this range.

Practice Problems

1. Find the frequency of a sound wave moving in air at room temperature with a wavelength of 0.667 m.
2. The human ear can detect sounds with frequencies between 20 Hz and 16 kHz. Find the largest and smallest wavelengths the ear can detect, assuming that the sound travels through air with a speed of 343 m/s at 20°C.
3. If you clap your hands and hear the echo from a distant wall 0.20 s later, how far away is the wall?
4. What is the frequency of sound in air at 20°C having a wavelength equal to the diameter of a 15 in. (38 cm) woofer loudspeaker? Of a 3.0 in. (7.6 cm) tweeter?

HELP WANTED

AUDIOLOGIST

A private practice seeks a certified audiologist to provide in-office services, as well as to deliver contract services to outside businesses and agencies. Patience, compassion, excellent communication skills, thorough knowledge of current equipment, strong concentration skills, and attention to detail are required for success in this career field. Only candidates with a master's degree need apply. For information contact:
American Speech-Language-Hearing Association
10801 Rockville Pike
Rockville, MD 20852

Loudness

The physical characteristics of sound waves are measured by frequency and wavelength, with which you are already familiar. Another physical characteristic of sound waves is amplitude. Amplitude is the measure of the variation in pressure along the wave. In humans, sound is detected by the ear and interpreted by the brain. The loudness of a sound, as perceived by our sense of hearing, depends primarily on the amplitude of the pressure wave.

The human ear is extremely sensitive to the variations in pressure waves, that is, the amplitude of the sound wave. The ear can detect pressure wave amplitudes of less than one billionth of an atmosphere, or 2×10^{-5} Pa. Recall from Chapter 13 that 1 atmosphere of pressure equals 10^5 pascals. At the other end of the audible range, the pressure variations that cause pain are a million times greater, about one thousandth of an atmosphere, or 20 Pa. Remember that the ear can detect only pressure variations. You can easily reduce the external pressure on your ear by thousands of pascals just by driving over a mountain pass, yet a difference of just a few pascals of pressure variation will cause pain.

Because of the wide range in pressure variations that humans can detect, these amplitudes are measured on a logarithmic scale called **sound level.** Sound level is measured in **decibels** (dB). The level depends on the ratio of the pressure variation of a given sound wave to the pressure variation in the most faintly heard sound, 2×10^{-5} Pa. Such an amplitude has a sound level of zero decibels (0 dB). A sound with a ten times larger pressure amplitude (2×10^{-4} Pa) is 20 dB. A pressure amplitude ten times larger than this is 40 dB. Most people perceive a 10 dB increase in sound level as about twice as loud as the original level. **Figure 15–3** shows the sound level in decibels for a variety of sounds.

F.Y.I.

The thermal energy equivalent of the sound emitted over a 90 minute period by a crowd of 50 000 at a football game is only enough to heat one cup of coffee.

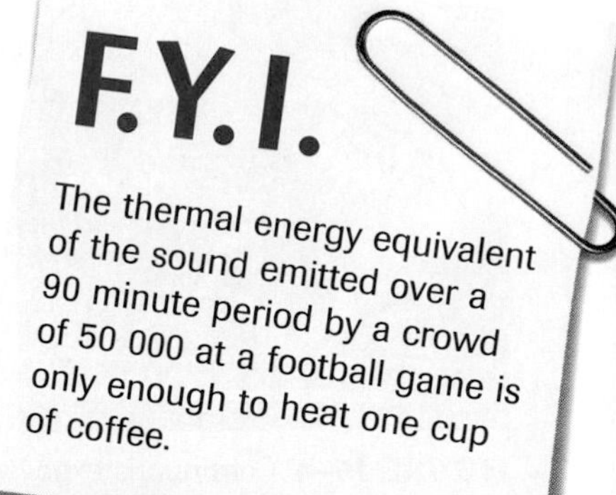

Sound Levels
110 dB
Painful
100 dB
60
70 dB
Noisy
50 dB
Moderate
30 dB
Quiet
10 dB
Barely
audible

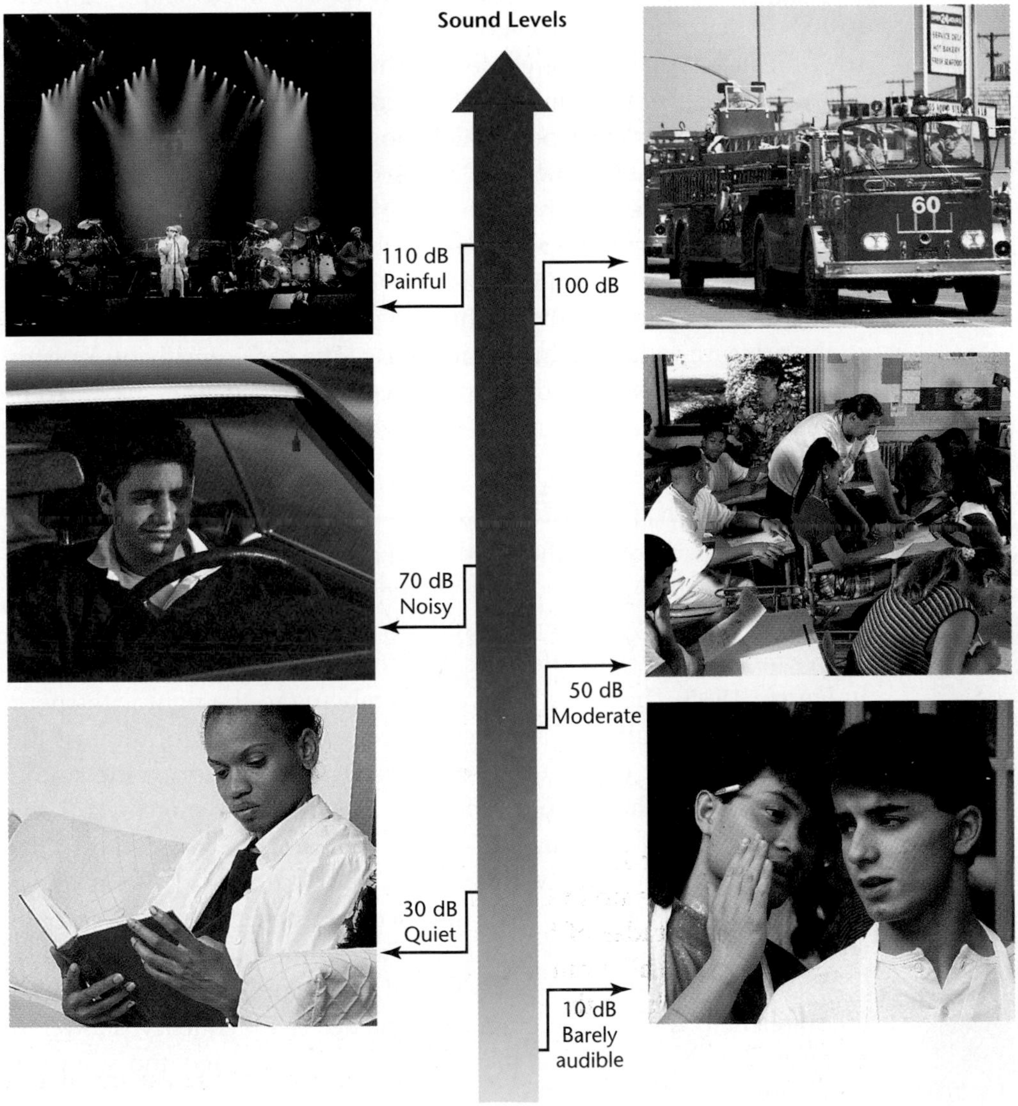

FIGURE 15–3 This decibel scale shows the sound level of some familiar sounds.

FIGURE 15–4 Continuous exposure to loud sounds can cause serious hearing loss. In many occupations, workers such as airline personnel and rock musicians wear ear protection.

Pitch

Marin Mersenne (1588-1648) and Galileo (1564-1642) first connected **pitch** with the frequency of vibration. Pitch also can be given a name on the musical scale. For instance, middle C has a frequency of 262 Hz and an E note has a frequency of 327 Hz. The ear is not equally sensitive to all frequencies. Most people cannot hear sounds with frequencies below 20 Hz or above 16 000 Hz. In general, people are most sensitive to sounds with frequencies between 1000 Hz and 5000 Hz. Older people are less sensitive to frequencies above 10 000 Hz than are young people. By age 70, most people can hear nothing above 8000 Hz. This loss affects the ability to understand speech.

Exposure to loud sounds, either noise or music, has been shown to cause the ear to lose its sensitivity, especially to high frequencies. The longer a person is exposed to loud sounds, the greater the effect. A person can recover from short-term exposure in a period of hours, but the effects of long-term exposure can last for days or weeks. Long exposure to 100 dB or greater sound levels can produce permanent damage. Many rock musicians have suffered serious hearing loss, some as much as 40%. Hearing loss also can result from loud music being transmitted to stereo headphones from personal radios and tape players. The wearer may be unaware just how high the sound level is. Cotton earplugs reduce the sound level only by about 10 dB. Special ear inserts can provide a 25-dB reduction. Specifically designed earmuffs and inserts can reduce the level by up to 45 dB, as shown in **Figure 15–4.**

Loudness, as perceived by the human ear, is not directly proportional to the pressure variations in a sound wave. The ear's sensitivity depends on both pitch and amplitude. Also, perception for pure tones is different than it is for mixtures of tones.

Bat Music

The Doppler Shift

Have you ever noticed the pitch of an ambulance, fire, or police siren as the vehicle sped past you? The frequency is higher when the vehicle is moving toward you, then it suddenly drops to a lower pitch as the source moves away. This effect is called the **Doppler shift** and is shown in **Figure 15–5.** The sound source, S, is moving to the right with speed v_S. The waves it emits spread in circles centered on the location of the source at the time it produced the wave. The frequency of the sound source does not change, but when the source is moving toward the sound detector, O_A in **Figure 15–5a,** more waves are crowded into the space between them. The wavelength is shortened to λ_A. Because the speed of sound is not changed, more crests reach the ear per second, which means the frequency of the detected sound increases.

When the source is moving away from the detector, O_B, in **Figure 15–5a,** the wavelength is lengthened to λ_B and the detected frequency is lower. A Doppler shift also occurs if the detector is moving and the source is stationary, as in **Figure 15–5b.**

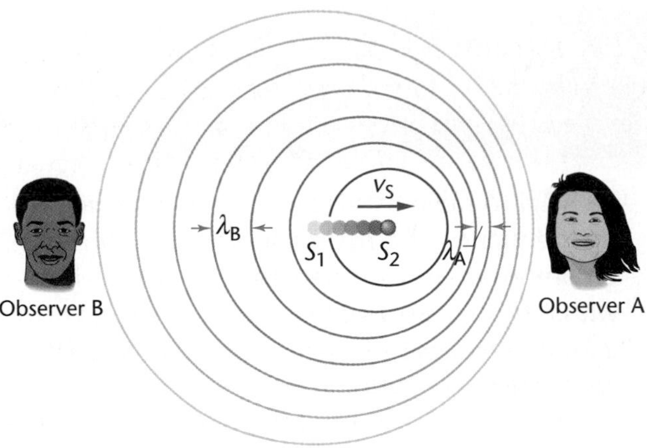

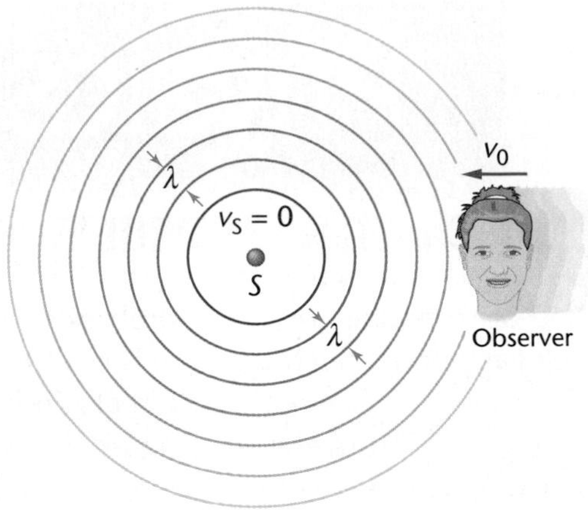

FIGURE 15–5 A Doppler shift can be produced either by the source moving **(a),** or by the receiver moving **(b).**

The Doppler shift occurs in all wave motion, both mechanical and electromagnetic. It has many applications. Radar detectors use the Doppler shift to measure the speed of baseballs and automobiles. Astronomers use the Doppler shift of light from distant galaxies to measure their speed and infer their distance. Physicians can detect the speed of the moving heart wall in a fetus by means of the Doppler shift in ultrasound. A bat uses the Doppler shift to detect and catch flying insects. When an insect is flying faster than the bat, the reflected frequency is lower, but when the bat is catching up to the insect, the reflected frequency is higher.

15.1 Section Review

1.1 The eardrum moves back and forth in response to the pressure variations of a sound wave. Sketch a graph of the displacement of the eardrum versus time for two cycles of a 1-kHz tone and for two cycles of a 2-kHz tone.

1.2 If you hear a train whistle pitch drop as the train passes you, can you tell from which direction the train was coming?

1.3 What physical characteristic of a wave would you change to increase the loudness of a sound? To change the pitch?

1.4 **Critical Thinking** To a person who is hard of hearing, normal conversation (60 dB) sounds like a soft whisper. What increase in sound levels (in dB) must a hearing aid provide? How many times must the sound wave pressure be increased? (Refer to **Figure 15–3.**)

Physics & Society

You *Can* Take It With You!

You or someone you know probably has a cellular phone. Such a device is essentially a sophisticated, two-way radio that sends and receives sound over radio frequencies. Cell phones, as they're often called, are similar to conventional wire-line phones, but a cell phone is mobile—you *can* take it with you to send or receive a phone call even in some of the most remote places on Earth.

The ever-increasing number of wireless communications devices is creating a dilemma between radio astronomers and the users of cellular phones, pagers, and other such devices. Because the wireless devices and radio telescopes operate on similar radio wave frequencies, the amount of radio frequency interference, or RFI, is increasing at an alarming rate.

What's out there?

Anything with a temperature above absolute zero emits radio waves. Stars, planets, and other objects in space give off these waves, which can be used to study many of these celestial objects' physical and chemical properties. Since their inception in the mid-1900s, radio telescopes have discovered clouds of water, ammonia, ethyl alcohol, and other complex molecules in interstellar space. These findings have led some astronomers to hypothesize that because the sun is a typical star, and Earth is a satellite of this star, intelligent life may exist on comparable planets orbiting similar stars somewhere in the more than 100 billion galaxies thought to make up the universe.

Some radio telescopes operate on frequencies ranging from 1610.6 to 1613.8 megahertz (MHz). These frequencies are vital to radio astronomy research because the frequency of a hydroxyl molecule, which is composed of hydrogen and oxygen, lies within this range. Hydroxyl molecules are common in areas of the universe where stars are forming or dying.

Another important frequency to radio astronomical research is 1420.406 MHz, the radio frequency of atomic hydrogen, which makes up over 90% of the universe.

Convenience at the price of research?

Hundreds of mobile communications satellites are currently orbiting Earth. More, no doubt, will soon be launched to keep up with the increasing demand for wireless communications devices. While some international laws and agreements have resulted in the allotment of some portions of the radio spectrum for astronomical research only, many radio astronomers believe that this is not enough. Many argue that their research is being hampered by idle chit-chat.

Investigating the Issue

1. **Acquiring Information** Find out more about how a radio telescope works. Then, suggest what could be done, from the radio astronomy perspective, to help alleviate this problem.
2. **Debating the Issue** Which side of this issue do you support? Explain your position.

Follow the link for this chapter on the Glencoe Homepage at **www.glencoe.com/sec/science** to find out more about radio astronomy and cell phones.

The Physics of Music

15.2

In the middle of the nineteenth century, German physicist Hermann Helmholtz studied sound production in musical instruments and the human voice. In the twentieth century, scientists and engineers developed electronic equipment that permits not only a detailed study of sound, but also the creation of electronic musical instruments and recording devices that allow us to have music whenever and wherever we wish.

OBJECTIVES

- **Describe** the origin of sound.
- **Demonstrate** an understanding of resonance, especially as applied to air columns.
- **Explain** why there is a variation among instruments and among voices using the terms *timbre, resonance, fundamental,* and *harmonic.*
- **Determine** why beats occur.

Sources of Sound

Sound is produced by a vibrating object. The vibrations of the object create molecular motions that cause pressure oscillations in the air. A loudspeaker has a cone that is made to vibrate by electrical currents. The surface of the cone creates the sound waves that travel to your ear and allow you to hear music. Musical instruments such as gongs, cymbals, and drums are other examples of vibrating surfaces that are sources of sound.

The human voice is produced by vibrations of the vocal cords, which are two membranes located in the throat. Air from the lungs rushing through the throat starts the vocal cords vibrating. The frequency of vibration is controlled by the muscular tension placed on the cords.

In brass instruments, such as the trumpet, trombone, and tuba, the lips of the performer vibrate, as shown in **Figure 15–6a.** Reed instruments, like the clarinet, saxophone, and oboe, have a thin wooden strip, or reed, that vibrates as a result of air blown across it, **Figure 15–6b.** In a flute, recorder, organ pipe, and whistle, air is forced across an opening in a pipe. Air moving past the opening sets the column of air in the instrument into vibration.

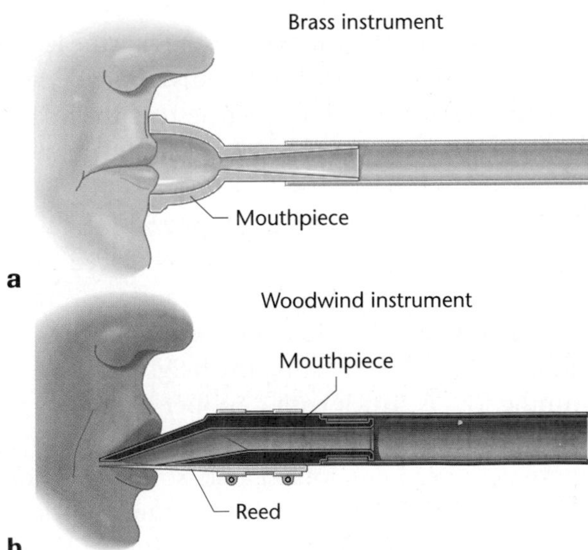

FIGURE 15–6 The shapes of the mouthpieces of a brass instrument **(a)** and a reed instrument **(b)** help determine the characteristics of the sound each instrument produces.

In stringed instruments, such as the piano, guitar, and violin, wires or strings are set into vibration. In the piano, the wires are struck; in the guitar, they are plucked. In the violin, the friction of the bow causes the strings to vibrate. Often, the strings are attached to a sounding board that vibrates with the strings. The vibrations of the sounding board cause the pressure oscillations in the air that we hear as sound. Electric guitars use electronic devices to detect and amplify the vibrations of the guitar strings.

FIGURE 15–7 In all the instruments pictured, changes in pitch are brought about by changing the length of the resonating column of air.

Resonance in Air Columns

If you have ever used just the mouthpiece of a brass or reed instrument, you know that the vibration of your lips or the reed alone does not make a sound with any particular pitch. The long tube that makes up the instrument must be attached if music is to result. When the instrument is played, the air within this tube vibrates at the same frequency, or in resonance, with a particular vibration of the lips or reed. Remember that resonance increases the amplitude of a vibration by repeatedly applying a small external force at the same natural frequency. The length of the air column determines the frequencies of the vibrating air that will be set into resonance. For the instruments shown in **Figure 15–7,** changing the length of the column of vibrating air varies the pitch of the instrument. The mouthpiece simply creates a mixture of different frequencies and the resonating air column acts on a particular set of frequencies to amplify a single note, turning noise into music.

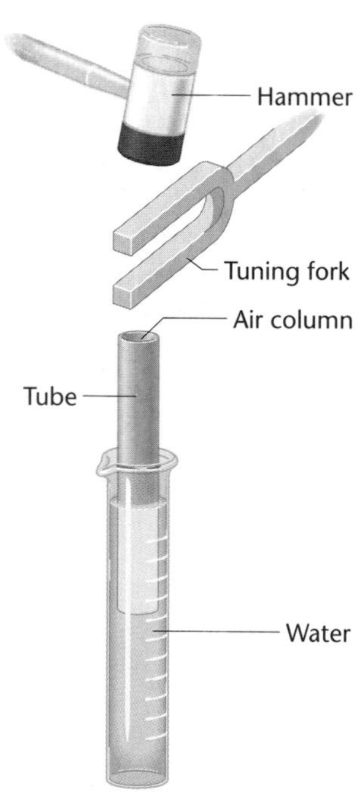

FIGURE 15–8 Raising and lowering the tube changes the length of the air column. When the air column is in resonance with the tuning fork, the sound is loudest.

A tuning fork above a hollow tube can provide resonance in an air column, as shown in **Figure 15–8.** The tube is placed in water so that the bottom end of the tube is below the water surface. A resonating tube with one end closed is called a **closed-pipe resonator.** The length of the air column is changed by adjusting the height of the tube above the water. If the tuning fork is struck with a rubber hammer and the length of the air column is varied as the tube is lifted up and down in the water, the sound alternately becomes louder and softer. The sound is loud when the air column is in resonance with the tuning fork. A resonating air column intensifies the sound of the tuning fork.

Standing pressure wave

How does resonance occur? The vibrating tuning fork produces a sound wave. This wave of alternate high- and low-pressure variations moves down the air column. When the wave hits the water surface, it is reflected back up to the tuning fork, as indicated in **Figure 15–9a.** If the reflected high-pressure wave reaches the tuning fork at the same moment that the fork produces another high-pressure wave, then the leaving and returning waves reinforce each other. This reinforcement of waves produces a standing wave, and resonance is achieved.

An **open-pipe resonator** is a resonating tube with both ends open that also will resonate with a sound source. In this case, the sound wave does not reflect off a closed end, but rather off an open end. The pressure of the reflected wave is inverted; for example, if a high-pressure wave strikes the open end, a low-pressure wave will rebound, as shown in **Figure 15–9b.**

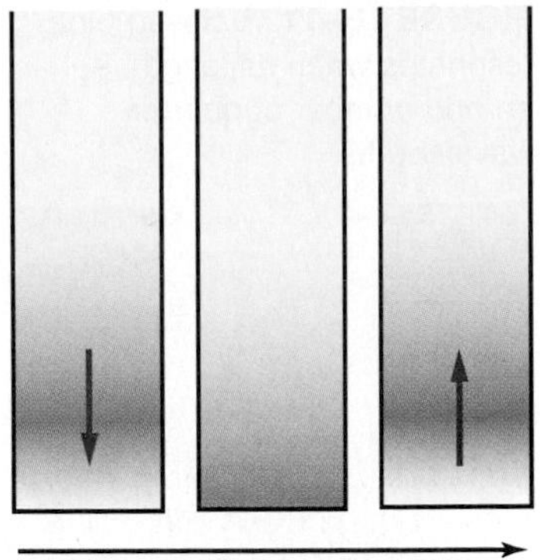

a

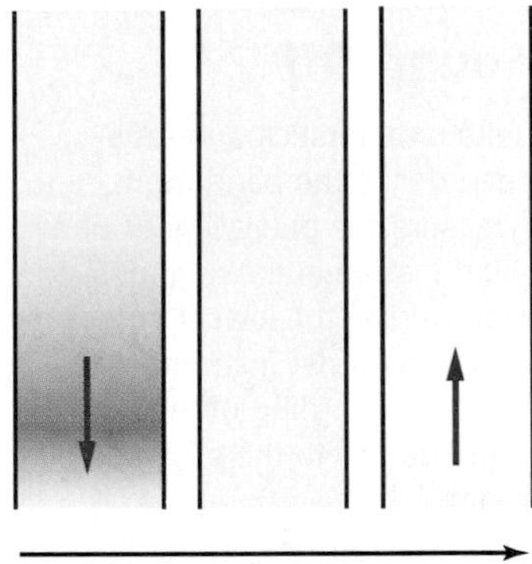

b

FIGURE 15–9 A wave of high pressure will reflect off the end of a pipe, whether the pipe is open or closed.

Resonance lengths

A standing sound wave in a pipe can be represented by a sine wave, as shown in **Figure 15–10.** Sine waves can represent either the air pressure or the displacement of the air molecules. You can see that standing waves have nodes and antinodes. In the pressure graphs, the nodes are regions of mean atmospheric pressure and at the antinodes, the pressure is at its maximum or minimum value. In the case of the displacement graph, the antinodes are regions of high displacement and the nodes are regions of low displacement. In both cases, two antinodes (or two nodes) are separated by one-half wavelength.

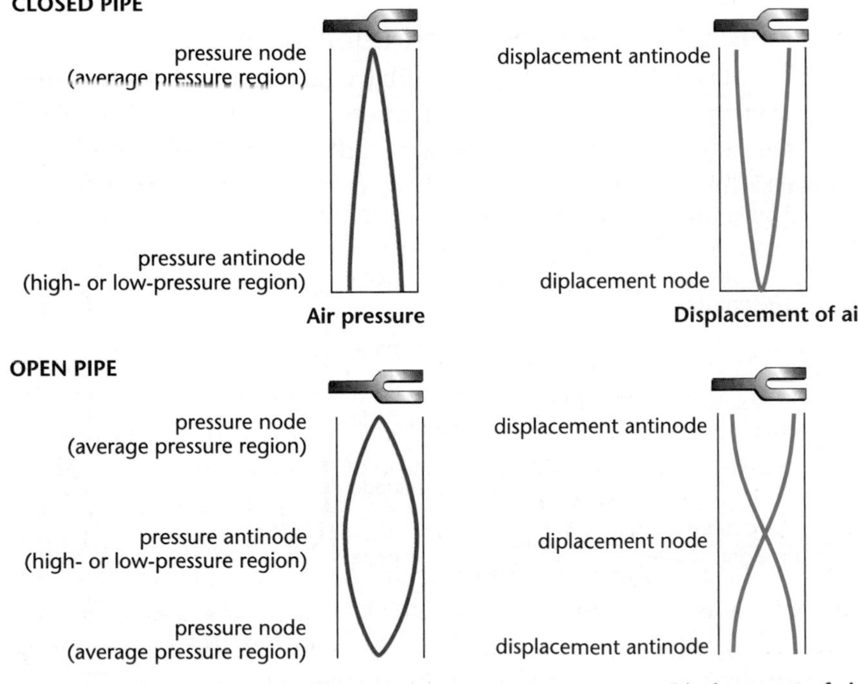

FIGURE 15–10 Sine waves represent standing waves in pipes.

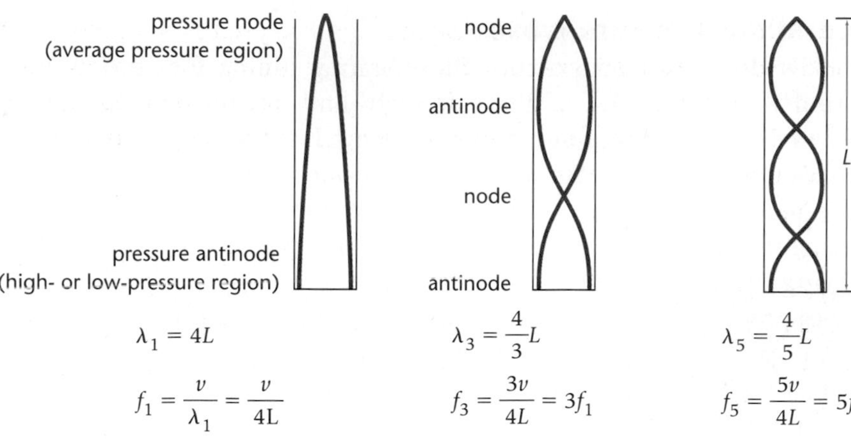

FIGURE 15–11 A closed pipe resonates when its length is an odd number of quarter wavelengths.

Sound Off

Take a meterstick and tape recorder to the band room. Measure the entire length of a wind instrument. Ask a musician to play the lowest note possible on her instrument. Make a recording of the lowest note. Return to the physics room.

Analyze and Conclude For the lowest note, $L = \lambda/2$, what is the wavelength played by the instrument? Use this estimate of the wavelength and the wave equation to predict the frequency. **Hint:** $v = \lambda f$. Use a frequency generator to try to match the recorded note. Read the value on the frequency generator. Is this reading close to your prediction?

Resonance frequencies in a closed pipe

The shortest column of air that can have an antinode at the closed end and a node at the open end is one-fourth wavelength long, as shown in **Figure 15–11.** As the frequency is increased, additional resonance lengths are found at half-wavelength intervals. Thus, columns of length $\lambda/4$, $3\lambda/4$, $5\lambda/4$, $7\lambda/4$, and so on will all be in resonance with a tuning fork.

In practice, the first resonance length is slightly longer than one-fourth wavelength. This is because the pressure variations do not drop to zero exactly at the open end of the pipe. Actually, the node is approximately 1.2 pipe diameters beyond the end. Each additional resonance length, however, is spaced by exactly one-half wavelength. Measurement of the spacings between resonances can be used to find the velocity of sound in air, as shown in the next example problem.

Resonance frequencies in an open pipe

The shortest column of air that can have nodes at both ends is one-half wavelength long, as shown in **Figure 15–12.** As the frequency is increased, additional resonance lengths are found at half-wavelength intervals. Thus columns of length $\lambda/2$, λ, $3\lambda/2$, 2λ, and so on will be in resonance with a tuning fork.

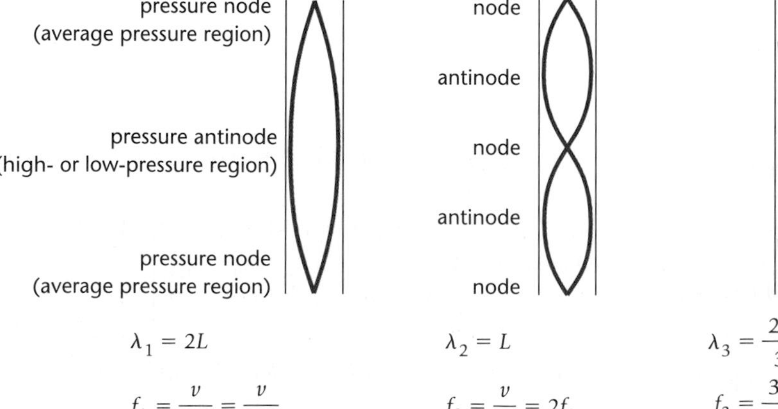

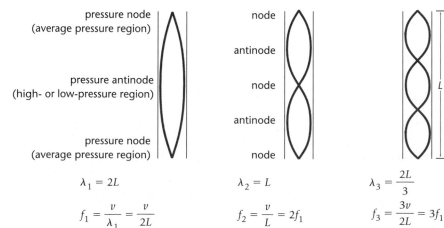

FIGURE 15–12 An open pipe resonates when its length is an even number of quarter wavelengths.

If open and closed pipes of the same length are used as resonators, the wavelength of the resonant sound for the open pipe will be half as long. Therefore, the frequency will be twice as high for the open pipe as for the closed pipe. For both pipes, resonance lengths are spaced by half-wavelength intervals.

Hearing resonance

Musical instruments use resonance to increase the loudness of particular notes. Open-pipe resonators include brass instruments, flutes, and saxophones. The hanging pipes under marimbas and xylophones are examples of closed-pipe resonators. If you shout into a long tunnel or underpass, the booming sound you hear is the tunnel acting as a resonator.

FIGURE 15–13 A shell acts as a closed-pipe resonator to magnify certain frequencies from the background noise.

Example Problem

Finding the Speed of Sound Using Resonance

When a tuning fork with a frequency of 392 Hz is used with a closed-pipe resonator, the loudest sound is heard when the column is 21.0 cm and 65.3 cm long. The air temperature is 27.0°C. What is the speed of sound at this temperature?

Sketch the Problem

- Draw the closed-pipe resonator.
- Mark the resonance lengths.

L_A L_B

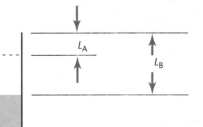

Calculate Your Answer

Known:	Strategy:	Calculations:
$f = 392$ Hz	In closed pipes, resonant lengths are spaced by one-half wavelength.	$L_B - L_A = \frac{1}{2}\lambda$
$L_A = 21.0$ cm		$\lambda = 2(L_B - L_A)$
$L_B = 65.3$ cm	Find wavelength from the difference of the two lengths.	$\lambda = 2(65.3\text{ cm} - 21.0\text{ cm}) = 88.6\text{ cm}$
$T = 27.0°\text{C}$		
Unknown:	Speed is found using frequency and wavelength.	$v = f\lambda$
$v = ?$		$v = (392\text{ Hz})(0.886\text{ m}) = 347\text{ m/s}$

Check Your Answer

- Are the units correct? (Hz)(m) = (1/s)(m) = m/s. The answer's units are correct.
- Does the sign make sense? Positive: speed, wavelength, and frequency are always positive.
- Is the magnitude realistic? Check using the relationship that v_{sound} increases approximately 0.6 m/s per °C. v_{sound} at 20°C is 343 m/s. (7°C)(0.6m/s·°C) = 4.2 m/s higher. The answer is reasonable.

Physics Lab

Speed of Sound

Problem

How can you measure the speed of sound?

Materials

tuning fork
hollow glass tube
1000-mL graduated cylinder
hot water
ice water
thermometer
tuning fork hammer
tape measure

Data and Observations					
Hot Water			Cold Water		
Known	f	=	Known	f	=
Measure	T	=	Measure	T	=
	L	=		L	=
Calculate	λ	=	Calculate	λ	=
	v	=		v	=

Procedure

1. Place cylinders with hot water on one side of the classroom and ice water on the other side of the classroom.
2. Record the value of the frequency that is stamped on the tuning fork and record the temperature of the water.
3. Wear goggles while using tuning forks next to the glass tubes. With the tube lowered in the cylinder, carefully strike the tuning fork with the rubber hammer.
4. Hold the tuning fork above the glass tube while you slowly raise the tube until the sound is amplified, and is loudest by the tube.
5. Measure L, the distance from the water to the top of the tube, to the nearest 0.5 cm.
6. Trade places with another group on the other side of the room and repeat steps 2-5 using the same tuning fork.
7. Repeat steps 2-6 using a different tuning fork.

Analyze and Conclude

1. **Calculating Results** Calculate the values for λ and v.
2. **Comparing Results** Were the values of v different for cold and hot air? How do the values of v compare for different tuning forks?
3. **Making Inferences** Write a general statement describing how the speed of sound depends on the variables tested in this experiment.
4. **Forming an Explanation** Describe a possible model of sound moving through air that will explain your results.

Apply

1. What would an orchestra sound like if the higher frequencies traveled faster than the lower frequencies?

Practice Problems

5. A 440-Hz tuning fork is held above a closed pipe. Find the spacings between the resonances when the air temperature is 20°C.
6. The 440-Hz tuning fork is used with a resonating column to determine the velocity of sound in helium gas. If the spacings between resonances are 110 cm, what is the velocity of sound in He?
7. The frequency of a tuning fork is unknown. A student uses an air column at 27°C and finds resonances spaced by 20.2 cm. What is the frequency of the tuning fork?
8. A bugle can be thought of as an open pipe. If a bugle were straightened out, it would be 2.65 m long.
   a. If the speed of sound is 343 m/s, find the lowest frequency that is resonant in a bugle (ignoring end corrections).
   b. Find the next two higher-resonant frequencies in the bugle.
9. A soprano saxophone is an open pipe. If all keys are closed, it is approximately 65 cm long. Using 343 m/s as the speed of sound, find the lowest frequency that can be played on this instrument (ignoring end corrections).

BIOLOGY CONNECTION

The human auditory canal acts as a closed-pipe resonator that increases the ear's sensitivity for frequencies between 2000 and 5000 Hz. In the middle ear, the three bones, or ossicles, act as a lever, a simple machine with a mechanical advantage of 1.5. By a feedback mechanism, the bones also help to protect the inner ear from loud sounds. Muscles, triggered by a loud noise, pull the third bone away from the oval window.

Detection of Pressure Waves

Sound detectors convert sound energy—the kinetic energy of the vibrating air molecules—into another form of energy. A common detector is a microphone, which converts sound waves into electrical energy. A microphone consists of a thin disk that vibrates in response to sound waves and produces an electrical signal. You'll learn about this transformation process in Chapter 25, after your study of electricity.

The human ear

The ear is an amazing sound detector. Not only can it detect sound waves over a very wide range of frequencies, but it also is sensitive to an enormous range of wave amplitudes. In addition, human hearing can distinguish many different qualities of sound. The ear is a complex sense organ. Knowledge of both physics and biology is required to understand it. The interpretation of sounds by the brain is even more complex, and it is not totally understood.

The human ear, shown in **Figure 15–14,** is a device that collects pressure waves and converts them to electrical impulses. Sound waves entering your auditory canal cause vibrations of the tympanic membrane. Three tiny bones then transfer this vibration to fluid in the cochlea. Tiny hairs lining the spiral-shaped cochlea pick up certain frequencies out of the fluid vibration. The hairs then stimulate nerve cells, which send an impulse to the brain, producing the sensation of sound.

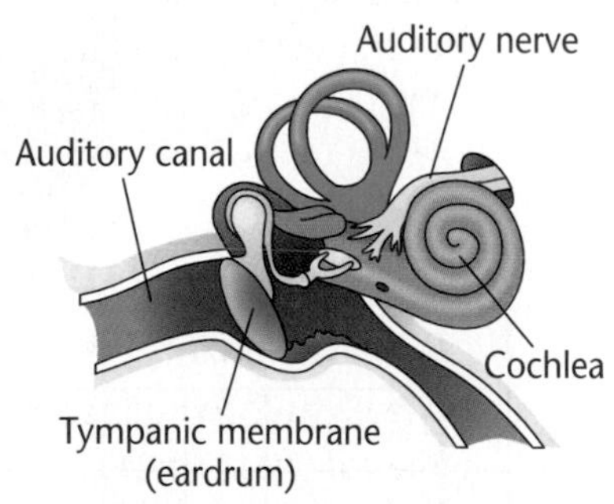

FIGURE 15–14 The human ear is a complex sense organ that translates sound vibrations into nerve impulses that are then sent to the brain for interpretation.

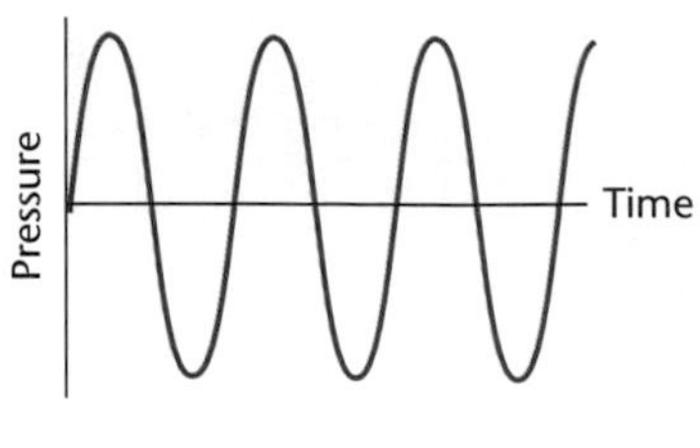

FIGURE 15–15 Graph of pure sound versus time **(a).** Graph of clarinet sound versus time **(b).**

Sound Quality

A tuning fork produces a soft and uninteresting sound. That's because its tines vibrate like simple harmonic oscillators, producing the simple sine wave shown in **Figure 15–15a.** Sounds made by the human voice and musical instruments are much more complex, like the wave in **Figure 15–15b.** Both waves have the same frequency or pitch, but they sound very different. The complex wave is produced by using the principle of superposition to add waves of many frequencies. The shape of the wave depends on the relative amplitudes of these frequencies. In musical terms, the difference between the two waves is called **timbre,** tone color, or tone quality.

The sound spectrum: fundamental and harmonics

The complex sound wave in **Figure 15–15b** was made by a clarinet. Why does an instrument such as a clarinet produce such a sound wave? The air column in a clarinet acts as a closed pipe. Look back at **Figure 15–11,** which shows three resonant frequencies for closed pipes. If the clarinet is of length L, then the lowest frequency, f_1, that will be resonant is $v/4L$. This lowest frequency is called the **fundamental.** But a closed pipe will also resonate at $3f_1$, $5f_1$, and so on. These higher frequencies, which are odd-number multiples of the fundamental frequency, are called **harmonics.** It is the addition of these harmonics that gives a clarinet its distinctive timbre.

Some instruments, such as an oboe, act as open-pipe resonators. Their fundamental frequency is $f_1 = v/2L$ with harmonics at frequencies of $2f_1$, $3f_1$, $4f_1$ and so on. Different combinations and amplitudes of these harmonics give each instrument its own unique timbre. A graph of the amplitude of a wave versus its frequency is called the sound spectrum. The spectra of three instruments are shown in **Figure 15–16.** Each spectrum is as different as is the timbre of the instrument.

FIGURE 15–16 A violin, clarinet, and piano produce characteristic sound spectra.

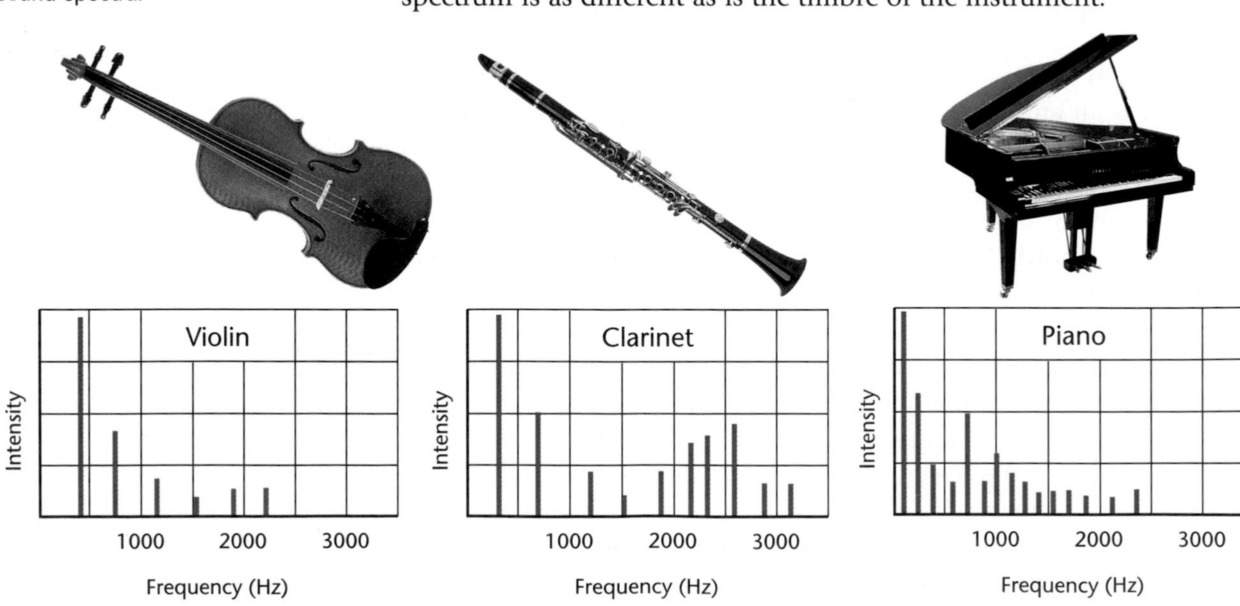

Sounds good? Consonance and dissonance

When sounds that have two different pitches are played at the same time, the resulting sound can be either pleasant or jarring. In musical terms, several pitches played together are called a chord. An unpleasant set of pitches is called **dissonance.** If the combination is pleasant, the sounds are said to be in **consonance.**

What makes a sound pleasant to listen to? Different cultures have different definitions, but most Western cultures accept the definitions of Pythagoras, who lived in ancient Greece. Pythagoras experimented by plucking two strings at the same time. He noted that pleasing sounds resulted when the strings had lengths in small, whole-number ratios, for example 1:2, 2:3, or 3:4. Musicians find that there are many such musical intervals that produce agreeable sounds.

F.Y.I.

A sonic boom will occur any time the speed of an aircraft goes above the speed of sound (Mach 1). Flying at supersonic speed causes the disturbed air to bunch together and collectively impact causing a thunder-like clap.

Musical intervals

Two notes with frequencies related by the ratio 1:2 are said to differ by an **octave.** For example, if a note has a frequency of 440 Hz, a note an octave higher has a frequency of 880 Hz. A note one octave lower has a frequency of 220 Hz. The fundamental and its harmonics are related by octaves; the first harmonic is one octave higher, the second is two octaves higher, and so on. The sum of the fundamental and the first harmonic is shown in **Figure 15–17a.** It is important to recognize that it is the ratio of two frequencies, not the size of the interval between them, that determines the musical interval.

In other common musical intervals, two pitches may be close together. For example, the ratio of frequencies for a "major third" is 4:5. A typical major third is made up of the notes C and E. The note C has a frequency of 262 Hz, so E has a frequency of (5/4)(262 Hz) = 327 Hz. In the same way, notes in a "fourth" (C and F) have a frequency ratio of 3:4, and those in a "fifth" (C and G) have a ratio of 2:3. Graphs of these pleasant sounds are shown in **Figure 15–17.** More than two notes sounded together also can produce consonance. You are probably familiar with the major chord made up of the three notes called *do, mi,* and *sol.* For at least 2500 years, this has been recognized as the sweetest of the three-note chords; it has the frequency ratio of 4:5:6.

Ring, Ring

How good is your hearing? Here is a simple test to find out. Find a penny, a nickel, a dime, and a quarter. Ask a lab partner to drop them in any order and listen closely. Can you identify the sound of each coin with your eyes closed?

Analyze and Conclude

Describe the differences in the sounds. What are the physical factors that cause the differences in the sounds? Can you suggest any patterns?

FIGURE 15–17 These time graphs show the superposition of two waves having the ratios of 1:2, 2:3, 3:4, and 4:5.

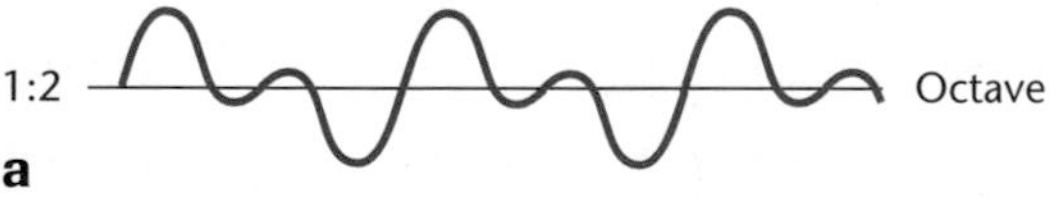

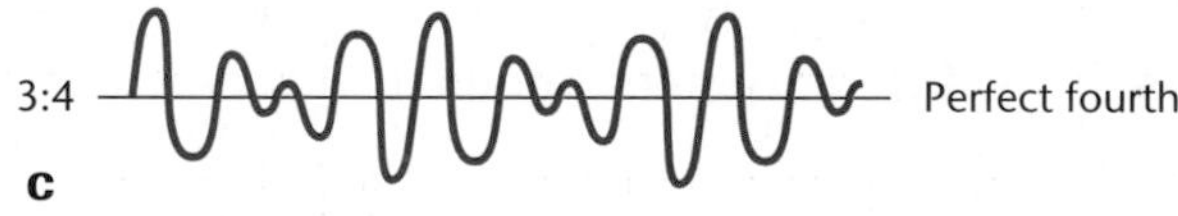

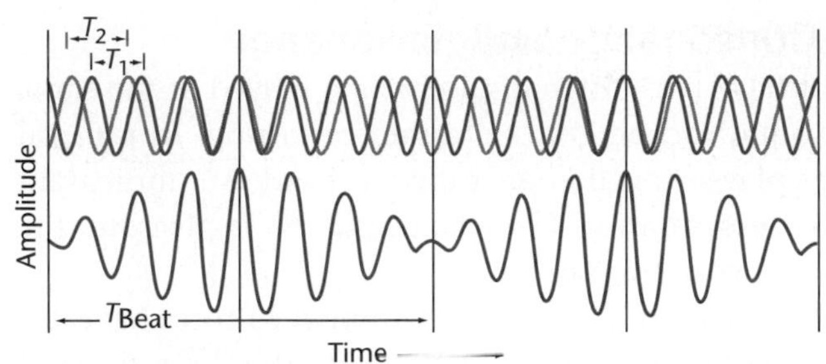

FIGURE 15–18 Beats occur as a result of the superposition of two sound waves of slightly different frequencies.

Beat Notes

You have seen that consonance is defined in terms of the ratio of frequencies. When the ratio becomes nearly 1:1, the frequencies become very close. Two frequencies that are nearly identical interfere to produce high and low sound levels, as illustrated in **Figure 15–18.** This oscillation of wave amplitude is called a **beat.** The frequency of a beat is the magnitude of difference between the frequencies of the two waves, $f_{beat} = |f_A - f_B|$. When the difference is less than 7 Hz, the ear detects this as a pulsation of loudness. Musical instruments often are tuned by sounding one against another and adjusting the frequency of one until the beat disappears.

Example Problem

Finding the Beat

Two tuning forks, one with frequency of 442 Hz, the other with frequency of 444 Hz, are struck at the same time. What beat frequency will result?

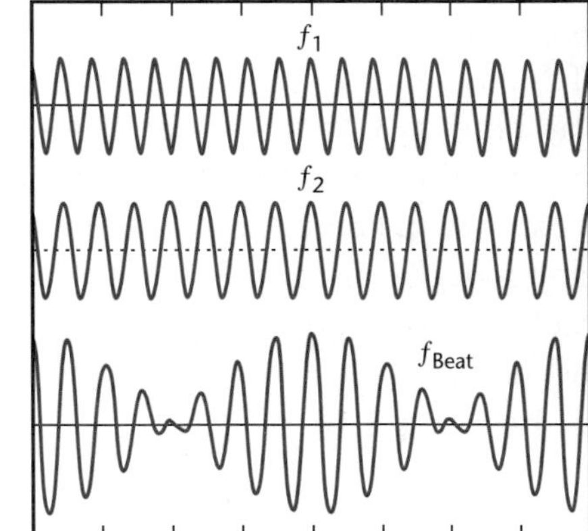

Sketch the Problem

- In beat problems, making a sketch isn't necessary. However, it is instructive to take the time to produce at least one sketch depicting the interference of two waves, such as the sketch to the right.

Calculate Your Answer

Known:	**Unknown:**
$f_A = 442$ Hz	$f_{beat} = ?$
$f_B = 444$ Hz	

Strategy:

The beat frequency is the magnitude of the difference between the two frequencies.

Calculations:

$f_{beat} = |f_A - f_B|$
$f_{beat} = |442 \text{ Hz} - 444 \text{ Hz}|$
$f_{beat} = 2$ Hz

Check Your Answer

- Are the units correct? Frequencies are measured in Hz.
- Does the sign make sense? Frequencies are always positive.

Practice Problems

10. A 330-Hz and a 333-Hz tuning fork are struck simultaneously. What will the beat frequency be?

Sound Reproduction and Noise

How often do you listen to sound produced directly by a human voice or musical instrument? Most of the time the sound has been recorded and played through electronic systems. To reproduce the sound faithfully, the system must accommodate all frequencies equally. A good stereo system keeps the amplitudes of all frequencies between 20 and 20 000 Hz the same to within 3 dB.

A telephone system, on the other hand, needs only to transmit the information in spoken language. Frequencies between 300 and 3000 Hz are sufficient. Reducing the number of frequencies present helps reduce the noise. A noise wave is shown in **Figure 15–19.** Many frequencies are present with approximately the same amplitude. While noise is not helpful in a telephone system, people claim that listening to noise has a calming effect, and some dentists use noise to help their patients relax.

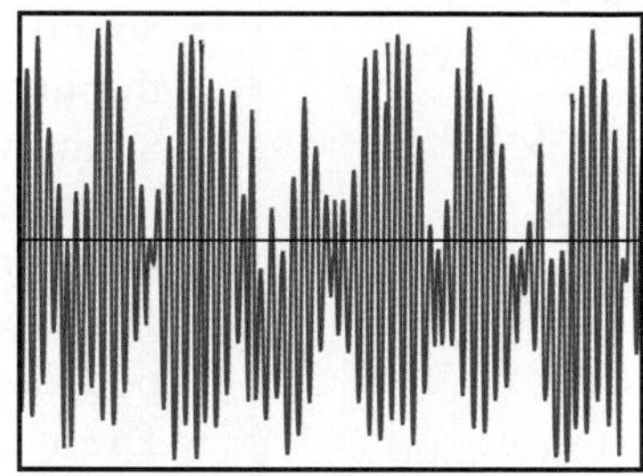

FIGURE 15–19 Noise is composed of several frequencies and involves random changes in frequency or amplitude.

15.2 Section Review

2.1 What is the vibrating object that produces sounds in

a. the human voice?

b. a clarinet?

c. a tuba?

2.2 Hold one end of a ruler against your desktop, with one-quarter of it extending over the desk. Pluck the extended end of the ruler.

a. Where does the noise you hear come from?

b. Test your answer by placing a towel between the ruler and desk. What do you hear and what does the towel do?

c. How does part **a** demonstrate sound production by a guitar?

2.3 The speech of a person with a head cold often sounds different from that person's normal speech. Explain.

2.4 **Critical Thinking** The end correction to a closed organ pipe increases the effective length by about 1.2 pipe diameters.

a. Are the frequencies of the fundamental and higher harmonics still related by the numbers 1, 3, 5, and so on?

b. Actually, the end correction depends slightly on wavelength. Does that change your answer to part **a**? Explain.

CHAPTER 15 REVIEW

Key Terms

15.1

- sound level
- decibel
- pitch
- Doppler shift

15.2

- closed-pipe resonator
- open-pipe resonator
- timbre
- fundamental
- harmonic
- dissonance
- consonance
- octave
- beat

Summary

15.1 Properties of Sound

- Sound is a pressure variation transmitted through matter as a longitudinal wave.
- Sound waves have frequency, wavelength, and speed. Sound waves reflect and interfere.
- The speed of sound in air at room temperature (20°C) is 343 m/s.
- The amplitude of a sound wave is measured in decibels (dB).
- The loudness of sound as perceived by the ear and brain depends mainly on its amplitude.
- The frequency of a sound wave is heard as its pitch.
- The Doppler shift is the change in frequency of sound caused by the motion of either the source or detector.

15.2 The Physics of Music

- Sound is produced by vibrating objects in matter.
- Most sounds are complex waves that are composed of more than one frequency.
- An air column can resonate with a sound source, increasing its amplitude.
- An open pipe resonates when its length is $\lambda/2$, $2\lambda/2$, $3\lambda/2$, and so on.
- A closed pipe resonates when its length is $\lambda/4$, $3\lambda/4$, $5\lambda/4$, and so on.
- Sound detectors convert the energy carried by a sound wave into another form of energy.
- The frequencies and intensities of the complex waves produced by a musical instrument determine the timbre that is characteristic of that instrument.
- The fundamental frequency and harmonics can be described in terms of resonance.
- The shape of the throat and mouth cavity determines the timbre of the human voice.
- Notes on a musical scale differ in frequency by small, whole-number ratios.
- Two waves with almost the same frequency interfere to produce a beat note.

Reviewing Concepts

Section 15.1

1. A firecracker is set off near a set of hanging ribbons. Describe how they might vibrate.
2. In the nineteenth century, people put their ears to a railroad track to get an early warning of an approaching train. Why did this work?
3. When timing the 100-m run, officials at the finish line are instructed to start their stopwatches at the sight of smoke from the starter's pistol and not at the sound of its firing. Explain. What would happen to the times for the runners if the timing started when sound was heard?
4. Does the Doppler shift occur for only some types of waves or for all types of waves?
5. Sound waves with frequencies higher than can be heard by humans, called ultrasound, can be transmitted through the human body. How could ultrasound be used to measure the speed of blood flowing in veins or arteries?

Section 15.2

6. How can a certain note sung by an opera singer cause a crystal glass to shatter?
7. In the military, as marching soldiers approach a bridge, the command "route step" is given. The soldiers then walk out-of-step with each other as they cross the bridge. Explain.

8. How must the length of an open tube compare to the wavelength of the sound to produce the strongest resonance?
9. Explain how the slide of a trombone changes the pitch of the sound in terms of a trombone being a resonance tube.
10. What property distinguishes notes played on both a trumpet and a clarinet if they have the same pitch and loudness?

Applying Concepts

11. A common method of estimating how far a lightning flash is from you is to count the seconds between the flash and the thunder, and then divide by 3. The result is the distance in kilometers. Explain how this rule works. Devise a similar rule for miles.
12. The speed of sound increases by about 0.6 m/s for each degree Celsius when the air temperature rises. For a given sound, as the temperature increases, what happens to
    a. the frequency?
    b. the wavelength?
13. In a *Star Trek* episode, a space station orbiting Tanuga IV blows up. The crew of the *Enterprise* immediately hears and sees the explosion; they realize that there is no chance for rescue. If you had been hired as an advisor, what two physics errors would you have found and corrected?
14. Suppose the horns of all cars emitted sound at the same pitch or frequency. What would be the change in the frequency of the horn of a car moving
    a. toward you?
    b. away from you?
15. A bat emits short pulses of high-frequency sound and detects the echoes.
    a. In what way would the echoes from large and small insects compare if they were the same distance from the bat?
    b. In what way would the echo from an insect flying toward the bat differ from that of an insect flying away from the bat?
16. If the pitch of sound is increased, what are the changes in
    a. the frequency?
    b. the wavelength?
    c. the wave velocity?
    d. the amplitude of the wave?
17. Does a sound of 40 dB have a factor of 100 (10^2) times greater pressure variation than the threshold of hearing, or a factor of 40 times greater?
18. The speed of sound increases with temperature. Would the pitch of a closed pipe increase or decrease when the temperature of the air rises? Assume that the length of the pipe does not change.
19. Two flutes are tuning up. If the conductor hears the beat frequency increasing, are the two flute frequencies getting closer together or farther apart?
20. A covered organ pipe plays a certain note. If the cover is removed to make it an open pipe, is the pitch increased or decreased?

Problems

Section 15.1

LEVEL 1

21. You hear the sound of the firing of a distant cannon 6.0 s after seeing the flash. How far are you from the cannon?
22. If you shout across a canyon and hear an echo 4.0 s later, how wide is the canyon?
23. A sound wave has a frequency of 9 800 Hz and travels along a steel rod. If the distance between compressions, or regions of high pressure, is 0.580 m, what is the speed of the wave?
24. A rifle is fired in a valley with parallel vertical walls. The echo from one wall is heard 2.0 s after the rifle was fired. The echo from the other wall is heard 2.0 s after the first echo. How wide is the valley?
25. A certain instant camera determines the distance to the subject by sending out a sound wave and measuring the time needed for the echo to return to the camera. How long would it take the sound wave to return to the camera if the subject were 3.00 m away?
26. Sound with a frequency of 261.6 Hz travels through water at a speed of 1435 m/s. Find the

sound's wavelength in water. Don't confuse sound waves moving through water with surface waves moving through water.

27. If the wavelength of a 4.40×10^2 Hz sound in freshwater is 3.30 m, what is the speed of sound in water?
28. Sound with a frequency of 442 Hz travels through steel. A wavelength of 11.66 m is measured. Find the speed of the sound in steel.
29. The sound emitted by bats has a wavelength of 3.5 mm. What is the sound's frequency in air?
30. Ultrasound with a frequency of 4.25 MHz can be used to produce images of the human body. If the speed of sound in the body is the same as in salt water, 1.50 km/s, what is the wavelength of the pressure wave in the body?
31. The equation for the Doppler shift of a sound wave of speed v reaching a moving detector, is $f_d = f_s(v + v_d)/(v - v_s)$, where v_d is the speed of the detector, v_s is the speed of the source, f_s is the frequency of the source, f_d is the frequency of the detector. If the detector moves toward the source, v_d is positive; if the source moves toward the detector, v_s is positive. A train moving toward a detector at 31 m/s blows a 305-Hz horn. What frequency is detected by a
    a. stationary train?
    b. train moving toward the first train at 21 m/s?
32. The train in problem 31 is moving away from the detector. Now what frequency is detected by
    a. a stationary train?
    b. a train moving away from the first train at a speed of 21 m/s?
33. Adam, an airport employee, is working near a jet plane taking off. He experiences a sound level of 150 dB.
    a. If Adam wears ear protectors that reduce the sound level to that of a chain saw (110 dB), what decrease in dB will be required?
    b. If Adam now hears something that sounds like a whisper, what will a person not wearing the protectors hear?
34. A rock band plays at an 80-dB sound level. How many times greater is the sound pressure from another rock band playing at
    a. 100 dB?
    b. 120 dB?

LEVEL 2

35. If you drop a stone into a mine shaft 122.5 m deep, how soon after you drop the stone do you hear it hit the bottom of the shaft?

Section 15.2

LEVEL 1

36. A slide whistle has a length of 27 cm. If you want to play a note one octave higher, the whistle should be how long?
37. An open vertical tube is filled with water, and a tuning fork vibrates over its mouth. As the water level is lowered in the tube, resonance is heard when the water level has dropped 17 cm, and again after 49 cm of distance exists from the water to the top of the tube. What is the frequency of the tuning fork?
38. The auditory canal, leading to the eardrum, is a closed pipe 3.0 cm long. Find the approximate value (ignoring end correction) of the lowest resonance frequency.
39. If you hold a 1.0-m aluminum rod in the center and hit one end with a hammer, it will oscillate like an open pipe. Antinodes of air pressure correspond to nodes of molecular motion, so there is a pressure antinode in the center of the bar. The speed of sound in aluminum is 5150 m/s. What would be the lowest frequency of oscillation?
40. The lowest note on an organ is 16.4 Hz.
    a. What is the shortest open organ pipe that will resonate at this frequency?
    b. What would be the pitch if the same organ pipe were closed?
41. One tuning fork has a 445-Hz pitch. When a second fork is struck, beat notes occur with a frequency of 3 Hz. What are the two possible frequencies of the second fork?
42. A flute acts as an open pipe and sounds a note with a 370-Hz pitch. What are the frequencies of the second, third, and fourth harmonics of this pitch?
43. A clarinet sounds the same note, with a pitch of 370 Hz , as in problem 42. The clarinet, however, produces harmonics that are only odd multiples of the fundamental frequency.

What are the frequencies of the lowest three harmonics produced by this instrument?

LEVEL 2

44. During normal conversation, the amplitude of a pressure wave is 0.020 Pa.
   a. If the area of the eardrum is 0.52 cm^2, what is the force on the eardrum?
   b. The mechanical advantage of the bones in the inner ear is 1.5. What force is exerted on the oval window?
   c. The area of the oval window is 0.026 cm^2. What is the pressure increase transmitted to the liquid in the cochlea?
45. One closed organ pipe has a length of 2.40 m.
   a. What is the frequency of the note played by this pipe?
   b. When a second pipe is played at the same time, a 1.40-Hz beat note is heard. By how much is the second pipe too long?
46. One organ pipe has a length of 836 mm. A second pipe should have a pitch one major third higher. The pipe should be how long?
47. In 1845, French scientist B. Ballot first tested the Doppler shift. He had a trumpet player sound an A, 440 Hz, while riding on a flatcar pulled by a locomotive. At the same time, a stationary trumpeter played the same note. Ballot heard 3.0 beats per second. How fast was the train moving toward him? (Refer to problem 31 for the Doppler shift equation.)
48. You try to repeat Ballot's experiment. You plan to have a trumpet played in a rapidly moving car. Rather than listening for beat notes, however, you want to have the car move fast enough so that the moving trumpet sounds a major third above a stationary trumpet. (Refer to problem 31 for the Doppler shift equation.)
   a. How fast would the car have to move?
   b. Should you try the experiment?

Critical Thinking Problems

49. Suppose that the frequency of a car horn (when not moving) is 300 Hz. What would the graph of the frequency versus time look like as the car approached and then moved past you? Complete a rough sketch.
50. Describe how you could use a stopwatch to estimate the speed of sound if you were near the green on a 200-m golf hole as another group of golfers were hitting their tee shots. Would your estimate of their velocities be too large or too small?
51. A light wave coming from a point on the left edge of the sun is found by astronomers to have a slightly higher frequency than light from the right side. What do these measurements tell you about the sun's motion?

Going Further

Computer Application Here is a way to measure the speed of sound with Calculator Based Lab equipment or a microcomputer based lab setup such as Vernier's universal lab interface (ULI). If you are using Vernier's ULI, load the sound software and connect the microphone directly to the ULI. You will be able to see the pattern of the pressure changes in the air for different sounds. Record the sound when you snap your fingers. How much time does this sound take to die out? Now snap your fingers near the end of a hollow tube (10-cm or 15-cm diameter and 1 to 3 m long) and you will be able to see the original sound and also the echo. Use twice the length of the tube and the time delay to calculate the speed of sound. Try different lengths of tubes if you have them.

interNET CONNECTION

Follow the link on the Glencoe Homepage at **www.glencoe.com/sec/science** to find out more about this chapter.

Colors to Mix and Match

Explain how each of the colors in the shadows is formed.

CHAPTER

16 Light

Light and sound are two ways that you receive information about the world. Of the two, light provides the greater variety of information. The eye can detect tiny changes in the size, position, brightness, and color of an object. Light is emitted by incandescent and fluorescent lamps; by television screens, lasers, and tiny light-emitting diodes (LEDs); and by flames, sparks, and even fireflies. Our major source of emitted light is the sun. However, our eyes receive light waves from most objects in the environment in the form of reflections. Light is reflected not only by mirrors and white paper, but also by the moon, trees, and even black cloth. In fact, it is difficult to find an object that does not reflect some light. Emission and reflection are just two of the ways that light interacts with matter. Although light is only a small portion of the entire range of electromagnetic waves, the study of light is, in many ways, a study of all electromagnetic radiation.

WHAT YOU'LL LEARN

- You will understand the fundamentals of light, including its speed, wavelength range, and intensity.
- You will describe the interactions between two or more light waves and between light waves and matter.

WHY IT'S IMPORTANT

- Light is a primary sensor to how the universe behaves. From learning about the biological patterns on Earth to discovering the astronomical rules of outer space, scientists rely upon detecting light waves.

interNET CONNECTION

Follow the link for this chapter on the Glencoe Homepage at **www.glencoe.com/sec/science** to find out more about light.

16.1 Light Fundamentals

OBJECTIVES

- **Recognize** that light is the visible portion of an entire range of electromagnetic frequencies.
- **Describe** the ray model of light.
- **Solve** problems involving the speed of light.
- **Define** *luminous intensity, luminous flux,* and *illuminance.*
- **Solve** illumination problems.

Color Conventions

- Light rays are **red.**

Early scientists considered light to be a stream of particles emitted by a light source. However, not all of the properties of light could be explained by this theory. Experiments showed that light also behaves like a wave. Today, the nature of light is explained in terms of both particles and waves. In this chapter, you will apply what you have learned about mechanical waves to the study of light.

The Facts of Light

What is light? **Light** is the range of frequencies of electromagnetic waves that stimulates the retina of the eye. Light waves have wavelengths from about 400 nm (4.00×10^{-7} m) to 700 nm (7.00×10^{-7} m). The shortest wavelengths are seen as violet light. As wavelength increases, the colors gradually change to indigo, blue, green, yellow, orange, and finally, red, as shown in **Figure 16–1.**

Light travels in a straight line in a vacuum or other uniform medium. How do you know this? If light from the sun or a flashlight is made visible by dust particles in the air, the path of the light is seen to be a straight line. When your body blocks sunlight, you see a sharp shadow. Also, our brains locate objects by automatically assuming that light travels from objects to our eyes along a straight path.

The straight-line path of light has led to the **ray model** of light. A ray is a straight line that represents the path of a narrow beam of light. The use of ray diagrams to study the travel of light is called ray optics or geometric optics. Even though ray optics ignores the wave nature of light, it is useful in describing how light is reflected and refracted.

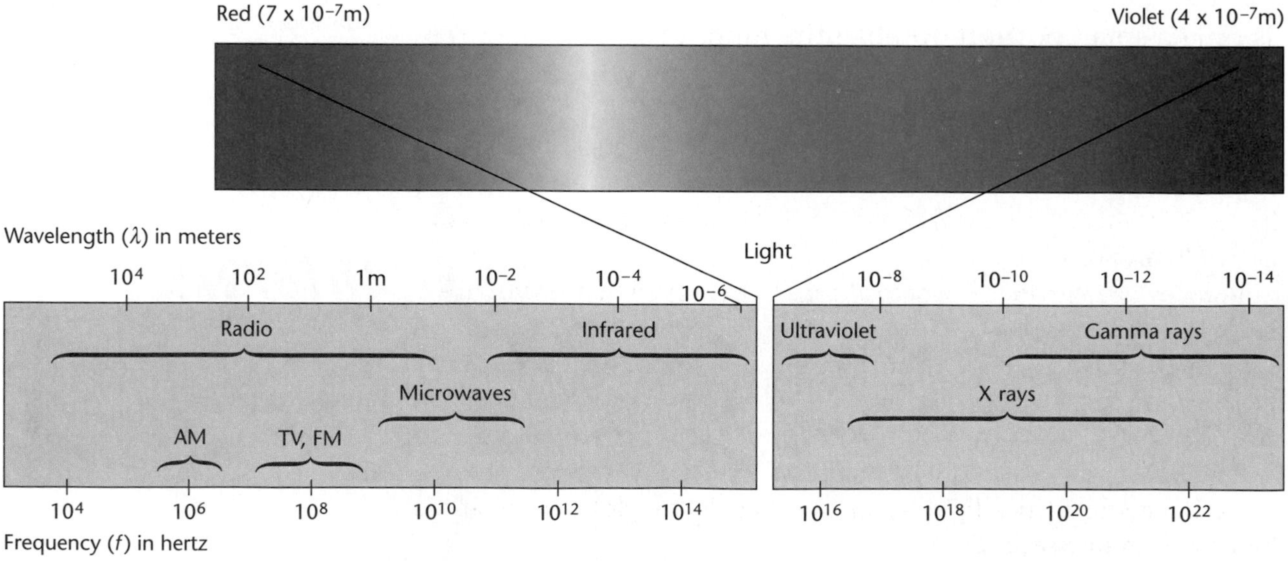

FIGURE 16–1 The visible spectrum is a very small portion of the whole electromagnetic spectrum.

Red (7 x 10⁻⁷m) — Violet (4 x 10⁻⁷m)

Wavelength (λ) in meters: 10^4, 10^2, 1m, 10^{-2}, 10^{-4}, 10^{-6}, Light, 10^{-8}, 10^{-10}, 10^{-12}, 10^{-14}

Radio, Infrared, Microwaves, AM, TV, FM, Ultraviolet, Gamma rays, X rays

Frequency (f) in hertz: 10^4, 10^6, 10^8, 10^{10}, 10^{12}, 10^{14}, 10^{16}, 10^{18}, 10^{20}, 10^{22}

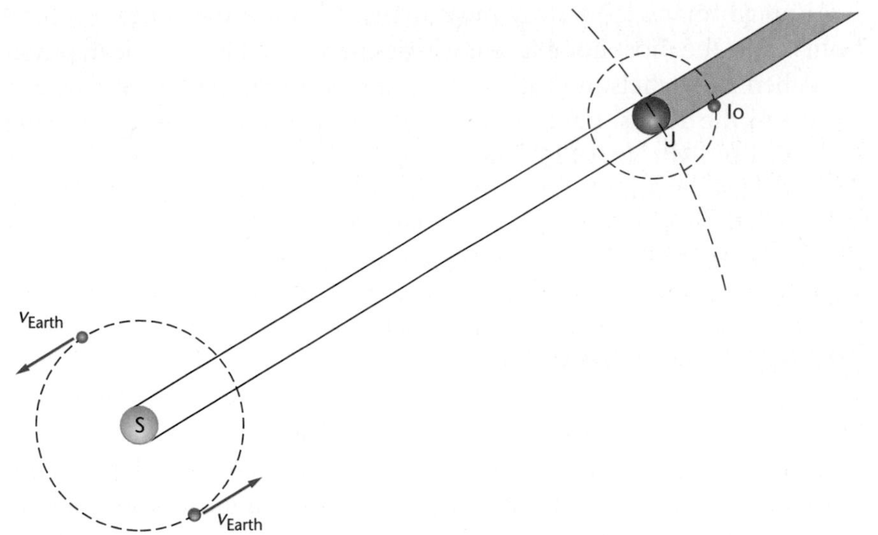

FIGURE 16–2 Roemer measured the time of an eclipse of one of Jupiter's moons, Io. During successive eclipses, the moon would come around faster or slower, depending on whether Earth was moving toward or away from Jupiter.

The Speed of Light

Before the 17th century, most people believed that light travels instantaneously. Galileo was the first to hypothesize that light has a finite speed and to suggest a method of determining it. His method, however, was not sensitive enough, and he was forced to conclude that the speed of light is too fast to be measured at all over a distance of a few kilometers. Danish astronomer Ole Roemer (1644–1710) was the first to determine that light does travel with a measurable speed. Between 1668 and 1674, Roemer made 70 careful measurements of the 42.5-hour orbital period of Io, one of the moons of Jupiter. He recorded the times when Io emerged from Jupiter's shadow, as shown in **Figure 16–2.**

He found that the period varied slightly depending on when the measurement was made. The variation was as much as 14 seconds longer when Earth was moving away from Jupiter and 14 seconds shorter when Earth was approaching Jupiter.

What might cause this discrepancy in Io's orbital period? Roemer concluded that as Earth moved away from Jupiter, the light from each new appearance of Io took longer to travel the increasing distance to Earth. Thus, the measured period increased. Likewise, as Earth approached Jupiter, Io's orbital period would seem to decrease. Based on these data, in 1676 Roemer calculated that light took 22 minutes to cross the diameter of Earth's orbit.

Roemer had successfully proved that light moved at a finite speed. Using the present value of the diameter of Earth's orbit, Roemer's value of 22 minutes gives a speed of light of about 220 million meters per second. This is only three quarters of what is now accepted as the correct value. Today we know that light takes 16 minutes, not 22, to cross Earth's orbit. Nevertheless, the speed of light was found to be finite but so fast that a light beam could circle the globe seven and a half times in one second.

F.Y.I.

Ole Roemer made his measurements in Paris as part of a project to improve maps by calculating the longitude of locations on Earth. This is an early example of the needs of technology resulting in scientific advances.

FIGURE 16–3 Albert A. Michelson became the first American to win a Nobel prize in science.

Although many laboratory measurements of the speed of light have been made, the most notable was a series performed by American physicist Albert A. Michelson (1852–1931), shown in **Figure 16–3.** Between 1880 and the 1920s, he developed Earth-based techniques to measure the speed of light. In 1926, Michelson measured the time required for light to make a round-trip between two California mountains 35 km apart. Michelson's best result was $2.997996 \pm 0.00004 \times 10^8$ m/s. For this work, he became the first American to receive a Nobel prize in science.

The speed of light defined

The development of the laser in the 1960s provided new methods of measuring the speed of light. As you learned in Chapter 14, the speed of a wave is equal to the product of its frequency and wavelength. The speed of light in a vacuum is such an important and universal value that it has its own special symbol, c. Thus, $c = \lambda f$. The frequency of light can be counted with extreme precision using lasers and the time standard provided by atomic clocks. Measurements of its wavelength, however, are much less precise. As a result, in 1983 the International Committee on Weights and Measurements decided to make the speed of light a defined quantity. In principle, an object's length is now measured in terms of the time required by light to travel from one end of the object to the other. The committee defined the speed of light in a vacuum to be exactly $c = 299\ 792\ 458$ m/s. For most calculations, however, it is sufficient to use $c = 3.00 \times 10^8$ m/s.

HELP WANTED

PHOTOGRAPHER

A large advertising agency is looking for a talented staff photographer. Your technological knowledge of cameras, lenses, sources of light, various types of film and the processes for developing them can be the result of formal or on-the-job training.
Because of the variety of our accounts, your subjects will range from products to people to places. Proof of your qualifications is in your portfolio, so send yours for our review. For information contact:
Professional Photographers of America, Inc.
57 Forsyth Street, N.W.
Suite 1600
Atlanta, GA 30303-2206

Practice Problems

1. What is the frequency of yellow light, $\lambda = 556$ nm?
2. One nanosecond (ns) is 10^{-9} s. Laboratory workers often estimate the distance light travels in a certain time by remembering the approximation "light goes one foot in one nanosecond." How far, in feet, does light actually travel in exactly 1 ns?
3. Modern lasers can create a pulse of light that lasts only a few femtoseconds.
   - **a.** What is the length of a pulse of violet light that lasts 6.0 fs?
   - **b.** How many wavelengths of violet light ($\lambda = 400$ nm) are included in such a pulse?
4. The distance to the moon can be found with the help of mirrors left on the moon by astronauts. A pulse of light is sent to the moon and returns to Earth in 2.562 s. Using the defined speed of light, calculate the distance from Earth to the moon.
5. Use the correct time taken for light to cross Earth's orbit, 16 minutes, and the diameter of the orbit, 3.0×10^{11} m, to calculate the speed of light using Roemer's method.

Physics Lab

Pinhole Camera

Problem

How do light rays travel?

Materials

large coffee can with translucent cover
5 cm of masking tape
40-watt lightbulb (nonfrosted) in a fixture
small and large nail

Procedure

1. Punch one hole in the bottom of the coffee can with each nail.
2. Place the masking tape over the larger hole.
3. Place the translucent top on the coffee can.
4. Turn on the 40-watt lightbulb. Turn off the lights in the room.
5. Point the hole at the light and note the image formed on the cover of the can.
6. Draw the path of the light to show the orientation of the image.

Data and Observations

1. Is the image reversed right to left? Design an activity to find out. Record your results.
2. Move the can farther away from the bulb. Note how the image changes.
3. For various positions, measure:
   distance from hole to object, d_o
   distance from hole to image, d_i
   height of object, h_o
   height of image, h_i.

Analyze and Conclude

1. **Analyzing Data** Make a drawing to show why the image gets smaller as the can is moved away from the light.
2. **Testing a Hypothesis** Predict how the image formed by the nail hole would compare to the image formed by a pinhole. List the similarities and differences. Test your hypothesis. Record your results.
3. **Inferring a Relationship** Try to determine a mathematical rule for the relationship among h_i, h_o, d_i, and d_o. Show your results.

Apply

1. Your eye is a form of pinhole camera. Would you expect the images to be upside-down? Explain.

FIGURE 16–4 The bridge is illuminated while the city lights are luminous.

Sources of Light

What's the difference between sunlight and moonlight? Sunlight, of course, is much, much brighter. But there is an important fundamental difference between the two. The sun is a luminous body, while the moon is an illuminated body. A **luminous** body emits light waves; an **illuminated** body simply reflects light waves produced by an outside source, as illustrated in **Figure 16–4.** An incandescent lamp, such as a common lightbulb, is luminous because electrical energy heats a thin tungsten wire in the bulb and causes it to glow. An incandescent object emits light as a result of its high temperature. A bicycle reflector, on the other hand, works as an illuminated body. It is designed to reflect automobile headlights.

Humans register the sensation of light when electromagnetic waves of the appropriate wavelength(s) reach our eyes. Our eyes have different sensitivities to different wavelengths.

Physics & Technology

Digital Versatile Discs

In 1982, when CDs (compact discs) were introduced, they revolutionized the audio electronics industry. A few years later, CD-ROMs (CD-read only memory) began to do the same thing for the personal computer industry. The high-quality sound and images, large storage capacity, durability, and ease of use made CDs and CD-ROMs popular with consumers.

CDs, CD-ROMS, and DVDs are all examples of optical storage technology. Information is stored on the disc in a spiral of microscopic pits. These pits store a digital code that is read by a laser. The primary difference between CD and DVDs is the amount of information that each can hold. Today's CD can store 0.68 gigabytes of data whereas DVDs have the ability to store from 4.7 to 17 gigabytes.

How are DVDs made to obtain a higher storage capacity? Storage capacity depends on the number of pits. Manufacturers of DVDs increase the number of pits by shrinking pit size and by recording data on as many as four layers. Reducing pit size allows the pits to be closer together and the spiral track to be tighter. Thus, more pits can fit on the surface providing DVDs with more than six times the storage capability of a CD. Because the smaller pits are shallower on a DVD than a CD, a shorter wavelength laser is required. DVDs use a red, 640-nm laser as opposed to CDs with an infrared, 780-nm laser.

Another boost to storage is layering technology. Advances in aiming and focusing of the laser allow data to be recorded on two layers. To read the second layer, the laser is simply focused a little deeper into the disc where the second layer of data is stored. Not only are the two layer discs possible, but so are double-sided discs. The possibility of four layers gives DVDs a storage capability of 17 gigabytes.

Thinking Critically What would be the result of using even smaller pits and even shorter-wavelength blue or green lasers to read optical storage discs?

Luminous flux

The rate at which visible light is emitted from a source is called the **luminous flux,** *P.* The unit of luminous flux is the **lumen,** lm. A typical 100-watt incandescent lightbulb emits approximately 1750 lm. Imagine placing the bulb at the center of a sphere, as shown in **Figure 16–5.** The bulb emits light in almost all directions. The 1750 lm of luminous flux characterize all of the light that strikes the inside surface of the sphere in a given unit of time.

Often, we may not be interested in the total amount of light emitted by a luminous object. We are more likely to be interested in the amount of illumination the object provides on a book, a sheet of paper, or a highway. The illumination of a surface is called the **illuminance,** *E,* and is the rate at which light falls on a surface. Illuminance is measured in lumens per square meter, lm/m^2, or **lux,** lx.

Consider the 100-watt lightbulb in the middle of the sphere. What is the illumination of the sphere's surface? The area of the surface of a sphere is $4\pi r^2$. **Figure 16–5** shows that the luminous flux striking each square meter of the sphere is as follows.

$$\frac{1750\text{ lm}}{4\pi r^2\text{ m}^2} = \frac{1750}{4\pi r^2}\text{ lx}$$

At a distance of 1 m from the bulb, the illumination is approximately 140 lx.

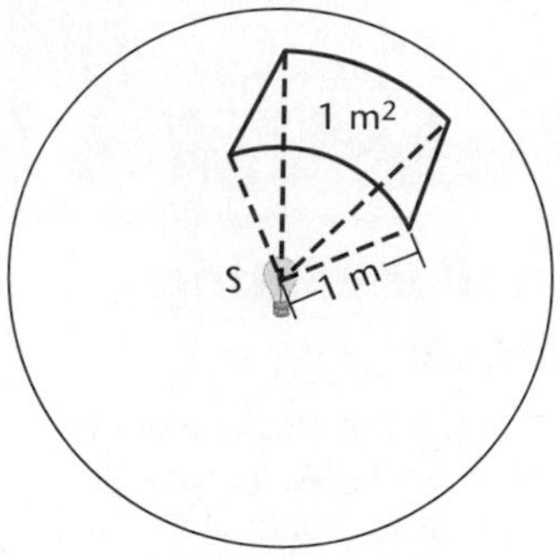

luminous flux
$P = 1750$ lm
1 m²
S
1 m
illuminance
$E = \dfrac{1750}{4\pi r^2}$ lx

FIGURE 16–5 Luminous flux is the rate that light is emitted from a bulb, whereas illuminance is the rate that light falls on some surface.

An inverse-square relationship

What would happen if the sphere surrounding the lamp were larger? If the sphere had a radius of 2 m, the luminous flux would still total 1750 lm, but the area of the sphere would then be $4\pi(2\text{ m})^2 = 16\pi\text{ m}^2$, four times larger. Consequently, the illumination on the surface would be reduced by a factor of four to 35 lx. Thus, if the distance of a surface from a point source of light is doubled, the illumination provided by the source on that surface is reduced by a factor of four. In the same way, if the distance is increased to 3 m, the illumination would be only $(1/3)^2$ or 1/9 as large as it was when the light source was 1 m away. Notice that illumination is proportional to $1/r^2$. This inverse-square relationship, as shown in **Figure 16–6,** is similar to that of gravitational force, which you studied in Chapter 8.

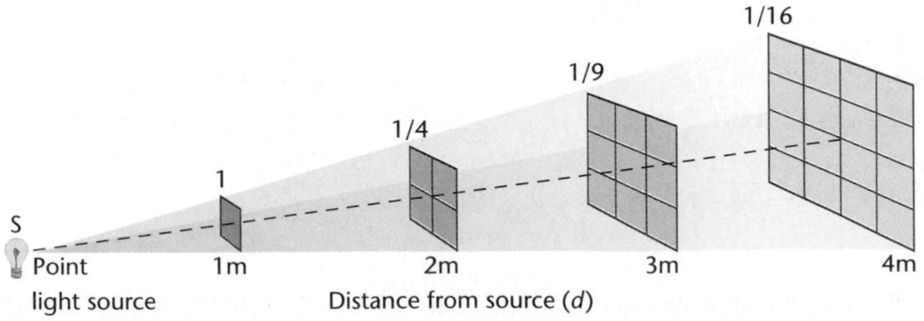

FIGURE 16–6 The illuminance of a surface varies inversely as the square of its distance from a light source.

Pocket Lab

An Illuminating Matter

Which is more efficient, or has the highest lm/W, a lower- or higher-power lightbulb? To find out, look at your lightbulbs at home and record the power and lumens for at least three different bulbs.

Graph Your Results Make a graph of power (horizontal axis) versus lumens (vertical axis). Summarize your results.

Luminous intensity

Some light sources are specified in **candela,** cd, or candle power. A candela is not a measure of luminous flux, but of luminous intensity. The **luminous intensity** of a point source is the luminous flux that falls on $1 m^2$ of a sphere 1m in radius. Thus, luminous intensity is luminous flux divided by 4π. A bulb with 1750 lm flux has an intensity of (1750 lm)/4π = 139 cd. A flashlight bulb labeled 1.5 cd emits a flux of 4π (1.5 cd) = 19 lm. The candela is the official SI unit from which all light intensity units are calculated.

How to illuminate a surface

There are two ways to increase the illumination on a surface. You can use a brighter bulb, which increases luminous flux, or you can move the surface closer to the bulb, decreasing the distance. Mathematically, the illuminance, E, directly under a small light source is represented by the following equation.

$$E = \frac{P}{4\pi d^2}$$

P represents the luminous flux of the source, and d represents its distance from the surface. This equation is valid only if the light from the source strikes the surface perpendicular to it. It is also valid only for sources that are small enough or far enough away to be considered point sources. Thus, the equation does not give accurate values with long fluorescent lamps, or with incandescent bulbs in large reflectors that are close to the illuminated surface.

Example Problem

Illumination of a Surface

What is the illumination on your desktop if it is lighted by a 1750-lm lamp that is 2.50 m above your desk?

Sketch the Problem

- Assume that the bulb is the point source.
- Diagram the position of the bulb and desktop. Label P and d.

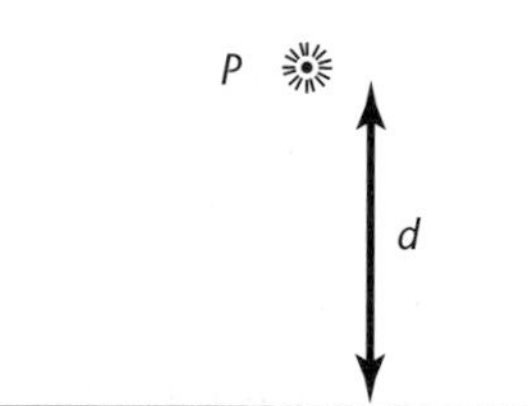

Calculate Your Answer

Known:

luminous flux, P = 1750 lm

d = 2.50 m

Unknown:

illuminance, E = ?

Strategy:

The surface is perpendicular to the direction the light ray is traveling, so you can use the illuminance equation.

Calculations:

$$E = \frac{P}{4\pi d^2}$$

$$E = \frac{1750 \text{ lm}}{4\pi(2.50 \text{ m})^2} = 22.3 \text{ lm/m}^2 = 22.3 \text{ lx}$$

Check Your Answer

- Are the units correct? $lm/m^2 = lx$, which the answer agrees with.
- Do the signs make sense? All quantities are positive, as they should be.
- Is the magnitude realistic? Answer agrees with quantities given.

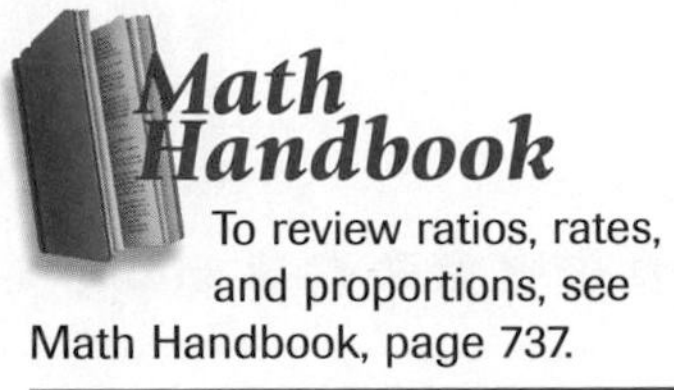

To review ratios, rates, and proportions, see Math Handbook, page 737.

Practice Problems

6. A lamp is moved from 30 cm to 90 cm above the pages of a book. Compare the illumination on the book before and after the lamp is moved.

7. What is the illumination on a surface 3.0 m below a 150-watt incandescent lamp that emits a luminous flux of 2275 lm?

8. Draw a graph of the illuminance from a 150-watt incandescent lamp between 0.50 m and 5.0 m.

9. A 64-cd point source of light is 3.0 m above the surface of a desk. What is the illumination on the desk's surface in lux?

10. The illumination on a tabletop is 2.0×10^1 lx. The lamp providing the illumination is 4.0 m above the table. What is the intensity of the lamp?

11. A public school law requires a minimum illumination of 160 lx on the surface of each student's desk. An architect's specifications call for classroom lights to be located 2.0 m above the desks. What is the minimum luminous flux the lights must deliver?

16.1 Section Review

1.1 How far does light travel in the time it takes sound to go 1 cm in air at 20°C?

1.2 The speed of light is slower in air and water than in a vacuum. The frequency, however, does not change when light enters water. Does the wavelength change? If so, in which direction?

1.3 Which provides greater illumination of a surface, placing two equal bulbs instead of one at a given distance or moving one bulb to half that distance?

1.4 **Critical Thinking** A bulb illuminating your desk provides only half the illumination it should. If it is currently 1.0 m away, how far should it be to provide the correct illumination?

16.2 Light and Matter

OBJECTIVES

- **Explain** the formation of color by light and by pigments or dyes.
- **Explain** the cause and **give examples** of interference in thin films.
- **Describe** methods of producing polarized light.

Objects can be seen clearly through air, glass, some plastics, and other materials. These materials, which transmit light waves without distorting images, are **transparent** materials. Materials that transmit light but do not permit objects to be seen clearly through them are **translucent** materials. Lamp shades and frosted lightbulbs are examples of translucent objects. Materials such as brick, which transmit no light but absorb or reflect all light incident upon them, are **opaque** materials. All three types of materials are illustrated in **Figure 16–7.**

FIGURE 16–7 Materials can be transparent, translucent, or opaque.

Color

One of the most beautiful phenomena in nature is a rainbow. Artificial rainbows can be produced when light passes through water or glass. How is the color pattern of a rainbow produced? In 1666, the 24-year-old Isaac Newton did his first scientific experiments on the colors produced when a narrow beam of sunlight passed through a prism, shown in **Figure 16–8.** Newton called the ordered arrangement of colors from violet to red a **spectrum.** He thought that some unevenness in the glass might be producing the spectrum.

FIGURE 16–8 White light, when passed through a prism, is separated into a spectrum of colors.

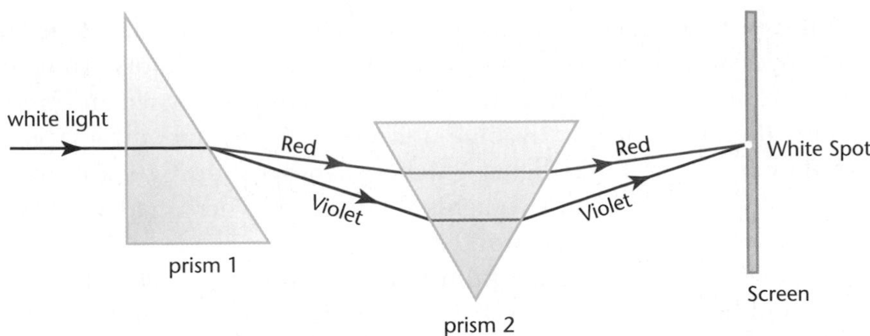

FIGURE 16–9 A second prism can recombine the colors separated by the first prism into white light again.

To test this assumption, he allowed the spectrum from one prism to fall on a second prism. If the spectrum were caused by irregularities in the glass, he reasoned, then the second prism should have increased the spread in colors. Instead, the second prism reversed the spreading of colors and recombined them to form white light, as shown in **Figure 16–9.** After more experiments, Newton concluded that white light is composed of colors. We now know that each color in the spectrum is associated with a specific wavelength of light, as represented in **Figure 16–1,** page 374.

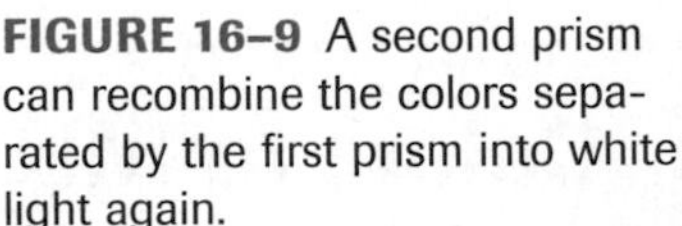

Hot and Cool Colors

Some artists refer to red and orange as hot colors and green and blue as cool colors. But does emitting red or orange light really indicate that an object is hotter than one emitting blue or green? Try this to find out. Obtain a pair of prism glasses or a piece of diffraction grating from your teacher. Find a lamp with a dimmer switch and turn off the light. Next, slowly turn the dimmer so that the light gets brighter and brighter. To get the best effect, turn off all the other lights in the room.

Analyze and Conclude Which colors appeared first when the light was dim? Which colors were the last to appear? How do these colors relate to the temperature of the filament?

Color by addition

White light can be formed from colored light in a variety of ways. For example, if correct intensities of red, green, and blue light are projected onto a white screen, as in **Figure 16–10,** the screen will appear to be white. Thus, red, green, and blue light added together form white light. This is called the additive color process. A color television tube uses the additive process. It has tiny dotlike sources of red, green, and blue light. When all have the correct intensities, the screen appears to be white. For this reason, red light, green light, and blue light are called the **primary colors** of light. The primary colors can be mixed by pairs to form three different colors. Red and green light together produce yellow light, blue and green light produce cyan, and red and blue light produce magenta. The three colors yellow, cyan, and magenta are called the **secondary colors** of light.

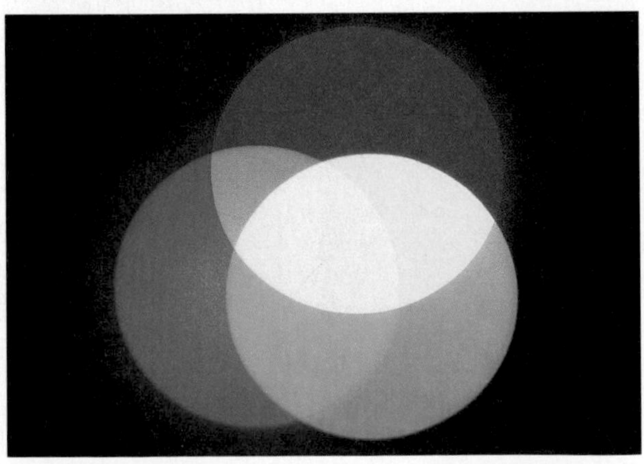

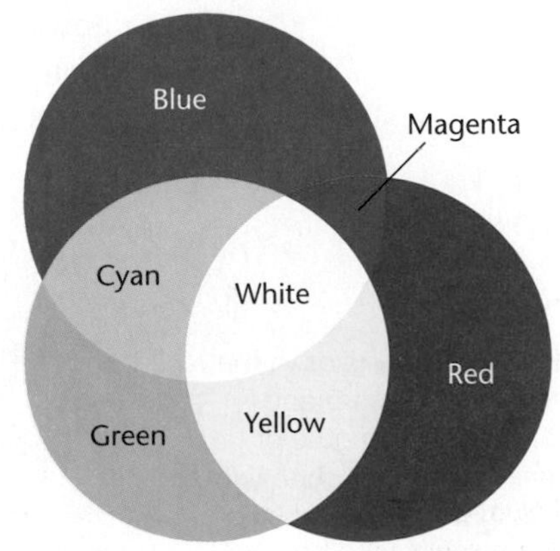

FIGURE 16–10 The additive mixture of blue, green, and red light produces white light.

Colors to Mix and Match

In the chapter-opening photograph, each shadow occurs when the girl blocks one color of light, leaving the secondary colors. Thus, in order from the right, the yellow shadow is illuminated by red and green lights, the cyan shadow by blue and green lights, and the magenta shadow by red and blue lights. Smaller shadows showing the primary light colors appear where two lights are blocked. Where there is a black shadow, all three lights are blocked.

Yellow light can be made from red light and green light. If yellow light and blue light are projected onto a white screen with the correct intensities, the surface will appear to be white. Thus, yellow and blue light combine to form white light, and consequently, yellow light is called the **complementary color** to blue light. Yellow light is made up of the two other primary colors. In the same way, cyan and red are complementary colors, as are magenta and green.

Colors by subtraction

A **dye** is a molecule that absorbs certain wavelengths of light and transmits or reflects others. A tomato is red because it reflects red light to our eyes. When white light falls on the red block in **Figure 16–11,** dye molecules in the red block absorb the blue and green light and reflect the red. When only blue light falls on the block, very little light is reflected and the block appears to be almost black.

CHEMISTRY CONNECTION

Like a dye, a **pigment** is a colored material that absorbs certain colors and transmits or reflects others. The difference is that a pigment particle is larger than a molecule and can be seen with a microscope. Often, a pigment is a finely ground inorganic compound such as titanium(IV) oxide (white), chromium(III) oxide (green), or cadmium sulfide (yellow). Pigments mix in a medium to form suspensions rather than solutions.

The absorption of light forms colors by the subtractive process. Pigments and dyes absorb certain colors from white light. A pigment that absorbs only one primary color from white light is called a **primary pigment.** Yellow pigment absorbs blue light and reflects red and green light. Yellow, cyan, and magenta are the primary pigments. A pigment that absorbs two primary colors and reflects one is a **secondary pigment.** The secondary pigments are red (which absorbs green and blue light), green (which absorbs red and blue light), and blue (which absorbs red and green light). Note that the primary pigment colors are the secondary light colors. In the same way, the secondary pigment colors are the primary light colors.

a

b

c

FIGURE 16–11 The dyes in the blocks selectively absorb and reflect various wavelengths of light. Illumination is by white light in **(a),** red light in **(b),** and blue light in **(c).**

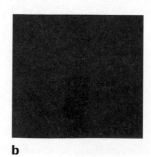

FIGURE 16–12 The primary pigment colors are yellow, cyan, and magenta. In each case, the pigment absorbs one of the primary light colors and reflects the other two.

The primary pigment yellow absorbs blue light. If it is mixed with the secondary pigment blue, which absorbs green and red light, all light will be absorbed. No light will be reflected, so the result will be black. Thus, yellow and blue are complementary pigments. Cyan and red, as well as magenta and green, are also complementary pigments. The primary pigments and their complementary pigments are shown in **Figure 16–12.**

Formation of Colors in Thin Films

Have you ever seen a spectrum of colors produced by a soap bubble or by the oily film on a water puddle in a parking lot? These colors are not the result of separation of white light by a prism or of absorption of colors in a pigment. In fact, the colors you see cannot be explained in terms of a ray model of light; they are a result of the constructive and destructive interference of light waves, or **thin-film interference.**

If a soap film is held vertically, as in **Figure 16–13,** its weight makes it thicker at the bottom than at the top. The thickness varies gradually from top to bottom. When a light wave strikes the film, part of it is reflected, as shown by R_1, and part is transmitted. The transmitted wave travels through the film to the back surface, where, again, part is reflected, as shown by R_2. If the thickness of the film is one fourth of the wavelength of the wave in the film ($\lambda/4$), the round-trip path length in

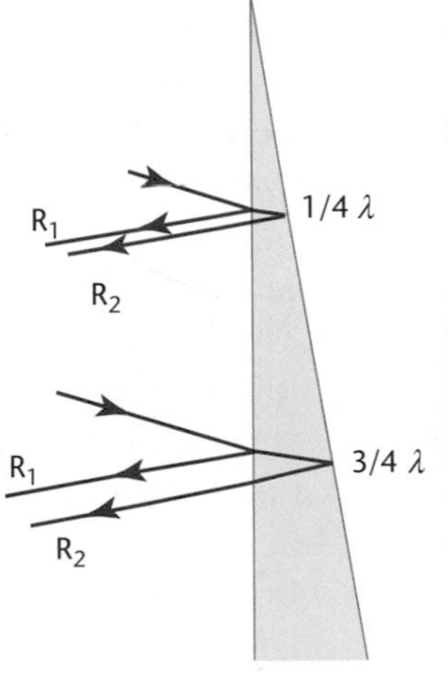

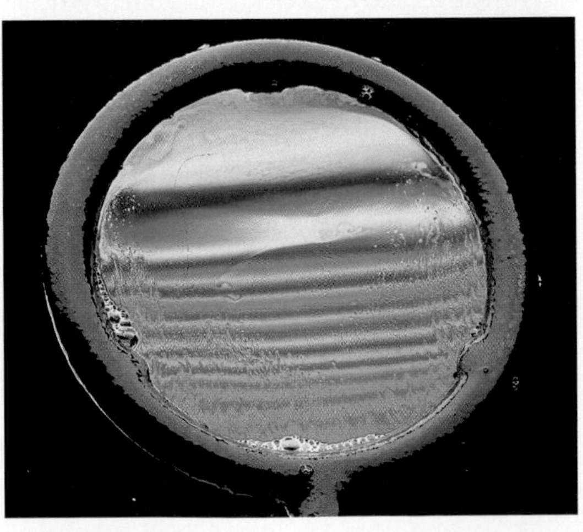

FIGURE 16–13 Each color is reinforced where the soap film is 1/4, 3/4, 5/4, and so on of the wavelength for that color. Because each color has a different wavelength, a series of color bands is reflected from the soap films.

Pocket Lab

Soap Solutions

Dip a ring into soap solution and hold it at a 45° angle to the horizontal. Look for color bands to form in horizontal stripes.

Analyze and Conclude Why do the bands move? Why are the bands horizontal? What type of pattern would you see if you looked through the soap with a red filter? Try it. Describe and explain your results.

the film is $\lambda/2$. In this case, it would appear that the wave returning from the back surface would reach the front surface one-half wavelength behind the first reflected wave and that the two waves would cancel by the superposition principle. But, as you learned in Chapter 14, when a transverse wave is reflected from a more optically dense medium, it is inverted. As a result, the first reflected wave, R_1, is inverted on reflection. The second reflected wave, R_2, is reflected from a less-dense medium and is not inverted. Thus, when the film has a thickness of $\lambda/4$, the wave reflected from the back surface returns to the front surface in sync with the first reflected wave. The two waves reinforce each other as they leave the film. Light with other wavelengths suffers partial or complete destructive interference. At any point on the film, the light most strongly reflected has a wavelength satisfying the requirement that the film thickness equals $\lambda/4$.

Different colors of light have different wavelengths. As the thickness of the film changes, the $\lambda/4$ requirement will be met at different locations for different colors. As the thickness increases, the light with the shortest wavelength, violet, will be most strongly reflected, then blue, green, yellow, orange, and finally red, which has the longest wavelength. A rainbow of color results.

Notice in **Figure 16–13** that the spectrum repeats. When the thickness is $3\lambda/4$, the round-trip distance is $3\lambda/2$, and constructive interference occurs again. Any thickness equal to an odd multiple of quarter wavelengths—$\lambda/4$, $3\lambda/4$, $5\lambda/4$, $7\lambda/4$, and so on—satisfies the conditions for reinforcement for a given color. At the top of the film, there is no color; the film appears to be black. Here, the film is too thin to produce constructive interference for any color. Shortly after the top of the film becomes thin enough to appear black, it breaks.

Polarization of Light

Have you ever looked at light reflected off a road through Polaroid sunglasses? As you rotate the glasses, the road first appears to be dark, then light, and then dark again. Light from a lamp, however, changes very little as the glasses are rotated. Why is there a difference? Part of the reason is that the light coming from the road is reflected. A second part is that the reflected light has become polarized.

Polarization can be understood by considering the rope model of light waves, as shown in **Figure 16–14.** The transverse mechanical waves in the rope represent the transverse electromagnetic waves of light. The slots represent what is referred to as the polarizing axis of the Polaroid material. When the rope waves are parallel to the slots, they pass through. When they are perpendicular to the slots, the waves are blocked. Polaroid material contains long molecules that allow electromagnetic waves of one direction to pass through while absorbing the waves vibrating in the other direction. One direction of the Polaroid material is called the polarizing axis. Only waves vibrating parallel to that axis can pass through.

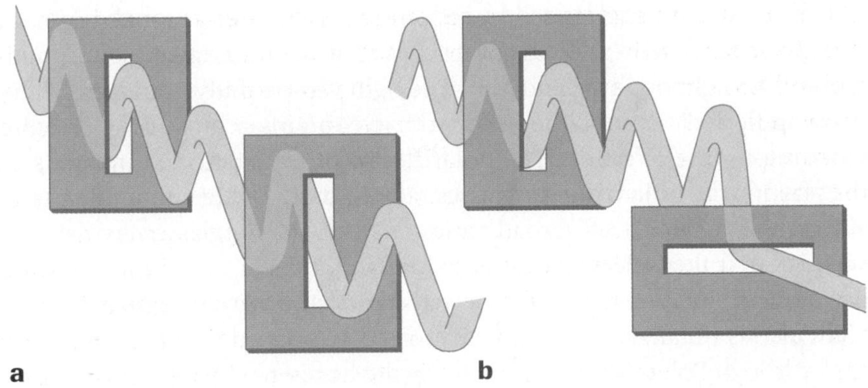

FIGURE 16–14 In the wave model of light, waves are polarized in relation to the vertical plane **(a).** Vertically polarized waves cannot pass through a horizontal polarizer **(b).**

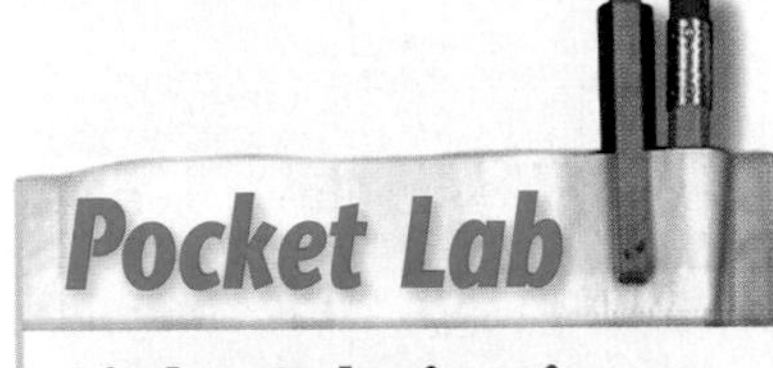

Light Polarization

Obtain a polarizing filter from your teacher to take home. Look through the filter at various objects as you rotate the filter. Make a record of those objects that seem to change in brightness as the filter is rotated.

Recognize Cause and Effect What seems to be the pattern?

Ordinary light contains electromagnetic waves vibrating in every direction perpendicular to its direction of travel. Each wave can be resolved into two perpendicular components in a manner similar to an acceleration or velocity vector. On the average, therefore, half the waves vibrate in one plane, while the other half vibrate in a plane perpendicular to the first. If polarizing material is placed in a beam of ordinary light, only those waves vibrating in one plane pass through. Half the light, passes through, and the intensity of the light is reduced by half. The polarizing material produces light that is **polarized** in a particular plane of vibration. The material is said to be a polarizer of light and is called a polarizing filter.

Suppose a second polarizing filter is placed in the path of the polarized light. If the polarizing axis of the second filter is perpendicular to the direction of vibration of the polarized light, no light will pass through, as shown in **Figure 16–15a.** If the filter, however, is at an angle, the component of light parallel to the polarizing axis of the filter will be transmitted, as shown in **Figure 16–15b.** Thus, a polarizing filter can determine the orientation of polarization of light and is often called an "analyzer."

FIGURE 16–15 The arrows show that unpolarized light vibrates in many planes. Polarized light from a polarizer is absorbed by an analyzer that is perpendicular to the plane of the polarized light **(a).** Polarized light from a polarizer is only partially absorbed by an analyzer that is at an angle to the plane of the polarized light **(b).**

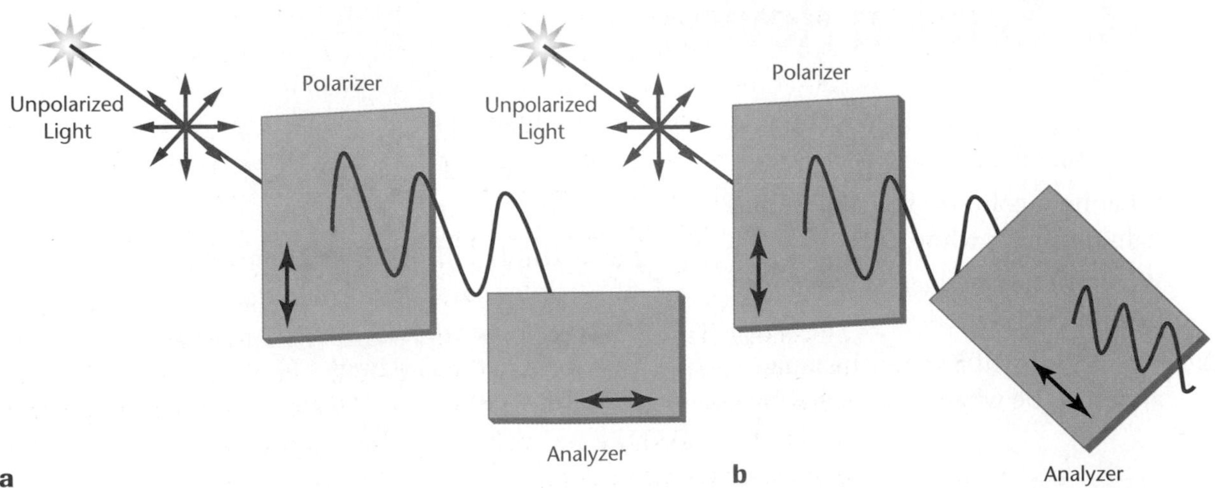

FIGURE 16–16 A polarizing filter over the lens of a camera can block the glare from reflecting surfaces.

Light also can be polarized by reflection. If you look through a polarizing filter at the light reflected by a sheet of glass and rotate the filter, you will see the light brighten and dim. The light was partially polarized when it was reflected. That is, the reflected ray contains a great deal of light vibrating in one direction. The polarization of light reflected by roads is the reason why polarizing sunglasses reduce glare. The fact that the intensity of light reflected off a road varies as Polaroid sunglasses are rotated suggests that the reflected light is partially polarized. Photographers can use polarizing filters to block reflected light, as shown in **Figure 16–16.** Light also is polarized when it is scattered by molecules in the air. If you look through Polaroid sunglasses along the horizon when the sun is overhead and rotate the glasses, you will see the brightness change, showing that the light is polarized.

The Ray and Wave Models of Light

You have learned that many characteristics of light can be explained with a simple ray model. An understanding of the interaction of light with thin films that produce colors, however, requires the use of a model of light that involves waves. This model also is used to explain polarization. In the next chapter, you'll find that the ray model is suitable for explaining how lenses and mirrors form images. In Chapter 18 you'll learn about other aspects of light that can be understood only through the use of the wave model. But both the ray and wave models of light have been found to be inadequate to explain some other interactions of light with matter. For such phenomena, we will need yet another model, which is much closer to the ray model than to the wave model. This model, often referred to as the particle theory of light, will be discussed in Chapter 27.

16.2 Section Review

2.1 Why might you choose a window shade that is translucent? Opaque?

2.2 What light color do you add to blue light to obtain white light?

2.3 What primary pigment colors must be mixed to get red?

2.4 What color will a yellow banana appear to be when illuminated by

a. white light?

b. green and red light?

c. blue light?

2.5 **Critical Thinking** Describe a simple experiment you could do to determine whether sunglasses in a store were polarizing.

CHAPTER 16 REVIEW

ey Terms

.1
ght
y model
minous
uminated
minous flux
men
uminance
x
ndela
minous tensity

2
nsparent
nslucent
aque
ectrum
imary color
condary color
mplementary lor
e
gment
imary pigment
condary gment
n-film erference
larized

Summary

16.1 Light Fundamentals

- Light is an electromagnetic wave that stimulates the retina of the eye. Its wavelengths are between 400 and 700 nm.
- Light travels in a straight line through any uniform medium.
- In a vacuum, light has a speed of 3.00×10^8 m/s.
- The luminous flux of a light source is the rate at which light is emitted. It is measured in lumens.
- Illuminance is the rate at which light falls on a unit area. It is measured in lux.

16.2 Light and Matter

- Materials may be characterized as being transparent, translucent, or opaque, depending on the amount of light they reflect, transmit, or absorb.
- White light is a combination of the spectrum of colors, each having different wavelengths.
- White light can be formed by adding together the primary light colors: red, blue, and green.
- The subtractive primary colors—cyan, magenta, and yellow—are used in pigments and dyes to produce a wide variety of colors.
- Colors in soap and oil films are caused by the interference of specific wavelengths of light reflected from the front and back surfaces of the thin films.
- Polarized light consists of waves vibrating in a particular plane.

Reviewing Concepts

Section 16.1

1. Sound does not travel through a vacuum. How do we know that light does?
2. What is the range of wavelength, from shortest to longest, that the human eye can detect?
3. What color of visible light has the shortest wavelength?
4. What was changed in the equation $v = \lambda f$ in this chapter?
5. Distinguish between a luminous body and an illuminated body.
6. Look carefully at an ordinary, frosted, incandescent bulb. Is it a luminous or an illuminated body?
7. Explain how we can see ordinary, nonluminous classroom objects.
8. What are the units used to measure each of the following?
   a. luminous intensity
   b. illuminance
   c. luminous flux
9. What is the symbol that represents each of the following?
   a. luminous intensity
   b. illuminance
   c. luminous flux

Section 16.2

10. Distinguish among transparent, translucent, and opaque objects.
11. Of what colors does white light consist?
12. Is black a color? Why does an object appear to be black?
13. Name each primary light color and its secondary light color.
14. Name each primary pigment and its secondary pigment.
15. Why can sound waves not be polarized?

Applying Concepts

16. What happens to the wavelength of light as the frequency increases?
17. To what is the illumination of a surface by a light source directly proportional? To what is it inversely proportional?
18. A point source of light is 2.0 m from screen A and 4.0 m from screen B. How does the illumination of screen B compare with the illumination of screen A?
19. You have a small reading lamp 35 cm from the pages of a book. You decide to double the distance. Is the illumination on the book the same? If not, how much more or less is it?
20. Why are the insides of binoculars and cameras painted black?
21. The eye is most sensitive to yellow-green light. Its sensitivity to red and blue light is less than ten percent as great. Based on this knowledge, what color would you recommend that fire trucks and ambulances be painted? Why?
22. Some very efficient streetlights contain sodium vapor under high pressure. They produce light that is mainly yellow with some red. Should a community having these lights buy dark-blue police cars? Why or why not?
23. Suppose astronauts made a soap film in the space shuttle. Would you expect an orderly set of colored lines, such as those in **Figure 16–13**? Explain.
24. Photographers often put polarizing filters over the camera lens to make clouds in the sky more visible. The clouds remain white while the sky looks darker. Explain this based on your knowledge of polarized light.
25. An apple is red because it reflects red light and absorbs blue and green light. Follow these steps to decide whether a piece of transparent red cellophane absorbs or transmits blue and green light:
    a. Explain why the red cellophane looks red in reflected light.
    b. When you hold it between your eye and a white light, it looks red. Explain.
    c. Now, what happens to the blue and green light?
26. A soap film is transparent and doesn't absorb any color. If such a film reflects blue light, what kind of light does it transmit?
27. You put a piece of red cellophane over one flashlight and a piece of green cellophane over another. You shine the light beams on a white wall. What color will you see where the two flashlight beams overlap?
28. You now put both the red and green cellophane pieces over one of the flashlights in problem 27. If you shine the flashlight beam on a white wall, what color will you see? Explain.
29. If you have yellow, cyan, and magenta pigments, how can you make a blue pigment? Explain.
30. Consider a thin film of gasoline floating on water. The speed of light is slower in gasoline than in air, and slower in water than in gasoline. Would you expect the $\lambda/4$ rule to hold in this case? Explain.

Problems

Section 16.1

LEVEL 1

31. Convert 700 nm, the wavelength of red light, to meters.
32. Light takes 1.28 s to travel from the moon to Earth. What is the distance between them?
33. The sun is 1.5×10^8 km from Earth. How long does it take for the sun's light to reach us?
34. Radio stations are usually identified by their frequency. One radio station in the middle of the FM band has a frequency of 99.0 MHz. What is its wavelength?
35. What is the frequency of a microwave that has a wavelength of 3.0 cm?
36. Find the illumination 4.0 m below a 405-lm lamp.
37. A screen is placed between two lamps so that they illuminate the screen equally. The first lamp emits a luminous flux of 1445 lm and is 2.5 m from the screen. What is the distance of the second lamp from the screen if the luminous flux is 2375 lm?

38. A three-way bulb uses 50, 100, or 150 W of electrical power to deliver 665, 1620, or 2285 lm in its three settings. The bulb is placed 80 cm above a sheet of paper. If an illumination of at least 175 lx is needed on the paper, what is the minimum setting that should be used?

39. Two lamps illuminate a screen equally. The first lamp has an intensity of 101 cd and is 5.0 m from the screen. The second lamp is 3.0 m from the screen. What is the intensity of the second lamp?

LEVEL 2

40. Ole Roemer found that the maximum increased delay in the disappearance of Io from one orbit to the next is 14 s.

a. How far does light travel in 14 s?

b. Each orbit of Io takes 42.5 h. Earth travels the distance calculated in **a** in 42.5 h. Find the speed of Earth in km/s.

c. See if your answer for **b** is reasonable. Calculate Earth's speed in orbit using the orbital radius, 1.5×10^8 km, and the period, one year.

41. Suppose you wanted to measure the speed of light by putting a mirror on a distant mountain, setting off a camera flash, and measuring the time it takes the flash to reflect off the mirror and return to you. Without instruments, a person can detect a time interval of about 0.1 s. How many kilometers away would the mirror have to be? Compare this distance with that of some known objects.

42. A streetlight contains two identical bulbs 3.3 m above the ground. If the community wants to save electrical energy by removing one bulb, how far from the ground should the streetlight be positioned to have the same illumination on the ground under the lamp?

43. A student wants to compare the luminous flux from a bulb with that of a 1750-lm lamp. The two bulbs illuminate a sheet of paper equally. The 1750-lm lamp is 1.25 m away; the unknown bulb is 1.08 m away. What is its luminous flux?

44. A 10-cd point source lamp and a 60-cd point source lamp cast equal intensities on a wall. If the 10-cd lamp is 6.0 m from the wall, how far is the 60-cd lamp?

Critical Thinking Problems

45. Suppose you illuminated a thin soap film with red light from a laser. What would you see?

46. If you were to drive at sunset in a city filled with buildings that have glass-covered walls, you might be temporarily blinded by the setting sun reflected off the building's walls. Would polarizing glasses solve this problem?

47. You use a small pinhole camera to form an image of the sun on the screen. Given that the diameter of the sun is approximately 1.4×10^6 km, use the relationship you developed in the Physics Lab between h_i, h_o, d_i, and d_o to estimate the distance to the sun.

Going Further

A hanging soap film **(Figure 16-13)** gets thicker at a rate of 150 nm for each centimeter from the top of the film. Use a calculator or computer to find the distances from the top of the film of the first three reflected fringes of each of the colors blue, green, yellow, and red.

The color is most strongly reflected when the thickness is an odd number of quarter wavelengths of that color ($\lambda/4$, $3\,\lambda/4$, $5\,\lambda/4$, etc.) The wavelength, however, is that of the light within the soap film. This wavelength is 3/4 of the wavelength in air. Use the following wavelengths in air: blue 460 nm, green 550 nm, yellow 600 nm, red 660 nm. Plot these locations on a sheet of paper and compare with **Figure 16-13.**

***inter*NET CONNECTION**

Follow the link on the Glencoe Homepage at **www.glencoe.com/sec/science** to find out more about this chapter.

A Smile in the Sky

Rainbows are created by a combination of refraction and reflection in raindrops. When you look at a rainbow, you see violet on the inside and red on the outside. How does the rainbow form, and why are the colors separated?

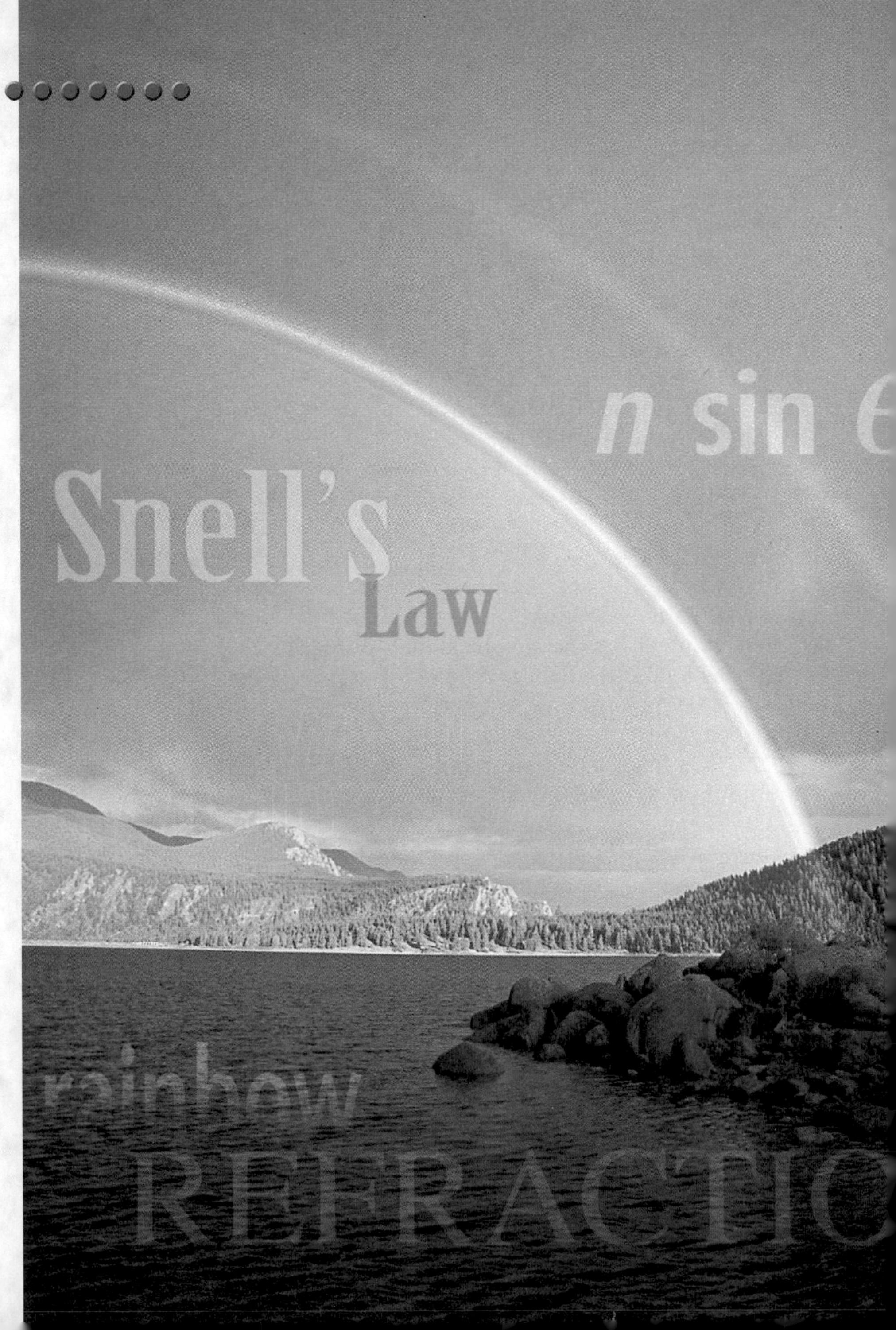

CHAPTER

17 Reflection and Refraction

Has anything in the sky ever captured the human imagination more than a rainbow? Yet all you need to produce a rainbow is simultaneous sunshine and rain. Even the drops from a hose or lawn sprinkler will do. Stand with the sun low and behind you and look into the water drops. Each drop of water separates sunlight into a spectrum: violet at the inside of the arc, then blue, green, yellow, and, at the outside, red. Look carefully at the photo, and you will see more. The sky is brighter inside the rainbow than it is outside. There is also a secondary rainbow, with the order of colors reversed. The same principles that define a rainbow are also responsible for some of the more mysterious observations you see on a daily basis. These include mirages, the illusion of a straw bending as it enters a glass filled with liquid, the appearance of multiple reflections when you look into a mirror or through a window, or the multiple lights and colors seen in the fiber optic bundle pictured above. Knowing how to explain these phenomena makes the rainbow no less beautiful and the other observations no less mysterious!

WHAT YOU'LL LEARN

- You will study how light is bent when it moves from one medium to another.
- You will understand why total internal reflection occurs.
- You will discover what effects are caused by changes in the index of refraction.

WHY IT'S IMPORTANT

- Your view of the world, from your reflection in a mirror to your use of a phone or computer, depends on the various ways light interacts with the matter around you.

interNET CONNECTION

Follow the link for this chapter on the Glencoe Homepage at **www.glencoe.com/sec/science** to find out more about reflection and refraction.

17.1 How Light Behaves at a Boundary

OBJECTIVES

- **Explain** the law of reflection.
- **Distinguish** between diffuse and regular reflection and **provide** examples.
- **Calculate** the index of refraction in a medium.

Reflections

Toss a tennis ball or a handball against a wall so that it will bounce to a lab partner, but first predict where the ball must hit on the wall to bounce in the right direction. If your partner moves closer (or farther) from the wall, does your rule still work?

Compare and Contrast Write a general rule that seems to work. Does your rule for the bouncing ball work for predicting the path of light? How is it similar?

Light travels in straight lines and at very high velocities. Its velocity varies however, depending on the medium through which it moves. In this sense, light acts just like any other wave moving from one medium to another. What happens to light striking a surface between air and glass?

The Law of Reflection

In **Figure 17–1** a mirror has been placed on the table in front of a protractor. A laser beam strikes the mirror and is reflected from it. What can you say about the angles the beams make with the mirror? If you look carefully, you can see that each makes an angle of 60° relative to a line perpendicular to the mirror. This line is called the normal to the surface, or normal. The angle that the incoming beam makes with the normal, the angle of incidence, is equal to the angle the outgoing beam makes, the angle of reflection.

When two parallel beams strike the mirror, as in **Figure 17–2a,** the two reflected beams also are parallel, which means that the angle of reflection was the same for the two beams, as shown in **Figure 17–2c.** A smooth surface such as the mirror causes **regular reflection,** in which light is reflected back to the observer in parallel beams.

What happens when light strikes another surface that seems to be smooth, such as the page of this book or a wall painted white? As in **Figure 17–2b,** there is no reflected beam. Rather, there is a fuzzy round dot because the light was reflected into a wide range of angles, as illustrated in **Figure 17–2d.** On the scale of the wavelength of light, even these seemingly smooth surfaces are rough, and therefore they result in **diffuse reflection.** Where there is regular reflection, as in a mirror, you can see your face. But no matter how much light is reflected off bright white paint, it will still result in diffuse reflection, and you will never be able to use the wall as a mirror.

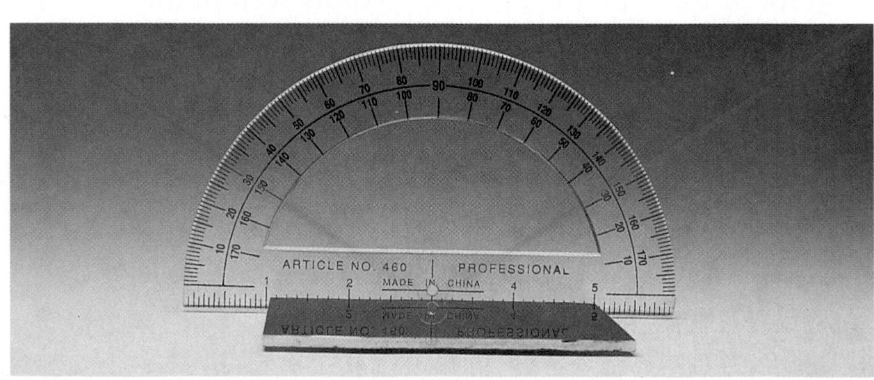

FIGURE 17–1 A light ray reflecting from a mirror shows that the angle of incidence equals the angle of reflection.

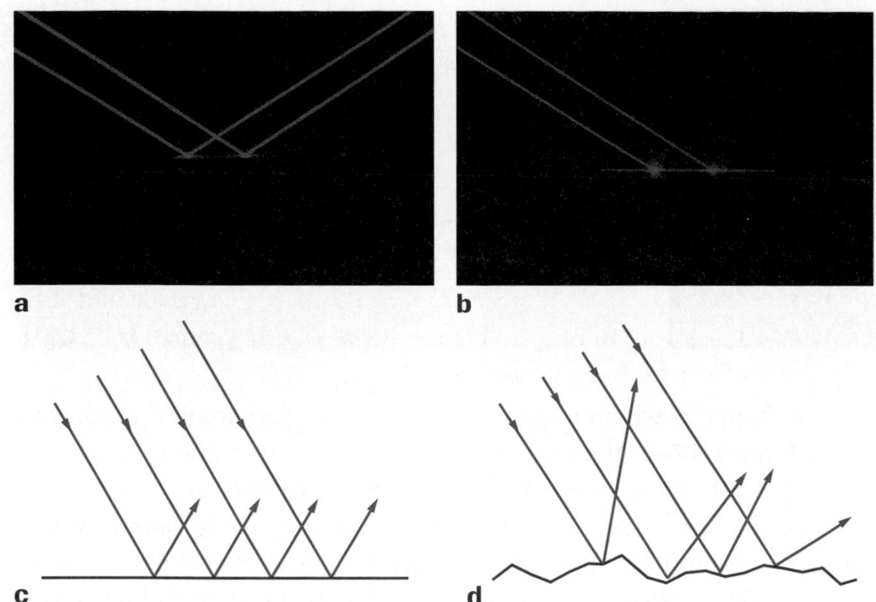

FIGURE 17–2 When parallel light rays strike a mirror surface, they reflect as parallel rays **(a) (c).** When parallel light rays strike a rough surface, they are randomly reflected **(b) (d).**

Refraction of Light

When light strikes the top of a block of plastic, as in **Figure 17–3a,** faint beams of light are reflected from the surface, but a bright beam goes into the block. It doesn't go in as a straight line, however; the beam is bent at the surface. Recall from Chapter 14 that this change in direction, or bending of a wave, at the boundary between two media is called refraction.

Note that when a light beam goes from air to glass at an angle, it is bent toward the normal, as shown in **Figure 17–3b.** The beam in the first medium is called the incident ray, and the beam in the second medium is called the refracted ray. In this case, the angle of incidence is larger than the **angle of refraction,** which is the angle that the refracted ray makes with the normal to the surface. If the angle of refraction is smaller than the angle of incidence, then the new medium is said to be more **optically dense.** Later in the chapter you'll learn that the speed of light is slower in more optically dense materials.

FIGURE 17–3 Light is refracted toward the normal as it enters denser medium. Compare the deflection of a set of wheels as it crosses a pavement-mud boundary. The first wheel that enters the mud is slowed, causing the wheels to change direction towards the perpendicular.

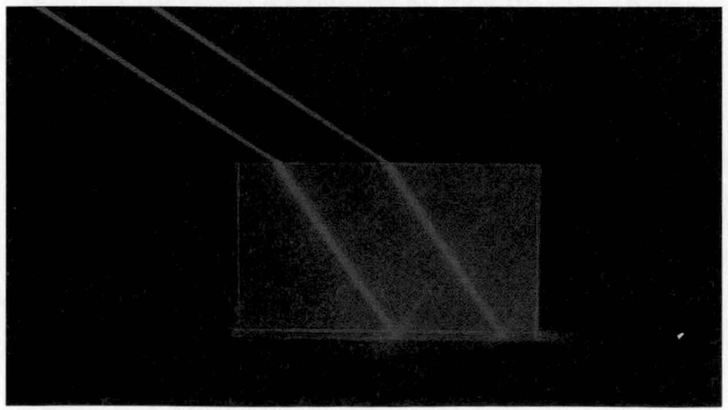

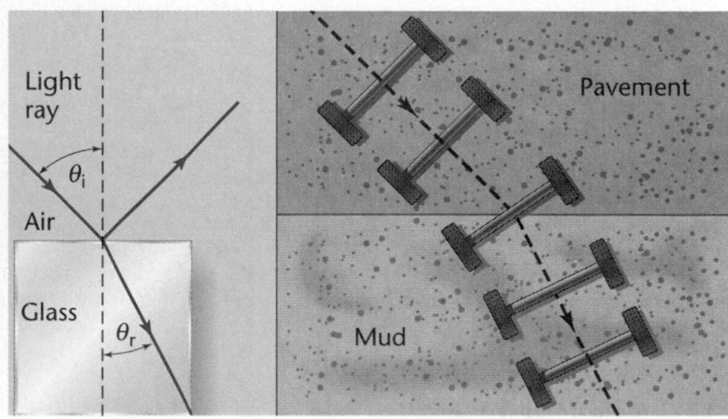

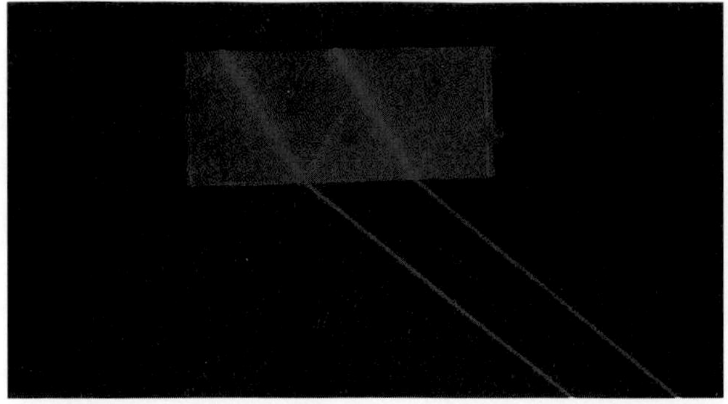

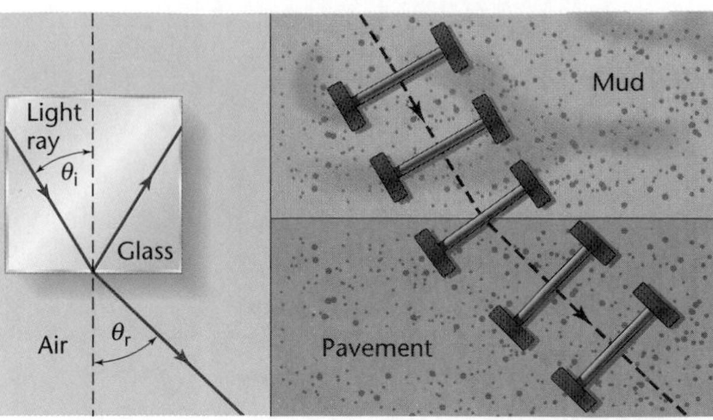

FIGURE 17–4 Light is refracted away from the normal as it enters a less-dense medium. Compare the deflection of a set of wheels as it crosses a mud-pavement boundary. The first wheel to leave the mud speeds up, and the direction of the wheels changes away from the perpendicular.

What happens when a light ray passes from glass to air? As you can see in **Figure 17–4,** the rays are refracted away from the normal. The angle of refraction is larger than the angle of incidence.

When light strikes a surface along the perpendicular, the angle of incidence is zero, and the angle of refraction will also be zero. The refracted ray leaves perpendicular to the surface and does not change direction.

Snell's Law

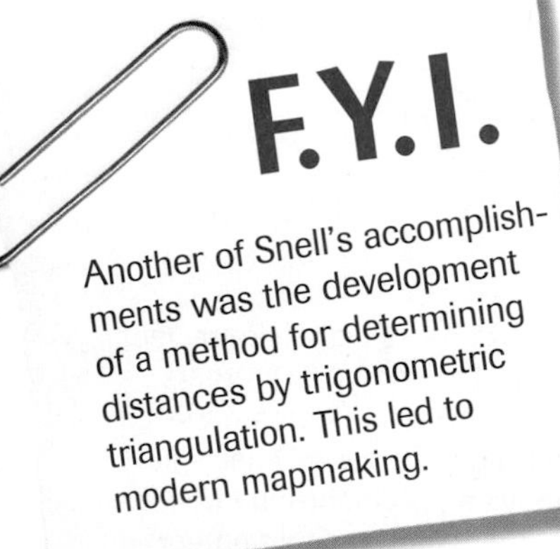

F.Y.I.
Another of Snell's accomplishments was the development of a method for determining distances by trigonometric triangulation. This led to modern mapmaking.

When light passes from one medium to another, it may be reflected and refracted. The degree to which it is bent depends on the angle of incidence, and the properties of the medium as shown in **Figure 17–5.** How does the angle of refraction depend on the angle of incidence? The answer to this question was found by Dutch scientist Willebrord Snell in 1621. **Snell's law** states that the ratio of the sine of the angle of incidence to the sine of the angle of refraction is a constant. For light going from the vacuum into another medium, the constant, n, is called the **index of refraction.** Snell's law, is written as

$$n = \frac{\sin \theta_i}{\sin \theta_r}$$

where θ_r is the angle of refraction, n the index of refraction, and θ_i is the angle of incidence. In the more general case, the relationship is written as

$$n_i \sin \theta_i = n_r \sin \theta_r.$$

FIGURE 17–5 When light passes from one medium to another, the angle of refraction depends upon the angle of incidence. This is shown very clearly by the rays of light leaving the glass prism.

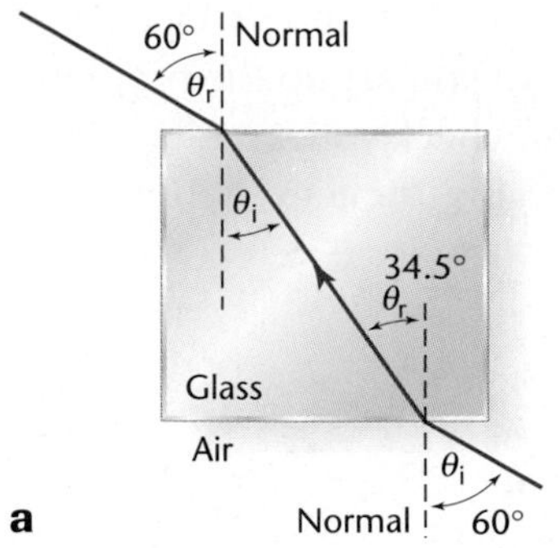

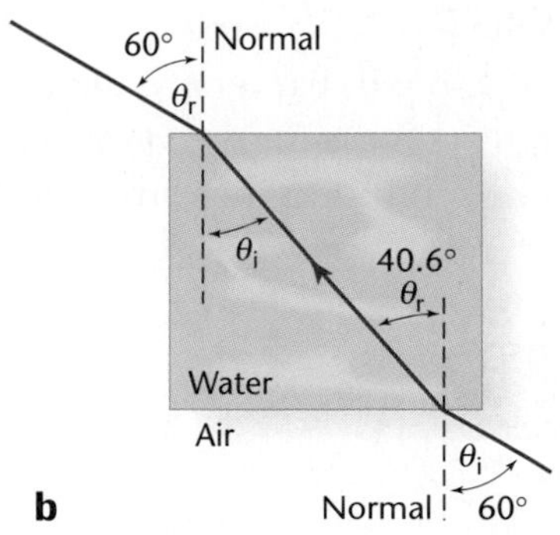

FIGURE 17–6 The index of refraction for glass is greater than that for water. As a result, the bending will be greater when light passes into or exits from the glass.

Here, n_i is the index of refraction of the medium in which the incident ray travels, the first medium, and n_r is the index of refraction of the medium in which the refracted ray moves, the second medium. **Figure 17–6** shows rays of light entering and leaving glass and water from air. Note how θ_i is always used for the angle the incident ray makes with the surface, regardless of the medium. From the angles of refraction, how would you expect the index of refraction of water to compare to that of glass?

The index of refraction of a substance often can be measured in the laboratory. To do this, direct a ray of light onto the substance's surface. Measure the angle of incidence and the angle of refraction. Then use Snell's law to find the index of refraction. **Table 17–1** presents indices of refraction for some common materials. Note that the index of refraction for air is only slightly larger than that of a vacuum. For all but the most precise measurements, you can set the index of refraction of air to 1.00.

TABLE 17–1
Indices of Refraction

Medium	n
vacuum	1.00
air	1.0003
water	1.33
ethanol	1.36
crown glass	1.52
quartz	1.54
flint glass	1.61
diamond	2.42

Problem Solving Strategy

Drawing Ray Diagrams

1. Draw a diagram showing the two media, as in **Figure 17–7.** Label the media, and indicate the two indices of refraction, n_i and n_r.
2. Draw the incident ray to the point where it hits the surface, then draw a normal to the surface at that point.
3. Use a protractor to measure the angle of incidence.
4. Use Snell's law to calculate the angle of refraction.
5. Use a protractor to draw the refracted ray leaving the surface at the point where the incident ray entered.
6. Evaluate your work. Make sure your answer agrees with the qualitative statement of Snell's law: light moving from a smaller n to a larger n is bent toward the normal; light moving from a larger n to a smaller n is bent away from the normal.

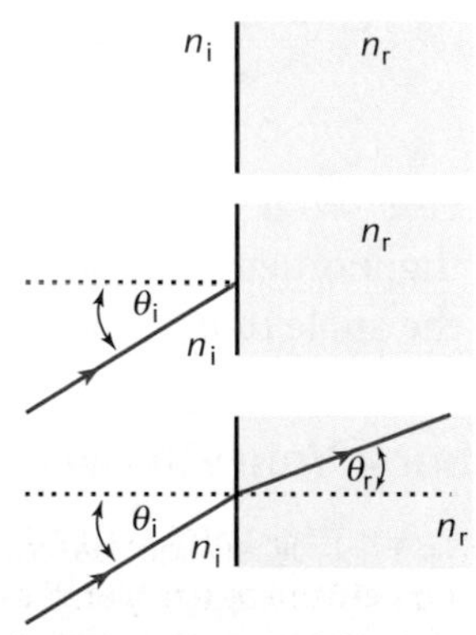

FIGURE 17–7 These are the steps to follow to draw a ray diagram.

When solving problems involving the reflection and refraction of light, you will use a ray diagram. This will help you set up the problem, assign symbols for the various quantities, and check your results. When drawing ray diagrams, use the Problem Solving Strategy outlined on page 397.

Example Problem

Snell's Law

A light beam in air hits a sheet of crown glass at an angle of 30.0°. At what angle is it refracted?

Sketch the Problem

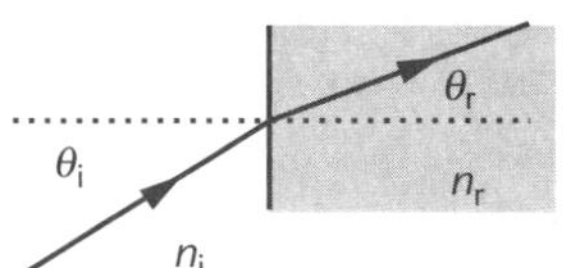

- Draw a picture of the beam moving from the air to the crown glass.
- Draw a ray diagram.
- Check to make sure that the angle in the medium with the larger n is smaller.

Calculate Your Answer

Known:

$\theta_i = 30.0°$

$n_i = 1.00$

$n_r = 1.52$

Unknown:

Angle of refraction $\theta_r = ?$

Strategy:

Use Snell's law written to be solved for the sine of the angle of refraction.

Calculations:

$$n_i \sin \theta_i = n_r \sin \theta_r$$

$$\sin \theta_r = \left(\frac{n_i}{n_r}\right) \sin \theta_i$$

$$= \left(\frac{1.00}{1.52}\right)(0.500)$$

$$= 0.329$$

Use your calculator or trigonometry tables to find the angle of refraction.

$$\theta_r = 19.2°$$

Check Your Answer

- Are the units correct? Angles are expressed in degrees.
- Is the magnitude realistic? The index of refraction, n_r, is greater than the index of refraction, n_i. Therefore, the angle of refraction, θ_r, must be less than the angle of incidence, θ_i.

Physics Lab

Bending of Light

Problem

How is the index of refraction of light in water determined?

Materials 

graph paper
felt-tip pen
ruler
semicircular plastic dish
water

Procedure

Part I

1. Draw a line dividing the graph paper in half.
2. Use the felt-tip pen to draw a vertical line at the center of the straight edge of the plastic dish. This line will be your object.
3. Place the edge of the dish along the straight line so that the dish is on the bottom half of the paper. Trace the outline of the dish on the paper.
4. Mark the position of the object on your paper.
5. Add water until the dish is 3/4 full.
6. Lay a ruler on the bottom half of the paper. Adjust the position until the edge of the ruler seems to point at the object when you look through the water.
7. Have a lab partner check to verify that the ruler position is accurate.
8. Draw a line along the ruler edge to the edge of the dish.
9. Repeat steps **6–8** for a different position of the ruler.

Part II

1. Wipe the vertical line from the dish and draw a vertical line at the center of the curved edge. This is your new object.
2. Repeat all steps from Part I, but this time sight the ruler on the top half of the paper.

Data and Observations

1. Look at the sight lines you drew in Part I. Did the light bend when moving from water to air?
2. For Part II, do the sight lines point directly toward the object?
3. For Part II, draw a line from the object position to the point where each sight line touches the dish.
4. Draw the normal at each point where the sight line touched the dish.
5. Measure the angles from the normal for the angles in air and water.

Analyze and Conclude

1. **Interpreting Data** Explain why the light did not bend in Part I. (**Hint:** Draw the normal to the surface.)
2. **Calculating Values** Calculate *n*, using Snell's law.

Apply

1. Could a flat piece of material be used for focusing light? Make a drawing to support your answer.

Practice Problems

1. Light in air is incident upon a piece of crown glass at an angle of 45.0°. What is the angle of refraction?
2. A ray of light passes from air into water at an angle of 30.0°. Find the angle of refraction.
3. A ray of light is incident upon a diamond at 45.0°.
   a. What is the angle of refraction?
   b. Compare your answer for 3a to your answer for problem 1. Does glass or diamond bend light more?
4. A block of unknown material is submerged in water. Light in the water is incident on the block at an angle of 31°. The angle of refraction in the block is 27°. What is the index of refraction of the unknown material?

Pocket Lab

Refraction 

Place a small hexagonal nut in the center of the bottom of a 1000-mL beaker. Pour water into the beaker until it is half full of water. Look through the sides of the beaker at the nut while placing a ruler along the tabletop so that the edge of the ruler appears to point to the center of the nut. Do you think that the ruler really points to the nut? Look from the top to see where the ruler points. Place a golf ball on the nut. Look through the sides of the beaker at the ball and adjust the edge of the ruler to point to the edge of the ball. Look from the top.

Analyze and Conclude
Describe your observations. Make a drawing to show why the ball appears to be so wide.

Index of Refraction and the Speed of Light

The index of refraction is a measure of how much light bends when it passes into a medium from a vacuum. But the index is also a measure of how fast light travels in the medium. To learn how these two are connected, you must explore the connections between the ray model and the wave model of light.

A ray, as you know, is the path of an extremely narrow beam of light. If you have a broader beam of light that is always the same width, then you can visualize this beam as a series of parallel, straight wave fronts. Each wave front represents the crest of the electromagnetic wave and is perpendicular to the direction of the ray. Therefore, the spacing between wavefronts is one wavelength. Recall from Chapter 14, that when a wave moves from one medium to another in which the wave speed is different, the frequency of the wave is unchanged, but the wavelength changes according to the equation $v = \lambda f$. Thus, in a vacuum, $v_{vacuum} = c$, and $\lambda_{vacuum} = c/f$. But if light travels at speed $v_{material}$ in a medium, then $\lambda_{material} = v_{material}/f$.

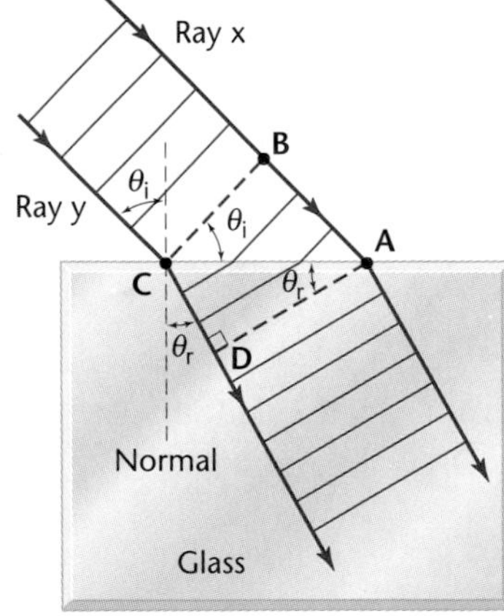

FIGURE 17–8 In this diagram, the refraction of two parallel light rays incident on a piece of glass is shown.

Figure 17–8 shows a beam of light, originally in a vacuum, that strikes a glass plate at an angle of incidence θ_i. The line BC represents the last wavefront totally in the vacuum, and the line AD represents the first wavefront entirely in the glass. In this example, these lines are three wavelengths apart. But the wavelength in glass is smaller than that in the vacuum, so the distance CD is shorter than the distance BA, and the wavefronts that are partially in the vacuum and partially in the glass are bent at the boundary. You can see that the direction of the ray is bent toward the normal when light moves from a vacuum into matter. But by how much is it bent?

First find the relative lengths of CD and BA, which are separated by three wavelengths. They are related in the following way.

$$\frac{BA}{CD} = \frac{3\lambda_{vacuum}}{3\lambda_{glass}} = \frac{3c/f}{3v_{glass}/f} = \frac{c}{v_{glass}}$$

Thus, these two lengths are related by the speed of light in the vacuum and in the glass.

Next find the relationship between the sines of the angles of incidence and refraction. Consider the two triangles, ABC and ADC. The sine of an angle is the length of the opposite side divided by the length of the hypotenuse. Thus, $\sin \theta_i = BA/CA$ and $\sin \theta_r = CD/CA$. Therefore, using Snell's law, the following is true.

$$\frac{\sin \theta_i}{\sin \theta_r} = \frac{BA/CA}{CD/CA} = \frac{BA}{CD}$$

You already know that $BA/CD = c/v_{glass}$, and, according to Snell's law, because the index of refraction of the vacuum is 1.00, the following is true.

$$\frac{\sin \theta_i}{\sin \theta_r} = n \text{ and } \frac{\sin \theta_i}{\sin \theta_r} = \frac{c}{v_{glass}}$$

$$\text{so, } n = \frac{c}{v_{glass}}$$

Therefore, it is possible to calculate the speed of light in many substances.

Example Problem

Speed of Light in Matter

Find the speed of light in water.

Calculate Your Answer

Known:

$n_{water} = 1.33$

$c = 3.00 \times 10^8$ m/s

Unknown:

Speed of light in water, $v_{water} = ?$

Strategy:

Use the relationship that the index of refraction of water equals the ratio of the speed of light in vacuum to the speed of light in water.

Calculations:

$n_{water} = c/v_{water}$

$v_{water} = c/n_{water}$

$= \frac{(3.00 \times 10^8 \text{ m/s})}{1.33}$

$= 2.26 \times 10^8$ m/s

Check Your Answer

- Is the magnitude realistic? Light slows as it passes through water. Therefore, the speed must be less than 3.00×10^8 m/s.

Practice Problems

5. Use **Table 17–1** to find the speed of light in the following.
   a. ethanol **b.** quartz **c.** flint glass
6. The speed of light in one type of plastic is 2.00×10^8 m/s. What is the index of refraction of the plastic?
7. What is the speed of light for the unknown material in problem 4?
8. Suppose two pulses of light were "racing" each other, one in air, the other in a vacuum. You could tell the winner if the time difference were 10 ns (10×10^{-9} s). How long would the race have to be to determine the winner?

The speed of light for some materials is listed in **Table 17–2**.

TABLE 17–2

Speed of Light in Various Materials

Material	v m/s
Vacuum	3.00×10^8
Air	3.00×10^8
Ice	2.29×10^8
Glycerine	2.04×10^8
Crown glass	1.97×10^8
Rock salt	1.95×10^8

17.1 Section Review

1.1 Give examples of diffuse and regular reflectors.

1.2 If you double the angle of incidence, the angle of reflection also doubles. Does the angle of refraction double as well?

1.3 You notice that when a light ray enters a certain liquid from water, it is bent toward the normal, but when it enters the same liquid from crown glass, it is bent away from the normal. What can you conclude about its index of refraction?

1.4 **Critical Thinking** Could an index of refraction ever be less than 1? What would that imply about the velocity of light in that medium?

Applications of Reflected and Refracted Light

17.2

OBJECTIVES

- **Explain** total internal reflection.
- **Define** the critical angle.
- **Explain** effects caused by the refraction of light in a medium with varying refractive indices.
- **Explain** dispersion of light in terms of the index of refraction.

Modern societies depend on communication systems. The telephone has become necessary for both homes and businesses. In many cities, however, the underground pipes containing telephone wires are so full that no new customers can be added. Now the old wires can be replaced by a bundle of optical fibers that can carry thousands of telephone conversations at once. Moreover, illegal tapping of optical fibers is almost impossible. This application of the reflection of light is revolutionizing our communication systems.

Total Internal Reflection

What happens when a ray of light passes from a more optically dense medium into air? You have learned that the light is bent away from the normal. In other words, the angle of refraction is larger than the angle of incidence. The fact that the angle of refraction is larger than the angle of incidence leads to an interesting phenomenon known as **total internal reflection,** which occurs when light passes from a more optically dense medium to a less optically dense medium at an angle so great that there is no refracted ray. **Figure 17–9** shows how this happens. Ray 1 is incident upon the surface of the water at angle θ_i. Ray 1 leaves by the angle of refraction, θ_r. Ray 2 is incident at such a large angle, θ_i, that the refracted ray lies along the surface of the water. The angle of refraction is 90°. What is θ_i?

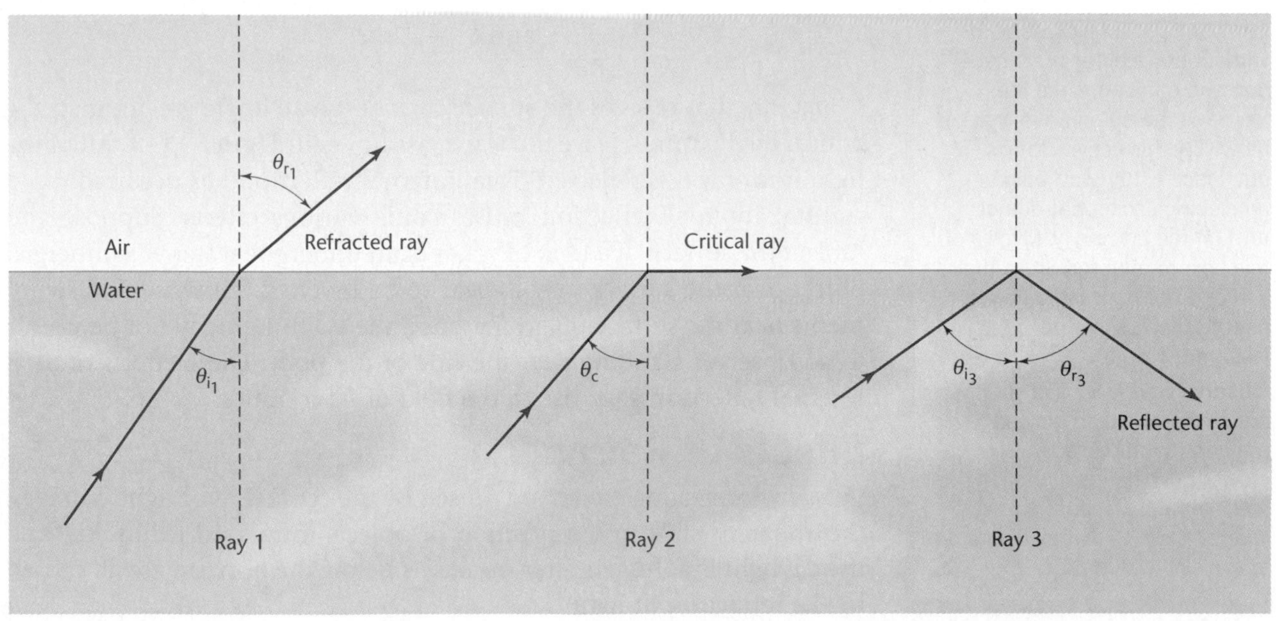

FIGURE 17–9 Ray 1 is refracted. Ray 2 is refracted along the boundary of the medium and forms the critical angle. An angle of incidence greater than the critical angle results in the total internal reflection of ray 3.

F.Y.I.

Most binoculars contain two prisms mounted at right angles to each other so that the light can be redirected from the object to the eyepiece by total internal reflection.

For light traveling from one medium into another, Snell's law is $n_i \sin \theta_i = n_r \sin \theta_r$. For this example, $n_{water} \sin \theta_i = n_{air} \sin \theta_r$ or $(1.33)(\sin \theta_i) = (1.00)(\sin 90°)$.

Solving the equation for $\sin \theta_i$,

$$\sin \theta_i = \frac{(1.00)(\sin 90°)}{1.33}$$

$$= 0.752$$

so $\theta_i = 48.8°$. When an incident ray of light passing from water to air makes an angle of 48.8°, the angle of refraction is 90°.

The incident angle that causes the refracted ray to lie right along the boundary of the substance, angle θ_c, is unique to the substance. It is known as the **critical angle** of the substance. The critical angle, θ_c, of any substance may be calculated as follows.

$$n_i \sin \theta_i = n_r \sin \theta_r$$

In this situation, $\theta_i = \theta_c$, $n_r = 1.00$, and $\theta_r = 90.0°$. Therefore, the following is true.

$$\sin \theta_c = \frac{(1.00)(\sin 90.0°)}{n_i}$$

$$= \frac{1.00}{n_i}$$

For crown glass, the critical angle can be calculated as follows.

$$\sin \theta_c = \frac{1.00}{1.52}$$

$$= 0.658$$

$$\text{and } \theta_c = 41.1°$$

Any ray that reaches the surface of water at an angle greater than the critical angle cannot leave the water, as shown in **Figure 17–9.** All of the light from ray 3 is reflected. Total internal reflection has occurred.

Total internal reflection causes some curious effects. Suppose you look at the surface of a tank of water from under the water. A submerged object near the surface may appear to be inverted. Likewise, if a swimmer is near the surface of a quiet pool, the swimmer may not be visible to an observer standing near the side of the pool. These effects of total internal reflection gave rise to the field of fiber optics.

Effects of Refraction

Many interesting effects are caused by the refraction of light. Mirages, the apparent shift in the position of objects immersed in liquids, and the daylight that lingers after the sun is below the horizon are all caused by the refraction of light.

Pocket Lab

Cool Images

Caution: *Avoid staring directly into the laser beam or at bright reflections.*

Can you light an electric bulb without any electrical connection and can you make the image of the bulb in a mirror glow? Try this activity to find out. Place a 100-watt bulb in an electric socket but do not turn on the electricity. Place this next to a mirror. Sit so that you can see both the bulb and its reflection. Aim a penlight laser at the bulb.

Observe and Infer Did you notice that the bulb glowed red and that the image also glowed? What would happen if you aimed the laser at the image? Try it. Use a ray diagram to explain your results.

How It Works

Optical Fibers

Fiber-optic technology has been applied to many fields. Physicians are able to use optical fibers to treat illnesses and as surgical tools to perform delicate operations. In the field of telecommunications, optical fibers can transmit audio and video information, as well as other data, as coded light signals. An information flow equivalent to 25 000 telephone conversations can be carried by a fiber the thickness of a human hair. Facsimile systems, oscilloscopes, photographic typesetting machines, and computer graphics systems also use optical fibers. For this reason, optical fibers are being used to transmit telephone, computer, and video signals within buildings; from city to city; and even across oceans. Even plants have been shown to use total internal reflection to transport light to cells that utilize light energy.

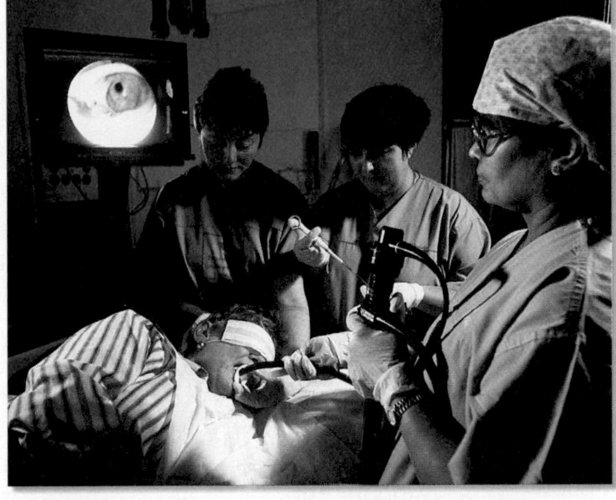

1. The core of an optical fiber is made of either glass or plastic. Because it transmits light over a variety of distances, this core must be highly transparent.

2. Light impulses from a source enter one end of the optical fiber. Each time the light strikes the surface, the angle of incidence is larger than the critical angle. The reflection is total, keeping the light within the fiber.

3. The outer covering of an optic fiber is called cladding. Cladding prevents light from scattering once it is in the core.

4. Light exits the cable at the opposite end, where a device, which could be the human eye, receives the light.

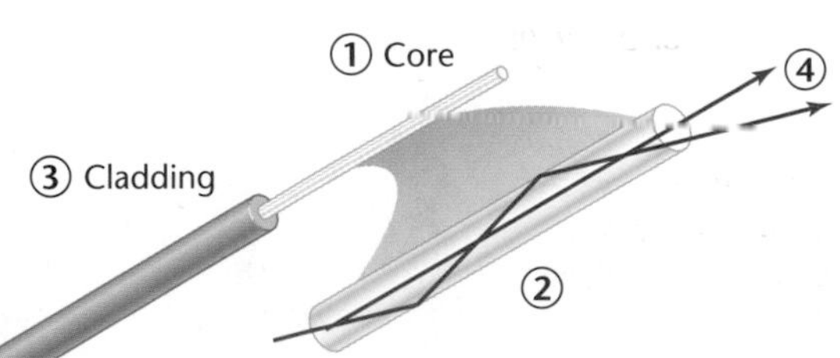

5. The thin, flexible optical fiber can be easily bent around corners or combined with many other fibers into a cable.

Thinking Critically

1. What advantages do fiber-optic cables have over their electrical counterparts?
2. Which part of a fiber-optic cable—the core or the cladding—has a higher index of refraction?

FIGURE 17–10 Mirages are caused by the refracting properties of a nonuniform atmosphere.

F.Y.I.

The blue color of the sky during the daylight hours results from the intense scattering of blue light from the sun. In the early morning and late evening, this blue light must travel through more air and the amount of blue you see is reduced by this scattering. This allows you to see the beautiful yellows, oranges, and reds of the sunrises and sunsets.

Have you ever seen the mirage effect shown in **Figure 17–10?** A driver looking almost parallel to the road sees what looks like a puddle of water. The puddle, however, disappears as the car approaches. The mirage is the result of the sun heating the road. The hot road, in turn, heats the air above it, while the air farther above the road remains cool. The index of refraction of air at 30°C, for example, is 1.00026, while that of air at 15°C is 1.00028. Thus, there is a continuous change in the index of refraction of the air. A ray of light aimed toward the road encounters the smaller index of refraction and is bent away from the normal, that is, it is bent to be more parallel to the road, as shown in **Figure 17–11.** The motorist actually sees light from the sky, which looks like light reflected from a puddle.

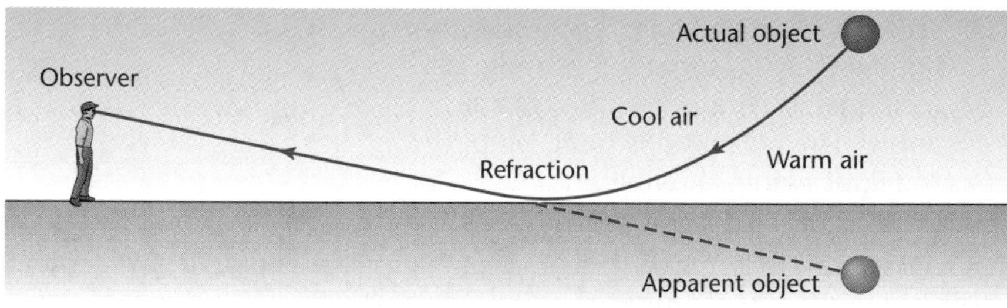

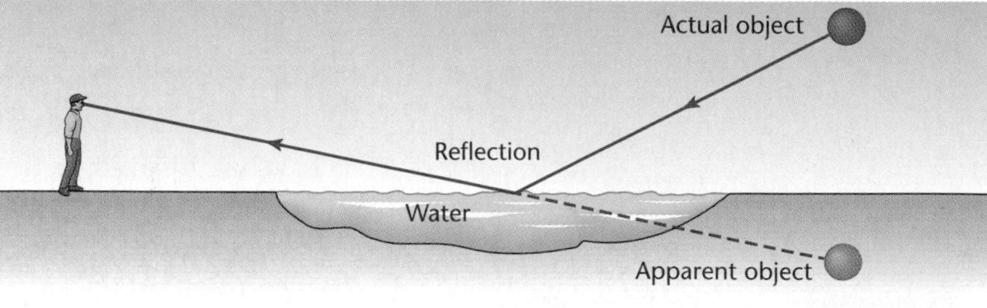

FIGURE 17–11 Refraction of light in air of different densities **(a)** produces an effect similar to the reflection of light off a pool of water **(b).**

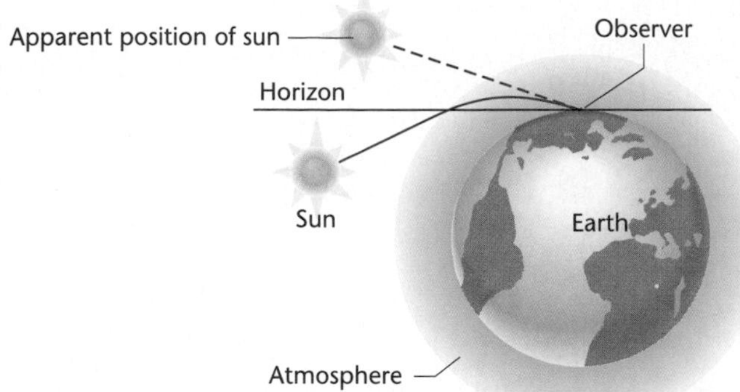

FIGURE 17–12 After the sun has actually set, it is still visible because of refraction of light over the horizon through the atmosphere.

An object submerged in a liquid is not where it appears to be. As a result of refraction, an object may appear to be much closer to the surface of the liquid than it actually is. Refraction also makes a spoon placed in a glass of water appear to be bent.

Light travels at a slightly slower speed in Earth's atmosphere than it does in outer space. As a result, sunlight is refracted by the atmosphere. In the morning, this refraction causes sunlight to reach us before the sun is actually above the horizon, as shown in **Figure 17–12.** In the evening, the sunlight is bent above the horizon after the sun has actually set. Thus, daylight is extended in the morning and evening because of the refraction of light.

LITERATURE CONNECTION

Rainbows result when sunlight is refracted and reflected by raindrops. To many writers, rainbows represent the mystery and beauty of nature. William Wordsworth wrote that he would rather be dead than unable to appreciate a rainbow. A rainbow in a story often signals that hard times are over. In folktales, rainbows point the way to pots of gold. Maybe rainbows seem miraculous because they are so rare. Why don't we see them more often?

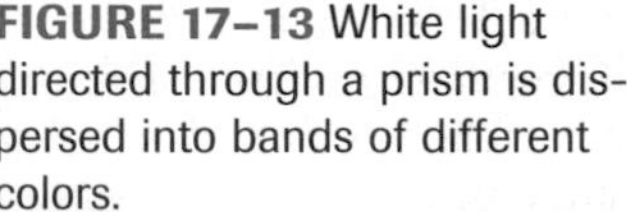

Dispersion of Light

How can a prism produce such beautiful colors? Did they come from the glass? Were they in the light itself? Although the formation of colors by light passing through glass, water, and other clear materials had been observed long before his time, Sir Isaac Newton was the first to discover how the colors were formed. He found that sunlight was separated into the spectrum of colors when it passed through a prism, as in **Figure 17–13.** To show that the colors weren't in the prism itself, Newton used the

FIGURE 17–13 White light directed through a prism is dispersed into bands of different colors.

FIGURE 17–14 This old sketch illustrates Sir Isaac Newton's demonstration with prisms. Prism ABC separates white light into colors. Prism EDG combines the colors back into white light.

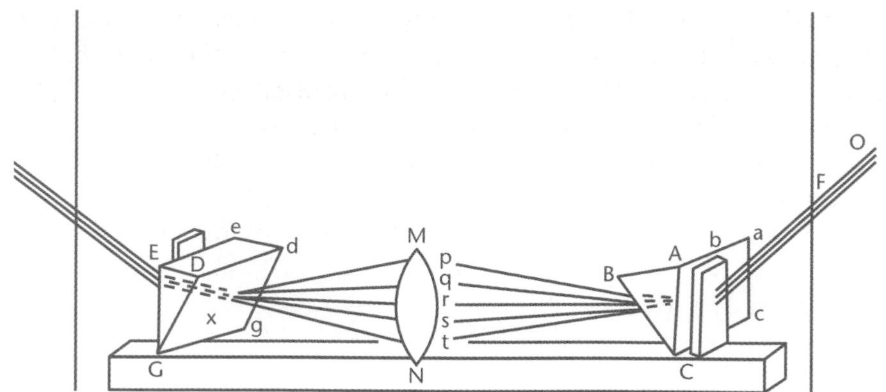

apparatus shown in **Figure 17–14.** When spectrum, p,q,r,s,t, produced by prism ABC, was directed through a lens onto a second prism, DEG, white light came out of the second prism. Thus, Newton showed that white light could be separated into all the colors of the rainbow, and that those colors could be added together again to produce white light.

The separation of light into its spectrum is called **dispersion.** Red light is bent the least as it goes through a prism, while violet is bent the most. This means that the index of refraction depends on the color, or the wavelength, of light. The index of refraction is smaller for red than it is for violet. Because the index of refraction and light speed are related, it further means that the speed of light in matter depends on the wavelength of the light.

Glass is not the only substance that disperses light. A diamond not only has one of the highest refractive indices of any material, but it also has one of the largest variations in the index. Thus, it disperses light more than most other materials. The intense colors visible when light is dispersed in a diamond are the reason why these gems are said to have "fire."

Different light sources have different spectra. A prism can be used to determine the spectrum of a source. Light from an incandescent lamp, for example, contains all visible wavelengths of light. When this light passes through a prism, a continuous band of color is seen. A fluorescent lamp produces both a continuous spectrum and light emitted at four individual wavelengths. Thus, its spectrum contains both a continuous band and bright lines at specific colors.

A prism is not the only means of dispersing light. A rainbow is a spectrum formed when sunlight is dispersed by water droplets in the atmosphere. Sunlight that falls on a water droplet is refracted. Because of dispersion, each color is refracted at a slightly different angle, as shown in **Figure 17–15a**. At the back surface of the droplet, the light undergoes total internal reflection. On the way out of the droplet, the light is once more refracted and dispersed. Although each droplet produces a complete spectrum, an observer will see only a certain wavelength of light

Personal Rainbow

You can make your own personal rainbow when the sun is out and low in the sky for easier viewing. Adjust a garden hose to produce a gentle spray. Face away from the sun so that you can see your shadow. Spray the water upwards above your shadow and watch closely until you see the colors. By moving the spray in an arc from side to side, you will produce your own personal rainbow.

Analyze and Conclude Did you notice the order of the colors in the spectrum of visible lights. Could you easily see each of the colors ROYGBIV? Which color was on the inside edge? Which color was on the outside edge?

from each droplet. The wavelength depends on the relative positions of sun, droplet, and observer. Because there are millions of droplets in the sky, a complete spectrum is visible. The droplets reflecting red light make an angle of 42° in relation to the direction of the sun's rays; the droplets reflecting blue light make an angle of 40°, as shown in **Figure 17–15b.**

Sometimes you can see a faint second-order rainbow. Light rays that are reflected twice inside the drop produce this sight. A third bow is possible but normally is too weak to observe.

A Smile in the Sky

FIGURE 17–15 Refraction occurs when rays pass into and out of a raindrop. Reflection occurs at the inside surface **(a).** The observer sees only certain wavelengths from each drop **(b).** (Ray angles have been exaggerated for clarity.)

17.2 Section Review

2.1 If you were to use quartz and crown glass to make an optical fiber, which would you use for the coating layer? Why?

2.2 Is there a critical angle for light going from glass to water? From water to glass?

2.3 Why can you see the image of the sun just above the horizon when the sun, itself, has already set?

2.4 **Critical Thinking** In what direction can you see a rainbow on a rainy late afternoon?

CHAPTER 17 REVIEW

Key Terms

17.1

- regular reflection
- diffuse reflection
- angle of refraction
- optically dense
- Snell's law
- index of refraction

17.2

- total internal reflection
- critical angle
- dispersion

Summary

17.1 How Light Behaves at a Boundary

- The law of reflection states that the angle of reflection is equal to the angle of incidence.
- Refraction is the bending of light rays at the boundary between two media. Refraction occurs only when the incident ray strikes the boundary at an angle.
- When light goes from a medium with a small n to one with a large n, it is bent toward the normal. Light going from materials with a large n to those with a small n is bent away from the normal.
- Snell's law states $n_i \sin \theta_i = n_r \sin \theta_r$.

17.2 Applications of Reflected and Refracted Light

- Total internal reflection occurs when light is incident on a boundary from the medium with the larger index of refraction. If the angle of incidence is greater than the critical angle, no light leaves; it is all reflected.
- Light waves of different wavelengths have slightly different refractive indices. Thus, they are refracted at different angles. Light falling on a prism is dispersed into a spectrum of colors.

Reviewing Concepts

Section 17.1

1. What is meant by the phrase "normal to the surface"?
2. How does regular reflection differ from diffuse reflection?
3. Does the law of reflection hold for diffuse reflection? Explain.
4. Copy **Figure 17–16** onto your paper. Draw a normal and label the angle of incidence and the angle of refraction for light moving from substance A to substance B.

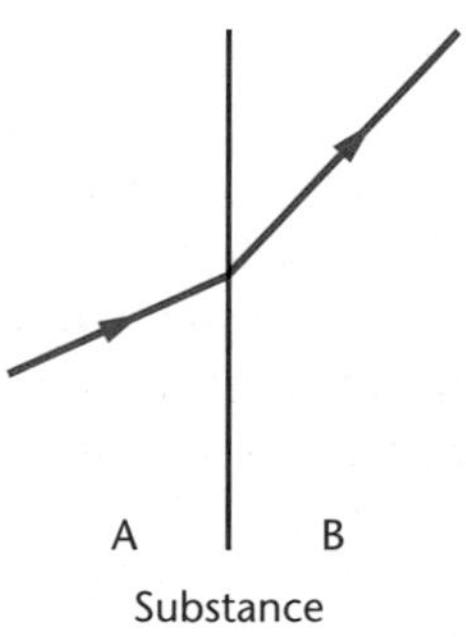

FIGURE 17–16

5. How does the angle of incidence compare with the angle of refraction when a light ray passes from air into glass at a non-zero angle?
6. How does the angle of incidence compare with the angle of refraction when a light ray leaves glass and enters air at a non-zero angle?
7. What are the units of the index of refraction?
8. State Snell's law in your own words.
9. Derive $n = \sin \theta_i / \sin \theta_r$ from the general form of Snell's law, $n_A \sin \theta_A = n_B \sin \theta_B$. State any assumptions and restrictions.

Section 17.2

10. What is the critical angle of incidence?
11. What happens to a ray of light with an angle of incidence greater than the critical angle?
12. Explain mirages.

13. Although the light coming from the sun is refracted while passing through Earth's atmosphere, the light is not separated into its spectrum. What does this tell us about the speeds of different colors of light traveling through air?
14. What evidence is there that diamonds have a slightly different index of refraction for each color of light?

Applying Concepts

15. A dry road is a diffuse reflector, while a wet road is not. Sketch a car with its headlights illuminating the road ahead. Show why the wet road would appear darker to the driver than the dry road would.
16. Why is it desirable that the pages of a book be rough rather than smooth and glossy?
17. Is it necessary to measure the volume of a glass block to find its optical density? Explain.
18. A light ray strikes the boundary between two transparent media. What is the angle of incidence for which there is no refraction?
19. In the example problem Snell's law, a ray of light is incident upon crown glass at 30.0°. The angle of refraction is 19.2°. Assume that the glass is rectangular in shape. Construct a diagram to show the incident ray, the refracted ray, and the normal. Continue the ray through the glass until it reaches the opposite edge.
    a. Construct a normal at this point. What is the angle at which the refracted ray is incident upon the opposite edge of the glass?
    b. Assume that the material outside the opposite edge is air. What is the angle at which the ray leaves the glass?
    c. As the ray leaves the glass, is it refracted away from or toward the normal?
    d. How is the orientation of the ray leaving the glass related to the ray entering the glass?
20. Assume that the angle of incidence in problem 19 remains the same. What happens to the angle of refraction as the index of refraction increases?
21. Which substance, A or B, in **Figure 17–16** has a larger index of refraction? Explain.
22. How does the speed of light change as the index of refraction increases?
23. How does the size of the critical angle change as the index of refraction increases?
24. Which two pairs of media, air and water or air and glass, have the smaller critical angle?
25. Examine **Figure 17–5.** Why do the two left-hand bottom rays that enter the prism exit vertically, while the two top rays exit horizontally? (**Hint:** If you look carefully, you will find that the middle ray has both vertical and horizontal intensity and that there is a trace of the other ray moving vertically.)
26. If you crack the windshield in your car, you will see a silvery line along the crack. The two pieces of glass have separated at the crack, and there is air between them. The silvery line indicates that light is reflecting off the crack. Draw a ray diagram to explain why this occurs. What phenomenon does this illustrate?
27. According to legend, Erik the Red sailed from Iceland and discovered Greenland after he had seen the land in a mirage. Describe how the mirage might have occurred.
28. A prism bends violet light more than it bends red light. Explain.
29. Which color of light travels fastest in glass: red, green, or blue?
30. Why would you never see a rainbow in the southern sky if you were in the northern hemisphere?

Problems

Section 17.1

LEVEL 1

31. A ray of light strikes a mirror at an angle of 53° to the normal.
    a. What is the angle of reflection?
    b. What is the angle between the incident ray and the reflected ray?
32. A ray of light incident upon a mirror makes an angle of 36.0° with the mirror. What is the angle between the incident ray and the reflected ray?
33. A ray of light has an angle of incidence of 30.0° on a block of quartz and an angle of refraction of 20.0°. What is the index of refraction for this block of quartz?

34. A ray of light travels from air into a liquid. The ray is incident upon the liquid at an angle of 30.0°. The angle of refraction is 22.0°.
a. What is the index of refraction of the liquid?
b. Refer to **Table 17–1.** What might the liquid be?

35. A ray of light is incident at an angle of 60.0° upon the surface of a piece of crown glass. What is the angle of refraction?

36. A light ray strikes the surface of a pond at an angle of incidence of 36.0°. At what angle is the ray refracted?

37. Light is incident at an angle of 60.0° on the surface of a diamond. Find the angle of refraction.

38. A ray of light has an angle of incidence of 33.0° on the surface of crown glass. What is the angle of refraction into the air?

39. A ray of light passes from water into crown glass at an angle of 23.2°. Find the angle of refraction.

40. Light goes from flint glass into ethanol. The angle of refraction in the ethanol is 25°. What is the angle of incidence in the glass?

41. A beam of light strikes the flat, glass side of a water-filled aquarium at an angle of 40.0° to the normal. For glass, $n = 1.50$. At what angle does the beam
a. enter the glass?
b. enter the water?

42. What is the speed of light in diamond?

43. The speed of light in chloroform is 1.99×10^8 m/s. What is its index of refraction?

LEVEL 2

44. A thick sheet of plastic, $n = 1.500$, is used as the side of an aquarium tank. Light reflected from a fish in the water has an angle of incidence of 35.0°. At what angle does the light enter the air?

45. A light source, S, is located 2.0 m below the surface of a swimming pool and 1.5 m from one edge of the pool. The pool is filled to the top with water.
a. At what angle does the light reaching the edge of the pool leave the water?
b. Does this cause the light viewed from this angle to appear deeper or shallower than it actually is?

46. A ray of light is incident upon a 60°–60°–60° glass prism, $n = 1.5$, **Figure 17–17.**
a. Using Snell's law determine the angle θ_r to the nearest degree.
b. Using elementary geometry, determine the values of angles A, B, and C.
c. Determine the angle, θ_{exit}.

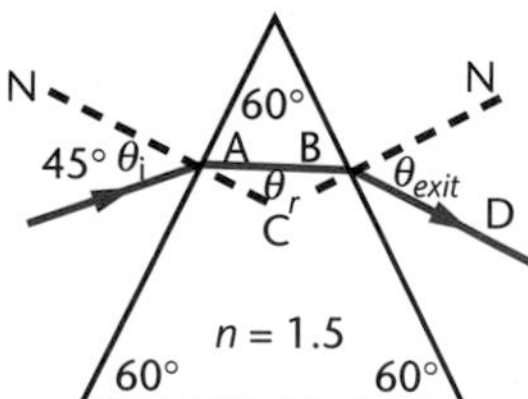

FIGURE 17–17

47. A sheet of plastic, $n = 1.5$, 25 mm thick is used in a bank teller's window. A ray of light strikes the sheet at an angle of 45°. The ray leaves the sheet at 45° but at a different location. Use a ray diagram to find the distance between the ray that leaves and the one that would have left if the plastic were not there.

48. The speed of light in a clear plastic is 1.90×10^8 m/s. A ray of light strikes the plastic at an angle of 22°. At what angle is the ray refracted?

49. How many more minutes would it take light from the sun to reach Earth if the space between them were filled with water rather than a vacuum? The sun is 1.5×10^8 km from Earth.

Section 17.2

LEVEL 1

50. Find the critical angle for diamond.

51. A block of glass has a critical angle of 45.0°. What is its index of refraction?

52. A ray of light in a tank of water has an angle of incidence of 55°. What is the angle of refraction in air?

53. The critical angle for a special glass in air is 41°. What is the critical angle if the glass is immersed in water?

54. A diamond's index of refraction for red light, 656 nm, is 2.410, while that for blue light, 434 nm, is 2.450. Suppose white light is incident on the diamond at 30.0°. Find the angles of refraction for these two colors.

55. The index of refraction of crown glass is 1.53 for violet light, and it is 1.51 for red light.
   a. What is the speed of violet light in crown glass?
   b. What is the speed of red light in crown glass?

56. Just before sunset, you see a rainbow in the water being sprayed from a lawn sprinkler. Carefully draw your location and the locations of the sun and the water coming from the sprinkler.

LEVEL 2

57. A light ray enters a rectangle of crown glass, as illustrated in **Figure 17–18.** Use a ray diagram to trace the path of the ray until it leaves the glass.

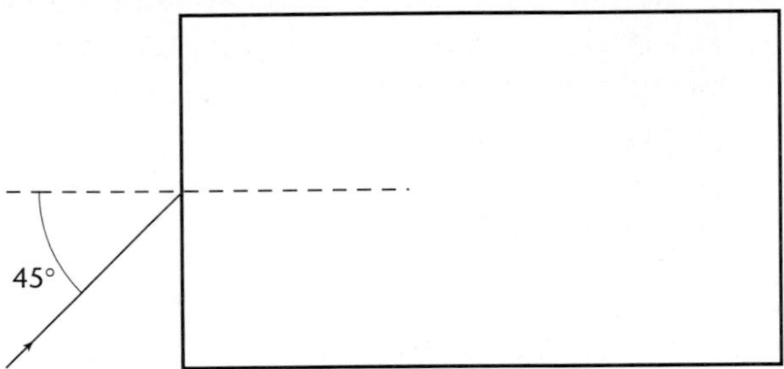

FIGURE 17–18

58. Crown glass's index of refraction for red light is 1.514, while that for blue light is 1.528. White light is incident on the glass at 30.0°.
   a. Find the angles of refraction for these two colors.
   b. Compare the difference between the angles of reflection and refraction for the crown glass and diamond. Angles for diamond were calculated in problem 54.
   c. Use the results to explain why diamonds are said to have "fire."

Critical Thinking Problems

59. How much dispersion is there when light goes through a slab of glass? For dense flint glass, $n = 1.7708$ for blue light ($\lambda = 435.8$ nm) and $n = 1.7273$ for red light ($\lambda = 643.8$ nm). White light in air ($n = 1.0003$) is incident at exactly 45°. Find the angles of refraction for the two colors, then find the difference in those angles in degrees. You should use five significant digits in all calculations.

60. Suppose the glass slab in problem 59 were rectangular. At what angles would the two colors leave the glass?

61. Find the critical angle for ice ($n = 1.31$). In a very cold world, would fiber-optic cables made of ice be better or worse than those made of glass? Explain.

Going Further

Using a Computer or Programmable Calculator Explore why the region inside the rainbow is brighter than the outside. First, **Figure 17–15** shows the path of a light ray that strikes a water drop of radius r and distance d from the center. Confirm that the angle through which the ray is bent is given by

$$\phi = 180° - 4\,\theta_r + 2\,\theta_i$$

where θ_i is the angle of incidence of the ray on the drop and θ_r is the angle of refraction. Next, demonstrate that $\sin\theta_i = d/r$, and that Snell's law then gives $\sin\theta_r = d/nr$. The index of refraction for water is 1.331 for red light and 1.337 for blue. Use your computer or calculator to find the angles θ_i, θ_r, and ϕ for at least ten values of d/r from 0 to 1. Plot ϕ versus d/r. You should find that the minimum angle through which the ray is bent is about 138°. Write a paragraph that shows how your results explain the brightness inside the rainbow.

interNET CONNECTION

Follow the link on the Glencoe Homepage at **www.glencoe.com/sec/science** to find out more about this chapter.

Four butterflies, but only one is real!

The others—small, large, upright, and turned upside down—are images that result from reflection and refraction in a single piece of glass. What must be the shape of the glass? How does the glass produce all those butterflies?

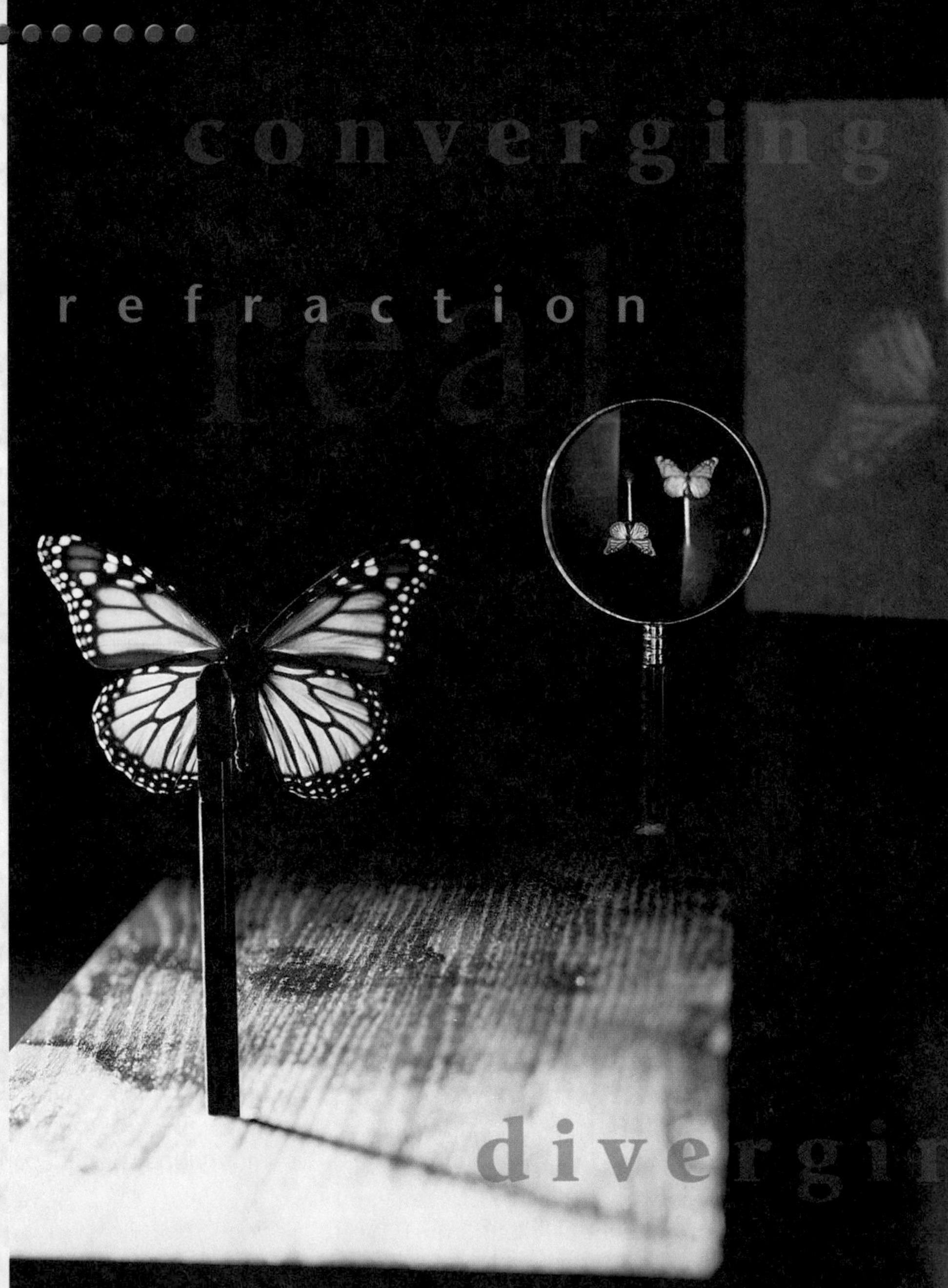

CHAPTER

18 Mirrors and Lenses

To answer the two questions, follow the paths taken by the rays of light that reflect from the real butterfly and enter the lens of the camera. The rays that formed the three images were shunted here and there before converging on the film to produce the photo. You will learn to follow light rays from the butterfly as they bend while passing through the glass or bounce off the polished glass surface.

Light rays may follow complex routes as they encounter mirrors or pass through lenses of different shapes. Yet, complex as a ray's route may be, you can use the laws of optics to trace its journey and discover where it joins other rays to form an image.

All the optical devices that are taken for granted in everyday life—eyeglasses, magnifying glasses, microscopes, cameras, camcorders—apply the laws of reflection and refraction to produce images. Moreover, our entire view of the world is the result of the optical images formed on our retinas by the lenses in our eyes.

WHAT YOU'LL LEARN

- You will locate real and virtual images produced by plane, concave, and convex mirrors.
- You will recognize the causes of aberrations in lenses and mirrors and how these can be minimized.

WHY IT'S IMPORTANT

- The laws of reflection and refraction that produce images on the retinas of your eyes are also the basis for optical instruments that allow scientists to see into outer space and into the world of subatomic particles.

interNET CONNECTION

Follow the link for this chapter on the Glencoe Homepage at **www.glencoe.com/sec/science** to find out more about mirrors and lenses.

18.1 Mirrors

OBJECTIVES

- **Explain** how concave, convex, and plane mirrors form images.
- **Locate** images using ray diagrams, and **calculate** image location and size using equations.
- **Explain** the cause of spherical aberration and how the effect may be overcome.
- **Describe** uses of parabolic mirrors.

Mirrors are the oldest optical instruments. Undoubtedly, prehistoric humans saw their faces reflected in the quiet water of lakes or ponds. Almost 4000 years ago, Egyptians used polished metal mirrors to view their images. But sharp, well-defined reflected images were not possible until 1857, when Jean Foucault, a French scientist, developed a method of coating glass with silver.

Objects and Images in Plane Mirrors

If you looked at yourself in a bathroom mirror this morning, you saw your image in a plane mirror. A **plane mirror** is a flat, smooth surface from which light is reflected by regular reflection rather than by diffuse reflection. This means that light rays are reflected with equal angles of incidence and reflection.

In describing mirrors and lenses, the word *object* is used in a new way. You were the object when you looked into the bathroom mirror. An **object** is a source of spreading, or diverging, light rays. Every point on an object is a source of diverging light rays. An object may be luminous, such as a candle and a lamp. But more often, an object, such as the moon or the page you are reading is illuminated. An illuminated object usually reflects light diffusely in all directions.

Figure 18–1 shows how some of the rays reflected off point O on the bill of a baseball cap strike a plane mirror. The equal angles of incidence and reflection are shown for three rays. Notice that they diverge when they leave the point of the cap, and they continue to diverge after they are reflected from the mirror. The person sees those rays that enter the pupil of his eye. The dashed lines are sight lines, the backward extensions of the rays leaving the mirror. They converge at point I. The eye and brain interpret the rays as having come from point I. This point is called the **image** of the bill of the cap. Because the rays do not actually converge on that point, this kind of image is called a **virtual image.**

Color Conventions

- Light rays are **red.**
- Lenses and mirrors are **light blue.**
- Objects are **indigo.**
- Images are **light violet.**

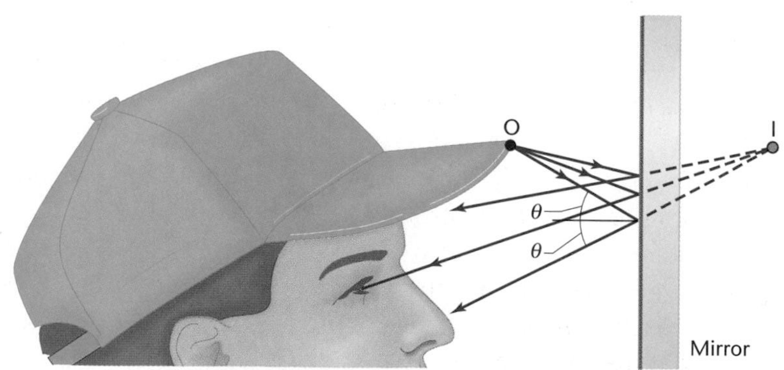

FIGURE 18–1 The reflected rays that enter the eye appear to originate at a point behind the mirror.

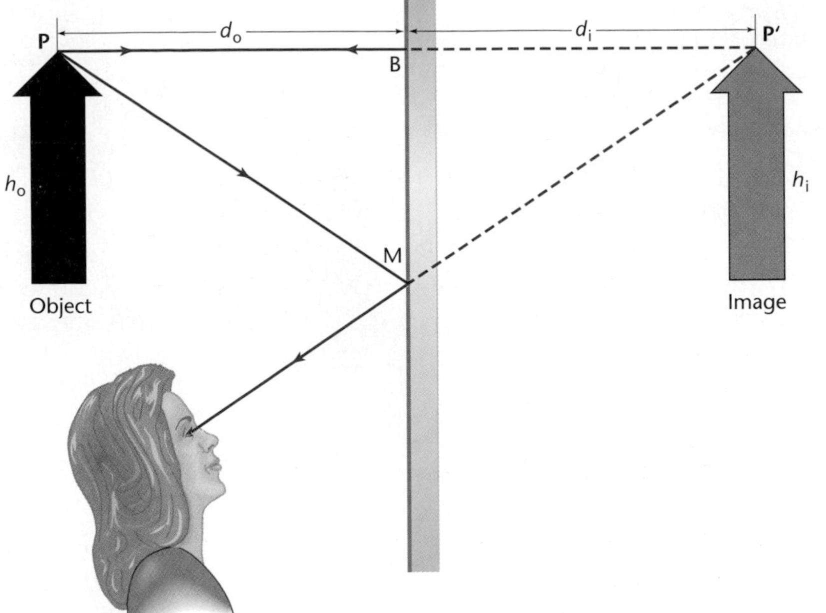

FIGURE 18–2 Light rays (two are shown) leave a point on the object. Some strike the mirror and are reflected into the eye. Sight lines, drawn as dashed lines, extend from the location on the mirror where the reflections occurred back to where they converge. The image is located where the sight lines converge. By geometry, $d_i = d_o$.

Where is the image located? **Figure 18–2** shows two of the rays that leave point P on the object. One ray strikes the mirror at B, the other at M. Both rays are reflected with equal angles of incidence and reflection. Ray PB, which strikes the mirror at an angle of 90°, is reflected back on itself. Ray PM is reflected into the observer's eye. Sight lines, shown in **Figure 18–2** as dashed lines, are extended back from B and M, the positions at which the two rays are reflected from the mirror. The sight lines converge at point P', which is the image of point P. The distance between the object and mirror, the object distance, is line PB, which has a length d_o. Similarly, the distance between the image and the mirror is the length of line P'B and is called the image distance, d_i. The object distance and the image distance, d_o and d_i respectively, are corresponding sides of the two congruent triangles PBM and P'BM. Therefore, $d_o = d_i$.

How large is the image? If you drew the paths and the sight lines of two rays originating from the bottom of the arrow, you would find that they converge at the bottom of the image. Therefore, the object and the image have the same size, or, as **Figure 18–2** shows, $h_o = h_i$. The image and the object are pointing in the same direction, so the image is called an **erect image.**

Is there a difference between you and your image in the mirror? Follow the rays and sight lines in **Figure 18–3a.** The ray that diverges from the right hand of the object converges at what appears to be the left hand of the image. You might ask why the top and bottom are not also reversed. If you look at the figure carefully, you'll see that the direction that is reversed is the one perpendicular to the surface of the mirror. Left and right are reversed, but in the same way that a right-hand glove can be worn on the left hand by turning it inside out. Thus, it is more correct to say that the front and back of an image are reversed.

Pocket Lab

Where's the image?

Suppose that you are standing directly in front of a mirror and see your image. Exactly where is the image? Here is a way to find out. Find a camera with a focusing ring that has distances marked on it. Stand 1.0 m from a mirror and focus on the edge of the mirror. Check the reading on the focusing ring. It should be 1.0 m. Now focus on your image. What is the reading on the ring now?

Analyze and Conclude Summarize your results and write a brief conclusion.

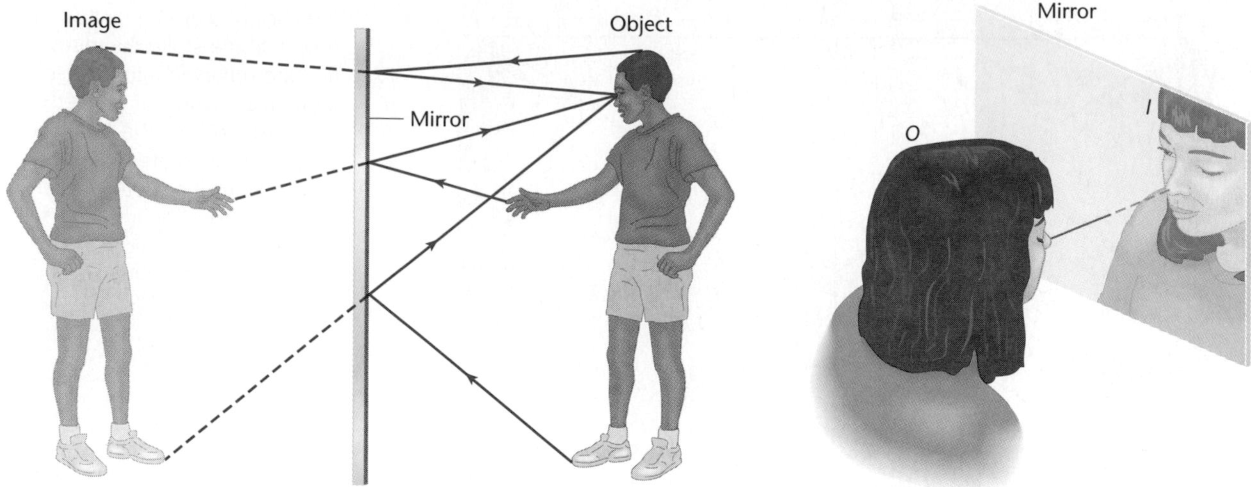

FIGURE 18–3 The image formed in a plane mirror is the same size as the object and is the same distance behind the mirror as the object is in front. If you blink your right eye, it looks as if your left eye blinks.

Concave Mirrors

Look at your reflection in the inside surface of a spoon. The spoon acts as a concave mirror. A **concave mirror** reflects light from its inner ("caved in") surface. In a spherical concave mirror, the mirror is part of the inner surface of a hollow sphere, as shown in **Figure 18–4.** The sphere of radius r has a geometric center, C. Point A is the center of the mirror, and the line CA is the **principal axis,** that is, the straight line perpendicular to the surface of the mirror at its center.

How does light reflect from a concave mirror? Think of a concave mirror as a large number of small plane mirrors arranged around the surface of a sphere, as shown in **Figure 18–5a.** Each mirror is perpendicular to a radius of the sphere. When a ray strikes a mirror, it is reflected with equal angles of incidence and reflection. **Figure 18–4** shows that a ray parallel to the principal axis is reflected at P and crosses the principal axis at some point, F. A parallel ray an equal distance below the principal axis would, by symmetry, also cross the principal axis at F. These parallel rays meet, or converge, at F, which is called the **focal point** of the mirror. The two sides FC and FP of the triangle CFP are equal in length. Thus, the focal point, F, is half the distance between the mirror and the center of curvature, C.

Pocket Lab

Real or Virtual?

Hold a small concave mirror at arm's length and look at your image. What do you see? Is the image in front or behind the mirror? What happens to the image as you slowly bring the mirror toward your face?

Analyze and Conclude Briefly summarize your observations and conclusions.

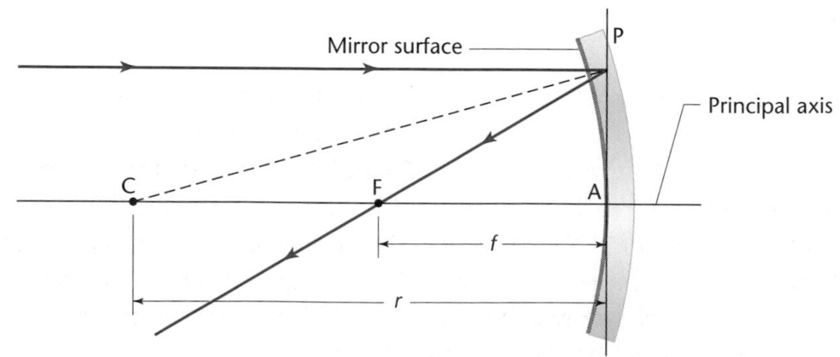

FIGURE 18–4 The focus of a spherical concave mirror is located halfway between the center of curvature and the mirror surface. Rays entering parallel to the principal axis are reflected to converge at the focal point, F.

a

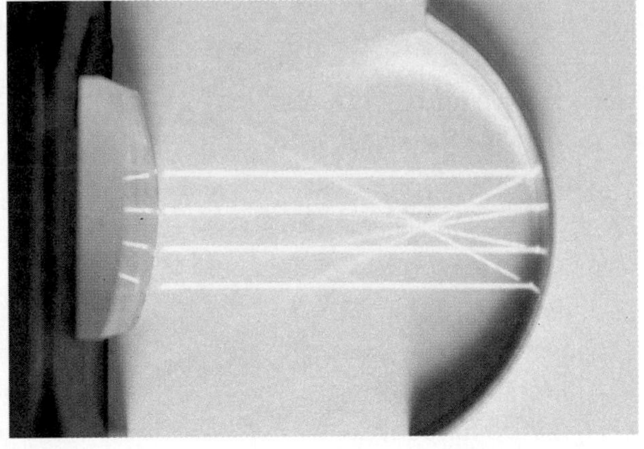

b

FIGURE 18–5 The surface of a concave mirror reflects light to a given point, as in **(b).** A solar furnace in the French Alps, shown in **(a),** reflects light in a similar way from a group of plane mirrors arranged in a curve.

How can you find the location of the focal point of a concave mirror? First you need parallel light rays, because only parallel rays will converge at the focal point. Because the sun is so far away, you can consider it a source of nearly parallel rays. If you point the principal axis of a concave mirror at the sun, all the rays will be reflected through the focal point. Hold a piece of paper near the mirror and move the paper toward and away from the mirror until the smallest and sharpest spot is formed. The spot must be at the focal point because, as was just discussed, the rays striking the mirror are, for all practical purposes, parallel. The distance from the focal point to the mirror along the principal axis is the **focal length,** *f*, of the mirror. In **Figure 18–4,** notice that the focal length is half the radius of curvature of the mirror, or $2f = r$.

Real versus virtual images

The bright spot that you see when you position a piece of paper at the focal point of a concave mirror as it reflects rays from the sun is an image of the sun. The image is a **real image** because rays actually converge and pass through the image. A real image can be seen on a piece of paper or projected onto a screen. In contrast, the image produced by a plane mirror is behind the mirror. The rays reflected from a plane mirror never actually converge but appear to diverge from a point behind the mirror. A **virtual image** cannot be projected onto a screen or captured on a piece of paper because light rays do not converge at a virtual image.

Real images formed by concave mirrors

To develop a graphical method of finding the image produced by a concave mirror, recall that every point on an object emits or reflects light rays in all possible directions. It's impossible and unnecessary to follow all those rays, but you can select just two rays and, for simplicity, draw them from only one point. You can also use a simplified model of the mirror in which all rays are reflected from a plane rather than from the curved surface of the mirror. That model will be explained shortly. Here is a set of rules to use in finding images.

Focal Points

Take a concave mirror into an area of direct sunlight. Use a piece of clay to hold the mirror steady so that the concave mirror directly faces the sun. Move your finger toward or away from the mirror in the area of reflected light to find the brightest spot (focal point). Turn the mirror so that the convex side faces the sun and repeat the experiment.

Analyze and Conclude Record and explain your results.

Lens/Mirror Equation Conventions Applied to Mirrors

d_o is positive for real objects.
d_o is negative for virtual objects.

d_i is positive for real images.
d_i is negative for virtual images.

f is positive for concave mirrors.
f is negative for convex mirrors.

Problem Solving Strategy

Locating Images in Mirrors by Ray Tracing

1. Choose a scale for your drawing such that the drawing is approximately the width of your paper, about 20 cm.
   a. If the object is beyond F, as shown in **Figure 18–6,** then the image will be on the object side of the mirror. Therefore, draw the mirror at the right edge of your paper.
   b. If the object is beyond C, the image distance will be smaller, so draw the object near the left edge of your paper.
   c. If the object is between C and F, the image will be beyond C. The closer the object is to F, the farther away the image will be, so leave room at the left side of your paper.
   d. Choose a scale such that the larger distance, that of the image or the object, is 15 to 20 cm on your paper. Let 1 cm on the paper represent 1, 2, 4, 5, or 10 actual centimeters.
2. Draw the principal axis. Draw a vertical line where the principal axis touches the mirror. If the focal point is known, indicate that position on the principal axis. Label it F. Locate and label the center of curvature, C, at twice the focal distance from the mirror.
3. Draw the object and label its top O_1. Choose a scale for the object that is different from that of the overall drawing because otherwise it may be too small to be seen.
4. Draw ray 1, the parallel ray. Ray 1 is parallel to the principal axis. All rays parallel to the principal axis are reflected through the focal point, F.
5. Draw ray 2, the focus ray. It passes through the focal point, F, on its way to the mirror and is reflected parallel to the principal axis.
6. The image is located where ray 1 and ray 2 cross after reflection. Label the image I_1. Draw a vertical line from I_1 to the principal axis to represent the image.

Makeup

Do you have a makeup mirror in your home? Does this mirror produce images that are larger or smaller than your face? What does this tell you about the curvature? Feel the surface of the mirror. Does this confirm your prediction about the curvature? Try to discover the focal length of this mirror.

Analyze and Conclude
Record your procedure and briefly explain your observations and results.

How would you describe the image in **Figure 18–6?** It is a real image because the rays actually converge at the point where the image is located. It is inverted. The object O_1 is above the axis, but the image point I_1 is below the axis. The image is reduced in size; it is smaller than the object. Thus, the image is real, inverted, and reduced.

Where is the image? If the object is beyond C, as it is in **Figure 18–6,** the image is between C and F. If the object is moved outward from C, the image moves inward toward F and shrinks in size. If the object is brought closer to C, the image moves outward from the mirror. If the object is at C, the image will be there also, and it will be the same size as the object. If the object is moved even closer to F, but not inside it, the image will move farther away and become larger.

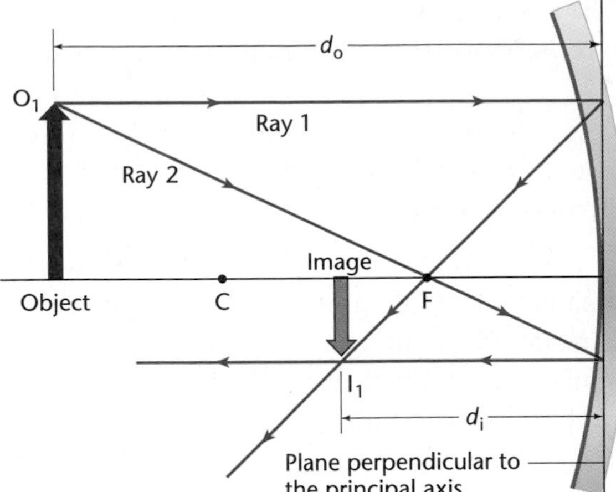

FIGURE 18–6 Note that both rays pass through the focal point, Ray 1 after it is reflected from the mirror and Ray 2 on its way to the mirror. The image is real, inverted, and reduced in size.

To review solving equations, see the Math Handbook, page 737.

All features of the image can be found mathematically. You can use geometry to relate the focal length of the mirror, f, to the distance from the object to the mirror, d_o, and to the distance from the image to the mirror, d_i. The equation for this is called the **lens/mirror equation:**

$$\frac{1}{f} = \frac{1}{d_i} + \frac{1}{d_o}$$

This is the first equation you have seen that contains the inverses of all quantities. You should first solve this equation for the quantity you are seeking. For example, if you are given the object and image distances and asked to find the focal length, you would first add the fractions on the right side of the equation using the least common denominator, $d_i d_o$.

$$\frac{1}{f} = \frac{d_o + d_i}{d_i d_o}$$

Then take the reciprocal of both sides.

$$f = \frac{d_i d_o}{d_o + d_i}$$

Another useful equation is the definition of magnification. **Magnification,** m, is the ratio of the size of the image, h_i, to the size of the object, h_o.

$$m = \frac{h_i}{h_o}$$

By using similar triangles in a ray diagram, you obtain the following.

$$m = \frac{-d_i}{d_o}$$

You can write a single equation for the image height in terms of the object height and the image and object distances by equating the two

HELP WANTED

OPTICIAN

Eye-care professionals are needed for customer assistance and behind-the-scenes positions. Interpersonal skills and skills in precision lab work are a must. Vocational/technical school preparation, including an associate's degree, is desired but extensive apprenticeship experience may be substituted. Excellent finger dexterity, flexibility to work irregular shifts, and appropriate certification are required. For information contact:

Opticians Association of America
10341 Democracy Lane
P.O. Box 10110
Fairfax, VA 22030

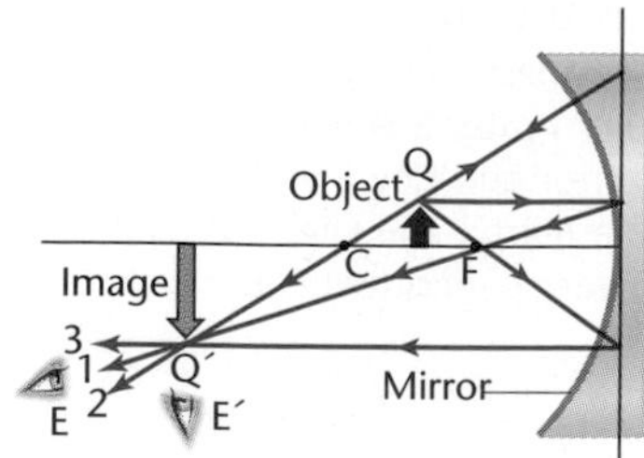

FIGURE 18–7 To the eye at E, it appears that there is an object at Q′ blocking the view of the mirror. However, if the eye moves to E′ and looks toward Q′, the object disappears because there is no light reflected from Q′ to E′.

preceding expressions. The right side of each equation equals the magnification, *m*, and therefore, they are equal to each other.

$$\frac{h_i}{h_o} = \frac{-d_i}{d_o}, \text{ or } h_i = -h_o\left(\frac{d_i}{d_o}\right)$$

Describing a real image

In the case of real images, d_i and d_o are both positive, so h_i will be negative. This means that the magnification is also negative. When the magnification is negative, the image is inverted. What is the magnification of a plane mirror? Recall that in that case, d_i and d_o have the same magnitude, but the image is behind the mirror. Therefore, d_i is negative and the magnification is $+1$, which means that the image and the object are the same size.

How can you tell if an image is real? If an image is real, the rays will converge on it in a ray diagram. In the lens/mirror equation, d_i will be positive. If you use an actual mirror, you can put a piece of paper at the location of the image and you'll see the image. You also can see the image floating in space if you place your eye so that the rays that form the image fall on your eye. But as **Figure 18–7** shows, you must stare at the location of the image and not at the mirror or object.

When solving problems involving mirrors, you may be asked to locate the image by means of a scale drawing using the methods of the problem solving strategy. At other times, when you are asked to find the image mathematically, you should also make a careful sketch to enable you to visualize the situation and check the reasonableness of your results.

Example Problem

Calculating a Real Image Formed by a Concave Mirror

A concave mirror has a radius of curvature of 20.0 cm. An object, 2.0 cm high, is placed 30.0 cm from the mirror.

a. Where is the image located?

b. How high is the image?

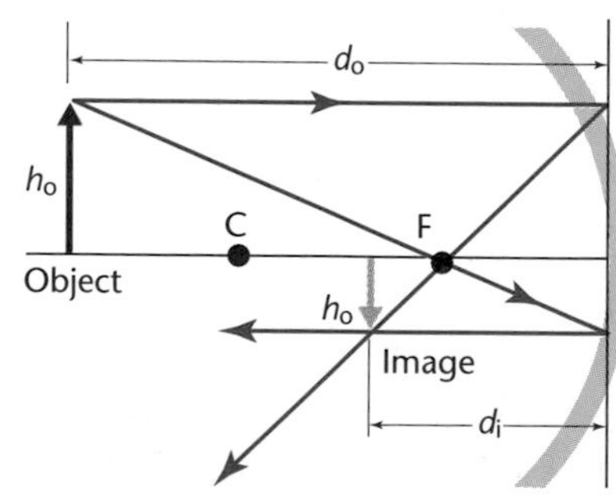

Sketch the Problem

- Sketch the situation; locate the object and mirror.
- Draw two principal rays.

Calculate Your Answer

Known:	**Unknown:**
$h_o = 2.0$ cm	$d_i = ?$
$d_o = 30.0$ cm	$h_i = ?$
$r = 20.0$ cm	

Strategy:

Focal length is half the radius of curvature. Use lens/mirror equation to find image location.

Calculations:

$f = r/2 = 10.0$ cm

$$\frac{1}{f} = \frac{1}{d_o} + \frac{1}{d_i}, \text{ so } d_i = \frac{fd_o}{(d_o - f)}$$

$$d_i = \frac{(10.0 \text{ cm})(30.0 \text{ cm})}{30.0 \text{ cm} - 10.0 \text{ cm}} = 15.0 \text{ cm}$$

Use magnification relations to find height of image.

$$m = \frac{h_i}{h_o} = \frac{-d_i}{d_o}, \text{ so } h_i = \frac{-d_i h_o}{d_o}$$

$$h_i = \frac{(-15.0 \text{ cm})(2.0 \text{ cm})}{30.0 \text{ cm}} = -1.0 \text{ cm}$$

Check Your Answer

- Are your units correct? All distances are in cm.
- Do the signs make sense? Positive location and negative height agree with the drawing.
- Are the magnitudes realistic? The magnitudes agree with the drawing.

Virtual images formed by concave mirrors

You have seen that as the object approaches the focal point, F, of a concave mirror, the image moves farther away from the mirror. If the object is at the focal point, all reflected rays are parallel. They never meet, and so the image is said to be at infinity. What happens if the object is moved even closer to the mirror, that is, between the focal point and the mirror? The ray diagram is shown in **Figure 18–8.** Again, two rays are drawn to locate the image of a point on an object. As before, ray 1 is drawn parallel to the principal axis and reflected through the focal point. To draw ray 2, first draw a dashed line from the focal point to the point on the object. Then draw ray 2 as an extension of the dashed line to the mirror where it is reflected parallel to the principal

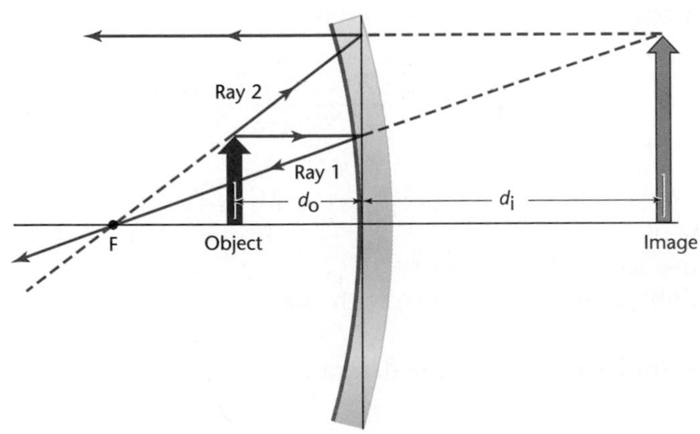

FIGURE 18–8 When an object is located between the focal point and a spherical concave mirror, an enlarged, upright, virtual image is formed behind the mirror.

FIGURE 18–9 Objects placed between the focal point and the surface of a concave mirror form enlarged, virtual images.

axis. Note that ray 1 and ray 2 diverge as they leave the mirror, so there can be no real image. However, the dashed lines behind the mirror are sight lines coming from an apparent origin behind the mirror. These sight lines converge to form a virtual image located behind the mirror. When you use the lens/mirror equation in solving problems involving concave mirrors, you will find that d_i is negative. The image is upright and enlarged like the statuette in **Figure 18–9.**

An upright, enlarged image is a feature of shaving and makeup mirrors, which are concave mirrors. When you use a shaving or makeup mirror, you must hold the mirror close to your face and in doing so, you are placing your face within the focal length of the mirror.

Example Problem

A Concave Mirror as a Magnifier

An object, 2.0 cm high, is placed 5.0 cm in front of a concave mirror with a focal length of 10.0 cm. How large is the image, and where is it located?

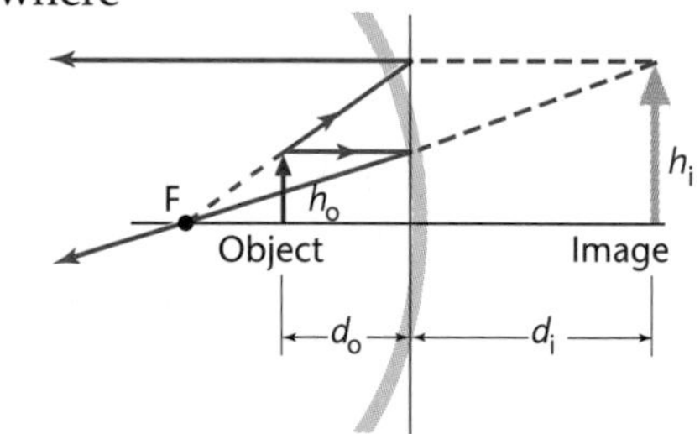

Sketch the Problem

- Sketch the situation; locate the object and mirror.
- Draw two principal rays.
- Extend the rays behind the mirror to locate the image.

Calculate Your Answer

Known:

$h_o = 2.0$ cm

$d_o = 5.0$ cm

$f = 10.0$ cm

Unknown:

$d_i = ?$

$h_i = ?$

Strategy:

Use lens/mirror equation to find image location.

Solve magnification relations to find height of image.

Calculations:

$$\frac{1}{f} = \frac{1}{d_o} + \frac{1}{d_i}, \text{ so } d_i = \frac{fd_o}{d_o - f}$$

$$d_i = \frac{(10.0 \text{ cm})(5.0 \text{ cm})}{5.0 \text{ cm} - 10.0 \text{ cm}} = -10.0 \text{ cm, virtual}$$

$$m = \frac{h_i}{h_o} = \frac{-d_i}{d_o}, \text{ so } h_i = \frac{-d_i h_o}{d_o}$$

$$h_i = \frac{-(-10.0 \text{ cm})(2.0 \text{ cm})}{5.0 \text{ cm}} = 4.0 \text{ cm, upright}$$

Check Your Answer

- Are your units correct? All are cm.
- Do your signs make sense? Negative location means virtual image; positive height means upright image. These agree with the ray diagram.
- Are the magnitudes realistic? Magnitudes agree with the diagram.

Practice Problems

Calculate a real image formed by a concave mirror.

1. Use a ray diagram drawn to scale to solve the Example Problem.
2. An object 3.0 mm high is 10.0 cm in front of a concave mirror having a 6.0-cm focal length. Find the image and its height by means of
   a. a ray diagram drawn to scale.
   b. the lens/mirror and magnification equations.
3. An object is 4.0 cm in front of a concave mirror having a 12.0-cm radius. Locate the image using the lens/mirror equation and a scale ray diagram.
4. A 4.0-cm-high candle is placed 10.0 cm from a concave mirror having a focal length of 16.0 cm. Find the location and height of the image.
5. What is the radius of curvature of a concave mirror that magnifies by a factor of +3.0 an object placed 25 cm from the mirror?

F.Y.I.

Sunlight can be concentrated at the focal point of a large concave mirror. If a cooking pot is placed at that point, the temperatures produced are high enough to cook the food.

Image defects in concave mirrors

In tracing rays, you have reflected the rays from a vertical line rather than the curved surface of the mirror. The mirror/lens equation also assumes that all reflections occur from a plane perpendicular to the principal axis that passes through the mirror. Real rays, however, are reflected off the mirror itself, so they will look like the drawing in **Figure 18–10a.** Notice that only parallel rays close to the principal axis are reflected through the focal point. Other rays converge at points closer to the mirror. The image formed by parallel rays in a large spherical mirror is a disk, not a point. This effect is called **spherical aberration.**

A mirror ground to the shape of a parabola, **Figure 18–10b,** suffers no spherical aberration; all parallel rays are reflected to a single spot. For that reason, parabolic mirrors have been used in telescopes. But many of the newest telescopes use spherical mirrors and specially shaped secondary mirrors or lenses to eliminate the aberration.

Burned Up

Convex (converging) lenses can be used as magnifying glasses. Use someone's eyeglasses to see if they magnify. Are the glasses converging? Can the lenses be used in sunlight to start a fire?

Analyze and Conclude Use your answers to describe the lens.

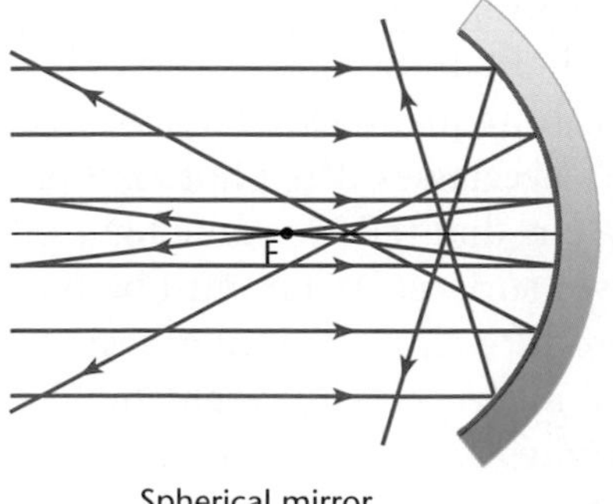

a

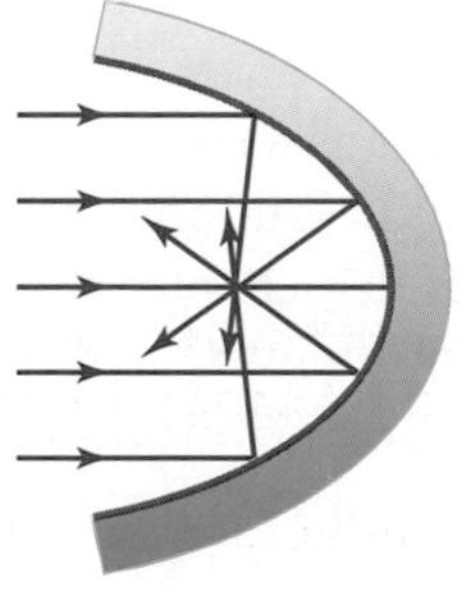

b

FIGURE 18–10 Some rays reflected from a concave spherical mirror converge at points other than the focus, as shown in **(a).** A parabolic mirror, such as the one shown in **(b),** focuses all parallel rays at a point.

FIGURE 18–11 No real image is formed by a convex mirror. An erect, virtual image, reduced in size, is formed at the apparent intersection of the extended rays. Convex mirrors are often used as wide-angle mirrors for safety and security.

Convex Mirrors

A **convex mirror** is a spherical mirror that reflects light from its outer surface. Rays reflected from a convex mirror always diverge. Thus, convex mirrors do not form real images. When drawing ray diagrams, the focal point, F, is placed behind the mirror, at a distance halfway between the mirror and the center of curvature. When using the lens/mirror equation, the focal length, f, of a convex mirror is a negative number, and d_i is negative because the image is behind the mirror.

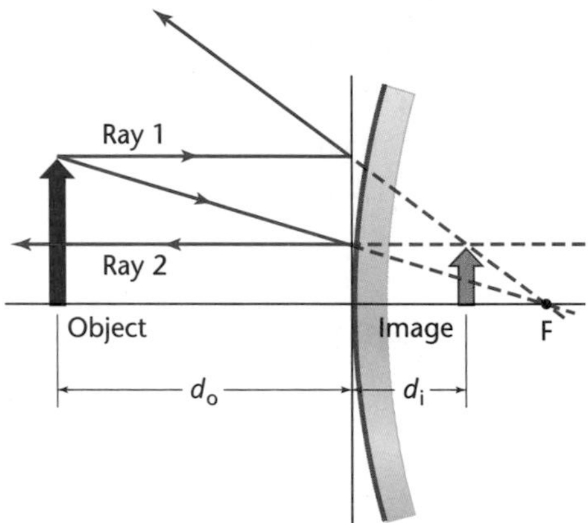

The ray diagram in **Figure 18–11** shows how an image is formed in a convex mirror. Ray 1 approaches the mirror parallel to the principal axis. To draw the reflected ray, draw a dashed line from the focal point, F, to the point where ray 1 strikes the mirror. The reflected ray is in the same direction as the dashed line. Ray 2 approaches the mirror on a path that, if extended behind the mirror, would pass through F. The reflected part of ray 2 is parallel to the principal axis. The two reflected rays diverge, as if coming from a point behind the mirror. The image, located at the apparent intersection of the extended rays behind the mirror, is virtual, erect, and reduced in size.

Convex mirrors form images reduced in size, and therefore, the images seem farther away. But convex mirrors also reflect an enlarged field of view. Rearview mirrors used in cars are often convex mirrors, as are mirrors used in stores to observe shoppers.

Four butterflies, but only one is real!

Ordinary glass also reflects some light. If the glass is curved outward, it will act as a convex mirror. You can frequently see reduced images of yourself if you look into someone's eyeglasses. The photo at the beginning of this chapter shows a glass lens that reflects some light off both its front (convex) and rear (concave) surfaces. What must be the shape of the glass? How does the glass produce all those butterflies?

Both reflected images are reduced in size; one is upright and the other is inverted. The upright image comes from the convex surface, the inverted one from the concave surface.

Example Problem

Image in a Security Mirror

A convex security mirror in a warehouse has a radius of curvature of 1.0 m. A 2.0-m-high forklift is 5.0 m from the mirror. What is the location and size of the image?

Sketch the Problem

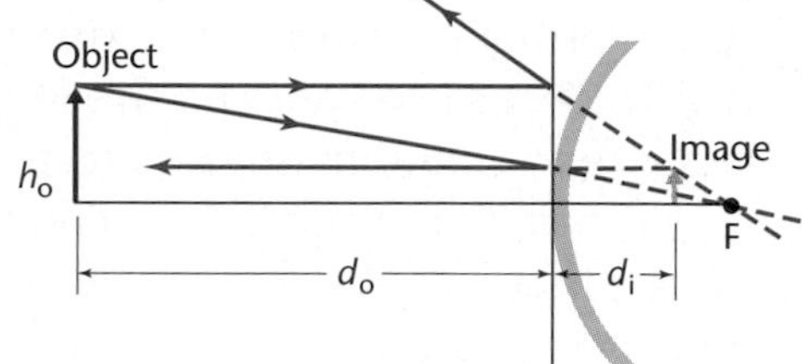

- Sketch the situation; locate the mirror and the object.
- Draw two principal rays.

Calculate Your Answer

Known:		**Unknown:**
$h_o = 2.0$ m	$r = 1.0$ m	$d_i = ?$
$d_o = 5.0$ m		$h_i = ?$

Strategy:	**Calculations:**
The focal length is negative for convex mirrors.	$f = (-1/2)r = -1.0 \text{ m}/2 = -0.5 \text{ m}$
Use the lens/mirror equation to find the location.	$d_i = fd_o/(d_o - f)$ $d_i = (-0.5 \text{ m})(5.0 \text{ m})/(5.0 \text{ m} - (-0.5 \text{ m}))$ $= -0.45$ m, virtual
Combine the magnification equations to determine height.	$h_i = -d_i h_o/d_o$ $h_i = -(-0.45 \text{ m})(2.0 \text{ m})/(5.0 \text{ m})$ $= 0.18$ m, upright, reduced

Check Your Answer

- Are your units correct? All distances are in meters.
- Do the signs make sense? Negative location means virtual image; positive height means upright image. These agree with the diagram.
- Are the magnitudes realistic? They agree with the diagram.

Practice Problems

6. An object is 20.0 cm in front of a convex mirror with a -15.0-cm focal length. Find the location of the image using
   a. a scale ray diagram. **b.** the lens/mirror equation.
7. A convex mirror has a focal length of -12 cm. A lightbulb with a diameter of 6.0 cm is placed 60.0 cm in front of the mirror. Locate the image of the lightbulb. What is its diameter?
8. A convex mirror is needed to produce an image three-fourths the size of the object and located 24 cm behind the mirror. What focal length should be specified?

F.Y.I.

In 1857, Jean Foucault developed a technique for silvering glass to make mirrors for telescopes. Silvered glass mirrors are lighter and less likely to tarnish than metal mirrors previously used.

Physics & Technology

The Hubble Space Telescope

Astronomers have known for decades that to see farther into space and time, telescopes must collect more light and this requires larger mirrors. But massive mirrors bend under their own weight, distorting the images being observed. Atmospheric distortions, temperature effects, and light pollution also limit the performance of Earth-based telescopes.

One approach scientists are using to overcome these difficulties is to place telescopes in Earth orbit. The Hubble Space Telescope, the first orbiting telescope, was launched in 1990. Among the revelations provided by the Hubble telescope are views of galaxies so distant that they show us what some parts of the universe looked like just a few hundred million years after its birth. Detailed images of the comet Hale-Bopp, which visited our region of the solar system in 1997, enabled astronomers to estimate the size of the comet's nucleus and observe violent eruptions that took place as different parts of the nucleus turned to face the sun. Hubble also has captured images of giant plumes of gas and dust produced by a volcanic eruption on Io, one of Jupiter's moons.

In 1997, Hubble produced spectacular images of jets of gas and dust blown into space by a massive black hole at the center of the Egg Nebula, about 3000 light-years from Earth.

Thinking Critically What are some of the advantages and disadvantages of placing a telescope in orbit?

18.1 Section Review

1.1 Draw a ray diagram showing your eye placed 12 cm from a plane mirror. Two rays leave a point on an eyelash and enter opposite sides of the pupil of your eye, 1 cm apart. Locate the image of the eyelash.

1.2 If a beam of parallel light rays is sent into a spherical concave mirror, do all the rays converge at the focal point?

1.3 If a mirror produces an erect, virtual image, can you immediately conclude that it is a plane mirror? Explain.

1.4 **Critical Thinking** A concave mirror is used to produce a real image of a distant object. A small plane mirror is put between the mirror and the image. The mirror is put at a 45° angle to the principal axis of the concave mirror.

a. Make a ray diagram. Is the image of the plane mirror real or virtual? Explain.

b. If the small mirror were a convex mirror, would the image be real or virtual? Explain.

Lenses 18.2

Eyeglasses were made from lenses as early as the thirteenth century. Around 1610, Galileo used two lenses as a telescope. With this instrument, he discovered the moons of Jupiter. Since Galileo's time, lenses have been used in many optical instruments such as microscopes and cameras. Lenses are probably the most useful and important of all optical devices.

OBJECTIVES

- **Describe** how real and virtual images are formed by convex and concave lenses.
- **Locate** the image with a ray diagram and find the image location and size using a mathematical model.
- **Define** chromatic aberration and **explain** how it can be reduced.
- **Explain** how optical instruments such as microscopes and telescopes work.

Types of Lenses

A **lens** is made of transparent material, such as glass or plastic, with a refractive index larger than that of air. Each of the lens's two faces is part of a sphere and can be convex, concave, or flat. A lens is called a **convex lens** if it is thicker at the center than at the edges. Convex lenses are converging lenses because they refract parallel light rays so that the light rays meet. A **concave lens** is thinner in the middle than at the edges and is called a diverging lens because rays passing through it spread out. Use **Figure 18–12** to compare the shapes of the two types of lenses and the paths of light rays as they pass through each lens.

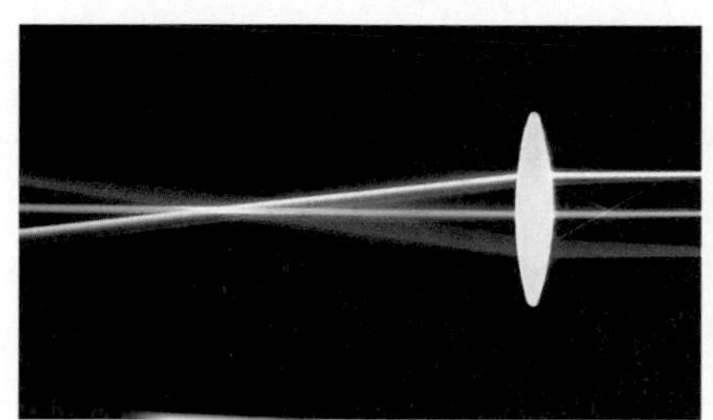

a

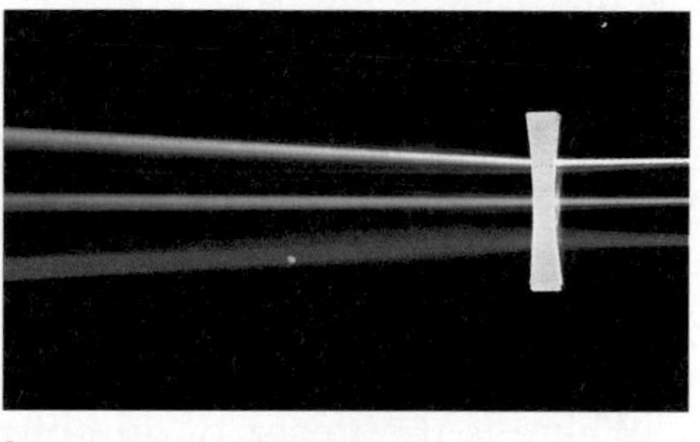

b

FIGURE 18–12 In **(a),** the refracted rays converge, while in **(b)** they diverge.

Convex Lenses

When light passes through a lens, refraction occurs at the two lens surfaces. In Chapter 17, you learned that Snell's law and geometry can be used to predict the paths of rays passing through a lens. To simplify your drawings and calculations, you will use the same approximation you used with mirrors, that is, that all refraction occurs on a plane, called the principal plane, that passes through the center of the lens. This approximation, called the thin lens model, applies to all the lenses you will learn about in this book.

Real images from convex lenses

Have you ever used a lens for the purpose shown in **Figure 18–13?** By positioning the lens so that the rays of the sun converge on the leaf, the camper produces the image of the sun on the leaf's surface. The image is real. Because the rays are converging on a small spot, enough energy is being concentrated there that it could set the leaf ablaze. The

FIGURE 18–13 A converging lens can be used to start a fire in a pile of leaves.

FIGURE 18–14 When an object is placed well beyond the principal focus of a convex lens, the image is real, inverted, and reduced in size. If the object were placed at the position of the image, you could locate the new image by tracing the same rays in the opposite direction.

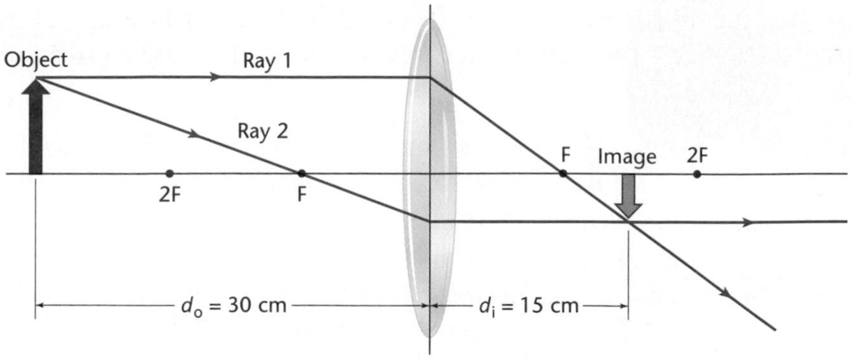

rays of the sun are examples of light rays that are almost exactly parallel to the principal axis because they have come from such a distant source. After being refracted in the lens, the rays converge at a point called the focal point, F, of the lens. **Figure 18–14** shows two focal points, one on each side of the lens. This is because the lens is symmetrical and light can pass through it in both directions. The two focal points are important in drawing rays, as you will see. The distance from the lens to a focal point is the focal length, f. The focal length depends upon the shape of the lens and the refractive index of the lens material.

In **Figure 18–14,** you can trace rays from an object located far from a convex lens. Ray 1 is parallel to the principal axis. It refracts and passes through F on the other side of the lens. Ray 2 passes through F on its way to the lens. After refraction, its path is parallel to the principal axis. The two rays intersect at a point beyond F and locate the image. Rays selected from other points on the object would converge at corresponding points on the image. Note that the image is real, inverted, and smaller than the object.

Lens/Mirror Equation Conventions Applied to Lenses

- f is positive for convex lenses.
- f is negative for concave lenses.
- d_o is positive on the object side of the lens.
- d_i is positive on the other side (image side) of the lens, where images are real.
- d_i is negative on the object side of the lens where images are virtual.

Where is the image of an object that is closer to the lens than the object in **Figure 18–14** is? You can find the location of the image without drawing another ray diagram. If you imagine the object in the position of the image in **Figure 18–14,** you can easily locate the new object by using a basic principle of optics that states that if a reflected or refracted ray is reversed in direction, it will follow its original path in the reverse direction. This means that the image and object may be interchanged by changing the direction of the rays. Imagine that the path of light through the lens in **Figure 18–14** is reversed and the object is at a distance of 15 cm from the right side of the lens. The new image, located at 30 cm from the left side of the lens, is again real and inverted, but it is now larger than the object.

If the object were placed at a distance twice the focal length from the lens, that is, at the point 2F on **Figure 18–14,** the image also would be found at 2F. Because of symmetry, the image and object would have the same size. Thus, you can conclude that if an object is more than twice the focal length from the lens, the image is reduced in size. If the object is between F and 2F, then the image is enlarged.

The lens/mirror equation can be used to find the location of an image, and the magnification equation can be used to find its size.

$$\frac{1}{f} = \frac{1}{d_i} + \frac{1}{d_o}$$

$$m = \frac{h_i}{h_o} = \frac{-d_i}{d_o}$$

Recall that you used the lens/mirror equation, as well as the equation for magnification, in solving problems involving mirrors. The following Example Problem will show you how to apply these equations to problems involving lenses.

Example Problem

An Image Formed by a Convex Lens

An object is placed 32.0 cm from a convex lens that has a focal length of 8.0 cm.

a. Where is the image?

b. If the object is 3.0 cm high, how high is the image?

c. Is the image inverted or upright?

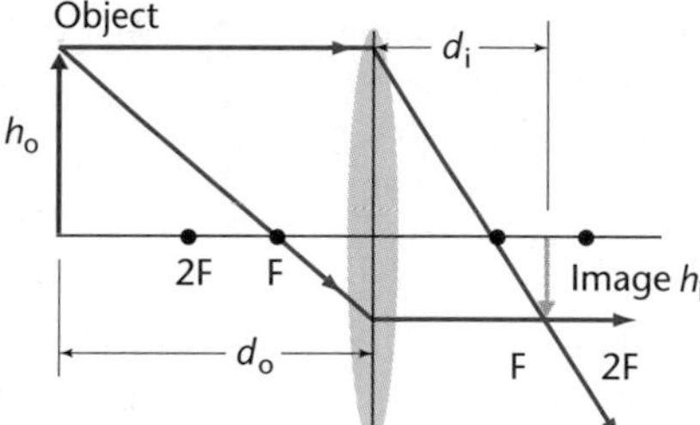

Sketch the Problem

- Sketch the situation, locating the object and lens.
- Draw two principal rays.

Calculate Your Answer

Known:

$d_o = 32.0$ cm

$h_o = 3.0$ cm

$f = 8.0$ cm

Unknown:

$d_i = ?$

$h_i = ?$

Strategy:

Use lens/mirror equation to determine d_i.

Solve magnification relations to find image height.

Calculations:

$$\frac{1}{f} = \frac{1}{d_o} + \frac{1}{d_i}, \text{ so } d_i = \frac{fd_o}{d_o - f}$$

$$d_i = \frac{(8.0 \text{ cm})(32.0 \text{ cm})}{32.0 \text{ cm} - 8.0 \text{ cm}}$$

$$= 11.0 \text{ cm, real}$$

$$m = \frac{h_i}{h_o} = \frac{-d_i}{d_o}, \text{ so } h_i = \frac{-d_i h_o}{d_o}$$

$$h_i = \frac{-(11.0 \text{ cm})(3.0 \text{ cm})}{32.0 \text{ cm}} = -1.0 \text{ cm, inverted}$$

Check Your Answer

- Are the units correct? All are in cm.
- Do the signs make sense? Location is positive (real); height is negative (inverted). These are in agreement with the diagram.
- Are the magnitudes realistic? Location and height agree with the drawing.

Practice Problems

9. Use a ray diagram to find the image position of an object 30 cm to the left of a convex lens with a 10-cm focal length. (Let 1 cm on the drawing represent 20 cm.)
10. An object, 2.25 mm high, is 8.5 cm to the left of a convex lens of 5.5-cm focal length. Find the image location and height.
11. An object is placed to the left of a 25-mm focal length convex lens so that its image is the same size as the object. What are the image and object locations?

Fish-Eye Lens

How can fish focus light with their eyes? The light from an object in the water goes from the water into the fish eye, which is also mostly water. Obtain a converging lens and observe that it can be used as a magnifying glass. Now hold the lens under water in an aquarium. Does the lens still magnify?

Analyze and Conclude Compare the magnifying ability of the glass lens when used under water and in air. Would a more curved lens bend the light more? Would you predict that the index of refraction of the material in a fish eye is the same as water? Defend your prediction.

Why use a large lens?

For simplicity, you have drawn ray diagrams as if only two rays formed the image. But, in reality, all the rays that leave a point on the object and pass through the lens converge and form an image at the same spot. **Figure 18–15a** shows more of the rays involved. Notice that only the rays that hit the lens are imaged at the same spot. If you put a piece of paper at the image location, the size of the spot will be smallest at that point. If you move the paper in either direction along the principal axis, the size of the spot gets bigger but fuzzier. You would say that the image is out of focus.

What would happen if you used a lens of larger diameter? More of the rays that miss the lens would now go through it, as you can see in **Figure 18–15b.** With more rays converging on the image, it would be brighter. Would the reverse be true if you used a smaller lens? Fewer rays would pass through the smaller lens and focus on the image, so the image would be dimmer. Cameras use this principle to allow the aperture to be adjusted for dimmer or brighter days.

Virtual images

If an object is placed at the focal point of a convex lens, the refracted rays will emerge in a parallel beam. If the object is brought closer to the lens, the rays do not converge on the opposite side of the lens. Instead, the image appears on the same side of the lens as the object. This image is virtual, erect, and enlarged.

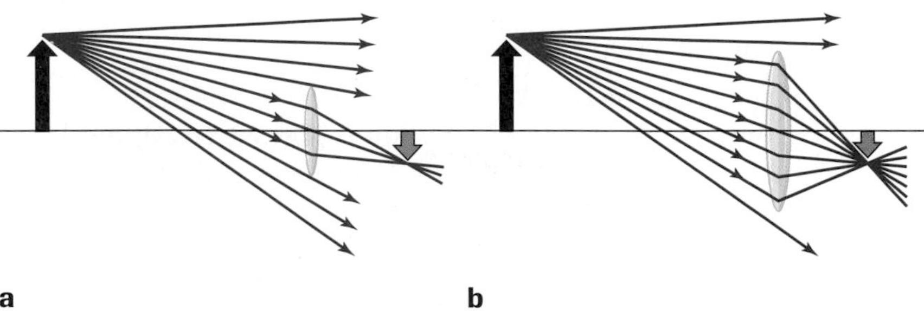

FIGURE 18–15 All the rays that pass through the lens focus at the same point. A larger lens allows more rays to pass through and thus produces a brighter image.

Physics Lab

Seeing Is Believing

Problem

Draw lines of sight to locate virtual images produced by lenses.

Materials  

large-diameter convex lens
large-diameter concave lens
2 small balls of clay
2 rulers
2- or 3-cm-long nail
2 pieces of paper

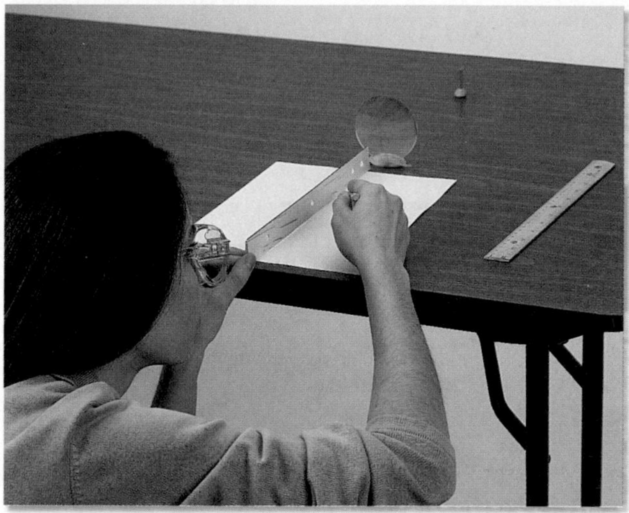

Procedure

1. Assemble the equipment as shown in the photo using the concave lens.
2. Look through the lens to make sure that you can see both ends of the nail. Move the nail closer or farther from the lens until both ends are visible.

Data and Observations

1. Mark the paper to show the tip of the nail, the head of the nail, and also the lens line.
2. Line up your straight edge to point to the head of the nail. Have your lab partner verify that the edge is accurate.
3. Draw the line of sight.
4. Move to another position and draw a second line of sight to the head of the nail.
5. Repeat steps 2–4, this time drawing two lines of sight to the tip of the nail.
6. Use a new sheet of paper and repeat steps 1–5 using the convex lens.

Analyze And Conclude

1. **Analyzing Data** The image position can be located by extending lines of sight until they intersect. Extend the two lines of sight that point to the image head. Extend the two lines of sight that point to the image tip. Describe the results.
2. **Analyzing Data** Repeat the analysis for the convex lens, and describe the results.
3. **Interpreting Data** Describe the image from the concave lens. What was surprising about the image?
4. **Interpreting Data** Describe the image from the convex lens. What was surprising about the image?

Apply

1. Describe an application of a similar arrangement for a convex lens.

Figure 18–16 shows how a convex lens forms a virtual image. The object is between F and the lens. Ray l, as usual, approaches the lens parallel to the principal axis and is refracted through the focal point, F. Ray 2 travels from the tip of the object, in the direction it would have if it had started at F on the object side of the lens. The dashed line from F to the object shows you how to draw ray 2. Ray 2 leaves the lens parallel to the principal axis. Rays 1 and 2 diverge as they leave the lens. Thus, no real image is possible. Drawing sight lines for the two rays back to their apparent intersection locates the virtual image. It is on the same side of the lens as the object, erect, and larger than the object.

Example Problem

A Magnifying Glass

A convex lens with a focal length of 6.0 cm is held 4.0 cm from an insect that is 0.50 cm long.

a. Where is the image located?

b. How large does the insect appear to be?

Sketch the Problem

- Sketch the situation, locating the lens and the object.
- Draw two principal rays.

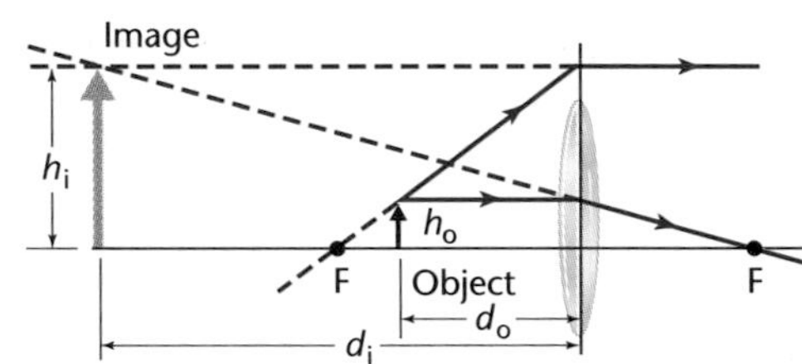

Calculate Your Answer

Known:	**Unknown:**
$d_o = 4.0$ cm	$d_i = ?$
$h_o = 0.50$ cm	$h_i = ?$
$f = 6.0$ cm	

Strategy:

Use lens/mirror equation to determine d_i.

Calculations:

$$\frac{1}{f} = \frac{1}{d_o} + \frac{1}{d_i}, \text{ so } d_i = \frac{fd_o}{d_o - f}$$

$$d_i = \frac{(6.0\text{ cm})(4.0\text{ cm})}{4.0\text{ cm} - 6.0\text{ cm}} = -12\text{ cm, virtual}$$

Solve magnification relations to find image height.

$$m = \frac{h_i}{h_o} = \frac{-d_i}{d_o}, \text{ so } h_i = \frac{-d_i h_o}{d_o}$$

$$h_i = \frac{-(-12\text{ cm})(0.50\text{ cm})}{4.0\text{ cm}} = 1.5\text{ cm, upright}$$

Check Your Answer

- Are your units correct? All are in cm.
- Do the signs make sense? Negative d_i means virtual image; positive h_i means upright image, as the diagram shows.
- Are the magnitudes realistic? They agree with the diagram.

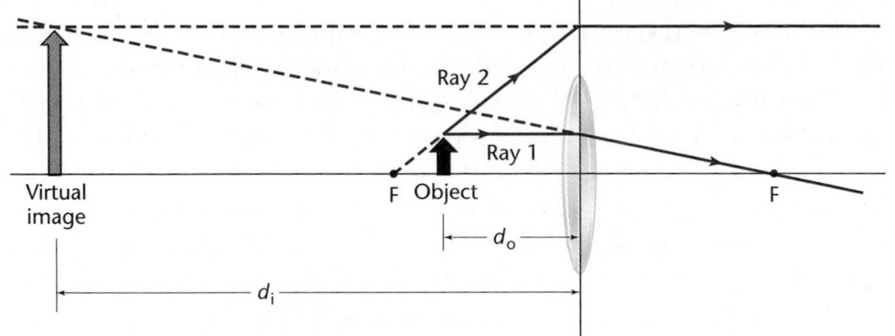

FIGURE 18–16 The two principal rays show that an upright, enlarged, virtual image, formed by a convex lens when the object is located between the lens and the principal focus. Drawing principal rays allows you to locate and construct the image of an object. These rays do not necessarily pass through the lens itself.

Practice Problems

12. A newspaper is held 6.0 cm from a convex lens of 20.0-cm focal length. Find the image distance of the newsprint image.

13. A magnifying glass has a focal length of 12.0 cm. A coin, 2.0 cm in diameter, is placed 3.4 cm from the lens. Locate the image of the coin. What is the diameter of the image?

14. A stamp collector wants to magnify an image by 4.0 when the stamp is 3.5 cm from the lens. What focal length is needed for the lens?

Pocket Lab

Bright Ideas

Stick the edge of a converging lens into a ball of clay and place the lens on a tabletop. Use a small lightbulb on one side and a screen on the other side to get a sharp image of the bulb. Predict what will happen to the image if you place your hand over the top half of the lens. Try it.

Analyze and Conclude What happened? How much of the lens is needed for a complete image?

Concave Lenses

In **Figure 18–17,** you can see how an image is formed by a concave lens. A concave lens causes all rays to diverge. Ray 1 leaves O_1 and approaches the lens parallel to the principal axis. It leaves the lens in the direction it would have if it had passed through the focal point. Ray 2 passes directly through the center of the lens without bending. Rays 1 and 2 diverge after passing through the lens. Their apparent intersection is *i*, on the same side of the lens as the object. The image is virtual, erect, and reduced in size. This is true no matter how far from the lens the object is located. The focal length of a concave lens is negative.

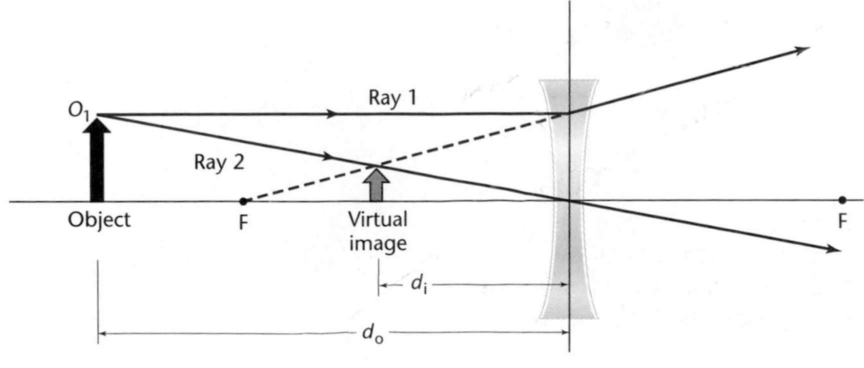

FIGURE 18–17
Concave lenses are used in eyeglasses to correct nearsightedness and in combination with convex lenses in cameras and telescopes.

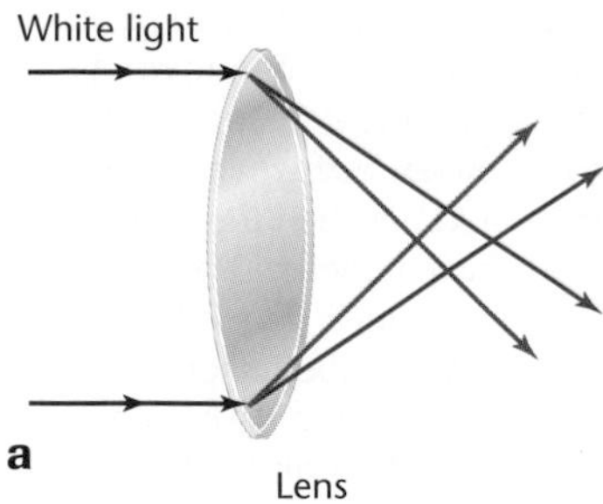

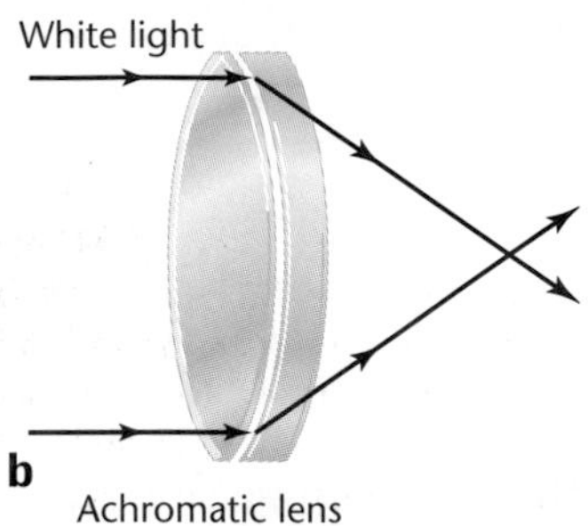

FIGURE 18–18
Chromatic aberration is depicted in **a,** which shows light of different wavelengths focused at different points. The achromatic lens in **b** is a combination of a convex and a concave lens, which minimizes the chromatic defect.

Defects of lenses

The model you have used for drawing rays through lenses suggests that all rays that pass through all parts of a lens focus at the same location. However, this is only an approximation. In real lenses, rays that pass through the extreme edges of the lens are focused at a location different from rays that pass through the center. This inability of the lens to focus all parallel rays to a single point is called spherical aberration. Lenses as well as mirrors have spherical aberration. Spherical aberration is eliminated in inexpensive cameras by using only the centers of lenses. In more expensive instruments, many lenses, often five or more, are used to form a sharp, well-defined image.

Lenses have a second defect that mirrors do not. The edges of a lens resemble a prism, and different wavelengths of light are bent at slightly different angles, as you can see in **Figure 18–18a.** Thus, the light that passes through a lens, especially near the edges, is slightly dispersed. An object viewed through a lens appears ringed with color. This effect is called **chromatic aberration.** The term *chromatic* comes from the Greek *chromo,* which means "related to color."

Chromatic aberration is always present when a single lens is used, but this defect can be greatly reduced by joining a convex lens with a concave lens that has a different index of refraction. Such a combination of lenses is shown in **Figure 18–18b.** Both lenses disperse light, but the dispersion caused by the converging lens is almost canceled by that caused by the diverging lens. The index of refraction of the diverging lens is chosen so that the combination lens still converges the light. A lens constructed in this way is called an **achromatic lens.** All precision optical instruments use achromatic lenses.

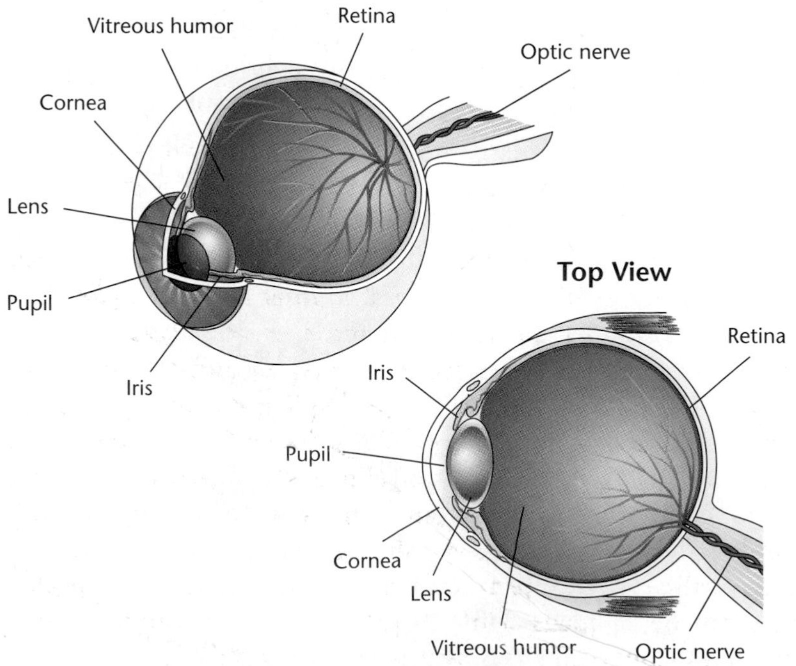

FIGURE 18–19
The curved surfaces of the cornea and lens refract light rays that enter the eye through the pupil to form an image on the retina.

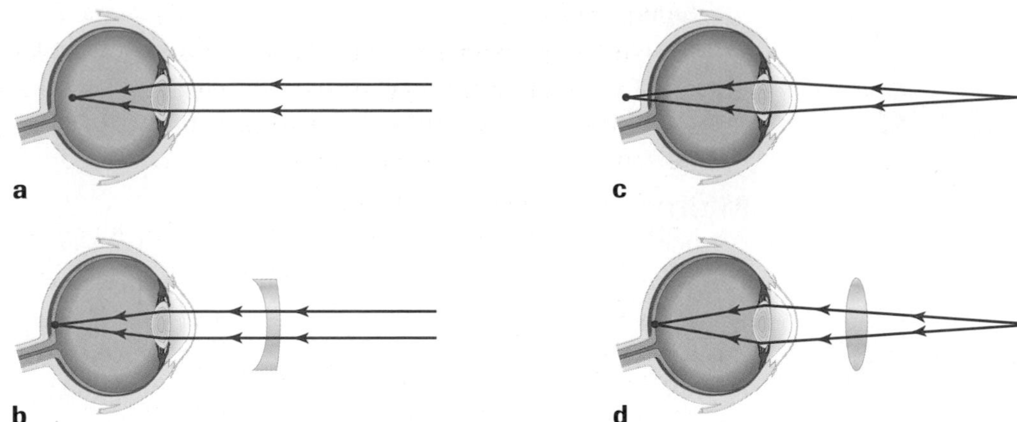

FIGURE 18–20 A nearsighted person cannot see distant objects because the image is focused in front of the retina as shown in **a**. The concave lens in **b** corrects this defect. A farsighted person cannot see close objects because the image is focused behind the retina as shown in **c**. The convex lens in **d** corrects this defect.

Optical Instruments that Use Lenses

Although the eye itself is a remarkable optical device, its abilities can be greatly extended by a wide variety of instruments composed of lenses and mirrors. The eye is a fluid-filled, almost spherical vessel that forms the image of an object on the retina, as shown in **Figure 18–19.** Most of the refraction occurs at the curved surface of the cornea. The eye lens is made of flexible material with a refractive index different from that of the fluid. Muscles can change the shape of the lens, thereby changing its focal length. When the muscles are relaxed, the image of distant objects is focused on the retina. When the muscles contract, the focal length is shortened, permitting images of objects 25 cm or closer to be focused on the retina.

The eyes of many people do not focus sharp images on the retina. Instead, images are found either in front of the retina or behind it. External lenses, in the form of eyeglasses or contact lenses, are needed to adjust the focal length and move the image to the retina. **Figure 18–20** shows that the nearsighted, or myopic, eye has too short a focal length. Images of distant objects are formed in front of the retina. Concave lenses correct this defect by diverging the light rays, thus increasing the image distance, and placing the image on the retina. You also can see in **Figure 18–20** that farsightedness, or hyperopia, is the result of too long a focal length, which results in the image falling behind the retina. A similar result is caused by the increasing rigidity of the lenses in the eyes of people more than about 45 years old. Their muscles cannot shorten the focal length enough to focus images of close objects on the retina. For either defect, convex lenses produce a virtual image farther from the eye than the object. This image then becomes the object for the eye lens and can be focused on the retina, thereby correcting the defect. Some people have lenses or eye shapes that are not spherical. This defect is called astigmatism, and the result is that vertical lines of images can be in focus while horizontal lines are not. Eyeglasses having a nonspherical shape can correct astigmatism.

BIOLOGY CONNECTION

F.Y.I.

The earliest eyeglasses were made of thick, convex lenses. These lenses reminded their makers of lentils. Hence the term *lens*, from the Latin for "lentil beans."

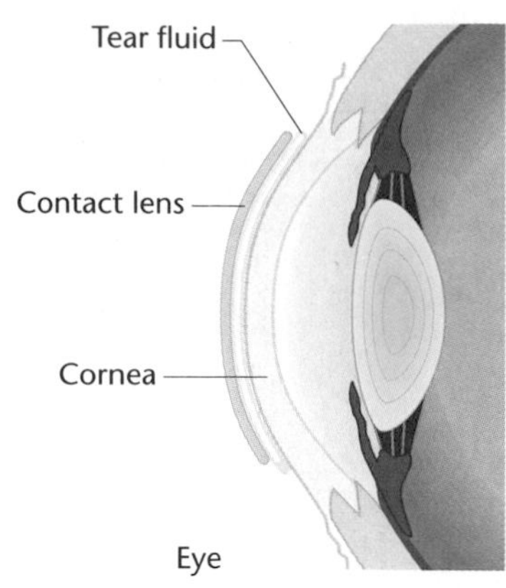

FIGURE 18–21 A contact lens rests on a layer of tears between it and the surface of the cornea.

Contact lenses produce the same results as eyeglasses. These very thin lenses are placed directly on the cornea, as shown in **Figure 18–21.** A thin layer of tears between the cornea and lens keeps the lens in place. Most of the refraction occurs at the air-lens surface, where the change in refractive index is greatest.

Microscopes and telescopes

Microscopes allow the eye to see extremely small objects. Most microscopes use at least two convex lenses. An object is placed very close to a lens with a very short focal length, the objective lens. This lens produces a real image located between the second lens, the ocular or eyepiece lens, and its focal point. The ocular produces a greatly magnified virtual image of the image formed by the objective lens.

An astronomical refracting telescope uses two convex lenses. The objective lens of a telescope has a long focal length. The parallel rays from a star or other distant object focus in a plane at the focal point of this lens. The eyepiece lens, with a short focal length, then refracts the rays into another parallel beam. The viewer sees a virtual, enlarged, inverted image. The primary purpose of a telescope is not to magnify the image. It is to increase the angle between the rays from two different stars and to collect more light than would strike the unaided eye.

18.2 Section Review

2.1 Which of the lenses whose cross sections are shown in **Figure 18–22** are convex or converging lenses? Which are concave or diverging lenses?

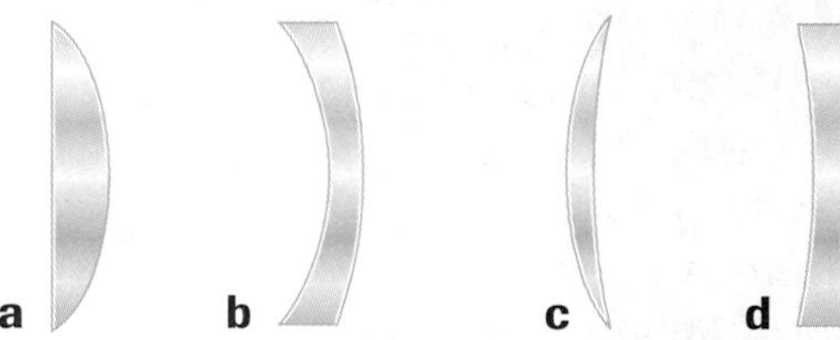

FIGURE 18–22

2.2 Suppose your camera was focused on a person 2 m away. You now want to focus it on a tree that is farther away. Should the lens be moved closer to the film or farther away?

2.3 You first focus white light through a single lens so that red is focused to the smallest point on a sheet of paper. Which direction should you move the paper to best focus blue?

2.4 **Critical Thinking** An air lens constructed of two watch glasses is placed in a tank of water. Copy **Figure 18–23** and draw the effect of this lens on parallel light rays incident on the lens.

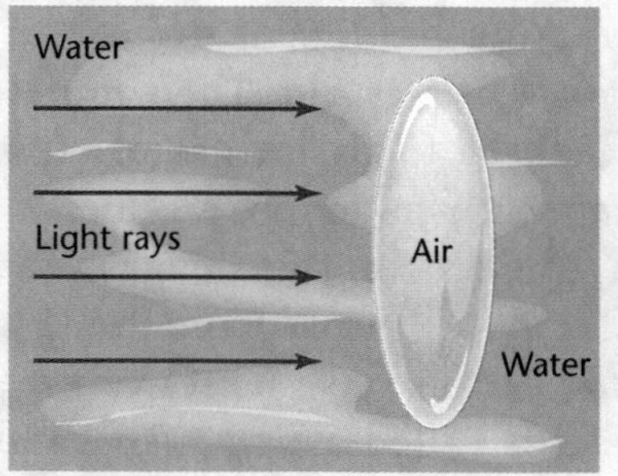

FIGURE 18–23

CHAPTER 18 REVIEW

Key Terms

18.1

- plane mirror
- object
- image
- virtual image
- erect image
- concave mirror
- principal axis
- focal point
- focal length
- real image
- lens/mirror equation
- magnification
- spherical aberration
- convex mirror

18.2

- lens
- convex lens
- concave lens
- chromatic aberration
- achromatic lens

Summary

18.1 Mirrors

- An object is a source of diverging light rays.
- Some mirrors reflect light rays that appear to diverge from a point on the other side of a mirror. The point from which they appear to diverge is called the virtual image.
- The image in a plane mirror is the same size as the object. It is as far behind the mirror as the object is in front of the mirror. The image is virtual and erect.
- The focal point of a convex or concave mirror is halfway between the center of curvature of the mirror and the mirror.
- Parallel rays striking a concave mirror converge at the focal point. Parallel rays striking a convex mirror appear to diverge from the focal point behind the mirror.
- Concave mirrors form real, inverted images if the object is farther from the mirror than the focal point, and virtual, upright images if the object is between the mirror and the focal point.
- Convex mirrors always produce virtual, upright, reduced images.
- Parallel light rays that are far from the principal axis are not reflected by spherical mirrors to converge at the focal point. This defect is called spherical aberration.

18.2 Lenses

- Convex lenses are thinner at their outer edges than at their centers. Concave lenses are thicker at their outer edges than at their centers.
- Convex lenses produce real, inverted images if the object is farther from the lens than the focal point. If the object is closer than the focal point, a virtual, upright, enlarged image is formed.
- Concave lenses produce virtual, upright, reduced images.
- Lenses have spherical aberrations because parallel rays striking a lens near its edge do not focus at one spot. Lenses also focus light of different wavelength (color) at different locations. This is called chromatic aberration.

Reviewing Concepts

Section 18.1

1. Describe the physical properties of the image seen in a plane mirror.
2. Where is the image of an object in a plane mirror?
3. Describe the physical properties of a virtual image.
4. A student believes that very sensitive photographic film can detect a virtual image. The student puts photographic film at the location of the image. Does this attempt succeed? Explain.
5. How can you prove to someone that an image is a real image?
6. What is the focal length of a plane mirror? Does the lens/mirror equation work for plane mirrors? Explain.
7. An object produces a virtual image in a concave mirror. Where is the object located?
8. Why are convex mirrors used as rearview mirrors?
9. What causes the defect that all concave spherical mirrors have?

Section 18.2

10. Locate and describe the physical properties of the image produced by a convex lens if an object is placed some distance beyond 2F.

11. What factor, other than the curvature of the surfaces of a lens, determines the location of the focal point of the lens?
12. To project an image from a movie projector onto a screen, the film is placed between F and 2F of a converging lens. This arrangement produces an inverted image. Why do the actors appear to be erect when the film is viewed?

Applying Concepts

13. Locate and describe the physical properties of the image produced by a concave mirror when the object is located at the center of curvature.
14. An object is located beyond the center of curvature of a spherical concave mirror. Locate and describe the physical properties of the image.
15. An object is located between the center of curvature and the focus of a concave mirror. Locate and describe the physical properties of the image of the object.
16. You have to order a large concave mirror for a piece of high-quality equipment. Should you order a spherical mirror or a parabolic mirror? Explain.
17. Describe the physical properties of the image seen in a convex mirror.
18. List all the possible arrangements in which you can use a spherical mirror, either concave or convex, to form a real image.
19. List all possible arrangements in which you can use a spherical mirror, either concave or convex, to form an image reduced in size.
20. The outside rearview mirrors of cars often carry the warning "Objects in the mirror are closer than they appear." What kind of mirror is this and what advantage does it have?
21. What physical characteristic of a lens distinguishes a converging lens from a diverging lens?
22. If you try to use a magnifying glass underwater, will its properties change? Explain.
23. Suppose **Figure 18–17** were redrawn with a lens of the same focal length but a larger diameter. How would the location of the image change?
24. Why is there chromatic aberration for light that goes through a lens but there is not chromatic aberration for light that reflects from a mirror?

Problems

Section 18.1

LEVEL 1

25. Penny wishes to take a picture of her image in a plane mirror. If the camera is 1.2 m in front of the mirror, at what distance should the camera lens be focused?
26. A concave mirror has a focal length of 10.0 cm. What is its radius of curvature?
27. Light from a star is collected by a concave mirror. How far from the mirror is the image of the star if the radius of curvature is 150 cm?
28. An object is 30.0 cm from a concave mirror of 15-cm focal length. The object is 1.8 cm high. Use the lens/mirror equation to find the image. How high is the image?
29. A jeweler inspects a watch with a diameter of 3.0 cm by placing it 8.0 cm in front of a concave mirror of 12.0-cm focal length.
    a. Where will the image of the watch appear?
    b. What will be the diameter of the image?
30. A dentist uses a small mirror of radius 40 mm to locate a cavity in a patient's tooth. If the mirror is concave and is held 16 mm from the tooth, what is the magnification of the image?

LEVEL 2

31. Draw a ray diagram of a plane mirror to show that if you want to see yourself from your feet to the top of your head, the mirror must be at least half your height.
32. Sunlight falls on a concave mirror and forms an image 3.0 cm from the mirror. If an object 24 mm high is placed 12.0 cm from the mirror, where will its image be formed?
    a. Use a ray diagram.
    b. Use the lens/mirror equation.
    c. How high is the image?
33. A production line inspector wants a mirror that produces an upright image with magnification of 7.5 when it is located 14.0 mm from a machine part.
    a. What kind of mirror would do this job?
    b. What is its radius of curvature?

34. Shiny lawn spheres placed on pedestals are convex mirrors. One such sphere has a diameter of 40 cm. A 12-cm robin sits in a tree 1.5 m from the sphere. Where is the image of the robin and how long is the image?

Section 18.2

LEVEL 1

35. The focal length of a convex lens is 17 cm. A candle is placed 34 cm in front of the lens. Make a ray diagram to locate the image.
36. The convex lens of a copy machine has a focal length of 25.0 cm. A letter to be copied is placed 40.0 cm from the lens.
    a. How far from the lens is the copy paper?
    b. The machine was adjusted to give an enlarged copy of the letter. How much larger will the copy be?
37. Camera lenses are described in terms of their focal length. A 50.0-mm lens has a focal length of 50.0 mm.
    a. A camera with a 50.0-mm lens is focused on an object 3.0 m away. Locate the image.
    b. A 1000-mm lens is focused on an object 125 m away. Locate the image.
38. A convex lens is needed to produce an image that is 0.75 times the size of the object and located 24 cm behind the lens. What focal length should be specified?
39. In order to clearly read a book 25 cm away, a farsighted person needs an image distance of −45 cm from the eye. What focal length is needed for the lens?

LEVEL 2

40. A slide of an onion cell is placed 12 mm from the objective lens of a microscope. The focal length of the objective lens is 10.0 mm.
    a. How far from the lens is the image formed?
    b. What is the magnification of this image?
    c. The real image formed is located 10.0 mm beneath the eyepiece lens. If the focal length of the eyepiece is 20.0 mm, where does the final image appear?
    d. What is the final magnification of this compound system?

Critical Thinking Problems

41. Your lab partner used a convex lens to produce an image with d_i = 25 cm and h_i = 4.0 cm. You are examining a concave lens with a focal length of −15 cm. You place the concave lens between the convex lens and the original image, 10 cm from the image. To your surprise, you see a real, enlarged image on the wall. You are told that the image from the convex lens is now the object for the concave lens, and because it is on the opposite side of the concave lens, it is a virtual object. Use these hints to find the location and size of the new image and to predict whether the concave lens changed the orientation of the original image.
42. What is responsible for the rainbow-colored fringe commonly seen at the edges of a spot of white light from a slide or overhead projector?
43. An overhead projector has a lens-mirror combination above the stage. If you move the screen farther from the projector, should you increase the distance between the stage and lens, keep it the same, or decrease it? Explain.
44. A lens is used to project the image of an object onto a screen. Suppose you cover the right half of the lens. What will happen to the image?

Going Further

Applying Calculators You are examining a 5-cm butterfly under a magnifying glass with a focal length of 15 cm. The equation for calculating how large the butterfly appears is $h_i = |(-h_o f)/(d_o - f)|$. Graph this equation on a graphing calculator with h_i on the y-axis (with a range of −50 cm to 50 cm) and d on the x-axis (with a range of 0 cm to 50 cm). How large does the butterfly appear at 5 cm? At 10 cm? 13 cm? 17 cm? 20 cm? 30 cm? 50 cm? For what distances is the image upright?

***inter*NET CONNECTION**

Follow the link on the Glencoe Homepage at **www.glencoe.com/sec/science** to find out more about this chapter.

On the Right Wave-length

Beetles can be annoying pests to gardeners, but it is still easy to admit how beautiful some of them can be. In daylight, the hard back of this ground beetle appears to be a mix of brilliant, metallic, iridescent colors. What characteristic of light could explain this unusual effect?

CHAPTER

19 Diffraction and Interference of Light

You have seen that dyes and pigments produce colors when they absorb some wavelengths of light while they transmit or reflect others. In raindrops and prisms, different wavelengths are bent through different angles. The shining colors you see in peacock tail-feathers, mother-of-pearl shells, and soap films all result from interference of light in thin films of matter. However, the colors you see on the backs of some beetles, which are some of the most beautiful in nature, are caused by a different light behavior.

From their observations of the behavior of light in nature, scientists have developed instruments that can accurately measure the wavelengths of light waves. With a knowledge of light behavior, scientists have also made it possible for you to observe microscopic organisms with ease. If you have examined microorganisms or other extremely small objects through optical microscopes, the sharp images you saw are made possible by the understanding and application of the principles governing light's behaviors.

WHAT YOU'LL LEARN

- You will define diffraction and relate it to the interference of light waves.
- You will describe the operation of a grating spectrometer.

WHY IT'S IMPORTANT

- By understanding diffraction, you can identify the resolving powers of microscopes and telescopes.
- It is possible to measure wavelengths of light accurately with a grating spectrometer.

*inter***NET**
CONNECTION

Follow the link for this chapter on the Glencoe Homepage at **www.glencoe.com/sec/science** to find out more about diffraction and interference of light.

19.1 When Light Waves Interfere

OBJECTIVES

- **Relate** the diffraction of light to its wave characteristics.
- **Explain** how light falling on two closely spaced slits produces an interference pattern, and use measurements to **calculate** wavelengths of light.
- **Apply** geometrical models to **explain** single-slit diffraction and two-slit interference patterns.

Sir Isaac Newton, whose laws of motion you studied in Chapter 5, believed that light was composed of fast-moving, unimaginably tiny particles, which he called corpuscles. He was aware that the Italian scientist Francesco Maria Grimaldi (1618–1663) had observed that the edges of shadows are not perfectly sharp. But Newton thought that Grimaldi's result was caused by the interaction of light corpuscles with the vibrating particles on the edges of openings. Newton probably never imagined that the wavelengths of visible light might be so tiny they could produce such small diffraction effects.

Diffraction

Grimaldi named the slight spreading of light around barriers diffraction. The Dutch scientist Christiaan Huygens (1629–1695) proposed a wave model to explain diffraction. According to Huygens, all the points of a wave front of light could be thought of as new sources of smaller waves. These wavelets expand in every direction and are in step with one another. A light source consists of an infinite number of point sources, which generate a plane wave front, as shown in **Figure 19–1.**

Much later, the English physician Thomas Young (1773–1829) read Newton's book on optics while studying the human eye. He became convinced that Newton's descriptions of light behavior in optics could be explained if light were a wave with an extremely small wavelength. In 1801, Young developed an experiment that allowed him to make a precise measurement of light's wavelength using diffraction.

Young's two-slit experiment

Young's experiment not only enabled him to measure light's wavelength, but also provided additional evidence of the wave nature of light. Young directed a beam of light at two closely spaced narrow slits in a barrier. The light was diffracted, and the rays from the two slits overlapped. When the overlapping light beams from the two slits fell on an observing screen on the other side of the barrier, the overlap did not

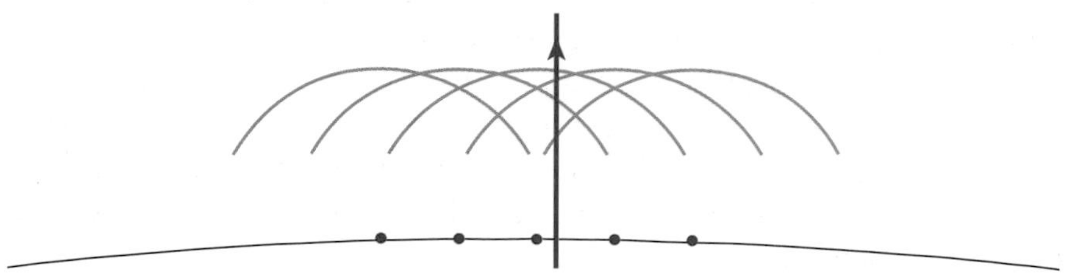

FIGURE 19–1 According to Huygens, the crest of each wave can be thought of as a series of point sources. Each point source creates a circular wavelet. All the wavelets add together. In the center of the beam, the wave front is flat, but at the edges the circular waves spread out. The beam no longer has sharp edges.

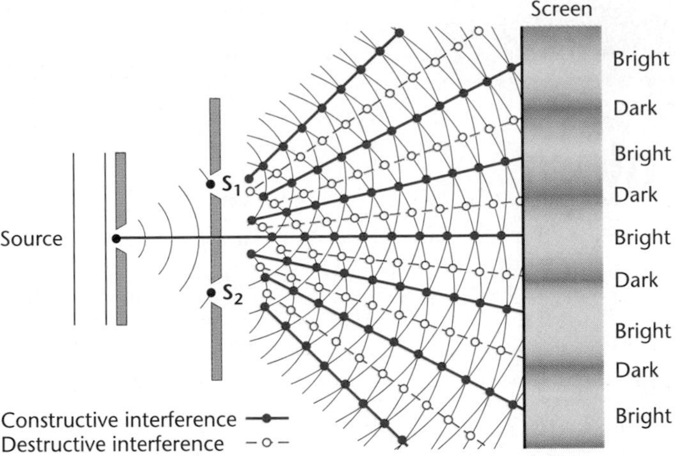

FIGURE 19–2 The diffraction of monochromatic light through a double slit produces bright and dark bands on a screen.

produce extra light, but a pattern of bright and dark bands, which Young called **interference fringes.** He explained that these bands must be the result of constructive and destructive interference of the light waves from the two slits.

Young placed a narrow slit in front of a **monochromatic light** source, one that emits light of only one wavelength. Only a small part of the light from the source passed through the slit, ensuring that the waves were in phase; that is, the waves' crests reached the same point at the same time—as did their troughs. Waves of this type are called **coherent waves.**

The waves spread out after passing through the single slit and fell on the double slit. The waves were again diffracted at the double slit, which acted as two sources of new circular waves spreading out on the far side of this second barrier, as shown in **Figure 19–2.** The semicircles represent wave crests moving outward from the slits. Midway between the crests are the troughs. At the points where the two crests overlap, the waves interfere constructively, and the light intensity increases creating a bright band on a screen. Where a crest and a trough meet, they interfere destructively, canceling each other out and creating a dark region.

Diffraction of white light

In a diffraction experiment that uses monochromatic light, constructive interference produces a bright central band on the screen, as well as other bright bands on either side, **Figure 19–3a** and **b.** Between the bright bands are dark areas located where destructive interference occurs. However, when white light is used in a double-slit experiment, diffraction causes the appearance of colored spectra instead of bright and dark bands, as shown in **Figure 19–3c.** The positions of the constructive and destructive interference bands depend on the wavelength of the light. All wavelengths interfere constructively in the central bright band, so that band is white. The positions of the other bands depend on the wavelength, so the light is separated by diffraction into a spectrum of color at each band.

FIGURE 19–3 The diffraction of a monochramatic light source produces interference on the screen resulting in a pattern, such as the one shown for blue light **(a)** and for red light **(b).** The diffraction of white light produces bands of different colors **(c).**

Physics Lab

Wavelengths of Colors

Problem

How can you accurately measure the wavelength of four colors of light?

Diffraction grating

Index card strip

x 90°

Clay

$L = 1.0$ m

Materials

meterstick
index card
40-W straight filament light
ball of clay
tape
diffraction grating

Data and Observations				
Color	x	d	L	λ

Procedure

1. Cut the index card lengthwise into four equal strips.
2. Write the letters "O" (orange), "Y" (yellow), "G" (green), and "B" (blue) on the strips.
3. Place the ball of clay 1.0 m on the bench in front of the lamp. Use the ball of clay to support the diffraction grating.
4. Plug in the lamp and turn off the room lights.
5. When you look through the diffraction grating, you should see bands of colors to the sides of the bulb. If you do not see the colors to the sides, then rotate the diffraction grating 90° until you do.
6. Have a lab partner stand behind the lamp and move the strip labeled "O" from side to side until you see it in place with the middle of its color. Ask your partner to tape the strip to the table at that point.
7. Repeat step 6 for each of the other colored strips.

Analyze and Conclude

1. **Observing and Inferring** What color is closest to the lamp? Suggest a reason and list the order that colors occur, beginning from red.
2. **Making and Using a Table** Make a data table like the one shown to record x, d, and L for each of the four colors. Measure and record x for each strip to the nearest 0.1 cm. Record the value of d provided by your teacher.
3. **Calculating** Use equation $\lambda = xd/L$ to calculate the wavelength for each color, and record this value in nanometers in your data table.

Apply

1. How could diffraction gratings be used in conjunction with telescopes?

Measuring the Wavelength of a Light Wave

Young used the double-slit experiment to make the first precise measurement of the wavelength of light. A diagram of this experiment is shown in **Figure 19–4,** which is not drawn to scale so that all points can be observed. Regardless of the wavelength of light used, light reaching point P_0 travels the same distance from each slit. Therefore, all wavelengths of light interact constructively. The first bright band on either side of the central band is called the first-order line. It falls on the screen at point P. The band is bright because light from the two slits, S_1 and S_2, interferes constructively. The two path lengths, which would be much larger in reality than is shown in the model, differ by one wavelength. That is, the distance PS_1 is one wavelength longer than PS_2.

To measure wavelength, Young first measured the distance between P_0 and P, labeled x in **Figure 19–4.** The distance between the screen and the slits is L, and the separation of the two slits is d. In the right triangle NS_1S_2, the side S_1N is the length difference of the two light paths. S_1N is one wavelength, λ, long. The lines from the slits to the screen are almost parallel because length L is so much larger than d. Thus, OP nearly equals the distance L, and the lines NS_2 and OP are nearly perpendicular to each other. Because the triangle NS_1S_2 is similar to triangle PP_0O, the ratio of the corresponding sides of these similar triangles is the same, as shown by the following equation.

$$\frac{x}{L} = \frac{\lambda}{d}$$

The equation to solve for λ is then given as follows.

$$\lambda = \frac{xd}{L}$$

The wavelengths of light waves can be measured with considerable precision using double-slit interference patterns. It is not unusual for wavelength measurements to be precise to four significant digits.

Hot Lights

Plug a 100-W clear lamp into a Variac (variable power supply). Turn off the room lights. Look through a diffraction grating at the lamp as you slowly increase the power.

Observing and Inferring Describe what you see. Which color appears first? What happens to the brightness of previous colors as new colors become visible? What is the order of the colors?

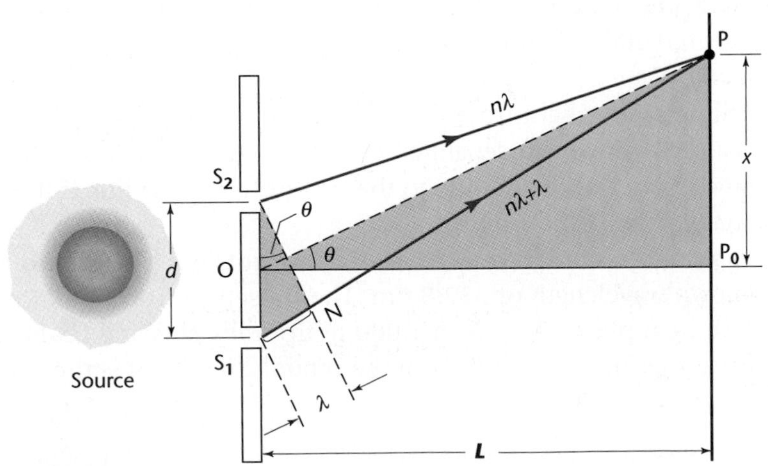

FIGURE 19–4 This diagram represents an analysis of the angles of light formed by double-slit interference. In reality, the distance, L, is about 10^5 times longer than the separation, d, between the two slits. It is necessary to distort the diagram so that the details close to the slit can be made clear.

Example Problem

Wavelength of Light

A two-slit experiment is performed to measure the wavelength of red light. The slits are 0.0190 mm apart. A screen is placed 0.600 m away and the separation between the central bright line and the first-order bright line is found to be 21.1 mm. What is the wavelength of the red light?

Sketch the Problem

- Sketch the experiment.
- Label knowns and unknowns.

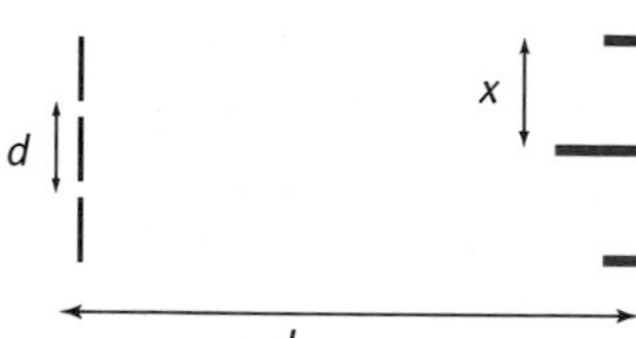

Calculate Your Answer

Known:

$d = 1.90 \times 10^{-5}$ m

$x = 2.11 \times 10^{-2}$ m

$L = 0.600$ m

Unknown:

$\lambda = ?$

Strategy:

Solve for the wavelength.

$$\lambda = \frac{xd}{L}$$

Calculations:

$$\lambda = \frac{(2.11 \times 10^{-2}\ \text{m})(1.90 \times 10^{-5}\ \text{m})}{0.600\ \text{m}}$$

$$= 668\ \text{nm}$$

Check Your Answer

- The answer is in m or nm, which are correct for wavelength.
- The wavelength of red light is near 700 nm; and that of blue is near 400 nm. So the answer is reasonable for red light.

Practice Problems

1. Violet light falls on two slits separated by 1.90×10^{-5} m. A first-order line appears 13.2 mm from the central bright line on a screen 0.600 m from the slits. What is the wavelength of the violet light?
2. Yellow-orange light from a sodium lamp of wavelength 596 nm is aimed at two slits separated by 1.90×10^{-5} m. What is the distance from the central line to the first-order yellow line if the screen is 0.600 m from the slits?
3. In a double-slit experiment, physics students use a laser with a known wavelength of 632.8 nm. The slit separation is unknown. A student places the screen 1.000 m from the slits and finds the first-order line 65.5 mm from the central line. What is the slit separation?

FIGURE 19–5 This diffraction pattern for red light was produced with a single slit having a width of 0.02 cm.

Single-Slit Diffraction

Suppose that you walk by the open door of the band rehearsal room at school. You hear the music as you walk toward the rehearsal room door long before you can see the players through the door. Sound seems to have reached you by bending around the edge of the door, whereas the light, which enables you to see the band players, has traveled only in a straight line. Both sound and light are composed of waves, so why don't they seem to act the same? In fact, they do behave in the same way. As Grimaldi first noted, the spreading of waves, or diffraction, occurs in both cases, but, because of light's much smaller wavelengths, the diffraction is much less obvious.

From one to many slits

When light passes through a single, small opening, light is diffracted, and a series of bright and dark bands appears. Instead of the equally spaced, bright bands you have seen produced by two slits, the pattern from a single slit has a wide, bright central band with dimmer bands on either side, as shown in **Figure 19–5.**

To understand single-slit diffraction, suppose that the single slit has a width w. Imagine the slit as being divided into a large number of even smaller slits of width dw. Just as in two-slit interference, a dark band is produced each time light passing through a pair of these smaller slits interferes destructively.

How can you choose pairs of tiny slits so that each pair has the same separation? Divide the single slit into two equal parts and choose one tiny slit from each part so that each pair will be separated by a distance $w/2$, as shown in **Figure 19–6a.** That is, for any tiny slit in the top half, there will be another tiny slit in the bottom half, a distance $w/2$ away.

FIGURE 19–6 A slit of width w is divided into pairs of tiny slits, each separated by $w/2$ **(a).** Light passing through the slit forms a diffraction pattern on the screen **(b).** By studying this diffraction pattern, it is possible to determine the slit width, w, if L and the wavelength of the light, λ, are known.

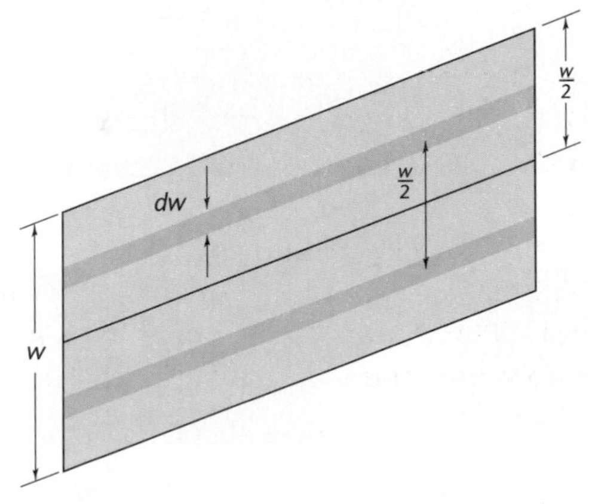

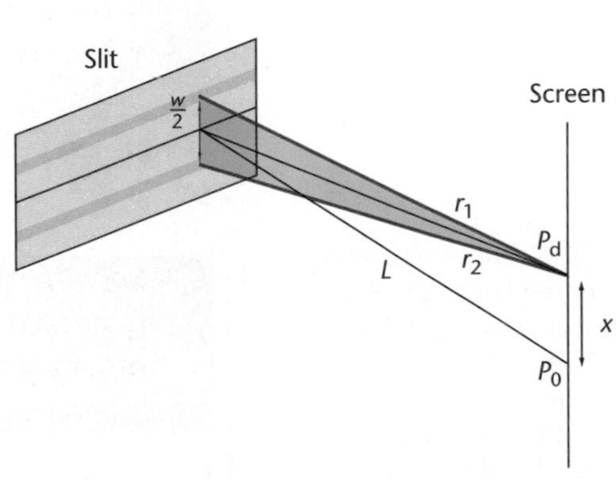

Pocket Lab

Laser Spots

Turn on a laser so that it makes a spot on the center of a movie screen. What would you expect to happen to the spot if you were to put a piece of window screening in the pathway of the beam? Explain your prediction.

Observing and Interpreting What really happened? Use the wave theory to explain your results.

Measuring a wavelength of light

If the slit is now illuminated, a central bright band appears at location P_0 on the screen, as shown in **Figure 19–6b.** But at position P_d, the path lengths r_2 and r_1 differ by one-half wavelength and produce a dark band. How far is the dark band from the central bright band? The situation is similar to that of double-slit interference, but the paths are now different by $\lambda/2$ and the separation between the slits is now $w/2$. The ratio of sides of the triangle can be shown in the following way.

$$\frac{x}{L} = \frac{\lambda/2}{w/2} = \frac{\lambda}{w}$$

The distance between the central bright band and the first dark band, x, can be determined by the following equation.

$$x = \frac{\lambda L}{w}$$

Additional dark bands occur where the path lengths differ by $3\lambda/2$, $5\lambda/2$, and so on. **Figure 19–7** shows examples of single-slit diffraction using different light sources.

It can be seen from this model that if you make the slit width smaller, you will make the bright band—that is, the distance between the dark bands—wider. If you use light with a longer wavelength, which is more toward the red end of the visible spectrum, you also increase the width of the bright band. Thus, the interference fringes that indicate the wave properties of light become noticeable when the light passes through small openings, which still are up to ten or 100 times the light's wavelength. Large openings, however, cast sharp shadows, as Newton first observed; thus, they do not as clearly reveal the wave nature of light.

a

b

c

FIGURE 19–7 These diffraction patterns for red light **(a),** blue light **(b),** and white light **(c)** were produced with a slit of width 0.02 cm. Note that the red light has a longer wavelength than that for the blue light.

Practice Problems

4. A double-slit apparatus, $d = 15\ \mu m$, is used to determine the wavelength of an unknown green light. The first-order line is 55.8 mm from the central line on a screen that is 1.6 m from the slits. What is the wavelength of the light?
5. Monochromatic green light of wavelength 546 nm falls on a single slit with width 0.095 mm. The slit is located 75 cm from a screen. How far from the center of the central band is the first dark band?
6. Light from a He-Ne laser ($\lambda = 632.8$ nm) falls on a slit of unknown width. A pattern is formed on a screen 1.15 m away on which the first dark band is 7.5 mm from the center of the central bright band. How wide is the slit?
7. Yellow light falls on a single slit 0.0295 mm wide. On a screen 60.0 cm away, there is a dark band 12.0 mm from the center of the bright central band. What is the wavelength of the light?
8. White light falls on a single slit 0.050 mm wide. A screen is placed 1.00 m away. A student first puts a blue-violet filter ($\lambda = 441$ nm) over the slit, then a red filter ($\lambda = 622$ nm). The student measures the width of the central peak, that is, the distance between the two dark bands.
   a. Which filter produced the wider band?
   b. Calculate the width of the central bright band for each of the two filters.

19.1 Section Review

1.1 Two very narrow slits are cut close to each other in a large piece of cardboard. They are illuminated by monochromatic red light. A sheet of white paper is placed far from the slits, and a pattern of bright and dark bands is seen on the paper. Explain why some regions are bright and others are dark.

1.2 Sketch the pattern described in question 1.1.

1.3 Sketch what happens to the pattern in question 1.1 if the red light is replaced by blue light.

1.4 **Critical Thinking** One of the slits in question 1.1 is covered so that no light can get through. What happens to the pattern?

19.2 Applications of Diffraction

OBJECTIVES

- **Explain** how diffraction gratings form interference patterns and how they are used in grating spectrometers.
- **Discuss** how diffraction limits the ability of a lens to distinguish two closely spaced objects.

On the Right Wavelength

The iridescent colors seen in many beetles are produced by diffraction. A beetle's hard back is covered with tiny ridges only a few hundred nanometers apart. Each space between the ridges acts as a slit and diffracts the light that hits it, thereby producing interference effects. The interference pattern from two slits is enhanced by this arrangement of many ridges and slits in series. In the same way, the spaces between the grooves on a compact disk diffract light and produce the familiar multicolored light reflected from a CD.

Diffraction Gratings

Although single-slit diffraction or two-slit interference can be used to measure the wavelength of light, diffraction gratings, such as those shown in **Figure 19–8,** are used in actual practice. A **diffraction grating** is a device that transmits or reflects light and forms an interference pattern in the same way that a double slit does. Diffraction gratings are made by scratching very fine lines with a diamond point on glass. The spaces between the scratched lines act like slits. Gratings can have as many as 10 000 lines per centimeter. That is, the spacing between the lines can be as small as 10^{-6} m, or 1000 nm. Less expensive replica gratings are made by pressing a thin plastic sheet onto a glass grating. When the plastic is pulled away, it contains an accurate imprint of the scratches. Jewelry made from replica gratings produces a spectrum just like that seen on the surface of a CD.

The gratings described above are called transmission gratings. Other gratings, called reflection gratings, are produced by scribing fine lines on metallic or reflective glass surfaces. Reflection gratings and interference gratings produce similar interference patterns, which can be analyzed in the same manner.

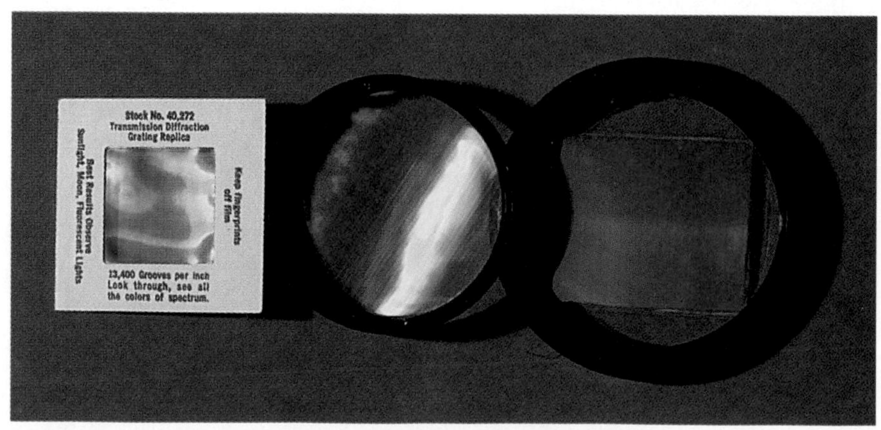

FIGURE 19–8 Diffraction gratings are used to create interference patterns for the analysis of light sources.

How It Works

Holograms

Holography is a form of photography that produces a three-dimensional image. Because they are difficult to reproduce, holograms are often placed on credit cards to help in the prevention of counterfeiting. In some manufacturing industries, "before" and "after" holograms are used to evaluate the effects of stress on various materials.

1 A hologram is made by first passing a coherent beam of light onto a semitransparent mirror.

2 The mirror then splits the beam into two beams: the object beam and the reference beam.

3 The object beam passes through a lens and is reflected from a mirror to illuminate the object. The beam in turn reflects from the object onto a photographic film or plate.

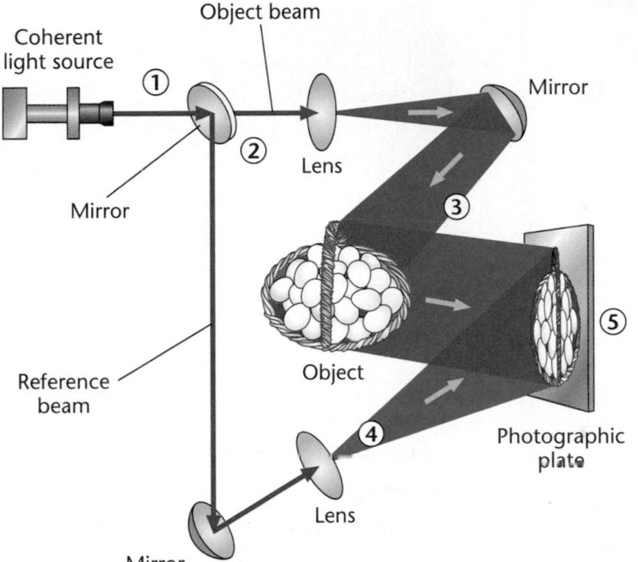

4 The reference beam is first reflected from a mirror, then it is spread by a lens and is directed over the object beam on the film or plate.

5 The superimposed beams on the plate form an interference pattern that allows the plate to record both the intensity and relative phase of the light from each point on the object.

6 When the photographic film or plate is developed, the resulting picture of the interference pattern becomes a hologram of the object. When the hologram is illuminated, a hovering image containing rainbowlike bands of color is visible.

Thinking Critically

1. Why can't a hologram be produced using a fluorescent light source?
2. Find out what the term *parallax* means. How can this term be used to describe a hologram?

FIGURE 19–9 A spectroscope **(a)** is used to measure the wavelengths of light emitted by a light source **(b).**

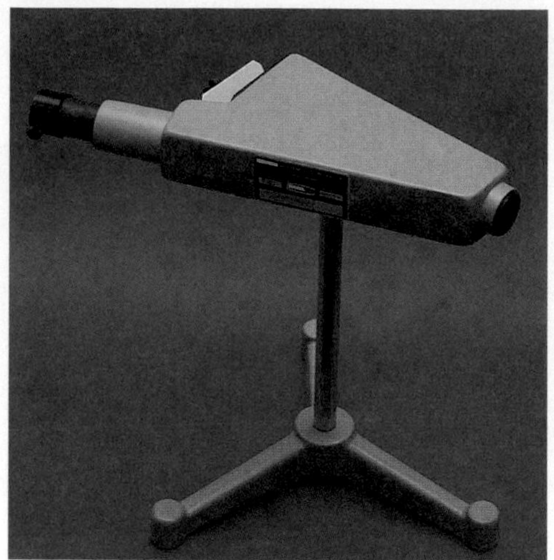

a

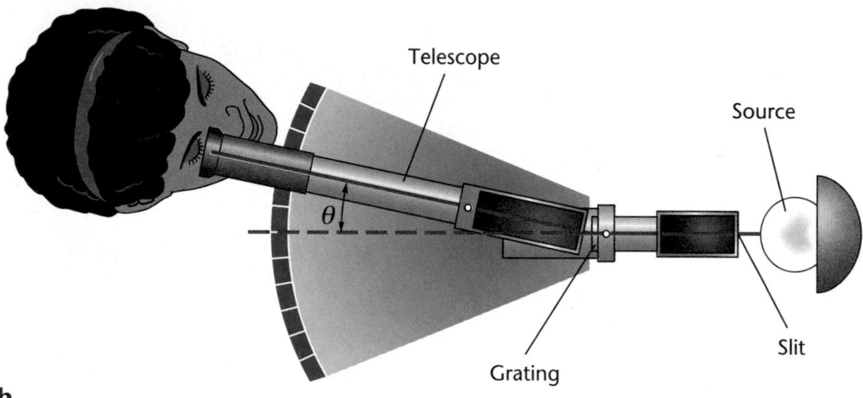

b

HELP WANTED

SPECTROSCOPIST

Position requires operating spectrometers, spectroscopes, other specialized laboratory instrumentation, and computers. Applicant with the appropriate skills will determine the fragmentation patterns of substances, and the absorption, reflection, and scattering of electromagnetic radiation with matter. A background in chemistry, metallurgy, and astrophysics is a must. Spectroscopists will have a doctorate degree in chemistry or physics. Candidates with a bachelor's degree in these areas may work as laboratory technicians. For more information, contact:

American Society of Mass Spectrometry
1202 Don Diego
Santa Fe, NM 87505 or
Society for Applied Spectroscopy
201 B Broadway Street
Frederick, MD 21701

The interference pattern produced by a diffraction grating has bright bands in the same locations caused by a double slit, but the bands are narrower and the dark regions are broader. As a result, individual colors can be distinguished more easily. Wavelengths can be measured more precisely with a diffraction grating than with double slits.

Earlier in this chapter, you used the following equation to calculate the wavelength of light using double-slit interference.

$$\frac{x}{L} = \frac{\lambda}{d}$$

The same equation holds for a diffraction grating, where d is the distance between the lines. Instead of measuring the distance from the central band to the first bright band, x, most laboratory instruments measure the angle θ, as indicated in **Figure 19–9.** Because x is so much smaller than L, the distance from the center of the slits to P, OP, is almost equal to the perpendicular distance L. Thus the ratio x/L can be replaced by $\sin\theta$. In equation form, this is shown as $\sin\theta \approx x/L$. Therefore, the wavelength can be found first by measuring the angle between

a

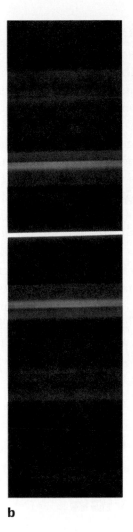

b

FIGURE 19–10 A grating was used to produce interference patterns for red light **(a)** and white light **(b)**.

the central bright band and the first-order line, and then by using the following equation.

$$\lambda = \frac{xd}{L} = d \sin \theta$$

The instrument used to measure light wavelengths produced by a diffraction grating is called a grating spectroscope, shown in **Figure 19–9a.** As you look through a telescope from one end, the source at the other end emits light that falls on a slit and then passes through a diffraction grating, **Figure 19–9b.** When monochromatic red light is used, you will see a series of bright bands to either side of the central bright line, as shown in **Figure 19–10a.** When white light falls on the instrument, each red band is replaced by a spectrum, as shown in **Figure 19–10b.** The red band in the spectrum is at the same location on the screen as it is for a monochromatic light. The telescope can be moved until the desired line appears in the middle of the viewer. The angle θ is then read directly from the calibrated base of the spectrometer. Because d is known, λ can be calculated.

Lights in the Night

Obtain small pieces of red and blue cellophane. When it is dark, find a long stretch of road and estimate the distance to cars when you can just barely tell that they have two headlights on. When a car is far away, its lights blend together. Look at these distant lights through the red cellophane and also through the blue cellophane. Which color makes it easier to resolve the two lights into separate images?

Determining Cause and Effect Explain why one color is more effective in separating the lights. Suggest how the use of blue filters might be useful for scientists working with telescopes or microscopes.

Resolving Power of Lenses

When light enters the lens of a telescope, it passes through a circular hole. The lens diffracts the light, just as a slit does. The smaller the lens, the wider the diffraction pattern. If the light comes from a star, the star will appear to be spread out. If two stars are close enough together, the images may be so blurred by diffraction that a viewer cannot tell whether there are two stars or only one.

ASTRONOMY CONNECTION

Some telescopes are not powerful enough to resolve the blurred images of the two stars. Lord Rayleigh (1842–1919) established the **Rayleigh criterion** for resolution. If the central bright band of one star falls on the first dark band of the second, the two stars will be just resolved. That is, a viewer will be able to tell that there are two stars and not just one. The effects of diffraction on the resolving power of the telescope can be reduced by increasing the size of the lens.

F.Y.I.

The Hubble Space Telescope was designed to resolve the images of stars. It can resolve the images of objects with spacing that is the equivalent to the spacing of car headlights 2500 miles away.

Diffraction limits the resolving power of microscopes as well as telescopes. The objective lens of a microscope cannot be enlarged, but the wavelength of light can be reduced. The diffraction pattern formed by blue light is narrower than that formed by red light. Thus, microscopes used by biologists often use blue or violet light to illuminate their objectives.

19.2 Section Review

2.1 Many narrow slits are close to each other and equally spaced in a large piece of cardboard. They are illuminated by monochromatic red light. A sheet of white paper is placed far from the slits, and a pattern of bright and dark bands is visible on the paper. Sketch the pattern that would be seen on the screen.

2.2 You shine a red laser light through one diffraction grating, forming a pattern of red dots on a screen. Then you substitute a second diffraction grating for the first one, forming a different pattern. The dots produced by one grating are spread more than those produced by the other. Which grating has more lines per millimeter?

2.3 An astronomer uses a telescope to view a number of closely spaced stars. Colored filters are available to select only certain colors from the starlight. Through which filter, red or blue, could the astronomer more easily count the stars? Explain.

2.4 **Critical Thinking** You are shown a spectrometer, but do not know whether it produces its spectrum with a prism or a grating. By looking at a white light spectrum, how could you tell?

CHAPTER 19 REVIEW

Key Terms

19.1
- interference fringe
- monochromatic light
- coherent wave

19.2
- diffraction grating
- Rayleigh criterion

Summary

19.1 When Light Waves Interfere

- Light has wave properties.
- Light passing through two closely spaced, narrow slits produces a pattern of dark and light bands on a screen called an interference pattern.
- Interference patterns can be used to measure the wavelength of light.
- Light passing through a narrow hole or slit is diffracted, or spread from a straight-line path, and produces a diffraction pattern on a screen.
- Both interference and diffraction patterns depend on the wavelength of light, the width or separation of the slits, and the distance to the screen.
- Interference patterns are narrower and sharper than diffraction patterns.

19.2 Applications of Diffraction

- Diffraction gratings consist of large numbers of slits and produce narrow interference patterns.
- Diffraction gratings can be used to measure the wavelength of light precisely or to separate light composed of different wavelengths.
- Diffraction limits the ability of a lens to distinguish two closely spaced objects.

Reviewing Concepts

Section 19.1

1. Why is it important that monochromatic light be used to make the interference pattern in Young's interference experiment?
2. Explain why the central bright line produced when light is diffracted by a double slit cannot be used to measure the wavelength of the light waves.
3. Describe how you could use light of a known wavelength to find the distance between two slits.
4. Why is the diffraction of sound waves more familiar in everyday experience than is the diffraction of light waves?
5. For each of the following examples, state whether the color is produced by diffraction, refraction, or the presence of pigments.
   a. soap bubbles
   b. rose petals
   c. mother of pearl
   d. oil films
   e. a rainbow

Section 19.2

6. As monochromatic light passes through a diffraction grating, what is the difference between the path lengths of light from two adjacent slits to a dark area on the screen?
7. When white light passes through a grating, what is visible on the screen? Why are no dark areas visible?
8. Why do diffraction gratings have large numbers of grooves? Why are these grooves so close together?
9. Why would a telescope with a small diameter not be able to resolve the images of two closely spaced stars?
10. Why is blue light used for illumination in an optical microscope?

Applying Concepts

11. How can you tell whether an interference pattern is from a single slit or a double slit?

12. Describe the changes in a single-slit pattern as slit width is decreased.
13. For a given diffraction grating, which color of visible light produces a bright line closest to the central bright line?
14. What are the differences in the characteristics of the interference patterns formed by diffraction gratings containing 10^4 lines/cm and 10^5 lines/cm?
15. Using **Figure 16–1,** decide for which part of the electromagnetic spectrum a picket fence could possibly be used as a diffraction grating.

Problems

Section 19.1

LEVEL 1

16. Light falls on a pair of slits 19.0 μm apart and 80.0 cm from the screen. The first-order bright line is 1.90 cm from the central bright line. What is the wavelength of the light?
17. Light of wavelength 542 nm falls on a double slit. First-order bright bands appear 4.00 cm from the central bright line. The screen is 1.20 m from the slits. How far apart are the slits?
18. Monochromatic light passes through a single slit with a width of 0.010 cm and falls on a screen 100 cm away. If the distance from the center of the pattern to the first band is 0.60 cm, what is the wavelength of the light?
19. Light with a wavelength of 4.5×10^{-5} cm passes through a single slit and falls on a screen 100 cm away. If the slit is 0.015 cm wide, what is the distance from the center of the pattern to the first dark band?
20. Monochromatic light with a wavelength of 400 nm passes through a single slit and falls on a screen 90 cm away. If the distance of the first-order dark band is 0.30 cm from the center of the pattern, what is the width of the slit?

LEVEL 2

21. Using a compass and ruler, construct a scale diagram of the interference pattern that results when waves 1 cm in length fall on two slits 2 cm apart. The slits may be represented by two dots spaced 2 cm apart and kept to one side of the paper. Draw a line through all points of reinforcement. Draw dotted lines through all nodal lines.

Section 19.2

LEVEL 1

22. A good diffraction grating has 2.5×10^3 lines per cm. What is the distance between two lines in the grating?
23. A spectrometer uses a grating with 12 000 lines/cm. Find the angles at which red light, 632 nm, and blue light, 421 nm, have first-order bright bands.
24. A camera with a 50-mm lens set at *f*/8 aperture has an opening 6.25 mm in diameter.
    a. Suppose this lens acts like a slit 6.25 mm wide. For light with $\lambda = 550$ nm, what is the resolution of the lens—the distance from the middle of the central bright band to the first-order dark band? The film is 50.0 mm from the lens.
    b. The owner of a camera needs to decide which film to buy for it. The expensive one, called fine-grained film, has 200 grains/mm. The less costly, coarse-grained film has only 50 grains/mm. If the owner wants a grain to be no smaller than the width of the central bright band calculated above, which film should be purchased?
25. Suppose the Hubble Space Telescope, 2.4 m in diameter, is in orbit 100 km above Earth and is turned to look at Earth, as in **Figure 19–11.** If you ignore the effect of the atmosphere, what is the resolution of this telescope? Use $\lambda = 500$ nm.

FIGURE 19–11

26. After passing through a grating with a spacing of 4.00×10^{-4} cm, a red line appears 16.5 cm from the central line on a screen. The screen is 1.00 m from the grating. What is the wavelength of the red light?

LEVEL 2

27. Marie uses an old 33-1/3 rpm record as a diffraction grating. She shines a laser, $\lambda =$ 632.8 nm, on the record. On a screen 4.0 m from the record, a series of red dots 21 mm apart are visible.
   a. How many ridges are there in a centimeter along the radius of the record?
   b. Marie checks her results by noting that the ridges came from a song that lasted 4.01 minutes and took up 16 mm on the record. How many ridges should there be in a centimeter?

Critical Thinking Problems

28. Yellow light falls on a diffraction grating. On a screen behind the grating you see three spots, one at zero degrees, where there is no diffraction, and one each at $+30°$ and $-30°$. You now add a blue light of equal intensity that is in the same direction as the yellow light. What pattern of spots will you now see on the screen?

29. Blue light of wavelength λ passes through a single-slit of width w. A diffraction pattern appears on a screen. If you now replace the blue light with a green light of wavelength 1.5λ, to what width should you change the slit in order to get the original pattern back?

30. At night, the pupil of a human eye can be considered to be a slit with a diameter of 8.0 mm. The diameter would be smaller in daylight. An automobile's headlights are separated by 1.8 m. How far away can the human eye distinguish the two headlights at night? **Hint:** Assume a wavelength of 500 nm and recall that Rayleigh's criterion stated that the peak of one image should be at the first minimum of the other.

Going Further

Team Project You and your team have been hired as consultants for a new sci-fi movie. The screenwriter is planning an attack by two groups of aliens. One group has eyes that can detect infrared wavelengths. The other has eyes sensitive to microwaves. The aliens can see (resolve) just as well as humans. The screenwriter asks you to decide whether or not this is reasonable.
   a. To determine the resolving power of humans, calculate the distance that a car is away from you so that you can still distinguish two headlights. **Hint:** Use yellow light to obtain λ, estimate the iris opening for humans, and estimate the separation of the car's headlights.
   b. Can you see a car's headlights at the distance calculated in **a?** Does diffraction limit your eyes' sensing ability? Hypothesize as to what might be the limiting factors.
   c. Determine the iris size for the alien that can detect the infrared wavelengths. Assume the same resolving ability as for humans and use a λ of 10 μm.
   d. Determine the iris size for the alien that can detect microwaves. Assume the same resolving ability as for humans and use a λ of 10 mm.
   e. Are the iris sizes of the two aliens reasonable? What would you tell the screenwriter concerning the design of his aliens?

interNET CONNECTION

Follow the link on the Glencoe Homepage at **www.glencoe.com/sec/science** to find out more about this chapter.

Sky Light

During a thunderstorm, Molly and Paresh were watching lightning bolts light up the sky. Paresh was impressed by the patterns of the lightning. Molly asked Paresh, "Why do you suppose lightning jumps between a cloud and Earth?" What explanation might Paresh give?

CHAPTER 20 Static Electricity

Nature provides few more awesome displays than lightning. Is there a way to experience lightning close up, perhaps safely at home or in the school lab? Is lightning related to some everyday occurrences? If you have ever scuffed your feet on the carpet so that you could create a small spark between your fingers and a friend's nose, then you can answer yes to these questions. The connection between rubbing surfaces together and sparks has been known for a long time. It was not until Benjamin Franklin, however, that the connection between lightning and sparks was established. In 1750, Franklin proposed his famous kite experiment. Two years later, he showed that "electrical" fire could be obtained from a cloud, setting off a flurry of research in the field of electricity. Over the next several chapters, you will investigate electrical phenomena and develop models to explain what you observe.

WHAT YOU'LL LEARN

- You will classify electrical charge and analyze how charge interacts with matter.
- You will infer the rules of how charge pushes and pulls on the world.

WHY IT'S IMPORTANT

- In this age of microprocessors and sensitive circuitry, a knowledge of static electrical charge may save your electronic components from damage.

interNET CONNECTION

Follow the link for this chapter on the Glencoe Homepage at **www.glencoe.com/sec/science** to find out more about static electricity.

20.1 Electrical Charge

You may have had the experience of rubbing your shoes on the carpet hard enough to create a lightning-like spark when you touched someone. Franklin's kite experiment showed that lightning is similar to electricity caused by friction. Electrical effects produced this way are called *static electricity.* In this chapter, you will investigate **electrostatics,** the study of electrical charges that can be collected and held in one place. Current electricity, produced by batteries and generators, will be explored in later chapters.

OBJECTIVES

- **Recognize** that objects that are charged exert forces, both attractive and repulsive.
- **Demonstrate** that charging is the separation, not the creation, of electrical charges.
- **Describe** the differences between conductors and insulators.

Charged Objects

Have you ever noticed the way your hair is attracted to the comb when you comb your hair on a dry day? Perhaps you also have found that socks sometimes stick together when you take them out of the clothes dryer. If so, you will recognize the attraction of the bits of paper to a comb shown in **Figure 20–1.** If the weather is dry, try this yourself now. Rub a plastic comb or ballpoint pen on your clothing. (Wool clothing is best.) Then, hold the pen or comb close to a pile of paper bits. Notice the way the paper pieces jump up toward the pen or comb. There must be a new, relatively strong force causing this upward acceleration because it is larger than the downward acceleration caused by the gravitational force of Earth.

There are other differences between this new force and gravity. Paper is attracted to a comb only after the comb has been rubbed. If you wait a while, the attractive property of the comb disappears. Gravity, on the other hand, does not require rubbing and does not disappear. The ancient Greeks noticed similar effects when they rubbed amber. The Greek word for amber is *elektron,* and today this attractive property is called "electrical." An object that exhibits electrical interaction after rubbing is said to be charged.

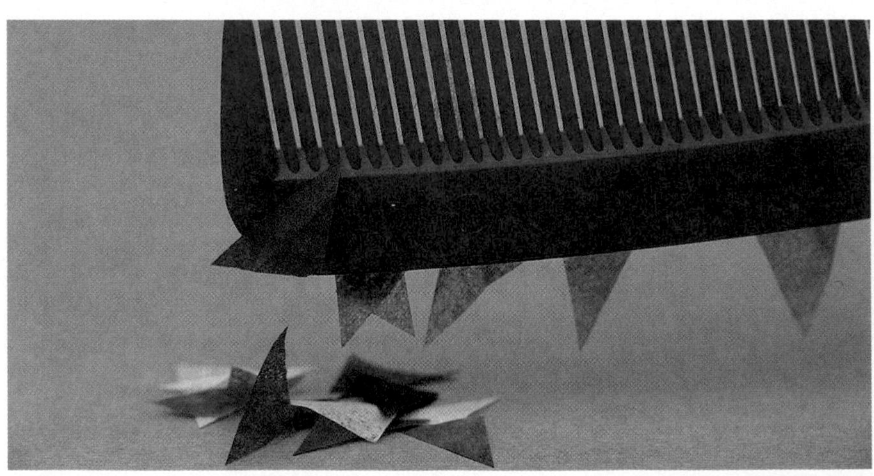

FIGURE 20–1 Running a comb through your hair transfers electrons to the comb, giving it a negative charge. When the comb is brought close to bits of paper, a charge separation is induced on the paper bits. The attractive electrical force accelerates the paper bits upward against the force of gravity.

Like charges

You can explore electrical interactions with very simple equipment such as transparent tape. Fold over about 5 mm at the end of a strip of tape for a handle and then stick the remaining 8- to 12-cm-long part of the strip on a dry, smooth surface such as your desktop. Then stick a second, similar piece next to the first. Quickly pull both strips off the desk and bring them near each other. What happens? The strips have a new property that causes them to repel each other. They are electrically charged. They were prepared in the same way, so they must have the same type of charge. Therefore, you have just demonstrated that two objects with the same type of charge repel each other.

You can learn about this charge by doing some simple experiments. You may have found that the tape is attracted to your hand. Are both sides attracted, or just one? If you wait a while, especially in humid weather, you'll find that the electrical charge disappears. You can restore it by again sticking the tape to the desk and pulling it off. You also can remove its charge by gently rubbing your fingers down both sides of the tape.

F.Y.I.

Charged objects will eventually return to their neutral state. The charge "leaks off" onto the water molecules in the air. On humid or rainy days it is difficult to make any object hold its charge for long.

Opposite charges

Now, stick one strip of tape on the desk and place the second strip on top of the first, as shown in **Figure 20–2a.** Use the handle of the bottom strip of tape to pull the two off the desk together. Rub them with your fingers until they are no longer attracted to your hand. You've now removed all the electrical charge. With one hand on the handle of one strip and the other on the handle of the second strip, quickly pull the two strips apart. You'll find that they're now both charged. They once again are attracted to your hands. But, do they still repel each other? No, they now attract each other. They are charged, but they are no longer charged alike. They have opposite charges and attract each other.

Is tape the only object that you can charge? Once again, stick one strip of tape to the desk and the second strip on top. Label the bottom strip B and the top strip T. Pull the pair off together. Discharge them, then pull them apart. Stick the handle end of each strip to the edge of a table, the bottom of a lamp shade, or some similar object. The two should hang down a short distance apart, as shown in **Figure 20–2b.** Finally, rub the comb or pen on your clothing and bring it near first one strip of tape and then the other. You will find that one strip will be attracted to the comb, the other repelled from it. You can now explore the interactions of charged objects with the strips of tape.

a

b

FIGURE 20–2 Strips of tape can be given opposite charges **(a),** then used to demonstrate the interactions of like and opposite charges **(b).**

Color Conventions

- Positive charges are **red.**
- Negative charges are **blue.**

Experimenting with charge

Try to charge other objects such as plates, glasses, and plastic bags. Rub them with different materials such as silk, wool, and plastic wrap. If the air is dry, scuff your shoes on the carpet and bring your finger near the strips of tape. You should find that most charged objects attract one strip and repel the other. To test silk or wool, slip a plastic bag over your hand before holding the cloth. After rubbing, take your hand out of the bag and bring both the bag and cloth near the strips of tape.

You will never find an object that repels both strips of tape, although you might find some that attract both. Bring your finger near first one strip, then the other. You will find that it attracts both. We will explore this effect later in this chapter.

Types of charge

From your experiments, you can make a list of objects labeled B for bottom, which have the same charge as the tape stuck on the desk. Another list can be made of objects labeled T, which have the same charge as the tape stuck on the top of the other tape. There are only two lists; thus, there are only two types of charge. You could give these types the names "yellow" and "green." Benjamin Franklin called them positive and negative charges. Using Franklin's convention, when hard rubber and plastic are rubbed, they become negatively charged. When materials such as glass and wool are rubbed, they become positively charged.

Just as you showed that an uncharged pair of strips of tape became oppositely charged, you were probably able to show that if you rubbed plastic with wool, the plastic became charged negatively, the wool positively. The two kinds of charges were not created alone, but in pairs. These experiments suggest that matter normally contains both charges, positive and negative. Friction in some way separates the two. To explore this further, you must consider the microscopic picture of matter.

A Microscopic View of Charge

Electrical charges exist within atoms. In 1890, J.J. Thomson discovered that all materials contain light, negatively charged particles he called electrons. Between 1909 and 1911, Ernest Rutherford, a New Zealander, discovered that atoms have a massive, positively charged nucleus. If the positive charge of the nucleus exactly balances the negative charge of the surrounding electrons, then the atom is **neutral.**

With the addition of energy, the outer electrons can be removed from atoms. An atom missing electrons has an overall positive charge, and consequently any matter made of these electron-deficient atoms is positively charged. The freed electrons can remain unattached or become attached to other atoms, resulting in negatively charged particles. From a microscopic viewpoint, acquiring charge is a process of transferring electrons.

HELP WANTED

OFFICE EQUIPMENT TECHNICIAN

Must be skilled in installing, servicing, and operating computers and other office equipment. Prefer graduate of two-year course at a technical institute or community college. High school courses should include algebra and other math, computer science, language arts, and physics or chemistry. Growing opportunity for those with computer skills. May work in a manufacturing plant, a corporation, or a repair shop. For more information, join a computer user group or contact:

Institute of Electrical and Electronics Engineers
345 East 47th Street
New York, NY 10017

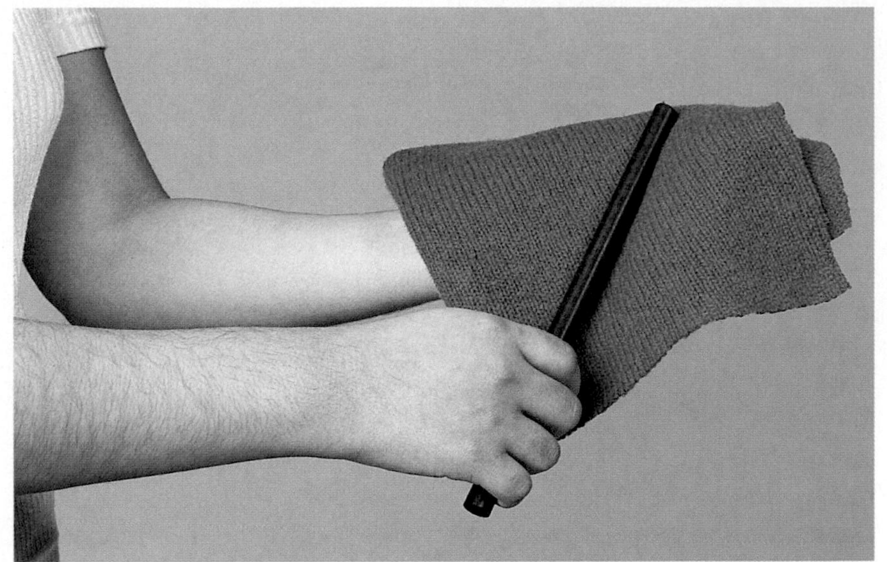

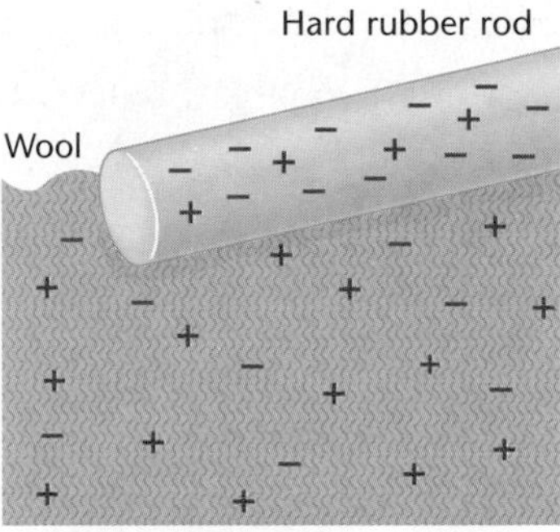

FIGURE 20–3 As the rubber rod strokes the wool, electrons are removed from the wool atoms and cling to the rubber atoms. In this way, both objects become charged.

Separation of charge

If two neutral objects are rubbed together, each can become charged. For instance, when rubber and wool are rubbed together, electrons from atoms on the wool are transferred to the rubber, as shown in **Figure 20–3.** The extra electrons on the rubber result in a net negative charge. The electrons missing from the wool result in a net positive charge. The combined total charge of the two objects remains the same. Charge is conserved, which is one way of saying that individual charges are never created or destroyed. All that happened was that the positive and negative charges were separated through a transfer of electrons.

Contact between the tires of a moving car or truck and the road can cause the tires to become charged. Processes inside a thundercloud can cause the cloud bottom to become negatively charged and the cloud top to become positively charged. In both these cases, no charge is made, only separated.

Conductors and Insulators

Hold a plastic rod or comb at its midpoint and rub only one end. You will find that only the rubbed end becomes charged. In other words, the charges you transferred to the plastic stayed where they were put, they did not move. **Figure 20–4** shows static charges—charges that are not moving—on an insulator. The strips of tape that you charged earlier in this chapter acted in the same way. Materials through which charges will not move easily are called electrical **insulators.** Glass, dry wood, most plastics, cloth, and dry air are all good insulators.

Suppose that you support a metal rod on an insulator so that it is isolated, or completely surrounded by insulators. If you then touch the charged comb to one end of the metal rod, you will find that the charge spreads very quickly over the entire rod. Materials such as metals that allow charges to move about easily are called electrical **conductors.** Electrons

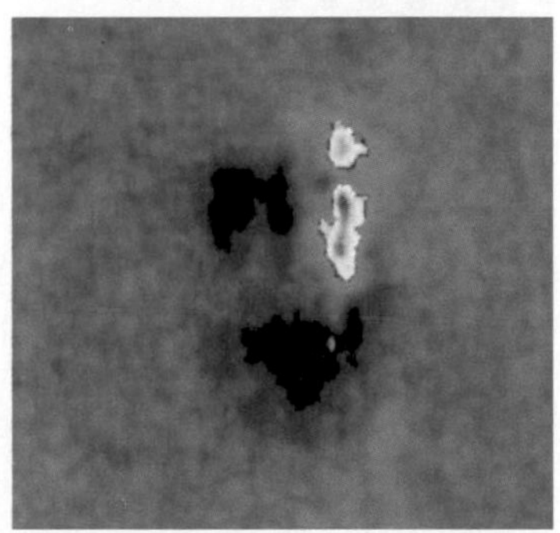

FIGURE 20–4 A piece of plastic 0.02 mm wide was given a net positive charge. Areas of negative charge are visible as dark regions. Areas of positive charge are visible as yellow regions.

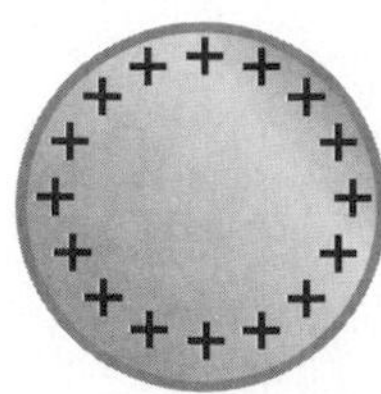

a Conductor

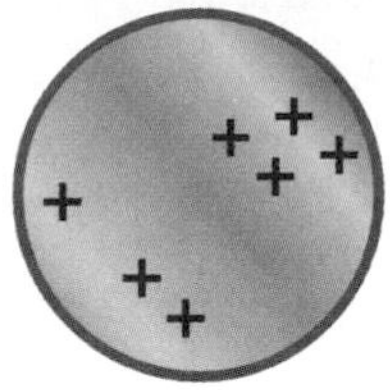

b Insulator

FIGURE 20–5 Charges placed on a conductor will spread over the entire surface **(a).** Charges placed on an insulator will remain where they are placed **(b).**

carry, or conduct, electric charge through the metal. Metals are good conductors because at least one electron on each atom of the metal can be removed easily. These electrons act as if they no longer belong to any one atom, but to the metal as a whole; consequently, they move freely throughout the piece of metal. **Figure 20–5** illustrates how charges behave when placed on a conductor or an insulator. Copper and aluminum are both excellent conductors and are used commercially to carry electricity. Plasma, a highly ionized gas, and graphite, the form of carbon used in pencils, also are good conductors of electrical charge.

When air becomes a conductor

Air is an insulator. However, under certain conditions, sparks or lightning occurs, allowing charge to move through air as if it were a conductor. The spark that jumps between your finger and a doorknob after you have rubbed your feet on the carpet discharges you. That is, you have become neutral because the excess charges have left you. Similarly, lightning discharges a thundercloud. In both these cases, for a brief moment, air became a conductor. Recall that conductors must have charges that are free to move. For a spark or lightning to occur, free moving charged particles must be formed in the normally neutral air. In the case of lightning, excess charges in the cloud and on the ground are great enough to remove electrons from the molecules in the air. The electrons and positively or negatively charged atoms become free to move. They form a conductor that is a plasma. The discharge of Earth and the thundercloud moving through this conductor forms a luminous arc called lightning. In the case of your finger and the doorknob, the discharge is called a spark.

20.1 Section Review

1.1 How could you find out which strip of tape, B or T, is positively charged?

1.2 Suppose you attach a long metal rod to a plastic handle so that the rod is isolated. You touch a charged glass rod to one end of the metal rod. Describe the charges on the metal rod.

1.3 In the 1730s, Stephan Gray tried to see how far metal rods could conduct electrical charge. He hung metal rods by thin silk cords from the ceiling. When the rods were longer than 293 feet, the silk broke. Gray replaced the silk with stronger wires made of brass, but then the experiments failed. The metal rods would no longer transmit a charge from one end to the other. Why not?

1.4 **Critical Thinking** Suppose there were a third type of charge. What experiments could you suggest to explore its properties?

Physics Lab

What's the Charge?

Problem

Can you see the effects of electrostatic charging? How can you increase the amount of charge on an object without discharging it?

Materials

30 cm × 30 cm block of polystyrene
22-cm aluminum pie pan
plastic cup
drinking straw
wool
transparent tape
thread
pith ball (or small piece of plastic foam packing material)
liquid graphite

Data and Observations	
Description of Event	Observations

Procedure

1. Paint the pith ball with graphite and allow it to dry.
2. Tape the inverted cup to the aluminum pie pan. Secure the straw to the top of the cup and use the thread to attach the ball as shown in the photograph.
3. Rub the foam with wool, then remove the wool.
4. Holding onto the plastic cup, lower the pie pan until it is about 3 cm above the foam block and then slowly lift it away.
5. Place the pie pan directly on the charged foam block and lift it away.
6. Bring your finger near the ball until they touch.
7. Place the pie pan on the foam block and touch the edge of the pie pan with your finger. Then remove the pie pan from the foam block and touch the ball again with your finger.
8. Repeat step 7 several times without recharging the foam block.

Analyze and Conclude

1. **Forming a Description** As the pie pan was brought near the charged block, could you detect a force between the neutral pie pan and the charged foam? Describe it
2. **Interpreting Observations** Explain what happened to the ball in step 4 and step 5.
3. **Analyzing Results** Make a drawing to show the distribution of charges on the neutral pie pan as it is lowered toward the charged foam block.
4. **Inferring Relationships** What was the reason for using the ball on a thread? Explain the back-and-forth motion of the ball in step 6.
5. **Interpreting Observations** Does the polystyrene block seem to run out of charges in step 8?

Apply

1. Clear plastic wrap is sold to seal up containers of food. Suggest a reason why it clings to itself.

20.2 Electrical Force

OBJECTIVES

- **Summarize** the relationship between forces and charges.
- **Describe** how an electroscope detects electric charge.
- **Explain** how to charge by conduction and induction.
- **Use** Coulomb's law to **solve** problems relating to electrical force.
- **Develop** a model of how charged objects can attract a neutral object.

Electrical forces must be strong because they can easily produce accelerations larger than the acceleration caused by gravity. We also have seen that they can be either repulsive or attractive while gravitational forces are always attractive. Many scientists made attempts to measure electrical force. Daniel Bernoulli, otherwise known for his work on fluids, made some crude measurements in 1760. In 1770, Henry Cavendish showed that electrical forces must obey an inverse square force law, but, being extremely shy, he did not publish his work. His manuscripts were discovered over a century later, after all his work had been duplicated by others.

Forces on Charged Bodies

The forces you observed on tape strips also can be demonstrated by suspending a negatively charged hard rubber rod so that it turns easily, as shown in **Figure 20–6.** If you bring another negatively charged rod near the suspended rod, it will turn away. The negative charges on the rods repel each other. It is not necessary for the rods to make contact; the force, called the electrical force, acts over a distance. If a positively charged glass rod is suspended and a similarly charged glass rod is brought close, the two positively charged rods also will repel each other. If a negatively charged rod is brought near the positively charged rod, however, the two will attract each other, and the suspended rod will turn toward the oppositely charged rod. The results of your tape experiments and these observations with charged rods can be summarized in this way:

- There are two kinds of electrical charges, positive and negative.
- Charges exert force on other charges over a distance.
- The force is stronger when the charges are closer together.
- Like charges repel; opposite charges attract.

FIGURE 20–6 A charged rod, when brought close to another suspended rod, will attract or repel the suspended rod.

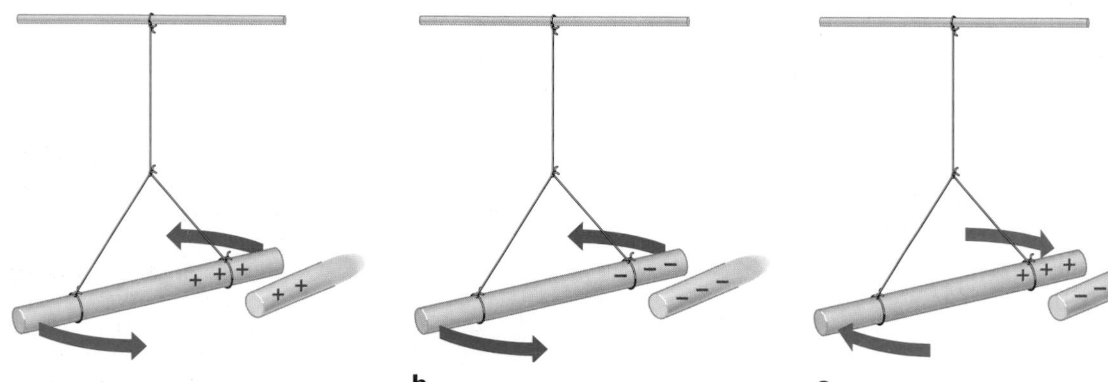

Neither a strip of tape nor a large rod hanging in open air is a very sensitive or convenient way of determining charge. Instead, a device called an **electroscope** is used. An electroscope consists of a metal knob connected by a metal stem to two thin, lightweight pieces of metal foil called leaves, as shown in **Figure 20–7.** Note that the leaves are enclosed to eliminate stray air currents.

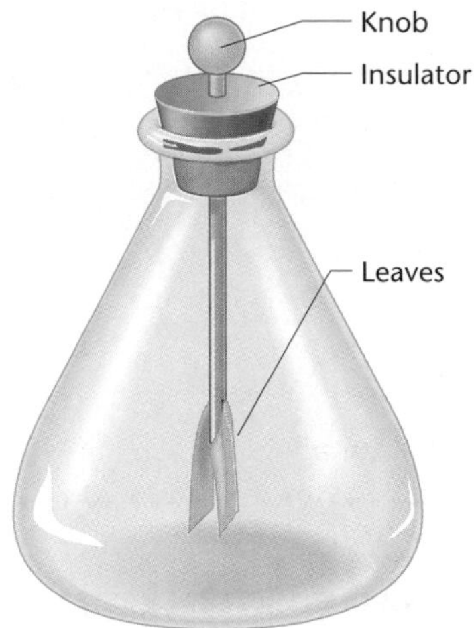

FIGURE 20–7 An electroscope is a device used for detecting electrical charges.

Charging by conduction

When a negatively charged rod is touched to the knob of an electroscope, negative charges (electrons) are added to the knob. The charges spread over all the metal surfaces. As shown in **Figure 20–8a,** the two leaves are charged negatively and repel each other, causing them to spread apart. The electroscope has been given a net charge. Charging a neutral body this way, by touching it with a charged body, is called **charging by conduction.**

The leaves also will spread if the electroscope is charged positively. How, then, can you find out whether the electroscope is charged positively or negatively? The type of charge can be determined by observing what happens to the spread leaves if a rod of known charge is brought close to the knob. The leaves will spread farther apart if the electroscope has the same charge as that of the rod, as shown in **Figure 20–8b.** The leaves will fall slightly if the electroscope has a charge opposite to that of the rod, as in **Figure 20–8c.**

Separation of charge on neutral objects

Earlier in the chapter, when you brought your finger near either charged strip of tape, the tape was attracted toward your finger. Your finger, however, was not charged; it was neutral and had equal amounts of positive and negative charge. You know that in materials that are conductors, charges can move easily, and that in the case of sparks, electric forces can change insulators into conductors. Given this information, you can develop a plausible model for the force your finger exerted on the charged objects.

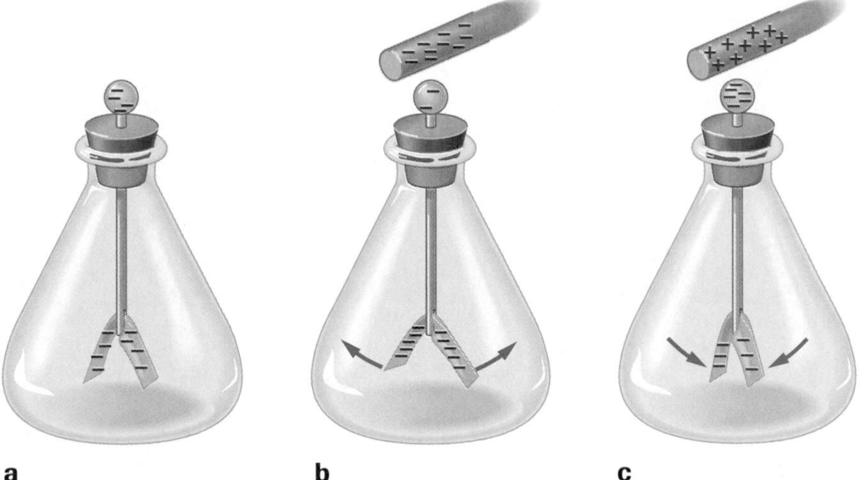

FIGURE 20–8 A negatively charged electroscope will have its leaves spread **(a).** A negatively charged rod pushes electrons down to the leaves, causing them to spread more **(b).** A positively charged rod attracts some of the electrons from the leaves, causing them to spread apart less **(c).**

Suppose you move your finger, or any other uncharged object, close to a positively charged object. The negative charges in your finger will be attracted to the positively charged object, and the positive charges in your finger will be repelled. Your finger will remain neutral, but the positive and negative charges will be separated. The electrical force is stronger for the charged objects that are closer together, therefore, the separation results in an attractive force between the neutral object and the charged object. The force of a charged comb on your hair or on neutral pieces of paper is the result of the same process, the separation of charge.

Sky Light

The negative charges at the bottom of thunderclouds also can cause charge separation in Earth. Positive charges in the ground are attracted to Earth's surface under the cloud. The forces of the charges in the cloud and those on Earth's surface can break molecules apart, separating the positively and negatively charged particles. These charged particles are free to move, and they establish a conducting path from the ground to the cloud. Travelling at 500 000 kilometers per hour, a lightning bolt courses through the sky to discharge the cloud.

Charging by induction

Charge separation can be used to charge an object without touching it. Suppose that a negatively charged rod is brought close to one of two identical insulated metal spheres that are touching, as in **Figure 20–9b.** Electrons from the first sphere will be forced onto the sphere farther from the rod and will make it negatively charged. The closer sphere is now positively charged. If the spheres are separated while the rod is nearby, each sphere will have a charge, and the charges will be equal but opposite, as shown in **Figure 20–9c.** This process is called **charging by induction.**

Coulomb's Law

You have seen that charges push and pull on each other. That is, a force acts between two or more charged objects. How does this force depend on the size of the charges and their separation? You demon-

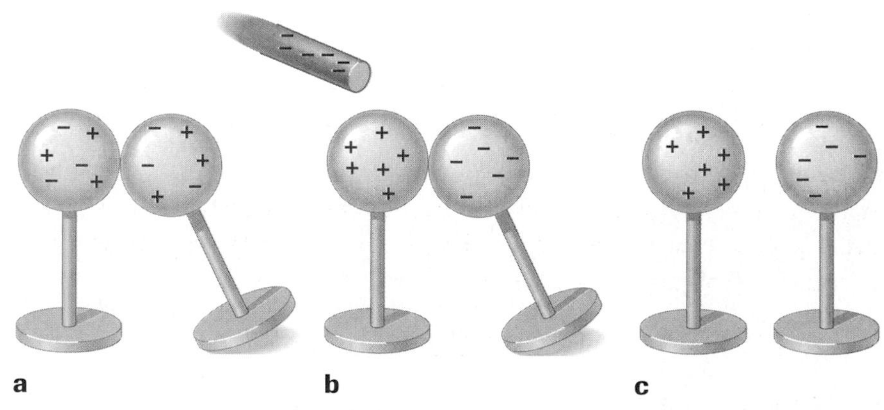

FIGURE 20–9 Start with neutral spheres that are touching **(a),** bring a charged rod near them **(b),** and then separate the spheres and remove the charged rod **(c).** This is one example of charging by induction.

strated in your experiments with tape some basic properties. You found that the force depends on distance. The closer you brought the charged rod to the tape, the stronger the force. You also found that the more you charged the rod, the stronger the force. But how can you vary the quantity of charge in a controlled way? This problem was solved in 1785 by French physicist Charles Coulomb (1736–1806). The type of apparatus used by Coulomb is shown in **Figure 20–10.** An insulating rod with small conducting spheres, A and A′, at each end was suspended by a thin wire. A similar sphere, B, was placed in contact with sphere A. When they were touched with a charged object, the charge spread evenly over the two spheres. Because they were the same size, they received equal amounts of charge. The symbol for charge is q. Therefore, the amount of charge on the spheres can be represented by the notation q_A and q_B.

Coulomb found how the force between the two charged spheres, A and B, depended on the distance. First, he carefully measured the amount of force needed to twist the suspending wire through a given angle. He then placed equal charges on spheres A and B and varied the distance, d, between them. The force moved A from its rest position, twisting the suspending wire. By measuring the deflection of A, Coulomb could calculate the force of repulsion. He showed that the force, F, varied inversely with the square of the distance between the centers of the spheres.

$$F \propto \frac{1}{d^2}$$

To investigate the way in which the force depended on the amount of charge, Coulomb had to change the charges on the spheres in a measured way. Coulomb first charged spheres A and B equally, as before. Then he selected an extra uncharged sphere, C, the same size as sphere B. When C was placed in contact with B, the spheres shared the charge that had been on B alone. Because the two were the same size, B now had only half its original charge. Therefore, the charge on B was only one half the charge on A. The extra sphere was then removed. After Coulomb adjusted the position of B so that the distance, d, between A and B was the same as before, he found that the force between A and B was half of its former value. That is, he found that the force varied directly with the charge of the bodies.

$$F \propto q_A q_B$$

After many similar measurements, Coulomb summarized the results in a law now known as **Coulomb's law:** the magnitude of the force between charge q_A and charge q_B, separated a distance d, is proportional to the magnitude of the charges and inversely proportional to the square of the distance.

$$F \propto \frac{q_A q_B}{d^2}$$

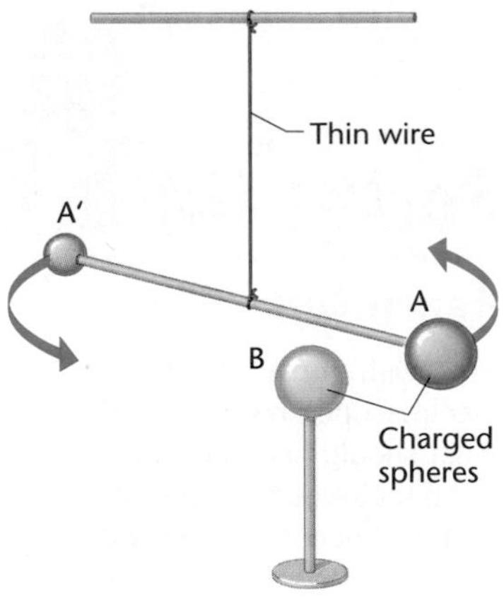

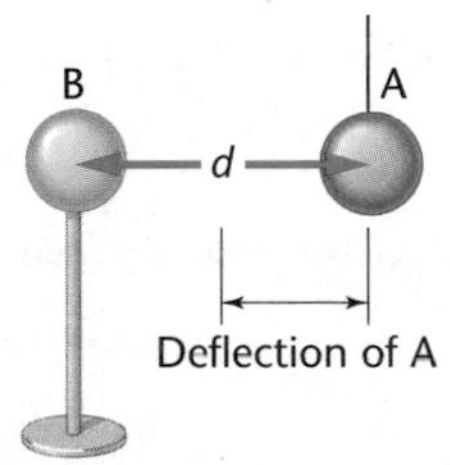

FIGURE 20–10 Coulomb used this type of apparatus to measure the force between two spheres, A and B. He observed the deflection of A while varying the distance between A and B.

Charged Up

Rub a balloon with wool. Touch the balloon to the knob of an electroscope and watch the leaves.

Analyze and Conclude
Describe the result. Make a drawing to explain the result. Touch the knob of the electroscope to make the leaves fall. Would you expect that the wool could move the leaves? Why? Try it. Explain your results.

Pocket Lab

Reach Out

Start with the leaves of an electroscope down. Predict what should happen if you bring a charged balloon near (but not touching) the top of the electroscope.

Analyze and Conclude
Explain your prediction. Try it. Describe and explain your results.

The unit of charge: The coulomb

The amount of charge an object has is difficult to measure directly. Coulomb, however, showed that the quantity of charge could be related to force. Thus, he could define a standard quantity of charge in terms of the amount of force it produces. The SI standard unit of charge is called the **coulomb** (C). One coulomb is the charge of 6.25×10^{18} electrons or protons. Recall that the charge of protons and electrons is equal. The charge that produces a large lightning bolt is about 10 coulombs. The charge on an individual electron is only 1.60×10^{-19} C. The magnitude of the charge of an electron is called the **elementary charge.** Thus, as you will calculate in problem 22 in the Chapter Review, even small pieces of matter, such as the coins in your pocket, contain up to 1 million coulombs of negative charge. This enormous amount of charge produces almost no external effects because it is balanced by an equal amount of positive charge. However, if the charge is unbalanced, even as small of a charge as 10^{-9} C can result in large forces.

According to Coulomb's law, the magnitude of the force on charge q_A caused by charge q_B a distance d away can be written as follows.

$$F = K\frac{q_A q_B}{d^2}$$

When the charges are measured in coulombs, the distance is measured in meters, and the force is measured in newtons, the constant, K, is 9.0×10^9 N·m²/C².

This equation gives the magnitude of the force that charge q_A exerts on q_B and also the force that q_B exerts on q_A. These two forces are equal in magnitude but opposite in direction. You can observe this example of Newton's third law of motion in action when you bring two strips of tape with like charges together. Each exerts forces on the other. If you bring a charged comb near either strip of tape, the strip, with its small mass, moves readily. The acceleration of the comb and you is, of course, much less because of the much greater mass.

The electrical force, like all other forces, is a vector quantity. Force vectors need both a magnitude and a direction. However, Coulomb's law will furnish only the magnitude of the force. To determine the direction, you need to draw a diagram and interpret charge relations carefully. Consider the direction of force on a positively charged object called A. If another positively charged object, B, is brought near, the force on A is repulsive. The force, $F_{B\text{ on }A}$, acts in the direction from B to A, as shown in **Figure 20–11a.** If, instead, B is negatively charged, the force on A is attractive and acts in the direction from A to B, as shown in **Figure 20–11b.**

FIGURE 20–11 The rule for determining direction of force is like charges repel, unlike charges attract.

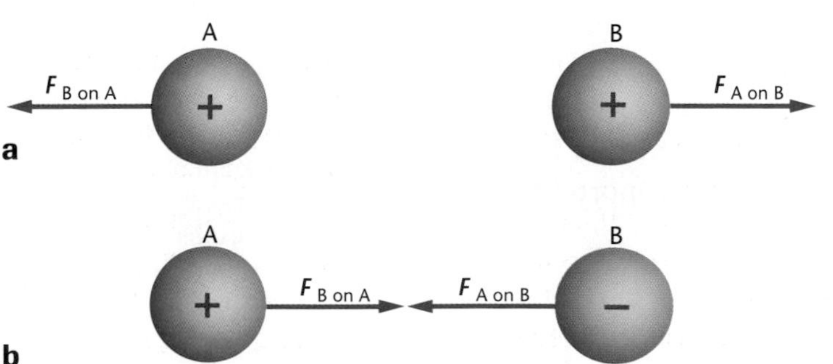

How It Works

Laser Printers

Static electricity is important to the operation of laser printers. Inside a laser printer, static electrical charges are used to transfer a black powder, commonly called toner, onto paper to create images or text. Letters and other shapes are actually composed of tiny dots of toner that are heat-bonded onto the paper. The dots are so numerous and placed so closely together that they merge to form very finely detailed images.

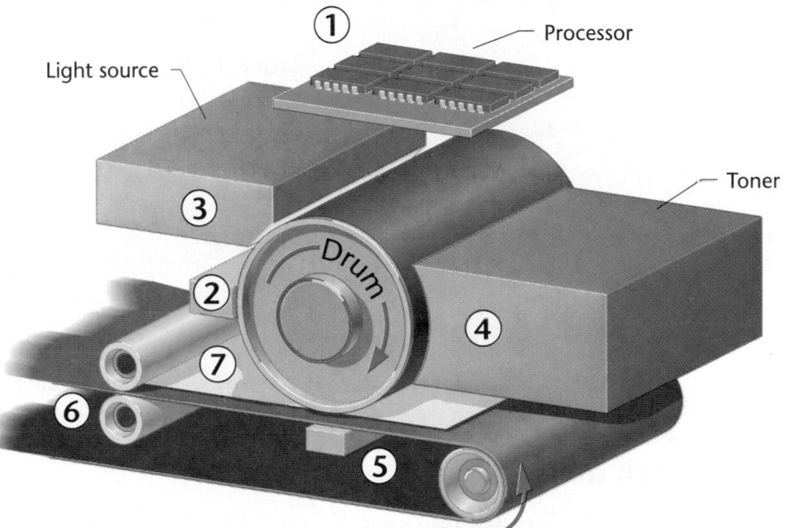

1. The computer sends instructions to the microprocessor that operates the laser printer. The microprocessor controls the aiming and timing of a light source, either a laser or a light-emitting diode (LED).

2. The surface of the metal drum is coated with a photosensitive semiconductor. In dark, the semiconductor is an insulator. As the drum rotates, it is sprayed with electrons, giving the surface a negative charge.

3. According to instructions received from the processor, a beam of light sweeps over the surface of the drum. Wherever the light strikes, the semiconductor becomes a conductor, and charges leak onto the supporting drum. These uncharged areas on the semiconductor correspond to the white portions of the page.

4. The drum is rotated so that it contacts uncharged toner particles. Toner contains small plastic beads coated in graphite. As the drum passes by the toner, the beads are attracted to the charged areas.

5. A sheet of paper that is being pushed by rollers through the printer's paper train receives a small opposite charge. The paper passing by the rotating drum electrically attracts the toner.

6. After picking up toner from the drum, the paper moves into the fusing system. Here the paper is exposed to heat and pressure that melts the grains of toner and binds them to the paper.

7. The drum is returned to the dark where it is an insulator. The surfaces are recharged and are ready to receive the image of another page.

Thinking Critically

1. Photocopiers also use toner, a photosensitive drum, and fusing system. Describe what you think might be some of the similarities and differences of photocopiers and laser printers?
2. In the process, notice that the areas exposed to laser light become the white part of a page. Design a process where the exposed areas become the black part of a page.

Problem Solving Strategy

Electrical Force Problems

1. Sketch the system showing all distances and angles to scale.
2. Diagram the vectors of the system; include derived vectors using dashed lines.
3. Use Coulomb's law to find the magnitude of the force.
4. Use your diagram along with trigonometric relations to find the direction of the force.
5. Perform all algebraic operations on units as well as the numbers. Make sure the units match the variable in question.
6. Consider the magnitude of your answer. Is it reasonable?

Notice in this problem solving strategy that Coulomb's law only is used to determine magnitudes. Therefore, it is unnecessary to include the sign of the charges or distance when evaluating Coulomb's law, because the answer is always positive.

Example Problem

Coulomb's Law

Two charges are separated by 3.0 cm. Object A has a charge of $+6.0\ \mu\text{C}$, while object B has a charge of $+3.0\ \mu\text{C}$. What is the force on object A?

Sketch the Problem

- Establish coordinate axes.
- Label spheres A and B.
- Draw distance d_{AB}.
- Diagram the force vectors.

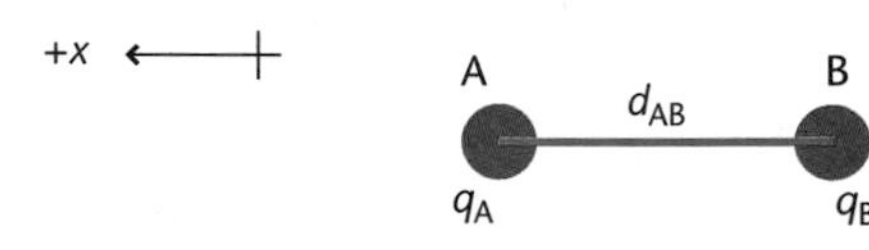

$F_{\text{B on A}}$ A

Calculate Your Answer

Known:

$q_A = +6.0\ \mu\text{C}$

$q_B = +3.0\ \mu\text{C}$

$d_{AB} = 0.030\ \text{m}$

Unknown:

$\mathbf{F}_{\text{B on A}} = ?$

Strategy:

Use Coulomb's law. Do not include signs when using Coulomb's law.

The direction of force is determined by the diagram.

Calculations:

$$F_{\text{B on A}} = K\frac{q_A q_B}{{d_{AB}}^2}$$

$$F_{\text{B on A}} = (9.0 \times 10^9\ \text{N}\cdot\text{m}^2/\text{C}^2)\frac{(6.0\ \mu\text{C})(3.0\ \mu\text{C})}{(3.0 \times 10^{-2}\ \text{m})^2}$$

$$\mathbf{F}_{\text{B on A}} = 1.8 \times 10^2\ \text{N},\ +x\ \text{direction}$$

Check Your Answer

- Are the units correct? Perform algebra on the units to ensure that your answer is in newtons.
- Does the direction make sense? It agrees with the coordinate axis and direction of push of charge B.

- Is the magnitude realistic? Look at the magnitudes in the equation $10^9 \times 10^{-6} \times 10^{-6} \div 10^{-2} \div 10^{-2} = 10^1$. This is close to the answer, so the magnitude checks.

Example Problem

Coulomb's Law with Three Charges

A sphere with charge 6.0 μC is located near two other charged spheres. A -3.0-μC sphere is located 4.00 cm to the right and a 1.5-μC sphere is located 3.00 cm directly underneath. Determine the net force on the 6.0-μC sphere.

Sketch the Problem

- Establish coordinate axes.
- Draw the displacement vectors.
- Diagram the force vectors.

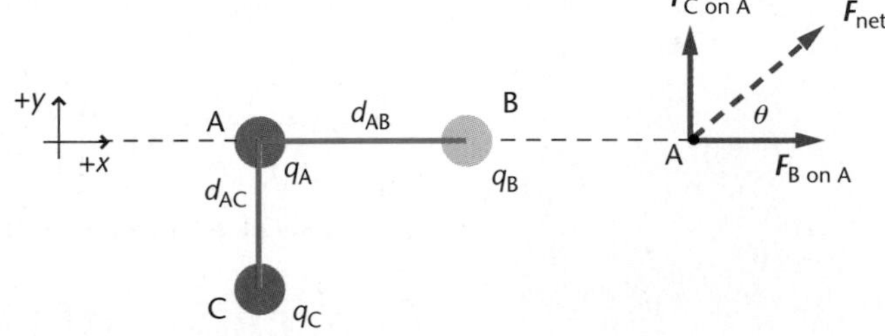

Calculate Your Answer

Known:

$q_A = 6.0\ \mu C$

$q_B = -3.0\ \mu C$

$q_C = 1.5\ \mu C$

$d_{AB} = 0.0400$ m

$d_{AC} = 0.0300$ m

Strategy:

Use Coulomb's law. Do not include signs when using Coulomb's law (refer to the problem-solving strategy). The direction of force is determined by the diagram.

Calculations:

$$F_{B\text{ on }A} = K\frac{q_A q_B}{d_{AB}^2} = (9.0\ \text{GN}\cdot\text{m}^2/\text{C}^2)\frac{(6.0\ \mu\text{C})(3.0\ \mu\text{C})}{(4.00 \times 10^{-2}\ \text{m})^2}$$

$$\mathbf{F}_{B\text{ on }A} = 1.0 \times 10^2\ \text{N, to the right}$$

$$F_{C\text{ on }A} = \frac{K q_A q_C}{d_{AC}^2} = (9.0\ \text{GN}\cdot\text{m}^2/\text{C}^2)\frac{(6.0\ \mu\text{C})(1.5\ \mu\text{C})}{(3.00 \times 10^{-2}\ \text{m})^2}$$

$$\mathbf{F}_{C\text{ on }A} = 9.0 \times 10^1\ \text{N, up}$$

Unknown:

$\mathbf{F}_{B\text{ on }A} = ?$

$\mathbf{F}_{C\text{ on }A} = ?$

$\mathbf{F}_{net} = ?$

Use the tangent function to find θ.

$$\tan\theta = \frac{F_{C\text{ on }A}}{F_{B\text{ on }A}} = \frac{9.0 \times 10^1\ \text{N}}{1.0 \times 10^{-2}\ \text{N}},\ \theta = 42$$

Use the Pythagorean theorem to find $\mathbf{F}_{net}$.

$$F_{net} = \sqrt{(1.0 \times 10^2\ \text{N})^2 + (9.0 \times 10^1\text{N})^2} = 130\ \text{N}$$

$$\mathbf{F}_{net} = 130\ \text{N, }42°\ \text{above } x\text{-axis}$$

Check Your Answer

- Are the units correct? Perform algebra on the units to ensure that your answer is in newtons.
- Does the direction make sense? It agrees with direction of force of the charge.
- Is the magnitude realistic? A magnitude of 130 N fits with the quantities given.

a

b

FIGURE 20–12 Static electricity precipitators are used to reduce the fly ash released into the environment. In **(a)** the device is off, in **(b)** it is on.

Practice Problems

1. A negative charge of -2.0×10^{-4} C and a positive charge of 8.0×10^{-4} C are separated by 0.30 m. What is the force between the two charges?
2. A negative charge of -6.0×10^{-6} C exerts an attractive force of 65 N on a second charge 0.050 m away. What is the magnitude of the second charge?
3. Two positive charges of 6.0 μC are separated by 0.50 m. What force exists between the charges?
4. An object with charge $+7.5 \times 10^{-7}$ C is placed at the origin. The position of a second object, charge $+1.5 \times 10^{-7}$ C, is varied from 1.0 cm to 5.0 cm. Draw a graph of the force on the object at the origin.
5. The charge on B in the second Example Problem is replaced by $+3.00$ μC. Use graphical methods to find the net force on A.

Application of Electrical Forces

There are many applications of electrical forces on neutral particles. For example, these forces can collect soot in smokestacks, thereby reducing air pollution, as shown in **Figure 20–12.** Tiny paint droplets, charged by induction, can be used to paint automobiles and other objects very uniformly. Photocopy machines use static electricity to place black toner on a page so that a precise reproduction of the original document is made.

20.2 Section Review

2.1 When an electroscope is charged, the leaves rise to a certain angle and remain at that angle. Why don't they rise farther?

2.2 Two charged spheres are on a frictionless horizontal surface. One has a $+3 \times 10^{-6}$ C charge, the other a $+6 \times 10^{-6}$ C charge. Sketch the two spheres, showing all forces on them. Make the length of your force arrows proportional to the strength of the forces.

2.3 Explain why a balloon that has been rubbed on a wool shirt sticks to the wall.

2.4 **Critical Thinking** Suppose you are testing Coulomb's law using a small, charged plastic sphere and a large, charged metal sphere. Both are charged positively. According to Coulomb's law, the force depends on $1/d^2$, where d is the distance between the centers of the spheres. As the two spheres get close together, the force is smaller than expected from Coulomb's law. Explain.

CHAPTER 20 REVIEW

Key Terms

20.1
- electrostatics
- neutral
- insulator
- conductor

20.2
- electroscope
- charging by conduction
- charging by induction
- Coulomb's law
- coulomb
- elementary charge

Summary

20.1 Electrical Charge

- There are two kinds of electrical charge, positive and negative.
- Electrical charge is not created or destroyed; it is conserved.
- Objects can be charged by the transfer of electrons.
- Charges added to one part of an insulator remain on that part.
- Charges added to a conductor quickly spread over the surface of the object.
- Charged objects exert forces on other charged objects. Like charges repel; unlike charges attract.

20.2 Electrical Force

- When an electroscope is charged, electrical forces cause its thin metal leaves to spread.
- An object can be charged by conduction by touching a charged object to it.
- To charge an object by induction, a charged object is first brought nearby, causing a separation of charges. Then the object to be charged is separated, trapping opposite charges on the two halves.
- Coulomb's law states that the force between two charges varies directly with the product of their charge and inversely with the square of the distance between them.
- The SI unit of charge is the coulomb. One coulomb (C) is the magnitude of the charge of 6.25×10^{18} electrons or protons. The elementary charge, the charge of the proton or electron, is 1.60×10^{-19} C.
- A charged object of either sign can produce separation of charge in a neutral body. Thus, a charged object attracts a neutral one.

Reviewing Concepts

Section 20.1

1. If you comb your hair on a dry day, the comb can become positively charged. Can your hair remain neutral? Explain.
2. List some insulators and conductors.
3. What property makes metal a good conductor and rubber a good insulator?

Section 20.2

4. Why do socks taken from a clothes dryer sometimes cling to other clothes?
5. If you wipe a stereo record with a clean cloth, why does the record then attract dust?
6. The combined charge of all electrons in a nickel is hundreds of thousands of coulombs. Does that imply anything about the net charge on the coin? Explain.
7. Name three methods of charging an object.
8. Explain how to charge a conductor negatively if you have only a positively charged rod.

Applying Concepts

9. How does the charge of an electron differ from the charge of a proton?
10. Using a charged rod and an electroscope, how can you find whether or not an object is a conductor?
11. A charged rod is brought near a pile of tiny plastic spheres. Some of the spheres are attracted to the rod, but as soon as they touch the rod, they fly away in different directions. Explain.
12. Lightning usually occurs when a negative charge in a cloud is transported to

Earth. If Earth is neutral, what provides the attractive force that pulls the electrons toward Earth?

13. Explain what happens to the leaves of a positively charged electroscope when rods with the following charges are nearby but not touching the electroscope.
a. positive
b. negative

14. Coulomb's law and Newton's law of universal gravitation appear to be similar. In what ways are the electrical and gravitational forces similar? How are they different?

15. The text describes Coulomb's method for obtaining two charged spheres, A and B, so that the charge on B was exactly half the charge on A. Suggest a way Coulomb could have placed a charge on sphere B that was exactly one third the charge on sphere A.

16. Coulomb measured the deflection of sphere A when spheres A and B had equal charges and were a distance *d* apart. He then made the charge on B one third the charge on A. How far apart would the two spheres then have to be for A to have the same deflection it had before?

17. Two charged bodies exert a force of 0.145 N on each other. If they are moved so that they are one fourth as far apart, what force is exerted?

18. The constant, *K*, in Coulomb's equation is much larger than the constant, *G*, in the universal gravitation equation. Of what significance is this?

19. Electrical forces between charges are enormous in comparison to gravitational forces. Yet we normally do not sense electrical forces between us and our surroundings, while we do sense gravitational interactions with Earth. Explain.

Problems

Section 20.2

LEVEL 1

20. Two charges, q_A and q_B, are separated by a distance, *d*, and exert a force, *F*, on each other. What new force will exist if
a. q_A is doubled?
b. q_A and q_B are cut in half?
c. *d* is tripled?
d. *d* is cut in half?
e. q_A is tripled and *d* is doubled?

21. How many excess electrons are on a ball with a charge of -4.00×10^{-17} C?

22. How many coulombs of charge are on the electrons in a nickel? Use the following method to find the answer.
a. Find the number of atoms in a nickel. A nickel has a mass of about 5 g. Each mole (6.02×10^{23} atoms) has a mass of about 58 g.
b. Find the number of electrons in the coin. A nickel is 75% Cu and 25% Ni, so each atom on average has 28.75 electrons.
c. Find how many coulombs of charge are on the electrons.

23. A strong lightning bolt transfers about 25 C to Earth.
a. How many electrons are transferred?
b. If each water molecule donates one electron, what mass of water lost an electron to the lightning? One mole of water has a mass of 18 g.

24. Two electrons in an atom are separated by 1.5×10^{-10} m, the typical size of an atom. What is the electrical force between them?

25. A positive and a negative charge, each of magnitude 1.5×10^{-5} C, are separated by a distance of 15 cm. Find the force on each of the particles.

26. Two negatively charged bodies, each charged with -5.0×10^{-5} C, are 0.20 m from each other. What force acts on each particle?

27. How far apart are two electrons if they exert a force of repulsion of 1.0 N on each other?

28. A force of -4.4×10^{3} N exists between a positive charge of 8.0×10^{-4} C and a negative charge of -3.0×10^{-4} C. What distance separates the charges?

29. Two identical positive charges exert a repulsive force of 6.4×10^{-9} N when separated by a distance of 3.8×10^{-10} m. Calculate the charge of each.

LEVEL 2

30. A positive charge of 3.0 μC is pulled on by two negative charges. One, -2.0 μC, is 0.050 m to the north and the other, $-4.0\ \mu C$, is 0.030 m to the south. What total force is exerted on the positive charge?

31. Three particles are placed in a line. The left particle has a charge of $-67\ \mu C$, the middle, $+45\ \mu C$, and the right, $-83\ \mu C$. The middle particle is 72 cm from each of the others, as shown in **Figure 20–13.**

a. Find the net force on the middle particle.

b. Find the net force on the right particle.

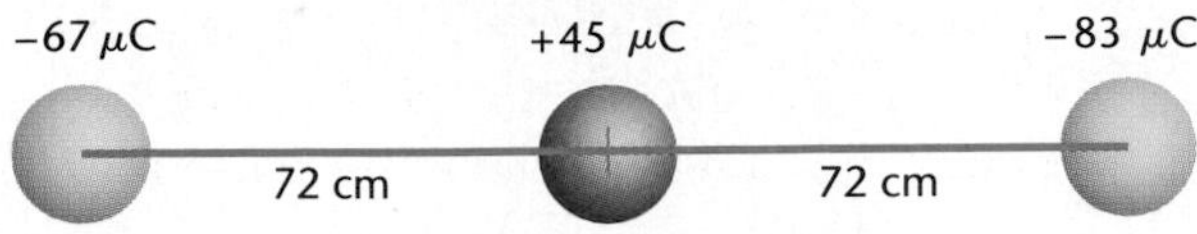

FIGURE 20–13

Critical Thinking Problems

32. Three charged spheres are located at the positions shown in **Figure 20–14.** Find the total force on sphere B.

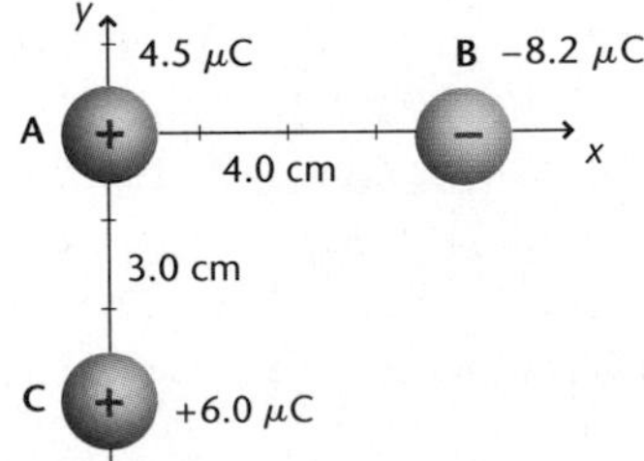

FIGURE 20–14

33. Two charges, q_A and q_B, are at rest near a positive test charge, q_T, of 7.2 μC. The first charge, q_A, is a positive charge of 3.6 μC, located 2.5 cm away from q_T at 35°; q_B is a negative charge of $-6.6\ \mu C$, located 6.8 cm away at 125°.

a. Determine the magnitude of each of the forces acting on q_T.

b. Sketch a force diagram.

c. Graphically determine the resultant force acting on q_T.

34. The two pith balls in **Figure 20–15** each have a mass of 1.0 g and equal charge. One pith ball is suspended by an insulating thread. The other is brought to 3.0 cm from the suspended ball. The suspended ball is now hanging with the thread forming an angle of 30.0° with the vertical. The ball is in equilibrium with F_E, F_g, and F_T. Calculate each of the following.

a. F_g

b. F_E

c. the charge on the balls

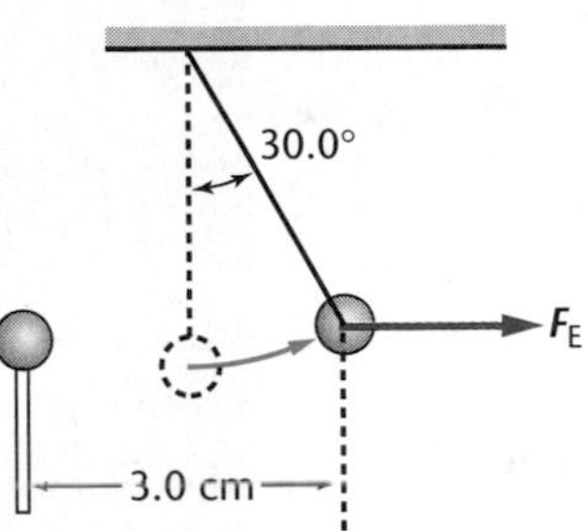

FIGURE 20–15

Going Further

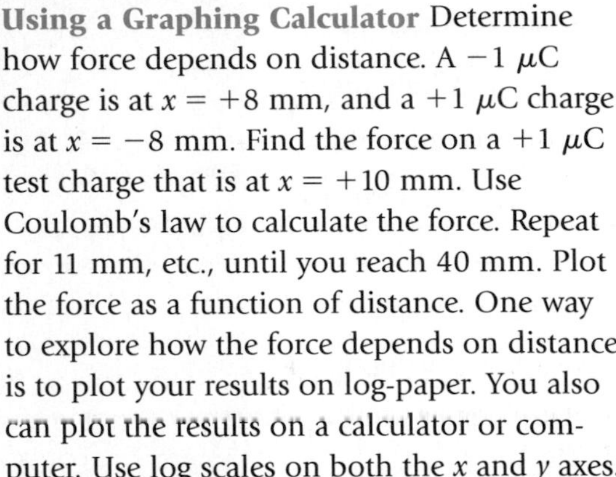

Using a Graphing Calculator Determine how force depends on distance. A $-1\ \mu C$ charge is at $x = +8$ mm, and a $+1\ \mu C$ charge is at $x = -8$ mm. Find the force on a $+1\ \mu C$ test charge that is at $x = +10$ mm. Use Coulomb's law to calculate the force. Repeat for 11 mm, etc., until you reach 40 mm. Plot the force as a function of distance. One way to explore how the force depends on distance is to plot your results on log-paper. You also can plot the results on a calculator or computer. Use log scales on both the x and y axes.

The slope of the line, m, indicates the power to which x is raised. For example, if x changes by a factor of 10 and y changes by a factor of 1/100, then the slope is $m = -2$, and the force is proportional to x^{-2}. If the slope is not constant, then you cannot write an equation for the force as x^{m}. Over what range of distances is the slope constant? What is its slope?

interNET CONNECTION

Follow the link on the Glencoe Homepage at **www.glencoe.com/sec/science** to find out more about this chapter.

High-Energy Halo

Research into the transmission of ultrahigh-voltage electricity measuring up to two million volts is essential for the development of future power technology. High-voltage air core resonant transformers can generate lightning-like discharges called coronas. Why can a corona be seen around a transformer or a power line when the effect is not seen in the electric wires in our homes?

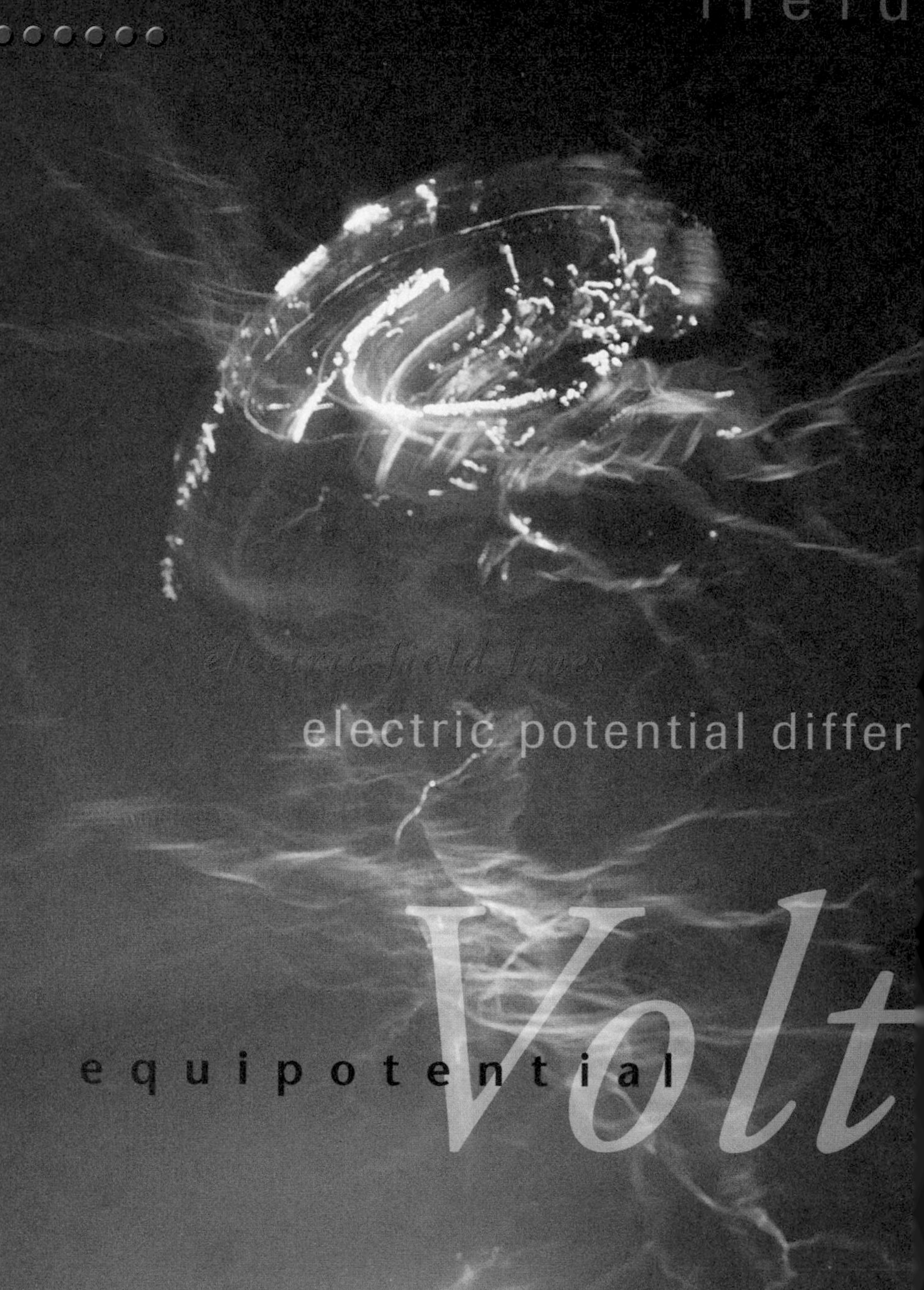

CHAPTER

21 Electric Fields

The blue glow in the photo is a modern version of St. Elmo's Fire. Sailors in the time of Columbus saw these ghostly streamers issuing from their ships' high masts. They recognized them as warning signs of an approaching lightning storm. The glow is indeed related to lightning. The "fire" in this photo is a corona discharge. Besides causing the eerie glow, it produces radio interference and can initiate sparking from the wire to another conductor, so designers of these experimental power lines try to reduce or eliminate it entirely.

Electric fields exist around any object that carries a charge. The potential difference between two charged objects allows you to transfer energy in the form of electricity between those objects. This, in turn, provides the power you use on a daily basis whether you plug an electrical cord into an outlet in your home or use a battery in a portable device.

WHAT YOU'LL LEARN

- You will distinguish between electric force and electric fields.
- You will understand how grounding is related to charge sharing.
- You will recognize the relationship between conductor shape and electric field strength.

WHY IT'S IMPORTANT

- You use electric power every day. Almost every appliance and many of the tools you use run on electricity. As the demand for electricity increases, you need to understand this powerful source of energy.

interNET CONNECTION

Follow the link for this chapter on the Glencoe Homepage at **www.glencoe.com/sec/science** to find out more about electric fields.

21.1 Creating and Measuring Electric Fields

OBJECTIVES

- **Define** and **measure** an electric field.
- **Solve** problems relating to charge, electric fields, and forces.
- **Diagram** electric field lines.

Charge and Field Conventions

- Positive charges are **red.**
- Negative charges are **blue.**
- Electric field lines are **indigo.**

Electric force, like the gravitational force you studied in Chapter 8, varies inversely as the square of the distance between two point objects. Both forces can act at a great distance. How can a force be exerted across what seems to be empty space? In trying to understand electric force, Michael Faraday developed the concept of an electric field. According to Faraday, a charge creates an electric field about it in all directions. If a second charge is placed at some point in the field, the second charge interacts with the field at that point. The resulting force is the result of a local interaction.

The Electric Field

The **electric field** is a vector quantity that relates the force exerted on a test charge to the size of the test charge. How does this work? An electric charge, q, produces an electric field that you can measure. This is shown in **Figure 21–1.** First, measure the field at a specific location. Call this point A. An electrical field can be observed only because it produces forces on other charges, so you must place a small positive test charge, q', at A. Then, measure the force exerted on the test charge, q', at this location.

According to Coulomb's law, the force is proportional to the test charge. If the size of the test charge is doubled, the force is doubled. Thus, the ratio of force to test charge is independent of the size of the test charge. If you divide the force, $\boldsymbol{F}$, on the test charge, measured at point A, by the size of the test charge, q', you obtain a vector quantity, $\boldsymbol{F}/q'$. This quantity does not depend on the test charge, only on the charge q and the location of point A. The electric field at point A, the location of q', is represented by the following equation.

$$\boldsymbol{E} = \frac{\boldsymbol{F}_{\text{on } q'}}{q'}$$

The direction of the electric field is the direction of the force on the positive test charge. The magnitude of the electric field is measured in newtons per coulomb, N/C.

A picture of an electric field can be made by using arrows to represent the field vectors at various locations, as shown in **Figure 21–1.** The length of the arrow will be used to show the strength of the field. The direction of the arrow shows the field direction. To find the field from two charges, the fields from the individual charges are added vectorially. A test charge can be used to map out the field resulting from any collection of charges. Typical electric fields produced by charge collections are shown in **Table 21-1.**

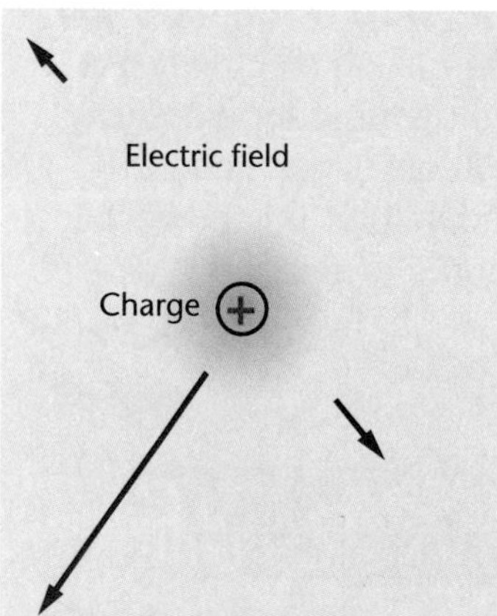

FIGURE 21–1 Arrows can be used to represent the magnitude and direction of the electric field about an electric charge at various locations.

TABLE 21-1	
Approximate Values of Typical Electric Fields	
Field	**Value (N/C)**
Nearby a charged hard rubber rod	1×10^3
In a television picture tube	1×10^5
Needed to create a spark in air	3×10^6
At an electron orbit in hydrogen atom	5×10^{11}

F.Y.I.

Robert Van de Graaff devised the high-voltage electrostatic generator in the 1930s. These generators can build up giant potentials that can accelerate particles to high energy levels.

An electric field should be measured only by a small test charge. Why? The test charge also exerts a force on q. It is important that the force exerted by the test charge doesn't move q to another location, and thus change the force on q' and the electric field being measured. A small test charge cannot move q.

Example Problem

Calculating an Electric Field

An electric field is to be measured using a positive test charge of 4.0×10^{-5} C. This test charge experiences a force of 0.60 N acting at an angle of 10°. What is the magnitude and direction of the electric field at the location of the test charge?

Sketch the Problem

- Draw the force vector on the test charge, q', on a two-dimensional grid at an angle of 10°.

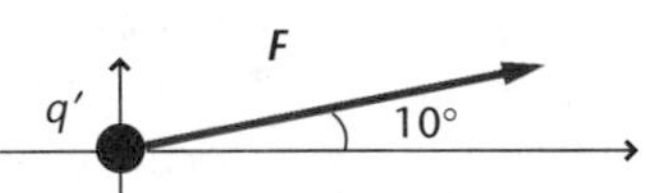

Calculate Your Answer

Known:

$q' = +4.0 \times 10^{-5}$ C

$\mathbf{F} = 0.60$ N at 10°

Unknown:

$\mathbf{E} = ?$ at 10°

Strategy:

Use the relationship $E = \frac{F}{q'}$

Calculations:

$$E = \frac{F}{q'} = \frac{0.60 \text{ N}}{4.0 \times 10^{-5} \text{ C}}$$

$\mathbf{E} = 1.5 \times 10^4$ N/C at 10°

Check Your Answer

- Are the units correct? The electric field is correctly measured in newtons per coulomb.
- Are the signs correct? The field direction is in the direction of the force because the test charge is positive, so **E** is positive.
- Is the magnitude correct? Electric fields can have values around 10^5 N/C as shown in **Table 21-1**.

Practice Problems

1. A negative charge of 2.0×10^{-8} C experiences a force of 0.060 N to the right in an electric field. What are the field magnitude and direction?
2. A positive test charge of 5.0×10^{-4} C is in an electric field that exerts a force of 2.5×10^{-4} N on it. What is the magnitude of the electric field at the location of the test charge?
3. Suppose the electric field in problem 2 was caused by a point charge. The test charge is moved to a distance twice as far from the charge. What is the magnitude of the force that the field exerts on the test charge now?
4. You are probing the field of a charge of unknown magnitude and sign. You first map the field with a 1.0×10^{-6} C test charge, then repeat your work with a 2.0×10^{-6} C test charge.
   a. Would you measure the same forces with the two test charges? Explain.
   b. Would you find the same fields? Explain.

Electric Fields

How does the electric field around a charged piece of plastic foam vary in strength and direction? Try this activity to find out. Tie a pith ball on the end of a 20-cm nylon thread and tie the other end to a plastic straw. When you hold the straw horizontally, notice that the ball hangs straight down on the thread. Now rub a piece of wool on a 30 cm × 30 cm square of plastic foam to charge both objects. Stand the foam in a vertical orientation. Hold the straw and touch the pith ball to the wool, then slowly bring the hanging ball towards the charged plastic foam. Move the pith ball to different regions and notice the angle of the thread.

Analyze and Conclude Why did the ball swing toward the charged plastic? Explain in terms of the electric field. Did the angle of the thread change? Why? Does the angle of the thread indicate the direction of the electric field? Explain.

So far you have measured the field at a single point. Now, imagine moving the test charge to another location. Measure the force on it again and calculate the electric field. Repeat this process again and again until you assign every location in space a measurement of the vector quantity of the electric field associated with it. The field is present even if there is no test charge to measure it. Any charge placed in an electric field experiences a force on it resulting from the electric field at that location. The strength of the force depends on the magnitude of the field, $\mathbf{E}$, and the size of the charge, q. Thus, $\mathbf{F} = \mathbf{E}q$. The direction of the force depends on the direction of the field and the sign of the charge.

Picturing the Electric Field

An alternative picture of an electric field is shown in **Figure 21–2.** The lines are called **electric field lines.** The direction of the field at any point is the tangent drawn to the field line at that point. The strength of the electric field is indicated by the spacing between the lines. The field is strong where the lines are close together. It is weaker where the lines are spaced farther apart. Although only two-dimensional models can be shown here, remember that electric fields exist in three dimensions.

The direction of the force on a positive test charge near another positive charge is away from the other charge. Thus, the field lines extend radially outward like the spokes of a wheel, as shown in **Figure 21–2a.** Near a negative charge, the direction of the force on the positive test charge is toward the negative charge, so the field lines point radially inward, as shown in **Figure 21–2b.** When there are two or more charges,

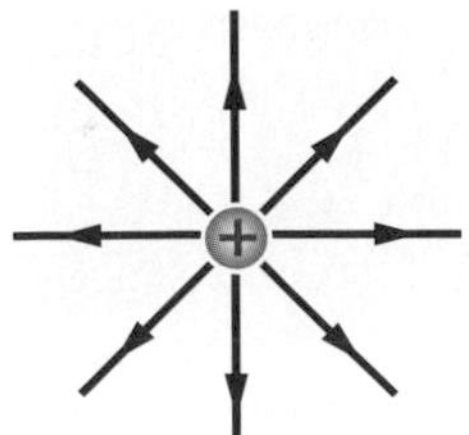

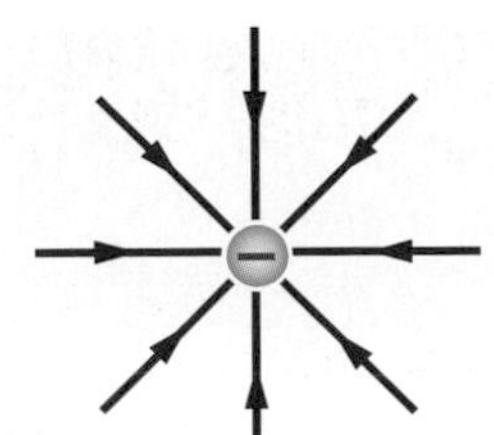

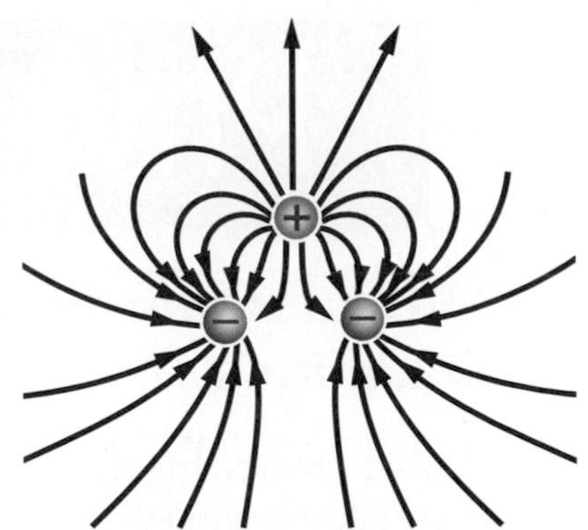

FIGURE 21–2 Lines of force are drawn perpendicularly away from the positively-charged object **(a)** and perpendicularly into the negatively-charged object **(b)**. Electric field lines between like charged and oppositely charged objects are shown in **(c)**.

the field is the vector sum of the fields resulting from the individual charges. The field lines become curved and the pattern is more complex, as shown in **Figure 21–2c.** Note that field lines always leave a positive charge and enter a negative charge.

The Van de Graaff machine is a device that transfers large amounts of charge from one part of the machine to the top metal terminal, as shown in **Figure 21–3a.** A person touching the terminal is charged electrically. The charges on the person's hairs repel each other, causing the hairs to follow the field lines. This transfer of charge is diagrammed in **Figure 21–3b.** Charge is transferred onto a moving belt at the base of the generator, Position A, and is transferred off the belt at the metal dome at the top, Position B. An electric motor does the work needed to increase the electric potential energy. Another method of visualizing field lines is to use grass seed in an insulating liquid such as mineral oil. The electric forces cause a separation of charge in each long, thin grass seed. The seeds then turn so that they line up along the direction of the electric field. The seeds thus form a pattern of the electric field lines. The patterns in **Figure 21–4** were made this way.

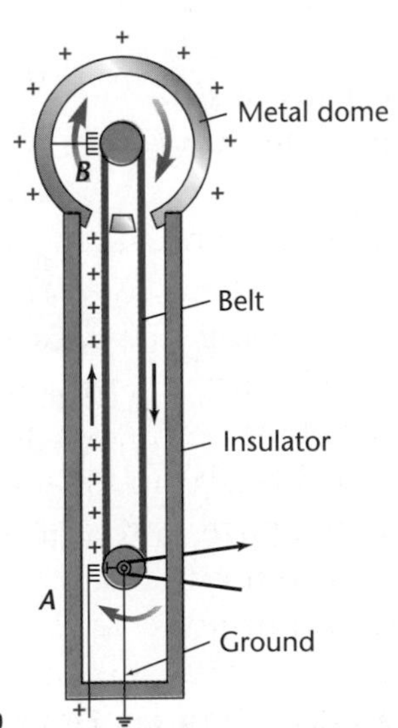

FIGURE 21–3 When a person touches a Van de Graaff generator, the results can be dramatic **(a).** In the Van de Graaff generator, charge is transferred onto a moving belt at A, and from the belt to the metal dome at B. An electric motor does the work needed to increase the electric potential energy **(b).**

Physics & Society

Computers for People with Disabilities

Studying physics, math, and other sciences can be difficult enough without the additional challenges presented by physical disabilities. People with impaired vision, hearing, or mobility require assistance that can be provided through the use of computers.

Computer technology that translates written text into Braille or speech provides people who are blind with ready access to textbooks and scientific journals without the need for another person to read the material to them. Speech also can be turned into written text by voice recognition software. Electronic readouts from laboratory instruments can be fed into computers programmed to produce Braille or speech. Also, devices have been developed to create textured charts and graphs meant to be felt with the fingers rather than viewed with the eyes.

People with hearing loss depend heavily on printed materials and on computerized translations and informational sources, including the Internet. Speech synthesis software enables those who cannot speak to participate in verbal discussions. Computer programs that teach sign language help facilitate communication between those who can hear and those who cannot.

Both speech synthesis and voice recognition software are valuable for people with mobility impairment, as are devices that replace or augment the conventional keyboard and mouse. Those who do not have use of their hands and arms can use mouth sticks or head sticks to press keys and to move the cursor around the screen by manipulating track balls or glide pads.

Another aid for people with mobility limitations is the sip-and-puff switch which is placed in the mouth. The user's inhalations (sips) and exhalations (puffs) create the dots and dashes of letters in Morse code, which are then translated into text or speech by a computer. Word-prediction software that completes words based on the first few letters helps to reduce the number of keystrokes required by a typist who is disabled.

Investigating the Issue

1. **Researching Information** Research the life and career of Stephen Hawking to find out more about his scientific contributions, his physical challenges, and the technologies that have helped him overcome some of those challenges.
2. **Debating the Issue** Computer aids for people with disabilities are expensive, and they often have to be paid for by the schools or employers who train or employ these people. Find out what kinds of tools are used by students and working adults with disabilities in your community, and how those tools are made available to them. Who decides how much money should be spent on these items? How do you think such decisions should be handled?

interNET CONNECTION

Follow the link on the Glencoe Homepage at **www.glencoe.com/sec/science** to find out more about computer applications for people with disabilities.

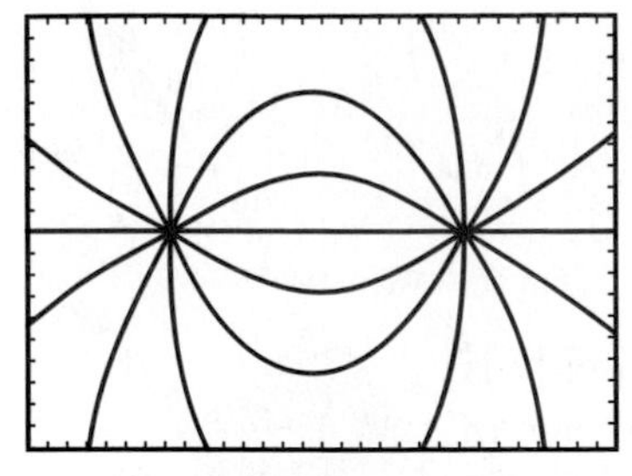

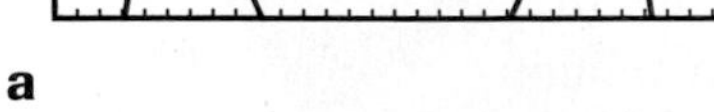

a

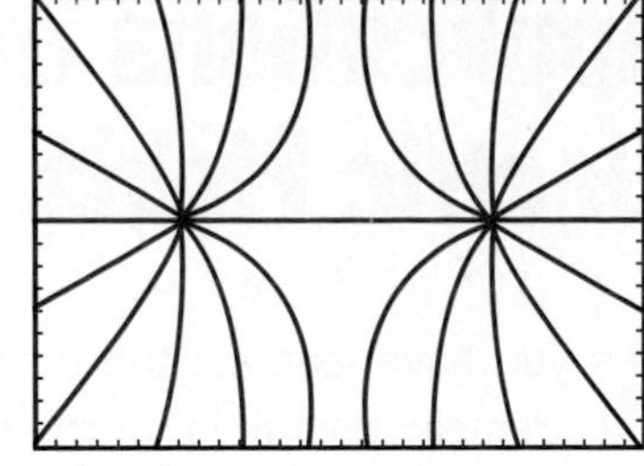

b

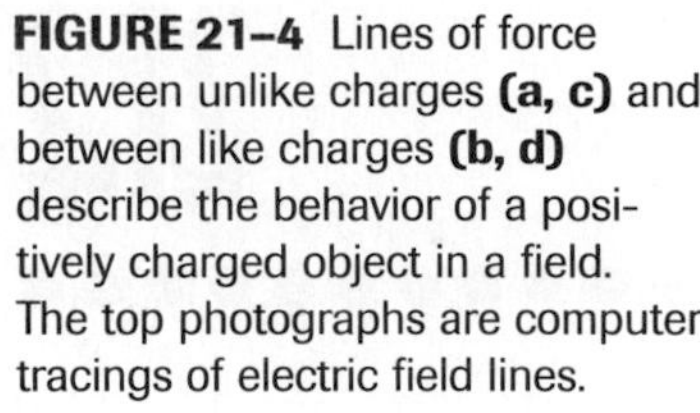

FIGURE 21–4 Lines of force between unlike charges **(a, c)** and between like charges **(b, d)** describe the behavior of a positively charged object in a field. The top photographs are computer tracings of electric field lines.

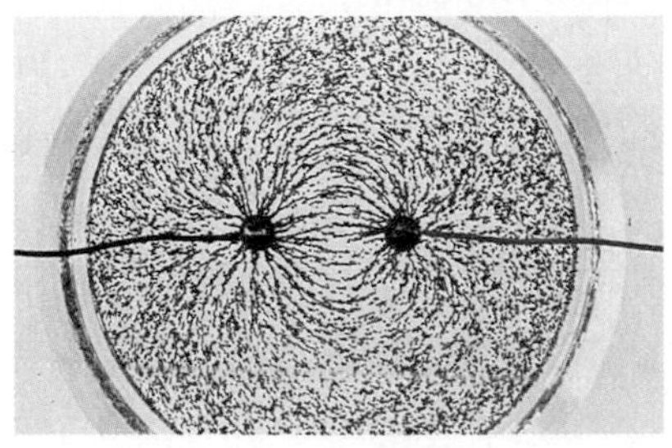

c

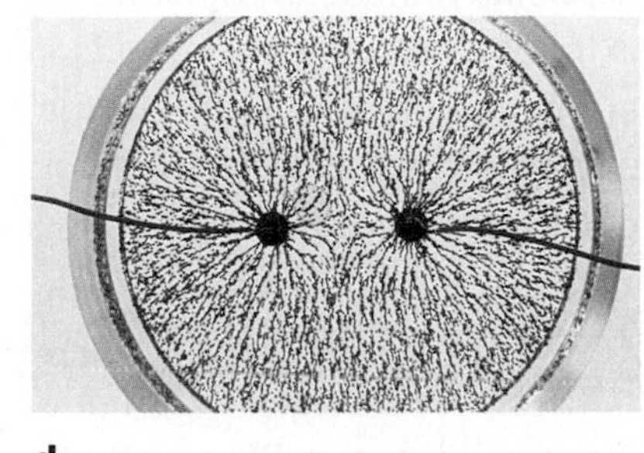

d

Field lines do not really exist. They are simply a means of providing a model of an electric field. Electric fields, on the other hand, do exist. They provide a method of calculating the force on a charged body, but do not explain, however, why charged bodies exert forces on each other.

21.1 Section Review

1.1 Suppose you are asked to measure the electric field in space. Answer the following questions about this step-by-step procedure.

- **a.** How do you detect the field at a point?
- **b.** How do you determine the magnitude of the field?
- **c.** How do you choose the size of the test charge?
- **d.** What do you do next?

1.2 Suppose you are given an electric field, but the charges that produce the field are hidden. If all the field lines point into the hidden region, what can you say about the sign of the charge in that region?

1.3 How does the electric field, ***E***, differ from the force, ***F***, on the test charge?

1.4 Critical Thinking **Figure 21–4b** shows the field from two like charges. The top positive charge in **Figure 21–2c** could be considered a test charge measuring the field resulting from the two negative charges. Is this positive charge small enough to produce an accurate measure of the field? Explain.

21.2 Applications of Electric Fields

OBJECTIVES

- **Define** and **calculate** electric potential difference.
- **Explain** how Millikan used electric fields to find the charge of the electron.
- **Determine** where charges reside on solid and hollow conductors.
- **Describe** capacitance and **solve** capacitor problems.

As you have learned, the concept of energy is extremely useful in mechanics. The law of conservation of energy allows us to solve motion problems without knowing the forces in detail. The same is true in the study of electrical interactions. The work performed moving a charged particle in an electric field can result in the particle gaining either potential or kinetic energy, or both. Because this chapter investigates charges at rest, only changes in potential energy will be discussed.

Energy and the Electric Potential

Recall the change in gravitational potential energy of a ball when it is lifted, as shown in **Figure 21–5.** Both the gravitational force, $\boldsymbol{F}$, and the gravitational field, $\boldsymbol{g} = \boldsymbol{F}/m$, point toward Earth. If you lift a ball against the force of gravity, you do work on it, increasing its potential energy.

The situation is similar with two unlike charges: they attract each other, and so you must do work to pull one charge away from the other. When you do the work, you store it as potential energy. The larger the test charge, the greater the increase in its potential energy, ΔU_e.

The force on the test charge depends on its magnitude, but it is convenient to define a quantity, the electric field, that does not depend on the magnitude of the test charge. Recall from Section 21.1 that $\boldsymbol{E} = \boldsymbol{F}/q'$, where q' is the magnitude of the test charge. The electric field is then the force per unit charge. In a similar way, the **electric potential difference,** ΔV, is defined as the work done moving a test charge in an electric field divided by the magnitude of the test charge.

$$\Delta V = \frac{W_{\text{on } q'}}{q'}$$

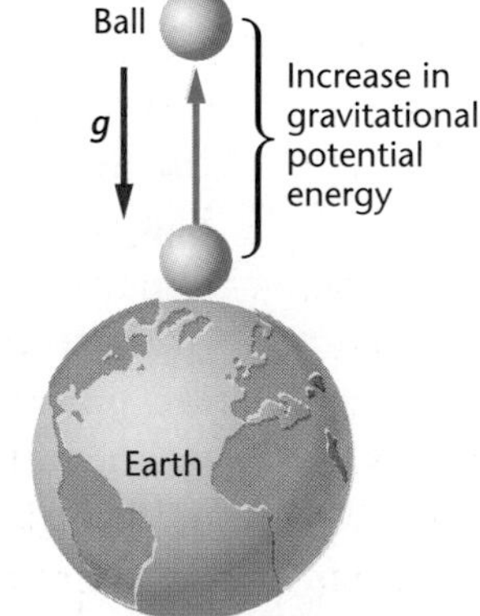

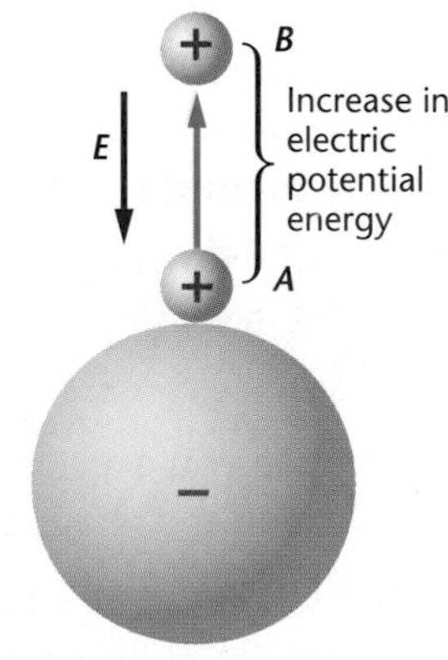

FIGURE 21–5 Work is needed to move an object against the force of gravity **(a)** and against the electric force **(b).** In both cases, the potential energy of the object is increased.

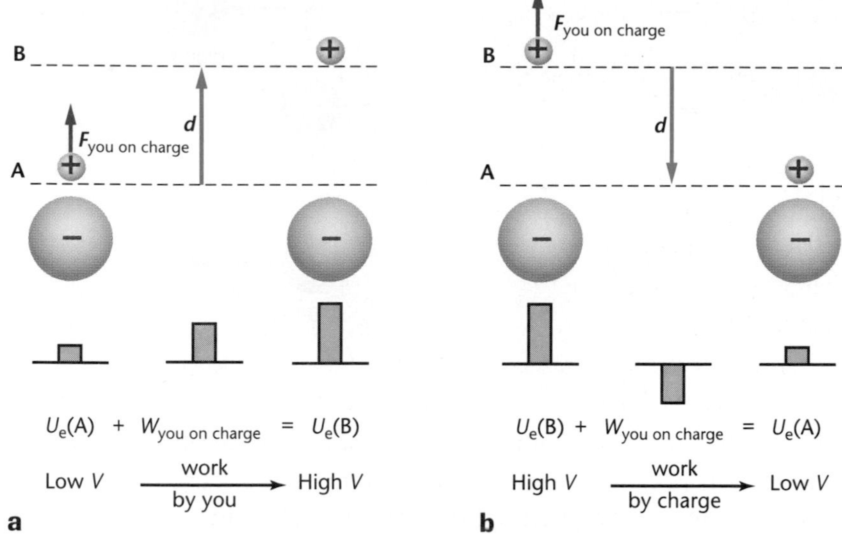

FIGURE 21–6 Electric potential difference is determined by measuring the force per unit charge. If you move the charges apart, you increase the electric potential difference **(a).** If you move the charges closer together, you reduce the electric potential difference **(b).**

Electric potential difference is measured in joules per coulomb. One joule per coulomb is called a **volt** (J/C = V).

Consider the situation shown in **Figure 21–6.** The negative charge creates an electric field, ***E,*** around itself. Suppose you placed a small positive test charge, q', in the field at position A. It will experience a force in the direction of the field. If you now move the test charge away from the negative charge to position B, you will have to exert a force, $\boldsymbol{F}_{\text{by you}}$, on the charge. Because the force you exert is in the same direction as the displacement, the work you do on the test charge is positive. Therefore there will also be a positive change in the electric potential difference. **Figure 21–6a** shows that the potential energy is raised by the amount of work done so, $U_e(A) + W_{\text{by you}} = U_e(B)$. The change does not depend on the size of the test charge. It only depends on the field and the displacement.

Suppose you now move the test charge back to position A from position B. The force you exert is now in the direction opposite the displacement, so the work you do is negative. The electric potential difference is also negative. In fact, it is equal and opposite to the potential difference for the move from position A to position B. **Figure 21–6b** shows that the potential energy is changed again by the amount of work you did, so $U_e(B) + W_{\text{by you}} = U_e(A)$. The electric potential difference does not depend on the path used to go from one position to another. It does depend on the two positions.

Is there always an electric potential difference between the two positions? Suppose you moved the test charge in a circle around the negative charge. The force the electric field exerts on the test charge is always perpendicular to the direction you moved it, so you do no work. Therefore, the electric potential difference is zero. Whenever the electric potential difference between two or more positions is zero, those positions are said to be at **equipotential.**

HELP WANTED

VENDING MACHINE REPAIRERS

If you have a knack for fixing things, and are able to learn quickly from our in-house training on specific equipment, we would like to hear from you. You must be able to read electrical diagrams and manuals and use sophisticated test equipment. Vocational school training is a plus, but experience will be considered. We need reliable, bondable, productive people who are willing to work when needed and are interested in advancement. For information contact:

National Automatic Merchandising Association
20 North Wacker Drive
Chicago, IL 60606

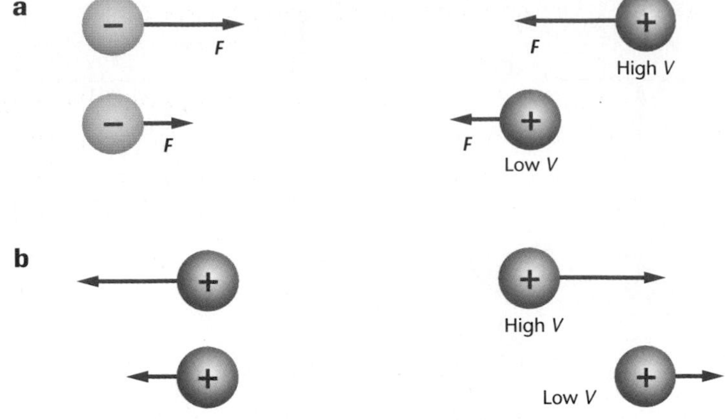

FIGURE 21–7 Electric potential is smaller when two unlike charges are closer together **(a)** and larger when two like charges are closer together **(b).**

F.Y.I.

Some manufacturers of computers recommend that you keep computers plugged in to maintain the ground even when they are not in use. It is also a good idea to never remove the grounding pin on electrical equipment.

Only differences in potential energy can be measured. The same is true of electric potential. Thus, only differences in electric potential are important. The electric potential difference from point A to point B is defined as $\Delta V = V_B - V_A$. Electric potential differences are measured with a voltmeter. Sometimes the electric potential difference is simply called the voltage. Do not confuse the electric potential difference, ΔV, with the volts, V.

As you learned in Chapter 11, the potential energy of a system can be defined as zero at any reference point. In the same way, the electric potential of any point can be defined as zero. Choose a point and label it point A. If $V_A = 0$, then $\Delta V = V_B$. If instead, $V_B = 0$, then $\Delta V = -V_A$. No matter what reference point is chosen, the value of the electric potential difference from point A to point B will always be the same.

You have seen that electric potential difference increases as a positive test charge is separated from a negative charge. What happens when a positive test charge is separated from a positive charge? There is a repulsive force between these two charges. Potential energy decreases as the two charges are moved farther apart. Therefore, the electric potential is smaller at points farther from the positive charge, as shown in **Figure 21–7.**

The Electric Potential in a Uniform Field

A uniform electric force and field can be made by placing two large, flat conducting plates parallel to each other. One is charged positively and the other negatively. The electric field between the plates is constant except at the edges of the plates. Its direction is from the positive to the negative plate. The grass seeds pictured in **Figure 21–8** represent the electric field between parallel plates.

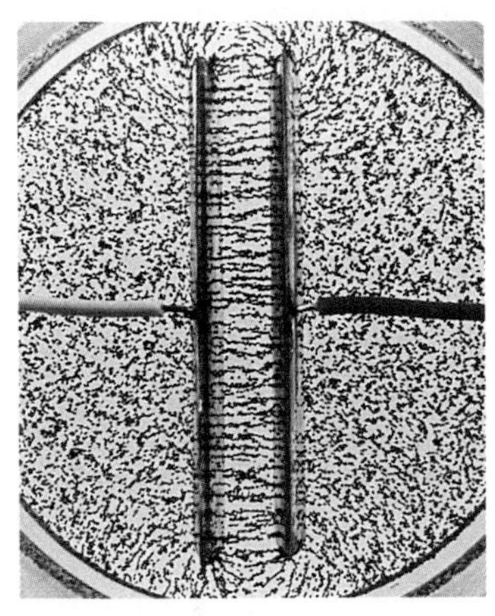

FIGURE 21–8 A representation of an electric field between parallel plates is shown.

If a positive test charge, q', is moved a distance d, in the direction opposite the electric field direction, the work done is found by the relationship $W_{\text{on } q'} = Fd$. Thus, the electric potential difference, the work done per unit charge, is $\Delta V = Fd/q' = (F/q')d$. Now, the electric field intensity is the force per unit charge, $E = F/q'$. Therefore, the electric potential difference, ΔV, between two points a distance d apart in a uniform field, **E,** is represented by the following equation.

$$\Delta V = Ed$$

The electric potential increases in the direction opposite the electric field direction. That is, the electric potential is higher near the positively charged plate. By dimensional analysis, the product of the units of E and d is (N/C)·(m). This is equivalent to one J/C, the definition of one volt.

Math Handbook

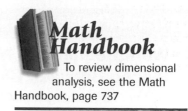

To review dimensional analysis, see the Math Handbook, page 737

Example Problem

Electric Field Between Two Parallel Plates

Two parallel plates are given opposite charges. A voltmeter measures the electric potential difference to be 60.0 V. The plates are 3.0 cm apart. What is the magnitude of the electric field between them?

Sketch the Problem

- Draw two parallel plates separated by 3.0 cm.
- Identify the lower plate as positively charged and the top plate as negatively charged.
- Indicate that the electric potential difference is located between the two plates.

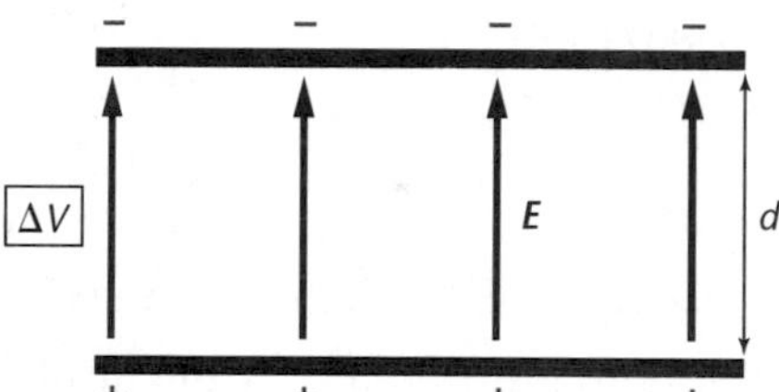

Calculate Your Answer

Known:

$\Delta V = 60.0$ V

$d = 0.030$ m

Unknown:

$E = ?$

Strategy:

For the uniform field between charged parallel plates, $\Delta V = Ed$

Therefore, $E = \frac{\Delta V}{d}$

Calculations:

$$E = \frac{\Delta V}{d}$$

$$= \frac{60.0 \text{ V}}{0.030 \text{ m}}$$

$$= 2.0 \times 10^3 \text{ J/C·m}$$

$$= 2.0 \times 10^3 \text{ N/C}$$

The lower plate is at a higher voltage, so the electric field points upward.

Check Your Answer

- Are the units correct? Values are expressed in newtons per coulomb.
- Are the signs correct? The electric field is represented by the electric potential difference divided by the plate separation. The potential is higher near the positively charged plate, and the electric field points away from the positively charged plate.
- Is the magnitude realistic? With ΔV of a few volts and a short distance, high electric field values are expected.

Example Problem

Work Required to Move a Proton Placed Between Charged Parallel Plates

Two large, charged parallel plates are 4.0 cm apart. The magnitude of the electric field between the plates is 625 N/C.

a. What is the electric potential difference between the plates?

b. What work will you do to move a charge equal to that of one proton from the negative to the positive plate?

Sketch the Problem

- Draw two parallel plates separated by 4.0 cm.
- Identify the lower plate as positively charged and the upper plate as negatively charged.
- Place a proton in the electric field.

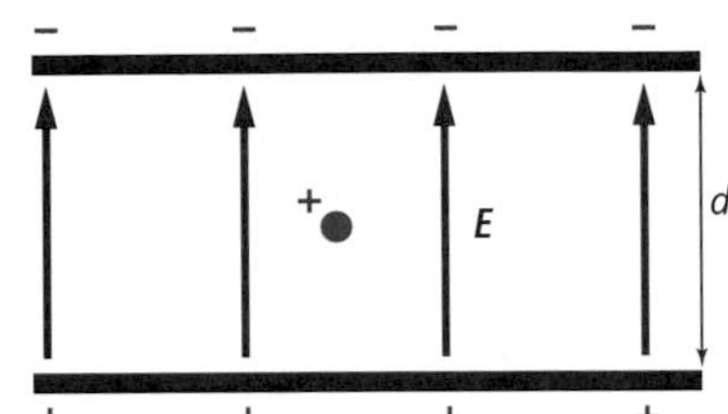

Calculate Your Answer

Known:

$E = 625$ N/C

$d = 0.040$ m

$q = 1.60 \times 10^{-19}$ C

Unknown:

$\Delta V = ?$

$W_{\text{by you}} = ?$

Strategy:

a. Use $\Delta V = Ed$ to determine the potential difference in the uniform field between parallel plates.

Calculations:

$$\begin{aligned}\Delta V &= Ed \\ &= (625\ \text{N/C})(0.040\ \text{m}) \\ &= 25\ \text{Nm/C} \\ &= 25\ \text{J/C} \\ &= 25\ \text{V}\end{aligned}$$

b. $W = q\Delta V$ to determine the work.

$$\begin{aligned}W_{\text{by you}} &= q\Delta V \\ &= (1.6 \times 10^{-19}\ \text{C})(25\ \text{J/C}) \\ &= 4.0 \times 10^{-18}\ \text{J}\end{aligned}$$

Check Your Answer

- Are the units correct? Work is in joules and potential is in volts.
- Are the signs correct? You do work to move a positive charge toward a positively charge plate, so sign of the electric potential difference is positive.
- Is the magnitude realistic? With a ΔV of a few volts and a small charge, the amount of work performed will be small.

Practice Problems

5. The electric field intensity between two large, charged, parallel metal plates is 8000 N/C. The plates are 0.05 m apart. What is the electric potential difference between them?
6. A voltmeter reads 500 V across two charged, parallel plates that are 0.020 m apart. What is the electric field between them?
7. What electric potential difference is applied to two metal plates 0.500 m apart if the electric field between them is 2.50×10^3 N/C?
8. What work is done when 5.0 C is moved through an electric potential difference of 1.5 V?

BIOLOGY CONNECTION

The electric eel shocks prey with groups of highly compacted nerve endings in its tail. The larger the eel, the larger the nerve-ending cells, and the stronger the voltage. A nine-foot eel can emit 650 volts, enough to stun a person or a large animal. After many discharges, eels must rest to build up voltage. They inhabit the freshwaters of South America.

Millikan's Oil-Drop Experiment

One important application of the uniform electric field between two parallel plates is the measurement of the charge of an electron. This was first determined by American physicist Robert A. Millikan in 1909. **Figure 21–9** shows the method used by Millikan to measure the charge carried by a single electron. Fine oil drops were sprayed from an atomizer into the air. These drops were charged by friction with the atomizer as they were sprayed. Gravity acting on the drops caused them to fall. A few entered the hole in the top plate of the apparatus. An electric potential difference was placed across the two plates. The resulting electric field between the plates exerted a force on the charged drops. When the top plate was made positive enough, the electric force caused negatively charged drops to rise. The electric potential difference between the plates was adjusted to suspend a charged drop between the plates. At this point, the downward force of Earth's gravitational field and the upward force of the electric field were equal in magnitude.

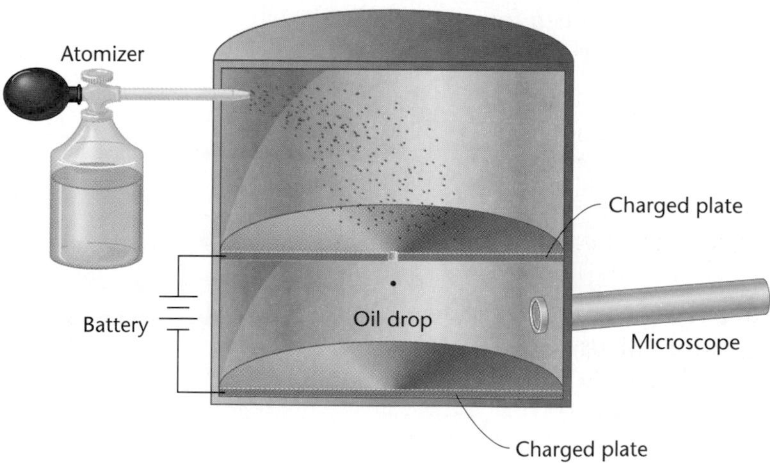

FIGURE 21–9 This illustration shows a cross-sectional view of the apparatus Millikan used to determine the charge on an electron.

F.Y.I.

Even if the inner surface of an object is pitted or bumpy, giving it a larger surface area than the outer surface, the charge still will be entirely on the outside.

The magnitude of the electric field, $\mathbf{E}$, was determined from the electric potential difference between the plates. A second measurement had to be made to find the weight of the drop using the relationship mg, which was too tiny to measure by ordinary methods. To make this measurement, a drop was first suspended. Then the electric field was turned off and the rate of the fall of the drop was measured. Because of friction with the air molecules, the oil drop quickly reached terminal velocity. This velocity was related to the mass of the drop by a complex equation. Using the measured terminal velocity to calculate mg, and knowing $\mathbf{E}$, the charge q could be calculated.

Millikan found that there was a great deal of variation in the charges of the drops. When he used X rays to ionize the air and add or remove electrons from the drops, he noted, however, that the changes in the charge were always a multiple of 1.60×10^{-19} C. The changes were caused by one or more electrons being added to or removed from the drops. He concluded that the smallest change in charge that could occur was the amount of charge of one electron. Therefore, Millikan proposed that each electron always carried the same charge, 1.60×10^{-19} C.

Millikan's experiment showed that charge is quantized. This means that an object can have only a charge with a magnitude that is some integral multiple of the charge of the electron.

The presently accepted theory of matter states that protons are made up of fundamental particles called quarks. The charge on a quark is either $+1/3$ or $-2/3$ the charge on an electron. One theory of quarks states that quarks can never be isolated. Many experimenters have used an updated Millikan apparatus to look for fractional charges on drops or tiny metal spheres. There have been no reproducible discoveries of fractional charges.

Example Problem

Finding the Charge on an Oil Drop

In a Millikan oil drop experiment, a drop has been found to weigh 1.9×10^{-14} N. When the electric field is 4.0×10^{4} N/C, the drop is suspended motionless.

a. What is the charge on the oil drop?

b. If the upper plate is positive, how many excess electrons does the oil drop have?

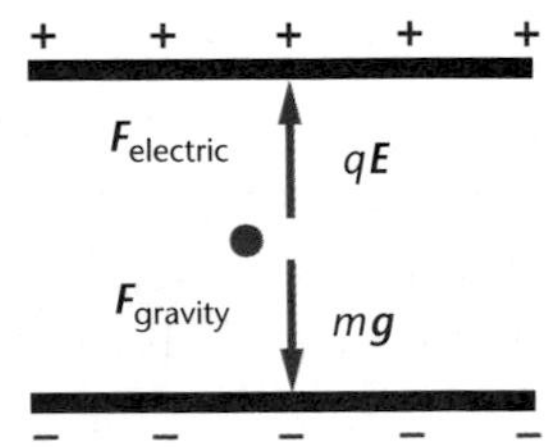

Sketch the Problem

- Draw two plates with the positive charge on the top plate and the negative charge on the bottom plate.
- Suspend an oil drop in the field and identify the gravitational force downward and the electric force upward.
- The oil drop will be motionless.

Calculate Your Answer

Known:

$mg = 1.9 \times 10^{-14}$ N

$E = 4.0 \times 10^4$ N/C

$e = 1.60 \times 10^{-19}$ C/electron

Unknown:

charge on drop $q = ?$

number of electrons, $n = ?$

Strategy:

a. When balanced, $F_{\text{electric}} = F_{\text{gravity}}$.
Then, $qE = mg$ so use $q = mg/E$

b. Determine the number of electrons by $n = q/e$

Calculations:

$$q = \frac{mg}{E}$$

$$q = \frac{1.9 \times 10^{-14}\ \text{N}}{4.0 \times 10^4\ \text{N/C}} = 4.8 \times 10^{-19}\ \text{C}$$

$$n = \frac{q}{e} = \frac{4.8 \times 10^{-19}\ \text{C}}{1.60 \times 10^{-19}\ \text{C/electron}}$$

$$= 3\ \text{electrons}$$

Check Your Answer

- Are the units correct? Charge is measured in coulombs.
- Are the signs correct? Charge is the product of the magnitude of the charge on one electron multiplied by the number of electrons. This is a positive number.
- Is the magnitude realistic? There is an excess of electrons, because the drop is attracted to the positively charged plate.

Practice Problems

9. A drop is falling in a Millikan oil-drop apparatus when the electric field is off.
   a. What are the forces acting on the oil drop, regardless of its acceleration?
   b. If the drop is falling at constant velocity, what can be said about the forces acting on it?
10. An oil drop weighs 1.9×10^{-15} N. It is suspended in an electric field of 6.0×10^3 N/C.
    a. What is the charge on the drop?
    b. How many excess electrons does it carry?
11. A positively charged oil drop weighs 6.4×10^{-13} N. An electric field of 4.0×10^6 N/C suspends the drop.
    a. What is the charge on the drop?
    b. How many electrons is the drop missing?
12. If three more electrons were removed from the drop in problem 11, what field would be needed to balance the drop?

Physics Lab

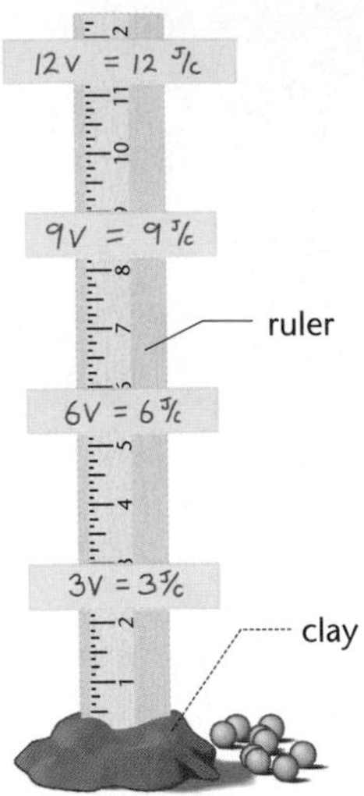

Charges, Energy, and Voltage

Problem

How can you make a model that demonstrates the relationship of charge, energy, and voltage?

Materials 

ball of clay
ruler
transparent tape
12 steel balls, 3-mm diameter
paper

Procedure

1. Use the clay to support the ruler vertically on the tabletop. The 0 end should be at the table.
2. Cut a 2 cm × 8 cm rectangular piece of paper and write on it "3 V = 3 J/C".
3. Cut three more rectangles and label them: 6 V = 6 J/C, 9 V = 9 J/C, and 12 V = 12 J/C.
4. Tape the 3-V rectangle to the 3" mark on the ruler, the 6-V to the 6" mark, and so on.
5. Let each steel ball represent 1 C of charge.
6. Lift and tape four steel balls to the 3-V rectangle, three to the 6-V rectangle, and so on.

Data and Observations

1. Make a data table with columns labeled "Level," "Charge," "Voltage," and "Energy."
2. Fill in the data table for your model for each level of the model.
3. The model shows different amounts of charges at different energy levels. Where should steel balls be placed to show a zero energy level? Explain.

Data and Observations

Level	Charge	Voltage	Energy

Analyze and Conclude

1. **Analyzing Data** How much energy is required to lift each coulomb of charge from the tabletop to the 9-V level?
2. **Analyzing Data** What is the total potential energy stored in the 9-V level?
3. **Relating Concepts** The total energy of the charges in the 6-V level is not 6 J. Explain this.
4. **Making Predictions** How much energy would be given off if the charges in the 9-V level fell to the 6-V level? Explain.

Apply

1. A 9-V battery is very small. A 12-V car battery is very big. Use your model to help explain why two 9-V batteries will not start your car.

Sharing of Charge

All systems come to equilibrium when the energy of the system is at a minimum. For example, if a ball is put on a hill it will finally come to rest in a valley, where its gravitational potential energy is smallest. This would also be the location where its gravitational potential has been reduced the largest amount. This same principle explains what happens when an insulated, negatively charged metal sphere, such as the one shown in **Figure 21–10,** touches a second, uncharged sphere.

The excess negative charges on sphere A repel each other, so when the neutral sphere B touches A, there is a net force on the charges on A toward B. If you were to move the first charge from A to B, you would have to exert a force opposing the force of the other charges. The force you exert on the first charge is in the direction opposite its displacement. Therefore, you do negative work on it, and the electric potential difference is negative. When the next few charges are moved, they feel a repulsive force from the charges already on B, so the negative work you do on these charge is smaller, that is, the electric potential difference is less negative. At some point, the force pushing the charge off A equals the repulsive force from the charges on B. No work is done moving that charge, and the electric potential difference is zero. You would have to do work to move the next charge to B, so the electric potential difference would now be higher. But, moving that charge would require an increase in energy, so it would not occur. Thus charges move until there is no electric potential difference between the two spheres.

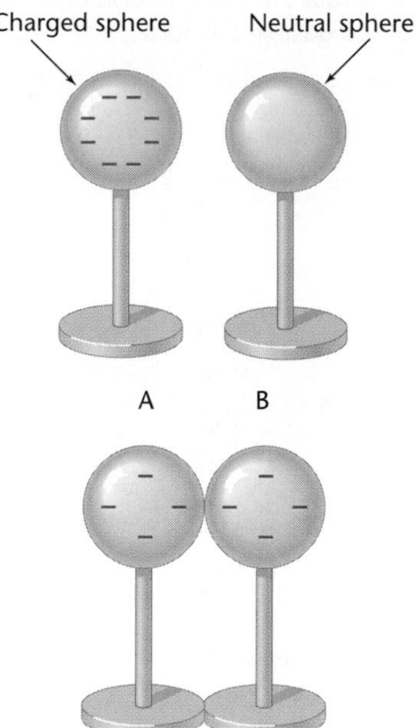

FIGURE 21–10 A charged sphere shares charge equally with a neutral sphere of equal size.

Suppose two spheres have different sizes, as in **Figure 21–11.** Although the total number of charges on the two spheres are the same, the larger sphere has a larger surface area, so charges can spread farther apart. With the charges farther apart, the repulsive force between them is reduced. So, if the two spheres are now touched together, there will be a net force that will move charges from the smaller to the larger sphere. Again, the charges will move to the sphere with the lower electric potential until there is no electric potential difference between the two spheres. In this case, the larger sphere will have a larger charge when equilibrium is reached.

Metal spheres of unequal size

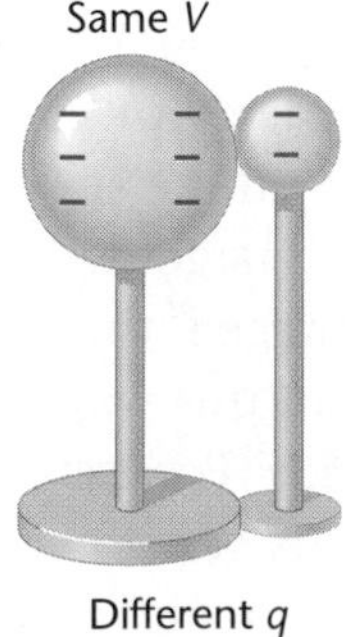

FIGURE 21–11 A charged sphere gives much of its charge to a larger sphere.

FIGURE 21–12 The ground wire on a fuel truck prevents explosion of the gasoline vapors.

The same principle explains how charges move on the individual spheres, or on any conductor. They distribute themselves so that the net force on each charge is zero. With no force, there is no electric field along the surface of the conductor. Therefore, there is no electric potential difference anywhere on the surface. The surface of a conductor is, therefore, an equipotential surface.

Earth is a very large sphere. If a charged body is touched to Earth, almost any amount of charge can flow to Earth until the electric potential difference between that body and Earth is reduced to zero. Thus, Earth can absorb all excess charge on a body. When this charge has flowed to Earth, the body becomes neutral. Touching a body to Earth to eliminate excess charge is called **grounding.** Gasoline trucks can become charged by friction. If that charge were to jump to Earth through gasoline vapor, it could cause an explosion. Instead, a metal wire on the truck safely conducts the charge to the ground, as shown in **Figure 21–12.** If a computer or other sensitive instrument were not grounded, static charges could accumulate, creating an electric potential difference between the computer and Earth. If a person touched the computer, charges could flow through the computer to the person, damaging the equipment or hurting the person.

F.Y.I.
Fewer modern cars and trucks are equipped with static strips because tires are now made to be slightly conductive.

Electric Fields Near Conductors

The charges on a conductor are spread as far apart as they can be to make the energy of the system as low as possible. The result is that all charges are on the surface of a solid conductor. If the conductor is hollow, excess charges will move to the outer surface. If a closed metal container is charged, there will be no charges on the inside surfaces of the container. In this way, a closed metal container shields the inside from electric fields. For example, people inside a car are protected from the electric fields generated by lightning. On an open coffee can there will be very few charges inside, and none near the bottom.

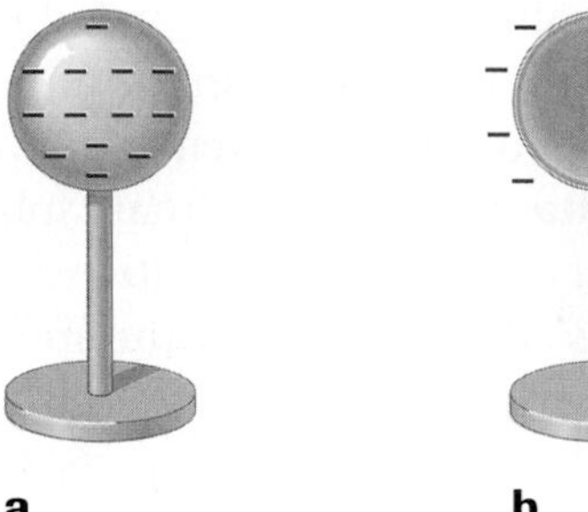
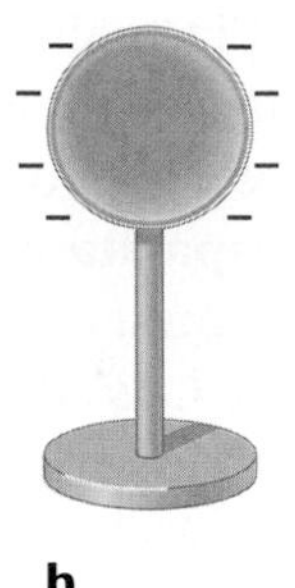
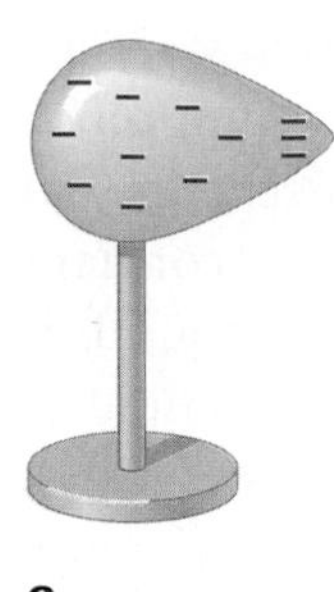

FIGURE 21–13 The electric field around a conducting body depends on the structure and shape (**b** is hollow) of the body.

So why is the electric charge visible as a corona on a power line but not in the electric wires in houses? Even though the surface of a conductor is at an equipotential, the electric field around the outside of it depends on the shape of the body as well as the electric potential difference between it and Earth. The charges are closer together at sharp points of a conductor. Therefore, the field lines are closer together; and the field is stronger, as indicated in **Figure 21–13.** This field can become so strong that nearby air molecules are separated into electrons and positive ions. As the electrons and ions recombine, energy is released and light is produced. The result is the blue glow of a corona. The electrons and ions are accelerated by the field. If the field is strong enough, when the particles hit other molecules, they will produce more ions and electrons. The stream of ions and electrons that results is a plasma, which is a conductor. The result is a spark, or, in extreme cases, lightning. To reduce coronae and sparking, conductors that are highly charged or operate at high potentials, especially those that service houses, are made smooth in shape to reduce the electric fields.

On the other hand, lightning rods are pointed so that the electric field will be strong near the end of the rod. Air molecules are pulled apart near the rod, forming the start of a conducting path from the rod to the clouds. As a result of the rod's sharply pointed shape, charges in the clouds spark to the rod rather than to a chimney or other high point on a house or other buildings. From the rod, a conductor takes the charges safely to the ground.

High-Energy Halo

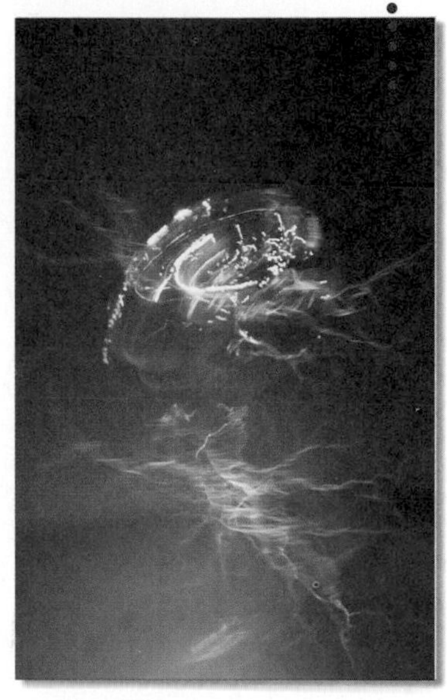

Storing Charges: The Capacitor

When you lift a book, you increase its gravitational potential energy. This can be interpreted as storing energy in a gravitational field. In a similar way, you can store energy in an electric field. In 1746, Dutch physician and physicist Pieter Van Musschenbroek invented a device that could store a large electric charge in a small device. In honor of the city in which he worked, it was called a Leyden jar. Benjamin Franklin used a Leyden jar to store the charge from lightning and in many other experiments. A version of the Leyden jar is still in use today in electric equipment. This version has a new form, a much smaller size, and a new name. It is called the capacitor.

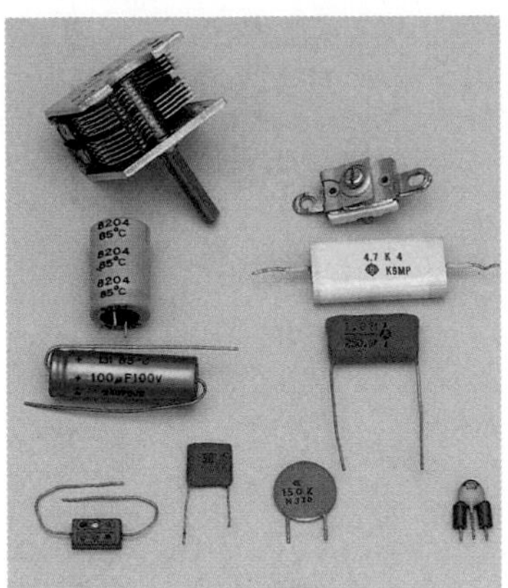
FIGURE 21-14 Various types of capacitors are pictured.

As charge is added to an object, the electric potential difference between that object and Earth increases. For a given shape and size of the object, the ratio of charge stored to electric potential difference, $q/\Delta V$, is a constant called the **capacitance,** C. For a small sphere far from the ground, even a small amount of added charge will increase the electric potential difference. Thus, C is small. The larger the sphere, the greater the charge that can be added for the same increase in electric potential difference, and thus the larger the capacitance.

The **capacitor** is a device that is designed to have a specific capacitance. All capacitors are made up of two conductors, separated by an insulator. The two conductors have equal and opposite charges. Capacitors are used today in electrical circuits to store charge. Commercial capacitors, such as those shown in **Figure 21–14,** contain strips of aluminum foil separated by thin plastic that are tightly rolled up to conserve space.

The capacitance of a capacitor is independent of the charge on it. Capacitance can be measured by first placing charge q on one plate and charge $-q$ on the other, and then measuring the electric potential difference, ΔV, that results. The capacitance is then found by using the following equation.

$$C = \frac{q}{\Delta V}$$

Capacitance is measured in farads, F, named after Michael Faraday. One farad is one coulomb per volt (C/V). Just as one coulomb is a large amount of charge, one farad is an enormous capacitance. Capacitors usually contain capacitances between 10 picofarads (10×10^{-12} F) and 500 microfarads (500×10^{-6} F). Note that if the charge is increased, the electric potential difference also increases. The capacitance depends only on the construction of the capacitor, not on the charge, q.

Example Problem

Finding the Capacitance

A sphere has an electric potential difference between it and Earth of 60.0 V when it has been charged to 3.0×10^{-6} C. What is its capacitance?

Sketch the Problem

- Draw a sphere above Earth identifying the electric potential difference and the charge.

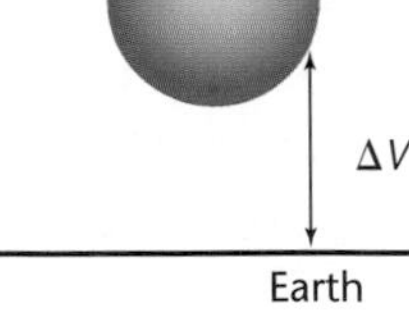

Calculate Your Answer

Known:

$\Delta V = 60.0$ V

$q = 3.0 \times 10^{-6}$ C

Unknown:

$C = ?$

Strategy:

Use the relationship $C = \frac{q}{\Delta V}$.

Calculations:

$$C = \frac{q}{\Delta V} = \frac{3.0 \times 10^{-6}\ \text{C}}{60.0\ \text{V}} = 5.0 \times 10^{-8}\ \text{C/V}$$
$$= 0.050\ \mu\text{F}$$

Check Your Answer

- Are the units correct? Capacitance is measured in farads.
- Are the signs correct? Values are positive.
- Is the magnitude realistic? A small capacitance has a moderate potential difference when a moderate charge is added.

Practice Problems

13. A 27-μF capacitor has an electric potential difference of 25 V across it. What is the charge on the capacitor?

14. Both a 3.3-μF and a 6.8-μF capacitor are connected across a 15-V electric potential difference. Which capacitor has a greater charge? What is it?

15. The same two capacitors are each charged to 2.5×10^{-4} C. Which has the larger electric potential difference? What is it?

16. A 2.2-μF capacitor is first charged so that the electric potential difference is 6.0 V. How much additional charge is needed to increase the electric potential difference to 15.0 V?

21.2 Section Review

2.1 An oil drop is motionless in a Millikan oil drop apparatus.

a. What is the direction of the electric field?

b. Which plate, upper or lower, is positively charged?

2.2 If the charge on a capacitor is changed, what is the effect on

a. the capacitance, C?

b. the electric potential difference, ΔV?

2.3 If a large, charged sphere is touched by a smaller, uncharged sphere, what can be said about

a. the potentials of the two spheres?

b. the charges on the two spheres?

2.4 **Critical Thinking** Suppose you have a large, hollow sphere that has been charged. Through a small hole in the sphere, you insert a small, uncharged sphere into the hollow interior. The two touch. What is the charge on the small sphere?

CHAPTER 21 REVIEW

Key Terms

21.1
- electric field
- electric field lines

21.2
- electric potential difference
- volt
- equipotential
- grounding
- capacitance
- capacitor

Summary

21.1 Creating and Measuring Electric Fields

- An electric field exists around any charged object. The field produces forces on other charged bodies.
- The electric field intensity is the force per unit charge. The direction of the electric field is the direction of the force on a tiny, positive test charge.
- Electric field lines provide a picture of the electric field. They are directed away from positive charges and toward negative charges.

21.2 Applications of Electric Fields

- Electric potential difference is the change in potential energy per unit charge in an electric field. Electric potential differences are measured in volts.
- The electric field between two parallel plates is uniform between the plates except near the edges.
- Robert Millikan's experiments showed that electric charge is quantized and that the negative charge carried by an electron is 1.60×10^{-19} C.
- Charges will move in conductors until the electric potential is the same everywhere on the conductor.
- A charged object can have its excess charge removed by touching it to Earth or to an object touching Earth. This is called grounding.
- Electric fields are strongest near sharply pointed conductors.
- Capacitance is the ratio of the charge on a body to its electric potential difference. It is independent of the charge on the body and the electric potential difference across it.

Reviewing Concepts

Section 21.1

1. Draw some of the electric field lines between
   a. two like charges.
   b. two unlike charges.
   c. two parallel plates of opposite charge.
2. What are the two properties a test charge must have?
3. How is the direction of an electric field defined?
4. What are electric field lines?
5. How is the strength of an electric field indicated with electric field lines?

Section 21.2

6. What SI unit is used to measure electric potential energy? What SI unit is used to measure electric potential difference?
7. Define *volt* in terms of the change in potential energy of a charge moving in an electric field.
8. Why does a charged object lose its charge when it is touched to the ground?
9. A charged rubber rod placed on a table maintains its charge for some time. Why is the charged rod not grounded immediately?
10. A metal box is charged. Compare the concentration of charge at the corners of the box to the charge concentration on the sides.
11. In your own words, describe a capacitor.

Applying Concepts

12. What happens to the size of the electric field if the charge on the test charge is halved?

13. Does it require more energy or less energy to move a fixed positive charge through an increasing electric field?

14. What will happen to the electric potential energy of a charged particle in an electric field when the particle is released and free to move?

15. **Figure 21–15** shows three spheres with charges of equal magnitude, with signs as shown. Spheres Y and Z are held in place but sphere X is free to move. Initially sphere X is equidistant from spheres Y and Z. Choose the path that sphere X will follow, assuming no other forces are acting.

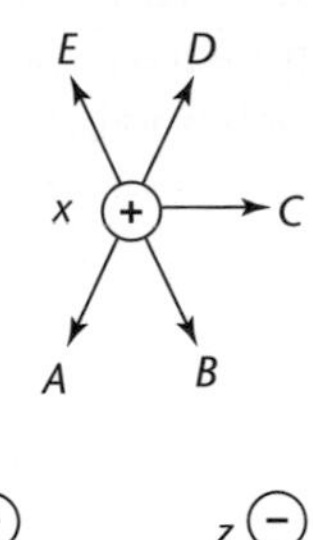

FIGURE 21–15

16. What is the unit of electric potential difference in terms of m, kg, s, and C?

17. What do the electric field lines look like when the electric field has the same strength at all points in a region?

18. When doing a Millikan oil-drop experiment, it is best to work with drops that have small charges. Therefore, when the electric field is turned on, should you try to find drops that are moving rapidly or slowly? Explain.

19. If two oil drops can be held motionless in a Millikan oil-drop experiment,

a. can you be sure that the charges are the same?

b. the ratios of what two properties of the drops would have to be equal?

20. Tim and Sue are standing on an insulating platform and holding hands when they are given a charge. Tim is larger than Sue. Who has the larger amount of charge, or do they both have the same amount?

21. Which has a larger capacitance, an aluminum sphere with a 1-cm diameter or one with a 10-cm diameter?

22. How can you store a different amount of charge in a capacitor?

Problems

Section 21.1

LEVEL 1

The charge of an electron is -1.60×10^{-19} C.

23. A positive charge of 1.0×10^{-5} C experiences a force of 0.20 N when located at a certain point. What is the electric field intensity at that point?

24. What charge exists on a test charge that experiences a force of 1.4×10^{-8} N at a point where the electric field intensity is 2.0×10^{-4} N/C?

25. A test charge experiences a force of 0.20 N on it when it is placed in an electric field intensity of 4.5×10^{5} N/C. What is the magnitude of the charge?

26. The electric field in the atmosphere is about 150 N/C, downward.

a. What is the direction of the force on a charged particle?

b. Find the electric force on a proton with charge $+1.6 \times 10^{-19}$ C.

c. Compare the force in ***b*** with the force of gravity on the same proton (mass = 1.7×10^{-27} kg).

27. Carefully sketch

a. the electric field produced by a +1.0-μC charge.

b. the electric field resulting from a +2.0-μC charge. Make the number of field lines proportional to the change in charge.

28. Charges X, Y, and Z are all equidistant from each other. X has a +1.0-μC charge, Y has a +2.0-μC charge, and Z has a small negative charge.

a. Draw an arrow showing the force on charge Z.

b. Charge Z now has a small positive charge on it. Draw an arrow showing the force on it.

29. A positive test charge of 8.0×10^{-5} C is placed in an electric field of 50.0-N/C intensity. What is the strength of the force exerted on the test charge?

LEVEL 2

30. Electrons are accelerated by the electric field in a television picture tube, whose value is given in **Table 21-1.**
a. Find the force on an electron.
b. If the field is constant, find the acceleration of the electron (mass = 9.11×10^{-31} kg).

31. A lead nucleus has the charge of 82 protons.
a. What are the direction and magnitude of the electric field at 1.0×10^{-10} m from the nucleus?
b. What are the direction and magnitude of the force exerted on an electron located at this distance?

Section 21.2

LEVEL 1

32. If 120 J of work are performed to move one coulomb of charge from a positive plate to a negative plate, what potential difference exists between the plates?

33. How much work is done to transfer 0.15 C of charge through an electric potential difference of 9.0 V?

34. An electron is moved through an electric potential difference of 500 V. How much work is done on the electron?

35. A 12-V battery does 1200 J of work transferring charge. How much charge is transferred?

36. The electric field intensity between two charged plates is 1.5×10^3 N/C. The plates are 0.080 m apart. What is the electric potential difference, in volts, between the plates?

37. A voltmeter indicates that the electric potential difference between two plates is 50.0 V. The plates are 0.020 m apart. What electric field intensity exists between them?

38. An oil drop is negatively charged and weighs 8.5×10^{-15} N. The drop is suspended in an electric field intensity of 5.3×10^3 N/C.
a. What is the charge on the drop?
b. How many electrons does it carry?

39. A capacitor that is connected to a 45.0-V source contains 90.0 μC of charge. What is the capacitor's capacitance?

40. What electric potential difference exists across a 5.4-μF capacitor that has a charge of 2.7×10^{-3} C?

41. What is the charge in a 15.0-pF capacitor when it is connected across a 75.0-V source?

LEVEL 2

42. A force of 0.053 N is required to move a charge of 37 μC a distance of 25 cm in an electric field. What is the size of the electric potential difference between the two points?

43. In an early set of experiments in 1911, Millikan observed that the following measured charges, among others, appeared at different times on a single oil drop. What value of elementary charge can be deduced from these data?
a. 6.563×10^{-19} C
b. 8.204×10^{-19} C
c. 11.50×10^{-19} C
d. 13.13×10^{-19} C
e. 16.48×10^{-19} C
f. 18.08×10^{-19} C
g. 19.71×10^{-19} C
h. 22.89×10^{-19} C
i. 26.13×10^{-19} C

44. The energy stored in a capacitor with capacitance C, having an electric potential difference ΔV, is represented by $W = 1/2C\,\Delta V^2$. One application of this is in the electronic photoflash of a strobe light. In such a unit, a capacitor of 10.0 μF is charged to 300 V. Find the energy stored.

45. Suppose it took 30 s to charge the capacitor in problem 44.
a. Find the power required to charge it in this time.
b. When this capacitor is discharged through the strobe lamp, it transfers all its energy in 1.0×10^{-4} s. Find the power delivered to the lamp.
c. How is such a large amount of power possible?

46. Lasers are used to try to produce controlled fusion reactions that might supply large amounts of electrical energy. The lasers require brief pulses of energy that are stored in large rooms filled with capacitors. One such room has a capacitance of 61×10^{-3} F charged to a potential difference of 10 kV.
   a. Find the energy stored in the capacitors, given that $W = 1/2\ \Delta V^2$.
   b. The capacitors are discharged in 10 ns (1.0×10^{-8} s). What power is produced?
   c. If the capacitors are charged by a generator with a power capacity of 1.0 kW, how many seconds will be required to charge the capacitors?

47. Two point charges, one at $+3.00 \times 10^{-5}$ C and the other at -5.00×10^{-5} C are placed at adjacent corners of a square 0.800 m on a side. A third charge of $+6.00 \times 10^{-5}$ C is placed at the corner diagonally opposite to the negative charge. Calculate the magnitude and direction of the force acting on the third charge. For the direction, determine the angle the force makes with the edge of the square parallel to the line joining the first two charges.

Critical Thinking Problems

48. In an ink-jet printer, drops of ink are given a certain amount of charge before they move between two large parallel plates. The purpose of the plates is to deflect the charges so that they are stopped by a gutter and do not reach the paper. This is shown in **Figure 21–16.** The plates are 1.5 cm long and have an electric field of $E = 1.2 \times 10^6$ N/C between them. Drops with a mass $m = 0.10$ ng and a charge $q = 1.0 \times 10^{-16}$ C are moving horizontally at a speed $v = 15$ m/s parallel to the plates. What is the vertical displacement of the drops when they leave the plates? To answer this question, go through the following steps.
   a. What is the vertical force on the drops?
   b. What is their vertical acceleration?
   c. How long are they between the plates?
   d. Given that acceleration and time, how far are they displaced?

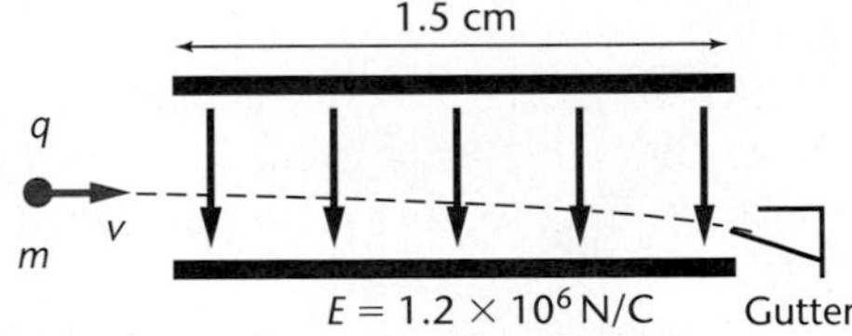

FIGURE 21–16

49. Suppose the moon had a net negative charge equal to $-q$ and Earth had a net positive charge equal to $+10q$. What value of q would yield the same magnitude force that you now attribute to gravity?

Going Further

Determining Unit Charge Millikan discovered that the electric charge always seemed to exist in multiples of a unit charge. How did he know this? What reasoning did he use to reach his conclusions? Design a modeling activity using a balance and several groups of different numbers of steel balls to determine a unit value of mass that can be used to determine the mass of a single ball. Make a bar graph of the masses versus the trials. Suggest other mass values that might result if more trial measurements were made. Consider whether the actual unit mass could be smaller or larger than your determination. Write a short summary of your findings.

interNET CONNECTION

Follow the link on the Glencoe Homepage at **www.glencoe.com/sec/science** to find out more about this chapter.

More for Less

High-voltage power transmission lines criss-cross our country. It would seem to be easier to simply transmit electrical energy at the lower voltages used in homes. Why is this not done?

CHAPTER

22 Current Electricity

How many ways do you use electric energy in your everyday life? Can you even count the ways? Where does this energy come from? Most likely, it comes from a power plant far away.

Electricity is such a common tool that you often take it for granted. Its importance is never truly considered until it is missing. The most important aspects of electric energy are its abilities to be changed into other forms of energy and to be transferred efficiently over long distances. The large amounts of natural potential and kinetic energy possessed by resources such as Niagara Falls are of little use to an industrial complex 100 km away unless that energy can be harnessed and transferred efficiently. Electric energy provides the means to transfer large quantities of energy great distances with little loss. This transfer is usually done at high potential differences, such as the power line shown here.

WHAT YOU'LL LEARN

- You will explain energy transfer in circuits.
- You will solve problems involving current, potential difference, and resistance.
- You will diagram simple electric circuits.
- You will solve problems involving the use and cost of electric energy

WHY IT'S IMPORTANT

- The electric tools and appliances you use are based upon the ability of electric circuits to transfer energy by potential difference and thus perform work.

interNET CONNECTION

Follow the link for this chapter on the Glencoe Homepage at **www.glencoe.com/sec/science** to find out more about current electricity.

22.1 Current and Circuits

You have had many experiences with electric circuits. Every time you turn on a light, radio, or television, or turn the ignition key to start a car or turn on a flashlight, you complete an electric circuit. Electric charges flow and energy is transferred. In this chapter you will learn how electric circuits work.

OBJECTIVES

- **Define** an electric current and the ampere.
- **Describe** conditions that create current in an electric circuit.
- **Draw** circuits and **recognize** that they are closed loops.
- **Define** power in electric circuits.
- **Define** *resistance* and **describe** Ohm's law.

Producing Electric Current

In Chapter 21 you learned that when two conducting spheres touch, charges flow from the sphere at a higher potential difference to the one at a lower potential difference. The flow continues until the potential differences of both spheres are equal.

A flow of charged particles is an **electric current.** In **Figure 22–1a,** two conductors, A and B, are connected by a wire conductor, C. Charges flow from the higher potential difference of B to A through C. This flow of positive charge is called **conventional current.** The flow stops when the potential differences of A, B, and C are equal. How could you keep the flow going? You would have to maintain a potential difference between B and A. This could be done by pumping charged particles from conductor A back to conductor B. The electric potential energy of the charges would have to be increased by this pump, so it would require external energy to run. The electric energy could come from one of several other forms of energy. One familiar source of electric energy, a voltaic or galvanic cell (a common dry cell), converts chemical energy to electric energy. Several of these cells connected together are called a **battery.** A second source of electric energy, the **photovoltaic cell,** or solar cell, changes light energy into electric energy. Yet another source of electric energy is a generator. A generator can be driven by moving water, rushing steam, or wind and converts kinetic energy into electric energy. One source of electric energy is pictured in **Figure 22–2.**

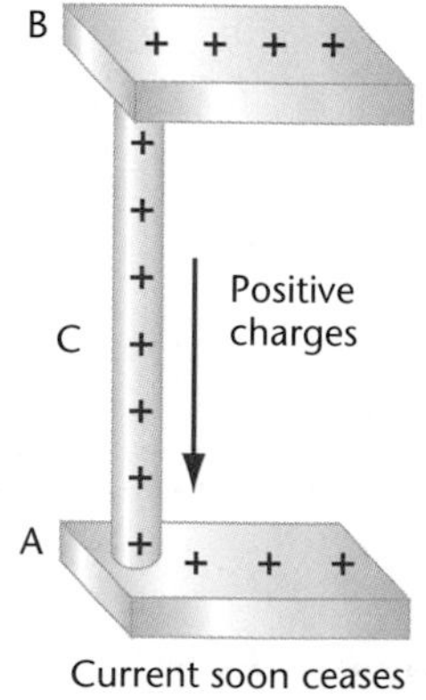

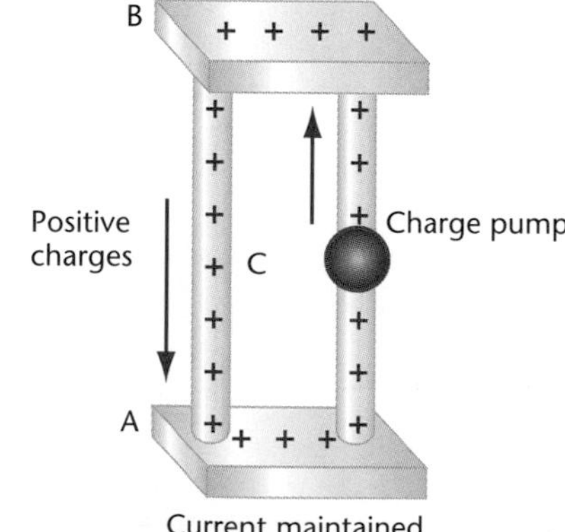

FIGURE 22–1 Conventional current is defined as positive charges flowing from the positive plate to the negative plate **(a).** A generator pumps the positive charges back to the positive plate, creating the current **(b).** In most metals, negatively-charged electrons actually flow from the negative to the positive plate. When the electrons move, they create positive charges which appear to move in the opposite direction from the positive plate to the negative plate.

Electric Circuits

The charges in **Figure 22–1b** move around a closed loop from the pump through B to A through C, and back to the pump. Such a closed loop is called an **electric circuit.** A circuit includes a charge pump, which increases the potential energy of the charges moving from A to B, and is connected to a device that reduces the potential energy of the charges moving from B to A. The potential energy lost by the charges, qV, in moving through the device is usually converted into some other form of energy. For example, a motor converts electric energy to kinetic energy, a lamp changes electric energy into light, and a heater converts electric energy into thermal energy. Note that the Δ is dropped from the V for potential difference. This is a historical convention rather than an appropriate one. For convenience and uniformity, this text will also use V rather than ΔV.

The charge pump creates the flow of charged particles, or current. Consider a generator driven by a waterwheel, such as the one pictured in **Figure 22–3a.** The water falls and rotates the waterwheel and generator. Thus, the kinetic energy of the water is converted to electric energy by the generator. The generator increases the electric potential difference, V, between B and A as it removes charges from wire B and adds them to wire A. Energy in the amount qV is needed to increase the potential difference of the charges. This energy comes from the change in energy of the water. No generator, however, is 100 percent efficient. Only 98 percent of the kinetic energy put into most generators is converted into electric energy. The remainder becomes thermal energy. The temperature of the generator increases, as shown in **Figure 22–3b.**

FIGURE 22–2 Sources of electric energy include chemical, solar, hydrodynamic, wind, and nuclear energies.

Charge and Field Conventions

- Positive charges are **red.**
- Negative charges are **blue.**
- Electric field lines are **indigo.**

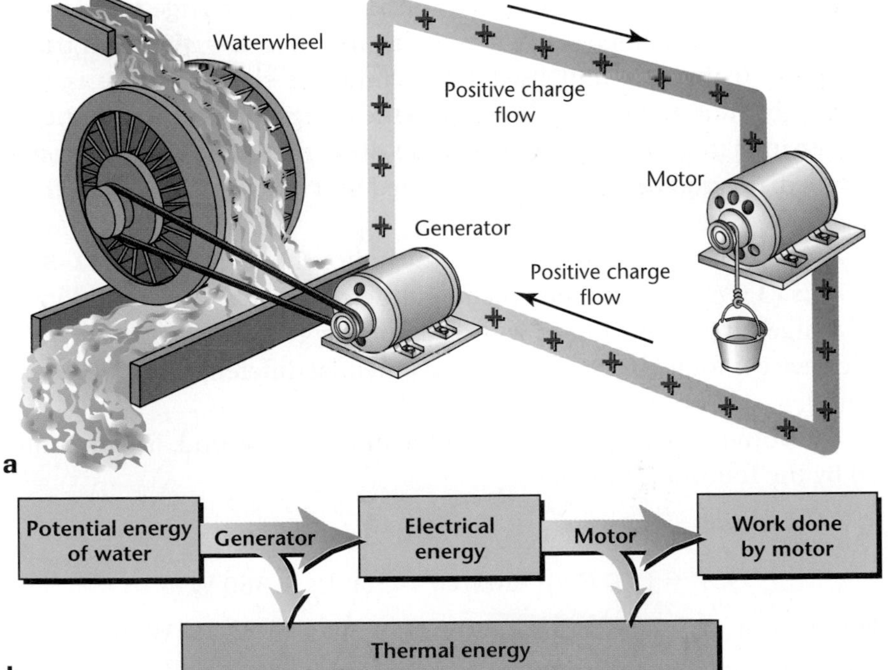

FIGURE 22–3 The potential energy of the waterfall is eventually converted into work done raising the bucket **(a).** The production and use of electric current is not 100 percent efficient. Some thermal energy is produced by the splashing water, friction, and electrical resistance **(b).**

If wires are connected to a motor, the charges in the wire flow into the motor. The flow continues through the circuit back to the generator. The motor converts electric energy to kinetic energy. Like generators, motors are not 100 percent efficient. Typically, only 90 percent of the electric energy converted by a motor is changed into kinetic energy.

You know that charges can't be created or destroyed, only separated. Thus, the total amount of charge (number of negative electrons and positive ions) in the circuit does not change. If one coulomb flows through the generator in one second, then one coulomb also will flow through the motor in one second. Thus, charge is a conserved quantity. Energy is also conserved. The change in electric energy, E, equals qV. Because q is conserved, the net change in potential energy of the charges going completely around the circuit must be zero. The potential difference increase produced by the generator equals the potential difference decrease across the motor.

If the potential difference between the two wires is 120 V, the generator must do 120 J of work on each coulomb of positive charge that it transfers from the more negative wire to the more positive wire. Every coulomb of positive charge that moves from the more positive wire through the motor and back to the more negative wire delivers 120 J of energy to the motor. Thus, electric energy serves as a way to transfer the energy of falling water to the kinetic energy of a turning motor.

Rates of Charge Flow and Energy Transfer

Power measures the rate at which energy is transferred. If a generator transfers one joule of kinetic energy to electric energy each second, it is transferring energy at the rate of one joule per second, or one watt. The energy carried by an electric current depends on the charge transferred and the potential difference across which it moves, $E = qV$. Recall from Chapter 20 that the unit used for quantity of electric charge is the coulomb. Thus, the rate of flow of electric charge, or electric current, I, is measured in coulombs per second. A flow of one coulomb per second is called an **ampere,** A, and is represented by 1 C/s = 1 A. A device that measures current is called an ammeter.

Suppose that the current through the motor shown in **Figure 22–3** is 3.0 C/s (3.0 A). Because the potential difference is 120V, each coulomb of charge supplies the motor with 120 J of energy. The power of an electric device is found by multiplying the potential difference, V, by the current, I.

The power, or energy delivered to the motor per second, is represented by the following equation

$$P = IV$$

$$P = (3.0\ \text{C/s})(120\ \text{J/C}) = 360\ \text{J/s} = 360\ \text{W}$$

Recall from Chapter 10 that power is defined in watts, W; thus, the power delivered by this motor is 360 watts.

Example Problem

Electric Power

A 6.0-V battery delivers a 0.50-A current to an electric motor that is connected across its terminals.

a. What power is consumed by the motor?

b. If the motor runs for 5.0 minutes, how much electric energy is delivered?

Sketch the Problem

- Draw a circuit showing the positive terminal of a battery connected to a motor and the return wire from the motor connected to the negative terminal of the battery.
- Show the conventional current.

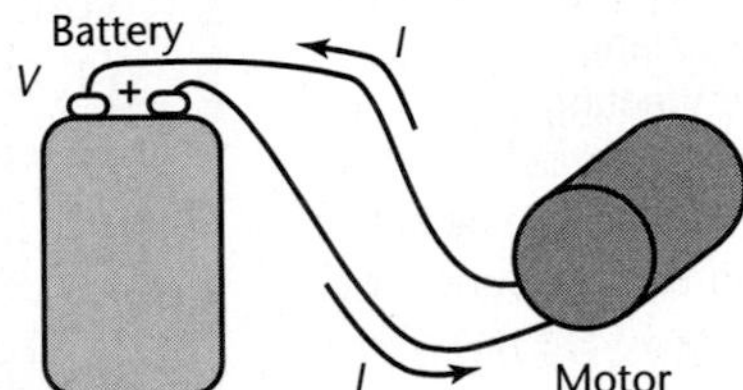

Calculate Your Answer

Known:

$V = 6.0$ V

$I = 0.50$ A

$t = 5.0$ min

Unknown:

$P = ?$

$E = ?$

Strategy:

a. Use $P = IV$ to find the power.

b. From Chapter 10 you learned that $P = E/t$, so use $E = Pt$ to find the energy.

Calculations:

a. $P = IV = (0.50\text{ A})(6.0\text{ V}) = 3.0\text{ W}$

b. $E = Pt = (3.0\text{ W})(5.0\text{ min})(60\text{ s}/1\text{ min}) = 9.0 \times 10^2\text{ J}$

Check Your Answer

- Are the units correct? Power is measured in watts and energy is measured in joules.
- Is the magnitude realistic? Small batteries deliver only a few watts of power.

Practice Problems

1. The current through a lightbulb connected across the terminals of a 120-V outlet is 0.50 A. At what rate does the bulb convert electric energy to light?
2. A car battery causes a current of 2.0 A through a lamp while 12 V is across it. What is the power used by the lamp?
3. What is the current through a 75-W lightbulb connected to a 120-V outlet?
4. The current through the starter motor of a car is 210 A. If the battery keeps 12 V across the motor, what electric energy is delivered to the starter in 10.0 s?

Lighting Up

Use a D cell and a 10-cm length of wire to light a miniature lamp. Make a sketch of two circuits that work and two circuits that do not work.

Analyze and Conclude Can you light the lamp without having two connecting points on the lamp? Does a lamp connected in your house have two connecting points? Suggest a reason why two are needed.

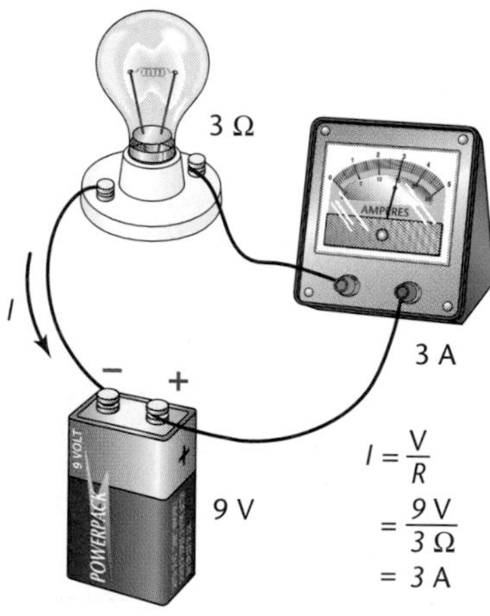

FIGURE 22–4 In a circuit having a 3-Ω resistance and a 9-V battery, there is a 3-A current.

Resistance and Ohm's Law

Suppose that two conductors have a potential difference between them. If you connect them with a copper rod, you will create a large current. If, on the other hand, you put a glass rod between them, there will be almost no current. The property that determines how much current will flow is called the **resistance.** Resistance is measured by placing a potential difference across two points on a conductor and measuring the current. The resistance, R, is defined to be the ratio of the potential difference, V, to the current, I.

$$R = \frac{V}{I}$$

The electric current, I, is in amperes. The potential difference, V, is in volts. The resistance of the conductor, R, is measured in ohms. One ohm (1 Ω) is the resistance that permits a current of 1 A to flow when a potential difference of 1 V is applied across the resistance. A simple circuit relating resistance, current, and voltage is shown in **Figure 22–4.** A 9-V battery is connected to a 3-Ω resistance lightbulb. The circuit is completed by connection to an ammeter. The current measures 3 A.

German scientist Georg Simon Ohm found that the ratio of the potential difference to the current is always a constant for a given conductor. Therefore, the resistance for most conductors does not vary as the magnitude or the direction of the potential applied to it changes. A device that has a constant resistance that is independent of the potential difference is said to obey Ohm's law.

Most metallic conductors obey Ohm's law, at least over a limited range of voltages. Many important devices, however, do not. A transistor radio or pocket calculator contains many devices, such as transistors and diodes, that do not obey Ohm's law. Even a lightbulb has a resistance that depends on the voltage and does not obey Ohm's law.

Wires used to connect electric devices have small resistance. One meter of a typical wire used in physics labs has a resistance of about 0.03 Ω. Wires used in house wiring offer as little as 0.004-Ω resistance for each meter of length. Because wires have so little resistance, there is almost no potential drop across them. To produce potential differences, you need large resistance concentrated into a small volume. **Resistors** are devices designed to have a specific resistance. They may be made of long, thin wires; graphite; or semiconductors.

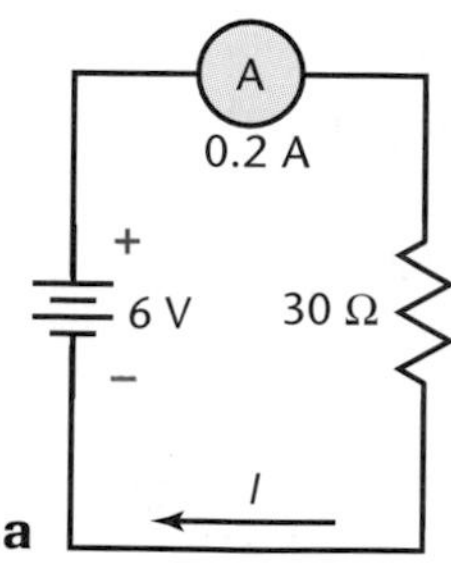

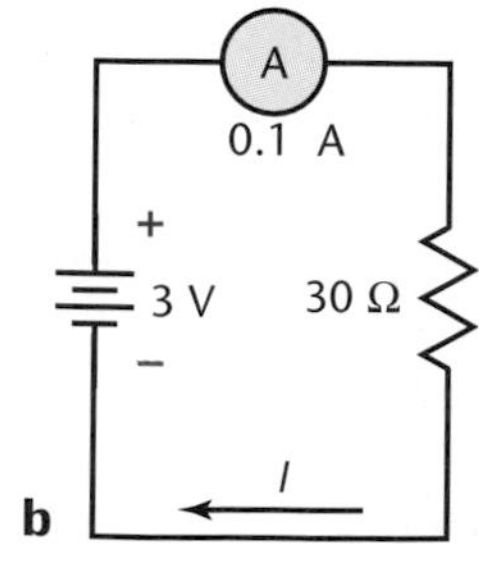

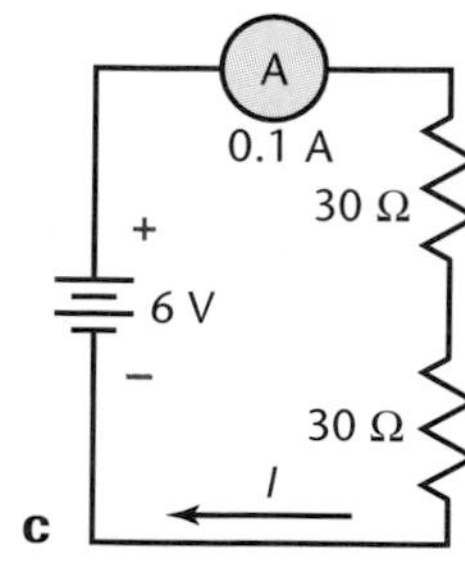

FIGURE 22–5 The current through a simple circuit **(a)** can be regulated by removing some of the dry cells **(b)** or increasing the resistance of the circuit **(c).**

Superconductors are materials that have zero resistance. There is no restriction of current in these materials, and so there is no potential difference, V, across a superconductor. Because the power dissipated in a conductor is given by the product IV, a superconductor can conduct electricity without loss of energy. The development of superconductors that can be cooled by relatively inexpensive liquid nitrogen may lead to more efficient transfer of electric energy.

There are two ways to control the current in a circuit. Because $I = V/R$, I can be changed by varying either V or R, or both. **Figure 22–5a** shows a simple circuit. When V is 6 V and R is 30 Ω, the current is 0.2 A. How could the current be reduced to 0.1 A?

The larger the potential difference, or voltage, placed across a resistor, the larger the current that passes through it. If the current through a resistor is to be cut in half, the potential difference will be cut in half. In **Figure 22–5b,** the voltage applied across the resistor has been reduced from 6 V to 3 V to reduce the current to 0.1 A. A second way to reduce the current to 0.1 A is to increase the resistance to 60 Ω by adding a 30-Ω resistor to the circuit, as shown in **Figure 22–5c,** or by replacing the 30-Ω resistor with a 60-Ω resistor. Both of these methods will reduce the current to 0.1 A. Resistors are often used to control the current in circuits or parts of circuits. Sometimes a smooth, continuous variation of the current is desired. A lamp dimmer switch allows continuous rather than step-by-step changes in light intensity. To achieve this kind of control, a variable resistor, called a rheostat, or **potentiometer,** is used. A circuit containing a potentiometer is shown in **Figure 22–6.** A variable resistor consists of a coil of resistance wire and a sliding contact point. By moving the contact point to various positions along the coil, the amount of wire added to the circuit is varied. As more wire is placed in the circuit, the resistance of the circuit increases; thus, the current

F.Y.I.

$V = IR$ is often taken as a statement of Ohm's law, but this is not true. The equation simply defines resistance and applies to all resistances whether or not they obey Ohm's law.

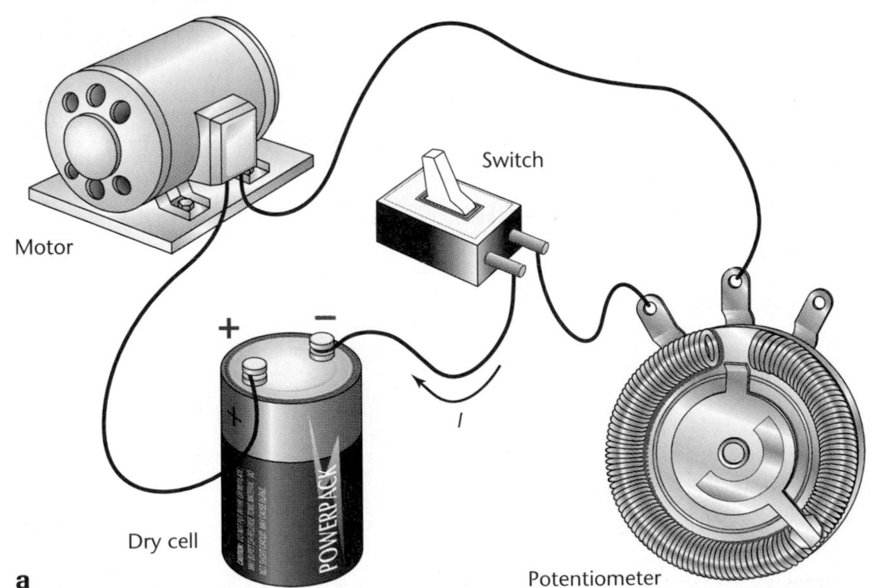

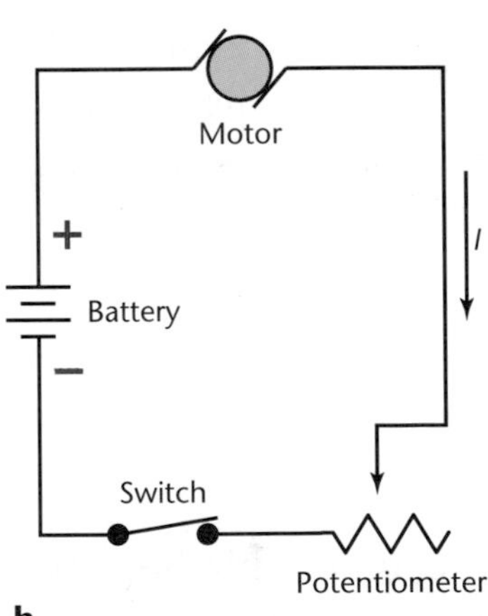

FIGURE 22–6 A potentiometer can be used to change current in an electric circuit.

FIGURE 22–7 An inside view of a potentiometer shows the coils and the sliding contact wire.

changes in accordance with the equation $I = V/R$. In this way, the light output of a lamp can be adjusted from bright with little wire in the circuit to dim with a lot of wire in the circuit. This type of device controls the speed of electric fans, electric mixers, and other appliances. To save space, the coil of wire is often bent into a circular shape and a sliding contact is moved by a knob, as shown in **Figure 22–7.**

Your body is a moderately good electrical conductor. Some physiological responses are documented in **Table 22–1.** If enough current is present through your body, your breathing or heart can stop. In addition, the energy transferred can burn you. If your skin is dry, its resistance is high enough to keep currents produced by small and moderate voltages low. If your skin becomes wet, however, its resistance is lower, and the currents can rise to dangerous levels, of tens of milliamps.

TABLE 22–1

The Damage Caused by Electric Shock

Current	Possible Effects
1 mA	mild shock can be felt
5 mA	shock is painful
15 mA	muscle control is lost
100 mA	death can occur

Example Problem

Current Through a Resistor

A 30.0-V battery is connected to a 10.0 Ω resistor. What is the current in the circuit?

Sketch the Problem

- Draw a circuit containing a battery, an ammeter, and a resistor.
- Show the conventional current.

Calculate Your Answer

Known:

$V = 30.0$ V

$R = 10.0$ Ω

Unknown:

$I = ?$

Strategy:

Use $I = V/R$ to determine the current.

Calculations:

$$I = \frac{V}{R} = \frac{30.0\ \text{V}}{10.0\ \Omega} = 3.00\ \text{A}$$

Check Your Answer

- Are the units correct? Current is measured in amperes.
- Is the magnitude realistic? Batteries deliver a few amperes.

Practice Problems

For all problems, you should assume that the battery voltage is constant, no matter what current is present.

5. An automobile headlight with a resistance of 30 Ω is placed across a 12-V battery. What is the current through the circuit?

6. A motor with an operating resistance of 32 Ω is connected to a voltage source. The current in the circuit is 3.8 A. What is the voltage of the source?

7. A transistor radio uses 2.0×10^{-4} A of current when it is operated by a 3.0-V battery. What is the resistance of the radio circuit?

8. A lamp draws a current of 0.50 A when it is connected to a 120-V source.

a. What is the resistance of the lamp?

b. What is the power consumption of the lamp?

9. A 75-W lamp is connected to 120 V.

a. What is the current through the lamp?

b. What is the resistance of the lamp?

10. A resistor is added in series with the lamp in problem 9 to reduce the current to half its original value.

a. What is the potential difference across the lamp? Assume that the lamp resistance is constant.

b. How much resistance was added to the circuit?

c. How much power is now dissipated in the lamp?

BIOLOGY CONNECTION

The resistance of the human body when the skin is dry is about 10^5 Ω. It drops to about 1500 Ω for wet hands. Therefore, the current through the body from 120 V is about 1.2 mA for dry skin and 80 mA for wet skin. A person will feel a tingle at about 1 mA, at 10 to 20 mA there will be muscular effects, and by 20 mA a person will not be able to let go of a conducting wire. Respiratory paralysis occurs between 20 and 100 mA, and currents above 100 mA can be fatal.

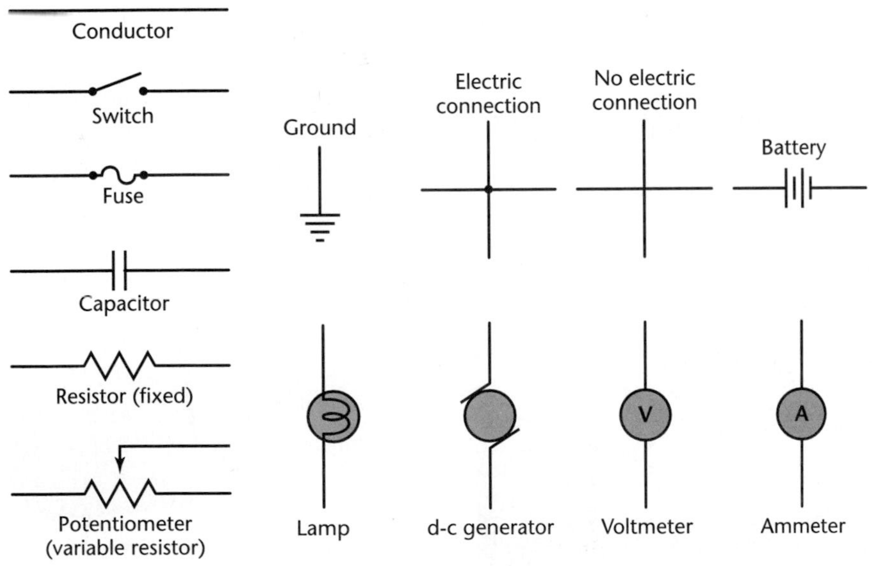

FIGURE 22–8 Symbols that are commonly used to design electric circuits are presented.

Pocket Lab

Running Out

Use the proper symbols and design a drawing that shows a power supply in a continuous circuit with two miniature lamps. Next, draw the circuit with an ammeter included to measure the electrical flow between the power supply and the bulbs. Make a third drawing to show the ammeter at a position to measure the electrical flow between the bulbs.

Test Your Prediction Would you predict the current between the lamps to be more or less than the current before the lamps? Why? Build the circuits to find out. Record your results.

Diagramming Circuits

A simple circuit can be described in words. It can also be depicted by photographs or artist's drawings of the parts. Most frequently, however, a diagram of an electric circuit is drawn using standard symbols for the circuit elements. Such a diagram is called a circuit **schematic.** Some of the symbols used in schematics are shown in **Figure 22–8.**

Both an artist's drawing and a schematic of the same circuit are shown in **Figure 22–9** and **Figure 22–10.** Notice that in both the artist's drawing and the schematic, current is shown out of the positive terminal of the battery. You learned earlier in this chapter that this is called the conventional current. To draw schematic diagrams, use the following strategy and always set up a conventional current.

Problem Solving Strategy

Drawing Schematic Diagrams

Follow these steps when drawing schematic diagrams.

1. Draw the symbol for the battery or other source of electric energy (such as a generator) at the left side of the page. Put the positive terminal on top.
2. Draw a wire coming out of the positive terminal. When you reach a resistor or other device, draw the symbol for it.
3. If you reach a point where there are two current paths, such as at a voltmeter, draw a ─┴─ in the diagram. Follow one path until the two current paths join again. Then draw the second path.
4. Follow the current path until you reach the negative terminal of the battery.
5. Check your work to make sure that you have included all parts and that there are complete paths for current to follow.

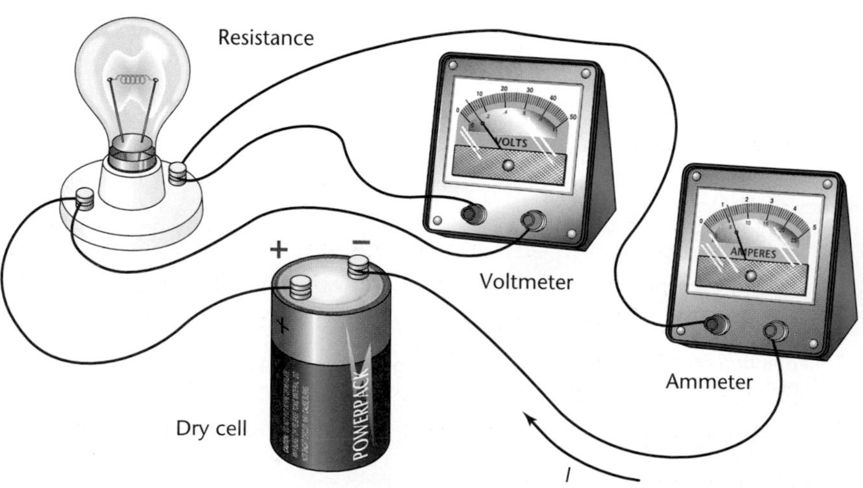

FIGURE 22–9 A simple electric circuit is represented pictorially.

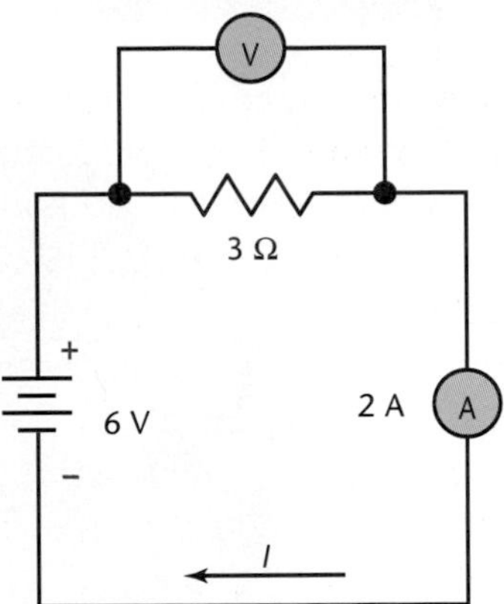

FIGURE 22–10 A simple electric circuit is represented schematically.

You have learned that an ammeter measures current and that a voltmeter measures potential differences. Each instrument has two terminals, usually labeled + and −. When it is in use, the voltmeter measures the potential difference across any component of a circuit. When connecting the voltmeter in a circuit, always connect the + terminal to the end of the circuit component closer to the positive terminal of the battery and connect the other terminal to the other side of the component. This kind of connection is called a **parallel connection** because the circuit component and the voltmeter are aligned parallel to each other in the circuit, as diagrammed in **Figure 22–9.** The potential difference across the voltmeter is equal to that across the circuit element. Always associate the two words *voltage across.*

The ammeter measures the current through a circuit component. The same current that goes through the component must go through the ammeter, so there can be only one current path. This connection is called a **series connection.** To add an ammeter to a circuit, you must remove the wire connected to the circuit component and connect it to the ammeter instead. Then connect another wire from the second terminal of the ammeter to the circuit component. Always associate the two words *current through.*

F.Y.I.

A fundamental rule of electric circuits states that electrical signals travel one foot in one nanosecond. This is one of the fundamental limitations on the possible speed of a computer. Signals in an electric circuit cannot move any faster.

Practice Problems

11. Draw a circuit diagram to include a 60-V battery, an ammeter, and a resistance of 12.5 Ω in series. Indicate the ammeter reading and the direction of current.

12. Draw a series-circuit diagram showing a 4.5-V battery, a resistor, and an ammeter reading 90 mA. Label the size of the resistor. Choose a direction for the conventional current and indicate the positive terminal of the battery.

13. Add a voltmeter that measures the potential difference across the resistors in problems 11 and 12 and repeat problems.

Physics Lab

Mystery Cans

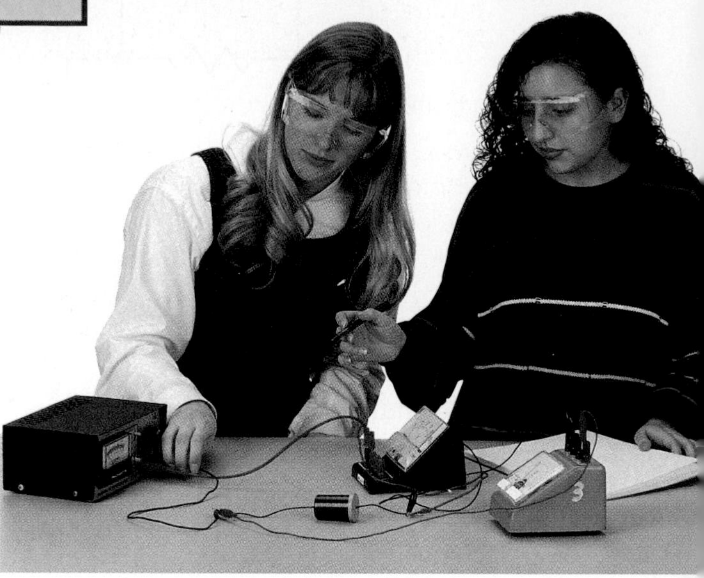

Problem

An electric device is inside each film can. How can you design and build a circuit to determine whether the resistance is constant for different voltages?

Materials

power supply with variable voltage
wires with clips
multimeter
ammeter
3 film cans for each group

Procedure

1. Identify the variables to be measured.
2. Design your circuit and label each component. Use the proper symbols to make your drawing of the set-up. Show your teacher your plan before proceeding further.
3. Build the circuit of your design and slowly increase the voltage on your power supply to make sure that your meters are working properly. Do not exceed one amp or the current limitation set by your teacher. Reverse connections as needed.
4. Make at least three measurements of voltage and current for each can.

Data and Observations

1. Make a data table with at least three places for measurements on each can.

Analyze and Conclude

1. **Calculating Results** Calculate *R* for each test.
2. **Graphing Data** Graph *V* versus *I* for all of your data. Draw a separate line for each can.
3. **Interpreting Graphs** Determine the slope for each of your lines.
4. **Comparing Values** Open each can to see the marked values of the resistors. Compare your predicted values to the actual values.

Apply

1. Most incandescent lamps burn out when they are switched on rather than when they have been on for a while. Predict what happens to the resistance and the current when a cold lamp is switched on. Make a graph of *R* versus *t* and also *I* versus *t* for the first few seconds. Calculate the resistance of an operating 60-W lamp at 120 V. Now use a multimeter as an ohmmeter to measure the resistance of a cold 60-W lamp. Describe your results.

Physics & Technology

Digital Systems

Digital systems transmit information using a string of signals in binary code—*0*s and *1*s. A signal, which can be light, sound, color, electric current, or some other quantity, is on when the code is *1* and off when the code is *0*. Unlike analog signals, which vary continuously in wave form, digital signals arrive almost exactly as they were sent. In other words, with analog technology, time is continuous; with digital technology, time is measured at discrete moments.

Digital technology has made its way into many businesses and homes worldwide. Most computers, for example, are digital. Such machines receive and process software much more easily than their analog counterparts. Digital photography, while still in its infancy, allows photographers to take pictures, instantly view the results, and save those that they want to keep. Some disadvantages of this filmless photography, however, include the lack of detail and somewhat fuzzy pictures.

Digital X rays allow dentists to better diagnose dental problems while exposing their patients to up to 90 percent less radiation than that used by conventional X-ray techniques. Compact discs employ digital technology to preserve your favorite tunes and computer games. Most telephone systems now use digital technology to transmit voices. High-definition television, or HDTV, uses digital technology to enhance your viewing pleasure by improving both the picture resolution and the sound quality.

Thinking Critically You've probably experienced surround sound at your local movie theater. Why do you think this use of digital technology seems to really draw you into the movie?

22.1 Section Review

1.1 Draw a schematic diagram of a circuit, containing a battery and a bulb, that will make the bulb light.

1.2 Joe argues that because $R = V/I$, if he increases the voltage, the resistance will increase. Is Joe correct? Explain.

1.3 You are asked to measure the resistance of a long piece of wire. Show how you would construct a circuit containing a battery, voltmeter, ammeter, and the wire to be tested to make the measurement. Specify what you would measure and how you would compute the resistance.

1.4 **Critical Thinking** We say that power is "dissipated" in a resistor. To *dissipate* is to use, or to waste, or to squander. What is "used" when charge flows through a resistor?

22.2 Using Electric Energy

OBJECTIVES

- **Explain** how electric energy is converted into thermal energy.
- **Determine** why high-voltage transmission lines are used to carry electric energy over long distances.
- **Define** *kilowatt-hour.*

F.Y.I.

Electricity is produced in hydroelectric power plants by the conversion of mechanical energy into electric energy.

Among the electrical appliances in your home, you probably have a hair dryer, several lamps, a television set, a stereo, a microwave oven, a refrigerator, and a stove. Each converts electrical energy into another form: light, kinetic, sound, or thermal energy. How much is converted, and at what rate?

Energy Transfer in Electric Circuits

Energy supplied to a circuit can be used in many different ways. A motor converts electric energy to mechanical energy. An electric lamp changes electric energy into light. Unfortunately, not all of the energy delivered to a motor or a light ends up in a useful form. Lights, especially incandescent bulbs, get hot. Motors are often too hot to touch. In each case, some energy is converted into thermal energy. Let's examine some devices designed to convert as much energy as possible into thermal energy.

Heating a Resistor

A space heater, a hot plate, and the heating element in a hair dryer are designed to convert almost all the electric energy into thermal energy. Household appliances, such as those pictured in **Figure 22–11,** act like resistors when these devices are in a circuit. When charge, q, moves through a resistor, its potential difference is reduced by an amount V. As you have learned, the energy change is represented by qV. In practical use, it is the rate at which energy is changed, that is, the power, $P = E/t$, that is important. You learned earlier that current is the rate at which charge flows, $I = q/t$, and that power dissipated in a resistor is represented by $P = IV$. For a resistor, $V = IR$. Substituting this expression into the equation for electric power, you obtain the following.

$$P = I^2R$$

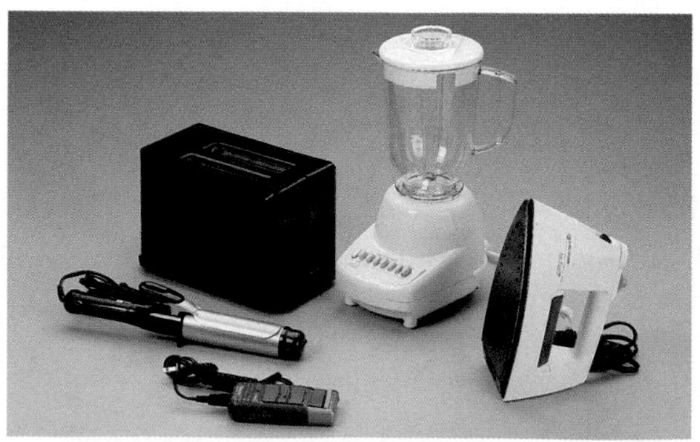

FIGURE 22–11 Which of these appliances are designed specifically to change electrical energy into thermal energy?

The power dissipated in a resistor is thus proportional to the square of the current that passes through it and to the resistance. The power is the rate at which energy is converted from one form to another. Energy is changed from electric to thermal energy, and the temperature of the resistor rises. That is, the resistor gets hot. If the resistor is an immersion heater or burner on an electric stove top for example, heat flows into cold water fast enough to bring the water to the boiling point in a few minutes.

If the power continues to be dissipated at this rate, then after a time, t, the energy converted to thermal energy, will be $E = Pt$. Because $P = I^2R$ the total energy that will be converted to thermal energy can be written in the following way.

$$E = I^2Rt$$

Appliances

Look closely at three electric appliances in your home. Record the power (watts) or the current (amps) of each appliance. Assume that each appliance operates at 120 V. Determine the resistance of each appliance.

Analyze and Conclude
Relate the power used in an appliance to its resistance.

Example Problem

Thermal Energy Produced by an Electric Current

A heater has a resistance of 10.0 Ω. It operates on 120.0 V.

a. What is the current through the resistance?

b. What thermal energy is supplied by the heater in 10.0 s?

Sketch the Problem

- Draw a circuit with a 120.0-V potential difference source and a 10.0 Ω resistor.

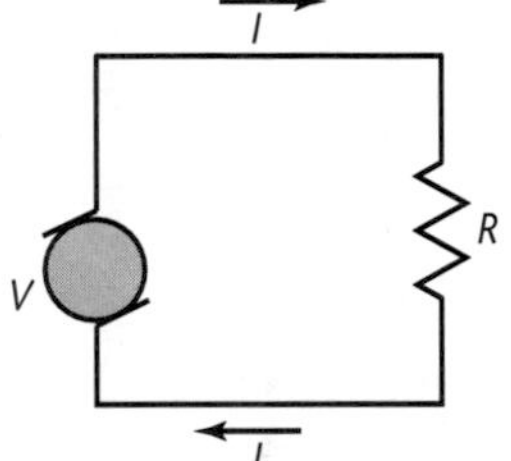

Calculate Your Answer

Known:

$R = 10.0\ \Omega$

$V = 120.0$ V

$t = 10.0$ s

Unknown:

$I = ?$

$E = ?$

Strategy:

a. Use $I = V/R$ to determine the current.

b. Use $E = I^2Rt$ to determine the energy.

Calculations:

$I = V/R = (120.0\text{ V})/(10.0\ \Omega) = 12.0$ A

$E = I^2Rt = (12.0\text{ A})^2(10.0\ \Omega)(10.0\text{ s}) = 14.4 \times 10^3$ J $= 14.4$ kJ

Check Your Answer

- Are the units correct? Current is measured in amperes, and energy is measured in joules.
- Is the magnitude realistic? Current produced by heating elements is about 10 A. Energy values of resistance heaters are large.

Practice Problems

14. A 15-Ω electric heater operates on a 120-V outlet.
- **a.** What is the current through the heater?
- **b.** How much energy is used by the heater in 30.0 s?
- **c.** How much thermal energy is liberated in this time?

15. A 30-Ω resistor is connected across a 60-V battery.
- **a.** What is the current in the circuit?
- **b.** How much energy is used by the resistor in 5.0 min?

16. A 100.0-W lightbulb is 20.0 percent efficient. This means that 20.0 percent of the electric energy is converted to light energy.
- **a.** How many joules does the lightbulb convert into light each minute it is in operation?
- **b.** How many joules of thermal energy does the lightbulb produce each minute?

17. The resistance of an electric stove element at operating temperature is 11 Ω.
- **a.** If 220 V are applied across it, what is the current through the stove element?
- **b.** How much energy does the element convert to thermal energy in 30.0 s?
- **c.** The element is being used to heat a kettle containing 1.20 kg of water. Assume that 70 percent of the heat is absorbed by the water. What is its increase in temperature during the 30.0s?

Pocket Lab

Heating Up

Look at the markings on three resistors. Predict which resistor would allow the most current through it using a constant voltage. Under the same conditions, predict which resistor will heat the most. Explain your prediction.

Test Your Predictions Tape a thermometer bulb to the resistor. Turn on the power for one minute. Measure the temperature. Allow the resistor to cool off and then repeat the procedure with the remaining two resistors.

CAUTION: *Do not touch the resistors with power supplied. They may be extremely hot. Wait two minutes after turning off the power to remove the thermometer.*

Transmission of Electric Energy

Niagara Falls and Hoover Dam can produce electric energy with little pollution. This hydroelectric energy, however, often must be transmitted long distances to reach homes and industries. How can the transmission take place with as little loss to thermal energy as possible?

Thermal energy is produced at a rate represented by $P = I^2R$. Electrical engineers call this unwanted thermal energy the joule heating or I^2R loss. To reduce this loss, either the current, I, or the resistance, R, must be reduced. All wires have some resistance, even though this resistance is small. For example, 1 km of the large wire used to carry electric current into a home has a resistance of 0.20 Ω.

Suppose that a farmhouse were connected directly to a power plant 3.5 km away, as depicted in **Figure 22–12.** The resistance in the wires needed to carry a current in a circuit to the home and back to the plant is 2(3.5 km)(0.2 Ω/km) = 1.4 Ω. An electric stove might cause a 41-A current through the wires. The power dissipated in the wires is represented by the following relationship.

$$P = I^2R = (41\ \text{A})^2 \times 1.4\ \Omega = 2400\ \text{W}$$

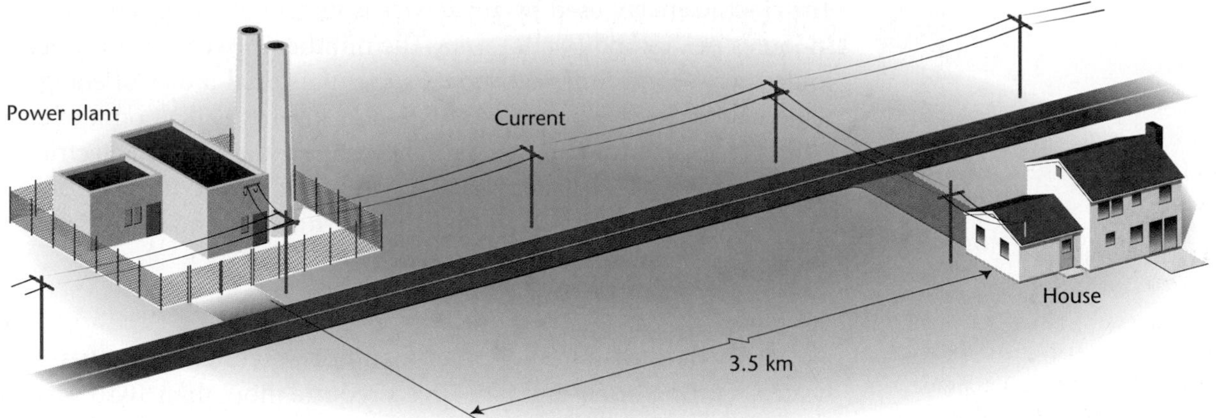

FIGURE 22–12 Electrical energy is transferred over long distances at high voltages to minimize I^2R losses.

All this power is converted to thermal energy and is wasted. This loss could be minimized by reducing the resistance. Cables of high conductivity and large diameter are currently used, but such cables are expensive and heavy. Because the loss is also proportional to the square of the current in the conductors, it is even more important to keep the current in the transmission lines low.

So why not just transmit the potential difference at household voltages directly instead of using the high-voltage transmission lines? The electrical energy per second (power) transferred over a long-distance transmission line is determined by the relationship $P = IV$. The current can be reduced without the power being reduced by increasing the voltage. Some long-distance lines use voltages of more than 500 000 volts. The resulting lower current reduces the I^2R loss in the lines by keeping the I^2 factor low. Long-distance transmission lines always operate at high voltage to reduce I^2R loss. The output voltage from the generating plant can be reduced upon arrival at electric substations to 2400 V and again to 240 V or 120 V before use at home.

More for Less

The Kilowatt-Hour

While electric companies often are called "power" companies, they really provide energy. When consumers pay their home electric bills, an example of which is shown in **Figure 22–13,** they actually pay for electric energy, not power.

a

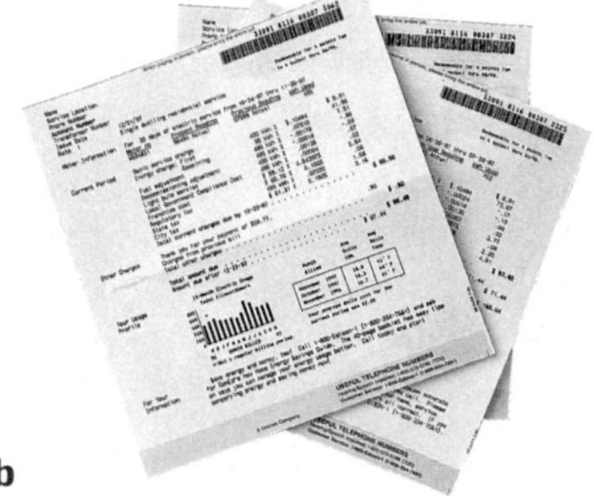

b

FIGURE 22–13 Watt-hour meters measure the amount of electric energy used by a consumer **(a).** The more current being used at a given time, the faster the horizontal disk in the center of the meter turns. Meter readings are then used in calculating the cost of energy **(b).**

The electric energy used by any device is its rate of energy consumption, in joules per second (watts) times the number of seconds it is operated. Joules per second times seconds, J·s/s, equals total joules of energy. The joule, also defined as a watt-second, is a relatively small amount of energy. This is too small for commercial sales use. For that reason, electric companies measure their energy sales in a unit of a large number of joules called a kilowatt-hour, kWh. A **kilowatt-hour** is equal to 1000 watts delivered continuously for 3600 seconds (1 hour).

$$1 \text{ kWh} = (1000 \text{ J/s})(3600 \text{ s}) = 3.6 \times 10^6 \text{ J}$$

Not many devices in homes other than hot-water heaters, stoves, heaters, curling irons, and hair dryers require more than 1000 watts of power. Ten 100-watt lightbulbs operating all at once would use one kilowatt hour of energy if they were left on for one full hour.

Example Problem

The Cost of Operating an Electric Device

A television set draws 2.0 A when operated on 120 V.

a. How much power does the set use?

b. If the set is operated for an average of 7.0 h/day, what energy in kWh does it consume per month (30 days)?

c. At 11¢ per kWh, what is the cost of operating the set per month?

Calculate Your Answer

Known:

$I = 2.0$ A

$V = 120.0$ V

$t = (7.0 \text{ h/day})(30 \text{ day})$

cost = 11¢/kWh

Unknown:

$E = ?$

total cost = ?

Strategy:

a. Use $P = IV$ to determine the power.

b. Use $E = Pt$ to determine the energy.

c. Use E/unit cost to determine the cost.

Calculations:

$P = IV = (2.0 \text{ A})(120.0 \text{ V}) = 2.4 \times 10^2 \text{ W}$

$E = Pt = (2.4 \times 10^2 \text{ W})(7.0 \text{ h/d})(30 \text{ d}) = 5.0 \times 10^4 \text{ Wh}$

$= 5.0 \times 10^1 \text{ kWh}$

cost = $(5.0 \times 10^1 \text{ kWh})(11¢\text{/kWh}) = \$5.50$

Check Your Answer:

- Are the units correct? Power is in watts, energy is in kilowatt-hours, and cost is in dollars.
- Is the magnitude realistic? A television does not require much power to operate. However, if you watch a lot of television, the cost of operation will be more than the cost for operating an appliance that requires more power.

Practice Problems

18. An electric space heater draws 15.0 A from a 120-V source. It is operated, on the average, for 5.0 h each day.
a. How much power does the heater use?
b. How much energy in kWh does it consume in 30 days?
c. At 11¢ per kWh, how much does it cost to operate the heater for 30 days?

19. A digital clock has a resistance of 12 000 Ω and is plugged into a 115-V outlet.
a. How much current does it draw?
b. How much power does it use?
c. If the owner of the clock pays 9¢ per kWh, how much does it cost to operate the clock for 30 days?

20. A four-slice toaster is rated at 1200 W and designed for use with 120-V circuits.
a. What is the resistance of the toaster?
b. How much current will flow when the toaster is turned on?
c. At what rate is heat generated in the toaster?
d. If all the heat generated were concentrated into 500 g of water at room temperature, at what rate would the temperature be rising?
e. The nichrome heating wires in the toaster total 2.00 m long if pulled straight. What is the electric field in the wire during operation if all the energy is converted in the nichrome wire?
f. If it takes 3 minutes to properly make toast and the cost per kilowatt-hour is 10 cents, how much does it cost to make one slice of toast?

F.Y.I.

In Ohm's day it was not possible to go to the hardware store and buy wire, something we take for granted today. Fortunately, he learned how to make wire while helping his father, who was a locksmith.

22.2 Section Review

2.1 A battery charges a capacitor. The capacitor is discharged through a photo flashlamp. List the forms of energy in these two operations.

2.2 A hair dryer operating from 120 V has two settings, hot and warm. In which setting is the resistance likely to be smaller? Why?

2.3 Why would a home using an electric range and hot-water heater have these appliances connected to 240 V rather than 120 V?

2.4 **Critical Thinking** When demand for electric power is high, power companies sometimes reduce the voltage, thereby producing a "brown out." What is being saved?

CHAPTER 22 REVIEW

Key Terms

22.1
- electric current
- conventional current
- battery
- photovoltaic cell
- electric circuit
- ampere
- resistance
- resistor
- potentiometer
- schematic
- parallel connection
- series connection

22.2
- kilowatt-hour

Summary

22.1 Current and Circuits

- Batteries, generators, and solar cells convert various forms of energy to electric energy.
- In an electric circuit, electric energy is transmitted from a device that produces electric energy to a resistor or other device that converts electric energy into the form needed.
- As a charge moves through resistors in a circuit, its potential energy is reduced. The energy released when the charge moves around the remainder of the circuit equals the work done to give the charge its initial potential energy.
- One ampere is one coulomb per second.
- Electric power is found by multiplying voltage by current.
- The resistance of a device is the ratio of the voltage across it divided by the current through it.
- In a device that obeys Ohm's law, the resistance remains constant as the voltage and current change.
- The current in a circuit can be varied by changing either the voltage or the resistance, or both.
- In a circuit diagram, conventional current is used. This is the direction in which a positive charge would move.

22.2 Using Electric Energy

- The thermal energy produced in a circuit from electric energy is equal to I^2Rt.
- In long-distance transmission, current is reduced without power being reduced by increasing the voltage.
- A kilowatt-hour, kWh, is an energy unit. It is equal to 3.6×10^6 J.

Reviewing Concepts

Section 22.1

1. Describe the energy conversions that occur in each of these devices.
   a. incandescent lightbulb
   b. clothes dryer
   c. digital clock radio
2. Define the unit of electric current in terms of fundamental MKS units.
3. Which wire conducts electricity with the least resistance: one with a large cross-sectional diameter or one with a small cross-sectional diameter?
4. How many electrons flow past a point in a wire each second if the wire has a current of 1 A?

Section 22.2

5. Why do lightbulbs burn out more frequently just as they are switched on rather than while they are operating?
6. A simple circuit consists of a battery, a resistor, and some connecting wires. Draw a circuit schematic of this simple circuit. Show the polarity of the battery and the direction of the conventional current.
7. A simple circuit consists of a resistor, a battery, and connecting wires.
   a. How must an ammeter be connected in a circuit to correctly read the current?
   b. How must a voltmeter be connected to a resistor in order to read the potential difference across it?
8. If a battery is short-circuited by a heavy copper wire being connected from one terminal to the other, the temperature of the copper wire rises.

Why does this happen?

9. Why does a wire become warmer as charges flow through it?
10. What electrical quantities must be kept small to transmit electric energy economically over long distances?

Applying Concepts

11. When a battery is connected to a complete circuit, charges flow in the circuit almost instantaneously. Explain.
12. Explain why a cow that touches an electric fence experiences a mild shock.
13. Why can birds perch on high-voltage lines without being injured?
14. Describe two ways to increase the current in a circuit.
15. You have two lightbulbs that work on a 120-V circuit. One is 50 W, the other is 100 W. Which bulb has a higher resistance? Explain.
16. If the voltage across a circuit is kept constant and the resistance is doubled, what effect does this have on the circuit's current?
17. What is the effect on the current in a circuit if both the voltage and the resistance are doubled? Explain.
18. Sue finds a device that looks like a resistor. When she connects it to a 1.5-V battery, only 45×10^{-6} A is measured, but when a 3.0-V battery is used, 25×10^{-3} A is measured. Does the device obey Ohm's law?
19. If the ammeter in **Figure 22–5** were moved to the bottom of the diagram, would the ammeter have the same reading? Explain.
20. Two wires can be placed across the terminals of a 6.0-V battery. One has a high resistance, and the other has a low resistance. Which wire will produce thermal energy at the faster rate? Why?

Problems

Section 22.1

LEVEL 1

21. The current through a toaster connected to a 120-V source is 8.0 A. What power is dissipated by the toaster?
22. A current of 1.2 A is measured through a lightbulb when it is connected across a 120-V source. What power is dissipated by the bulb?
23. A lamp draws 0.50 A from a 120-V generator.
    a. How much power is delivered?
    b. How much energy does the lamp convert in 5.0 min?
24. A 12-V automobile battery is connected to an electric starter motor. The current through the motor is 210 A.
    a. How many joules of energy does the battery deliver to the motor each second?
    b. What power, in watts, does the motor use?
25. A 4000-W clothes dryer is connected to a 220-V circuit. How much current does the dryer draw?
26. A flashlight bulb is connected across a 3.0-V potential difference. The current through the lamp is 1.5 A.
    a. What is the power rating of the lamp?
    b. How much electric energy does the lamp convert in 11 min?
27. A resistance of 60 Ω has a current of 0.40 A through it when it is connected to the terminals of a battery. What is the voltage of the battery?
28. What voltage is applied to a 4.0-Ω resistor if the current is 1.5 A?
29. What voltage is placed across a motor of 15-Ω operating resistance if there is 8.0 A of current?
30. A voltage of 75 V is placed across a 15-Ω resistor. What is the current through the resistor?
31. A 20.0-Ω resistor is connected to a 30.0-V battery. What is the current through the resistor?
32. A 12-V battery is connected to a device and 24 mA of current is measured. If the device obeys Ohm's law, how much current is present when a 24-V battery is used?
33. A person with dry skin has a resistance from one arm to the other of about 1×10^5 Ω. When skin is wet, resistance drops to about 1.5×10^3 Ω. (refer to **Table 22–1**).
    a. What is the minimum voltage placed across the arms that would produce a current that could be felt by a person with dry skin?
    b. What effect would the same voltage have if the person had wet skin?

c. What would be the minimum voltage that would produce a current that could be felt when the skin is wet?

34. A lamp draws a 66-mA current when connected to a 6.0-V battery. When a 9.0-V battery is used, the lamp draws 75 mA.
   a. Does the lamp obey Ohm's law?
   b. How much power does the lamp dissipate at 6.0 V?
   c. How much power does it dissipate at 9.0 V?

LEVEL 2

35. How much energy does a 60-W lightbulb use in half an hour? If the lightbulb is 12 percent efficient, how much thermal energy does it generate during the half hour?

36. Some students connected a length of nichrome wire to a variable power supply that could produce from 0.0 V to 10.0 V across the wire. They then measured the current through the wire for several voltages. They recorded the data showing the voltages used and currents measured. These are presented in **Table 22–2.**

TABLE 22–2

Voltage V (volts)	Current I (amps)	Resistance $R = V/I$ (ohms)
2.00	0.014	
4.00	0.027	
6.00	0.040	
8.00	0.052	
10.00	0.065	
- 2.00	- 0.014	
- 4.00	- 0.028	
- 6.00	- 0.039	
- 8.00	- 0.051	
- 10.00	- 0.064	

   a. For each measurement, calculate the resistance.
   b. Graph I versus V.
   c. Does the nichrome wire obey Ohm's law? If not, for all the voltages, specify the voltage range for which Ohm's law holds.

37. The current through a lamp connected across 120 V is 0.40 A when the lamp is on.
   a. What is the lamp's resistance when it is on?
   b. When the lamp is cold, its resistance is one fifth as large as it is when the lamp is hot. What is its cold resistance?
   c. What is the current through the lamp as it is turned on if it is connected to a potential difference of 120 V?

38. The graph in **Figure 22–14** shows the current through a device called a silicon diode.
   a. A potential difference of +0.70 V is placed across the diode. What resistance would be calculated?
   b. What resistance would be calculated if a +0.60-V potential difference were used?
   c. Does the diode obey Ohm's law?

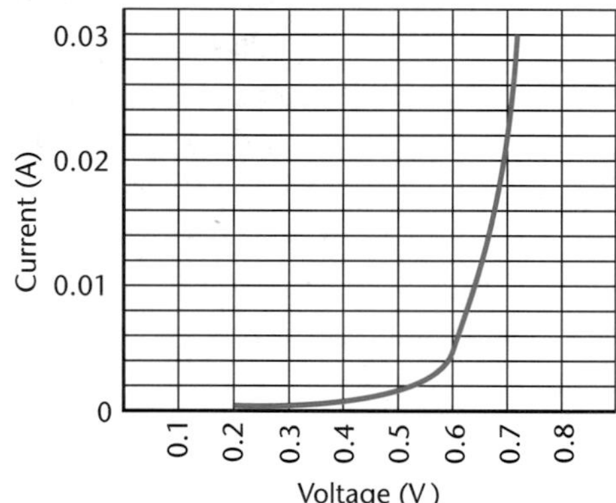

FIGURE 22–14

39. Draw a schematic diagram to show a circuit that includes a 90-V battery, an ammeter, and a resistance of 45 Ω connected in series. What is the ammeter reading? Draw arrows showing the direction of conventional current.

40. Draw a series circuit diagram to include a 16-Ω resistor, a battery, and an ammeter that reads 1.75 A. Conventional current is measured through the meter from left to right. Indicate the positive terminal and the voltage of the battery.

Section 22.2

LEVEL 1

41. What is the maximum current that should be allowed in a 5.0-W, 220-Ω resistor?

42. The wire in a house circuit is rated at 15.0 A and has a resistance of 0.15 Ω.
  a. What is its power rating?
  b. How much heat does the wire give off in 10.0 min?

43. A current of 1.2 A is measured through a 50-Ω resistor for 5.0 min. How much heat is generated by the resistor?

44. A 6.0-Ω resistor is connected to a 15-V battery.
  a. What is the current in the circuit?
  b. How much thermal energy is produced in 10 min?

45. A 110-V electric iron draws 3.0 A of current. How much thermal energy is developed each hour?

LEVEL 2

46. An electric motor operates a pump that irrigates a farmer's crop by pumping 10 000 L of water a vertical distance of 8.0 m into a field each hour. The motor has an operating resistance of 22.0 Ω and is connected across a 110-V source.
  a. What current does it draw?
  b. How efficient is the motor?

47. A transistor radio operates by means of a 9.0-V battery that supplies it with a 50-mA current.
  a. If the cost of the battery is \$0.90 and it lasts for 300 h, what is the cost per kWh to operate the radio in this manner?
  b. The same radio, by means of a converter, is plugged into a household circuit by a homeowner who pays 8¢ per kWh. What does it now cost to operate the radio for 300 h?

Critical Thinking Problems

48. A heating coil has a resistance of 4.0 Ω and operates on 120 V.
  a. What is the current in the coil while it is operating?
  b. What energy is supplied to the coil in 5.0 min?
  c. If the coil is immersed in an insulated container holding 20.0 kg of water, what will be the increase in the temperature of the water? Assume that 100 percent of the heat is absorbed by the water.
  d. At 8¢ per kWh, how much does it cost to operate the heating coil 30 min per day for 30 days?

49. An electric heater is rated at 500 W.
  a. How much energy is delivered to the heater in half an hour?
  b. The heater is being used to heat a room containing 50 kg of air. If the specific heat of air is 1.10 kJ/kg·°C, and 50 percent of the thermal energy heats the air in the room, what is the change in air temperature in half an hour?
  c. At 8¢ per kWh, how much does it cost to run the heater 6.0 h per day for 30 days?

Going Further

Formulating Models How much energy is stored in a capacitor? The energy needed to increase the potential difference of a charge, q is represented by $E = qV$. But in a capacitor, $V = q/C$. Thus, as charge is added, the potential difference increases. But as more charge is added, it takes more energy to add the additional charge.

Consider a 1.0-F "supercap" used as an energy storage device in a personal computer. Plot a graph of V as the capacitor is charged by adding 5.0 C to it. What is the voltage across the capacitor? The area under the curve is the energy stored in the capacitor. Find the energy in joules. Is it equal to the total charge times the final potential difference? Explain.

*inter***NET**
CONNECTION

Follow the link on the Glencoe Homepage at **www.glencoe.com/sec/science** to find out more about this chapter.

Which Bulb Burns Brighter?

One is a 60-watt bulb and the other a 100-watt bulb, and they are connected in an electric circuit.

CHAPTER

23 Series and Parallel Circuits

The answer may surprise you. You know which of the two lightbulbs you would choose for your desk lamp if you needed more light. A watt is a unit of power, so you might expect that a 100-W lightbulb would produce more light than a 60-W bulb. But the electric circuit that connects these bulbs is not like the parallel circuit that would connect them in your home. In the photo, the bulbs are connected in series. Does that change your answer? Why or why not? To answer these questions, you need to know how a series electric circuit differs from a parallel circuit.

In Chapter 22, you studied circuits that have one source of electrical energy, for example, a battery and one device such as a motor or a lamp that converted the energy to another form. But circuits can be much more versatile than that. Now you'll explore several ways in which devices may be connected in electric circuits. By the end of this chapter, you'll have no trouble deciding which of the two bulbs is brighter, and you'll be able to explain why.

WHAT YOU'LL LEARN

- You will distinguish between parallel and series circuits and series-parallel combinations and solve problems dealing with them.
- You will explain the function of fuses, circuit breakers, and ground fault interrupters, and describe ammeters and voltmeters.

WHY IT'S IMPORTANT

- Electrical circuits are the basis of every electrical device, from electric lights to microwave ovens to computers. Understanding circuits helps you to use them, and to use them safely.

interNET CONNECTION

Follow the link for this chapter on the Glencoe Homepage at **www.glencoe.com/sec/science** to find out more about electrical circuits.

23.1 Simple Circuits

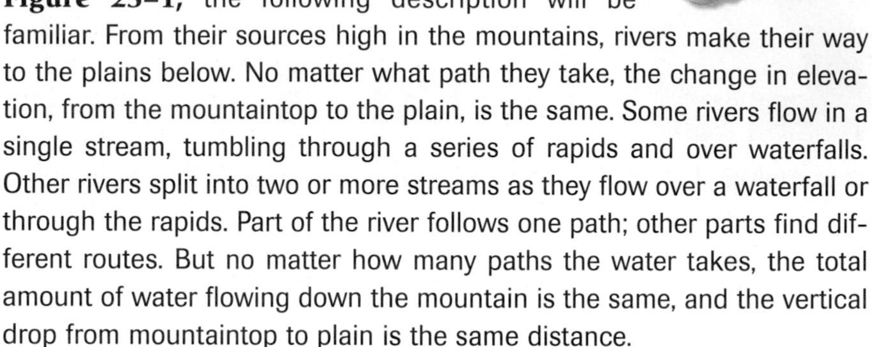

OBJECTIVES

- **Describe** both a series connection and a parallel connection and **state** the important characteristics of each.
- **Calculate** current, voltage drops, and equivalent resistance for devices connected in series and in parallel.
- **Describe** a voltage divider and **solve** problems involving one.

If you have ever had the chance to explore rivers in the mountains, such as the river shown below in **Figure 23–1,** the following description will be familiar. From their sources high in the mountains, rivers make their way to the plains below. No matter what path they take, the change in elevation, from the mountaintop to the plain, is the same. Some rivers flow in a single stream, tumbling through a series of rapids and over waterfalls. Other rivers split into two or more streams as they flow over a waterfall or through the rapids. Part of the river follows one path; other parts find different routes. But no matter how many paths the water takes, the total amount of water flowing down the mountain is the same, and the vertical drop from mountaintop to plain is the same distance.

Mountain rivers can serve as a model for electric circuits. The distance the water drops is similar to potential difference. The amount of water flowing through the river each second is similar to current. Narrow rapids are like a large resistance. But what is similar to a battery or generator? Just as in an electrical circuit, an energy source is needed to raise water to the top of the mountain. As you learned in Earth science, the source of energy is the sun. Solar energy evaporates water from lakes and seas and forms clouds that release rain or snow to fall on the tops of mountains. Think about the mountain river model as you read about the current in electrical circuits.

FIGURE 23–1 No matter what path the water of a river takes down a mountain, the amount of water and the drop in elevation are the same.

Series Circuits

Pat, Chris, and Ali were connecting two identical lamps to a battery as illustrated in **Figure 23–2.** Before making the final connection to the battery, their teacher asked them to predict the brightness of the two lamps. They knew that the brightness of a lamp depends on the current flowing through it. Pat said that only the lamp closer to the + terminal of the battery would light because all the current would be converted into light. Chris said that the second lamp would light, but it would be dimmer than the other one because some electrical energy would be changed into thermal and light energy. Consequently, there would be less electrical energy left for the second lamp. Ali said that both lamps would be equally bright because current is a flow of charge, and because the charge leaving the first lamp had nowhere else to go in the circuit except through the second lamp, the current would be the same in the two lamps. Who do you think is right?

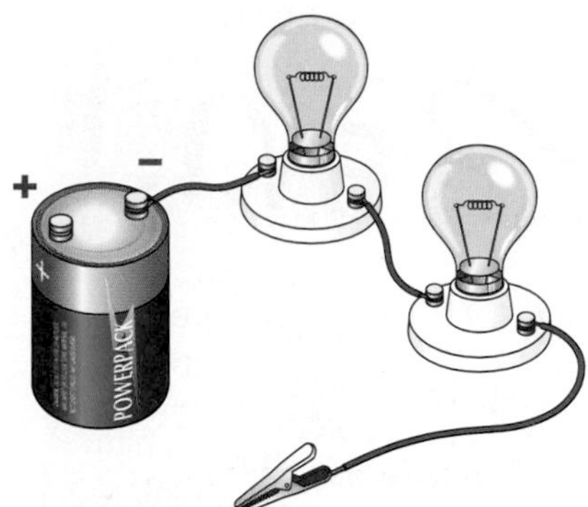

FIGURE 23–2 What is your prediction about the brightness of the two lightbulbs?

If you consider the mountain river model for this circuit, you'll see that Ali is correct. Charge has only one path to follow. Recall from Chapter 20 that charge cannot be created or destroyed, so the same amount of charge must leave a circuit as enters the circuit. This means that the current is the same everywhere in the circuit. If you connect three ammeters into a circuit as shown in **Figure 23–3,** they all have the same value. A circuit such as this, in which all current travels through each device, is called a **series circuit.**

But how could you answer Chris? If the current is the same, what changes in the lamp to produce the thermal and light energy? Recall that power, the rate at which electrical energy is converted, is represented by $P = IV$. Thus, if there is a potential difference or voltage drop across the lamp, then electrical energy is being converted into another form. The resistance of the lamp is defined as $R = V/I$. Thus, the potential difference, also called the voltage drop, is $V = IR$.

What is the current in the series circuit?

From the river model, you know that the sum of the drops in height at each rapid is equal to the total drop from the top of the mountain to sea level. In the electrical circuit, the increase in voltage provided by the generator or other energy source, V_{source}, is equal to the sum of voltage drops across the lamps A and B.

$$V_{source} = V_A + V_B$$

Because the current, I, through the lamps is the same, $V_A = IR_A$ and $V_B = IR_B$. Therefore, $V_{source} = IR_A + IR_B$ or $V_{source} = I(R_A + R_B)$. The current through the circuit is represented by the following.

$$I = \frac{V_{source}}{R_A + R_B}$$

This equation applies to any number of resistances in series, not just two. The same current would exist with a single resistor, R, that has a

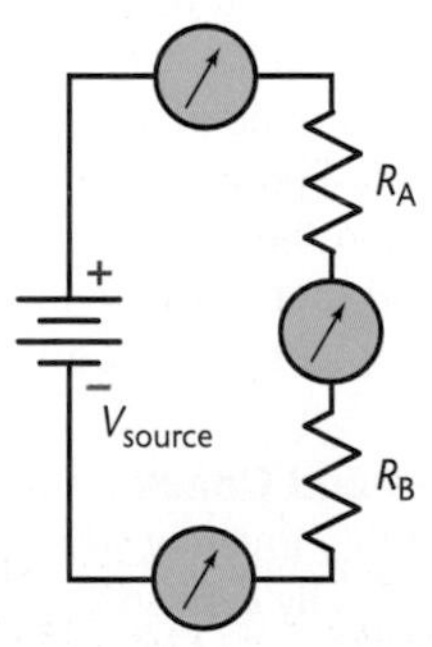

FIGURE 23–3 The ammeters show that in a series circuit the current is the same everywhere.

resistance equal to the sum of the resistances of the two lamps. Such a resistance is called the equivalent resistance of the circuit. For resistors in series, the **equivalent resistance** is the sum of all the individual resistances.

$$R = R_A + R_B + \ldots$$

Notice that the equivalent resistance is larger than any single resistance. Therefore, if the battery voltage doesn't change, adding more devices in series always decreases the current. To find the current, I, through a series circuit, first calculate the equivalent resistance, R, and then use the following equation to calculate I.

$$I = \frac{V_{source}}{R}$$

Practice Problems

1. Three 20-Ω resistors are connected in series across a 120-V generator. What is the equivalent resistance of the circuit? What is the current in the circuit?
2. A 10-Ω resistor, a 15-Ω resistor, and a 5-Ω resistor are connected in series across a 90-V battery. What is the equivalent resistance of the circuit? What is the current in the circuit?
3. Consider a 9-V battery in a circuit with three resistors connected in series.
   a. If the resistance of one of the resistors increases, how will the series resistance change?
   b. What will happen to the current?
   c. Will there be any change in the battery voltage?
4. A string of holiday lights has ten bulbs with equal resistances connected in series. When the string of lights is connected to a 120-V outlet, the current through the bulbs is 0.06 A.
   a. What is the equivalent resistance of the circuit?
   b. What is the resistance of each bulb?
5. Calculate the voltage drops across the three resistors in problem 2, and check to see that their sum equals the voltage of the battery.

Voltage drops in a series circuit

In any circuit, the net change in potential as current moves through it must be zero. This is because the electrical energy source in the circuit, the battery or generator, raises the potential. As current passes through the resistors, the potential drops an amount equal to the increase, and, therefore, the net change is zero.

The potential drop across each resistor in a series circuit can be calculated by rearranging the equation that defines resistance, $R = V/I$, to

F.Y.I.

Henry Cavendish used the direct approach to measure the strength of an electric current. Lacking the appropriate instruments, he instead shocked himself with the current and then estimated the pain.

Series Resistance

Hook up a power supply, a resistor, and an ammeter in a series circuit. Predict what will happen to the current in the circuit when a second, identical resistor is added in series to the circuit. Predict the new currents when the circuit contains three and four resistors in series. Explain your prediction. Try it.

Analyze and Conclude Make a data table to show your results. Briefly explain your results. (**Hint:** Include the idea of resistance.)

solve for V, $V = IR$. First, find the equivalent resistance, R, in the circuit by calculating the sum of all the individual resistances. Then, to find the current, which is the same everywhere in the circuit, use the equivalent resistance and the equation $I = V/R$, where V is the potential drop. Having determined the current in the circuit, multiply I by the resistance of the individual resistor to find the potential drop across that resistor.

An important application of series resistors is the voltage divider. A **voltage divider** is a series circuit used to produce a voltage source of desired magnitude from a higher-voltage battery. Suppose you have a 9-V battery but need a 5-V potential source. A voltage divider can supply this voltage. Consider the circuit shown in **Figure 23–4**. Two resistors, R_A and R_B, are connected in series across a battery of magnitude V. The equivalent resistance of the circuit is $R = R_A + R_B$. The current, I, is represented by the following equation.

$$I = \frac{V}{R} = \frac{V}{R_A + R_B}$$

The desired voltage, 5 V, is the voltage drop, V_B, across resistor R_B.

$$V_B = IR_B.$$

I is replaced by the preceding equation.

$$V_B = IR_B = \left(\frac{V}{R_A + R_B}\right) R_B$$

$$V_B = \frac{VR_B}{R_A + R_B}$$

Voltage dividers are often used with sensors such as photoresistors. The resistance of a photoresistor depends upon the amount of light that strikes it. Photoresistors are made of semiconductors such as silicon, selenium, and cadmium sulfide. A typical photoresistor can have a resistance of 400 Ω when light strikes it, but 400 000 Ω when in the dark. The output voltage of a voltage divider that uses a photoresistor depends upon the amount of light striking the photoresistor sensor. This circuit can be used as a light meter, such as the one in **Figure 23–5.** In this device, an electronic circuit detects the potential difference and converts it to a measurement of illuminance that can be read on the digital display.

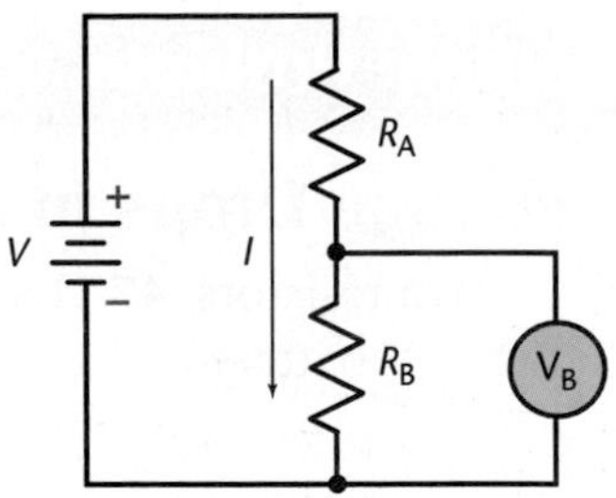

FIGURE 23–4 The values of R_A and R_B are chosen such that the voltage drop across R_B is the desired voltage.

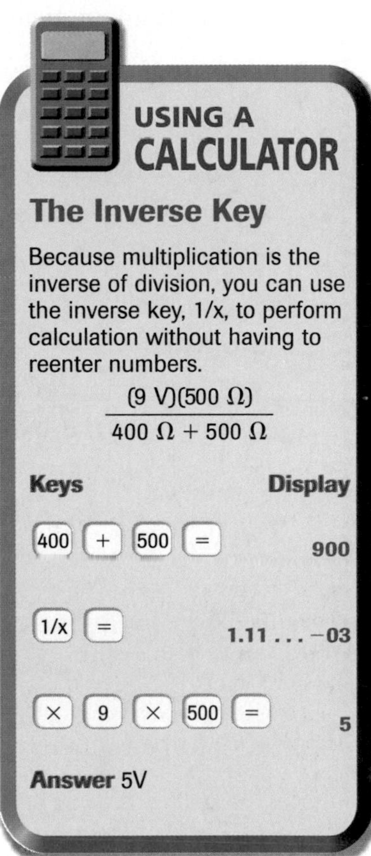

The Inverse Key

Because multiplication is the inverse of division, you can use the inverse key, 1/x, to perform calculation without having to reenter numbers.

$$\frac{(9\ \text{V})(500\ \Omega)}{400\ \Omega + 500\ \Omega}$$

Keys	**Display**
400 + 500 =	900
1/x =	1.11 . . . −03
× 9 × 500 =	5

Answer 5V

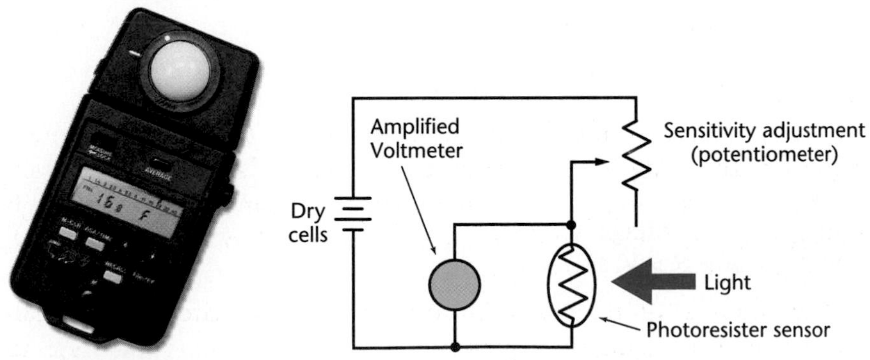

FIGURE 23–5 The output voltage of this voltage divider depends upon the amount of light striking the photoresister sensor. This is the basis for the light meter shown at left.

Example Problem

Voltage Drops in a Series Circuit

Two resistors, 47-Ω and 82-Ω, are connected in series across a 45.0-V battery.

a. What is the current in the circuit?

b. What is the voltage drop across each resistor?

c. The 47-Ω resistor is replaced by a 39-Ω resistor. Will the current increase, decrease, or remain the same?

d. What will happen to the voltage drop across the 82-Ω resistor?

Sketch the Problem

- Draw a schematic of the circuit.
- Include an ammeter and voltmeters.

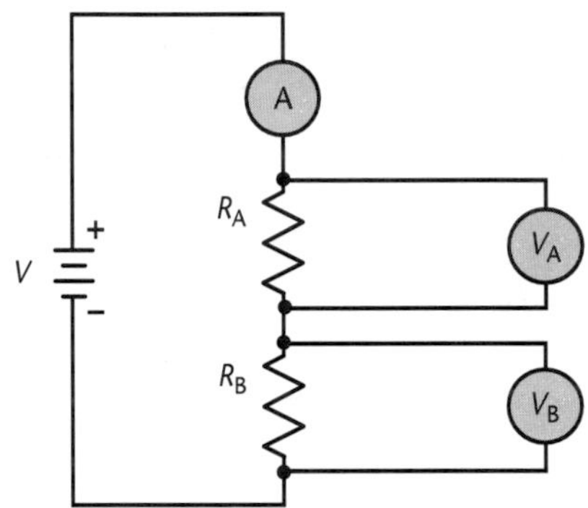

Calculate Your Answer

Known:	Unknown:
$V_{source} = 45.0$ V	$I = ?$
$R_A = 47\ \Omega$	$V_A = ?$
$R_B = 82\ \Omega$	$V_B = ?$
	effects of changing R_A

Strategy:

a. To determine the current, first find the equivalent resistance.

Calculations:

$$R = R_A + R_B$$

$$I = \frac{V_{source}}{R} = \frac{V_{source}}{R_A + R_B}$$

$$I = \frac{45.0\text{ V}}{47\ \Omega + 82\ \Omega} = 0.349\text{ A}$$

b. Use $V = IR$ for each resistor.

$$V_A = IR_A = (0.349)(47\ \Omega) = 16\text{ V}$$

$$V_B = IR_B = (0.349)(82\ \Omega) = 29\text{ V}$$

c. Calculate current again using R_A as 39 Ω.

$$I = \frac{V_{source}}{R_A + R_B} = \frac{45.0\text{ V}}{39\ \Omega + 82\ \Omega} = 0.372\text{ A}$$

The current will increase.

d. Determine new voltage drop.

$$V_B = IR_B = (0.372)(82\ \Omega) = 31\text{ V}$$

The voltage will increase.

Check Your Answer

- Are the units correct? Current, A = V/Ω, voltage, V = AΩ.
- Is the magnitude realistic? Numerically, $R > V$, so $I < 1$. R decreases so I increases. V_B changes because it depends on I.

Example Problem

Voltage Divider

A 9.0-V battery and two resistors, 400 Ω and 500 Ω, are connected as a voltage divider. What is the voltage across the 500-Ω resistor?

Sketch the Problem

- Draw the battery and resistors in a series circuit.

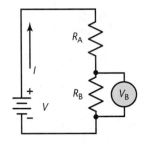

Calculate Your Answer

Known:	Strategy:	Calculations:
$V_{source} = 9.0$ V	Write the expression for the current through the circuit.	$I = \frac{V_{source}}{R}$
$R_A = 400$ Ω		
$R_B = 500$ Ω	Determine equivalent resistance, R.	$R = R_A + R_B$
Unknown:	Use voltage drop equation to determine V_B.	$V_B = IR_B = \frac{(V_{source})(R_B)}{R_A + R_B}$
$V_B = ?$		$V_B = \frac{(9.0\text{ V})(500\ \Omega)}{400\ \Omega + 500\ \Omega} = 5\text{ V}$

Check Your Answer

- Are the units correct? V = V Ω/Ω.
- Is the magnitude realistic? The voltage drop is less than the battery voltage. Because 500 Ω is more than half of the equivalent resistance, the voltage drop is more than half the battery voltage.

Practice Problems

6. A 20.0-Ω resistor and a 30.0-Ω resistor are connected in series and placed across a 120-V potential difference.
   a. What is the equivalent resistance of the circuit?
   b. What is the current in the circuit?
   c. What is the voltage drop across each resistor?
   d. What is the voltage drop across the two resistors together?
7. Three resistors of 3.0 kΩ, 5.0 kΩ, and 4.0 kΩ are connected in series across a 12-V battery.
   a. What is the equivalent resistance?
   b. What is the current through the resistors?
   c. What is the voltage drop across each resistor?
   d. Find the total voltage drop across the three resistors.

The largest voltages in your home are in your television set, where 15 000 V to 20 000 V are common. The largest currents are likely to be the 40 A in an electric range.

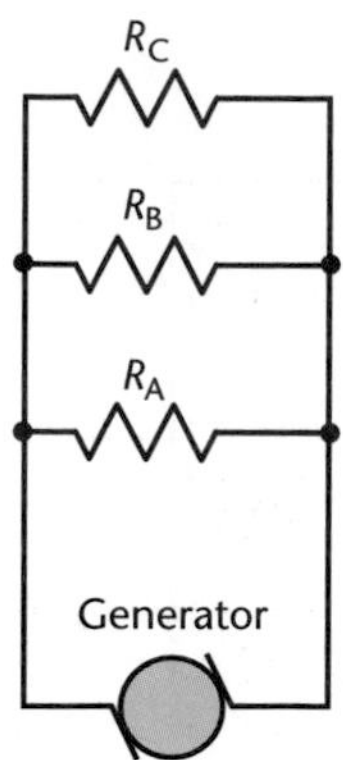

FIGURE 23–6 The parallel paths for current in this diagram are analogous to the paths that a river may take down the mountain.

8. A photoresistor is used in a voltage divider as R_B. $V = 9.0$ V and $R_A = 500\ \Omega$.

a. What is the output voltage, V_B, across R_B, when a bright light strikes the photoresistor and $R_B = 475\ \Omega$?

b. When the light is dim, $R_B = 4.0$ kΩ. What is V_B?

c. When the photoresistor is in total darkness, $R_B = 0.40$ MΩ ($0.40 \times 10^6\ \Omega$). What is V_B?

9. A student makes a voltage divider from a 45-V battery, a 475-kΩ ($475 \times 10^3\ \Omega$) resistor, and a 235-kΩ resistor. The output is measured across the smaller resistor. What is the voltage?

Parallel Circuits

Look at the circuit shown in **Figure 23–6.** How many current paths are there? The current from the generator can go through any of the three resistors. A circuit in which there are several current paths is called a **parallel circuit**. The three resistors are connected in parallel; both ends of the three paths are connected together. In the mountain river model for circuits, such a circuit is illustrated by three paths for the water over a waterfall. Some paths may have a large flow of water, others a small flow. The sum of the flows, however, is equal to the total flow of water over the falls. In addition, it doesn't matter which channel the water flows through because the drop in height is the same. Similarly, in a parallel electrical circuit, the total current is the sum of the currents through each path, and the potential difference across each path is the same.

What is the current through each resistor? It depends upon the individual resistance. For example, in **Figure 23–7,** the potential difference across each resistor is 120 V. The current through a resistor is given by $I = V/R$, so you can calculate the current through the 24-Ω resistor as $I = (120\text{V})/(24\ \Omega) = 5$ A. Calculate the currents through the other two resistors. The total current through the generator is the sum of the currents through the three paths, in this case, 38 A.

What would happen if the 6-Ω resistor were removed from the circuit? Would the current through the 24-Ω resistor change? That current depends only upon the potential difference across it and its resistance, and neither has changed, so the current is unchanged. The same is true

F.Y.I.

André Marie Ampère was a physicist, chemist, and mathematician. His discovery that magnetic fields are produced by the flow of current through a wire led to the invention of the galvanometer.

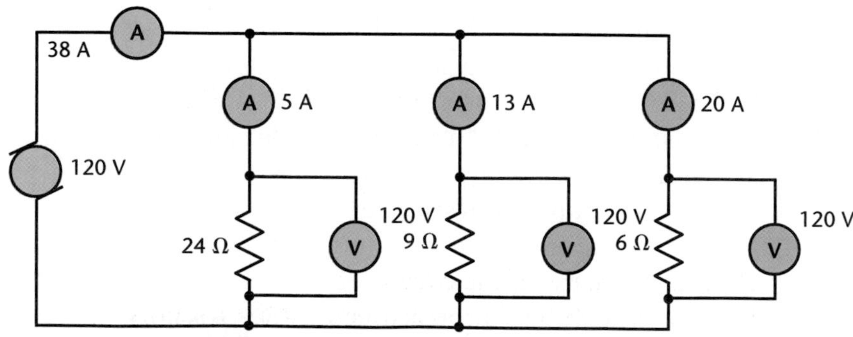

FIGURE 23–7 In a parallel circuit, the reciprocal of the total resistance is equal to the sum of the reciprocals of the individual resistances.

for the current through the 9-Ω resistor. The branches of a parallel circuit are independent of each other. The total current through the generator, however, would change, and the sum of the currents in the branches would then be 18 A.

How can you find the equivalent resistance of a parallel circuit? In **Figure 23–7**, the total current through the generator is 38 A. A single resistor that would have a 38-A current when 120 V were placed across it would be represented by the following equation.

$$R = \frac{V}{I} = \frac{120\ \text{V}}{38\ \text{A}} = 3.2\ \Omega$$

Notice that this resistance is smaller than that of any of the three resistors in parallel. Placing two or more resistors in parallel always decreases the equivalent resistance of a circuit. The resistance decreases because each new resistor provides an additional path for current, increasing the total current while the potential difference remains unchanged.

To calculate the equivalent resistance of a parallel circuit, first note that the total current is the sum of the currents through the branches. If I_A, I_B, and I_C are the currents through the branches and I is the total current, then $I = I_A + I_B + I_C$.

The potential difference across each resistor is the same, so the current through each resistor, for example, R_A, can be found from $I_A = V/R_A$. Therefore, this becomes the equation for the sum of the currents:

$$\frac{V}{R} = \frac{V}{R_A} + \frac{V}{R_B} + \frac{V}{R_C}.$$

Dividing both sides of the equation by V provides an equation for the equivalent resistance of the three parallel resistors.

$$\frac{1}{R} = \frac{1}{R_A} + \frac{1}{R_B} + \frac{1}{R_C}$$

This equation can be used for any number of resistors in parallel.

Parallel Resistance

Hook up a power supply, a resistor, and an ammeter in a series circuit. Predict what will happen to the current in the circuit when a second, identical resistor is added in parallel to the first. Predict the new currents when the circuit contains three and four resistors in parallel. Explain your prediction. Try it.

Analyze and Conclude Make a data table to show your results. Briefly explain your results. (**Hint:** Include the idea of resistance.)

Example Problem

Equivalent Resistance and Current in a Parallel Circuit

Three resistors, 60.0 Ω, 30.0 Ω, and 20.0 Ω, are connected in parallel across a 90.0-V battery.

a. Find the current through each branch of the circuit.

b. Find the equivalent resistance of the circuit.

c. Find the current through the battery.

Sketch the Problem

- Draw a schematic of the circuit.
- Include ammeters to show the paths of the currents.

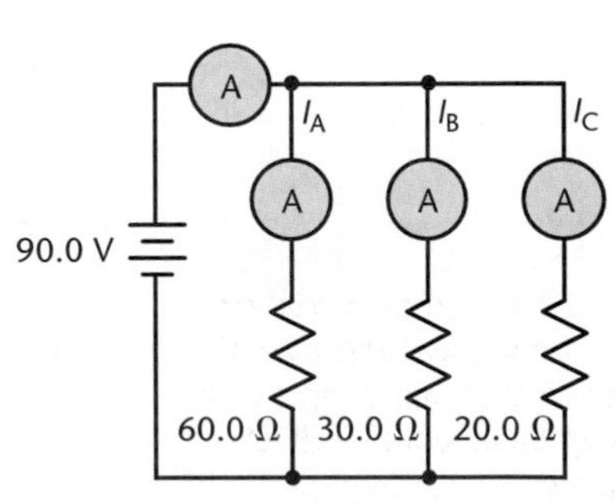

Calculate Your Answer

Known:

$R_A = 60.0\ \Omega$ $R_C = 20.0\ \Omega$

$R_B = 30.0\ \Omega$ $V = 90.0\ V$

Unknown:

$I_A = ?$ $I_C = ?$ $R = ?$

$I_B = ?$ $I = ?$

Strategy:

a. The voltage across each resistor is the same, so use $I = \frac{V}{R}$ for each branch.

Calculations:

$$I_A = \frac{V}{R_A} = \frac{90.0\ V}{60.0\ \Omega} = 1.50\ A$$

$$I_B = \frac{V}{R_B} = \frac{90.0\ V}{30.0\ \Omega} = 3.00\ A$$

$$I_C = \frac{V}{R_C} = \frac{90.0\ V}{20.0\ \Omega} = 4.50\ A$$

b. Use the equivalent resistance equation for parallel circuits.

$$\frac{1}{R} = \frac{1}{R_A} + \frac{1}{R_B} + \frac{1}{R_C}$$

$$\frac{1}{R} = \frac{1}{60\ \Omega} + \frac{1}{30\ \Omega} + \frac{1}{20\ \Omega} = 0.100\ \Omega^{-1}$$

$$R = 10.0\ \Omega$$

c. Use $I = \frac{V}{R}$ to find the total current.

$$I = \frac{V}{R} = \frac{90.0\ V}{10.0\ \Omega} = 9.00\ A$$

Check Your Answer

- Are your units correct? Currents are in amps, resistances are in ohms.
- Do the signs make sense? All are positive, as they should be.
- Is the magnitude realistic? $I_A + I_B + I_C = I$

Practice Problems

10. Three 15-Ω resistors are connected in parallel and placed across a 30-V battery.
  a. What is the equivalent resistance of the parallel circuit?
  b. What is the current through the entire circuit?
  c. What is the current through each branch of the circuit?

11. A 120.0-Ω resistor, a 60.0-Ω resistor, and a 40.0-Ω resistor are connected in parallel and placed across a 12.0-V battery.
  a. What is the equivalent resistance of the parallel circuit?
  b. What is the current through the entire circuit?
  c. What is the current through each branch of the circuit?

12. Suppose one of the 15.0-Ω resistors in problem 10 is replaced by a 10.0-Ω resistor.
  a. Does the equivalent resistance change? If so, how?

b. Does the amount of current through the entire circuit change? In what way?

c. Does the amount of current through the other 15.0-Ω resistors change? In what way?

Now that you have learned about both parallel and series circuits, you can analyze the brightness of the 60-W and 100-W lamps shown in the photo at the beginning of this chapter. The brightness of a lightbulb is proportional to the power dissipated by it. Used in the normal way, each bulb would be connected across 120 V. Based on what you learned in Chapter 22, the resistance of the 60-W bulb is higher than that of the 100-W bulb. But when the bulbs are connected in series, the current through the two bulbs is the same, so $P = I^2R$. The higher-resistance lamp, the 60-W lamp, now dissipates more power and glows brighter than the 100-W lamp.

Which Bulb Burns Brighter?

23.1 Section Review

1.1 Are car headlights connected in series or parallel? Draw on your experience.

1.2 Lamp dimmers often contain variable resistors.

- **a.** Would a dimmer be hooked in series or in parallel with the lamp to be controlled? Why?
- **b.** Should the resistance of the dimmer be increased or decreased to dim the lamp?

1.3 A switch is connected in series with a 75-W bulb to a source of 120 V.

- **a.** What is the potential difference across the switch when it is closed or on?
- **b.** What is the potential difference across the switch when it is open, or off? Explain.

1.4 **Critical Thinking** The circuit in **Figure 23–8** has four identical resistors. Suppose a wire is added to connect points A and B. Answer the following questions, explaining your reasoning.

- **a.** What is the current through the wire?
- **b.** What happens to the current through each resistor?
- **c.** What happens to the current drawn from the battery?
- **d.** What happens to the potential difference across each resistor?

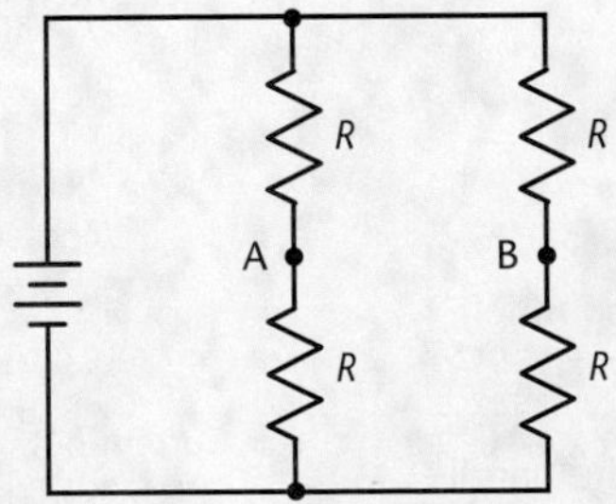

FIGURE 23–8

23.2 Applications of Circuits

You have already learned some of the elements of household wiring circuits. It's important to understand the requirements and limitations of these systems. Above all, it is important to be aware of the safety measures that must be practiced to prevent accidents.

OBJECTIVES

- **Explain** how fuses, circuit breakers, and ground-fault interrupters protect household wiring.
- **Analyze** combined series-parallel circuits and **calculate** the equivalent resistance of such circuits.

  State the important characteristics of voltmeters and ammeters, and **explain** how each is used in circuits.

Safety Devices

In an electric circuit, fuses and circuit breakers are switches that act as safety devices. They prevent circuit overloads that can occur when too many appliances are turned on at the same time or a short circuit occurs in one appliance. When appliances are connected in parallel, each additional appliance placed in operation reduces the equivalent resistance in the circuit and causes more current through the wires. The additional current may produce enough thermal energy ($P = I^2R$) to melt insulation on the wires and cause a short circuit in the wires or even a fire.

A **fuse** is a short piece of metal that melts if too large a current passes through it. The thickness of the metal to be used is determined by the amount of current that can be safely handled by the circuit. Should there be a larger current through the circuit, the fuse will melt and break the circuit. A **circuit breaker,** shown in **Figure 23–9,** is an automatic switch that opens when the current reaches some set value. If current greater than the set value flows in the circuit, the circuit is overloaded. The circuit breaker will open and thereby stop all current.

A **ground-fault interrupter** is often required by law in electrical outlets in bathrooms and kitchens. Current follows a single path from the power source into the electrical outlet and back to the source. Sometimes, when an appliance such as a hair dryer is used, the appliance or user might touch a cold water pipe or a sink full of water and in this way create another current path through the user. If a current as small as 5 mA should follow this path through a person, it could result in serious injury. The ground-fault interrupter contains an electronic circuit

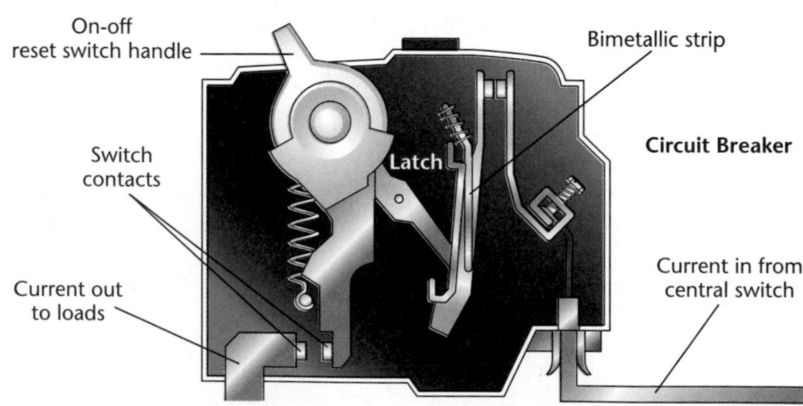

FIGURE 23–9 When too much current flows through the bimetallic strip, it will bend down and release the latch. The handle moves to the off position causing the switch to open, and that breaks the circuit.

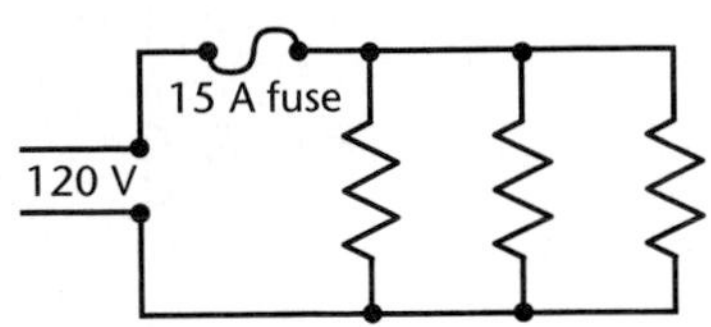

FIGURE 23–10 This parallel wiring arrangement permits the use of more than one appliance simultaneously, but if all three appliances are used at once, the fuse could melt.

that detects small differences in current caused by an extra current path and opens the circuit, thereby preventing dangerous shocks.

Electric wiring in homes uses parallel circuits, such as the one diagrammed in **Figure 23–10,** so that the current in any one circuit does not depend upon the current in the other circuits. The current in a device that dissipates power, P, when connected to a voltage source, V, is represented by $I = P/V$. Suppose, that in the schematic diagram shown in **Figure 23–10,** a 240-W television is plugged into a 120-V outlet. The current that flows is represented by $I = (240\ \text{W})/(120\ \text{V}) = 2\ \text{A}$. Then, a 720-W curling iron is plugged in. The current through the curling iron is $I = (720\ \text{W})/(120\ \text{V}) = 6\ \text{A}$. Finally, a 1440-W hair dryer is plugged in. The current through the hair dryer is $I = (1440\ \text{W})/(120\ \text{V}) = 12\ \text{A}$.

The current through these three appliances can be found by considering them as resistors in a parallel circuit in which the current through each appliance is independent of the others. The value of the resistance is found by calculating the current the appliance draws and then using the equation $R = V/I$. The equivalent resistance of the three appliances is

$$\frac{1}{R} = \frac{1}{10\ \Omega} + \frac{1}{20\ \Omega} + \frac{1}{60\ \Omega} = \frac{1}{6\ \Omega}$$

$$R = 6\ \Omega.$$

The 15-A fuse is connected in series with the power source so the entire current passes through it. The current through the fuse is

$$I = \frac{V}{R} = \frac{120\ \text{V}}{6\ \Omega} = 20\ \text{A}.$$

The 20-A current exceeds the rating of the 15-A fuse, so that the fuse will melt, or blow, cutting off current to the entire circuit.

A **short circuit** occurs when a circuit is formed that has a very low resistance. The low resistance causes the current to be very large. If there were no fuse or circuit breaker, such a large current could easily start a fire. A short circuit can occur if the insulation on a lamp cord becomes old and brittle. The two wires in the cord could accidentally touch. The resistance of the wire might be only 0.010 Ω. When placed across 120 V, this resistance would result in the following current.

$$I = \frac{V}{R} = \frac{120\ \text{V}}{0.010\ \Omega} = 12\ 000\ \text{A}$$

HELP WANTED

ELECTRICIAN

Electrical contractor needs electricians who have successfully completed a 5-year apprenticeship program conducted by a union or professional builder's association. You must be a high school grad, be in good physical condition, have excellent dexterity and color vision, and be willing to work when and where there is work. You will do all aspects of the job, including reading blueprints, dealing with all types of wires, conduits, and equipment. Safety and quality work must be your highest priorities. For information contact:

International Brotherhood of Electrical Workers
1125 15th Street, N.W.
Washington, DC 20005

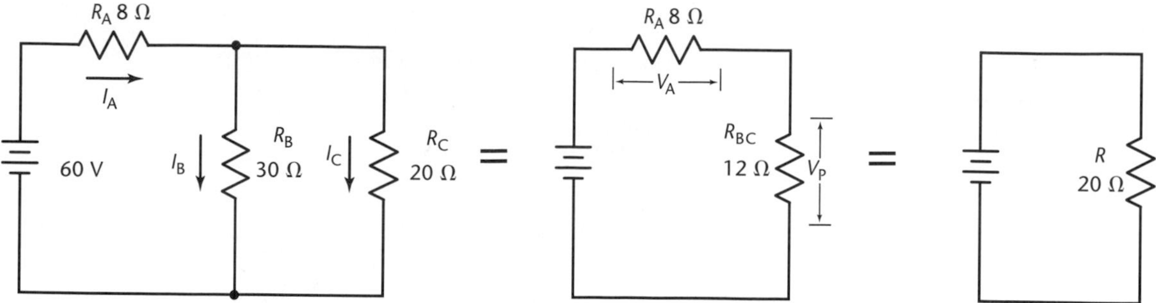

FIGURE 23–11 Use these diagrams as you study the following Problem Solving Strategy.

Such a current would cause a fuse or a circuit breaker to open the circuit immediately, thereby preventing the wires from becoming hot enough to start a fire.

Combined Series-Parallel Circuits

Have you ever noticed the light in your bathroom dim when you turned on a hair dryer? The light and the hair dryer were connected in parallel across 120 V. This means that the current through the lamp should not have changed when you plugged in the dryer. Yet the light dimmed, so the current must have changed. The dimming occurred because the house wiring had a small resistance in series with the parallel circuit. This is a **combination series-parallel circuit.** The following is a strategy for analyzing such circuits. Refer to **Figure 23–11** which illustrates the procedure described in steps 1, 2, and 3 of the Problem Solving Strategy.

ECONOMICS CONNECTION

Thomas Edison developed the first lightbulb that could be used in homes and a commercially viable distribution system for electricity. He started with the basic assumption that customers should be able to turn on and off their own lightbulbs without affecting the other bulbs in the circuit. Therefore, bulbs had to be connected in parallel. Because only part of the circuit current flowed through each bulb, Edison had to design a high-resistance filament that required only a small current for a large heating effect.

Problem Solving Strategy

Series-Parallel Circuits

1. Draw a schematic diagram of the circuit.
2. Find any parallel resistors. Resistors in parallel have separate current paths. They must have the same potential differences across them. Calculate the single equivalent resistance that can replace them. Draw a new schematic using that resistor.
3. Are any resistors (including the equivalent resistor) now in series? Resistors in series have one and only one current path through them. Calculate a single new equivalent resistance that can replace them. Draw a new schematic diagram using that resistor.
4. Repeat steps 2 and 3 until you can reduce the circuit to a single resistor. Find the total circuit current. Then go backwards through the circuits to find the currents through and the voltages across individual resistors.

Design Your Own **Physics Lab**

Circuits Lab

Problem

Suppose that three identical lamps are connected to the same power supply. Can a circuit be made such that one lamp is brighter than the others and stays on if either of the others is loosened in its socket?

Hypothesis

One lamp should be brighter than the other two and remain at the same brightness when either of the other two lamps is loosened in its socket so that it goes out.

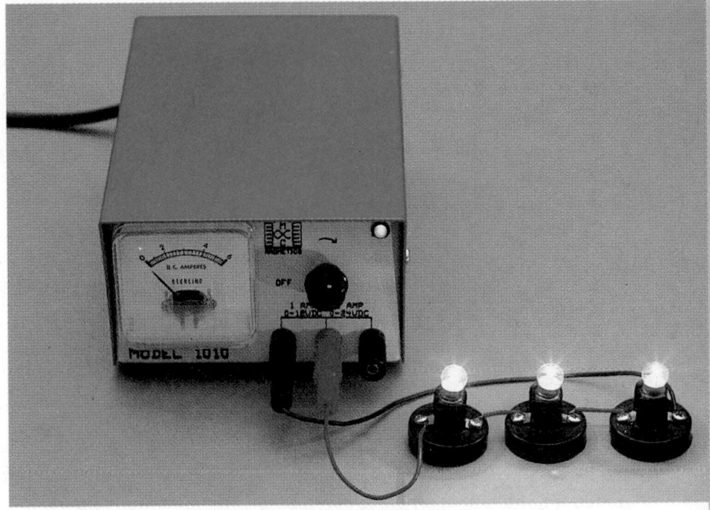

Possible Materials

power supply with variable voltage
wires with clips
3 identical lamps and sockets

Plan the Experiment

1. Sketch a series circuit and predict the relative brightness of each lamp. Predict what would happen to the other lamps when one is loosened so that it goes out.
2. Sketch a parallel circuit and predict the relative brightness of each lamp. Predict what would happen to the other lamps when one is loosened so that it goes out.
3. Draw a combination circuit. Label the lamps A, B, and C. Would the bulbs have the same brightness? Predict what would happen to the other two lamps when each lamp in turn is loosened so that it goes out.
4. **Check the Plan** Show your circuits and predictions to your teacher before starting to build the circuits.

Analyze and Conclude

1. **Interpreting Data** Did the series circuit meet the requirements? Explain.
2. **Interpreting Data** Did the parallel circuit meet either of the requirements? Explain.
3. **Formulating Hypotheses** Explain the circuit that solved the problem in terms of current.
4. **Formulating Hypotheses** Use the definition of *resistance* to explain why one lamp was brighter and the other two were equally dim.
5. **Making Predictions** Predict how the voltages would compare when measured across each lamp in the correct circuit.
6. **Testing Conclusions** Use a voltmeter to check your prediction.

Apply

1. Can one wall switch control several lights in the same room? Are the lamps in parallel or series? Are the switches in parallel or series with the lamps? Explain.

Example Problem

Series-Parallel Circuit

A hair dryer with a resistance of 12 Ω and a lamp with a resistance of 125 Ω are connected in parallel to a 125-V source through a 1.5-Ω resistor in series.

a. Find the current through the lamp when the hair dryer is off.

b. Find the current when the hair dryer is on.

c. Does the 1.5-Ω resistance explain why the lamp dims? Explain.

Sketch the Problem

- Draw a diagram of the simple series circuit when the dryer is off.
- Draw the series-parallel circuit including the dryer and lamp.
- Replace R_A and R_B with a single equivalent resistance.

Calculate Your Answer

Known:	Unknown:
$R_A = 125\ \Omega$	$I_{A1} = ?$
$R_B = 12\ \Omega$	$I_{A2} = ?$
$R_C = 1.5\ \Omega$	
$V_{source} = 125\ V$	

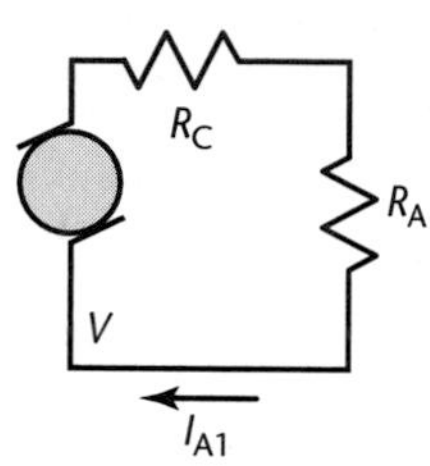

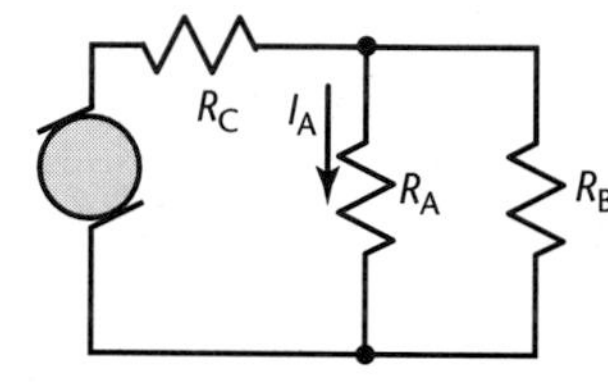

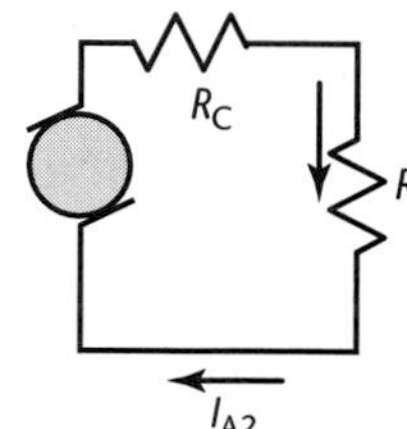

Strategy:

a. When the hairdryer is off, the circuit is a simple series circuit.

Calculations:

$$I_{A1} = \frac{V}{R} = \frac{V}{R_A + R_C} = \frac{125V}{125\ \Omega + 1.5\ \Omega} = 0.99A$$

b. Find the equivalent resistance for parallel circuit.

$$\frac{1}{R_p} = \frac{1}{R_A} + \frac{1}{R_B},\ \text{so } R_p = \frac{R_A R_B}{R_A + R_B}$$

Find the equivalent resistance for entire circuit.

$$R = R_C + R_p = R_C + \frac{R_A R_B}{R_A + R_B}$$

$$R = 1.5\ \Omega + \frac{(125\ \Omega)(12\ \Omega)}{125\ \Omega + 12\ \Omega} = 12.5\ \Omega$$

Use equivalent resistance to determine the current.

$$I_2 = \frac{V_{source}}{R} = \frac{125\ V}{12.5\ \Omega} = 1.0 \times 10^1\ A$$

c. The greater current means a greater voltage drop across R_C, which causes less voltage across R_A. A decrease in voltage means current decreases, which is less power.

$$V_C = IR_C = (1.0 \times 10^1\ A)(1.5\ \Omega) = 15\ V$$

$$V_A = V_{source} - V_C = 125\ V - 15\ V = 110V$$

Current drops from 0.99 A to 0.880 A. Power, $P = I^2R$, is smaller, consequently the light dims.

$$I_{A2} = \frac{V_A}{R_A} = \frac{110\ V}{125\ \Omega} = 0.88\ A$$

Check Your Answer

- Are the units correct? Current is in amps, potential drops are in volts.
- Is the magnitude realistic? Decreased parallel resistance increases the current, causing a voltage drop in the series resistor. This leaves less voltage across the combination, so the current is smaller.

Practice Problems

13. Two 60-Ω resistors are connected in parallel. This parallel arrangement is connected in series with a 30-Ω resistor. The combination is then placed across a 120-V battery.

a. Draw a diagram of the circuit.

b. What is the equivalent resistance of the parallel portion of the circuit?

c. What single resistance could replace the three original resistors?

d. What is the current in the circuit?

e. What is the voltage drop across the 30-Ω resistor?

f. What is the voltage drop across the parallel portion of the circuit?

g. What is the current in each branch of the parallel portion of the circuit?

Pocket Lab

Ammeter Resistance

Design an experiment using a power supply, a voltmeter, an ammeter, a resistor, and some wires to determine the resistance of the ammeter. Make a sketch of your setup and include measurements and equations.

Communicating Results
What is the resistance of the ammeter? Be prepared to present your experiment to the class.

Ammeters and Voltmeters

An **ammeter** is used to measure the current in any branch or part of a circuit. If, for example, you want to measure the current through a resistor, you would place an ammeter in series with the resistor. This requires opening a current path and inserting an ammeter. The use of an ammeter should not change the current in the circuit you wish to measure. Because current would decrease if the ammeter increased the resistance in the circuit, the resistance of an ammeter should be as low as possible. **Figure 23–12** shows a real ammeter as an ideal, zero-resistance meter placed in series with a 0.01-Ω resistor. The ammeter resistance is much smaller than the values of the resistors. The current decrease would be from 1.0 A to 0.9995 A, too small to notice.

Another instrument, called a **voltmeter,** is used to measure the voltage drop across some part of a circuit. To measure the potential drop across a resistor, connect the voltmeter in parallel with the resistor. A voltmeter should have a very high resistance so that it causes the smallest possible change in currents or voltages in the circuit. Consider the circuit shown in **Figure 23–13.** A typical inexpensive voltmeter consists of an ideal, zero-resistance meter in series with a 10-kΩ resistor. When it is connected in parallel with R_B, the equivalent resistance of the combination is smaller than R_B alone. Thus, the total resistance of the circuit decreases, increasing the

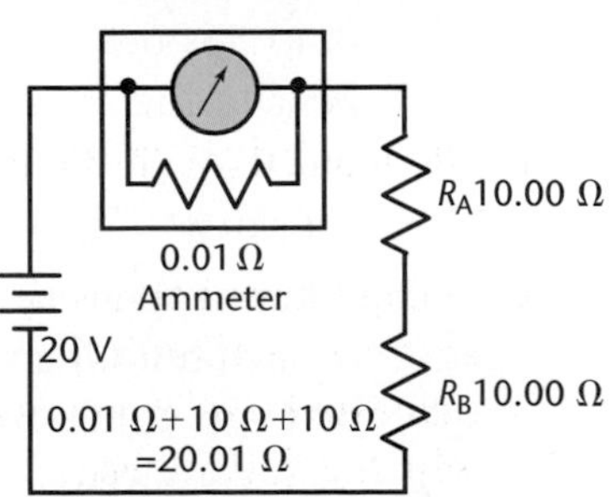

FIGURE 23–12 An ammeter measures current and so it is always placed in series in a circuit.

FIGURE 23–13 A laboratory voltmeter such as this one measures potential difference. Voltmeters are placed in parallel.

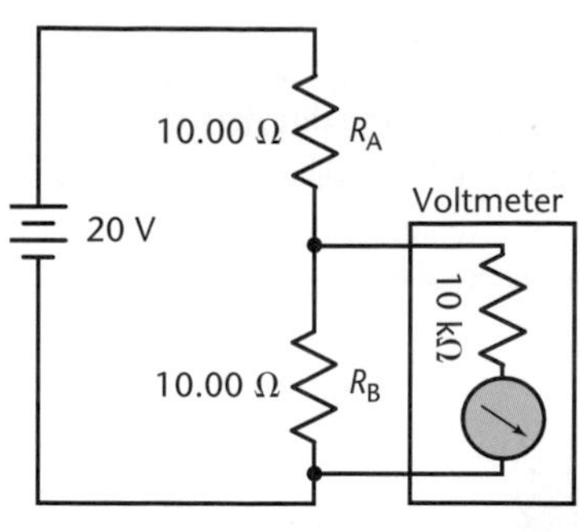

current. R_A has not changed, but the current through it has increased, increasing the potential drop across it. The battery, however, holds the potential drop across R_A and R_B constant. Thus, the potential drop across R_B must decrease. The result of connecting a voltmeter across a resistor is to lower the potential drop across it. The higher the resistance of the voltmeter, the smaller the voltage change. Using a voltmeter with a 10 000-Ω resistance changes the voltage across R_B from 10 V to 9.9975 V, too small a change to detect. Modern electronic multimeters have even higher resistances, 10^7 Ω, and so produce even smaller changes.

23.2 Section Review

2.1 Consider the circuit in **Figure 23–14** made with identical bulbs.

a. Compare the brightness of the three bulbs.

b. What happens to the brightness of each bulb when bulb 1 is unscrewed from its socket? What happens to the three currents?

c. Bulb 1 is screwed in again and bulb 3 is unscrewed. What happens to the brightness of each bulb? What happens to the three currents?

d. What happens to the brightness of each bulb if a wire is connected between points B and C?

e. A fourth bulb is connected in parallel with bulb 3 alone. What happens to the brightness of each bulb?

2.2 **Critical Thinking** In the circuit in **Figure 23–14,** the wire at point C is broken and a small resistor is inserted in series with bulbs 2 and 3. What happens to the brightness of the two bulbs? Explain.

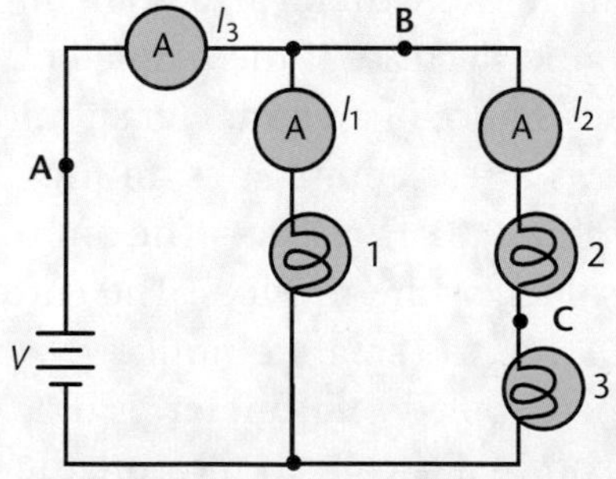

FIGURE 23–14

How It Works

Electric Switch

An electric switch is a device that is used to interrupt, complete, or divert an electrical current in a circuit. Switches are found on everything from hair dryers and toaster ovens, to calculators and video games, to computers and airplane instrument panels. Some switches are simple mechanical switches. In certain devices, such as computers, however, mechanical switches are too slow and are replaced by electronic switches made from semiconducting materials.

Probably the most common type of switch is the mechanical switch you use to operate the small appliances and lights in your home or school. The switch shown below is a snap-action toggle switch typically used to turn lights off and on.

1. The insulated handle of a snap-action toggle switch can be flipped in either of two positions-off or on.

2. When the handle is in the off position, as it is in this diagram, metal contacts within the switch are separated, interrupting the path of the current.

3. When the handle is in the on position, the metal contacts, which are linked by wires, come together to complete the circuit.

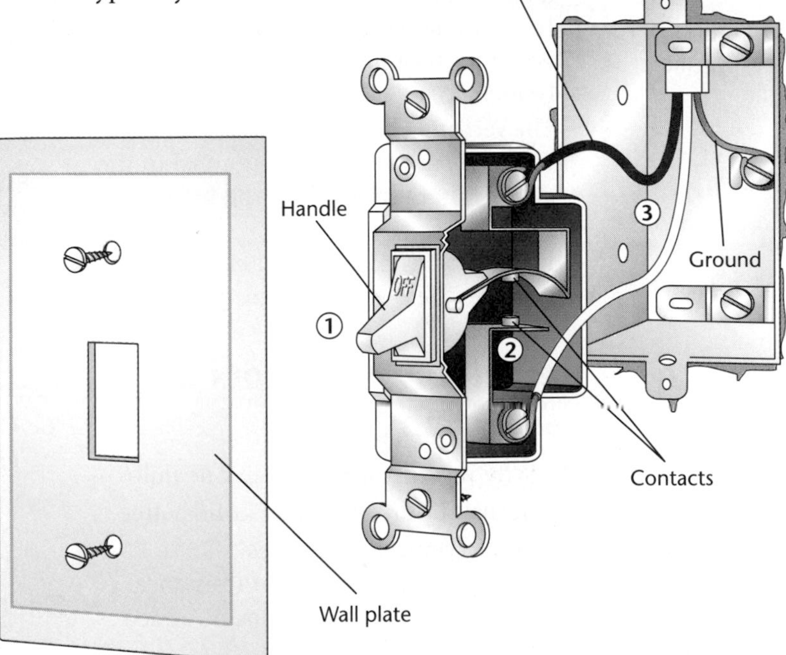

Thinking Critically

1. Circuit breakers are automatic switches. Explain why this is so.
2. Why must the contacts in a switch be metal? Why are switch handles often plastic?

CHAPTER 23 REVIEW

Key Terms

23.1

- series circuit
- equivalent resistance
- voltage divider
- parallel circuit

23.2

- fuse
- circuit breaker
- ground-fault interrupter
- short circuit
- combination series-parallel circuit
- ammeter
- voltmeter

Summary

23.1 Simple Circuits

- The current is the same everywhere in a simple series circuit.
- The equivalent resistance of a series circuit is the sum of the resistances of its parts.
- The sum of the voltage drops across resistors in series is equal to the potential difference applied across the combination.
- A voltage divider is a series circuit used to produce a voltage source from a higher-voltage battery.
- The voltage drops across all branches of a parallel circuit are the same.
- In a parallel circuit, the total current is equal to the sum of the currents in the branches.
- The reciprocal of the equivalent resistance of parallel resistors is equal to the sum of the reciprocals of the individual resistances.
- If any branch of a parallel circuit is opened, there is no current in that branch. The current in the other branches is unchanged.

23.2 Applications of Circuits

- A fuse or circuit breaker, placed in series with appliances, creates an open circuit when dangerously high currents flow.
- A complex circuit is often a combination of series and parallel branches. Any parallel branch is first reduced to a single equivalent resistance. Then any resistors in series are replaced by a single resistance.
- An ammeter is used to measure the current in a branch or part of a circuit. An ammeter always has a low resistance and is connected in series.
- A voltmeter measures the potential difference (voltage) across any part or combination of parts of a circuit. A voltmeter always has a high resistance and is connected in parallel with the part of the circuit being measured.

Reviewing Concepts

Section 23.1

1. Why is it frustrating when one bulb burns out on a string of holiday tree lights connected in series?
2. Why does the equivalent resistance decrease as more resistors are added to a parallel circuit?
3. Several resistors with different values are connected in parallel. How do the values of the individual resistors compare with the equivalent resistance?
4. Why is household wiring done in parallel instead of in series?
5. Why is there a difference in total resistance between three 60-Ω resistors connected in series and three 60-Ω resistors connected in parallel?
6. Compare the amount of current entering a junction in a parallel circuit with that leaving the junction. *Note:* A junction is a point where three or more conductors are joined.

Section 23.2

7. Explain the function of a fuse in an electric circuit.
8. What is a short circuit? Why is a short circuit dangerous?
9. Why does an ammeter have a very low resistance?
10. Why does a voltmeter have a very high resistance?
11. How does the way in which an ammeter is connected in a circuit differ from the way a voltmeter is connected?

Applying Concepts

12. What happens to the current in the other two lamps if one lamp in a three-lamp series circuit burns out?

13. Suppose that in the voltage divider in **Figure 23–4,** the resistor R_A is made to be a variable resistor. What happens to the voltage output, V_B, of the voltage divider if the resistance of the variable resistor is increased?

14. Circuit A contains three 60-Ω resistors in series. Circuit B contains three 60-Ω resistors in parallel. How does the current in the second 60-Ω resistor change if a switch cuts off the current to the first 60-Ω resistor in

a. circuit A?

b. circuit B?

15. What happens to the current in the other two lamps if one lamp in a three-lamp parallel circuit burns out?

16. An engineer needs a 10-Ω resistor and a 15-Ω resistor. But there are only 30-Ω resistors in stock. Must new resistors be purchased? Explain.

17. If you have a 6-V battery and many 1.5-V bulbs, how could you connect them so that they light but do not have more than 1.5 V across each bulb?

18. Two lamps have different resistances, one larger than the other.

a. If they are connected in parallel, which is brighter (dissipates more power)?

b. When connected in series, which is brighter?

19. For each of the following, write the form of circuit that applies: series or parallel.

a. The current is the same throughout.

b. The total resistance is equal to the sum of the individual resistances.

c. The voltage drop is the same across each resistor.

d. The voltage drop is proportional to the resistance.

e. Adding a resistor decreases the total resistance.

f. Adding a resistor increases the total resistance.

g. If the current through one resistor goes to zero, there is no current in the entire circuit.

h. If the current through one resistor goes to zero, the current through all other resistors remains the same.

i. This form is suitable for house wiring.

20. Why is it dangerous to replace a 15-A fuse in a circuit with a fuse of 30 A?

Problems

Section 23.1

LEVEL 1

21. A 20.0-Ω lamp and a 5.0-Ω lamp are connected in series and placed across a potential difference of 50.0 V. What is

a. the equivalent resistance of the circuit?

b. the current in the circuit?

c. the voltage drop across each lamp?

d. the power dissipated in each lamp?

22. The load across a battery consists of two resistors, with values of 15 Ω and 45 Ω, connected in series.

a. What is the total resistance of the load?

b. What is the voltage of the battery if the current in the circuit is 0.10 A?

23. A lamp having a resistance of 10 Ω is connected across a 15-V battery.

a. What is the current through the lamp?

b. What resistance must be connected in series with the lamp to reduce the current to 0.50 A?

24. A string of 18 identical holiday tree lights is connected in series to a 120-V source. The string dissipates 64.0 W.

a. What is the equivalent resistance of the light string?

b. What is the resistance of a single light?

c. What power is dissipated by each lamp?

25. One of the bulbs in problem 24 burns out. The lamp has a wire that shorts out the lamp filament when it burns out. This drops the resistance of the lamp to zero.

a. What is the resistance of the light string now?

b. Find the power dissipated by the string.

c. Did the power go up or down when a bulb burned out?

26. A 75.0-W bulb is connected to a 120-V source.
  a. What is the current through the bulb?
  b. What is the resistance of the bulb?
  c. A lamp dimmer puts a resistance in series with the bulb. What resistance would be needed to reduce the current to 0.300 A?

27. In problem 26, you found the resistance of a lamp and a dimmer resistor.
  a. Assuming that the resistances are constant, find the voltage drops across the lamp and the resistor.
  b. Find the power dissipated by the lamp.
  c. Find the power dissipated by the dimmer resistor.

28. A 16.0-Ω and a 20.0-Ω resistor are connected in parallel. A difference in potential of 40.0 V is applied to the combination.
  a. Compute the equivalent resistance of the parallel circuit.
  b. What is the current in the circuit?
  c. How large is the current through the 16.0-Ω resistor?

LEVEL 2

29. Amy needs 5.0 V for some integrated circuit experiments. She uses a 6.0-V battery and two resistors to make a voltage divider. One resistor is 330 Ω. She decides to make the other resistor smaller. What value should it have?

30. Pete is designing a voltage divider using a 12.0-V battery and a 100.0-Ω resistor as R_B. What resistor should be used as R_A if the output voltage across R_B is to be 4.00 V?

31. A typical television dissipates 275 W when it is plugged into a 120-V outlet.
  a. Find the resistance of the television.
  b. The television and 2.5-Ω wires connecting the outlet to the fuse form a series circuit that works like a voltage divider. Find the voltage drop across the television.
  c. A 12-Ω hair dryer is plugged into the same outlet. Find the equivalent resistance of the two appliances.
  d. Find the voltage drop across the television and hairdryer. The lower voltage explains why the television picture sometimes shrinks when another appliance is turned on.

Section 23.2

LEVEL 1

32. A circuit contains six 240-Ω lamps (60-W bulbs) and a 10.0-Ω heater connected in parallel. The voltage across the circuit is 120 V. What is the current in the circuit
  a. when four lamps are turned on?
  b. when all lamps are on?
  c. when six lamps and the heater are operating?

33. If the circuit in problem 32 has a fuse rated at 12 A, will the fuse melt if everything is on?

34. Determine the reading of each ammeter and each voltmeter in **Figure 23–15.**

35. Determine the power used by each resistance shown in **Figure 23–15.**

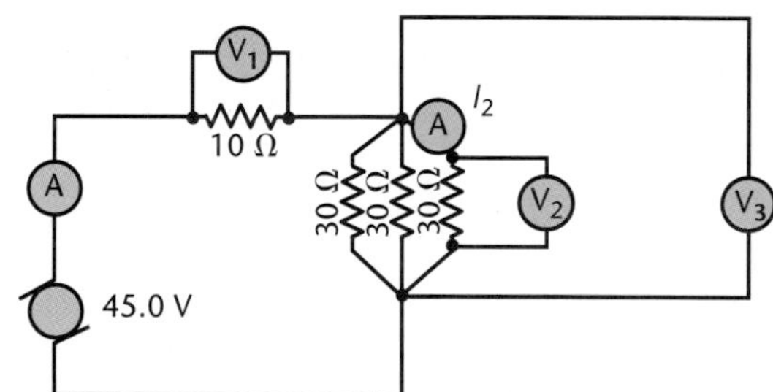

FIGURE 23–15

LEVEL 2

36. During a laboratory exercise, you are supplied with a battery of potential difference *V*, two heating elements of low resistance that can be placed in water, an ammeter of very small resistance, a voltmeter of extremely high resistance, wires of negligible resistance, a beaker that is well insulated and has negligible heat capacity, and 100.0 g of water at 25°C.
  a. By means of a diagram and standard symbols, show how these components should be connected to heat the water as rapidly as possible.
  b. If the voltmeter reading holds steady at 50.0 V and the ammeter reading holds steady at 5.0 A, estimate the time in seconds required to completely vaporize the water in the beaker. Use 4200 J/kg·°C as the specific heat of water and 2.3×10^6 J/kg as the heat of vaporization of water.

37. A typical home circuit is diagrammed in **Figure 23–16.** Note that the lead lines to the kitchen lamp each has a very low resistance of 0.25 Ω. The lamp has a resistance of 240.0 Ω. Although the circuit is a parallel circuit, the lead lines are in series with each of the components of the circuit.

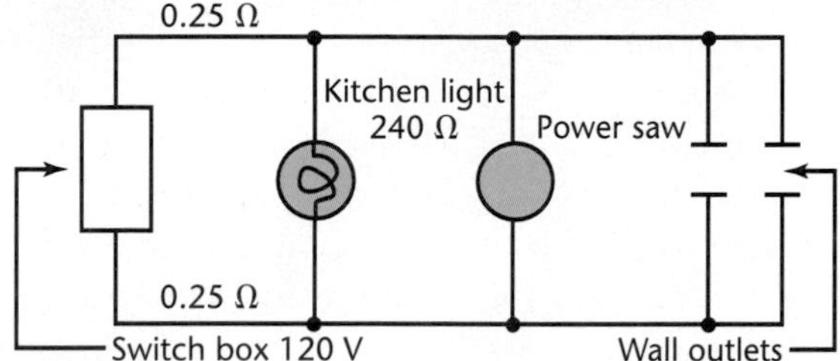

FIGURE 23–16

a. Compute the equivalent resistance of the circuit consisting of just the light and the lead lines to and from the light.
b. Find the current to the bulb.
c. Find the power dissipated in the bulb.
d. Because the current in the bulb is 0.50 A, the current in the lead lines also must be 0.50 A. Calculate the voltage drop across each of the two leads.

38. A power saw is operated by an electric motor. When electric motors are first turned on, they have a very low resistance. Suppose that a kitchen light in problem 37 is on and a power saw is turned on. The saw and lead lines have an initial total resistance of 6.0 Ω.
a. Compute the equivalent resistance of the light-saw parallel circuit.
b. What is the total current flowing in the circuit?
c. What is the total voltage drop across the two leads to the light?
d. What voltage remains to operate the light? Will this cause the light to dim temporarily?

Critical Thinking Problems

39. A 50-200-250-W three-way bulb has three terminals on its base. Sketch how these terminals could be connected inside the bulb to provide the three brightnesses. Explain how to connect 120 V across two terminals at a time to obtain a low, medium, and high level of brightness.

40. Batteries consist of an ideal source of potential difference in series with a small resistance. The electrical energy of the battery is produced by chemical reactions that occur in the battery. However, these reactions also result in a small resistance that, unfortunately, cannot be completely eliminated. A flashlight contains two batteries in series. Each has a potential difference of 1.50 V and an internal resistance of 0.20 Ω. The bulb has a resistance of 22.0 Ω.
a. What is the current through the bulb?
b. How much power does the bulb dissipate?
c. How much greater would the power be if the batteries had no internal resistance?

Going Further

Using What You've Learned An ohmmeter is made by connecting a 6.0-V battery in series with an adjustable resistor and an ideal ammeter. The ammeter deflects full-scale with a current of 1.0 mA. The two leads are touched together and the resistance is adjusted so that 1.0 mA flows.
a. What is the resistance of the adjustable resistor?
b. The leads are now connected to an unknown resistance. What resistance would produce a current of half-scale, 0.50 mA?
c. What resistance would produce a reading of quarter-scale, 0.25 mA?
d. What resistance would produce a reading of three-quarters-scale, 0.75 mA?
e. To make a usable ohmmeter, a scale reading in ohms is attached to the ammeter. Using your answers to the preceding questions, show what resistances would appear at various places on the ammeter dial.

interNET CONNECTION

Follow the link for this chapter on the Glencoe Homepage at **www.glencoe.com/sec/science** to find out more about this chapter.

Northern Light Show

The displays of the northern lights, also known as the aurora borealis, have fascinated and mystified people for centuries. Storytellers and poets have found inspiration in the luminous streaks and patches of color in the night sky of the northern hemisphere. What causes these unusual light shows of the northern sky?

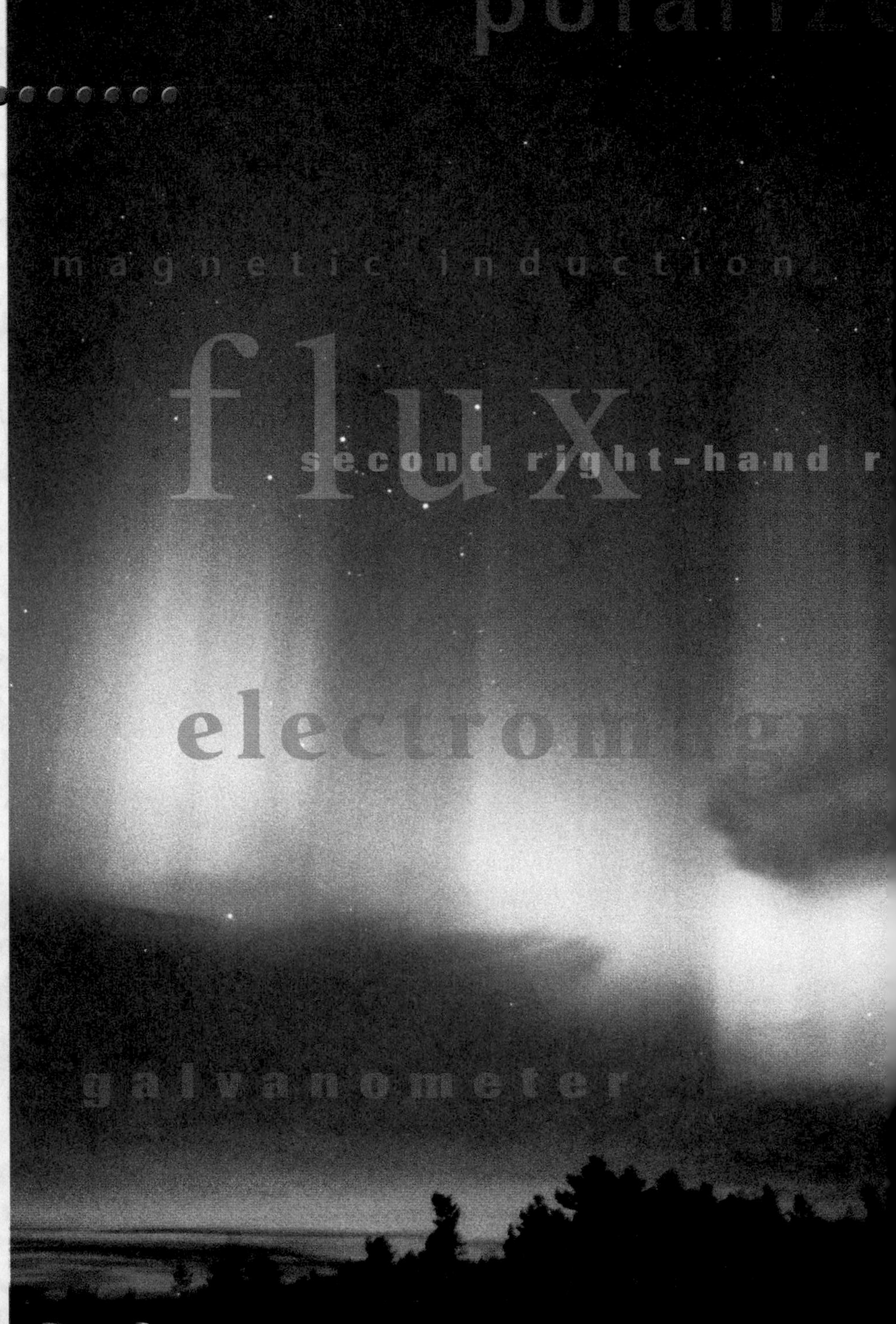

CHAPTER

24 Magnetic Fields

According to ancient myths, the northern lights were thought to be caused by ghostly spirits. Today's scientific knowledge tells us that the aurora borealis results from the interaction between the solar wind and Earth's magnetic field. To fully understand how the aurora borealis occurs, you will need to know more about these two phenomena.

The existence of magnets and magnetic fields has been known for more than 2000 years. The Chinese were using magnets as compasses when the first European explorers reached China in the 1500s. Magnetic rocks, called lodestones, were studied by the earliest scientists. Today, magnets are used in the generators that supply electricity. In addition, motors, television sets, and tape recorders depend on the magnetic effects of electric currents. Thus, the study of magnetism is an important part of your investigation of electricity as well as your understanding of many common electronic devices.

WHAT YOU'LL LEARN

- You will relate magnetism to electric charge and electricity.
- You will describe how electromagnetism is harnessed to produce mechanical work.

WHY IT'S IMPORTANT

- Using electromagnetism in electric motors, you can convert electrical energy to mechanical energy.
- Every day, you apply mechanical energy produced from electrical energy.

*inter*NET CONNECTION

Follow the link for this chapter on the Glencoe Homepage at **www.glencoe.com/sec/science** to find out more about magnetic fields.

24.1 Magnets: Permanent and Temporary

OBJECTIVES

- **Describe** the properties of magnets and the origin of magnetism in materials.
- **Compare** various magnetic fields.

If you have ever used a compass to tell direction or picked up tacks or paper clips with a magnet, you have observed some effects of magnetism. You might even have made an electromagnet by winding wire around a nail and connecting it to a battery. The properties of magnets become most obvious when you experiment with two of them.

General Properties of Magnets

To enhance your study of magnetism, you can experiment with two bar or ceramic magnets such as those shown in **Figure 24–1.**

Magnetic poles

Suspend a magnet from a thread, as in **Figure 24–2a.** If it is a bar magnet, you may have to tie a yoke to keep it horizontal. Note that when the magnet comes to rest, it has lined up in a north-south direction. Put a mark on the magnet end that points north. If you rotate it away from that direction, it will return. From this simple experiment, you can conclude that a magnet is **polarized,** that is, it has two ends, one of which is the north-seeking end, or north pole; the other is the south-seeking end, or south pole. A compass is nothing more than a small magnet mounted so that it is free to turn.

Suspend another magnet and mark the end that points north. While one magnet is suspended, observe the interaction of two magnets by bringing the second magnet near, as in **Figure 24–2b.** Note that the two ends that pointed north, the north poles, repel each other, as do the two south poles. The north pole of one magnet, however, will attract the south pole of the other magnet. That is, like poles repel; unlike poles attract. Magnets always have two opposite magnetic poles. If you break a magnet in half, you create two smaller magnets, but each still has two poles. Scientists have tried breaking magnets into separate north and south poles, or "monopoles," but no one has succeeded, not even on the microscopic level.

Knowing that magnets always orient themselves in a north-south direction, it may occur to you that Earth itself is probably a giant magnet. Because opposite poles attract and the north pole of a compass magnet points north, the south pole of the Earth-magnet must be near Earth's geographic north pole.

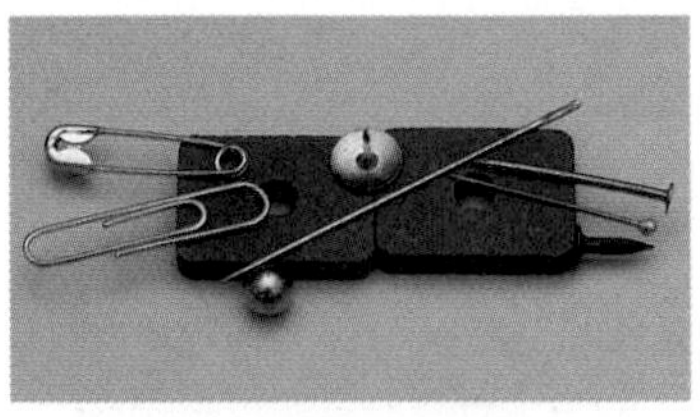

FIGURE 24–1 Ceramic magnets are commonly available in most hardware stores.

How do magnets affect other materials?

As you probably discovered as a child, magnets attract things besides other magnets—things like nails, tacks, paper clips, and many other

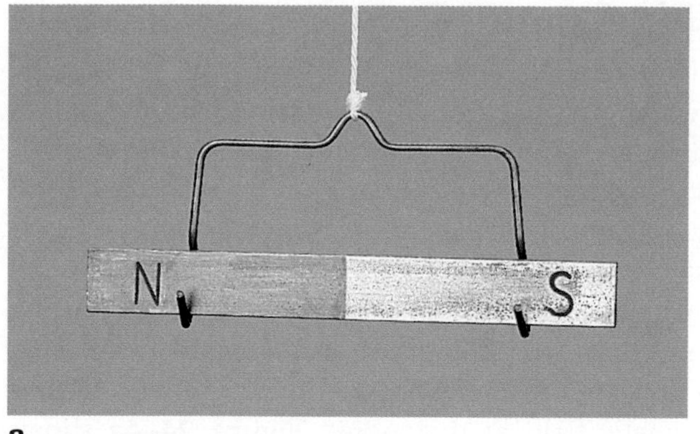

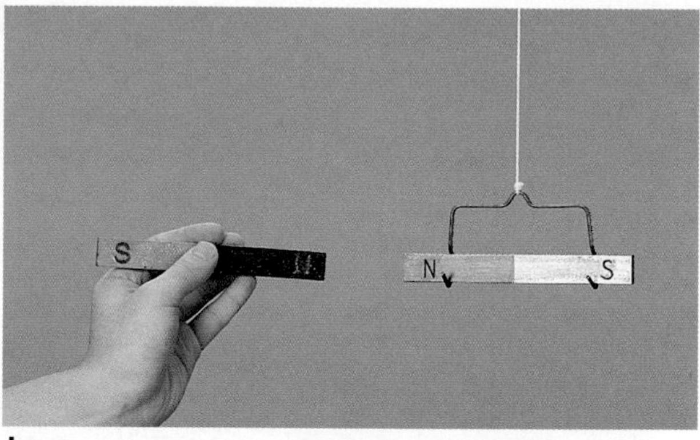

FIGURE 24-2 If you suspend a magnet by a thread, it will align itself with Earth's magnetic field **(a).** Its north pole will point north. If you then move the north pole of a second magnet toward the north pole of the suspended magnet, the suspended magnet will move away **(b).**

metal objects. Unlike the interaction between two magnets, however, either end of a magnet will attract either end of a piece of metal. How can you explain this behavior? First, you can touch a magnet to a nail and then touch the nail to smaller metal pieces. The nail itself becomes a magnet, as shown in **Figure 24–3.** The magnet causes the nail to become polarized. The direction of polarization of the nail depends on the polarization of the magnet. The nail is only temporarily magnetized; if you pull away the magnet, the nail's magnetism disappears. The polarization induced in the nail is similar to the polarization induced in a conductor by a nearby charged object, which you learned about in Chapter 20.

Permanent magnets

The magnetism of permanent magnets is produced in the same way that you created the magnetism of the nail. But because of the microscopic structure of the magnet material, the induced magnetism becomes permanent. Many permanent magnets are made of ALNICO V, an iron alloy containing 8% **Al**uminum, 14% **Ni**ckel, and 3% **Co**balt. A variety of rare earth elements, such as neodymium and gadolinium, produce permanent magnets that are extremely strong for their size.

Monopoles?

Place a disk magnet flat on the center of your paper. Place another disk magnet flat at the top of your paper and slowly slide it toward the center magnet.

Observing and Inferring Does the first magnet attract or repel the second magnet? Rotate one of the magnets and note the effect on the other. Does each magnet have only one pole?

FIGURE 24-3 If you touch an iron nail with a magnet, the nail will become magnetized and will in turn attract other iron objects. However, as soon as the magnet is removed, the nail will lose its magnetism.

Color Conventions

- Positive charges are **red.**
- Negative charges are **blue.**
- Electric field lines are **indigo.**
- Magnetic field lines are **green.**

Magnetic Fields Around Permanent Magnets

When you experimented with two magnets, you noticed that the forces between magnets, both attraction and repulsion, occur not only when the magnets touch each other, but also when they are held apart. In the same way that long-range electric and gravitational forces can be described by electric and gravitational fields, magnetic forces can be described by the existence of **magnetic fields** around magnets.

The presence of a magnetic field around a magnet can be shown using iron filings. Each long, thin, iron filing becomes a small magnet by induction. Just like a tiny compass needle, the iron filing rotates until it is parallel to the magnetic field at that point. **Figure 24–4a** shows filings in a glycerol solution surrounding a bar magnet. The three-dimensional shape of the field is visible. In **Figure 24–4b,** the filings make up a two-dimensional plot of the field. These lines of filings can help you to visualize magnetic field lines.

Magnetic field lines

Note that magnetic field lines, like electric field lines, are imaginary. Not only do field lines help us visualize the field, but they also provide a measure of its strength. The number of magnetic field lines passing through a surface is called the **magnetic flux.** The flux per unit area is proportional to the strength of the magnetic field. As you can see in **Figure 24–4,** the magnetic flux is most concentrated at the poles, and this is where the magnetic field strength is the greatest.

The direction of a magnetic field line is defined as the direction in which the N-pole of a compass points when it is placed in the magnetic field. Outside the magnet, the field lines come out of the magnet at its N-pole and enter the magnet at its S-pole, as illustrated in **Figure 24–5.** What happens inside the magnet? There are no isolated poles on which field lines can start or stop, so magnetic field lines always travel inside the magnet from the south pole to the north pole to form closed loops.

FIGURE 24–4 The magnetic field of a bar magnet shows up clearly in three dimensions when the magnet is suspended in glycerol with iron filings **(a).** It is, however, easier to set up a magnet beneath a sheet of paper covered with iron filings to see the pattern of magnetic fields in two dimensions **(b).**

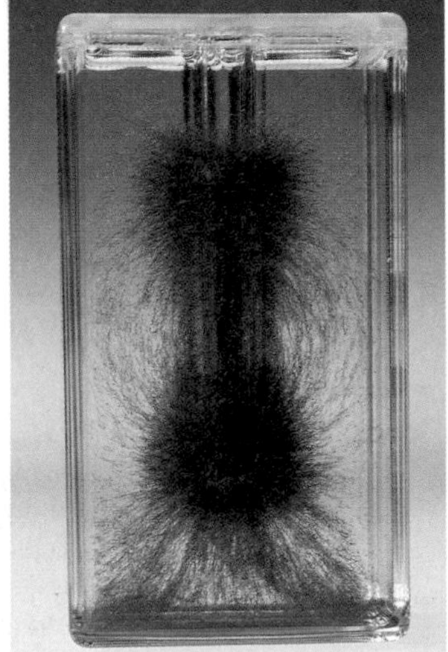

a

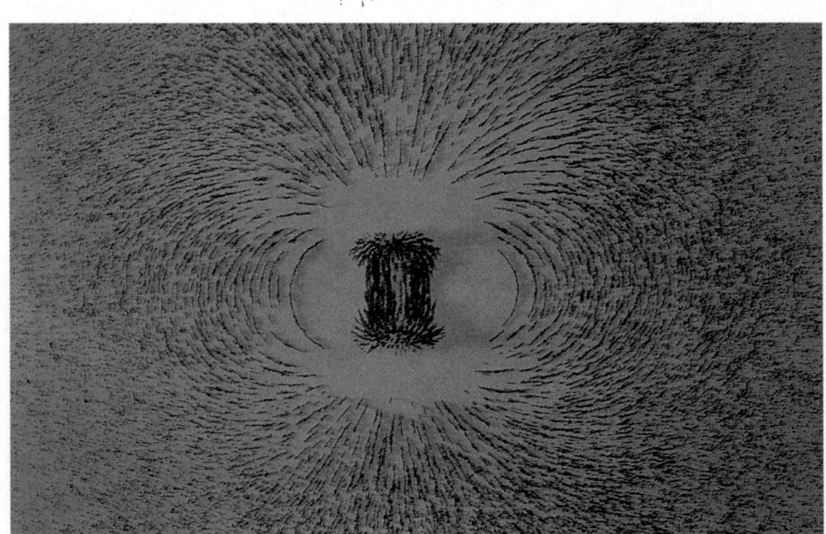

b

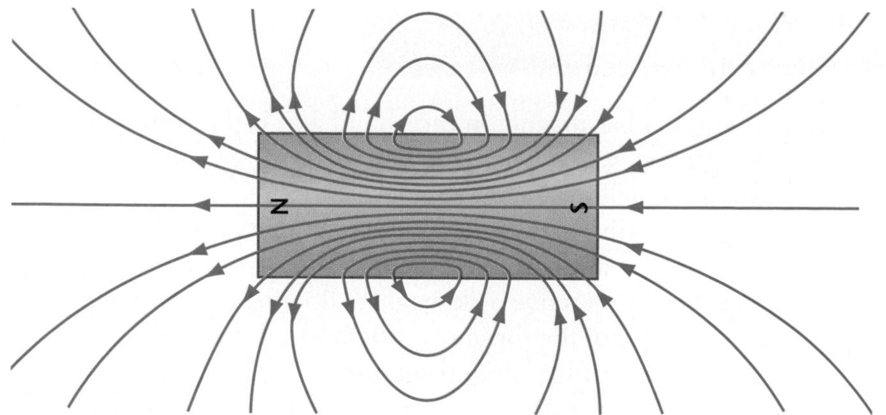

FIGURE 24–5 Magnetic field lines can be visualized as closed loops that leave the north pole of a magnet and enter the south pole.

What kinds of magnetic fields are produced by pairs of bar magnets? You can visualize the fields by placing a sheet of paper over the poles of two magnets. Then sprinkle the paper with iron filings. **Figure 24–6a** shows the field lines between two like poles. By contrast, two unlike poles (N and S) placed close together produce the pattern shown in **Figure 24–6b.** The filings show that the field lines between two unlike poles run directly from one magnet to the other.

Forces on objects in magnetic fields

Magnetic fields exert forces on other magnets. The field produced by the N-pole of one magnet pushes the N-pole of a second magnet away in the direction of the field line. The force exerted by the same field on the S-pole of the second magnet is attractive, in a direction opposite the field lines. The second magnet attempts to line up with the field, like a compass needle.

When a sample made of iron, cobalt, or nickel is placed in the magnetic field of a permanent magnet, the field lines become concentrated within the sample. Lines leaving the N pole of the magnet enter one end of the sample, pass through it, and leave the other end. Thus, the end of the sample closest to the magnet's N-pole becomes the sample's S-pole, and the sample is attracted to the magnet.

Pocket Lab

Funny Balls

Place a disk magnet flat on your paper. Roll a 3-mm steel ball at the magnet. Place a second steel ball on the paper, touching the magnet and the first steel ball.

Hypothesizing What happens? Why? Make a sketch to help explain your hypothesis. Devise a procedure to test your hypothesis.

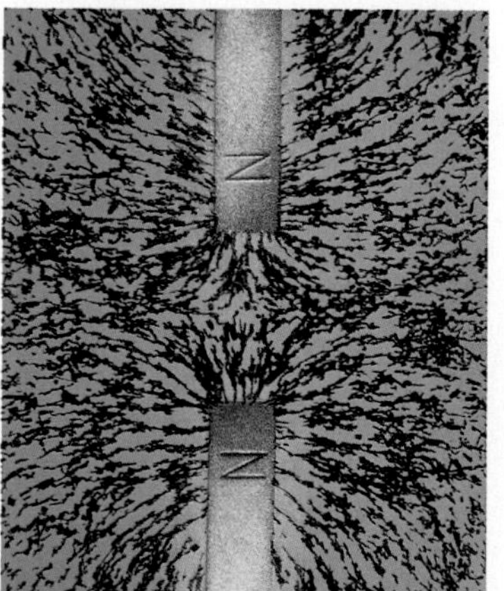

a

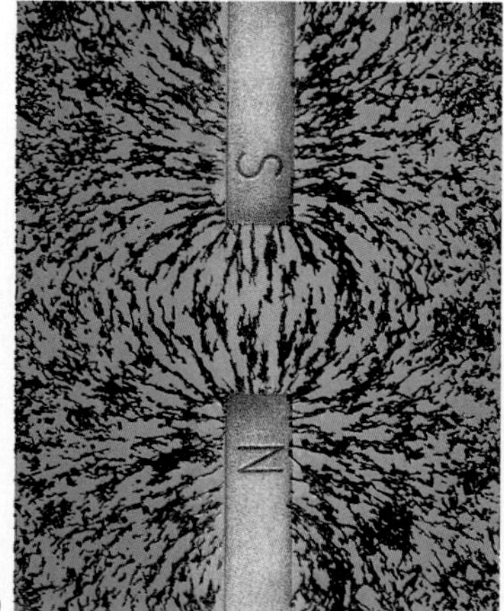

b

FIGURE 24–6 The magnetic field lines indicated by iron filings on paper over two magnets clearly show that like poles repel **(a)** and unlike poles attract **(b).** The iron filings do not form continuous lines between like poles. Between a north and a south pole, however, the iron filings show that field lines run directly between the two magnets.

FIGURE 24–7

Practice Problems

1. If you hold a bar magnet in each hand and bring your hands close together, will the force be attractive or repulsive if the magnets are held so that
   a. the two north poles are brought close together?
   b. a north pole and a south pole are brought together?
2. **Figure 24–7** shows five disk magnets floating above each other. The north pole of the top-most disk faces up. Which poles are on the top side of the other magnets?
3. A magnet attracts a nail, which, in turn, attracts many small tacks, as shown in **Figure 24–3.** If the N-pole of the permanent magnet is the top face, which end of the nail is the N-pole?

Electromagnetism

How does electric charge affect a magnet? When you bring a magnet near a charged strip of transparent tape, there is no effect on the magnet or the tape. However, there is a marked effect on the magnet when the charge moves as an electrical current.

Magnetic field near a current-carrying wire

In 1820, Danish physicist Hans Christian Oersted (1777–1851) was experimenting with electric currents in wires. Oersted laid a wire across the top of a small compass and connected the ends of the wire to complete an electrical circuit, as shown in **Figure 24–8.** He had expected the needle to point toward the wire or in the same direction as the current in the wire. Instead, he was amazed to see that the needle rotated until it pointed perpendicular to the wire. The forces on the compass magnet's poles were perpendicular to the direction of current in the wire. Oersted also found that when there was no current in the wire, no magnetic forces existed.

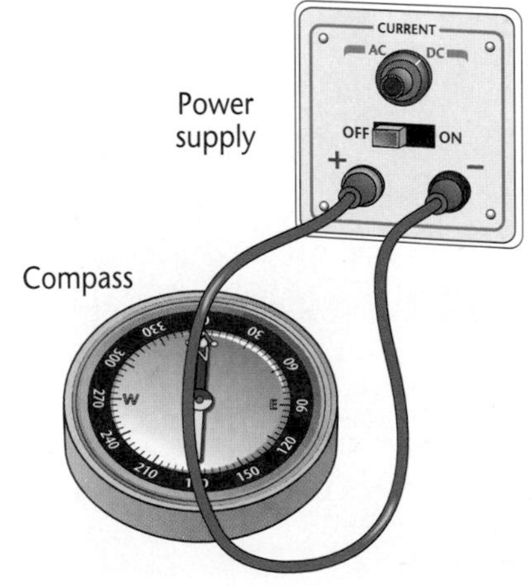

FIGURE 24–8 Using an apparatus similar to the one shown here, Oersted was able to demonstrate a connection between magnetism and electricity.

If a compass needle turns when placed near a wire carrying an electric current, it must be the result of a magnetic field created by the current. You can easily show the magnetic field around a current-carrying wire by placing a wire vertically through a horizontal piece of cardboard on which iron filings are sprinkled. When there is a current, tap the cardboard. The filings will form a pattern of concentric circles around the wire, as shown in **Figure 24–9.**

FIGURE 24–9 The magnetic field produced by current in a straight wire through a cardboard disc shows up as concentric circles of iron filings around the wire.

The circular lines indicate that magnetic field lines around current-carrying wires form closed loops in the same way that field lines about permanent magnets form closed loops. The strength of the magnetic field around a long, straight wire is proportional to the current in the wire. The strength of the field also varies inversely with the distance from the wire.

A compass shows the direction of the field lines. If you reverse the direction of current, the compass needle also reverses its direction, as shown in **Figure 24–10a.** You can find the direction of the field around a wire using the **first right-hand rule.** Imagine holding a length of insulated wire with your right hand. Keep your thumb pointed in the direction of the conventional (positive) current. The fingers of your hand circle the wire and point in the direction of the magnetic field, as illustrated in **Figure 24–10b.**

Magnetic field near a coil

An electric current in a single circular loop of wire forms a magnetic field all around the loop. Applying the right-hand rule to any part of the wire loop, it can be shown that the direction of the field inside the loop is always the same. In **Figure 24–11a,** the field is always up, or out of the page. Outside the loop, it is always down, or into the page.

When a wire is looped several times to form a coil and a current is allowed to flow through the coil, the field around all the loops is always in the same direction, as shown in **Figure 24–11b.** A long coil of wire consisting of many loops is called a **solenoid.** The field from each loop in a solenoid adds to the fields of the other loops.

When there is an electric current in a coil of wire, the coil has a field like that of a permanent magnet. When this current-carrying coil is brought close to a suspended bar magnet, one end of the coil repels the

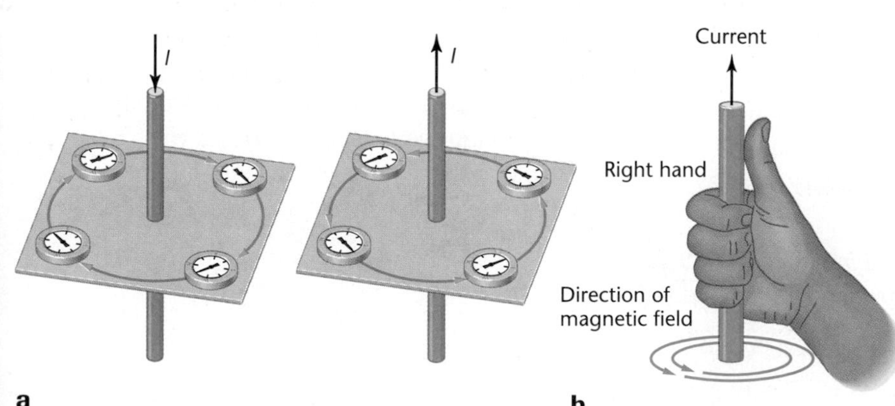

FIGURE 24–10 The magnetic field produced by current in a straight wire conductor reverses if the current in the wire is reversed **(a).** The right-hand rule for a straight, current-carrying wire shows the direction of the magnetic field **(b).**

Design Your Own **Physics Lab**

Coils and Currents

Problem

You have seen that an electric current affects a magnetic compass needle. What happens to pieces of iron located inside a coil that carries a current? What is the effect of changing the magnitude of the current? Does an alternating current produce a different effect from that of a direct current?

Hypothesis  

Write a testable hypothesis that addresses the questions posed in the problem.

Possible Materials

a ring stand with crossbar and clamp

two 20-cm lengths of thick, insulated iron wire

75 cm of thread

magnetic compass

miniature lamp with socket

500-turn, air-core solenoid

a variable power supply that can produce AC and DC voltages and currents

electrical leads and alligator clips

Plan the Experiment

1. Develop a plan and design a circuit you can use to test your hypothesis.
2. **Check the Plan** Show your teacher your plan before you start to build the circuit. **CAUTION:** *Be sure the power supply is off as you build the circuit.*
3. **CAUTION:** *Your teacher must inspect your setup before you turn the power on and begin your investigation.*

Analyze and Conclude

1. **Making Observations** Describe your observations as you increased the direct current produced by the power supply.
2. **Drawing Conclusions** What conclusion can you make regarding the strength of the magnetic field as you increased the current?
3. **Interpreting Results** What can you conclude from the results of your experimentation comparing the effects of direct and alternating currents?

Apply

1. Large and powerful electromagnets are often used at scrap metal facilities. Would you expect that these magnets use AC current or DC current? Explain why.

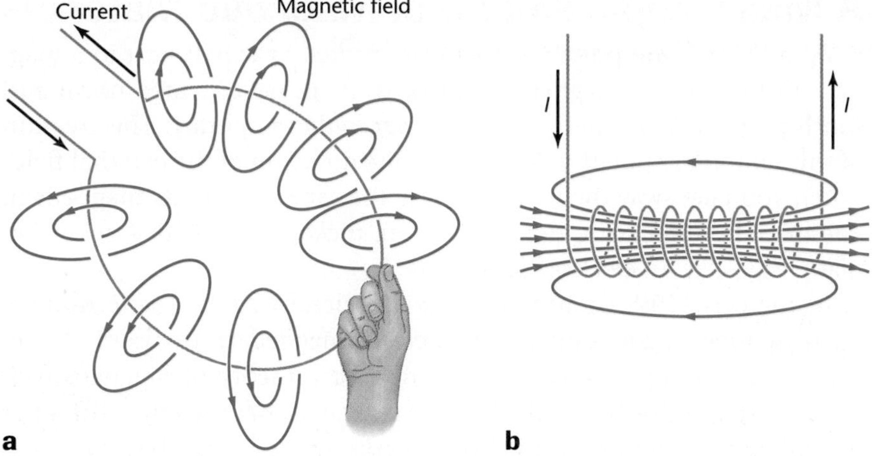

FIGURE 24–11 The magnetic field around a circular loop of current-carrying wire can be modeled with the aid of the right-hand rule **(a).** A current in a solenoid creates a magnetic field with the field from each coil adding to all the others **(b).**

north pole of the magnet. Thus, the current-carrying coil has a north and a south pole and is itself a magnet. This type of magnet is called an **electromagnet.** The strength of the field of an electromagnet is proportional to the current in the coil. The magnetic field produced by each loop of a coil is the same as that produced by any other loop. Because these fields are in the same direction, increasing the number of loops in an electromagnet increases the strength of the magnetic field.

The strength of an electromagnet can also be increased by placing an iron rod or core inside the coil, because the field inside the coil magnetizes the core. The magnetic strength of the core adds to that of the coil to produce a much stronger magnet.

The direction of the field produced by an electromagnet can be found by using the **second right-hand rule.** Imagine holding an insulated coil with your right hand. Curl your fingers around the loops in the direction of the conventional (positive) current, as in **Figure 24–12.** Your thumb points toward the N-pole of the electromagnet.

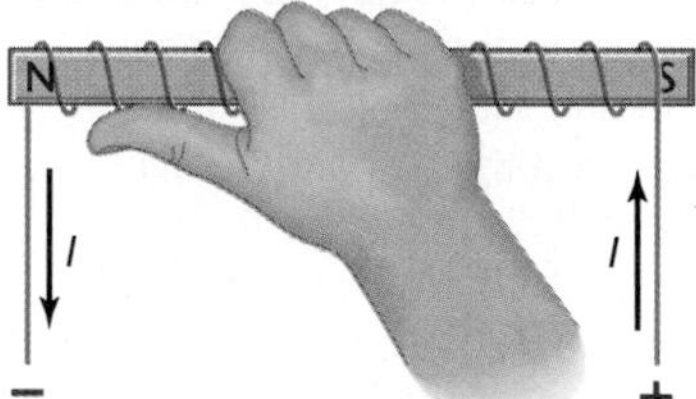

FIGURE 24–12 The second right-hand rule can be used to determine the polarity of an electromagnet.

Practice Problems

4. A long, straight, current-carrying wire runs from north to south.
   a. A compass needle placed above the wire points with its N-pole toward the east. In what direction is the current flowing?
   b. If a compass is put underneath the wire, in which direction will the compass needle point?
5. How does the strength of the magnetic field 1 cm from a current-carrying wire compare with
   a. the strength of the field 2 cm from the wire?
   b. the strength of the field 3 cm from the wire?
6. A student makes a magnet by winding wire around a nail and connecting it to a battery, as shown in **Figure 24–13.** Which end of the nail, the pointed end or the head, will be the north pole?

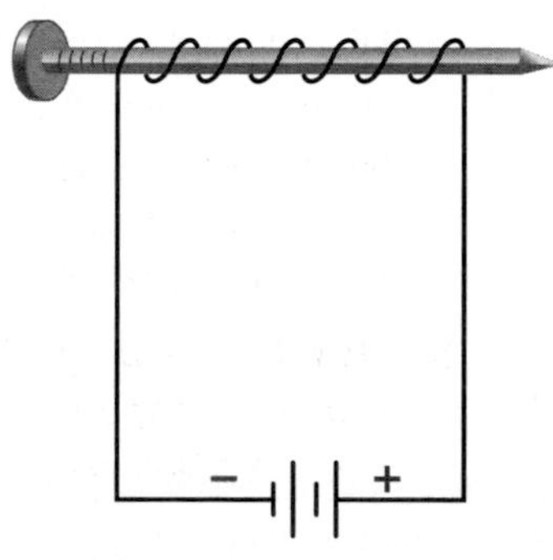

FIGURE 24–13

Pocket Lab

3-D Magnetic Fields

Most illustrations are able to show the shape of the magnetic field around a magnet only two-dimensionally. Try this activity to see the shape of a magnetic field in 3-D. Tie a string to the middle of a nail so that the nail will hang horizontally. Put a small piece of tape around the string where it wraps around the nail so that the string will not slip. Insert the nail into a coil and apply a voltage to the coil. This will magnetize the nail. Turn off the power to the coil and remove the nail. Now hold the string to suspend the nail and slowly move it close to a permanent magnet. Try this for magnets of various shapes.

Analyze What evidence do you have that the nail became magnetized? Using your results, make a 3-D drawing that shows the magnetic field around the nail.

A Microscopic Picture of Magnetic Materials

Recall that if you put a piece of iron, nickel, or cobalt next to a magnet, it too becomes magnetic. That is, you will have created north and south poles. The magnetism is, however, only temporary. The creation of this temporary polarity depends on the direction of the external field. When you take away the external field, the sample loses its magnetism. The three ferromagnetic elements—iron, nickel, and cobalt—behave in many ways like an electromagnet.

In the early 19th century, the French scientist André-Marie Ampère knew that the magnetic effects of an electromagnet are the result of electric current through its loops. He proposed a theory of magnetism in iron to explain this behavior. Ampère reasoned that the effects of a bar magnet must result from tiny loops of current within the bar.

Magnetic domains

Although the details of Ampère's reasoning were wrong, his basic idea was correct. Each electron in an atom acts like a tiny electromagnet. The magnetic fields of the electrons in a group of neighboring atoms can combine together. Such a group is called a **domain.** Although they may contain 10^{20} individual atoms, domains are still very small—usually from 10 to 1000 microns. Thus, even a small sample of iron contains a huge number of domains.

When a piece of iron is not in a magnetic field, the domains point in random directions. Their magnetic fields cancel one another. If, however, the iron is placed in a magnetic field, the domains tend to align with the external field, as shown in **Figure 24–14.** In the case of a temporary magnet, after the external field is removed, the domains return to their random arrangement. In permanent magnets, the iron has been alloyed with other substances that keep the domains aligned after the external magnetic field is removed.

Electromagnets make up the recording heads of audiocassette and videotape recorders. Recorders create electrical signals that represent the sounds or pictures being recorded. The electric signals produce currents in the recording head. When magnetic recording tape, which has many tiny bits of magnetic material bonded to thin plastic, passes over the

a

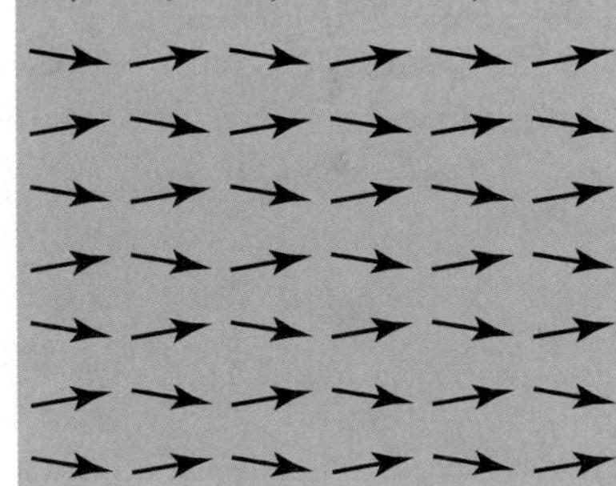

b

FIGURE 24–14 A piece of iron **(a)** becomes a magnet only when its domains align **(b)**.

How It Works

Computer Storage Disks

Magnetic fields are essential to the operation of computer storage disks. Data and software commands for computers are processed digitally in bits. Each bit is identified as either a 0 or a 1. How are these bits stored?

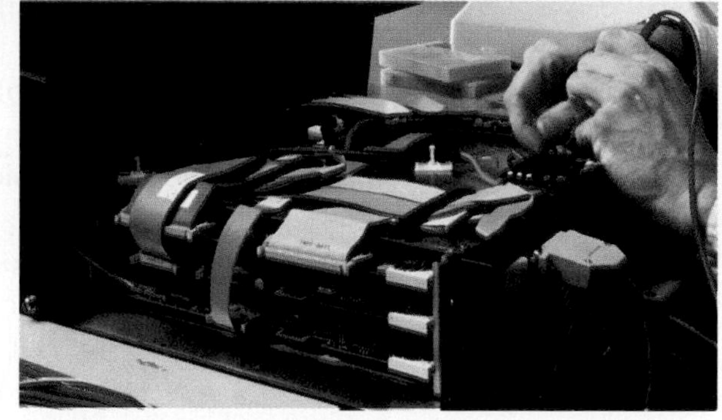

1 The surface of a computer storage disk is covered with an even distribution of magnetic particles within a film. The direction of the particles' domains changes in response to a magnetic field. Formatting organizes the disk's surface into sectors and tracks.

2 During recording onto the disk, current is routed to the disk drive's read/write head, which is an electromagnet composed of a wire-wrapped iron core. The current through the wire induces a magnetic field in the core.

3 As the read/write head passes over the spinning storage disk, the domains of atoms in the magnetic film line up in bands. The orientation of the domains depends on the direction of the current.

4 Two bands code for one bit. Two bands magnetized with the poles oriented in the same direction represent 0. Two bands represent 1 with poles oriented in opposite directions. The recording current always reverses when the read/write head begins recording the next data bit.

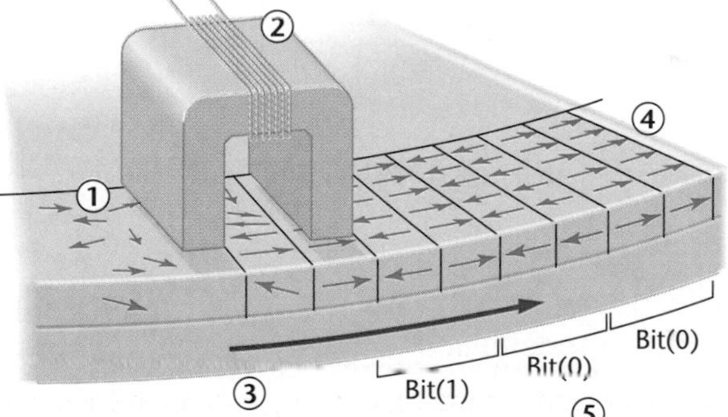

5 To retrieve data, no current is sent to the read/write head. Rather, the magnetized bands in the disk induce current in the coil as the disk spins beneath the head. Changes in the direction of the induced current are sensed by the computer and interpreted as 0s and 1s.

Thinking Critically

1. Why do manufacturers recommend keeping floppy disks away from objects such as electric motors, computer and television screens, and audio speakers?
2. After a data bit has been stored, the direction of current through the read/write head is automatically reversed to begin the next data bit. Explain why.

recording head, the domains of the bits are aligned by the magnetic fields of the head. The directions of the domains' alignments depend on the direction of the current in the head and become a magnetic record of the sounds or pictures being recorded. The material on the tape is chosen so that the domains can keep their alignments permanently. On playback of the tape, a pair of signals produced by the magnetic particles goes to an amplifier and a pair of loudspeakers or earphones. When a previously recorded tape is used to record new sounds, an erase head produces a rapidly alternating magnetic field that disorients the magnetic particles on the tape.

EARTH SCIENCE CONNECTION

Rocks that contain iron have recorded the history of the direction of Earth's magnetic field. Rocks on the seafloor were produced when molten rock poured out of cracks in the bottom of the oceans. As they cooled, they were magnetized in the direction of Earth's field at that time. The seafloor spreads, so rocks farther from the crack are older than those near the crack. Scientists examining seafloor rocks were surprised to find that the direction of the magnetization in different rocks varied. They concluded from their data that the north and south magnetic poles of Earth have exchanged places many times in Earth's history. The origin of Earth's magnetic field is not well understood. How this field might reverse direction periodically is even more of a mystery.

24.1 Section Review

1.1 Is a magnetic field real, or is it just a means of scientific modeling?

1.2 A wire is passed through a card on which iron filings are sprinkled. The filings show the magnetic field around the wire. A second wire is close to the first wire and parallel to it. There is an identical current in the second wire. If the two currents are in the same direction, how would the first magnetic field be affected? What if the two currents are in opposite directions?

1.3 Describe the right-hand rule used to determine the direction of a magnetic field around a straight, current-carrying wire.

1.4 **Critical Thinking** Imagine a toy containing two parallel, horizontal metal rods.

a. The top rod floats above the lower one. If the top rod's direction is reversed, however, it falls down onto the lower rod. Explain why the rods could behave in this way.

b. Assume that the top rod was lost and replaced with another one. In this case, the top rod falls down no matter what its orientation is. What type of replacement rod must have been used?

Forces Caused by Magnetic Fields

24.2

While studying the behaviors of magnets, Ampère noted that an electric current produces a magnetic field like that of a permanent magnet. Because a magnetic field exerts forces on permanent magnets, Ampère hypothesized that there is also a force on a current-carrying wire that is placed in a magnetic field.

OBJECTIVES

- **Relate** magnetic induction to the direction of the force on a current-carrying wire in a magnetic field.
- **Solve** problems involving magnetic field strength and the forces on current-carrying wires, and on moving, charged particles in magnetic fields.
- **Describe** the design and operation of an electric motor.

Forces on Currents in Magnetic Fields

The force on a wire in a magnetic field can be demonstrated using the arrangement shown in **Figure 24–15.** A battery produces the current in a wire that passes directly between two bar magnets. Recall that the direction of the magnetic field between two magnets is from the N-pole of one magnet to the S-pole of a second magnet. When there is a current in the wire, a force is exerted on the wire. As you can see, depending on the direction of the current, the force on the wire either pushes it down, as shown in **Figure 24–15a,** or pulls it up, as shown in **Figure 24–15b.** Michael Faraday (1791–1867) discovered that the force on the wire is at right angles to both the direction of the magnetic field and the direction of the current.

Faraday's description of the force on a current-carrying wire does not completely describe the direction. The force can be up or down. The direction of the force on a current-carrying wire in a magnetic field can be found by using the **third right-hand rule,** which is illustrated in **Figure 24–16.** The magnetic field can be indicated by the symbol ***B.*** Its direction is represented by a series of arrows, but when a field is directly into or out of the page, its direction is indicated by crosses or dots. The crosses suggest the feathers at the end of an archery arrow, and the dots suggest the point. To use the third right-hand rule, point the fingers of your right hand in the direction of the magnetic field. Point your thumb in the direction of the conventional (positive) current in the wire. The palm of your hand then faces in the direction of the force acting on the wire.

Soon after Oersted announced his discovery that the direction of the magnetic field in a wire is perpendicular to the flow of electric current in the wire, Ampère was able to demonstrate the forces that current-carrying wires exert on each other. **Figure 24–17a** shows the direction of the magnetic field around each of the current-carrying wires, which you recall is determined by the first right-hand rule. By applying the third right-hand rule to either wire, you can show why the wires attract each other. **Figure 24–17b** demonstrates the opposite situation. That is, when currents are in opposite directions, the forces push the wires apart.

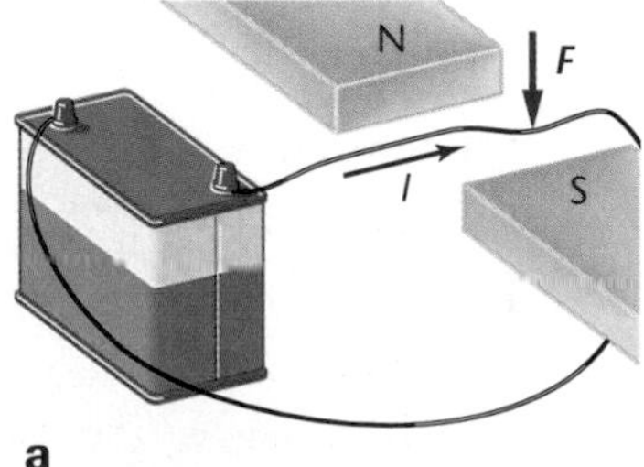

a

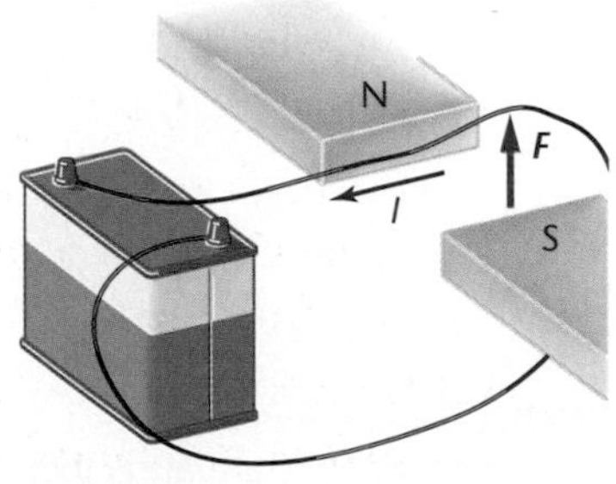

b

FIGURE 24–15 Current-carrying wires experience forces when they are placed in magnetic fields.

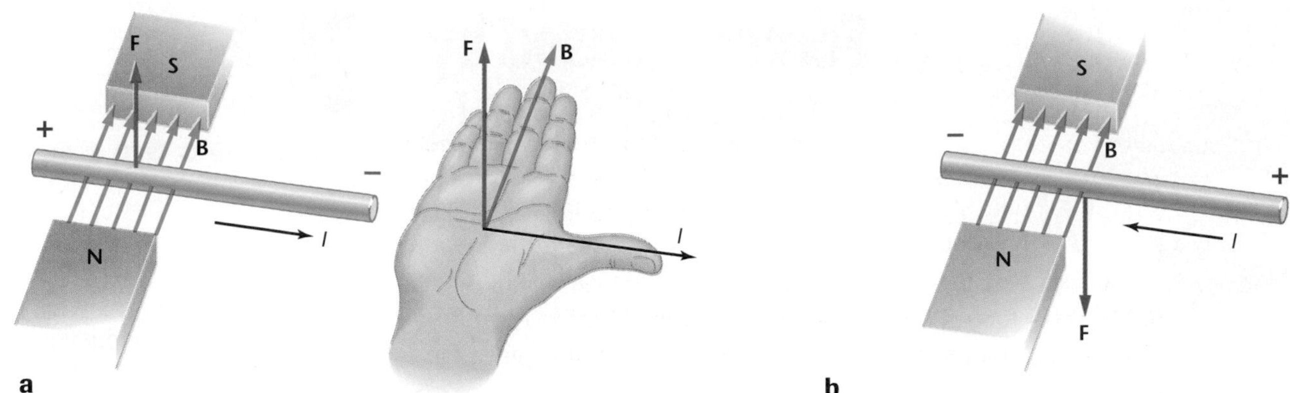

FIGURE 24–16 The third right-hand rule can be used to determine the direction of force when the current and magnetic field are known **(a).** Use your own hand to demonstrate the direction of force for the setup in **(b).**

Force on a wire resulting from a magnetic field

It is possible to determine the force of magnetism that is exerted on a current-carrying wire passing through a magnetic field at right angles to the wire. Experiments show that the magnitude of the force, F, on the wire is proportional to three factors: the strength of the field, B, the current, I, in the wire, and the length, L, of the wire that lies in the magnetic field. The relationship of these four factors is as follows.

$$F = BIL$$

The strength of a magnetic field, B, is measured in teslas, T. The strengths of some typical magnetic fields are provided in **Table 24–1.** A magnetic field having a strength of one tesla causes a force of one newton to be exerted on a 1-m length of straight wire carrying one ampere of current. Based on $B = F/IL$, the following is obtained.

$$1\ \text{T} = 1\ \text{N/A·m}$$

TABLE 24–1
Typical Magnetic Field

Source and Location	Strength (T)
Surface of neutron star (predicted)	10^8
Strong laboratory electromagnet	10
Small bar magnet	0.01
Earth's magnetic field	5×10^{-5}

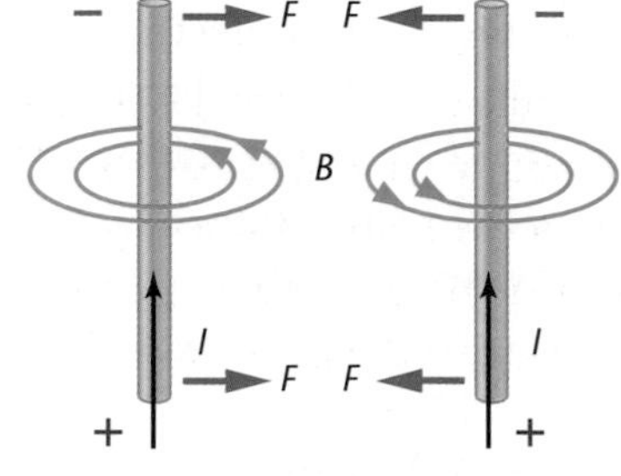

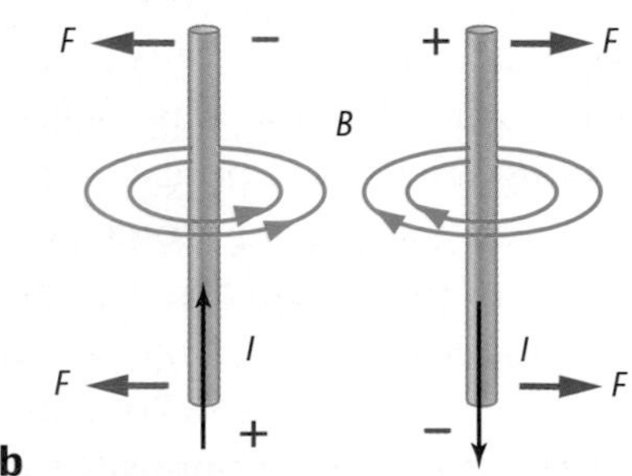

FIGURE 24–17 Two current-carrying conductors are attracted when the currents are in the same direction **(a),** and are repelled when the currents are in opposite directions **(b).**

Example Problem

Calculating the Strength of a Magnetic Field

A straight wire that carries a 5.0-A current is in a uniform magnetic field oriented at right angles to the wire. When 0.10 m of the wire is in the field, the force on the wire is 0.20 N. What is the strength of the magnetic field, *B*?

Sketch the Problem

- Sketch the wire and show the direction of the current with an arrow; the magnetic field lines, labeled ***B***; and the force on the wire, ***F***.
- Determine the direction of the force using the third right-hand rule.

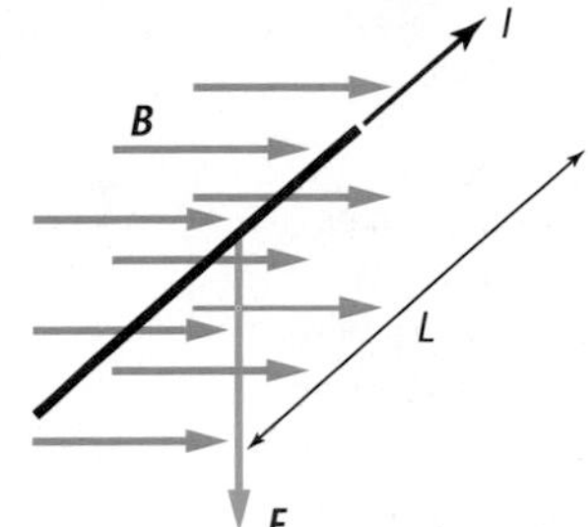

Calculate Your Answer

Known:

$I = 5.0$ A

$L = 0.10$ m

$F = 0.20$ N

All are at right angles.

Unknown:

$B = ?$

Strategy:

Use the equation $F = BIL$ because B is uniform and because B and I are perpendicular to each other. Calculate B.

Calculations:

$F = BIL$, so $B = F/IL$

$$B = \frac{0.20 \text{ N}}{(5.0 \text{ A})(0.10 \text{ m})} = 0.40 \text{ N/A·m} = 0.40 \text{ T}$$

Check Your Answer

- Are the units correct? The answer is in teslas, the correct unit for magnetic field.
- Is the magnitude realistic? The force is large for the current and length.

Practice Problems

7. A wire 0.50 m long carrying a current of 8.0 A is at right angles to a 0.40-T magnetic field. How strong a force acts on the wire?
8. A wire 75 cm long carrying a current of 6.0 A is at right angles to a uniform magnetic field. The magnitude of the force acting on the wire is 0.60 N. What is the strength of the magnetic field?
9. A copper wire 40 cm long carries a current of 6.0 A and weighs 0.35 N. A certain magnetic field is strong enough to balance the force of gravity on the wire. What is the strength of the magnetic field?

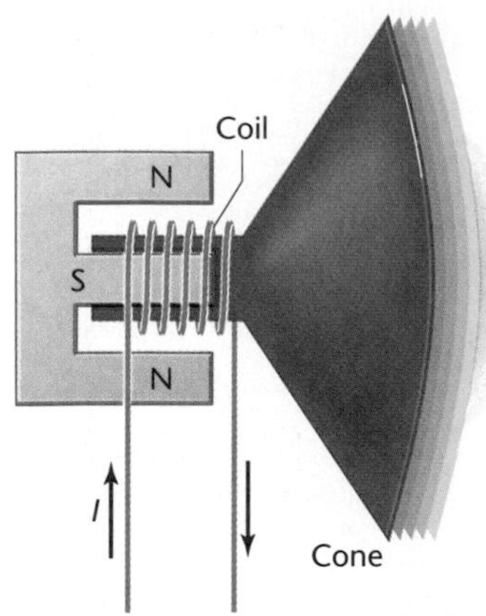

FIGURE 24–18 Sound waves can be created by exerting a force on a current-carrying wire in a magnetic field. This diagram of a loudspeaker shows how the coil can be pushed into and out of the magnetic field with changes in direction of the current.

Loudspeakers

One use of the force on a current-carrying wire in a magnetic field is in a loudspeaker. A loudspeaker changes electrical energy to sound energy using a coil of fine wire mounted on a paper cone and placed in a magnetic field, as shown in **Figure 24–18.** The amplifier driving the loudspeaker sends a current through the coil. The current changes direction between 20 and 20 000 times each second, depending on the pitch of the tone it represents. A force, exerted on the coil because it is in a magnetic field, pushes the coil either into or out of the field, depending on the direction of the current. The motion of the coil causes the cone to vibrate, creating sound waves in the air.

Galvanometers

The forces exerted on a loop of wire in a magnetic field can be used to measure current. If a small loop of current-carrying wire is placed in the strong magnetic field of a permanent magnet, as in **Figure 24–19a,** it is possible to measure very small currents. The current passing through the loop goes in one end of the loop and out the other end. Applying the third right-hand rule to each side of the loop, note that one side of the loop is forced down, while the other side of the loop is forced up. The resulting torque rotates the loop. The magnitude of the torque acting on the loop is proportional to the magnitude of the current. This principle of measuring small currents is used in a galvanometer. A **galvanometer** is a device used to measure very small currents. For this reason, a galvanometer can be used as a voltmeter or an ammeter.

A small spring in the galvanometer exerts a torque that opposes the torque resulting from the current; thus, the amount of rotation is proportional to the current. The meter is calibrated by finding out how much the coil turns when a known current is sent through it, as shown in **Figure 24–19b.** The galvanometer can then be used to measure unknown currents.

FIGURE 24–19 If a wire loop is placed in a magnetic field when there is a current, the loop will rotate because of the torque exerted by the field **(a).** An unknown current passing through a galvanometer can be metered, because the coil rotates in proportion to the magnitude of the current **(b).**

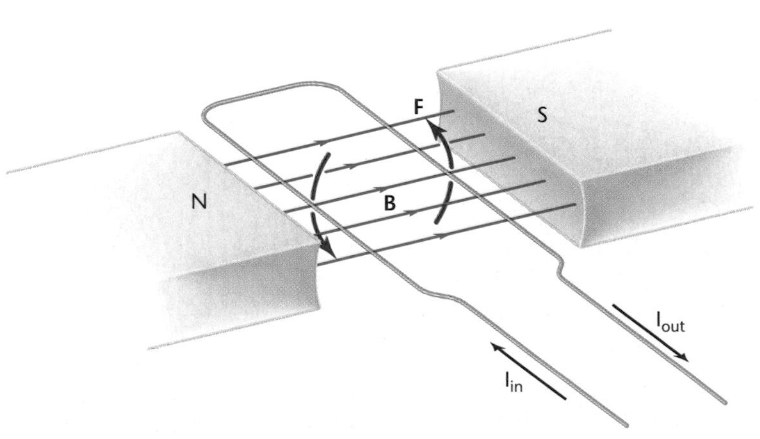

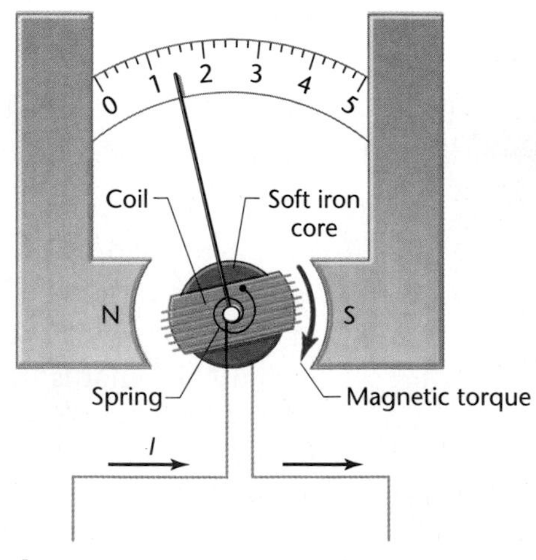

Many galvanometers produce full-scale deflections with as little as 50 μA (50×10^{-6} A) of current. The resistance of the coil of wire in a sensitive galvanometer is about 1000 ohms. In order to measure larger currents, such a galvanometer can be converted into an ammeter by placing a resistor with resistance smaller than that of the galvanometer in parallel with the meter, as shown in **Figure 24–20a.** Most of the current, I_s, passes through the resistor, called the shunt, because the current is inversely proportional to resistance, whereas only a few microamps, I_m, flow through the galvanometer. The resistance of the shunt is chosen according to the desired deflection scale.

A galvanometer also can be connected as a voltmeter. To make a voltmeter, a resistor, called the multiplier, is placed in series with the meter, as shown in **Figure 24–20b.** The galvanometer measures the current through the multiplier. The current is represented by $I = V/R$, where V is the voltage across the voltmeter and R is the effective resistance of the galvanometer and the multiplier resistor. Suppose you want a voltmeter that reads full-scale when 10 V is placed across it. The resistor is chosen so that at 10 V the meter is deflected full-scale by the current through the meter and resistor.

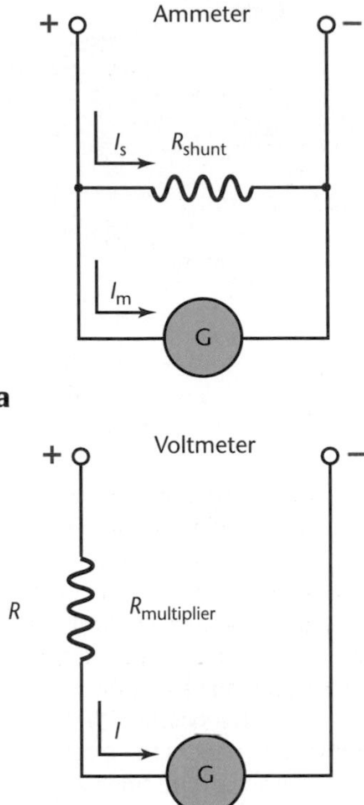

FIGURE 24–20 A galvanometer can be connected as either an ammeter **(a)** or a voltmeter **(b).**

Electric motors

You have seen how the simple loop of wire used in a galvanometer cannot rotate more than 180°. The forces push the right side of the loop up and the left side of the loop down until the loop reaches the vertical position. The loop will not continue to turn because the forces are still up and down, now parallel to the loop, and can cause no further rotation.

In an **electric motor,** an apparatus that converts electrical energy to kinetic energy, the loop must rotate a full 360° in the field; thus, the current running through the loop must reverse direction just as the loop reaches its vertical position. This reversal allows the loop to continue rotating, as illustrated in **Figure 24–21.** To reverse current direction, a split-ring commutator is used. Brushes, pieces of graphite that make contact with the commutator, allow current to flow into the loop. The split ring is arranged so that each half of the commutator changes brushes just as the loop reaches the vertical position. Changing brushes reverses the current in the loop. As a result, the direction of the force on each side of the loop is reversed, and the loop continues to rotate. This process repeats each half-turn, causing the loop to spin in the magnetic field.

Although only one loop is indicated in **Figure 24–21,** in an electric motor, the loop of wire, called the **armature,** is made of several loops mounted on a shaft or axle. The total force acting on the armature is proportional to $nBIL$, where n is the total number of turns on the armature, B is the strength of the magnetic field, I is the current, and L is the length of wire in each turn that moves through the magnetic field. The magnetic field is produced either by permanent magnets or by an electromagnet called a field coil. The torque on the armature, and, as a result, the speed of the motor, is controlled by varying the current through the motor.

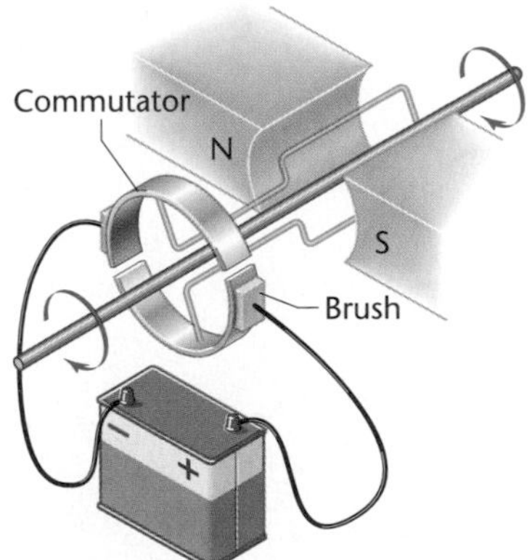

FIGURE 24–21 In an electric motor, split-ring commutators allow the wire loops in the motor to rotate 360°.

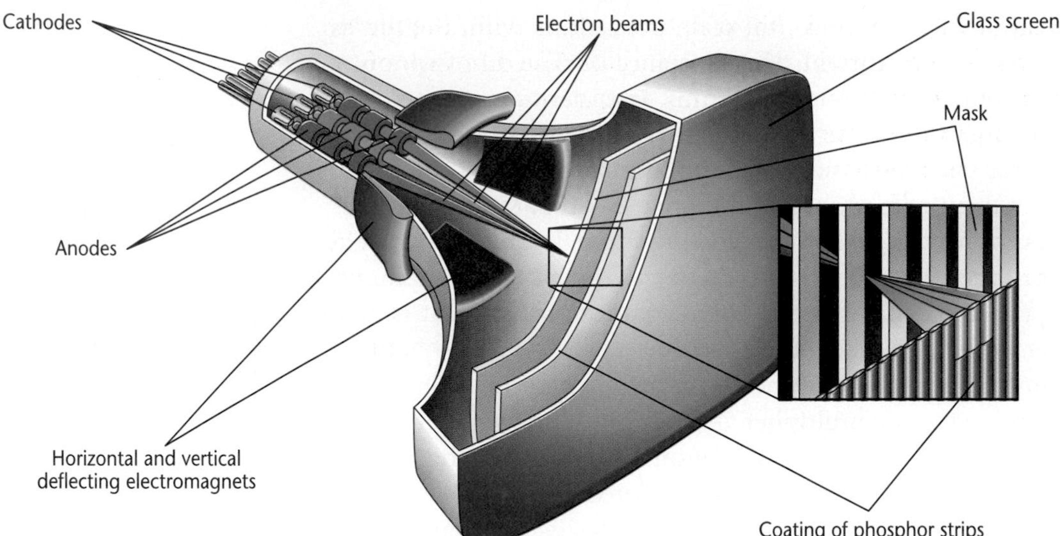

FIGURE 24–22 A computer monitor and a television use a cathode-ray tube to form pictures for viewing. Notice that the pairs of magnets deflect the electron beam vertically and horizontally.

The Force on a Single Charged Particle

Charged particles do not have to be confined to a wire, but can move across any region as long as the air has been removed to prevent collisions with air molecules. The picture tube, also called a cathode-ray tube, in computer monitors or television sets uses electrons deflected by magnetic fields to form the pictures on the screen, as illustrated in **Figure 24–22.** In a cathode-ray tube, electric fields pull electrons off atoms in the negative electrode, or cathode. Other electric fields gather, accelerate, and focus the electrons into a narrow beam. Magnetic fields are used to control the motion of the beam back and forth and up and down across the screen of the tube. The screen is coated with a phosphor that glows when it is struck by the electrons, thereby producing the picture.

The force produced by a magnetic field on a single electron depends on the velocity of the electron, the strength of the field, and the angle between directions of the velocity and the field. Consider a single electron moving in a wire of length L. The electron is moving perpendicular to the magnetic field. The current, I, is equal to the charge per unit time entering the wire, $I = q/t$. In this case, q is the charge of the electron and t is the time it takes to move the length of the wire, L. The time required for a particle with speed, v, to travel distance, L, is found by using the equation of motion, $d = vt$, or, in this case, $t = L/v$. As a result, the equation for the current, $I = q/t$, can be replaced by $I = qv/L$. Therefore, the force on a single electron moving perpendicular to a magnetic field of strength, B, can be found.

$$F = BIL = B\left(\frac{qv}{L}\right)L = Bqv$$

The particle's charge is measured in coulombs, its velocity in m/s, and the strength of the magnetic field in teslas, T.

The direction of the force is perpendicular to both the velocity of the particle and the magnetic field. Note, however, that the direction of the force is *opposite* that given by the third right-hand rule with the thumb pointed along the velocity of the positive particle. The direction of the force is opposite because the electron has a negative charge, and conventional current has a positive charge.

Electrons and positive ions trapped in the magnetic field of Earth form the Van Allen radiation belts. Solar storms send tremendous numbers of high-energy charged particles toward Earth. They disturb Earth's magnetic field, dumping electrons out of the Van Allen belts. These electrons excite atoms of nitrogen and oxygen in Earth's atmosphere and cause them to emit the red, green, and blue colors called the aurora borealis, or northern lights, that circle the north magnetic pole.

Northern Light Show

Example Problem

Force on a Charged Particle in a Magnetic Field

A beam of electrons travels at 3.0×10^6 m/s through a uniform magnetic field of 4.0×10^{-2} T at right angles to the field. How strong is the force that acts on each electron?

Sketch the Problem

- Represent the beam of electrons and its direction of motion; the magnetic field of lines, labeled *B*; and the force on the electron beam, *F*. Remember that the force is opposite that given by the third right-hand rule because of the electron's negative, elementary charge.

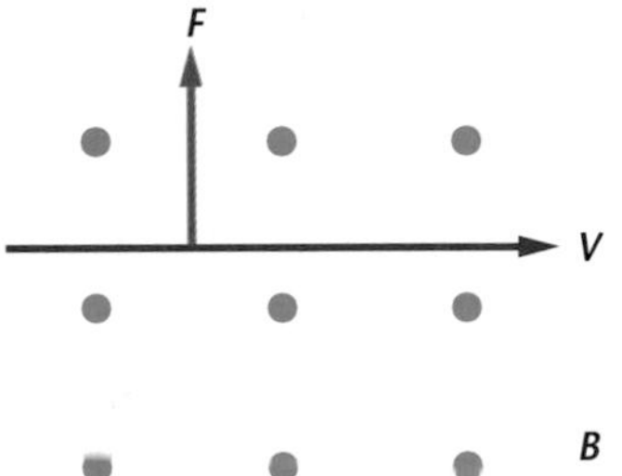

Calculate Your Answer

Known:	**Unknown:**
$v = 3.0 \times 10^6$ m/s	$F = ?$
$B = 4.0 \times 10^{-2}$ T	
$q = -1.60 \times 10^{-19}$ C	

Strategy:

Substitute the knowns, along with their respective units, into the equation $F = Bqv$. Calculate *F*.

Calculations:

$$F = (4.0 \times 10^{-2}\ \text{T})(-1.60 \times 10^{-19}\ \text{C})(3.0 \times 10^6\ \text{m/s})$$
$$= -1.9 \times 10^{-14}\ \text{T·C·m/s}$$
$$= -1.9 \times 10^{-14}\ \text{N}$$

Check Your Answer

- Are the units correct? T = N/A·m, and A = C/s; so T = N·s/C·m. Thus, T·C·m/s = N, the unit for force.
- Does the direction of the force make sense? Use the third right-hand rule to verify that the directions of the forces are correct, recalling that the force on the electron is opposite that given by the third right-hand rule.
- Is the magnitude realistic? Yes, forces on electrons and protons are always small fractions of a newton.

HELP WANTED

METEOROLOGIST

Do you have a sincere interest in the thorough study of data? Are you willing to make predictions, even when the data provide much less than a 100% probability? Join us and prepare for a highly respected profession. A bachelor's degree from an accredited meteorology program is required. Willingness to work your way up the ladder and take your turn with irregular shifts and possible relocation will help your career climb. For information contact:

American Meteorological Society, 45 Beacon Street, Boston, MA 02108

Practice Problems

10. An electron passes through a magnetic field at right angles to the field at a velocity of 4.0×10^6 m/s. The strength of the magnetic field is 0.50 T. What is the magnitude of the force acting on the electron?

11. A stream of doubly ionized particles (missing two electrons and thus carrying a net charge of two elementary charges) moves at a velocity of 3.0×10^4 m/s perpendicular to a magnetic field of 9.0×10^{-2} T. What is the magnitude of the force acting on each ion?

12. Triply ionized particles in a beam carry a net positive charge of three elementary charge units. The beam enters a magnetic field of 4.0×10^{-2} T. The particles have a speed of 9.0×10^6 m/s. What is the magnitude of the force acting on each particle?

13. Doubly ionized helium atoms (alpha particles) are traveling at right angles to a magnetic field at a speed of 4.0×10^{-2} m/s. The field strength is 5.0×10^{-2} T. What force acts on each particle?

24.2 Section Review

2.1 A horizontal, current-carrying wire runs north-south through Earth's magnetic field. If the current flows north, in which direction is the force on the wire?

2.2 A beam of electrons in a cathode-ray tube approaches the deflecting magnets. The north pole is at the top of the tube; the south pole is on the bottom. If you are looking at the tube from the direction of the phosphor screen, in which direction are the electrons deflected?

2.3 Compare the diagram of a galvanometer in **Figure 24–19** with the electric motor in **Figure 24–21.** How is the galvanometer similar to an electric motor? How are they different?

2.4 When the plane of the coil in a motor is perpendicular to the magnetic field, the forces do not exert a torque on the coil. Does this mean the coil doesn't rotate? Explain.

2.5 **Critical Thinking** How do you know that the forces on parallel current-carrying wires aren't a result of electrostatics? **Hint:** Consider what the charges would be like when the force is attractive. Then consider what the forces are if three wires carry currents in the same direction.

CHAPTER 24 REVIEW

Key Terms

24.1

- polarized
- magnetic field
- magnetic flux
- first right-hand rule
- solenoid
- electromagnet
- second right-hand rule
- domain

24.2

- third right-hand rule
- galvanometer
- electric motor
- armature

Summary

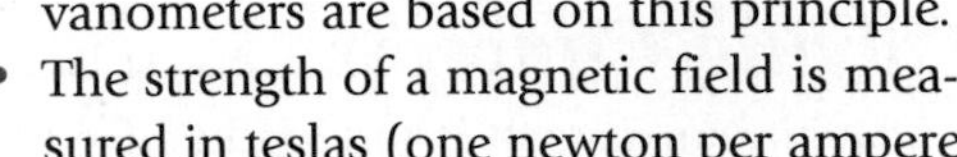

24.1 Magnets: Permanent and Temporary

- Like magnetic poles repel; unlike magnetic poles attract.
- Magnetic fields exit from the north pole of a magnet and enter its south pole.
- Magnetic field lines always form closed loops.
- A magnetic field exists around any wire that carries current.
- A coil of wire that carries a current has a magnetic field. The field about the coil is like the field about a permanent magnet.

24.2 Forces Caused by Magnetic Fields

- When a current-carrying wire is placed in a magnetic field, there exists a force on the wire that is perpendicular to both the field and the wire. Galvanometers are based on this principle.
- The strength of a magnetic field is measured in teslas (one newton per ampere per meter).
- An electric motor consists of a coil of wire placed in a magnetic field. When there is a current in the coil, the coil rotates as a result of the force on the wire in the magnetic field.
- The force a magnetic field exerts on a charged particle depends on the velocity and charge of the particle and the strength of the field. The direction of the force is perpendicular to both the field and the particle's velocity.

Reviewing Concepts

Section 24.1

1. State the rule for magnetic attraction and repulsion.
2. Describe how a temporary magnet differs from a permanent magnet.
3. Name the three most important common magnetic elements.
4. Draw a small bar magnet and show the magnetic field lines as they appear around the magnet. Use arrows to show the direction of the field lines.
5. Draw the magnetic field between two like magnetic poles and then between two unlike magnetic poles. Show the directions of the fields.
6. If you broke a magnet in two, would you have isolated north and south poles? Explain.
7. Describe how to use the right-hand rule to determine the direction of a magnetic field around a straight current-carrying wire.
8. If a current-carrying wire is bent into a loop, why is the magnetic field inside the loop stronger than the magnetic field outside?
9. Describe how to use the right-hand rule to determine the polarity of an electromagnet.
10. Each electron in a piece of iron is like a tiny magnet. The iron, however, may not be a magnet. Explain.
11. Why will dropping or heating a magnet weaken it?

Section 24.2

12. Describe how to use the right-hand rule to determine the direction of force on a current-carrying wire placed in a magnetic field.

13. A strong current is suddenly switched on in a wire. No force acts on the wire, however. Can you conclude that there is no magnetic field at the location of the wire? Explain.
14. What kind of meter is created when a shunt is added to a galvanometer?

Applying Concepts

15. A small bar magnet is hidden in a fixed position inside a tennis ball. Describe an experiment you could do to find the location of the N-pole and the S-pole of the magnet.
16. A piece of metal is attracted to one pole of a large magnet. Describe how you could tell whether the metal is a temporary magnet or a permanent magnet.
17. Is the magnetic force that Earth exerts on a compass needle less than, equal to, or greater than the force the compass needle exerts on Earth? Explain.
18. You are lost in the woods but have a compass with you. Unfortunately, the red paint marking the N-pole has worn off. You do have a flashlight with a battery and a length of wire. How could you identify the N-pole?
19. A magnet can attract a piece of iron that is not a permanent magnet. A charged rubber rod can attract an uncharged insulator. Describe the different microscopic processes that produce these similar phenomena.
20. A current-carrying wire runs across a laboratory bench. Describe at least two ways you could find the direction of the current.
21. In what direction in relation to a magnetic field would you run a current-carrying wire so that the force on it resulting from the field is minimized or even made to be zero?
22. Two wires carry equal currents and run parallel to each other.
    a. If the two currents are in opposite directions, where will the magnetic field from the two wires be larger than the field from either wire alone?
    b. Where will the magnetic field be exactly twice as large as that of either wire?
    c. If the two currents are in the same direction, where will the magnetic field be exactly zero?
23. How is the range of a voltmeter changed when the resistor's resistance is increased?
24. A magnetic field can exert a force on a charged particle. Can the field change the particle's kinetic energy? Explain.
25. A beam of protons is moving from the back to the front of a room. It is deflected upward by a magnetic field. What is the direction of the field causing the deflection?
26. Earth's magnetic field lines are shown in **Figure 24–23.** At what location, poles or equator, is the magnetic field strength greatest? Explain.

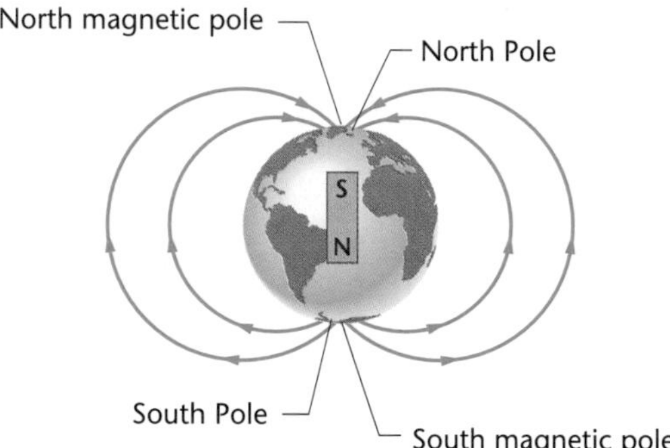

FIGURE 24–23

Problems

Section 24.1

LEVEL 1

27. A wire 1.50 m long carrying a current of 10.0 A is at right angles to a uniform magnetic field. The force acting on the wire is 0.60 N. What is the strength of the magnetic field?
28. A conventional current is in a wire as shown in **Figure 24–24.** Copy the wire segment and sketch the magnetic field that the current generates.

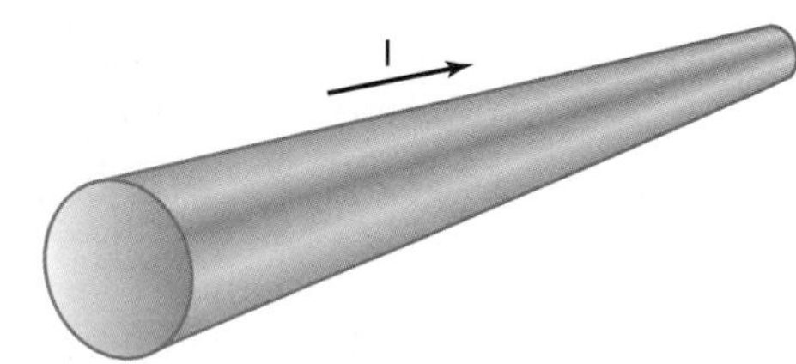

FIGURE 24–24

29. The current is coming straight out of the page in **Figure 24–25.** Copy the figure and sketch the magnetic field that the current generates.

FIGURE 24–25

30. **Figure 24–26** shows the end view of an electromagnet with the current as shown.
 a. What is the direction of the magnetic field inside the loop?
 b. What is the direction of the magnetic field outside the loop?

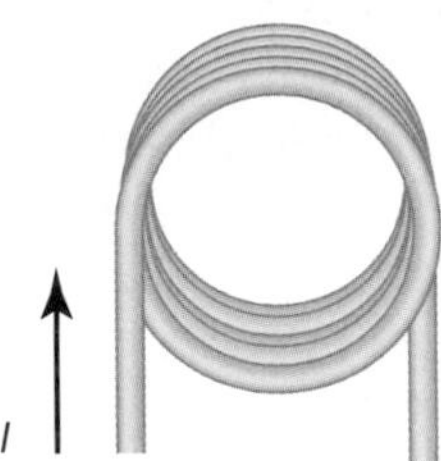

FIGURE 24–26

LEVEL 2

31. The repulsive force between two ceramic magnets was measured and found to depend on distance, as given in **Table 24–2.**

TABLE 24–2

Separation, *d* (mm)	Force, *F* (N)
10	3.93
12	0.40
14	0.13
16	0.057
18	0.030
20	0.018
22	0.011
24	0.0076
26	0.0053
28	0.0038
30	0.0028

 a. Plot the force as a function of distance.
 b. Does this force follow an inverse square law?

Section 24.2

LEVEL 1

32. A current-carrying wire is placed between the poles of a magnet, as shown in **Figure 24–27.** What is the direction of the force on the wire?

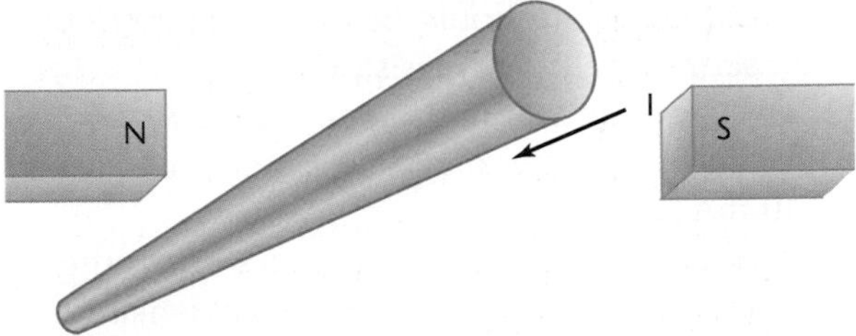

FIGURE 24–27

33. A wire 0.50 m long carrying a current of 8.0 A is at right angles to a uniform magnetic field. The force on the wire is 0.40 N. What is the strength of the magnetic field?
34. The current through a wire 0.80 m long is 5.0 A. The wire is perpendicular to a 0.60-T magnetic field. What is the magnitude of the force on the wire?
35. A wire 25 cm long is at right angles to a 0.30-T uniform magnetic field. The current through the wire is 6.0 A. What is the magnitude of the force on the wire?
36. A wire 35 cm long is parallel to a 0.53-T uniform magnetic field. The current through the wire is 4.5 A. What force acts on the wire?
37. A wire 625 m long is in a 0.40-T magnetic field. A 1.8-N force acts on the wire. What current is in the wire?
38. The force on a 0.80 m wire that is perpendicular to Earth's magnetic field is 0.12 N. What is the current in the wire?
39. The force acting on a wire at right angles to a 0.80-T magnetic field is 3.6 N. The current in the wire is 7.5 A. How long is the wire?
40. A power line carries a 225-A current from east to west parallel to the surface of Earth.
 a. What is the magnitude of the force resulting from Earth's magnetic field acting on each meter of the wire?

b. What is the direction of the force?

c. In your judgment, would this force be important in designing towers to hold these power lines?

41. A galvanometer deflects full-scale for a 50.0-μA current.

a. What must be the total resistance of the series resistor and the galvanometer to make a voltmeter with 10.0-V full-scale deflection?

b. If the galvanometer has a resistance of 1.0 kΩ, what should be the resistance of the series (multiplier) resistor?

42. The galvanometer in problem 41 is used to make an ammeter that deflects full-scale for 10 mA.

a. What is the potential difference across the galvanometer (1.0 kΩ resistance) when a current of 50 μA passes through it?

b. What is the equivalent resistance of parallel resistors that have the potential difference calculated in **a** for a circuit with a total current of 10 mA?

c. What resistor should be placed in parallel with the galvanometer to make the resistance calculated in **b?**

43. A beam of electrons moves at right angles to a magnetic field of 6.0×10^{-2} T. The electrons have a velocity of 2.5×10^{6} m/s. What is the magnitude of the force on each electron?

44. A beta particle (high-speed electron) is traveling at right angles to a 0.60-T magnetic field. It has a speed of 2.5×10^{7} m/s. What size force acts on the particle?

45. The mass of an electron is 9.11×10^{-31} kg. What is the acceleration of the beta particle described in problem 44?

46. A magnetic field of 16 T acts in a direction due west. An electron is traveling due south at 8.1×10^{5} m/s. What are the magnitude and direction of the force acting on the electron?

47. A muon (a particle with the same charge as an electron) is traveling at 4.21×10^{7} m/s at right angles to a magnetic field. The muon experiences a force of 5.00×10^{-12} N. How strong is the field?

48. The mass of a muon is 1.88×10^{-28} kg. What acceleration does the muon described in problem 47 experience?

49. A singly ionized particle experiences a force of 4.1×10^{-13} N when it travels at right angles through a 0.61-T magnetic field. What is the velocity of the particle?

LEVEL 2

50. A room contains a strong, uniform magnetic field. A loop of fine wire in the room has current flowing through it. Assuming you rotate the loop until there is no tendency for it to rotate as a result of the field, what is the direction of the magnetic field relative to the plane of the coil?

51. The magnetic field in a loudspeaker is 0.15 T. The wire consists of 250 turns wound on a 2.5-cm diameter cylindrical form. The resistance of the wire is 8.0 Ω. Find the force exerted on the wire when 15 V is placed across the wire.

52. A wire carrying 15 A of current has a length of 25 cm in a magnetic field of 0.85 T. The force on a current-carrying wire in a uniform magnetic field can be found using the equation $F = BIL \sin \theta$. Find the force on the wire if it makes an angle with the magnetic field lines of

a. 90°. **b.** 45°. **c.** 0°.

53. An electron is accelerated from rest through a potential difference of 20 000 V, which exists between plates P_1 and P_2, shown in **Figure 24–28.** The electron then passes through a small opening into a magnetic field of uniform field strength, *B*. As indicated, the magnetic field is directed into the page.

a. State the direction of the electric field between the plates as either P_1 to P_2 or P_2 to P_1.

b. In terms of the information given, calculate the electron's speed at plate P_2.

c. Describe the motion of the electron through the magnetic field.

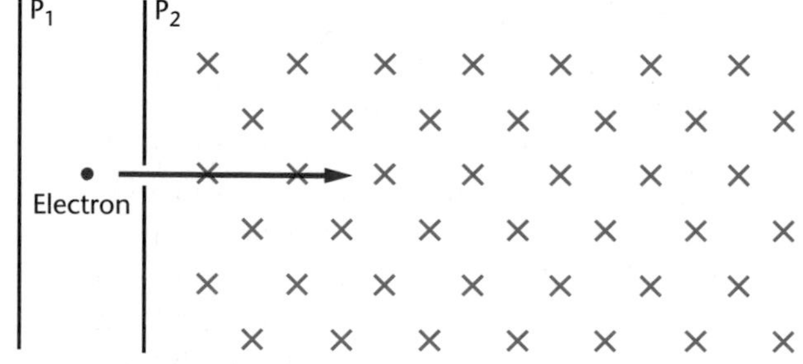

FIGURE 24–28

54. A force of 5.78×10^{-16} N acts on an unknown particle traveling at a 90° angle through a magnetic field. If the velocity of the particle is 5.65×10^{4} m/s and the field is 3.20×10^{-2} T, how many elementary charges does the particle carry?

Critical Thinking Problems

55. A current is sent through a vertical spring as shown in **Figure 24–29.** The end of the spring is in a cup filled with mercury. What will happen? Why?

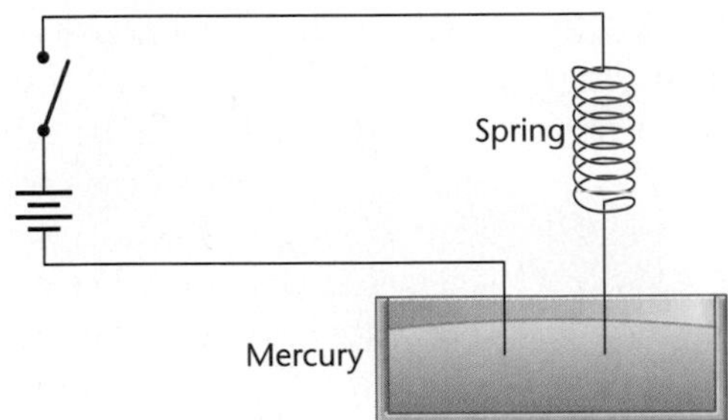

FIGURE 24–29

56. The magnetic field produced by a long, current-carrying wire is represented by $B = 2 \times 10^{-7}(\text{T}\cdot\text{m/A})I/d$, where B is the field strength in teslas, I is the current in amps, and d is the distance from the wire in meters. Use this equation to estimate some magnetic fields that you encounter in everyday life.

a. The wiring in your home seldom carries more than 10 A. How does the field 0.5 m from such a wire compare to Earth's magnetic field?

b. High-voltage power transmission lines often carry 200 A at voltages as high as 765 kV. Estimate the magnetic field on the ground under such a line, assuming that it is about 20 m high. How does this field compare with that in your home?

c. Some consumer groups have recommended that pregnant women not use electric blankets in case the magnetic fields cause health problems. Blankets typically carry currents of about 1 A. Estimate the distance a fetus might be from such a wire, clearly stating your assumptions, and find the magnetic field at the location of the fetus. Compare this with Earth's magnetic field.

Going Further

Adding Vectors In almost all cases described in problem 56, a second wire carries the same current in the opposite direction. Find the net magnetic field a distance 0.10 m from each wire that carries 10 A. The wires are 0.01 m apart. Make a scale drawing of the situation. Calculate the magnitude of the field from each wire and use the right-hand rule to draw vectors showing the direction of the fields. Finally, find the vector sum of the two fields. State its magnitude and direction.

*inter***NET**
CONNECTION

Follow the link on the Glencoe Homepage at **www.glencoe.com/sec/science** to find out more about this chapter.

Go with the Flow

Two aluminum rings, one with a slit and one a continuous ring, are placed over a magnetic field generator that is producing a constantly changing magnetic field. Why does one ring float while the other does not?

CHAPTER

25 Electromagnetic Induction

What magic levitates the ring? Why isn't the other ring also floating? Here are the facts: the rings are made of aluminum, and the coil of wire around the central rod produces a continually changing magnetic field. However, aluminum is a nonmagnetic substance. In other words, a magnet will not attract a piece of aluminum much less push it up. Yet the photo clearly shows an upward force being exerted on the top, uncut ring. What force could be pushing the ring up to counterbalance the downward force of gravity? And why doesn't this upward force act on the cut ring? You've learned that superconductors and permanent magnets can cause objects to float. There is, however, no superconductor or permanent magnet in this photograph. You may never have seen such a floating ring, but the physical principle that explains why it is levitated also explains how electricity is produced and delivered to your home and school.

WHAT YOU'LL LEARN

- You will describe how changing magnetic fields can generate electric current and potential difference.
- You will apply this phenomenon to the construction of generators and transformers.

WHY IT'S IMPORTANT

- The relationship between magnetic fields and currents makes possible the three cornerstones of electrical technology: motors, generators, and transformers.

interNET CONNECTION

Follow the link for this chapter on the Glencoe Homepage at **www.glencoe.com/sec/science** to find out more about electromagnetic induction.

25.1 Creating Electric Current from Changing Magnetic Fields

OBJECTIVES

- **Explain** how a changing magnetic field produces an electric current.
- **Define** electromotive force, and **solve** problems involving wires moving in a magnetic field.
- **Describe** how an electric generator works and how it differs from a motor.
- **Recognize** the difference between peak and effective voltage and current.

In 1822, Michael Faraday wrote a goal in his notebook: "Convert Magnetism into Electricity." After nearly ten years of unsuccessful experiments, he was able to show that a changing magnetic field could produce electric current. In the same year, Joseph Henry, an American high school teacher, made the same discovery.

Faraday's Discovery

In Chapter 24 you read about how Hans Christian Oersted discovered that an electric current produces a magnetic field. Michael Faraday thought that the reverse must also be true, that a magnetic field produces an electric current. Faraday tried many combinations of magnetic fields and wires without success, until he found that he could induce current by moving a wire through a magnetic field. **Figure 25–1** shows one of Faraday's experiments. A wire loop that is part of a closed circuit is placed in a magnetic field. When the wire moves up through the field, the current is in one direction. When the wire moves down through the field, the current is in the opposite direction. When the wire is held stationary or is moved parallel to the magnetic field, there is no current. An electric current is generated in a wire only when the wire cuts magnetic field lines.

Creating current

Faraday found that to generate current, either the conductor can move through a magnetic field or the magnetic field can move past a conductor. It is the relative motion between the wire and the magnetic field that produces the current. The process of generating a current through a circuit in this way is called **electromagnetic induction.**

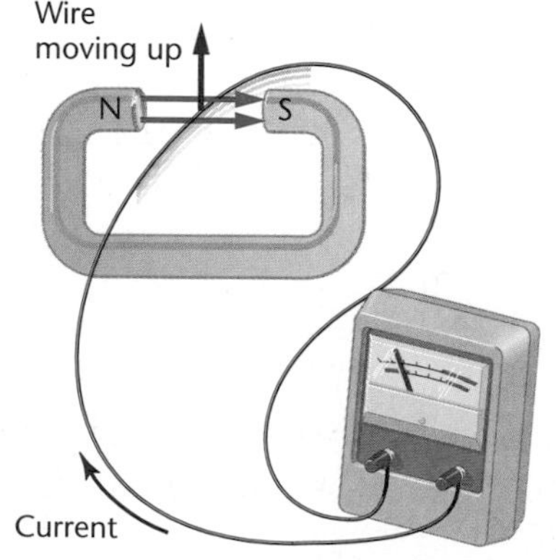

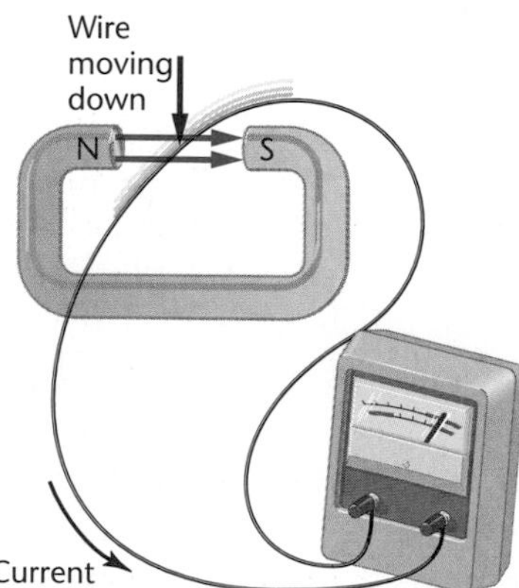

FIGURE 25–1 When a wire is moved in a magnetic field, there is an electric current in the wire, but only while the wire is moving. The direction of the current depends on the direction the wire is moving through the field. The arrows indicate the direction of conventional current.

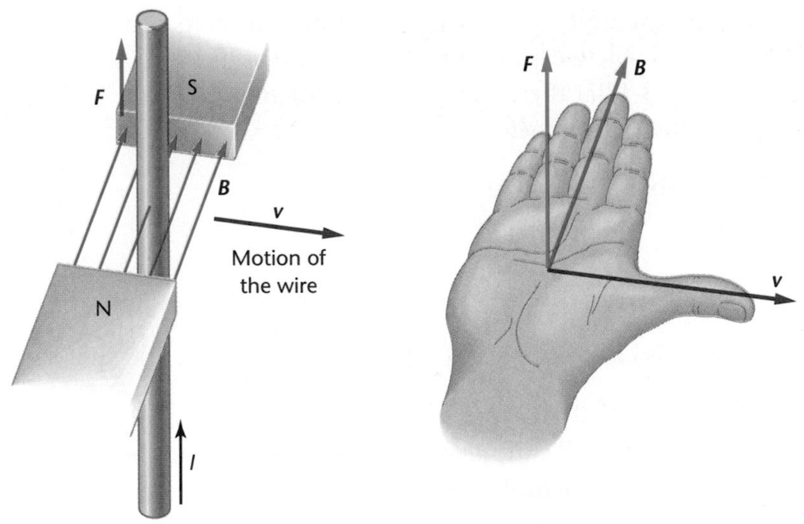

FIGURE 25–2 The right-hand rule can be used to find the direction of the forces on the charges in a conductor that is moving in a magnetic field.

In what direction is the current? To find the force on the charges in the wire, use the third right-hand rule described in Chapter 24. Hold your right hand so that your thumb points in the direction in which the wire is moving and your fingers point in the direction of the magnetic field. The palm of your hand will point in the direction of the conventional (positive) current, as illustrated in **Figure 25–2.**

Electromotive Force

When you studied electric circuits, you learned that a source of electrical energy, such as a battery, is needed to produce a continuous current. The potential difference, or voltage, given to the charges by a battery is called the **electromotive force,** or *EMF.* Electromotive force, however, is not a force; it is a potential difference and is measured in volts. Thus, the term *EMF* is misleading. Like many other historical terms still in use, it originated before electricity was well understood.

What created the potential difference that caused an induced current in Faraday's experiment? When you move a wire through a magnetic field, you exert a force on the charges and they move in the direction of the force. Work is done on the charges. Their electrical potential energy, and thus their potential, is increased. The difference in potential is called the induced *EMF.* The *EMF,* measured in volts, depends on the magnetic field, $\mathbf{B}$, the length of the wire in the magnetic field, L, and the velocity of the wire in the field, $\mathbf{v}$. If $\mathbf{B}$, $\mathbf{v}$, and the direction of the length of the wire are mutually perpendicular, then *EMF* is the product of the three.

$$EMF = BLv$$

If a wire moves through a magnetic field at an angle to the field, only the component of the wire's velocity that is perpendicular to the direction of the magnetic field generates *EMF.*

HISTORY CONNECTION

In the late 1800s, there was much debate in the United States over the best way to transmit power from power plants to consumers. The existing plants transmitted direct current (DC), but DC had serious limitations. A key decision was made to use alternating current (AC) in the new hydroelectric power plant at Niagara Falls. With the first successful transmission of AC to Buffalo, New York in 1896, the Niagara Falls plant paved the way for the development of AC power plants across the country.

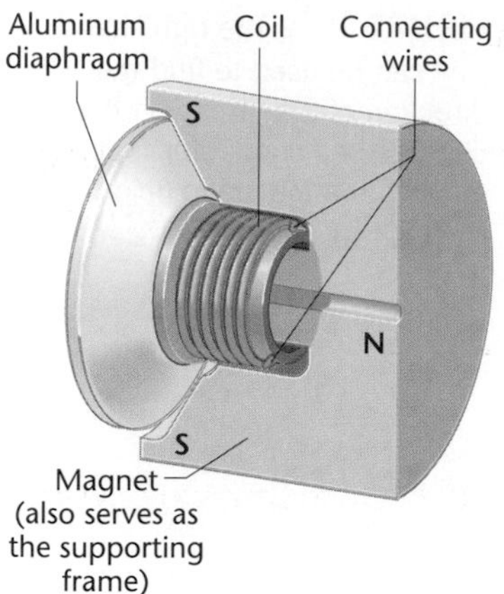

FIGURE 25–3 In this schematic of a moving coil microphone, the aluminum diaphragm is connected to a coil in a magnetic field. When sound waves vibrate the diaphragm, the coil moves in the magnetic field, generating a current proportional to the sound wave.

Checking the units of the *EMF* equation will help you work algebra correctly in problems. The units for *EMF* are volts, V. In Chapter 24, B was defined as F/IL; therefore, the units for B are N/A·m. Following is the unit equation for *EMF*.

$$\text{variables: } EMF = BLv$$

$$\text{units: V} = \left(\frac{\text{N}}{\text{A}\cdot\text{m}}\right)(\text{m})(\text{m/s}) = \frac{\text{N}\cdot\text{m}}{\text{A}\cdot\text{s}} = \frac{\text{J}}{\text{C}} = \text{V}$$

From previous chapters, recall that J = N·m, A = C/s, and that V = J/C.

Application of induced *EMF*

A microphone is a simple application that depends on an induced *EMF*. A dynamic microphone is similar in construction to a loudspeaker. The microphone in **Figure 25–3** has a diaphragm attached to a coil of wire that is free to move in a magnetic field. Sound waves vibrate the diaphragm, which moves the coil in the magnetic field. The motion of the coil, in turn, induces an *EMF* across the ends of the coil. The induced *EMF* varies as the frequency of the sound varies. In this way, the sound wave is converted to an electrical signal. The voltage generated is small, typically 10^{-3} V, but it can be increased, or amplified, by electronic devices.

Example Problem

Induced *EMF*

A straight wire, 0.20 m long, moves at a constant speed of 7.0 m/s perpendicular to magnetic field of strength 8.0×10^{-2} T.

a. What *EMF* is induced in the wire?

b. The wire is part of a circuit that has a resistance of 0.50 Ω. What is the current through the wire?

Sketch the Problem

- Establish a coordinate system.
- Draw a straight wire of length L. Connect an ammeter to the wire to represent a current measurement.
- Choose a direction for the magnetic field that is perpendicular to the length of the wire.
- Choose a direction for the velocity that is perpendicular to both the length and the magnetic field.

Calculate Your Answer

Known:		**Unknown:**
$v = 7.0$ m/s	$B = 8.0 \times 10^{-2}$ T	$EMF = ?$
$L = 0.20$ m	$R = 0.50$ Ω	$I = ?$

Strategy:

a. For motion at a right angle through a field, start with the *EMF* equation. Keep track of units. It is helpful to remember that $B = F/IL$, $energy = Fd$, $I = q/t$, and $V = energy/q$.

b. Use the current equation. Recall that voltage and *EMF* are equivalent. Use the right-hand rule to determine current direction. The thumb is $\boldsymbol{v}(+x)$, fingers are $\boldsymbol{B}(+z)$, and palm is current, which points down $(-y)$. Downward current corresponds to a counterclockwise current in the loop.

Calculations:

$$EMF = BLv$$

$$EMF = (8.0 \times 10^{-2}\text{T})(0.20\text{ m})(7.0\text{ m/s})$$

$$EMF = 0.11\text{ T·m}^2\text{/s} = 0.11\text{ J/C} = 0.11\text{ V}$$

$$I = \frac{V}{R} = \frac{EMF}{R}$$

$$I = \frac{0.11\text{ V}}{0.50\ \Omega} = 0.22\text{ A, counterclockwise}$$

Check Your Answer

- Are the units correct? **a.** The answer is in volts, which is the correct unit for electromotive force. **b.** Ohms are defined as V/A. Perform the algebra with the units to confirm that I is measured in A.
- Does the direction make sense? The direction obeys the right-hand rule: $\boldsymbol{v}$ is the thumb, $\boldsymbol{B}$ is the fingers, $\boldsymbol{F}$ is the palm. Current is in the direction of the force.
- Is the magnitude realistic? The answers are near 10^{-1}. This agrees with the quantities given and the algebra performed.

Practice Problems

1. A straight wire, 0.5 m long, is moved straight up at a speed of 20 m/s through a 0.4-T magnetic field pointed in the horizontal direction.
   a. What *EMF* is induced in the wire?
   b. The wire is part of a circuit of total resistance of 6.0 Ω. What is the current in the circuit?
2. A straight wire, 25 m long, is mounted on an airplane flying at 125 m/s. The wire moves in a perpendicular direction through Earth's magnetic field ($B = 5.0 \times 10^{-5}$ T). What *EMF* is induced in the wire?
3. A straight wire, 30.0 m long, moves at 2.0 m/s in a perpendicular direction through a 1.0-T magnetic field.
   a. What *EMF* is induced in the wire?
   b. The total resistance of the circuit of which the wire is a part is 15.0 Ω. What is the current?
4. A permanent horseshoe magnet is mounted so that the magnetic field lines are vertical. If a student passes a straight wire between the poles and pulls it toward herself, the current flow through the wire is from right to left. Which is the N-pole of the magnet?

Making Currents

Hook the ends of a 1-m length of wire to the binding posts of a galvanometer (or microammeter). Make several overlapping loops in the wire. Watch the readings on the wire as you move a pair of neodymium magnets (or a strong horseshoe magnet) near the loops. Record your observations.

Analyze and Conclude What can you do to increase the current? Replace the 1-m length of wire with a preformed coil and see how much current you can produce. Describe your results.

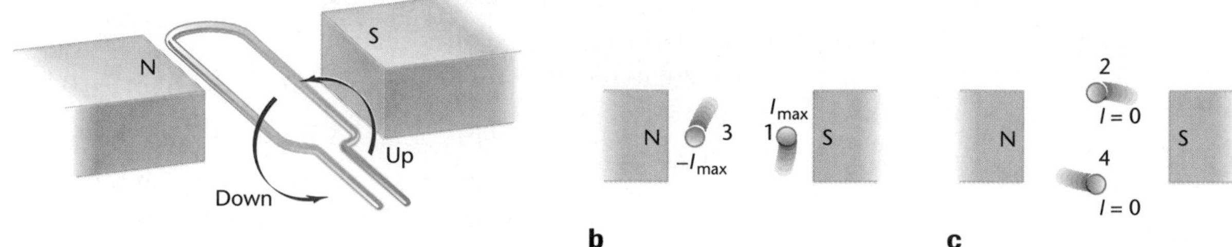

FIGURE 25–4 An electric current is generated in a wire loop as the loop rotates **(a).** The cross-sectional view shows the position of the loop when maximum current is generated **(b).** The numbered positions correspond to the numbered points on the graph in **Figure 25–5.**

Electric Generators

The **electric generator,** invented by Michael Faraday, converts mechanical energy to electrical energy. An electric generator consists of a number of wire loops placed in a strong magnetic field. The wire is wound around an iron form to increase the strength of the magnetic field. The iron and wires are called the armature, which is similar to that of an electric motor.

The armature is mounted so that it can rotate freely in the magnetic field. As the armature turns, the wire loops cut through the magnetic field lines, inducing an *EMF.* Commonly called the voltage, the *EMF* developed by the generator depends on the length of wire rotating in the field. Increasing the number of loops in the armature increases the wire length, thereby increasing the induced *EMF.*

Current from a generator

When a generator is connected in a closed circuit, the induced *EMF* produces an electric current. **Figure 25–4** shows a single-loop generator. The direction of the induced current can be found from the third right-hand rule described in Chapter 24. As the loop rotates, the strength and direction of the current change. The current is greatest when the motion of the loop is perpendicular to the magnetic field—when the loop is in the horizontal position. In this position, the component of the loop's velocity perpendicular to the magnetic field is greatest. As the loop rotates from the horizontal to the vertical position, it moves through the magnetic field lines at an ever-increasing angle. Thus, it cuts through fewer magnetic field lines per unit time, and the current decreases. When the loop is in the vertical position, the wire segments move parallel to the field and the current is zero. As the loop continues to turn, the segment that was moving up begins to move down, reversing the direction of the current in the loop. This change in direction takes place each time the loop turns through 180°. The current changes smoothly from zero to some maximum value and back to zero during each half-turn of the loop. Then it reverses direction. The graph of current versus time is shown in **Figure 25–5.**

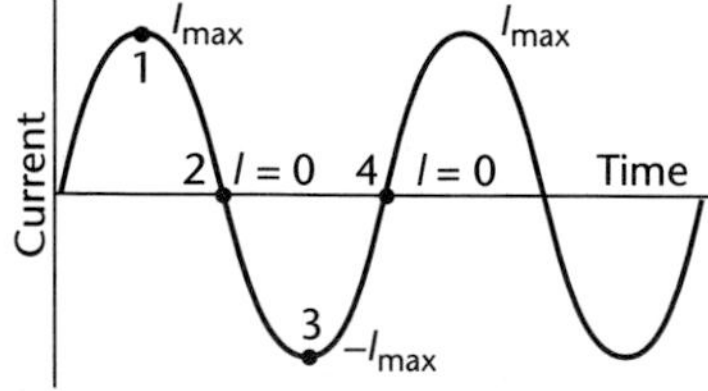

FIGURE 25–5 This graph shows the variation of current with time as the loop in **Figure 25–4** rotates. The variation of *EMF* with time can be shown with a similar graph.

Generators and motors are almost identical in construction, but they convert energy in opposite directions. A generator converts mechanical energy to electrical energy, while a motor converts electrical energy to mechanical energy.

Physics & Society

Electromagnetic Fields (EMFs)

Alternating currents produce both electric and magnetic fields, EMFs. These low-frequency fields, 60 Hz, are generated by nearly all electrical appliances as well as by outdoor power lines and indoor electrical wiring. About two decades ago, some doctors suggested that there was a direct link between EMFs and childhood leukemia, a rare form of cancer. This possibility spurred many studies on the topic both in physics and biology. In addition to the cost of the studies, concern about EMF health risks are costing the U.S. society an estimated $1 billion a year in lawsuits, rerouting of power lines, and redesigning of products such as computer monitors and household appliances. In 1995, the American Physical Society issued a statement that its review of scientific literature found "no consistent, significant link between cancer and power line fields." But there are still questions to be resolved because of the subtle effects of EMFs on cell membranes.

Cells, Tissues, and EMFs

To date, studies on human cells and tissues and on laboratory animals have shown that exposure to EMFs at the low frequencies common in most households does not alter cell functions. Research also has shown that EMFs tens or hundreds of times stronger than those in residential structures do cause changes in the chemical signals that cells send to each other. This correlation, however, doesn't seem to have adverse effects on health.

Furthermore, current research indicates that exposure to even very high EMFs does not affect a cell's DNA. Damaged DNA is currently thought to be the cause of most cancers. In 1997, the National Cancer Institute finished an exhaustive, seven-year study with the conclusion "that if there is any link at all, it's far too weak to be concerned about."

Toss the toaster and bag the blanket?

Although current research indicates that the EMF issue is no cause for alarm, some people (including some scientists and physicians) are not convinced that their toasters, electric blankets, alarm clocks, computer terminals, televisions, and other commonly used household items are truly safe.

Investigating the Issue

1. **Acquiring Information** Read current articles about EMFs. Note the major points made in each article. While reading, look for bias and evidence of factual documentation.
2. **Debating the Issue** What do you think should be done? Do you think there is sufficient and conclusive evidence that links health hazards and EMFs? Should people limit their exposure to EMFs? Should companies, including government, continue to spend millions of dollars to research the bioeffects of EMFs, redesign products to emit less, and move or bury power lines?
3. **Recognizing Cause and Effect** While there appears to be no correlation between exposure to EMFs and certain cancers, some studies have suggested a possible link between the two. What might be the reasons for this inconsistency?

interNET CONNECTION

Follow the link for this chapter on the Glencoe Homepage at **www.glencoe.com/sec/science** to find out more about EMFs.

FIGURE 25–6 Alternating-current generators transmit current to an external circuit by way of a brush-slip-ring arrangement **(a)**. The alternating current produced varies with time **(b)**. The resulting power is always positive and also is sinusoidal **(c)**.

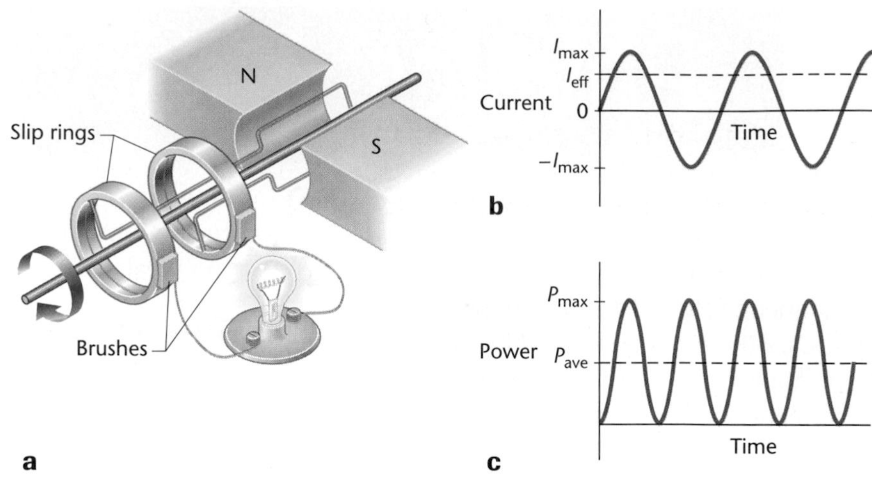

Alternating-Current Generator

An energy source turns the armature of a generator in a magnetic field at a fixed number of revolutions per second. In the United States, electric utilities use a 60-Hz frequency. The current goes from one direction to the other and back to the first, 60 times a second. **Figure 25–6a** shows how an alternating current, AC, in an armature is transmitted to the rest of the circuit. The brush-slip-ring arrangement permits the armature to turn freely while still allowing the current to pass into the external circuit. As the armature turns, the alternating current varies between some maximum value and zero, as shown in the graph in **Figure 25–6b.**

Average power

The power produced by a generator is the product of the current and the voltage. Because both I and V vary, the power associated with an alternating current varies. **Figure 25–6c** shows a graph of the power produced by an AC generator. Note that power is always positive because I and V are either both positive or both negative. Average power, P_{AC}, is defined as half the maximum power, $P_{AC} = 1/2\ P_{AC\ max}$.

Effective voltage and currents

It is common to describe alternating current and voltage in terms of effective current and voltage, rather than referring to their maximum values. Recall from Chapter 22 that $P = I^2R$. Thus, you can relate effective current, I_{eff}, in terms of the average AC power.

$$P_{AC} = I_{eff}^2R$$

To determine I_{eff} in terms of maximum current, I_{max}, start with the power relationship and substitute in I^2R, then solve for I_{eff}.

$$P_{AC} = 1/2\ P_{AC\ max}$$

$$I_{eff}^2R = 1/2\ I_{max}^2R$$

Motor and Generator

Make a series circuit with a Genecon (or efficient DC motor), a miniature lamp, and an ammeter. Rotate the handle (or motor shaft) to try to light the lamp.

Analyze and Conclude
Describe your results. Predict what might happen if you connect your Genecon to the Genecon from another lab group and crank yours. Try it. Describe what happens. Can more than two be connected?

$$I_{eff} = \sqrt{1/2\ I_{max}^2}$$

$$I_{eff} = 0.707\ I_{max}$$

Similarly, it can be shown that $V_{eff} = 0.707\ V_{max}$.

In the United States, the voltage generally available at wall outlets is described as 120 V, where 120 V is the magnitude of the effective voltage, not the maximum voltage.

Practice Problems

5. A generator develops a maximum voltage of 170 V.
   - **a.** What is the effective voltage?
   - **b.** A 60-W lightbulb is placed across the generator with an I_{max} of 0.70 A. What is the effective current through the bulb?
   - **c.** What is the resistance of the lightbulb when it is working?

6. The effective voltage of an AC household outlet is 117 V.
   - **a.** What is the maximum voltage across a lamp connected to the outlet?
   - **b.** The effective current through the lamp is 5.5 A. What is the maximum current in the lamp?

7. An AC generator delivers a peak voltage of 425 V.
   - **a.** What is the V_{eff} in a circuit placed across the generator?
   - **b.** The resistance of the circuit is $5.0 \times 10^2\ \Omega$. What is the effective current?

8. If the average power dissipated by an electric light is 100 W, what is the peak power?

HELP WANTED
POWER PLANT OPERATIONS TRAINEE

A utility company needs high school graduates who have the interest and ability to learn detailed information about the production and distribution of electrical power. Your high school record of reliability, effort, and cooperation are most important.

This entry-level position includes extensive training. Success in this field is dependent on quality work, continued training, and the ability to contribute positively to the work environment. For information contact:

Utility Workers of America
815 16th Street, NW
Washington, DC 20006

25.1 Section Review

1.1 Could you make a generator by mounting permanent magnets on a rotating shaft and keeping the coil stationary? Explain.

1.2 A bike generator lights the headlamp. What is the source of the energy for the bulb when the rider travels along a flat road?

1.3 Consider the microphone shown in **Figure 25–3.** When the diaphragm is pushed in, what is the direction of the current in the coil?

1.4 **Critical Thinking** A student asks: "Why does AC dissipate any power? The energy going into the lamp when the current is positive is removed when the current is negative. The net is zero." Explain why this reasoning is wrong.

25.2 Changing Magnetic Fields Induce *EMF*

OBJECTIVES

- **State** Lenz's law, and **explain** back-*EMF* and how it affects the operation of motors and generators.
- **Explain** self-inductance and how it affects circuits.
- **Describe** a transformer and **solve** problems involving voltage, current, and turn ratios.

In a generator, current begins when the armature turns through a magnetic field. You learned in Chapter 24 that when there is a current through a wire in a magnetic field, a force is exerted on the wire. Thus, the act of generating current produces a force on the wires in the armature.

Lenz's Law

In what direction is the force on the wires of the armature? To determine the direction, consider a section of one loop that moves through a magnetic field, as shown in **Figure 25–7.** In the last section, you learned that an *EMF* will be induced in the wire governed by the equation $EMF = BLv$. If the magnetic field is out of the page and velocity is to the right, then the right-hand rule shows a downward *EMF*, and consequently a downward current is produced. In Chapter 24, you learned that a wire carrying a current through a magnetic field will experience a force acting on it. This force results from the interaction between the existing magnetic field and the magnetic field generated around all currents. To determine the direction of the force, use the third right-hand rule: if current I is down and magnetic field B is out, then the resulting force is to the left. This means that the direction of the force on the wire opposes the original motion of the wire, v. That is, the force acts to slow down the rotation of the armature. The method of determining the direction of a force was first demonstrated in 1834 by H.F.E. Lenz and is therefore called Lenz's law.

Lenz's law states that the direction of the induced current is such that the magnetic field resulting from the induced current opposes the change in the field that caused the induced current. Note that it is the change in the field and not the field itself that is opposed by the induced magnetic effects.

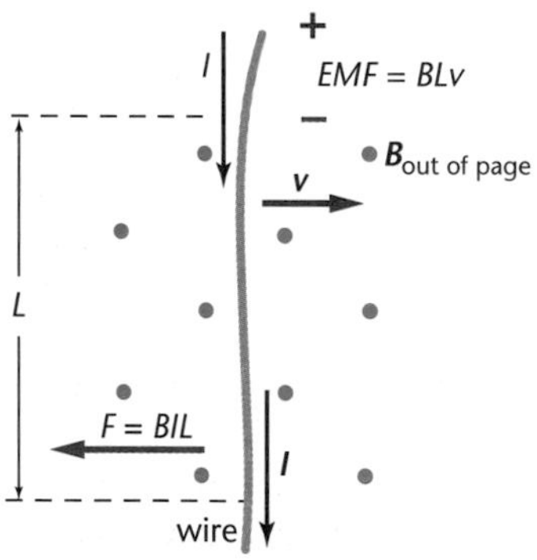

+
I
$EMF = BLv$
–
$B_{\text{out of page}}$
v
L
$F = BIL$
I
wire

FIGURE 25–7 A wire, length L, moving through a magnetic field, ***B,*** induces an electromotive force. If the wire is part of a circuit, then there will be a current, I. This current will now interact with the magnetic field producing force, ***F.*** Notice that the resulting force opposes the motion, ***v,*** of the wire.

Opposing change

Figure 25–8 is an example of how Lenz's law works. The N-pole of a magnet is moved toward the left end of a coil. To oppose the approach of the N-pole, the left end of the coil also must become an N-pole. In other words, the magnetic field lines must emerge from the left end of the coil. Using the second right-hand rule you learned in Chapter 24, you will see that if Lenz's law is correct, the induced current must be in a counterclockwise direction. Experiments have shown that this is so. If the magnet is turned so that an S-pole approaches the coil, the induced current will flow in a clockwise direction.

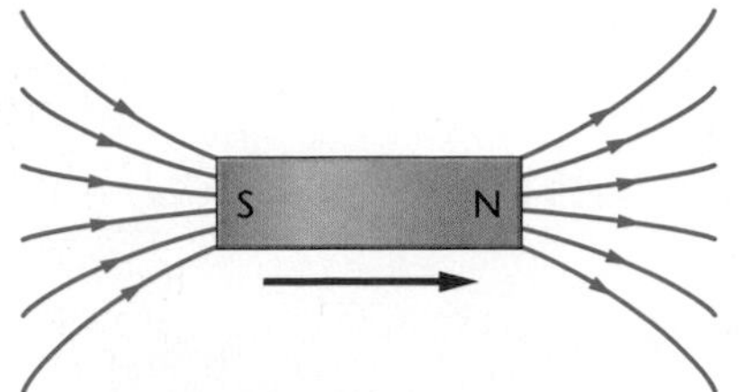

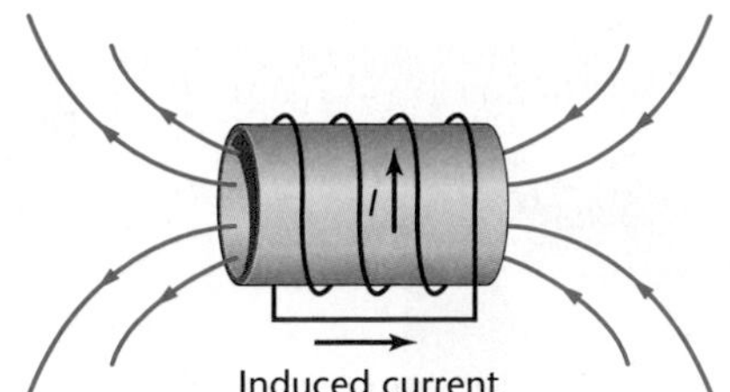

FIGURE 25–8 The magnet approaching the coil causes an induced current to flow. Lenz's law predicts the direction of flow shown.

If a generator produces only a small current, then the opposing force on the armature will be small, and the armature will be easy to turn. If the generator produces a larger current, the force on the larger current will be greater, and the armature will be more difficult to turn. A generator supplying a large current is producing a large amount of electrical energy. The opposing force on the armature means that mechanical energy must be supplied to the generator to produce the electrical energy, consistent with the law of conservation of energy.

Motors and Lenz's law

Lenz's law also applies to motors. When a current-carrying wire moves in a magnetic field, an *EMF* is generated. This *EMF*, called the back-*EMF*, is in a direction that opposes the current. When a motor is first turned on, there is a large current because of the low resistance of the motor. As the motor begins to turn, the motion of the wires across the magnetic field induces a back-*EMF* that opposes the current. Therefore, the net current through the motor is reduced. If a mechanical load is placed on the motor, as in a situation in which work is being done to lift a weight, the rotation of the motor will slow. This slowing down will decrease the back-*EMF*, which will allow more current through the motor. Note that this is consistent with the law of conservation of energy: if current increases, so does the rate at which electric power is being sent to the motor. This power is delivered in mechanical form to the load. If the mechanical load stops the motor, current can be so high that wires overheat.

The heavy current required when a motor is started can cause voltage drops across the resistance of the wires that carry current to the motor. The voltage drop across the wires reduces the voltage across the motor. If a second device, such as a lightbulb, is in a parallel circuit with the motor, the voltage at the bulb also will drop when the motor is started. The bulb will dim. As the motor picks up speed, the voltage will rise again and the bulb will brighten.

When the current to the motor is interrupted by a switch in the circuit being turned off or by the motor's plug being pulled from a wall outlet, the sudden change in the magnetic field generates a back-*EMF*. This reverse voltage can be large enough to cause a spark across the switch or between the plug and the wall outlet.

Slow Motor

Make a series circuit with a miniature DC motor, an ammeter, and a DC power supply. Hook up a voltmeter in parallel across the motor. Adjust the setting on the power supply so that the motor is running at medium speed. Make a data table to show the readings on the ammeter and voltmeter.

Analyze and Conclude Predict what will happen to the readings on the circuit when you hold the shaft and keep it from turning. Try it. Explain the results.

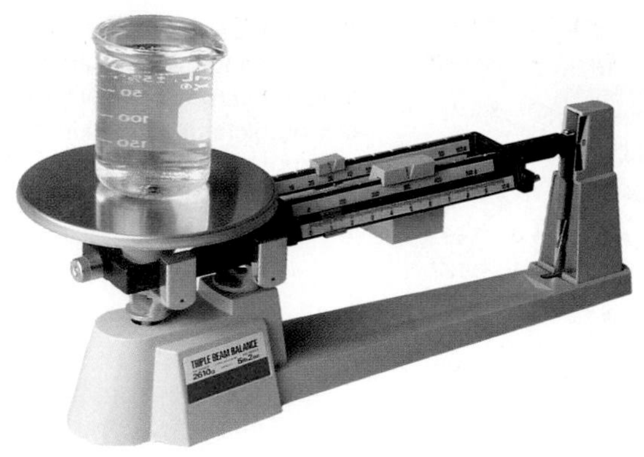

a

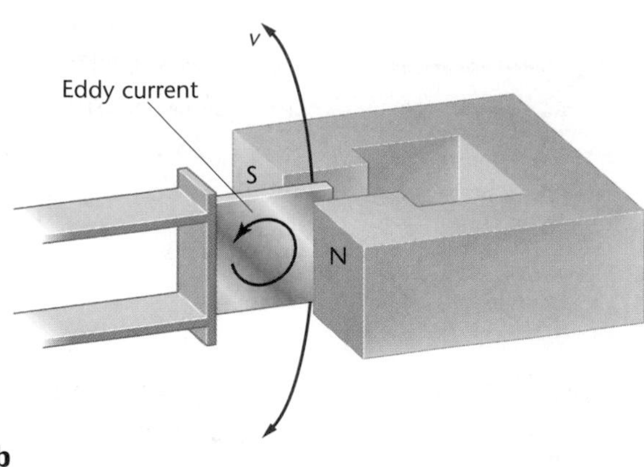

b

FIGURE 25–9 Sensitive balances use eddy-current damping to control oscillations of the balance beam **(a).** As the metal plate on the end of the beam moves through the magnetic field, a current is generated in the metal. This current, in turn, produces a magnetic field that opposes the motion that caused it, and the motion of the beam is dampened **(b).**

Application of Lenz's law

A sensitive balance, such as the kind used in chemistry laboratories, shown in **Figure 25–9,** uses Lenz's law to stop its oscillation when an object is placed on the pan. A piece of metal attached to the balance arm is located between the poles of a horseshoe magnet. When the balance arm swings, the metal moves through the magnetic field. Currents called **eddy currents** are generated in the metal. These currents produce a magnetic field that acts to oppose the motion that caused the currents. Thus, the metal piece is slowed down. The force opposes the motion of the metal in either direction but does not act if the metal is still. Thus, it does not change the mass read by the balance. This effect is called eddy current damping.

Eddy currents are generated when a piece of metal moves through a magnetic field. The reverse is also true: a current is generated when a metal loop is placed in a changing magnetic field. According to Lenz's law, the current generated will oppose the changing magnetic field. How does the current do the opposing? It generates a magnetic field of its own in the opposite direction. This induced magnetic field is what causes the uncut, aluminum ring in the chapter-opening photo to float. An AC current is in the coil, so a constantly changing magnetic field is generated. This changing magnetic field induces an *EMF* in the rings. For the uncut ring, the *EMF* causes a current that produces a magnetic field. This magnetic field will oppose the change in the generating magnetic field. The interaction of these two magnetic fields causes the ring to push away from the coil similar to how the north poles of two magnets push away from each other. For the lower ring, which has been sawed through, an *EMF* is generated, but no current can result. Hence, no opposing magnetic field is produced by the ring.

Go with the Flow

Self-Inductance

Back-*EMF* can be explained another way. As Faraday showed, *EMF* is induced whenever a wire cuts lines of magnetic field. Consider the coil of wire shown in **Figure 25–10.** The current through the wire increases

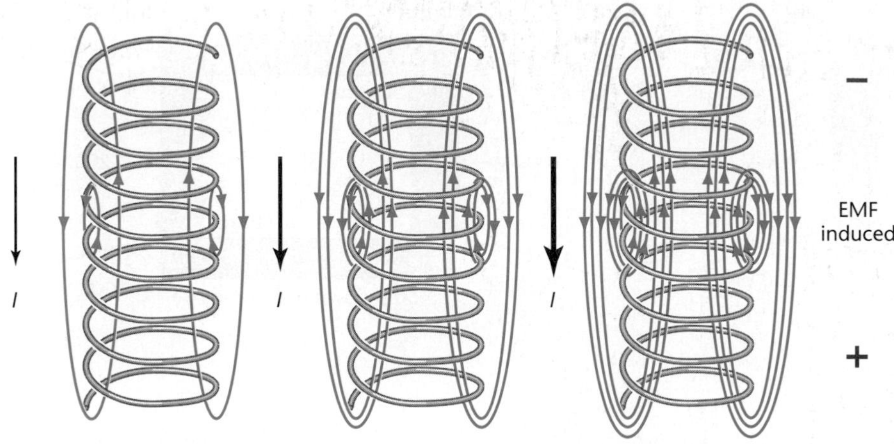

FIGURE 25–10 As the current in the coil increases from left to right, the magnetic field generated by the current also increases. This increase in magnetic field produces an *EMF* that opposes the current direction.

from left to right. The current generates a magnetic field, shown by magnetic field lines. As the current and magnetic field increase, you can imagine that new lines are created. As the lines expand, they cut through the coil wires, generating an *EMF* to oppose the current changes. The *EMF* will make the potential of the top of the coil more negative than the bottom. This induction of *EMF* in a wire carrying changing current is called **self-inductance.** The size of the induced *EMF* is proportional to the rate at which field lines cut through the wires. The faster the current is changed, the larger the opposing *EMF*. If the current reaches a steady value, the magnetic field is constant, and the *EMF* is zero. When the current is decreased, an EMF is generated that tends to prevent the reduction in magnetic field and current.

Because of self-inductance, work has to be done to increase the current flowing through the coil. Energy is stored in the magnetic field. This is similar to the way a charged capacitor stores energy in the electric field between its plates.

Transformers

Inductance between coils is the basis for the operation of a transformer. A **transformer** is a device used to increase or decrease AC voltages. Transformers are widely used because they change voltages with relatively little loss of energy.

How transformers work

Self-inductance produces an *EMF* when current changes in a single coil. A transformer has two coils, electrically insulated from each other, but wound around the same iron core. One coil is called the **primary coil.** The other coil is called the **secondary coil.** When the primary coil is connected to a source of AC voltage, the changing current creates a varying magnetic field. The varying magnetic field is carried through the core to the secondary coil. In the secondary coil, the varying field induces a varying *EMF*. This effect is called **mutual inductance.**

Slow Magnet

Lay a 1-m length of copper tube on the lab table. Try to pull the copper with a pair of neodymium magnets. Can you feel any force on the copper? Hold the tube by one end so that it hangs straight down. Drop a small steel marble through the tube. Use a stopwatch to measure the time needed for first the marble and then for the pair of magnets to fall through the tube. Catch the magnets in your hand. If they hit the table or floor, they will break.

Analyze and Conclude
Devise a hypothesis that would explain the strange behavior of the falling magnets and suggest a method of testing your hypothesis.

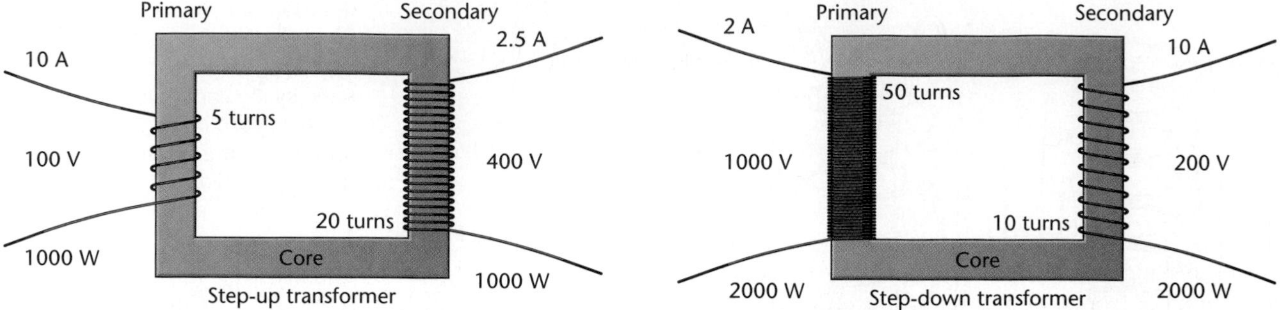

FIGURE 25–11 In a transformer, the ratio of input voltage to output voltage depends upon the ratio of the number of turns on the primary coil to the number of turns on the secondary coil.

The *EMF* induced in the secondary coil, called the secondary voltage, is proportional to the primary voltage. The secondary voltage also depends on the ratio of turns on the secondary coil to turns on the primary coil.

$$\frac{\text{secondary voltage}}{\text{primary voltage}} = \frac{\text{number of turns on secondary coil}}{\text{number of turns on primary coil}}$$

$$\frac{V_s}{V_p} = \frac{N_s}{N_p}$$

If the secondary voltage is larger than the primary voltage, the transformer is called a **step-up transformer.** If the voltage coming out of the transformer is smaller than the voltage put in, then it is called a **step-down transformer.**

In an ideal transformer, the electric power delivered to the secondary circuit equals the power supplied to the primary circuit. An ideal transformer dissipates no power itself.

$$P_p = P_s$$

$$V_pI_p = V_sI_s$$

Rearrange the equation and you find the current in the primary circuit depends on how much current is required by the secondary circuit.

$$\frac{I_s}{I_p} = \frac{V_p}{V_s} = \frac{N_p}{N_s}$$

FIGURE 25–12 If the input voltage is connected to the coils on the left, where there is a larger number of turns, the transformer functions as a step-down transformer. If the input voltage is connected at the right, the transformer functions as a step-up transformer.

A step-up transformer increases voltage. Because transformers cannot increase the power output, there must be a corresponding decrease in current through the secondary circuit. Similarly, in a step-down transformer, the current is greater in the secondary circuit than it is in the primary circuit. A voltage decrease corresponds to a current increase. **Figure 25–11** illustrates the principles of step-up and step-down transformers.

Some transformers can function either as step-up transformers or step-down transformers depending on how they are hooked up. **Figure 25–12** shows an example of such a transformer.

Design Your Own **Physics Lab**

Swinging Coils

Problem

Electricity that you use in your everyday life comes from the wall socket or from chemical batteries. Modern theory suggests that current can be caused by the interactions of wires and magnets. Exactly how do coils and magnets interact?

Hypothesis

Form a testable hypothesis that relates to the interaction of magnets and coils. Be sure to include some symmetry tests in your hypothesis. Try to design a system of coils and magnets so that you can use one pair as a generator and one pair as a motor.

Possible Materials

coils of enameled wire
identical sets of magnets
masking tape
supports and bars

Plan the Experiment

1. Devise a means to test stationary effects: those that occur when the magnet and coils are not moving.
2. Consider how to test moving effects: those that occur when the magnet moves in various directions in relation to the coil.
3. Include different combinations of connecting, or not connecting, the ends of the wires.
4. Consider polarity, magnetic strength, and any other variables that might influence the interaction of the coils and magnet.
5. **Check the Plan** Make sure that your teacher has approved your final plan before you proceed with your experiment.

Analyze and Conclude

1. **Organizing Results** Construct a list of tests that you performed and their results.
2. **Analyzing Data** Summarize the effects of the stationary magnet and the moving magnet. Explain how connecting the wires influenced your results.
3. **Relating Concepts** Describe and explain the effects of changing polarity, direction, number of coils, and any other variables you used.
4. **Checking Your Hypothesis** Did the experiment yield expected results? Did you determine any new interactions?

Apply

1. The current that you generated in this activity was quite small. List several factors that you could change to generate more current. (**Hint:** Think of a commercial generator.)

Example Problem

Step-Up Transformer

A step-up transformer has a primary coil consisting of 200 turns and a secondary coil that has 3000 turns. The primary coil is supplied with an effective AC voltage of 90.0 V.

a. What is the voltage in the secondary circuit?

b. The current in the secondary circuit is 2.00 A. What is the current in the primary circuit?

c. What is the power in the primary circuit?

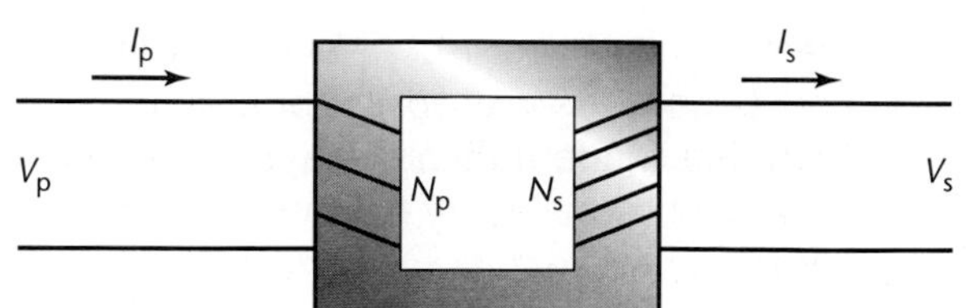

Sketch the Problem

- Draw an iron core with turns of wire.
- Label the variables *I*, *V*, and *N*.

Calculate Your Answer

Known:	Strategy:	Calculations:
$N_p = 200$ $N_s = 3000$ $V_p = 90.0$ V $I_s = 2.00$ A	**a.** Voltage and turn ratios are equal. Solve for V_s.	$\frac{V_s}{V_p} = \frac{N_s}{N_p}$ $V_s = \frac{N_s}{N_p} V_p = \frac{3000}{200}(90.0\ \text{V}) = 1.35\ \text{kV}$
Unknown: $V_s = ?$ $I_p = ?$ $P_p = ?$	**b.** Assuming that the transformer is perfectly efficient, the power in the primary and secondary circuits is equal.	$P_p = P_s$ $V_p I_p = V_s I_s$ $I_p = \frac{V_s}{V_p} I_s = \frac{1350\ \text{V}}{90.0\ \text{V}}(2.00\ \text{A}) = 30.0\ \text{A}$
	c. Use the power relation to solve for P_p.	$P_p = V_p I_p = (90.0\ \text{V})(30.0\ \text{A}) = 2.70\ \text{kW}$

Check Your Answer

- Are the units correct? Check the units with algebra. Voltage: V; Current: A; and Power: W.
- Do the signs make sense? All numbers are positive.
- Is the magnitude realistic? A large step-up ratio of turns results in a large secondary voltage yet a smaller secondary current. Answers agree.

Practice Problems

For all problems, effective currents and voltages are indicated.

9. A step-down transformer has 7500 turns on its primary coil and 125 turns on its secondary coil. The voltage across the primary circuit is 7.2 kV.

a. What voltage is across the secondary circuit?

b. The current in the secondary circuit is 36 A. What is the current in the primary circuit?

10. A step-up transformer's primary coil has 500 turns. Its secondary coil has 15 000 turns. The primary circuit is connected to an AC generator having an *EMF* of 120 V.

a. Calculate the *EMF* of the secondary circuit.

b. Find the current in the primary circuit if the current in the secondary circuit is 3.0 A.

c. What power is drawn by the primary circuit? What power is supplied by the secondary circuit?

11. A step-up transformer has 300 turns on its primary coil and 90 000 turns on its secondary coil. The *EMF* of the generator to which the primary circuit is attached is 60.0 V.

a. What is the *EMF* in the secondary circuit?

b. The current in the secondary circuit is 0.50 A. What current is in the primary circuit?

FIGURE 25–13 Step-down transformers are used to reduce the high voltage in transmission lines to levels appropriate for consumers at the points of use.

Everyday uses of transformers

As you learned in Chapter 22, long-distance transmission of electrical energy is economical only if low currents and very high voltages are used. Step-up transformers are used at power sources to develop voltages as high as 480 000 V. The high voltage reduces the current required in the transmission lines, keeping the energy lost to resistance low. When the energy reaches the consumer, step-down transformers, such as the one shown in **Figure 25–13,** provide appropriately low voltages for consumer use. Transformers in your appliances further adjust voltages to useable levels.

25.2 Section Review

2.1 You hang a coil of wire with its ends joined so it can swing easily. If you now plunge a magnet into the coil, the coil will swing. Which way will it swing with respect to the magnet and why?

2.2 If you unplugged a running vacuum cleaner from the wall outlet, you would be much more likely to see a spark than you would be if you unplugged a lighted lamp from the wall. Why?

2.3 Frequently, transformer windings that have only a few turns are made of very thick (low-resistance) wire, while those with many turns are made of thin wire. Why?

2.4 **Critical Thinking** Would permanent magnets make good transformer cores? Explain.

CHAPTER 25 REVIEW

Key Terms

25.1
- electromagnetic induction
- electromotive force
- electric generator

25.2
- Lenz's law
- eddy current
- self-inductance
- transformer
- primary coil
- secondary coil
- mutual inductance
- step-up transformer
- step-down transformer

Summary

25.1 Creating Electric Current from Changing Magnetic Fields

- Michael Faraday discovered that if a wire moves through a magnetic field, an electric current can flow.
- The current produced depends upon the angle between the velocity of the wire and the magnetic field. Maximum current occurs when the wire is moving at right angles to the field.
- Electromotive force, *EMF*, is the potential difference created across the moving wire. *EMF* is measured in volts.
- The *EMF* in a straight length of wire moving through a uniform magnetic field is the product of the magnetic field, B, the length of the wire, L, and the component of the velocity of the moving wire, v, perpendicular to the field.
- A generator and a motor are similar devices. A generator converts mechanical energy to electrical energy; a motor converts electrical energy to mechanical energy.

25.2 Changing Magnetic Fields Induce *EMF*

- Lenz's law states that an induced current is always produced in a direction such that the magnetic field resulting from the induced current opposes the change in the magnetic field that is causing the induced current.
- Self-inductance is a property of a wire carrying a changing current. The faster the current is changing, the greater the induced *EMF* that opposes that change.
- A transformer has two coils wound about the same core. An AC current through the primary coil induces an alternating *EMF* in the secondary coil. The voltages in alternating-current circuits may be increased or decreased by transformers.

Reviewing Concepts

Section 25.1

1. How are Oersted's and Faraday's results similar? How are they different?
2. You have a coil of wire and a bar magnet. Describe how you could use them to generate an electric current.
3. What does *EMF* stand for? Why is the name inaccurate?
4. What is the armature of an electric generator?
5. Why is iron used in an armature?
6. What is the difference between a generator and a motor?
7. List the major parts of an AC generator.
8. Why is the effective value of an AC current less than its maximum value?
9. Water trapped behind a dam turns turbines that rotate generators. List all the forms of energy that take part in the cycle that includes the stored water and the electricity produced.

Section 25.2

10. State Lenz's law.
11. What produces the back-*EMF* of an electric motor?
12. Why is there no spark when you close a switch, putting current through an inductor, but there is a spark when you open the switch?
13. Why is the self-inductance of a coil a major factor when the coil is in an AC circuit but a minor factor when the coil is in a DC circuit?
14. Explain why the word *change* appears so often in this chapter.
15. Upon what does the ratio of the *EMF* in the primary circuit of a transformer to the *EMF* in the secondary circuit of the transformer depend?

Applying Concepts

16. Substitute units to show that the units of BLv are volts.
17. When a wire is moved through a magnetic field, resistance of the closed circuit affects
    a. current only.
    b. *EMF* only.
    c. both.
    d. neither.
18. As Logan slows his bike, what happens to the *EMF* produced by his bike's generator? Use the term *armature* in your explanation.
19. The direction of AC voltage changes 120 times each second. Does that mean that a device connected to an AC voltage alternately delivers and accepts energy?
20. A wire is moved horizontally between the poles of a magnet, as shown in **Figure 25–14.** What is the direction of the induced current?

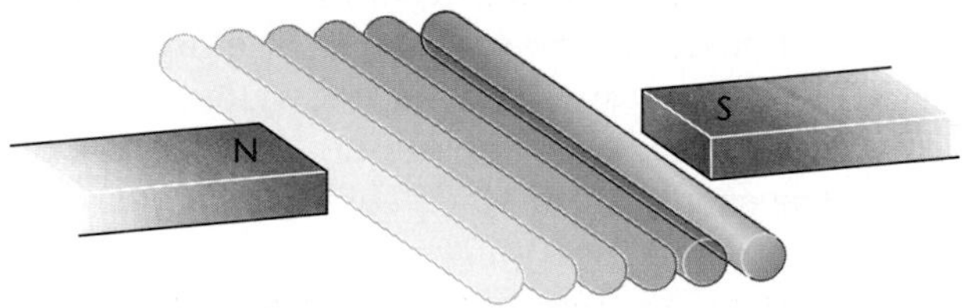

FIGURE 25–14

21. You make an electromagnet by winding wire around a large nail. If you connect the magnet to a battery, is the current larger just after you make the connection or several tenths of a second after the connection is made? Or is it always the same? Explain.
22. A segment of a wire loop is moving downward through the poles of a magnet, as shown in **Figure 25–15.** What is the direction of the induced current?

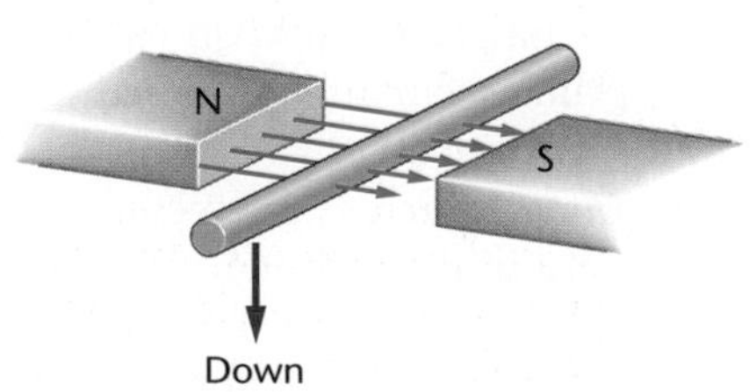

FIGURE 25–15

23. A transformer is connected to a battery through a switch. The secondary circuit contains a lightbulb. Which of the following statements best describes when the lamp will be lighted? Explain.
    a. as long as the switch is closed
    b. only the moment the switch is closed
    c. only the moment the switch is opened
24. The direction of Earth's magnetic field in the northern hemisphere is downward and to the north. If an east-west wire moves from north to south, in which direction is the current?
25. You move a length of copper wire down through a magnetic field B, as shown in **Figure 25–15.**
    a. Will the induced current move to the right or left in the wire segment in the diagram?
    b. As soon as the wire is moved in the field, a current appears in it. Thus, the wire segment is a current-carrying wire located in a magnetic field. A force must act on the wire. What will be the direction of the force acting on the wire as a result of the induced current?
26. A physics instructor drops a magnet through a copper pipe, as illustrated in **Figure 25–16.** The magnet falls very slowly, and the class concludes that there must be some force opposing gravity.
    a. What is the direction of the current induced in the pipe by the falling magnet if the S pole is toward the bottom?
    b. The induced current produces a magnetic field. What is the direction of the field?
    c. How does this field reduce the acceleration of the falling magnet?

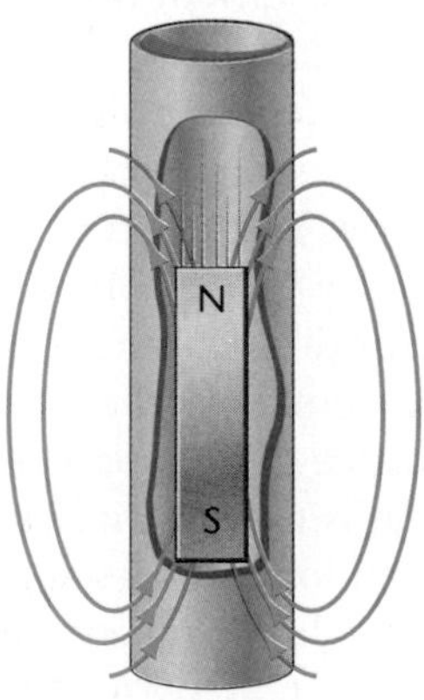

FIGURE 25–16

27. Why is a generator more difficult to rotate when it is connected to a circuit and supplying current than it is when it is standing alone?

Problems

Section 25.1

LEVEL 1

28. A wire, 20.0 m long, moves at 4.0 m/s perpendicularly through a magnetic field. An *EMF* of 40 V is induced in the wire. What is the strength of the magnetic field?
29. An airplane traveling at 950 km/h passes over a region where Earth's magnetic field is 4.5×10^{-5} T and is nearly vertical. What voltage is induced between the plane's wing tips, which are 75 m apart?
30. A straight wire, 0.75 m long, moves upward through a horizontal 0.30-T magnetic field at a speed of 16 m/s.
    a. What *EMF* is induced in the wire?
    b. The wire is part of a circuit with a total resistance of 11 Ω. What is the current?
31. At what speed would a 0.20-m length of wire have to move across a 2.5-T magnetic field to induce an *EMF* of 10 V?
32. An AC generator develops a maximum *EMF* of 565 V. What effective *EMF* does the generator deliver to an external circuit?
33. An AC generator develops a maximum voltage of 150 V. It delivers a maximum current of 30.0 A to an external circuit.
    a. What is the effective voltage of the generator?
    b. What effective current does it deliver to the external circuit?
    c. What is the effective power dissipated in the circuit?
34. An electric stove is connected to an AC source with an effective voltage of 240 V.
    a. Find the maximum voltage across one of the stove's elements when it is operating.
    b. The resistance of the operating element is 11 Ω. What is the effective current?
35. A wire, 20.0 m long, moves at 4.0 m/s perpendicularly through a magnetic field. An *EMF* of 40 V is induced in the wire. What is the strength of the magnetic field?

LEVEL 2

36. A 40-cm wire is moved perpendicularly through a magnetic field of 0.32 T with a velocity of 1.3 m/s. If this wire is connected into a circuit of 10-Ω resistance, what is the current?
37. You connect both ends of a copper wire, total resistance 0.10 Ω, to the terminals of a galvanometer. The galvanometer has a resistance of 875 Ω. You then move a 10.0-cm segment of the wire upward at 1.0 m/s through a 2.0×10^{-2}-T magnetic field. What current will the galvanometer indicate?
38. The direction of a 0.045-T magnetic field is 60° above the horizontal. A wire, 2.5 m long, moves horizontally at 2.4 m/s.
    a. What is the vertical component of the magnetic field?
    b. What *EMF* is induced in the wire?
39. A generator at a dam can supply 375 MW (375×10^6 W) of electrical power. Assume that the turbine and generator are 85% efficient.
    a. Find the rate at which falling water must supply energy to the turbine.
    b. The energy of the water comes from a change in potential energy, $U = mgh$. What is the change in U needed each second?
    c. If the water falls 22 m, what is the mass of the water that must pass through the turbine each second to supply this power?

Section 25.2

LEVEL 1

40. The primary coil of a transformer has 150 turns. It is connected to a 120-V source. Calculate the number of turns on the secondary coil needed to supply these voltages.
    a. 625 V **b.** 35 V **c.** 6.0 V
41. A step-up transformer has 80 turns on its primary coil. It has 1200 turns on its secondary coil. The primary circuit is supplied with an alternating current at 120 V.
    a. What voltage is across the secondary circuit?
    b. The current in the secondary circuit is 2.0 A. What current is in the primary circuit?
    c. What is the power input and output of the transformer?

42. A laptop computer requires an effective voltage of 9.0 volts from the 120-V line.
   a. If the primary coil has 475 turns, how many does the secondary coil have?
   b. A 125-mA current is in the computer. What current is in the primary circuit?
43. A hair dryer uses 10 A at 120 V. It is used with a transformer in England, where the line voltage is 240 V.
   a. What should be the ratio of the turns of the transformer?
   b. What current will the hair dryer now draw?

LEVEL 2

44. A 150-W transformer has an input voltage of 9.0 V and an output current of 5.0 A.
   a. Is this a step-up or step-down transformer?
   b. What is the ratio of V_{output} to V_{input}?
45. A transformer has input voltage and current of 12 V and 3.0 A, respectively, and an output current of 0.75 A. If there are 1200 turns on the secondary side of the transformer, how many turns are on the primary side?
46. Scott connects a transformer to a 24-V source and measures 8.0 V at the secondary circuit. If the primary and secondary circuits were reversed, what would the new output voltage be?

Critical Thinking Problems

47. Suppose an "anti-Lenz's law" existed that meant a force was exerted to increase the change in magnetic field. Thus, when more energy was demanded, the force needed to turn the generator would be reduced. What conservation law would be violated by this new "law"? Explain.
48. Real transformers are not 100% efficient. That is, the efficiency, in percent, is represented by $e = (100\%)P_s/P_p$. A step-down transformer that has an efficiency of 92.5% is used to obtain 28.0 V from the 125-V household voltage. The current in the secondary circuit is 25.0 A. What is the current in the primary circuit?
49. A transformer that supplies eight homes has an efficiency of 95%. All eight homes have electric ovens running that draw 35 A from 240 V lines. How much power is supplied to the ovens in the eight homes? How much power is dissipated as heat in the transformer?

Going Further

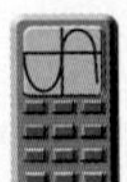

Graphing Calculator Show that the average power in an AC circuit is half the peak power. **Figure 25–5** shows how the current produced by a generator varies in time. The equation that describes this variation is $I = I_{max}\sin(2\pi ft)$. In the U.S., $f = 60$ Hz. If a resistor is connected across a generator, then the voltage drop across the resistor will be given by $V = I_{max}R\sin(2\pi ft)$. The power dissipated by the resistor is $P = I_{max}^2R \sin^2(2\pi ft)$. Suppose that $R = 10\ \Omega$, $I_{max} = 1$ A.

a. Plot the power as a function of time from $t = 0$ to $t = 1/60$ s (one complete cycle).

b. Determine the energy transfer to the resistor. When the power is constant, the energy is the product of the power and the time interval. When the power varies, the energy can be calculated as the area under the curve of the graph of power versus time. There are different ways you can find this, depending on the tools you have. You could transfer the plot to a large piece of graph paper and count squares under the curve. Or, you could calculate the power every 1/1200 s and multiply by the time interval (1/1200 s) to find the incremental energy transfer, then add up all the small increments of energy. Or, you might use a computer.

c. The average power is given by the total energy transferred divided by the time interval. Find the average power from your result above and compare with the peak power, 10 W.

***inter*NET CONNECTION**

Follow the link on the Glencoe Homepage at **www.glencoe.com/sec/science** to find out more about this chapter.

Big Ears

This parabolic dish antenna is designed to receive television signals from satellites orbiting hundreds of kilometers above Earth's surface. Why are dish-style antennas rather than more familiar television or radio antennas used for detecting radio signals from deep space?

CHAPTER

26 Electro-magnetism

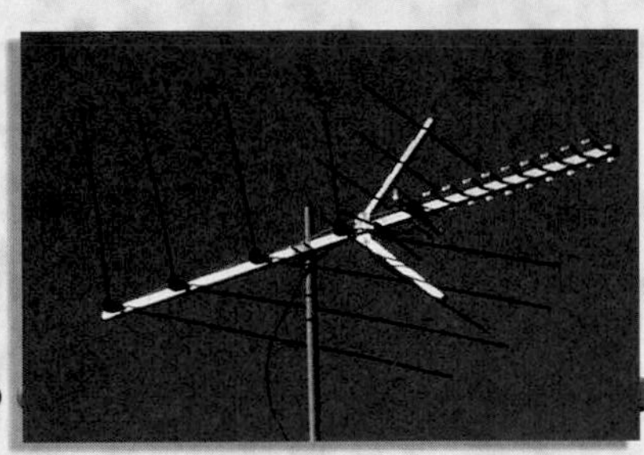

If there are intelligent beings on another planet, either in our galaxy or in a far-off galaxy, how could they communicate with people on Earth? They probably would send a message in the form of electromagnetic waves—most likely, as microwaves. Electromagnetic waves travel at the fastest speed possible—the speed of light. They can carry information, are easily generated and detected, and are undeflected by the galactic magnetic field. Only in the microwave region of the electromagnetic spectrum is the universe fairly quiet. Thus, microwaves are ideal for transmitting messages over long distances.

Could we receive such messages? Huge parabolic dish antennas detect electromagnetic waves from space. They are presently used to receive information sent from space probes in our solar system and to pick up signals from distant galaxies, but they also could receive messages from hypothetical extraterrestrial beings. These dish antennas have the same function as the more familiar television antennas on the roofs of homes, but they are very different in form.

WHAT YOU'LL LEARN

- You will learn how combined electric and magnetic fields can be used to find the masses of electrons, atoms, and molecules.
- You will explain how electromagnetic waves are created, travel through empty space, and are detected.

WHY IT'S IMPORTANT

- Mass spectrometers are widely used in many branches of science.
- Radio and TV, light and X rays—all these electromagnetic waves play vital roles in our lives.

interNET CONNECTION

Follow the link for this chapter on the Glencoe Homepage at **www.glencoe.com/sec/science** to find out more about electromagnetism.

26.1 Interaction Between Electric and Magnetic Fields and Matter

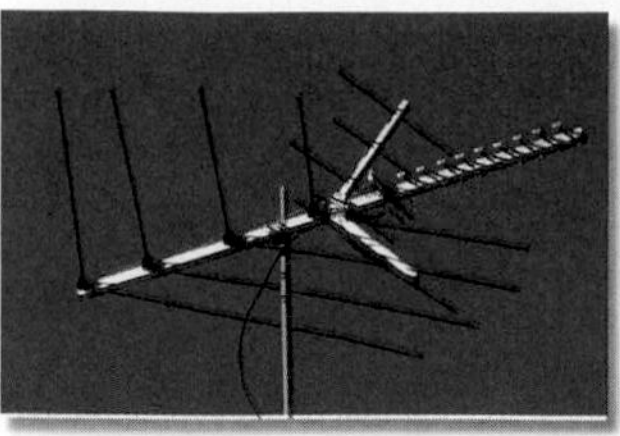

OBJECTIVES

- **Describe** the measurement of the charge-to-mass ratio of the electron and **solve** problems related to this measurement.
- **Explain** how a mass spectrometer separates ions of different masses and **solve** problems involving this instrument.

F.Y.I.

Sir Joseph John Thomson was awarded the Nobel Prize in Physics in 1906 for his work on the electron. In 1908, he was knighted. Eventually, seven of his former research assistants won Nobel prizes.

The source of most radio and television waves is accelerating electrons. It is the electrons' charge that results in electric fields, and the electrons' motion that produces magnetic fields. Electrons are also part of every atom that makes up the dish antennas that receive the waves. Therefore, it is important to understand some of the properties of electrons. The same techniques used to study electrons can be extended to study the properties of positive ions, which are atoms stripped of one or more electrons.

Mass of the Electron

The charge of an electron was first measured by Robert Millikan, using the force of an electric field on the charge on an oil drop, as described in Chapter 21. The mass of an electron, however, is far too small to measure directly on an ordinary balance. But it is possible to find the charge-to-mass ratio of an electron, q/m, by utilizing the forces of electric and magnetic fields acting on a moving electron. This can be accomplished by measuring deflections of electrons in cathode-ray tubes. From this ratio and the charge, the mass can be found.

Thomson's experiments with electrons

The ratio of charge to mass of an electron was first measured in 1897 by British physicist J. J. Thomson (1856–1940). For his experiment, he used a cathode-ray tube similar to the one shown in **Figure 26–1.** In this experiment, all air is removed from the glass tube. An electric field accelerates electrons off the negatively charged cathode and toward the positively charged anode. Some of the electrons pass through a slit in the anode and travel in a narrow beam toward a fluorescent coating. When the electrons hit the coating, they cause it to glow at the point where the electrons hit.

Electric and magnetic fields in the center of the tube exert forces on

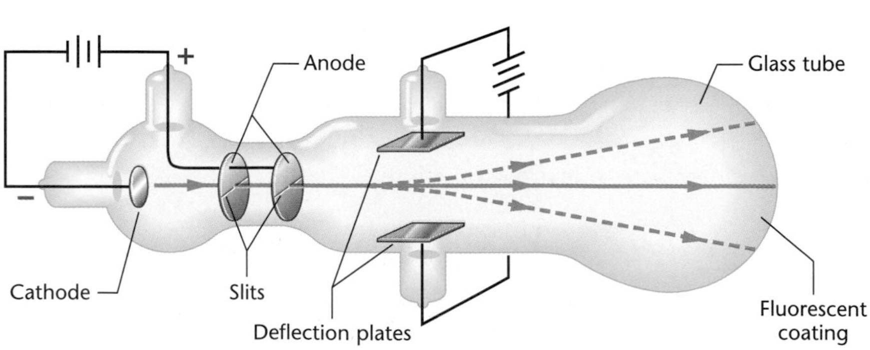

FIGURE 26–1 The charge-to-mass ratio of an electron was first measured with the Thomson adaptation of a cathode-ray tube. In the diagram, the electromagnets have been removed to show the deflection plates. When the tube is in use, the electromagnets and the deflection plates lie in the same plane.

the electrons. The electric field is produced by charged parallel plates and is perpendicular to the beam. The electric field, of intensity E, produces a force, qE, on the electrons that deflects the electrons upward. The magnetic field is produced by two coils and is at right angles to both the beam and the electric field.

Recall from Chapter 24 that the force exerted by a magnetic field is perpendicular to the field and to the direction of motion of the electrons. The magnitude of the force exerted on the electrons by the magnetic field is equal to Bqv. Here, B is the magnetic field strength, and v is the electron velocity. The magnetic force depicted in **Figure 26–1** acts downward.

The electric and magnetic fields may be adjusted until the beam of electrons follows a straight, or undeflected, path. Then the forces due to the two fields are equal in magnitude and opposite in direction, and the following equation is true.

$$Bqv = Eq$$

Solving this equation for v, the following expression is obtained.

$$v = \frac{Eq}{Bq} = \frac{E}{B}$$

This equation reveals that the forces are balanced only for electrons that have a specific velocity, v. If the electric field is turned off, only the force due to the magnetic field remains. The magnetic force is perpendicular to the direction of motion of the electrons, causing a centripetal acceleration of the electrons. They follow a circular path with radius r. Using Newton's second law of motion, the following equation can be obtained.

$$Bqv = \frac{mv^2}{r}$$

Solving for q/m results in the following equation.

$$\frac{q}{m} = \frac{v}{Br}$$

Thomson calculated the straight trajectory velocity, v, using measured values of E and B. Next, he measured the distance between the undeflected spot and the position of the spot when only the magnetic field acted on the electrons. Using this distance, he calculated the radius of the circular path of the electron, r. This allowed Thomson to calculate q/m. The average of many experimental trials produced the value $q/m = 1.759 \times 10^{11}$ C/kg. Based on the value $q = 1.602 \times 10^{-19}$ C, the mass of the electron, m, can be calculated.

$$m = \left(\frac{q}{q/m}\right) = \frac{1.602 \times 10^{-19}\ \text{C}}{1.759 \times 10^{11}\ \text{C/kg}} = 9.107 \times 10^{-31}\ \text{kg}$$

$$m \cong 9.11 \times 10^{-31}\ \text{kg}$$

HELP WANTED

BROADCAST TECHNICIAN

An electronics and computer whiz is needed to be the technical expert of a large TV station to monitor and adjust equipment, as well as to troubleshoot and repair it in high-pressure situations. An associate degree or a vocational/technical school background, with knowledge of current FCC rules and appropriate licensure certification, is necessary. For information on careers and schools offering preparation in this field, contact:

National Association of Broadcasters
1771 N. Street, N.W.
Washington, DC 20036

FIGURE 26–2 This photograph shows the circular tracks of electrons (e^-) and positrons (e^+) moving through the magnetic field in a bubble chamber. Note that the electrons and positrons curve in opposite directions.

Thomson's experiments with protons

Thomson put his apparatus to an additional use by finding q/m for positive ions in the same way that he measured the quantity for electrons. He took advantage of the fact that positively charged particles, when they are in either an electric field or a magnetic field, bend the opposite way from electrons. This is shown in **Figure 26–2.** To accelerate positively charged particles into the deflection region, he reversed the direction of the field between the cathode and anode. A small amount of hydrogen gas was then put into the tube. The field pulled the electrons off the hydrogen atoms, leaving them with a net positive charge, and accelerated these positively charged protons through a tiny slit in the negatively charged anode. The proton beam then passed through the electric and magnetic deflecting fields to the fluorescent coating of the Thomson cathode-ray tube. The mass of the proton was determined in the same manner as was the mass of the electron. The mass of the proton was found to be 1.67×10^{-27} kg. Heavier ions produced by stripping an electron from gases such as helium, neon, and argon were measured by a similar method.

Example Problem

Path of an Electron in a Magnetic Field

An electron of mass 9.11×10^{-31} kg moves through a cathode-ray tube with a speed of 2.0×10^{5} m/s across and perpendicular to a magnetic field of 3.5×10^{-2} T. The electric field is turned off. What is the radius of the circular path followed by the electron?

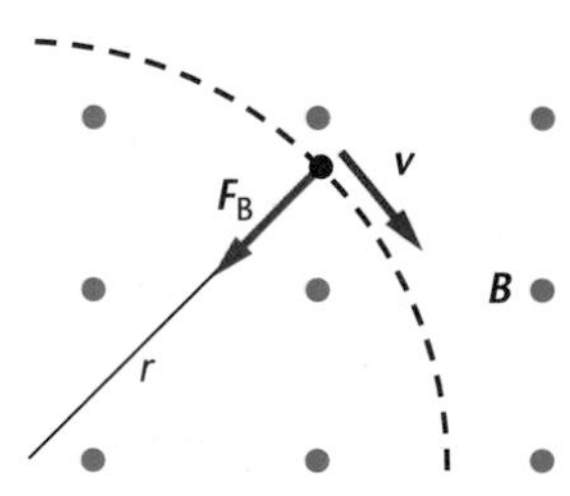

Sketch the Problem

- Draw the path of the electron, and label the velocity, $\boldsymbol{v}$.
- Sketch the magnetic field perpendicular to the velocity.
- Diagram the force acting on the electron. Add the radius of the electron's path to your sketch.

Calculate Your Answer

Known:

$v = 2.0 \times 10^5$ m/s

$B = 3.5 \times 10^{-2}$ T

$m = 9.11 \times 10^{-31}$ kg

$q = 1.60 \times 10^{-19}$ C

Unknown:

$r = ?$

Strategy:

Solve for the radius using the equation obtained from Newton's second law of motion.

Calculations:

$$Bqv = \frac{mv^2}{r}$$

$$r = \frac{mv}{Bq} = \frac{(9.11 \times 10^{-31}\ \text{kg})(2.0 \times 10^5\ \text{m/s})}{(3.5 \times 10^{-2}\ \text{T})(1.60 \times 10^{-19}\ \text{C})}$$

$$r = 3.3 \times 10^{-5}\ \text{m}$$

Check Your Answer

- Are the units correct? The radius of the circular path is a length measurement, the meter.

Practice Problems

Assume that the direction of all moving charged particles is perpendicular to the uniform magnetic field.

1. Protons passing without deflection through a magnetic field of 0.60 T are balanced by a 4.5×10^3-N/C electric field. What is the speed of the moving protons?
2. A proton moves at a speed of 7.5×10^3 m/s as it passes through a 0.60-T magnetic field. Find the radius of the circular path. The charge carried by the proton is equal to that of the electron, but it is positive.
3. Electrons move through a 6.0×10^{-2}-T magnetic field balanced by a 3.0×10^3-N/C electric field. What is the speed of the electrons?
4. Calculate the radius of the circular path that the electrons in Practice Problem 3 follow in the absence of the electric field.

Rolling Along

Place a small ball of clay under one end of a grooved ruler to make a ramp. Roll a 6-mm-diameter steel ball down the ramp and along the tabletop. Place a strong magnet near the path of the ball so that the ball will curve, but not hit the magnet. Predict what will happen to the path when the ball is started higher or lower on the ramp. Try it.

Analyze and Conclude Is this consistent for a charged particle moving through a magnetic field?

The Mass Spectrometer

When Thomson put neon gas into his tube, he found two dots on the screen instead of one, and thus two values for q/m. He ultimately concluded that atoms of the same element could have the same chemical properties but different masses; he had shown the existence of **isotopes,** a possibility first proposed by chemist Frederick Soddy.

The masses of positive ions can be measured precisely by using a **mass spectrometer,** an adaptation of the Thomson tube. It is used to determine the charge-to-mass ratios of gases and of materials that can be heated to form gases. The material under investigation is called the ion source. In the ion source, accelerated electrons strike the gas atoms, knocking off electrons and thus forming positive gas ions. A potential difference, V, between the electrodes produces an electric field that accelerates the ions. One type of mass spectrometer is shown in **Figure 26–3.**

To select ions with a specific velocity, the ions first are passed through electric and magnetic deflecting fields, as in the Thomson tube. The ions that go through undeflected move into a region with a uniform magnetic field. There they follow a circular path, which has a radius that can be obtained from Newton's second law: $Bqv = mv^2/r$. Solving for r yields the following equation.

$$r = \frac{mv}{qB}$$

The velocity of the undeflected ion can be found from the equation for the kinetic energy of ions accelerated from rest through a known potential difference, V.

$$K = 1/2mv^2 = qV$$

$$v = \sqrt{\frac{2qV}{m}}$$

Substituting this expression for v in the equation $r = mv/qB$ gives the radius of the circular path.

$$r = \frac{mv}{qB} = \frac{m}{qB}\sqrt{\frac{2qV}{m}} = \frac{1}{B}\sqrt{\frac{2Vm}{q}}$$

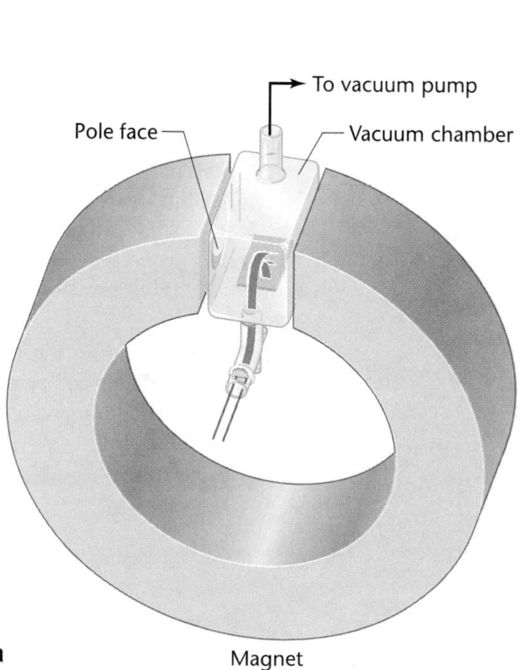

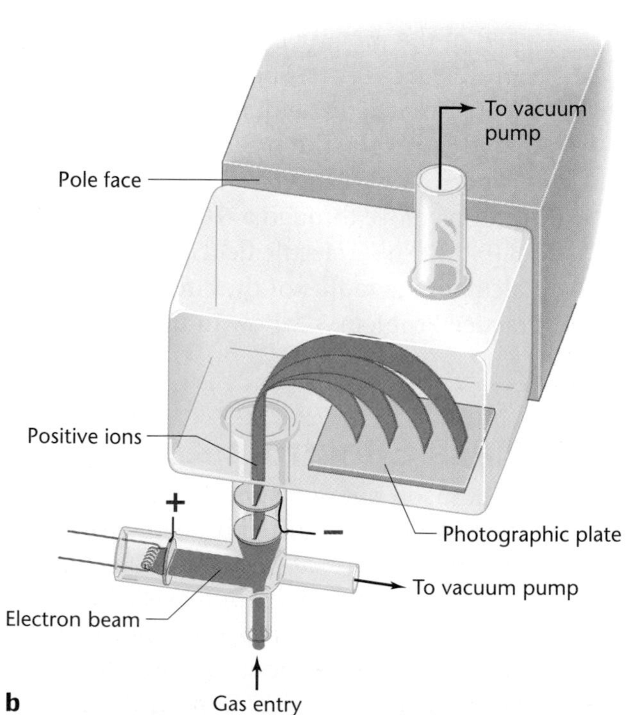

FIGURE 26–3 The mass spectrometer is used extensively to analyze isotopes of an element. Inside the spectrometer, a magnet causes the positive ions to be deflected according to their mass **(a).** In the vacuum chamber, the process is recorded on a photographic plate or a solid-state detector **(b).**

From this equation, the charge-to-mass ratio of the ion is determined.

$$Br = \sqrt{\frac{2mV}{q}}, \text{ so } \frac{q}{m} = \frac{2V}{B^2 r^2}$$

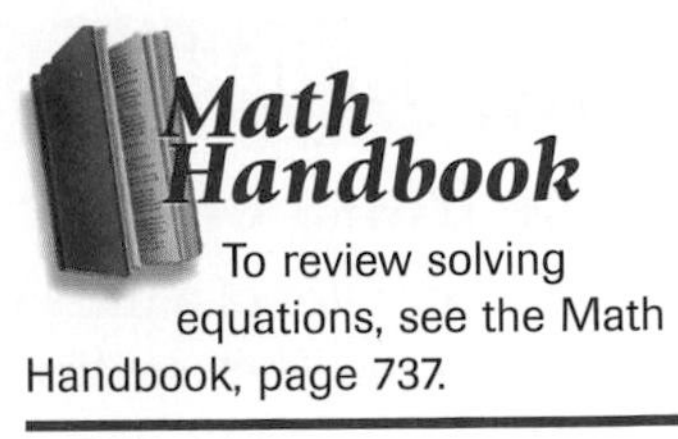

To review solving equations, see the Math Handbook, page 737.

In one type of mass spectrometer, the ions hit a photographic film, where they leave a mark. The radius, *r*, is found by measuring the distance between the mark and the slit in the electrode. This distance is twice the radius of the circular path. **Figure 26–4** shows marks on film from the four isotopes of the element chromium. The isotope with mass number 52 makes the darkest mark, showing that most chromium atoms have this mass.

All of the chromium ions that hit the film have the same charge. The charge depends on how many electrons were removed in the ion source. It takes more-energetic electrons to remove a second electron from the gas atoms. For low electron energies, only one electron is removed from an atom. When the energy is increased, however, both singly and doubly charged ions are produced. In this way, the operator of the mass spectrometer can choose the charge on the ion.

Mass spectrometers are extremely versatile tools. For example, they can be used to separate isotopes of atoms such as uranium. Instead of film, cups are used to collect the separated isotopes. In another application, chemists use a mass spectrometer (often called an MS) as a very sensitive tool to find small amounts of molecules in a sample. Amounts as small as one molecule in 10 billion molecules can be identified. Investigators detect the ions using electronic devices and are able to separate ions with mass differences of one ten-thousandth of one percent.

FIGURE 26–4 The mass spectrometer **(a)** is widely used to determine relative concentrations of various isotopes of an element. This spectrometer's computer display is used to check sensitivity and resolution prior to analysis. During analysis, marks are left on a film **(b)** by ^{50}Cr, ^{52}Cr, ^{53}Cr, and ^{54}Cr. Note that the weight of the mark is proportional to the percentage of the isotope in the element.

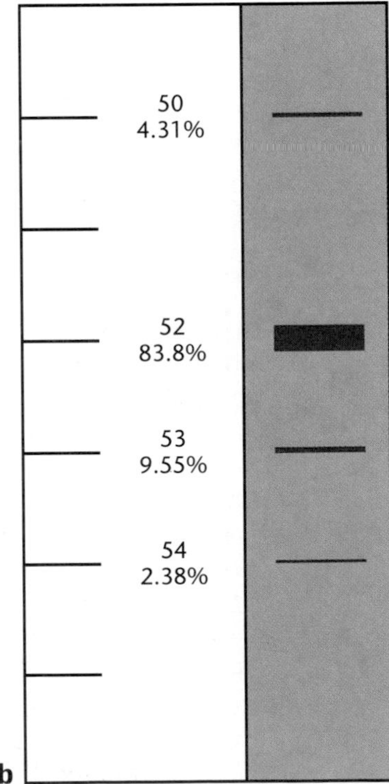

Example Problem

The Mass of a Neon Atom

The operator of a mass spectrometer produces a beam of doubly ionized neon atoms. They are first accelerated by a potential difference of 34 V. In a 0.050-T magnetic field, the radius of the path of the ions is 53 mm. Find the mass of the neon atom as a whole number of proton masses.

Sketch the Problem

- Draw the circular path of the ions. Label the radius.
- Draw and label the potential difference between the electrodes.

Calculate Your Answer

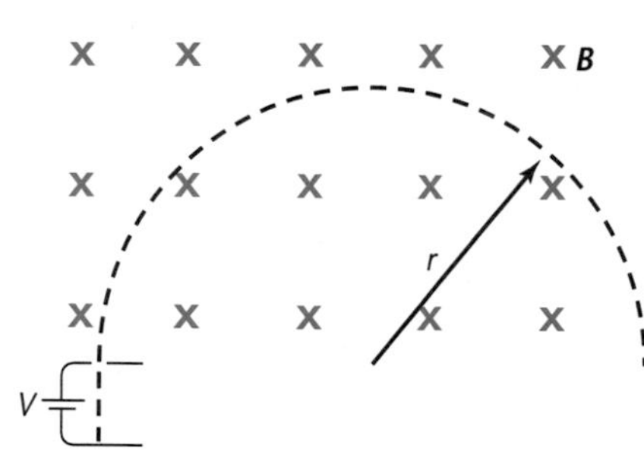

Known:

$V = 34\text{ V}$

$B = 0.050\text{ T}$

$r = 0.053\text{ m}$

$q = 2(1.60 \times 10^{-19}\text{ C}) = 3.20 \times 10^{-19}\text{ C}$

$m_{\text{proton}} = 1.67 \times 10^{-27}\text{ kg}$

Unknown:

$m_{\text{neon}} = ?$

$N_{\text{protons}} = ?$

Strategy:

Find the charge-to-mass ratio using the equation at the right.

Calculations:

$$\frac{q}{m} = \frac{2V}{B^2r^2}, \text{ so } m = \frac{qB^2r^2}{2V}$$

Because the charge is known, the mass can be found.

$$m_{\text{neon}} = \frac{(3.20 \times 10^{-19}\text{ C})(0.050\text{ T})^2(0.053\text{ m})^2}{2(34\text{ V})}$$

$$= 3.3 \times 10^{-26}\text{ kg}$$

Divide by the mass of a proton to find the number of proton masses.

$$N_{\text{protons}} = \frac{m_{\text{neon}}}{m_{\text{proton}}} = \frac{(3.3 \times 10^{-26}\text{ kg})}{(1.67 \times 10^{-27}\text{ kg/proton})}$$

$$\cong 20\text{ protons}$$

Check Your Answer

- Are the units correct? Mass should be measured in grams or kilograms. The number of protons should not have any units.
- Is the magnitude realistic? Yes, neon has two isotopes, with masses of approximately 20 and 22 proton masses.

Practice Problems

5. A stream of singly ionized lithium atoms is not deflected as it passes through a 1.5×10^{-3}-T magnetic field perpendicular to a 6.0×10^2-V/m electric field.

a. What is the speed of the lithium atoms as they pass through the crossed fields?

b. The lithium atoms move into a magnetic field of 0.18 T. They follow a circular path of radius 0.165 m. What is the mass of a lithium atom?

6. A mass spectrometer analyzes and gives data for a beam of doubly ionized argon atoms. The values are $q = 2(1.60 \times 10^{-19}\ \text{C})$, $B = 5.0 \times 10^{-2}\ \text{T}$, $r = 0.106\ \text{m}$, and $V = 66.0\ \text{V}$. Find the mass of an argon atom.

7. A beam of singly ionized oxygen atoms is sent through a mass spectrometer. The values are $B = 7.2 \times 10^{-2}\ \text{T}$, $q = 1.60 \times 10^{-19}\ \text{C}$, $r = 0.085\ \text{m}$, and $V = 110\ \text{V}$. Find the mass of an oxygen atom.

8. You found the mass of a neon isotope in the last Example Problem. Another neon isotope has a mass of 22 proton masses. How far from the first isotope would these ions land on the photographic film?

The Federal Communications Commission assigns each radio and TV station a carrier wave with a specific frequency. A station broadcasts by varying its carrier wave. When the wave is received by your radio or TV set, the carrier wave is stripped away and the information from the wave is processed so that you can see or hear it.

26.1 Section Review

1.1 Consider what changes Thomson would have had to make to accelerate protons rather than electrons in his cathode-ray tube.

a. To select particles of the same velocity, would the ratio E/B have to be changed?

b. For the deflection caused by the magnetic field alone to remain the same, would the B field have to be made smaller or larger? Explain.

1.2 As Thomson raised the energy of the electrons producing the ions in his tube, he found ions with two positive elementary charges rather than just one. How would he have recognized this?

1.3 A modern mass spectrometer can analyze molecules having masses of hundreds of proton masses. If the singly charged ions of these molecules are produced using the same accelerating voltage, how would the magnetic field have to be changed for them to hit the film?

1.4 **Critical Thinking** Thomson did not know the number of electrons in the atoms. With most atoms, he found that, as he raised the energy of the electrons that produced ions, he would first get ions with one electron missing, then ions with two electrons missing, and so on. With hydrogen, however, he could never remove more than one electron. What could he then conclude about the positive charge of the hydrogen atom?

Physics Lab

Simulating a Mass Spectrometer

Problem

How can you simulate the working parts of a mass spectrometer?

Materials

2 balls of clay
6-mm steel ball
graph paper
masking tape
2 permanent magnets
cafeteria tray or glass wave tank
grooved ruler
glass marble

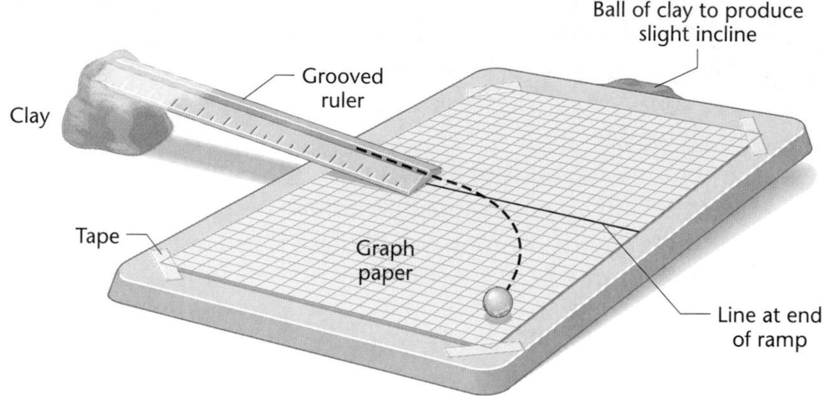

Procedure

1. Build the apparatus as shown in the diagram. Place a ball of clay under one side of the wave tank so that the tank is slightly sloped.
2. Make a test trial, allowing the steel ball to roll down the track. The ball should follow a curved path similar to the one shown in the diagram when it is started halfway up the ruler.
3. Starting from the same spot on the ruler, roll the steel ball down the track three times. Mark the positions where the ball crosses the far side of the graph paper.
4. Place the permanent magnets on the paper so that they pull the ball slightly upward. Adjust the magnets so that the ball follows a straight path across the graph paper, as shown by the solid line in the diagram.

Data and Observations

1. Describe the path of the ball in step 3.
2. Describe the path of the ball in step 4.

Analyze and Conclude

1. **Thinking Critically** In this model, you used gravity to simulate the electric field of a mass spectrometer. How could the electric field in this model be varied?
2. **Analyzing Data** What happens to the path as the magnet is brought closer to the path of the ball? Why?

Apply

1. Predict what would happen to a 6-mm ball that had the same mass, but less or no iron content. Explain your prediction. Test it.

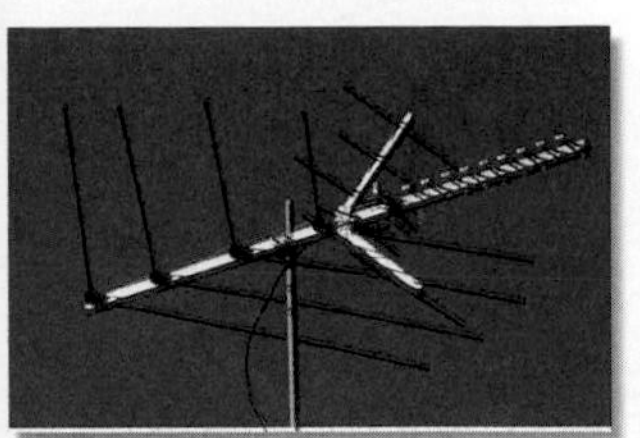

Electric and Magnetic Fields in Space

26.2

Signals emanating from galaxies, satellites, and television stations are electromagnetic waves. The properties of the electric and magnetic fields that constitute these waves were studied during most of the nineteenth century. In 1820, Oersted discovered that currents produce magnetic fields, and 11 years later, Faraday discovered induction. In the 1860s, Maxwell predicted that even without wires, electric fields changing in time cause magnetic fields, and that the magnetic fields changing in time produce electric fields. The result of this coupling is energy transmitted across empty space in the form of electromagnetic waves. Maxwell's theory led to a complete description of electricity and magnetism. It also gave us radio, television, and many other devices that have become part of our daily lives.

OBJECTIVES

- **Describe** how electric and magnetic fields can produce more electric and magnetic fields.
- **Explain** how accelerated charges produce electromagnetic waves.
- **Explain** the process by which electromagnetic waves are dctected.

Electromagnetic Waves

Oersted found that an electric current in a conductor produces a magnetic field. Changing the current changes the magnetic field, and, as Faraday discovered, changing the magnetic field can induce an electric current in a wire. Furthermore, the current-producing electric fields exist even without a wire, as illustrated in **Figure 26–5a.** Thus, a changing magnetic field produces a changing electric field. The field lines of the induced electric field will be closed loops, because unlike an electrostatic field, there are no charges on which the lines begin or end.

In 1860, Maxwell postulated that the opposite also is true. A changing electric field produces a changing magnetic field, as shown in **Figure 26–5b.** Maxwell suggested that charges were not necessary; the changing electric field alone would produce the magnetic field.

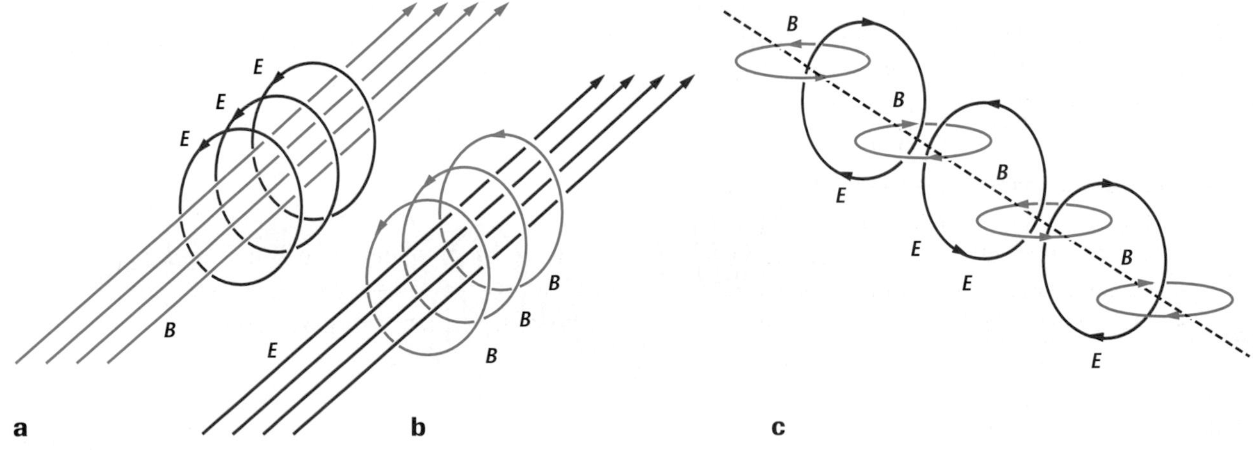

FIGURE 26–5 These diagrams represent an induced electric field **(a)**, a magnetic field **(b)**, and both electric and magnetic fields **(c)**.

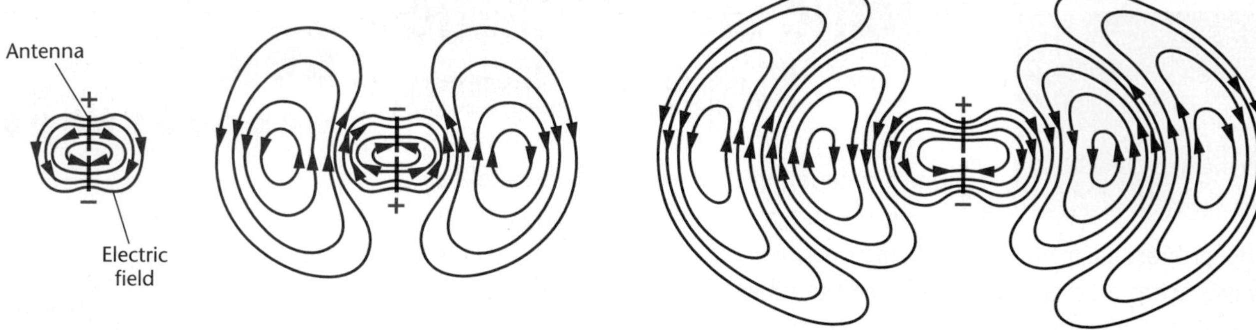

FIGURE 26–6 The changing current in the antenna generates a changing electric field. The changing electric field generates a changing magnetic field (not shown).

Maxwell then predicted that either accelerating charges or changing magnetic fields would produce electric and magnetic fields that move through space, **Figure 26–5c**. The combined fields are called an **electromagnetic wave.** The speed at which the wave moves, calculated Maxwell, was the speed of light, 3.00×10^8 m/s, as measured by Fizeau in 1849. Not only were electricity and magnetism linked, but also, optics, the study of light, became a branch of the study of electricity and magnetism. Heinrich Hertz (1857–1894), a German physicist, demonstrated experimentally in 1887 that Maxwell's theory was correct.

Figure 26–6 shows the formation of an electromagnetic wave. A wire, called an **antenna,** is connected to an alternating current (AC) source. The source produces changing currents in the antenna that alternate at the frequency of the AC source. The changing currents generate a changing electric field that moves outward from the antenna. There is also a changing magnetic field perpendicular to the page that is generated by the changing electric field, although the magnetic field is not shown in the figure. The electromagnetic waves spread out in space, moving at the speed of light.

If you stood to the right of the antenna as the waves approached, you could imagine the electric and magnetic fields changing in time, as in **Figure 26–7.** The electric field oscillates, first up, then down. The magnetic field oscillates at right angles to the electric field. The two fields are also at right angles to the direction of the motion of the wave. An electromagnetic wave produced by an antenna such as the one shown in **Figure 26–6** is polarized; that is, the electric field is always parallel to the direction of the antenna wires.

Catching the Wave

When you listen to your radio, you are hearing the information that is carried by electromagnetic waves. Many electronic and electrical devices produce low-frequency electromagnetic waves. Use a telephone pickup coil along with an amplifier to try to pick up signals from such devices as a television, computer, light, burning candle, coffee maker, or vacuum cleaner.

Analyzing Data Describe and interpret your results.

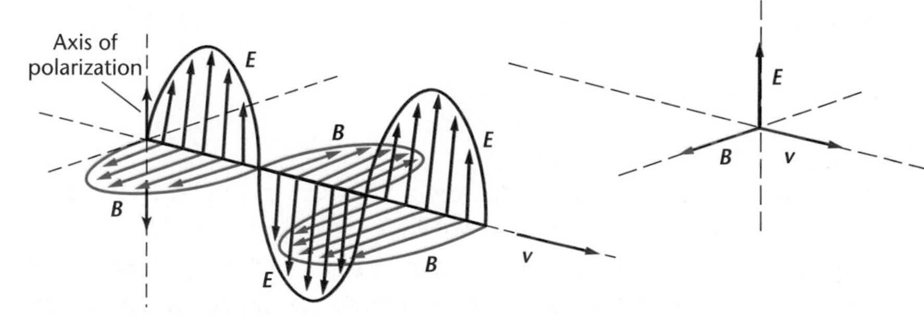

FIGURE 26–7 Portions of the electric and magnetic fields generated by an antenna might look like this at an instant in time.

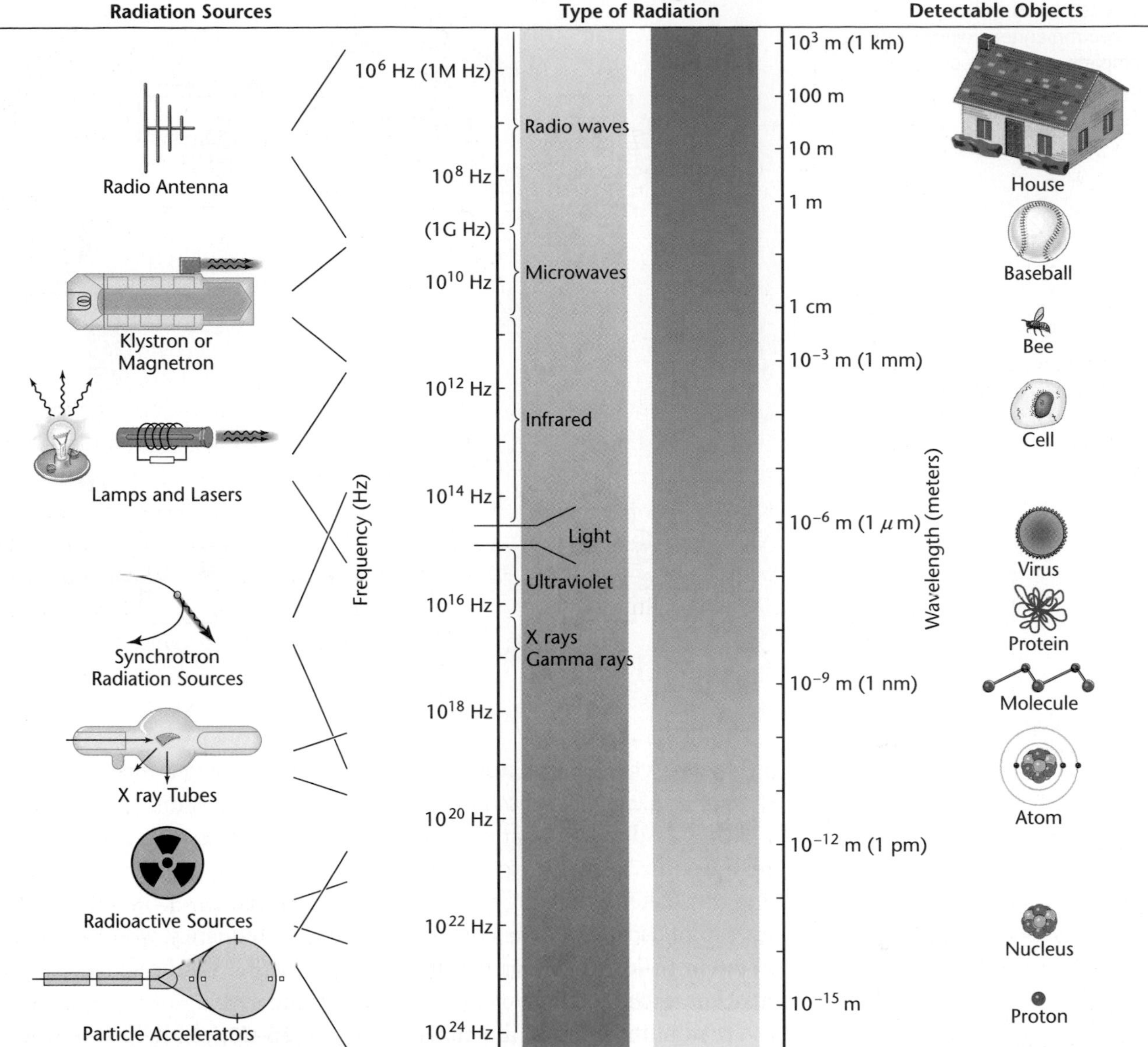

FIGURE 26–8 Representative examples of various types of electromagnetic radiation and their wavelengths are shown here.

Production of Electromagnetic Waves

Electromagnetic waves can be generated over a wide range of frequencies. **Figure 26–8** shows the electromagnetic spectrum. As you have learned, the AC generator is one method of creating the oscillating fields in the antenna. The frequency of the wave can be changed by varying the speed at which the generator is rotated. The highest frequency that can be generated in this way is about 1000 Hz.

Using a coil and capacitor

The most common method of generating waves of higher frequencies is to use a coil and capacitor connected in a series circuit. If the capacitor is charged by a battery, the potential difference across the capacitor creates an electric field. When the battery is removed, the capacitor discharges,

FIGURE 26–9 In the production of electromagnetic waves, one complete oscillation cycle is shown.

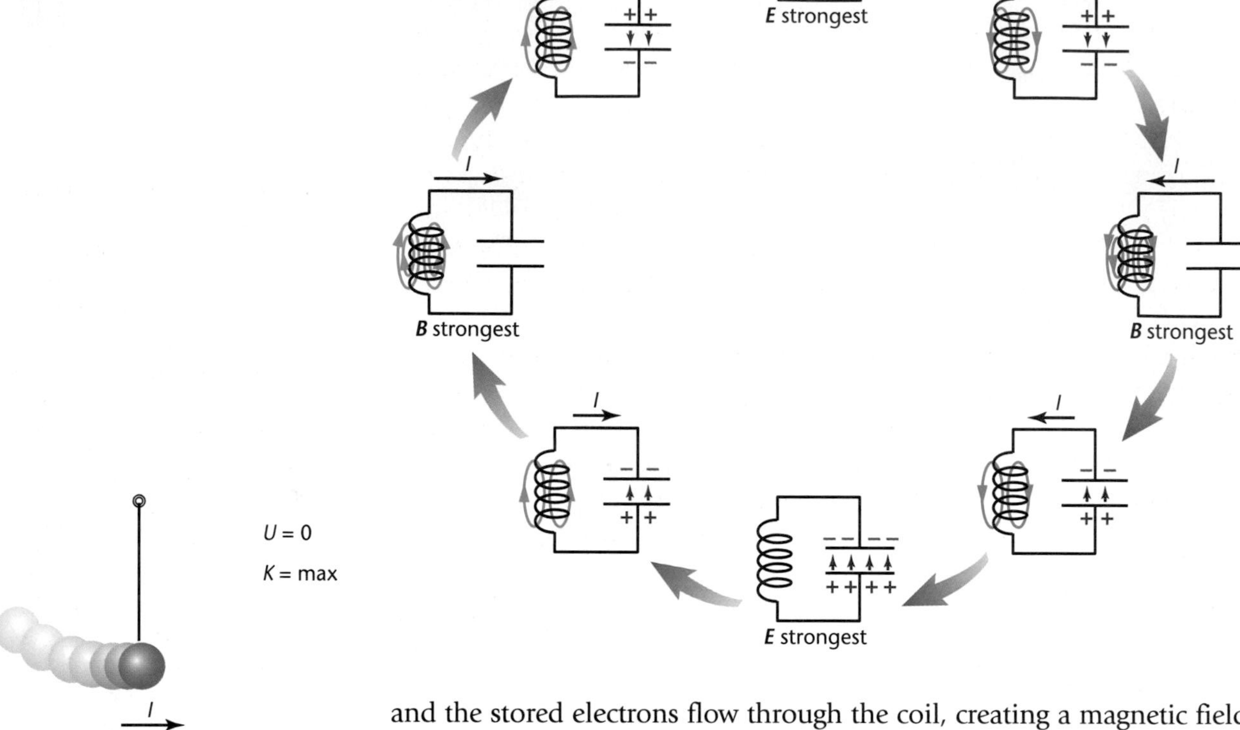

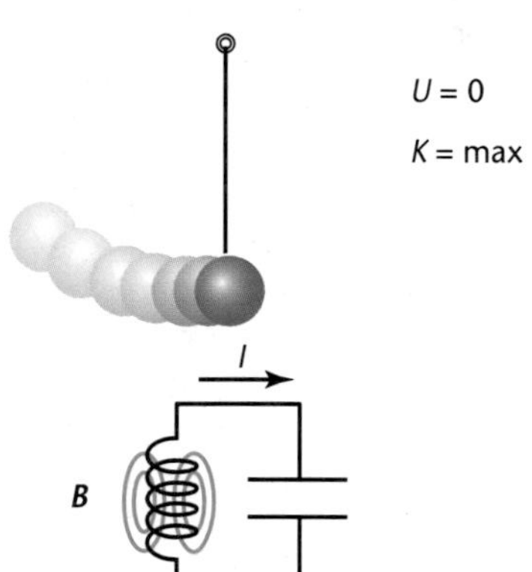

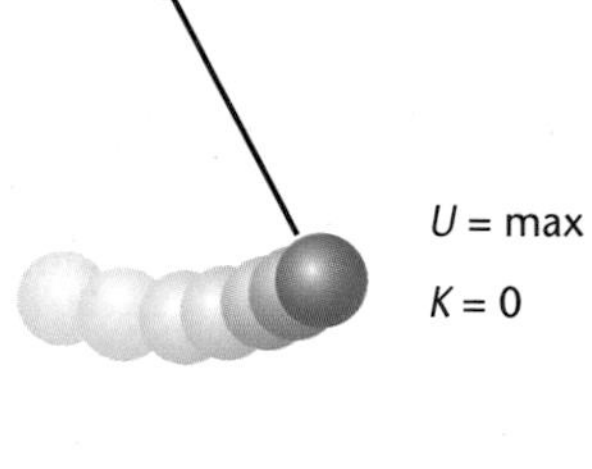

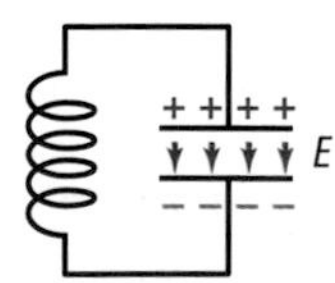

FIGURE 26–10 A pendulum is analogous to the action of electrons in a coil and capacitor combination.

and the stored electrons flow through the coil, creating a magnetic field. After the capacitor has discharged, the magnetic field of the coil collapses. A back-*EMF* develops that recharges the capacitor, this time in the opposite direction. The capacitor again discharges, and the process is repeated. One complete oscillation cycle is shown in **Figure 26–9.** Recall that the number of oscillations each second is called the frequency, which depends on the size of the capacitor and the coil. The antenna, connected across the capacitor, extends the fields of the capacitor into space.

A pendulum analogy, illustrated in **Figure 26–10,** can help you understand the coil and capacitor circuit. The electrons in the coil and capacitor are represented by the pendulum bob. The bob moves fastest when its displacement from vertical is zero. This is similar to the largest current flowing in the coil when the charge on the capacitor is zero. When the bob is at its greatest angle, its displacement from the vertical is largest, and it has zero velocity. This position is like the instant when the capacitor holds the largest charge and the current through the coil is zero.

Energy considerations

The pendulum model also can be used to describe energy. The potential energy of the pendulum is largest when its displacement is greatest. The kinetic energy is largest when the velocity is greatest. The sum of the potential and kinetic energies—the total energy—is constant. Both the magnetic field produced by the coil and the electric field in the capacitor contain energy. When the current is largest, the energy stored in the magnetic field is greatest. When the current is zero, the electric field of the capacitor is largest, and all the energy is in the electric field. The total energy—the sum of the magnetic field energy, the electric field energy, the thermal losses, and the energy carried away by the electromagnetic waves being generated—is constant. Energy carried, or radiated, in the form of electromagnetic waves is frequently called **electromagnetic radiation.**

Just as the pendulum will eventually stop swinging if it is left alone, the oscillations in a coil and capacitor also will die out because of resistance in the circuit, unless energy is added to the circuit. Gentle pushes, applied at the correct times, will keep a pendulum moving. The swing of largest amplitude occurs when the frequency of pushing is the same as the frequency of swinging. This is the condition of resonance, which was discussed in Chapter 6. Similarly, voltage pulses applied to the coil and capacitor circuit at the right frequency keep the oscillations going. One way of doing this is to add a second coil to form a transformer, as in **Figure 26–11.** The AC induced in the secondary coil is increased by an amplifier and added back to the coil and capacitor. This type of circuit can produce frequencies up to approximately 400 MHz.

Increasing the oscillation frequency

To increase the oscillation frequency, the size of the coil and capacitor must be made smaller. Above 1000 MHz, individual coils and capacitors will not work. For these electromagnetic waves, called microwaves, a rectangular box called a resonant cavity acts as both a coil and a capacitor. The size of the box determines the frequency of oscillation. Such a cavity is found in every microwave oven.

At frequencies of infrared waves, the size of resonant cavities would have to be reduced to the size of molecules. The oscillating electrons that produce infrared waves are, in fact, within the molecules. Visible and ultraviolet waves are generated by electrons within atoms.

High-frequency waves, such as X rays and gamma waves, are the result of accelerating charges in the nucleus of an atom. Most electromagnetic waves arise from accelerated charges, and all travel at the speed of light.

Piezoelectricity

Coils and capacitors are not the only method of generating oscillation voltages. Quartz crystals have a property called **piezoelectricity:** they bend or deform when a voltage is applied across them. Just as a piece of metal will vibrate at a specific frequency when it is bent and released, so too will a quartz crystal. A crystal can be cut so that it will vibrate at a specific desired frequency. An applied voltage bends it so that it starts vibrating. The piezoelectric property also generates an *EMF* when the crystal is bent. Because this *EMF* is produced at the vibrating frequency of the crystal, it can be amplified and returned to the crystal to keep it vibrating. Quartz crystals are used in wristwatches because the frequency of vibration is so constant.

More Radio Stuff

Radio stations broadcast with tremendous power from tall towers. Would it be possible to make your own radio transmitter and receiver? Try this activity to find out. Pop your favorite tape into a tape player, or tune your portable radio to a strong local AM station. Next, put a miniplug into the earphone jack, and attach the two leads to the ends of an air core solenoid. Put a second miniplug into a miniamp speaker, and attach the leads to the ends of a second air core solenoid. Turn the power on for the miniamp speaker, and listen for the sound.

Make a Hypothesis Make a hypothesis explaining why this lab works. Try changing the distance between the solenoids. Explain your results.

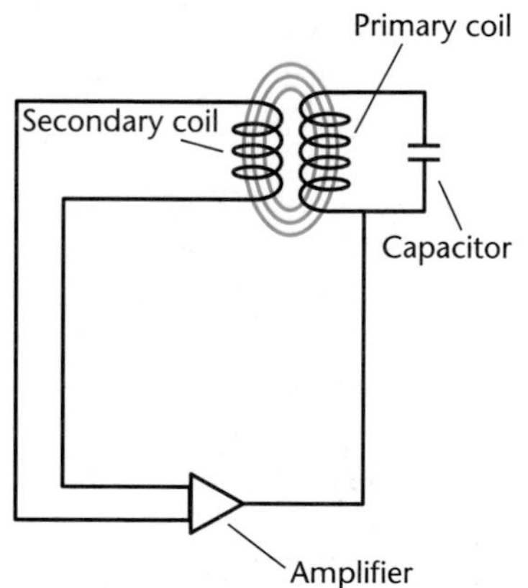

FIGURE 26–11 The amplified pulse from the secondary coil is in resonance with the coil and capacitor circuit and keeps the oscillations going.

How It Works

Bar-Code Scanners

A bar-code scanner uses light to "read" a code made up of bands of black and white bars. The computer links the code with data about the bar-coded item. In a supermarket, for example, the computer can display and print the name and price of the item, record the sale for the store's records, and even keep track of how many of those items remain in stock.

1 Bar codes consist of a series of alternating black bars and white spaces. There are many different bar codes. Each one uses a specific arrangement of bars and spaces of different widths to stand for a letter, number, or other character. Bands that indicate the beginning and end of the code enable the scanner to read either forward or backward.

2 Most supermarket checkout counters use laser scanners to read bar codes. The laser, housed beneath the clear glass window on the checkout counter, produces a beam of light that shines through a beam spreader, then onto a partially silvered, tilted mirror.

3 The mirror reflects the light up through a rotating disc. The disc focuses and directs the beam through the scanner window.

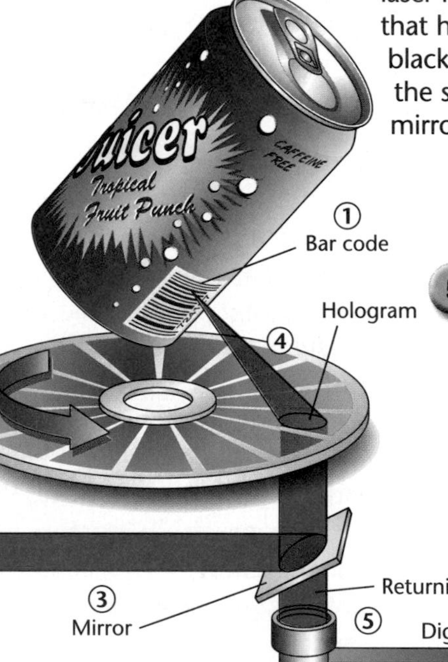

4 As the cashier drags each bar-coded item across the scanner window, the laser light scans the bar code. Light that hits the spaces between the black bars is reflected back through the scanner window, through the mirror, and onto a detector below.

5 The bursts of light striking the detector correspond to the width of the black bars and white spaces of the bar code. The detector changes these bursts into a digital signal that is sent to the computer for processing.

Thinking Critically

1. Review the descriptions of digital versatile discs in Chapter 16 and laser printers in Chapter 20. Compare and contrast the operation of a bar code scanner with these two devices.
2. Why is the scanner able to read the bar code when the bar-coded item is held in almost any orientation?

Reception of Electromagnetic Waves

Now let's examine how electromagnetic waves can be detected. When the electric fields in these waves strike another antenna, as shown in **Figure 26–12,** they accelerate the electrons in it. The acceleration is largest when the antenna is turned in the direction of the polarization of the wave; that is, when it is parallel to the direction of the electric fields in the wave. An *EMF* across the terminals of the antenna oscillates at the frequency of the electromagnetic wave. The *EMF* is largest if the length of the antenna is one-half the wavelength of the wave. The antenna then resonates in the same way an open pipe one-half wavelength long resonates with sound waves. For that reason, an antenna designed to receive radio waves is much longer than one designed to receive microwaves.

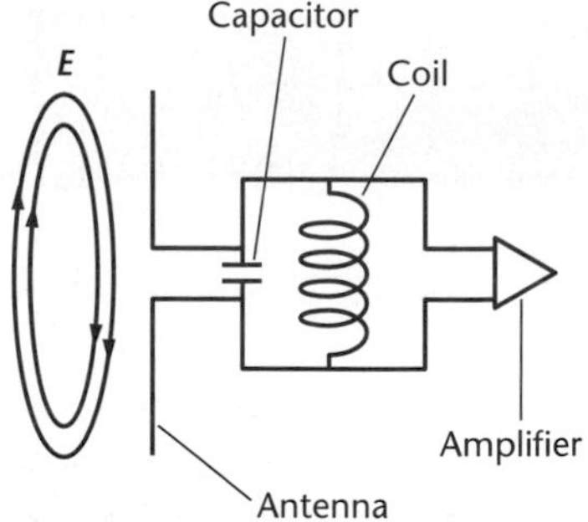

FIGURE 26–12 The changing electric fields from a radio station cause electrons in the antenna to accelerate.

While a simple wire antenna can detect electromagnetic waves, several wires can be used to increase the detected *EMF.* A television antenna often consists of two or more wires spaced about one-quarter wavelength apart. Electric fields generated in the individual wires form constructive interference patterns that increase the strength of the signal. At very short wavelengths, parabolic dishes reflect the waves, just as parabolic mirrors reflect light waves. Giant parabolic dishes focus waves with wavelengths of 2 to 6 cm on the antennas held by the tripod above the dish.

Big Ears

Selection of waves

Radio and television waves transmit information across space. Many different radio and television stations produce electromagnetic waves at the same time. If the information being broadcast is to be understood, the waves of a particular station must be selected. To select waves of a particular frequency and reject the others, a coil and capacitor circuit is connected to the antenna. The capacitance is adjusted until the oscillation frequency of the circuit equals the frequency of the desired wave. Only this frequency can cause significant oscillations of the electrons in the circuit. The information carried by the oscillations is then amplified and ultimately drives a loudspeaker. The combination of antenna, coil and capacitor circuit, and amplifier is called a **receiver.**

Energy from waves

Waves carry energy as well as information. At microwave and infrared frequencies, the electromagnetic waves accelerate electrons in molecules. The energy of the electromagnetic waves is converted to thermal energy in the molecules. Microwaves cook foods in this way. Infrared waves from the sun produce the warmth you feel on a bright, sunny day.

Light waves can transfer energy to electrons in atoms. In photographic film, this energy causes a chemical reaction. The result is a permanent record of the light reaching the camera from the subject. In the eye, the energy produces a chemical reaction that stimulates a nerve, resulting in a response in the brain that we call vision. At higher frequencies, UV radiation causes many chemical reactions to occur, including those in living cells that produce sunburn and tanning.

MEDICINE CONNECTION

X Rays

In 1895, German physicist Wilhelm Roentgen (1845–1923) sent electrons through an evacuated glass tube similar to the one shown in **Figure 26–13.** Roentgen used a very high voltage across the tube to give the electrons a large kinetic energy. The electrons struck the metal anode of the tube. When this happened, Roentgen noted a glow on a phosphorescent screen a short distance away. The glow continued even when a piece of wood was placed between the tube and the screen. He concluded that some kind of highly penetrating rays were coming from the tube.

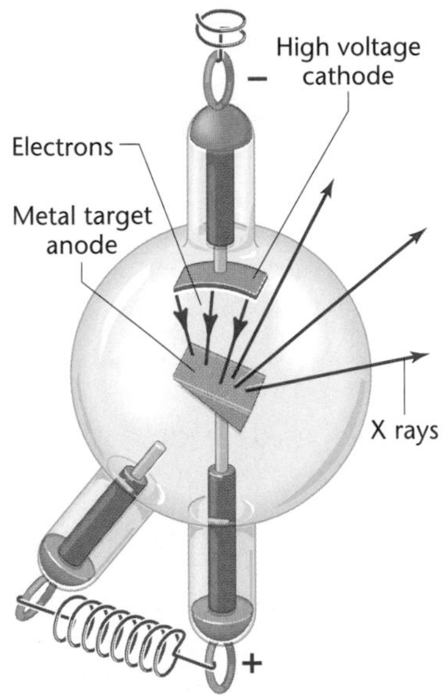

FIGURE 26–13 The metal target in this X-ray tube can be changed to produce X rays of different wavelengths.

Because Roentgen did not know what these strange rays were, he called them X rays. A few weeks later, Roentgen found that photographic plates were darkened by X rays. He also discovered that soft body tissue was transparent to the rays, but that bone blocked them. He produced an X-ray picture of his wife's hand. Within months, doctors recognized the valuable medical uses of this phenomenon.

It is now known that **X rays** are high-frequency electromagnetic waves. They are produced when electrons are accelerated to high speeds by means of potential differences of 20 000 or more volts. When the electrons crash into matter, their kinetic energies are converted into the very high-frequency electromagnetic waves called X rays.

Electrons are accelerated to these speeds in cathode-ray tubes, such as the picture tube in a television. When the electrons hit the face plate, they cause the colored phosphors to glow. The sudden stopping of the electrons also can produce X rays. The face-plate glass in television screens contains lead to stop the X rays and protect viewers.

26.2 Section Review

2.1 What was Maxwell's contribution to electromagnetism?

2.2 Television antennas normally have the metal rod elements in a horizontal position. From that, what can you deduce about the directions of the electric fields in television signals?

2.3 Television channels 2 through 6 have frequencies just below the FM radio band, while channels 7 through 13 have much higher frequencies. Which signals would require a longer antenna, those of channel 7 or those of channel 6?

2.4 **Critical Thinking** Most of the UV radiation from the sun is blocked by the ozone layer in Earth's atmosphere. Scientists have found a thinning of the ozone layer over both Antarctica and the Arctic Ocean. Should we be concerned about this?

CHAPTER 26 REVIEW

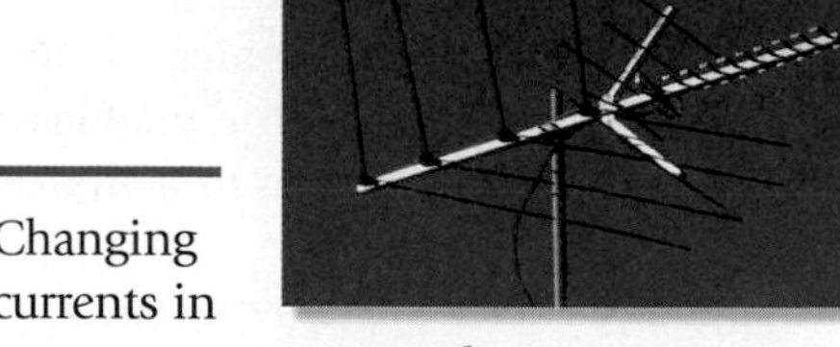

Key Terms

26.1

- isotope
- mass spectrometer

26.2

- electromagnetic wave
- antenna
- electromagnetic radiation
- piezoelectricity
- receiver
- X ray

Summary

26.1 Interaction Between Electric and Magnetic Fields and Matter

- The ratio of charge to mass of the electron was measured by J. J. Thomson using balanced electric and magnetic fields in a cathode-ray tube.
- An electron's mass can be found by combining Thomson's result with Millikan's measurement of the electron's charge.
- The mass spectrometer uses both electric and magnetic fields to measure the masses of ionized atoms and molecules.

26.2 Electric and Magnetic Fields in Space

- Electromagnetic waves are coupled, changing electric and magnetic fields that move through space.
- Changing currents in an antenna generate electromagnetic waves.
- The frequency of oscillating currents can be selected by a resonating coil and capacitor circuit.
- Electromagnetic waves can be detected by the *EMF* they produce in an antenna. The length of the most efficient antenna is one-half the wavelength of the wave to be detected.
- Microwave and infrared waves can accelerate electrons in molecules, producing thermal energy.
- When high-energy electrons strike an anode in an evacuated tube, their kinetic energies are converted to electromagnetic waves of very high energy called X rays.

Reviewing Concepts

Section 26.1

1. What is the mass of an electron? What is its charge?
2. What are isotopes?

Section 26.2

3. The direction of an induced magnetic field is always at what angle to the changing electric field?
4. Like all waves, microwaves can be transmitted, reflected, and absorbed. Why can soup be heated in a ceramic mug but not in a metal pan in a microwave oven? Why does the mug's handle not get as hot as the soup?
5. Why must an AC generator be used to propagate electromagnetic waves? If a DC generator were used, when would it create electromagnetic waves?
6. A vertical antenna wire transmits radio waves. Sketch the antenna and the electric and magnetic fields it creates.
7. What happens to quartz crystals when a voltage is placed across them?
8. Car radio antennas are vertical. What is the direction of the electric fields they detect?
9. How does an antenna receiving circuit select electromagnetic radio waves of a certain frequency and reject all others?

Applying Concepts

10. The electrons in a Thomson tube travel from left to right. Which deflection plate should be charged positively to bend the electron beam upward?
11. The electron beam in question 10 has a magnetic field to make the beam path straight. What would be the direction of the magnetic field needed to bend the beam downward?
12. Show that the units of E/B are the same as the units for velocity.

13. A mass spectrometer operates on neon ions. What is the direction of the magnetic field needed to bend the beam in a clockwise semicircle?
14. Charged particles are moving through an electric field and a magnetic field that are perpendicular to each other. Suppose you adjust the fields so that a certain ion, with the correct velocity, passes without deflection. Then, another ion with the same velocity but a different mass enters the fields. Describe the path of the second ion.
15. If the sign of the charge on the particle in question 14 is changed from positive to negative, do the directions of either or both of the two fields have to be changed to keep the particle undeflected? Explain.
16. Do radio waves, light, or X rays have the largest **a.** wavelength? **b.** frequency? **c.** velocity?
17. The frequency of television waves broadcast on channel 2 is about 58 MHz. The waves broadcast on channel 7 are about 180 MHz. Which channel requires a longer antenna?
18. Suppose the eyes of an alien being are sensitive to microwaves. Would you expect such a being to have larger or smaller eyes than yours? Why?

Problems

Section 26.1

LEVEL 1

19. A beam of ions passes undeflected through a pair of crossed electric and magnetic fields. E is 6.0×10^5 N/C and B is 3.0×10^{-3} T. What is the ions' speed?
20. Electrons moving at 3.6×10^4 m/s pass through an electric field with an intensity of 5.8×10^3 N/C. How large a magnetic field must the electrons also experience for their path to be undeflected?
21. The electrons in a beam move at 2.8×10^8 m/s in an electric field of 1.4×10^4 N/C. What value must the magnetic field have if the electrons pass through the crossed fields undeflected?
22. A proton moves across a 0.36-T magnetic field in a circular path of radius 0.20 m. What is the speed of the proton?
23. Electrons move across a 4.0-mT magnetic field. They follow a circular path with radius 2.0 cm.
    a. What is their speed?
    b. An electric field is applied perpendicularly to the magnetic field. The electrons then follow a straight-line path. Find the magnitude of the electric field.
24. A proton enters a 6.0×10^{-2}-T magnetic field with a speed of 5.4×10^4 m/s. What is the radius of the circular path it follows?
25. A proton enters a magnetic field of 6.4×10^{-2} T with a speed of 4.5×10^4 m/s. What is the circumference of its circular path?
26. A 3.0×10^{-2}-T magnetic field in a mass spectrometer causes an isotope of sodium to move in a circular path with a radius of 0.081 m. If the ions have a single positive charge and are moving with a speed of 1.0×10^4 m/s, what is the isotope's mass?
27. An alpha particle, a doubly ionized helium atom, has a mass of 6.7×10^{-27} kg and is accelerated by a voltage of 1.0 kV. If a uniform magnetic field of 6.5×10^{-2} T is maintained on the alpha particle, what will be the particle's radius of curvature?
28. An electron is accelerated by a 4.5-kV potential difference. How strong a magnetic field must be experienced by the electron if its path is a circle of radius 5.0 cm?
29. A mass spectrometer yields the following data for a beam of doubly ionized sodium atoms: $B = 8.0 \times 10^{-2}$ T, $q = 2(1.60 \times 10^{-19}$ C), $r = 0.077$ m, and $V = 156$ V. Calculate the mass of a sodium atom.

LEVEL 2

30. An alpha particle has a mass of approximately 6.6×10^{-27} kg and bears a double elementary positive charge. Such a particle is observed to move through a 2.0-T magnetic field along a path of radius 0.15 m.
    a. What speed does the particle have?
    b. What is its kinetic energy?
    c. What potential difference would be required to give it this kinetic energy?

31. In a mass spectrometer, ionized silicon atoms have curvatures with radii of 16.23 cm and 17.97 cm. If the smaller radius corresponds to a mass of 28 proton masses, what is the mass of the other silicon isotope?
32. A mass spectrometer analyzes carbon-containing molecules with a mass of 175×10^3 proton masses. What percent differentiation is needed to produce a sample of molecules in which only carbon isotopes of mass 12, and none of mass 13, are present?

Section 26.2

LEVEL 1

33. The radio waves reflected by a parabolic dish are 2.0 cm long. How long should the antenna be that detects the waves?

LEVEL 2

34. The difference in potential between the cathode and anode of a spark plug is 1.0×10^4 V.
    a. What energy does an electron give up as it passes between the electrodes?
    b. One-fourth of the energy given up by the electron is converted to electromagnetic radiation. The frequency of the wave is related to the energy by the equation $E = hf$, where h is Planck's constant, 6.6×10^{-34} J/Hz. What is the frequency of the waves?
35. Channel 6 broadcasts on a frequency of 85 MHz.
    a. What is the wavelength of the electromagnetic wave broadcast on channel 6?
    b. What is the length of an antenna that will detect channel 6 most easily?

Critical Thinking Problems

36. H.G. Wells wrote a science fiction book called *The Invisible Man,* in which a man drinks a potion and becomes invisible, although he retains all of his other faculties. Explain why it wouldn't be possible for an invisible person to be able to see.
37. You are designing a mass spectrometer using the principles discussed in this chapter, but with an electronic detector replacing the photographic film. You want to distinguish singly ionized molecules of 175 proton masses from those with 176 proton masses, but the spacing between adjacent cells in your detector is 0.1 mm. The molecules must have been accelerated by a potential difference of at least 500 volts to be detected. What are some of the values of V, B, and r that your apparatus should have?

Going Further

Project Experiment with the production and detection of electromagnetic waves using two boom boxes or one radio or tape player with an earphone jack and another with a microphone input jack. You will need two plugs fitting the jacks that have two wires to which you can make contact. The boom box with the plug into the earphone jack will generate electromagnetic waves; the boom box with the plug into the microphone jack will detect them. To create and detect electric fields, connect each wire coming from the earphone plug to a soda can. Stand the soda cans side-by-side about 10 cm apart. **CAUTION:** *Do not allow any of the cans to touch.* Use a second pair of cans connected to the microphone plug. Turn both volume controls up high. See how far you can transmit electromagnetic waves from one pair of cans to the other. You can replace one or both pairs of cans with a coil of wire of at least 100 turns and about 20 cm in diameter. You then will be generating and/or detecting magnetic fields. These electromagnetic waves have frequencies equal to those of the sound waves you generated. What are their wavelengths?

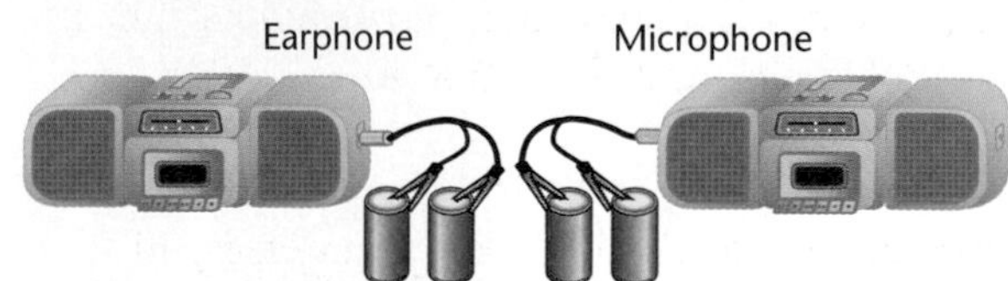

FIGURE 26–14

interNET CONNECTION

Follow the link on the Glencoe Homepage at **www.glencoe.com/sec/science** to find out more about this chapter.

Electron Corral

Look at these extraordinary images. Would you believe that they are corrals for electrons? The peaks are individual iron atoms that IBM scientists placed, one by one, on a copper surface using a scanning tunneling microscope (STM). How does quantum theory permit the STM to work?

CHAPTER

27 Quantum Theory

The startling ideas of quantum theory are the basis for the amazing scanning tunneling microscope. The STM allows scientists to penetrate deeper into the world of atoms and to manipulate matter in ways that were inconceivable only a short time ago. Not surprisingly, the Nobel prize in physics was awarded in 1986 to Gerd Binnig and Heinrich Rohrer, the inventors of the STM, in recognition of the important contribution this instrument is making to the study of matter at the atomic level.

The STM and its cousins can produce pictures of the surfaces of metals, insulators, and other materials including strands of DNA. They can be used to study friction on an atomic scale, and to create new materials. They can even assemble "designer" molecules one atom at a time.

All of this is possible and can be explained by quantum theory. But the STM is not the only phenomenon that has its roots in quantum theory. You have seen more familiar quantum effects in the bright lights of neon signs, in the colors of chemical flame tests, and in fireworks displays. As you learn more about the wave and particle nature of light and matter, you will begin to identify many other ways in which quantum ideas play a part in your everyday life.

WHAT YOU'LL LEARN

- You will describe light as a discrete, or quantized, bundle of momentum and energy.
- You will recognize that atom-sized particles of matter behave like waves, showing diffraction and interference effects.

WHY IT'S IMPORTANT

- Microwave ovens, lasers, televisions, computer monitors, and home security systems are a few of the many devices that depend upon the quantum nature of light and matter.

*inter*NET CONNECTION

Follow the link for this chapter on the Glencoe Homepage at **www.glencoe.com/sec/science** to find out more about the duality of waves and particles.

27.1 Waves Behave Like Particles

OBJECTIVES

- **Describe** the spectrum emitted by a hot body and **explain** the basic theory that underlies the emission of hot-body radiation.
- **Explain** the photoelectric effect and **recognize** that quantum theory can explain it, whereas the wave theory cannot.
- **Explain** the Compton effect and **describe** it in terms of the momentum and energy of the photon.
- **Describe** experiments that demonstrate the particle-like properties of electromagnetic radiation.

In 1889, the experiments of Heinrich Hertz confirmed the predictions of Maxwell's theory, which you learned about in Chapter 26. All of optics seemed to be explainable in terms of electromagnetic theory. Only two small problems remained. Wave theory could not describe the spectrum of light emitted by a hot body such as molten steel or an incandescent lightbulb. Also, as discovered by Hertz himself, ultraviolet light discharged electrically charged metal plates. This effect, called the photoelectric effect, could not be explained by Maxwell's wave theory.

Radiation from Incandescent Bodies

Why was radiation from hot bodies a puzzle? Hot bodies contain vibrating particles, which radiate electromagnetic waves. Maxwell's theory should have had no conflict with this, but its prediction was wrong. Why? What radiation do hot bodies emit?

If you look through a prism at the light coming from an incandescent lightbulb, you will see all the colors of the rainbow. The bulb also emits infrared radiation, that you cannot see. **Figure 27–1** shows the spectra of incandescent bodies at three different temperatures: 4000 K, 5800 K, and 8000 K. A spectrum is a plot of the intensity of radiation emitted at various frequencies. Light and infrared radiation are produced by the vibration of the charged particles within the atoms of a body that is so hot it glows, or is incandescent. The shapes of the curves in **Figure 27–1** show that energy is emitted at a variety of frequencies that depend on temperature. At each temperature, there is a frequency at which the maximum amount of energy is emitted. By comparing the three curves, you can see that as the temperature increases, the frequency at which the maximum energy is emitted also increases.

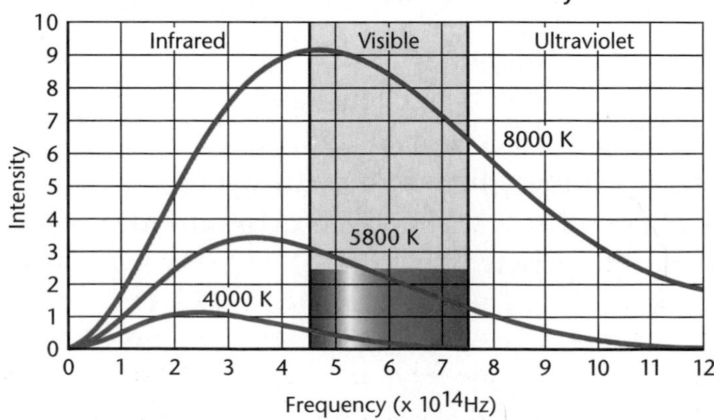

FIGURE 27–1 This graph shows the spectra of incandescent bodies at three temperatures.

Suppose you put a lightbulb on a dimmer control. You gradually turn up the voltage, increasing the temperature of the glowing filament. The color changes from deep red through orange to yellow and finally, to white. Because higher temperature results in radiation of higher frequency, more radiation at the higher-frequency end of the visible spectrum, the violet end, is produced and the body appears to be whiter. The colors you see depend upon the relative amounts of emission at various frequencies and the sensitivity of your eyes to those colors. Compare the position of the maximum of each curve in **Figure 27–1** with the visible spectrum.

The total power emitted also increases with temperature. The amount of energy emitted every second in electromagnetic waves is proportional to the temperature in kelvins raised to the fourth power, T^4. Thus, hotter sources radiate considerably more power than cooler bodies do. The sun, for example, is a dense ball of gases heated to incandescence by the energy produced within it. It has a surface temperature of 5800 K and a yellow color. The sun radiates 4×10^{26} W, an enormous amount of power. On average, every square meter on Earth's surface receives about 1000 J of energy each second.

Why does the spectrum have the shape shown in **Figure 27–1?** Maxwell's theory could not account for it. Between 1887 and 1900, many physicists tried to predict the shape of this spectrum using existing physical theories, but all failed. In 1900, the German physicist Max Planck (1858–1947), shown in **Figure 27–2,** found that he could calculate the spectrum only if he introduced a revolutionary hypothesis—that energy is not continuous. Planck assumed that the energy of vibration of the atoms in a solid could have only specific frequencies as shown by the following equation.

$$E = nhf$$

In the equation, f is the frequency of vibration of the atom, h is a constant, and n is an integer such as 0, 1, 2, or 3. The energy, E, could have the values hf, $2hf$, $3hf$, and so on, but never, for example, $2/3hf$. This behavior is described by saying that energy is **quantized.** Quantized energy comes only in packages of specific amounts.

Planck also proposed that atoms do not always radiate electromagnetic waves when they are vibrating, as predicted by Maxwell. Instead, he proposed that they emit radiation only when their vibration energy changes. For example, if the energy of an atom changes from $3hf$ to $2hf$, the atom emits radiation. The energy radiated is equal to the change in energy of the atom, in this case hf.

Planck found that the constant h was extremely small, about 7×10^{-34} J/Hz. This means that the energy-changing steps are too small to be noticeable in ordinary bodies. Still, the introduction of quantized energy was extremely troubling to physicists, especially Planck himself. It was the first hint that the physics of Newton and Maxwell might be valid only under certain conditions.

FIGURE 27–2 Max Planck (1858–1947) was awarded the Nobel prize in 1918 for his discovery of the quantized nature of energy.

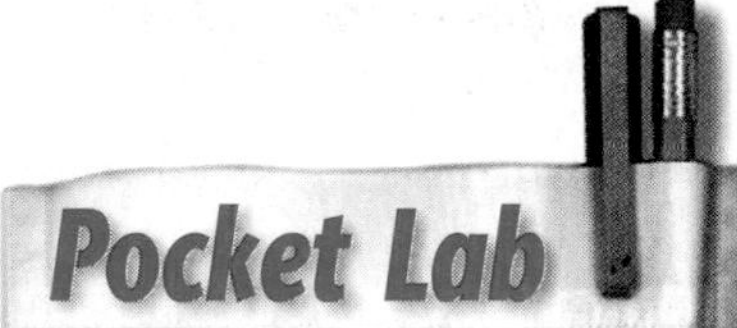

Glows in the Dark

Close the shades and turn off the lights in the room. Shine a flashlight at a beaker that contains fluorescein. Now place a red filter over the flashlight so that only red light hits the beaker. Describe the results. Repeat the experiment using a green filter. Explain the results. Would you expect the fluorescein to glow when a blue filter is used? Explain your prediction. Try it.

Analyze and Conclude Write a brief explanation of your observations.

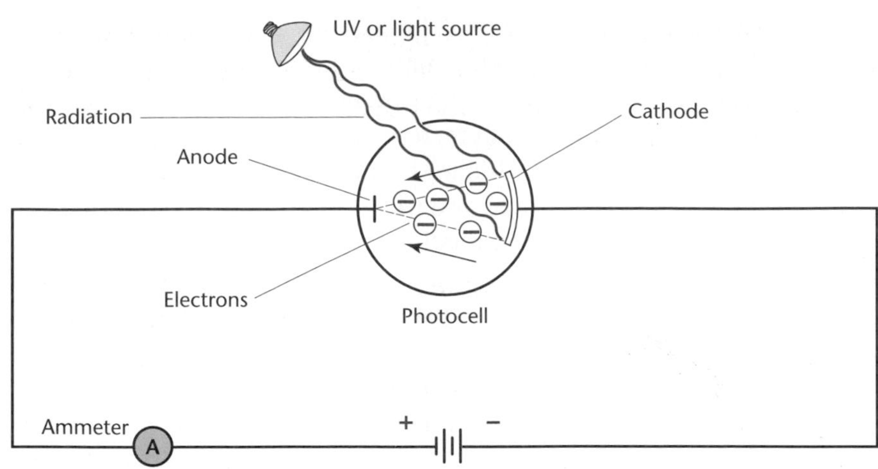

FIGURE 27–3 Electrons ejected from the cathode flow to the anode and thus complete the circuit.

The Photoelectric Effect

There was a second troubling experimental result unexplained by Maxwell. A negatively charged zinc plate was discharged when ultraviolet radiation fell on it, but it remained charged when ordinary visible light fell on it. Both ultraviolet light and visible light are electromagnetic radiation, so why would the zinc plate be discharged by one and not by the other? And why is a positively charged zinc plate not similarly discharged? Further study showed that the negatively charged zinc plate was discharged by losing or emitting electrons. The emission of electrons when electromagnetic radiation falls on an object is called the **photoelectric effect.**

The photoelectric effect can be studied in a photocell like the one in **Figure 27–3.** The cell contains two metal electrodes sealed in an evacuated tube. The air has been removed to keep the metal surface clean and to keep electrons from being stopped by air molecules. The large electrode, the cathode, is usually coated with cesium or another alkali metal. The second electrode, the anode, is made of a thin wire so that it blocks only the smallest amount of radiation. The tube is often made of quartz to permit ultraviolet wavelengths to pass through. A potential difference that attracts electrons to the anode is placed across the electrodes.

When no radiation falls on the cathode, there is no current in the circuit. But when radiation does fall on the cathode, there is a current, as shown by the meter in **Figure 27–3.** The current results from the ejection of electrons from the cathode by the radiation. Hence, these electrons are called photoelectrons. The electrons travel to the anode, the positive electrode.

Not all radiation results in a current. Electrons are ejected only if the frequency of the radiation is above a certain minimum value, called the **threshold frequency,** f_0. The threshold frequency varies with the metal. All wavelengths of visible light except red will eject electrons from

SOCIOLOGY CONNECTION

Many public buildings have doors that open automatically as people approach. These doors are often operated by so-called electric eyes. A light beam shines across the door opening onto a photovoltaic cell, causing electrons to be given off and a current to flow in a circuit. When the beam is broken, the current stops, and a mechanism is triggered that opens the door. These devices have allowed easier access to many people who would otherwise have difficulty opening a door.

cesium, but no wavelength of visible light will eject electrons from zinc. Ultraviolet light is needed for zinc. Radiation of a frequency below f_0 does not eject any electrons from the metal, no matter how intense the light is. However, even if the incident light is very dim, radiation at or above the threshold frequency causes electrons to leave the metal immediately; the greater the intensity of the incident radiation, the larger the flow of photoelectrons.

The electromagnetic wave theory cannot explain all of these facts. According to the wave theory, it is the intensity of the radiation, not the frequency, that determines the strength of the electric and magnetic fields. A more intense radiation, regardless of frequency, has stronger electric and magnetic fields. According to wave theory, the electric field accelerates and ejects the electrons from the metal. With very faint light shining on the metal, electrons would need to absorb energy for a very long time before they gained enough to be ejected. But, as you have learned, electrons are ejected immediately even in dim light if the frequency of the radiation is at or above the threshold frequency.

In 1905, Albert Einstein published a revolutionary theory that explained the photoelectric effect. According to Einstein, light and other forms of radiation consist of discrete bundles of energy, which were later called **photons.** The energy of each photon depends on the frequency of the light. The energy is represented by the equation $E = hf$, where h is Planck's constant, 6.63×10^{-34} J/Hz. Because the unit Hz = 1/s or s^{-1}, the unit of Planck's constant also can be expressed as J·s.

It is important to note that Einstein's theory of the photon goes further than Planck's theory of hot bodies. While Planck had proposed that vibrating atoms emitted radiation with energy equal to hf, he did not suggest that light and other forms of radiation acted like particles. Einstein's theory of the photon reinterpreted and extended Planck's theory of hot bodies.

Einstein's photoelectric-effect theory explains the existence of a threshold frequency. A photon with a minimum energy, hf_0, is needed to eject an electron from the metal. If the photon has a frequency below f_0, the photon does not have the energy needed to eject an electron. Light with a frequency greater than f_0 has more energy than is needed to eject an electron. The excess energy, $hf - hf_0$, becomes the kinetic energy of the electron.

$$K = hf - hf_0$$

Einstein's equation is a statement of conservation of energy. The incoming photon has energy hf. An amount of energy, hf_0, is needed to free the electron from the metal. The remainder becomes the kinetic energy of the electron. Note that an electron cannot simply accumulate photons until it has enough energy; only one photon interacts with one electron. In addition, hf_0 is the minimum energy needed to free an electron. The minimum energy is the amount of energy needed to release

HELP WANTED

PARTICLE PHYSICIST

The successful candidate will analyze elementary particles (quarks, leptons, and bosons) and must be skilled in using accelerators to study these particles and their interactions. A knowlege of computers and specialized cameras is also required. Most positions in high-energy physics are in government or university research facilities and require a doctorate in physics or a master's degree and extensive experience. For more information, contact one of the following organizations at the same address.
American Institute of Physics or the American Physical Society
One Physics Ellipse
College Park, MD 20740

FIGURE 27–4 The kinetic energy of the ejected electrons can be measured using this apparatus. An ammeter measures the current through the circuit.

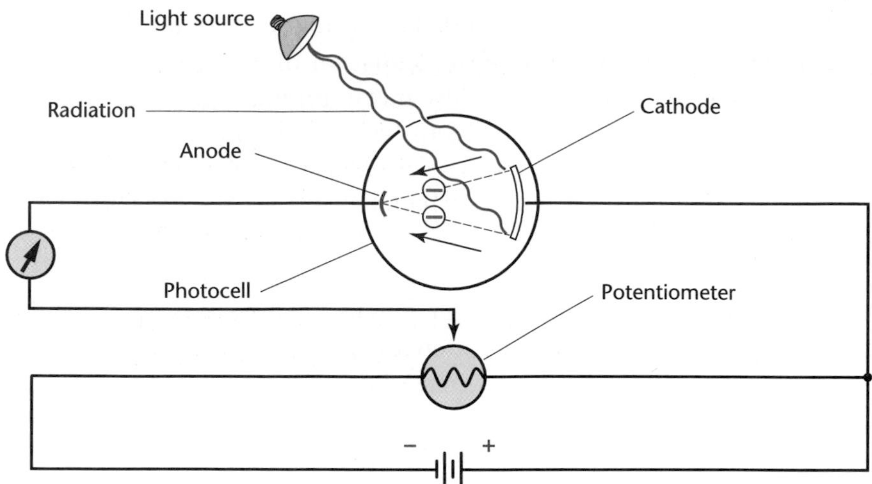

the most loosely held electron in an atom. Because not all electrons in an atom have the same energy, some need more than this minimum to escape, and as a result, they will have differing kinetic energies. Thus, the expression *kinetic energy of the ejected electrons* refers to the maximum kinetic energy an ejected electron could have. Some electrons will have less kinetic energy.

How can Einstein's theory be tested? The kinetic energy of the ejected electrons can be measured indirectly by a device like the one pictured in **Figure 27–4.** A variable electric potential difference across the tube makes the anode negative. By analogy to gravity, the electrons must expend energy, in effect climb a hill, to reach the anode. Only if they have enough kinetic energy when they leave the cathode will they reach the anode before being turned back. Light of the chosen frequency illuminates the cathode. An ammeter measures the current flowing through the circuit. Gradually, the experimenter increases the opposing potential difference, making the anode more negative. As the opposing potential difference increases, more and more kinetic energy is needed for the electrons to reach the anode, and fewer and fewer electrons arrive there to complete the circuit. At some voltage, called the stopping potential, no electrons have enough kinetic energy to reach the anode, and the current falls to zero. The maximum kinetic energy, K, at the cathode equals the work done by the electric field in stopping them. That is, $K = -qV_0$. In the equation, V_0 is the magnitude of the stopping potential in volts (J/C), and q is the charge of the electron (-1.60×10^{-19} C).

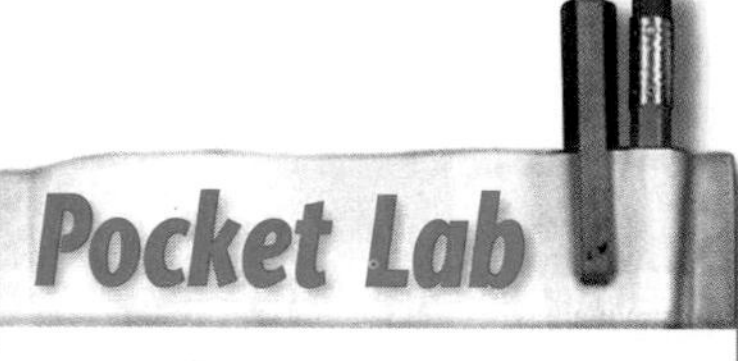

See the Light

Close the shades and turn off all the lights in the room. Look at a 120-W lamp as you use a dimmer switch to slowly increase and decrease the voltage across the lamp. Describe what you observe. What would you expect to see if you repeated the experiment while looking through a diffraction grating? Why? Try it. Describe your results.

Analyze and Conclude Describe your observations.

The joule is too large a unit of energy to use with atomic systems. A more convenient energy unit is the electron volt (eV). One electron volt is the energy of an electron accelerated across a potential difference of one volt.

$$1 \text{ eV} = (1.60 \times 10^{-19} \text{ C})(1 \text{ V}) = 1.60 \times 10^{-19} \text{ C·V}$$

$$1 \text{ eV} = 1.60 \times 10^{-19} \text{ J}$$

Example Problem

The Kinetic Energy of a Photoelectron

The stopping potential, V_0, that prevents electrons from flowing across a certain photocell is 4.0 V. What is the kinetic energy given to the electrons by the incident light? Give your answer in both J and eV.

Sketch the Problem

- Draw the cathode and anode, the incident radiation, and the direction of the path of the ejected electron.

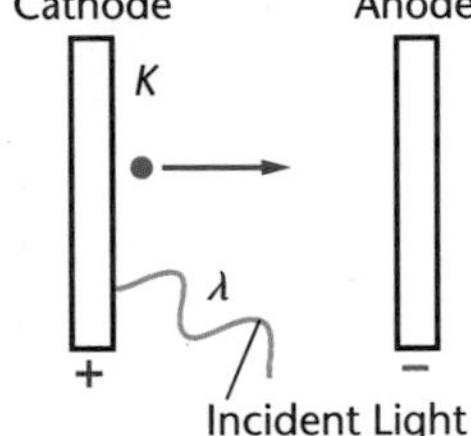

Calculate Your Answer

Known:	**Unknown:**
$V_0 = 4.0$ V	K (in J and eV) = ?
$q = -1.60 \times 10^{-19}$ C	

Strategy:

The electric field does work on the electrons. When the work done, W, equals the negative of the initial kinetic energy, K, electrons no longer flow across the photocell. Use V_0 to find the work done, which equals the kinetic energy.

Calculations:

$K + W = 0; K = -W$

$W = qV_0$, so $K = -qV_0$

$K = -(-1.60 \times 10^{-19}\text{ C})(4.0\text{ V})$

$= +6.4 \times 10^{-19}$ J

$1\text{ eV} = 1.60 \times 10^{-19}$ J

so $K = (6.4 \times 10^{-19}\text{ J})(1\text{ eV}/1.60 \times 10^{-19}\text{ J})$

$= 4.0$ eV

Check Your Answer

- Are the units correct? 1 V = 1 J/C, so CV = J.
- Does the sign make sense? Kinetic energy is always positive.
- Is the magnitude realistic? Energy in electron volts is equal in magnitude to stopping potential difference in volts.

Problem Solving Strategy

A Useful Unit for *hc*.

The energy of a photon of wavelength λ is given by $E = hf$. But, $f = c/\lambda$, so $E = hc/\lambda$. Thus, it's helpful to know the value of hc in eV·nm so that when you divide by λ in nm, you obtain the energy in eV.

Convert the constant hc to the unit eV·nm as follows.

$hc = (6.626 \times 10^{-34}\text{ J/Hz})(2.998 \times 10^{8}\text{ m/s})(1\text{ eV}/1.602 \times 10^{-19}\text{ J})(10^{9}\text{ nm/m})$

$= 1240$ eV·nm

Thus, $E = hc/\lambda = (1240\text{ eV·nm})/\lambda$,

where λ is in nm and E is in eV.

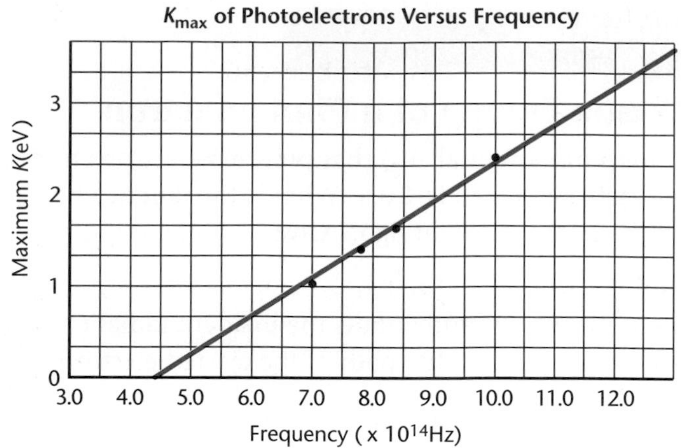

Figure 27–5 The graph shows that as the frequency of the incident radiation increases, the kinetic energy of the ejected electrons increases proportionally.

A graph of the kinetic energies of the electrons ejected from a metal versus the frequencies of the incident photons is a straight line, as you can see in **Figure 27–5.** All metals have similar graphs with the same slope. The slope of all of the lines is Planck's constant, h.

$$h = \frac{\Delta K}{\Delta f} = \frac{\text{change in maximum kinetic energy of ejected electrons}}{\text{change in frequency of incident radiation}}$$

The graphs of various metals differ only in the threshold frequency that is needed to free electrons. In **Figure 27–5,** the threshold frequency, f_0, is the point at which $K = 0$. In this case, f_0, located at the intersection of the curve with the x-axis, is approximately 4.4×10^{14} Hz. The threshold frequency is related to the energy needed to free the most weakly bound electron from a metal. This is called the **work function** of the metal. The work function is thus measured by hf_0. When a photon of frequency f_0 is incident on a metal, the energy of the photon is sufficient to release the electron, but not sufficient to provide the electron with any kinetic energy.

F.Y.I.

How are waves and particles alike? They both carry momentum and energy. Particles carry these quantities in their mass, waves in their fields.

Example Problem

Finding the Energy of Photoelectrons

Sodium has a threshold wavelength of 536 nm.

a. Find the work function of sodium in eV.

b. If ultraviolet radiation with a wavelength of 348 nm falls on sodium, what is the energy of the ejected electrons in eV?

Sketch the Problem

- In your drawing, include the positive cathode and negative anode.

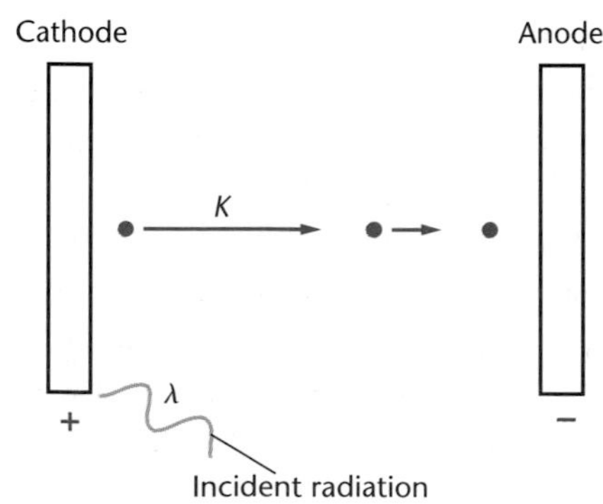

Calculate Your Answer

Known:	**Unknown:**
$\lambda_0 = 536$ nm	$W = ?$
$\lambda = 348$ nm	$K = ?$
$hc = 1240$ eV·nm	

Strategy:	**Calculations:**
a. Find the work function using Planck's constant and the threshold wavelength.	$W = hf_0 = hc/\lambda_0$ $= (1240 \text{ eV·nm})/(536 \text{ nm})$ $= 2.31$ eV
b. Use Einstein's photoelectric-effect equation to determine the energy of the ejected electron.	Photon energy $= hc/\lambda$ $= (1240 \text{ eV·nm})/(348 \text{ nm})$ $= 3.56$ eV
Subtract the work function to find the kinetic energy of the electrons.	$K = hf - hf_0 = hc/\lambda - hc/\lambda_0$ $= 3.56 \text{ eV} - 2.31 \text{ eV}$ $= 1.25$ eV

Check Your Answer

- Are the units correct? Performing algebra on the units verifies K in eV.
- Do the signs make sense? K should be positive.
- Are the magnitudes realistic? Energies should be a few electron volts.

Practice Problems

1. The stopping potential required to prevent current through a photocell is 3.2 V. Calculate the kinetic energy in joules of the photoelectrons as they are emitted.
2. The stopping potential for a photoelectric cell is 5.7 V. Calculate the kinetic energy of the emitted photoelectrons in eV.
3. The threshold wavelength of zinc is 310 nm.
   a. Find the threshold frequency of zinc.
   b. What is the work function in eV of zinc?
   c. Zinc in a photocell is irradiated by ultraviolet light of 240 nm wavelength. What is the kinetic energy of the photoelectrons in eV?
4. The work function for cesium is 1.96 eV.
   a. Find the threshold wavelength for cesium.
   b. What is the kinetic energy in eV of photoelectrons ejected when 425-nm violet light falls on the cesium?

F.Y.I.

A dictionary defines the word *quantum* as 1: a quantity or an amount and 2: the smallest amount of a physical quantity that can exist independently.

Physics Lab

Red Hot or Not?

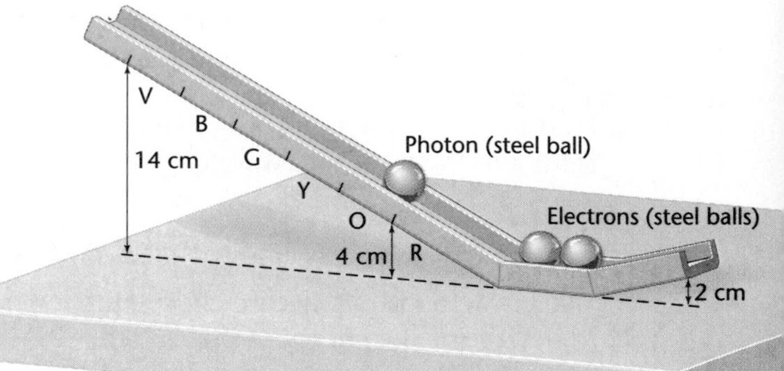

Problem

How well do steel balls simulate the photo-electric effect?

Materials

2-cm steel balls

grooved channel (U-channel or shelf bracket)

red, orange, yellow, green, blue, and violet marking pens or colored stickers

Procedure

1. Shape the grooved channel as shown in the diagram. Mark a point on the channel, 4 cm above the table, with R for red.
2. Mark a point on the channel, 14 cm above the table, with V for violet. Place marks for blue, green, yellow, and orange uniformly between R and V.
3. Place two steel balls at the lowest point on the channel. These steel balls represent valence electrons in the atom.
4. Place a steel ball on the channel at the red mark. This represents a photon of red light which has the lowest energy of the six colors of light being modeled.
5. Release the photon and see if the electrons are removed from the atom, that is, see if either steel ball escapes from the channel.
6. Remove the steel ball that represents the photon from the lower part of the channel.
7. Repeat steps 4–6 for each color's mark on the channel. **Note:** Always start with two electrons at the low point in the channel. Record your observations.

Data and Observations

1. Identify the photons by the color mark from which they were released. Which color of photons was able to remove the electrons?
2. Did one photon ever remove more than one electron? If so, what was its color?
3. Summarize your observations in terms of the energies of the photons.

Analyze and Conclude

1. **Making Predictions** Predict what would happen if two red photons could hit the electrons at the same time.
2. **Testing Predictions** Start two steel balls (photons) at the red mark on the channel and see what happens. Describe the results.
3. **Making Inferences** Some materials hold their valence electrons tighter than others. How could the model be modified to show this?

Apply

1. Photographers often have red lights in their darkrooms. Explain why they use red light but not blue light.

The Compton Effect

The photoelectric effect demonstrates that a photon, even though it has no mass, has kinetic energy just as a particle does. In 1916, Einstein predicted that the photon should have another particle property, momentum. He showed that the momentum of a photon should be hf/c. Because $f/c = 1/\lambda$, the photon's momentum is represented by the following equation.

$$p = \frac{hf}{c} = \frac{h}{\lambda}$$

Experiments done by an American physicist, Arthur Holly Compton, in 1922 tested Einstein's theory. Compton directed X rays of known wavelength at a graphite target, as shown in **Figure 27–6a,** and measured the wavelengths of the X rays scattered by the target. He found that some of the X rays were scattered without change in wavelength. Other scattered X rays, however, had a longer wavelength than the original radiation, as shown in **Figure 27–6b.** The wavelength of the first maximum corresponds to the wavelength of the original incident X rays; the maximum at longer wavelength results from the scattered X rays.

Recall that the energy of a photon is hf, and that $f = c/\lambda$. Thus, the energy of the photon is expressed by the following equation.

$$E = \frac{hc}{\lambda}$$

The equation tells you that the energy of a photon is inversely proportional to its wavelength. The increase in wavelength that Compton observed meant that the X-ray photons had lost both energy and momentum. The shift in the energy of scattered photons is called the **Compton effect.** It is a tiny shift, about 10^{-3} nm, and so is measurable

F.Y.I.

A photovoltaic cell can act like an on-off switch. There is current in a circuit when light of a given frequency falls upon the photovoltaic cell. When the light beam is broken, the current ceases.

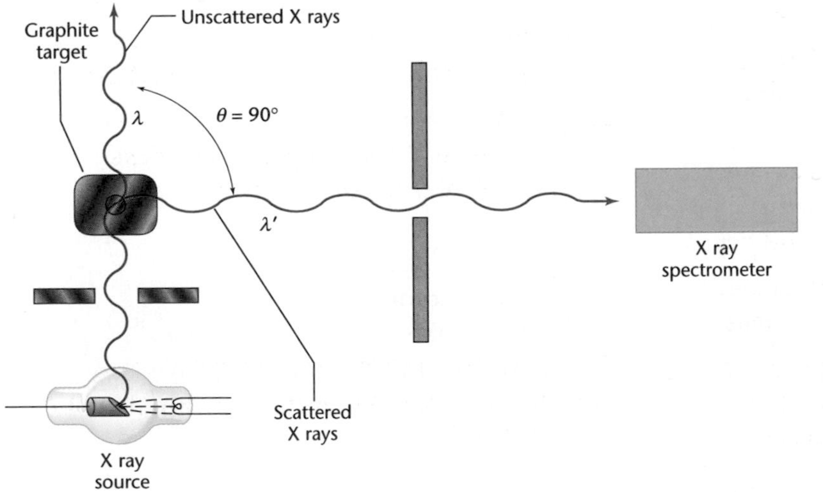

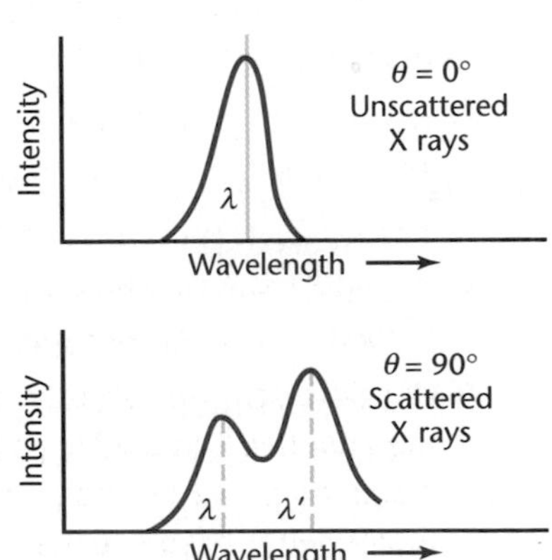

FIGURE 27–6 Compton used an apparatus similar to that shown in **(a)** to study the nature of photons. In **(b),** the increased wavelength of the scattered photons is evidence that the X-ray photons have lost energy.

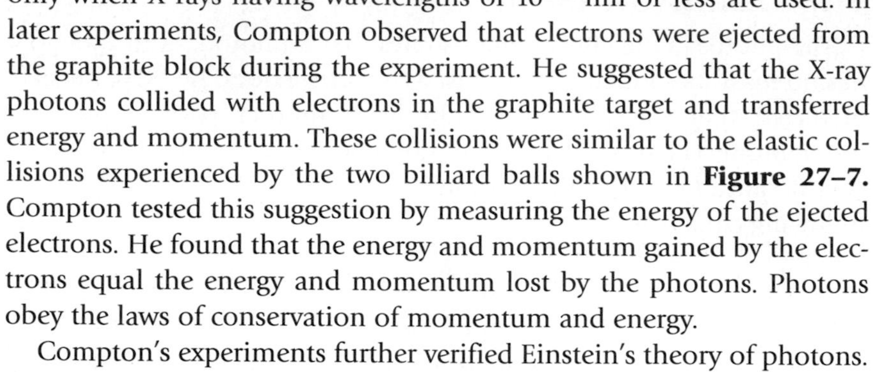

only when X rays having wavelengths of 10^{-2} nm or less are used. In later experiments, Compton observed that electrons were ejected from the graphite block during the experiment. He suggested that the X-ray photons collided with electrons in the graphite target and transferred energy and momentum. These collisions were similar to the elastic collisions experienced by the two billiard balls shown in **Figure 27–7.** Compton tested this suggestion by measuring the energy of the ejected electrons. He found that the energy and momentum gained by the electrons equal the energy and momentum lost by the photons. Photons obey the laws of conservation of momentum and energy.

Compton's experiments further verified Einstein's theory of photons. A photon is a particle that has energy and momentum. Unlike matter, however, a photon has no mass and travels at the speed of light.

FIGURE 27–7 When a photon strikes an electron, the energy and momentum gained by the electron equal the energy and momentum lost by the photon.

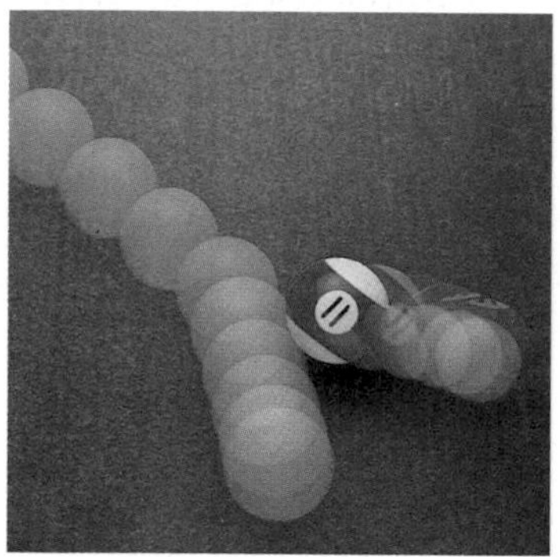

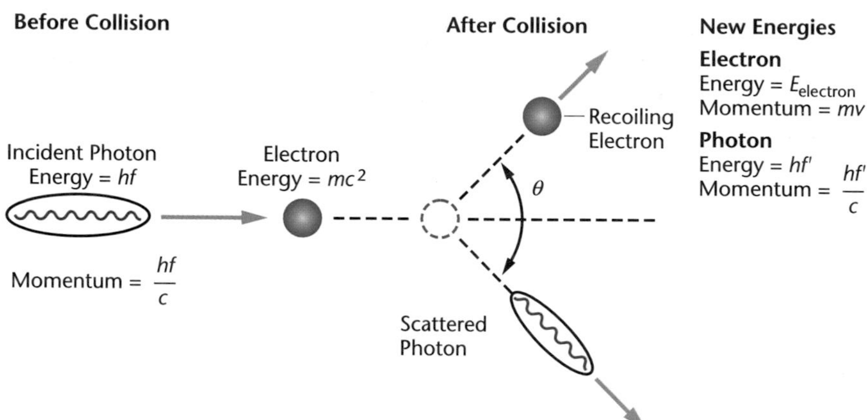

27.1 Section Review

1.1 Describe, in general, how the intensity of radiation from a hot body at a fixed temperature varies with frequency of radiation.

1.2 As the temperature of a body is increased, how does the frequency of peak intensity change? How does the total energy radiated change?

1.3 An experimenter sends an X ray into a target. An electron, but no other radiation, comes out. Was the event the photoelectric or Compton effect?

1.4 **Critical Thinking** In the Compton effect, the collision of two billiard balls serves as a model for the interaction of a photon and an electron. Suppose the electron were replaced by the much more massive proton. Would the proton gain as much energy from the collision as the electron did? Would the photon lose as much energy as it did colliding with the electron?

Particles Behave Like Waves 27.2

The photoelectric effect and Compton scattering showed that an electromagnetic wave has the properties of a particle. If a wave behaves like a particle, could a particle behave like a wave? French physicist Louis-Victor de Broglie (1892–1987) suggested in 1923 that material particles do have wave properties. This proposal was so extraordinary that it was ignored by other scientists until Einstein read de Broglie's papers and supported his ideas.

OBJECTIVES

- **Describe** evidence of the wave nature of matter and **solve** problems relating wavelength to particle momentum.
- **Recognize** the dual nature of both waves and particles and the importance of the Heisenberg uncertainty principle.

Matter Waves

Recall that the momentum of an object is equal to its mass times its velocity, $p = mv$. By analogy with the momentum of the photon, $p = h/\lambda$, de Broglie proposed that the momentum of a particle is represented by the following equation.

$$p = mv = \frac{h}{\lambda}$$

If you solve this equation for wavelength, you will find that the **de Broglie wavelength** of the particle is represented by

$$\lambda = \frac{h}{p} = \frac{h}{mv}.$$

According to de Broglie, particles such as electrons and protons should show wavelike properties. Effects such as diffraction and interference had never been observed for particles, so de Broglie's work was greeted with considerable doubt. However, in 1927, the results of two different experiments showed that electrons are diffracted just as light is. In one experiment, English physicist G. P. Thomson aimed a beam of electrons at a very thin crystal. The atoms in crystals are arrayed in a regular pattern that acts as a diffraction grating. Electrons diffracted from the crystal formed the same patterns that X rays of a similar wavelength formed. **Figure 27–8** shows the pattern made by diffracting electrons. The two experiments proved that material particles have wave properties.

The wave nature of objects of ordinary size is not observable because the wavelengths are extremely short. Consider the de Broglie wavelength of a 0.25-kg baseball when it leaves a bat with a speed of 21 m/s.

$$\lambda = \frac{h}{mv} = \frac{6.63 \times 10^{-34}\ \text{J·s}}{(0.25\ \text{kg})(20\ \text{m/s})} = 1.3 \times 10^{-34}\ \text{m}$$

The wavelength is far too small to have observable effects. However, you will see in the following example problem that an object as small as an electron has a wavelength that can be observed and measured.

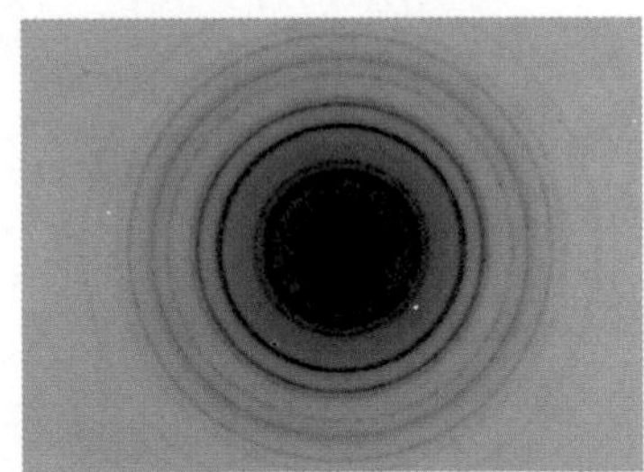

FIGURE 27–8 Electron diffraction patterns, such as this one for which a cubic zirconium crystal was used, demonstrate the wave properties of particles.

Example Problem

The de Broglie Wavelength of an Electron

An electron is accelerated by a potential difference of 75 V. What is its de Broglie wavelength?

Calculate Your Answer

Known:

$V = 75$ V

$q = 1.60 \times 10^{-19}$ C

$m = 9.11 \times 10^{-31}$ kg

$h = 6.63 \times 10^{-34}$ J·s

Unknown:

λ (de Broglie) = ?

Strategy:

Use the kinetic energy provided by the acceleration across a potential difference to find velocity.

Calculations:

$$K = qV, \text{ so } 1/2mv^2 = qV$$

$$v = \sqrt{\frac{2qV}{m}} = \sqrt{\frac{2(1.60 \times 10^{-19}\text{ C})(75\text{ V})}{9.11 \times 10^{-31}\text{ kg}}} = 5.1 \times 10^6\text{ m/s}$$

Find p.

$$p = mv = (9.11 \times 10^{-31}\text{ kg})(5.1 \times 10^6\text{ m/s}) = 4.6 \times 10^{-24}\text{ kg·m/s}$$

Use p to find the de Broglie wavelength.

$$\lambda = \frac{h}{p} = \frac{6.63 \times 10^{-34}\text{ J·s}}{4.6 \times 10^{-24}\text{ kg·m/s}} = 1.4 \times 10^{-10}\text{ m} = 0.14\text{ nm}$$

Check Your Answer

- Are the units correct? Performing algebra on the units verifies m/s for v and nm for λ.
- Do the signs make sense? Positive values are expected for both v and λ.
- Are the magnitudes realistic? λ is close to 0.1 nanometer, the spacing between the atoms in a solid.

Practice Problems

5. An electron is accelerated by a potential difference of 250 V.
   a. What is the speed of the electron?
   b. What is the de Broglie wavelength of this electron?
6. A 7.0-kg bowling ball rolls with a velocity of 8.5 m/s.
   a. What is the de Broglie wavelength of the bowling ball?
   b. Why does the bowling ball exhibit no observable wave behavior?
7. An X ray with a wavelength 5.0×10^{-12} m is traveling in a vacuum.
   a. Calculate the momentum associated with this X ray.
   b. Why does the X ray exhibit little particle behavior?

Particles and Waves

When you think of a particle, you think of mass, size, kinetic energy, and momentum. The properties of a wave, on the other hand, are frequency, wavelength, and amplitude. Is light a particle or a wave? Many physicists and philosophers have tried to work out an answer to this question. Most share a belief that the particle and wave aspects of light show complementary views of the true nature of light and must be taken together. Either model alone is incomplete.

How does quantum theory permit the scanning tunneling microscope to work? The STM works because electrons behave not only like particles, but also like waves. An STM has a fine needle-like tip positioned only a few nanometers above the surface to be studied. A potential difference is applied between the tip and surface. Because the tip and surface do not touch, you might think there would be no current. But electrons, acting like waves, can "tunnel" through the electric potential barrier created by the gap. The tiny current they create is very sensitive to the width of the gap. As the needle moves across the surface, the size of the current changes as the height of the surface changes. A computer creates an image of the shape of the surface.

Electron Corral

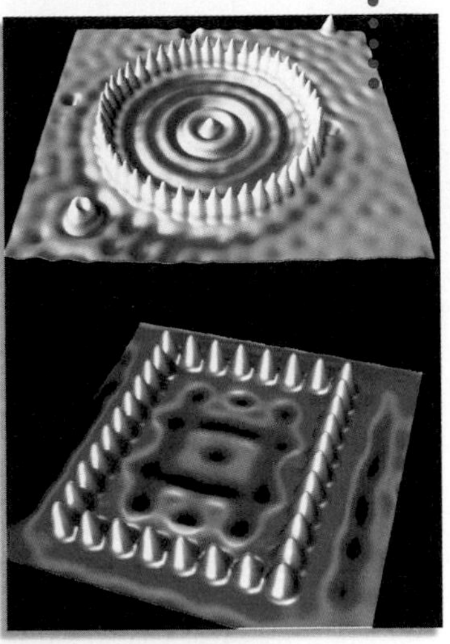

Physics & Technology

Say "Cheese"!

Look carefully at the photograph. Could you guess that this creature is a diatom that normally inhabits ocean waters? This breathtaking picture was taken by an electron microscope, a device that uses lenses, an electron beam, and often a computer to produce amazingly detailed images of objects that are as small as a few tenths of a nanometer.

A transmission electron microscope (TEM) uses electrons, electromagnetic lenses, and a vacuum to produce images that vary in contrast depending upon the heavy-metal stains used in specimen preparation. These metals prevent the electron beam from penetrating certain areas of the specimen while allowing electrons to flow freely through other sections, thereby creating contrast.

A scanning electron microscope, SEM, is identical to a TEM except that an SEM contains a deflection coil. This enables the electron beam to sweep across the sample and produce an image of electrons reflected from it.

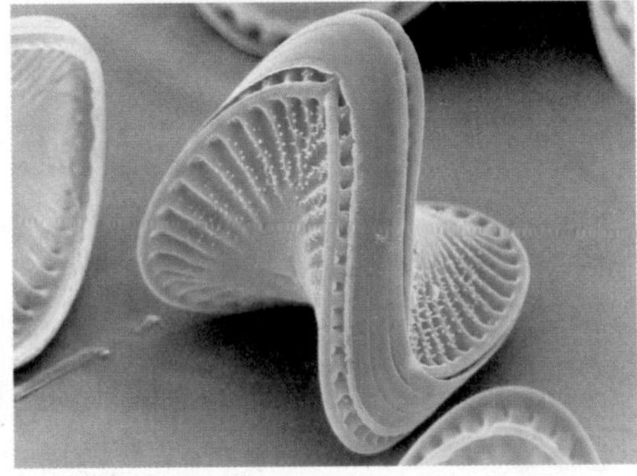

Thinking Critically Like the SEM, the scanning transmission electron microscope, (STEM), contains a deflection coil. What advantages might the STEM have over either of its counterparts for the viewing of biological specimens?

Determining location and momentum

Many physicists hold that the properties of an object can be defined only by devising an experiment that can measure them. One cannot simply say that a particle is at a certain location moving with a specific speed. Rather, an experiment must be described that will locate the particle and measure its speed.

How can you find the location of a particle? You must touch it or reflect light from it, as illustrated in **Figure 27-9**. Then the reflected light must be collected by an instrument or the human eye. Because of diffraction effects, light spreads out, making it impossible to locate the particle exactly. The spreading can be reduced by decreasing the wavelength of the light; the shorter the wavelength, the more precisely the location can be measured.

The Compton effect, however, means that when light of short wavelengths (high energy) strikes a particle, the momentum of the particle is changed. Therefore, the act of precisely measuring the location of a particle has the effect of changing the particle's momentum. The more precise the effort to measure the particle's location, the larger the likely change in its momentum.

In the same way, if the momentum of the particle is measured, the position of the particle changes. The conclusion is that it is impossible to measure precisely both the position and momentum of a particle at the same time. This fact is called the **Heisenberg uncertainty principle,** named for German physicist Werner Heisenberg. It is the result of the dual wave and particle description of light and matter.

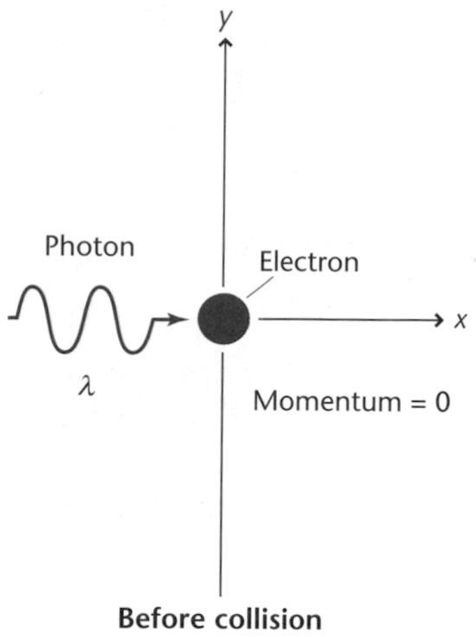

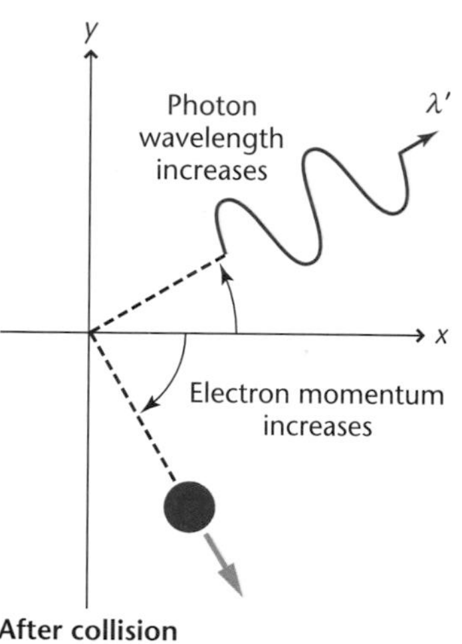

FIGURE 27–9 A particle can be seen only when light is scattered from it, but the scattering changes the momentum of the electron.

27.2 Section Review

2.1 If you want to increase the wavelength of a proton, you slow it down. What could you do to increase the wavelength of a photon?

2.2 If you try to find the location of a beam of light by having it pass through a narrow hole, why can you not tell much about the direction of the beam?

2.3 **Critical Thinking** Physicists recently made a diffraction grating of standing waves of light. They sent atoms through this grating and observed interference. If the spacing of the slits in the grating were half a wavelength, 250 nm, roughly what would you expect that the de Broglie wavelength of the atoms should be?

CHAPTER 27 REVIEW

Key Terms

27.1
- quantized
- photoelectric effect
- threshold frequency
- photon
- work function
- Compton effect

27.2
- de Broglie wavelength
- Heisenberg uncertainty principle

Summary

27.1 Waves Behave Like Particles

- Objects hot enough to be incandescent emit light because of the vibrations of the charged particles inside their atoms.
- The spectrum of incandescent objects covers a broad range of wavelengths. The spectrum depends upon the temperature of the incandescent objects.
- Planck explained the spectrum of an incandescent object by supposing that a particle can have only certain energies that are multiples of a constant now called Planck's constant.
- The photoelectric effect is the emission of electrons by certain metals when they are exposed to electromagnetic radiation.
- Einstein explained the photoelectric effect by postulating that light comes in bundles of energy called photons.
- The photoelectric effect allows the measurement of Planck's constant, h.
- The work function, the energy with which electrons are held inside metals, is measured by the threshold frequency in the photoelectric effect.

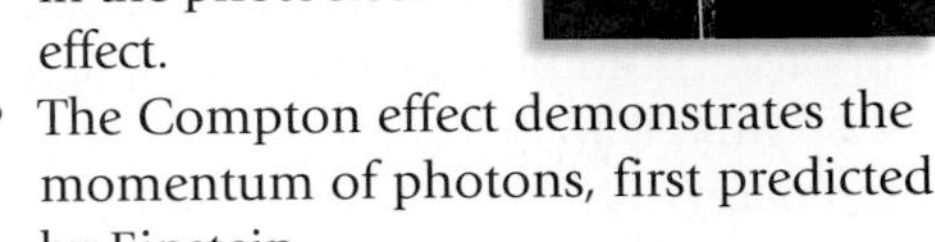

- The Compton effect demonstrates the momentum of photons, first predicted by Einstein.
- Photons, or light quanta, are massless and travel at the speed of light. Yet they have energy, hf, and momentum, $p = h/\lambda$.

27.2 Particles Behave Like Waves

- The wave nature of material particles was suggested by de Broglie and verified experimentally by diffracting electrons off crystals.
- The particle and wave aspects are complementary parts of the complete nature of both matter and light.
- The Heisenberg uncertainty principle states that it is not possible to measure precisely the position and momentum of a particle (light or matter) at the same time.

Reviewing Concepts

Section 27.1

1. Explain the concept of quantized energy.
2. In Max Planck's interpretation of the radiation of incandescent bodies, what is quantized?
3. What is a quantum of light called?
4. Light above the threshold frequency shines on the metal cathode in a photocell. How does Einstein's theory explain the fact that as the light intensity is increased, the current of photoelectrons increases?
5. Explain how Einstein's theory accounts for the fact that light below the threshold frequency of a metal produces no photoelectrons, regardless of the intensity of the light.
6. Certain types of black-and-white film are not sensitive to red light; they can be developed with a red safelight on. Explain this on the basis of the photon theory of light.
7. How does the Compton effect demonstrate that photons have momentum as well as energy?

Section 27.2

8. The momentum of a material particle is mv. Can you calculate the momentum of a photon using mv? Explain.

9. Describe how the following properties of the electron could be measured. Explain in each case what is to be done.
   a. charge
   b. mass
   c. wavelength

10. Describe how the following properties of a photon could be measured. Explain in each case what is to be done.
   a. energy
   b. momentum
   c. wavelength

Applying Concepts

11. What is the change in the intensity of red light given off by an incandescent body if the temperature increases from 4000 K to 8000 K?

12. Two iron rods are held in a fire. One glows dark red while the other glows bright orange.
   a. Which rod is hotter?
   b. Which rod is radiating more energy?

13. Will high-frequency light eject a greater number of electrons from a photosensitive surface than low-frequency light, assuming that both frequencies are above the threshold frequency?

14. Potassium in a photocell emits photoelectrons when struck by blue light. Tungsten emits them only when ultraviolet light is used.
   a. Which metal has a higher threshold frequency?
   b. Which metal has a larger work function?

15. Compare the de Broglie wavelength of a baseball moving 21 m/s with the size of the baseball.

Problems

Section 27.1

LEVEL 1

16. The stopping potential of a certain metal is 5.0 V. What is the maximum kinetic energy of the photoelectrons in
   a. electron volts?
   b. joules?

17. What potential difference is needed to stop electrons having a maximum kinetic energy of 4.8×10^{-19} J?

18. The threshold frequency of sodium is 4.4×10^{14} Hz. How much work must be done to free an electron from the surface of sodium?

19. If light with a frequency of 1.00×10^{15} Hz falls on the sodium in the previous problem, what is the maximum kinetic energy of the photoelectrons?

20. Barium has a work function of 2.48 eV. What is the longest wavelength of light that will cause electrons to be emitted from barium?

21. A photocell is used by a photographer to measure the light falling on the subject to be photographed. What should be the work function of the cathode if the photocell is to be sensitive to red light ($\lambda = 680$ nm) as well as the other colors?

22. The threshold frequency of tin is 1.2×10^{15} Hz.
   a. What is the threshold wavelength?
   b. What is the work function of tin?
   c. Electromagnetic radiation with a wavelength of 167 nm falls on tin. What is the kinetic energy of the ejected electrons in eV?

23. What is the momentum of a photon of yellow light whose wavelength is 600 nm?

LEVEL 2

24. A home uses about 4×10^{11} J of energy each year. In many parts of the United States, there are about 3000 h of sunlight each year.
   a. How much energy from the sun falls on one square meter each year?
   b. If this solar energy can be converted to useful energy with an efficiency of 20 percent, how large an area of converters would produce the energy needed by the home?

25. The work function of iron is 4.7 eV.
   a. What is the threshold wavelength of iron?
   b. Iron is exposed to radiation of wavelength 150 nm. What is the maximum kinetic energy of the ejected electrons in eV?

26. Suppose a 5.0-g object, such as a nickel, vibrates while connected to a spring. Its maximum velocity is 1.0 cm/s.
   a. Find the maximum kinetic energy of the vibrating object.
   b. The object emits energy in the form of light

of frequency 5.0×10^{14} Hz and its energy is reduced by one step. Find the energy lost by the object.

c. How many step reductions would this object have to make to lose all its energy?

Section 27.2

LEVEL 1

27. Find the de Broglie wavelength of a deuteron of mass 3.3×10^{-27} kg that moves with a speed of 2.5×10^4 m/s.

28. An electron is accelerated across a potential difference of 54 V.

a. Find the maximum velocity of the electron.

b. Calculate the de Broglie wavelength of the electron.

29. A neutron is held in a trap with a kinetic energy of only 0.025 eV.

a. What is the velocity of the neutron?

b. Find the de Broglie wavelength of the neutron.

30. The kinetic energy of the hydrogen atom electron is 13.65 eV.

a. Find the velocity of the electron.

b. Calculate its de Broglie wavelength.

c. Compare your answer with the radius of the hydrogen atom, 5.19 nm.

LEVEL 2

31. An electron has a de Broglie wavelength of 400 nm, the shortest wavelength of visible light.

a. Find the velocity of the electron.

b. Calculate the energy of the electron in eV.

32. An electron microscope is useful because the de Broglie wavelength of electrons can be made smaller than the wavelength of visible light. What energy in eV has to be given to an electron for it to have a de Broglie wavelength of 20 nm?

33. An electron has a de Broglie wavelength of 0.18 nm.

a. How large a potential difference did it experience if it started from rest?

b. If a proton has a de Broglie wavelength of 0.18 nm, how large is the potential difference it experienced if it started from rest?

Critical Thinking Problems

34. A HeNe laser emits photons with a wavelength of 632.8 nm.

a. Find the energy, in joules, of each photon.

b. A typical small laser has a power of 0.5 mW = 5×10^{-4} J/s. How many photons are emitted each second by the laser?

35. The intensity of a light that is just barely visible is 1.5×10^{-11} W/m^2.

a. If this light shines into your eye, passing through the pupil with a diameter of 7.0 mm, what is the power, in watts, that enters your eye?

b. If the light has a wavelength of 550 nm, how many photons per second enter your eye?

Going Further

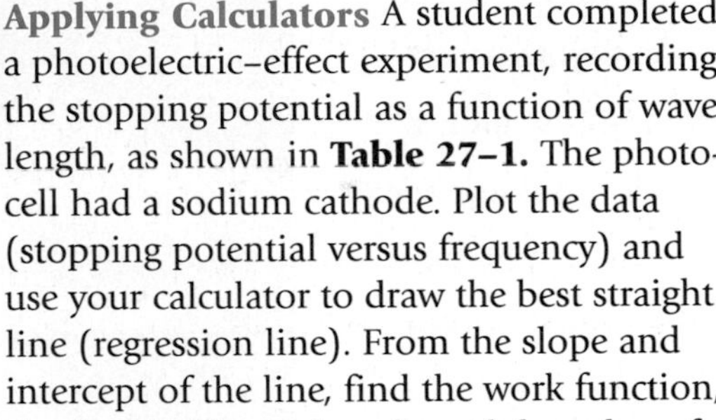

Applying Calculators A student completed a photoelectric–effect experiment, recording the stopping potential as a function of wavelength, as shown in **Table 27–1.** The photocell had a sodium cathode. Plot the data (stopping potential versus frequency) and use your calculator to draw the best straight line (regression line). From the slope and intercept of the line, find the work function, the threshold wavelength, and the value of h/q from this experiment. Compare the value of h/q to the accepted value.

TABLE 27–1

Stopping Potential Versus Wavelength

λ(nm)	V_0(eV)
200	4.20
300	2.06
400	1.05
500	0.41
600	0.03

*inter***NET** CONNECTION

Follow the link on the Glencoe Homepage at **www.glencoe.com/sec/science** to find out more about this chapter.

Atomic Secrets

The street lamps, the building lights, and the car headlights were photographed with a diffraction grating over the camera lens. Why are the images of the lights different, and how could these differences be used to identify the types of lights?

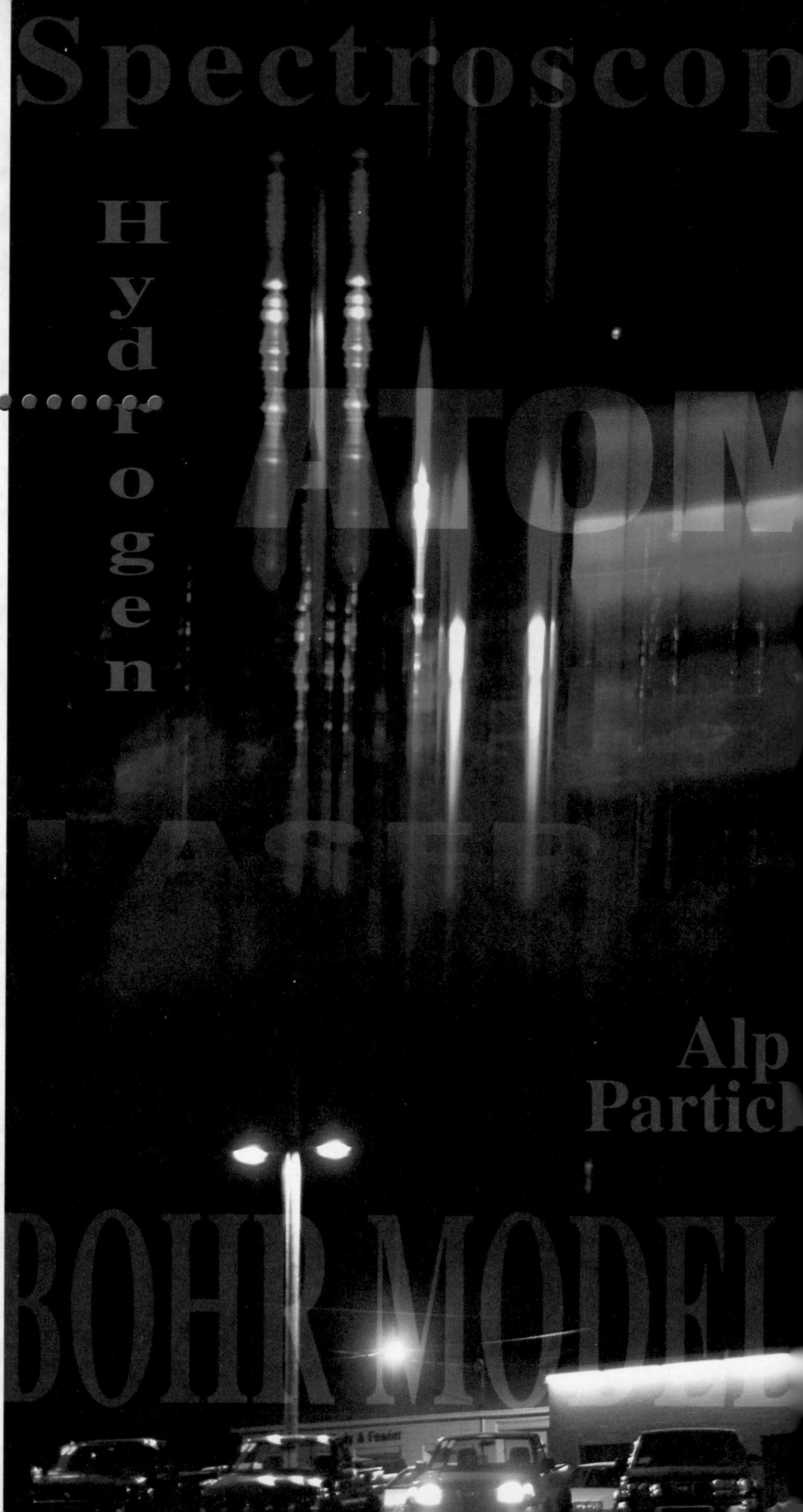

CHAPTER 28 The Atom

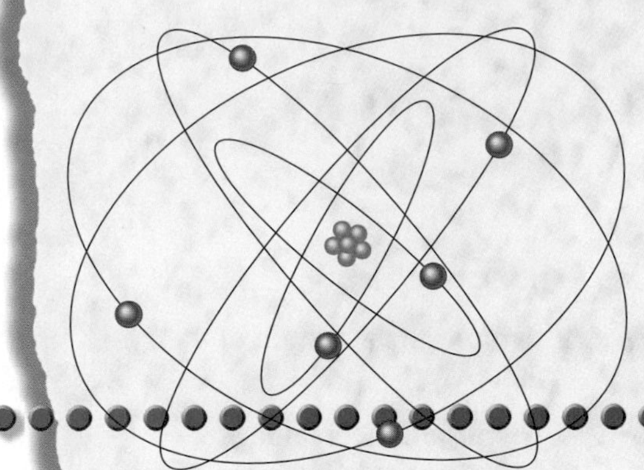

As you have learned, both prisms and diffraction gratings separate light into its component colors. But do all sources of light have the same spectrum, that is, the same mix of colors? Look at the light sources in the photograph. You can see headlights, street lamps, and building lights. The spectra were produced by placing a diffraction grating over the camera lens. Note the differences among the spectra. In some, every color is present; in others, there are brighter bands at specific colors.

Scientists have studied spectra like these to identify the sources of light for more than 100 years. Because the materials that make up these sources are all different, scientists have determined that different spectra can be obtained from every type of atom. This information has been used to determine the number and kind of atoms present in the sun and in distant stars. Scientists also have used spectral analysis to characterize matter and investigate impurities in manufactured items, such as molten iron and glass.

When the study of spectra began, the structure of atoms was unknown. It was the study of the spectrum produced by the simplest atom, hydrogen, that led to discovery of the structure of not only hydrogen, but also all other atoms.

WHAT YOU'LL LEARN

- You will examine the composition of the atom.
- You will determine specific energies of electrons.
- You will explore the probability of finding the electrons at specific locations or with specific momenta.
- You will study lasers.

WHY IT'S IMPORTANT

- The quantum model of the atom and the transition of electrons between orbitals are responsible for the ability to produce artificial lighting, determine the composition of materials, operate electronic equipment, and produce special effects for movies.

interNET CONNECTION

Follow the link for this chapter on the Glencoe Homepage at **www.glencoe.com/sec/science** to find out more about the atom.

28.1 The Bohr Model of the Atom

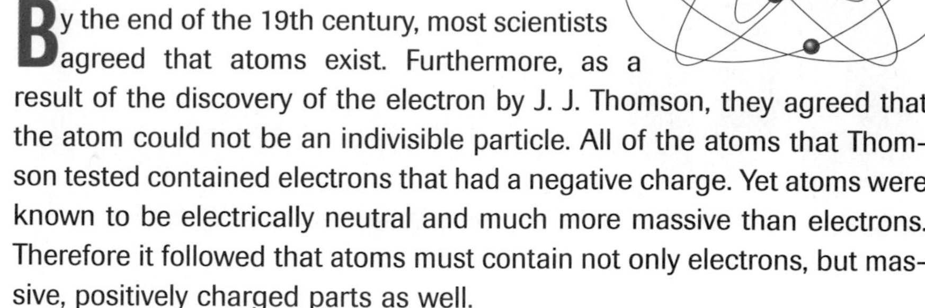

OBJECTIVES

- **Explain** the structure of the atom.
- **Distinguish** continuous spectra from line spectra.
- **Contrast** emission and absorption spectra.
- **Solve** problems using the orbital radius and energy-level equations.

By the end of the 19th century, most scientists agreed that atoms exist. Furthermore, as a result of the discovery of the electron by J. J. Thomson, they agreed that the atom could not be an indivisible particle. All of the atoms that Thomson tested contained electrons that had a negative charge. Yet atoms were known to be electrically neutral and much more massive than electrons. Therefore it followed that atoms must contain not only electrons, but massive, positively charged parts as well.

The Nuclear Model

Discovering the nature of the massive part of the atom and the arrangement of the electrons was a major challenge. Physicists and chemists from many countries both cooperated and competed in searching for the solution to this puzzle. The result provided not only knowledge of the structure of the atom, but also a totally new approach to both physics and chemistry. The history of this work is one of the most exciting stories of the twentieth century.

J. J. Thomson believed that a massive, positively charged substance filled the atom. He pictured the electrons arranged within this substance like raisins in a muffin. However, Ernest Rutherford, who was working in England at the same time, performed a series of brilliant experiments showing that the atom had a very different structure.

Compounds containing uranium had been found to emit penetrating rays. Some of these emissions were found to be massive, positively charged particles moving at high speed. These were later named **alpha (α) particles.** The α particles could be detected by a screen coated with zinc sulfide that emitted a small flash of light, or **scintillation,** each time an α particle hit it, as shown in **Figure 28–1.**

F.Y.I.

J. J. Thomson, the discoverer of the electron, first called these subatomic particles "corpuscles."

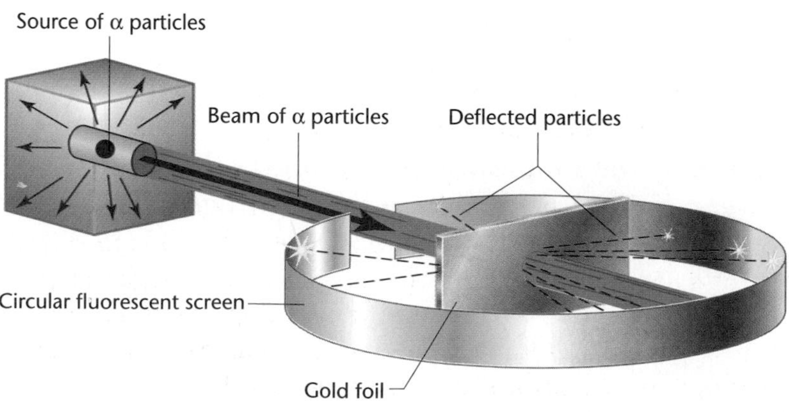

FIGURE 28–1 After bombarding metal foil with alpha particles, Rutherford's team concluded that most of the mass of the atom was concentrated in the nucleus.

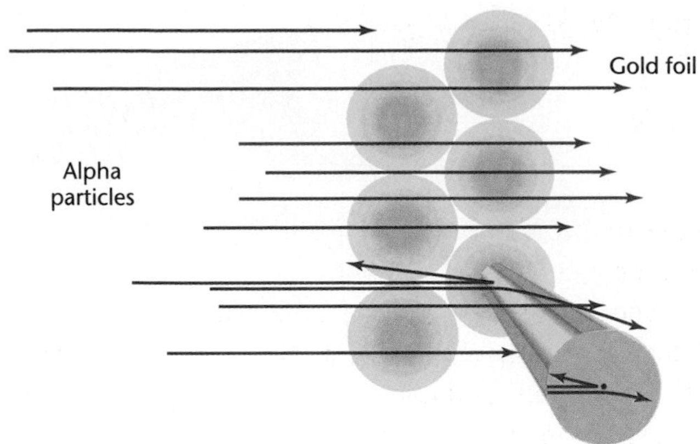

FIGURE 28–2 Most of the alpha particles that Rutherford directed at a thin sheet of gold foil went through it without deflection. A few, however, were deflected at large angles.

Rutherford directed a beam of α particles at a thin sheet of gold foil only a few atoms thick. He observed that while most of the α particles passed through the sheet, the beam was spread slightly by the foil. Detailed studies of the deflection of α particles showed that a few of the particles were deflected at large angles, even larger than 90°. A diagram of this is shown in **Figure 28–2.** Rutherford was amazed. He had assumed that the mass was spread more or less evenly throughout the atom. He compared his amazement to that of firing a 15-inch cannon shell at tissue paper and then having the shell bounce back and hit him.

Using Coulomb's force law and Newton's laws of motion, Rutherford concluded that the results could be explained only if all the positive charge of the atom were concentrated in a tiny, massive central core, now called the nucleus. Rutherford's model is therefore called the **nuclear model** of the atom. All the positive charge and more than 99.9 percent of the mass of the atom are in its nucleus. Electrons are outside and far away from the nucleus, and they do not contribute a significant amount of mass. The diameter of the atom is 10 000 times larger than the diameter of the nucleus; thus, the atom is mostly empty space. This discovery accounted for the previous observations that almost all of the α particles had passed through the gold foil, and that some of the particles had been deflected at large angles.

Atomic spectra

How are the electrons arranged around the nucleus of the atom? One of the clues scientists used to answer this question came from studying the light emitted by atoms. The set of light wavelengths emitted by an atom is called the atom's **emission spectrum.**

Atoms of gases can be made to emit their characteristic colors by using a gas discharge tube apparatus. A glass tube containing a gas at low pressure has metal electrodes at each end. When a high voltage of electricity is applied across the tube, electrons pass through the gas. The electrons collide with the gas atoms, transferring energy to them. The atoms then give up this extra energy, emitting it in the form of light. The light emitted by mercury is shown in **Figure 28–3.**

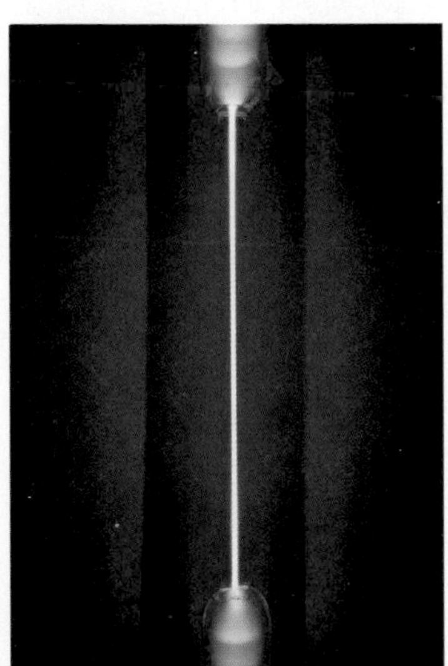

FIGURE 28–3 In a gas discharge tube apparatus, mercury gas glows when high voltage is applied.

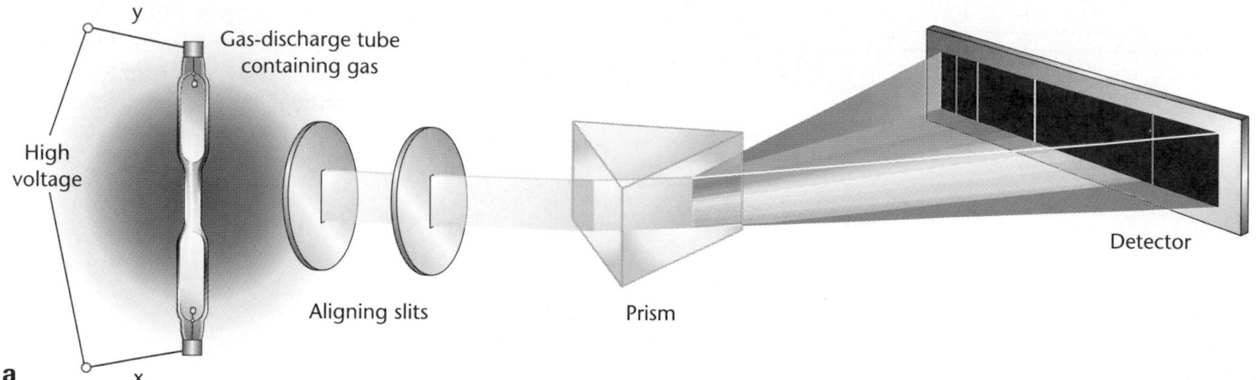

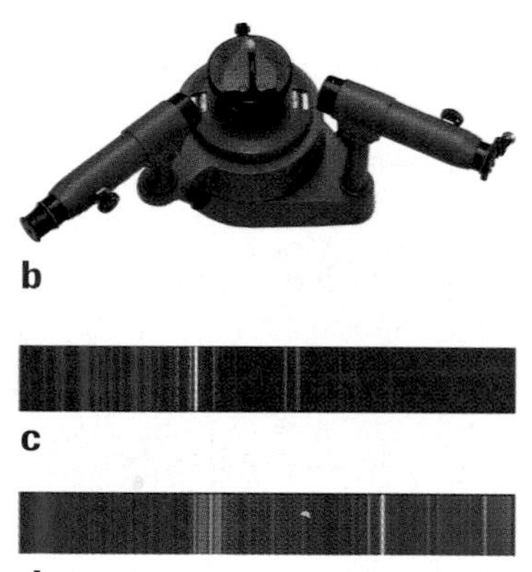

FIGURE 28–4 A prism spectroscope can be used to observe emission spectra **(a, b).** The emission spectra of neon **(c)** and molecular hydrogen **(d)** show characteristic lines.

The emission spectrum of an atom can be seen by looking at the light through a prism or a diffraction grating or by putting such a grating in front of a camera lens, as was done in the opening photo. The spectrum can be studied in greater detail using the instrument diagrammed in **Figure 28–4a** and pictured in **Figure 28–4b.** In this **spectroscope,** the light passes through a slit and is then dispersed as it travels through a prism or diffraction grating. A lens system collects the dispersed light for viewing through a telescope or for recording on a photographic plate. Each wavelength of light forms an image of the slit at different positions. The spectrum of an incandescent solid is a continuous band of colors from red through violet. The spectrum of a gas, however, is a series of lines of different colors. Each line corresponds to a particular wavelength of light emitted by the atoms of the gas. Suppose an unidentified gas, perhaps neon or hydrogen, is contained in a tube. When it is excited, the gas will emit light at wavelengths characteristic of the atoms of that gas. Thus, the gas can be identified by comparing its wavelengths with the lines present in the spectrum of a known sample. The emission spectra for neon and hydrogen are shown in **Figure 28–4c,** and **d,** respectively.

How can the differences in these emissions be used to identify the atoms involved? When the emission spectrum of a combination of elements is photographed, analysis of the lines on the photograph can indicate the identities and the relative concentrations of the elements present. If the material being examined contains a large amount of any particular element, the lines for that element are more intense on the photograph. Through comparison of the intensities of the lines, the percentage composition of the material can be determined. Thus, an emission spectrum is a useful analytic tool.

FIGURE 28-5 This apparatus is used to produce the absorption spectrum of sodium.

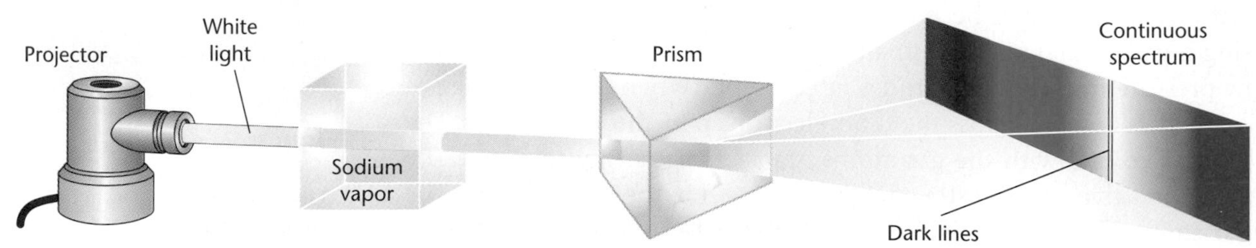

FIGURE 28–6 The emission spectrum **(a)** and the absorption spectrum **(b)** of sodium are pictured.

A gas that is cool and does not emit light will absorb light at characteristic wavelengths. This set of wavelengths is called an **absorption spectrum.** To obtain an absorption spectrum, white light is sent through a sample of gas and then through a spectroscope, as shown in **Figure 28–5.** The normally continuous spectrum of the white light then has dark lines in it. These lines show that light of some wavelengths has been absorbed. Often, the bright lines of the emission spectrum and the dark lines of the absorption spectrum of a gas occur at the same wavelengths. Thus, cool gaseous elements absorb the same wavelengths that they emit when excited, as shown in **Figure 28–6** for the emission spectrum and the absorption spectrum of sodium. Analysis of the wavelengths of the dark lines in an absorption spectrum also can be used to indicate the composition of the gas.

In 1814, while examining the spectrum of sunlight, Josef von Fraunhofer noticed some dark lines. These dark lines, now called Fraunhofer lines, are shown in **Figure 28–7.** To account for these lines, Fraunhofer assumed that the sun has a relatively cool atmosphere of gaseous elements. He reasoned that as light leaves the sun, it passes through these gases, which absorb light at their characteristic wavelengths. As a result, these wavelengths are missing from the sun's absorption spectrum. Through a comparison of the missing lines with the known lines of the various elements, the composition of the atmosphere of the sun was determined. In this manner, the element helium was discovered in the sun before it was found on Earth. Spectrographic analysis also has made it possible to determine the composition of stars.

Both emission and absorption spectra are valuable scientific tools. As a result of the elements' characteristic spectra, chemists are able to analyze , identify, and quantify unknown materials by observing the spectra they emit or absorb. The emission and absorption spectra of elements are important in industry as well as in scientific research. For example, steel mills reprocess large quantities of scrap iron of varying compositions. The exact composition of a sample of scrap iron can be determined in minutes by spectrographic analysis. The composition of the steel can then be adjusted to suit commercial specifications. Aluminum, zinc, and other metal processing plants employ the same method.

Atomic Secrets

ASTRONOMY CONNECTION

FIGURE 28–7 Fraunhofer lines appear in the absorption spectrum of the sun.

FIGURE 28–8 The emission spectrum of hydrogen in the visible range has four lines.

The study of spectra is a branch of science known as **spectroscopy.** Spectroscopists are employed throughout research and industrial communities. Spectroscopy has proven an effective tool to analyze materials on Earth, and it is the only currently available tool to study the composition of stars over the vast expanse of space.

The Bohr Model of the Atom

In the nineteenth century, many physicists tried to use atomic spectra to determine the structure of the atom. Hydrogen was studied extensively because it is the lightest element and has the simplest spectrum. The visible spectrum of hydrogen consists of four lines: red, green, blue, and violet, as shown in **Figure 28–8.**

Any theory that explained the structure of the atom would have to account for these wavelengths, and it would also have to fit Rutherford's nuclear model. Rutherford had suggested that electrons orbited the nucleus much as the planets orbit the sun. There was, however, a serious problem with this planetary model.

An electron in an orbit is constantly accelerated toward the nucleus. As you learned in Chapter 26, electrons that have been accelerated will radiate energy by emitting electromagnetic waves. At the rate that an orbiting electron would lose energy, it would spiral into the nucleus in only 10^{-9} second. But, atoms are known to be stable. Thus, the planetary model was not consistent with the laws of electromagnetism. In addition, if the planetary theory were true, the accelerated electron should radiate energy at all wavelengths. But, as you have seen, the light emitted by atoms is radiated only at specific wavelengths.

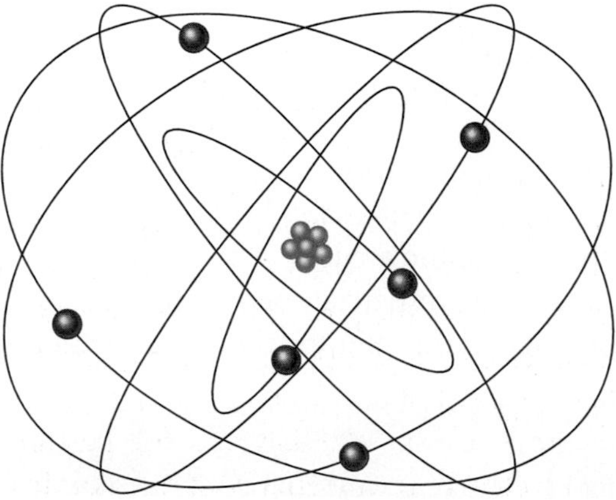

FIGURE 28–9 Bohr's planetary model of the atom postulated that electrons moved in fixed orbits around the nucleus.

Danish physicist Niels Bohr went to England in 1911 and soon joined Rutherford's group to work on determining the structure of the atom. He tried to unite the nuclear model with Einstein's theory of light. This was a courageous idea, because in 1911 Einstein's revolutionary theory of the photoelectric effect had not yet been confirmed by experiments and was not widely accepted. Recall from Chapter 27 that according to Einstein, light and other forms of electromagnetic radiation consist of discrete bundles of energy called photons. That is, light seems to act like a stream of tiny particles.

F.Y.I.

The proton was observed in 1919 as a particle that is emitted when the nucleus of an atom is bombarded by alpha particles.

Bohr energy is quantized

Bohr started with the planetary arrangement of electrons, diagrammed in **Figure 28–9,** but he made the bold hypothesis that the laws of electromagnetism do not operate inside atoms. He hypothesized that an electron in a stable orbit does not radiate energy, even though it is accelerating.

If energy was not radiated when electrons were in stable orbits but elements could emit a characteristic energy spectrum, when was the energy radiated? Bohr suggested that light is emitted when the electron's energy changes, as shown in **Figure 28–10.** According to Einstein, the energy of a photon of light is represented by the equation $E = hf = hc/\lambda$. Bohr reasoned that if the emission spectrum contains only certain wavelengths, then an electron can emit or absorb only specific amounts of energy. Therefore, the atomic electrons can have only certain amounts of energy. Recall from Chapter 27 that when energy is found only in certain amounts, it is said to be quantized.

The quantization of energy in atoms is unlike everyday experience. For example, if the energy of a pendulum were quantized, it could only oscillate with certain amplitudes, such as 10 cm or 20 cm, but not for example, 11.3 cm. Electrons in an atom have different quantized amounts of energy that are called **energy levels.** When an electron has the smallest allowable amount of energy, it is in the lowest energy level, called the **ground state.** If an electron absorbs energy, it can make a transition to a higher energy level, called an **excited state.** Atomic electrons usually remain in excited states for only a few billionths of a second before returning to the ground state and emitting energy.

The energy of an orbiting electron in an atom is the sum of the kinetic energy of the electron and the potential energy resulting from the attractive force between the electron and the nucleus. The energy of an electron in an orbit near the nucleus is less than that of an electron in an orbit farther away because work must be done to move an electron to orbits farther away from the nucleus. The electrons in excited states have larger orbits and correspondingly higher energies. The model of an atom having a central nucleus with its electrons occupying specific quantized energy levels is known as the **Bohr model** of the atom.

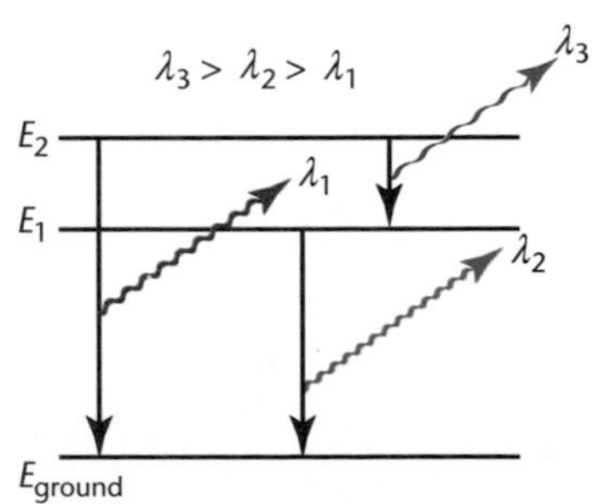

FIGURE 28–10 The energy of the emitted photon is equal to the difference in energy between two energy levels.

Einstein's theory holds that the light photon has an energy hf. Bohr postulated that the change in the energy of an atomic electron when a

Nuclear Bouncing

Place a 9″ aluminum pie pan on a table. Gently press four glass marbles (protons) into the pie pan so that they sit in small indentations near the center of the pan. Roll a 12-mm steel ball (alpha particle) down a grooved ruler to see if you can hit the marbles. Then remove the marbles and put the steel balls into the indentations (each steel ball represents a nucleus) in the pie pan and roll a marble (alpha particle) down the grooved ruler.

Analyze and Conclude When you roll the steel ball at the marbles, does it change its path? Does the steel ball ever bounce back? When you switch the balls and marbles, how are the results different? Why are the results different? Hypothesize what will happen when an alpha particle hits a proton.

photon is absorbed is equal to the energy of that photon. When the electron makes the return transition to the ground state, a photon is emitted. The energy of the photon is equal to the energy difference between the excited and ground states as defined by the following relationship.

$$hf = E_{excited} - E_{ground}$$

Molecules have additional discrete energy levels. For example, they can rotate and vibrate, which individual atoms cannot do. As a result, molecules can emit a much wider variety of light frequencies than can individual atoms.

Predictions of the Bohr Model

A scientific theory must do more than present postulates; it must allow predictions to be made that can be checked against experimental data. A good theory also can be applied to many different problems, and it ultimately provides a simple, unified explanation of some part of the physical world.

Bohr was able to calculate the wavelengths of light emitted by hydrogen. The calculations were in excellent agreement with the values measured by other scientists. As a result, Bohr's model was widely accepted. Unfortunately, the model could not predict the spectrum of the next simplest element, helium. In addition, there was no reason to suggest that the laws of electromagnetism should work everywhere but inside the atom. Not even Bohr believed that his model was a complete theory of the structure of the atom.

Despite its shortcomings, the Bohr model describes the energy levels and wavelengths of light emitted and absorbed by hydrogen atoms remarkably well. You can use this model to calculate the wavelengths of light emitted by an atom. Bohr's calculations start with Newton's law, $F = ma$, applied to an electron of mass m and charge $-q$ in a circular orbit of radius r about a massive particle, a proton, of charge q. Remember that $F = Kq^2/r^2$ and that ma can be represented as mv^2/r. Thus, the following is true.

$$F = ma$$

$$\frac{Kq^2}{r^2} = \frac{mv^2}{r}$$

Here, K is the constant, $9.0 \times 10^9\ \text{N}\cdot\text{m}^2/\text{C}^2$, from Coulomb's law.

FIGURE 28–11 Ernest Rutherford and Niels Bohr devised the planetary model of the atom.

Bohr proposed that the angular momentum, which is the product of the momentum of the electron and the radius of its circular orbit, mvr, can have only certain values. These values are represented by the equation $mvr = nh/2\pi$. Here, h is Planck's constant and n is an integer. Because the quantity mvr can have only certain values, the angular momentum is said to be quantized.

Because $Kq^2/r^2 = mv^2/r$, and $v = nh/2m\pi r$, Bohr substituted for v and predicted the radii of the orbits of the electrons in the hydrogen atom by using the following equation.

$$r_n = \frac{h^2n^2}{4\pi^2Kmq^2}$$

By substituting SI values for the quantities into the equation, you can calculate the radius of the innermost orbit of the hydrogen atom, where $n = 1$, by using the following relationship.

$$r_n = \frac{(6.626 \times 10^{-34}\ \text{J·s})^2\,(1)^2}{4\pi^2(9.00 \times 10^9\ \text{N·m}^2/\text{C}^2)(9.11 \times 10^{-31}\ \text{kg})(1.60 \times 10^{-19}\ \text{C})^2}$$

$$= 5.30 \times 10^{-11}\ \text{J}^2\text{·s}^2/\text{N·m}^2\text{·kg}$$

$$= 5.30 \times 10^{-11}\ \text{m, or } 0.053\ \text{nm}$$

A little more algebra shows that the total energy of the electron in its orbit, which is the sum of the potential and kinetic energies of the electron and is defined by $-1/2\ Kq^2/r$, is represented by the following equation.

$$E_n = \frac{-2\pi^2K^2mq^4}{h^2} \times \frac{1}{n^2}$$

By substituting numerical values for the constants, you can calculate the energy of the electron in joules, which yields the following equation.

$$E_n = -2.17 \times 10^{-18}\ \text{J} \times \frac{1}{n^2}$$

This can be written in electron volts by the following equation.

$$E_n = -13.6\ \text{eV} \times \frac{1}{n^2}$$

Both the radius of an orbit and the energy of the electron can have only certain values. That is, both are quantized. The integer n is called the principal **quantum number.** The number n determines the values of r and E. **Figure 28–12** shows that the radius increases as the square of n. The energy depends on $1/n^2$. The energy is negative to indicate the amount of energy that must to be added to the electron in the energy level to free it from the attractive force of the nucleus, that is, to ionize the atom. When an electron moves from a lower energy level to a higher energy level, it gains energy. The value of the energy difference is positive. The energy levels of hydrogen are shown in **Figure 28–13.**

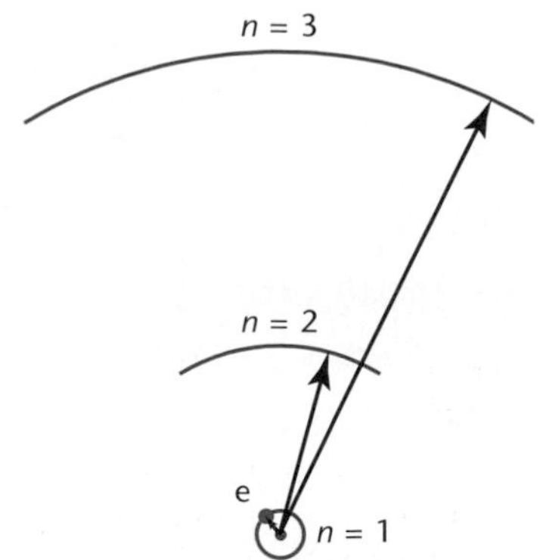

FIGURE 28–12 Radii of electron orbits for the first three energy levels of hydrogen according to the Bohr model.

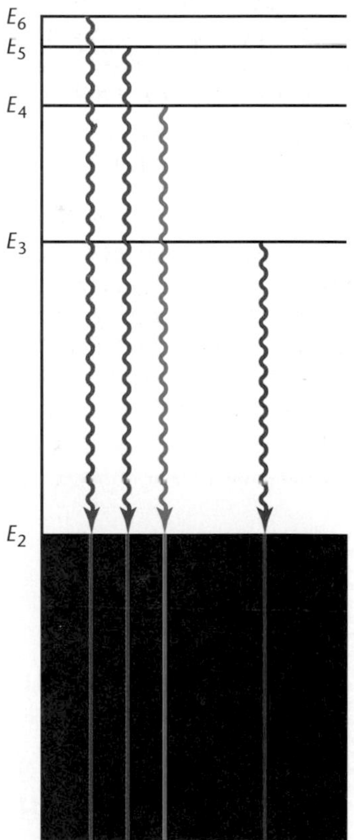

FIGURE 28–13 Bohr's model of the hydrogen atom showed that a definite amount of energy is released when an electron moves from a higher to a lower energy level. The energy released in each transition corresponds to a definite line in the hydrogen spectrum.

The light emitted by hydrogen when the atom drops into its ground state from any excited state is in the ultraviolet range. The four visible lines in the hydrogen spectrum are all produced when the atom drops from the $n = 3$ or higher state into the $n = 2$ state.

Example Problem

Orbital Energy of Electrons in the Hydrogen Atom

For the hydrogen atom, determine the energy of the innermost energy level ($n = 1$), the energy of the second energy level ($n = 2$), and the energy difference between the first and second energy levels.

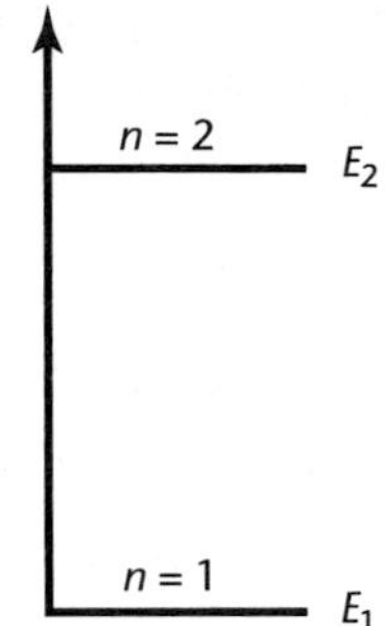

Sketch the Problem

- Diagram two energy levels 1 and 2. Indicate increasing energy.

Calculate Your Answer

Known:

innermost energy level, $n = 1$

next energy level, $n = 2$

Unknown:

energy of level 1, $E_1 = ?$

energy of level 2, $E_2 = ?$

difference in energy, $\Delta E = ?$

Strategy:

Use $E_n = -13.6 \text{ eV} \times 1/n^2$ for each energy level and calculate the difference between the energy levels.

Calculations:

$$E_n = -13.6 \text{ eV} \times \frac{1}{n^2}$$

$$n = 1, E_1 = -13.6 \text{ eV} \times \frac{1}{1^2} = -13.6 \text{ eV}$$

$$n = 2, E_2 = -13.6 \text{ eV} \times \frac{1}{2^2} = -3.40 \text{ eV}$$

$$\Delta E = E_2 - E_1$$

$$= -3.40 \text{ eV} - (-13.6 \text{ eV})$$

$$= 10.2 \text{ eV} = 10.2 \text{ eV of energy is absorbed.}$$

Check Your Answer

- Are the units correct? Energy values are in electron volts.
- Is the sign correct? The energy difference is positive when electrons move from lower energy levels to higher energy levels.
- Is the magnitude realistic? The energy needed to move an electron from the first energy level to the second energy level should be approximately 10 eV.

Example Problem

Frequency and Wavelength of Emitted Photons

An electron in an excited hydrogen atom drops from the second energy level to the first energy level. Calculate the energy, the frequency, and the wavelength of the photon emitted.

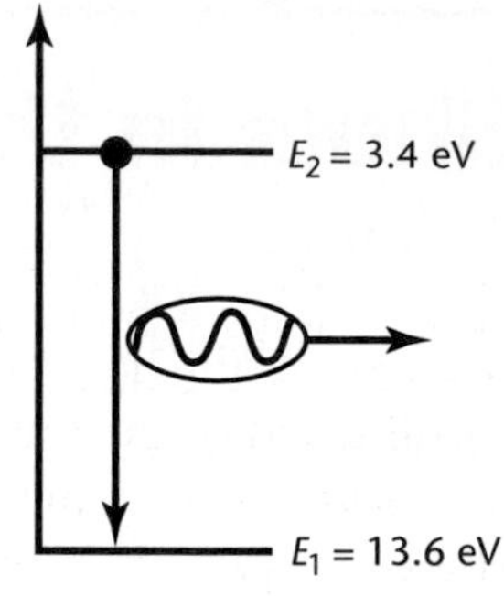

Sketch the Problem

- Diagram two energy levels showing the electron moving from level 2 to level 1.

Calculate Your Answer

Known:

second energy level, $n = 2$

innermost energy level, $n = 1$

Planck's constant, $h = 6.63 \times 10^{-34}$ J·s

speed of light, $c = 3.00 \times 10^{8}$ m/s

energy of 1 eV $= 1.60 \times 10^{-19}$ J

Unknown:

energy of level 2, $E_2 = ?$

energy of level 1, $E_1 = ?$

difference in energy, $\Delta E = hf = ?$

frequency, $f = ?$

wavelength, $\lambda = ?$

Strategy:

Use $E_n = -13.6 \text{ eV} \times 1/n^2$ to determine the energy of each level.

To determine the energy difference, use the following equations.

$\Delta E = hf$, so $f = \dfrac{\Delta E}{h}$

and $c = \lambda f$, so $\lambda = \dfrac{c}{f}$

Calculations:

for $n = 2$, $E_2 = -13.6 \text{ eV} \times \dfrac{1}{2^2} = -3.40 \text{ eV}$

for $n = 1$, $E_1 = -13.6 \text{ eV} \times \dfrac{1}{1^2} = -13.6 \text{ eV}$

$$\Delta E = E_1 - E_2 = -13.6 \text{ eV} - (-3.40 \text{ eV})$$

$$= -10.2 \text{ eV Energy is emitted.}$$

$$f = \frac{\Delta E}{h} = \frac{(10.2 \text{ eV})(1.60 \times 10^{-19} \text{ J/eV})}{6.63 \times 10^{-34} \text{ J·s}}$$

$$= 2.46 \times 10^{15} \text{ s}^{-1}$$

$$\lambda = \frac{c}{f} = \frac{3.00 \times 10^{8} \text{ m/s}}{2.46 \times 10^{15} \text{ s}^{-1}} = 1.22 \times 10^{-7} \text{ m}$$

$$= 122 \text{ nm}$$

Check Your Answer

- Are the units correct? Wavelengths are measured in multiples of meters. Energy is measured in electron volts.
- Are the signs correct? The energy is released as the electron moves from the second energy level to the first. The energy difference is thus negative.
- Is the magnitude realistic? Energy released in this transition produces light in the ultraviolet region below 400 nm.

Physics Lab

Shots in the Dark

Problem

Given that the atom is mostly empty space, how easy is it to hit a nucleus and cause atomic scattering?

Materials

3 dozen rubber stoppers
bedsheet (or blanket)
blindfold (or darkened goggles)
6 9-inch aluminum pie pans
4 1.5-meter 1 × 2 wood pieces
fishing line

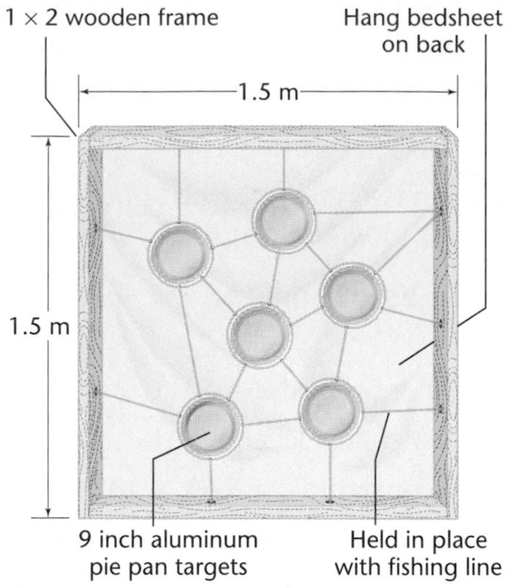

Procedure

1. Construct the model according to the diagram.
2. Each student will be blindfolded, led to a position 3 m directly in front of the target area, and allowed to toss ten rubber stoppers (one at a time) into the target area. If a rubber stopper does not strike within the target area, the shooter should be told "too high," "too low," and so on and be given an extra rubber stopper.
3. Students will be able to hear the nuclear "hit" when the rubber stopper hits the target area. Only one hit will be counted on a single target.

Data and Observations		
Student's Name	Number of Shots	Number of Hits

Analyze and Conclude

There are six circular targets within the target area. The ratio of hits to shots is represented by the following:

$$\frac{\text{hits}}{\text{shots}} = \frac{\text{total target area}}{\text{total model area}} = \frac{6\pi r^2}{\text{width} \times \text{height}}$$

1. **Analyzing Data** Use the class totals for shots and hits to calculate the total area for the six targets. Estimate the area for each target. Then calculate the radius for each target.
2. **Relating Concepts** The uncertainty for this experiment decreases with more shots. The percentage uncertainty is represented by the following:

$$\%\ \text{uncertainty} = \frac{(\text{shots})^{1/2}}{\text{shots}} \times 100\%$$

Find the uncertainty for your class.

Apply

1. A recent phone poll sampled 800 people. Estimate the uncertainty in the poll.

Practice Problems

1. According to the Bohr model, how many times larger is the orbit of a hydrogen electron in the second level than in the first?
2. You learned how to calculate the radius of the innermost orbit of the hydrogen atom. Note that all factors in the equation are constants with the exception of n^2. Use the solution to the Example Problem "Orbital Energy of Electrons in the Hydrogen Atom" to find the radius of the orbit of the second, third, and fourth allowable energy levels in the hydrogen atom.
3. Calculate the energies of the second, third, and fourth energy levels in the hydrogen atom.
4. Calculate the energy difference between E_3 and E_2 in the hydrogen atom. Find the wavelength of the light emitted. Which line in **Figure 28–8** is the result of this transmission?
5. The diameter of the hydrogen nucleus is 2.5×10^{-15} m and the distance between the nucleus and the first electron is about 5×10^{-9} m. If you use a baseball with diameter of 7.5 cm to represent the nucleus, how far away would the electron be?
6. A mercury atom drops from 8.82 eV to 6.67 eV.
   a. What is the energy of the photon emitted by the mercury atom?
   b. What is the frequency of the photon emitted by the mercury atom?
   c. What is the wavelength of the photon emitted by the mercury atom?

28.1 Section Review

1.1 Which of these quantities is quantized: your height, the number of your siblings, or the mass of a sample of gas?

1.2 Why don't the electrons in Rutherford's nuclear model fly away from the nucleus?

1.3 Explain how energy is conserved when an atom absorbs a photon of light.

1.4 How does the Bohr model differ from the Rutherford nuclear model?

1.5 **Critical Thinking** An emission spectrum can contain wavelengths produced when an electron moves from the third to the second level. Could you see this line in the absorption spectrum with your eyes or will you need a special detector? Explain.

28.2 The Quantum Model of the Atom

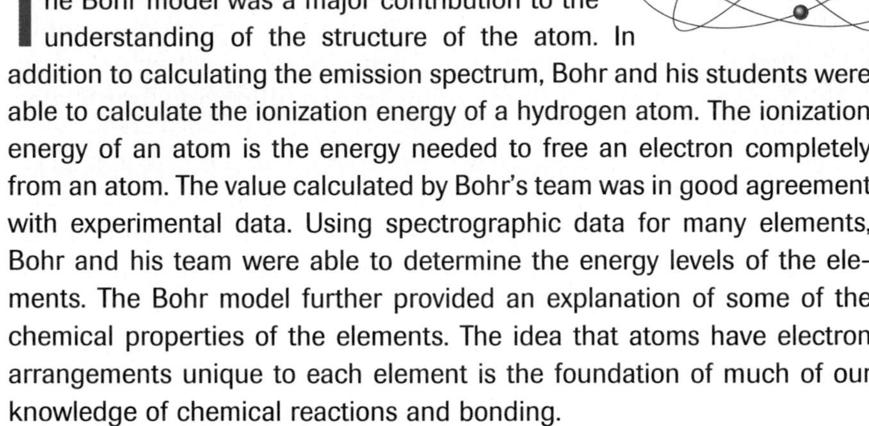

OBJECTIVES

- **Describe** the shortcomings of the Bohr model of the atom.
- **Describe** the quantum model of the atom.
- **Explain** how a laser works and **describe** the properties of laser light.

The Bohr model was a major contribution to the understanding of the structure of the atom. In addition to calculating the emission spectrum, Bohr and his students were able to calculate the ionization energy of a hydrogen atom. The ionization energy of an atom is the energy needed to free an electron completely from an atom. The value calculated by Bohr's team was in good agreement with experimental data. Using spectrographic data for many elements, Bohr and his team were able to determine the energy levels of the elements. The Bohr model further provided an explanation of some of the chemical properties of the elements. The idea that atoms have electron arrangements unique to each element is the foundation of much of our knowledge of chemical reactions and bonding.

The postulates that Bohr made could not be explained on the basis of known physics. For example, the theories of electromagnetism required that the accelerated particles radiate energy, which would cause the rapid collapse of the atom. In addition, Bohr could not explain the reason for the quantization of angular momentum. How could Bohr's work be put on a firm foundation?

From Orbits to an Electron Cloud

The first hint to the solution of these problems was provided by Louis de Broglie. Recall from Chapter 27 that he proposed that particles have wave properties just as light has particle properties. The wavelength of a particle with momentum mv was defined earlier to be $\lambda = h/mv$, and the angular momentum was defined to be $mvr = hr/\lambda$. The Bohr quantization condition, $mvr = nh/2\pi$, can be written in the following way.

$$\frac{hr}{\lambda} = \frac{nh}{2\pi} \text{ or } n\lambda = 2\pi r$$

Note that the circumference of the Bohr orbit, $2\pi r$, is equal to a whole number multiple, n, of the wavelength of the electron, λ.

In 1926, Austrian physicist Erwin Schroedinger used de Broglie's wave model to create a quantum theory of the atom based on waves. This theory did not propose a simple planetary picture of an atom, as Bohr's model had. In particular, the radius of the electron orbit was not likened to the radius of the orbit of a planet about the sun.

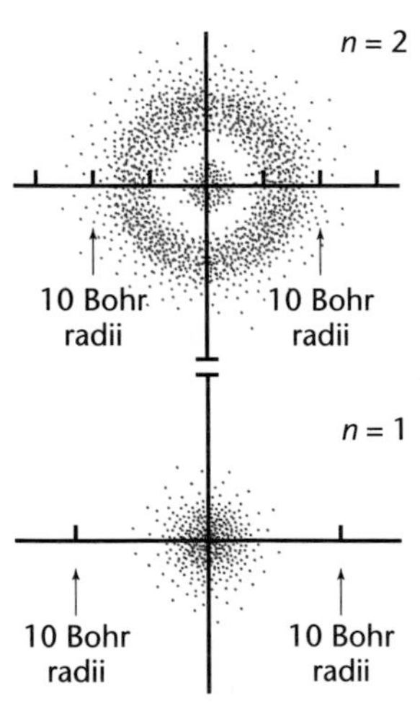

FIGURE 28–14 These plots show the probability of finding the electron in a hydrogen atom at any given location. The denser the points, the higher the probability of finding the electron.

The wave-particle nature of matter means that it is impossible to know both the position and momentum of an electron at the same time. Thus, the modern **quantum model** of the atom predicts only the probability that an electron is at a specific location. The most probable distance of the electron from the nucleus in hydrogen is found to be the same as the radius of the Bohr orbit. The probability that the electron is at any radius can be calculated, and a three-dimensional plot can be constructed that shows regions of equal probability. The region in which there is a high probability of finding the electron is called the **electron cloud. Figure 28–14** shows a slice through the electron cloud for the two lowest states of hydrogen.

Even though the quantum model of the atom is difficult to visualize, **quantum mechanics,** the study of the properties of matter using its wave properties, uses this model and has been extremely successful in predicting many details of the structure of the atom. These details are very difficult to calculate precisely for all but the simplest atoms. Only very sophisticated computers can make highly accurate approximations for the heavier atoms. Quantum mechanics also enables the structure of many molecules to be calculated, allowing chemists to determine the arrangement of atoms in the molecules. Guided by quantum mechanics, chemists have been able to create new and useful molecules that are not otherwise available.

Quantum mechanics is also used to analyze the details of the emission and absorption of light by atoms. As a result of this theory, a new source of light was developed.

Pocket Lab

Bright Lines

Turn on a gas-discharge tube power supply attached to a gas tube so that the tube glows. **CAUTION:** *Do not touch any exposed metal when the power supply is turned on. Dangerous high voltages are present. Always turn off the power supply before changing gas tubes.* Turn off the room lights. Describe the color. Now look through a diffraction grating at the tube.

Analyze and Conclude Make a sketch of the results. Repeat this activity with a different gas tube. Explain the differences.

Lasers

Light emitted by an incandescent source has many wavelengths and travels in all directions. Light produced by an atomic gas consists of only a few different wavelengths, but it also is emitted in all directions. The light waves emitted by atoms at one end of a discharge tube are not necessarily in step with the waves from the other end. That is, the waves are not necessarily all at the same point in their cycle. Some will be in step, with the minima and maxima of the waves coinciding. Such light is called **coherent light**. Others will be out of step. Such light is called **incoherent light.** This is shown in **Figure 28–15.**

Light is emitted by atoms that have been excited. So far, you have learned about two ways in which atoms can be excited: thermal excitation and electron collision. Atoms also can be excited by collisions with photons of exactly the right energy.

FIGURE 28–15 Waves of incoherent light **(a)** and coherent light **(b)** are shown.

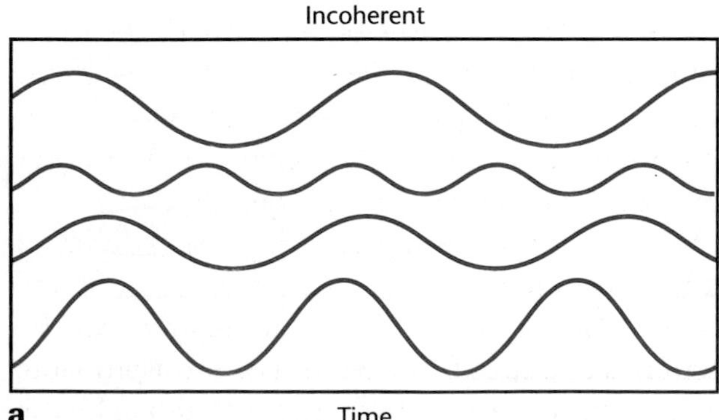

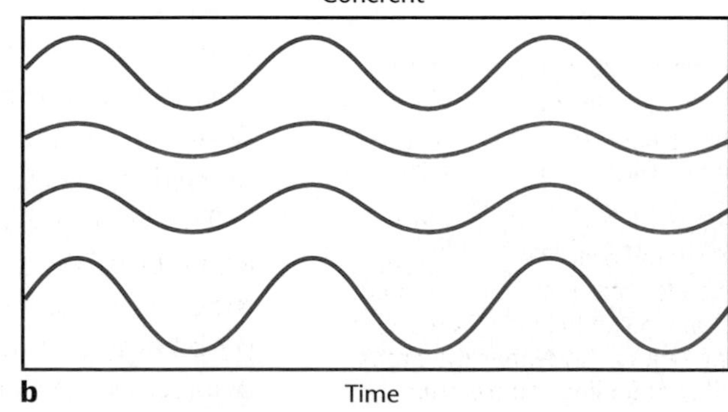

F.Y.I.

If enough atoms could be squeezed together to fill one cubic centimeter with each nucleus touching another nucleus, the total weight would be over 250 million metric tons.

What happens when an atom is in an excited state? After a very short time, it normally returns to the ground state, giving off a photon of the same energy that it absorbed, as shown in **Figure 28–16a.** This is called spontaneous emission.

In 1917, Einstein considered what would happen to an atom already in an excited state that is struck by another photon of the same energy as the original photon. He showed that the atom will emit a photon of the same energy and move to a lower state. The photon that caused, or stimulated, the emission will not be affected. This process is called **stimulated emission.** The two photons leaving the atom not only will have the same wavelength, but also they will be in step, as shown in **Figure 28–16b.**

Either of the two photons can now strike other excited atoms, producing additional photons that are in step with the original photons. This process can continue, producing an avalanche of photons, all of the same wavelength and all having their maxima and minima at the same times.

To make this process happen, certain conditions must be met. First, of course, there must be other atoms in the excited state. Second, the photons must be collected so that they strike the excited atoms. A device that fulfills both these conditions was invented in 1959 and is called a **laser.** The word *laser* is an acronym. It stands for **l**ight **a**mplification by **s**timulated **e**mission of **r**adiation. An atom that emits light when it is stimulated in a laser is said to lase.

The atoms in a laser can be put into the excited state, or pumped, in different ways, as outlined in **Figure 28–17.** An intense flash of light with a wavelength shorter than that of the laser can pump the atoms. The more energetic photons produced by the flash collide with and excite the lasing atoms. One atom decays to a lower energy state, starting the avalanche. As a result, a brief flash or pulse of laser light is emitted. Alternatively, a continuous electric discharge such as that in a neon sign can be used to put atoms in the excited state. The laser light resulting from this process is continuous rather than pulsed. The helium-

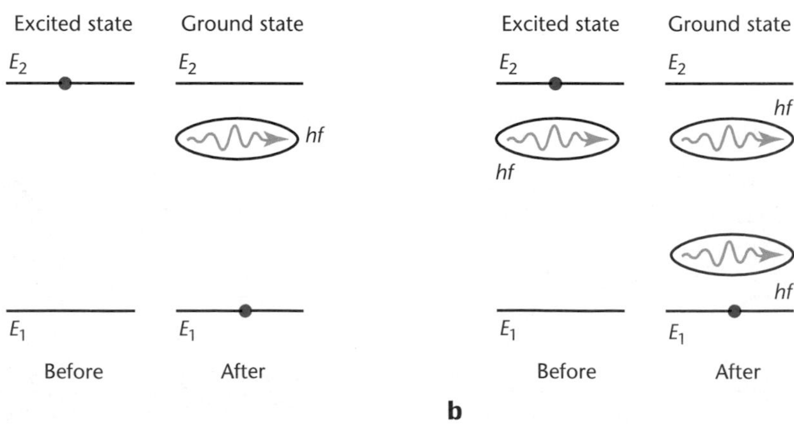

FIGURE 28–16 The spontaneous emission of a photon with energy hf when an electron in an atom drops from an excited state E_2 to the ground state E_1 is shown **(a).** The stimulated emission of a photon when an excited atom is struck by a photon with energy hf is shown **(b).** For both, $hf = E_2 - E_1$.

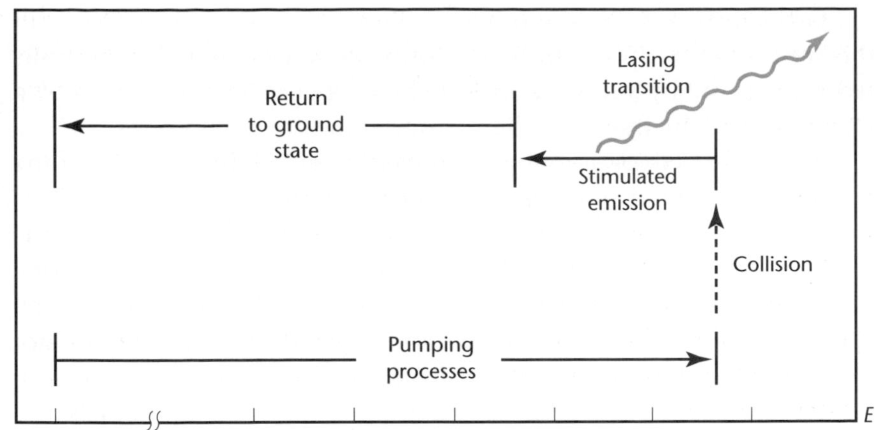

FIGURE 28–17 A photon striking an atom in the excited state stimulates it to make a transition to a lower state and emit a second coherent photon.

neon lasers often seen in science classrooms are continuous lasers. An electric discharge excites the helium atoms. They collide with the neon atoms, pumping them to an excited state and causing them to lase.

The photons emitted by atoms are collected by placing a glass tube containing the atoms between two parallel mirrors. One mirror is 100 percent reflective and will reflect all the light hitting it while the other is only partially reflective and will allow only about one percent of the light to pass through. When a photon strikes an atom in the excited state, it stimulates the atom to make a transition to the lower state. Thus, two photons leave the atom. These photons can strike other atoms and produce more photons, thereby starting the avalanche. Photons that are directed toward the ends of the tube will be reflected back into the gas by the mirrors. The reflected photons reinforce one another with each pass between the mirror and build to a high intensity. The photons that exit the tube through the partially reflecting mirror produce the laser beam. This is shown in **Figure 28–18.**

Laser light is highly directional because of the parallel mirrors. The light beam is very small, typically only about 1/2 mm in diameter, so the light is very intense. The light is all of one wavelength, or monochromatic, because the transition of electrons between only one pair of energy levels in one type of atom is involved. Because all the stimulated photons are emitted in step with the photons that struck the atoms, laser light is coherent light.

Pocket Lab

Laser Diffraction

Caution: *Avoid staring directly into the laser beam or at bright reflections.* A diffraction grating separates light from discharge tubes into individual wavelengths. Predict what will happen when you shine a laser light through a diffraction grating. Shine the laser at a sheet of white paper about 1 foot away. Then place the diffraction grating next to the laser to see what happens.

Analyze and Conclude
Describe and explain your results. Predict how your results would be similar and different with a green laser light.

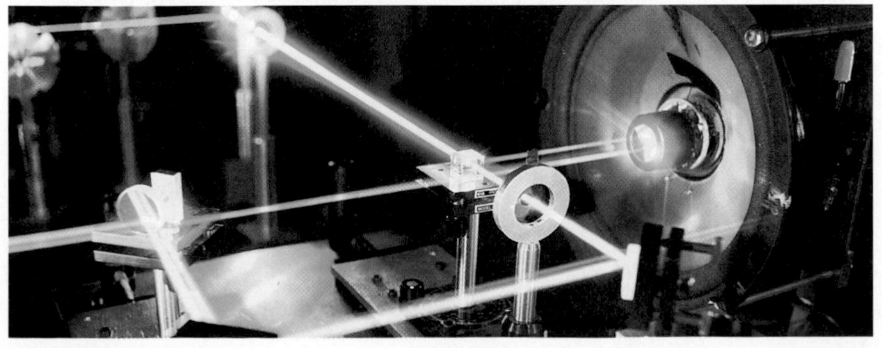

FIGURE 28–18 A laser produces a beam of coherent light.

HELP WANTED

LASER TECHNICIAN

Will produce, test, operate, and/or repair lasers. May work in a hospital, the fiber optics industry, a research lab, a manufacturing plant, the military, or on space projects or construction sites. Some laser technicians create holograms for charge cards, diagnosis of medical problems, or identification of flaws in machine parts. Position requires a two-year degree in laser technology. Four-year degree preferred. For more information, contact:

The Laser Institute of America
12424 Research Parkway
Suite 125
Orlando, FL 32826-3274

Many substances—solids, liquids, and gases—can be made to lase in this way. Most produce laser light at only one wavelength. For example, red is produced by a neon laser, blue by an argon laser, and green by a helium-cadmium laser. The light from some lasers, however, can be tuned, or adjusted, over a range of wavelengths. **Table 28–1** shows the types of wavelengths produced by some common lasers.

All lasers are very inefficient. No more than one percent of the electrical energy delivered to a laser is converted to light energy. Despite this inefficiency, the unique properties of laser light have led to many applications. Laser beams are narrow and highly directional. They do not spread out over long distances. Surveyors use laser beams for this reason. Laser beams are also used to check the straightness of long tunnels and pipes. When astronauts visited the moon, they left a mirror which was used by scientists to reflect a laser beam from Earth. The distance between Earth and the moon was thus accurately determined.

Laser light is used in fiber optics as well. A fiber uses total internal reflection to transmit light over many kilometers with little loss. The laser is switched on and off rapidly, transmitting information through the fiber. In many cities, optical fibers have replaced copper wires for the transmission of telephone calls, computer data, and even television pictures.

The single wavelength of light emitted by lasers makes lasers valuable in spectroscopy. Laser light is used to excite other atoms. The atoms then return to the ground state, emitting characteristic spectra. Samples with extremely small numbers of atoms can be analyzed in this way. In fact, single atoms have been detected by means of laser excitation and even have been held almost motionless by laser beams.

The concentrated power of laser light is used in a variety of ways. In medicine, lasers can be used to repair the retina in an eye. Lasers also can be used in surgery in place of a knife to cut flesh with little loss of

TABLE 28–1

Common Lasers

Medium	Wavelength (nm)	Type
Nitrogen	337 (UV*)	Pulsed
Helium-cadmium	441.6	Continuous
Argon ion	476.5, 488.0	Continuous
Krypton ion	524.5	Continuous
Neon	632.8	Continuous
Gallium aluminum arsenide	680	Continuous
Ruby	694.3	Pulsed
Gallium arsenide$^+$	840-1350 (IR*)	Continuous
Neodymium	1040 (IR*)	Pulsed
Carbon dioxide	10 600 (IR*)	Continuous

* UV means ultraviolet, IR means infrared.

$^+$ The wavelength of gallium arsenide depends on temperature.

FIGURE 28–19 Because laser light is so concentrated and directed, lasers can be used in industry to cut and weld steel.

Physics & Technology

All Aglow!

Fluorescence is a phenomenon that occurs when certain substances are exposed to radiation. As a result of their being stimulated by the radiation, fluorescent substances reemit the radiation to produce that glow-in-the dark quality found on various decals and stickers; T-shirts, tennis shoes and other pieces of clothing; light-switch plates; and children's books, puzzles, and toys; just to name a few examples.

In addition to those just named above, fluorescent substances have many practical applications. In medicine, for example, the fluorescent protein found in jellyfish can be used to mark and subsequently change, or mutate, the amino acids that make up a protein. Staining chromosomes with fluorescent dyes allows physicians to identify DNA abnormalities that cause diseases such as Down's syndrome and leukemia. Another medical application of fluorescence involves using the glowing substances to count proteins and other chemicals to provide early warning signs of certain types of cancers.

Fluorescence also has its importance in the art world. Jewelers and gemologists have used fluorescent techniques to study the Hope diamond, the largest known gem of its kind at 45.5 carats. Determining the conditions under which this diamond formed could lead to the discovery of other such diamonds. Diamonds, the hardest known natural substances, have uses in dentistry as polishers, and in jewelry and industry as industrial drills and abrasives.

Fluorescence can sometimes solve or prevent certain crimes. Forensic labs, for example, use fluorescent powders to lift fingerprints. Glow-in-the-dark decals are used by some cities to deter auto theft, which generally occurs between midnight and 6:00 A.M. Users of the decals sign a statement saying that they normally don't drive between those hours. Thus, if a car with these special fluorescent stickers is seen on the road, police are alerted to a possible theft.

Thinking Critically Think about what you've just read. Compile a list of at least five other practical applications of fluorescence. Explain the usefulness of each of your ideas.

FIGURE 28–20 When a hologram is made on film, a laser beam is split into two parts by a half-silvered mirror. Interference occurs on the film as the direct laser light meets laser light reflected off an object. The interference of both beams of light allows the film to record both the intensity and phase of light from the object.

blood. In industry, lasers are used to cut materials such as steel, as shown in **Figure 28–19,** and to weld materials together. In the future, lasers may be able to produce nuclear fusion to create an almost inexhaustible energy source.

Holograms are made possible by the coherent nature of laser light. A hologram, shown in **Figure 28–20,** is a photographic recording of both the phase and the intensity of light. Holograms form realistic three-dimensional images and can be used, among other applications, in industry to study the vibration of sensitive equipment and components.

28.2 Section Review

2.1 Which of the lasers in **Table 28–1** emits the most red light (visible light with the longest wavelength)? Which emits in the blue region of the spectrum?

2.2 Could green light be used to pump a red laser? Why could red light not be used to pump a green laser?

2.3 Why does the Bohr model of the atom conflict with the uncertainty principle while the quantum model does not?

2.4 **Critical Thinking** Suppose that an electron cloud were to get so small that the atom was almost the size of the nucleus. Use the uncertainty principle to explain why this would take a tremendous amount of energy.

CHAPTER 28 REVIEW

Key Terms

28.1
- alpha (α) particle
- scintillation
- nuclear model
- emission spectrum
- spectroscope
- absorption spectrum
- spectroscopy
- energy level
- ground state
- excited state
- Bohr model
- quantum number

28.2
- quantum model
- electron cloud
- quantum mechanics
- coherent light
- incoherent light
- stimulated emission
- laser

Summary

28.1 The Bohr Model of the Atom

- Ernest Rutherford directed positively charged, high-speed alpha particles at thin metal foils. By studying the paths of the reflected particles, he showed that atoms are mostly empty space with a tiny, massive, positively-charged nucleus at the center.
- The spectra produced by atoms of an element can be used to identify that element.
- If white light passes through a gas, the gas absorbs the same wavelengths that it would emit if it were excited. If light leaving the gas goes through a prism, an absorption spectrum is visible.
- In the model of the atom developed by Niels Bohr, the electrons can have only certain energy levels.
- In the Bohr model of the atom, electrons can make transitions between energy levels. As they do, they emit or absorb electromagnetic radiation.
- The frequency and wavelength of absorbed and emitted radiation can be calculated for the hydrogen atom using the Bohr model.

28.2 The Quantum Model of the Atom

- The quantum mechanical model of the atom cannot be visualized easily. Only the probability that an electron is at a specific location can be calculated.
- Quantum mechanics is extremely successful in calculating the properties of atoms, molecules, and solids.
- Lasers produce light that is directional, powerful, monochromatic, and coherent. Each property gives the laser useful applications.

Reviewing Concepts

Section 28.1

1. Describe how Rutherford determined that the positive charge in an atom is concentrated in a tiny region rather than spread throughout the atom.
2. How does the Bohr model explain why the absorption spectrum of hydrogen contains exactly the same frequencies as its emission spectrum?
3. What are some of the problems with a planetary model of the atom?
4. What three assumptions did Bohr make in developing his model of the atom?
5. How does the Bohr model account for the spectra emitted by atoms?

Section 28.2

6. A laboratory laser has a power of only 0.8 mW (8×10^{-4} W). Why does it seem stronger than the light of a 100 W lamp?
7. A device similar to a laser that emits microwave radiation is called a maser. What words are likely to make up this acronym?
8. What properties of laser light led to its use in light shows?

Applying Concepts

9. The northern lights are the result of high-energy particles coming from the sun and striking atoms high in Earth's atmosphere. If you looked at these lights through a spectrometer, would you expect to see a continuous or line spectrum? Explain.

10. If white light were emitted from Earth's surface and observed by someone in space, would its spectrum appear to be continuous? Explain.
11. Suppose you wanted to explain quantization to a younger brother or sister. Would you use money or water as an example? Explain.
12. A photon with energy of 6.2 eV enters a mercury atom in the ground state. Will it be absorbed by the atom? See **Figure 28–21.** Explain.

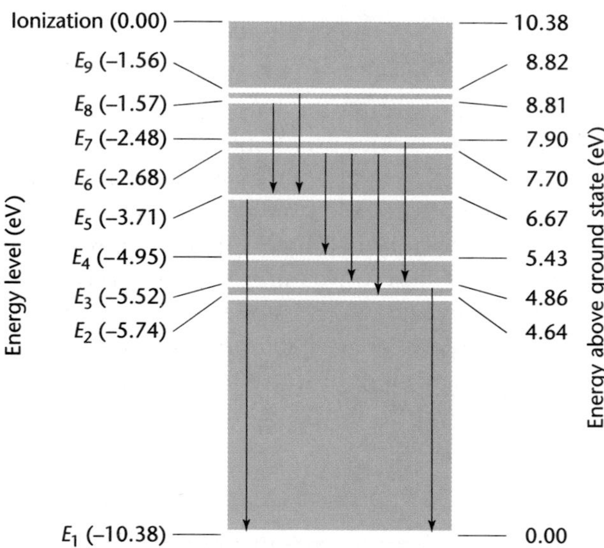

FIGURE 28–21

13. **Figure 28–22** shows the energy levels of a certain atom. If an electron can make transitions between any two levels, how many spectral lines can the atom emit? Which transition produces the photon with the highest energy?

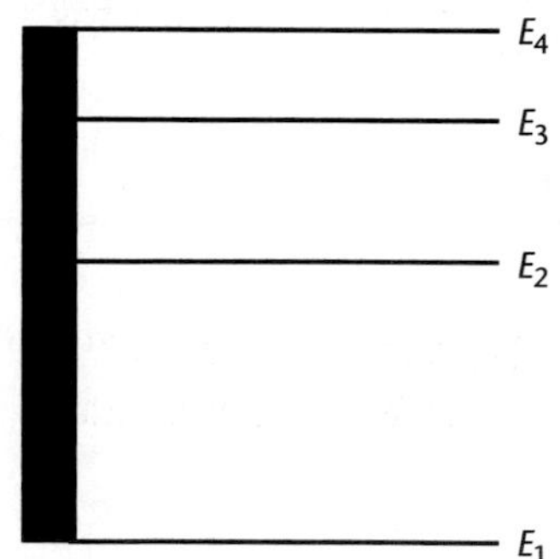

FIGURE 28–22

14. A photon is emitted when an electron drops through energy levels within an excited hydrogen atom. What is the maximum energy the photon can have? If this same amount of energy were given to an electron in the ground state in a hydrogen atom, what would happen?
15. When electrons fall from higher energy levels to the third energy level within hydrogen atoms, are the photons that are emitted infrared, visible, or ultraviolet light? Explain.
16. Compare the quantum mechanical theory of the atom with the Bohr model.
17. You have a laser that emits a red light, a laser that emits a green light, and a laser that emits a blue light. Which laser, produces photons with the highest energy?

Problems

Section 28.1

LEVEL 1

*See **Figure 28–21** for Problems 18, 19, and 20.*

18. A mercury atom is in an excited state when its energy level is 6.67 eV above the ground state. A photon of energy 2.15 eV strikes the mercury atom and is absorbed by it. To what energy level is the mercury atom raised?
19. A mercury atom is in an excited state at the E_6 energy level.
   a. How much energy would be needed to ionize the atom?
   b. How much energy would be released if the electron dropped down to the E_2 energy level instead?
20. A mercury atom in an excited state has an energy of -4.95 eV. It absorbs a photon that raises it to the next-higher energy level.
   a. What is the energy of the photon?
   b. What is the photon's frequency?
21. A photon with an energy of 14.0 eV enters a hydrogen atom in the ground state and ionizes it. With what kinetic energy will the electron be ejected from the atom?
22. Calculate the radius of the orbital associated with the energy levels E_5 and E_6 of the hydrogen atom.

23. What energies are associated with a hydrogen atom's energy levels E_2, E_3, E_4, E_5, and E_6?
24. Using the values that are calculated in problem 23, calculate the following energy differences for a hydrogen atom.
 a. $E_6 - E_5$
 b. $E_6 - E_3$
 c. $E_4 - E_2$
 d. $E_5 - E_2$
 e. $E_5 - E_3$
25. Use the values from problem 24 to determine the frequencies of the photons emitted when an electron in a hydrogen atom makes the level changes listed.
26. Determine the wavelengths of the photons of the frequencies that you calculated in problem 25.
27. Determine the frequency and wavelength of the photon emitted when an electron drops
 a. from E_3 to E_2 in an excited hydrogen atom.
 b. from E_4 to E_3 in an excited hydrogen atom.
28. What is the difference between the energies of the E_4 and E_1 energy levels of the hydrogen atom?

LEVEL 2

29. From what energy level did an electron fall if it emits a photon of 94.3 nm wavelength when it reaches ground state within a hydrogen atom?
30. For a hydrogen atom in the $n = 3$ Bohr orbital, find
 a. the radius of the orbital.
 b. the electric force acting between the proton and the electron.
 c. the centripetal acceleration of the electron.
 d. the orbital speed of the electron. Compare this speed with the speed of light.
31. A hydrogen atom has its electron in the $n = 2$ level.
 a. If a photon with a wavelength of 332 nm strikes the atom, show that the atom will be ionized.
 b. When the atom is ionized, assume that the electron receives the excess energy from the ionization. What will be the kinetic energy of the electron in joules?

Section 28.2

LEVEL 2

32. Gallium arsenide lasers are used in CD players. If such a laser emits at 840 nm, what is the difference in eV between the two lasing energy levels?
33. A carbon dioxide laser emits very high-power infrared radiation. What is the energy difference in eV between the two lasing energy levels?

Critical Thinking Problems

34. The four brightest lines in the mercury spectrum have wavelengths of 405 nm, 436 nm, 546 nm, and 578 nm. What are the differences in energy levels for each of these lines?
35. After the emission of these visible photons, the mercury atom continues to emit photons until it reaches the ground state. From inspection of **Figure 28–21,** determine whether or not any of these photons would be visible. Explain.

Going Further

Measuring Energy Levels A positronium atom consists of an electron and its antimatter relative, the positron bound together. Although the lifetime of this "atom" is very short—on the average it lives 1/7 of a microsecond—its energy levels can be measured. The Bohr model can be used to calculate energies with the mass of the electron replaced by one-half its mass. Describe how the radii of the orbits and the energy of each level would be affected. What would be the wavelength of the E_2 to E_1 transition?

*inter***NET**
CONNECTION

Follow the link on the Glencoe Homepage at **www.glencoe.com/sec/science** to find out more about this chapter.

The Inside Dope

How are conductors, resistors, diodes, and transistors, which are an essential part of this microchip, produced from a pure crystal of silicon?

CHAPTER

29 Solid State Electronics

In our society, most people use electronic devices every day. The principles of solid-state physics and electrical engineering surround us in the form of radios, televisions, calculators, microcomputer-controlled microwave ovens, and CD players.

In the 1950s, transistors the size of a pencil eraser began to replace the much larger vacuum tubes that had given birth to the field of electronics in the 1920s. By the 1960s, when engineers had realized the potential for smaller transistors, they found ways of putting many individual transistors on a single crystal of silicon. This integrated circuit, or microchip, was the key invention that led to many of today's technological achievements. Modern circuits containing millions of transistors are about the same size and cost as a single early transistor. Thus, it is now possible for designers to put systems as complex as those used in today's computers on your wrist.

WHAT YOU'LL LEARN

- You will be able to distinguish among electric conductors, semiconductors, and insulators.
- You will examine how pure semiconductors are modified to produce desired electrical properties.
- You will compare how diodes and transistors are made and used.

WHY IT'S IMPORTANT

- Semiconductor electronics is used in many aspects of your daily life.

interNET CONNECTION

Follow the link for this chapter on the Glencoe Homepage at **www.glencoe.com/sec/science** to find out more about solid state electronics.

29.1 Conduction in Solids

Electronic devices depend not only on natural conductors and insulators but also on materials that have been designed and produced by many scientists and engineers working together. This brief investigation into electronics begins with a study of how materials conduct electricity.

OBJECTIVES

- **Describe** electron motion in conductors and semiconductors.
- **Compare** and **contrast** *n*-type and *p*-type semiconductors.

Band Theory of Solids

Recall that materials can be either electrical conductors or insulators. In conductors, electrical charges can move easily, but in insulators, charges are much more difficult to displace. When you examine these two types of materials at the atomic level, the difference in the way they are able to carry charges becomes apparent.

You learned in Chapter 13 that crystalline solids consist of atoms bound together in regular arrangements. You also know from Chapters 27 and 28 that an atom consists of a dense, positively charged nucleus surrounded by a cloud of negatively charged electrons. These electrons can occupy only certain allowed energy levels. Under most conditions, the electrons in an atom occupy the lowest possible energy levels. This condition is referred to as the ground state. Because the atoms can have only certain energies, any energy changes that occur are quantized; that is, the energy changes occur in specific amounts.

Suppose you could construct a solid by assembling atoms together, one by one. You would start with all the atoms in the ground state. If you brought two atoms close together, the electric field of each atom would affect the other. The energy levels of one atom would be raised slightly, while the energy levels of the other atom would be lowered, as illustrated in **Figure 29–1a.** Note that there are two atoms, and there are two different sets of energy levels.

FIGURE 29–1 Atomic energy levels are split apart when two atoms are brought together **(a)** and when four atoms are brought together **(b).** An energy band is formed when many atoms are brought together **(c).**

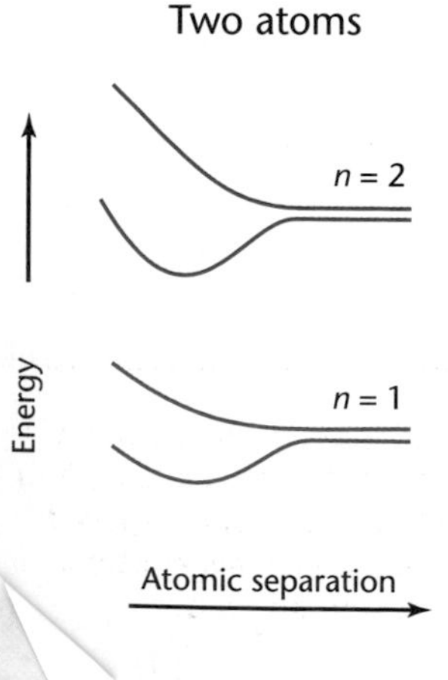

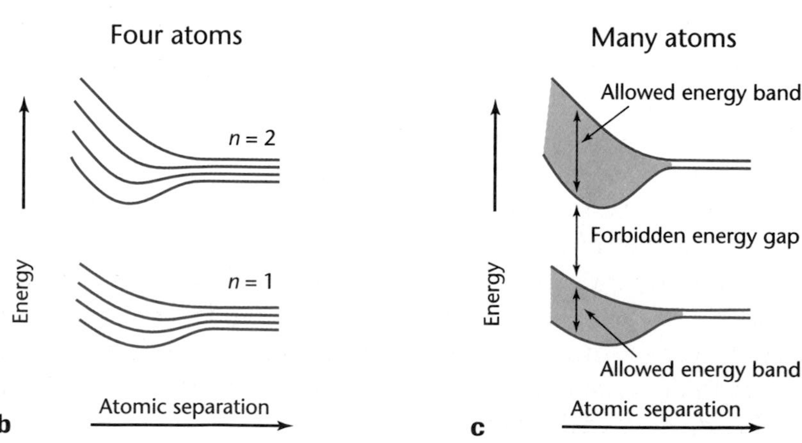

Now consider what happens when you bring many atoms close together. The original energy levels in the atoms split into energy levels so close together that they can no longer be easily identified as distinct, as indicated in **Figure 29–1b.** Rather, the levels are spread into broad bands, as shown in **Figure 29–1c.** The bands are separated by values of energy that no electrons possess. These energies are called **forbidden gaps.** Electrical conduction in solids explained in terms of these energy bands and forbidden gaps is called the **band theory** of solids.

Recall that any system will adjust itself until its energy is minimized. Therefore, electrons fill the energy levels of an atom beginning with the level of lowest energy and continuing to the highest; no two electrons can have the same energy at the same time. For atoms in the ground state, the lower energy levels are completely full. The outermost band that contains electrons is called the **valence band.** The lowest band that is not filled to capacity with electrons is called the **conduction band.**

Materials with partially filled bands are conductors, as indicated in **Figure 29–2.** The size of the forbidden gap between the valence band and the conduction band determines whether a solid is an insulator or a semiconductor.

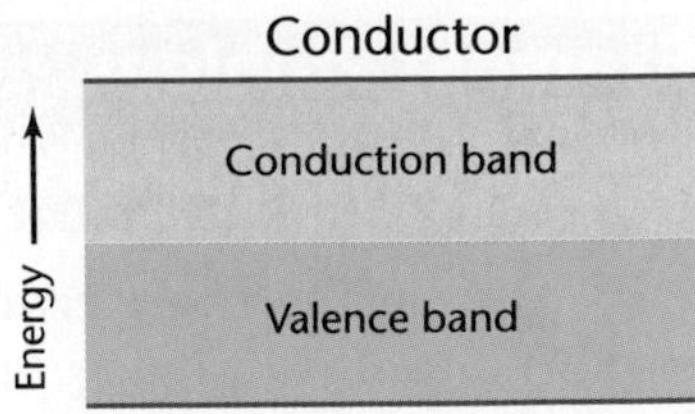

FIGURE 29–2 In a material that is a good conductor, the valence band is partially filled. The blue-shaded area shows energies occupied by electrons.

Conductors

When a potential difference is placed across a material, the resulting electric field exerts a force on the electrons. The electrons accelerate and gain energy; the field does work on them. If there are bands within the material that are only partially filled, then there are energy levels available that are only slightly higher than the electrons' present energy levels. As a result, the electrons that gain energy from the field can move from one atom to the next. Such movement of electrons from one atom to the next is an electric current, and the entire process is known as electrical conduction. Materials with partially filled bands, such as the metals aluminum and copper, conduct electricity easily.

The free electrons in conductors act like atoms in a gas or water molecules in the sea. The electrons move about rapidly in a random way, changing directions only when they collide with the cores of the atoms. However, if an electric field is put across a length of wire, there will be a net force pushing the electrons in one direction. Although their motion is not greatly affected, they have a slow overall movement dictated by the electric field. **Figure 29–3** shows a model of how electrons continue to move rapidly with speeds of 10^6 m/s in random directions, but how they also drift very slowly at speeds of 10^{-5} m/s or slower toward the positive end of the wire. This model of conductors is called the electron gas model. If the temperature is increased, the speeds of the electrons increase, and, consequently, they collide more frequently with atomic cores. Thus, as the temperature rises, the conductivity of metals is reduced. Conductivity is the reciprocal of resistivity. As conductivity is reduced, a material's resistance rises.

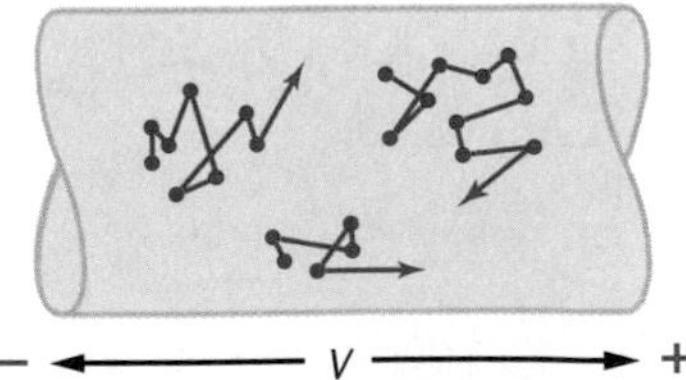

FIGURE 29–3 The electrons move rapidly and randomly in a conductor. If a field is applied across the wire, the electrons drift toward one end.

Example Problem

The Free-Electron Density of a Conductor

How many free electrons exist in a cubic centimeter of copper? Each atom contributes one electron. The density, atomic mass, and number of atoms per mole of copper can be found in Appendix D.

Calculate Your Answer

Known:

For copper: 1 free e^- per atom

$\rho = 8.96\ \text{g/cm}^3$

$M = 63.54\ \text{g/mole}$

$N_A = 6.02 \times 10^{23}$ atoms/mole

Unknown:

Free $e^-/\text{cm}^3 = ?$

Strategy:

Use Avogadro's number, the atomic mass (g/mole), and density (g/cm^3) to find atoms/cm^3.

Calculations:

$$\frac{\text{free } e^-}{\text{cm}^3} = \left(\frac{1 \text{ free } e^-}{1 \text{ atom}}\right)\left(\frac{6.02 \times 10^{23} \text{ atoms}}{1 \text{ mole}}\right)\left(\frac{1 \text{ mole}}{63.54 \text{ g}}\right)\left(\frac{8.96 \text{ g}}{1 \text{ cm}^3}\right) = 8.49 \times 10^{22} \frac{\text{free } e^-}{\text{cm}^3 \text{ Cu}}$$

Check Your Answer

- Are the units correct? The number of free electrons per cubic centimeter answers the question.
- Is the magnitude reasonable? Yes, you would expect a large number of electrons in a cubic centimeter.

Practice Problems

1. Zinc, density $7.13\ \text{g/cm}^3$, atomic mass 65.37 g/mole, has two free electrons per atom. How many free electrons are there in each cubic centimeter of zinc?

Insulators

In an insulating material such as sulfur, table salt, or glass, the valence band is filled to capacity and the conduction band is empty. As shown in **Figure 29–4,** the valence band and the conduction band are separated by a forbidden gap. An electron must gain a large amount of energy to go to the next energy level. Recall that an electron volt (eV) is a convenient energy unit for energy changes in atomic systems. In an insulator, the lowest energy level in the conduction band is 5 to 10 eV above the highest energy level in the valence band, as shown in **Figure 29–4a.** There is at least a 5-eV gap of energies that no electrons can possess.

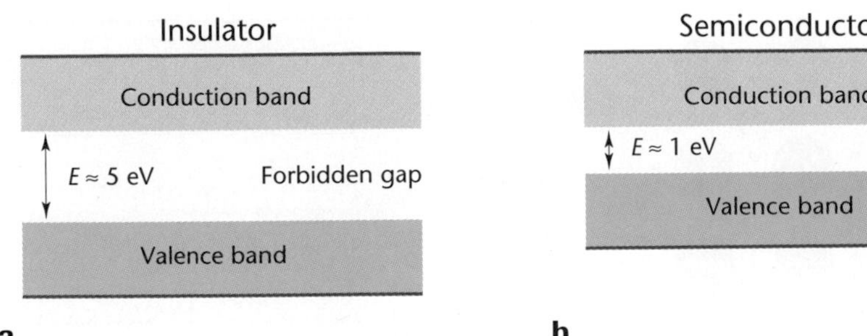

FIGURE 29–4 Compare the valence and conduction bands in an insulator **(a)** and in a semiconductor **(b).** Notice that the forbidden gap is wider in **a** than in **b.** Compare these diagrams with that shown in **Figure 29–2.**

Although electrons have some kinetic energy as a result of their thermal energy, the average kinetic energy of electrons at room temperature is not sufficient for them to jump the forbidden gap. If a small electric field is placed across an insulator, almost no electrons gain enough energy to reach the conduction band, so there is no current. Electrons in an insulator must be given a large amount of energy to be pulled free from one atom and moved to the next. As a result, the electrons in an insulator tend to remain in place, and the material does not conduct electricity.

Semiconductors

Electrons can move more freely in semiconductors than in insulators, but not as easily as in conductors, as shown in **Figure 29–4b.** The energy gap between the valence band and the conduction band is 1 eV or less. How does the structure of a semiconductor explain its electronic characteristics? Atoms of the most common semiconductors, silicon (Si) and germanium (Ge), each have four valence electrons. These four electrons are involved in binding the atoms together into the solid crystal. The valence electrons form a filled band, as in an insulator, but the forbidden gap between the valence and conduction bands is much smaller than in an insulator. Not much energy is needed to pull one of the electrons from a silicon atom and put it into the conduction band, as illustrated in **Figure 29–5a.** Indeed, the gap is so small that some electrons reach the conduction band as a result of their thermal kinetic energy alone. That is, the random motion of atoms and electrons gives some electrons enough energy to break free of their home atoms and wander around the silicon crystal. If an electric field is applied to a semiconductor, even more electrons are moved into the conduction band and move through the solid according to the direction of the applied electric field. In contrast to the effect in a metal, the higher the temperature of a semiconductor, the more electrons are able to reach the conduction band, and the higher the conductivity.

An atom from which an electron has broken free is missing an electron and is said to contain a hole. As shown in **Figure 29–5b,** a **hole** is an empty energy level in the valence band. The atom now has a net positive charge. If an electron breaks free from another atom, it can land on the hole and become bound to an atom once again. When a hole and a free electron recombine, their opposite charges cancel each other. The

HELP WANTED

SYSTEMS ANALYST

If you can analyze a problem; select the best hardware, software, and support system to solve it; and then work with other staff members to implement that solution, we are looking for you to fill a systems analyst position.

A B.S. or an advanced degree in related areas, such as business or scientific engineering, will determine your position within this booming computer service company. Your sense of logic, ability to keep up with advances in our industry, communication skills, and willingness to work hard will determine your success.

For information contact:
Association for Systems Management
24587 Bagley Road
Cleveland, OH 44138

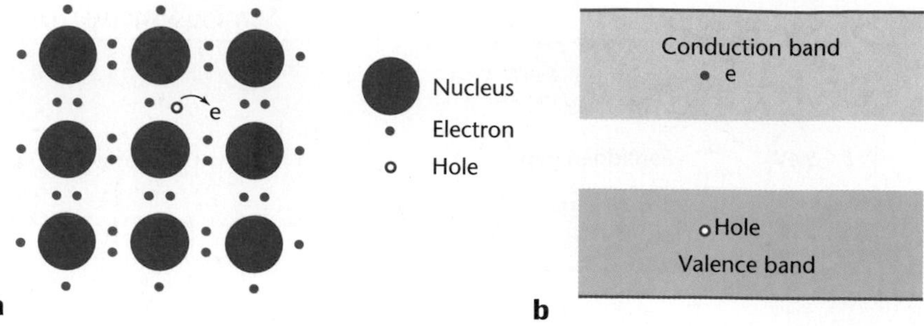

FIGURE 29–5 Some electrons in semiconductors have enough thermal kinetic energy to break free and wander through the crystal, as shown in the crystal structure **(a)** and in the bands **(b)**.

electron, however, has left behind a hole on its previous atom. Thus, as in a game of musical chairs, the negatively charged, free electrons move in one direction and the positively charged holes move in the opposite direction. Pure semiconductors that conduct as a result of thermally freed electrons and holes are called **intrinsic semiconductors.** Because so few electrons or holes are available to carry charge, conduction in intrinsic semiconductors is very small; thus, their resistance is very large.

Example Problem

Fraction of Free Electrons in an Intrinsic Semiconductor

Find the number of atoms in silicon that have free electrons. That is, find the ratio of the number of free e^-/cm^3 to the number of $atoms/cm^3$. Because of the thermal kinetic energy of the solid at room temperature, there are 1×10^{13} free electrons per cm^3.

Calculate Your Answer

Known:

$\rho = 2.33\ g/cm^3$

$M = 28.09\ g/mole$

Unknown:

free e^-/ atoms of Si = ?

Strategy:

Use Avogadro's number, the atomic mass (g/mole), and density (g/cm^3) to find $atoms/cm^3$.

Calculations:

$$\frac{\text{atoms}}{\text{cm}^3} = \left(\frac{2.33\ \text{g}}{\text{cm}^3}\right)\left(\frac{1\ \text{mole}}{28.09\ \text{g}}\right)\left(\frac{6.02 \times 10^{23}\ \text{atoms}}{\text{mole}}\right) = 4.99 \times 10^{22}\ \text{atoms/cm}^3$$

$$\frac{\text{free e}^-}{\text{atom}} = \left(\frac{1 \times 10^{13}\ \text{free e}^-}{\text{cm}^3}\right)\left(\frac{1\ \text{cm}^3}{4.99 \times 10^{22}\ \text{atoms}}\right) = 2 \times 10^{-10}\ \text{free e}^-/\text{atom}$$

or, 1 out of 5 billion Si atoms has a free electron

Check Your Answer

- Using the factor-label method confirms the correct units.
- Is the magnitude reasonable? In an intrinsic semiconductor at room temperature, few atoms have free electrons.

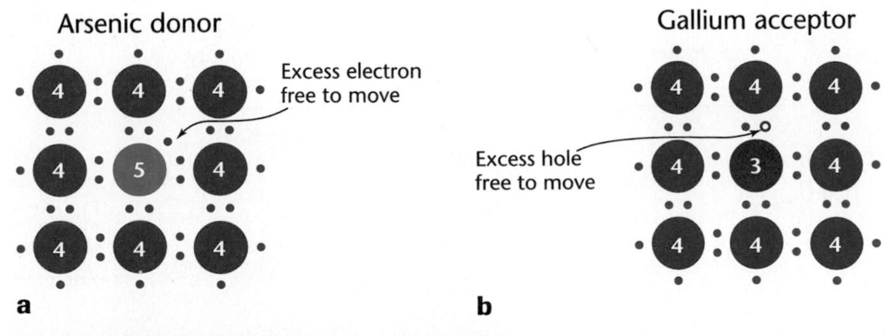

FIGURE 29–6 Donor atoms of arsenic with five valence electrons replace acceptor atoms and provide excess electrons in the silicon crystal **(a).** Acceptor atoms of gallium with three valence electrons create holes in the crystal **(b).**

Practice Problems

2. In pure germanium, density 5.23 g/cm^3, atomic mass 72.6 g/mole, there are 2×10^{16} free electrons/cm^3 at room temperature. How many free electrons are there per atom?

Doped Semiconductors

Although conductivity does not depend only on the number of free electrons, materials with fewer than one free electron per million atoms will not conduct electricity very well. To make a practical device, the conductivity of semiconductors must be increased greatly. This can be done by adding certain other atoms, or impurities, which will create **extrinsic semiconductors.** Impurity atoms, often called **dopants,** increase conductivity by adding either electrons or holes to a semiconductor.

n-type semiconductors

There are two kinds of semiconductors: those that conduct by means of electrons, and those that conduct by means of holes. The type of semiconductor that conducts by means of electrons is called an ***n*-type semiconductor** because conduction is by means of negatively charged particles. **Figure 29–6a** shows how a few dopant atoms replace silicon atoms in the crystal. Arsenic (As) atoms that have five valence electrons can be used as a dopant. Four of the five electrons bind to neighboring silicon atoms because silicon needs electrons to fill its valence band. The fifth electron is not needed in bonding and so can move relatively freely. This is called the donor electron. The energy of this donor electron is so close to the conduction band that thermal energy can easily remove it from the impurity atom, putting an electron in the conduction band, as shown in **Figure 29–7a.**

p-type semiconductors

The type of semiconductor that conducts by means of holes is called a ***p*-type semiconductor.** What kind of dopant atom can be used to create holes? A gallium (Ga) atom, for example, has only three valence electrons. If a gallium atom replaces a silicon atom, one binding electron is missing, as shown in **Figure 29–6b.** The gallium atom is called

All Aboard! 

Metals become better conductors when they are cooled. Semiconductors become better conductors when they are heated. Does a thermistor act like a metal or a semiconductor?

Make a series circuit with a low-voltage DC power supply, a thermistor, and an ammeter (0-100 mA scale). Slowly turn up the power supply until the needle is in the middle of the scale (50 mA). The voltage will be about 0.6 V. Watch what happens to the current when you hold the thermistor between your fingers. Describe the results.

Comparing and Contrasting List several possible advantages of thermistors over standard thermometers.

FIGURE 29–7 In an *n*-type semiconductor, donor energy levels place electrons in the conduction band **(a).** In a *p*-type semiconductor, acceptor energy levels result in holes in the valence band **(b).**

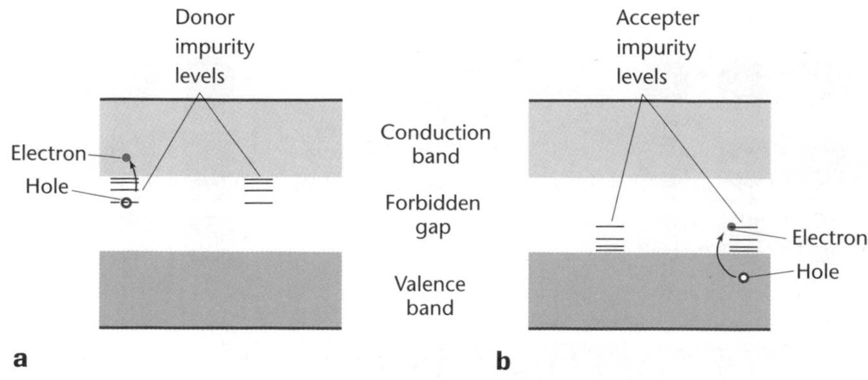

an electron acceptor. That is, the gallium atom creates a hole in the silicon semiconductor. Only thermal energy is needed to excite electrons from the valence band into this hole creating a hole on a silicon atom that is free to move through the crystal. Conduction is the result of the motion of positively charged holes in the valence band, as shown in **Figure 29–7b.** Both *p*-type and *n*-type semiconductors are electrically neutral. Adding dopant atoms of either type does not add any net charge to a semiconductor. If there are free electrons, then there are the same number of positively charged atoms. When a semiconductor conducts electricity by means of holes, there is a corresponding number of negatively charged atoms.

Silicon is doped by putting a silicon crystal in a vacuum with a sample of the impurity material. The impurity material is heated until it is vaporized, and the atoms condense on the cold silicon. When the silicon is warmed gently, the impurities diffuse into the material. Only a few impurity atoms per million silicon atoms are needed to increase the conductivity of the semiconductor by a factor of 1000 or more. Thus, the electrical properties of a semiconductor can be determined by controlling the number of impurity atoms doped into it. Finally, a thin layer of aluminum or gold is evaporated on the crystal, and a wire is welded to the conductor. The wire allows the user to put current into and bring it out of the doped silicon.

F.Y.I.

Silicon circuits are now being built on a layer of sapphire crystal. Unlike semiconductor bases that allow some current to pass through, the sapphire layer is nonconductive. Engineers may now be able to put several integrated chips for a cellular phone on a single chip.

Thermistors

The electrical conductivity of intrinsic and extrinsic semiconductors is sensitive to both temperature and light. Unlike metals in which conductivity is reduced when the temperature rises, an increase in temperature of a semiconductor allows more electrons to reach the conduction band. Thus, conductivity increases and resistance decreases. One semiconductor device, the thermistor, is designed so that its resistance depends very strongly on temperature. Thus, the thermistor can be used as a sensitive thermometer and to compensate for temperature variations of other components in an electrical circuit. Thermistors can also be used to measure the power of radio-frequency, infrared, and visible light sources.

Example Problem

The Conductivity of Doped Silicon

Silicon is doped with arsenic so that one in every million Si atoms is replaced by an arsenic atom. Each As atom donates one electron to the conduction band.

a. What is the density of free electrons?

b. By what ratio is this density greater than that of intrinsic silicon with 1×10^{13} free e^-/cm^3?

c. Is conduction mainly by the electrons of the silicon or the arsenic?

Calculate Your Answer

Known:

1 As atom/10^6 Si atoms

1 free e^-/As atom

4.99×10^{22} Si atoms/cm^3

1×10^{13} free e^-/cm^3 in intrinsic Si

Unknown:

a. free e^-/cm^3 donated by As

b. ratio of As-donated free e^- to intrinsic free e^-

Strategy:

From the density of Si atoms, find the density of As atoms. Because each As atom donates one e^-, this number is the density of free electrons.

Calculations:

a. $$\frac{\text{free e}^-}{\text{cm}^3} = \left(\frac{1 \text{ free e}^-}{1 \text{ As atom}}\right)\left(\frac{1 \text{ As atom}}{1 \times 10^6 \text{ Si atoms}}\right)\left(\frac{4.99 \times 10^{22} \text{ Si atoms}}{\text{cm}^3}\right) = 4.99 \times 10^{16} \text{ free e}^-/\text{cm}^3$$

b. Ratio is $$\frac{4.99 \times 10^{16} \text{ free e}^-/\text{cm}^3 \text{ in doped Si}}{1 \times 10^{13} \text{ free e}^-/\text{cm}^3 \text{ in intrinsic Si}} = 4.99 \times 10^3$$

c. Because there are 5000 arsenic-donated electrons for every intrinsic electron, conduction is mainly by the arsenic-donated electrons.

Check Your Answer

- Are the units correct? Using the factor-label method confirms the correct units.
- Is the magnitude reasonable? The ratio is large enough so that intrinsic electrons make almost no contribution to conductivity.

Practice Problems

3. If you wanted to have 5×10^3 as many electrons from As doping as thermally free electrons in the germanium semiconductor described in Practice Problem 2, how many As atoms should there be per Ge atom?

Light meters

Other useful applications of semiconductors depend on their light sensitivity. When light falls on a semiconductor, the light can excite electrons from the valence band to the conduction band, in the same way that light excites atoms. Thus, the resistance decreases as the light intensity increases. Materials such as silicon and cadmium sulfide are used as light-dependent resistors in light meters used by astronomers to measure the brightness of stars; by lighting engineers to design the illumination of stores, offices, and homes; and by photographers to capture the best image, as shown in **Figure 29–8.**

FIGURE 29–8 Photographers sometimes use a light meter to measure the intensity of incident light on an object. The meter converts the measurement into a unit that tells the photographer what exposure to set on the camera.

29.1 Section Review

1.1 In which type of material, a conductor, a semiconductor, or an insulator, are electrons most likely to remain with the same atom?

1.2 Magnesium oxide has a forbidden gap of 8 eV. Is this material a conductor, an insulator, or a semiconductor?

1.3 You are designing an integrated circuit using a single crystal of silicon. You want to have a region with relatively good insulating properties. Should you dope this region or leave it as an intrinsic semiconductor?

1.4 **Critical Thinking** If the temperature increases, the number of free electrons in an intrinsic semiconductor increases. For example, raising the temperature by 10°C doubles the number of free electrons. Is it more likely that an intrinsic semiconductor or a doped semiconductor will have a conductivity that depends on temperature? Explain.

Electronic Devices 29.2

Today's electronic instruments, such as radios, televisions, CD players, and microcomputers, often use semiconductor devices that are combined on chips of semiconducting silicon a few millimeters wide. The chips contain not only regions of doped silicon that act as wires or resistors, but also areas where two or three differently doped regions are in contact. In these devices, current and voltage vary in more complex ways than are described by Ohm's law. Because the variation is not linear, the devices can change current from AC to DC and amplify voltages.

OBJECTIVES:

- **Describe** how diodes limit current to motion in only one direction.
- **Explain** how a transistor can amplify or increase voltage changes.

Diodes

The simplest semiconductor device is the **diode.** A diode consists of joined regions of *p*-type and *n*-type semiconductors. Rather than two separate pieces of doped silicon being joined, a single sample of intrinsic silicon is treated first with a *p*-dopant, then with an *n*-dopant. Metal contacts are coated on each region so that wires can be attached, as shown in **Figure 29–9.** The boundary between the *p*-type and *n*-type regions is called the junction. The resulting device therefore is called a ***pn*-junction diode.**

The holes and electrons in the *p*- and *n*-regions are affected by the junction. There are forces on the free-charge carriers in the two regions near the junction. The free electrons on the *n*-side are attracted to the positive holes on the *p*-side. The electrons readily move into the *p*-side and recombine with the holes. Holes from the *p*-side similarly move into the *n*-side, where they recombine with electrons. As a result of this flow, the *n*-side has a net positive charge, and the *p*-side has a net negative charge. These charges produce forces in the opposite direction that stop further movement of charge carriers. The region around the junction is left with neither holes nor free electrons. This region, depleted of charge carriers, is called the **depletion layer.** Because it has no charge carriers, it is a poor conductor of electricity. Thus, a junction diode consists of relatively good conductors at the ends that surround a poor conductor.

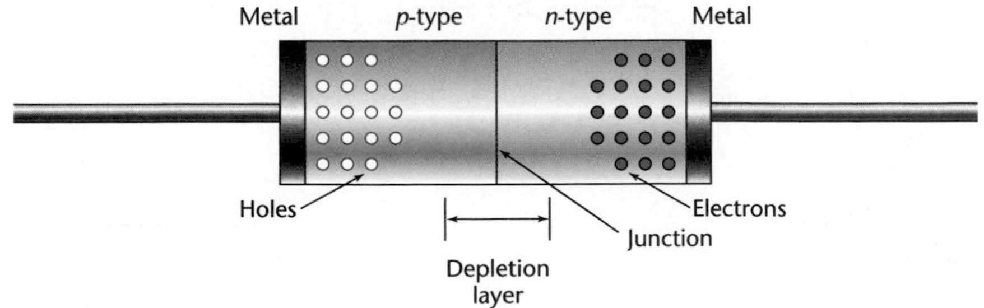

FIGURE 29–9 A diagram of the *pn* junction diode shows the depletion layer, where there are no charge carriers.

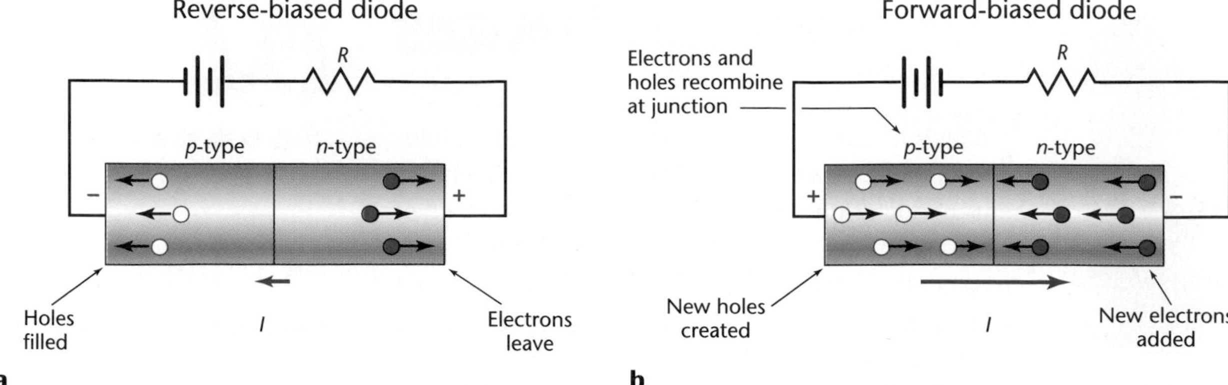

FIGURE 29–10 Compare the direction of current in a reverse-biased diode **(a)** and a forward-biased diode **(b).**

When a diode is connected into a circuit in the way shown in **Figure 29–10a,** both the free electrons in the *n*-type semiconductor and the holes in the *p*-type semiconductor are attracted toward the battery. The width of the depletion layer is increased, and no charge carriers meet. Almost no current flows through the diode: it acts like a very large resistor, almost an insulator. A diode oriented in this manner is a reverse-biased diode.

If the battery is connected in the opposite direction, as shown in **Figure 29–10b,** charge carriers are pushed toward the junction. If the voltage of the battery is large enough, 0.6 V for a silicon diode, electrons reach the *p*-end and fill the holes. The depletion layer is eliminated, and a current flows. The battery continues to supply electrons for the *n*-end. It removes electrons from the *p*-end, which is the same as supplying holes. With further increases in voltage from the battery, the current increases. A diode in this kind of circuit is a forward-biased diode.

The graph shown in **Figure 29–11** indicates the current through a silicon diode as a function of voltage across it. If the applied voltage is negative, the reverse-biased diode acts like a very high resistor; only a tiny current flows (about 10^{-11} A for a silicon diode). If the voltage is positive, the diode is forward-biased and acts like a small resistor, but not, however, one that obeys Ohm's law. One major use of a diode is to convert AC voltage to a voltage that has only one polarity. When a diode is used in a circuit such as that illustrated in **Figure 29–12,** it is called a rectifier. The arrow in the symbol for the diode shows the direction of conventional current.

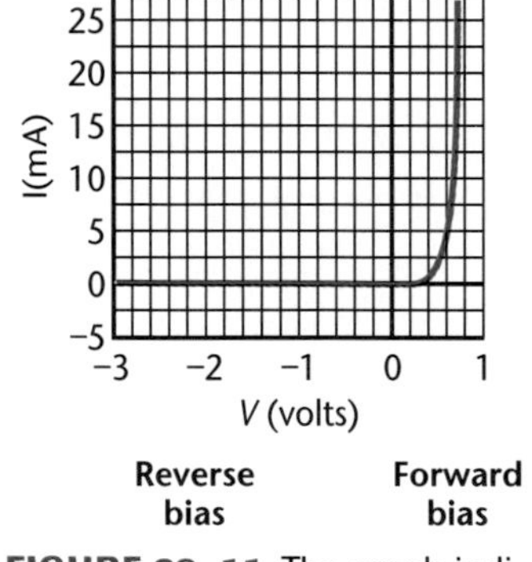

FIGURE 29–11 The graph indicates current-voltage characteristics for a silicon junction diode.

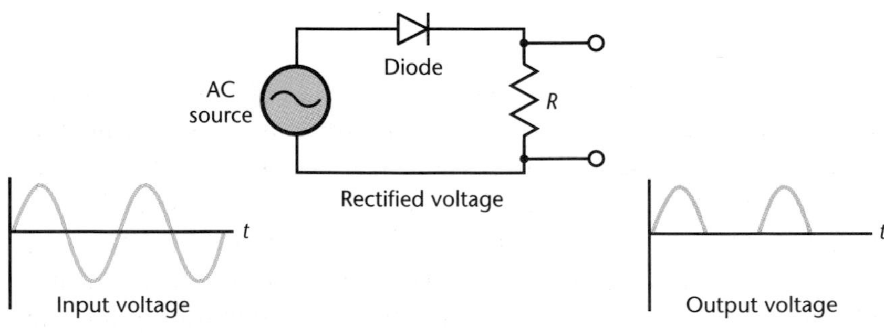

FIGURE 29–12 A diode can be used as a rectifier in a circuit.

Example Problem

A Diode in a Simple Circuit

A silicon diode, with *I/V* characteristics like those shown in **Figure 29–11,** is connected to a power supply through a 470-Ω resistor. The power supply forward-biases the diode, and its voltage is adjusted until the diode current is 12 mA. What is the battery voltage?

Sketch the Problem

- Draw a circuit diagram indicating the direction of current.

Calculate Your Answer

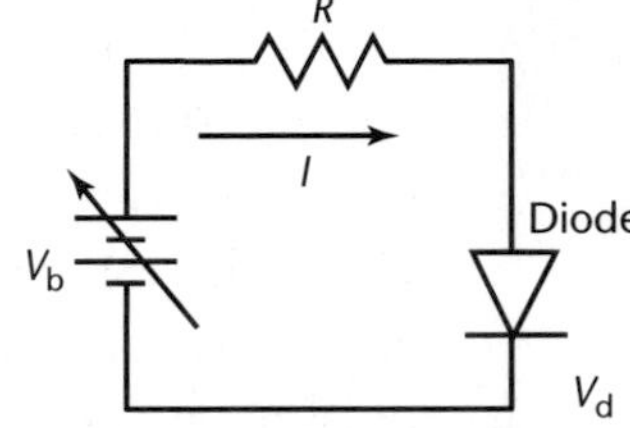

Known:	**Unknown:**
$I = 0.012$ A	$V_b = ?$
$V_d = 0.7$ V	
$R = 470\ \Omega$	

Strategy:

The voltage drop across the resistor is known from $V = IR$, and this is the difference between the battery voltage and the diode voltage drop.

Calculations:

$V_b = IR + V_d$

$V_b = (0.012\ \text{A})(470\ \Omega) + 0.7\ \text{V}$

$= 6.3\ \text{V}$

Check Your Answer

- Are the units correct? The battery's potential difference is in volts.
- Is the magnitude reasonable? Yes, it is larger than the diode voltage drop, but less than 12 V, which is typical of batteries.

Practice Problems

4. What battery voltage would be needed to produce a current of 2.5 mA in the diode in the preceding Example Problem?
5. A Ge diode has a voltage drop of 0.4 V when 12 mA flow through it. If the same 470-Ω resistor is used, what battery voltage is needed?

Diodes can do more than provide one-way paths for current. Diodes made from combinations of gallium and aluminum with arsenic and phosphorus emit light when they are forward-biased. When electrons reach the holes in the junction, they recombine and release the excess energy at the wavelengths of light. These diodes are called light-emitting diodes, or LEDs. Certain semiconductor crystals can be cut with parallel faces so that the light waves reflect back and forth in the crystal. The

FIGURE 29–13 Compare the circuit symbols used to represent a *pnp* transistor **(a)** and an *npn* transistor **(b)**.

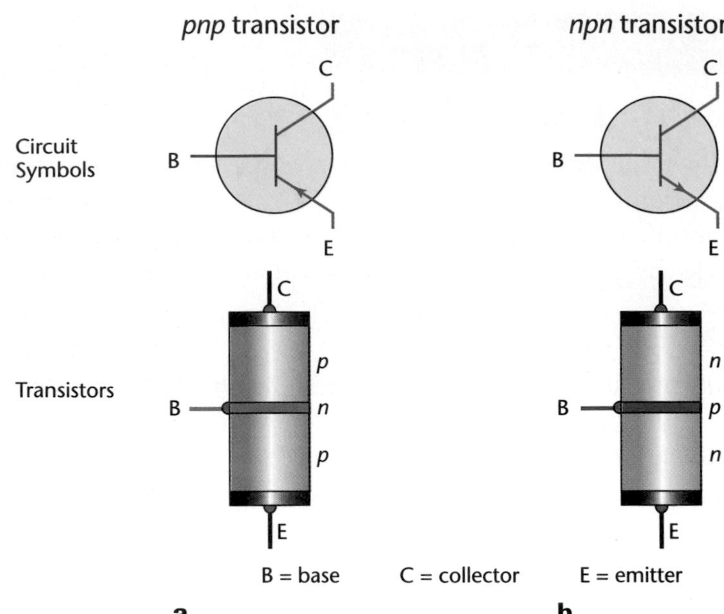

Red Light

Make a series circuit with a power supply, a 470-Ω resistor, and a red LED. Connect the short lead of the LED to the negative side of the power supply. Attach the other lead to the resistor. Hook the remaining resistor lead to the positive side of the power supply. Slowly increase the voltage until the LED glows. Note the voltage setting on the power supply.

Hypothesize What will happen if you reverse the direction of current? Why? Try it and explain what happens.

result is a diode laser that emits a narrow beam of coherent, monochromatic light, or infrared radiation. Diode lasers are used in CD players and supermarket bar-code scanners. They are compact, powerful light sources.

Both CD players and supermarket scanners must detect the laser light reflected from the CD or bar code. Diodes can detect light as well as emit it. A reverse-biased *pn*-junction diode is usually used as a light detector. Light falling on the junction creates pairs of electrons and holes. These are pulled toward the ends of the diode, resulting in a current that depends on the light intensity.

Transistors and Integrated Circuits

A **transistor** is a simple device made of doped semiconducting material that is used in most electronic circuits. One example of a transistor consists of a region of one type of doped semiconductor sandwiched between layers of the opposite type. An *npn* transistor consists of *n*-type semiconductors surrounding a thin *p*-type layer. If the center is an *n*-type region, then the device is called a *pnp* transistor. In either case, the central layer is called the base. The two surrounding regions are the emitter and the collector. The schematic symbols for the two transistor types are shown in **Figure 29–13.** The arrow is on the emitter and shows the direction of conventional current.

The operation of an *npn* transistor is illustrated in **Figure 29–14.** The *pn*-junctions in the transistor can be thought of as two back-to-back diodes. The battery on the right keeps the potential difference between collector and emitter, V_{CE}, positive. The base-collector diode is reverse-biased, so there is no current. The battery on the left is connected so that the base is more positive than the emitter. That is, the base-emitter diode is forward-biased.

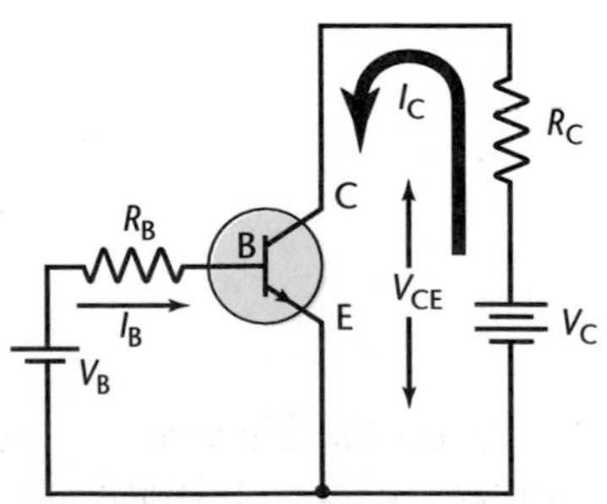

FIGURE 29–14 The circuit of an *npn* transistor demonstrates how voltage can be amplified.

Physics & Society

A Revolution in Robots

As our knowledge of solid state electronics advances, so does our ability to create smaller, more powerful computers called robots. The essential characteristics of a robot include the ability to be programmed to carry out a sequence of actions and to repeat those actions over and over again, as instructed.

Robots for hazardous tasks

Robots were first used for monotonous assembly-line tasks such as bolting together engine parts and handling molten metal. Now, they are used for work in hazardous or unusual environments. A robot rover called Pioneer, developed during the 1990s to clean toxic-waste storage tanks, helped decontaminate the damaged Chernobyl nuclear reactor in Russia by locating radioactive hot spots and providing details of structural damage.

Robots as explorers

Unlike humans, robots require no life-support systems, so they are less expensive to send on interplanetary expeditions. The *Mars Sojourner* rover was equipped with solar panels for power, wheels for travel, sensors and analyzers for examining rocks and soil, and communications equipment for receiving instructions from and sending data back to human operators on Earth.

Microrobots

Engineers have discovered that tiny robots are better than larger ones at handling tiny components, such as those used in the manufacture of electronic circuits. Microrobots have also become very important in medicine. Because tiny instruments can be inserted into the body through small incisions, they reduce the pain and trauma of surgery. A surgeon, viewing the body's interior with a remote camera, can control the robot to remove a gall bladder or repair knee ligaments. Some medical equipment is so precise that it can more accurately manipulate tissues than the larger hand of a skilled human surgeon.

Robots even smaller than those currently in use are in the works. Microtechnologists have created a working helicopter the size of a peanut. Some of its parts are a hundred times smaller than the diameter of a human hair. Medical technologists and computer makers are among those keeping a close watch on the microtechnology industry—an industry that develops the microscopic structures used in microdevices.

Investigating the Issue

1. **Using the Internet** Search for information about recent advances in artificial intelligence, then prepare to discuss the following questions. Has anyone invented a robot that can make human decisions? Do you think robots will ever be capable of human thought and emotions?
2. **Thinking Critically** Some researchers are trying to develop robots that look and act like humans. Would you prefer to work alongside a robot that looks like a machine or one that looks like a person? Why?

interNET CONNECTION

Follow the link for this chapter on the Glencoe Homepage at **www.glencoe.com/sec/science** to find out more about robots.

Physics Lab

The Stoplight

Problem

How can you design a circuit so that changing the direction of the current changes the LED that lights up?

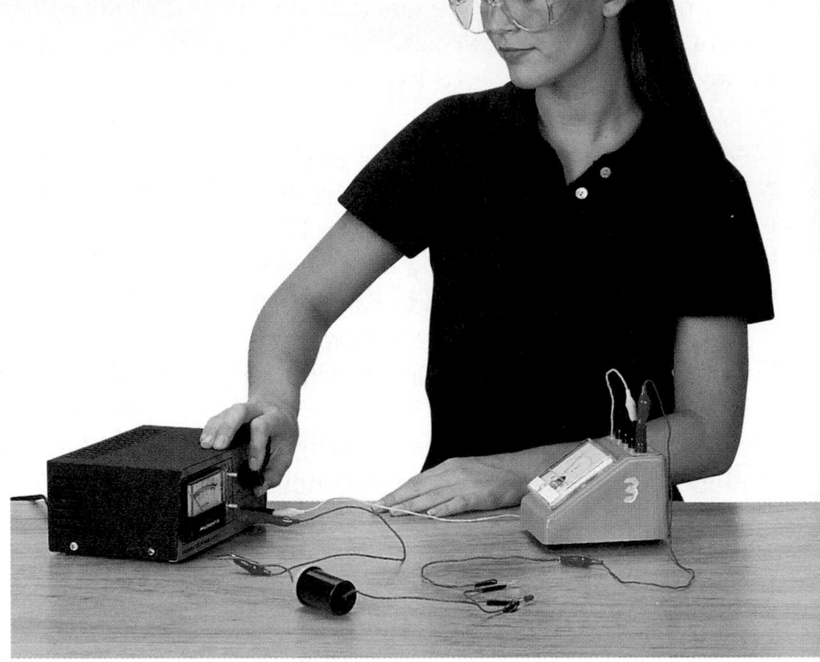

Materials

0- to 12-V variable power supply
red LED
green LED
bicolored LED
wires
470-Ω resistor
voltmeter

Procedure

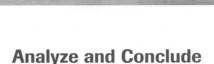

1. Connect a series circuit with the power supply, the resistor, and the red and green LEDs to light them both. Do not bypass or omit the resistor with an LED. Always have the resistor between an LED and one side of the power supply.
2. Reverse the direction of the current in the circuit and note the result. Measure the voltage across an LED.
3. Design a circuit so that changing the direction of the current will change the color that lights up.
4. Test your circuit.

Data and Observations

1. What voltage was needed to light the LEDs?
2. Describe what happened when the current was reversed.

Analyze and Conclude

1. **Diagramming a Circuit** Make a drawing to show your stoplight circuit (red on, green off; then green on, red off).
2. **Explaining Results** Why does your stoplight circuit work?
3. **Analyzing Results** Is your circuit a series or parallel circuit?
4. **Making Predictions** What change would you observe if you replaced the resistor with a 330-Ω resistor?

Apply

1. Design and conduct experiments to discover what type of LED the bicolored LED is. Remember to leave the resistor connected to the power supply.
2. How does an LED differ from a 60-W lightbulb?

There is conventional current in the direction of the arrow from the base into the emitter. Thus, electrons flow from the emitter into the base. But the base layer is very thin, often less than 10^{-6} m wide. As a result, most of the electrons pass through the base to the collector. The current through the collector, I_C, is much larger than the current through the base, I_B. The collector current causes a voltage drop across resistor R_C. Small changes in the voltage on the base produce large changes in the collector current, and thus changes in the voltage drop across R_C. As a result, the transistor amplifies small voltage changes into much larger changes. A *pnp* transistor works the same way, except that the potentials of both batteries are reversed, and holes carry the current in the transistor. The energy for amplification comes from the battery.

In a tape player, the small voltage variations from the voltage induced in a coil by magnetized regions on the tape are amplified to move the speaker coil. In computers, small currents in the base-emitter circuits can turn on or turn off large currents in the collector-emitter circuits. In addition, several transistors can be connected together to perform logic operations or to add numbers together. In this case, they act as fast switches rather than as amplifiers.

F.Y.I.

Materials using microelectromechanical systems are almost sentient. They can "sense" changes in temperature, pressure, and motion. They are able to actuate control mechanisms and communicate. Researchers foresee clouds of these tiny machines monitoring hurricanes, performing medical procedures, and controlling machine parts.

Microchips

An integrated circuit, or a microchip, consists of thousands of transistors, diodes, resistors, and conductors, each no more than a few micrometers across. All these components can be made by doping silicon, as shown in **Figure 29–15.** A microchip begins as an extremely pure single crystal of silicon, 10–20 cm in diameter and 1/2 m long. The silicon is sliced by a diamond-coated saw into wafers less than 1 mm thick. The circuit is then built layer by layer on the surface of this wafer.

The Inside Dope

FIGURE 29–15 A piece of ultrapure silicon forms the basis for an integrated circuit chip.

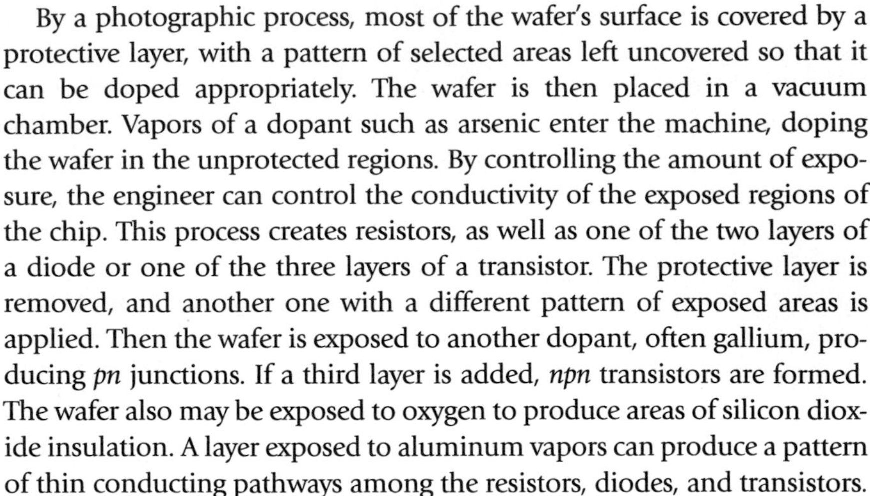

By a photographic process, most of the wafer's surface is covered by a protective layer, with a pattern of selected areas left uncovered so that it can be doped appropriately. The wafer is then placed in a vacuum chamber. Vapors of a dopant such as arsenic enter the machine, doping the wafer in the unprotected regions. By controlling the amount of exposure, the engineer can control the conductivity of the exposed regions of the chip. This process creates resistors, as well as one of the two layers of a diode or one of the three layers of a transistor. The protective layer is removed, and another one with a different pattern of exposed areas is applied. Then the wafer is exposed to another dopant, often gallium, producing *pn* junctions. If a third layer is added, *npn* transistors are formed. The wafer also may be exposed to oxygen to produce areas of silicon dioxide insulation. A layer exposed to aluminum vapors can produce a pattern of thin conducting pathways among the resistors, diodes, and transistors.

Hundreds of identical circuits, usually called chips, are produced at one time on a single wafer with a 10- to 20-cm diameter. The chips are then sliced apart and tested, attached to wires, and mounted in a plastic protective body. The tiny size of integrated circuits allows the placement of complicated circuits in a small space. Because electronic signals need to travel tiny distances, this miniaturization has increased the speed of computers.

Semiconductor electronics requires that physicists, chemists, and engineers work together. Physicists contribute their understanding of the motion of electrons and holes in semiconductors. Physicists and chemists together add precisely controlled amounts of impurities to extremely pure silicon. Engineers develop the means of mass-producing chips containing thousands of miniaturized diodes and transistors. Together, their efforts have brought our world into this electronic age.

FINE ARTS CONNECTION

Magic eye pictures, or stereograms, allow viewers to see three-dimensional images without using a viewing device. Computer software creates vertical strips of nearly identical dot patterns and combines them. Viewers must decouple their eyes, allowing each eye to see a different image. The brain matches the pattern one eye sees in a strip with the pattern the other eye sees in the next strip. The brain then fuses the two patterns into a three-dimensional image. Could you see a three-dimensional image with one eye closed?

29.2 Section Review

2.1 Compare the resistance of a *pn*-junction diode when it is forward-biased and reverse-biased.

2.2 In a light-emitting diode, which terminal should be connected to the *p*-end to make the diode light?

2.3 If the diode shown in **Figure 29–11** is forward-biased by a battery and a series resistor so that there is more than 10 mA of current, the voltage drop is always about 0.7 V. Assume that the battery voltage is increased by 1 V.

a. By how much does the voltage across the diode or the voltage across the resistor increase?

b. By how much does the current through the resistor increase?

2.4 **Critical Thinking** Could you replace an *npn* transistor with two separate diodes connected by their *p*-terminals? Explain.

CHAPTER 29 REVIEW

ey Terms

.1

rbidden gap

and theory

lence band

nduction band

ole

trinsic
miconductor

trinsic
miconductor

pant

type
miconductor

type
miconductor

2

ode

-junction
ode

pletion layer

nsistor

Summary

29.1 Conduction in Solids

- Electrical conduction may be explained by the band theory of solids.
- In solids, the allowed energy levels are spread into broad bands. The bands are separated by regions called the forbidden gaps, that is, by values of energies that electrons may not possess.
- In conductors, electrons can move because the valence band is only partially filled.
- Electrons in metals have a very fast random motion. A potential difference across the metal causes a very slow drift of the electrons.
- In insulators, more energy is needed to move electrons than is generally available.
- Conduction in semiconductors is usually the result of doping pure crystals with impurity atoms.
- *n*-type semiconductors conduct by means of free electrons, and *p*-type semiconductors conduct by means of holes.

29.2 Electronic Devices

- Diodes conduct charges in one direction only and can be used to produce current in one direction only.
- A transistor has alternate layers of *n*-type and *p*-type semiconductors and can amplify voltage changes.

Reviewing Concepts

Section 29.1

1. How do the energy levels in a crystal of an element differ from the energy levels in a single atom of that element?
2. Why does heating a semiconductor increase its conductivity?
3. What is the main current carrier in a *p*-type semiconductor?

Section 29.2

4. An ohmmeter is an instrument that places a potential difference across a device to be tested, measures the current, and displays the resistance of the device. If you connect an ohmmeter across a diode, will the current you measure depend on which end of the diode was connected to the positive terminal of the ohmmeter? Explain.
5. What is the significance of the arrowhead at the emitter in a transistor circuit symbol?
6. Redraw **Figure 29–14** as a *pnp* transistor.

Applying Concepts

7. The resistance of graphite decreases as temperature rises. Does graphite conduct electricity like copper or like silicon?
8. Which of the following materials would make a better insulator: one with a forbidden gap 8 eV wide, one with a forbidden gap 3 eV wide, or one with no forbidden gap?
9. Consider atoms of the three materials in question 8. From which material would it be most difficult to remove an electron?
10. How does the size of the gap between the conduction band and the valence band differ in semiconductors and insulators?
11. Which would make a better insulator, an extrinsic or an intrinsic semiconductor? Explain.
12. Silicon is doped with phosphorus. Will the dopant atoms be donors or acceptors? Will the semiconductor conduct with holes or electrons?

13. Doping silicon with gallium produces a *p*-type semiconductor. Why are the holes mainly on the silicon atoms rather than on the gallium atoms that caused the holes to be produced?
14. You use an ohmmeter to measure the resistance of a *pn*-junction diode. Would the meter show a higher resistance when the diode is forward-biased or reverse-biased?
15. If the ohmmeter in problem 14 shows the lower resistance, is the ohmmeter lead connected to the arrow side of the diode at a higher or lower potential than the lead connected to the other side?
16. If you dope pure germanium with gallium alone, do you produce a resistor, a diode, or a transistor?

Problems

Section 29.1

LEVEL 1

17. The forbidden gap in silicon is 1.1 eV. Electromagnetic waves striking the silicon cause electrons to move from the valence band to the conduction band. What is the longest wavelength of radiation that could excite an electron in this way? Recall that

$$E = \frac{1240 \text{ eV}\cdot\text{nm}}{\lambda}$$

18. A light-emitting diode (LED) produces green light with a wavelength of 550 nm when an electron moves from the conduction band to the valence band. Find the width of the forbidden gap in eV in this diode.
19. How many free electrons exist in a cubic centimeter of sodium? Its density is 0.971 g/cm^3, its atomic mass is 22.99 g/mole, and there is one free electron per atom.

LEVEL 2

20. At a temperature of 0°C, thermal energy frees 1.1×10^{12} e$^-$/cm^3 in pure silicon. The density of silicon is 2.33 g/cm^3, and the atomic mass of silicon is 28.09 g/mole. What is the fraction of atoms that have free electrons?
21. Use the periodic table to determine which of the following elements could be added to germanium to make a *p*-type semiconductor: B, C, N, P, Si, Al, Ge, Ga, As, In, Sn, and Sb.
22. Which of the elements listed in problem 21 would produce an *n*-type semiconductor?

Section 29.2

LEVEL 1

23. The potential drop across a glowing LED is about 1.2 V. In **Figure 29–16,** the potential drop across the resistor is the difference between the battery voltage and the LED's potential drop, 6.0 V − 1.2 V = 4.8 V. What is the current through
    a. the LED?
    b. the resistor?

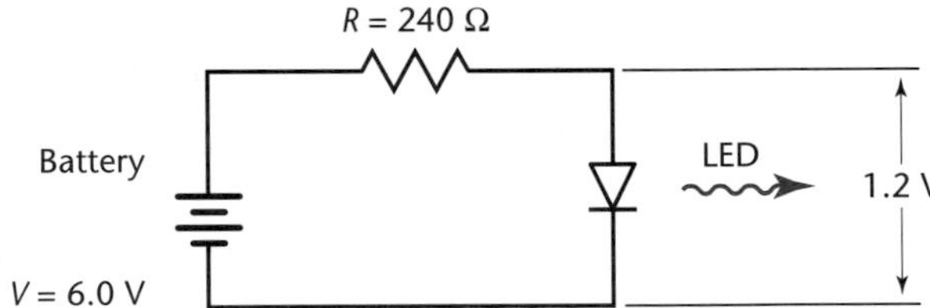

FIGURE 29–16

24. Jon wanted to raise the current through the LED in problem 23 up to 30 mA so that it would glow brighter. Assume that the potential drop across the LED is still 1.2 V. What resistor should be used?
25. **Figure 29–17** shows a battery, diode, and bulb connected in series so that the bulb lights. Note that the diode is forward-biased. State whether the bulb in each of the pictured circuits, 1, 2, and 3, is lighted.

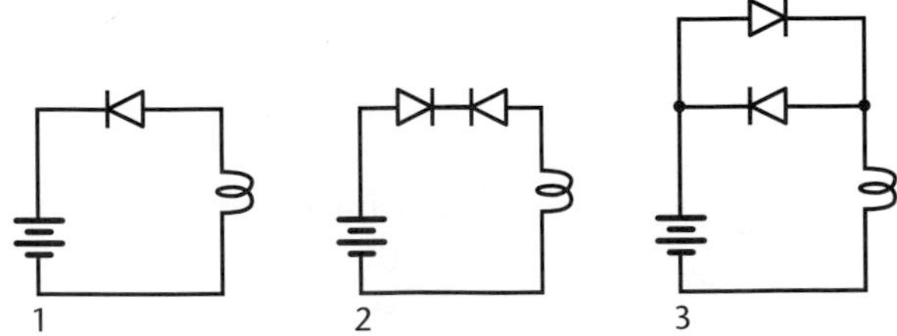

FIGURE 29–17

26. In the circuit shown in **Figure 29–18,** tell whether lamp L_1, lamp L_2, both, or neither is lighted.

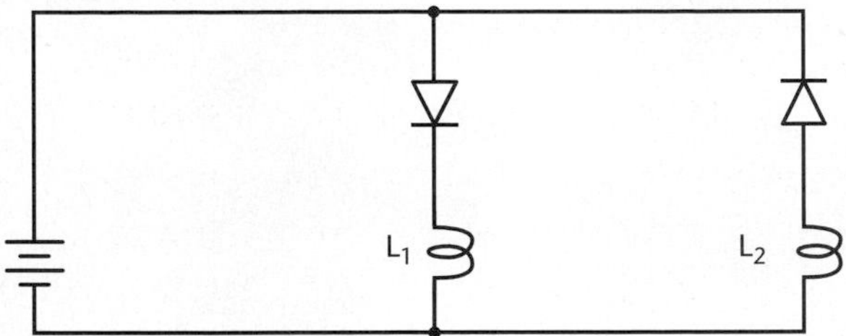

FIGURE 29–18

27. A silicon diode whose I/V characteristics are shown in **Figure 29–11** is connected to a battery through a 270-Ω resistor. The battery forward-biases the diode, and its voltage is adjusted until the diode current is 15 mA. What is the battery voltage?

28. What bulbs are lighted in the circuit shown in **Figure 29–19** when
a. switch 1 is closed and switch 2 is open?
b. switch 2 is closed and switch 1 is open?

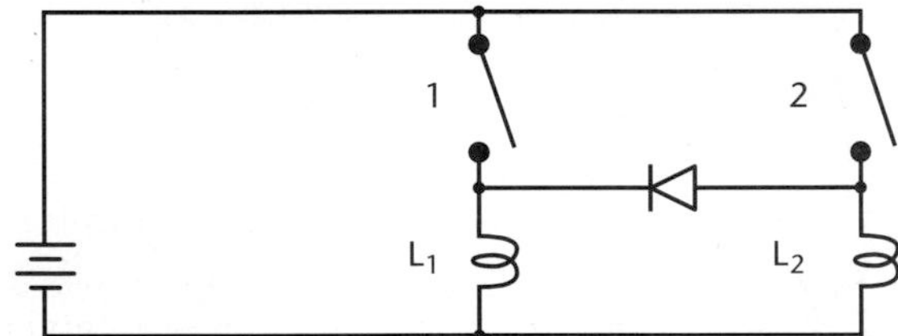

FIGURE 29–19

LEVEL 2

29. Which element or elements could be used as the second dopant used to make a diode, if the first dopant were boron?

Critical Thinking Problems

30. The I/V characteristics of two LEDs that glow with different colors are shown in **Figure 29–20.** Each is to be connected through a resistor to a 9.0-V battery. If each is to be run at a current of 0.040 A, what resistors should be chosen for each?

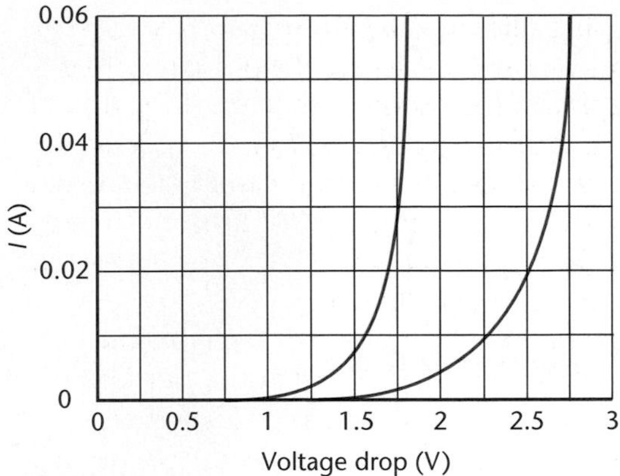

FIGURE 29–20

31. Suppose that the two LEDs in problem 30 are now connected in series. If the same battery is to be used and a current of 0.035 A is desired, what resistor should be used?

Going Further

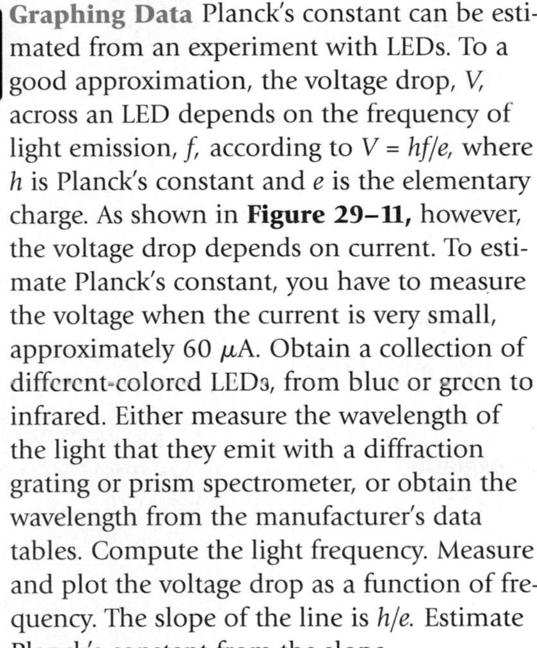

Graphing Data Planck's constant can be estimated from an experiment with LEDs. To a good approximation, the voltage drop, V, across an LED depends on the frequency of light emission, f, according to $V = hf/e$, where h is Planck's constant and e is the elementary charge. As shown in **Figure 29–11,** however, the voltage drop depends on current. To estimate Planck's constant, you have to measure the voltage when the current is very small, approximately 60 μA. Obtain a collection of different-colored LEDs, from blue or green to infrared. Either measure the wavelength of the light that they emit with a diffraction grating or prism spectrometer, or obtain the wavelength from the manufacturer's data tables. Compute the light frequency. Measure and plot the voltage drop as a function of frequency. The slope of the line is h/e. Estimate Planck's constant from the slope.

*inter***NET** CONNECTION

Follow the link on the Glencoe Homepage at **www.glencoe.com/sec/science** to find out more about this chapter.

Phantom Tracks

This technician's work seeks clues into the nature of the atomic nucleus. In the highlighted box are two particles moving in opposite spiral paths. The curvatures are about the same, so their momenta must be equal. What might cause this pair of paths?

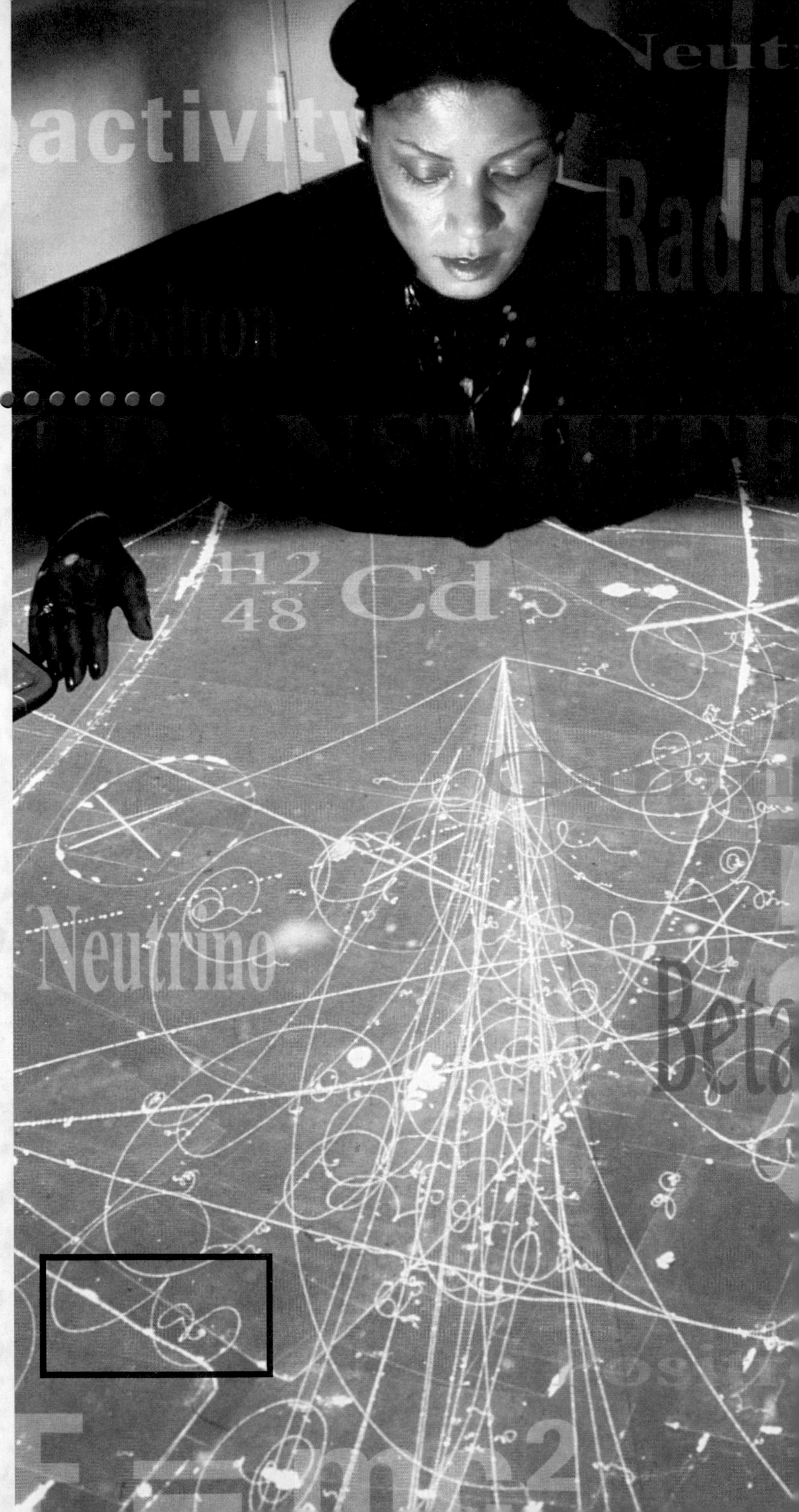

CHAPTER

30 The Nucleus

Ernest Rutherford not only established the existence of the nucleus, but he also conducted some of the early experiments to discover its structure. Later, scientists looked at the products that result when a nucleus breaks apart in radioactive decay. Today, modern accelerators and detectors have given physicists the ability to study nuclei and the particles that compose them with much greater precision than was possible during Rutherford's time. This photo, for example, shows a technician measuring the tracks of subatomic particles moving through a bubble chamber. The charged particles are bent by a magnetic field. The direction of the curve shows their charge. The faster they move, the less they bend. Thus, their momenta can be determined as well. We will see how bubble chambers and other tools have enlarged our knowledge of not only the nucleus, but of a whole new class of particles that may be the ultimate building blocks of matter.

WHAT YOU'LL LEARN

- You will describe the components of a nucleus and how radioactive decay affects these components.
- You will compare the various tools that physicists use to study the nucleus and nuclear reactions.

WHY IT'S IMPORTANT

- Atomic nuclei are used in various ways to determine what is happening inside the human body, from utilizing nuclei in magnetic resonance imaging (MRI) to using radioactive particles in positron emission tomography (PET).

interNET CONNECTION

Follow the link for this chapter on the Glencoe Homepage at **www.glencoe.com/sec/science** to find out more about the nucleus.

30.1 Radioactivity

A fter the discovery of radioactivity by Becquerel in 1896, many scientists studied this new phenomenon. French scientists Marie and Pierre Curie discovered the new elements polonium and radium in samples of radioactive uranium. In Canada, Ernest Rutherford and Fredrick Soddy used radioactivity to probe the center of the atom, the nucleus.

OBJECTIVES

- **Determine** the number of neutrons and protons in nuclides.
- **Describe** three forms of radioactive decay and **solve** nuclear equations.
- **Define** half-life and **calculate** the amount of material and its activity remaining after a given number of half-lives.

Description of the Nucleus

Recall from Chapter 28 how the nucleus was discovered. Ernest Rutherford directed a beam of α particles at metal foil and noticed that a few α particles were scattered at large angles. To explain the results, he hypothesized that the nucleus consisted of massive, positively charged particles. Around 1921, the name proton was adopted for these particles and each was defined as possessing one unit of elementary charge, e. In Chapter 20, you learned that elementary charge is the magnitude of charge existing on one electron, $e = 1.60 \times 10^{-19}$ C.

How was the other component of the nucleus, the neutron, discovered? Again, the scattering experiment played a key role. In 1909, only the mass of the nucleus was known. The charge of the nucleus was found as a result of X-ray scattering experiments done by Moseley, a member of Rutherford's team. The results showed that the positively charged protons accounted for roughly half the mass of the nucleus. One hypothesis was that the extra mass was the result of protons, and that electrons in the nucleus reduced the charge to the observed value. This hypothesis had some fundamental problems, however. In 1932, English physicist James Chadwick solved the problem when he discovered a neutral particle that had a mass approximately that of the proton. This particle, the neutron, accounted for the missing mass of the nucleus without increasing its charge.

FIGURE 30–1 Nuclear physicist Dr. Sekazi Mtingwa served as president of the National Society of Black Physicists from 1992 to 1994.

Mass and charge of the nucleus

The only charged particle in the nucleus is the proton. Therefore, the total charge of the nucleus is the number of protons, Z, times the elementary charge, e.

$$\textit{nuclear charge} = Z \times e$$

Both the proton and the neutron have mass that is approximately equal to 1u, where u is the **atomic mass unit,** 1.66×10^{-27} kg. To determine the approximate mass of the nucleus, multiply the number of neutrons and protons, A, by u.

$$\textit{nuclear mass} \approx A \times \mathrm{u}$$

The symbols Z and A are **atomic number** and **mass number,** respectively.

Size of the nucleus

As you know, the electrical force between positively charged protons is repulsive. The neutrons are not affected by the electrical force. What, then, holds the nucleus together? An attractive **strong nuclear force** is exerted by one proton on any other proton or neutron that is near it. This force, which also exists between neutrons, is stronger than the electrostatic repulsing force. Because of the strong nuclear force, the nucleus is held together at a diameter of about 10 fm (10^{-14} m). Typical atomic radii are on the order of 0.1 nm, which is 10 000 times larger than the size of the nucleus.

F.Y.I.

Science is awash with serendipity; science is hard work when done properly, but in the hard work there is joy and in the discovery there is abundant reward.

Sherwin B. Nuland
Doctors: The Biography of Medicine, 1988

Isotopes

Looking at the periodic table, you might notice that the first four elements all have mass number near a whole number. Boron, on the other hand, has a mass of 10.8 u. If, as was thought, the nucleus is made up of only protons and neutrons, each with a mass of approximately 1u, then the total mass of any atom should near a whole number.

The puzzle of atomic masses that were not whole numbers was solved with the mass spectrometer. You learned in Chapter 26 how the mass spectrometer demonstrated that an element could have atoms with different masses. For example, in an analysis of a pure sample of neon, not one, but two spots appeared on the film of the spectrometer. The two spots were produced by neon atoms of different masses. One variety of neon atom was found to have a mass of 20 u, the second type a mass of 22 u. All neutral neon atoms have ten protons in the nucleus and ten electrons in the atom. One kind of neon atom, however, has 10 neutrons in its nucleus, while the other has 12 neutrons. The two kinds of atoms are called isotopes of neon. The nucleus of an isotope is called a **nuclide.** All nuclides of an element have the same number of protons but have different numbers of neutrons, as illustrated by the hydrogen and helium nuclides shown in **Figure 30–2**. All isotopes of a neutral element have the same number of electrons around the nucleus and chemically behave the same way.

FIGURE 30–2 The nuclides of hydrogen **(a)** and helium **(b)** illustrate that all the nuclides of an element have the same numbers of protons but have different numbers of neutrons. Protons are red and neutrons are gray.

Hydrogen isotopes

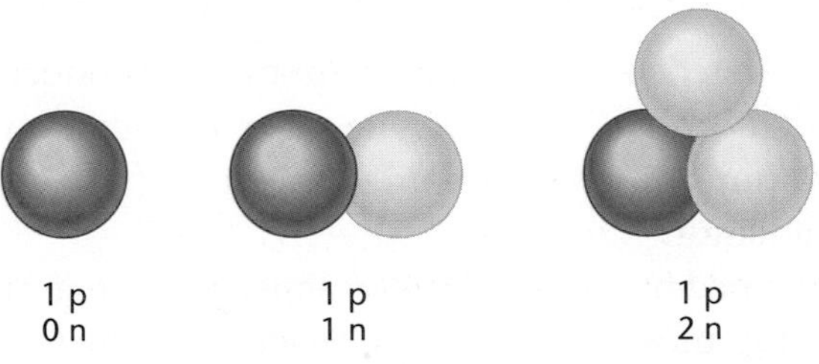

a

Helium isotopes

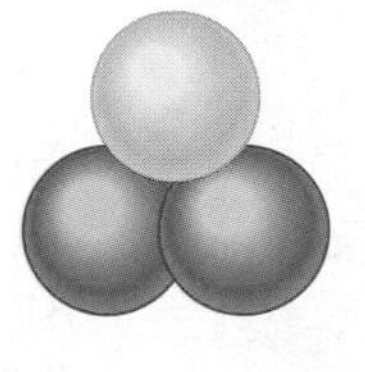

2 p
1 n

2 p
2 n

b

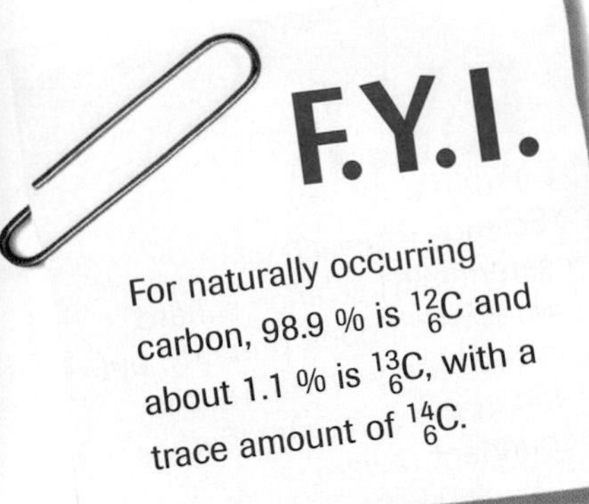

The measured mass of neon gas is 20.183 u. This figure is now understood to be the average mass of the naturally occurring isotopes of neon. Thus, while the mass of an individual atom of neon is close to a whole number of mass units, the atomic mass of an average sample of neon atoms is not. Most elements have several isotopic forms that occur naturally. The mass of one isotope of carbon, $^{12}_{6}C$, is now used to define the mass unit. One u is defined to be 1/12 the mass of the $^{12}_{6}C$ isotope.

A special method of notation is used to describe an isotope. A subscript representing the atomic number, Z, is written to the left of the symbol for the element. A superscript written to the left of the symbol is the mass number, A. This notation takes the form $^{A}_{Z}X$, where X is any element. For example, the two isotopes of neon, with atomic number 10, are written as $^{20}_{10}Ne$ and $^{22}_{10}Ne$.

Practice Problems

1. Three isotopes of uranium have mass numbers of 234, 235, and 238. The atomic number of uranium is 92. How many neutrons are in the nuclei of each of these isotopes?
2. An isotope of oxygen has a mass number of 15. How many neutrons are in the nuclei of this isotope?
3. How many neutrons are in the mercury isotope $^{200}_{80}Hg$?
4. Write the symbols for the three isotopes of hydrogen which have zero, one, and two neutrons in the nucleus.

HELP WANTED

NUCLEAR ENGINEER

Several federal departments need bright, conscientious nuclear engineers to continue to work on refining the production and distribution process of nuclear energy to make it the world's best and safest energy source. The key word is safety. If you have an advanced degree, a creative yet logical approach to problem solving, a desire to be part of an exciting team, and the ability to help convince the public about the future of our industry, Uncle Sam wants you. For information contact:
American Nuclear Society
555 North Kensington Ave.
LaGrange Park, IL 60525

Radioactive Decay

In 1896, Henri Becquerel was working with compounds containing the element uranium. To his surprise, he found that photographic plates that had been covered to keep out light became fogged, or partially exposed, when these uranium compounds were anywhere near the plates. This fogging suggested that some kind of ray had passed through the plate coverings. Several materials other than uranium or its compounds also were found to emit these penetrating rays. Materials that emit this kind of radiation are now said to be **radioactive** and to undergo radioactive decay.

In 1899, Rutherford discovered that uranium compounds produce three different kinds of radiation. He separated the types of radiation according to their penetrating ability and named them α (alpha), β (beta), and γ (gamma) radiation.

Alpha radiation can be stopped by a thick sheet of paper, while 6 mm of aluminum is needed to stop most beta particles. Several centimeters of lead are required to stop gamma rays, which have proved to be high-energy photons. Gamma rays are the most penetrating of the three because they have no charge. Rutherford determined that an alpha particle is the nucleus of helium atoms, ${}^{4}_{2}\text{He}$. By having a +2 charge, alpha particles interact strongly with matter, in fact, just a few centimeters of air are enough to absorb them. Beta particles were later identified as high-speed electrons. Moving faster and having less charge allows beta particles a greater penetrating ability. **Figure 30–3** illustrates the difference in charge of these three particles.

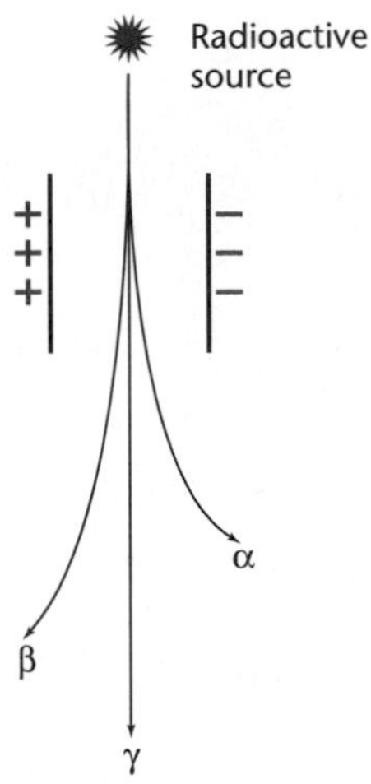

FIGURE 30–3 Alpha, beta, and gamma emissions behave differently in an electric field. Alpha and beta particles are deflected because of their charge.

Alpha decay

The emission of an α particle is a process called **alpha decay.** Because α particles consist of two protons and two neutrons, they must come from the nucleus of an atom. The nucleus that results from α decay will have a mass and charge different from those of the original nucleus. A change in nuclear charge means that the element has been changed, or **transmuted,** into a different element. The mass number of an α particle, ${}^{4}_{2}\text{He}$, is 4, so the mass number, A, of the decaying nucleus is reduced by 4. The atomic number of ${}^{4}_{2}\text{He}$ is 2, and therefore the atomic number of the nucleus, Z, is reduced by 2.

For example, when ${}^{238}_{92}\text{U}$ emits an α particle, the atomic number, Z, changes from 92 to 90. From **Table D–6** of the Appendix, we find that $Z = 90$ is thorium. The mass number of the newly formed nucleus is $A = 238 - 4 = 234$. A thorium isotope, ${}^{234}_{90}\text{Th}$, is formed. The uranium isotope has been transmuted into thorium.

Beta decay

Beta particles are electrons emitted by the nucleus. However, the nucleus contains no electrons, so where do the electrons come from? **Beta decay** occurs when a neutron is changed to a proton within the nucleus. In all reactions, charge must be conserved. That is, the charge before the reaction must equal the charge after the reaction. In beta decay, when a neutron, charge 0, changes to a proton, charge +1, an electron charge, −1, also appears. Charge conservation is satisfied. Consequently, as a result of beta decay, a nucleus with N neutrons and Z protons ends up with a nucleus of N−1 neutrons and Z+1 protons. Another particle, an antineutrino, is also emitted in beta decay. The reason for the appearance of this small, massless, chargeless particle will be discussed in the next section. The symbol for an antineutrino is the Greek letter nu with a bar over it, ${}^{0}_{0}\bar{\nu}$.

Gamma decay

Gamma radiation results from the redistribution of the charge within the nucleus. The γ ray is a high-energy photon. Neither the mass number nor the atomic number is changed when a nucleus emits a γ ray in **gamma decay.** Gamma radiation often accompanies alpha and beta decay.

Radioactive elements often go through a series of successive decays, or transmutations, until they form a stable nucleus. For example, ${}^{238}_{92}\text{U}$ undergoes 14 separate transmutations before the stable lead isotope, ${}^{206}_{82}\text{Pb}$, is produced.

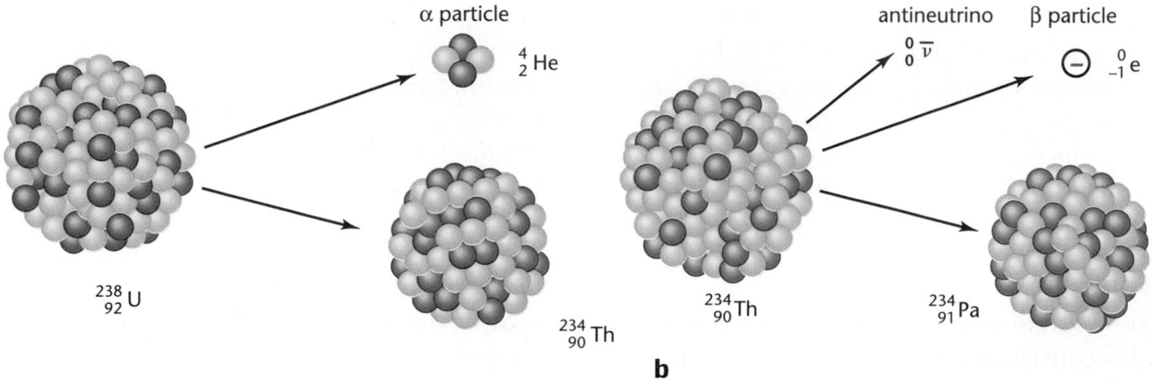

FIGURE 30–4 The emission of an alpha particle by uranium-238 results in the formation of thorium-234 **(a)**. The emission of a beta particle by thorium-234 results in the formation of protactinium-234 **(b)**.

Nuclear Reactions and Equations

A **nuclear reaction** occurs whenever the number of neutrons or protons in a nucleus changes. Just as in chemical reactions, some nuclear reactions occur with a release of energy; others occur only when energy is added to a nucleus.

One form of nuclear reaction is the emission of particles by radioactive nuclei. The reaction releases excess energy in the form of the kinetic energy of the emitted particles. One such reaction is the thorium-234 decaying into protactinium-234 by the emission of a high-speed electron and antineutrino, as shown in **Figure 30–4b.**

While nuclear reactions can be described in words or in pictures, as in **Figure 30–4,** they can be written more easily in equation form. The symbols used for the nuclei in nuclear equations make the calculation of atomic number and mass number in nuclear reactions simpler. For example, this is the word equation for the change of uranium to thorium resulting from α decay: uranium-238 yields thorium-234 plus an α particle. The nuclear equation for this reaction is as follows.

$$^{238}_{92}\text{U} \rightarrow {}^{234}_{90}\text{Th} + {}^{4}_{2}\text{He}$$

The total number of nuclear particles stays the same during the nuclear reaction. Thus, the sum of the superscripts on the right side of the equation must equal the sum of the superscripts on the left side of the equation. The sum of the superscripts on both sides of the equation is 238. Electric charge also is conserved. Thus, the sum of the subscripts on the right is equal to the sum of the subscripts on the left.

A β particle is an electron and is represented by the symbol $^{0}_{-1}$e. This indicates that the electron has one negative charge and a mass number of 0. The transmutation of a thorium atom by the emission of a β particle is shown in **Figure 30–4b.**

$$^{234}_{90}\text{Th} \rightarrow {}^{234}_{91}\text{Pa} + {}^{0}_{-1}\text{e} + {}^{0}_{0}\bar{\nu}$$

Note that the sum of the left-side superscripts equals the sum of the right-side superscripts. Equality must also exist between the left-side subscripts and the right-side subscripts.

Example Problem

Nuclear Equations: Alpha Decay

Write the equation for the process by which a radioactive radium isotope, ${}^{226}_{88}\text{Ra}$, emits an α particle and becomes the radon isotope ${}^{222}_{86}\text{Rn}$.

Calculate your Answer

Strategy:

The sum of the superscripts, A, and the sum of the subscripts, Z, must be equal on the two sides of the equation.

Calculations:

${}^{226}_{88}\text{Ra} \rightarrow {}^{222}_{86}\text{Rn} + {}^{4}_{2}\text{He}$

Check your Answer

- Sum of A: $226 = 222 + 4$
- Sum of Z: $88 = 86 + 2$

Example Problem

Nuclear Equations: Beta Decay

Write the equation for the decay of the radioactive lead isotope, ${}^{209}_{82}\text{Pb}$, into the bismuth isotope, ${}^{209}_{83}\text{Bi}$, by the emission of a β particle and an antineutrino.

Calculate your Answer

Strategy:

The sum of the superscripts, A, and the sum of the subscripts, Z must be equal on the two sides of the equation.

Calculations:

${}^{209}_{82}\text{Pb} \rightarrow {}^{209}_{83}\text{Bi} + {}^{0}_{-1}\text{e} + {}^{0}_{0}\bar{\nu}$

Check your Answer

- Sum of A: $209 = 209 + 0 + 0$
- Sum of Z: $82 = 83 - 1 + 0$

Practice Problems

5. Write the nuclear equation for the transmutation of a radioactive uranium isotope, ${}^{234}_{92}\text{U}$, into a thorium isotope, ${}^{230}_{90}\text{Th}$, by the emission of an α particle.
6. Write the nuclear equation for the transmutation of a radioactive thorium isotope, ${}^{230}_{90}\text{Th}$, into a radioactive radium isotope, ${}^{226}_{88}\text{Ra}$, by the emission of an α particle.

7. Write the nuclear equation for the transmutation of a radioactive radium isotope, $^{226}_{88}Ra$, into a radon isotope, $^{222}_{86}Rn$, by α decay.
8. A radioactive lead isotope, $^{214}_{82}Pb$, can change to a radioactive bismuth isotope, $^{214}_{83}Bi$, by the emission of a β particle and an antineutrino. Write the nuclear equation.

Pocket Lab

Background Radiation

Place a Geiger counter on the lab table far away from any sources of radiation. Turn the counter on and record the number of counts for a three-minute interval. Tape a piece of paper around the tube to cover the window and repeat the measurements.

Analyze and Conclude Did the count go down? What type of radiation could the counter be receiving? Explain.

Half-Life

The time required for half of the atoms in any given quantity of a radioactive isotope to decay is the **half-life** of that element. Each particular isotope has its own half-life. For example, the half-life of the radium isotope $^{226}_{88}Ra$ is 1600 years. That is, in 1600 years, half of a given quantity of $^{226}_{88}Ra$ decays into another element. In a second 1600 years, half of the remaining sample will have decayed. In other words, one fourth of the original amount still will remain after 3200 years.

The decay rate, or number of decays per second, of a radioactive substance is called its **activity.** Activity is proportional to the number of radioactive atoms present. Therefore, the activity of a particular sample is also reduced by one half in one half-life. Consider $^{131}_{53}I$, with a half-life of 8.07 days. If the activity of a certain sample is 8×10^5 decays per second when the $^{131}_{53}I$ is produced, then 8.07 days later its activity will be 4×10^5 decays per second. After another 8.07 days, its activity will be 2×10^5 decays per second. The activity of a sample is also related to its half-life. The shorter the half-life, the higher the activity. Consequently, if you know the activity of a substance and the amount of that substance, you can determine its half-life. The SI unit for decays per second is a Bequerel, Bq.

TABLE 30–1

Half-Life of Selected Isotopes

Element	Isotope	Half-Life	Radiation Produced
hydrogen	$^{3}_{1}H$	12.3 years	β
carbon	$^{14}_{6}C$	5730 years	β
cobalt	$^{60}_{27}Co$	30 years	β,γ
iodine	$^{131}_{53}I$	8.07 days	β,γ
lead	$^{212}_{82}Pb$	10.6 hours	β
polonium	$^{194}_{84}Po$	0.7 seconds	α
polonium	$^{210}_{84}Po$	138 days	α,γ
uranium	$^{235}_{92}U$	7.1×10^8 years	α,γ
uranium	$^{238}_{92}U$	4.51×10^9 years	α,γ
plutonium	$^{236}_{94}Pu$	2.85 years	α
plutonium	$^{242}_{94}Pu$	3.79×10^5 years	α,γ

Practice Problems

Refer to ***Figure 30–5*** *and* ***Table 30–1*** *to solve the following problems.*

9. A sample of 1.0 g of tritium, ${}^{3}_{1}H$, is produced. What will be the mass of the tritium remaining after 24.6 years?

10. The isotope ${}^{238}_{93}Np$ has a half-life of 2.0 days. If 4.0 g of neptunium is produced on Monday, what will be the mass of neptunium remaining on Tuesday of the next week?

11. A sample of polonium-210 is purchased for a physics class on September 1. Its activity is 2×10^{6} Bq. The sample is used in an experiment on June 1. What activity can be expected?

12. Tritium, ${}^{3}_{1}H$, was once used in some watches to produce a fluorescent glow so that the watches could be read in the dark. If the brightness of the glow is proportional to the activity of the tritium, what would be the brightness of such a watch, in comparison to its original brightness, when the watch is six years old?

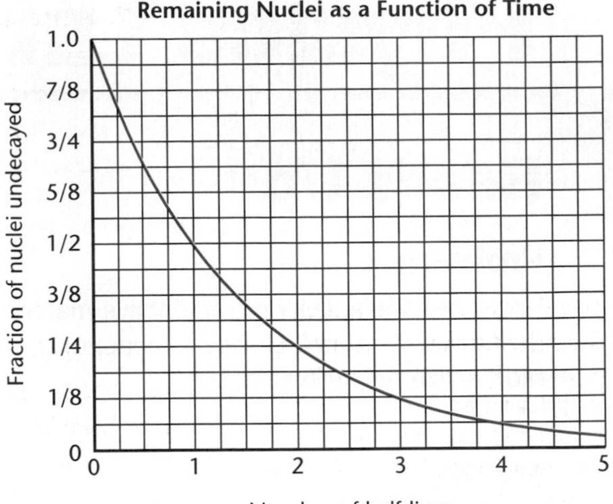

FIGURE 30–5

30.1 Section Review

1.1 Consider these two pairs of nuclei: ${}^{12}_{6}C$ and ${}^{13}_{6}C$ and ${}^{11}_{5}B$ and ${}^{12}_{6}C$. In which way are the two alike? In which way are they different?

1.2 How can an electron be expelled from a nucleus in β decay if the nucleus has no electrons?

1.3 Use **Figure 30–5** and **Table 30–1** to estimate in how many days a sample of ${}^{131}_{53}I$ would have 3/8 its original activity.

1.4 **Critical Thinking** Alpha emitters are used in smoke detectors. An emitter is mounted on one plate of a capacitor, and the α particles strike the other plate. As a result, there is a potential difference across the plates. Explain and predict which plate has the more positive potential.

Physics Lab

Heads Up

Problem

How does the activity of radioactive materials decrease over time? Devise a model of the radioactive decay system.

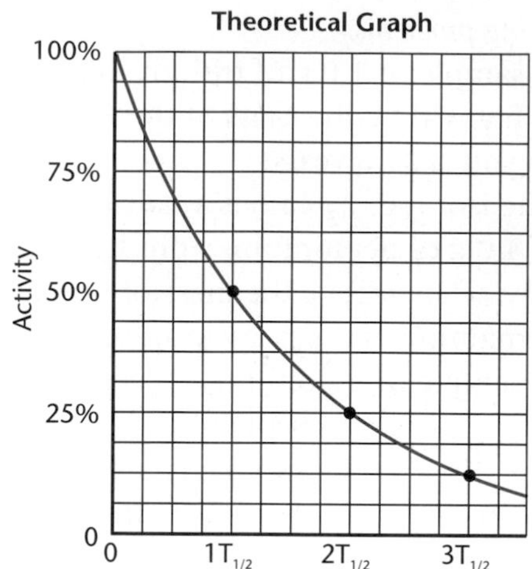

Materials

20 pennies
graph paper

Procedure

1. Set up a data table as shown. Turn the pennies so that they are all heads. In this simulation, a heads indicates that the nucleus has not decayed.
2. Flip each coin separately and put the heads and tails into separate piles.
3. Record the number of heads on your data sheet. Remove the pennies that came up tails.
4. Flip all remaining coins and separate the heads and tails. Count the number of heads and record the value.
5. Repeat steps 2-4 one more time.
6. Share your data with four other students and copy their data onto your data sheet.

Data and Observations						
	you	other students				tot
Begin	20	20	20	20	20	10
Trial 1						
Trial 2						
Trial 3						

Analyze and Conclude

1. **Comparing Data** Did each person have the same number of heads after each trial?
2. **Analyzing Data** Is the number of heads close to what you expected?
3. **Graphing Results** Total the number of heads remaining for each trial. Make a graph of the number of heads (vertical) verses the trial (horizontal).
4. **Interpreting Graphs** Compare your results to the theoretical graph shown in the lab. Propose explanations for any differences you notice.
5. **Understanding Procedures** Explain the rationale for collecting the results from other students and using the sum of all the results for graphing and analysis.

Apply

1. Laws mandate that hospitals keep radioactive materials for 10 half-lives before disposing of them. Calculate the fraction of the original activity left at the end of 10 half-lives.

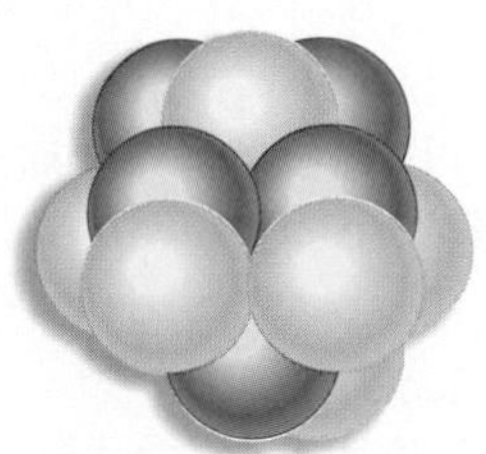

The Building Blocks of Matter

30.2

Why are some isotopes radioactive while others are stable? What holds the nucleus together against the repulsive force of the charged protons? These questions and many others motivated some of the best physicists to study the nucleus. The tiny size of the nucleus meant that new tools had to be developed for this study. Studies of nuclei have also led to an understanding of the structure of the particles found in the nucleus, the proton and the neutron, and the nature of the forces that hold the nucleus together.

OBJECTIVES

- **Describe** the operation of particle detectors and particle accelerators.
- **Define** antiparticles and **calculate** the energy of γ rays emitted when particles and their antiparticles annihilate one another.
- **Describe** the quark and lepton model of matter and **explain** the role of force carriers.

Nuclear Bombardment

The first tool used to study the nucleus was the product of radioactivity. Rutherford bombarded many elements with α particles, using them to cause a nuclear reaction. For example, when nitrogen gas was bombarded, Rutherford noted that high-energy protons were emitted from the gas. A proton has a charge of 1, while an α particle has a charge of 2. Rutherford hypothesized that the nitrogen had been artificially transmuted by the α particles. The unknown results of the transmutation can be written ${}^{A}_{Z}X$, and the nuclear reaction can be written as follows.

$$ {}^{4}_{2}\text{He} + {}^{14}_{7}\text{N} \rightarrow {}^{1}_{1}\text{H} + {}^{A}_{Z}\text{X} $$

Simple arithmetic shows that the atomic number of the unknown isotope is $Z = 2 + 7 - 1 = 8$. The mass number is $A = 4 + 14 - 1 = 17$. From **Table D–6** in the Appendix, you can see that the isotope must be ${}^{17}_{8}\text{O}$. The transmutation of nitrogen to oxygen is shown in **Figure 30–6.** The identity of this ${}^{17}_{8}\text{O}$ isotope was confirmed with a mass spectrometer several years after Rutherford's experiment.

Bombarding ${}^{9}_{4}\text{Be}$ with α particles produced a radiation more penetrating than any that had been discovered previously. In 1932, Irene Joliot-Curie (daughter of Marie and Pierre Curie) and her husband, Frederic Joliot, discovered that high-speed protons were expelled from paraffin wax that was exposed to this new radiation from beryllium. In the same year, James Chadwick showed that the particles emitted from

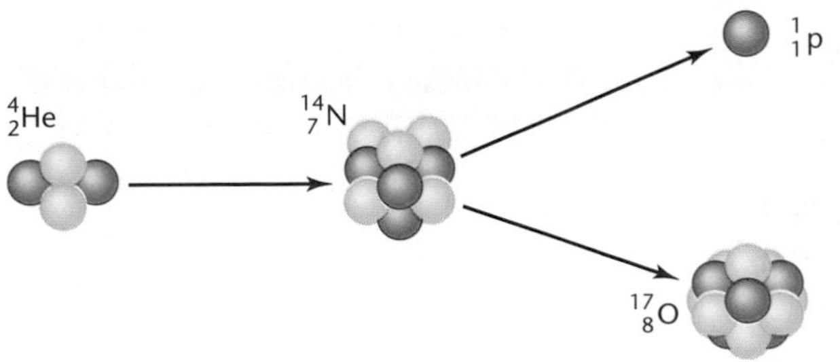

FIGURE 30–6
Oxygen-17 can be produced from the artificial transmutation of nitrogen.

beryllium were uncharged, but they had approximately the same mass as protons. This is how he discovered the neutron. The reaction can be written using the symbol for the neutron, ${}_{0}^{1}\text{n}$.

$$ {}_{2}^{4}\text{He} + {}_{4}^{9}\text{Be} \rightarrow {}_{6}^{12}\text{C} + {}_{0}^{1}\text{n} $$

Neutrons, being uncharged, are not repelled by the nucleus. As a result, neutrons often are used to bombard nuclei.

As Rutherford had showed, alpha particles are useful in producing nuclear reactions. Alpha particles from radioactive materials, however, have fixed energies. In addition, sources that emit a large number of particles per second are difficult to produce. Thus, methods of artificially accelerating particles to high energies are needed. Energies of several million electron volts are required to produce nuclear reactions. Consequently, several types of particle accelerators have been developed. The linear accelerator, the cyclotron, and the synchrotron are the accelerator types in greatest use today.

Linear Accelerators

A linear accelerator consists of a series of hollow tubes within a long evacuated chamber. The tubes are connected to a source of high-frequency alternating voltage, as illustrated in **Figure 30–7b.** Protons are produced in an ion source similar to that described in Chapter 26. When the first tube has a negative potential, protons are accelerated into it. There is no electric field within the tube, so the protons move at constant velocity. The length of the tube and the frequency of the voltage are adjusted so that when the protons have reached the far end of the tube, the potential of the second tube is negative in relation to that of the first. The resulting electric field in the gap between the tubes accelerates the protons into the second tube. This process continues, with the protons receiving an acceleration between each pair of tubes. The energy of the protons is increased by 10^5 eV with each acceleration. The protons ride along the crest of an electric field wave much as a surfboard moves on the ocean. At the end of the accelerator, the protons can have energies of many millions or billions of electron volts.

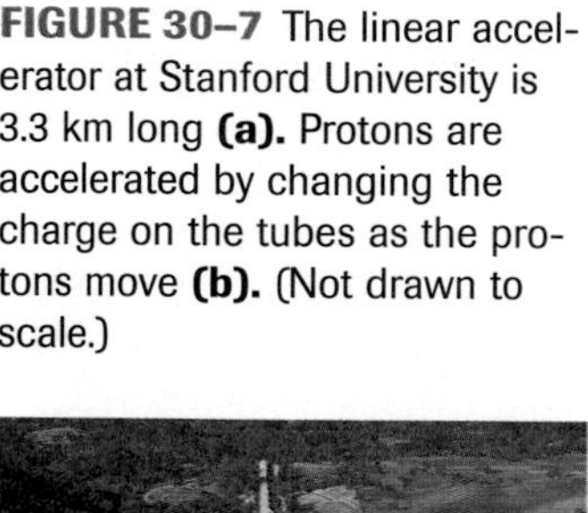

FIGURE 30–7 The linear accelerator at Stanford University is 3.3 km long **(a).** Protons are accelerated by changing the charge on the tubes as the protons move **(b).** (Not drawn to scale.)

a

b

Linear accelerators can be used with both electrons and protons. The largest linear accelerator, at Stanford University in California, is shown in **Figure 30–7a.** It is 3.3 km long and accelerates electrons to energies of 20 GeV (2×10^{10} eV).

The Synchrotron

An accelerator may be made smaller by using a magnetic field to bend the path of the particles into a circle. In a device known as a synchrotron, the bending magnets are separated by accelerating regions, as shown in **Figure 30–8a.** In the straight regions, high-frequency alternating voltage accelerates the particles. The strength of the magnetic field and the length of the path are chosen so that the particles reach the location of the alternating electric field precisely when the field's polarity will accelerate them. One of the largest synchrotrons in operation is at the Fermi National Accelerator Laboratory near Chicago, shown in **Figure 30–8b.** Protons there reach energies of 1 TeV (10^{12} eV). Two proton beams travel the circle in opposite directions. The beams collide in an interaction region and the results are studied.

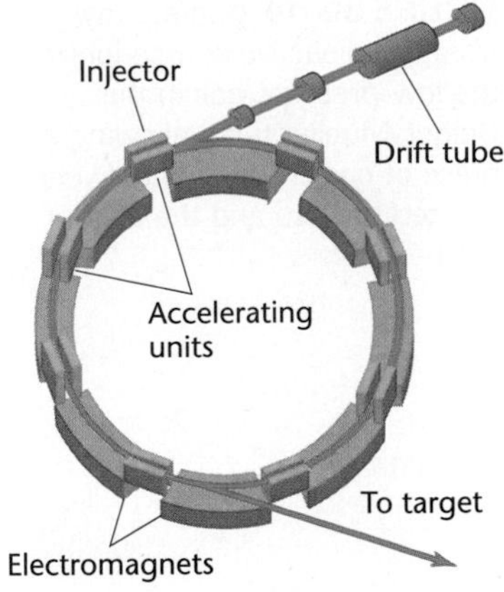

FIGURE 30–8 The synchrotron is a circular accelerator. Magnets are used to control the path and acceleration of the particles **(a).** Fermi Laboratory's synchrotron has a diameter of 2 km **(b).**

Particle Detectors

Once particles are produced, they need to be detected. In other words, they need to interact with matter in such a way that we can sense them with our relatively weak human senses. Your hand will stop an alpha particle, yet you will have no idea that the particle struck you. And as you read this sentence, billions of solar neutrinos pass through your body without so much as a twitch from you. Fortunately, scientists have devised tools to detect and distingwuish the products of nuclear reactions.

In the last section you read how uranium samples fogged photographic plates. When α particles, β particles, or γ particles strike photographic film, the film becomes fogged, or exposed. Thus, photographic film can be used to detect these particles and rays. Many other devices are used to detect charged particles and γ rays. Most of these devices make use of the fact that a collision with a high-speed particle will remove electrons from atoms. That is, the high-speed particles ionize the matter that they bombard. In addition, some substances fluoresce, or emit photons, when they are exposed to certain types of radiation. Thus, fluorescent substances also can be used to detect radiation. These three means of detecting radiation are illustrated in **Figure 30–9.**

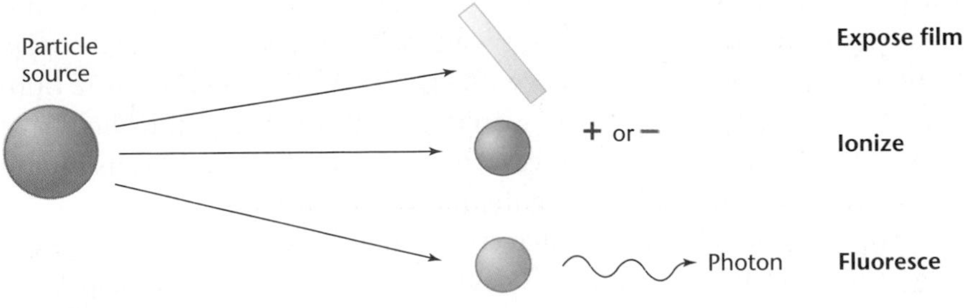

FIGURE 30–9 Particles can be detected when they interact with matter, exposing film, charging the matter, or causing the matter to emit a photon.

FIGURE 30–10 Gamma rays from a radioactive source ionize the low-pressure gas in the Geiger-Mueller tube, allowing a pulse of current to flow between the central wire and the copper tube.

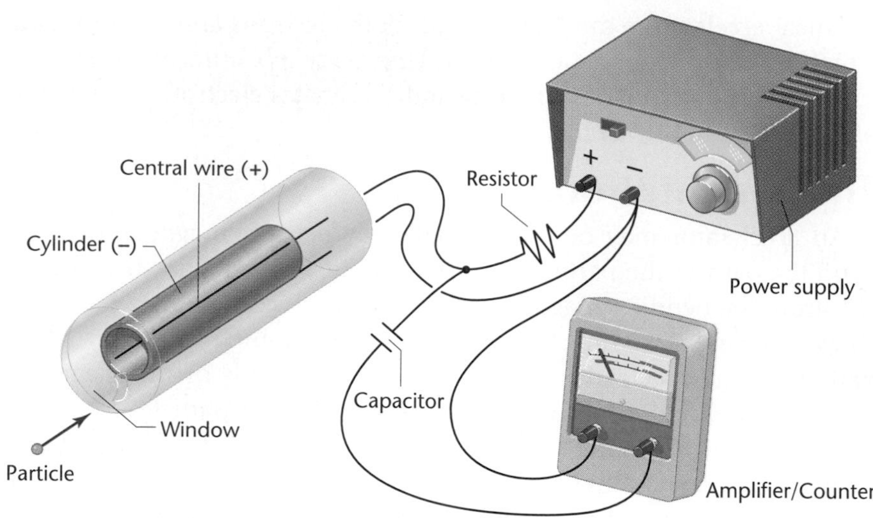

In the Geiger-Mueller tube, particles ionize gas atoms, as illustrated in **Figure 30–10.**The tube contains a gas at low pressure (10 kPa). At one end of the tube is a very thin window through which charged particles or gamma rays pass. Inside the tube is a copper cylinder with a negative charge. A rigid wire with a positive charge runs down the center of this cylinder. The voltage across the wire and cylinder is kept just below the point at which a spontaneous discharge, or spark, occurs. When a charged particle or gamma ray enters the tube, it ionizes a gas atom between the copper cylinder and the wire. The positive ion produced is accelerated toward the copper cylinder by the potential difference. The electron is accelerated toward the positive wire. As these charged particles move toward the electrodes, they strike other atoms and form even more ions in their path.

Thus, an avalanche of charged particles is created and a pulse of current flows through the tube. The current causes a potential difference across a resistor in the circuit. The voltage is amplified and registers the arrival of a particle by advancing a counter or producing an audible signal, such as a click. The potential difference across the resistor then lowers the voltage across the tube so that the current flow stops. Thus, the tube is ready for the beginning of a new avalanche when another particle or gamma ray enters it.

A device once used to detect particles was the Wilson cloud chamber. The chamber contained an area supersaturated with water vapor or ethanol vapor. When charged particles traveled through the chamber, leaving a trail of ions in their paths, the vapor tended to condense into small droplets on the ions. In this way, visible trails of droplets, or fog, were formed, as shown in **Figure 30–11.** In another detector, the bubble chamber, charged particles would pass through a liquid held just above the boiling point. In this case, the trails of ions would cause small vapor bubbles to form, marking the particle's trajectory.

Modern experiments use detection chambers that are like giant Geiger-Mueller tubes. Huge plates are separated by a small gap filled

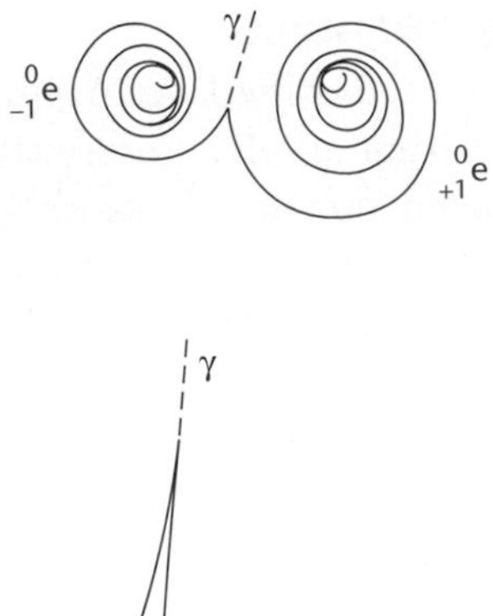

FIGURE 30–11 False-color bubble chamber photograph shows the symmetrical production of matter and antimatter.

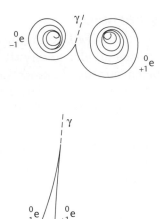

with a low-pressure gas. A discharge is produced in the path of a particle passing through the chamber. A computer locates the discharge and records its position for later analysis. Neutral particles do not produce discharges, and thus they do not leave tracks. The laws of conservation of energy and momentum in collisions can be used to tell if any neutral particles were produced. Other detectors measure the energy of the particles. The entire array of detectors used in high-energy accelerator experiments, such as the Collider Detector at Fermilab (CDF), can be up to three stories high, as shown in **Figure 30–12a.** The CDF is designed to monitor a quarter-million particle collisions each second, as though the detector functioned as a 5,000-ton camera, creating a computer picture of the collision events, as shown in **Figure 30–12b.**

The Elementary Particles

The model of the hydrogen atoms in 1930 was fairly simple: a proton, a neutron, and orbiting electrons. There also was a particle with no mass called a gamma particle (photon). Different combinations of these four **elementary particles** seemed to describe the contents of the universe quite well. Yet there were other elementary particles that needed to be explained, such as the antineutrino and a positive electron that was found in 1932. Physicists tried to answer the questions: Of what is matter composed? What holds matter together? by devising more sophisticated particle detectors and accelerators.

In 1935, a remarkable hypothesis by Japanese physicist Hideki Yukawa spurred much research in the years to follow. Recall from Chapter 27 how the electric field propagates through space by means of an electromagnetic wave, that is, a photon. The interaction of the field on a charge produces an electrical force. In essence, the electrical force is carried through space by a photon. Could a particle carry the nuclear force as well? Yukawa hypothesized the existence of a new particle that could carry the nuclear force through space. He predicted that this new particle would have a mass intermediate between the electron and the proton; hence, it was called the mesotron, later shortened to meson. In 1947 this particle was discovered and is called π-meson or simply pion.

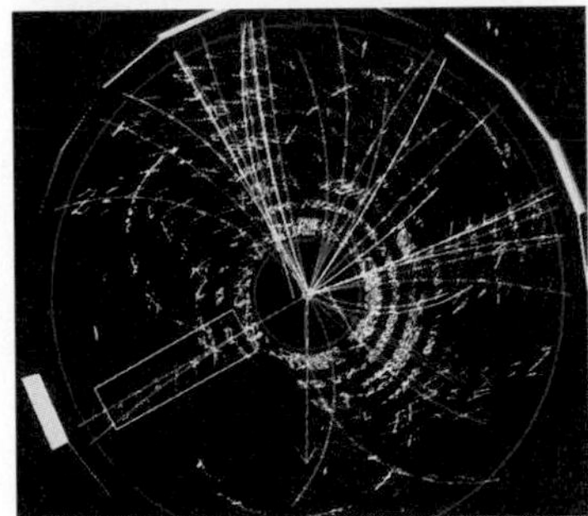

FIGURE 30–12 The Collider Detector at Fermilab (CDF) records the tracks from billions of collisions **(a).** A CDF computer image of a top quark event **(b).**

Follow the Tracks

Prepare a cloud chamber by soaking the cloth ring in alcohol. Place the radioactive needle into the side of the cloud chamber and then place the chamber on a block of dry ice. After the chamber cools down, you should be able to observe the tracks of the radiation.

Analyze and Conclude Predict what might happen when you place a small neodymium magnet in the bottom of the center of the chamber. Try it. Describe the results.

Fundamental types of elementary particles

Can the elementary particles be grouped or classified according to shared characteristics? Physicists now believe that all elementary particles can be grouped into three families labeled **quarks, leptons,** and **force carriers.** Quarks make up protons, neutrons, and mesons. Leptons are particles like electrons and antineutrinos. Quarks and leptons make up all the matter in the universe, whereas force carriers are particles that carry, or transmit, forces between matter. For example, photons carry the electromagnetic interaction. Eight particles, called gluons, carry the strong nuclear interaction that binds quarks into protons and the protons and neutrons into nuclei. Three particles, the weak bosons, are involved in the weak nuclear interaction, which operates in beta decay. The graviton is the name given to the yet-undetected carrier of the gravitational interaction. The properties of particles are summarized in **Table 30–2.**

Charges are given in units of the elementary charge, and masses are stated as energy equivalents, given by Einstein's formula, $E = mc^2$. The energy equivalent of these particles is much larger than the electron volt, and so is shown in MeV (mega-electron volts, or 10^6 eV) and GeV (giga-electron volts, or 10^9 eV).

Particles and Antiparticles

The α particles and γ rays emitted by radioactive nuclei have single energies that depend on the decaying nucleus. For example, the energy of the α particle emitted by thorium-234 is always 4.2 MeV. Beta particles, however, are emitted with a wide range of energies. One might expect the energy of the β particles to be equal to the difference between the energy of the nucleus before decay and the energy of the nucleus produced by the decay. In fact, the wide range of energies of electrons emitted during β decay suggested to Niels Bohr that energy might not be conserved in

TABLE 30–2

Quarks				Leptons			
Name	**Symbol**	**Mass**	**Charge**	**Name**	**Symbol**	**Mass**	**Charge**
down	d	333 MeV	$-1/3\ e$	electron	e	0.511 MeV	$-e$
up	u	330 MeV	$2/3\ e$	neutrino	ν_e	0	0

Force Carriers				
Force	**Name**	**Symbol**	**Mass**	**Charge**
Electromagnetic	photon	γ	0	0
Weak	Weak boson	W^+	80.2 GeV	$+e$
		W^-	80.2 GeV	$-e$
		Z^0	91.2 GeV	0
Strong	Gluon (8)	g	0	0
Gravitational	graviton (?)	G	0	0

nuclear reactions. Wolfgang Pauli in 1931 and Enrico Fermi in 1934 suggested that an unseen neutral particle was emitted with the β particle. Named the **neutrino** ("little neutral one" in Italian) by Fermi, the particle, which is actually an antineutrino, was not directly observed until 1956.

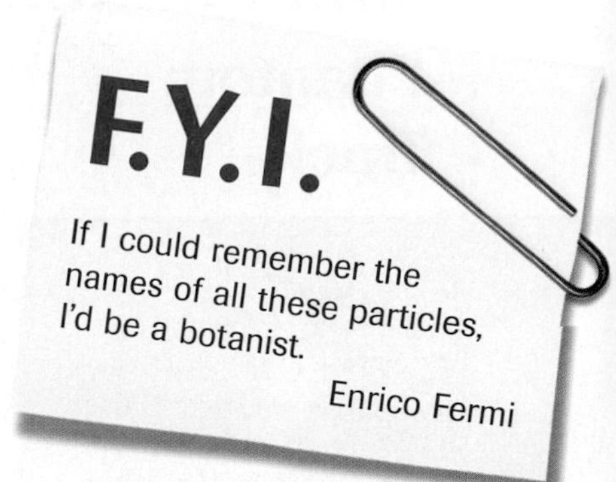

In a stable nucleus, the neutron does not decay. A free neutron, or one in an unstable nucleus, can decay into a proton by emitting a β particle. Sharing the outgoing energy with the proton and β particle is an antineutrino, ${}^{0}_{0}\overline{\nu}$. The antineutrino has zero mass and is uncharged, but like the photon, it carries momentum and energy. The neutron decay equation is written as follows.

$${}^{1}_{0}\text{n} \rightarrow {}^{1}_{1}\text{p} + {}^{0}_{-1}\text{e} + {}^{0}_{0}\overline{\nu}$$

When an isotope decays by emission of a **positron,** or antielectron, a process like β decay occurs. A proton within the nucleus changes into a neutron with the emission of a positron, ${}^{0}_{1}\text{e}$, and a neutrino, ${}^{0}_{0}\nu$

$${}^{1}_{1}\text{p} \rightarrow {}^{1}_{0}\text{n} + {}^{0}_{1}\text{e} + {}^{0}_{0}\nu$$

The weak nuclear interaction

The decay of neutrons into protons and protons into neutrons cannot be explained by the strong force. The existence of β decay indicates that there must be another interaction, the **weak nuclear force,** acting in the nucleus. This second type of nuclear force is much weaker than the strong nuclear force, and it can be detected only in certain types of radioactive decay.

Annihilation and production

The positron is an example of an **antiparticle,** a particle of antimatter. The electron and positron have the same mass and charge magnitude; however, the sign of their charge is opposite. When a positron and an electron collide, the two can annihilate each other, resulting in energy in the form of γ rays, as shown in **Figure 30–13.** Matter is converted directly into energy. The amount of energy can be calculated using Einstein's equation for the energy equivalent of mass.

$$E = mc^2$$

The mass of the electron is 9.11×10^{-31} kg. The mass of the positron is the same. Therefore, the energy equivalent of the positron and the electron together can be calculated as follows.

$$E = 2(9.11 \times 10^{-31}\text{ kg})(3.00 \times 10^{8}\text{ m/s})^2$$

$$E = 1.64 \times 10^{-13}\text{ J}\left(\frac{1\text{eV}}{1.60 \times 10^{-19}\text{J}}\right)$$

$$E = 1.02 \times 10^{6}\text{ eV or } 1.02\text{ MeV}$$

Recall from Chapter 27 that $1\text{eV} = 1.60 \times 10^{-19}$ J. When a positron and an electron at rest annihilate each other, the sum of the energies of the γ rays emitted is 1.02 MeV.

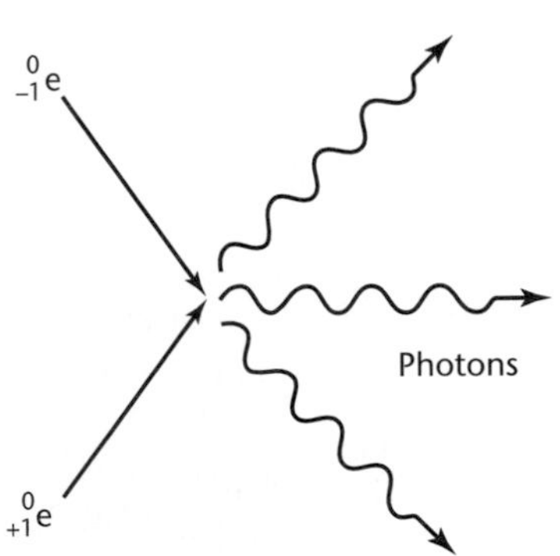

FIGURE 30–13 The collision of a positron and an electron results in gamma ray production.

Phantom Tracks

The inverse of annihilation also can occur. That is, energy can be converted directly into matter. If a γ ray with at least 1.02 MeV energy passes close by a nucleus, a positron and electron pair can be produced.

$$\gamma \rightarrow e^- + e^+$$

This is called **pair production.** Individual reactions such as $\gamma \rightarrow e^-$ and $\gamma \rightarrow e^+$, however, cannot occur, because such events would violate the law of conservation of charge. Matter and antimatter particles must always be produced in pairs.

The production of a positron-electron pair is shown in the chapter-opening photograph of bubble chamber tracks. A magnetic field around the bubble chamber causes the oppositely charged particles to curve in opposite directions. The γ ray that produced the pair produced no track. If the energy of the γ ray is larger than 1.02 MeV, the excess energy goes into kinetic energy of the positron and electron. The positron soon collides with another electron and they are both annihilated, resulting in the production of two or three γ rays with a total energy of no less than 1.02 MeV.

Particle conservation

Each quark particle and each lepton particle also has its antiparticle. The antiparticles are identical to the particles except that for charged particles, an antiparticle will have the opposite charge. When a particle and its antiparticle collide, they annihilate each other and are transformed into photons, or lighter particle-antiparticle pairs and energy. The total number of quarks and the total number of leptons in the universe is constant. That is, quarks and leptons are created or destroyed only in particle-antiparticle pairs. Consequently, the number of charge carriers is not conserved; the total charge, however, is conserved. On the other hand, force carriers such as gravitons, photons, gluons, and weak bosons can be created or destroyed if there is enough energy.

Antiprotons also can be created. An antiproton has a mass equal to that of the proton but is negatively charged. Protons have 1836 times as much mass as electrons. Thus, the energy needed to create proton-antiproton pairs is comparably larger. The first proton-antiproton pair was produced and observed at Berkeley, California in 1955. Neutrons also have an antiparticle, aptly named an antineutron.

F.Y.I.

The electromagnetic attraction of an electron and a positron is 4.2×10^{42} times stronger than their gravitational attraction.
-Isaac Asimov's *Book of Facts*

Practice Problems

13. The mass of a proton is 1.67×10^{-27} kg.

- **a.** Find the energy equivalent of the proton's mass in joules.
- **b.** Convert this value to eV.
- **c.** Find the smallest total γ ray energy that could result in a proton-antiproton pair.

How It Works

Smoke Detectors

A smoke detector is a sensing device that sounds an alarm when it detects the presence of smoke particles in the air. Two types are commonly available. One uses photons to detect the smoke; the other uses alpha particles and is more popular because it is less expensive and more sensitive.

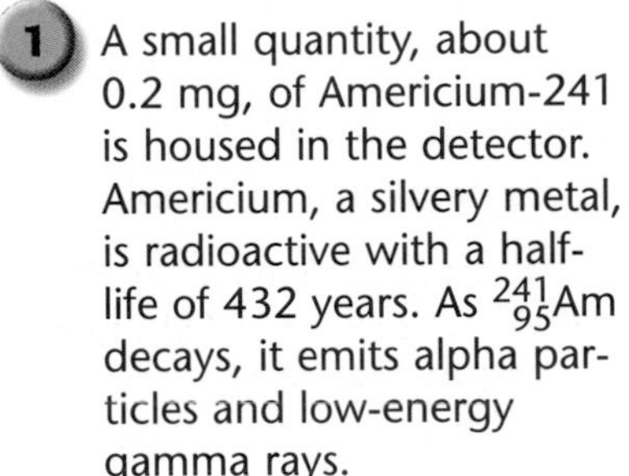

1. A small quantity, about 0.2 mg, of Americium-241 is housed in the detector. Americium, a silvery metal, is radioactive with a half-life of 432 years. As $^{241}_{95}Am$ decays, it emits alpha particles and low-energy gamma rays.

2. Alpha particles emitted by Am-241 collide with the oxygen and nitrogen in the air in the detector's chamber to produce ions.

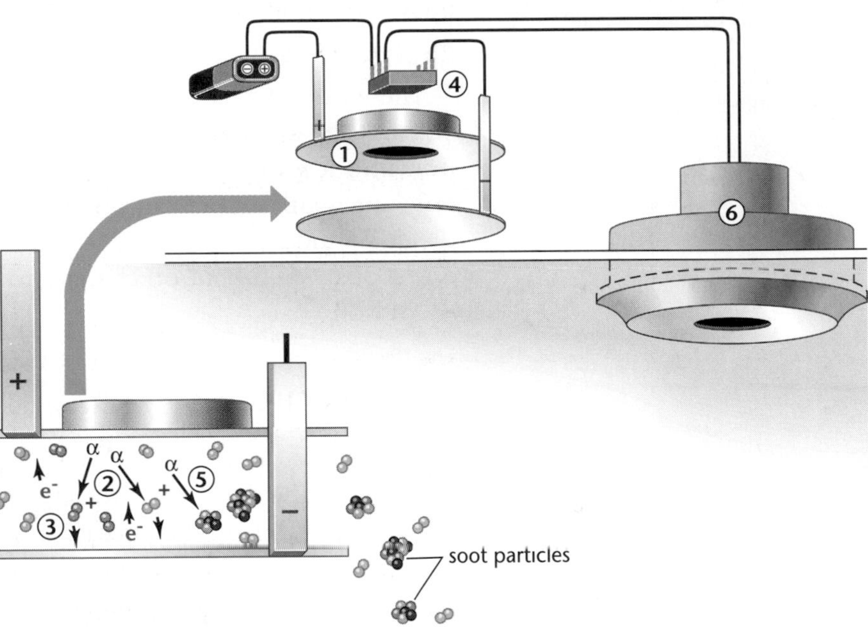

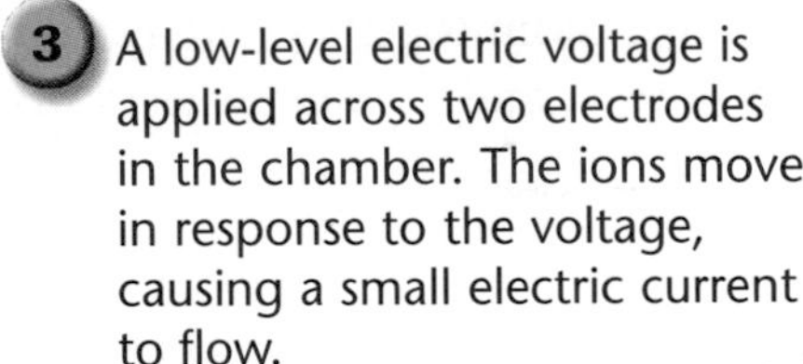

3. A low-level electric voltage is applied across two electrodes in the chamber. The ions move in response to the voltage, causing a small electric current to flow.

4. A battery-operated microchip monitors this current. Any decrease in the current is detected by the microchip.

5. As smoke enters the detector's chamber, soot particles absorb some of the alpha particles.

6. Fewer alpha particles means fewer ionized air molecules, and consequently there is less current across the electrodes. The microchip senses this change and activates the alarm.

Thinking Critically

1. Why do ionizing smoke detectors contain an isotope that undergoes alpha decay rather than beta or gamma decay?
2. Explain how ions permit an electric current in the detection chamber.

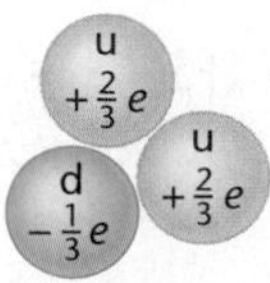

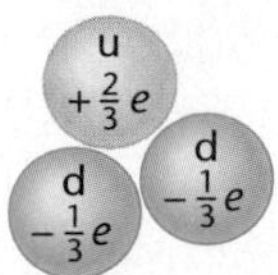

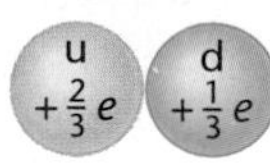

FIGURE 30–14 Even though quarks have fractional charge, all the particles they make have whole-number charge.

The Quark Model of Nucleons

The quark model describes nucleons, the proton and the neutron, as an assembly of quarks. The nucleons are each made up of three quarks. The proton has two up quarks, u, (charge $+2/3\ e$) and one down quark, d, (charge $-1/3\ e$), as shown in **Figure 30–14.** A proton is described as p = uud. The charge on the proton is the sum of the charges of the three quarks, $(2/3 + 2/3 + -1/3)e = +e$. The neutron is made up of one up quark and two down quarks, n = udd. The charge of the neutron is zero, $(2/3 + -1/3 + -1/3)e = 0$.

Individual quarks cannot be observed, because the strong force that holds them together becomes larger as the quarks are pulled farther apart. In this sense, the strong force acts like the force of a spring. It is unlike the electric force, which becomes weaker as charged particles are moved farther apart. In the quark model, the strong force is the result of the emission and absorption of gluons that carry the force.

Quark model of beta decay

The weak interaction involves three force carriers: W^+, W^-, and Z^0 bosons. The weak force exhibits itself in beta decay, the decay of a neutron into a proton, electron, and antineutrino. As was shown before, only one quark in the neutron and the proton is different. Beta decay in the quark model occurs in two steps, as shown in **Figure 30–15b.** First, one d quark in a neutron changes to a u quark with the emission of a W^- boson.

$$\mathrm{d} \rightarrow \mathrm{u} + \mathrm{W}^-$$

Then the W^- boson decays into an electron and an antineutrino.

$$\mathrm{W}^- \rightarrow \mathrm{e}^- + \overline{\nu}$$

Similarly, in the decay of a proton, a neutron and a W^+ boson are emitted. The weak boson then decays into a positron and a neutrino.

The emission of a Z^0 boson is not accompanied by a change from one quark to another. The Z^0 boson produces an interaction between the nucleons and the electrons in atoms that is similar to, but much weaker than, the electromagnetic force holding the atom together. The interaction was first detected in 1979. The W^+, W^-, and Z^0 bosons were first observed directly in 1983.

FIGURE 30–15 Beta decay can be represented by two different models, proton-neutron or quark.

Models of beta decay

a proton-neutron model: $^1_0\mathrm{n} \rightarrow {}^1_1\mathrm{p} + {}^0_{-1}\mathrm{e} + {}^0_0\overline{\nu}$

b quark model: udd $\rightarrow$ uud + W^- $\rightarrow$ uud + ${}^0_{-1}\mathrm{e} + {}^0_0\overline{\nu}$

Unification theories

The differences between the four fundamental interactions are evident: the forces may act on different quantities such as charge or mass, they may have different dependencies on distance, and the force carriers have different properties. However, there are some similarities among the interaction. For instance, the force between charged particles, the electromagnetic interaction, is carried by photons in much the same way as weak bosons carry the weak interaction. The electric force acts over a long range because the photon has zero mass, while the weak force acts over short distances because the W and Z bosons are relatively massive. The mathematical structures of the theories of the weak interaction and electromagnetic interaction, however, are similar. In the high-energy collisions produced in accelerators, the electromagnetic and weak interactions have the same strength and range.

Astrophysical theories of supernovae indicate that during massive stellar explosions, the two interactions are identical. Present theories of the origin of the universe suggest that the two forces were identical during the early moments of the cosmos as well. For this reason, the electromagnetic and weak forces are said to be unified into a single force, called the electroweak force.

In the same way that the electromagnetic and weak forces were unified into the electroweak force during the 1970s, physicists are presently trying to create a **grand unification theory** that includes the strong force as well. Work is still incomplete. Theories are being improved and experiments to test these theories are being planned. A fully unified theory that includes gravitation will require even more work.

LITERATURE CONNECTION

In 1964, an American physicist, Murray Gell-Mann, suggested that particles with a charge equal to 1/3*e* or 2/3*e* might exist. He named these particles "quarks," a word that comes from James Joyce's novel *Finnegan's Wake.*

Additional quarks and leptons

Bombardment of particles at high energies creates many particles of medium and large mass that have very short lifetimes. Some of these particles are combinations of two or three u or d quarks or a quark-antiquark pair. Combinations of the u and d quarks and antiquarks, however, cannot account for all the particles produced. Combinations of four other quarks are necessary to form all known particles. Two additional pairs of leptons also are produced in high-energy collisions. The additional quarks and leptons are listed in **Table 30–3.**

No one knows why there are six quarks and six leptons or whether there are still other undiscovered particles. Physicists are planning to build higher-energy accelerators to attempt to answer this question. **Figure 30–16** describes the fundamental building blocks of matter and where they might be found.

The branch of physics that is involved in the study of these particles is called elementary particle physics. The field is very exciting because new discoveries occur almost every week. Each new discovery, however, seems to raise as many questions as it answers. The question of what makes up the universe does not yet have a complete answer.

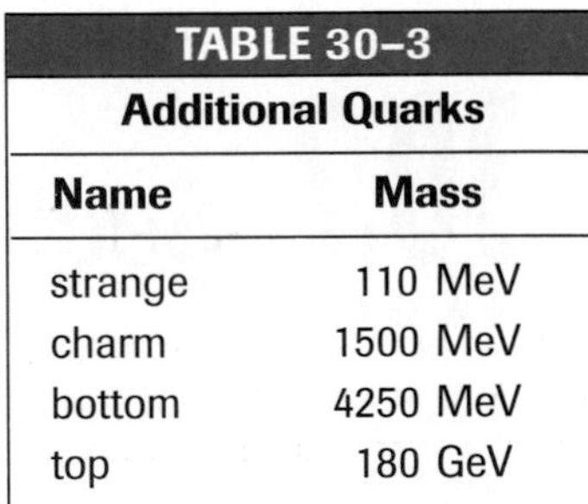

TABLE 30–3

Additional Quarks

Name	Mass
strange	110 MeV
charm	1500 MeV
bottom	4250 MeV
top	180 GeV

Additional Leptons

Name	Mass
muon	105.7 MeV
muon neutrino	0
tau	1777 MeV
tau neutrino	0

FIGURE 30–16 The known quarks and leptons are divided into three families. The everyday world is made from particles in the bottom family. Particles in the middle group are found in cosmic rays and are routinely produced in particle accelerators. Particles in the top family are believed to have existed briefly during the earliest moments of the Big Bang and are created in high-energy collisions.

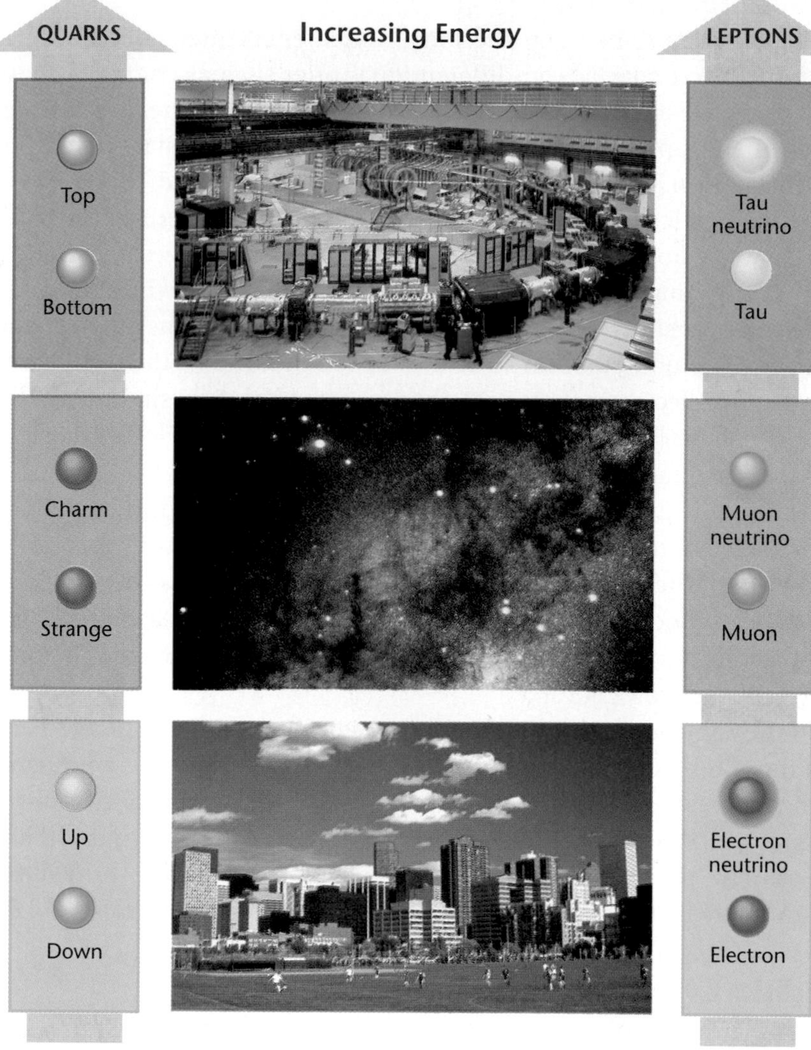

30.2 Section Review

2.1 Why would a more energetic proton than a neutron be required to produce a nuclear reaction?

2.2 Protons in the Fermi Laboratory accelerator, **Figure 30–8,** move counterclockwise. In what direction is the magnetic field of the bending magnets?

2.3 **Figure 30–11** shows the production of two electron/positron pairs. Why does the bottom set of tracks curve less than the top pair of tracks?

2.4 **Critical Thinking** Write an equation for beta decay in which you write the quarks comprising both the neutron and the proton. Include both steps in the decay.

CHAPTER 30 REVIEW

Key Terms

30.1

- atomic mass unit
- atomic number
- mass number
- strong nuclear force
- nuclide
- radioactive
- alpha decay
- transmute
- beta decay
- gamma decay
- nuclear reaction
- half-life
- activity

30.2

- elementary particle
- quark
- lepton
- force carrier
- neutrino
- positron
- weak nuclear force
- antiparticle
- pair production
- grand unification theory

Summary

30.1 Radioactivity

- The number of protons in a nucleus is given by the atomic number.
- The sum of the numbers of protons and neutrons in a nucleus is equal to the mass number.
- Atoms having nuclei with the same number of protons but different numbers of neutrons are called isotopes.
- An unstable nucleus undergoes radioactive decay, transmuting into another element.
- Radioactive decay produces three kinds of particles. Alpha (α) particles are helium nuclei, beta (β) particles are high-speed electrons, and gamma (γ) rays are high-energy photons.
- In nuclear reactions, the sum of the mass number, A, and the total charge, Z, are not changed.
- The half-life of a radioactive isotope is the time required for half of the nuclei to decay.
- The number of decays of a radioactive sample per second is the activity.

30.2 The Building Blocks of Matter

- Bombardment of nuclei by protons, neutrons, alpha particles, electrons, gamma rays, or other nuclei can produce a nuclear reaction.
- Linear accelerators and synchrotrons produce high-energy particles.
- The Geiger-Mueller counter and other particle detectors use the ionization caused by charged particles passing through matter.
- When antimatter and matter combine, all mass is converted into energy or lighter matter-antimatter particle pairs.
- By pair production, energy is transformed into a matter-antimatter particle pair.
- The weak interaction, or weak force, operates in beta decay. The strong force binds the nucleus together.
- All matter appears to be made up of two families of particles, quarks and leptons.
- Matter interacts with other matter through a family of particles called force carriers.

Reviewing Concepts

Section 30.1

1. Define the term *transmutation* as used in nuclear physics and give an example.
2. What are the common names for an α particle, β particle, and γ radiation?
3. What happens to the atomic number and mass number of a nucleus that emits an alpha particle?
4. What happens to the atomic number and mass number of a nucleus that emits a beta particle?
5. What two quantities must always be conserved in any nuclear equation?

Section 30.2

6. Why would a linear accelerator not work with a neutron?
7. In which of the four interactions (strong, weak, electromagnetic, and gravitational) do the following particles take part?
   a. electron **b.** proton **c.** neutrino

8. What happens to the atomic number and mass number of a nucleus that emits a positron?
9. Give the symbol, mass, and charge of the following particles.
   a. proton
   b. positron
   c. α particle
   d. neutron
   e. electron

Applying Concepts

10. Which are generally more unstable, small or large nuclei?
11. Which isotope has the greater number of protons, uranium-235 or uranium-238?
12. Which is usually larger, *A* or *Z*? Explain.
13. Which is most like an X ray: alpha particles, beta particles, or gamma particles?
14. Could a deuteron, ${}^{2}_{1}\mathrm{H}$, decay through alpha decay? Explain.
15. Which will give a higher reading on a radiation detector: a radioactive substance that has a short half-life or an equal amount of a radioactive substance that has a long half-life?
16. Why is carbon dating useful in establishing the age of campfires but not the age of a set of knight's armor?
17. What would happen if a meteorite made of antiprotons, antineutrons, and positrons landed on Earth?

Problems

Section 30.1

LEVEL 1

18. An atom of an isotope of magnesium has an atomic mass of about 24 u. The atomic number of magnesium is 12. How many neutrons are in the nucleus of this atom?
19. An atom of an isotope of nitrogen has an atomic mass of about 15 u. The atomic number of nitrogen is 7. How many neutrons are in the nucleus of this isotope?
20. List the number of neutrons in an atom of each of the following isotopes.
   a. ${}^{112}_{48}\mathrm{Cd}$
   b. ${}^{209}_{83}\mathrm{Bi}$
   c. ${}^{208}_{83}\mathrm{Bi}$
   d. ${}^{80}_{35}\mathrm{Br}$
   e. ${}^{1}_{1}\mathrm{H}$
   f. ${}^{40}_{18}\mathrm{Ar}$
21. Find the symbols for the elements that are shown by the following symbols, where *X* replaces the symbol for the element.
   a. ${}^{18}_{9}X$
   b. ${}^{241}_{95}X$
   c. ${}^{21}_{10}X$
   d. ${}^{7}_{3}X$
22. A radioactive bismuth isotope, ${}^{214}_{83}\mathrm{Bi}$, emits a β particle. Write the complete nuclear equation, showing the element formed.
23. A radioactive polonium isotope, ${}^{210}_{84}\mathrm{Po}$, emits an α particle. Write the complete nuclear equation, showing the element formed.
24. An unstable chromium isotope, ${}^{56}_{24}\mathrm{Cr}$, emits a β particle. Write a complete equation showing the element formed.
25. During a reaction, two deuterons, ${}^{2}_{1}\mathrm{H}$, combine to form a helium isotope, ${}^{3}_{2}\mathrm{He}$. What other particle is produced?
26. On the sun, the nuclei of four ordinary hydrogen atoms combine to form a helium isotope, ${}^{4}_{2}\mathrm{He}$. What particles are missing from the following equation for this reaction?
$$4{}^{1}_{1}\mathrm{H} \rightarrow {}^{4}_{2}\mathrm{He} + ?$$
27. Write a complete nuclear equation for the transmutation of a uranium isotope, ${}^{227}_{92}\mathrm{U}$, into a thorium isotope, ${}^{223}_{90}\mathrm{Th}$.
28. In an accident in a research laboratory, a radioactive isotope with a half-life of three days is spilled. As a result, the radiation is eight times the maximum permissible amount. How long must workers wait before they can enter the room?
29. If the half-life of an isotope is two years, what fraction of the isotope remains after six years?
30. The half-life of strontium-90 is 28 years. After 280 years, how would the intensity of a sample of strontium-90 compare to the original intensity of the sample?

LEVEL 2

31. $^{238}_{92}U$ decays by α emission and two successive β emissions back into uranium again. Show the three nuclear decay equations and predict the atomic mass number of the uranium formed.

32. A Geiger-Mueller counter registers an initial reading of 3200 counts while measuring a radioactive substance. It registers 100 counts 30 hours later. What is the half-life of this substance?

33. A 14-g sample of $^{14}_{6}C$ contains Avogadro's number, 6.02×10^{23}, of nuclei. A 5.0 g-sample of C-14 will have how many nondecayed nuclei after 11 460 years?

34. A 1.00-μg sample of a radioactive material contains 6.0×10^{14} nuclei. After 48 hours, 0.25 μg of the material remains.
  a. What is the half-life of the material?
  b. How could one determine the activity of the sample at 24 hours using this information?

Section 30.2

LEVEL 1

35. What would be the charge of a particle composed of three u quarks?

36. The charge of an antiquark is opposite that of a quark. A pion is composed of a u quark and an anti-d quark. What would be the charge of this pion?

37. Find the charge of a pion made up of
  a. u and anti-u quark pair.
  b. d and anti-u quarks.
  c. d and anti-d quarks.

LEVEL 2

38. The synchrotron at the Fermi Laboratory has a diameter of 2.0 km. Protons circling in it move at approximately the speed of light.
  a. How long does it take a proton to complete one revolution?
  b. The protons enter the ring at an energy of 8.0 GeV. They gain 2.5 MeV each revolution. How many revolutions must they travel before they reach 400 GeV of energy?
  c. How long does it take the protons to be accelerated to 400 GeV?
  d. How far do the protons travel during this acceleration?

Critical Thinking Problems

39. Gamma rays carry momentum. The momentum of a gamma ray of energy E_γ is equal to E_γ/c, where c is the speed of light. When an electron-positron pair decays into two gamma rays, both momentum and energy must be conserved. The sum of the energies of the gamma rays is 1.02 MeV. If the positron and electron are initially at rest, what must be the magnitude and direction of the momentum of the two gamma rays?

40. An electron-positron pair, initially at rest, also can decay into three gamma rays. If all three gamma rays have equal energies, what must be their relative directions?

Going Further

Using a Graphing Calculator An archaeological expedition group in Jerusalem finds three wooden bowls, each in a different part of the city. An analysis of the levels of carbon-14 reveals that the three bowls have 90%, 75%, and 60% of the carbon-14 one would expect to find in a similar bowl made today. The equation for the decay of carbon-14 is $y = 100 \times 2^{(-x/5730)}$ where y is the percentage of carbon-14 remaining and x is the age of the object in years. Graph this equation on a graphing calculator with axes scaled from 0 to 5000 years and from 100% to 50%. Trace along the equation to find the age of the three objects.

interNET CONNECTION

Follow the link on the Glencoe Homepage at **www.glencoe.com/sec/science** to find out more about this chapter.

Shock Waves

The photo of a bundle of uranium-filled fuel rods in water may be familiar to you. What causes the blue glow surrounding the assembly?

CHAPTER

31 Nuclear Applications

No area of physics has stirred more controversy than nuclear physics. It is also an area with the greatest potential for developments to help people. The nuclear force of the atom has been used as weapons and as a potent force for destruction. However, the application of nuclear physics to other areas is more important. Nuclear physics in medicine has provided tools to treat cancer and diagnose illness and disease. Nuclear power has provided electric energy without the vast pollution caused by the burning of fossil fuels. This has helped to ease the problems of smog and has relieved the suffering of many who are afflicted with breathing disorders. Nuclear power cells have provided the power for space exploration and for tiny pacemakers to regulate the beating of the human heart. Smoke detectors rely on nuclear physics to protect people from fires. In many ways, your modern existence is not possible without nuclear physics. An informed citizen must understand some of the science and technology to fully appreciate its impact on society.

WHAT YOU'LL LEARN

- You will calculate the energy released in nuclear reactions.
- You will examine how radioactive isotopes are produced and used.
- You will examine how fission and fusion occur.
- You will study how nuclear reactors work.

WHY IT'S IMPORTANT

- Fission reactors are controversial methods of producing electric power, and a knowledge of how they work makes a person better equipped to discuss their strengths and weaknesses.

interNET CONNECTION

Follow the link for this chapter on the Glencoe Homepage at **www.glencoe.com/sec/science** to find out more about nuclear applications.

31.1 Holding the Nucleus Together

What holds the nucleus together? The negatively charged electrons that surround the positively charged nucleus of an atom are held in place by the attractive electromagnetic force. But the nucleus consists of positively charged protons and neutral neutrons. The repulsive electromagnetic force between the protons might be expected to cause them to fly apart. This does not happen because an even stronger attractive force exists within the nucleus.

OBJECTIVES

- **Define** binding energy of the nucleus.
- **Relate** the energy released in a nuclear reaction to the change in binding energy during the reaction.

The Strong Nuclear Force

You learned in Chapter 30 that the force that overcomes the mutual repulsion of the charged protons is called the strong nuclear force. The strong force acts between protons and neutrons that are close together, as they are in a nucleus. This force is more than 100 times as intense as the electromagnetic force. The range of the strong force is short, only about the radius of a proton, 1.3×10^{-15} m. It is attractive and is of the same strength between protons and protons, protons and neutrons, and neutrons and neutrons. As a result of this equivalence, both neutrons and protons are called **nucleons.**

The strong force holds the nucleons in the nucleus. If a nucleon were to be pulled out of a nucleus, work would have to be done to overcome the attractive force. Doing work adds energy to the system. Thus, the assembled nucleus has less energy than the separate protons and neutrons that make it up. The difference is the **binding energy** of the nucleus. Because the assembled nucleus has less energy, the binding energy is identified as a negative value.

Binding Energy of the Nucleus

The binding energy comes from the nucleus converting some of its mass to hold the nucleons together, and it can be expressed in the form of an equivalent amount of mass, according to the equation $E = mc^2$. The unit of mass used in nuclear physics is the atomic mass unit, u. One atomic mass unit is 1/12 the mass of the ${}^{12}_{6}C$ nucleus.

Because energy has to be added to take a nucleus apart, the mass of the assembled nucleus is less than the sum of the masses of the nucleons that compose it. For example, the helium nucleus, ${}^{4}_{2}He$, consists of two protons and two neutrons. The mass of a proton is 1.007825 u. The mass of a neutron is 1.008665 u. If the mass of the helium nucleus were equal to the sum of the masses of the two protons and the two neutrons, you would expect that the mass of the nucleus would be 4.032980 u. Careful measurement, however, shows that the mass of a helium nucleus is only 4.002603 u. The actual mass of the helium nucleus is less than

F.Y.I.

In the 1950s, Ford Motors designed the Ford Nucleon, which was to be powered by a small atomic reactor core located under a circular cover at the rear of the car. It was to be recharged with nuclear fuel. The car was never built.

the mass of its constituent parts by 0.0030377 u. The difference between the sum of the masses of the individual nucleons and the actual mass is called the **mass defect.** The energy equivalent of the missing mass is the binding energy, the energy that holds the nucleus together. The binding energy can be calculated from the experimentally determined mass defect by using $E = mc^2$.

Masses are normally measured in atomic mass units. It will be useful, then, to determine the energy equivalent of 1 u (1.6605×10^{-27} kg). To determine the energy, you must multiply the mass by the square of the speed of light in air (2.9979×10^8 m/s). This is expressed to five significant digits.

$$E = mc^2$$

$$= (1.6605 \times 10^{-27}\ \text{kg})(2.9979 \times 10^8\ \text{m/s})^2$$

$$= 14.924 \times 10^{-11}\ \text{kg·m}^2/\text{s}^2 = 14.924 \times 10^{-11}\ \text{J}$$

Remember that 1 kg·m^2/s^2 equals 1 joule. The most convenient unit of energy to use is the electron volt. To express the energy in electron volts, you must convert from joules to electron volts.

$$E = (14.923 \times 10^{-11}\ \text{J})(1\ \text{eV}/1.6022 \times 10^{-19}\ \text{J})$$

$$= 9.3149 \times 10^8\ \text{eV}$$

$$= 931.49\ \text{MeV}$$

Hence, 1 u of nuclear mass is equivalent to 931.49 MeV of binding energy. **Figure 31–1** shows how the binding energy per nucleon depends on the size of the nucleus. Heavier nuclei are bound more strongly than lighter nuclei. Except for a few nuclei, the binding energy per nucleon becomes more negative as the mass number, *A*, increases to a value of 56, that of iron, Fe. $^{56}_{26}$Fe is the most tightly bound nucleus; thus, nuclei become more stable as their mass numbers approach that of iron. Nuclei whose mass numbers are larger than that of iron are less strongly bound and are therefore less stable.

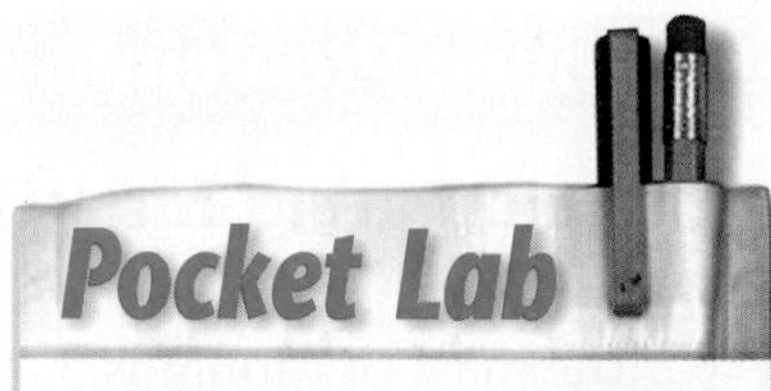

Binding Energy 

Particles within the nucleus are strongly bonded. Place two disk magnets together to represent a proton and neutron within a nucleus. Slowly pull them apart. Feel how the force changes with separation.

Analyze and Conclude
Describe how this analogy could be extended for a nucleus that contains several protons and neutrons.

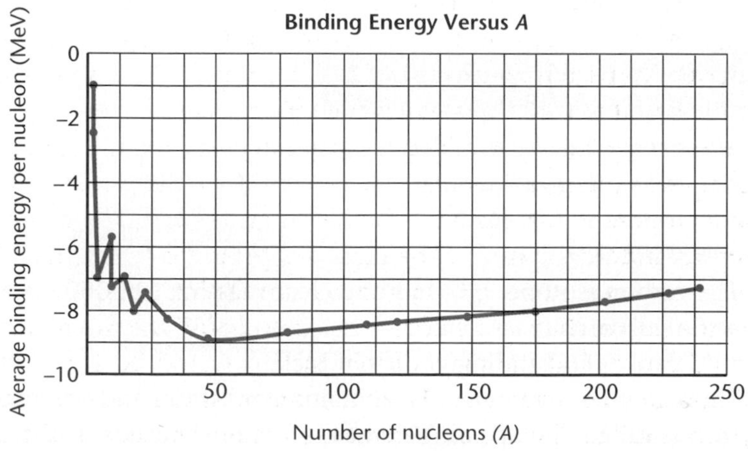

FIGURE 31–1 A graph of the binding energy per nucleon is shown.

Example Problem

Calculating Mass Defect and Nuclear Binding Energy

Find the nuclear mass defect and binding energy of tritium, ${}^{3}_{1}H$. The mass of tritium is 3.016049 u, the mass of a proton is 1.007825 u, and the mass of a neutron is 1.008665 u.

Calculate Your Answer

Known:

mass of 1 proton = 1.007825 u
mass of 2 neutrons = 2.017330 u
mass of tritium = 3.016049 u
binding energy of 1 u = 931.49 MeV

Unknown:

total mass of nucleons = ?
mass defect = ?

Strategy:

Add the masses of the nucleons.
The mass defect is equal to the actual mass of tritium less the mass of the one proton and two neutrons that comprise it.

Calculations:

Mass of 1 proton	1.007825 u
Plus mass of 2 neutrons	+2.017330 u
Total mass of nucleons	3.025155 u

Mass of tritium	3.016049 u
Less mass of nucleons	−3.025155 u
Mass defect	−0.009106 u

The binding energy is the energy equivalent of the mass defect.

Binding energy: 1 u = 931.49 MeV,
so $E = (-0.009106\ \text{u})(931.49\ \text{MeV/u})$
$= -8.482\ \text{MeV}$

Check Your Answer

- Are the units correct? Mass is measured in u, and energy is measured in MeV.
- Does the sign make sense? Binding energy should be negative.
- Is the magnitude realistic? According to **Figure 31–1,** binding energies per nucleon are between −1 MeV and −2 MeV, so the answer for three nucleons is reasonable.

Practice Problems

Use these values to solve the following problems:
mass of proton = 1.007825 u
mass of neutron = 1.008665 u
1 u = 931.49 MeV.

1. The carbon isotope, ${}^{12}_{6}C$, has a nuclear mass of 12.0000 u.
   a. Calculate its mass defect.
   b. Calculate its binding energy in MeV.
2. The isotope of hydrogen that contains one proton and one neutron is called deuterium. The mass of its nucleus is 2.014102 u.

a. What is its mass defect?

b. What is the binding energy of deuterium in MeV?

3. A nitrogen isotope, $^{15}_{7}N$, has seven protons and eight neutrons. Its nucleus has a mass of 15.00011 u.

a. Calculate the mass defect of this nucleus.

b. Calculate the binding energy of the nucleus.

4. An oxygen isotope, $^{16}_{8}O$, has a nuclear mass of 15.99491 u.

a. What is the mass defect of this isotope?

b. What is the binding energy of its nucleus?

HELP WANTED

RADIOLOGIC TECHNOLOGIST

Do you have an associate degree and professional certification in radiological technology? We are looking for a person with both of the above who is compassionate, conscientious, and cooperative. You must have a penchant for perfection when it comes to safety regulations and procedures. A large hospital offers an excellent opportunity for advancement with increased training. For information contact:
American Society of Radiologic Technologists
15000 Central Avenue, S.E.
Albuquerque, NM 87123

A nuclear reaction will occur naturally if energy is released by the reaction. Energy will be released if the nucleus that results from the reaction is more tightly bound than the original nucleus. When a heavy nucleus, such as $^{238}_{92}U$, decays by releasing an alpha particle, the binding energy per nucleon of the resulting $^{234}_{90}Th$ has a more negative value than that of the uranium. This means that the excess energy of the $^{238}_{92}U$ nucleus is transferred into the kinetic energy of the alpha particle and that the thorium nucleus is more stable than the uranium nucleus. At low atomic numbers, those below $Z = 26$, reactions that add nucleons to a nucleus make the binding energy of the nucleus more negative and increase the stability of the nucleus. Thus, the binding energy of the larger nucleus is less than the sum of the energies of the two smaller ones.

Energy is released when a spontaneous nuclear reaction occurs. In the sun and other stars, the production of heavier nuclei such as helium and carbon from hydrogen releases energy that will become the electromagnetic radiation that you see as visible light from the stars.

31.1 Section Review

1.1 When tritium, $^{3}_{1}H$, decays, it emits a beta particle and becomes $^{3}_{2}He$. Which nucleus would you expect to have a more negative binding energy?

1.2 Which of those two nuclei in 1.1 would have the larger mass defect?

1.3 The range of the strong force is so short that only nucleons that touch each other feel the force. Use this fact to explain why, in large nuclei, the repulsive electromagnetic force can overcome the strong attractive force and make the nucleus unstable.

1.4 **Critical Thinking** In old stars, not only are helium and carbon produced by joining more tightly bound nuclei, but so are oxygen ($Z = 8$) and silicon ($Z = 14$). What would be the atomic number of the heaviest nucleus that could be formed in this way? Explain.

31.2 Using Nuclear Energy

OBJECTIVES

- **Define** how radioactive isotopes can be artificially produced and used.
- **Solve** nuclear equations.
- **Define** *nuclear fission* and *chain reaction.*
- **Describe** the operation of one or more types of nuclear reactors.
- **Describe** the fusion process.

In no other area of physics has basic knowledge led to applications as quickly as in the field of nuclear physics. The medical use of the radioactive element radium began within 20 years of its discovery. Proton accelerators were tested for medical applications less than one year after being invented. In the case of nuclear fission, the military application was under development before the basic physics was even known. Peaceful applications followed in less than ten years. Questions surrounding the uses of nuclear science in our society are important for all citizens today.

Artificial Radioactivity

Marie and Pierre Curie noted as early as 1899 that substances placed close to radioactive uranium became radioactive themselves. In 1934, Irene Joliot-Curie and Frederic Joliot bombarded aluminum with alpha particles, which produced phosphorus atoms and neutrons by the following reaction (alpha particles can be represented by ${}^{4}_{2}\text{He}$).

$${}^{4}_{2}\text{He} + {}^{27}_{13}\text{Al} \rightarrow {}^{30}_{15}\text{P} + {}^{1}_{0}\text{n}$$

In addition to neutrons, the Joliot-Curies found another particle coming from the reaction, a positively charged electron, or positron. Remember from Chapter 30 that the positron is a particle with the same mass as the electron but with a positive charge. The most interesting result of the Joliot-Curies' experiment was that positrons continued to be emitted after the alpha bombardment stopped. The positrons were found to come from the phosphorus isotope ${}^{30}_{15}\text{P}$. The Joliot-Curies had produced a radioactive isotope not previously known. The decay of the new isotope occurs via the following reaction.

$${}^{30}_{15}\text{P} \rightarrow {}^{0}_{+1}\text{e} + {}^{30}_{14}\text{Si} \qquad ({}^{0}_{+1}\text{e is a positron})$$

Radioactive isotopes can be formed from stable isotopes by bombardment with alpha particles, protons, neutrons, electrons, or gamma rays. The resulting unstable nuclei emit radiation until they are converted into stable isotopes. Recall from Chapter 30 that this process is known as transmutation. The ${}^{30}_{14}\text{Si}$ isotope of silicon from the reaction above is stable. The radioactive nuclei may emit alpha, beta, and gamma radiation as well as positrons. **Figure 31–2** shows an example of a use of gamma radiation for food preservation.

Artificially produced radioactive isotopes have many other uses, especially in medicine. In many medical applications, patients are given radioactive isotopes that are absorbed by specific parts of the body. The detection of the decay products of these isotopes allows doctors to trace the movement of the isotopes and of the molecules to which they are attached through the body. Iodine, for example, is primarily used in the

FIGURE 31–2 The strawberries in the two cartons are identical except that those in the carton on the left have been treated with gamma radiation to destroy the organisms that cause spoilage. The berries do not absorb the radiation and are not radioactive.

MEDICINE CONNECTION

thyroid gland. The thyroid gland uses iodine obtained from food and water to produce thyroid hormone, which controls metabolism. When a gland produces too much hormone, the condition is called hyperthyroidism. When the patient is given radioactive iodine, ${}^{131}_{53}\mathrm{I}$, it is concentrated in the thyroid gland. Excess energy in the nucleus of radioactive iodine slows the production of the thyroid hormone. A physician uses a radiation detector counter to monitor the activity of ${}^{131}_{53}\mathrm{I}$ in the region of the thyroid. The amount of iodine taken up by this gland is a measure of its ability to function. **Figure 31–3** shows an image of a healthy thyroid gland superimposed on the area of the throat where it is found.

The Positron Emission Tomography Scanner, or PET scanner, is another medical application of isotopes. A positron-emitting isotope is included in a solution injected into a patient's body. In the body, the isotope decays, releasing a positron. The positron annihilates an electron, and two gamma rays are emitted. The PET scanner detects the gamma rays and pinpoints the location of the positron-emitting isotope. A computer is then used to make a three-dimensional map of the isotope distribution, which is affected by the density of the surrounding tissues. This information can then be used to determine the location of tumors and other anomalies. Details such as the use of nutrients in particular regions of the brain also can be traced. For example, if a person in a PET scanner were solving a physics problem, more nutrients would flow to the part of the brain being used to solve the problem. The decay of the positrons in this part of the brain would increase, and the PET scanner could map this area.

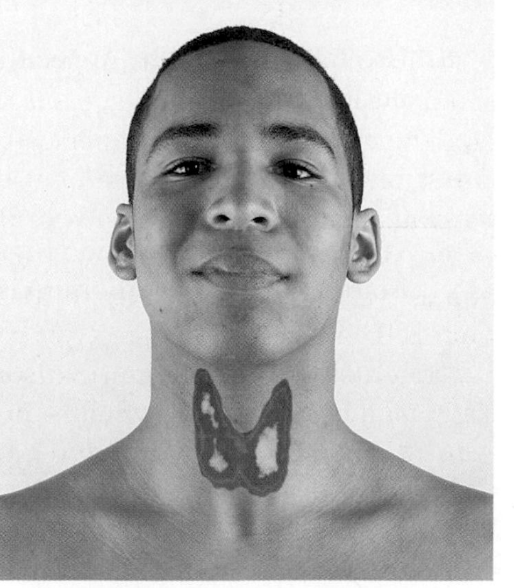

FIGURE 31–3 The photo shows an image of a normal thyroid gland obtained by using radioactive iodine-131. This image is superimposed on the area of the throat where it is found.

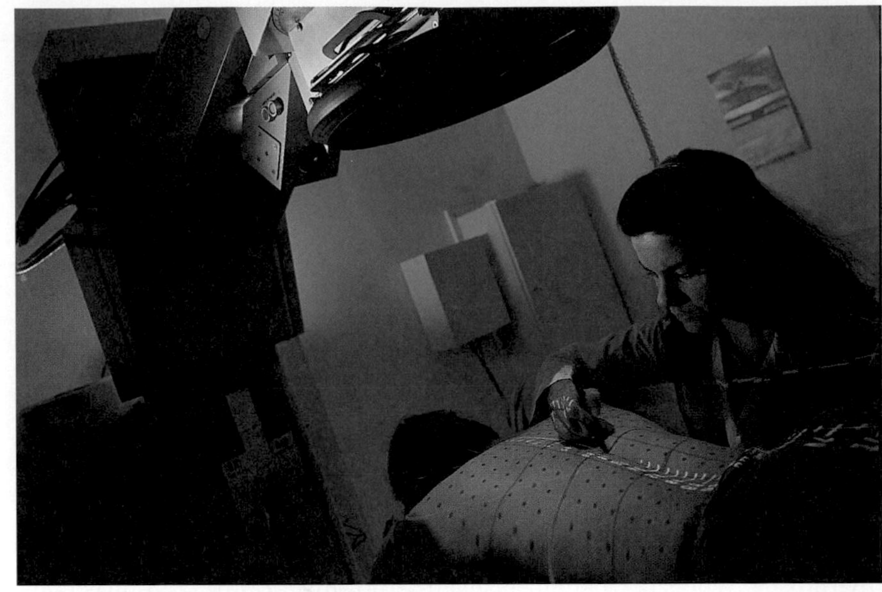

FIGURE 31–4 Radiation produced by cobalt-60 is carefully aimed at tumor cells. These cells are more sensitive to the damaging effects of radiation because they will divide more often than normal cells.

F.Y.I.

Gamma rays destroy both cancerous and healthy cells; thus, the beams of radiation must be directed only to the cancerous cells.

Another use of radioactivity in medicine is the destruction of cancerous cells. Often, gamma rays from the isotope $^{60}_{27}Co$ are used to treat cancer patients as is shown in **Figure 31–4.** The ionizing radiation produced by radioactive iodine can be localized and used to destroy cells in a diseased thyroid gland with minimal harm to the rest of the body. Another method of reducing damage to healthy cells is to use particles such as pions produced by particle accelerators such as the synchrotron. The physician adjusts the accelerator so that the particles decay only in the cancerous tissue. These unstable particles pass through body tissue without doing damage and are trapped in cancerous tissues. When they decay, however, the emitted particles destroy cells.

Practice Problems

5. Use **Table D–6** of the Appendix to complete the following nuclear equations.

a. $^{14}_{6}C \rightarrow ? + {}^{0}_{-1}e$

b. $^{55}_{24}Cr \rightarrow ? + {}^{0}_{-1}e$

6. Write the nuclear equation for the transmutation of a uranium isotope, $^{238}_{92}U$, into a thorium isotope, $^{234}_{90}Th$, by the emission of an alpha particle.

7. A radioactive polonium isotope, $^{214}_{84}Po$, undergoes alpha decay and becomes lead. Write the nuclear equation.

8. Write the nuclear equations for the beta decay of these isotopes.

a. $^{210}_{80}Pb$ **b.** $^{210}_{83}Bi$ **c.** $^{234}_{90}Th$ **d.** $^{239}_{93}Np$

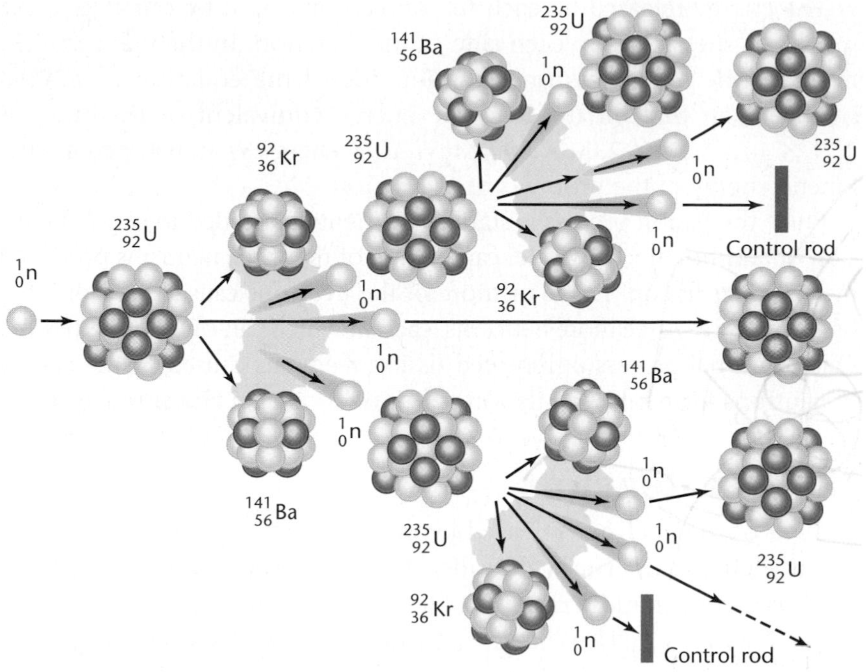

FIGURE 31–5 The nuclear fission chain reaction of uranium-235 takes place in the core of a nuclear reactor.

Nuclear Fission

The possibility of obtaining useful forms of energy from nuclear reactions was discussed in the 1930s. The most promising results came from bombarding substances with neutrons. In Italy, in 1934, Enrico Fermi and Emilio Segre produced many new radioactive isotopes by bombarding uranium with neutrons. They believed that they had formed new elements with atomic numbers larger than 92, the atomic number of uranium.

German chemists Otto Hahn and Fritz Strassmann made careful chemical studies of the results of bombarding uranium with neutrons. In 1939, their analyses showed that the resulting atoms acted, chemically, like barium. The two chemists could not understand how barium, with an atomic number of 56, could be produced from uranium.

One week later, Lise Meitner and Otto Frisch proposed that the neutrons had caused a division of the uranium into two smaller nuclei, resulting in a large release of energy. Such a division of a nucleus into two or more fragments is called **fission.** The possibility that fission could be not only a source of energy, but also an explosive weapon, was immediately realized by many scientists.

The uranium isotope, $^{235}_{92}U$, undergoes fission when it is bombarded with neutrons. The elements barium and krypton are typical results of fission, as shown in **Figure 31–5.** The reaction is defined by the following equation.

$$^{1}_{0}n + ^{235}_{92}U \rightarrow ^{92}_{36}Kr + ^{141}_{56}Ba + 3\ ^{1}_{0}n + 200\text{ MeV}$$

The energy released by each fission can be found by calculating the masses of the atoms on each side of the equation. In the uranium-235 reaction, the total mass on the right side of the equation is 0.215 u smaller than that on the left. The energy equivalent of this mass is 3.21×10^{-11} J, or 2.00×10^{2} MeV. This energy is transferred to the kinetic energy of the products of the fission.

Once the fission process is started, the neutron needed to cause the fission of additional $^{235}_{92}U$ nuclei can be one of the three neutrons produced by an earlier fission. If one or more of the neutrons causes a fission, that fission releases three more neutrons, each of which can cause more fission. This continual process of repeated fission reactions caused by the release of neutrons from previous fission reactions is called a **chain reaction.**

F.Y.I.

A 1000-MW nuclear power plant uses 25 metric tons of fuel per year and produces 25 metric tons of waste. An equivalent coal-burning power plant uses 2.5 million metric tons of coal per year and produces more than 15 million metric tons of waste.

Nuclear Reactors

Most of the neutrons released by the fission of $^{235}_{92}U$ atoms are moving at high speeds. These are called **fast neutrons.** In addition, naturally occurring uranium consists of less than one percent $^{235}_{92}U$ and more than 99 percent $^{238}_{92}U$. When a $^{238}_{92}U$ nucleus absorbs a fast neutron, it does not undergo fission, but becomes a new isotope, $^{239}_{92}U$. The absorption of neutrons by $^{238}_{92}U$ keeps most of the neutrons from reaching the fissionable $^{235}_{92}U$ atoms. Thus, most neutrons released by the fission of $^{235}_{92}U$ are unable to cause the fission of another $^{235}_{92}U$ atom.

Fermi suggested that a chain reaction would occur if the uranium were broken up into small pieces and placed in a **moderator,** a material that can slow down, or moderate, the fast neutrons. When a neutron collides with a light atom, such as carbon, it transfers momentum and energy to the atom. In this way, the neutron loses energy. The moderator creates many **slow neutrons,** which are more likely to be absorbed by $^{235}_{92}U$ than by $^{238}_{92}U$. The larger number of slow neutrons greatly increases the probability that a neutron released by the fission of a $^{235}_{92}U$ nucleus will cause another $^{235}_{92}U$ nucleus to fission. If there is enough $^{235}_{92}U$ in the sample, a chain reaction can occur.

A neutron loses the most energy when it strikes a hydrogen nucleus. Thus, hydrogen is an ideal moderator. Fast neutrons, however, cause a nuclear reaction with normal hydrogen nuclei, $^{1}_{1}H$. For this reason, when Fermi produced the first controlled chain reaction on December 2, 1942, he used graphite (carbon) as a moderator.

Heavy water, in which the hydrogen, $^{1}_{1}H$, is replaced by the isotope deuterium, $^{2}_{1}H$, does not react with fast neutrons. Thus, the problem of the unwanted nuclear reaction with hydrogen is avoided. As a result, heavy water is used as a moderator with natural uranium in the Canadian CANDU reactors.

The process that increases the number of fissionable nuclei is called **enrichment.** Enrichment of uranium is difficult and requires large, expensive equipment. The U.S. government operates the plants that produce enriched uranium for most of the world.

Physics Lab

Solar Power

Problem

How can you measure the local power output from the nearest continuous running fusion reactor, the sun?

Materials

solar cell
voltmeter
ammeter
electrical leads
ruler

Procedure

1. With no load attached, measure the voltage output of a solar cell when the cell is outdoors and directly facing the sun.
2. Measure the current from the solar cell when the cell is outdoors and directly facing the sun.
3. Measure the length and width of the solar cell and determine its surface area.
4. Remeasure the voltage and current when the sunlight passes through a window.

Data and Observations			
Voltage indoor	Voltage outdoor	Current indoor	Current outdoor

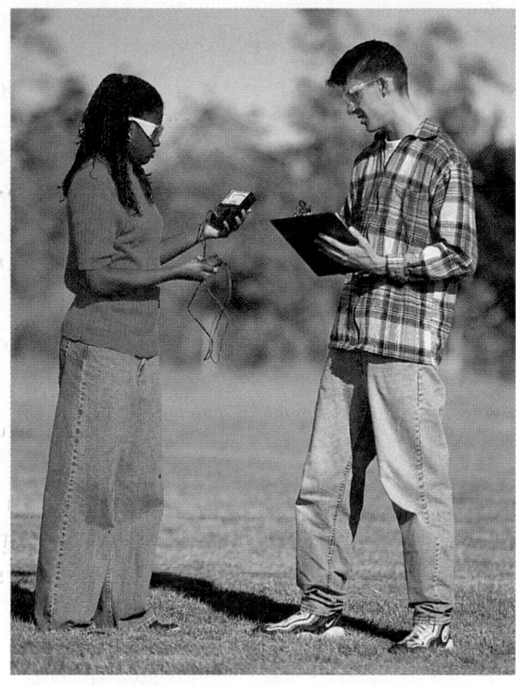

Analyze and Conclude

1. **Calculating Results** Calculate the power, *IV*, for the solar cell outdoors and indoors. What percentage of power did the window stop?
2. **Calculating Results** Calculate the amount of power that could be produced by a cell that has an area of 1.0 square meter.
3. **Calculating Efficiency** The sun supplies about 1000 W of power per square meter to Earth. Calculate the efficiency of your solar cell.

Apply

1. You are planning to install 15 square meters of solar cells on your roof. How much power will you expect them to produce?

Pocket Lab

Power Plant

Call your local electric company and ask the following questions.

1. Where is the nearest nuclear power plant?
2. What fraction (or percentage) of electricity is supplied by nuclear power in your area?

Analyze and Conclude How much electrical energy in your neighborhood is provided by nuclear power and how much is provided by other sources?

Shock Waves

The type of nuclear reactor used in the United States, the pressurized water reactor, contains about 200 metric tons of uranium sealed in hundreds of metal rods. The rods are immersed in water. Water not only is the moderator, but also transfers thermal energy away from the fission of uranium. Rods of cadmium metal are placed between the uranium rods. Cadmium absorbs neutrons easily and also acts as a moderator. The cadmium rods are moved in and out of the reactor to control the rate of the chain reaction. Thus, the rods are called **control rods.** When the control rods are inserted completely into the reactor, they absorb enough of the neutrons released by the fission reactions to prevent any further chain reaction. As the control rods are removed from the reactor, the rate of energy release increases, with more free neutrons available to continue the chain reaction.

Energy released by the fission heats the water surrounding the uranium rods. The water itself doesn't boil because it is under high pressure, which increases its boiling point. As shown in **Figure 31–6,** this water is pumped to a heat exchanger, where it causes other water to boil, producing steam that turns turbines. The turbines are connected to generators that produce electrical energy.

Some of the fission energy goes to increase the kinetic energy of electrons, giving these particles speeds near the speed of light in a vacuum. You learned in Chapter 14 that when light enters a medium of greater density, the speed of the light in that medium is reduced. In a similar way, when these energized electrons moving at almost the speed of light in a vacuum enter the water, their speed exceeds the speed of light in the water. As a result, a blue glow is emitted when fuel rods are placed in water. The glow is called the Cerenkov effect. It is not the result of radioactivity in the water; radioactive objects do not emit a blue glow.

Fission of ${}^{235}_{92}\text{U}$ nuclei produces Kr, Ba, and other atoms in the fuel rods. Most of these atoms are radioactive. About once a year, some of the uranium fuel rods must be replaced. The old rods no longer can be used in the reactor, but they are still extremely radioactive and must be stored in a location that can be secured. Methods of permanently storing these radioactive waste products are currently being developed.

Among the products of fission is an isotope of plutonium, ${}^{239}_{94}\text{Pu}$. This isotope is fissionable when it absorbs neutrons and can be used in nuclear weapons. It is also toxic. As a result, plutonium-containing materials are all stored in secured locations, never in waste dumps. There is hope that, in the future, this fissionable isotope might be removed from radioactive waste and recycled to fuel other reactors.

The world's supply of uranium is limited. If nuclear reactors are used to supply a large fraction of the world's energy, uranium will become scarce. Even though the plutonium produced by normal reactors might be recovered to fuel other reactors, there is still a net loss of fuel. To extend the supply of uranium, **breeder reactors** have been developed.

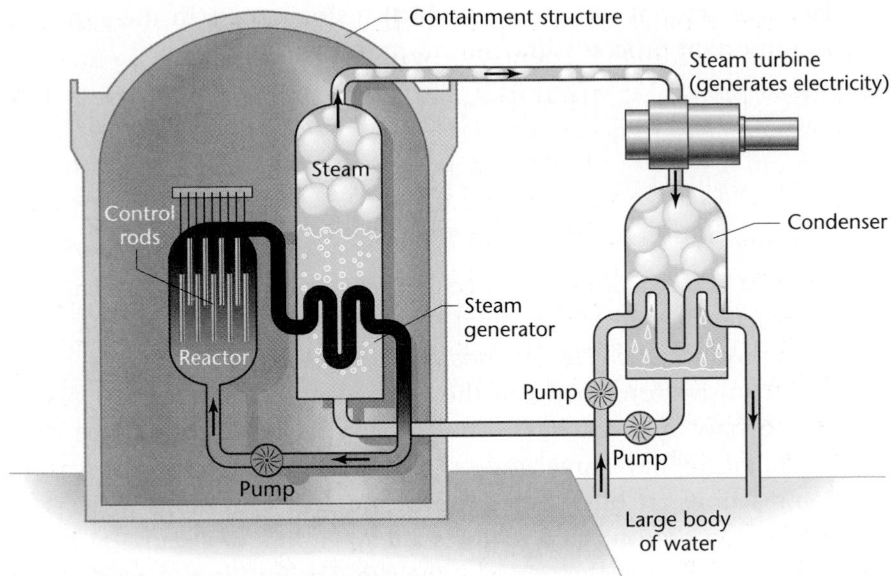

FIGURE 31–6 In a nuclear power plant, the thermal energy released in nuclear reactions is converted to electric energy.

When a reactor contains both plutonium and $^{238}_{92}U$, the plutonium will undergo fission just as $^{235}_{92}U$ does. Many of the free neutrons from the fission are absorbed by the $^{238}_{92}U$ to produce additional $^{239}_{94}Pu$. For every two plutonium atoms that undergo fission, three new ones are formed. More fissionable fuel can be recovered from a breeder reactor than was originally present.

Nuclear Fusion

In nuclear **fusion,** nuclei with small masses combine to form a nucleus with a larger mass, shown in **Figure 31–7.** In the process, energy is released. You learned earlier in this chapter that the larger nucleus is more tightly bound, so its mass is less than the sum of the masses of the smaller nuclei. A typical example of fusion is the process that occurs in the sun. Four hydrogen nuclei (protons) fuse in several steps to form one helium nucleus. The mass of the four protons is greater than the mass of the helium nucleus that is produced. The energy equivalent of this mass difference is transferred to the kinetic energy of the resultant particles. The energy released by the fusion of one helium nucleus is 25 MeV. In comparison, the energy released when one dynamite molecule reacts chemically is about 20 eV, almost 1 million times smaller.

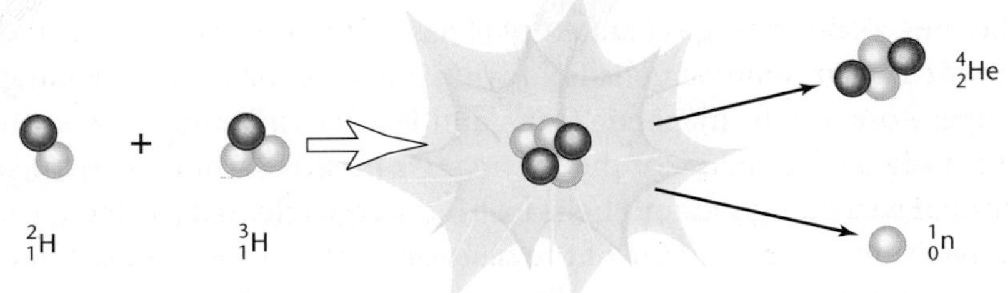

FIGURE 31–7 The fusion of deuterium and tritium produces helium. Protons are red and neutrons are gray in the figure.

F.Y.I.

The sun is a giant nuclear fusion reactor that supplies Earth with energy from more than 150 million km away.

There are several processes by which fusion occurs in the sun. The most important process is the proton-proton chain.

$$^{1}_{1}H + ^{1}_{1}H \rightarrow ^{2}_{1}H + ^{0}_{+1}e + ^{0}_{0}\nu$$

$$^{1}_{1}H + ^{2}_{1}H \rightarrow ^{3}_{2}He + \gamma$$

$$^{3}_{2}He + ^{3}_{2}He \rightarrow ^{4}_{2}He + 2\,^{1}_{1}H$$

The first two reactions must occur twice in order to produce the two $^{3}_{2}He$ particles needed for the final reaction. The net result is that four protons produce one $^{4}_{2}He$, two positrons, and two neutrinos.

The repulsive force between the charged nuclei requires the fusing nuclei to have high energies. Thus, fusion reactions take place only when the nuclei have large amounts of thermal energy. For this reason, fusion reactions are often called **thermonuclear reactions.**

The proton-proton chain requires a temperature of about 2×10^7 K, such as that found in the center of the sun. Fusion reactions also occur in a hydrogen, or thermonuclear, bomb. In this device, the high temperature necessary to produce the fusion reaction is produced by exploding a uranium fission, or atomic, bomb.

Controlled Fusion

Could the huge energy available from fusion be used safely on Earth? Safe energy requires control of the fusion reaction. One reaction that might produce **controlled fusion** is the following.

$$^{2}_{1}H + ^{3}_{1}H \rightarrow ^{4}_{2}He + ^{1}_{0}n + 17.6 \text{ MeV}$$

Here, one deuterium atom fuses with one tritium atom to form a helium atom with a resulting release of a neutron and 17.6 million electron volts of energy.

Deuterium, $^{2}_{1}H$, is available in large quantities in seawater, and tritium, $^{3}_{1}H$, is easily produced from deuterium. Therefore, controlled fusion would give the world an almost limitless source of energy without the formation of radioactive wastes. In order to control fusion, however, some difficult problems must be solved.

Fusion reactions require that the atoms be raised to temperatures of millions of degrees. No material now in existence can withstand temperatures even as high as 5000 K. In addition, the atoms would be cooled if they touched confining material. Magnetic fields, however, can confine charged particles. Energy is added to the atoms, stripping away electrons and forming separated plasmas of electrons and ions. A sudden increase in the magnetic field will compress the plasma, raising its temperature. Electromagnetic fields and fast-moving neutral atoms can also increase the energy of the plasma. Using this technique, hydrogen nuclei have been fused into helium. The energy released by the reaction becomes the kinetic energy of the neutron and helium ion. This energy would be used to heat some other material, possibly liquefied lithium.

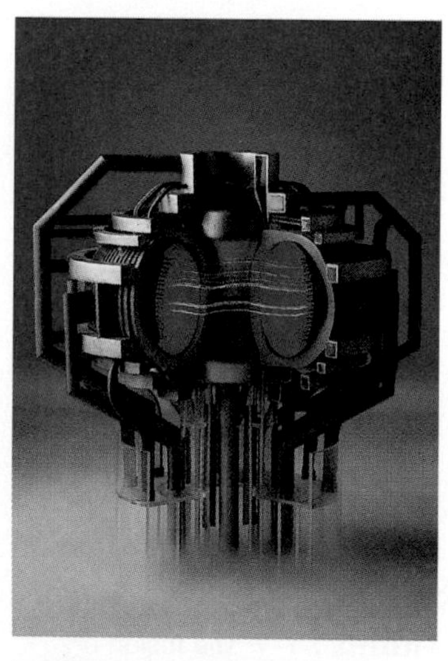

FIGURE 31–8 The Tokamak is an experimental controlled fusion reactor.

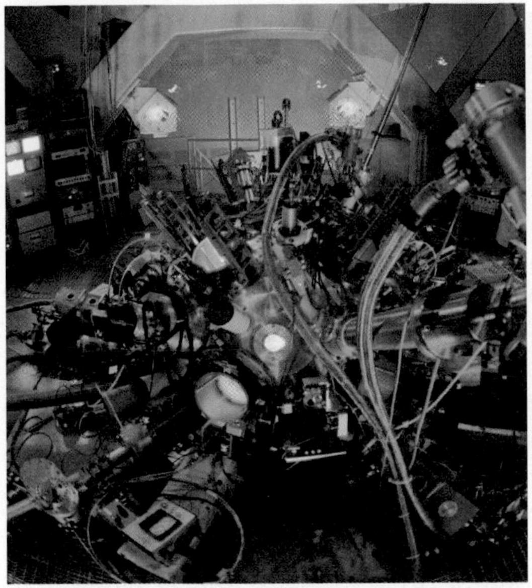

FIGURE 31–9 In laser confinement, pellets containing deuterium and tritium are imploded by many giant lasers, producing helium and large amounts of thermal energy. Electrical discharges are visible on the surface of the water covering the Particle Beam Fusion Accelerator II when a pulse of ions is fired.

The lithium, in turn, would boil water, producing steam to turn electric generators.

A useful reactor must produce more energy than it consumes. So far, the energy produced by fusion has been only a tiny fraction of the energy required to create and hold the plasma. The confinement of plasma is a difficult problem because instabilities in the magnetic field allow the plasma to escape. The Tokamak reactor, shown in **Figure 31–8,** provides a doughnut-shaped magnetic field in which the plasma is confined.

A second approach to controlled fusion is called **inertial confinement fusion.** Deuterium and tritium are liquefied under high pressure and confined in tiny glass spheres. Multiple laser beams are directed at the spheres, as shown in **Figure 31–9.** The energy deposited by the lasers results in forces that make the pellets implode, squeezing their contents. The tremendous compression of the hydrogen that results raises the temperature to levels needed for fusion.

Practice Problems

9. Calculate the mass defect and the energy released for the deuterium-tritium fusion reaction used in the Tokamak, defined by the following reaction.

$$^{2}_{1}\mathrm{H} + ^{3}_{1}\mathrm{H} \rightarrow ^{4}_{2}\mathrm{He} + ^{1}_{0}\mathrm{n}$$

10. Calculate the energy released for the overall reaction in the sun where four protons produce one $^{4}_{2}\mathrm{He}$, two positrons, and two neutrinos.

Physics & Technology

Radioactive Tracers

Radioactive isotopes are extremely useful in many scientific and industrial applications. Ecologists use radioactive tracers to follow the movement of pesticides or pollutants through ecosystems. For example, many pesticides contain sulfur (S). Replacing some of the S with radioactive sulfur-35 makes it possible to follow the pesticide as it moves through soil and into lakes, streams, or groundwater. Biologists can tag insects, bats, and other small animals with tiny amounts of a radioisotope such as cobalt-60. This enables the researchers to follow the animals' movements at night or in areas not accessible to humans.

Geologists and petrochemical engineers use radiation to learn about underground rock formations. Radioisotopes are lowered into a test well. The characteristics of the radiation that is reflected back from underground rocks and fluids reveal details about the density and composition of the rock. These characteristics also can be used to tell whether water, hydrocarbon deposits, or salt beds are present.

Testing the strength of a material without destroying it is important in industry. Carbon-containing engine parts can be exposed to radiation that turns some of the carbon (C) to carbon-14. After the treated part is installed in the engine and used, lubricating fluid from the engine is tested for carbon-14 content. Extremely small amounts of wear can be measured this way, enabling engineers to determine the rate of wear and predict how long a part can be expected to last.

Phosphorus (P) is a mineral important to plant growth. Plants grown in soil tagged with the radioisotope phosphorus-32 reveal how much of the mineral is absorbed and where it is located in the plant. The amount and location of the radiation emitted by the plant is detected by placing the plant next to a sheet of photographic film.

Thinking Critically Suppose you conducted an experiment with phosphorus-32 tagged soil. After developing the photographic film, what would you expect to see? Explain your reasoning.

31.2 Section Review

2.1 What happens to the energy released in a fusion reaction in the sun?

2.2 One fusion reaction involves two deuterium nuclei. A deuterium molecule contains two deuterium atoms. Why doesn't this molecule undergo fusion?

2.3 How does a breeder reactor differ from a normal reactor?

2.4 **Critical Thinking** Fusion powers the sun. The temperatures are hottest, and the number of fusion reactions greatest, in the center of the sun. What contains the fusion reaction?

CHAPTER 31 REVIEW

Key Terms

31.1
- nucleon
- binding energy
- mass defect

31.2
- fission
- chain reaction
- fast neutron
- moderator
- slow neutron
- enrichment
- control rod
- breeder reactor
- fusion
- thermonuclear reaction
- controlled fusion
- inertial confinement fusion

Summary

31.1 Holding the Nucleus Together

- The strong force binds the nucleus together.
- The energy released in a nuclear reaction can be calculated by finding the mass defect, the difference in mass of the particles before and after the reaction.
- The binding energy is the energy equivalent of the mass defect.

31.2 Using Nuclear Energy

- Bombardment can produce radioactive isotopes not found in nature. These are called artificial radioactive nuclei and are often used in medicine.
- In nuclear fission, the uranium nucleus is split into two smaller nuclei with a release of neutrons and energy.
- Nuclear reactors use the energy released in fission to generate electrical energy.
- The fusion of hydrogen nuclei into a helium nucleus releases the energy that causes stars to shine.
- Development of a process for controlling fusion for use on Earth might provide large amounts of energy safely.

Reviewing Concepts

Section 31.1

1. What force inside a nucleus acts to push the nucleus apart? What force inside the nucleus acts in a way to hold the nucleus together?
2. Define the mass defect of a nucleus. To what is it related?

Section 31.2

3. List three medical uses of radioactivity.
4. What sequence of events must occur for a chain reaction to take place?
5. In a fission reaction, binding energy is converted into thermal energy. Objects with thermal energy have random kinetic energy. What objects have kinetic energy after fission?
6. A newspaper claims that scientists have been able to cause iron nuclei to undergo fission. Is the claim likely to be true? Explain.
7. What role does a moderator play in a fission reactor?
8. The reactor at the Chernobyl power station that exploded and burned used blocks of graphite. What was the purpose of the graphite blocks?
9. Breeder reactors generate more fuel than they consume. Is this a violation of the law of conservation of energy? Explain.
10. Scientists think that Jupiter might have become a star if the temperatures inside the planet had not been so low. Why must stars have a high internal temperature?
11. Fission and fusion are opposite processes. How can each release energy?
12. What two processes are being studied to control the fusion process?

Applying Concepts

13. What is the relationship between the average binding energy per nucleon and the degree of stability of a nucleus?
14. Use the graph of binding energy per nucleon in **Figure 31–1** to determine whether the reaction ${}^{2}_{1}H + {}^{1}_{1}H \rightarrow {}^{3}_{2}He$ is energetically possible.
15. Give an example of a naturally and an artificially produced radioactive isotope. Explain the difference.

16. In a nuclear reactor, water that passes through the core of the reactor flows through one loop while the water that produces steam for the turbines flows through a second loop. Why are there two loops?
17. The fission of a uranium nucleus and the fusion of four hydrogen nuclei both produce energy.
    a. Which produces more energy?
    b. Does the fission of a kilogram of uranium nuclei or the fusion of a kilogram of deuterium produce more energy?
    c. Why are your answers to parts **a** and **b** different?
18. Explain how it might be possible for some fission reactors to produce more fissionable fuel than they consume. What are such reactors called?
19. What is the difference between the fission process in an atomic bomb and in a reactor?
20. Why might a fusion reactor be safer than a fission reactor?

Problems

Section 31.1

LEVEL 1

21. A carbon isotope, ${}^{13}_{6}C$, has a nuclear mass of 13.00335 u.
    a. What is the mass defect of this isotope?
    b. What is the binding energy of its nucleus?
22. A nitrogen isotope, ${}^{12}_{7}N$, has a nuclear mass of 12.0188 u.
    a. What is the binding energy per nucleon?
    b. Does it require more energy to separate a nucleon from a ${}^{14}_{7}N$ nucleus or from a ${}^{12}_{7}N$ nucleus? ${}^{14}_{7}N$ has a mass of 14.00307 u.
23. The two positively charged protons in a helium nucleus are separated by about 2.0×10^{-15} m. Use Coulomb's law to find the electric force of repulsion between the two protons. The result will give you an indication of the strength of the strong nuclear force.

LEVEL 2

24. A ${}^{232}_{92}U$ nucleus, mass = 232.0372 u, decays to ${}^{228}_{90}Th$, mass = 228.0287 u, by emitting an α particle, mass = 4.0026 u, with a kinetic energy of 5.3 MeV. What must be the kinetic energy of the recoiling thorium nucleus?
25. The binding energy for ${}^{4}_{2}He$ is 28.3 MeV. Calculate the mass of a helium nucleus in atomic mass units.

Section 31.2

LEVEL 1

26. The radioactive nucleus indicated in each of the following equations disintegrates by emitting a positron. Complete each nuclear equation.
    a. ${}^{21}_{11}Na \rightarrow ? \; {}^{0}_{+1}e + ?$
    b. ${}^{49}_{24}Cr \rightarrow ? \; {}^{0}_{+1}e + ?$
27. A mercury isotope, ${}^{200}_{80}Hg$, is bombarded with deuterons, ${}^{2}_{1}H$. The mercury nucleus absorbs the deuterons and then emits an α particle.
    a. What element is formed by this reaction?
    b. Write the nuclear equation for the reaction.
28. When bombarded by protons, a lithium isotope, ${}^{7}_{3}Li$, absorbs a proton and then ejects two α particles. Write the nuclear equation for this reaction.
29. Each of the following nuclei can absorb an α particle, assuming that no secondary particles are emitted by the nucleus. Complete each equation.
    a. ${}^{14}_{7}N + {}^{4}_{2}He \rightarrow ?$
    b. ${}^{27}_{13}Al + {}^{4}_{2}He \rightarrow ?$
30. When a boron isotope, ${}^{10}_{5}B$, is bombarded with neutrons, it absorbs a neutron and then emits an α particle.
    a. What element is also formed?
    b. Write the nuclear equation for this reaction.
31. When a boron isotope, ${}^{11}_{5}B$, is bombarded with protons, it absorbs a proton and emits a neutron.
    a. What element is formed?
    b. Write the nuclear equation for this reaction.
    c. The isotope formed is radioactive and decays by emitting a positron. Write the complete nuclear equation for this reaction.

LEVEL 2

32. The isotope most commonly used in PET scanners is $^{18}_{9}$F.

a. What element is formed by the positron emission of this element?

b. Write the equation for this reaction.

c. The half-life of $^{18}_{9}$F is 110 min. A solution containing 10.0 mg of this isotope is injected into a patient at 8:00 A.M. How much remains in the patient's body at 3:30 P.M.?

33. The first atomic bomb released an energy equivalent of 2.0×10^{1} kilotons of TNT. One kiloton of TNT is equivalent to 5.0×10^{12} J. What was the mass of the uranium-235 that underwent fission to produce this energy?

34. Complete the following fission reaction.

$$^{239}_{94}\text{Pu} + ^{1}_{0}\text{n} \rightarrow ^{137}_{52}\text{Te} + ? + 3\ ^{1}_{0}\text{n}$$

35. Complete the following fission reaction.

$$^{235}_{92}\text{U} + ^{1}_{0}\text{n} \rightarrow ^{92}_{36}\text{Kr} + ? + 3\ ^{1}_{0}\text{n}$$

36. Complete each of the following fusion reactions.

a. $^{2}_{1}\text{H} + ^{2}_{1}\text{H} \rightarrow ? + ^{1}_{0}\text{n}$

b. $^{2}_{1}\text{H} + ^{2}_{1}\text{H} \rightarrow ? + ^{1}_{1}\text{H}$

c. $^{2}_{1}\text{H} + ^{3}_{1}\text{H} \rightarrow ? + ^{1}_{0}\text{n}$

37. One fusion reaction is $^{2}_{1}\text{H} + ^{2}_{1}\text{H} \rightarrow ^{4}_{2}\text{He}$.

a. What energy is released in this reaction?

b. Deuterium exists as a diatomic, two-atom molecule. One mole of deuterium contains 6.022×10^{23} molecules. Find the amount of energy released, in joules, in the fusion of one mole of deuterium molecules.

c. When one mole of deuterium burns, it releases 2.9×10^{6} J. How many moles of deuterium molecules would have to burn to release just the energy released by the fusion of one mole of deuterium molecules?

Critical Thinking Problems

38. One fusion reaction in the sun releases about 25 MeV of energy. Estimate the number of such reactions that occur each second from the luminosity of the sun, which is the rate at which it releases energy, 4×10^{26} W.

39. The mass of the sun is 2×10^{30} kg. If 90 percent of the sun's mass is hydrogen, find the number of hydrogen nuclei in the sun. From the number of fusion reactions each second that you calculated in problem 38, estimate the number of years the sun could continue to "burn" its hydrogen.

40. If a uranium nucleus were to split into three pieces of approximately the same size instead of two, would more or less energy be released?

Going Further

Data Analysis An isotope undergoing radioactive decay is monitored by a radiation detector. The number of counts in each five-minute interval is recorded. The results are shown in **Table 31–1.** The sample is then removed and the radiation detector records 20 counts resulting from cosmic rays in five minutes. Find the half-life of the isotope. Note that you should first subtract the 20-count background reading from each result. Then plot the counts as a function of time. From your graph, determine the half-life.

TABLE 31–1

Radioactive Decay Measurements

Time (min)	Counts (per 5 minutes)
0	987
5	375
10	150
15	70
20	40
25	25
30	18

*inter***NET**
CONNECTION

Follow the link on the Glencoe Homepage at **www.glencoe.com/sec/science** to find out more about this chapter.

Appendices Contents

Appendix A
Math Handbook

Basic Math Calculations

Fractions, decimals, and percents To express a fraction as a decimal, divide the numerator by the denominator. The resulting number, the quotient, will be the decimal equivalent of the fraction. To express a fraction as a percent, multiply the quotient by 100%. Round the result to the correct number of significant digits.

Example: Express $\frac{33}{59}$ as a decimal and as a percent.

Strategy:

- Divide 33 by 59, expressing the answer in two significant digits.
- Multiply the decimal by 100%.

Solution: $\frac{33}{59} = 0.5593 = 0.56$

$$0.56 \times 100\% = 56\%$$

To express a percent as a decimal, write the percent in the form of a fraction, $\frac{x}{100}$, and then find the quotient. A shortcut is to simply move the decimal point two places left and add the percent symbol.

Example: Express 91.6% as a decimal.

Strategy:

- Move the decimal point two places to the left.

Solution: $91.6\% \Rightarrow 0.916$

Calculating relative uncertainty and relative error Whenever you measure a physical quantity, there is some degree of uncertainty in the measurement. The type of measuring device chosen, as shown by the two metersticks in **Figure 1**, and how carefully the measuring device was used affect precision and accuracy.

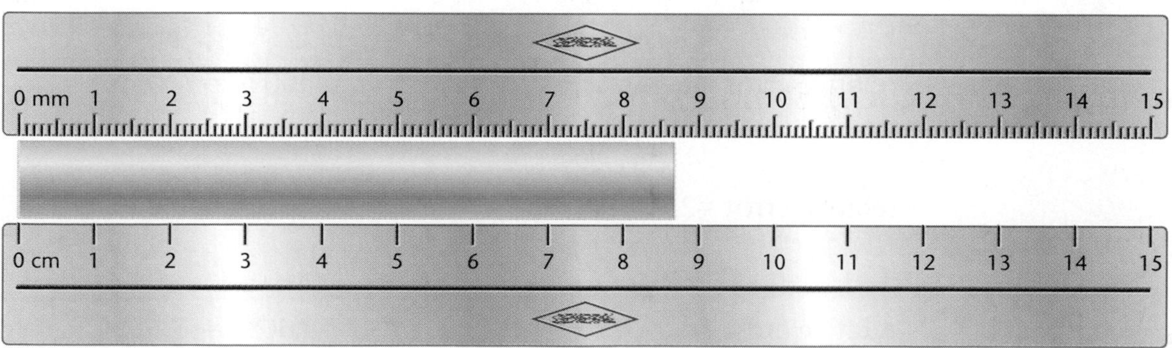

FIGURE 1

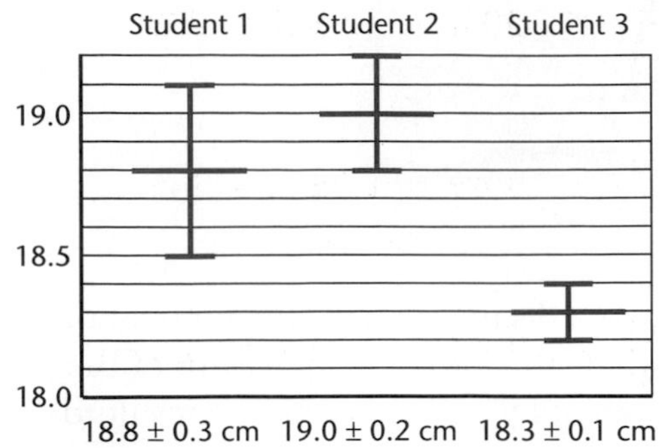

FIGURE 2

The precision of an experimental result can be expressed as estimated uncertainty. Examine the experimental results in **Figure 2** reported by the three students in Chapter 2. Each student measured the length of a block of wood. Student 1's result was reported as (18.8 ± 0.3) cm. The estimated uncertainty in that measurement is represented by ± 0.3. Notice that each student reported a different estimated uncertainty.

The students also could have reported each result using relative uncertainty.

$$\text{relative uncertainty} = \frac{\text{estimated uncertainty}}{\text{actual measurement}} \times 100\%$$

Often, experimental data are compared to accepted values. Relative error is the percent deviation from an accepted value, that is, the uncertainty of a measurement in terms of accuracy. The relative error is calculated according to the following formula.

$$\text{relative error} = \frac{|\text{accepted value} - \text{experimental value}|}{\text{accepted value}} \times 100\%$$

Example: Compare the relative error and relative uncertainty of each student's measurement shown in **Figure 2.** The actual length of the block of wood is 19.0 cm.

Strategy:

- Identify the experimental value and estimated uncertainty measured by each student.
- Use the formulas above to calculate both unknown quantities.
- Round the answers to the correct number of significant digits.

Solution:

$$\text{relative error \#1} = \frac{|19.0\text{ cm} - 18.8\text{ cm}|}{19.0\text{ cm}} \times 100\% = 1.05\%$$

$$\text{relative error \#2} = \frac{|19.0\text{ cm} - 19.0\text{ cm}|}{19.0\text{ cm}} \times 100\% = 0.00\%$$

$$\text{relative error \#3} = \frac{|19.0\text{ cm} - 18.3\text{ cm}|}{19.0\text{ cm}} \times 100\% = 3.68\%$$

$$\text{relative uncertainty \#1} = \frac{0.3 \text{ cm}}{18.8 \text{ cm}} \times 100\% = 1.59\% = 2\%$$

$$\text{relative uncertainty \#2} = \frac{0.2 \text{ cm}}{19.0 \text{ cm}} \times 100\% = 1.05\% = 1\%$$

$$\text{relative uncertainty \#3} = \frac{0.1 \text{ cm}}{18.3 \text{ cm}} \times 100\% = 0.54\% = 0.5\%$$

Student 3 reported the smallest relative uncertainty. His measurement was the most precise. Student 2's measurement was the most accurate. She had the smallest relative error.

Ratios, rates, and proportions A ratio is a comparison between two numbers by division. Ratios are often expressed as fractions. A rate is a ratio between two measurements with different units. For example, the ratio, $\frac{\text{meters}}{\text{second}}$, compares the distance traveled to a period of time. In physics, you will need to solve problems that relate ratios. A proportion is a statement of equality of two or more ratios. To solve for the unknown quantity in a proportion, cross multiply the terms in the ratios and solve for the unknown. Notice that the cross products *ad* and *cb* are equal.

$$\text{If } \frac{a}{b} = \frac{c}{d}, \text{ then } ad = cb.$$

Example: $\frac{1.0 \text{ in.}}{2.54 \text{ cm}} = \frac{3.5 \text{ in.}}{x}$

Strategy:

- Cross multiply.
- Solve for x.
- Round the answer to two significant digits.

Solution: $\frac{1.0 \text{ in.}}{2.54 \text{ cm}} = \frac{3.5 \text{ in.}}{x}$

$$(1.0 \text{ in.})x = (2.54 \text{ cm})(3.5 \text{ in.})$$

$$x = \frac{(2.54 \text{ cm})(3.5 \text{ in.})}{1.0 \text{ in.}}$$

$$x = 8.89 \text{ cm}$$

$$x = 8.9 \text{ cm}$$

Algebra

Solving equations To solve for one unknown, perform arithmetic operations on both sides of the equal sign until the unknown is by itself on one side of the equation.

Example: Solve the following equation for x.

$$\frac{ay}{x} = cb + 5$$

Strategy:

- Multiply both sides by x.
- Divide both sides by the term $cb + 5$.

Solution:

$$\frac{ay}{x} = cb + 5$$

$$x\left(\frac{ay}{x}\right) = x(cb + 5)$$

$$ay = x(cb + 5)$$

$$x = \frac{ay}{cb + 5}$$

Unit operations/dimensional analysis Most physical quantities have units as well as numerical values. When you substitute a value into an equation, you must write both the value and the unit. You have learned in the factor-label method of unit conversion that, when a term has several units, you can operate on the units like any other mathematical quantity. You will often be able to tell when you have set up the equation incorrectly by inspecting the units. This procedure is often called dimensional analysis. If your answer has the wrong units, you have made an error in the calculation of your answer.

Example: Find d when $v = 67$ meters/second and $t = 5.0$ minutes.

Strategy:

- v, t, and d are related by the equation $d = vt$.
- Set up the equation and operate on the units.
- Be sure the resulting unit is correct for d.

Solution: $d = vt$

$$d = \frac{67 \text{ meters}}{\text{second}} \times \frac{60 \text{ seconds}}{1 \text{ minute}} \times 5.0 \text{ minutes}$$

$$d = \frac{67 \text{ meters}}{\cancel{\text{second}}} \times \frac{60 \cancel{\text{seconds}}}{1 \cancel{\text{minute}}} \times 5.0 \cancel{\text{minutes}}$$

$$d = 2.0 \times 10^4 \text{ meters}$$

d is measured in units of length. The solution is correct.

Properties of exponents An exponent tells how many times a number, called a base, is used as a factor. In the example, $a \times a \times a = a^3$, a is raised to the third power.

For any nonzero number a and any integer n, the following properties apply.

- Exponent of Zero: $a^0 = 1$
- Exponent of One: $a^1 = \text{a}$
- Negative Exponents: $a^{-n} = \dfrac{1}{a^n}$

For all integers a and b and all integers m, n, and p, the following properties apply.

- Product of Powers: $a^m \times a^n = a^{m+n}$
- Power of Powers: $(a^m)^n = a^{mn}$
- Quotient of Powers: $\dfrac{a^m}{a^n} = a^{m-n}$
- The n-Root of Powers: $\sqrt[n]{a^m} = a^{m/n}$
- Power of a Product: $(ab)^m = a^m b^m$
- Power of a Monomial: $(a^m b^n)^p = a^{mp} b^{np}$

Example: Simplify $(2a^4b)^3[(-2b)^3]^2$

Strategy:
- Use the power of powers property.
- Use the power of a monomial property.
- Use the product of powers property.

Solution:

$$\begin{aligned}(2a^4b)^3[(-2b)^3]^2 &= (2a^4b)^3(-2b)^6\\ &- 2^3(u^4)^3b^3(-2)^6b^6\\ &= 8a^{12}b^3(64)b^6\\ &= 512a^{12}b^9\end{aligned}$$

Example: Simplify $\sqrt[4]{\dfrac{1}{a^2}}$

Strategy:
- Use the negative exponents property.
- Use the n-root of powers property.

Solution:

$$\begin{aligned}\sqrt[4]{\frac{1}{a^2}} &= \sqrt[4]{a^{-2}}\\ &= a^{-2/4}\\ &= a^{-1/2}\end{aligned}$$

The quadratic formula Any equation in one variable, where the highest power is two, is a quadratic equation. The graph of a quadratic equation is a parabola. The roots of a quadratic equation in the form $ax^2 + bx + c = 0$, where $a \neq 0$, are given by the quadratic formula.

$$x = \frac{-b \pm \sqrt{b^2 - 4ac}}{2a}$$

The quantities a, b and c typically are given.

The expression $b^2 - 4ac$ is called the discriminant. The discriminant tells us the nature of the roots of the quadratic equation.

Discriminant	Nature of the Roots
$b^2 - 4ac > 0$	two distinct real roots
$b^2 - 4ac = 0$	exactly one real root
$b^2 - 4ac < 0$	two distinct imaginary roots

Example: Solve $x^2 - 6x - 40 = 0$.

Strategy:

- Substitute the values in the quadratic formula.
- $a = 1$, $b = -6$, and $c = -40$.

Solution:

$$x = \frac{-(-6) \pm \sqrt{(-6)^2 - 4(1)(-40)}}{2(1)}$$

$$x = \frac{6 \pm \sqrt{36 + 160}}{2}$$

$$x = \frac{6 \pm \sqrt{196}}{2}$$

$$x = \frac{6 \pm 14}{2}$$

$$x = \frac{6 + 14}{2} \quad \text{or} \quad x = \frac{6 - 14}{2}$$

$$= 10 \qquad\qquad = -4$$

Notice in the example that there are two solutions, $x = 10$ and $x = -4$. Sometimes in physics problems, only one solution corresponds to a real-life situation. In that case, one of the solutions would be discarded.

Geometry and Trigonometry

Perimeter, area, and volume Use the following table to solve problems involving perimeter, circumference, area, and volume.

	Perimeter/ Circumference	Area	Surface Area	Volume
Circle radius r	$C = 2\pi r$	$A = \pi r^2$		
Square side a	$P = 4a$	$A = a^2$		
Rectangle length l width w	$P = 2l + 2w$	$A = lw$		
Triangle base b height h		$A = \frac{1}{2}bh$		
Cylinder radius r height h			$SA = 2\pi rh + 2\pi r^2$	$V = \pi r^2 h$
Sphere radius r			$SA = 4\pi r^2$	$V = \frac{4}{3}(\pi r^3)$
Cube side a			$SA = 6a^2$	$V = a^3$

Calculating the area under a graph The calculation of the area under a graph, as shown in **Figure 3-a** and **b,** can often yield useful information. When you do not know the formula for the area of a figure with a curved edge, you can approximate the area by drawing rectangles at small intervals, as shown in **Figure 3-b.** The smaller the intervals, the closer the sum of the areas of the rectangles will be to the actual area under the curve.

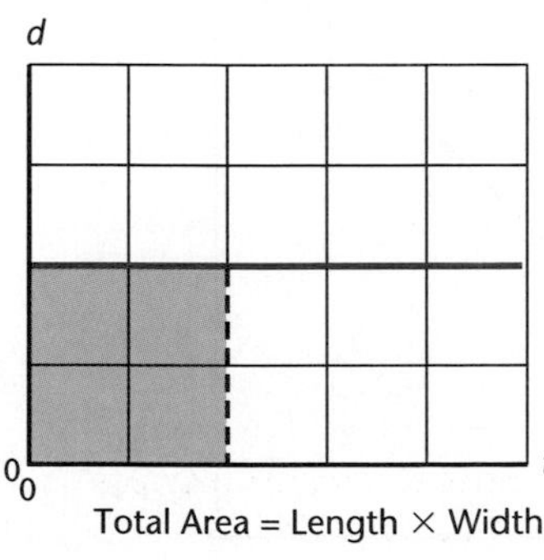

a

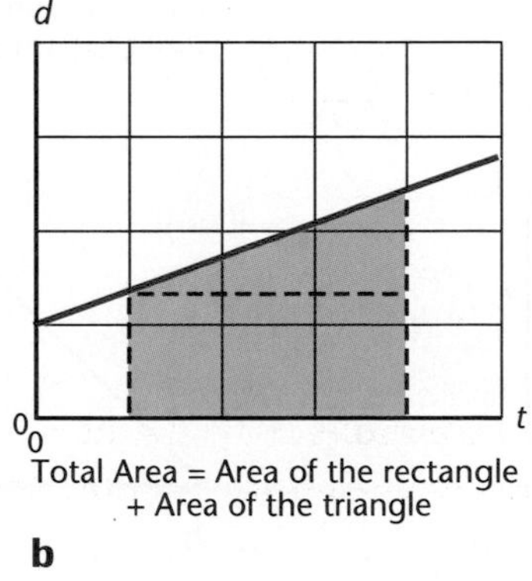

b

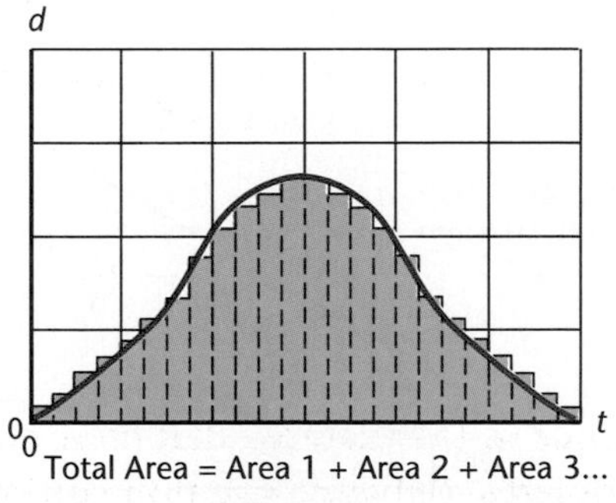

c

FIGURE 3

Pythagorean Theorem If a and b represent the measures of the legs of a right triangle and c represents the measure of the hypotenuse, then $c^2 = a^2 + b^2$, or $c = \sqrt{a^2 + b^2}$.

Example: Find the distance c from A to B in **Figure 4.**

Strategy:

- Use the graph to determine a and b.
- Use the Pythagorean Theorem to find c.

Solution: Distance between B and $C = a = |4 - 1| = 3$

Distance between A and $C = b = |1 - 5| = 4$

$$c = \sqrt{4^2 + 3^2}$$

$$= \sqrt{16 + 9}$$

$$= 5$$

The distance from A to B is 5.

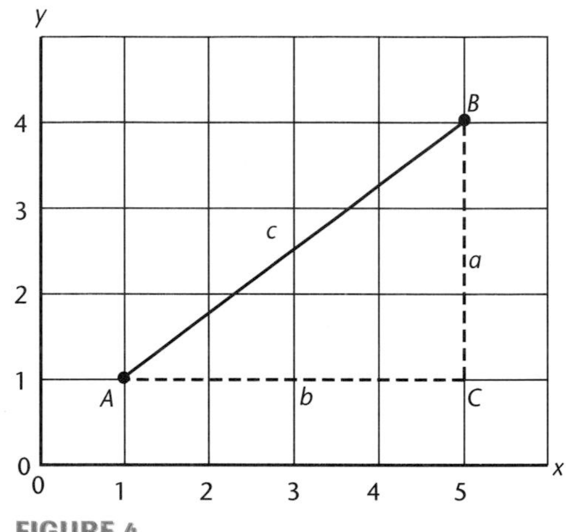

FIGURE 4

Special triangles In physics, it is useful to know the relationships between the sides of a 30°-60°-90° right triangle and the sides of a 45°-45°-90° right triangle, shown in **Figure 5.** If the length of one side of the triangle is known, the unknown sides can be easily calculated.

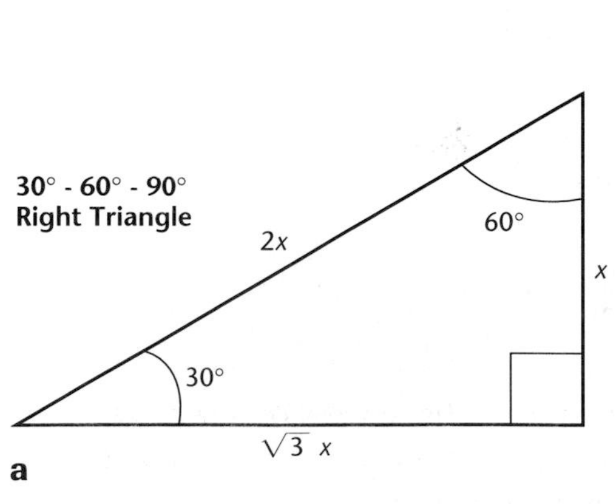

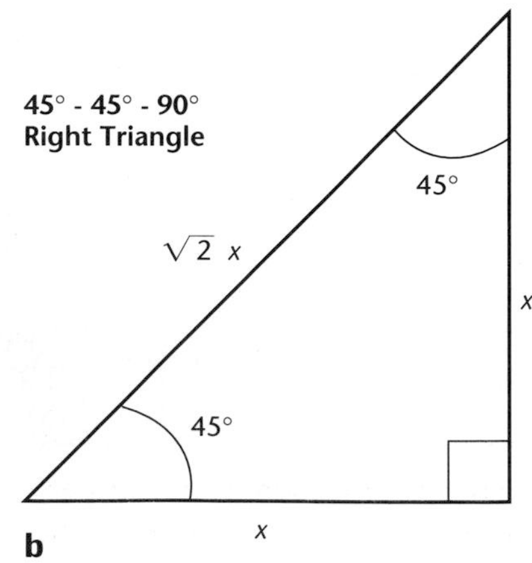

FIGURE 5

The trigonometric ratios The ratios of the lengths of the sides of a right triangle can be used to define the basic trigonometric functions, sine (sin), cosine (cos), and tangent (tan). The sides a and b form the right angle, $\angle C$. The angle θ is formed by sides b and c. Side a is opposite angle θ. Side c, opposite the right angle, is called the hypotenuse.

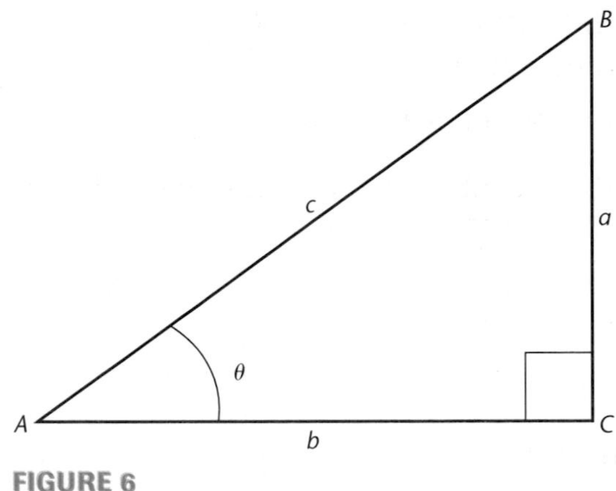

FIGURE 6

$$\sin\theta = \frac{\text{opposite}}{\text{hypotenuse}} = \frac{a}{c} \qquad \cos\theta = \frac{\text{adjacent}}{\text{hypotenuse}} = \frac{b}{c} \qquad \tan\theta = \frac{\text{opposite}}{\text{adjacent}} = \frac{a}{b}$$

Use the first letter from the terms in each relationship to form the acronym SOH-CAH-TOA, an easy way to remember the trigonometric ratios.

Example: For the triangle ABC in **Figure 6,** find $\sin\theta$, $\cos\theta$, and $\tan\theta$, if $a = 48$ cm, $b = 55$ cm, and $c = 73$ cm.

Strategy:

- Use the trigonometric ratios, SOH-CAH-TOA.

Solution:

$$\text{SOH: } \sin\theta = \frac{\text{opposite}}{\text{hypotenuse}} = \frac{48\text{ cm}}{73\text{ cm}} = 0.66$$

$$\text{CAH: } \cos\theta = \frac{\text{adjacent}}{\text{hypotenuse}} = \frac{55\text{ cm}}{73\text{ cm}} = 0.75$$

$$\text{TOA: } \tan\theta = \frac{\text{opposite}}{\text{adjacent}} = \frac{48\text{ cm}}{55\text{ cm}} = 0.87$$

If the value of the sine, cosine, or tangent can be determined from the lengths of two sides of a triangle, the corresponding angle can be found by using a trigonometric functions table or by using the inverse function ($\sin^{-1}$, $\cos^{-1}$, or $\tan^{-1}$) on a calculator.

The Law of Cosines and Law of Sines Sometimes, you will need to work with a triangle that is not a right triangle. The Law of Cosines and Law of Sines apply to all triangles.

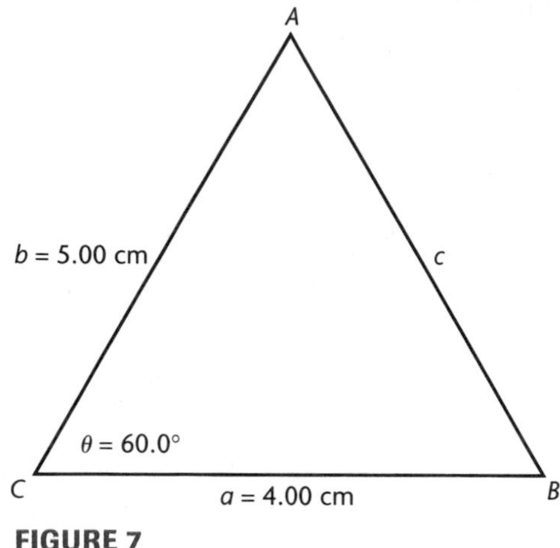

FIGURE 7

The Law of Cosines is useful when you know the measure of two sides and the angle formed by them or the measures of all three sides of a triangle.

$$c^2 = a^2 + b^2 - 2ab \cos \theta$$

Example: For triangle *ABC* in **Figure 7,** find the length of side *c*.

Strategy:

- Substitute the known values into the Law of Cosines.
- $a = 4.00$ cm, $b = 5.00$ cm, and $\theta = 60.0°$.

Solution:

$$c = \sqrt{a^2 + b^2 - 2ab \cos\theta}$$

$$c = \sqrt{(4.00 \text{ cm})^2 + (5.00 \text{ cm})^2 - 2(4.00 \text{ cm})(5.00 \text{ cm})\cos 60.0°}$$

$$c = \sqrt{16.0 \text{ cm}^2 + 25.0 \text{ cm}^2 - 40.0 \text{ cm}^2(0.500)}$$

$$c = \sqrt{21.0 \text{ cm}^2}$$

$$c = 4.58 \text{ cm}$$

Similarly, it is true for any triangle such as *ABC* in **Figure 7** that

$$a^2 = b^2 + c^2 - 2bc \cos A$$

$$b^2 = a^2 + c^2 - 2ac \cos B.$$

If an angle is larger than 90°, its cosine is negative and is numerically equal to the cosine of its supplement. In the triangle *DEF* below, angle *F* is 120.0°. Therefore, its cosine is the negative of the cosine of (180.0° − 120.0°) or 60.0°. The cosine of 60.0° is 0.500. Thus, the cosine of 120.0° is −0.500.

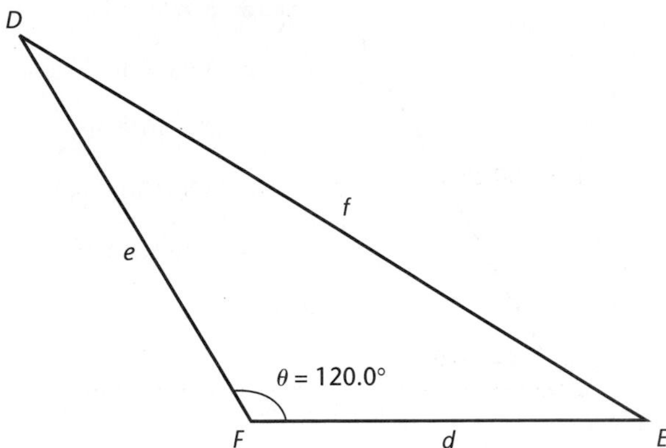

The Law of Sines is useful when you know the measures of two angles and any side of a triangle or the measures of two sides of a triangle and an angle opposite one of these sides.

$$\frac{\sin A}{a} = \frac{\sin B}{b} = \frac{\sin C}{c}$$

Example: For the triangle *ABC* in **Figure 7,** find the measure of angle *A*.

Strategy:

* Substitute the known values into the Law of Sines.
* Use a calculator or a trig table to go from sin *A* to *A*.

Solution: $$\frac{\sin A}{a} = \frac{\sin C}{c}$$

$$\sin A = \frac{a \sin C}{c}$$

$$\sin A = \frac{4.00 \text{ cm}(\sin 60.0°)}{4.58 \text{ cm}}$$

$$\sin A = \frac{(4.00 \text{ cm})(0.867)}{4.58 \text{ cm}}$$

$$\sin A = 0.757$$

$$A = 49.2°$$

Appendix B
Solutions for Practice Problems

CHAPTER 1

No practice problems.

CHAPTER 2

1. a. 5.8×10^3 m; **b.** 4.5×10^5 m

c. 3.02×10^8 m; **d.** 8.6×10^{10} m

2. a. 5.08×10^{-4} kg; **b.** 4.5×10^{-7} kg

c. 3.600×10^{-4} kg; **d.** 4×10^{-3} kg

3. a. 3×10^5 s; **b.** 1.86×10^5 s

c. 9.3×10^7 s

4. a. $(1.1 \text{ cm})\dfrac{(1 \times 10^{-2} \text{ m})}{(1 \text{ cm})} = 1.1 \times 10^{-2} \text{ m}$

b. $(76.2 \text{ pm})\dfrac{(1 \times 10^{-12} \text{ m})}{(1 \text{ pm})}\left(\dfrac{1 \times 10^3 \text{ mm}}{\text{m}}\right)$

$= 7.62 \times 10^{-8} \text{ mm}$

c. $(2.1 \text{ km})\dfrac{(1 \times 10^3 \text{ m})}{(1 \text{ km})} = 2.1 \times 10^3 \text{ m}$

d. $(2.278 \times 10^{11} \text{ m})\left(\dfrac{1 \text{ km}}{1 \times 10^3 \text{ m}}\right)$

$= 2.278 \times 10^8 \text{ km}$

5. a. $1 \text{ kg} = 1 \times 10^3 \text{ g}$ so $147\text{g}\left[\dfrac{1 \text{ kg}}{1 \times 10^3 \text{ g}}\right]$

$= 147 \times 10^{-3} \text{ kg}$

$= 1.47 \times 10^{-1} \text{ kg}$

b. $1 \text{ Mg} = 1 \times 10^6 \text{ g}$ and $1 \text{ kg} = 1 \times 10^3 \text{ g}$

so $11 \text{ Mg}\left(\dfrac{1 \times 10^6 \text{ g}}{\text{Mg}}\right)\left(\dfrac{1 \text{ kg}}{1 \times 10^3 \text{ g}}\right)$

$= 1.1 \times 10^4 \text{ kg}$

c. $1 \ \mu g = 1 \times 10^{-6} \text{ g}$

$7.23 \ \mu g\left(\dfrac{1 \text{g}}{1 \times 10^6 \ \mu g}\right)\left(\dfrac{1 \text{ kg}}{1 \times 10^3 \text{ g}}\right)$

$= 7.23 \times 10^{-9} \text{ kg}$

d. $478 \text{ mg}\left[\dfrac{1.00 \times 10^{-3} \text{ g}}{1.00 \text{ mg}}\right]\left[\dfrac{1.00 \text{ kg}}{1.00 \times 10^3 \text{ g}}\right]$

$= 4.78 \times 10^{-4} \text{ kg}$

6. a. 8×10^{-7} kg; **b.** 7×10^{-3} kg

c. 3.96×10^{-19} kg; **d.** 4.6×10^{-12} kg

7. a. $2 \times 10^{-8} \text{ m}^2$; **b.** $-1.52 \times 10^{-11} \text{ m}^2$

c. $3.0 \times 10^{-9} \text{ m}^2$

d. $0.46 \times 10^{-18} \text{ m}^2 = 4.6 \times 10^{-19} \text{ m}^2$

8. a. $5.0 \times 10^{-7} \text{ mg} + 4 \times 10^{-8} \text{ mg}$

$= 5.0 \times 10^{-7} \text{ mg} + 0.4 \times 10^{-7} \text{ mg}$

$= 5.4 \times 10^{-7} \text{ mg}$

b. $6.0 \times 10^{-3} \text{ mg} + 2 \times 10^{-4} \text{ mg}$

$= 6.0 \times 10^{-3} \text{ mg} + 0.2 \times 10^{-3} \text{ mg}$

$= 6.2 \times 10^{-3} \text{ mg}$

c. $3.0 \times 10^{-2} \text{ pg} - 2 \times 10^{-6} \text{ ng}$

$= 3.0 \times 10^{-2} \times 10^{-12} \text{ g} - 2 \times 10^{-6} \times 10^{-9} \text{ g}$

$= 3.0 \times 10^{-14} \text{ g} - 0.2 \times 10^{-14} \text{ g}$

$= 2.8 \times 10^{-14} \text{ g}$

d. $8.2 \text{ km} - 3 \times 10^2 \text{ m}$

$= 8.2 \times 10^3 \text{ m} - 0.3 \times 10^3 \text{ m}$

$= 7.9 \times 10^3 \text{ m}$

9. a. $(2 \times 10^4 \text{ m})(4 \times 10^8 \text{ m}) = 8 \times 10^{4+8} \text{ m}^2$

$= 8 \times 10^{12} \text{ m}^2$

b. $(3 \times 10^4 \text{ m})(2 \times 10^6 \text{ m}) = 6 \times 10^{4+6} \text{ m}^2$

$= 6 \times 10^{10} \text{ m}^2$

c. $(6 \times 10^{-4} \text{ m})(5 \times 10^{-8} \text{ m})$

$= 30 \times 10^{-4-8} \text{ m}^2$

$= 3 \times 10^{-11} \text{ m}^2$

d. $(2.5 \times 10^{-7} \text{ m})(2.5 \times 10^{16} \text{ m})$

$= 6.25 \times 10^{-7+16} \text{ m}^2$

$= 6.3 \times 10^9 \text{ m}^2$

10. a. $\dfrac{6 \times 10^8 \text{ kg}}{2 \times 10^4 \text{ m}^3} = 3 \times 10^{8-4} \text{ kg/m}^3$

$= 3 \times 10^4 \text{ kg/m}^3$

b. $\dfrac{6 \times 10^8 \text{ kg}}{2 \times 10^{-4} \text{ m}^3} = 3 \times 10^{8-(-4)} \text{ kg/m}^3$

$= 3 \times 10^{12} \text{ kg/m}^3$

c. $\dfrac{6 \times 10^{-8}\text{ m}}{2 \times 10^{4}\text{ s}} = 3 \times 10^{-8-4}\text{ m/s}$

$= 3 \times 10^{-12}\text{ m/s}$

d. $\dfrac{6 \times 10^{-8}\text{ m}}{2 \times 10^{-4}\text{ s}} = 3 \times 10^{-8-(-4)}\text{ m/s}$

$= 3 \times 10^{-4}\text{ m/s}$

11. a. $\dfrac{(3 \times 10^{4}\text{ kg})(4 \times 10^{4}\text{ m})}{6 \times 10^{4}\text{ s}}$

$= \dfrac{12 \times 10^{4+4}\text{ kg} \cdot \text{m}}{6 \times 10^{4}\text{ s}}$

$= 2 \times 10^{8-4}\text{ kg} \cdot \text{m/s} = 2 \times 10^{4}\text{ kg} \cdot \text{m/s}$

The evaluation may be done in several other ways. For example

$(3 \times 10^{4}\text{ kg})(4 \times 10^{4}\text{ m})/(6 \times 10^{4}\text{ s})$

$= (0.5 \times 10^{4-4}\text{ kg/s})(4 \times 10^{4}\text{ m})$

$= (0.5\text{ kg/s})(4 \times 10^{4}\text{ m})$

$= 2 \times 10^{4}\text{ kg} \cdot \text{m/s}$

b. $(2.5 \times 10^{6}\text{ kg})(6 \times 10^{4}\text{ m})/(5 \times 10^{-2}\text{ s}^2)$

$= 15 \times 10^{6+4}\text{ kg} \cdot \text{m}/(5 \times 10^{-2}\text{ s}^2)$

$= 3 \times 10^{10-(-2)}\text{ kg} \cdot \text{m/s}^2$

$= 3 \times 10^{12}\text{ kg} \cdot \text{m/s}^2$

12. a. $(4 \times 10^{3}\text{ mg})(5 \times 10^{4}\text{ kg})$

$= (4 \times 10^{3} \times 10^{-3}\text{ g})(5 \times 10^{4} \times 10^{3}\text{ g})$

$= 20 \times 10^{7}\text{ g}^2$

$= 2 \times 10^{8}\text{ g}^2$

b. $(6.5 \times 10^{-2}\text{ m})(4.0 \times 10^{3}\text{ km})$

$= (6.5 \times 10^{-2}\text{ m})(4.0 \times 10^{3} \times 10^{3}\text{ m})$

$= 26 \times 10^{4}\text{ m}^2$

$= 2.6 \times 10^{5}\text{ m}^2$

c. $(2 \times 10^{3}\text{ ms})(5 \times 10^{-2}\text{ ns})$

$= (2 \times 10^{3} \times 10^{-3}\text{ s})(5 \times 10^{-2} \times 10^{-9}\text{ s})$

$= 10 \times 10^{-11}\text{ s}^2$

$= 1 \times 10^{-10}\text{ s}^2$

13. a. $\dfrac{2.8 \times 10^{-2}\text{ mg}}{2.0 \times 10^{4}\text{ g}} = \dfrac{2.8 \times 10^{-2} \times 10^{-3}\text{ g}}{2.0 \times 10^{4}\text{ g}}$

$= 1.4 \times 10^{-9}$

b. $\dfrac{(6 \times 10^{2}\text{ kg})(9 \times 10^{3}\text{ m})}{(2 \times 10^{4}\text{ s})(3 \times 10^{6}\text{ ms})}$

$= \dfrac{(6 \times 10^{2}\text{ kg})(9 \times 10^{3}\text{ m})}{(2 \times 10^{4}\text{ s})(3 \times 10^{6} \times 10^{-3}\text{ s})}$

$= \dfrac{54 \times 10^{5}\text{ kg} \cdot \text{m}}{6 \times 10^{7}\text{ s}^2}$

$= 9 \times 10^{-2}\text{ kg} \cdot \text{m/s}^2$

14. $\dfrac{(7 \times 10^{-3}\text{ m}) + (5 \times 10^{-3}\text{ m})}{(9 \times 10^{7}\text{ km}) + (3 \times 10^{7}\text{ km})}$

$= \dfrac{12 \times 10^{-3}\text{ m}}{12 \times 10^{7}\text{ km}}$

$= \dfrac{12 \times 10^{-3}\text{ m}}{12 \times 10^{7} \times 10^{3}\text{ m}} = \dfrac{12 \times 10^{-3}\text{ m}}{12 \times 10^{10}\text{ m}}$

$= 1 \times 10^{-13}$

15. a. 4 **b.** 3 **c.** 2

d. 4 **e.** 2 **f.** 3

16. a. 2 **b.** 4 **c.** 4

d. 3 **e.** 4 **f.** 3

17. a. 26.3 cm (round from 26.281 cm)

b. 1600 m or 1.6 km

18. a. 2.5 g (rounded from 2.536 g)

b. 475 m (rounded from 474.5832 m)

19. a. $3.0 \times 10^2\text{ cm}^2$ (the result 301.3 cm^2 expressed to two significant digits. Note that the expression in the form 300 cm^2 would not indicate how many of the digits are significant.)

b. 13.6 km^2 (the result 13.597335 expressed to three significant digits)

c. $35.7\text{ N} \cdot \text{m}$

20. a. 2.73 cm/s (the result 2.726045 . . . cm/s expressed to three significant digits)

b. 0.253 cm/s (the result 0.253354 . . . cm/s expressed to three significant digits)

c. 1.22×10^3 (the result $1.219469 \ldots \times 10^3$ expressed to three significant digits)

d. 4.1 g/cm^3 (the result $4.138636 \ldots \text{g/cm}^3$ expressed to two significant digits)

21. a.

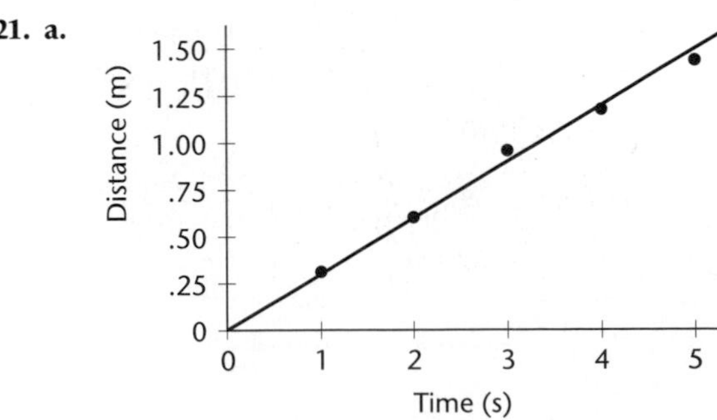

b. straight line **c.** Linear relationship

d. $M = \frac{\Delta y}{\Delta x} = \frac{1.5 - 0.60}{5 - 2} = \frac{0.90}{3} = 0.30 \text{ m/s}$

e. $d = 0.30(t)$

CHAPTER 3

No practice problems.

CHAPTER 4

1.

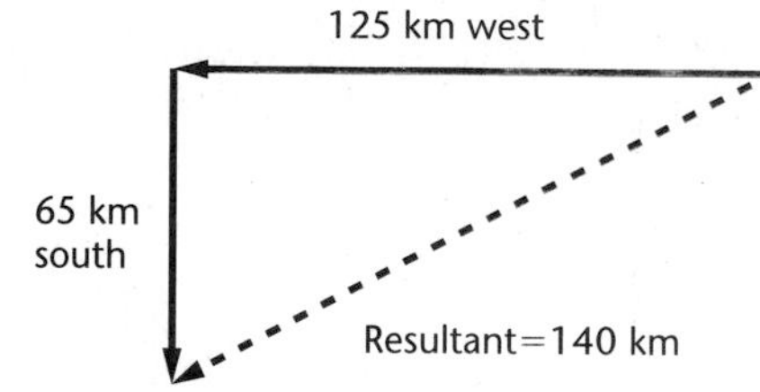

$R^2 = A^2 + B^2$

$R^2 = 65 \text{ km}^2 + 125 \text{ km}^2$

$R^2 = 19\,850 \text{ km}^2$

$R = 140 \text{ km}$

2.

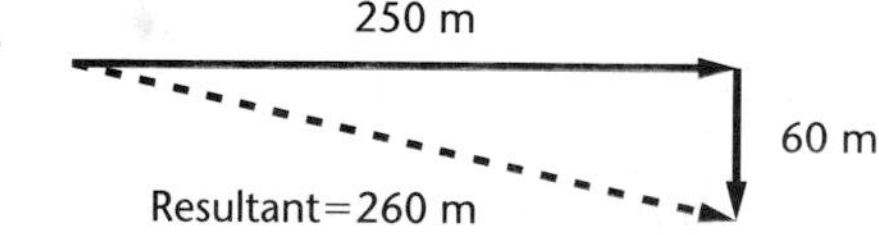

$R^2 = (250 \text{ m})^2 + (60 \text{ m})^2 = 66\,100 \text{ m}^2$

$R = 260 \text{ m}$

3.

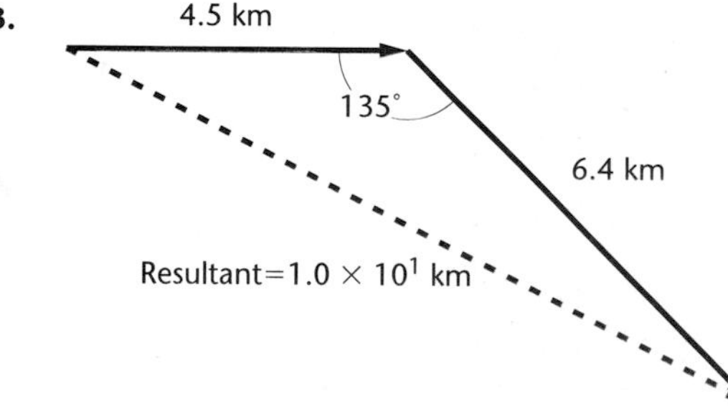

$R^2 = A^2 + B^2 - 2AB \cos\theta$

$R = [(4.5 \text{ km})^2 + (6.4 \text{ km})^2 - (2)(4.5 \text{ km})(6.4 \text{ km})(\cos 135°)]^{1/2}$

$R = 1.0 \times 10^1 \text{ km}$

4.

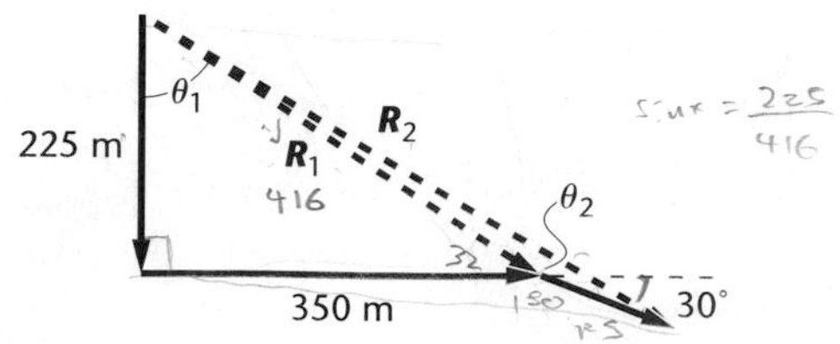

$R_1 = [(225 \text{ m})^2 + (350 \text{ m})^2]^{1/2} = 416 \text{ m}$

$\theta_1 = \tan^{-1} \frac{350 \text{ m}}{225 \text{ m}} = 57.3$

$\theta_2 = 180 - (60 - 57.3) = 177°$

$R_2 = [(416 \text{ m})^2 + (125 \text{ m})^2 - 2(416 \text{ m})(125 \text{ m})(\cos 177°)]^{1/2}$

$R_2 = 540 \text{ m}$

5. Magnitude of change in velocity

$= 45 - (-30) = 75 \text{ km/h}$

direction of change is from east to west

6. $+2.0 \text{ m/s} + 4.0 \text{ m/s} = 6.0 \text{ m/s}$ relative to street

7. $v_{result} = [v_b^2 + v_r^2]^{1/2}$

$= [(11 \text{ m/s})^2 + (5.0 \text{ m/s})^2]^{1/2} = 12 \text{ m/s}$

$\theta = \tan^{-1} \frac{5.0 \text{ m/s}}{11 \text{ m/s}} = 24°$

$v_{result} = 12$ m/s, 66° East of north

8. 2.5 m/s

⟶ boat

2.0 m/s River

⟵

→

0.5 m/s Resultant

2.5 m/s − 0.5 m/s = 2.0 m/s against the boat

9. $v = [v_p^2 + v_w^2]^{1/2} = [(150 \text{ km/h})^2 + (75 \text{ km/h})^2]^{1/2}$

$= 170 \text{ km/h}$

10.

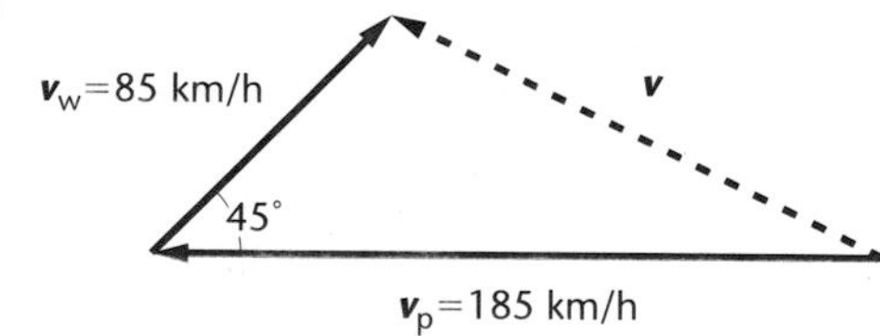

$v = [v_p^2 + v_w^2 - 2v_p v_w \cos\theta]^{1/2}$

$= [(185 \text{ km/h})^2 + (85 \text{ km/h})^2 - (2)(185 \text{ km/h})(85 \text{ km/h})(\cos 45°)]^{1/2}$

$= 140 \text{ km/h}$

11.

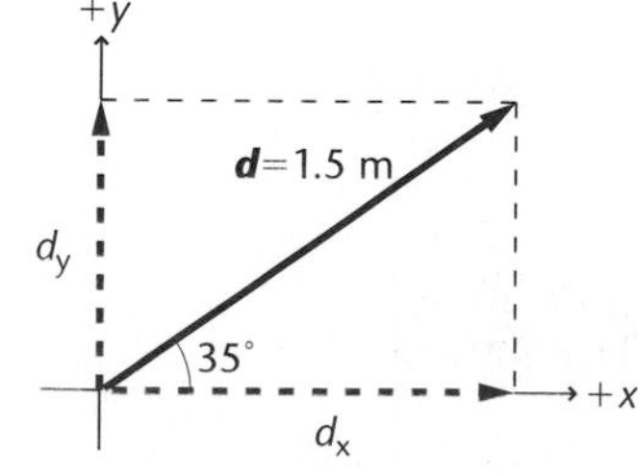

$d_x = 1.5 \text{ m} \cos 35° = 1.2 \text{ m}$

$d_y = 1.5 \text{ m} \sin 35° = 0.86 \text{ m}$

12.

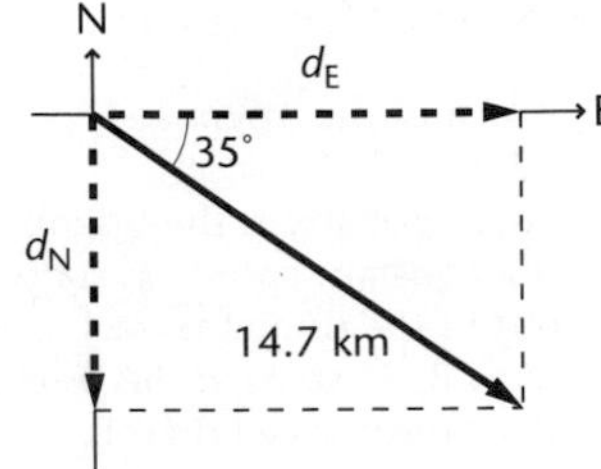

$d_E = 14.7 \text{ km} \cos 35° = 12 \text{ km}$

$d_N = -14.7 \text{ km} \sin 35° = -8.4 \text{ km}$

13.

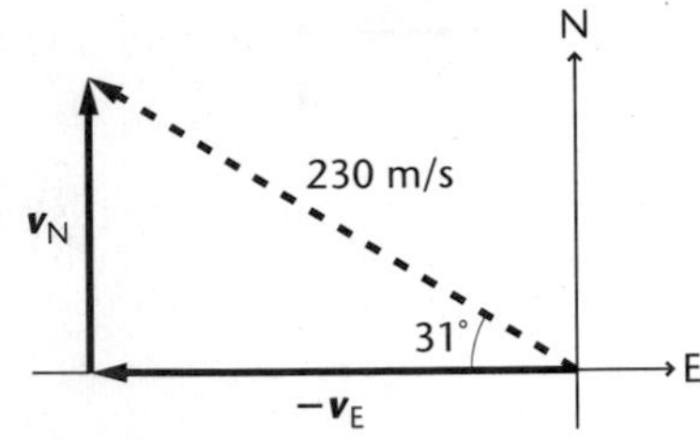

$v_E = -230 \text{ m/s} \cos 31° = -200 \text{ m/s}$

$v_N = 230 \text{ m/s} \sin 31° = 120 \text{ m/s}$

14.

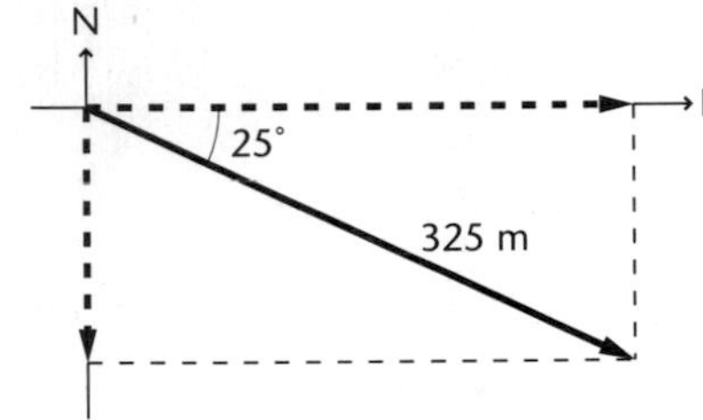

$d_E = 325 \text{ m} \cos 25° = 290 \text{ m}$

$d_N = -325 \text{ m} \sin 25° = -140 \text{ m}$

15.

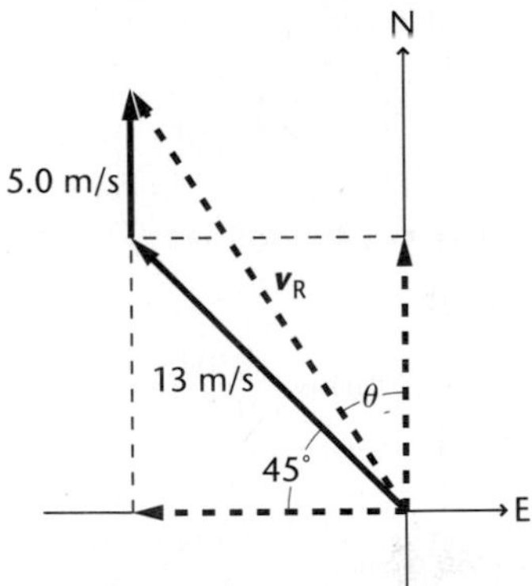

$v_{bW} = (13 \text{ m/s}) \cos 45° = 9.2 \text{ m/s}$

$v_{bN} = (13 \text{ m/s}) \sin 45° = 9.2 \text{ m/s}$

$v_{rN} = 5.0 \text{ m/s}$

$v_{rW} = 0$

$v_{RW} = 9.2 \text{ m/s} + 0 = 9.2 \text{ m/s}$

$v_{RN} = 9.2 \text{ m/s} + 5.0 \text{ m/s} = 14.2 \text{ m/s}$

$v_R = [(9.2 \text{ m/s})^2 + (14.2 \text{ m/s})^2]^{1/2} = 17 \text{ m/s}$

$\theta = \tan^{-1} \dfrac{9.2 \text{ m/s}}{14.2 \text{ m/s}} = \tan^{-1} 0.648 = 33°$

$v_R = 17$ m/s, 33° west of north

16.

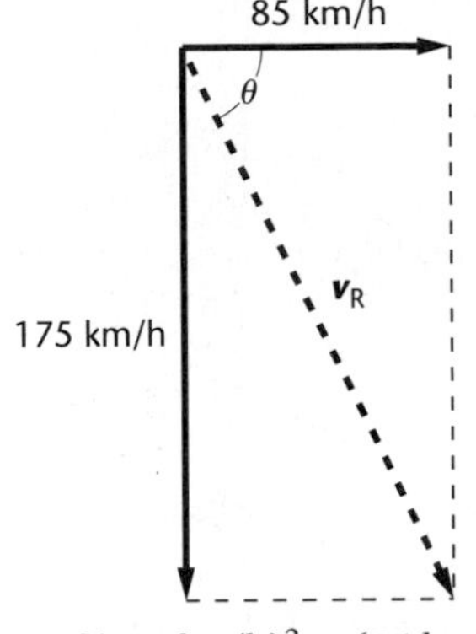

$v_R = [(175 \text{ km/h})^2 + (85 \text{ km/h})^2]^{1/2} = 195 \text{ km/hr}$

$\theta = \tan^{-1} \dfrac{175 \text{ km/h}}{85 \text{ km/h}} = \tan^{-1} 2.06 = 64°$

$v_R = 195$ km/h, 64° south of east

17.

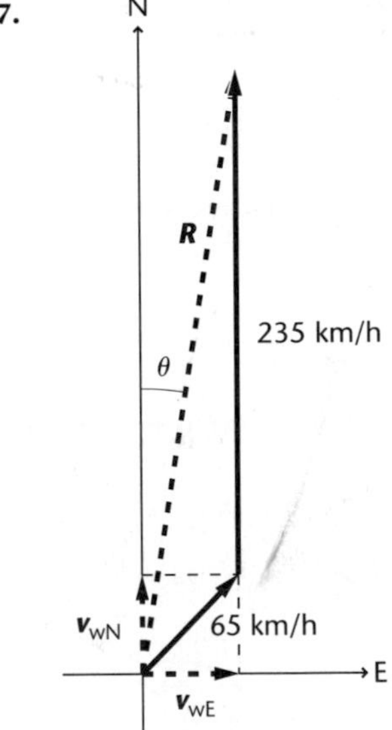

$v_{wN} = 65 \text{ km/h} \sin 45° = 46 \text{ km/h}$

$v_{wE} = 65 \text{ km/h} \cos 45° = 46 \text{ km/h}$

$R_N = 46 \text{ km/h} + 235 \text{ km/h} = 281 \text{ km/h}$

$R_E = 46 \text{ km/h}$

$R = [(281 \text{ km/h})^2 + (46 \text{ km/h})^2]^{1/2} = 285 \text{ km/hr}$

$\theta = \tan^{-1} \dfrac{46 \text{ km/h}}{281 \text{ km/h}} = 9.3°$ east of north

18.

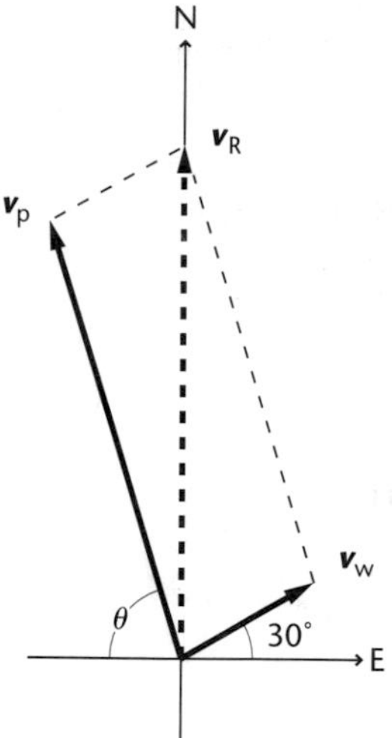

To travel north, the east components must be equal and opposite.

$v_{pE} = v_{wE} = 95$ km/h cos 30° = 82 km/h

$\theta = \cos^{-1} \dfrac{82 \text{ km/h}}{285 \text{ km/h}} = 73°$

$v_{pN} = 285$ km/h sin 73° = 273 km/h

$v_{wN} = 95$ km/h sin 30° = 47.5

$v_R = 320$ km/h north

CHAPTER 5

1. A starts at High St., walking east at constant velocity.

   B starts west of High St., walking east at slower constant velocity.

   C walks west from High St., first fast, but slowing to a stop.

   D starts east of High St., walking west at constant velocity.

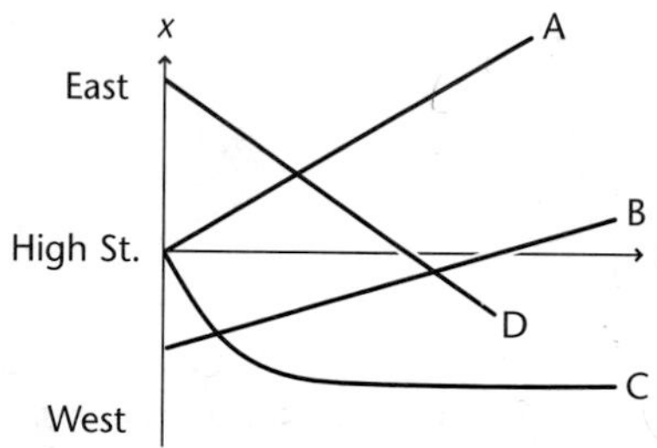

2. The car starts at the origin, moves backward (selected to be the negative direction) at a constant speed of 2 m/s for 10 s, then stops and stays at that location (−20 m) for 20 seconds. It then moves forward at 2.5 m/s for 20 seconds when it is at +30 m. It immediately goes backward at a speed of 1.5 m/s for 20 s, when it has returned to the origin.

3. **a.** Between 10 and 30 s.

   b. 30 m east of the origin

   c. At point D, 30 m east of the origin at 50 s.

4. **a.** A remains stationary. B starts at the origin; moves forward at a constant speed. C starts east (positive direction) of the origin, moves forward at the same speed as B. D starts at the origin, moves forward at a slower speed than B.

   b.

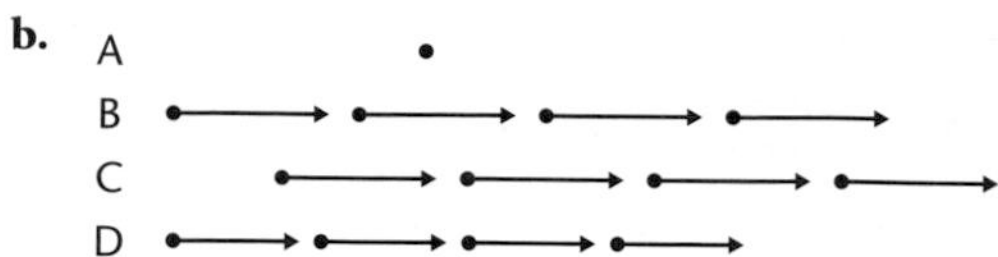

   c. B = C > D > A

5.

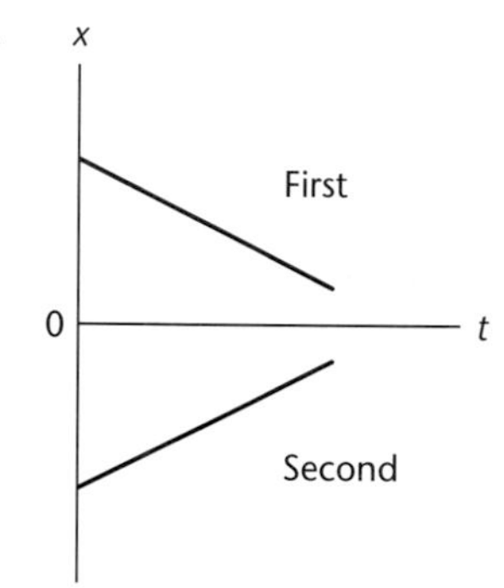

6. Average velocity is 75 m/s. At one second

   $\dfrac{d}{t} = \dfrac{(115 \text{ m})}{(1 \text{ s})} = 115 \text{ m/s}$

   while at 3 seconds, $\dfrac{d}{t} = 88$ m/s.

7. **a.** Into mph:

   10 m/s × (3600 s/h) × (0.6214 mi/km) × (0.001 km/m)

   = 22 mph

   Into km/h:

   10 m/s × (3600 s/h) × (0.001 km/m) = 36 km/h

   b. Into km/h:

   $65 \text{ mph} \times (5280 \text{ ft/mi}) \times \left(\dfrac{0.3048 \text{ m/ft}}{1000 \text{ m/km}}\right)$

   $= 1.0 \times 10^2$ km/h

   Into m/s:

   $65 \text{ mph} \times (5280 \text{ ft/mi}) \times \dfrac{(0.3048 \text{ m/ft})}{(3600 \text{ s/h})}$

   = 29 m/s

   c. Into km/h:

   $4 \text{ mph} \times (5280 \text{ ft/mi}) \times \left(\dfrac{0.3048 \text{ m/ft}}{1000 \text{ m/km}}\right) = 6.4 \text{ km/h}$

   Into m/s:

   $4 \text{ mph} \times (5280 \text{ ft/mi}) \times \dfrac{(0.3048 \text{ m/ft})}{(3600 \text{ s/h})} = 2 \text{ m/s}$

8.

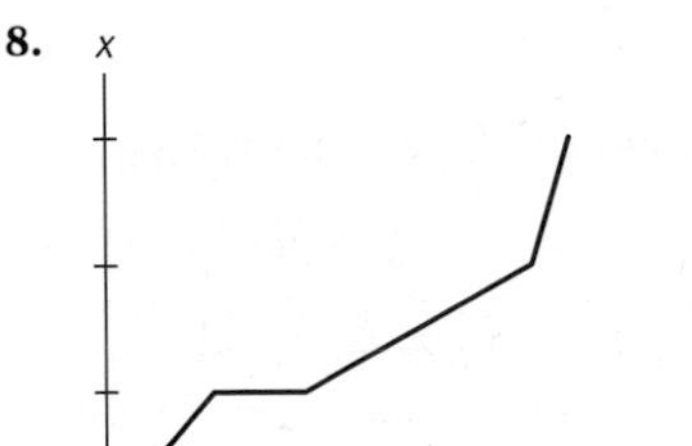

9. a. $v = \frac{8.0\text{ m}}{0.80\text{ s}} = 10\text{ m/s}$

so $x = (-2.0\text{ m}) + (10\text{ m/s})t$

b. At +8.0 m.

10. a. $v = \frac{-4.0\text{ m}}{0.60\text{ s}} = -6.7\text{ m/s}$

so $x = (-2.0\text{ m}) - (6.7\text{ m/s})t$

b. At 1.2 s.

11. a. $x = -(200\text{ m}) + (15\text{ m/s})t$

b. x (at 600 s) = 8800 m

c. The time at which $x = 0$ is given by $t = \frac{200\text{ m}}{15\text{ m/s}} = 13\text{ s}.$

12. a.

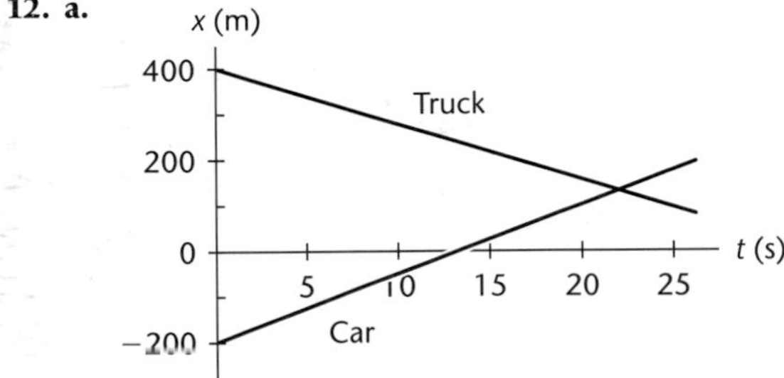

b. Equation for truck:
$x_T = (400\text{ m}) - (12\text{ m/s})t.$
They pass each other when

$-(200\text{ m}) + (15\text{ m/s})t = (400\text{ m}) - (12\text{ m/s})t$

or $-600\text{ m} = -(27\text{ m/s})t$

That is, $t = 22$ s. $x_T = 133$ m.

13. a. At 1.0 s, $v = 74$ m/s.

b. At 2.0 s, $v = 78$ m/s.

c. At 2.5 s, $v = 80$ m/s.

14. $\frac{(75\text{ m/s}) \times (3600\text{ s/h})}{1000\text{ m/km}} = 270\text{ km/h}$

15.

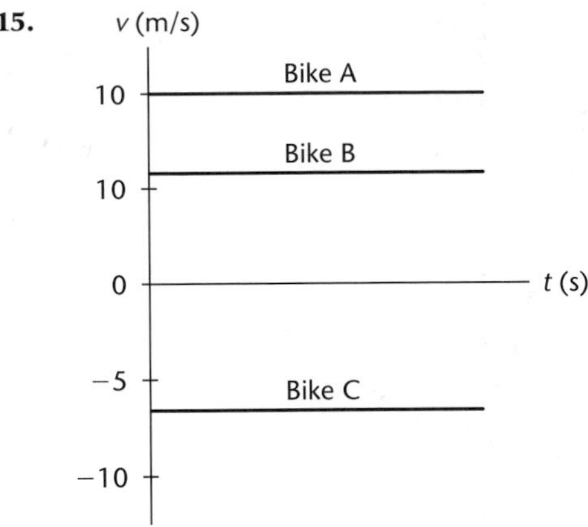

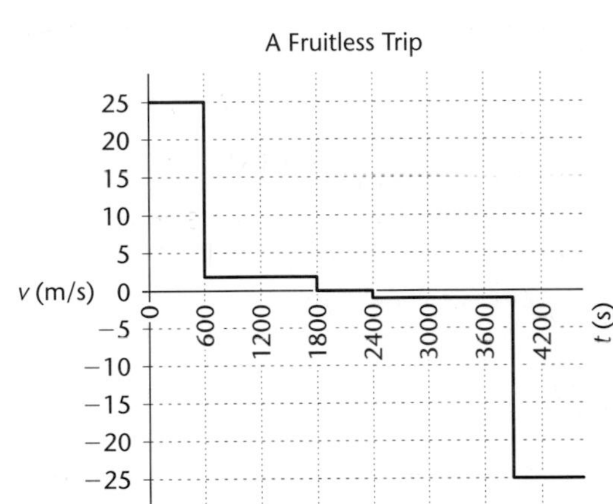

16. a.

b.

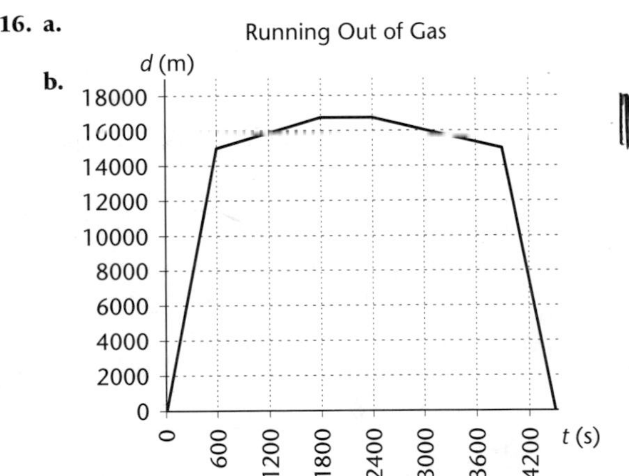

17. $\bar{a} = \frac{\Delta v}{\Delta t} = \frac{36\text{ m/s} - 4.0\text{ m/s}}{4.0\text{ s}} = 8.0\text{ m/s}^2$

18. $\bar{a} = \frac{v_2 - v_1}{t_2 - t_1} = \frac{15\text{ m/s} - 36\text{ m/s}}{3.0\text{ s}} = -7.0\text{ m/s}^2$

19. $\bar{a} = \frac{v_2 - v_1}{t_2 - t_1} = \frac{4.5\text{ m/s} - (-3.0\text{ m/s})}{2.5\text{ s}} = 3.0\text{ m/s}^2$

20. a. $\bar{a} = \frac{v_2 - v_1}{t_2 - t_1} = \frac{0\text{ m/s} - 25.0\text{ m/s}}{3.0\text{ s}} = -8.3\text{ m/s}^2$

b. Half as great (-4.2 m/s^2).

21. a. 5 to 15 s and 21 to 28 s

b. 0 to 6 s

c. 15 to 20 s

22. a. 2 m/s^2

b. -1.2 m/s^2

c. 0 m/s^2

23. a. $v = v_o + at = 2.0\text{ m/s} + (-0.50\text{ m/s}^2)(2.0\text{ s})$

$= 1.0\text{ m/s}$

b. $v = v_o + at = 2.0\text{ m/s} + (-0.50\text{ m/s}^2)(6.0\text{ s})$

$= -1.0\text{ m/s}$

c. The ball's velocity simply decreased in the first case. In the second case the ball slowed to a stop and then began rolling back down the hill.

1st case: v

2nd case: v

24. $a = (3.5\text{ m/s}^2)(1\text{ km}/1000\text{ m})(3600\text{ s/h})$

$= 12.6\text{ (km/h)/s}$

$v = v_o + at = 30\text{ km/h} + (12.6\text{(km/h)/s})(6.8\text{ s})$

$= 30\text{ km/h} + 86\text{ km/h}$

$= 116\text{ km/h}$

25. $v = v_o + at$

so $t = \frac{v - v_o}{a} = \frac{28\text{ m/s} - 0\text{ m/s}}{5.5\text{ m/s}^2} = 5.1\text{ s}$

26. $v = v_o + at$

so $t = \frac{v - v_o}{a} = \frac{3\text{ m/s} - 22\text{ m/s}}{-2.1\text{ m/s}^2} = 9.0\text{ s}$

27. $d = \frac{1}{2}(v + v_o)t = \frac{1}{2}(22\text{ m/s} + 44\text{ m/s})(11\text{ s})$

$= 3.6 \times 10^2\text{ m}$

28. $d = \frac{1}{2}(v - v_o)t$

so $t = \frac{2d}{v + v_o} = \frac{2(125\text{ m})}{25\text{ m/s} + 15\text{ m/s}} = 6.3\text{ s}$

29. $d = \frac{1}{2}(v - v_o)t$

so $v_o = \frac{2d}{t} - v = \frac{2(19\text{ m})}{4.5\text{ s}} - 7.5\text{ m/s} = 0.94\text{ m/s}$

30. a. $d = v_o t + \frac{1}{2}at^2$

$= (0\text{ m/s})(30.0\text{ s}) + \frac{1}{2}(3.00\text{ m/s}^2)(30.0\text{ s})^2$

$= 0\text{ m} + 1350\text{ m} = 1.35 \times 10^3\text{ m}$

b. $v = v_o + at = 0\text{ m/s} + (3.00\text{ m/s}^2)(30.0\text{ s})$

$= 90.0\text{ m/s}$

31. a. $v = v_o + at,\ a = -g = -9.80\text{ m/s}^2$

$v = 0\text{ m/s} + (-9.80\text{ m/s}^2)(4.0\text{ s})$

$v = -39\text{ m/s}$ (downward)

b. $d = v_o t + \frac{1}{2}at^2$

$= 0 + \frac{1}{2}(-9.80\text{ m/s}^2)(4.0\text{ s})^2$

$= \frac{1}{2}(-9.80\text{ m/s}^2)(16\text{ s}^2)$

$d = -78\text{ m}$ (downward)

32. a. Since $a = -g$, and, at the maximum height, $v = 0$, using $v^2 = v_0^2 + 2a(d - d_0)$, gives

$v_0^2 = 2gd$

or $d = \frac{v_0^2}{2g} = \frac{(22.5\text{ m/s})^2}{2(9.80\text{ m/s}^2)} = 25.8\text{ m}$

b. Time to rise: use $v = v_0 + at$, giving

$t = \frac{v_0}{g} = \frac{22.5\text{ m/s}}{9.80\text{ m/s}^2} = 2.30\text{ s}$

So, it is in the air for 4.6 s. To show that the time to rise equals the time to fall, when $d = d_0$

$v^2 = v_0^2 + 2a(d - d_0)$

gives $v^2 = v_0^2$ or $v = -v_0$. Now, using $v = v_0 + at$ where, for the fall, $v_0 = 0$ and $v = -v_0$, we get

$t = \frac{v_0}{g}$.

33. Given $v_0 = 65.0\text{ m/s}$, $v = 162.0\text{ m/s}$, and $t = 10.0\text{ s}$ and needing d, we use

$d = d_0 + \frac{1}{2}(v_0 + v)t$

or $d = \frac{1}{2}(65.0\text{ m/s} + 162.0\text{ m/s})(10.0\text{ s})$

$= 1.14 \times 10^3\text{ m}$

CHAPTER 6

1. a.

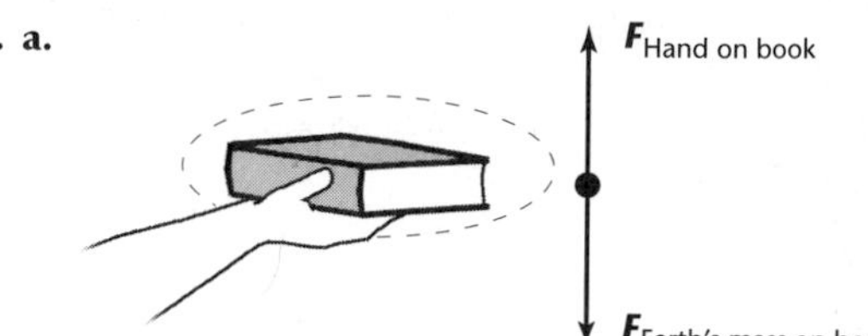

b.

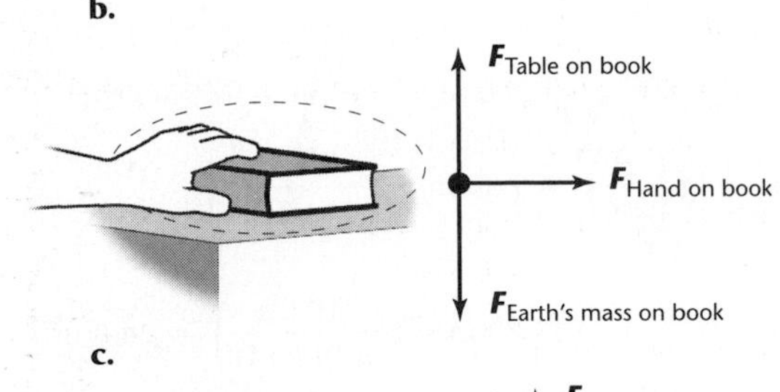

c.

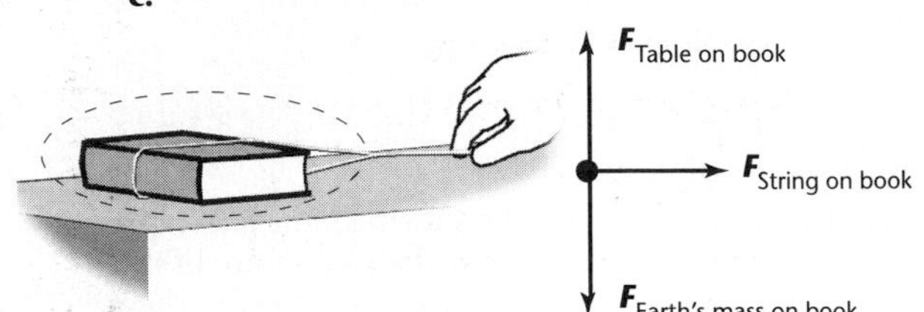

d.

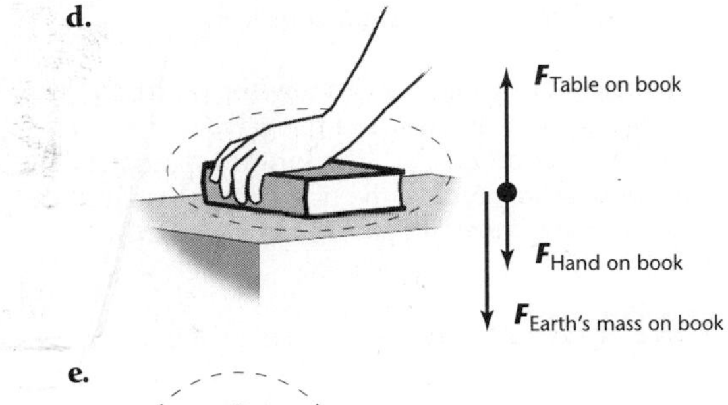

e.

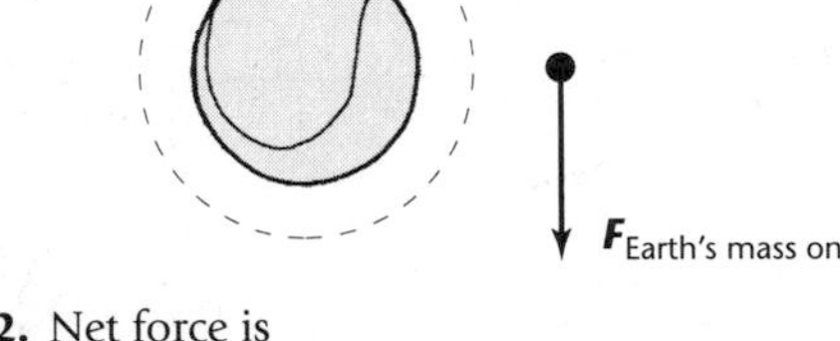

2. Net force is

$$225 \text{ N} + 165 \text{ N} = 3.90 \times 10^2 \text{ N}$$

in the direction of the two forces.

3. Net force is

$$225 \text{ N} - 165 \text{ N} = 6.0 \times 10^1 \text{ N}$$

in the direction of the larger force.

4. Magnitude and direction

$$F = \sqrt{(225 \text{ N})^2 + (165 \text{ N})^2} = 279 \text{ N}$$

$$\tan\theta = \frac{225}{165} = 1.36$$

$$\theta = 53.7° \text{ N of E}$$

5. The downward force is one pound, or 4.5 N. The force is

$$6.5 \text{ N} - 4.5 \text{ N} = 2.0 \text{ N upward}$$

6. $F = mg = (0.454 \text{ kg/lb})(9.80 \text{ m/s}^2) = 4.45 \text{ N/lb}$

Same force if you lie on the floor.

7.

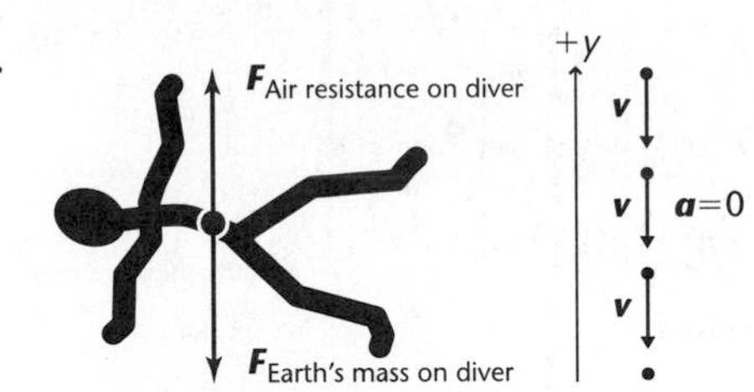

8.

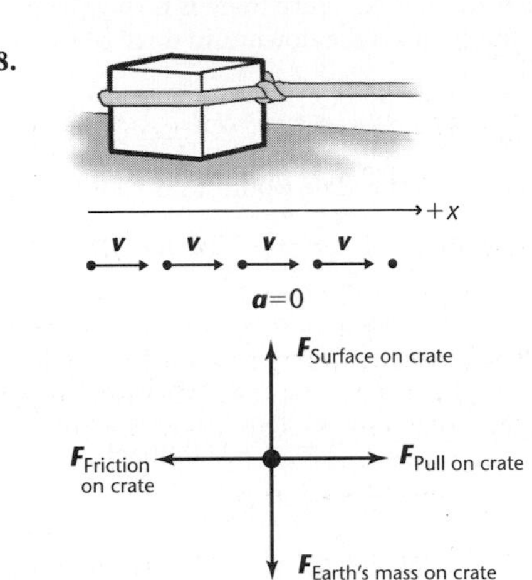

9.

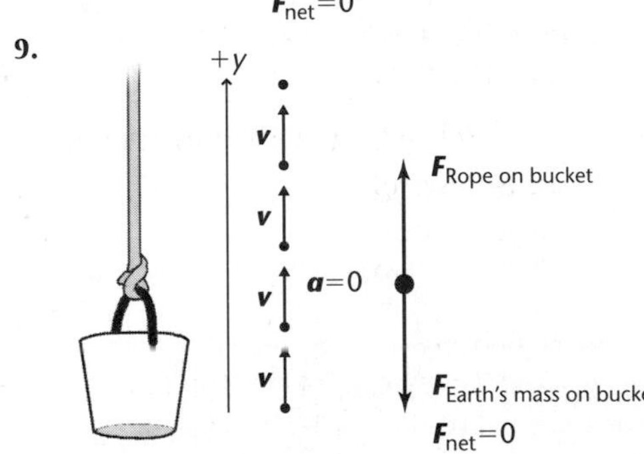

10.

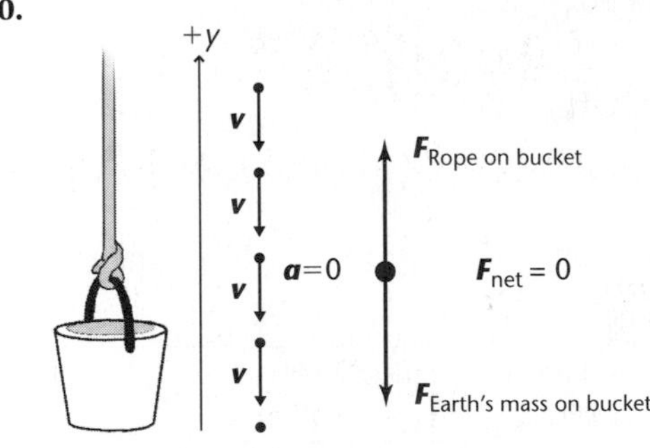

11.

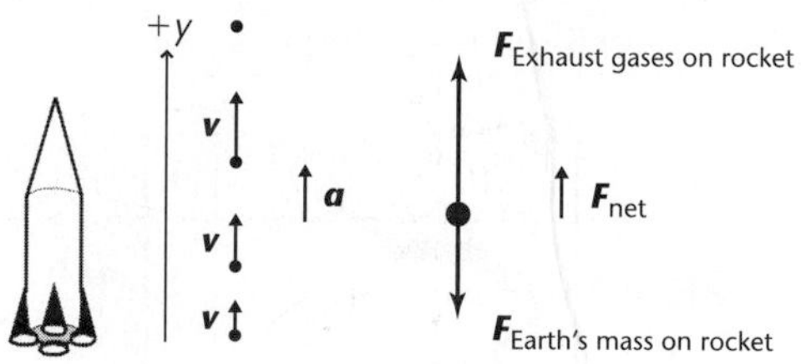

12. a. Scale reads 585 N. Since there is no acceleration your force equals the downward force of gravity. Mass

$$m = \frac{F_g}{g} = 59.7 \text{ kg}$$

b. On the moon the scale would read 95.5 N.

13. a. Mass = 75 kg

b. Slows while moving up or speeds up while moving down,

$$F_{scale} = m(g + a)$$
$$= (75 \text{ kg})(9.80 \text{ m/s}^2 - 2.0 \text{ m/s}^2)$$
$$= 5.9 \times 10^2 \text{ N}$$

c. Slows while moving up or speeds up while moving down,

$$F_{scale} = m(g + a)$$
$$= (75 \text{ kg})(9.80 \text{ m/s}^2 - 2.0 \text{ m/s}^2)$$
$$= 5.9 \times 10^2 \text{ N}$$

d. $F_{scale} = 7.4 \times 10^2$ N

e. Depends on the magnitude of the acceleration.

14. $F_N = mg = 52$ N

Since the speed is constant, the friction force equals the force exerted by the boy, 36 N. But,

$$F_f = \mu_k F_N$$

$$\text{so } \mu_k = \frac{F_f}{F_N} = \frac{(36 \text{ N})}{(52 \text{ N})} = 0.69$$

15. At constant speed, applied force equals friction force, so

$$F_f = \mu F_N = (0.12)(52 \text{ N} + 650 \text{ N}) = 84 \text{ N}$$

16. The initial velocity is 1.0 m/s, the final velocity 2.0 m/s, and the acceleration 2.0 m/s^2, so

$$t = \frac{(v - v_0)}{a} = \frac{(1.0 \text{ m/s})}{(2.0 \text{ m/s}^2)} = 0.50 \text{ s}$$

17. For a pendulum

$$T = 2\pi\sqrt{\frac{l}{g}}$$

$$\text{so } l = g\left(\frac{T}{2\pi}\right)^2 = 9.80 \text{ m/s}^2\left[\frac{1.00 \text{ s}}{(2)(3.14)}\right]^2$$
$$= 0.248 \text{ m}$$

18. $l = g\left(\frac{T}{2\pi}\right)^2 = (9.80 \text{ m/s}^2)\left(\frac{10.0 \text{ s}}{(2)(3.14)}\right)^2 = 24.8 \text{ m}$

No. This is over 75 feet long!

19. $g = l\left(\frac{2\pi}{T}\right)^2 = (0.65 \text{ m})\left(\frac{(2)(3.14)}{(2.8 \text{ s})}\right)^2 = 3.3 \text{ m/s}^2$

20. The force of your hand on the ball, the gravitational force of Earth's mass on the ball. The force of the ball on your hand, the gravitational force of the ball's mass on Earth. The force of your feet on Earth, the force of Earth on your feet.

21. The backward (friction) and upward (normal) force of the road on the tires and the gravitational force of Earth's mass on the car. The forward (friction) and the downward force of the tires on the road and the gravitational force of the car's mass on Earth.

CHAPTER 7

1. $F_A = F_B$

$$F_A = \frac{F_g}{2 \sin \theta} = \frac{168 \text{ N}}{2 \times \sin 42°} = 130 \text{ N}$$

2.

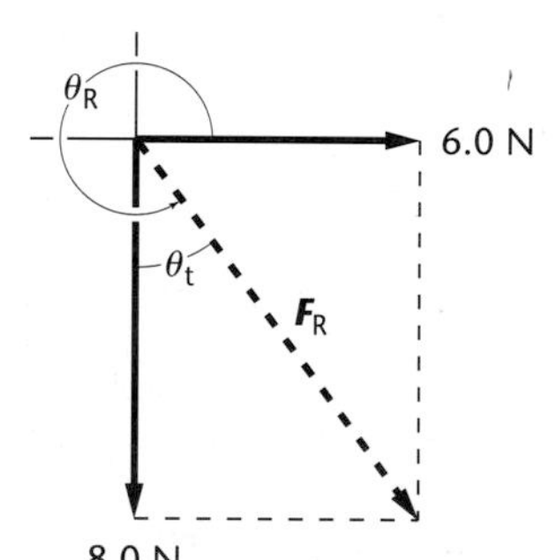

a. $F_R = \sqrt{(6.0 \text{ N})^2 + (8.0 \text{ N})^2} = 1.0 \times 10^1$ N

$$\theta_t = \tan^{-1}\left(\frac{6.0}{8.0}\right) = 37°$$
$$\theta_R = 270° + \theta_t = 307° = 310°$$
$$F_R = 1.0 \times 10^1 \text{ N at } 310°$$

b. $F_E = 1.0 \times 10^1$ N at 310° − 180° = 130°

3. 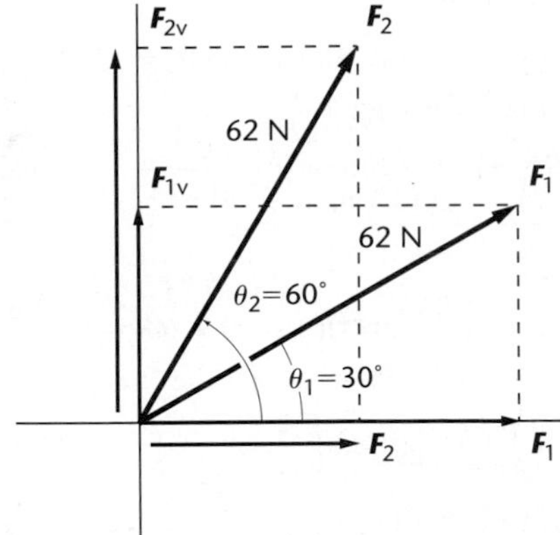

a. Vector addition is most easily carried out by using the method of addition by components. The first step in this method is the resolution of the given vectors into their horizontal and vertical components.

$F_{1h} = F_1 \cos \theta_1 = (62 \text{ N}) \cos 30° = 54 \text{ N}$

$F_{1v} = F_1 \sin \theta_1 = (62 \text{ N}) \sin 30° = 31 \text{ N}$

$F_{2h} = F_2 \cos \theta_2 = (62 \text{ N}) \cos 60° = 31 \text{ N}$

$F_{2v} = F_2 \sin \theta_2 = (62 \text{ N}) \sin 60° = 54 \text{ N}$

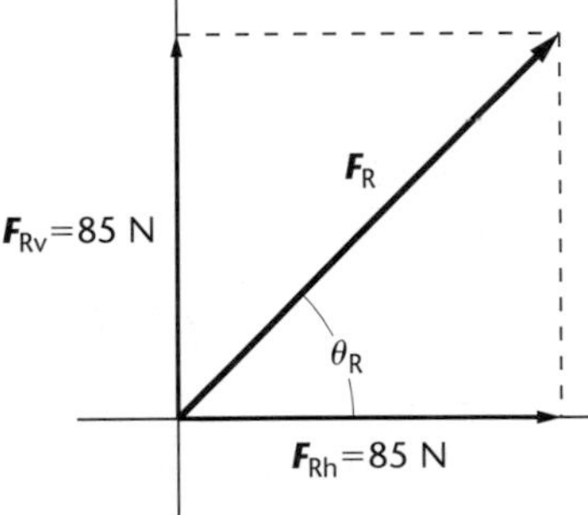

At this point, the two original vectors have been replaced by four components, vectors that are much easier to add. The horizontal and vertical components of the resultant vector are found by simple addition.

$F_{Rh} = F_{1h} + F_{2h} = 54 \text{ N} + 31 \text{ N} = 85 \text{ N}$

$F_{Rv} = F_{1v} + F_{2v} = 31 \text{ N} + 54 \text{ N} = 85 \text{ N}$

The magnitude and direction of the resultant vector are found by the usual method.

$F_R = \sqrt{(F_{Rh})^2 + (F_{Rv})^2}$

$= \sqrt{(85 \text{ N})^2 + (85 \text{ N})^2} = 120 \text{ N}$

$\tan \theta_R = \dfrac{F_{Rv}}{F_{Rh}} = \dfrac{85 \text{ N}}{85 \text{ N}} = 1$

$\theta_R = 45°$

$F_R = 120 \text{ N at } 45°$

b. $F_E = 120$ N, at $45° + 180° = 225°$

4. 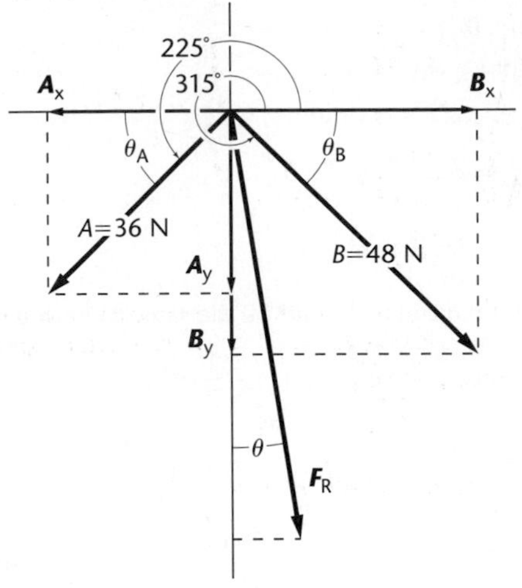

$\theta_A = 225° - 180° = 45°$

$\theta_B = 360° - 315° = 45°$

$A_x = -A \cos \theta_A = -(36 \text{ N}) \cos 45° = -25 \text{ N}$

$A_y = -A \sin \theta_A = (-36 \text{ N}) \sin 45° = -25 \text{ N}$

$B_x = B \cos \theta_B = (48 \text{ N}) \cos 45° = 34 \text{ N}$

$B_y = -B \sin \theta_B = -(48 \text{ N}) \sin 45° = -34 \text{ N}$

$F_x = A_x + B_x = -25 \text{ N} + 34 \text{ N} = 9 \text{ N}$

$F_y = A_y + B_y = -25 \text{ N} - 34 \text{ N} = -59 \text{ N}$

$F_R = \sqrt{F_x^2 + F_y^2} = \sqrt{(+9 \text{ N})^2 + (-59 \text{ N})^2}$

$= 60 \text{ N or } 6.0 \times 10^1 \text{ N}$

$\tan \theta = \dfrac{9}{59} = 0.153 \quad \theta = 9°$

$\theta_R = 270° + 9° = 279°$

$F_R = 60 \text{ N at } 279°$

$F_E = 60 \text{ N}$

$\theta_E = 279° - 180° = 99°$

5. a. $a = \dfrac{F}{m} = \dfrac{+mg \sin \theta}{m}$

$= +g \sin \theta = +(9.80 \text{ m/s}^2)(\sin 30.0°)$

$= 4.90 \text{ m/s}^2$

b. $v = v_0 + at = (4.90 \text{ m/s}^2)(4.0 \text{ s}) = 19.6 \text{ m/s}$

6. $F_{gx} = mg \sin \theta$

$= (62 \text{ kg})(9.80 \text{ m/s}^2)(0.60) = 3.6 \times 10^2 \text{ N}$

$F_{gy} = mg \cos \theta$

$= (62 \text{ kg})(9.80 \text{ m/s}^2)(0.80) = 4.9 \times 10^2 \text{ N}$

7. Since $a = g(\sin \theta - \mu \cos \theta)$,

$a = 9.80 \text{ m/s}^2(0.50 - (0.15)(0.866)) = 4.0 \text{ m/s}^2$

8. $a = g(\sin\theta - \mu\cos\theta)$

$a = g\sin\theta - g\mu\cos\theta$

If $a = 0$,

$$0 = g\sin\theta - g\mu\cos\theta$$

$$g\mu\cos\theta = g\sin\theta$$

$$\mu = \frac{g\sin\theta}{g\cos\theta} = \frac{\sin\theta}{\cos\theta}$$

$$\mu = \frac{\sin 37°}{\cos 37°} = 0.75$$

If $a = 0$, velocity would be the same as before.

9. **a.** Since $v_y = 0$, $y - v_y t = -\frac{1}{2}gt^2$ becomes

$$y = -\frac{1}{2}gt^2$$

or $$t^2 = -\frac{2y}{g} = \frac{-2(-78.4\text{ m})}{9.80\text{ m/s}^2} = 16\text{ s}^2$$

$$t = \sqrt{16\text{ s}^2} = 4.0\text{ s}$$

b. $x = v_x t = (5.0\text{ m/s})(4.0\text{ s}) = 2.0 \times 10^1\text{ m}$

c. $v_x = 5.0$ m/s. This is the same as the initial horizontal speed because the acceleration of gravity influences only the vertical motion. For the vertical component, use $v = v_o + gt$ with $v = v_y$ and v_o, the initial vertical component of velocity zero.

At $t = 4.0$ s

$$v_y = gt = (9.80\text{ m/s}^2)(4.0\text{ s}) = 39\text{ m/s}$$

10. **a.** (a) no change; 4.0 s

(b) twice the previous distance; 4.0×10^1 m

(c) v_x doubles; 1.0×10^1 m/s no change in v_y; 39 m/s

b. (a) increases by $\sqrt{2}$, since $t = \sqrt{\frac{-2y}{g}}$ and y doubles; 5.7 s

(b) increases by $\sqrt{2}$, since t increases by $\sqrt{2}$; 28 m

(c) no change in v_x; 5.0 m/s
v_y increases by $\sqrt{2}$, since t increases by $\sqrt{2}$; 55 m/s

11. Since $v_y = 0$, $y = -\frac{1}{2}gt^2$ and the time to reach the ground is

$$t = \sqrt{\frac{-2y}{g}} = \sqrt{\frac{-2(-0.950\text{ m})}{9.80\text{ m/s}^2}} = 0.440\text{ s}$$

From $x = v_x t$,

$$v_x = \frac{x}{t} = \frac{0.352\text{ m}}{0.440\text{ s}} = 0.800\text{ m/s}$$

12. $v_x = v_i\cos\theta = (27.0\text{ m/s})\cos 30.0° = 23.4\text{ m/s}$

$v_y = v_i\sin\theta = (27.0\text{ m/s})\sin 30.0° = 13.5\text{ m/s}$

When it lands, $y = v_y t - \frac{1}{2}gt^2 = 0$.

Therefore,

$$t = \frac{2v_y}{g} = \frac{2(13.5\text{ m/s})}{9.80\text{ m/s}^2} = 2.76\text{ s}$$

Distance:

$$x = v_x t = (23.4\text{ m/s})(2.76\text{ s}) = 64.6\text{ m}$$

Maximum height occurs at half the "hang time," or 1.38 s. Thus,

$$y = v_y t - \frac{1}{2}gt^2$$
$$= (13.5\text{ m/s})(1.38\text{ s}) - \frac{1}{2}(+9.80\text{ m/s}^2)(1.38\text{ s})^2$$
$$= 18.6\text{ m} - 9.33\text{ m} = 9.27\text{ m}$$

13. Following the method of Practice Problem 5,

$$v_x = v_o\cos\theta = (27.0\text{ m/s})\cos 60.0° = 13.5\text{ m/s}$$

$$v_y = v_o\sin\theta = (27.0\text{ m/s})\sin 60.0° = 23.4\text{ m/s}$$

$$t = \frac{2v_y}{g} = \frac{2(23.4\text{ m/s})}{9.80\text{ m/s}^2} = 4.78\text{ s}$$

Distance:

$$x = v_x t = (13.5\text{ m/s})(4.78\text{ s}) = 64.5\text{ m}$$

Maximum height:

at $t = \frac{1}{2}(4.78\text{ s}) = 2.39\text{ s}$

$$y = v_y t - \frac{1}{2}gt^2$$
$$= (23.4\text{ m/s})(2.39\text{ s}) - \frac{1}{2}(+9.80\text{ m/s}^2)(2.39\text{ s})^2$$
$$= 27.9\text{ m}$$

14. **a.** Since r and T remain the same,

$$v = \frac{2\pi r}{T} \quad\text{and}\quad a = \frac{v^2}{r}$$

remain the same. The new value of the mass is $m_2 = 2m_1$. The new force is $F_2 = m_2 a = 2m_1 a = 2F_1$, double the original force.

b. The new radius is $r_2 = 2r_1$, so the new velocity is

$$v_2 = \frac{2\pi r_2}{T} = \frac{2\pi(2r_1)}{T} = 2v_1$$

twice the original velocity. The new acceleration is

$$a_2 = \frac{(v_2)^2}{r_2} = \frac{(2v_1)^2}{2r_1} = 2a_1$$

twice the original. The new force is

$$F_2 = ma_2 = m(2a_1) = 2F_1$$

twice the original.

c. new velocity,

$$v_2 = \frac{2\pi r}{T_2} = \frac{2\pi r}{\left(\frac{1}{2}T\right)} = 2v_1$$

twice the original ; new acceleration

$$a_2 = \frac{(v_2)^2}{r} = \frac{(2v_1)^2}{r} = 4a_1$$

four times original; new force,

$$F_2 = ma_2 = m(4a_1) = 4F_1$$

four times original

15. a. $a_c = \frac{v^2}{r} = \frac{(8.8 \text{ m/s})^2}{25 \text{ m}} = 3.1 \text{ m/s}^2$

b. The frictional force of the track acting on the runner's shoes exerts the force on the runner.

16. a. $a_c = \frac{v^2}{r} = \frac{(32 \text{ m/s})^2}{56 \text{ m}} = 18 \text{ m/s}^2$

b. Recall $F_f = \mu F_N$. The friction force must supply the centripetal force so $F_f = ma_c$. The normal force is $F_N = -mg$. The coefficient of friction must be at least

$$\mu = \frac{F_f}{F_N} = \frac{ma_c}{mg} = \frac{a_c}{g} = \frac{18 \text{ m/s}^2}{9.80 \text{ m/s}^2} = 1.8$$

CHAPTER 8

1. $\left[\frac{T_a}{T_E}\right]^2 = \left[\frac{r_a}{r_E}\right]^3$ with $r_a = 2r_E$

Thus, $T_a = \left[\left(\frac{r_a}{r_E}\right)^3 T_E^2\right]^{1/2}$

$= \left[\left(\frac{2r_E}{r_E}\right)^3 (1 \text{ yr})^2\right]^{1/2} = 2.8 \text{ yr}$

2. $\left[\frac{T_M}{T_E}\right] = \left[\frac{r_M}{r_E}\right]^3$ with $r_M = 1.52r_E$

Thus, $T_M^2 = \left[\frac{r_M}{r_E}\right]^3 T_E^2 = \left[\frac{1.52r_E}{r_E}\right]^3 (365 \text{ days})^2$

$= 4.68 \times 10^5 \text{ days}^2$

$T_M = 684 \text{ days}$

3. $\left[\frac{T_s}{T_m}\right]^2 = \left[\frac{r_s}{r_m}\right]^3$

$T_s^2 = \left[\frac{r_s}{r_m}\right]^3 T_m^2 = \left[\frac{6.70 \times 10^2 \text{ km}}{3.90 \times 10^5 \text{ km}}\right]^3 (27.3 \text{ days})^2$

$= 3.78 \times 10^{-3} \text{ days}^2$

$T_s = 6.15 \times 10^{-2} \text{ days} = 88.6 \text{ min}$

4. $\left[\frac{T_s}{T_m}\right]^2 = \left[\frac{r_s}{r_m}\right]^3$ so $r_s^3 = r_m^3\left[\frac{T_s}{T_m}\right]^2$

$= (3.90 \times 10^5 \text{ km})^3 \left[\frac{1.00}{27.3}\right]^2$

$= 7.96 \times 10^{13} \text{ km}^3$

so $r_s = 4.30 \times 10^4 \text{ km}$

5. a. $v = \sqrt{\frac{Gm_E}{r}}$

$= \sqrt{\frac{(6.67 \times 10^{-11} \text{ N} \cdot \text{m}^2/\text{kg}^2)(5.97 \times 10^{24} \text{ kg})}{6.52 \times 10^6}}$

$= 7.81 \times 10^3 \text{ m/s}$

b. $T = 2\pi \sqrt{\frac{r^3}{Gm_E}}$

$= 2\pi \sqrt{\frac{(6.52 \times 10^6 \text{ m})^3}{(6.67 \times 10^{-11} \text{ N} \cdot \text{m}^2/\text{kg}^2)(5.97 \times 10^{24} \text{ kg})}}$

$= 5.24 \times 10^3 \text{ s} = 87.3 \text{ min}$

6. a. $v = \sqrt{\frac{Gm_M}{r}}$ with $r = r_M + 265$ km

$r = 2.44 \times 10^6 \text{ m} + 0.265 \times 10^6 \text{ m}$

$= 2.71 \times 10^6 \text{ m}$

$v = \sqrt{\frac{(6.67 \times 10^{-11} \text{ N} \cdot \text{m}^2/\text{kg}^2)(3.30 \times 10^{23} \text{ kg})}{2.71 \times 10^6 \text{ m}}}$

$= 2.85 \times 10^3 \text{ m/s}$

b. $T = 2\pi \sqrt{\frac{r^3}{Gm_M}}$

$= 2\pi \sqrt{\frac{(2.71 \times 10^6 \text{ m})^3}{(6.67 \times 10^{-11} \text{ N} \cdot \text{m}^2/\text{kg}^2)(3.30 \times 10^{23} \text{ kg})}}$

$= 5.97 \times 10^3 \text{ s} = 1.66 \text{ h}$

7. $v = \sqrt{\frac{Gm}{r}}$, where here m is the mass of the sun.

$v_M = \sqrt{\frac{(6.67 \times 10^{-11} \text{ N} \cdot \text{m}^2/\text{kg}^2)(1.99 \times 10^{30} \text{ kg})}{5.80 \times 10^{10} \text{ m}}}$

$= 4.78 \times 10^4 \text{ m/s}$

$v_S = \sqrt{\frac{(6.67 \times 10^{-11} \text{ N} \cdot \text{m}^2/\text{kg}^2)(1.99 \times 10^{30} \text{ kg})}{1.427 \times 10^{12} \text{ m}}}$

$= 9.64 \times 10^3$ m/s, about 1/5 as fast as Mercury

8. a. Use $T = 2\pi \sqrt{\frac{r^3}{Gm}}$, with

$T = 2.5 \times 10^8 \text{ y} = 7.9 \times 10^{15} \text{ s}$

$m = \frac{4\pi^2 r^3}{GT^2}$

$= \frac{4\pi^2(2.2 \times 10^{20} \text{ m})^3}{(6.67 \times 10^{-11} \text{ N} \cdot \text{m}^2/\text{kg}^2)(7.9 \times 10^{15} \text{ s})^2}$

$= 1.0 \times 10^{41} \text{ kg}$

b. number of stars $= \frac{\text{total galaxy mass}}{\text{mass per star}}$

$= \frac{1.0 \times 10^{41} \text{ kg}}{2.0 \times 10^{30} \text{ kg}} = 5.0 \times 10^{10}$

c. $v = \sqrt{\frac{Gm}{r}}$

$= \sqrt{\frac{(6.67 \times 10^{-11} \text{ N} \cdot \text{m}^2/\text{kg}^2)(1.0 \times 10^{41} \text{ kg})}{2.2 \times 10^{20} \text{ m}}}$

$= 1.7 \times 10^5 \text{ m/s} = 6.1 \times 10^5 \text{ km/h}$

CHAPTER 9

1. a. 100 km/h = 27.8 m/s

$p = mv = (725\text{ kg})(27.8\text{ m/s})$

$= 2.01 \times 10^4\text{ kg}\cdot\text{m/s eastward}$

b. $v = \frac{p}{m} = \frac{(2.01 \times 10^4\text{ kg}\cdot\text{m/s})}{(2175\text{ kg})}$

$= 9.24\text{ m/s} = 33.3\text{ km/h eastward}$

2. a. Impulse $= F\,\Delta t = (-5.0 \times 10^3\text{ N})(2.0\text{ s})$

$= -1.0 \times 10^4\text{ kg}\cdot\text{m/s westward}$

The impulse is directed westward and has a magnitude of 1.0×10^4 kg · m/s.

b. $p_1 = mv_1$

$= (725\text{ kg})(27.8\text{ m/s})$

$= 2.01 \times 10^4\text{ kg}\cdot\text{m/s eastward}$

$F\,\Delta t = \Delta p = p_2 - p_1$

$p_2 = F\,\Delta t + p_1$

$= -1.0 \times 10^4\text{ kg}\cdot\text{m/s} + 2.01 \times 10^4\text{ kg}\cdot\text{m/s}$

$p_2 = 1.0 \times 10^4\text{ kg}\cdot\text{m/s eastward}$

c. $p_2 = mv_2$

$v_2 = \frac{p_2}{m} = \frac{1.0 \times 10^4\text{ kg}\cdot\text{m/s}}{725\text{ kg}}$

$= 14\text{ m/s} = 50\text{ km/h eastward}$

3. a. $\text{impulse}_A = (5.0\text{ N})(2.0\text{ s} - 1.0\text{ s})$

$= 5.0\text{ N}\cdot\text{s} = 5.0\text{ kg}\cdot\text{m/s}$

$F\,\Delta t = \Delta p = p_2 - p_1$

$p_2 = F\,\Delta t + p_1$

$p_2 = 5.0\text{ kg}\cdot\text{m/s} + 14\text{ kg}\cdot\text{m/s}$

$= 19\text{ kg}\cdot\text{m/s}$

$p_2 = mv_2$

$v_2 = \frac{p}{m} = \frac{19\text{ kg}\cdot\text{m/s}}{7.0\text{ kg}}$

$= 2.7$ m/s in the same direction

b. impulse $= F\,\Delta t$

$\text{impulse}_B = (-5.0\text{ N})(2.0\text{ s} - 1.0\text{ s})$

$= -5.0\text{ N}\cdot\text{s} = -5.0\text{ kg}\cdot\text{m/s}$

$F\,\Delta t = \Delta p = p_2 - p_1$

$p_2 = F\,\Delta t + p_1$

$p_2 = -5.0\text{ kg}\cdot\text{m/s} + 14\text{ kg}\cdot\text{m/s} = 9.0\text{ kg}\cdot\text{m/s}$

$p_2 = mv_2$

$v_2 = \frac{p_2}{m} = \frac{9.0\text{ kg}\cdot\text{m/s}}{7.0\text{ kg}}$

$= 1.3$ m/s in the same direction

4. a.

6.00 m/s

28.0 m/s

b. $\Delta p = F\,\Delta t$

$= m(v_2 - v_1) = 240.0\text{ kg}(28.0\text{ m/s} - 6.00\text{ m/s})$

$= 5.28 \times 10^3\text{ kg}\cdot\text{m/s}$

c. $F = \frac{\Delta p}{\Delta t} = \frac{(5.28 \times 10^3\text{ kg}\cdot\text{m/s})}{(60.0\text{ s})} = 88.0\text{ N}$

5. a. Given: $m = 0.144$ kg
initial velocity,
$v_1 = +38.0$ m/s
final velocity,
$v_2 = -38.0$ m/s

Unknown: impulse

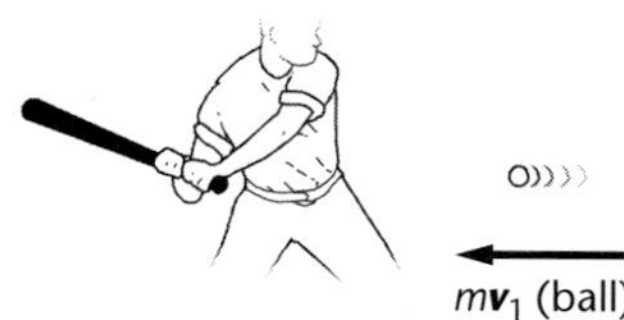

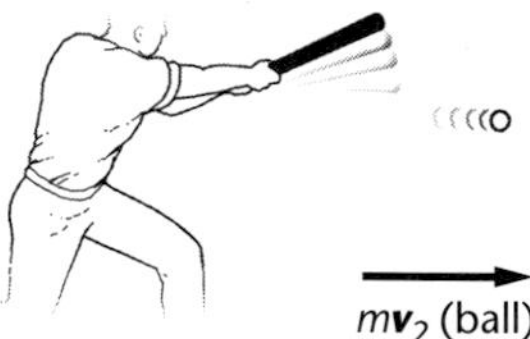

Basic equation: $F\,\Delta t = \Delta p$

b. Take the positive direction to be the direction of the ball after it leaves the bat.

$\Delta p = mv_2 - mv_1 = m(v_2 - v_1)$

$= (0.144\text{ kg})(+38.0\text{ m/s} - (-38.0\text{ m/s}))$

$= (0.144\text{ kg})(76.0\text{ m/s}) = 10.9\text{ kg}\cdot\text{m/s}$

c. $F\Delta t = \Delta p = 10.9 \text{ kg}\cdot\text{m/s}$

d. $F\Delta t = \Delta p$

so $F = \frac{\Delta p}{\Delta t} = \frac{10.9 \text{ kg}\cdot\text{m/s}}{8.0 \times 10^{-4} \text{ s}} = 1.4 \times 10^4 \text{ N}$

6. a. $p_1 = mv_1 \quad p_2 = 0$

$p_1 = (60 \text{ kg})(26 \text{ m/s}) = 1.6 \times 10^3 \text{ kg}\cdot\text{m/s}$

$F\Delta t = \Delta p = p_2 - p_1$

$F = \frac{0 - 1.6 \times 10^3 \text{ kg}\cdot\text{m/s}}{0.20 \text{ s}}$

$= 8 \times 10^3$ N opposite to the direction of motion

b. $F_g = mg$

$m = \frac{F_g}{g} = \frac{8 \times 10^3 \text{ N}}{9.80 \text{ m/s}^2} = 800 \text{ kg}$

Such a mass is too heavy to lift. You cannot safely stop yourself with your arms.

7. $p_1 = p_2$

$(3.0 \times 10^5 \text{ kg})(2.2 \text{ m/s}) = (2)(3.0 \times 10^5 \text{ kg})(v)$

$v = 1.1 \text{ m/s}$

8. $p_{h1} + p_{g1} = p_{h2} + p_{g2}$

$m_h v_{h1} + m_g v_{g1} = m_h v_{h2} + m_g v_{g2}$

Since $v_{g1} = 0$, $m_h v_{h1} = (m_h + m_g)v_2$

where $v_2 = v_{h2} = v_{g2}$ is the common final speed of goalie and puck.

$v_2 = \frac{m_h v_{h1}}{(m_h + m_g)} = \frac{(0.105 \text{ kg})(24 \text{ m/s})}{(0.105 \text{ kg} + 75 \text{ kg})} = 0.034 \text{ m/s}$

9. $m_b v_{b1} + m_w v_{w1} = (m_b + m_w)v_2$

where v_2 is the common final velocity of bullet and wooden block.

Since $v_{w1} = 0$,

$v_{b1} = \frac{(m_b + m_w)v_2}{m_b}$

$= \frac{(0.035 \text{ kg} + 5.0 \text{ kg})(8.6 \text{ m/s})}{(0.035 \text{ kg})}$

$= 1.2 \times 10^3 \text{ m/s}$

10. $m_b v_{b1} + m_w v_{w1} = m_b v_{b2} + m_w v_{w2}$

with $v_{w1} = 0$

$v_{w2} = \frac{(m_b v_{b1} - m_b v_{b2})}{m_w} = \frac{m_b(v_{b1} - v_{b2})}{m_w}$

$= \frac{(0.035 \text{ kg})(475 \text{ m/s} - 275 \text{ m/s})}{(2.5 \text{ kg})} = 2.8 \text{ m/s}$

11. $p_{A1} + p_{B1} = p_{A2} + p_{B2}$

so $p_{B2} = p_{B1} + p_{A1} - p_{A2}$

$m_B v_{B2} = m_B v_{B1} + m_A v_{A1} - m_A v_{A2}$

or $v_{B2} = \frac{m_B v_{B1} + m_A v_{A1} - m_A v_{A2}}{m_B}$

$= \frac{(0.710 \text{ kg})(+0.045 \text{ m/s}) + (0.355 \text{ kg})(+0.095 \text{ m/s})}{0.710 \text{ kg}}$

$- \frac{(0.355 \text{ kg})(+0.035 \text{ m/s})}{0.710 \text{ kg}}$

$= 0.075$ m/s in the initial direction

12. $m_A v_{A1} + m_B v_{B1} = m_A v_{A2} + m_B v_{B2}$

so v_{B2}

$= \frac{m_A v_{A1} + m_B v_{B1} - m_A v_{A2}}{m_B}$

$= \frac{(0.50 \text{ kg})(6.0 \text{ m/s}) + (1.00 \text{ kg})(-12.0 \text{ m/s})}{1.00 \text{ kg}}$

$- \frac{(0.50 \text{ kg})(-14 \text{ m/s})}{1.00 \text{ kg}}$

$= 2.0$ m/s, in opposite direction

13. $p_{r1} + p_{f1} = p_{r2} + p_{f2}$ where $p_{r1} + p_{f1} = 0$

If the initial mass of the rocket (including fuel) is $m_r = 4.00$ kg, then the final mass of the rocket is

$m_{r2} = 4.00 \text{ kg} - 0.0500 \text{ kg} = 3.95 \text{ kg}$

$0 = m_{r2}v_{r2} + m_f v_{f2}$

$v_{r2} = \frac{-m_f v_{f2}}{m_{r2}}$

$= \frac{-(0.0500 \text{ kg})(-625 \text{ m/s})}{(3.95 \text{ kg})} = 7.91 \text{ m/s}$

14. $p_{A1} + p_{B1} = p_{A2} + p_{B2}$ with $p_{A1} = p_{B1} = 0$

$m_B v_{B2} = -m_A v_{A2}$

so $v_{B2} = \frac{-m_A v_{A2}}{m_B} = \frac{-(1.5 \text{ kg})(-27 \text{ cm/s})}{(4.5 \text{ kg})}$

$= 9.0$ cm/s to the right

15. $p_{A1} + p_{B1} = p_{A2} + p_{B2}$ with $p_{A1} = p_{B1} = 0$

$m_A v_{A2} = -m_B v_{B2}$

so $v_{B2} = \frac{-m_A v_{A2}}{m_B} = \frac{-(80.0 \text{ kg})(4.0 \text{ m/s})}{(115 \text{ kg})}$

$= 2.8$ m/s in the opposite direction

16. a. Both the cannon and the ball fall to the ground in the same time from the same height. In that fall time, the ball moves 215 m, the cannon an unknown distance we will call x. Now

$t = \frac{d}{v}$

so $\frac{(215 \text{ m})}{v_{ball}} = \frac{x}{v_{cannon}}$

so $x = 215 \text{ m} \left[\frac{v_{cannon}}{v_{ball}}\right]$

related by conservation of momentum;

$$(4.5\text{ kg})v_{\text{ball}} = -(225\text{ kg})v_{\text{cannon}}$$

so $\left[\dfrac{-v_{\text{cannon}}}{v_{\text{ball}}}\right] = \dfrac{4.5\text{ kg}}{225\text{ kg}}$

Thus $x = -\left[\dfrac{4.5}{225}(215\text{ m})\right] = -4.3\text{ m}$

b. While on top, the cannon moves with no friction, and its velocity doesn't change, so it can take any amount of time to reach the back edge.

17.

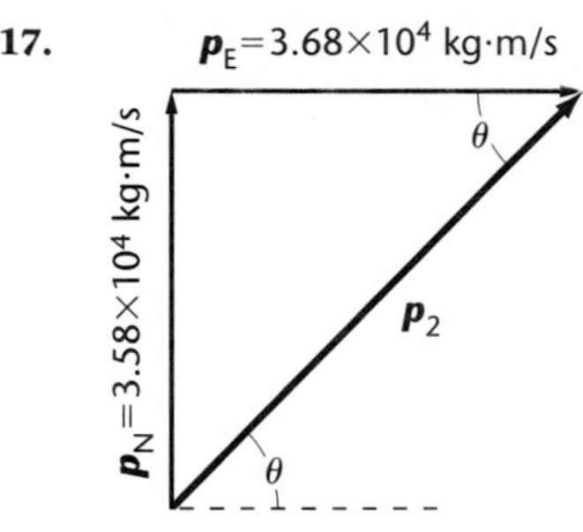

$\boldsymbol{p}_N + \boldsymbol{p}_E = \boldsymbol{p}_2$ (vector sum)

$$p_N = m_N v_N = (1325\text{ kg})(27.0\text{ m/s})$$
$$= 3.58 \times 10^4\text{ kg}\cdot\text{m/s}$$

$$p_E = m_E v_E = (2165\text{ kg})(17.0\text{ m/s})$$
$$= 3.68 \times 10^4\text{ kg}\cdot\text{m/s}$$

$$\tan\theta = \frac{p_N}{p_E} = \frac{3.58 \times 10^4\text{ kg}\cdot\text{m/s}}{3.68 \times 10^4\text{ kg}\cdot\text{m/s}} = 0.973$$

$\theta = 44.2°$, north of east

$$(p_2)^2 = (p_N)^2 + (p_E)^2$$
$$= (3.58 \times 10^4\text{ kg}\cdot\text{m/s})^2 + (3.68 \times 10^4\text{ kg}\cdot\text{m/s})^2$$
$$= 2.64 \times 10^9\text{ kg}^2\text{ m}^2/\text{s}^2$$

$$p_2 = 5.13 \times 10^4\text{ kg}\cdot\text{m/s}$$

$$p_2 = m_2 v_2 = (m_N + m_E)v_2$$

$$v_2 = \frac{p_2}{(m_N + m_E)} = \frac{(5.13 \times 10^4\text{ kg}\cdot\text{m/s})}{(1325\text{ kg} + 2165\text{ kg})}$$
$$= 14.7\text{ m/s}$$

18.

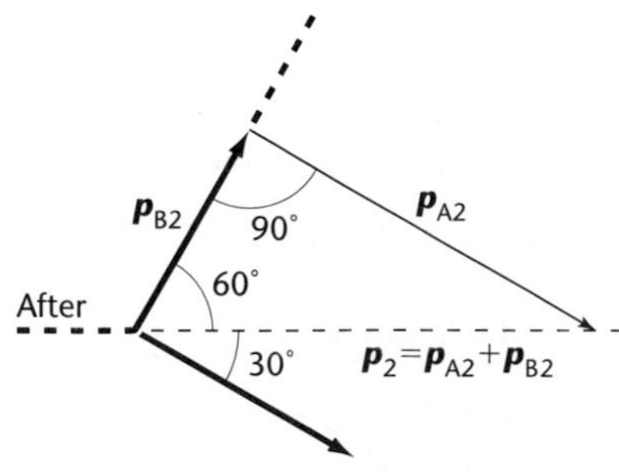

$\boldsymbol{p}_{A1} + \boldsymbol{p}_{B1} = \boldsymbol{p}_{A2} + \boldsymbol{p}_{B2}$ (vector sum) with $p_{A1} = 0$

$$m_1 = m_2 = m = 0.17\text{ kg}$$

$$p_{B1} = m_{B1}v_{B1} = (0.17\text{ kg})(4.0\text{ m/s})$$
$$= 0.68\text{ kg}\cdot\text{m/s}$$

$p_{A2} = p_{B1}\sin 60.0°$ $\quad mv_{A2} = mv_{B1}\sin 60.0°$

$$v_{A2} = v_{B1}\sin 60.0° = (4.0\text{ m/s})\sin 60.0°$$
$= 3.5$ m/s, 30.0° to right

$p_{B2} = p_{B1}\cos 60.0°$ $\quad mv_{B2} = mv_{B1}\cos 60.0°$

$$v_{B2} = v_{B1}\cos 60.0° = (4.0\text{ m/s})\cos 60.0°$$
$= 2.0$ m/s, 60.0° to left

CHAPTER 10

1. $W = Fd = (185\text{ N})(0.800\text{ m}) = 148$ joules

2. a. $W = Fd = (825\text{ N})(35\text{ m}) = 2.9 \times 10^4\text{ J}$

b. $W = Fd$
$$= (2)(825\text{ N})(35\text{ m}) = 5.8 \times 10^4\text{ J}$$

The amount of work doubles.

3. $F_g = mg = (0.180\text{ kg})(9.80\text{ m/s}^2) = 1.76\text{ N}$

$W = Fd = (1.76\text{ N})(2.5\text{ m}) = 4.4\text{ J}$

4. $W = Fd = mgd$

so $m = \dfrac{W}{gd} = \dfrac{7.0 \times 10^3\text{ J}}{(9.80\text{ m/s}^2)(1.2\text{ m})} = 6.0 \times 10^2\text{ kg}$

5. Both do the same amount of work. Only the height lifted and the vertical force exerted count.

6. Both the force and displacement are in the same direction, so

$W = Fd = (25\text{ N})(3.5\text{ m}) = 88\text{ J}$

7. a. Since gravity acts vertically, only the vertical displacement needs to be considered.

$W = Fd = (215\text{ N})(4.20\text{ m}) = 903\text{ J}$

b. Force is upward, but vertical displacement is downward, so

$$W = Fd\cos\theta = Fd\cos 180°$$
$$= (215\text{ N})(4.20\text{ m})(-1.00) = -903\text{ J}$$

8. $W = Fd\cos\theta = (628\text{ N})(15.0\text{ m})(\cos 46.0°)$
$$= 6.54 \times 10^3\text{ J}$$

9. $P = \dfrac{W}{t} = \dfrac{Fd}{t} = \dfrac{(575\text{ N})(20.0\text{ m})}{10.0\text{ s}}$
$$= 1.15 \times 10^3\text{ W} = 1.15\text{ kW}$$

10. a. $W = mgd = (7.5\text{ kg})(9.80\text{ m/s}^2)(8.2\text{ m})$
$$= 6.0 \times 10^2\text{ J}$$

b. $W = Fd + 6.0 \times 10^2\ \text{J}$

$= (645\ \text{N})(8.2\ \text{m}) + 6.0 \times 10^2\ \text{J} = 5.9 \times 10^3\ \text{J}$

c. $P = \frac{W}{t} = \frac{5.9 \times 10^3\ \text{J}}{(30\ \text{min})(60\ \text{s/min})} = 3\ \text{W}$

11. $P = \frac{W}{t}$ and $W = Fd$

so $F = \frac{Pt}{d} = \frac{(65 \times 10^3\ \text{W})(35\ \text{s})}{17.5\ \text{m}} = 1.3 \times 10^5\ \text{N}$

12. The work done is the area of the trapezoid under the solid line:

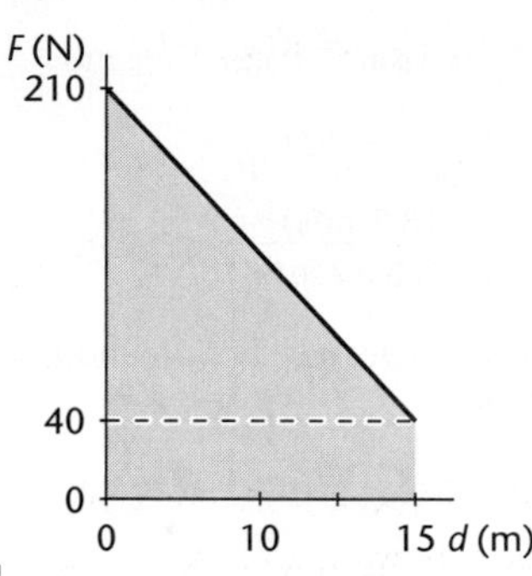

$W = \frac{1}{2}d(F_1 + F_2)$

$= \frac{1}{2}(15\ \text{m})(210\ \text{N} + 40\ \text{N}) = 1.9 \times 10^3\ \text{J}$

13. a. $IMA = \frac{d_e}{d_r} = \frac{20\ \text{cm}}{5.0\ \text{cm}} = 4.0$

b. $MA = \frac{F_r}{F_e} = \frac{1.9 \times 10^4\ \text{N}}{9.8 \times 10^3\ \text{N}} = 1.9$

c. efficiency $= \left[\frac{MA}{IMA}\right] \times 100\%$

$= \left[\frac{1.9}{4.0}\right] \times 100\% = 48\%$

14. a. $F_r = mg = (24.0\ \text{kg})(9.80\ \text{m/s}^2) = 235\ \text{N}$

$MA = \frac{F_r}{F_e} = \frac{235\ \text{N}}{129\ \text{N}} = 1.82$

b. efficiency $= \left[\frac{MA}{IMA}\right] \times 100\%$ where

$IMA = \frac{d_e}{d_r} = \frac{33.0\ \text{m}}{16.5\ \text{m}} = 2.00$

so efficiency $= \frac{1.82}{2.00} \times 100\% = 91.0\%$

15. efficiency $= \frac{W_o}{W_i} \times 100\% = \frac{F_r d_r}{F_e d_e} \times 100\%$

so $d_e = \frac{F_r d_r (100\%)}{F_e(\text{efficiency})}$

$= \frac{(1.25 \times 10^3\ \text{N})(0.13\ \text{m})(100\%)}{(225\ \text{N})(88.7\%)} = 0.81\ \text{m}$

16. $IMA = \frac{8.00\ \text{cm}}{35.6\ \text{cm}} = 0.225$

$MA = (95.0\%)\frac{0.225}{100\%} = 0.214$

$F_r = (MA)(F_e) = (0.214)(155\ \text{N}) = 33.2\ \text{N}$

$d_e = (IMA)(d_r) = (0.225)(14.0\ \text{cm}) = 3.15\ \text{cm}$

All of the above quantities are doubled.

CHAPTER 11

1. a. $\frac{22\text{m}}{\text{s}} \times \frac{3600\ \text{s}}{1\ \text{h}} \times \frac{1\ \text{km}}{1000\ \text{m}} = 79\ \text{km/h}$

b. $W = \Delta K = K_f - K_i$

$= 2.12 \times 10^5\ \text{J} - 8.47 \times 10^5\ \text{J}$

$= -6.35 \times 10^5\ \text{J}$

c. $W = \Delta K = 0 - 8.47 \times 10^5\ \text{J}$

$= -8.47 \times 10^5\ \text{J}$

d. $W = Fd$, so distance is proportional to work. The ratio is

$\frac{-6.35 \times 10^5\ \text{J}}{-2.12 \times 10^5\ \text{J}} = 3$

It takes three times the distance to slow the car to half its speed than it does to slow it to a complete stop.

2. a.

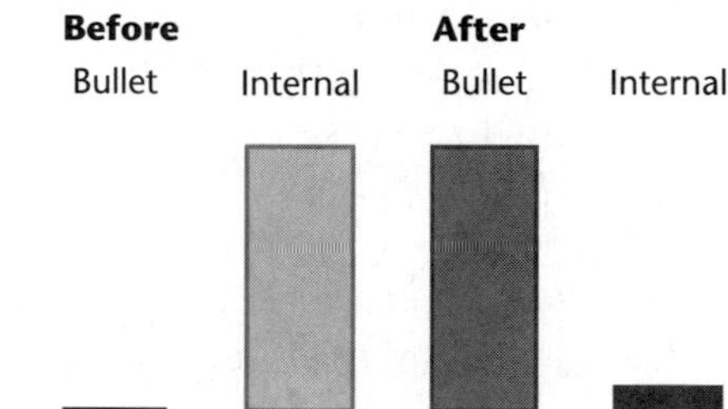

b. $K = \frac{1}{2}mv^2 = \frac{1}{2}(0.00420\ \text{kg})(965\ \text{m/s})^2$

$= 1.96 \times 10^3\ \text{J}$

c. $W = \Delta K = 1.96 \times 10^3\ \text{J}$

d. $W = Fd$

so $F = \frac{W}{d} = \frac{1.96 \times 10^3\ \text{J}}{0.75\ \text{m}} = 2.6 \times 10^3\ \text{N}$

e. $F = \frac{W}{d} = \frac{\Delta K}{d} = \frac{1.96 \times 10^3\ \text{J}}{0.015\ \text{m}}$

$= 1.3 \times 10^5\ \text{N}$, forward

3. a. $K = \frac{1}{2}mv^2$

$= \frac{1}{2}(7.85 \times 10^{11}\ \text{kg})(2.50 \times 10^4\ \text{m/s})^2$

$= 2.45 \times 10^{20}\ \text{J}$

b. $\frac{K_{comet}}{K_{bomb}} = \frac{2.45 \times 10^{20}\ \text{J}}{4.2 \times 10^{15}\text{J}} = 5.8 \times 10^4$
bombs would be required

4. a. Since $W_A = \Delta K = \frac{1}{2}mv_A^2$, then $v_A = \sqrt{\frac{2W_A}{m}}$

If $W_B = \frac{1}{2}W_A$,

$v_B = \sqrt{\frac{2W_B}{m}}$

$= \sqrt{2\left[\frac{1}{2}W_A\right]/m} = \sqrt{\frac{1}{2}}\,v_A$

$= (0.707)(100\ \text{km/h}) = 71\ \text{km/h}$

b. If $W_C = 2W_A$,

$v_{BC} = \sqrt{2}\,(100\ \text{km/h}) = 140\ \text{km/h}$

5. a. $U_g = mgh$

$U_g = (2.00\ \text{kg})(9.80\ \text{m/s}^2)$
$(0\ \text{m} - 2.10\ \text{m} + 1.65\ \text{m})$

$U_g = -8.8\ \text{J}$

b. $U_g = mgh$

$U_g = (2.00\ \text{kg})(9.80\ \text{m/s}^2)(0\ \text{m} - 2.10\ \text{m})$

$U_g = -41.2\ \text{J}$

6. $U_g = mgh$

At the edge,

$U_g = (90\ \text{kg})(9.80\ \text{m/s}^2)(+45\ \text{m}) = 4 \times 10^4\ \text{J}$

At the bottom,

$U_g = (90\ \text{kg})(9.80\ \text{m/s}^2)(+45\ \text{m} - 85\ \text{m})$

$= -4 \times 10^4\ \text{J}$

7. a. $U_g = mgh = (50.0\ \text{kg})(9.80\ \text{m/s}^2)(425\ \text{m})$

$= 2.08 \times 10^5\ \text{J}$

b. $\Delta U_g = mgh_f - mgh_i = mg(h_f - h_i)$

$= (50.0\ \text{kg})(9.80\ \text{m/s}^2)(225\ \text{m} - 425\ \text{m})$

$= -9.80 \times 10^4\ \text{J}$

8.

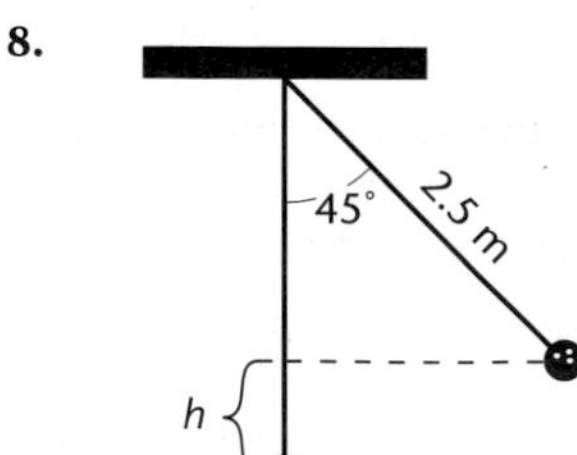

a. $h = (2.5\ \text{m})(1 - \cos\theta) = 0.73\ \text{m}$

$U_g = mgh = (7.26\ \text{kg})(9.80\ \text{m/s}^2)(0.73\ \text{m}) = 52\ \text{J}$

b. the height of the ball when the rope was vertical

9. a. The system is the bike + rider + Earth. No external forces, so total energy is conserved.

b. $K = \frac{1}{2}mv^2$

$= \frac{1}{2}(85\ \text{kg})(8.5\ \text{m/s})^2 = 3.1 \times 10^3\ \text{J}$

c. $K_{before} + U_{g\ before} = K_{after} + U_{g\ after}$

$\frac{1}{2}mv^2 + 0 = 0 + mgh,$

$h = \frac{v^2}{2g} = \frac{(8.5\ \text{m/s})^2}{(2)(9.80\ \text{m/s}^2)} = 3.7\ \text{m}$

d. No. It cancels because both K and U_g are proportional to m.

10. a. $K_{before} + U_{g\ before} = K_{after} + U_{g\ after}$

$0 + mgh = \frac{1}{2}mv^2 + 0$

$v^2 = 2gh = 2(9.80\ \text{m/s}^2)(4.0\ \text{m}) = 78.4\ \text{m}^2/\text{s}^2$

$v = 8.9\ \text{m/s}$

b. No **c.** No

11. a. $K_{before} + U_{g\ before} = K_{after} + U_{g\ after}$

$0 + mgh = \frac{1}{2}mv^2 + 0$

$v^2 = 2gh = 2(9.80\ \text{m/s}^2)(45\ \text{m}) = 880\ \text{m}^2/\text{s}^2$

$v = 30\ \text{m/s}$

b. $K_{before} + U_{g\ before} = K_{after} + U_{g\ after}$

$0 + mgh_i = \frac{1}{2}\text{mv}^2 + mgh_f$

$v^2 = 2g(h_i - h_f)$

$= 2(9.80\ \text{m/s}^2)(45\ \text{m} - 40\ \text{m}) = 98\ \text{m}^2/\text{s}^2$

$v = 10\ \text{m/s}$

c. No

12. a. The system of Earth, bike, and rider remains the same, but now the energy involved is not mechanical energy alone. The rider must be considered as having stored energy, some of which is converted to mechanical energy.

b. Energy came from the chemical potential energy stored in the rider's body.

13. a. 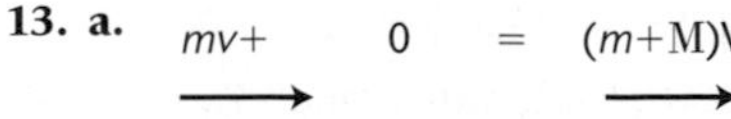

b. From the conservation of momentum,

$$mv = (m + M)V$$

so $$V = \frac{mv}{m + M}$$

$$= \frac{(0.00200 \text{ kg})(538 \text{ m/s})}{0.00200 \text{ kg} + 0.250 \text{ kg}} = 4.27 \text{ m/s}$$

c. $K = \frac{1}{2} mv^2 = \frac{1}{2} (0.00200 \text{ kg})(538 \text{ m/s})^2 = 289 \text{ J}$

d. $$K_f = \frac{1}{2} (m + M)V^2$$

$$= \frac{1}{2} (0.00200 \text{ kg} + 0.250 \text{ kg})(4.27 \text{ m/s})^2$$

$$= 2.30 \text{ J}$$

e. $$\%K \text{ lost} = \left(\frac{\Delta K}{K_i}\right) \times 100$$

$$= \left(\frac{287 \text{ J}}{289 \text{ J}}\right) \times 100 = 99.3\%$$

14. Conservation of momentum $mv = (m + M)V$, or

$$v = \frac{(m + M)V}{m}$$

$$= \frac{(0.00800 \text{ kg} + 9.00 \text{ kg})(0.10 \text{ m/s})}{0.00800 \text{ kg}}$$

$$= 113 \text{ m/s}$$

15. We have used conservation of momentum

$$mv_i + MV_i = mv_f + MV_f$$

and conservation of energy

$$\frac{1}{2} mv_i^2 + \frac{1}{2} MV_i^2 = \frac{1}{2} mv_f^2 + \frac{1}{2} MV_f^2$$

where m, v_i, v_f refer to the bullet, M, V_i, V_f to Superman, and $V_i = 0$. v_i may be eliminated from these equations by solving the momentum equation for $v_f = \frac{(mv_i - MV_f)}{m}$ and substituting this into the energy equation

$$mv_i^2 = mv_f^2 + MV_f^2$$

This gives a quadratic equation for V_f which, in factored form, is

$$MV_f\,[(M + m)V_f - 2mv_i] = 0$$

We are not interested in the solution $V_f = 0$ which corresponds to the case where the bullet does not hit Superman. We want the other,

$$V_f = \frac{2mv_i}{M + m} = \frac{2(0.00420 \text{ kg})(835 \text{ m/s})}{104 \text{ kg} + 0.00420 \text{ kg}}$$

$$= 6.74 \times 10^{-2} \text{ m/s}$$

16. a. 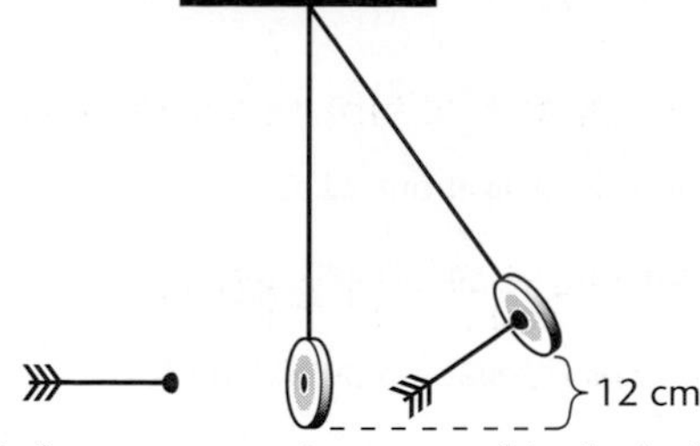

b. Only momentum is conserved in the inelastic dart-target collision, so

$$mv_i + MV_i = (m + M)V_f$$

where $V_i = 0$ since the target is initially at rest and V_f is the common velocity just after impact. As the dart-target combination swings upward, energy is conserved so $\Delta U_g = \Delta K$ or, at the top of the swing,

$$(m + M)gh = \frac{1}{2} (m + M)V_f^2$$

c. Solving this for V_f and inserting into the momentum equation gives

$$v_i = (m + M) \frac{\sqrt{2gh_f}}{m}$$

$$= \frac{(0.025 \text{ kg} + 0.73 \text{ kg}) \sqrt{2(9.8 \text{ m/s}^2)(0.12 \text{ m})}}{0.025 \text{ kg}}$$

$$= 46 \text{ m/s}$$

CHAPTER 12

1. a. $T_K = T_C + 273 = 0 + 273 = 273 \text{ K}$

b. $T_C = T_K - 273 = 0 - 273 = -273°\text{C}$

c. $T_K = T_C + 273 = 273 + 273 = 546 \text{ K}$

d. $T_C = T_K - 273 = 273 - 273 = 0°\text{C}$

2. a. $T_K = T_C + 273 = 27 + 273 = 3.00 \times 10^2 \text{ K}$

b. $T_K = T_C + 273 = 150 + 273 = 4.23 \times 10^2 \text{ K}$

c. $T_K = T_C + 273 = 560 + 273 = 8.33 \times 10^2 \text{ K}$

d. $T_K = T_C + 273 = -50 + 273 = 2.23 \times 10^2 \text{ K}$

e. $T_K = T_C + 273 = -184 + 273 = 89 \text{ K}$

f. $T_K = T_C + 273 = -300 + 273 = -27 \text{ K}$

impossible temperature—below absolute zero

3. a. $T_C = T_K - 273 = 110 - 273 = -163 \text{ °C}$

b. $T_C = T_K - 273 = 70 - 273 = -203 \text{ °C}$

c. $T_C = T_K - 273 = 22 - 273 = -251°\text{C}$

d. $T_C = T_K - 273 = 402 - 273 = 129°\text{C}$

e. $T_C = T_K - 273 = 323 - 273 = 5.0 \times 10^1\ °C$

f. $T_C = T_K - 273 = 212 - 273 = -61°C$

4. a. about 72°F is about 22°C, 295 K

b. about 40°F is about 4°C, 277 K

c. about 86°F is about 30°C, 303 K

d. about 0°F is about –18°C, 255 K

5. $Q = mC\Delta T$

$= (0.0600\ \text{kg})(385\ \text{J/kg}\cdot°\text{C})(80.0°\text{C} - 20.0°\text{C})$

$= 1.39 \times 10^3\ \text{J}$

6. a. $Q = mC\Delta T$

$$\Delta T = \frac{Q}{mC} = \frac{836.0 \times 10^3\ \text{J}}{(20.0\ \text{kg})(4180\ \text{J/kg}\cdot°\text{C})} = 10.0°\text{C}$$

b. Using $1\ \text{L} = 1000\ \text{cm}^3$, the mass of methanol required is

$m = \rho V = (0.80\ \text{g/cm}^3)(20.0\ \text{L})(1000\ \text{cm}^3/\text{L})$

$= 16{,}000\ \text{g or } 16\ \text{kg}$

$$\Delta T = \frac{Q}{mC} = \frac{836.0 \times 10^3\ \text{J}}{(16\ \text{kg})(2450\ \text{J/kg}\cdot°\text{C})} = 21°\text{C}$$

c. Water is the better coolant since its temperature increase is less than half that of methanol when absorbing the same amount of heat.

7. $m_A C_A(T_f - T_{Ai}) + m_B C_B(T_f - T_{Bi}) = 0$

Since $m_A = m_B$ and $C_A = C_B$, there is cancellation in this particular case so that

$$T_f = \frac{(T_{Ai} + T_{Bi})}{2} = \frac{(80.0°\text{C} + 10.0°\text{C})}{2} = 45.0°\text{C}$$

8. $m_A C_A(T_f - T_{Ai}) + m_W C_W(T_f - T_{Wi}) = 0$

Since, in this particular case, $m_A = m_W$, the masses cancel and

$$T_f = \frac{C_A T_{Ai} + C_W T_{Wi}}{C_A + C_W}$$

$$= \frac{(2450\ \text{J/kg}\cdot\text{K})(16.0°\text{C}) + (4180\ \text{J/kg}\cdot\text{K})(85.0°\text{C})}{2450\ \text{J/kg}\cdot\text{K} + 4180\ \text{J/kg}\cdot\text{K}}$$

$= 59.5°\text{C}$

9. $m_B C_B(T_f - T_{Bi}) + m_W C_W(T_f - T_{Wi}) = 0$

$$T_f = \frac{m_B C_B T_{Bi} + m_W C_W T_{Wi}}{m_B C_B + m_W C_W}$$

$$= \frac{(0.100\ \text{kg})(376\ \text{J/kg}\cdot\text{K})(90.0°\text{C})}{(0.100\ \text{kg})(376\ \text{J/kg}\cdot\text{K}) + (0.200\ \text{kg})(4180\ \text{J/kg}\cdot\text{K})}$$

$$+ \frac{(0.200\ \text{kg})(4180\ \text{J/kg}\cdot\text{K})(20.0°\text{C})}{(0.100\ \text{kg})(376\ \text{J/kg}\cdot\text{K}) + (0.200\ \text{kg})(4180\ \text{J/kg}\cdot\text{K})}$$

$= 23.0°\text{C}$

10. $m_A C_A(T_f - T_{Ai}) + m_W C_W(T_f - T_{Wi}) = 0$

Since $m_A = m_W$, the masses cancel and

$$C_A = \frac{-C_W(T_f - T_{Wi})}{(T_f - T_{Ai})}$$

$$= \frac{-(4180\ \text{J/kg}\cdot\text{K})(25.0°\text{C} - 10.0°\text{C})}{(25.0°\text{C} - 100.0°\text{C})}$$

$= 836\ \text{J/kg}\cdot\text{K}$

11. To warm the ice to 0°C:

$Q_W = mC\,\Delta T$

$= (0.100\ \text{kg})(2060\ \text{J/kg}\cdot°\text{C})(0.0° - (-20.0°\text{C}))$

$= 4120\ \text{J} = 0.41 \times 10^5\ \text{J}$

To melt the ice:

$Q_M = mH_f = (0.100\ \text{kg})(3.34 \times 10^5\ \text{J/kg})$

$= 3.34 \times 10^4\ \text{J}$

Total heat required:

$Q = Q_W + Q_M = 0.41 \times 10^4\ \text{J} + 3.34 \times 10^4\ \text{J}$

$= 3.75 \times 10^4\ \text{J}$

12. To heat the water from 60.0°C to 100.0°C:

$Q_1 = mC\,\Delta T$

$= (0.200\ \text{kg})(4180\ \text{J/kg}\cdot°\text{C})(40.0°\text{C})$

$= 0.334 \times 10^5\ \text{J}$

To change the water to steam:

$Q_2 = mH_v = (0.200\ \text{kg})(2.26 \times 10^6\ \text{J/kg})$

$= 4.52 \times 10^5\ \text{J}$

To heat the steam from 100.0°C to 140.0°C:

$Q_3 = mC\,\Delta T$

$= (0.200\ \text{kg})(2020\ \text{J/kg}\cdot°\text{C})(40.0°\text{C})$

$= 0.162 \times 10^5\ \text{J}$

$Q_{total} = Q_1 + Q_2 + Q_3 = 5.02 \times 10^5\ \text{J}$

13. Warm ice from –30.0°C to 0.0°C:

$Q_1 = mC\Delta T$

$= (0.300\ \text{kg})(2060\ \text{J/kg}\cdot°\text{C})(30.0°\text{C})$

$= 0.185 \times 10^5\ \text{J}$

Melt ice:

$Q_2 = mH_f = (0.300\ \text{kg})(3.34 \times 10^5\ \text{J/kg})$

$= 1.00 \times 10^5\ \text{J}$

Heat water 0.0°C to 100.0°C:

$Q_3 = mC\Delta T$

$= (0.300\ \text{kg})(4180\ \text{J/kg}\cdot°\text{C})(100.0°\text{C})$

$= 1.25 \times 10^5\ \text{J}$

Vaporize water:

$Q_4 = mH_v = (0.300\ \text{kg})(2.26 \times 10^6\ \text{J/kg})$

$= 6.78 \times 10^5\ \text{J}$

Heat steam 100.0°C to 130.0°C:

$$Q_5 = mC\Delta T$$
$$= (0.300 \text{ kg})(2020 \text{ J/kg} \cdot \text{°C})(30.0\text{°C})$$
$$= 0.182 \times 10^5 \text{ J}$$
$$Q_{total} = Q_1 + Q_2 + Q_3 + Q_4 + Q_5 = 9.40 \times 10^5 \text{ J}$$

14. a. To freeze, lead must absorb

$$Q = -mH_f = -(0.175 \text{ kg})(2.04 \times 10^4 \text{ J/kg})$$
$$= -3.57 \times 10^3 \text{ J}$$

This will heat the water

$$\Delta T = \frac{Q}{mC} = \frac{3.57 \times 10^3 \text{ J}}{(0.055 \text{ kg})(4180 \text{ J/kg} \cdot \text{°C})} = 16\text{°C}$$
$$T = T_i + \Delta T = 20.0\text{°C} + 16\text{°C} = 36\text{°C}$$

b. Now, $T_f = (m_AC_AT_{Ai} + m_BC_BT_{Bi})/(m_AC_A + m_BC_B)$

$$= \frac{(0.175 \text{ kg})(130 \text{ J/kg} \cdot \text{K})(327\text{°C})}{(0.175 \text{ kg})(130 \text{ J/kg} \cdot \text{K}) + (0.055 \text{ kg})(4180 \text{ J/kg} \cdot \text{K})}$$
$$+ \frac{(0.055 \text{ kg})(4180 \text{ J/kg} \cdot \text{K})(36.0\text{°C})}{(0.175 \text{ kg})(130 \text{ J/kg} \cdot \text{K}) + (0.055 \text{ kg})(4180 \text{ J/kg} \cdot \text{K})}$$
$$= 62\text{°C}$$

CHAPTER 13

1. $P = \dfrac{F}{A}$

so $F = PA = (1.0 \times 10^5 \text{ Pa})(1.52 \text{ m})(0.76 \text{ m})$
$$= 1.2 \times 10^5 \text{ N}$$

2. $F = mg$

$$A = 4(l \times w)$$
$$P = \frac{F}{A} = \frac{(925 \text{ kg})(9.80 \text{ m/s}^2)}{(4)(0.12 \text{ m})(0.18 \text{ m})}$$
$$= 1.05 \times 10^5 \text{ N/m}^2 = 1.0 \times 10^5 \text{ Pa}$$

3. $F_g = (11.8 \text{ g/cm}^3)(10^{-3} \text{ kg/g})(5.0 \text{ cm}) \times (10.0 \text{ cm})(20.0 \text{ cm})(9.80 \text{ m/s}^2)$

$$= 116 \text{ N}$$
$$A = (0.050 \text{ m})(0.100 \text{ m}) = 0.0050 \text{ m}^2$$
$$P = \frac{F}{A} = \frac{116 \text{ N}}{0.0050 \text{ m}^2} = 23 \text{ kPa}$$

4. $F_{net} = F_{outside} - F_{inside}$

$$= (P_{outside} - P_{inside})A$$
$$= (0.85 \times 10^5 \text{ Pa} - 1.00 \times 10^5 \text{ Pa}) \times (1.82 \text{ m})(0.91 \text{ m})$$
$$= -2.5 \times 10^4 \text{ N (toward the outside)}$$

5. $\dfrac{F_1}{A_1} = \dfrac{F_2}{A_2}$

$$F_1 = \frac{F_2A_1}{A_2} = \frac{(1600 \text{ N})(72 \text{ cm}^2)}{(1440 \text{ cm}^2)} = 8.0 \times 10^1 \text{ N}$$

6. $F_g = F_{buoyant} = \rho_{water} Vg$

$$V = \frac{F_g}{\rho_{water}\, g}$$
$$= \frac{600 \text{ N}}{(1000 \text{ kg/m}^3)(9.80 \text{ m/s}^2)} = 0.06 \text{ m}^3$$

This volume does not include that portion of her head that is above the water.

7. $F_T + F_{buoyant} = F_g$ where F_g is the air weight of the camera.

$$F_T = F_g - F_{buoyant} = F_g - \rho_{water} Vg$$
$$= 1250 \text{ N} - (1000 \text{ kg/m}^3)(0.083 \text{ m}^3)(9.80 \text{ m/s}^2)$$
$$= 4.4 \times 10^2 \text{ N}$$

8. $\Delta L = \alpha L_i \Delta T$

$$= [25 \times 10^{-6}(\text{°C})^{-1}](3.66 \text{ m})(67\text{°C})$$
$$= 6.1 \times 10^{-3} \text{ m, or } 6.1 \text{ mm}$$

9. $L_2 = L_1 + \alpha L_1(T_2 - T_1)$

$$= (11.5 \text{ m}) + [12 \times 10^{-6}(\text{°C})^{-1}](11.5 \text{ m}) \times (1221\text{°C} - 22\text{°C})$$
$$= 12 \text{ m}$$

10. a. For water $\beta = 210 \times 10^{-6}(\text{°C})^{-1}$, so

$$\Delta V = \beta V \Delta T$$
$$= [210 \times 10^{-6}(\text{°C})^{-1}](354 \text{ mL})(30.1\text{°C})$$
$$= 2.2 \text{ mL}$$
$$V = 354 \text{ mL} + 2.2 \text{ mL} = 356 \text{ mL}$$

b. For Al $\beta = 75 \times 10^{-6}(\text{°C})^{-1}$, so

$$\Delta V = \beta V \Delta T$$
$$= [75 \times 10^{-6}(\text{°C})^{-1}](354 \text{ mL})(30.1\text{°C})$$
$$= 0.80 \text{ mL}$$
$$V = 354 \text{ mL} + 0.80 \text{ mL} = 355 \text{ mL}$$

c. The difference will spill,

$$2.2 \text{ mL} - 0.80 \text{ mL} = 1.4 \text{ mL}$$

11. a. $V_2 = V_1 + \beta V_1(T_2 - T_1)$

$$= 45\,725 \text{ L} + [950 \times 10^{-6}(\text{°C})^{-1}] \times (45\,725 \text{ L})(-18.0\text{°C} - 32.0\text{°C})$$
$$= 43\,553 \text{ L} = 43\,600 \text{ L}$$

b. Its volume has decreased because of a temperature decrease.

CHAPTER 14

1. a. $v = \dfrac{d}{t} = \dfrac{515 \text{ m}}{1.50 \text{ s}} = 343 \text{ m/s}$

b. $T = \dfrac{1}{f} = \dfrac{1}{436 \text{ Hz}} = 2.29 \text{ ms}$

c. $\lambda = \frac{v}{f} = \frac{d}{ft}$

$\lambda = \frac{515 \text{ m}}{(436 \text{ Hz})(1.50 \text{ s})} = 0.787 \text{ m}$

2. a. $v = \frac{d}{t} = \frac{685 \text{ m}}{2.00 \text{ s}} = 343 \text{ m/s}$

b. $v = \lambda f$

$f = \frac{v}{\lambda} = \frac{343 \text{ m/s}}{0.750 \text{ m}} = 457 \text{ s}^{-1}$, or 457 Hz

c. $T = \frac{1}{f} = \frac{1}{457 \text{ s}^{-1}} = 2.19 \times 10^{-3}$ s, or 2.19 ms

3. At a lower frequency, because wavelength varies inversely with frequency

4. $v = \lambda f = (0.600 \text{ m})(2.50 \text{ Hz}) = 1.50 \text{ m/s}$

5. $\lambda = \frac{v}{f} = \frac{15.0 \text{ m/s}}{5.00 \text{ Hz}} = 3.00 \text{ m}$

6. $\frac{0.100 \text{ s}}{5 \text{ pulses}} = 0.0200$ s/pulse, so $T = 0.0200$ s

$v = \frac{\lambda}{T} = \frac{1.20 \text{ cm}}{0.0200 \text{ s}} = 60.0 \text{ cm/s} = 0.600 \text{ m/s}$

7. $v = \lambda f = (0.400 \text{ m})(20.0 \text{ Hz}) = 8.00 \text{ m/s}$

8. a. The pulse is partially reflected, partially transmitted.

b. Erect, because reflection is from a less dense medium.

c. It is almost totally reflected from the wall.

d. Inverted, because reflection is from a more dense medium.

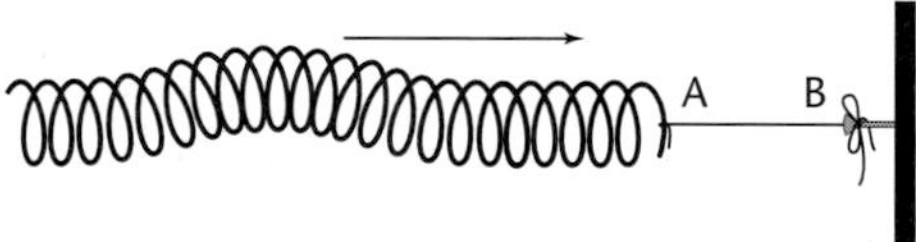

9. Pulse inversion means rigid boundary; attached to wall.

10. a. The pulse is partially reflected, partially transmitted; it is almost totally reflected from the wall.

b. Inverted, because reflection is from a more dense medium; inverted, because reflection is from a more dense medium.

CHAPTER 15

1. $v = \lambda f$

so $f = \frac{v}{\lambda} = \frac{343 \text{ m/s}}{0.667 \text{ m}} = 514 \text{ Hz}$

2. From $v = \lambda f$ the largest wavelength is

$\lambda = \frac{v}{f} = \frac{343 \text{ m/s}}{2.0 \times 10^1 \text{ Hz}} = 17 \text{ m}$

the smallest is

$\lambda = \frac{v}{f} = \frac{343 \text{ m/s}}{16000 \text{ Hz}} = 0.021 \text{ m}$

3. Assume that $v = 343$ m/s

$2d = vt = (343 \text{ m/s})(0.20 \text{ s}) = 68.6 \text{ m}$

$d = \frac{68.6 \text{ m}}{2} = 34 \text{ m}$

4. Woofer diameter 38 cm:

$f = \frac{v}{\lambda} = \frac{343 \text{ m/s}}{0.38 \text{ m}} = 0.90 \text{ kHz}$

Tweeter diameter 7.6 cm:

$f = \frac{v}{\lambda} = \frac{343 \text{ m/s}}{0.076 \text{ m}} = 4.5 \text{ kHz}$

5. Resonance spacing is $\frac{\lambda}{2}$ so using $v = \lambda f$ the resonance spacing is

$\frac{\lambda}{2} = \frac{v}{2f} = \frac{343 \text{ m/s}}{2(440 \text{ Hz})} = 0.39 \text{ m}$

6. Resonance spacing $= \frac{\lambda}{2} = 1.10$ m so

$\lambda = 2.20$ m

and $v = f\lambda = (440 \text{ Hz})(2.20 \text{ m}) = 970 \text{ m/s}$

7. From the previous example problem $v = 347$ m/s at 27°C and the resonance spacing gives

$\frac{\lambda}{2} = 0.202 \text{ m}$

or $\lambda = 0.404$ m

Using $v = \lambda f$,

$f = \frac{v}{\lambda} = \frac{347 \text{ m/s}}{0.404 \text{ m}} = 859 \text{ Hz}$

8. a. $\lambda_1 = 2L = 2(2.65 \text{ m}) = 5.30 \text{ m}$

so that the lowest frequency is

$f_1 = \frac{v}{\lambda_1} = \frac{343 \text{ m/s}}{5.30 \text{ m}} = 64.7 \text{ Hz}$

b. $f_2 = \frac{v}{\lambda_2} = \frac{v}{L} = \frac{343 \text{ m/s}}{2.65 \text{ m}} = 129 \text{ Hz}$

$f_3 = \frac{v}{\lambda_3} = \frac{3v}{2L} = \frac{3(343 \text{ m/s})}{2(2.65 \text{ m})} = 194 \text{ Hz}$

9. The lowest resonant frequency corresponds to the wavelength given by $\frac{\lambda}{2} = L$, the length of the pipe.

$\lambda = 2L = 2(0.65 \text{ m}) = 1.3 \text{ m}$

so $f = \frac{v}{\lambda} = \frac{343 \text{ m/s}}{1.3 \text{ m}} = 260 \text{ Hz}$

10. Beat frequency $= |f_2 - f_1|$

$$= |333\text{ Hz} - 330\text{ Hz}| = 3\text{ Hz}$$

CHAPTER 16

1. $c = \lambda f$

$$\text{so } f = \frac{c}{\lambda} = \frac{(3.00 \times 10^8\text{ m/s})}{(556 \times 10^{-9}\text{ m})} = 5.40 \times 10^{14}\text{ Hz}$$

2. $d = ct = (3.00 \times 10^8\text{ m/s})(1.00 \times 10^{-9}\text{ s}) \times (3.28\text{ ft/1 m})$

$$= 0.984\text{ ft}$$

3. a. $d = ct = (3.00 \times 10^8\text{ m/s})(6.0 \times 10^{-15}\text{ s})$

$$= 1.8 \times 10^{-6}\text{ m}$$

b. Number of wavelengths $= \dfrac{\text{pulse length}}{\lambda_{\text{violet}}}$

$$= \frac{1.8 \times 10^{-6}\text{ m}}{4.0 \times 10^{-7}\text{ m}}$$

$$= 4.5$$

4. $d = ct = (299\ 792\ 458\text{ m/s})\left(\frac{1}{2}\right)(2.562\text{ s})$

$$= 3.840 \times 10^8\text{ m}$$

5. $v = \dfrac{d}{t} = \dfrac{(3.0 \times 10^{11}\text{ m})}{(16\text{ min})(60\text{ s/min})} = 3.1 \times 10^8\text{ m/s}$

6. $\dfrac{E_{\text{after}}}{E_{\text{before}}} = \dfrac{P/4\pi d^2_{\text{after}}}{P/4\pi d^2_{\text{before}}} = \dfrac{d^2_{\text{before}}}{d^2_{\text{after}}} = \dfrac{(30\text{ cm})^2}{(90\text{ cm})^2} = \dfrac{1}{9}$

7. $E = \dfrac{P}{4\pi d^2} = \dfrac{2275\text{ lm}}{4\pi(3.0\text{ m})^2} = 2.0 \times 10^1\text{ lx} = 20\text{ lx}$

8. Illuminance of a 150-watt bulb

$P = 2275$, $d = 0.5, 0.75, \ldots, 5$

$$E(d) = \frac{P}{4\pi d^2}$$

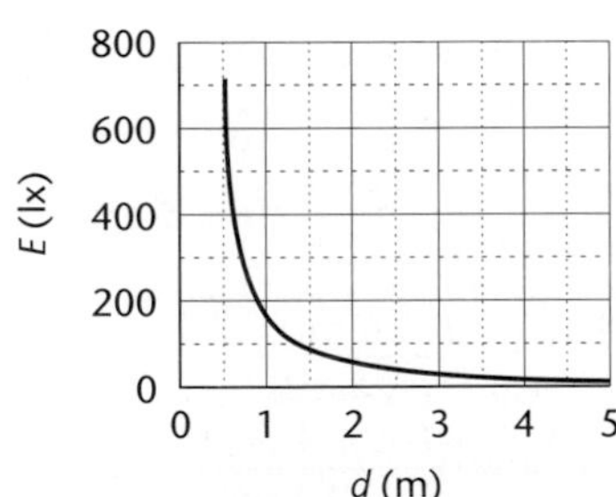

9. $P = 4\pi I = 4\pi(64\text{ cd}) = 256\pi\text{ lm}$

$$\text{so } E = \frac{P}{4\pi d^2} = \frac{256\pi\text{ lm}}{4\pi(3.0\text{ m})^2} = 7.1\text{ lx}$$

10. From $E = \dfrac{P}{4\pi d^2}$

$$P = 4\pi d^2 E = 4\pi(4.0\text{ m})^2(2.0 \times 10^1\text{ lx})$$

$$= 1.3 \times 10^3\ \pi\text{ lm} = 1300\pi\text{ lm}$$

$$\text{so } I = \frac{P}{4\pi d^2} = \frac{1280\pi\text{ lm}}{4\pi} = 3.2 \times 10^2\text{ cd} = 320\text{ cd}$$

11. $E = \dfrac{P}{4\pi d^2}$

$$P = 4\pi E d^2 = 4\pi(160\text{ lm/m}^2)(2.0\text{ m})^2$$

$$= 8.0 \times 10^3\text{ lm}$$

CHAPTER 17

1. The light is incident from air. From $n_i \sin\theta_i = n_r \sin\theta_r$,

$$\sin\theta_r = \frac{n_i \sin\theta_i}{n_r} = \frac{(1.00)\sin 45.0^\circ}{1.52}$$

$$= 0.465, \text{ or } \theta_r = 27.7^\circ$$

2. $n_i \sin\theta_i = n_r \sin\theta_r$

$$\text{so } \sin\theta_r = \frac{n_i \sin\theta_i}{n_r} = \frac{(1.00)\sin 30.0^\circ}{1.33} = 0.376$$

or $\theta_r = 22.1^\circ$

3. a. Assume the light is incident from air.

$n_i \sin\theta_i = n_r \sin\theta_r$ gives

$$\sin\theta_r = \frac{n_i \sin\theta_i}{n_r} = \frac{(1.00)\sin 45.0^\circ}{2.42} = 0.292$$

or $\theta_r = 17.0^\circ$

b. Diamond bends the light more.

4. $n_1 \sin\theta_1 = n_2 \sin\theta_2$

$$\text{so } n_2 = \frac{n_1 \sin\theta_1}{\sin\theta_2} = \frac{(1.33)(0.515)}{0.454} = 1.5$$

5. a. $v_{\text{ethanol}} = \dfrac{c}{n_{\text{ethanol}}} = \dfrac{3.00 \times 10^8\text{ m/s}}{1.36}$

$$= 2.21 \times 10^8\text{ m/s}$$

b. $v_{\text{quartz}} = \dfrac{c}{n_{\text{quartz}}} = \dfrac{3.00 \times 10^8\text{ m/s}}{1.54}$

$$= 1.95 \times 10^8\text{ m/s}$$

c. $v_{\text{flint glass}} = \dfrac{c}{n_{\text{flint glass}}} = \dfrac{3.00 \times 10^8\text{ m/s}}{1.61}$

$$= 1.86 \times 10^8\text{ m/s}$$

6. $n = \dfrac{c}{v} = \dfrac{3.00 \times 10^8\text{ m/s}}{2.00 \times 10^8\text{ m/s}} = 1.50$

7. $n = 1.5$

$$\text{so } v = \frac{c}{n} = \frac{3.00 \times 10^8\text{ m/s}}{1.5} = 2.0 \times 10^8\text{ m/s}$$

8. $t = \dfrac{d}{v} = \dfrac{dn}{c}$

$$\Delta t = \frac{d(n_{\text{air}} - n_{\text{vacuum}})}{c}$$

$$= \frac{d(1.0003 - 1.0000)}{3.00 \times 10^8\text{ m/s}} = d(1 \times 10^{-12}\text{ s/m})$$

Thus, $d = \dfrac{\Delta t}{1 \times 10^{-12}\ \text{s/m}}$

$= \dfrac{1 \times 10^{-8}\ \text{s}}{1 \times 10^{-12}\ \text{s/m}} = 10^4\ \text{m} = 10\ \text{km}$

CHAPTER 18

1.

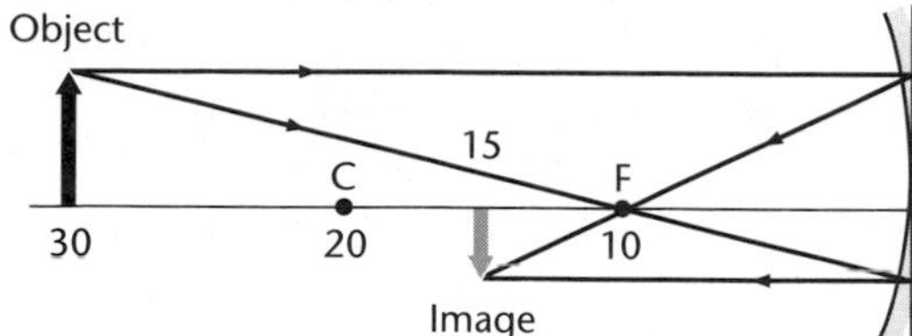

2. a.

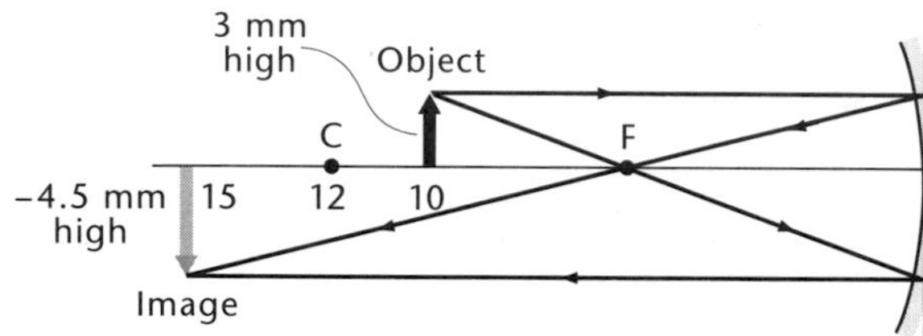

b. $\dfrac{1}{f} = \dfrac{1}{d_i} + \dfrac{1}{d_o}$

$\dfrac{1}{6.0} = \dfrac{1}{d_i} + \dfrac{1}{10.0}$

$\dfrac{1}{d_i} = \dfrac{1}{6.0} - \dfrac{1}{10.0}$

$d_i = \dfrac{1}{\dfrac{1}{6.0} - \dfrac{1}{10.0}} = 15\ \text{cm}$

$m = \dfrac{-d_i}{d_o} = \dfrac{-(15\ \text{cm})}{10.0\ \text{cm}} = -1.5$

$m = \dfrac{h_i}{h_o}$

so $h_i = mh_o = (-1.5)(3.0\ \text{mm}) = -4.5\ \text{mm}$

3.

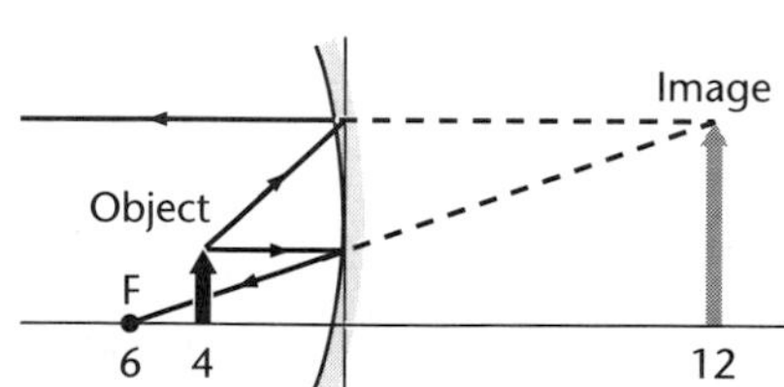

$f = \dfrac{r}{2} = \dfrac{(12.0\ \text{cm})}{2} = 6.0\ \text{cm}$

$\dfrac{1}{d_o} + \dfrac{1}{d_i} = \dfrac{1}{f}$

so $d_i = \dfrac{fd_o}{d_o - f} = \dfrac{(6.0\ \text{cm})(4.0\ \text{cm})}{(4.0\ \text{cm} - 6.0\ \text{cm})} = -12\ \text{cm}$

4. $\dfrac{1}{d_o} + \dfrac{1}{d_i} = \dfrac{1}{f}$

so $d_i = \dfrac{fd_o}{d_o - f}$

$= \dfrac{(16.0\ \text{cm})(10.0\ \text{cm})}{(10.0\ \text{cm} - 16.0\ \text{cm})} = -26.7\ \text{cm}$

$m = \dfrac{h_i}{h_o} = -\dfrac{d_i}{d_o} = \dfrac{-(-26.7\ \text{cm})}{(10.0\ \text{cm})} = +2.67$

so $h_i = mh_o = (2.67)(4.0\ \text{cm}) = 11\ \text{cm}$

5. $m = \dfrac{-d_i}{d_o} = 3$

so $d_i = -md_o = -3(25\ \text{cm}) = -75\ \text{cm}$

$\dfrac{1}{f} = \dfrac{1}{d_o} + \dfrac{1}{d_i}$

so $f = \dfrac{d_o d_i}{d_o + d_i} = \dfrac{(25\ \text{cm})(-75\ \text{cm})}{25\ \text{cm} + (-75\ \text{cm})}$

$= 37.5\ \text{cm}$, and $r = 2f = 75\ \text{cm}$

6. a.

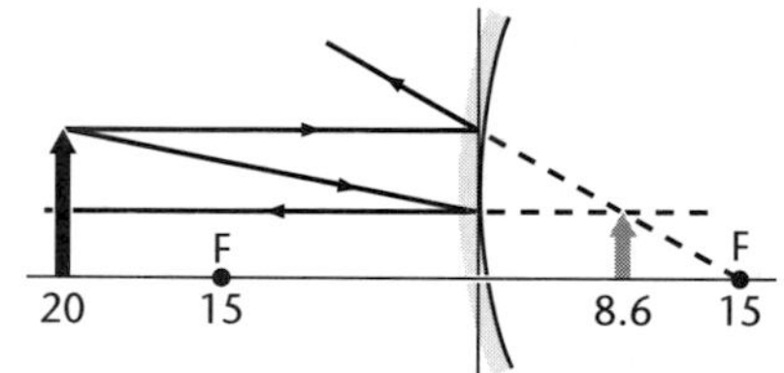

b. $\dfrac{1}{d_o} + \dfrac{1}{d_i} = \dfrac{1}{f}$

so $d_i = \dfrac{fd_o}{d_o - f}$

$= \dfrac{(-15.0\ \text{cm})(20.0\ \text{cm})}{20.0\ \text{cm} - (-15.0\ \text{cm})} = -8.57\ \text{cm}$

7. $\dfrac{1}{d_o} + \dfrac{1}{d_i} = \dfrac{1}{f}$

so $d_i = \dfrac{fd_o}{d_o - f}$

$= \dfrac{(-12\ \text{cm})(60.0\ \text{cm})}{60.0\ \text{cm} - (-12\ \text{cm})} = -1.0 \times 10^1\ \text{cm}$

$m = \dfrac{h_i}{h_o} = \dfrac{-d_i}{d_o} = \dfrac{-(-1.0 \times 10^1\ \text{cm})}{60.0\ \text{cm}} = +0.17$

so $mh_o = (0.17)(6.0\ \text{cm}) = 1.0\ \text{cm}$

8. $\dfrac{1}{f} = \dfrac{1}{d_o} + \dfrac{1}{d_i}$

so $f = \dfrac{d_o d_i}{d_o + d_i}$

and $m = \dfrac{-d_i}{d_o}$ so $d_o = \dfrac{-d_i}{m}$

Since $d_i = -24$ cm and $m = 0.75$,

$$d_o = \frac{-(-24\text{ cm})}{0.75} = 32\text{ cm}$$

$$\text{and } f = \frac{(32\text{ cm})(-24\text{ cm})}{32\text{ cm} + (-24\text{ cm})} = -96\text{ cm}$$

9.

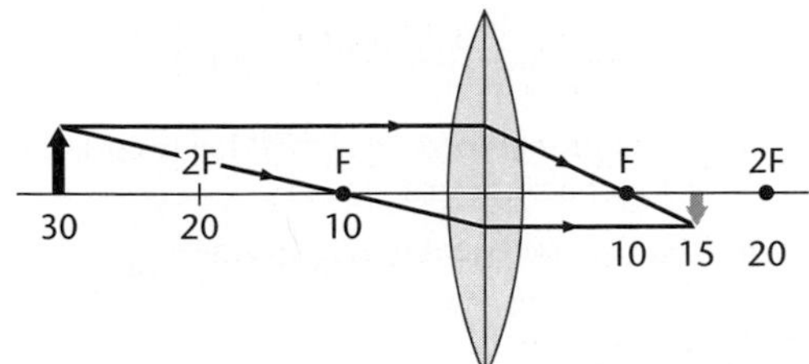

10. $\frac{1}{d_o} + \frac{1}{d_i} = \frac{1}{f}$

$$\text{so } d_i = \frac{fd_o}{d_o - f} = \frac{(5.5\text{ cm})(8.5\text{ cm})}{8.5\text{ cm} - 5.5\text{ cm}} = 16\text{ cm}$$

$$h_i = \frac{-d_i h_o}{d_o} = \frac{-(16\text{ cm})(2.25\text{ mm})}{8.5\text{ cm}} = -4.2\text{ mm}$$

11. $\frac{1}{d_o} + \frac{1}{d_i} = \frac{1}{f}$

with $d_o = d_i$ since

$$m = \frac{-d_i}{d_o} \quad \text{and} \quad m = -1$$

Therefore,

$$\frac{2}{d_i} = \frac{1}{f} \text{ and } \frac{2}{d_o} = \frac{1}{f}$$

$$d_i = 2f = 5.0 \times 10^1\text{ mm and}$$

$$d_o = 2f = 5.0 \times 10^1\text{ mm}$$

12. $\frac{1}{d_o} + \frac{1}{d_i} = \frac{1}{f}$

$$\text{so } d_i = \frac{f\,d_o}{d_o - f} = \frac{(20.0\text{ cm})(6.0\text{ cm})}{6.0\text{ cm} - 20.0\text{ cm}} = -8.6\text{ cm}$$

13. $\frac{1}{d_o} + \frac{1}{d_i} = \frac{1}{f}$

$$\text{so } d_i = \frac{f\,d_o}{d_o - f} = \frac{(12.0\text{ cm})(3.4\text{ cm})}{3.4\text{ cm} - 12.0\text{ cm}} = -4.7\text{ cm}$$

$$h_i = \frac{-h_o d_i}{d_o} = \frac{-(2.0\text{ cm})(-4.7\text{ cm})}{(3.4\text{ cm})} = 2.8\text{ cm}$$

14. $m = \frac{-d_i}{d_o}$

$$\text{so } d_i = -md_o = -(4.0)(3.5\text{ cm}) = -14\text{ cm}$$

$$\frac{1}{f} = \frac{1}{d_o} + \frac{1}{d_i}$$

$$\text{so } f = \frac{d_o d_i}{d_o + d_i} = \frac{(3.5\text{ cm})(-14\text{ cm})}{3.5\text{ cm} + (-14\text{ cm})} = 4.7\text{ cm}$$

CHAPTER 19

1. $\lambda = \frac{xd}{L} = \frac{(13.2 \times 10^{-3}\text{ m})(1.90 \times 10^{-5}\text{ m})}{(0.600\text{ m})}$

$= 418$ nm

2. $x = \frac{\lambda L}{d} = \frac{(5.96 \times 10^{-7}\text{ m})(0.600\text{ m})}{(1.90 \times 10^{-5}\text{ m})} = 18.8\text{ mm}$

3. $d = \frac{\lambda L}{x} = \frac{(6.328 \times 10^{-7}\text{ m})(1.000\text{ m})}{(65.5 \times 10^{-3}\text{ m})} = 9.66\ \mu\text{m}$

4. $\lambda = \frac{xd}{L} = \frac{(55.8 \times 10^{-3}\text{ m})(15 \times 10^{-6}\text{ m})}{(1.6\text{ m})}$

$= 520$ nm

5. $x = \frac{\lambda L}{w} = \frac{(5.46 \times 10^{-7}\text{ m})(0.75\text{ m})}{(9.5 \times 10^{-5}\text{ m})} = 4.3\text{ mm}$

6. $w = \frac{\lambda L}{x} = \frac{(6.328 \times 10^{-7}\text{ m})(1.15\text{ m})}{(7.5 \times 10^{-3}\text{ m})} = 97\ \mu\text{m}$

7. $\lambda = \frac{wx}{L} = \frac{(2.95 \times 10^{-5}\text{ m})(1.20 \times 10^{-2}\text{ m})}{(0.600\text{ m})}$

$= 590$ nm

8. a. Red, because central peak width is proportional to wavelength.

b. Width $= 2x = \frac{2\lambda L}{w}$

For blue,

$$2x = \frac{2(4.41 \times 10^{-7}\text{ m})(1.00\text{ m})}{(5.0 \times 10^{-5}\text{ m})} = 18\text{ mm}$$

For red,

$$2x = \frac{2(6.22 \times 10^{-7}\text{ m})(1.00\text{ m})}{(5.0 \times 10^{-5}\text{ m})} = 25\text{ mm}$$

CHAPTER 20

1. $F = \frac{Kq_Aq_B}{d_{AB}^2}$

$$= \frac{(9.0 \times 10^9\text{ N} \cdot \text{m}^2/\text{C}^2)(2.0 \times 10^{-4}\text{ C})(8.0 \times 10^{-4}\text{ C})}{(0.30\text{ m})^2}$$

$= 1.6 \times 10^4$ N

2. $F = \frac{Kq_Aq_B}{d_{AB}^2}$

$$q_B = \frac{Fd_{AB}^2}{Kq_A} = \frac{(65\text{ N})(0.050\text{ m})^2}{(9.0 \times 10^9\text{ N} \cdot \text{m}^2/\text{C}^2)(6.0 \times 10^{-6}\text{ C})}$$

$= 3.0 \times 10^{-6}$ C

3. $F = \frac{Kq_Aq_B}{d_{AB}^2}$

$$= \frac{(9.0 \times 10^9\text{ N} \cdot \text{m}^2/\text{C}^2)(6.0 \times 10^{-6}\text{ C})(6.0 \times 10^{-6}\text{ C})}{(0.5\text{ m})^2}$$

$= 1.3$ N

4. At $d = 1.0$ cm,

$$F = \frac{Kq_Aq_B}{d_{AB}^2}$$

$$= \frac{(9.0 \times 10^9\text{ N} \cdot \text{m}^2/\text{C}^2)(7.5 \times 10^{-7}\text{ C})(1.5 \times 10^{-7}\text{ C})}{(1.0 \times 10^{-2}\text{ m})^2}$$

$= 1.0 \times 10^1$ N

Since force varies as distance squared, the force at $d = 5.0$ cm is $\frac{1}{25}$ the force at $d = 1.0$ cm, or 4.1×10^{-2} N.

The force varies as $\frac{1}{d^2}$, so the graph looks like

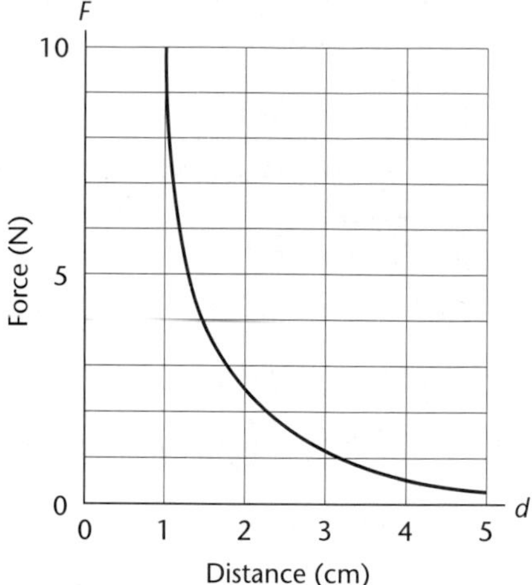

5. Magnitudes of all forces remain the same. The direction changes to 42° above the $-x$ axis, or 138°.

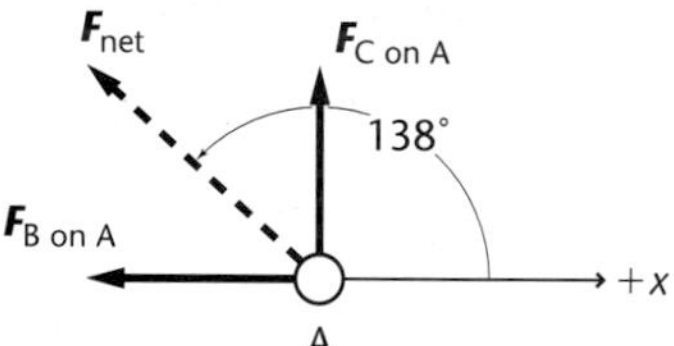

CHAPTER 21

1. $E = \frac{F}{q} = \frac{0.060 \text{ N}}{2.0 \times 10^{-8} \text{ C}}$
$= 3.0 \times 10^6$ N/C directed to the left

2. $E = \frac{F}{q} = \frac{2.5 \times 10^{-4} \text{ N}}{5.0 \times 10^{-4} \text{ C}} = 0.50$ N/C

3. $\frac{F_2}{F_1} = \frac{(Kq_Aq_B/d_2{}^2)}{(Kq_Aq_B/d_1{}^2)} = \left(\frac{d_1}{d_2}\right)^2$ with $d_2 = 2d_1$

$F_2 = \left(\frac{d_1}{d_2}\right)^2 F_1 = \left(\frac{d_1}{2d_1}\right)^2 (2.5 \times 10^{-4} \text{ N})$
$= 6.3 \times 10^{-5}$ N

4. **a.** No. The force on the 2.0 μC charge would be twice that on the 1.0 μC charge.

b. Yes. You would divide the force by the strength of the test charge, so the results would be the same.

5. $\Delta V = Ed = (8000 \text{ N/C})(0.05 \text{ m}) = 400$ J/C
$= 4 \times 10^2$ V

6. $\Delta V = Ed$

$E = \frac{\Delta V}{d} = \frac{500 \text{ V}}{0.020 \text{ m}} = 3 \times 10^4$ N/C

7. $\Delta V = Ed = (2.50 \times 10^3 \text{ N/C})(0.500 \text{ m})$
$= 1.25 \times 10^3$ V

8. $W = q\Delta V = (5.0 \text{ C})(1.5 \text{ V}) = 7.5$ J

9. **a.** Gravitational force (weight) downward, frictional force of air upward.

b. The two are equal in magnitude.

10. **a.** $F = Eq$

$q = \frac{F}{E} = \frac{1.9 \times 10^{-15} \text{ N}}{6.0 \times 10^3 \text{ N/C}} = 3.2 \times 10^{-19}$ C

b. # electrons $= \frac{q}{q_e} = \frac{3.2 \times 10^{-19} \text{ C}}{1.6 \times 10^{-19} \text{ C/electron}}$
$= 2$ electrons

11. **a.** $F = Eq$

$q = \frac{F}{E} = \frac{6.4 \times 10^{-13} \text{ N}}{4.0 \times 10^6 \text{ N/C}} = 1.6 \times 10^{-19}$ C

b. # electrons $= \frac{q}{1.6 \times 10^{-19} \text{ C/electron}}$
$= 1$ electron

12. $E = \frac{F}{q} = \frac{6.4 \times 10^{-13} \text{ N}}{(4)(1.6 \times 10^{-19} \text{ C})} = 1.0 \times 10^6$ N/C

13. $q = C\Delta V = (27\ \mu\text{F})(25 \text{ V}) = 6.8 \times 10^{-4}$ C

14. $q = C\Delta V$, so the larger capacitor has a greater charge.
$q = (6.8 \times 10^{-6} \text{ F})(15 \text{ V}) = 1.0 \times 10^{-4}$ C

15. $\Delta V = q/C$, so the smaller capacitor has the larger potential difference.

$\Delta V = \frac{(2.5 \times 10^{-4} \text{ C})}{(3.3 \times 10^{-6} \text{ F})} = 76$ V

16. $q = C\Delta V$ so $\Delta q = C(V_2 - V_1)$
$\Delta q = (2.2\ \mu\text{F})(15.0 \text{ V} - 6.0 \text{ V}) = 2.0 \times 10^{-5}$ C

CHAPTER 22

1. $P = VI = (120 \text{ V})(0.5 \text{ A}) = 60$ J/s $= 60$ W

2. $P = VI = (12 \text{ V})(2.0 \text{ A}) = 24$ W

3. $P = VI$
$I = \frac{P}{V} = \frac{75 \text{ W}}{120 \text{ V}} = 0.63$ A

4. $P = VI = (12 \text{ V})(210 \text{ A}) = 2500$ W
In 10 s,
$E = Pt = (2500 \text{ J/s})(10 \text{ s})$
$= 25000 \text{ J} = 2.5 \times 10^4$ J

5. $I = \frac{V}{R} = \frac{12\text{ V}}{30\ \Omega} = 0.40\text{ A}$

6. $V = IR = (3.8\text{ A})(32\ \Omega) = 1.2 \times 10^2\text{ V}$

7. $R = \frac{V}{I} = \frac{3.0\text{ V}}{2.0 \times 10^{-4}\text{ A}} = 2.0 \times 10^4\ \Omega$

8. a. $R = \frac{V}{I} = \frac{120\text{ V}}{0.5\text{ A}} = 2 \times 10^2\ \Omega$

b. $P = VI = (120\text{ V})(0.5\text{ A}) = 60\text{ W}$

9. a. $I = \frac{P}{V} = \frac{75\text{ W}}{120\text{ V}} = 0.63\text{ A}$

b. $R = \frac{V}{I} = \frac{120\text{ V}}{0.63\text{ A}} = 190\ \Omega$

10. a. The new value of the current is

$$\frac{0.63\text{ A}}{2} = 0.315\text{ A}$$

so $V = IR = (0.315\text{ A})(190\text{ W}) = 60\text{ V}$

b. The total resistance of the circuit is now

$$R_{\text{total}} = \frac{V}{I} = \frac{(120\text{ V})}{(0.315\text{ A})} = 380\ \Omega$$

Therefore,

$$R_{\text{res}} = R_{\text{total}} - R_{\text{lamp}} = 380\ \Omega - 190\ \Omega = 190\ \Omega$$

c. $P = VI = (60\text{ V})(0.315\text{ A}) = 19\text{ W}$

11. $I = \frac{V}{R} = \frac{60\text{ V}}{12.5\ \Omega} = 4.8\text{ A}$

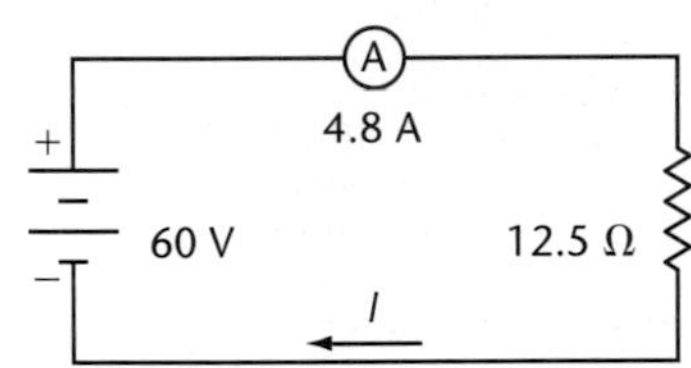

12. $R = \frac{V}{I} = \frac{4.5\text{ V}}{0.090\text{ A}} = 50\ \Omega$

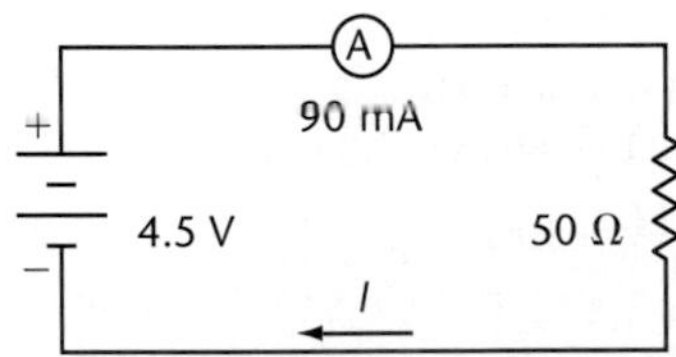

13. Both circuits will take the form

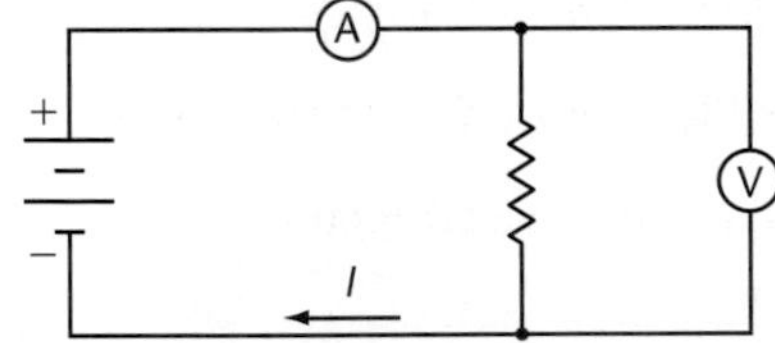

Since the ammeter resistance is assumed zero, the voltmeter readings will be

practice problem 11	60 V
practice problem 12	4.5 V

14. a. $I = \frac{V}{R} = \frac{120\text{ V}}{15\ \Omega} = 8.0\text{ A}$

b. $E = I^2Rt = (8.0\text{ A})^2(15\ \Omega)(30.0\text{ s}) = 2.9 \times 10^4\text{ J}$

c. 2.9×10^4 J, since all electrical energy is converted to thermal energy.

15. a. $I = \frac{V}{R} = \frac{60\text{ V}}{30\ \Omega} = 2\text{ A}$

b. $E = I^2Rt = (2\text{ A})^2(30\ \Omega)(5\text{ min})(60\text{ s/min})$
$= 4 \times 10^4\text{ J}$

16. a. $E = (0.200)(100.0\text{ J/s})(60.0\text{ s}) = 1.20 \times 10^3\text{ J}$

b. $E = (0.800)(100.0\text{ J/s})(60.0\text{ s}) = 4.80 \times 10^3\text{ J}$

17. a. $I = \frac{V}{R} = \frac{220\text{ V}}{11\ \Omega} = 20\text{ A}$

b. $E = I^2Rt = (20\text{ A})^2(11\ \Omega)(30.0\text{ s}) = 1.3 \times 10^5\text{ J}$

c. $Q = mC\,\Delta T$ with $Q = 0.70E$

$$\Delta T = \frac{0.70E}{mC} = \frac{(0.70)(1.3 \times 10^5\text{ J})}{(1.20\text{ kg})(4180\text{ J/kg}\cdot\text{C°})} = 18\text{°C}$$

18. a. $P = IV = (15.0\text{ A})(120\text{ V}) = 1800\text{ W} = 1.80\text{ kW}$

b. $E = Pt = (1.8\text{ kW})(5.0\text{ h/day})(30\text{ days})$
$= 270\text{ kWh}$

c. $\text{Cost} = (0.11\text{ \$/kWh})(270\text{ kWh}) = \$30$

19. a. $I = \frac{V}{R} = \frac{115\text{ V}}{12000\ \Omega} = 9.6 \times 10^{-3}\text{ A}$

b. $P = VI = (115\text{ V})(9.6 \times 10^{-3}\text{ A}) = 1.1\text{ W}$

c. $\text{Cost} = (1.1 \times 10^{-3}\text{ kW})(\$0.09\text{/kWh}) \times (30\text{ days})(24\text{ h/day})$
$= \$0.07$

20. a. $I = \frac{P}{V} = \frac{1200\text{ W}}{120\text{ V}} = 1.0 \times 10^1\text{ A}$

$R = \frac{V}{I} = \frac{120\text{ V}}{10\text{ A}} = 12\text{ W}$

b. 1.0×10^1 A or 10 A

c. $P = IV = (1.0 \times 10^1\text{ A})(120\text{ V}) = 1200\text{ W}$
$= 1.2 \times 10^3\text{ J/s}$

d. $Q = mC\,\Delta T$

In one s,

$$\Delta T = \frac{Q}{mC} = \frac{1.20 \times 10^3\text{ J/s}}{(0.500\text{ kg})(4180\text{ J/kg}\cdot\text{K})}$$
$$= 0.574\text{°C/s}$$

e. $\frac{120\text{ V}}{2.00\text{ m}} = 60.0\text{ V/m}$

f. $P = 1.2 \times 10^3\text{ W} = 1.2\text{ kW}$

$\text{Cost/3 min} = (1.2\text{ kW})(\$0.10)\left(\frac{3\text{ min}}{60\text{ min/h}}\right)$
$= \$0.006$ or 0.6 cents

If only one slice is made, 0.6 cents; if four slices are made, 0.15 cents per slice.

CHAPTER 23

1. $R = R_1 + R_2 + R_3 = 20\ \Omega + 20\ \Omega + 20\ \Omega = 60\ \Omega$

$$I = \frac{V}{R} = \frac{120\ \text{V}}{60\ \Omega} = 2\text{A}$$

2. $R = 10\ \Omega + 15\ \Omega + 5\ \Omega = 30\ \Omega$

$$I = \frac{V}{R} = \frac{90\ \text{V}}{30\ \Omega} = 3\ \text{A}$$

3. **a.** It will increase.

b. $I = \frac{V}{R}$, so it will decrease.

c. No. It does not depend on the resistance.

4. **a.** $R = \frac{V}{I} = \frac{120\ \text{V}}{0.06\ \text{A}} = 2000\ \Omega$

b. $\frac{2000\ \Omega}{10} = 200\ \Omega$

5. $V = IR = 3\ \text{A}(10\ \Omega + 15\ \Omega + 5\ \Omega)$

$= 30\ \text{V} + 45\ \text{V} + 15\ \text{V}$

$= 90\ \text{V} =$ voltage of battery

6. **a.** $R = 20.0\ \Omega + 30.0\ \Omega = 50.0\ \Omega$

b. $I = \frac{V}{R} = \frac{120\ \text{V}}{50.0\ \Omega} = 2.4\ \text{A}$

c. $V = IR$

Across 20.0-Ω resistor,

$V = (2.4\ \text{A})(20.0\ \Omega) = 48\ \text{V}$

Across 30.0-Ω resistor,

$V = (2.4\ \text{A})(30.0\ \Omega) = 72\ \text{V}$

d. $V = 48\ \text{V} + 72\ \text{V} = 120\ \text{V}$

7. **a.** $R = 3.0\ \text{k}\Omega + 5.0\ \text{k}\Omega + 4.0\ \text{k}\Omega = 12.0\ \text{k}\Omega$

b. $I = \frac{V}{R} = \frac{12\ \text{V}}{12.0\ \text{k}\Omega} = 1.0\ \text{mA} = 1.0 \times 10^{-3}\ \text{A}$

c. $V = IR$

so $V = 3.0$ V, 5.0 V, and 4.0 V

d. $V = 3.0\ \text{V} + 5.0\ \text{V} + 4.0\ \text{V} = 12.0\ \text{V}$

8. **a.** $V_B = \frac{VR_B}{R_A + R_B} = \frac{(9.0\ \text{V})(475\ \Omega)}{500\ \Omega + 475\ \Omega} = 4\ \text{V}$

b. $V_B = \frac{VR_B}{R_A + R_B} = \frac{(9.0\ \text{V})(4.0\ \text{k}\Omega)}{0.5\ \text{k}\Omega + 4.0\ \text{k}\Omega} = 8\ \text{V}$

c. $V_B = \frac{VR_B}{R_A + R_B} = \frac{(9.0\ \text{V})(4.0 \times 10^5\ \Omega)}{0.005 \times 10^5\ \Omega + 4.0 \times 10^5\ \Omega}$

$= 9\ \text{V}$

9. $V_2 = \frac{VR_2}{R_1 + R_2} = \frac{(45\ \text{V})(235\ \text{k}\Omega)}{475\ \text{k}\Omega + 235\ \text{k}\Omega} = 15\ \text{V}$

10. **a.** $\frac{1}{R} = \frac{1}{R_1} + \frac{1}{R_2} + \frac{1}{R_3} = \frac{3}{15\ \Omega}$

$R = 5.0\ \Omega$

b. $I = \frac{V}{R} = \frac{30\ \text{V}}{5.0\ \Omega} = 6\ \text{A}$

c. $I = \frac{V}{R} = \frac{30\ \text{V}}{15.0\ \Omega} = 2\ \text{A}$

11. **a.** $\frac{1}{R} = \frac{1}{120.0\ \Omega} + \frac{1}{60.0\ \Omega} + \frac{1}{40.0\ \Omega}$

$R = 20.0\ \Omega$

b. $I = \frac{V}{R} = \frac{12.0\ \text{V}}{20.0\ \Omega} = 0.600\ \text{A}$

c. $I_1 = \frac{V_1}{R_1} = \frac{12.0\ \text{V}}{120.0\ \Omega} = 0.100\ \text{A}$

$I_2 = \frac{V}{R_2} = \frac{12.0\ \text{V}}{60.0\ \Omega} = 0.200\ \text{A}$

$I_3 = \frac{V}{R_3} = \frac{12.0\ \text{V}}{40.0\ \Omega} = 0.300\ \text{A}$

12. **a.** Yes. Smaller.

b. Yes. Gets larger.

c. No. It remains the same. Currents are independent.

13. **a.**

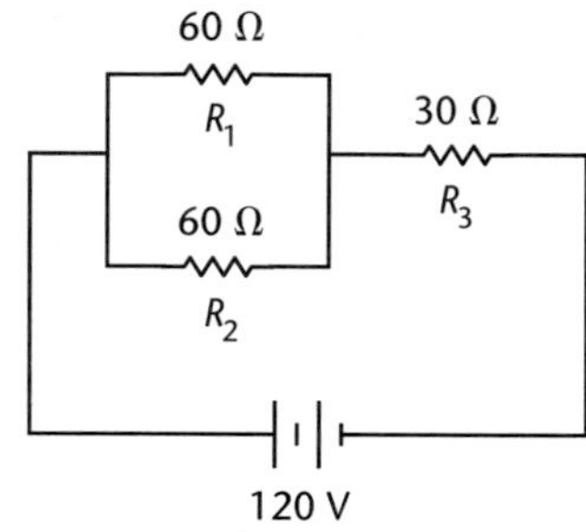

b. $\frac{1}{R} = \frac{1}{60\ \Omega} + \frac{1}{60\ \Omega} = \frac{2}{60\ \Omega}$

$R = \frac{60\ \Omega}{2} = 30\ \Omega$

c. $R_E = 30\ \Omega + 30\ \Omega = 60\ \Omega$

d. $I = \frac{V}{R} = \frac{120\ \text{V}}{60\ \Omega} = 2\ \text{A}$

e. $V_3 = IR_3 = (2\ \text{A})(30\ \Omega) = 60\ \text{V}$

f. $V = IR = (2\ \text{A})(30\ \Omega) = 60\ \text{V}$

g. $I = \frac{V}{R_1} = \frac{V}{R_2} = \frac{60\text{V}}{60\ \Omega} = 1\ \text{A}$

CHAPTER 24

1. a. repulsive b. attractive

2. south, north, south, north

3. the bottom (the point)

4. a. from south to north b. west

5. a. Since magnetic field strength varies inversely with the distance from the wire, it will be half as strong.

   b. It is one-third as strong.

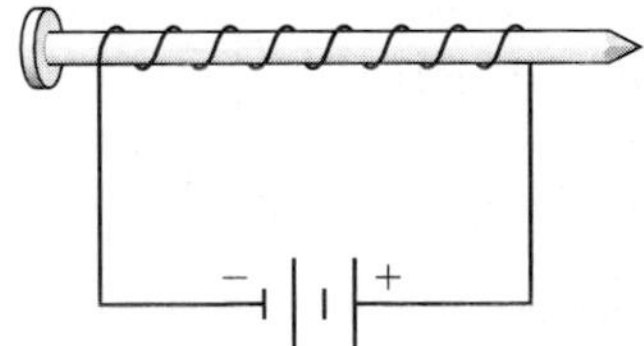

6. the pointed end

7. $F = BIL = (0.40\ \text{N/A}\cdot\text{m})(8.0\ \text{A})(0.50\ \text{m}) - 1.6\ \text{N}$

8. $B = \frac{F}{IL} = \frac{0.60\ \text{N}}{(6.0\ \text{A})(0.75\ \text{m})} = 0.13\ \text{T}$

9. $F = BIL$, $F =$ weight of wire.

   $B = \frac{F}{IL} = \frac{0.35\ \text{N}}{(6.0\ \text{A})(0.4\ \text{m})} = 0.1\ \text{T}$

10. $F = Bqv$

    $= (0.50\ \text{T})(1.6 \times 10^{-19}\ \text{C})(4.0 \times 10^{6}\ \text{m/s})$

    $= 3.2 \times 10^{-13}\ \text{N}$

11. $F = Bqv$

    $= (9.0 \times 10^{-2}\ \text{T})(2)(1.60 \times 10^{-19}\ \text{C}) \times (3.0 \times 10^{4}\ \text{m/s})$

    $= 8.6 \times 10^{-16}\ \text{N}$

12. $F = Bqv$

    $= (4.0 \times 10^{-2}\ \text{T})(3)(1.60 \times 10^{-19}\ \text{C}) \times (9.0 \times 10^{6}\ \text{m/s})$

    $= 1.7 \times 10^{-13}\ \text{N}$

13. $F = Bqv$

    $= (5.0 \times 10^{-2}\ \text{T})(2)(1.60 \times 10^{-19}\ \text{C}) \times (4.0 \times 10^{-2}\ \text{m/s})$

    $= 6.4 \times 10^{-22}\ \text{N}$

CHAPTER 25

1. a. $EMF = BLv$

   $= (0.4\ \text{N/A}\cdot\text{m})(0.5\ \text{m})(20\ \text{m/s}) = 4\ \text{V}$

   b. $I = \frac{V}{R} = \frac{4\ \text{V}}{6.0\ \Omega} = 0.7\ \text{A}$

2. $EMF = BLv$

   $= (5.0 \times 10^{-5}\ \text{T})(25\ \text{m})(125\ \text{m/s}) = 0.16\ \text{V}$

3. a. $EMF = BLv = (1.0\ \text{T})(30.0\ \text{m})(2.0\ \text{m/s})$

   $= 6.0 \times 10^{1}\ \text{V}$

   b. $I = \frac{V}{R} = \frac{BLv}{R}$

   $I = \frac{(1.0\ \text{T})(30.0\ \text{m})(2.0\ \text{m/s})}{15.0\ \Omega} = 4.0\ \text{A}$

4. Using the right-hand rule, the north pole is at the bottom.

5. a. $V_{eff} = (0.707)V_{max} = (0.707)(170\ \text{V}) = 120\ \text{V}$

   b. $I_{eff} = (0.707)I_{max} = (0.707)(0.70\ \text{A}) = 0.49\ \text{A}$

   c. $R = \frac{V_{eff}}{I_{eff}} = \frac{120\ \text{V}}{0.49\ \text{A}} = 240\ \text{ohms}$

6. a. $V_{max} = \frac{V_{eff}}{0.707} = \frac{117\ \text{V}}{0.707} = 165\ \text{V}$

   b. $I_{max} = \frac{I_{eff}}{0.707} = \frac{5.5\ \text{A}}{0.707} = 7.8\ \text{A}$

7. a. $V_{eff} = (0.707)(425\ \text{V}) = 3.00 \times 10^{2}\ \text{V}$

   b. $I_{eff} = \frac{V_{eff}}{R} = \frac{3.00 \times 10^{2}\ \text{V}}{5.0 \times 10^{2}\ \Omega} = 0.60\ \text{A}$

8. Since $P = V_{eff}I_{eff}$

   $= (0.707\ V_{max})(0.707\ I_{max}) = \frac{1}{2}P_{max}$

   $P_{max} = 2P = 2(100\ \text{W}) = 200\ \text{W}$

9. a. $\frac{V_S}{V_P} = \frac{N_S}{N_P}$

   $V_S = \frac{V_PN_S}{N_P} = \frac{(7200\ \text{V})(125)}{7500} = 120\ \text{V}$

   b. $V_PI_P = V_SI_S$

   $I_P = \frac{V_SI_S}{V_P} = \frac{(120\ \text{V})(36\ \text{A})}{7200\ \text{V}} = 0.60\ \text{A}$

10. a. $\frac{V_P}{V_S} = \frac{N_P}{N_S}$

    $V_S = \frac{V_PN_S}{N_P} = \frac{(120\ \text{V})(15\ 000)}{500} = 3.6 \times 10^{3}\ \text{V}$

    b. $V_PI_P = V_SI_S$

    $I_P = \frac{V_SI_S}{V_P} = \frac{(3600\ \text{V})(3.0\ \text{A})}{120\ \text{V}} = 9.0 \times 10^{1}\ \text{A}$

    c. $V_PI_P = (120\ \text{V})(90\ \text{A}) = 1.1 \times 10^{4}\ \text{W}$

    $V_SI_S = (3600\ \text{V})(3.0\ \text{A}) = 1.1 \times 10^{4}\ \text{W}$

11. a. $V_S = \frac{V_PN_S}{N_P} = \frac{(60.0\ \text{V})(90\ 000)}{300} = 1.80 \times 10^{4}\ \text{V}$

    b. $I_P = \frac{V_SI_S}{V_P} = \frac{(1.80 \times 10^{4}\ \text{V})(0.50\ \text{A})}{60.0\ \text{V}}$

    $= 1.5 \times 10^{2}\ \text{A}$

CHAPTER 26

1. $Bqv = Eq$

$$v = \frac{E}{B} = \frac{4.5 \times 10^3\ \text{N/C}}{0.60\ \text{T}} = 7.5 \times 10^3\ \text{m/s}$$

2. $Bqv = \frac{mv^2}{r}$

$$r = \frac{mv}{Bq} = \frac{(1.67 \times 10^{-27}\ \text{kg})(7.5 \times 10^3\ \text{m/s})}{(0.60\ \text{T})(1.60 \times 10^{-19}\ \text{C})}$$
$$= 1.3 \times 10^{-4}\ \text{m}$$

3. $Bqv = Eq$

$$v = \frac{E}{B} = \frac{3.0 \times 10^3\ \text{N/C}}{6.0 \times 10^{-2}\ \text{T}} = 5.0 \times 10^4\ \text{m/s}$$

4. $Bqv = \frac{mv^2}{r}$

$$r = \frac{mv}{Bq} = \frac{(9.11 \times 10^{-31}\ \text{kg})(5.0 \times 10^4\ \text{m/s})}{(6.0 \times 10^{-2}\ \text{T})(1.60 \times 10^{-19}\ \text{C})}$$
$$= 4.7 \times 10^{-6}\ \text{m}$$

5. a. $Bqv = Eq$

$$v = \frac{E}{B} = \frac{6.0 \times 10^2\ \text{N/C}}{1.5 \times 10^{-3}\ \text{T}} = 4.0 \times 10^5\ \text{m/s}$$

b. $Bqv = \frac{mv^2}{r}$

$$m = \frac{Bqr}{v} = \frac{(0.18\ \text{T})(1.60 \times 10^{-19}\ \text{C})(0.165\ \text{m})}{4.0 \times 10^5\ \text{m/s}}$$
$$= 1.2 \times 10^{-26}\ \text{kg}$$

6. $m = \frac{B^2r^2q}{2V}$

$$= \frac{(5.0 \times 10^{-2}\ \text{T})^2(0.106\ \text{m})^2(2)(1.60 \times 10^{-19}\ \text{C})}{2(66.0\ \text{V})}$$
$$= 6.8 \times 10^{-26}\ \text{kg}$$

7. $m = \frac{B^2r^2q}{2V}$

$$= \frac{(7.2 \times 10^{-2}\ \text{T})^2(0.085\ \text{m})^2(1.60 \times 10^{-19}\ \text{C})}{2(110\ \text{V})}$$
$$= 2.7 \times 10^{-26}\ \text{kg}$$

8. Use $r = \frac{1}{B}\sqrt{\frac{2Vm}{q}}$ to find the ratio of radii of the two isotopes. If M represents the number of proton masses, then $\frac{r_{22}}{r_{20}} = \sqrt{\frac{M_{22}}{M_{20}}}$, so

$$r_{22} = r_{20}\left[\frac{22}{20}\right]^{1/2} = 0.056\ \text{m}$$

Separation then is

$$2(0.056\ \text{m} - 0.053\ \text{m}) = 6\ \text{mm}$$

CHAPTER 27

1. $K = -qV_0 = -(-1.60 \times 10^{-19}\ \text{C})(3.2\ \text{J/C})$
$= 5.1 \times 10^{-19}\ \text{J}$

2. $K = -qV_0 = \frac{-(-1.60 \times 10^{-19}\ \text{C})(5.7\ \text{J/C})}{1.60 \times 10^{-19}\ \text{J/eV}} = 5.7\ \text{eV}$

3. a. $c = f_0\lambda_0$

$$f_0 = \frac{c}{\lambda_0} = \frac{3.00 \times 10^8\ \text{m/s}}{310 \times 10^{-9}\ \text{m}} = 9.7 \times 10^{14}\ \text{Hz}$$

b. $hf_0 = (6.63 \times 10^{-34}\ \text{J/Hz})(9.7 \times 10^{14}\ \text{Hz})$

$$= (6.4 \times 10^{-19}\ \text{J})\left[\frac{\text{eV}}{1.60 \times 10^{-19}\ \text{J}}\right] = 4.0\ \text{eV}$$

c. $K_{max} = \frac{hc}{\lambda} - hf_0$

$$= \frac{\left[(6.63 \times 10^{-34}\ \text{J/Hz})(3.00 \times 10^8\ \text{m/s})\left(\frac{\text{eV}}{1.60 \times 10^{-19}\ \text{J}}\right)\right]}{(240 \times 10^{-9}\ \text{m})} - 4.0\ \text{eV}$$

$= 5.2\ \text{eV} - 4.0\ \text{eV} = 1.2\ \text{eV}$

4. a. $W = \text{work function} = hf_0 = \frac{1240\ \text{eV}\cdot\text{nm}}{\lambda_0}$

where λ_0 has units of nm and W has units of eV.

$$\lambda_0 = \frac{1240\ \text{eV}\cdot\text{nm}}{W} = \frac{1240\ \text{eV}\cdot\text{nm}}{1.96\ \text{eV}} = 633\ \text{nm}$$

b. $K_{max} = hf - hf_0 = E_{photon} - hf_0$

$$= \frac{1240\ \text{eV}\cdot\text{nm}}{\lambda} - hf_0$$
$$= \frac{1240\ \text{eV}\cdot\text{nm}}{425\ \text{nm}} - 1.96\ \text{eV}$$
$$= 2.92\ \text{eV} - 1.96\ \text{eV} = 0.96\ \text{eV}$$

5. a. $\frac{1}{2}mv^2 = qV_0$

$$v^2 = \frac{2qV}{m} = \frac{2(1.60 \times 10^{-19}\ \text{C})(250\ \text{J/C})}{9.11 \times 10^{-31}\ \text{kg}}$$
$$= 8.8 \times 10^{13}\ \text{m}^2/\text{s}^2$$

$v = 9.4 \times 10^6\ \text{m/s}$

b. $\lambda = \frac{h}{mv}$

$$= \frac{6.63 \times 10^{-34}\ \text{J}\cdot\text{s}}{(9.11 \times 10^{-31}\ \text{kg})(9.4 \times 10^6\ \text{m/s})}$$
$$= 7.7 \times 10^{-11}\ \text{m}$$

6. a. $\lambda = \frac{h}{mv} = \frac{6.63 \times 10^{-34}\ \text{J}\cdot\text{s}}{(7.0\ \text{kg})(8.5\ \text{m/s})} = 1.1 \times 10^{-35}\ \text{m}$

b. The wavelength is too small to show observable effects.

7. a. $p = \frac{h}{\lambda} = \frac{6.63 \times 10^{-34}\ \text{J} \cdot \text{s}}{5.0 \times 10^{-12}\ \text{m}}$
$= 1.3 \times 10^{-22}\ \text{kg} \cdot \text{m/s}$

b. Its momentum is too small to affect objects of ordinary size.

CHAPTER 28

1. Four times as large since orbit radius is proportional to n^2, where n is the integer labeling the level.

2. $r_n = n^2k$, where $k = 5.30 \times 10^{-11}$ m

$r_2 = (2)^2(5.30 \times 10^{-11}\ \text{m}) = 2.12 \times 10^{-10}$ m

$r_3 = (3)^2(5.30 \times 10^{-11}\ \text{m}) = 4.77 \times 10^{-10}$ m

$r_4 = (4)^2(5.30 \times 10^{-11}\ \text{m}) = 8.48 \times 10^{-10}$ m

3. $E_n = \frac{-13.6\ \text{eV}}{n^2}$

$E_2 = \frac{-13.6\ \text{eV}}{(2)^2} = -3.40$ eV

$E_3 = \frac{-13.6\ \text{eV}}{(3)^2} = -1.51$ eV

$E_4 = \frac{-13.6\ \text{eV}}{(4)^2} = -0.850$ eV

4. Using the results of Practice Exercise 3,

$E_3 - E_2 = (-1.51\ \text{eV}) - (-3.40\ \text{eV}) = 1.89$ eV

$\lambda = \frac{hc}{\Delta E} = \frac{(6.63 \times 10^{-34}\ \text{J} \cdot \text{s})(3.00 \times 10^8\ \text{m/s})}{(1.89\ \text{eV})(1.61 \times 10^{-19}\ \text{J/eV})}$

$= 6.54 \times 10^{-7}\ \text{m} = 654$ nm

5. $\frac{x}{0.075\ \text{cm}} = \frac{5 \times 10^{-9}\ \text{m}}{2.5 \times 10^{-15}\ \text{m}}$

$x = 200\ 000$ m or 200 km

6. a. $\Delta E = 8.82\ \text{eV} - 6.67\ \text{eV} = 2.15$ eV

b. $\Delta E = hf = 2.15\ \text{eV} \left[\frac{1.60 \times 10^{-19}\ \text{J}}{\text{eV}}\right]$

$= 3.44 \times 10^{-19}$ J

so $f = \frac{E}{h} = \frac{3.44 \times 10^{-19}\ \text{J}}{6.63 \times 10^{-34}\ \text{J} \cdot \text{s}} = 5.19 \times 10^{14}$ Hz

c. $c = f\lambda$, so

$\lambda = \frac{c}{f} = \frac{3.00 \times 10^8\ \text{m/s}}{5.19 \times 10^{14}/\text{s}}$

$= 5.78 \times 10^{-7}$ m, or 578 nm

CHAPTER 29

1. $\frac{\text{free e}^-}{\text{cm}^3}$

$= \frac{(2\ \text{e}^-/\text{atom})(6.02 \times 10^{23}\ \text{atoms/mol})(7.13\ \text{g/cm}^3)}{65.37\ \text{g/mol}}$

$= 1.31 \times 10^{23}$ free e^-/cm^3

2. atoms/cm^3

$= \frac{(6.02 \times 10^{23}\ \text{atoms/mol})(5.23\ \text{g/cm}^3)}{72.6\ \text{g/mol}}$

$= 4.34 \times 10^{22}$ atoms/cm^3

free e^-/atom $= \frac{(2 \times 10^{16}\ \text{free e}^-/\text{cm}^3)}{(4.34 \times 10^{22}\ \text{atoms/cm}^3)}$

$= 5 \times 10^{-7}$

3. There were 5×10^{-7} free e^-/Ge atom, so we need 5×10^3 as many As dopant atoms, or 3×10^{-3} As atom/Ge atom.

4. At $I = 2.5$ mA, $V_d = 0.7$ V, so

$V = V_d + IR = 0.7\ \text{V} + (2.5 \times 10^{-3}\ \text{A})(470\ \Omega)$

$= 1.9$ V

5. $V = V_d + IR = 0.4\ \text{V} + (1.2 \times 10^{-2}\ \text{A})(470\ \Omega)$

$= 6.0$ V

CHAPTER 30

1. $A - Z =$ neutrons

$234 - 92 = 142$ neutrons

$235 - 92 = 143$ neutrons

$238 - 92 = 146$ neutrons

2. $A - Z = 15 - 8 = 7$ neutrons

3. $A - Z = 200 - 80 = 120$ neutrons

4. ${}^1_1\text{H}$, ${}^2_1\text{H}$, ${}^3_1\text{H}$

5. ${}^{234}_{92}\text{U} \rightarrow {}^{230}_{90}\text{Th} + {}^4_2\text{He}$

6. ${}^{230}_{90}\text{Th} \rightarrow {}^{226}_{88}\text{Ra} + {}^4_2\text{He}$

7. ${}^{226}_{88}\text{Ra} \rightarrow {}^{222}_{86}\text{Rn} + {}^4_2\text{He}$

8. ${}^{214}_{82}\text{Pb} \rightarrow {}^{214}_{83}\text{Bi} + {}^{0}_{-1}\text{e} + {}^0_0\bar{\nu}$

9.

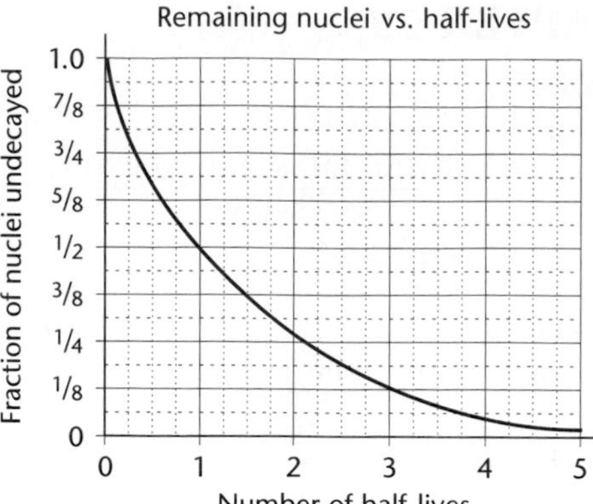

24.6 years = 2(12.3 years)

which is 2 half-lives. Since $\frac{1}{2} \times \frac{1}{2} = \frac{1}{4}$ there will be $(1.0 \text{ g})\left[\frac{1}{4}\right] = 0.25$ g remaining

10. Amount remaining = (original amount) $\left[\frac{1}{2}\right]^N$ where

N is the number of half-lives elapsed. Since

$$N = \frac{8 \text{ days}}{2.0 \text{ days}} = 4$$

$$\text{Amount remaining} = (4.0 \text{ g})\left[\frac{1}{2}\right]^4 = 0.25 \text{ g}$$

11. The half-life of $^{210}_{84}\text{Po}$ is 138 days.

There are 273 days or about 2 half-lives between September 1 and June 1. So the activity

$$= \left[2 \times 10^6 \frac{\text{decays}}{\text{s}}\right]\left[\frac{1}{2}\right]\left[\frac{1}{2}\right] = 5 \times 10^5 \text{ Bq}$$

12. From Table 30-1, 6 years is approximately 0.5 half-life for tritium. Since Figure 30-5 indicates that approximately $\frac{11}{16}$ of the original nuclei remain after 0.5 half-life, the brightness will be about $\frac{11}{16}$ of the original.

13. a. $E = mc^2 = (1.67 \times 10^{-27} \text{ kg})(3.00 \times 10^8 \text{ m/s})^2$

$= 1.50 \times 10^{-10}$ J

b. $E = \dfrac{1.50 \times 10^{-10} \text{ J}}{1.60 \times 10^{-19} \text{ J/eV}} = 9.38 \times 10^8 \text{ eV}$

$= 938$ MeV

c. The energy will be

(2)(938 MeV) = 1.88 GeV

CHAPTER 31

1. a.

6 protons = (6)(1.007825 u)	=	6.046950	u
6 neutrons = (6)(1.008665 u)	=	6.051990	u
total		12.098940	u
mass of carbon nucleus		−12.000000	u
mass defect		−0.098940	u

b. −(0.098940 u)(931.49 MeV/u) = −92.162 MeV

2. a. What is its mass defect?

1 proton	=	1.007825	u
1 neutron	=	1.008665	u
total		2.016490	u
mass of deuterium nucleus	=	−2.014102	u
mass defect		−0.002388	u

b. −(0.002388 u)(931.49 MeV/u) = −2.2 MeV

3. a.

7 protons = 7(1.007825 u)	=	7.054775	u
8 neutrons = 8(1.008665 u)	=	8.069320	u
total		15.124095	u
mass of nitrogen nucleus	=	−15.00011	u
mass defect of nitrogen nucleus	=	−0.12399	u

b. −(0.12399 u)(931.49 MeV/u) = −115.50 MeV

4. a.

8 protons = (8)(1.007825 u)	=	8.062600	u
8 neutrons = (8)(1.008665 u)	=	8.069320	u
total		16.131920	u
mass of oxygen nucleus		−15.99491	u
mass defect		−0.13701	u

b. −(0.13701 u)(931.49 MeV/u)

= −127.62 MeV

5. a. $^{14}_{6}\text{C} \rightarrow {}^{14}_{7}\text{N} + {}^{0}_{-1}\text{e}$

b. $^{55}_{24}\text{Cr} \rightarrow {}^{55}_{25}\text{Mn} + {}^{0}_{-1}\text{e}$

6. $^{238}_{92}\text{U} \rightarrow {}^{234}_{90}\text{Th} + {}^{4}_{2}\text{He}$

7. $^{214}_{84}\text{Po} \rightarrow {}^{210}_{82}\text{Pb} + {}^{4}_{2}\text{He}$

8. a. $^{210}_{82}\text{Pb} \rightarrow {}^{210}_{83}\text{Bi} + {}^{0}_{-1}\text{e}$

b. $^{210}_{83}\text{Bi} \rightarrow {}^{210}_{84}\text{Po} + {}^{0}_{-1}\text{e}$

c. $^{234}_{90}\text{Th} \rightarrow {}^{234}_{91}\text{Pa} + {}^{0}_{-1}\text{e}$

d. $^{239}_{93}\text{Np} \rightarrow ^{239}_{94}\text{Pu} + ^{0}_{-1}\text{e}$

9. Input masses
2.014102 u + 3.016049 u = 5.030151 u.

Output masses
4.002603 u + 1.008665 u = 5.011268 u.

Difference is –0.018883 u

Mass defect is –0.018883 u

Energy equivalent = –(0.0188883 u)(931.49 meV/u)
= –17.589 MeV

10. Positron mass

$$= (9.109 \times 10^{-31}\ \text{kg}) \left[\frac{1\ \text{u}}{1.6605 \times 10^{-27}\ \text{kg}}\right]$$
$$= 0.0005486\ \text{u}$$

Input mass: 4 protons = 4(1.007825 u)
= 4.031300 u

Output mass: $^{4}_{2}\text{He}$ + 2 positrons
= 4.002603 u + 2(0.0005486 u)
= 4.003700

Mass difference = 0.027600 u

Energy released = (0.027600 u)(931.49 MeV/u)
= 25.709 MeV

Appendix C
Equations

Chapter 4

$R^2 = A^2 + B^2 - 2AB\cos\theta$

$A_x = A\cos\theta$

$A_y = A\sin\theta$

$A = A_x + A_y$

Chapter 5

$\bar{v} = \frac{\Delta d}{\Delta t}$

$v = v_0 + at$

$d = d_0 + 1/2(v_0 + v)t$

$d = d_0 + v_0 t + 1/2at^2$

$v^2 = v_0{}^2 + 2a(d - d_0)$

Chapter 6

$F_{net} = ma$

$F_f = \mu F_N$

$T = 2\pi\sqrt{\frac{l}{g}}$

Chapter 7

$x = v_x t$

$y = v_y t + 1/2gt^2$

$a_c = \frac{v^2}{r} = \frac{4\pi^2 r}{T^2}$

Chapter 8

$\left(\frac{T_A}{T_B}\right)^2 = \left(\frac{r_A}{r_B}\right)^3$

$F = G\frac{m_A m_B}{d^2}$

Chapter 9

$p = mv$

$F\Delta t = \Delta p$

$p_{A1} + p_{B1} = p_{A2} + p_{B2}$

Chapter 10

$K = 1/2mv^2$

$W = \Delta K = Fd\cos\theta$

$P = \frac{W}{t}$

$MA = \frac{F_r}{F_e}$

$IMA = \frac{d_e}{d_r}$

$efficiency = \frac{MA}{IMA} \times 100\%$

Chapter 11

$U_g = mgh$

$E = K + U_g$

$E_{before} + U_{g\ before} = E_{after} + U_{g\ after}$

Chapter 12

$Q = mC\Delta T$

$Q = mH_f$

$Q = mH_v$

Chapter 13

$\frac{F_1}{A_1} = \frac{F_2}{A_2}$

$P = \frac{F}{A} = \rho hg$

$F_{buoyant} = \rho Vg$

$\alpha = \frac{\Delta L}{L_i \Delta T}$

Chapter 14

$v = \lambda f$

Chapter 16

$E = \frac{P}{4\pi d^2}$

Chapter 17

$n_i \sin\theta = n_r \sin\theta$

$n = \frac{c}{v}$

Chapter 18

$\frac{1}{f} = \frac{1}{d_i} + \frac{1}{d_o}$

$m = \frac{h_i}{h_o} = \frac{-d_i}{d_o}$

Chapter 19

$\lambda = \frac{xd}{L} = d\sin\theta$

Chapter 20

$F = K\frac{q_A q_B}{d^2}$

Chapter 21

$E = \frac{F_{on\ q'}}{q'}$

$\Delta V = Ed$

$W = q\Delta V$

$C = \frac{q}{\Delta V}$

Chapter 22

$P = IV$

$E = Pt$

$V = IR$

Chapter 23

$R = R_A + R_B + \ldots$

$\frac{1}{R} = \frac{1}{R_A} + \frac{1}{R_B} + \ldots$

Chapter 24

$F = BIL = Bqv$

Chapter 25

$EMF = BLv$

$I_{eff} = 0.707 I_{max}$

$V_{eff} = 0.707 V_{max}$

$\frac{I_s}{I_p} = \frac{V_p}{V_s} = \frac{N_p}{N_s}$

Chapter 26

$\frac{q}{m} = \frac{v}{Br} = \frac{2V}{B^2 r^2}$

Chapter 27

$E = nhf$

$K = hf - hf_0$

$p = \frac{h}{\lambda}$

Chapter 28

$r_n = \frac{h^2 n^2}{4\pi^2 Kmq^2}$

$E_n = -13.6\text{eV} \times \frac{1}{n^2}$

$\Delta E = hf$

$n\lambda = 2\pi r$

Chapter 31

$E = mc^2$

Appendix D
Tables

TABLE D-1
Color Conventions

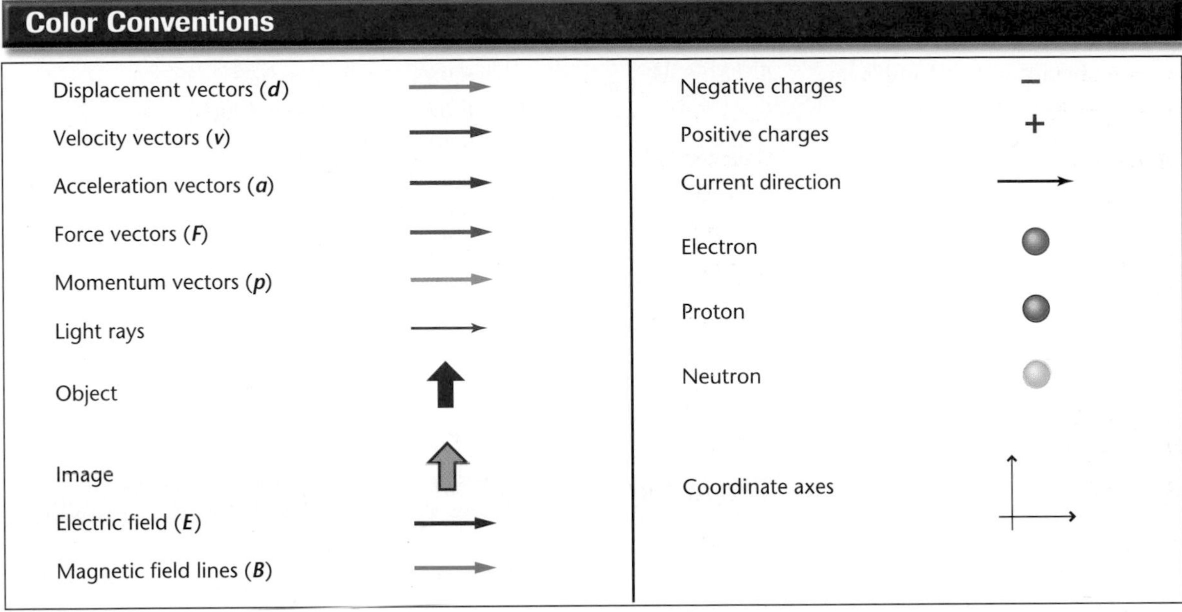

Displacement vectors (*d*)
Velocity vectors (*v*)
Acceleration vectors (*a*)
Force vectors (*F*)
Momentum vectors (*p*)
Light rays
Object
Image
Electric field (*E*)
Magnetic field lines (*B*)
Negative charges
Positive charges
Current direction
Electron
Proton
Neutron
Coordinate axes

TABLE D-2
Electric Circuit Symbols

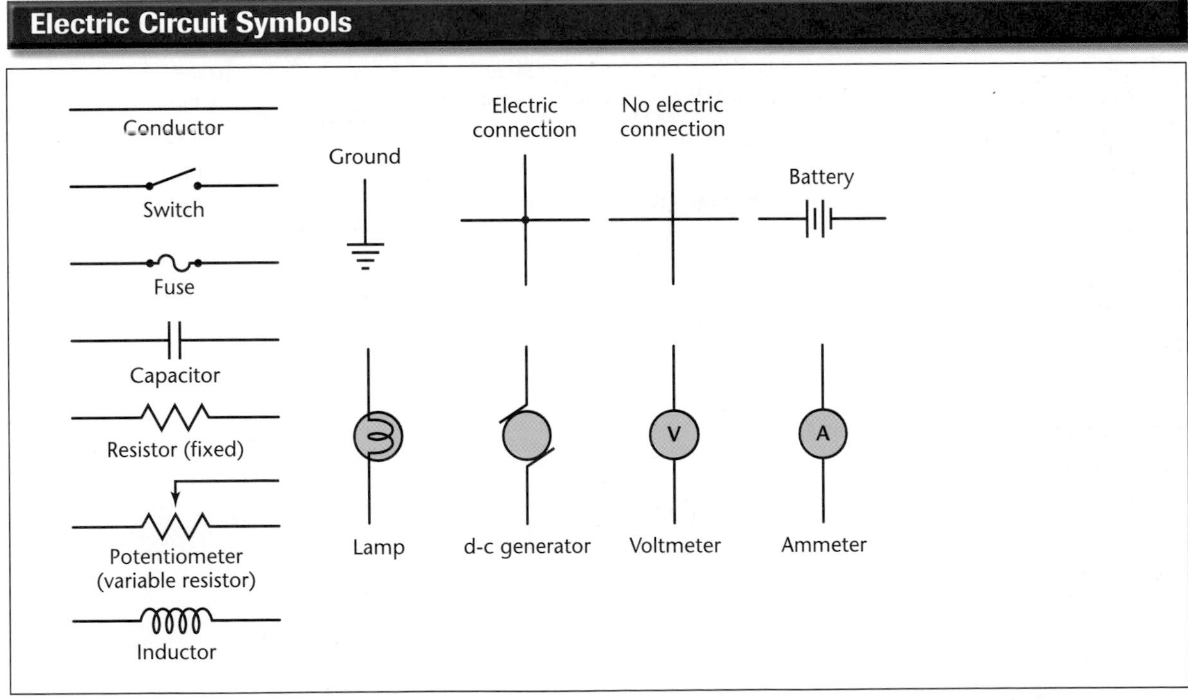

TABLE D-3

Physical Constants

Quantity	Symbol	Value
Atomic mass unit	u	1.661×10^{-27} kg
Avogadro's number	N_A	6.022×10^{23} particles/mole
Elementary charge	e	-1.602×10^{-19} C
Constant in Coulomb's law	K	8.988×10^{9} N·m^2/C^2
Gravitational constant	G	6.673×10^{-11} N·m^2/kg^2
Mass of an electron	m_e	9.109×10^{-31} kg
Mass of a proton	m_p	1.673×10^{-27} kg
Planck's constant	h	6.626×10^{-34} J·s
Speed of light in a vacuum	c	2.998×10^{8} m/s

TABLE D-4

Physical Data

Quantity	Value
Absolute zero temperature	−273.15° C
Acceleration due to gravity	9.801 m/s^2
Earth-moon distance	384×10^{3} km
Earth-sun distance (mean)	149.6×10^{6} km
Earth radius (mean)	6.38×10^{3} km
Heat of fusion of water	3.34×10^{5} J/kg
Heat of vaporization of water	2.26×10^{6} J/kg
Mass: Earth	5.97×10^{24} kg
Moon	7.36×10^{22} kg
Sun	1.99×10^{30} kg
Specific heat of water	4180 J/kg · K
Speed of sound in air (20°C)	343 m/s

TABLE D-5

Densities of Some Common Substances

Substance	Density at 20° C (g/cm^3)
Aluminum	2.70
Copper	8.96
Gold	19.32
Ice	0.99987
Iron	7.86
Lead	11.34
Silver	10.49
Ethyl alcohol	0.789
Mercury	13.534
Water (4°C)	1.000
Hydrogen	0.0000837
Oxygen	0.001331

TABLE D-6

The Elements

Element	Symbol	Atomic number	Atomic mass
Actinium	Ac	89	227.028
Aluminum	Al	13	26.982
Americium	Am	95	243.061
Antimony	Sb	51	121.757
Argon	Ar	18	39.948
Arsenic	As	33	74.922
Astatine	At	85	209.987
Barium	Ba	56	137.327
Berkelium	Bk	97	247.070
Beryllium	Be	4	9.012
Bismuth	Bi	83	208.980
Bohrium	Bh	107	(262)
Boron	B	5	10.811
Bromine	Br	35	79.904
Cadmium	Cd	48	112.411
Calcium	Ca	20	40.078
Californium	Cf	98	251.080
Carbon	C	6	12.011
Cerium	Ce	58	140.115
Cesium	Cs	55	132.905
Chlorine	Cl	17	35.453
Chromium	Cr	24	51.996
Cobalt	Co	27	58.933
Copper	Cu	29	63.546
Curium	Cm	96	247.070
Dubnium	Db	105	(262)
Dysprosium	Dy	66	162.50
Einsteinium	Es	99	252.083
Erbium	Er	68	167.26
Europium	Eu	63	151.965
Fermium	Fm	100	257.095
Fluorine	F	9	18.998
Francium	Fr	87	223.020
Gadolinium	Gd	64	157.25
Gallium	Ga	31	69.723
Germanium	Ge	32	72.61
Gold	Au	79	196.967
Hafnium	Hf	72	178.49
Hassium	Hs	108	(265)
Helium	He	2	4.003
Holmium	Ho	67	164.930
Hydrogen	H	1	1.008
Indium	In	49	114.82
Iodine	I	53	126.904
Iridium	Ir	77	192.22
Iron	Fe	26	55.847
Krypton	Kr	36	83.80
Lanthanum	La	57	138.906
Lawrencium	Lr	103	260.105
Lead	Pb	82	207.2
Lithium	Li	3	6.941
Lutetium	Lu	71	174.967
Magnesium	Mg	12	24.305
Manganese	Mn	25	54.938
Meitnerium	Mt	109	(266)
Mendelevium	Md	101	258.099
Mercury	Hg	80	200.59
Molybdenum	Mo	42	95.94
Neodymium	Nd	60	144.24
Neon	Ne	10	20.180
Neptunium	Np	93	237.048
Nickel	Ni	28	58.693
Niobium	Nb	41	92.906
Nitrogen	N	7	14.007
Nobelium	No	102	259.101
Osmium	Os	76	190.2
Oxygen	O	8	15.999
Palladium	Pd	46	106.42
Phosphorus	P	15	30.974
Platinum	Pt	78	195.08
Plutonium	Pu	94	244.064
Polonium	Po	84	208.982
Potassium	K	19	39.098
Praseodymium	Pr	59	140.908
Promethium	Pm	61	144.913
Protactinium	Pa	91	231.036
Radium	Ra	88	226.025
Radon	Rn	86	222.018
Rhenium	Re	75	186.207
Rhodium	Rh	45	102.906
Rubidium	Rb	37	85.468
Ruthenium	Ru	44	101.07
Rutherfordium	Rf	104	(261)
Samarium	Sm	62	150.36
Scandium	Sc	21	44.956
Seaborgium	Sg	106	(263)
Selenium	Se	34	78.96
Silicon	Si	14	28.086
Silver	Ag	47	107.868
Sodium	Na	11	22.990
Strontium	Sr	38	87.62
Sulfur	S	16	32.066
Tantalum	Ta	73	180.948
Technetium	Tc	43	97.907
Tellurium	Te	52	127.60
Terbium	Tb	65	158.925
Thallium	Tl	81	204.383
Thorium	Th	90	232.038
Thulium	Tm	69	168.934
Tin	Sn	50	118.710
Titanium	Ti	22	47.88
Tungsten	W	74	183.85
Uranium	U	92	238.029
Vanadium	V	23	50.942
Xenon	Xe	54	131.290
Ytterbium	Yb	70	173.04
Yttrium	Y	39	88.906
Zinc	Zn	30	65.39
Zirconium	Zr	40	91.224
Element 110*	Uun		
Element 111*	Uuu		
Element 112*	Uub		

*Names for elements 110-112 have not yet been approved by the IUPAC.

Appendix E
Safety Symbols

SAFETY SYMBOLS	HAZARD	EXAMPLES	PRECAUTION	REMEDY
Disposal	Special disposal considerations required	chemicals, broken glass, living organisms such as bacterial cultures, protists, etc.	Plan to dispose of wastes as directed by your teacher.	Ask your teacher how to dispose of laboratory materials
Biological	Organisms or organic materials that can harm humans	bacteria, fungus, blood, raw organs, plant material	Avoid skin contact with organisms or material. Wear dust mask or gloves. Wash hands thoroughly.	Notify your teacher if you suspect contact.
Extreme Temperature	Objects that can burn skin by being too cold or too hot	boiling liquids, hot plates, liquid nitrogen, dry ice, all burners	Use proper protection when handling. Remove flammables from area around open flames or spark sources.	Go to your teacher for first aid.
Sharp Object	Use of tools or glass-ware that can easily puncture or slice skin	razor blade, scalpel, awl, nails, push pins	Practice common sense behavior and follow guidelines for use of the tool.	Go to your teacher for first aid.
Fume	Potential danger to olfactory tract from fumes	ammonia, heating sulfur, moth balls, nail polish remover, acetone	Make sure there is good ventilation and never smell fumes directly.	Leave foul area and notify your teacher immediately.
Electrical	Possible danger from electrical shock or burn	improper grounding, liquid spills, short circuits	Double-check setup with instructor. Check condition of wires and apparatus.	Do not attempt to fix electrical problems. Notify your teacher immediately.
Irritant	Substances that can irritate the skin or mucus membranes	pollen, mothballs, steel wool, potassium permanganate	Dust mask or gloves are advisable. Practice extra care when handling these materials.	Go to your teacher for first aid.
Corrosive	Substances (Acids and Bases) that can react with and destroy tissue and other materials	acids such as vinegar, hydrochloric acid, hydrogen peroxide; bases such as bleach, soap, sodium hydroxide	Wear goggles and an apron.	Immediately begin to flush with water and notify your teacher.
Toxic	Poisonous substance that can be acquired through skin absorption, inhalation, or ingestion	mercury, many metal compounds, iodine, Poinsettia leaves	Follow your teacher's instructions. Always wash hands thoroughly after use.	Go to your teacher for first aid.
Flammable	Flammable and Combustible materials that may ignite if exposed to an open flame or spark	alcohol, powders, kerosene, potassium permanganate	Avoid flames and heat sources. Be aware of locations of fire safety equipment.	Notify your teacher immediately. Use fire safety equipment if applicable.

Eye Safety
This symbol appears when a danger to eyes exists.

Clothing Protection
This symbol appears when substances could stain or burn clothing.

Animal Safety
This symbol appears when safety of live animals and students must be ensured.

Glossary

absolute zero (12.1): The temperature at which no further thermal energy can be removed from an object; usually shown as –273° C.
absorption spectrum (28.1): A characteristic set of light wavelengths absorbed by a material, which can be used to indicate the composition of the material.
acceleration due to gravity (5.4): The acceleration of an object in free fall resulting from Earth's gravity.
accuracy (2.2): How well the results of an experiment agree with the measured and accepted value.
achromatic lens (18.2): A lens for which all light colors have the same focal length.
activity (30.1): The number of decays per second, or decay rate, of a radioactive substance.
adhesion (13.1): The electromagnetic force of attraction that acts between particles of different substances.
agent (6.1): A specific, immediate, identifiable cause of a force.
algebraic representation (4.1): Representation of a vector with an italicized letter in boldface type, which is often used in printed materials.
alpha decay (30.1): The radioactive decay process in which the nucleus of an atom emits an alpha particle.
alpha (α) particle (28.1): Positively-charged particles consisting of two protons and two neutrons emitted by radioactive materials.
ammeter (23.2): A device that measures current.
amorphous solids (13.2): Solids with definite volume and shape but no regular crystal structure.
ampere (A) (22.1): The base SI unit of current; one coulomb per second.
amplitude (6.2): The maximum distance an object moves from equilibrium in any periodic motion.
angle of refraction (17.1): The angle the refracted ray makes with the normal to the surface.
angular momentum (9.1): Quantity of motion used with objects rotating about a fixed axis.
antenna (26.2): A device, usually a wire or rod, that generates electromagnetic waves when connected to an alternating current (AC) source.
antinode (14.2): The point of largest amplitude when two wave pulses meet of two superimposed waves.
antiparticle (30.2): A particle of antimatter.
apparent weight (6.2): The weight of an object that is sensed as a result of contact forces on it.
Archimedes' principle (13.1): The buoyant force on an object immersed in a fluid is equal to the weight of the fluid displaced by the object.
armature (24.2): A loop of wire in an electric motor that is mounted on a shaft or axle.
atomic mass unit (u) (30.1): Unit of mass equal to 1/12 the mass of the carbon-12 isotope.
atomic number (Z) (30.1): The number of protons in the nucleus of an atom.
average acceleration (3.3): The change in average velocity divided by time.
average speed (3.3): The ratio of the total distance traveled to the time interval.
average velocity (3.3): The ratio of the change of position to the time interval during which that change occurred.

band theory (29.1): The theory that electrical conduction in solids can be explained in terms of energy bands and forbidden gaps.
base units (2.1): The seven fundamental SI units of measure.
battery (22.1): A group of several voltaic or galvanic cells connected together to convert chemical energy to electric energy.
beat (15.2): Oscillation of wave amplitude that occurs as a result of the superposition of two sound waves having nearly identical frequencies.
Bernoulli's principle (13.1): The pressure exerted by a fluid decreases as its velocity increases.
beta decay (30.1): The process of radioactive decay that occurs when a neutron is changed to a proton within the nucleus of an atom, and a beta particle and an antineutrino are emitted.
binding energy (31.1): The energy equivalent of the mass defect; it is converted to thermal energy in a fission reaction.
Bohr model (28.1): A model of an atom with a central nucleus and electrons in specific quantized energy levels from which they can make transitions, emitting or absorbing electromagnetic radiation.
boiling point (12.2): The specific temperature at which added thermal energy causes a substance to change from a liquid to a gaseous state.
breeder reactor (31.2): A reactor that generates more fissionable fuel than it consumes.
buoyant force (13.1): The upward force exerted on an object immersed in fluid.

calorimeter (12.1): A device that provides a closed, isolated system with which to measure changes in the thermal energy of a substance.
candela (cd) (16.1): The SI base unit of luminous intensity; candle power.
capacitance (21.2): The ratio of an object's stored charge to its potential difference; measured in farads.
capacitor (21.2): A device with a specific capacitance that is used in electrical circuits to store charge.
capillary action (13.1): The rise of liquid in a narrow tube that occurs because the adhesive forces between glass and liquid molecules are stronger than the cohesive forces between liquid molecules.
centripetal acceleration (7.3): The center-directed acceleration of an object in uniform circular motion.
centripetal force (7.3): The necessary net force exerted in the centripetal direction to cause centripetal acceleration.
chain reaction (31.2): Continual fission reactions that result from the release of neutrons from previous fission reactions.
charging by conduction (20.2): Charging a neutral object by touching it with a charged object.
charging by induction (20.2): Charging a neutral object by bringing it close to a charged object, causing a separation of charges, then removing the object to be charged, trapping opposite but equal charges.
chromatic aberration (18.2): Variation in focal length of a lens with the wavelength of light.
circuit breaker (23.2): An automatic switch that opens, stopping all current, when the circuit is overloaded.
closed system (9.2): A collection of objects that does not gain or lose mass.
closed-pipe resonator (15.2): A hollow tube with one closed end and a sound source at the other end.
coefficient of linear expansion (13.2): The length change divided by the original length and by temperature change.
coefficient of volume expansion (13.2): The volume change divided by original volume and by temperature change.

Glossary

coherent light (28.2): Light with the minima and maxima of the waves coinciding (in step).
coherent waves (19.1): Waves that are in phase—the wave crests and troughs reach the same point at the same time.
cohesive forces (13.1): The electromagnetic forces of attraction that like particles exert on one another.
combination series-parallel circuit (23.2): A complex circuit that is a combination of series and parallel branches.
complementary color (16.2): Two colors of light which when added together produce white light.
components (4.2): Scalar projections of the component vectors, with positive or negative signs, indicating their directions.
compound machine (10.2): A machine consisting of two or more simple machines.
Compton effect (27.1): The shift in the energy of scattered photons.
concave lens (18.2): A diverging lens, thicker at its outer edge than at its center.
concave mirror (18.1): A mirror that reflects light from its inwardly curved surface and produces either inverted, real images or upright, virtual images.
condensation (13.1): The process where particles of liquid that have evaporated into the air return to the liquid phase due to a decrease in kinetic energy or temperature.
conduction (12.1): The process that transfers kinetic energy when particles collide.
conduction band (29.1): In the atom of a solid, the lowest band that is not filled to capacity with electrons.
conductor (20.1): A material that allows electrical charges to move about easily. Charges added to a conductor spread over its entire surface.
consonance (15.2): A pleasant set of pitches.
constant acceleration (5.3): Acceleration that does not change over time.
constructive interference (14.2): Superposition of waves resulting in increased wave displacement.
contact force (6.1): A force exerted on an object only when touching it.
continuous wave (14.1): A regularly repeating sequence of wave pulses.
control rod (31.2): A cadmium rod that is moved in and out of a pressurized water reactor to control the rate of chain reaction.
controlled fusion (31.2): A type of fusion that could safely provide Earth with a nearly limitless energy source and no radioactive waste formation.
convection (12.1): The transfer of heat by means of motion in a fluid.
conventional current (22.1): The flow of positive charges.
convex lens (18.2): A converging lens, thinner at its outer edge than at its center.
convex mirror (18.1): A spherical mirror that reflects light from its outer surface and produces virtual, reduced, upright images.
coordinate system (3.2): A system used to describe motion that indicates where the zero point of the variable being studied is located and the direction in which the values of the variable increase.
coulomb (C) (20.2): The SI unit of electrical charge.
Coulomb's law (20.2): The magnitude of a force between two charges varies directly with the magnitude of the charges and inversely with the square of the distance between them.
crest (14.1): The high point of a wave.
critical angle (17.2): The incident angle unique to a substance that causes the refracted ray to lie along the boundary of the substance.
crystal lattice (13.2): A fixed, regular arrangement of atoms.

de Broglie wavelength (27.2): The length of de Broglie wave of a particle.
decibel (dB) (15.1): Unit of sound level.
depletion layer (29.2): In a *pn*-junction diode, the area that is lacking charge carriers and thus is a poor conductor of electricity.
derived units (2.1): Unit of a quantity, that is a combination of base units.
destructive interference (14.2): Superposition of waves with opposite but equal amplitudes.
diffraction (14.2): The bending of waves around a barrier.
diffraction grating (19.2): A device with parallel ridges that reflects light and forms an interference pattern.
diffuse reflection (17.1): The scattered, fuzzy reflection from a rough surface where light is randomly reflected.
diode (29.2): Semiconductor device used to produce current in only one direction.
dispersion (17.2): Variation of the speed of light through matter resulting in separation of light into a spectrum.
displacement (3.2): The vector quantity that defines the distance and direction between two positions.
dissonance (15.2): An unpleasant set of pitches.
distance (3.2): A scalar quantity that is the length, or size, of the displacement vector.
domain (24.1): A small group, usually 10 to 1000 microns, that is formed when the magnetic fields of the electrons in a group of neighboring atoms combine together.
dopant (29.1): An impurity atom that increases the conductivity of a semiconductor by adding either holes or electrons.
Doppler shift (15.1): A change in sound frequency due to the relative motion of either the source or the detector.
dye (16.2): A molecule that absorbs some light wavelengths and reflects or transmits others.

eddy currents (25.2): Currents that produce a magnetic field that opposes the motion that caused the currents. They are generated when a piece of metal moves through a magnetic field or when a metal loop is placed in a changing magnetic field.
efficiency (10.2): The ratio of output work to input work.
effort force (F_e) (10.2): The force exerted on a machine.
elastic collision (11.2): A collision in which the kinetic energy remains unchanged.
elastic potential energy (11.1): The potential energy stored in an object that is released as kinetic energy when the object undergoes a change in form or shape.
elasticity (13.2): The ability of a solid object to return to its original form after external forces are exerted on it.
electric circuit (22.1): A continuous path through which electric charges can flow.
electric current (22.1): The flow of charged particles.
electric field (21.1): A vector quantity that relates the force exerted on a charge to the size of the charge..
electric field lines (21.1): The lines providing a picture of the electric field, showing direction and strength.
electric generator (25.1): A device that converts mechanical energy to electrical energy.
electric motor (24.2): A device that converts electrical energy to kinetic energy.
electric potential difference (21.2): In an electric field, the change in potential energy per unit charge.

electromagnet (24.1): A current-carrying coil with a north and south pole that is itself a magnet.
electromagnetic induction (25.1): The generation of current through a circuit due to the relative motion between a wire and a magnetic field.
electromagnetic radiation (26.2): Energy carried through space in the form of electromagnetic waves.
electromagnetic wave (26.2): A wave consisting of coupled changing magnetic and electric fields that moves through space at the speed of light.
electromotive force (*EMF*) (25.1): The potential difference that is produced by electromagnetic induction.
electron cloud (28.2): The region in which there is a high probability of finding an electron.
electroscope (20.2): A device used to determine electrical force.
electrostatics (20.1): The study of electrical charges at rest.
elementary charge (20.2): The magnitude of the charge of an electron or proton, 1.60×10^{-19} C.
elementary particles (30.2): The three families of particles (quarks, leptons, and force carriers) that appear to make up all matter in the universe and through which matter interacts with other matter.
emission spectrum (28.1): The characteristic set of light wavelengths emitted by an atom.
energy (10.1): The property of an object that allows it to produce change in the environment or in itself.
energy level (28.1): The quantized amount of energy that an electron may have.
enrichment (31.2): The process that increases the number of fissionable nuclei.
entropy (12.2): A measure of the disorder of a system.
environment (6.1): The world around an object that exerts contact forces or long-range forces on it.
equilibrant (7.1): A single, additional force that is exerted on an object to produce equilibrium, which is the same magnitude as the resultant force but opposite in direction.
equilibrium (6.1): The condition where the net force on an object is zero.
equipotential (21.2): The potential difference of zero between two or more positions in an electric field.
equivalent resistance (23.1): In a series circuit, the sum of individual resistances.
erect image (18.1): A mirror image that points in the same direction as the reflected object.
evaporation (13.1): The change from liquid to vapor state.
excited state (28.1): A higher energy level reached by an electron when it absorbs energy.
external forces (9.2): All the forces outside a closed system.
extrinsic semiconductor (29.1): A semiconductor with greatly enhanced conductivity due to the addition of impurity atoms (dopants).

F

factor-label method (2.1): The method of converting a quantity expressed in one unit to that quantity in another unit.
fast neutron (31.2): A neutron that is emitted by the fission of uranium-235 atoms.
first law of thermodynamics (12.2): The total increase in the thermal energy of a system is the sum of the heat added to it and the work done on it.
first right-hand rule (24.1): The method used to determine the direction of the magnetic field around a current-carrying wire.
fission (31.2): Division of a nucleus into two or more fragments, resulting in a release of neutrons and energy.
flight time (7.2): The amount of time that a projectile is in the air.
fluid (13.1): A material that flows and has no definite shape.
focal length (18.1): The distance from the focal point to the mirror along the principal axis.
focal point (18.1): The point where parallel light rays converge or appear to diverge after reflecting from a mirror or refracting from a lens.
forbidden gap (29.1): The region that separates bands of energy levels in solids by values of energy that no electrons possess.
force (6.1): A push or pull exerted on an object having magnitude and direction; it may be either a contact or long-range force.
force carriers (30.2): Elementary particles—photons, gluons, weak bosons, or yet-undetected gravitrons—that carry forces between matter.
force of gravity (6.1): An attractive, long-range force that exists between all masses.
free-body diagram (6.1): A diagram that shows the direction of the force operating on each object.
frequency (f) (14.1): In any periodic motion, the number of complete oscillations measured in hertz.
fundamental (15.2): The lowest frequency of sound produced by a musical instrument.
fuse (23.2): A short piece of metal that will melt, or blow, if too large a current passes through it and that acts as a safety device by stopping current to the entire circuit.
fusion (31.2): The process in which nuclei with small masses are combined to form a nucleus with a larger mass.

G

galvanometer (24.2): A device used to measure very small currents.
gamma decay (30.1): The radioactive process of decay that takes place when the nucleus of an atom emits a gamma ray.
grand unification theory (30.2): A theory in the process of development to unify strong force and electroweak force into one force.
graphical representation (4.1): An arrow or arrow-tipped line segment that symbolizes a vector quantity with a specified length and direction.
gravitational force (8.1): The attractive force between all masses.
gravitational mass (8.2): Ratio of gravitational mass to an objects acceleration.
gravitational potential energy (11.1): The stored energy in a system resulting from gravitational interaction between masses.
ground state (28.1): The lowest energy level occupied by an electron.
ground-fault interrupter (23.2): A device containing an electronic circuit that detects current differences caused by an extra current path; it opens the circuit and prevents electric shocks.
grounding (21.2): Removing excess charge from a charged body by connecting it to Earth.

H

half-life (30.1): The amount of time required for half the nuclei of a given quantity of a radioactive isotope to decay.
harmonics (15.2): Higher frequencies of sound that are multiples of the fundamental frequency produced by a musical instrument.
heat (12.1): The energy transferred between two objects because of a difference in temperature.

Glossary

heat engine (12.2): A device that converts thermal energy to mechanical energy.
heat of fusion (12.2): The amount of energy required to melt one kilogram of a substance.
heat of vaporization (12.2): The amount of thermal energy required to vaporize one kilogram of a liquid.
Heisenberg uncertainty principle (27.2): It is not possible to precisely measure both the position and momentum of a particle of light or matter at the same time.
hole (29.1): Absence of an electron in a semiconductor.

ideal mechanical advantage (*IMA*) (10.2): The ratio of the effort distance to the resistance distance.
illuminance (16.1): The rate at which light falls on a surface; measured in lux.
illuminated (16.1): A body that reflects light waves produced by an outside source.
image (18.1): Reproduction of object formed with mirrors or lenses.
impulse (9.1): The product of the average net force exerted on an object and the time interval over which the force acts.
impulse-momentum theorem (9.1): The impulse given an object equals its change in momentum.
incident wave (14.2): A wave that strikes the boundary between two media.
incoherent light (28.2): Light with the minima and maxima of the waves not necessarily coinciding.
index of refraction (17.1): The ratio of the speed of light in a vacuum to its speed in a material.
inelastic collision (11.2): A collision in which the kinetic energy decreases.
inertia (6.1): The tendency of an object either at rest or moving at a constant speed to resist changing velocity.
inertial confinement fusion (31.2): A controlled fusion that uses deuterium, tritium, compression, and lasers.
inertial mass (8.2): The ratio of the net force exerted on an object to its acceleration.
instantaneous acceleration (5.3): The acceleration of an object at a particular instant of time.
instantaneous velocity (3.3): The speed and direction of an object at a particular instant in time.
insulator (20.1): A material through which electrical charges do not easily move.
interaction pair (6.3): A force pair composed of two forces that are opposite in direction and equal in magnitude.
interference (14.2): The interaction of two or more waves.
interference fringes (19.1): The pattern of dark and light bands on a screen due to constructive and destructive interference of light waves passing through two narrow, closely spaced slits.
internal forces (9.2): All the forces within a closed system.
intrinsic semiconductor (29.1): A pure semiconductor that carries charge as a result of thermally freed electrons and holes.
inverse relationship (2.3): Mathematical relationship between two variables, x and y, summarized by the equation $xy = k$, where k is a constant.
isolated system (9.2): A closed system on which the net external force is zero.
isotopes (26.1): Atomic nuclei having the same number of protons but different numbers of neutrons.

joule (10.1): SI unit of energy equal to one newton-meter.

kelvin (12.1): An interval on the Kelvin scale, equal to the size of one Celsius degree.
Kepler's laws of planetary motion (8.1): Three mathematical laws describing the behavior of all planets and satellites that state that the planets move in elliptical orbits, that they sweep out equal areas in equal time intervals, and that the square of the ratio of the periods of any two planets equals the cube of the ratio of their average distances from the sun.
kilogram (kg) (2.1): The SI base unit of mass.
kilowatt-hour (kWh) (22.2): Energy unit; the amount of energy equal to 1000 watts delivered continuously for one hour (3600 seconds).
kinetic energy (10.1): Energy of an object due to motion.
kinetic friction force (6.2): The force that opposes relative motion between surfaces in contact.
kinetic-molecular theory (12.1): Description of matter as being made up of tiny particles in random, constant motion.

laser (28.2): A device that generates coherent, directional, monochromatic light.
law of conservation of energy (11.2): The energy in a closed, isolated system is constant.
law of conservation of momentum (9.2): The momentum remains the same for any closed system upon which there is no net external force.
law of reflection (14.2): The angle of incidence is equal to the angle of reflection.
law of universal gravitation (8.1): Gravitational force between any two objects is directly proportional to the product of their masses and inversely proportional to the square of the distance between their centers.
lens (18.2): A transparent optical device, with a larger refractive index than air, used to converge or diverge light.
lens/mirror equation (18.1): $1/f = 1/d_i + 1/d_o$, where f is the focal length, d_i is the image distance, and d_o is the object distance.
Lenz's law (25.2): The magnetic field resulting from induced current opposes the change in the field that caused the induced current.
leptons (30.2): Elementary particles like electrons or antineutrinos; leptons and quarks appear to compose all the matter in the universe.
lever arm (7.3): The perpendicular distance from the axis of rotation to a line along which the force acts.
light (16.1): Electromagnetic radiation with wavelengths from 400 to 700 nm.
linear momentum (9.1): The product of mass and velocity of an object.
linear relationship (2.3): Relationship between two variables, x and y, summarized by the equation $y = ax + b$, where a and b are constant.
longitudinal wave (14.1): A wave that displaces matter parallel to the direction of wave motion.
long-range force (6.1): A force that acts on an object without touching it.
lumen (16.1): Unit of luminous flux.
luminous (16.1): A body that emits light waves.
luminous flux (16.1): The rate at which light is emitted from a light source; measured in lumens.
luminous intensity (16.1): The luminous flux that falls on one square meter of a sphere one meter in radius.
lux (16.1): Unit of illuminance; lumens per square meter.

machine (10.2): A device that changes the magnitude or the direction of a force needed to do work, making the task easier to accomplish.
magnetic field (24.1): A vector quantity that relates the force exerted on a magnet or current-carrying wire to the strength of the magnet or the size of the current.
magnetic flux (24.1): The number of magnetic field lines that pass through a surface.
magnification (18.1): The optical enlargement of an object; the ratio of the size of the image to the size of the object.
mass defect (31.1): The difference between the sum of the masses of individual nucleons and the actual mass.
mass number (*A*) (30.1): The sum of neutrons and protons in the nucleus of an atom.
mass spectrometer (26.1): A device that uses both magnetic and electric fields to precisely measure the masses of ionized atoms and molecules.
maximum height (7.2): The height of a projectile when the vertical velocity is zero.
mechanical advantage (*MA*) (10.2): The ratio of resistance force to effort force.
mechanical energy (11.2): The sum of kinetic energy and gravitational potential energy in a given system.
mechanical resonance (6.2): Condition at which the natural oscillation frequency equals the frequency of the driving force.
melting point (12.2): The temperature at which added thermal energy overcomes the forces holding the particles of a substance together, causing the substance to change from a solid to a liquid state.
meter (m) (2.1): The SI base unit of length.
metric system (2.1): A set of standards of measurement where units of different sizes are related by powers of 10.
moderator (31.2): A material that causes fast neutrons to lose speed when the neutrons collide with it.
monochromatic light (19.1): Light of only one wavelength.
motion diagram (3.1): A tool for the study of motion that uses a series of images to show the position of a moving object after equal time intervals.
mutual inductance (25.2): Effect that occurs in a transformer when a varying magnetic field created in the primary coil is carried through the iron core to the secondary coil, where the varying field induces a varying *EMF.*

***n*-type semiconductor** (29.1): A semiconductor that conducts by means of free electrons.
net force (6.1): The vector sum of all of the forces on an object.
neutral (20.1): An atom whose positively charged nucleus exactly balances the negatively charged surrounding electrons; no net electrical charge.
neutrino (30.2): A neutral, massless particle that carries momentum and energy and is emitted with the beta particle during beta decay.
Newton's first law (6.1): If a system has no net force on it, then its velocity will not change.
Newton's second law (6.1): Acceleration of an object equals the net force on that object divided by its mass.
Newton's third law (6.3): All forces come in pairs that are equal in magnitude and opposite in direction.
node (14.2): The stationary point where two equal wave pulses meet and displacement is zero.
nuclear model (28.1): An atom is mostly empty space with a tiny, massive positively-charged central nucleus.
nuclear reaction (30.1): Reaction in which the number of protons or neutrons in the nucleus of an atom changes.
nucleon (31.1): Either a proton or a neutron.
nuclide (30.1): The nucleus of an isotope.

object (optics) (18.1): A source of diverging light rays; may be luminous or illuminated.
octave (15.2): Two musical notes with frequencies related by a 1:2 ratio.
opaque (16.2): A material that absorbs or reflects light, not allowing objects to be seen through it.
open-pipe resonator (15.2): A hollow tube, with both ends open, that resonates with a sound source.
operational definition (3.1): Defines a concept in terms of the procedure or operation used.
optically dense (17.1): Materials with larger indices of refraction with respect to other materials.
origin (3.2): The point in a coordinate system at which the variables have a zero value.

***p*-type semiconductor** (29.1): A semiconductor that conducts by means of holes.
pair production (30.2): The conversion of energy into matter-antimatter particle pairs.
parallax (2.2): The apparent shift in position of an object when it is seen from different angles.
parallel circuit (23.1): A circuit with several current paths, whose total current equals the sum of the currents in its branches.
parallel connection (22.1): The arrangement of electric devices in a circuit where there is more than one current path.
particle model (3.1): A simplified version of a motion diagram in which the object in motion is replaced by a series of single points.
pascal (Pa) (13.1): The SI unit of pressure; one newton per square meter.
Pascal's principle (13.1): Any change in applied pressure on a confined fluid is transmitted undiminished through the fluid.
period (6.2): In any periodic motion, the time needed to repeat one complete cycle of motion.
photoelectric effect (27.1): The emission of electrons produced when electromagnetic radiation falls on certain metals.
photon (27.1): A light quantum that is massless, has energy and momentum, and travels at the speed of light.
photovoltaic cell (22.1): A solar cell that changes light energy into electric energy.
physics (1): The study of matter and energy and their relationships.
piezoelectricity (26.2): Property of quartz crystals, which, when bent or deformed by an applied voltage, will vibrate at a specific frequency and generate an *EMF*.
pigment (16.2): A colored material that absorbs certain colors and reflects or transmits other colors.
pitch (15.1): The frequency of a sound wave.
plane mirror (18.1): A flat, smooth surface that reflects light rays by regular reflection, not by diffuse reflection. Forms a virtual, erect image the same size as the object and the same distance behind the mirror as the object is in front.
plasma (13.2): The gaslike state of matter made up of positively charged ions or negatively charged electrons or a mixture of them.

Glossary

***pn*-junction diode** (29.2): A semiconductor device having an *n*-side with a net positive charge, a *p*-side with a net negative charge, and a region around the junction without charge carriers.
polarized (16.2): Light consisting of waves that vibrate on a specific plane. (24.1): The quality of having two opposite magnetic poles, one south-seeking and one north-seeking.
position vector (3.2): The arrow on a motion diagram that is drawn from the origin to the moving object.
positron (30.2): An antielectron.
potentiometer (22.1): A variable resistor that allows continuous, rather than step-by-step, changes in current in an electric circuit; also called a rheostat.
power (10.1): The rate of doing work; the rate at which energy is transferred.
precision (2.2): The degree of exactness with which a quantity is measured using a given instrument.
pressure (13.1): The force applied to a surface.
primary coil (25.2): An insulated transformer coil that creates a varying magnetic field when it is connected to an alternating-current (AC) voltage source.
primary color (16.2): Color from which other colors can be derived.
primary pigment (16.2): A pigment that absorbs only one color from white light.
principal axis (18.1): A straight line perpendicular to the surface of a spherical mirror at its center. (18.2) Line perpendicular to the plane of a lens passing through its center.
principle of superposition (14.2): The displacement of a medium caused by two or more waves is the algebraic sum of the displacements of the individual waves.
problem solving strategy (3.3): A stepwise procedure for solving problems that can involve sketching the problem, calculating the answer, and checking the answer.
projectile (7.2): An object with independent vertical and horizontal motions that moves through the air only under the force of gravity after an initial thrust.

Q

quadratic relationship (2.3): Parabolic relationship between two variables that exists where one variable depends on the square of another.
quantized (27.1): Small but measurable increments.
quantum mechanics (28.2): The study of the properties of matter using its wave properties.
quantum model (28.2): A model that predicts only the probability that an electron is at a specific location.
quantum number (28.1): The integer ratio of energy to its quantum increment.
quarks (30.2): Elementary particles that make up protons, mesons, and neutrons, and that together with leptons appear to make up all matter in the universe.

R

radiation (12.1): The transfer of energy by electromagnetic waves.
radioactive (30.1): A material that undergoes radioactive decay and emits radiation.
range (7.2): The horizontal distance traveled by a projectile.
ray model (16.1): Light moves in a straight-line path through a medium.
Rayleigh criterion (19.2): When the central bright band of one star falls on the first dark band of the second, the two stars will be just resolved.
real image (18.1): An optical image formed when light rays converge and pass through the image, producing an image that can be viewed on paper or projected onto a screen.
receiver (26.2): A device used for reception of electromagnetic waves.
reference level (11.1): An arbitrary position selected to solve a problem.
reflected wave (14.2): A returning wave that is either inverted or displaced in the same direction as the incident wave.
refraction (14.2): A change in the direction of waves crossing a boundary between two different media.
regular reflection (17.1): Reflection off a smooth surface, such as a mirror, where light is reflected back to the observer in parallel beams, producing a clear image.
resistance (22.1): The ratio of the voltage across a device to the current through it.
resistance force (F_r) (10.2): The force exerted by a machine.
resistors (22.1): Devices with a specific resistance.
resultant vector (4.1): The sum of two or more vectors.
rigid rotating object (7.3): A mass that rotates around its own axis.

S

scalar quantity (3.2): A quantity that has only magnitude.
schematic (22.1): A diagram of an electric circuit that uses standard symbols for circuit elements.
scientific method (1): Systematic problem-solving method of observing, experimenting, and analyzing.
scientific notation (2.1): The expression of numbers as powers of 10.
scintillation (28.1): The small flash of light emitted when a substance is struck by a radioactive particle.
second (s) (2.1): The SI base unit of time.
second law of thermodynamics (12.2): States that the entropy of the universe is always maintained or increased.
second right-hand rule (24.1): The method used to determine the direction of the field produced by an electromagnet.
secondary coil (25.2): An insulated transformer coil in which varying *EMF* is induced.
secondary color (16.2): Color formed by a pair of primary colors.
secondary pigment (16.2): A pigment that absorbs two primary colors from white light and reflects one.
self-inductance (25.2): The induction of *EMF* in a wire carrying a changing current.
series circuit (23.1): A circuit in which current passes through each device, one after another.
series connection (22.1): The connection of electric devices in a circuit where there is only one current path.
short circuit (23.2): Low-resistance connection between two points.
SI (2.1): (Système Internationale d'Unités) International standards of measurement adapted from the metric system.
significant digits (2.2): All valid digits in a measurement.
simple harmonic motion (6.2): Motion that returns an object to its equilibrium position as a result of a restoring force that is directly proportional to the object's displacement.
slope (2.3): The ratio of vertical change to horizontal change of a graph.
slow neutron (31.2): A lower-energy neutron created when a fast neutron collides with a moderator.
Snell's law (17.1): The ratio of the sine of the angle of incidence to the sine of the angle of refraction is a constant.
solenoid (24.1): A coil of wire with many loops that acts like a magnet when a current is allowed to flow through it.
sound level (15.1): A logarithmic scale that measures amplitude in decibels.

specific heat (12.1): The amount of energy that must be added to a material to raise the temperature of a unit mass one temperature unit.
spectroscope (28.1): Instrument used to study spectra.
spectroscopy (28.1): The study of spectra.
spectrum (16.2): The ordered arrangement of wavelengths.
spherical aberration (18.1): The image defect of a spherical mirror that does not allow parallel light rays far from the principal axis to converge at the focal point.
standing wave (14.2): A wave with stationary nodes and antinodes.
static friction force (6.2): The force that opposes the start of relative motion between the two surfaces in contact.
step-down transformer (25.2): A transformer in which the output voltage is smaller than the input voltage.
step-up transformer (25.2): A transformer in which the output voltage is larger than the input voltage.
stimulated emission (28.2): Process that occurs when an excited atom is struck by a photon and releases a photon of equal energy.
strong nuclear force (30.1): Force that holds the protons and neutrons together in the nucleus of an atom.
surface tension (13.1): The tendency of the surface of a liquid to contract to the smallest area possible.
surface wave (14.1): A wave that displaces matter both parallel and perpendicular to the direction of wave motion.
system (6.1): A defined collection of objects.

temperature (12.1): The "hotness" of an object as measured using a specific scale.
terminal velocity (6.2): The constant velocity of a moving object that is achieved when the drag force equals the force of gravity.
thermal energy (12.1): The overall energy of motion of the particles of an object.
thermal equilibrium (12.1): The state where the rate of energy transfer between bodies becomes equal and the bodies will be at the same temperature.
thermal expansion (13.2): The increase in length or volume of a material when heated.
thermodynamics (12.1): The study of heat.
thermometer (12.1): A device that measures temperature.
thermonuclear reaction (31.2): A nuclear fusion reaction.
thin-film interference (16.2): Light interference caused by reflection from the front and back surface of a thin layer of liquid or solid.
third right-hand rule (24.2): The method used to determine the direction of the force on a current-carrying wire in a magnetic field.
threshold frequency (27.1): The certain minimum frequency at which radiation causes the ejection of electrons from a metal.
timbre (15.2): Tone color, or sound quality determined by the frequencies and intensities of complex waves.
time interval (Δt) (3.2): Difference in time between two clock readings.
torque (7.3): The product of the force and the lever arm.
total internal reflection (17.2): Occurs when light is incident on the boundary to a less optically dense medium at an angle so large there is no refracted ray.
trajectory (7.2): The path of a projectile through space.
transformer (25.2): A device used to increase or decrease AC voltages with little loss of energy.
transistor (29.2): A semiconductor device used in electronic circuits to amplify voltage changes.
translucent (16.2): A material that transmits light but distorts its path.
transmuted (30.1): The change of an element into a different element by means of a change in nuclear charge.
transparent (16.2): A material that transmits light without distorting images.
transverse wave (14.1): A wave that displaces matter perpendicular to the direction of wave motion.
trough (14.1): The low point of a wave.

uniform circular motion (7.3): Motion at constant speed around a circle with a fixed radius.
uniform motion (5.1): Motion where equal displacements occur during successive equal time intervals.

valence band (29.1): In a material, the outermost band that contains electrons.
vector quantity (3.2): A quantity that has both magnitude and a direction.
vector resolution (4.2): The process of breaking a vector into its components.
virtual image (18.1): The point from which light rays appear to diverge without actually doing so.
volatile (13.1): Describes a liquid that evaporates quickly at a relatively low temperature.
volt (21.2): Electric potential difference measured in joules per coulomb.
voltage divider (23.1): A series circuit used to produce a voltage source from a higher-voltage battery.
voltmeter (23.2): A device that measures voltage.

W

watt (W) (10.1): Unit of power; one joule per second.
wave (14.1): A rhythmic disturbance that carries energy through matter or space.
wave pulse (14.1): A single disturbance traveling through a medium.
wavelength (λ) (14.1): The shortest distance between points where the wave pattern repeats itself.
weak nuclear force (30.2): A weak force acting in the nucleus apparent during radioactive decay.
weightlessness (6.2): An apparent weight of zero that results when there are no contact forces pushing against an object.
work (10.1): The process of changing the energy of a system by means of forces.
work-energy theorem (10.1): Work done on an object results in a change in kinetic energy.
work function (27.1): The energy with which electrons are held inside a metal.

X ray (26.2): High-frequency, short wave length electromagnetic waves.

***y*-intercept** (2.3): The point in a linear relationship where the line crosses the y-axis.

Index

Index

Index

Photo Credits

Cover Paul & Lindamarie Ambrose/FPG; **iv-v** Holway & Lobel Globos/Stock Market, **vi (t)**Alfred Pasieka/Peter Arnold, Inc, **(b)**NASA; **vii (tl)**StudiOhio, **(tr)**Morrison Photography, **(b)**Steve Lissau; **ix** Richard Megna/Fundamental Photograph; **x** John Feingerish/Stock Market, **xi** Index Stock Photography; **xii** Roger Du Buisson/Stock Market; **xiii (t)**Matt Meadows, **(b)**Mark C. Burnett/Photo Researchers; **xv** Jeff Blaton/Tony Stone Images: **xvi** Hank Morgan/Rainbow; **1** Michael Tamborrino/FPG; **2 3 4** NASA; **5 (l)**Scala/Art Resource, **(r)**Gianni Tortoli/Photo Researchers; **6** Stock Montage; **9 10** NASA; **11 (tl)**Bob Masini/PhotoTake, **(tr)**Morrison Photography, **(cl cr)**Michael Newman/PhotoEdit, **(bl)**Martin Bond/Photo Researchers, **(bc)**Krasowitz Michael/FPG, **(br)**Dustin W. Carr/ Harold G. Craighead/Physics News Graphics/American Institute of Physics; **12** Matt Meadows; **13 14** NASA, **15** Tom Pantages; **16 (t)**Tom Pantages, **(b)**David R. Frazier/Tony Stone Images; **17 (t)**National Bureau of Weights and Measures, **(b)**Tami Levinson; **19 (tl)**Omicron/Science Source/Photo Researchers, **(tc)**SW Productions, **(tr)**Jerry Rodriguss/Photo Researchers, **(bl)**R. Williams/NASA, **(br)**NASA, **24 25 29 30** Tom Pantages; **32** Matt Meadows; **33** NASA; **35** Tony Freeman/PhotoEdit; **37** Tom Pantages; **42** Thomas Zimmerman/Tony Stone Images; **43 44** Tom Pantages; **44 (t)** Tom Pantages, **(ctl)**Index Stock Photography, **(ctr)**Tony Freeman/PhotoEdit, **(cbl)**E. R. Degginger; **(cbr)**David Madison, **(b)**Jack Demuth; **45** Jack Demuth; **47 (t)**Tom Pantages, **(c b)**Jack Demuth; **48** Jack Demuth; **49** Tom Pantages; **53 (t)**Tom Pantages, **(b)**Jack Demuth; **57** Thomas Zimmerman/Tony Stone Images; **58** Jack Demuth; **60** Tom Pantages; **62** SuperStock; **63** Jim Sugar/Corbis; **64 (t)**Jim Sugar/Corbis, **(b)**color artwork by Folon/The New Yorker Magazine, Inc. Reprinted by special permission. All rights reserved.; **69** IndexStock; **72** Jim Sugar/Corbis; **75** SuperStock; **77** Jim Sugar/Corbis; **80** Dan Feicht/Cedar Point; **81 82** NASA; **84** Tony Stone Images; **86** David Madison; **90 94** NASA, **100** J & M Studios/Liaison International; **103** Skip Peticolas/ Fundamental Photographs; **104 (t)**NASA, **(b)**Loren Winters; **105** Dan Feicht/Cedar Point; **107** NASA, 115 David Barnes/Tony Stone Images; **116** Keith Kent/Peter Arnold, Inc.; **117 118** Richard Megna/Fundamental Photographs; **120** Matt Meadows; **126** Richard Megna/ Fundamental Photographs; **127** Morrison Photography; **133** courtesy Lawrence Livermore National Lab; **134 (t)**Keith Kent/Peter Arnold, Inc., **(b)**Ray Ellis/ Photo Researchers; **137** Matt Meadows; **138** Richard Megna/ Fundamental Photographs; **139** Yellow Dog Productions/Image Bank, **142** SuperStock, **144** Richard Megna/Fundamental Photographs; **148** Bob Daemmrich; **149 150** Photo Researchers; **152** Walter Bibikow/FPG; **155 (t)**Photo Researchers, **(b)**Richard Megna/Fundamental Photographs; **158** Bob Daemmrich; **159** Kodansha; **160** Richard Mackson/FPG; **162** SW Productions; **163 (t)**Photo Researchers, **(b)**Renee Lynn/Photo Researchers; **164** DUOMO/Image Bank, **169** Photo Researchers; **173** David R. Frazier Photolibrary; **174 175** NASA, **176 (t)**NASA, **(b)**The Newberry Library/Stock Montage; **179** Matt Meadows; **182 185 189** NASA; **191 (t)**Tom Pantages, **(b)**Arthur Selby; **193** NASA, **197** Craig Lovell/Corbis; **198** Donald Johnston/Tony Stone Images; **199** SuperStock; **200 (t)**SuperStock, **(b)**Loren Winters; **202** Tom Pantages; **203** Donald Johnston/Tony Stone Images; **204** Tony Freeman/PhotoEdit; **206** NASA/TSADO/Tom Stack & Associates; **207** SuperStock; **210** courtesy Daedalon Corp.; **211** SW Productions; **213** Matt Meadows; **217** SuperStock; **222** Karl Weatherly/Corbis; **223** Matt Meadows; **224 (t)**Matt Meadows, **(b)**George Disario/The Stock Market; **227** David Young-Wolff/PhotoEdit; **230 232** SW Productions; **233 (t)**Matt Meadows, **(b)**Alan Oddie/PhotoEdit; **238** Karl Weatherly/Corbis; **240** Morrison Photography; **241** Matt Meadows; **246** Rafael Macia/Photo Researchers; **247 248** Charles D. Winters/Photo Researchers; 251 Morrison Photography; **253** Deborah Gilbert/Image Bank; **255 (l)**Armwell/Tony Stone Images, **(r)**Dennis O'Clair/Tony Stone Images; **257** Matt Meadows; **258** Charles D. Winters/Photo Researchers; **259** Rafael Macia/Photo Researchers; **267** Charles D. Winters/Photo Researchers; 272 Eric Meola/Image Bank 273 **274** SuperStock; **276** Tom Pantages; **277 (l)**Dennis D. Cicco/Peter Arnold Inc., **(cl)**Jeff Greenberg/PhotoEdit, **(cr)** European Space Agency/Science Photo Library/Photo Researchers, **(r)**Rob Atkins/Image Bank; 281 Tom Pantages; **285 (t)**SuperStock, **(b)**David R. Frazier Photolibrary; **290** Eric Meola/Image Bank **293 (t)**Morrison Photography, **(b)**David Young-Wolff/PhotoEdit; **295** SuperStock; **298** David Madison; **299** Bob Firth/International Stock, **300 (t)**Bob Firth/International Stock, **(b)**Morrison Photography; **301 (t)**Don C. Nieman, **(c)** Morrison Photography, **(b)**David Frazier Photolibrary, **304 308** Tom Pantages; **309** Morrison Photography; **310 (t)**Dale E. Boyer/NASA/Photo Researchers, **(b)**Trent Steffler/David Frazier Photolibrary, **311** Herman Eisenbeiss/Photo Researchers; **312** Matt Meadows; **313** Mark E. Gibson; **314 (t)**Bob Firth/ International Stock, **(b)**From *Computers in Physics "How Do Crystals Melt?"*, courtesy Argonne National Lab/U.S. DOE/S.R Phillpot, S.Yip, D.Wolf; **315 (t)**SuperStock, **(b)**David Madison; **316** courtesy Dr. Andrew Kish/US DOT; **320** Tom Pantages; **321** Deborah Davis/PhotoEdit; **322** Bob Firth/International Stock, **323** Tim Courlas; **326** SuperStock, **327 328** Morrison Photography; **329** SuperStock, **330** Morrison Photography, **331** Tom Pantages; **336 (t)**Morrison Photography, **(b)**Tom Pantages; **337** Tom Pantages; **340** Kodansha; **341** Runk/Schoenberger from Grant Heilman; **342 (tl)** Runk/Schoenberger from Grant Heilman, **(tr)**Joe Cornish/Tony Stone Images, **(b)** Kodansha; **343** Kodansha; **344** Morrison Photography; **345** Kenji Kerins; **348** Erick and David Hosking/Corbis; **349 350** E.E. Kingsley/Photo Researchers; **353 (tl)**Paul Natkin/Photo Reserve, **(tr)**David Young-Wolff/ PhotoEdit, **(cl)**Michael Newman/PhotoEdit, **(cr)**Mary Kate Denny/ PhotoEdit, **(bl)**Aaron Haupt, **(br)**Richard Hutchings/Photo Researchers; **354(t)** Jamie Villaseca/Image Bank, **(b)**Erick and David Hosking/Corbis, **357** E.E. Kingsley/Photo Researchers; **358** Morrison Photography, **361** Michael Newman/PhotoEdit; **362** Matt Meadows; **364 (l)**SuperStock, **(c)**Morrison Photography, **(r)** Spencer Grant/ PhotoEdit; **368** E.E. Kingsley/Photo Researchers; **372** Matt Meadows; **373 374** Tom Pantages; **376** Yerkes Observatory/ Photo Researchers; **377** Matt Meadows; **378** D.W. Hamilton/ Image Bank; **382 (t)**Tom Pantages, **(c)**Beverly Carter, **(b)**David Parker/Photo Researchers; **383** Richard Megna/ Fundamental Photographs; **384 (t)**Matt Meadows, **(b)**Fundamental Photographs; **385 (t)**Matt Meadows, **(b)**Richard Menga/Fundamental Photographs; **388** William H. Mullins; **389** Tom Pantages; **392** Frank Siteman/PhotoEdit, **393** Alfred Pasieka/Peter Arnold, Inc.; **394 (t)**Alfred Pasieka/Peter Arnold, Inc., **(b)**Doug Martin; **395 396** Henry Leap/James Lehman; **399** Matt Meadows; **403** Alfred Pasieka/Peter Arnold, Inc.; **405** Deep Light Productions/Science Photo Library/Photo Researchers; **406** Kent Wood/Photo Researchers; **407** Jim Hamilton; **409** Frank Siteman/PhotoEdit, **410** Alfred Pasieka/Peter Arnold, Inc.; **414** Matt Meadows; **415 416** PhotoDisc; **419 (l)**AGE/Peter Arnold, Inc., **(r)**David Parker/Photo Researchers; **424** Morrison Photography, **426 (t)**Leonard Lessin/Peter Arnold, Inc., **(b)**Matt Meadows; **429 (t)**PhotoDisc, **(c)**David Parker/ Photo Researchers, **(b)**Matt Meadows; **433** Matt Meadows; **439** PhotoDisc, **442** Robert and

Photo Credits

Linda Mitchell; **443 444** Kristen Brochmann/Fundamental Photographs; **445 449 450** Kodansha; **452 (t)**Kristen Brochmann/Fundamental Photographs, **(c)**Robert and Linda Mitchell, **(b)**Tom Pantages; **453** courtesy Museum of Holography, Chicago; **454** Tom Pantages; **455** Kodansha; **457** Kristen Brochmann/Fundamental Photographs; **458** Mark Marten/NASA/Photo Researchers; **460** Keith Kent/Peter Arnold, Inc.; **461 462** Maisonneuve/Publiphoto/Photo Researchers; **463** Tom Pantages; **465 (t)**Morrison Photography, **(b)**courtesy B.D. Terris/IBM Almaden Research Center; **467** Matt Meadows; **468** Maisonneuve/Publiphoto/Photo Researchers; **470** Keith Kent/Peter Arnold, Inc.; **473** Gary Buss/ FPG; **476 (t)**Alan Oddie/PhotoEdit, **(b)**David R. Frazier Photolibrary; **477** Maisonneuve/Publiphoto/Photo Researchers; **480** Patrick Lambke; **481 482** Tom Pantages; **485** Mark C. Burnett/Photo Researchers; **487** Kodansha; **488** Tom Pantages; **490** Kodansha; **498** Tim Courlas; **499** Patrick Lambke; **500 502** Tom Pantages; **506** E. R. Degginger; **507 508** Tom Pantages; **509** Jeff Greenberg/Peter Arnold, Inc.; **514 518** Matt Meadows; **520 (t)**Tom Pantages, **(b)**Matt Meadows; **523 (l)**Tom Pantages, **(c)**Morrison Photography, **(r)**E.R. Degginger; **526** Tom Pantages; **530** Matt Meadows; **531** Morrison Photography; **532 (t)**Morrison Photography, **(b)**Glencoe Photo; **535** Mark C. Burnett; **541** Matt Meadows; **542 543** Morrison Photography; **545 547 548** Tom Pantages; **549** Doug Martin; **550** Morrison Photography, **554** Pekka Parviainen/Science Photo Library/Photo Researchers; **555** Tom Pantages; **557** Matt Meadows; **558 (l)**Phil Degginger/Color-Pic Inc., **(r)** Vaughan Fleming/Science Photo Library/Photo Researchers; **559** Richard Megna/Fundamental Photographs; **560** Tom Pantages; **561** Kodansha; **562** Matt Meadows; **565** Morton Beebe/Corbis; **567** Tom Pantages; **573** Pekka Parviainen/Science Photo Library/Photo Researchers; **575** Tom Pantages; **579** Phil Degginger/Color-Pic Inc.; **580** Richard Megna/ Fundamental Photographs; **581 582 590** Barros & Barros/Image Bank; **592 (t)**Tom Pantages, **(b)**Richard Menga/Fundamental Photographs; **594** courtesy Central Scientific Company; **595** Matt Meadows; **597** David R. Frazier Photolibrary; **598** Barros & Barros/Image Bank, **602** Richard Megna/Fundamental Photographs; **603 604** Beverly Carter; **606** Patrice Loiez, CERN/Science Photo library/Photo Researchers; **609** James Holmes/Oxford Centre for Molecular Science/Science Photo Library/Photo Researchers; **613** Beverly Carter; **618** Patti McConville/Image Bank; **619** Richard Megna/Fundamental Photographs; **621** Beverly Carter; **624** courtesy IBM Corporation, Research Division, Almaden Research Center, **625 626** Gregg Adams/Tony Stone Images; **627** FPG; **636** Richard Megna/Fundamental Photographs; **637 (t)** Gregg Adams/Tony Stone Images, **(b)**E.R. Degginger; **639 (t)**courtesy IBM Corporation, Research Division, Almaden Research Center, **(b)**Andrew Syred/Science Photo Library/Photo Researchers; **641** Gregg Adams/Tony Stone Images; **644** Tom Pantages; **647** Rich Treptow/Photo Researchers; **648 (t)**Tom Pantages, **(b)**Kodansha; **649 (t)**Physics Dept., Imperial College/Science Photo Library/Photo Researchers, **(c)**Tom Pantages, **(b)**Kodansha; **652** Science Photo Library/ Photo Researchers; **661** Hank Morgan/VHSID Lab ECE Dept. at U. MA. Amhurst/Science Source/Photo Researchers; **663** Lawrence Livermore National Laboratory/Science Photo Library/Photo Researchers; **664** Hank Morgan/Science Source/Photo Researchers; **668** Milan Chuckovich/Tony Stone Images; 669 **670** Jeffrey Sylvester/FPG; **678** Matt Meadows; **679** Jeffrey Sylvester/FPG; **684** Matt Meadows; **685 (t)**Milan Chuckovich /Tony Stone Images, **(b)**Richard Megna/Fundamental Photographs; **687** Jeffery Sylvester/FPG; **690** David Parker/Science Photo Library/Photo Researchers; **692** Courtesy Dr. Sekazi Mtingwa; **702** Bill W. Marsh/Photo Researchers; **703** Fermilab National Accelerator Laboratory/Science Photo Library/Photo Researchers; **705 (t)**Lawrence Berkley Laboratory/Science Photo Library/ Photo Researchers, **(c b)**Fermilab/Photo Researchers; 708 David Parker/Science Photo Library/Photo Researchers; 709 Morrison Photography; **712 (t)**CERN/Science Photo Library/Photo Researchers, **(c)**Luke Dodd/Science Photo Library/Photo Researchers, **(b)**Vic Bider/PhotoEdit; **716** Mitch Kezar/PhotoTake; **723 (t)**Tom Pantages, **(b)**Matt Meadows, **(inset)** Chris Priest/Science Photo Library/Photo Researchers; **724** Yoav Levy/PhotoTake; **727** SW Productions; **728** Mitch Kezar/PhotoTake; **730** Gordon Gahar/Lawrence Livermore National Laboratory; **731 (1)**Roger Ressmeyer/Corbis, **(r)**Ray Nelson/ PhotoTake.

Illustration Credits

Dartmouth Publishing, Inc. pages: 24, 31, 33, 34, 36, 40, 64, 65, 68, 72, 74, 85, 87, 97, 99, 105(t), 108, 109, 110, 111, 112, 113, 150, 154, 156, 159(t), 163, 164, 167, 170, 229, 233, 235, 236, 248, 249, 252, 253, 258, 259, 260, 262, 263, 374, 375, 379, 383(t), 385, 387, 508, 512(b), 528, 559, 563(br), 564, 568(b), 571(t), 576(br), 577, 578, 609, 613, 614, 616, 617, 619, 626, 628, 630, 632(t), 636, 640, 651, 653, 659, 660, 661, 666, 670, 671, 673, 674, 675, 676, 679, 680, 688, 689(tl)(c), 695, 702, 703, 738.

Thomas J. Gagliano/Gagliano Graphics pages: 142, 168, 185, 240, 363, 409, 436(b), 453, 454, 509, 512(l), 513, 516, 517, 533(t), 542, 549, 560, 572, 618, 648.

Precision Graphics pages: 25, 27, 35, 49, 52, 56, 59, 119, 127(tc), 131, 135, 136, 139, 143, 147, 160, 171, 172, 177, 182, 183, 190, 192, 196, 200, 207, 208, 214, 238, 243(br), 274, 275, 276, 282, 289, 290, 291(t), 297, 304, 309, 311, 315, 317, 321, 328, 329, 332, 337,338, 339, 341, 343(tr), 346, 355, 357, 358, 395(br), 396, 397(t), 400, 403, 405, 406, 407, 416, 417, 418, 421, 422(tl), 423, 425, 426, 430, 432, 435, 436(tl), 437, 438, 444, 445, 446, 447, 449, 465, 468, 469, 470, 471, 473, 488, 493, 496, 515, 523, 561, 563(t)(cr), 565, 567, 568(t), 570, 571(b), 576(cr), 579, 582, 583, 584(t), 586(t), 588, 591, 592, 593, 594, 599, 604, 608, 612, 615, 620, 634, 635, 645, 646, 647, 650(b), 656, 658, 665, 691, 692, 693, 696, 701, 704, 709, 710, 713, 725, 729, 737.

Preface, Inc. pages: 19, 45, 46, 47, 48, 50, 51, 54, 55, 57, 66, 67, 73, 75, 78, 82, 83, 84, 86, 90, 91, 92, 94, 95, 96, 98, 101, 102, 105(bc), 120, 121, 124, 126, 127(tr), 128, 129, 132, 138, 140, 151, 152, 153, 157, 159(br), 165, 180, 187, 194, 201, 203, 205, 209, 212, 215, 218, 219, 220, 221, 225, 226, 227, 228, 230, 237, 243(l), 244, 245, 250, 254, 261, 264, 277, 280, 283, 286, 288, 291(br), 302, 305, 306, 307, 318, 324, 331, 333, 343(b), 347, 350, 351, 353, 359, 360, 361, 364, 365, 366, 367, 380, 383(br), 395(c), 397(br), 398, 408, 410, 412, 413, 422(br), 424, 427, 431, 434, 448, 466, 472, 474, 475, 479, 482, 483, 485, 487, 489, 490, 491, 492, 494, 497, 499, 500, 503, 505, 511, 514, 521, 533(b), 535, 536, 537, 538, 539, 541, 543, 544, 546, 547, 548, 552, 553, 569, 573, 584(r), 586(bl), 590, 596, 606, 610, 623, 631, 632(b), 650(t), 654, 655, 681, 682, 689(tr), 699, 700, 705,707, 712, 717, 718, 719, 722, 733, 743, 744, 745, 746, 747, 781.